# Laser Spectroscopy and Laser Imaging

## An Introduction

# Series in Optics and Optoelectronics

Handbook of 3D Machine Vision: Optical Metrology and Imaging
**Song Zhang**

Handbook of Optoelectronics, Second Edition: Enabling Technologies (Volume Two)
**John P. Dakin, Robert G. W. Brown**

Handbook of Optoelectronic Device Modeling and Simulation: Fundamentals, Materials, Nanostructures, LEDs, and Amplifiers, Vol. 1
**Joachim Piprek**

Handbook of Optoelectronics, Second Edition: Applied Optical Electronics (Volume Three)
**John P. Dakin, Robert G. W. Brown**

Handbook of Optoelectronic Device Modeling and Simulation: Lasers, Modulators, Photodetectors, Solar Cells, and Numerical Methods, Vol. 2
**Joachim Piprek**

Handbook of Optoelectronics, Second Edition: Concepts, Devices, and Techniques (Volume One)
**John P. Dakin, Robert Brown**

Handbook of GaN Semiconductor Materials and Devices
**Wengang (Wayne) Bi, Haochung (Henry) Kuo, Peicheng Ku, Bo Shen**

Handbook of Optoelectronic Device Modeling and Simulation (Two-Volume Set)
**Joachim Piprek**

Handbook of Optoelectronics, Second Edition (Three-Volume Set)
**John P. Dakin, Robert G. W. Brown**

Optical MEMS, Nanophotonics, and Their Applications
**Guangya Zhou, Chengkuo Lee**

Thin-Film Optical Filters, Fifth Edition
**H. Angus Macleod**

Laser Spectroscopy and Laser Imaging: An Introduction
**Helmut H. Telle, Ángel González Ureña**

For more information about this series, please visit:
*https://www.crcpress.com/Series-in-Optics-and-Optoelectronics/book-series/TFOPTICSOPT*

# Laser Spectroscopy and Laser Imaging
## An Introduction

Helmut H. Telle

Instituto Pluridisciplinar - Universidad Complutense de Madrid
Madrid, Spain

Ángel González Ureña

Instituto Pluridisciplinar - Universidad Complutense de Madrid
Madrid, Spain

CRC Press
Taylor & Francis Group
Boca Raton  London  New York

CRC Press is an imprint of the
Taylor & Francis Group, an **informa** business

CRC Press
Taylor & Francis Group
6000 Broken Sound Parkway NW, Suite 300
Boca Raton, FL 33487-2742

First issued in paperback 2019

© 2018 by Taylor & Francis Group, LLC
CRC Press is an imprint of Taylor & Francis Group, an Informa business

No claim to original U.S. Government works

ISBN-13: 978-1-4665-8822-6 (hbk)
ISBN-13: 978-0-367-86821-5 (pbk)

**Visit the Taylor & Francis Web site at**
**http://www.taylorandfrancis.com**

**and the CRC Press Web site at**
**http://www.crcpress.com**

# Contents

# Detailed Contents

## Chapter 13
## Enhancement Techniques in Raman
## Spectroscopy                                  **333**

# Series Preface

This textbook series offers pedagogical resources for the physical sciences. It publishes high-quality, high-impact texts to improve understanding of fundamental and cutting-edge topics, as well as to facilitate instruction. The authors are encouraged to incorporate numerous problems and worked examples, as well as making available solutions manuals for undergraduate- and graduate-level course adoptions. The format makes these texts useful as professional self-study and refresher guides as well. Subject areas covered in this series include condensed matter physics, quantum sciences, atomic, molecular, and plasma physics, energy science, nanoscience, spectroscopy, mathematical physics, geophysics, environmental physics, and so on, in terms of both theory and experiment.

New books in the series are commissioned by invitation. Authors are also welcome to contact the publisher (Lou M. Han, Executive Editor: lou.han@ taylorandfrancis.com) to discuss new title ideas.

# Preface

Scientific and real-life applications of spectroscopy and imaging, which use a laser in one form or another, are abundant and a plethora of books have been written about particular, individual methodologies. To a lesser degree have general overviews been given, and if so these have been published many years, if not decades, ago.

In this book, we cover as wide a range of laser spectroscopic and imaging techniques as possible, and we tried to put these into perspective of use in the laboratory and in mainstream, real-world applications. While we attempted to be as inclusive as feasible, we are almost certain that we may have missed one or the other technique or application. But the book was never meant to be an encyclopedia. Rather, we intended to provide the reader with an overview of widely used techniques, including (1) their basic concepts; (2) an overview of the underlying theory and common experimental setups; (3) a handful of representative key applications that highlight the advantages and caveats of particular techniques; and (4) a comparison of the merits of certain implementations with those of other techniques. We end each chapter with brief snapshots of key events in the evolution of the technique, and cutting-edge developments. Of course, because of the necessary brevity, their selection is probably rather subjective, and in particular what is judged to be cutting-edge today may become routine tomorrow.

Overall, the text has been written—as far as possible—at a level suitable for advanced students and researchers in all fields of science, without delving into the depths of the underlying theoretical framework, but at the same time providing sufficiently detailed formulas for the reader to gauge the usefulness of a particular technique for his or her specific application.

**Helmut H. Telle**
**Ángel González Ureña**
*Madrid*

# Acknowledgments

We take this opportunity to acknowledge the support of many of our friends, students, postdocs, and colleagues; their efforts in reading through pages of the manuscript are gratefully recognized. They all provided invaluable input to improve the text, by pointing out where we went astray or by suggesting scientific aspects we had overlooked or forgotten.

In particular, we are indebted to our Executive Editor, Lou M. Han, for his guidance in polishing the layout and content of the book; but also for his skills in prodding us when we slacked, in giving encouragement when we were down, and—foremost—in being extremely patient when once again we did not keep to our schedule in reaching a promised milestone.

Last but not least, without the goodwill of our families, who supported us through long night hours and weekends, we might have thrown in the towel well before finishing this book; our gratitude and appreciation goes to them.

# Authors

The authors have known each other well for more than 30 years; they love and share many research interests in fundamental and applied physics and chemistry, all involving lasers in one form or another.

**Helmut H. Telle** earned his degrees in physics from the University of Cologne, Germany, in 1972 (BSc), 1974 (MSc), and 1979 (PhD), respectively. He exploited his newly gained experience in and passion for laser spectroscopy during an extensive postdoctoral research period, which found him expanding his horizons at universities and research institutions in Canada and France, at physics and chemistry departments. In 1984, he settled in Wales, United Kingdom, to embrace a career in teaching and research in laser physics at Swansea University. His research activities—both at Swansea and within the framework of numerous international collaborations—encompass a wide range of laser-spectroscopic techniques. He used these predominantly for trace detection of atomic and molecular species, and applied them to analytical problems in industry, biomedicine, and the environment on the one hand, but also to various fundamental aspects in science on the other hand. After nearly 30 years in Wales, he relocated to Spain to join the Instituto Pluridisciplinar of Madrid's Universidad Complutense. Here he pursues new frontiers in laser spectroscopy of exotic species of interest to astroparticle physics and astronomy.

**Ángel González Ureña** graduated in chemistry from the University of Granada (Spain) in 1968, and then earned his PhD in physical chemistry at the Universidad Complutense de Madrid in 1972. During the period 1972–1974, he carried out postdoctoral research at the Universities of Madison (Wisconsin, USA) and Austin (Texas, USA), embracing reaction dynamics in molecular beams. On his return to Spain, he took up the position of associate professor in chemical physics at the Universidad Complutense de Madrid, and was promoted to full professor in 1983.

The focus of his research activities was mainly on gas-phase, cluster, and surface reaction dynamics, mostly utilizing molecular beam and laser spectroscopic techniques. In said work, he was one of the pioneers in measuring threshold energies in chemical reactivity when changing the translational and electronic energy of the reactants. In recent years, his interests branched out into the application of laser technologies to analytical chemistry, environmental chemistry, biology, and food science. He is heading the Department of Molecular Beams and Lasers at the Instituto Pluridisciplinar, associated with Madrid's Universidad Complutense; for the first 10 years of the institute's existence, he also was its first director. In 2016 he received the *Richard B. Bernstein Medal* for "... lifetime achievements in the field of chemical stereo-dynamics."

# CHAPTER 1
# Introduction

Humans have always been fascinated with color and imaging of nature, and semiscientific or philosophical descriptions of these optical phenomena can be traced back more than two millennia. It is difficult to gauge which of the two was really understood first in a rigorous manner since scientific investigations in the modern sense—merging experimental observations with theoretical models—were not always the stated goal of the early "scientists." It is probably fair to say that imaging might have been the topic easier to understand from the simple, observational standpoint lending itself to conceptual sketching of the phenomenon. Color would have been more elusive to understanding; while its existence was easy to realize, to explain its origins was largely beyond the understanding of physics and chemistry before the Middle Ages.

Here we provide only a few historical remarks regarding the development of spectroscopy and imaging toward the modern-age revolution that the advent of the laser brought to these two fields of science.

Imaging of "reality" in the form of paintings can be traced back to prehistoric times and is evidenced by numerous cave drawings. Such imaging became more sophisticated with time, and different techniques evolved, including among others mosaic laying—which in a certain way may be seen as being akin to modern, pixilated imaging using digital camera sensors. The use of a real imaging instrument in the form of the so-called *camera obscura* goes back to "before common era" (BC). The first written reference to the *camera obscura* can be found in the works of the Chinese philosopher Mozi (about 470–390 BC) who correctly noted that the inverted upside-down image was due to light traveling in straight lines from its source. The *camera obscura* was used by many natural philosophers who experimented with different aspects of pinhole imaging in ever more sophisticated ways.

**Figure 1.1** First known drawing of the *camera obscura*; translation of the inscription: "Observing the solar eclipse of 24 January 1544." (From Gemma Frisius, R. "De radio astronomico et geométrico." Antwerp: apud. Greg. Brontium, 1545.)

Probably the first clear description of its workings is by Leonardo da Vinci in his *Codex Atlanticus*, published in 1502. And the first known drawing of a *camera obscura* is that published in the works of Rainer Gemma Frisius (Gemma Frisius 1545); he used a *camera obscura* in a "scientific" way, namely to observe and monitor the solar eclipse of 1544 (see Figure 1.1).

Color and its relation to "spectroscopy" proved to be more elusive to understanding and applications of scientific significance. It is probably the spectrum of the rainbow (see the top panel of Figure 1.2) that inspired not only artists but also many natural philosophers to try to explain the phenomenon behind the generation of a color-dispersed spectrum.

The earliest natural philosopher—at least according to written records—to seriously investigate the rainbow phenomenon was the Greek scholar Aristotle (384–322 BC). Although some of his reasoning was flawed albeit relatively consistent, his qualitative explanations remained unrivaled until the Middle Ages. Then, for the first time, the Persian astronomer Qutb al-Din al-Shirazi (1236–1311 AD) and his student Kamāl al-Dīn al-Fārisī (1267–1319 AD) provided a fairly accurate explanation for the rainbow phenomenon; the latter developed a mathematically satisfactory explanation, proposing a model of refraction and linking refraction to the decomposition of light. Incidentally, in his experiments, al-Fārisī made use of a *camera obscura*.

The understanding of color as part of the (rainbow) spectrum evolved from Newton's prism experiments in the seventeenth century. He discovered that a prism can "disassemble and reassemble white light," and in his publication on the topic (see Newton 1671), the word "spectrum" is used for the first time in the context of decomposition of white light into its components. The concept of his experiments is given in the bottom part of Figure 1.2. He observed that when a beam of white (sun) light struck the entrance face of a glass prism at an angle, that (1) some fraction of the light was reflected and that (2) another fraction of the light passed through the glass to emerge from the opposite prism face as a band of light with different (continuously changing) colors. Newton hypothesized light to constitute "particles" of different colors, and that these moved at different speeds in transparent matter, with red light particles moving faster than blue light particles in glass, but with red light being bent ("refracted") to a lesser degree than blue light.

It took until the early nineteenth century that the concepts of spectra and spectroscopy became more palatable, in particular with the discovery of light outside the visible range. And for the first time, the wavelengths of different colors of light were measured (Young 1802). From then on, different spectroscopic phenomena and techniques were developed, hand in hand with the discovery and evolution of photon and image recording devices. With the evolution of quantum physics in the early part of the twentieth century, spectroscopy became one of the indispensable cornerstones for the measurement and understanding of many physical, chemical, and biological processes.

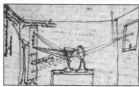

**Figure 1.2**   Origin of spectra and their explanation. (Top) Dispersion of white light in a rainbow. (Middle) Sketch of Newton's prism experiment, from his notes (see also Newton 1671). (Bottom) Nineteenth century color engraving of Newton's prism experiment.

## 1.1 LASERS AND THEIR IMPACT ON SPECTROSCOPY AND IMAGING

It is probably fair to state that the advent of the laser has by and large revolutionized spectroscopy and spectroscopic imaging. In particular, it is the specific properties of laser light sources that make all the difference in comparison to other light sources. Namely, these properties include (1) spectral intensity and monochromaticity; (2) wavelength tunability, often over broad intervals; (3) coherence, both temporally and spatially; and (4) temporal variation from continuous operation to ultrashort pulses.

### 1.1.1 Laser properties of importance to spectroscopy

It is most certainly the *intensity* within *narrow spectral intervals* that sets the laser apart from conventional light sources. This is because, in general, response of a system to a light stimulus can be described by the relation $I_{\text{signal}} \propto N \cdot S_\lambda \cdot I_\lambda$, where $N$ stands for the number of atoms or molecules taking part in the process, $S_\lambda$ incorporates the transition probability for the photon interaction, and $I_\lambda$ is the spectral intensity of the light source, optimally matched to the bandwidth of the transition, if possible. The spectral intensity can be orders of magnitude larger than for ordinary light sources. This means many things, among others that a much better signal-to-noise ratio is achievable. But more importantly, it becomes much easier to probe processes in which only very small numbers of particles participate. For example, in chemistry, one can trace transient species, reaction intermediates, photodissociation fragments, and so on; and in (analytical) measurements, one is able to record signals at extremely low particle densities, ultimately investigating individual particles of a species.

The *tunability* of the laser, in conjunction with high *monochromaticity*, lends itself for high-resolution spectral measurements. This means that the excitation by laser light can be molecule-specific—a great advantage in chemical reaction dynamics or compositional analysis—and individual quantum states can be selected, even resolving quantum features like hyperfine structure, or targeting a specific isotope of a species.

The extraordinary *coherence* properties of laser light allow for numerous applications that require, e.g., tight focusing or time-delayed overlap between two light beams. In this context, high temporal coherence will allow for tailoring of short laser pulses, which can be utilized in well-controlled pump—probe experiments to study ultrafast processes in real time. The property of spatial coherence can be exploited to focus a laser beam to diameters of the order of the light wavelength itself, thus reaching extremely high power densities (to initiate, e.g., nonlinear processes), or to single out individual atoms or molecules.

Finally, the exceptional control over the *time duration* of the laser radiation opens up many applications that would be inaccessible to traditional light sources. While light pulses down to about 1µs are feasible, by and large, any shorter pulse duration (nano-, pico-, femto-, and attosecond) is the domain of lasers. It is in particular the realm of ultrashort pulses with $t_{\text{p}} \sim 10^{-15}$–$10^{-18}$ s that has given researchers access to physical and chemical processes that could only be postulated theoretically but eluded experimental proof. For example, the full knowledge of chemical reactivity requires the full understanding of an elementary chemical reaction occurring in just one single, short time-scale event. The intermediate steps in such a reaction can be studied spectroscopically (energy regime) and dynamically (time regime), giving rise to the exciting field of femtochemistry.

One extraordinary example of the usefulness of ultrashort attosecond pulses, reaching beyond the limit of fundamental chemistry, is found in the imaging of the quantum motion of electrons not only in real time, but also in real space (see, e.g., Dixit et al. 2012).

### 1.1.2 Concepts of laser spectroscopy and imaging

The spectroscopy—and imaging—of atoms, molecules, and chemical reactions of materials and samples depends significantly on the state of matter (thermodynamically distinct phases that can occur at equilibrium) and the degree (size) of aggregation. Optimized for particular cases, laser spectroscopic techniques have been developed to study processes in the gas phase (mainly isolated particles), clusters (small-size aggregates), solutions, and surfaces. The concepts for laser spectroscopy and imaging are shown schematically in Figure 1.3, from the laser source through to the detection and interpretation of the measurement results.

## 1.2 ORGANIZATION OF THE BOOK

It is evident from Figure 1.3 that *en route* from the laser source to the "finished" spectrum or image, a number of steps are involved, and that information about the light–matter interactions needs to be "known" or to be answered. The aim of this book is not so much to provide a gapless summary of all and every methodology ever devised but to detail the underlying fundamental aspects of the most common (and some not-so-common) techniques of spectroscopy and spectral imaging, and to highlight weaknesses and strengths through a wealth of representative, practical examples.

Broadly, the book is organized into three main segments. First, we describe the basic photon–matter interaction processes, together with excitation sources and detection devices. Then the major part of the book deals with the individual spectroscopic techniques and subgroups thereof whenever required. And finally, applications of spectrally resolved imaging are summarized.

Throughout the book, we have put the emphasis on the understanding of the fundamental principles, underpinned by suitable examples, but at the same time we have made an effort to convey modern trends in the respective topics as well.

### 1.2.1 Introduction to photon–matter interaction processes, laser sources, and detection methodologies

Conceptually, all laser spectroscopy experiments are made up of the same general setup scheme: central to any experiment or measurement is the interaction zone with the sample to be probed, into which the excitation laser radiation is transferred, and out of which signal response is collected. The actual configurations of the interaction region and the experimental apparatus will differ vastly, depending on the nature of the sample under investigation and the selected spectroscopic technique, and on what answers are sought in a particular investigation. Accordingly, the laser source and the detection equipment may have to be tailored to the selected spectroscopic of imaging methodology (see Figure 1.4 for some selected aspects to be considered).

Therefore, the book begins with a brief summary on photon–matter interactions, in which we introduce the phenomena of absorption, emission, and scattering. While it is well beyond the scope of this book to describe all aspects in depth, we provide the relevant theoretical framework that is necessary to gauge how they are linked to qualitative and quantitative laser spectroscopy and spectroscopic

LASER SOURCE
*Continuous-wave/pulsed*
*fixed/variable/broadband* λ

SAMPLE
*Gaseous, liquid, solid*

INTERACTION OUTCOME
*Photon effects*
*(Absorption/emission/scattering)*
*Charged-particle effects*
*(Ionization into electrons/ions)*

DETECTION
*Point/1D/2D*

DATA ANALYSIS/EVALUATION
*Spectrum/image visualization*

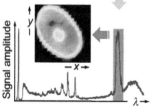

**Figure 1.3** Concept of laser spectroscopy and imaging, from the laser source through to the sample, its interaction with the laser photons, its detectable outcome, photon or charged-particle detection, until the data analysis provides the result in the form of spectra or spectral images. Here, this is exemplified for the spectral imaging of a single cell.

imaging. Of course, references are given to textbooks on the topics where appropriate or required, and some details relevant to a particular spectroscopic methodology are picked up again in the chapters that discuss them.

Lasers and laser radiation are the key ingredients in any of the techniques described later in the book. We first summarize the most fundamental aspects of lasers that impact on laser spectroscopy and spectrally resolved imaging, and in particular address the aforementioned aspects of laser wavelength and monochromaticity, bandwidth, tenability, and temporal behavior (from continuous-wave operation to ultrashort pulses). In addition, where required, laser cavities and constituent components are described in an extended fashion, to give the reader a feeling for which laser properties or configurations may be appropriate for a specific task. Individual laser classes and particular devices are collated in two chapters (Chapters 4 and 5); instead of listing all and every possible laser, by and large, we have focused our description on systems that are most commonly encountered in laser spectroscopy experiments, and that are commercially available (an example of a narrow-bandwidth, low-power laser source is shown in the top part of Figure 1.4).

Finally, we provide a summary of detection techniques and devices that are commonly encountered to measure the response to the laser stimulus: photon detection, (charged) particle detection, and detection by indirect phenomena are included. An example of a photon-detecting device is given in the bottom part of Figure 1.4.

### 1.2.2 Spectroscopic techniques and their applications

A galore of laser spectroscopic techniques has been developed over the past 50 years, but the underlying principles of fundamental spectroscopy have not changed much—it is mostly the light source that has been replaced. Thus, it is photon absorption, emission, and scattering phenomena and their consequences for the analyzed specimen that are at the heart of spectroscopic techniques. On the other hand, the laser enabled researchers to exploit processes that are difficult or impossible to realize using conventional light sources, and consequently new methodologies could be developed albeit still relying on the three fundamental photon processes. Examples are nonlinear spectroscopic techniques and femtosecond spectroscopy, to name but two.

While we have tried to paint as complete a picture as possible of the range of available methodologies and techniques, we had to be somewhat selective. Therefore, emphasis is placed on those technique that nowadays are the most common in scientific research and practical applications. And by and large we describe methods that rely on photon detection and charged particle (electrons or ions) detection. Out of the many possible laser spectroscopic methods, three (randomly selected) cases are collated in Figure 1.5, trying to highlight the universality of the approach to atomic and molecular samples, and to address aspects that emphasize the strength of lasers as the light source.

We begin with *absorption spectroscopy* for which the laser has revolutionized proceedings. Utilizing tunable laser sources, in most cases a spectrometer is no longer required—the laser itself provides the spectral selectivity. In particular, since laser sources with extremely narrow bandwidth (single-longitudinal mode with band widths down to less than a few kilohertz) are commercially available,

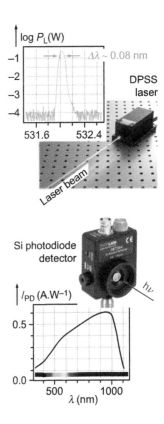

**Figure 1.4** Examples of an excitation laser and its associated bandwidth (top), and a photodiode detector and its spectral response (bottom).

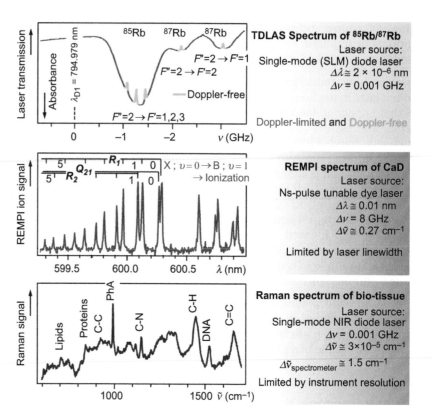

**Figure 1.5** Examples of laser-induced spectra. (Top) Doppler-free spectrum of the D1-line of $^{85}$Rb/$^{87}$Rb, using a single-mode diode laser and the technique of tunable diode laser absorption spectroscopy (or TDLAS) in saturated absorption configuration. (Middle) Resonant-multiphoton ionization (or REMPI) spectrum of CaD in a molecular beam, using a medium-resolution pulsed dye laser. (Bottom) Raman spectrum of a biological tissue sample, with the spectral resolution limited by spectrometer and molecular-band properties.

extremely high resolution can be achieved, being able to resolve hyperfine and isotopic details that were impossible to resolve with standard spectrometers (see the top panel of Figure 1.5). In addition to the general discussion on absorption spectroscopy (Chapter 7), very common implementations and some derivative techniques are described in a separate chapter (Chapter 8), including tunable diode laser absorption spectroscopy (TDLAS) and cavity ring-down spectroscopy. Selected examples of the many laboratory and practical applications are discussed.

In contrast to absorption spectroscopy, *laser-induced fluorescence spectroscopy* (LIFS) has a lesser penetration in commercial applications; this is mainly due to the fact that quantification of the observed signals is extremely difficult, and thus its use in quantitative analysis (which commercial users strive for) is relatively limited. On the other hand, LIFS is still often the method of choice for many scientific problems where the narrow bandwidth of the laser allows access to individual quantum states. This is because it combines the selectivity of the absorption transition with the nearly background-free observation of fluorescence photons, which allows for extreme sensitivities, ultimately involving only a single atom or molecule. Two chapters are dedicated to LIFS, covering its general principles (Chapter 9) and specific high-resolution and high-sensitivity applications (Chapter 10).

Despite being confined for a long time to scientific laboratory use, *Raman spectroscopy* has mostly benefited from the advent of the laser. Today, the technique has evolved as one of the cornerstones in (commercial) analytic spectroscopy because of its universality: all aggregate states, i.e., gaseous, liquid, and solid samples, can be probed; and it lends itself to qualitative and quantitative analysis of multispecies samples, which in particular for biological and pharmaceutical applications has proven to be invaluable, if not indispensable (an example of this is shown in the bottom panel of Figure 1.5). After the usual introduction to the topic (Chapter 11), we tried to convey the versatility and power of Raman spectroscopy in three additional chapters. These encompass (1) "conventional" linear Raman spectroscopy in Chapter 12, with examples selected to demonstrate its universality; (2) "enhancement" techniques in Chapter 13, covering, among others, resonance Raman spectroscopy; and (3) nonlinear Raman spectroscopy in Chapter 14, where we specifically highlight the techniques of surface-enhanced Raman spectroscopy and coherent anti-Stokes Raman scattering. One may perceive the coverage of Raman spectroscopy as "imbalanced" with respect to absorption and fluorescence spectroscopies; however, we think that this is justifiable since some modalities have evolved into nearly independent fields of spectroscopy.

The laser spectroscopy part concludes with brief surveys of techniques in which the laser is used to "destroy" the sample, for example, in *plasma generation and ablation* (nonresonant laser excitation) and in *ionization* (resonant laser interaction). The former group of techniques is exemplified by laser-induced breakdown spectroscopy (Chapter 15), while the latter encompasses resonance-enhanced multiphoton ionization (REMPI) and zero-electron kinetic energy spectroscopy (Chapter 16). Note that an example of a REMPI spectrum is included in the center panel of the overview in Figure 1.5.

### 1.2.3 Laser-spectroscopic imaging

The book concludes with a summary on laser (spectroscopic) imaging. We like to note here that we only included techniques that also incorporate the use of spectroscopic information, meaning that techniques like, e.g., holography or ellipsometry have been excluded on purpose.

The proceedings begin with a survey (Chapter 17) that highlights the general features and procedures of laser-spectroscopic imaging—whether globally or "pixel-by-pixel" (for the concept, see Figure 1.6). This is followed by three chapters addressing specific methodologies. In Chapters 18 and 19, techniques involving fluorescence and Raman spectroscopy are discussed; Chapter 20 is dedicated to diffuse optical imaging; and we finish in Chapter 21 with a collection of imaging techniques relying on absorption of the laser radiation (detection either directly or indirectly, e.g., photoacoustic detection), or on the "destructive" action of the laser on the specimen under investigation (e.g., ion imaging).

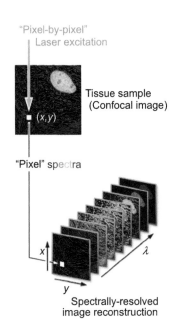

**Figure 1.6**  Concept of laser imaging, exemplified for the reconstruction of spectrally resolved pixel maps ($\lambda$-planes) from a spatial scan of a specimen by an excitation laser; full spectra are sampled for each interaction region.

# CHAPTER 2

# Interaction of Light with Matter

The interaction of light with matter takes place anytime, anywhere in the Universe. It explains why the sky is blue, why water is colorless, and why we see ourselves in a mirror. Indeed, colors are due to the interaction of charged particles (electrons and nuclei) with photons; therefore, colors are associated with quantum effects.

In this chapter, we briefly describe how the light and matter interaction depends on the light properties, like its frequency; on the optical properties of the matter, like their absorption; or on scattering coefficients, to mention just a few key examples. The aim is not to provide a full, comprehensive picture of all light–matter interactions but rather to concentrate on those main aspects that are underlying the spectroscopic and imaging techniques studied in subsequent chapters. For a more thorough description of light–matter interactions, the reader is referred to standard textbooks dealing with the topic, including, e.g., Atkins (2010), Demtröder (2010), or Hollas (2004).

## 2.1 ABSORPTION AND EMISSION OF RADIATION

As illustrated in Figure 2.1, let us consider an atom that is exposed to light and whose frequency is such that the photon energy coincides with the energy difference of two energy levels, i.e., $h\nu = E_2 - E_1$, where $E_2$, $E_1$, and $h$ stand for the higher and lower level energies, and Planck's constant, respectively. Under these conditions, the atom may absorb the photon energy transferring a (valence) electron from the lower to the higher state, a process called (induced) *absorption*.

The atom not only can absorb energy but also can emit light by the two other processes shown in Figure 2.1, provided that the excited state is populated initially. First, the excited atoms can undergo a process called *spontaneous emission* (see the middle panel in the figure); the phase of the light is randomly distributed when linked to the initial, exciting (absorbed) photon. Secondly, an incident photon initiates the transition, i.e., the incident photon triggers the emission of a further photon with the atom undergoing a transition from the excited (2) to the ground (1) state, a process called *stimulated emission* (see the bottom panel in the figure); the process was first proposed and correctly described by Einstein in his 1916/1917 publications on the quantum theory of light.

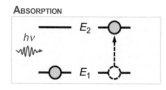

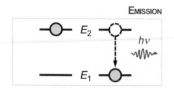

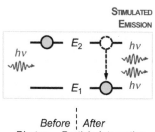

*Before | After*
*Photon ↔ Particle interaction*

$$E_2 - E_1 = \Delta E \equiv h\nu$$

**Figure 2.1** Illustration of the three most important light–matter interaction modes: absorption, spontaneous emission, and stimulated emission. For simplicity, a two-level system with only the ground state (1) and one excited (2) state is assumed.

Note that the frequency and phase of the stimulated emitted light are identical to those of the incident light, and that—as discussed in Chapter 3—this process is the key to laser emission. Note also that because of the random/equal phase conditions for the two emission processes, they are inherently different: while the stimulated process is a coherent phenomenon, spontaneous emission is not.

In the above context, the traditional explanation of color is associated to light absorption of organic molecular compounds and therefore to electronic transitions between their molecular orbital. Within this picture, the colors of the leaves, flowers, clothes, and countless other objects seen in our living environment are due to preferential absorption. For example, a substance of green appearance absorbs light of all visible frequencies, except the green. Thus, the absorption of a colorful substance covers the visible spectrum with the exception of the actual color of the substance.

But what about transparent substances? Take for example a window pane or the surface of water. One observes that a minute part of the incident light is reflected but that the rest is transmitted. If there is no resonant absorption, the substance looks (nearly) colorless, even though their outlines are visible because of the wavelength independent light reflection at the surface of standard window glass or water.

The fact that water is transparent in the visible region does not prove it may absorb in other regions of the spectrum. In fact, it is well known how water absorbs in the UV and the IR. More specifically, water molecules exhibit a slight absorption in the red near 750 nm, which explains the greenish-blue color of sea water; however, this should not be confused with the intense blue often observed in association with the reflection of the blue sky. It should be noted that the mechanisms responsible for the color in substances are not unique, and the reader is referred to, e.g., Juster (1962) and Schiller (2014) for a detailed description of them.

### 2.1.1 Einstein coefficients and transition probabilities

The framework for the modern (quantum) treatment of interaction of light with matter—in the form of atoms, molecules, and macroscopic solids—was laid just about 100 years ago by Einstein (see Figure 2.2) in his famous publication "On the Quantum Theory of Radiation" (Einstein 1917), which, however, at the time was probably overshadowed by his controversial work on relativity. Regardless, his pioneering description of the interaction processes and rate equations has proven to be indispensable, and is used in one form or another in most quantum treatments related to photon interactions.

Sort of following in the steps of Einstein, in this section, the absorption and emission of radiation is treated in a more quantitative manner, rather than the hand-waving arguments used in the previous, introductory section. To this end, let us consider an atom or molecule with only two stationary states denoted 1 and 2, as illustrated in Figure 2.1; these states can be electronic or, in the case of a molecule, rotational or vibrational. With reference to this figure, the three processes described in the previous section are represented by vertical up or down arrows, and the state energies (and populations) are denoted by $E_1$ and $E_2$ (and $N_1$ and $N_2$), respectively.

In the presence of a radiation field with a spectral density $\rho(v)$, given by Planck's distribution

$$\rho(v) = \frac{8\pi h v^3/c_0^3}{e^{hv/kT} - 1}, \qquad (2.1)$$

where $c_0$ is the speed of light. The change of population $N_2$ of state 2 can be due to the following three contributions:

- Induced absorption, given by the rate $(dN_2/dt) = N_1 B_{12}\rho(v)$
- Induced emission, given by the rate $(dN_2/dt) = -N_2 B_{21}\rho(v)$
- Spontaneous emission, given by the rate $(dN_2/dt) = -N_2 A_{21}$

In these rate equations, $B_{21}$ is the Einstein coefficient for the induced $(2 \to 1)$ process that is (without state degeneracy) equal to $B_{12}$. Likewise, $A_{21}$ is the Einstein coefficient for the spontaneous $(2 \to 1)$ emission.

When the populations have reached their equilibrium (i.e., no temporal variation is observed any longer), one can write

$$(dN_2/dt) = 0 \equiv (N_1 - N_2)B_{21}\rho(v) - N_2 A_{21}.$$

At equilibrium, the populations are related to the Boltzmann distribution law according to

$$(N_2/N_1) = (g_2/g_1) \cdot e^{-(E_2-E_1)/kT} \equiv e^{-\Delta E/kT}, \qquad (2.2)$$

where it is assumed that the degeneracies are identical ($g_2 = g_1$), and $k$ and $T$ stand for Boltzmann's constant and the absolute temperature, respectively. Introducing this $N_2/N_1$ ratio and the radiation density function $\rho(v)$ in the equilibrium condition (Equation 2.2), one obtains after some algebra the relationship

$$A_{21} = \left(8\pi h v^3/c_0^3\right) \cdot B_{21}, \qquad (2.3)$$

which is a useful equation allowing one to calculate $A_{21}$ if $B_{21}$ is known or vice versa. A relevant point of Equation 2.3 concerns the $v^3$ dependence between $A_{21}$ and $B_{21}$. It shows that the spontaneous emission increases with the third power of the (photon's) frequency with respect to the induced emission (or absorption). Also, since the lifetime of the excited state is related to the Einstein coefficient for spontaneous emission by $\tau_{21} = 1/A_{21}$, one can understand why the lifetime of an excited molecular electronic state is much shorter than that of its excited vibrational state.

It is thus clear that if one knows one Einstein coefficient, one can calculate the other; but how does one know the first one?

To answer this question, one can carry out an experiment, or one can calculate the so-called *transition moment* $\boldsymbol{\mu}_{21}$, given by

$$\boldsymbol{\mu}_{21} = \int \psi_2^* \hat{\boldsymbol{\mu}} \psi_1 dV = <2|\boldsymbol{\mu}|1>. \qquad (2.4)$$

**Figure 2.2** Einstein playing the violin. (From E.O. Hoppe (1878–1972). http://en.wikipedia.org/wiki/File:Albert_Einstein_violin.jpg, public domain.)

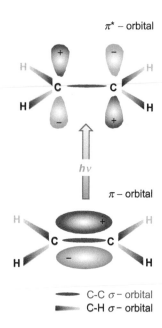

$\pi^*$ – orbital

C-C $\sigma$ – orbital
C-H $\sigma$ – orbital

$hv$

$\pi$ – orbital

**Figure 2.3** Illustration of the $\pi$ and $\pi^*$ molecular orbitals associated with the >C=C< chromophore electronic transition.

Here $\psi_2$ and $\psi_1$ are the wave functions of the two states, and $\hat{\mu}$ is the electric dipole moment operator with magnitude $\mu = \sum_j q_j|r_j|$, where $q_j$ and $r_j$ are the charge and position vector of the $j$th particle, respectively.

To understand the meaning of the transitions moment, Figure 2.3 shows, as an example, the $\pi$ and $\pi^*$ molecular orbitals associated with the >C=C< chromophore (note that both orbitals lie in the plane of the paper).

Imagine an electron is excited to $\pi^*$ from $\pi$ in an electronic transition; it becomes clear from the figure that during the transitions, a process of electron displacement takes place. Consequently, a transition moment arises that lasts the duration of the electronic excitation, because neither before nor after is there a permanent dipole moment, due to the orbital symmetry of both states.

It can be shown, using quantum mechanics, that the intensity of the absorption is proportional to the square of the transition dipole moment. Furthermore, the $B_{21}$ coefficient is given by

$$B_{21} = (1/6\varepsilon_0\hbar^2) \cdot |\mu_{21}|^2, \tag{2.5}$$

where the vector quantity $\mu$ represents the transition moment operator, and $\varepsilon_0$ is the vacuum permittivity. The transition probability $|\mu_{21}|^2$ is related to the transition moment defined in Equation 2.4 and underlies spectroscopic selection rules. Note also that if $\mu_{21} \neq 0$, the transition is allowed (always, of course, in the framework of the electronic selection rules); when $\mu_{21} = 0$, then the transition is said to be forbidden. For more details on transition moments, Einstein coefficients, and selection rules, see the standard texts covering quantum light–matter interaction, like Atkins (2010), Demtröder (2010), or Hollas (2004).

## 2.1.2 Quantitative description of light absorption— The Beer–Lambert law

One particular question one may ask is, can the Einstein coefficient for induced absorption, $B_{12}$, be determined experimentally? The simple answer to this question is yes; one can estimate this coefficient by measuring the absorbance of the respective substance.

Imagine a container filled with a gas (or liquid) of a given substance, with molar concentration, $c$. If one denotes the incident light intensity with $I_0$, at frequency $v$ (matching that of a transition of the substance), it is obvious that the transmitted light intensity should be less that $I_0$ due to the absorption of light.

Experimentally, it is found that there is an exponential relationship between the ratio $I/I_0$, and the concentration $c$ and path length $\ell$ through the absorbing substance (see Figure 2.4); this relationship is given by

$$I/I_0 = \exp(-\varepsilon c\ell) = 2.3 \cdot 10^{-\varepsilon c\ell}, \tag{2.6}$$

depending on whether one utilizes a functional representation in base-$e$ (predominantly in physics) or base-10 (normally in analytical chemistry) exponentials. Taking the latter representation, one obtains

$$\log_{10}(I/I_0) = -\varepsilon c\ell \text{ or } \log_{10}(I_0/I) = \varepsilon c\ell. \tag{2.7}$$

Here $\varepsilon$ is the so-called molar absorption coefficient and is proportional to the probability of absorption of the specific substance under consideration. This quantity was originally known as the molar extinction coefficient, though; this name is no longer in use.

The quantity $\log_{10}(I_0/I)$ noted in Equation 2.7 is called *absorbance A*, or sometimes *optical density* (OD). The linear relation $A = \varepsilon c \ell$ is known as the Beer–Lambert law. This law can be applied except where high concentrations or high-intensity light beams (lasers) are employed. In the first case, molecular complex association may take place; in the second case, a significant fraction of the molecules is in the excited rather than in the ground state. In both cases, the absorbing (molecule) concentration does not correspond to the initial monomer or ground state concentration.

The value of molar absorption coefficient $\varepsilon$ used in Equations 2.6 and 2.7 depends on the wavelength $\lambda$ of the interacting light, i.e., one needs to investigate $\varepsilon(\lambda) = A(\lambda)/c\ell$. In fact, a plot of $\varepsilon$ (or $\log_{10}\varepsilon$) as a function of wavelength (or alternatively in units of wavenumber) is called *absorption spectrum*. Note that, for historical reasons, the units of $\varepsilon$ are not expressed in SI units, but in $M^{-1}\cdot cm^{-1}$, or equivalently in $l\cdot mol^{-1}\cdot cm^{-1}$.

Often, one measures the so-called *transmittance T* $[ = (I/I_0)]$, with the relationship between this quantity and the absorbance given by $A = -\log_{10}T$.

It has to be noted that absorption normally does not only occur for a singular wavelength $\lambda$ (or frequency $\nu$), as the schematic in Figure 2.1 might suggest, but that the transition normally exhibits a line shape related to the actual nature of the atomic or molecular levels involved in and contributing to the transition. Thus, using only the maximum value of the molar absorption coefficient as an indicator of the transition intensities may lead to an erroneous result. In order to avoid this deficiency, one should use the band-integrated absorption coefficient $\boldsymbol{\mathcal{A}}$, given by

$$\boldsymbol{\mathcal{A}} = \int_{band} \varepsilon(\lambda)d\lambda \;\; or \;\; \boldsymbol{\mathcal{A}} = \int_{band} \varepsilon(\nu)d\nu.$$

In Figure 2.5, two absorption spectra are shown whose transition centers are equal, as are their respective values $\varepsilon_{max}$, but whose bandwidth is different. Clearly, the band-integrated absorption coefficient $\boldsymbol{\mathcal{A}}$ is different.

Note that—as hinted at further above—in some cases the Beer–Lambert law is expressed in the form $I = I_0 \cdot e^{-kcl}$, where $k = \ln(10)\cdot\varepsilon \cong 2.3\cdot\varepsilon$. Normally the concentration $c$ is given in units of $mol\cdot l^{-1}$, but in certain cases, one uses the alternative notation of $C$ in units of $molecules\cdot cm^{-3}$. In this latter case, the Beer–Lambert equation is written in the form $I = I_0 \cdot e^{-\sigma Cl}$. Here, $\sigma$ is the absorption cross section whose units are $cm^2\cdot molecule^{-1}$ (often abbreviated as $cm^2$ only). The relationship between $\sigma$ and $\varepsilon$ is given by

$$\sigma = \left(10^3/N_A\right) \cdot 2.3 \cdot \varepsilon = 3.82236 \cdot 10^{-21}\varepsilon,$$

where $N_A$ stands for the Avogadro's constant. In Chapter 9, line profiles and lifetimes of excited atoms or molecules are addressed in more detail.

In general, the absorption spectra of organic compounds in nonpolar solvents are rather broad and often featureless mainly due to the large number of vibrational

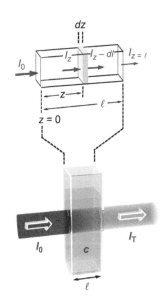

**Figure 2.4** Attenuation of light passing through a cuvette containing an absorbing solution, and the principle geometry to derive the Beer–Lambert law.

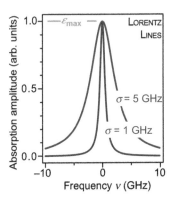

**Figure 2.5** Absorption line spectra for a relative transition center ($\nu = 0$ GHz), with the same maximum value $\varepsilon_{max}$ but with different widths, $\sigma$, yielding different integrated absorption coefficients, $A$.

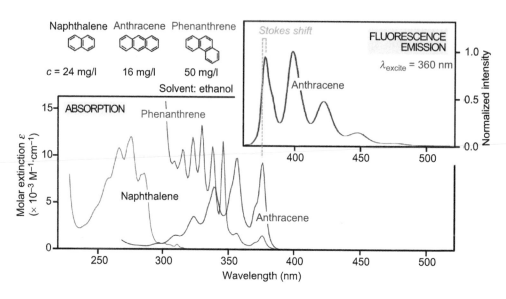

**Figure 2.6**  Absorption spectra of the naphthalene, anthracene, and phenanthrene diluted in ethanol; notice the structure due to vibronic bands. On the right, the fluorescence spectrum is shown for anthracene, excited near the absorption maximum at ~360 nm; note that the emission resembles a mirror-like image of the absorption spectrum.

levels but also due to the "blurring" effect associated with the interaction of the organic molecules with the solvent. On the other hand, in some favorable cases, clear structures are observed, which manifest vibronic transition features and allow for the identification of species. An example is shown in Figure 2.6 for a number of selected aromatic compounds, diluted in ethanol. In addition, the figure demonstrates the chromophoric shift associated with molecules comprising a multitude of similar building blocks (here benzene rings) or varying steric configurations.

## 2.2 FLUORESCENCE AND PHOSPHORESCENCE

Once an (organic) molecule is excited after absorption of a photon, different relaxation processes may take place. These excited-state photophysical processes are classified into radiative and nonradiative transitions. The two radiative processes are fluorescence and phosphorescence, whose main features are as follows:

- *Fluorescence*: radiative transition between states of the same multiplicity that occurs typically from the lowest vibrational level of the lowest excited singlet state $S_1$

- *Phosphorescence*: spin-forbidden radiative transition between states of different multiplicity, normally from the lowest vibrational level ($v = 0$) of the lowest triplet state, $T_1$, back into $S_0$

Fluorescence is a faster process than phosphorescence; while typical timescales for fluorescence emission are in the picosecond to microsecond region, i.e., $10^{-12}$–$10^{-6}$ s, phosphorescence emission is of the order of $10^{-3}$–$10^2$ s. This means that spontaneous fluorescent emission terminates right after the exciting

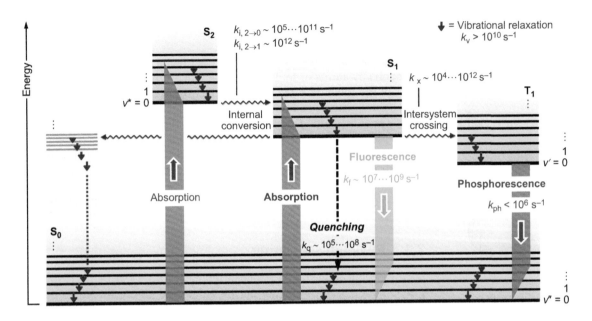

**Figure 2.7** Simplified Jablonsky diagram, showing the ground ($S_0$) and the excited ($S_1$ and $S_2$) singlet electronic states, and the first triplet state ($T_1$) of an organic molecule. The radiative processes absorption, fluorescence, and phosphorescence, and the nonradiative processes vibrational relaxation, internal conversion, intersystem crossing, and quenching, are indicated; typical rate constants for the various processes are included.

radiation is off. In contrast, for phosphorescence, the emission may last for up to many seconds after excitation ceased. These significantly different timescales suggest the existence of a different number of photophysical steps leading to both radiative processes.

Figure 2.7 illustrates a typical so-called Jablonsky diagram with the specific sequences of steps, and their timescales, leading to fluorescence and phosphorescence. Indeed, while both processes have in common the initial photon absorption from the ground to the excited electronic state, they markedly differ in the subsequent processes. For fluorescence, the radiative process is a transition from the first excited singlet state (for organic molecules normally designated as $S_1$) to all vibrational states of the initial ground state, $S_0$. For phosphorescence, first an intersystem crossing has to take place from $S_1$ to the first triplet state, $T_1$:

$$S_1 \xrightarrow{k_i} T_1.$$

Only when the triplet state is populated and after the (radiationless) vibrational relaxation reaches the lowest vibrational level in the triplet system $T_1(v = 0)$ does the slow, spin-forbidden phosphorescence (radiative) process commence.

Further above the absorption spectrum of anthracene was shown, together with a related fluorescence spectrum, both exhibiting evident rovibronic structure (see Figure 2.6). The "red shift" (Stokes shift) can be understood from the schematic energy level and transition structure shown here in Figure 2.7. The appearance and details of the observed mirror-like structure between absorption and emission will be discussed later in Chapter 9.

## 2.3 LIGHT SCATTERING

Light scattering is defined as the process in which light is redirected in different directions by atoms or molecules. When an atom is exposed to light, the electric fields of the incident wave force the electron to oscillate back and forth from its initial equilibrium positions. According to electromagnetic (EM) theory, when a charge is accelerated (in an oscillating field), it emits radiation in all directions in the plane perpendicular to the electron oscillations. In other words, the incident electric field induces a dipole moment in the particle that radiates light in all directions. This re-radiated light is called *scattered* radiation. In principle, this scattered radiation can be elastic (at the same) or inelastic (at different) frequency than that of incident radiation. Elastic scattering is also known as *Rayleigh scattering*; the underlying principles are briefly outlined next.

### 2.3.1 Rayleigh scattering

Lord Rayleigh published a paper in the early 1870s (Strutt 1871) describing the phenomenon of light scattering; he calculated the scattered intensity $I_s$ of a dipole whose dimension is significantly smaller than the incident wavelength (Cox et al. 2002). This scattered intensity is given by

$$I_s = I_0 \cdot \left(8\pi^4 \alpha^2 / R^2 \lambda^4\right) \cdot \left(1 + \cos^2 \theta\right); \tag{2.8}$$

where $R$ is the distance to the observed, $\alpha$ is the molecular polarizability, $\lambda$ is the wavelength, $I_0$ is the incident light intensity, and $\theta$ is the scattering angle. For $N$ dipole scatterers, the total scattered intensity is just $N$ times that of a single dipole scatterer. It should be noted that this formula was deduced for unpolarized incident light. From Equation 2.8, one finds that the intensity of the scattered light depends on the scattering angle being equal at $\theta = 0°$ and $\theta = 180°$, and that at $\theta = 90°$ is half of the forward/backward maximum values; this is also illustrated in Figure 2.8. For a derivation of Equation 2.8, see Bohren et al. (1983).

When the light is polarized (in a plane), the angular distribution of the scattered light is different; one finds

$$I_{\parallel} = I_0 \cdot \left(16\pi^4 \alpha^2 / R^2 \lambda^4\right) \cdot \left(\cos^2 \theta\right) \ \text{and} \ I_{\perp} = I_0 \cdot \left(16\pi^4 \alpha^2 / R^2 \lambda^4\right)$$

for incident light polarized parallel and perpendicular to the scattering plane, respectively. It should be noted that the scattering signal for unpolarized light $I_s$ is given by

$$I_s = \left(I_{\parallel} + I_{\perp}\right)/2.$$

It is also interesting to note that the scattering intensity depends roughly on $1/\lambda^4$. This means that the shorter the wavelength, the higher the scattering intensity; in other words, blue light scatters more than red light.

As shown in Figure 2.9, for high values of the scattering angle $\theta$ (i.e., looking upward), one observes that the sky appears to be blue since the blue rays are scattered toward the observer. On the other hand, at sunset and looking directly at the sun, blue light is mostly scattered out of the narrow range of scattering angle samples by the observer, and only the low-$\theta$ scattered light remains—the sky appears to be red.

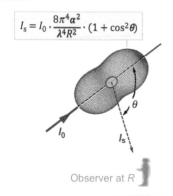

$$I_s = I_0 \cdot \frac{8\pi^4 \alpha^2}{\lambda^4 R^2} \cdot (1 + \cos^2\theta)$$

$I_0$   $I_s$   $\theta$

Observer at $R$

**Figure 2.8** Angular distribution of Rayleigh scattering intensity for unpolarized light; see text for mathematical details.

Blue sky

Red sunset

Short wavelengths
Scatter stronger
$\propto 1/\lambda^4$

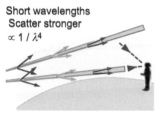

**Figure 2.9** Schematic illustration showing the enhanced blue scattering responsible for the sky's blue color and that of red scattering predominant at sunset; see text for comments.

For small particle sizes, which for the sake of simplicity are assumed to be spherical, the Rayleigh scattering intensity is often expressed as a function of the particle's diameter, $d$, and its refractive index, $n$. Thus, the scattering intensity adopts the form

$$I_s = I_0 \cdot \left(\frac{2\pi}{\lambda}\right)^4 \cdot \frac{(1 + \cos^2\theta)}{2} \cdot \left(\frac{(n^2 - 1)}{(n^2 + 1)}\right)^2 \cdot \left(\frac{d}{2}\right)^6; \qquad (2.9)$$

this equation gives the light intensity scattered in the fractional direction between $\theta$ and $\theta + d\theta$. Equation 2.9 can be rewritten in terms of the so-called differential scattering cross section $\sigma' = d\sigma_s(\theta)/d\Omega$, where $d\Omega$ is the differential element of solid angle. This leads to the expression

$$I_s = I_0 \cdot \left(\sigma'/R^2\right) \text{ with } \sigma' = I_0 \cdot \left(\frac{2\pi}{\lambda}\right)^4 \cdot \frac{(1 + \cos^2\theta)}{2} \cdot \left(\frac{(n^2 - 1)}{(n^2 + 1)}\right)^2 \cdot \left(\frac{d}{2}\right)^6.$$

If one wishes to obtain the total scattered intensity, one has to integrate $I_s$ over the full range of scattering directions. To do so, one should recall that the area $dA$ subtended by the solid angle $d\Omega$ is given by $dA = R^2 d\Omega = R^2 \sin\theta d\theta \cdot d\varphi$. Thus, the total scattering cross section $\sigma_s$ becomes the integral

$$\sigma_s = \int_A (I_s/I_0)dA.$$

After the integration, one obtains

$$\sigma_s = \left(\frac{2\pi^5 d^6}{3\,\lambda^4}\right) \cdot \left(\frac{(n^2 - 1)}{(n^2 + 1)}\right)^2. \qquad (2.10)$$

Thus, the integrated irradiance over the whole scattering area, that is, *the rate of the total incident energy scattered by the particle* $P_s$ (measured in W), is given by

$$P_s\,(\text{W}) = I_0(\text{W m}^{-2}) \cdot \sigma_s(\text{m}^2).$$

Frequently, the size of the scattering particles is parametrized by the dimensionless ratio $x = 2\pi r/\lambda$, with $r$ being the particle radius. Thus, Rayleigh scattering applies when $x \ll 1$, which typically is the case for particle sizes $r < \lambda/10$ and $m \cdot x < 1$, where $m = n_p/n_m$, i.e., the ratio of the refractive index of the particle to that of the scattering medium.

It is interesting to note that the scattering cross section depends on the sixth power of the particle size (see Equation 2.10). This explains, for example, why light traversing through pure water hardly experiences any scattering. On the other hand, when light passes through a colloidal suspension, such as, e.g., milk or fog, substantial scattering occurs. The phenomenon was discovered by John Tyndall in the nineteenth century and is now widely known as the Tyndall effect. In fact, measuring the Tyndall effect is a direct way to determine whether a mixture is colloidal or not, i.e., if it is a mixture in which at least one of its dimensions lies within the range of 1–1000 nm.

In principle, there is no rigorous mathematical theory to estimate the Tyndall scattering, but if the colloidal particles are spheroid, the scattering can be described by the Mie theory, outlined next.

Small particle MIE SCATTERING

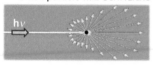

Large particle MIE SCATTERING

**Figure 2.10** Comparison between the angular distribution of the Rayleigh (top) and Mie scattering from small (middle) and large (bottom) particles. Note the enhanced forward scattering in the Mie distribution.

## 2.3.2 Mie scattering

For particle size such that $x \approx 1$, Rayleigh scattering is no longer valid, and the most used treatment is due to Gustav Mie (Mie 1908). This type of scattering takes place mainly in the forward direction, as illustrated in Figure 2.10, and typically it is not so wavelength-dependent as Rayleigh scattering (Caruthers 2011).

Basically, the Mie theory of scattering describes the interaction of EM waves with spherical particles. The mathematical description is rather more complicated than that of Rayleigh since the total scattering cross section is expressed as a function of series of spherical Bessel functions of $x$ and $mx$, the so-called Mie coefficients. The actual Mie scattering cross sections are often further classified as the cross sections for (1) total scattering $\sigma_s$, (2) extinction $\sigma_e$, (3) back scattering $\sigma_b$, and (4) absorption $\sigma_a$. All depend on the scattering angle, the particle size, and various optical parameters of the scattering medium (see Mie 1908 or, more accessibly, Kolwas 2010); the relevant expressions are listed in Table 2.1.

In practical terms, rather than describing or measuring the process by its cross sections, the results for Mie scattering—normally only calculated or measured in forward or backward direction rather than the full angular pattern distribution—are often displayed in dimensionless quantities. Not only does one use the dimensionless size parameter $x = 2\pi r/\lambda$ (as defined further above) but also the scattering effect itself is expressed dimensionless in the form of the so-called scattering efficiencies, $Q_i = \sigma_i/\pi r^2$, i.e., the scattering cross section is normalized to the geometric cross section of the particle.

In order to avoid the full mathematical framework of unwieldy Bessel functions and summation over the (complex) Mie coefficients, in a first—but often already quite accurate—approximation, the scattering particles are treated as ideal transparent spheres without absorption, i.e., the refractive index only constitutes its real part. This is the so-called *anomalous diffraction approximation* (see, e.g., van de Hulst 2003), which is given by

$$Q(x) = 2 - (4/p) \cdot \sin p + \left(4/p^2\right) \cdot (1 - \cos p), \qquad (2.11)$$

with the size parameter $x$ appearing in the form of the variable

$$p = 2 \cdot \left(n_{\text{particle}}/n_{\text{air}} - 1\right) \cdot x;$$

the physical meaning of $p$ is simply associated with the phase delay of the wave passing through the center of the sphere.

Without going into too fine details, simply inspecting the functional behavior of Equation 2.11, one finds the following:

- $Q$ has its largest value when the particle size is of the order of the wavelength of the scattering light.

- For $x \to \infty$, the scattering efficiency tends to $Q(x) \to 2$.

- For $x \ll 1$, one has $Q(x) \propto x^4 \propto \lambda^{-4}$, i.e., one reaches the Rayleigh regime.

- The oscillations associated with the sin/cos functions depend on the size of the refractive index; the larger the value of $n$, the faster the oscillation.

Note that physically the oscillations stem from the interference of the transmitted and diffracted waves. Simplistically, the two interfering contributions are due to

**Table 2.1**  Mie Cross-Section Formulas with $a_j$ and $b_j$, the so-called Mie Coefficients Representing the Magnetic and Electric Multipoles of Order $j$, respectively

| Process | Formula for Cross Section |
| --- | --- |
| Scattering | $\sigma_s = \dfrac{2\pi r^2}{x^2} \cdot \displaystyle\sum_{j=1}^{\infty}(2j+1)\cdot(|a_j|^2 + |b_j|^2)$ |
| Extinction | $\sigma_e = \dfrac{2\pi r^2}{x^2} \cdot \displaystyle\sum_{j=1}^{\infty}(2j+1)\cdot(-Re(a_j + b_j))$ |
| Backscattering | $\sigma_b = \dfrac{\pi r^2}{x^2} \cdot \left|\displaystyle\sum_{j=1}^{\infty}(2j+1)\cdot(-1)^j\cdot(a_j - b_j)\right|$ |
| Absorption | $\sigma_a = \sigma_e - \sigma_s$ |

specular reflection from the sphere on the one hand and to a creeping wave that skirts the shadowed side on the other hand. As the size parameter increases, the two contributions go in and out of phase leading to the interference pattern.

However, it should be kept in mind that Equation 2.11 constitutes an oversimplification since only the real part of the refractive index is taken into account. In reality, one has to take into account the complex refractive index $\tilde{n} = n + i\kappa$, where the complex part is associated with absorption through the absorption coefficient $\alpha = 4\pi\kappa/\lambda$. In that case, one will find that with increasing absorption, i.e., larger $\kappa$, the amplitude of the oscillation diminishes. Furthermore, for $x \to \infty$, the scattering efficiency now tends to $Q(x) \to 1$.

In Figure 2.11, the scattering (extinction) efficiency $Q_e$ is plotted as a function of the size parameter $x$ for three transparent spherical particles of different refractive index within the practically most important (aerosol) range $n = 1.33...2.0$, namely for $n \sim 1.33$ (pure-water droplets, in scattering theory often addressed as "oceanic water"); for $n \sim 1.5$ (dust-like aerosols, with a refractive index close to that of, e.g., BK7 glass, which is often used for modeling); and for $n \sim 2.0$ (soot-type aerosols, with a refractive index close to that of $Sc_2O_3$ often used for modeling). The scattering light is at wavelength $\lambda = 532$ nm. Note that for all three materials, a very small but nonzero absorption contribution has been included. Typical $n$- and $\kappa$-values for a range of frequently observed scattering particles can be found in, e.g., Levoni et al. (1997).

The various features mentioned above are clearly visible. In addition, one notices a ripple structure on top of the oscillatory functions; these are due to surface waves (resonance effects) for the scattering particles.

On the right-hand part of the figure, the extinction efficiency for water droplets is plotted on a log–log scale, which nicely allows one to "classify" the scattering into the normally utilized regimes of Rayleigh (reaching the $\lambda^{-4}$ regime, with slope 4 in the plot), Mie (oscillatory behavior), and geometric (asymptotic "constant") scattering. It should be noted that the transition between the three regimes is somewhat arbitrary, depending on which actual numerical value one chooses to mean $x \ll 1$ or $x \gg 1$.

In Chapter 20, light scattering through diffuse media, and in particular imaging techniques, will be studied to investigate the optical properties of human and biological tissues, mostly for medical image and diagnosis (see, e.g., Yodh and

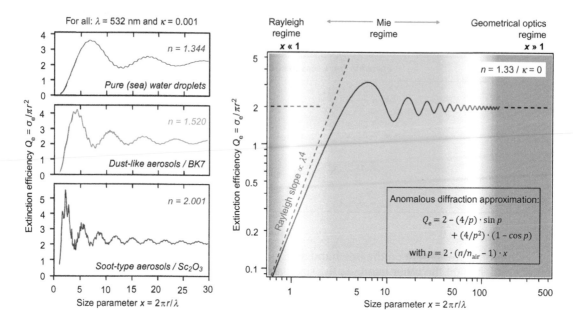

**Figure 2.11**  Extinction (scattering) efficiency $Q_s \cong Q_e$ for transparent spherical particles, as a function of size (normalized size parameter $x = 2\pi r/\lambda$); the incident light is at $\lambda = 532$ nm. (Left) $Q_e$ plots for water droplets, dust-like aerosols (refractive index similar to BK7 glass) and soot-type aerosols (refractive index similar to *Scandia*, $Sc_2O_3$). (Right) Log–log $Q_e$ plot for ideal (no absorption) water droplets; the calculation is based on the anomalous diffraction approximation (see, e.g., van de Hulst 2003). The different scattering regimes are indicated; for further details, see text.

Chance 1995). There the most recent techniques for diffuse optical imaging will be described using the most advanced theoretical treatments, which, however, in one form or another, are all based on fundamental Mie scattering.

### 2.3.3 Reflection and refraction

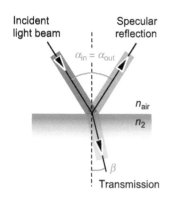

**Figure 2.12**  Illustration of incident, reflected, and refracted light rays between two media of different refractive index (for example, air and water). The respective propagation vectors and angles are indicated.

Two important types of light interaction with matter directly related to light scattering are reflection and refraction. Figure 2.12 shows a typical illustration of these phenomena in which a beam of light is reflected and refracted at the interface between two media as, for example, air and water, with refractive indices $n_{air}$ and $n_2$, respectively. The three (directional) beams of light—incident, reflected, and transmitted (refracted)—have wave vectors $k_i$, $k_r$, and $k_t$, respectively, and are shown together with their respective angles ($\alpha_{in} = \alpha_{out}$ and $\beta$) as well as the unit vector $\hat{n}$ normal to the interface located in the x–y plane.

Within the picture of matter–light wave interaction, one should recall that the incident light induces dipole oscillations that then reemit light at the same frequency as the incident one. The reemitted light may interfere with the incident light constructively or destructively, depending on their phase difference: if the interference is destructive, the light is attenuated and absorption takes place; if the interference is constructive, reflection or refraction occurs.

Optics theory for electromagnetic (light) waves tells us that the phases of the three waves (incident, reflected, and refracted) must match at the interface, a condition that leads to the so-called Snell's formulas for reflection and refraction at the said interface:

$$n_1 \cdot \sin|\alpha_{\text{in}}| = n_1 \cdot \sin|\alpha_{\text{out}}| \text{ and } n_1 \cdot \sin|\alpha_{\text{in}}| = n_2 \cdot \sin|\beta|, \qquad (2.12)$$

where the respective angles are indicated in Figure 2.12; in the specific case shown there, $n_1 \equiv n_{\text{air}}$. The reflection case in Equation 2.12 demands that the condition $|\alpha_{\text{in}}| = |\alpha_{\text{out}}|$ holds; this is called *specular reflection* and normally occurs when a light beam impinges in a smooth (flat) surface, like a mirror. Should the surface be not smooth any longer but exhibit irregularities that are of the order or larger than the light wavelength, then an extended beam experiences reflected beam rays into different directions, depending on the local surface normal for individual beam segments. Overall, the reflected rays add up to ray distribution patterns as illustrated in Figure 2.13. The result is an effect known as *diffuse reflection*, which gradually deteriorates from a semispecular distribution (top) into a near-$\cos^2$ (bottom) intensity distribution the "rougher" the surface becomes. Note that the majority of objects that do not emit light by themselves are seen because of diffuse refection.

At this point, one remark on the propagation of EM waves in media should be added. The index of refraction of a given medium is equal to the ratio between the speed of light in vacuum, $c$, and the medium, $c_m$, that is, $n = c/c_m$. Here one should bear in mind that the frequency of the light (wave) does not change when it passes from one medium to another; thus, it is the speed of light and consequently the wavelength that changes. In this context, one therefore finds that $v \cdot \lambda = c \to v \cdot \lambda' = n \cdot c_m \to \lambda' = \lambda/n$.

We wish to conclude this section on elastic light scattering with another remark on color that touches on a fundamental aspect of the relation between elastic scattering and reflection/refraction. Further above, it was described in a hand-waving way (although a rigorous derivation is possible as well) that the apparent blue color of the sky is a consequence of Rayleigh scattering of light in the atmosphere. In the same context of observed phenomena, one may now pose another similar probing question, namely, what process is responsible for the effect that snow or a cloud appears to be white. In short, this effect can be explained by a detailed consideration of the refraction and reflection of light from interfaces of condensed media (both snow and water droplets in a cloud resemble microcondensates).

When light impinges on a surface of a liquid or solid, in the EM framework, the electrons of the atoms or molecules "vibrate" under the interaction of the light electric filed and reemit light waves, as already mentioned further above. In the "bulk" of the sample, there is no incoherent scattering as would occur in low-density gases. Instead, one finds that, due to the close arrangement of the atoms or molecules, the great majority of the reemitted light waves add up coherently to form the refracted (transmitted) wave.

This is not the case for the atoms or molecules that are located near the surface, which start to "see" the discontinuity of the air/medium interface. As a consequence, now the scattering is mostly incoherent except from a thin layer (about half of the wavelength of the light) closest to the surface. The back-radiation from the oscillating dipoles in this thin layer is not cancelled by interferences but coherently superimposes to become the reflected part of the wave an observer sees. The area $A_{\text{F1}}$ of the said layer—in general, known as the first Fresnel zone (which is of size $A_{\text{F1}} = \pi\lambda r$, provided that the distance to the observer is much larger than the radius $r$ of the said first Fresnel zone)—from which the reflected wave arrives in phase to a given location is also proportional to the wavelength.

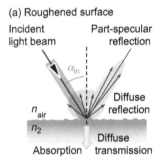

(a) Roughened surface

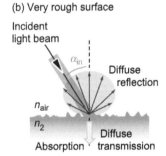

(b) Very rough surface

**Figure 2.13** Interaction of an incident light beam with surfaces of increasing roughness, increasingly morphing from (a) specular reflection to (b) full diffusive reflection.

Utilizing this knowledge, one can show (see, e.g., Weisskopf 1968) that the number of participating oscillators, $N$, which emit light coherently, is proportional to $\lambda^2$: one $\lambda$ dependence is related to the layer thickness, and the other one is linked to the surface area related to the Fresnel zone. It should be stressed that any light rays originating outside the Fresnel zone will interfere destructively with each other and will thus not contribute to the radiation amplitude at the observer location. This interpretation, put forward in the Weisskopf discourse, is shown schematically in Figure 2.14.

From quantum mechanics, one also knows that the intensity of the light emitted by $N$ coherent oscillators is $N^2$ times that of an individual oscillator. This $\lambda^4$ dependence associated with the $N^2$ proportionality cancels out the inverse $1/\lambda^4$ dependence of the scattered (Rayleigh) light, as noted further above. As a result, the reflected light intensity shows no frequency dependence; this then means that clouds appear as white because the water droplets reflect the sunlight with no change in the wavelength. The same arguments hold for a layer of snow that is irradiated with white (sun) light.

## 2.4 LIGHT SCATTERING: INELASTIC PROCESSES

In Rayleigh and Mie scattering, only the direction and intensity of an incident light beam were affected but not the photon energy. In addition to these elastic processes, there are two inelastic processes—*Brillouin* and *Raman* scattering—in which, as the name inelastic suggests, energy is transferred from the light beam to the scatterer but without actual absorption of photons from the beam.

### 2.4.1 Brillouin scattering

From a quantum physics point of view, Brillouin scattering describes the scattering of photons from collective, large-scale vibrational modes in "bulk" solids and liquids; in the former case, the interaction is with vibrational quanta of lattice vibrations (or acoustic phonons), while in the latter case, the interaction is with elastic waves. The effect was first predicted by Léon Brillouin in 1922. It should be noted here that Brillouin scattering differs starkly from Raman scattering, which describes—as discussed further below—the inelastic interaction of photons with individual molecules and molecule-internal quantum levels.

Because the scattering is inelastic, the energy (or wave frequency, $\omega$) and the momentum (or wave vector, $\boldsymbol{k}$) of the incident photon are not conserved. The photon can either lose or gain energy by creating or absorbing the energy of a (acoustic) phonon. Energy and momentum conservation for the complete scattering process lead to the relations

$$\omega_{S,AS} = \omega_0 \pm \omega_q \text{ and } \boldsymbol{k}_{S,AS} = \boldsymbol{k}_0 \pm \boldsymbol{q}, \tag{2.13}$$

where the indices "0" and "S,AS" refer to the incoming and scattered photon, respectively, and "q" identifies the phonon quantum. The plus/minus refers to the loss (–) or gain (+) of a transferred phonon, and the associated processes are named Stokes (S) or anti-Stokes (AS).

According to relations 2.13, the frequency and path of the scattered photon differ from those of the incident photon. The observed Brillouin shift $\omega_q$ depends on the wavelength of the incident photon, $\lambda_0$; the refractive index $n$ of the scattering

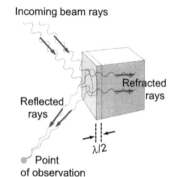

Incoming beam rays

Refracted rays

Reflected rays

$\lambda/2$

Point of observation

**Figure 2.14**  Reflection and refraction of light rays at the air/solid (liquid) interface, showing the onset of constructive interference generating the reflected wave. (Adapted from Weisskopf, V.F. *Sci. Am.* 219, no. 3 (1968): 60–71.)

material; the angle $\theta$ between incident and scattered light waves; and the phase velocity $\upsilon$ of the acoustic phonon wave:

$$\omega_q = \pm(4n\pi/\lambda_0) \cdot \upsilon \cdot \sin(\theta/2).$$

The observed shift frequencies are typically in the range 1–100 GHz.

Brillouin spectroscopy is mainly used for the measurement of certain (bulk) material properties, specifically elastic properties that describe temporary, normally reversible volume and shape changes that occur when stresses (extreme pressure and/or temperature) are applied to the sample. As such, while a valuable analytical tool, Brillouin spectroscopy falls slightly outside the framework of this book, and thus will not be discussed further.

### 2.4.2 Raman scattering

In Section 2.3, the interaction of light with matter, and how this interaction produces the elastic (Rayleigh and Mie) scattering from molecules, was discussed. Interestingly, it is possible that a small fraction of the incident photons scatter inelastically, i.e., some of the scattered photons emerge with different frequency than the incident ones—as in the case of Brillouin scattering but now associated with the energy transfer to molecules rather than phonon waves. This type of inelastic scattering is known as *Raman scattering*, named after C.V. Raman who discovered the phenomenon in 1928, and for which he was awarded the Nobel Prize in physics in 1930.

If the frequencies of the Raman scattered and the incident photons are $\nu_s$ and $\nu_0$, respectively, and if $E_i$ and $E_f$ are the energies of the molecule before and after the photon is scattered, then energy conservation implies that for the inelastic scattering part

$$h\nu_0 + E_i = h\nu_s + E_f \quad \text{or} \quad \Delta E = E_f - E_i = h(\nu_0 - \nu_s).$$

The energy difference $\Delta E$ is equal to the energy difference between the two molecular energies prior and after the scattering event; therefore, by measuring the so-called Raman shift $\Delta\nu = \nu_f - \nu_i$ (in units of frequency, or more often in units of $cm^{-1}$) gives direct access to molecular energy level differences, both for vibrational and rotational levels.

As illustrated in Figure 2.15, the Raman effect is a two-photon process and, like Rayleigh scattering, depends on the molecular polarizability. The spectral analysis of the Raman lines reveals the existence of lines with lesser energy than the Rayleigh line; these are called *Stokes lines*. When there is significant population of the vibrational states, Raman lines will also be observed whose frequency is higher than that of Rayleigh line; these lines are called *anti-Stokes lines*. It is clear from the ro-vibrational level structure shown schematically in the figure that not only vibrational transitions will contribute but that rotational substructures in the spectra will be observed as well. The latter gives rise to a band structure, governed by the selection rules associated with the two-photon Raman process (see standard text books covering Raman spectroscopy).

The availability of intense lasers sources has enabled the development of qualitative and quantitative Raman spectroscopy in many areas of physics and

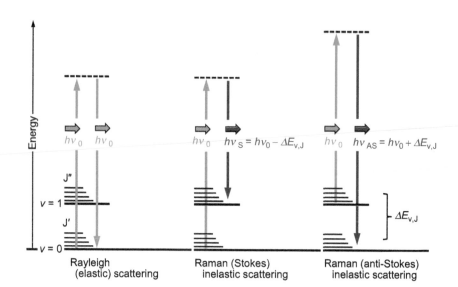

**Figure 2.15** Illustration of elastic (Rayleigh) and (inelastic) Raman scattering; both Stokes and anti-Stokes lines are indicated.

chemistry, in particular in the field of analytical chemistry. A detailed account of Raman spectroscopy and its applications is given in Chapters 11 through 14.

As will be shown in the Raman spectroscopy chapters, the efficiency of the Raman processes depicted in Figure 2.15 is normally very low, often only $10^{-6}$ or less. Without going into too much detail at this point, an attractive and now of increasingly wide use is a technique known as resonance Raman spectroscopy. Here the laser excitation frequency is selected to be in or near resonance with an electronic transition of the substance under study. It can be shown that this resonant effect significantly enhances the intensity of the vibrational modes (up to five to six orders of magnitude), whose motion are coupled with the electronic absorption, i.e., those vibrational modes localized in the molecular configuration involved by the electronic transition.

Said electronic-vibrational conditions are unique for many large (organic) molecules; thus, one observes a clear selectivity since the resonance only affects a few vibrational modes, simplifying the spectrum of, e.g., large biological molecules with normally a multitude of vibrational eigenfrequencies. Since the resonance is also only favoring the selected molecule, it also makes it easier to pick out the said molecule even in the presence of a vast amount of other molecular compounds. As an example of such improvement in Raman signal amplitude and species selectivity, in Figure 2.16, the Raman spectrum of a carrot sample is shown, which was excited by a green laser ($\lambda$ = 532 nm) under resonant conditions for the $\pi^* \leftarrow \pi$ transition of the pigment $\beta$-carotene. In contrast to the usual crowded spectrum obtained in ordinary Raman spectroscopy of biological samples, here the spectrum is very sparse: out of the many-compound vegetable matrix, only this pigment contributes significantly to the spectrum; moreover, the four main characteristic Raman bands of $\beta$-carotene dominate the spectrum (for further details, see Gonzálvez and Ureña 2012).

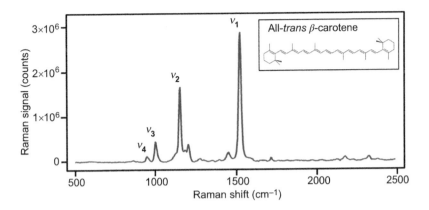

**Figure 2.16** Transmission resonance Raman (TRR) spectrum of $\beta$-carotene in a carrot slice (thickness = 400 µm); the most relevant vibrational bands ($v_1$ to $v_4$) are marked (for details, see Gonzálvez Ureña 2012).

## 2.5 BREAKTHROUGHS AND THE CUTTING EDGE

Our understanding of colors as being based on quantum effects is a modern, twentieth-century vision. Today "we know all about" molecular energy levels and photon transitions, and how these give rise to the different spectral colors. On the other hand, colors always have fascinated humans since the beginning of time. In that respect, the technology of how to make and exploit pigments probably belongs to the earliest developed by mankind: "At first man only saw the quality and variety of colors in nature, but later tried to make them himself" (Berke 2002).

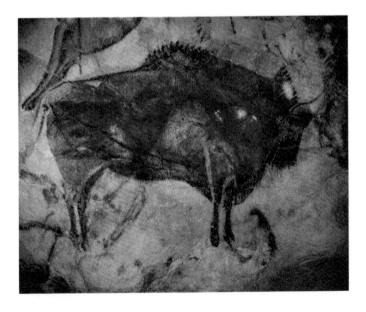

**Figure 2.17** Bison painting from Altamira Cave National Museum. (From Rammssos. "Reproduction of a bison of the cave of Altamira" (2008). http://commons.wikimedia.org/wiki/File:AltamiraBison.jpg, public domain.)

### 2.5.1 Breakthrough: Color in prehistoric times

The cave of Altamira in Cantabria, in the north of Spain, was the first cave in which colored prehistoric, Paleolithic cave paintings were discovered in 1879; the site has been declared a World Heritage Site by UNESCO in 1985. Recent investigations using uranium–thorium dating revealed that some of the earliest paintings go back even to 35,000 BC (Pike et al. 2012). In Figure 2.17, a bison is shown, which constitutes one of the most beautiful paintings of Altamira cave, dating from the *Lower Magdalenian* period (16,500–14,000 BC).

Those animal paintings, of which the one shown in the figure is only one example, demonstrate that prehistoric humans knew how to use different-colored earth or how to grind rocks to make powders of different color, an "empirical science" whose know-how continued through time and was exploited by the alchemists many millennia later. But even today, with all our knowledge and spectroscopic analysis based on quantum physics and chemistry, it still often remains an art to make and reproduce particular, subtle tones of color.

### 2.5.2 At the cutting edge: Single-photon spectroscopy of a single molecule

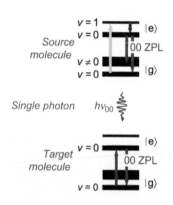

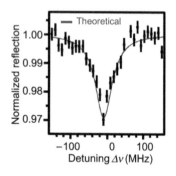

**Figure 2.18** (Top) Schematic of the free-space connection between two molecules via a single photon. (Bottom) Reflection of the single photon emitted by the source molecule from the target molecule, as a function of the frequency detuning between their mutual (zero-phonon) resonance; the solid red line represents the theoretical calculation. (Data adapted from Rezus, Y.L.A. et al., *Phys. Rev. Lett.* 108, no. 9 (2012).)

Nowadays, the interaction of light and matter at the limit of single photons and single emitters is of great relevance from a fundamental point of view, but also in relation with new quantum engineering applications, like, for example, the development of transistors or memory at the single-photon level.

Recently, work carried out by Rezus et al. (2012) demonstrated how a single photon can be created from a single (source) molecular emitter and, subsequently, the said single photon can be used to perform coherent spectroscopy on a second (target) emitter placed at a considerable distance. For this, the authors employed organic-dye molecules of dibenzanthanthrene (DBATT), imbedded in a tetradecane organic matrix at $T = 1.4$ K for both the source and the target samples. Light coupling between the source and the target substrate was achieved using a polarization-maintaining single-mode optical fiber.

Under these experimental connections, the researchers were able to produce photons, emitted by a selected source molecule in the transition from the $v' = 0$ level of the excited to the $v'' = 0$ level of the ground electronic states (the narrow zero-phonon line, 00-ZPL), as illustrated in the top of Figure 2.18, with a coherence length of several meters. The matrix isolation of both the source and target molecules, at the same low temperature, ensured that individual dye molecules could be excited selectively, in order to produce photons that matched the absorption frequency of a single, selected target molecule.

In order to perform the actual (transition line) spectroscopy on the target molecule, utilizing the stream of single photons emitted from the source molecule, the transition of either the source or the target molecule needed to be "tuned." In their experiment, the authors tuned the 00-ZPL frequency of the target molecule, exploiting Stark tuning by applying a voltage to microelectrodes attached to the target substrate. Essentially, the experiment measured the reflection of the single photon, emitted from the DBATT source molecule by another DBATT target molecule, as a function of frequency detuning with respect to the (zero-phonon) resonance transition. The main result obtained by the researchers is depicted in the bottom panel of Figure 2.18; a clear attenuation of about 3% was found, centered at the transition resonance frequency.

These findings not only illustrate the state of the art of spectroscopy of a single molecule using only single photons, but also reveal the strong coherent coupling of two molecules (emitter and target) separated at a distance smaller than the coherence length of the employed photon. Indeed, this constitutes a new type of molecular coupling that resembles the well-studied dipole–dipole coupling; this may pave the way for new quantum engineering applications such as, for example, the design and use of nanoguides.

# CHAPTER 3

# The Basics of Lasers

Since its invention in 1960, the laser has become one of the cornerstones in spectroscopy and imaging, primarily due to its phenomenal technological advances as a versatile and reliable light source for a plethora of applications. No wonder then that many often voluminous books have been written about lasers and the theory underlying their construction and operation. It is way beyond the scope of this book to cover the wealth of detailed knowledge accumulated over the years. But since lasers have become in many cases "black boxes," nevertheless some understanding of the laser system itself—including sometimes detailed insight into the underlying theory—is often essential if the user wishes to apply his or her laser system in the most efficient way.

So, rather than delving into the depths of laser theory, we wish to concentrate just on a few general aspects that are key to the understanding of a range of important issues in laser spectroscopy and imaging. Those who want or need to acquire deeper knowledge about laser theory and practical implementation are referred to standard textbooks (e.g., Siegman 1990; Silfvast 2008; Renk 2012; Eichhorn 2014, to name but a few) or to more specialized texts on the various laser effects or laser types (the most important of those will be referenced at the appropriate location).

With reference to Figure 3.1, there are three key aspects to lasers and laser radiation:

- The quantum processes and interactions that lead to the necessary population inversion between the upper and lower laser levels; in particular, pumping of the laser-active state and the internal transitions in the active medium are of the highest importance (some are undesirable and compete with the all-important laser transition itself).

- The actual laser cavity that converts the population inversion into useful laser radiation via gain (stimulated emission) on repeated passage through the laser-active medium; in addition to the actual mirror selection and cavity configuration, further optical elements in the cavity can be used to tailor the various properties of the laser output.

- The properties of the laser radiation propagating from the laser to the location of the experiment, including its wavelength, linewidth, polarization, pulse duration, beam profile, and light power/energy contained in the beam.

These will be discussed in the following sections. It should be noted that in the coverage we will not always follow the order in which the topics may appear in standard laser textbooks. We have opted for an approach that we thought to be

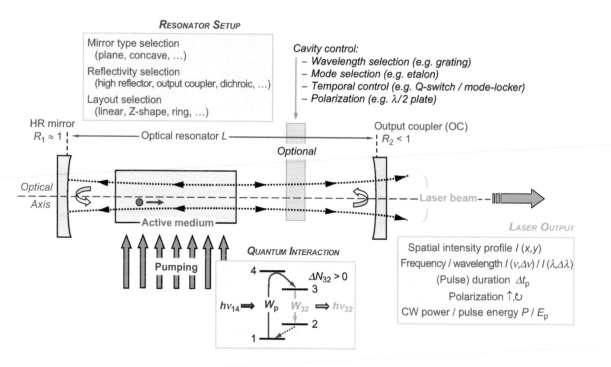

**Figure 3.1**    Conceptual realization of a laser; the key features, processes, and parameters are indicated.

the most convenient for linking fundamental aspects of laser theory and characterization with those of general relevance in laser spectroscopy and laser imaging.

# 3.1 FRAMEWORK FOR LASER ACTION

As has just been pointed out, while not being the center point of this book, some brief summary of the underlying physics describing the processes leading to and exploiting population inversion is certainly useful. Here we give a short overview of the rate equations describing the photon transitions between energy levels and their population underlying the four-level laser scheme; for the other schemes, the reader is referred to any of the standard laser textbooks. In addition to the quantum theoretical treatment, a few remarks are made about the key features of optical (laser) cavities, and how laser gain is achieved (treating both the steady-state and time-evolution scenarios).

## 3.1.1 Rate equations

The most common approach to describe the interplay of pump light photons with the population of the various energy levels involved in the overall scheme is the framework of coupled, time-dependent rate equations.

In standard textbooks, the discussion normally follows the sequence of two-, three-, and four-level systems. The underlying rationale behind this philosophy is to demonstrate that (1) in an isolated system with only two levels, no population inversion (required for laser action) can be realized; (2) the addition of a

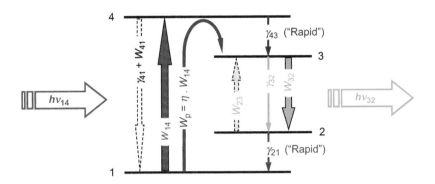

**Figure 3.2** Simplified four-level laser scheme; all relevant excitation and emission probabilities are indicated.

third level makes it possible to achieve population inversion, although with some difficulty; and (3) with four levels involved in the scheme, it becomes relatively easy to reach population inversion. For the description of these general scenarios, the reader may consult common textbooks covering basic laser theory in varying depth.

Here we restrict ourselves to a brief summary related to the four-level case, including various comments on laser-related issues beyond the actual level population rate equations.

A simplified term scheme related to four-level lasers is shown in Figure 3.2. With reference to the sketch, it should be noted that in very basic discussions of four-level laser schemes, even further simplifications are introduced. It also becomes clear that, in principle, two main radiation fields need to be considered, namely the pump radiation from the ground level 1 to the (intermediate) pump level 4, and the photon flux originating from the transition between the laser level 3 and level 2.

In quite a few texts, the stimulated transition probability is ignored when discussing only the issue of population inversion in the laser medium *per se*, rather than the generation of laser radiation. It also should be noted that, at times, the sequence of excitation from level 1 into level 4 followed by the relaxation into the upper laser level 3 is lumped together into an overall pumping rate $W_\mathrm{p}$, although strictly speaking, with "pump radiation," one actually means photons with energy $h\nu_{14}$.

The rate equations associated with the scheme depicted in Figure 3.2 are collected in Table 3.1. Note that in the notation of the table, rate constants $\gamma_{ij}$ are used rather than the Einstein coefficients $A_{ij}$ (both in units of $s^{-1}$). This allows one to include, in principle, also nonradiative relaxation processes. On the other hand, said nonradiative processes may lead to the violation of particle number conservation, i.e., the overall number of ground and excited particles, $N_{tot}$, does not remain constant, which is assumed in most basic theoretical descriptions, but which is encountered in a number of real laser systems.

In order to follow the dynamic evolution of the level populations, the four time-dependent coupled rate equation system needs to be solved as a function of time. Of course, this is straightforward in principle, since most mathematical software packages these days incorporate the relevant algorithms for this task. In fact, plotting the time-dependent evolution of the level populations may help in the understanding how to configure the resonator around the optical medium in order to optimize laser action.

**Table 3.1**   Simplified Rate Equations for a Pure Four-Level Laser System, under the Assumption of Particle Conservation

| | **Rate Equation** |
|---|---|
| Pump level | $dN_4/dt = -\gamma_4 N_4 - W_p \cdot (N_4 - N_1)$ with $\tau_4^{-1} \equiv \gamma_4 = \gamma_{41} + \gamma_{42} + \gamma_{43}$ and $W_p = B_{41} \cdot I_p$ |
| Upper laser level | $dN_3/dt = +\gamma_{43} N_4 - \gamma_3 N_3 - \boxed{W_L \cdot (N_3 - N_2)}$ with $\tau_3^{-1} \equiv \gamma_3 = \gamma_{31} + \gamma_{32}$ and $W_L = B_{32} \cdot I_L \equiv B_{32}^{ph} \cdot N_{Lph}$ |
| Lower laser level | $dN_2/dt = +\gamma_{42} N_4 + \gamma_{32} N_3 - \gamma_{21} N_2 + \boxed{W_L \cdot (N_3 - N_2)}$ with $\tau_2^{-1} \equiv \gamma_2 = \gamma_{21}$ |
| Ground level | $dN_1/dt = +\gamma_{41} N_4 + \gamma_{31} N_3 + \gamma_{21} N_2 + W_p \cdot (N_4 - N_1)$ |
| Laser photons | $\boxed{dN_{Lph}/dt = W_L W_3 - (N_{Lph} - \tau_c)}$ |
| Particle conservation | $N_{tot} = \sum\limits_{i=1-4} N_i$ |

*Note:*   If the active medium is positioned within a laser cavity, the highlighted terms containing the probability $W_L$ for laser photons leaving the cavity need to be included.

### 3.1.2 Population inversion in the steady-state limit

From a conceptual point of view, it is often intuitive to solve the equation system for the steady state, i.e., for all levels $i$, one has $dN_i/dt = 0$. This reduces the differential-equation system to an algebraic system, which can easily be solved (normally with a few further simplifying assumptions). The main additional assumption often is that the pump power flux $I_p$ does not deplete the ground level 1, i.e., at all times $N_1 \gg N_4$ and that therefore because of the in general low Boltzmann population of excited states at room temperature, one also has $N_1 \cong N_{tot}$. Under these circumstances, one finds, using the level 4 rate equation from Table 3.1, that in the steady state

$$(dN_4/dt) = 0 = -\gamma_4 N_4 - W_p \cdot (-N_{tot}) \Rightarrow N_4 \cong (W_p/\gamma_4) \cdot N_{tot}. \tag{3.1}$$

The knowledge about the population of level 4 can be introduced into the rate equations for laser levels 3 and 2. This simplifies the algebra significantly, and after some reshuffling of terms, one finds for the population inversion between the upper and lower laser levels

$$\Delta N = N_3 - N_2 = \frac{(1 - \gamma_{32}/\gamma_{21}) \cdot R_{p3} - (\gamma_3/\gamma_{21}) \cdot R_{p2}}{\gamma_2 + (1 + \gamma_{31}/\gamma_{21}) \cdot W_L}, \tag{3.2}$$

where $R_{pi}$ are the radiative population rates down from the pumped level 4, given by

$$R_{p3} = (\gamma_{43}/\gamma_3) \cdot W_p \quad \text{and} \quad R_{p2} = (\gamma_{42}/\gamma_3) \cdot W_p.$$

According to Equation 3.2, population inversion between the two laser levels ensues provided that (1) optical pumping into level 4 takes place and (2) the radiative relaxation from level 4 into level 3 (the upper laser level) is greater than that into level 2 (the lower laser level). And, of course, the inversion is reduced with an increase in stimulated emission on the laser transition (the second term in the denominator of Equation 3.2).

The population difference $\Delta N$ according to Equation 3.2 is plotted conceptually in Figure 3.3 as a function of the normalized pump rate $W_p/\gamma_{ul}$ (i.e., the ratio of population and depopulation rates of the upper laser level). It is clear from the arguments above and the trace in the figure that population inversion is reached

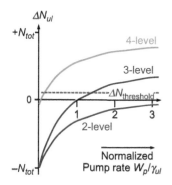

**Figure 3.3**   Population inversion between the upper and lower (laser) levels, $\Delta N_{ul}$, in dependence of the normalized pump rate, for two-, three-, and four-level systems.

as soon as the (normally extremely small) thermal population in the laser levels has been exceeded, thereafter the inversion increases rapidly. The rise is nonlinear because of the combination of the following: (1) an increase in $W_L$, which is functionally dependent on $N_3$, and (2) because for larger pump power flux, the assumption of nondepletion of level 1 does not hold any longer.

Similar treatments as the one outlined here for the four-level scheme can be applied to three- and two-level schemes; the formalism is not shown here, but the conceptual results for their population difference $\Delta N$ are also included in Figure 3.3. From the related traces, it becomes visually clear that the realization of lasers based on three-level schemes is normally much harder than those based on four-level schemes. No population inversion is achieved in the two-level system, irrespective of how hard the system is pumped.

### 3.1.3 Laser cavities

Above, the conceptual treatment of optical pumping and level populations in the laser-active medium was restricted to its intrinsic quantum properties and to photon transition rates. In order to utilize any of the population inversion between the prospective laser levels, feedback has to be applied to amplify, exploiting the stimulated emission process, some of the initial spontaneous photons to become the practicably useful laser light wave. For this, an optical cavity configuration has to be set up to resonate and amplify light waves within said cavity. Ultimately, a standing wave pattern is produced, which has to fulfill a range of boundary conditions in space and frequency; the associated frequency and intensity patterns are called modes and will be described in the following sections.

In the simplest case, an optical cavity consists of two mirrors facing each other. The most fundamental and frequently utilized configuration is that of two plane mirrors, which result in a so-called *plane-parallel* or *Fabry–Perot* cavity. While fundamentally simple (from the point of view of calculation), flat mirrors are rarely used because they are difficult to align to the required precision to support a standing wave pattern, unless the cavity length is very short (as a rule of thumb, $L < 1$ cm). Such an exception is found in semiconductor diode lasers, which possess an intrinsic, perfectly aligned plane-parallel cavity (see Chapter 5.1). When curved mirrors (both concave and convex) are incorporated into the setup, different resonator types ensue. These are distinguished by the focal lengths of the mirror(s), with radii of curvature $R_1$ and $R_2$, and the distance between them, the cavity length $L$.

However, whatever the cavity geometry, it is important that it is chosen so that the oscillating wave remains "stable," i.e., it supports a sufficient number of oscillations so that the light wave is amplified to useful power levels: only certain ranges of values for $R_1$, $R_2$, and $L$ result in stable resonators. Based on mathematical analysis of the resonator, by methods such as, e.g., ray transfer matrix analysis (see, e.g., Smith et al. 2007), it is possible to characterize the resonator with respect to the so-called *stability criterion*:

$$0 \leq \left(1 - \frac{L}{R_1}\right) \cdot \left(1 - \frac{L}{R_2}\right) = g_1 \cdot g_2 \leq 1. \tag{3.3}$$

A diagram including key information about cavity stability is shown in Figure 3.4, in which the stability parameters $g_1$ and $g_2$ are plotted against each other; the blue

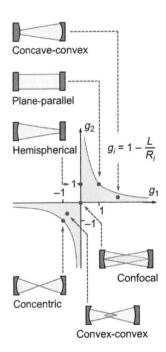

**Figure 3.4** Conditions for stable and unstable laser cavities, as a function of the geometrical factors $g_i = 1 - L/R_i$, with $L$ the spacing between the optical mirrors (cavity length) and $R_i$ the curvature radii of the mirrors. The stability regime $0 \leq g_1 \cdot g_2 \leq 1$ is indicated in blue. Some common cavity configurations are sketched.

area bounded by the hyperbolas $g_1 \cdot g_2 = 1$ indicates the stability regime for which Equation 3.3 is satisfied.

Any configuration which results in a stability parameter product <0 or >1 is unstable. Note that any cavity for which $g_1 = g_2$ is symmetric. Note also that cavities whose parameter values coincide with any of the boundary values of the stability region, i.e., the axes or the hyperbola lines, are only marginally stable. Any small variation in cavity length or misalignment may cause the resonator to become unstable. A few commonly known cavity types are sketched in the figure, indicated by their conceptual mirror configuration; these include, for example, plane-parallel (1,1), confocal (0,0), concentric (–1,–1), and hemispherical (–1,0); in brackets, the $(g_1, g_2)$ combinations are given.

The treatment of laser cavities described here is rather simplistic; the formalism was developed under the assumption that the cavity is empty. For a laser cavity, this is clearly incorrect: as a minimum additional entity, the cavity contains the gain medium. As a consequence, the value of $L$ is not the physical separation of the mirrors, but the *optical* path length between the mirrors, i.e., the effective lengths governed by the refractive indices of the media have to be taken into account.

Many modern, commercial laser systems comprise more mirrors than the two-mirror cavity just discussed; numerous three- and four-mirror configurations are commonly encountered, including V-type and ring cavities for the former and Z-shaped cavities for the latter. Often a pair of curved mirrors is arranged to form a confocal section, with the laser-active medium at the focus location; the remainder of the cavity is set to be quasi-collimated using plane mirrors (the shape of the beam depends on the resonator type). Specific cavity configurations are discussed further in conjunction with particular lasers in Chapter 4. In addition to the gain medium, often further elements are placed inside the cavity. In confocal sections of the cavity, these may be, for example, spatial filters for transverse mode control or modulators for Q-switching and mode-locking (for this see further below); other components—like filters, prisms, or diffraction gratings—in general, require that the beam passing through them is quasi-collimated.

### 3.1.4 Laser gain

For the discussion of laser gain, the conceptual laser cavity in Figure 3.5 is taken for reference; here the length of the optical medium, $L_m$, is shorter than the actual cavity length, $L$; the cavity is terminated by a high-reflective end mirror with reflectivity $RM_1 \cong 1$ and an output coupler with reflectivity $MR_2 < 1$—as is typical for common lasers. Note that here we use the notation MR for the mirror reflectivities, instead of R, to distinguish from the later-used mirror curvature radius, $R$. For simplicity, here we treat the problem in the steady-state regime. Then, for a circulating wave of frequency $\omega = 2\pi\nu$ to be maintained (i.e., not decreasing in amplitude after a round trip through the cavity), the amplitudes of the initial wave's EM field $U_i$ and the field $U_f$ after the round trip needs to satisfy the condition

$$U_f / U_i \equiv a_1 \cdot a_2 \cdot \exp\left(2g_m L_m - \frac{i2\omega L}{c}\right) = 1 \qquad (3.4)$$

In the equation, the coefficients $a_i$ (with $|a_i| \leq 1$) stand for the normalized fraction of the wave reflected from the two mirrors (these are related to the wave amplitudes $A(\mathbf{r})$—see Section 3.2—and the mirror reflectivites $MR_i = |a_i|^2$); exp $(2g_m L_m)$ stands for the round-trip amplification on passage through the gain

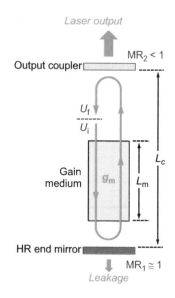

**Figure 3.5** Concept for a circulating wave to experience gain in a balance of amplification in passing through the gain medium, with gain coefficient $g_m$, and losses through the cavity mirrors, with reflectivities $MR_i$. For gain in a full round trip, $U_f \geq U_i$ is required.

medium of length $L_m$ and with gain coefficient $g_m$; and $\exp(-i2\omega L/c)$ represents the round-trip phase through the full cavity length $L$.

For the real part of Equation 3.4, i.e., the magnitude, one therefore has

$$a_1 \cdot a_2 \cdot \exp(2g_m L_m) = 1 \quad \text{or} \quad g_m = \frac{1}{4L_m} \cdot \ln\left(\frac{1}{MR_1 MR_2}\right), \qquad (3.5)$$

exploiting that $MR_i = |a_i|^2$. This is the minimum (net) gain required for the wave to be amplified and not to wither. It should be noted that Equations 3.4 and 3.5 are derived for the ideal case that the mirror transmissions are the only cavity losses. However, in real laser systems, additional losses are often encountered in the form of losses inside the cavity, for example, by inserted optical-control components or apertures, and losses in the active medium through scattering and absorption. Then the gain coefficient in Equation 3.4 has to include these and takes the form

$$g_m = \frac{1}{4L_m} \cdot \ln\left(\frac{1}{MR_1 MR_2(1-\alpha_1)(1-\alpha_2)}\right) + \alpha_m, \qquad (3.6)$$

where $\alpha_1$ and $\alpha_2$ stand for the cavity losses on the mirror-1 and mirror-2 sides of the gain medium, respectively, and $\alpha_m$ denotes the losses within the gain medium.

This gain coefficient is linked to the minimum, or threshold population inversion, $\Delta N_{th}$ to achieve laser action in a given system. One finds (see laser textbooks for a derivation)

$$\Delta N_{ul} = N_u - N_l \geq \Delta N_{th} \equiv \frac{\pi \, \Delta\omega_a}{\lambda^2 \gamma_{ul} L_m} \cdot \ln\left(\frac{1}{MR_1 MR_2}\right). \qquad (3.7)$$

Here $\lambda$ is the wavelength of the laser transition and $\Delta\omega_{ul}$ is its linewidth. Equation 3.7 constitutes the steady-state threshold condition for laser oscillation; it is clear from this equation that this threshold can be reached easiest (1) if the laser has a narrow transition linewidth and (2) if the external cavity losses through the mirrors are small, meaning basically that $MR_1, MR_2 \to 1$.

### 3.1.5 Cavity dynamics and the evolution of laser photons

Most of the above discussion on level populations and cavity behavior was based on the assumption that the steady state had been reached, i.e., no temporal change in any of the properties is seen any longer. While certainly steady-state conditions are desirable—in general, one wants stable, nonvarying laser output—the evolution toward the steady state is time-dependent, as the rate equations collected in Table 3.1 clearly indicate. In particular, temporal changes will be expected on switching on the laser, or for pulsed lasers. Here we provide a simplified view of the interplay between pumping of the upper laser level, evolution of its population, and the generation of laser photons.

Again, only the four-level system utilized throughout this section is considered. To start with, for the discussion below, it is assumed that the whole system is operated in the so-called small-signal gain regime. This means that (1) the population in the ground state is not depleted and thus is approximately constant and (2) the pumping level 4 and the lower laser level 2 are never populated significantly, i.e., $N_4 \approx N_2 \approx 0$. Then one can reduce the coupled equation system of the rate equations from Table 3.1 to just the equation for the upper laser level population and that for the laser photon generation.

With these two assumptions, and renaming levels 3 and 2 to upper and lower laser levels, respectively, said two rate equations can be written in terms of the population difference $\Delta N_{ul} \cong N_u$

$$dN_u/dt \cong \Delta N_{ul}/dt = P_u - \gamma_u N_u - K \cdot \Delta N_{ul} \cdot N_{Lph}$$
$$\cong P_u - (\gamma_u + K) \cdot N_u \cdot N_{Lph}$$

(3.8a)

$$dN_{Lph}/dt = K \cdot \Delta N_{ul} N_{Lph} + \gamma_{ul} N_u - (N_{Lph}/\tau_c)$$
$$\cong \left(K \cdot N_u - \tau_c^{-1}\right) \cdot N_{Lph} + \gamma_{ul} N_u .$$

(3.8b)

Here the rate parameter for laser photon generation is $K \cong V_a \cdot B_{32}^{ph}$, incorporating information about the laser mode volume and the Einstein coefficient for stimulated emission. Note the difference between $\gamma_{ul}$ and $\gamma_u$: the spontaneous radiation loss of the upper laser levels is to all lower levels, while the addition of spontaneously emitted photons $h\nu_{ul}$ to the number of stimulated laser photons only constitutes the fraction associated with the rate $\gamma_{ul}$. The term $P_u$ denotes the pumping into the upper laser level, which depends on the total number of particles available for the process and the power of the pump radiation; for the sake of simplicity, it is assumed to be constant here. The residence time for photons in the cavity is $\tau_c$; note that this can be approximated by

$$\tau_c \cong \frac{(2/c) \cdot (n_m L_m + n(L - L_m))}{2\alpha_i L - \ln(\mathrm{MR}_1 \mathrm{MR}_2)},$$

(3.9)

where $n_m$ and $n$ are the refractive indices of the medium and the rest of the cavity, respectively; and $\alpha_i$ stands for all cavity-internal losses.

This system of coupled, time-dependent differential Equation 3.8 is easy to solve (for example, using built-in capabilities in Mathematica™), integrating over time. The plot of data from a model system, based on parameters for a (diode-pumped) Yb:YAG laser, is shown in Figure 3.6a, together with measurement data from a pulsed DPSS Nd:YAG laser system in Figure 3.6b. For relevant data, see, e.g., Demirkhanyan (2006) and Barnes and Walsh (2003), respectively.

The figure clearly reveals the relaxation oscillation, which commences with a short delay after "switch on"; both the upper state level population and the number of generated laser photons exhibit this linked behavior. Here only one example combination of numerical values for the parameters $P_u$, $\gamma_u$, $\gamma_{ul}$, and $K$ is displayed; the choice of these parameters influences the buildup to population inversion (and hence the delay until laser action commences), the oscillation frequency, and the final steady-state limit (see, e.g., Daraei et al. 2013). A very intuitive, interactive visualization of the interplay between these parameters can be found, e.g., in a Mathematica™ demonstration project (Brosson 2005).

Of course, when waiting long enough, the rates described in Equations 3.8a and 3.8b will reach the steady state (provided the pumping is sufficient to support

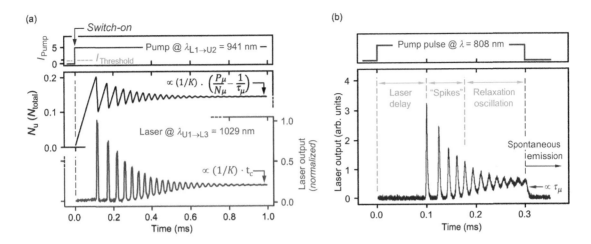

**Figure 3.6** Relaxation oscillations in free-running four-level laser systems. (a) Conceptual calculations for the upper laser level population and a selected, single laser transition of a Yb:YAG laser. (Energy levels and transition probabilities adapted from Demirkhanyan, G.G., *Laser Phys.* 16, no. 7 (2006): 1054–1057.) (b) Experimental measurement of the laser output relaxation oscillation of a pulse-pumped DPSS Nd:YAG laser. (Data adapted from Barnes, N.P. and B.M. Walsh. "Relaxation oscillation suppression in Nd: YAG lasers using intra-resonator harmonic generation". In *Conference on Lasers and Electro-Optics/Quantum Electronics and Laser Science Conference*, Technical Digest (Optical Society of America, 2003): paper CFM5.) For details see text.

both the population values). Thus, setting the left-hand sides in Equations 3.8a and 3.8b to zero, and after some algebra, one finds for the steady-state regime above threshold

$$N_u = (1/Kt_c) \cdot \left( \frac{N_{Lph}}{N_{Lph}+1} \right) \cong (1/Kt_c) \qquad (3.10a)$$

and

$$N_{Lph} = (1/K) \cdot \left( \frac{P_u}{N_u} - \frac{1}{\tau_u} \right). \qquad (3.10b)$$

Note that the upper level population to a good approximation only depends on the pump rate and the photon residence lifetime in the cavity; but it also should be kept in mind that all the simplified equations only hold in the small-signal gain regime, and assume that coupling to other levels can be neglected. For example, the conceptual simulation of population and laser output data were calculated for the pump transition $^2F_{7/2}(L_1) \rightarrow ^2F_{5/2}(U_2)$ at $\lambda = 941$ nm and the subsequent laser transition $^2F_{5/2}(U_1) \rightarrow ^2F_{7/2}(L_3)$ at $\lambda = 1029$ nm. In actual Yb-doped host materials, laser action is also observed for the neighboring $^2F_{5/2}$ $(U_1) \rightarrow ^2F_{7/2}(L_4,L_2)$ transition at $\lambda = 1047$ nm,1024 nm (for the level scheme, see Figure 3.7). These laser transitions compete for the population in the upper laser level $U_1$, leading to coupled-oscillation behavior (see, e.g., Pan et al. 2009, for the same laser transitions in an Yb-doped fiber laser).

Note also that after termination of the pump pulse, the laser output decays exponentially, since the upper level is longer replenished. This is clearly visible for the pulsed excitation of the DPSS Nd:YAG laser depicted in Figure 3.6b. The laser output data also show that, other than for the Yb-doped fiber laser, full steady state has not been reached yet at the end of the pump pulse.

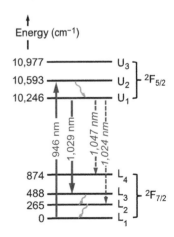

**Figure 3.7** Energy levels of Yb$^{3+}$ in a YAG crystal and relevant transitions for the model calculations in Figure 3.6.

## 3.2 LASER CAVITIES: SPATIAL FIELD DISTRIBUTIONS AND LASER BEAMS

The propagating electromagnetic wave of a laser beam, whether oscillating within the cavity or evolving outside away from the cavity, must satisfy the complex wave equation

$$\nabla^2 U - \left(\frac{1}{c^2}\right) \cdot \left(\frac{\partial^2 U}{\partial t^2}\right) = 0$$

The (complex amplitude) wave function is of the form

$$U(\boldsymbol{r}, t) = [A(\boldsymbol{r}) \cdot \exp(i\varphi(\boldsymbol{r}))] \cdot [\exp(i\omega t)] \equiv U(\boldsymbol{r}) \cdot U(t), \qquad (3.11)$$

where $\boldsymbol{r}$ incorporates the spatial dependence, $A(\boldsymbol{r})$ and $\varphi(\boldsymbol{r})$ are the amplitude and phase factors, respectively, and $\omega = 2\pi\nu$ is the oscillation frequency of the wave. As Equation 3.11 suggests, the overall wave function can be separated into its spatial and temporal dependent parts. For the $\boldsymbol{r}$-dependence, one arrives at the so-called paraxial wave equation; this can be written in either Cartesian or cylindrical coordinates, i.e., $U(x,y,z)$ or $U(r,\phi,z)$. The latter representation is only used when complete radial symmetry in the laser cavity is encountered. However, for the majority of practical lasers, it is rarely the case (note that the symmetry of the optical resonator is often restricted by optical components, such as, e.g., polarizing elements). Assuming that the laser radiation propagates in the $z$-direction, one finds for the two representations

$$\text{for Cartesian coordinates}: \left(\frac{\partial^2}{\partial x^2} + \frac{\partial^2}{\partial y^2} + 2ik\frac{\partial}{\partial z}\right) \times U(x, y, z) = 0; \qquad (3.12a)$$

$$\text{for cylindrical coordinates}: \left(\frac{1}{r}\frac{\partial}{\partial r}\left(r\frac{\partial}{\partial r}\right) + \frac{1}{r^2}\frac{\partial^2}{\partial \phi^2} + 2ik\frac{\partial}{\partial z}\right) \qquad (3.12b)$$
$$\times U(r, \phi, z) = 0.$$

### 3.2.1 Transverse mode structure

Solving these equations with the boundary conditions for the specific geometries of a laser cavity is, in principle, straightforward but requires sophisticated algebraic and numerical procedures. It was first successfully executed by Fox and Li (1961) for Fabry–Perot and confocal laser cavities; their approach is still widely used today. In particular, the solutions provide the lateral field distribution patterns of various orders, known as transverse electromagnetic modes (TEMs).

For Cartesian coordinate symmetry, these are designated $\text{TEM}_{mn}$ with $m$ and $n$ being the horizontal (normally $x$-direction) and vertical (normally $y$-direction) orders of the pattern, respectively, and for cylindrical symmetry, they are denoted $\text{TEM}_{pl}$, where $p$ and $l$ label the radial and angular mode orders, respectively.

The solutions for the electric field mode patterns $U$, the related intensity mode patterns $I = U^2$, and the definition of relevant parameters are collated in Table 3.2. Note that because of their functional structure, the modes described in Cartesian coordinate symmetry are often named Hermite–Gaussian (HG) modes ($\text{TEM}_{mn}^{\text{HG}}$ or $\text{HG}_{mn}$), while those in cylindrical coordinate symmetry are called

**Table 3.2** Solutions for the Paraxial Wave Equations in Cartesian (HG) and Cylindrical (LG) Representation

| | **Mathematical Expression** |
|---|---|
| **HG modes** | |
| Complex wave function | $U_{m,n}^{HG}(x, y, z) = C_{mn}^{HG} \cdot w^{-1} \cdot H_m(x\sqrt{2}/w) \cdot H_n(y\sqrt{2}/w)$ |
| | $\cdot \exp\left(-ik\dfrac{x^2 + y^2}{2R}\right) \cdot \exp(-i\Phi) \cdot \exp\left(-\dfrac{x^2 + y^2}{w^2}\right)$ |
| Normalization factor | $C_{mn}^{HG} = \sqrt{(2/\pi m! \, n!)} \cdot 2^{-(m+n)/2}$ |
| Hermite polynomial | $H_j(u) = (-1)^j \cdot \exp(u^2) \cdot \dfrac{d^j}{du^j}(\exp(-u^2))$ |
| | $\equiv \left(2u - \dfrac{d}{du}\right)^j \cdot 1$ |
| Guoy phase | $\Phi = (m + n + 1) \cdot \psi(z) = (m + n + 1) \cdot \tan^{-1}(z/z_R)$ |
| **LG-modes** | |
| Complex wave function | $U_{pl}^{LG}(r, \phi, z) = C_{mn}^{LG} \cdot w^{-1} \cdot L_{min(p,l)}^{|p-l|} \cdot (2r^2/w^2) \cdot \left(\dfrac{r\sqrt{2}}{w}\right)^{|p-l|}$ |
| | $\cdot \exp\left(-ik\dfrac{r^2}{2R}\right) \cdot \exp(-i\Phi) \cdot \exp(-i(p-l)\phi) \cdot \exp\left(-\dfrac{r^2}{w^2}\right)$ |
| Normalization factor | $C_{pl}^{LG} = \sqrt{(2/\pi p! \, l!)} \cdot min(p, l)!$ |
| Associated Laguerre polynomial | $L_n^m(u) = \left(\dfrac{1}{n!}\right) \cdot \left(\dfrac{\exp(u)}{u^m}\right) \cdot \dfrac{d^n}{du^n}(\exp(-u) \cdot u^{n+m})$ |
| | $\equiv \left(\dfrac{1}{n!}\right) \cdot \left(\dfrac{1}{u^m}\right) \cdot \left(\dfrac{d}{du} - 1\right)^n \cdot u^{n+m}$ |
| Gouy phase | $\Phi = (p + l + 1) \cdot \psi(z) = (p + l + 1) \cdot \tan^{-1}(z/z_R)$ |

$z$ = direction of wave propagation; $w$ = beam waist parameter; $R$ = mirror curvature.

Laguerre–Gaussian (LG) modes ($\text{TEM}_{pl}^{LG}$ or $LG_{pl}$). Some of the low-order mode patterns are visualized in Figure 3.8. Note that in Cartesian coordinate symmetry, intuitively the mode order numbers $m$ and $n$ refer to the number of amplitude minima in the direction of the electric field oscillation, $x$ or $y$, respectively. The LG modes exhibit an increasing number of concentric rings with increasing radial order $p$, while with increasing angular mode order $l$, additional angularly distributed lobes appear. Note that the $LG_{01*}$ mode, the so-called *doughnut mode*, is a special case comprising the superposition of two $LG_{0i}$ modes ($i = 1,2,3$), rotated $360°/4i$ with respect to each other.

It is noteworthy that besides the apparent difference in the lateral intensity pattern, there is a further distinction between the two mode types. In contrast to HG beams, LG beams exhibit rotational symmetry along their propagation axis and carry an intrinsic rotational orbital angular momentum of $|L_{ph}| = i\hbar$ per photon. Note that this intrinsic rotational momentum should not be confused with the angular momentum due to the polarization of light! The presence of this

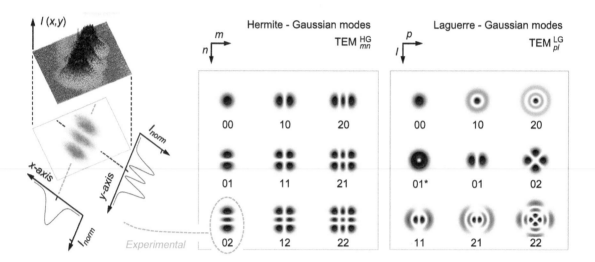

**Figure 3.8**   Theoretical TEM patterns for a selection of low-order transverse spatial HG (Cartesian coordinates) and LG (cylindrical coordinates) modes. A measured intensity distribution pattern for the $TEM_{02}$ HG mode and its evaluation is shown on the left.

rotational angular momentum gives rise to a torque on any refractive object placed in the beam, along the propagation axis. This feature is of considerable practical importance, e.g., in optical trapping, or transfer of this momentum to a particle thus making it spin; for such applications, great efforts are made to realize laser output with LG-mode structure.

## 3.2.2 Gaussian beams and their propagation

For all the beauty of mode patterns, in many practical applications a uniform intensity distribution across the beam is desirable, which in addition has an easy-to-handle mathematical description. The overall output intensity profile may constitute a superposition of any of the transverse modes allowed for a specific laser cavity, which would make the intensity distribution rather messy and in most cases undesirable. It is often straightforward to achieve that only the $TEM_{00}$ mode can oscillate. This is linked to the fact that the overall lateral extension of order modes is relatively larger in comparison to the fundamental mode; thus, placing an aperture of suitable diameter into the cavity will increase the losses of the higher modes and therefore suppress their oscillation.

Looking at the mode examples shown in Figure 3.8, both the lowest order HG and LG modes with $m = n = 0$ or $p = l = 0$ seem to meet the above requirement of simplicity; one finds that the intensity distribution for both is exactly equal and axis-symmetric. For this common $TEM_{00}$ beam with (monochromatic) wavelength $\lambda$, propagating in the $z$-direction, the complex (time-invariant) electric field amplitude—often called phasor—is given by (compare to the entry in Table 3.2)

$$U(\rho, z) = U_0 \cdot \frac{w_0}{w(z)} \cdot \exp\left(-\frac{\rho^2}{w(z)^2}\right) \cdot \exp\left(-i\left[kz - \tan^{-1}\left(\frac{z}{z_R}\right) + \frac{k\rho^2}{2R(z)}\right]\right), \quad (3.13)$$

where $\rho \equiv r \equiv (x^2 + y^2)^{1/2}$ is the lateral (axis-symmetric) distance parameter; $U_0$ and $w_0$ stand for the peak amplitude beam radius at the beam waist position $z = 0$; $k = 2\pi/\lambda$ is related to the wave propagation vector (along the $z$-direction); $z_R$ stands for the so-called Rayleigh length (for a definition, see below); and $R(z)$ describes the radius of curvature of the wave front. Note that in order to obtain the oscillating real electric field, one has to multiply phasor equation 3.13 with $\exp(i\omega t) = \exp(i2\pi ct/\lambda)$ and take the real part of the resulting product.

The related TEM$_{00}$ mode intensity distribution ($I = U^2$) can be written as

$$I(\rho, z) = I_0 \cdot \left(\frac{w_0}{w(z)}\right)^2 \cdot \exp\left(-\frac{2\rho^2}{w(z)^2}\right). \tag{3.14}$$

An example of the intensity distribution of a TEM$_{00}$ beam, measured relatively close to the laser aperture, is shown in Figure 3.9.

From Equations 3.13 and 3.14, three practically important parameters emerge, namely the beam radius $w(z)$, the wave front curvature $R(z)$, and the Rayleigh length $z_R$. The Rayleigh length is defined as

$$z_R = \pi w_0^2/\lambda; \tag{3.15a}$$

it determines the distance over which the beam propagates without diverging significantly. The beam radius varies along the propagation direction according to

$$w(z) = w_0 \cdot \left[1 + (z/z_R)^2\right]^{1/2}. \tag{3.15b}$$

And finally, the radius of curvature $R$ of the wave fronts evolves according to

$$R(z) = z \cdot \left[1 + (z_R/z)^2\right]. \tag{3.15c}$$

Note that the position $z = 0$ in the equation above corresponds to the so-called *beam waist focus*, where the beam radius is at its minimum, and the phase profile is flat. Note also that for a so-called *collimated beam*, i.e., a beam with (approximately) constant beam radius, the Rayleigh length has to be large in comparison to the envisaged propagation distance. Finally, at a distance from the waist equal to the Rayleigh range $z_R$, the width $w$ of the beam is $w(\pm z_R) = \sqrt{2} \cdot w_0$. The distance between these two points is often called the *confocal parameter* or *depth of focus* of the beam, $b = 2z_R = 2\pi w_0^2/\lambda$.

For very large distances from the origin ($z \gg z_R$), the parameter $w(z)$ increases linearly with $z$: the geometrical optics limit has been reached. The angle between the straight line $\rho = w(z)$ and the central (symmetry) axis of the beam ($\rho = 0$) is called the *divergence* of the beam; it is given by $\theta \cong \lambda/\pi w_0$ ($\theta$ in radian). Note that laser manufacturers often quote the "total" angular spread, $2\theta$, for the divergence but confusingly use the same symbol ($\theta$) for this. When tracing the propagation of a Gaussian beam away from the laser (and its original reference point $z = 0$), it is common to plot the beam width parameter $w(z)$ as a function of $z$. This is shown schematically in Figure 3.10, together with an indication of the aforementioned key parameters.

Referring back to the introductory remarks, laser power often is one of the key entities in spectroscopy and imaging. The power $P$ (in units of W) from a Gaussian TEM$_{00}$ beam passing through a circular aperture (of radius $\rho$), oriented vertically to the beam propagation direction position $z$, is given by

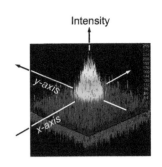

Intensity

**Figure 3.9** TEM$_{00}^{HG}$ beam intensity profile from a 488 nm Ar$^+$ laser, measured close to the laser output aperture. The laser beam was attenuated not to saturate the CCD webcam detector.

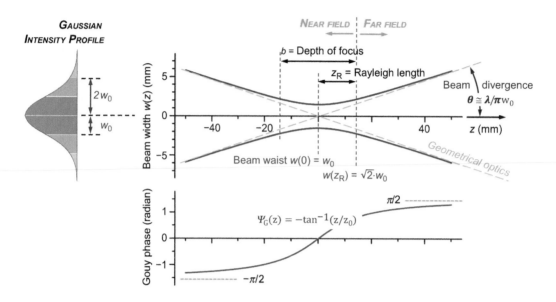

**Figure 3.10** Gaussian beam width $w(z)$ as a function of the distance $z$ along the beam, for beam waist $w_0 = 1.5$ mm and wavelength $\lambda = 532$ nm. In the bottom trace, the related Gouy phase is plotted.

$$P(\rho, z) = P_0 \cdot \left[1 - \exp\left(-\frac{2\rho^2}{w(z)^2}\right)\right], \text{ with } P_0 = \left(\frac{1}{2}\pi w_0^2\right) \cdot I_0. \tag{3.16}$$

For a radius $\rho = n \cdot w(z)$ with $n = 1, 2, 3$, the fractional transmitted power, $P(z)/P_0$, is equal to $[1 - e^{-2n}]$. These points are related to the $1/e$, $1/e^2$, and $1/e^3$ points in the intensity distribution, and one finds that 86.5%, 98.2%, and 99.8% of the total power are transmitted, respectively.

We like to conclude this section on Gaussian beam propagation with a remark on phase. In Table 3.2, the Gouy phase shift was mentioned, which for a TEM$_{00}$ beam is given as

$$\psi_G(z) = -\tan^{-1}(z/z_0). \tag{3.17}$$

Basically, the Gouy phase shift—which is named after C.R. Gouy who observed the effect in 1890—means that a Gaussian beam, when passing through a beam waist (focus), acquires a phase shift $\psi_G(z)$ which is in addition to the usual exp$(-ikz)$ phase shift inherent to a plane wave, with the same optical frequency. The Gouy phase shift behavior is schematically included in the bottom of Figure 3.10.

As such, Gouy phase shift is an inevitable consequence of beam focusing; its physical origin has been attributed to be due to the spread in transverse momentum in the focused laser beam (see, e.g., Feng and Winful 2001). To name but one example, its influence can clearly be seen experimentally as an alteration in phase matching conditions in nonlinear media, when comparing the plane-wave and optimally focused case. Essentially this means that care has to be taken when designing, e.g., second-harmonic generation in nonlinear optical crystals for which one has to rely on strong focusing of the input beam to achieve satisfactory conversion efficiencies (for nonlinear frequency conversion, see Chapter 5.3).

# 3.3 LASER CAVITIES: MODE FREQUENCIES, LINE SHAPES, AND SPECTRA

In Section 3.2, the complex wave equation and its related space- and time-dependent wave function $U(\mathbf{r},t) = U(\mathbf{r}) \cdot U(t)$ were introduced; for the $\mathbf{r}$-dependence, one arrived at the paraxial wave equation, which provided information about the spatial distribution of the electromagnetic field distribution, inside and outside of the laser cavity, and which resulted the transverse $TEM_{mn}$ mode structure description. The boundary conditions for a wave to be resonant along the propagation direction dictate that a "standing-wave" solution is only found for certain wave frequencies. This information is contained in the spatial phase shift expression (predominantly in the $z$-propagation direction) and is linked to the $\exp(i\omega t)$ part of the wave function.

## 3.3.1 Frequency mode structure

A wave (mode) resonates in the cavity if the EM-field phase is the same after a round trip inside the cavity, i.e., the total phase variation has to be equal to a multiple of $2\pi$. The phase term for a (Gaussian) wave is associated with the dual-exponential $\exp(ikz) \cdot \exp(\psi(z))$, as listed in Table 3.2. The first exponential term constitutes the phase shift due to (temporal) wave propagation, while the second term incorporates the Gouy phase contribution $\psi$, a specificity of Gaussian beams.

If $\phi(z)$ denotes the phase along the $z$-axis, one finds for the phase difference, at the locations of the two cavity mirrors at $z_1$ and $z_2$,

$$\Delta\phi = \phi(z_2) - \phi(z_1) = -k \cdot (z_2 - z_1) + |\psi(z_2) - \psi(z_1)| = -q\pi, \qquad (3.18)$$

in order to fulfil the round-trip standing-wave condition; $q$ is an integer equal to the number of half-wavelengths over the (longitudinal) cavity dimension $L \equiv z_2 - z_1 \equiv q \cdot (\lambda/2)$. This resonance condition described by Equation 3.18 needs to be taken into account for the characterization of the TEMs inside the cavity, and accordingly an additional descriptive index has to be added yielding $TEM_{mnq}$ rather than the $TEM_{nm}$ discussed in the previous section.

Using the well-known relation $k = 2\pi v/c$, one obtains the resonant frequencies of $TEM_{nmq}$ Gaussian modes in the cavity

$$v_{mnq} = \frac{c}{2L} \cdot \left[ q + \frac{1}{\pi} \cdot (m + n + 1) \cdot \left( \tan^{-1}(z_2/z_R) - \tan^{-1}(z_1/z_R) \right) \right], \qquad (3.19a)$$

which after some calculations yields

$$v_{mnq} = \frac{c}{2L} \cdot \left[ q + \frac{1}{\pi} \cdot (m + n + 1) \cdot \cos^{-1}\left( \pm\sqrt{g_1 g_2} \right) \right]. \qquad (3.19b)$$

Note that $g_i = 1 - L/R_i$ have the same sign because of the stability criterion (see Figure 3.4 and the related discussion above); if their signs are positive, one should take the + sign in Equation 3.19b, and vice versa. Note that these frequencies depend on the values of the mirror curvature radii, and therefore the numerical values of mode frequencies and their pattern are not necessarily easy to jot down.

For some intuitive visualization, some limiting cases may be considered, such as, e.g., an almost plane–plane mirror configuration, i.e., a near-Fabry–Perot cavity. For such a resonator cavity, one therefore has $R_1 = R_2 \ll L$, which implies that the $\cos^{-1}$ term in Equations 3.19a and 3.19b becomes $\cos^{-1}(g)$, and with $g \cong 1$, one obtains

$$\cos^{-1}(g) \approx \sqrt{2L/R} \ll 1 \tag{3.20}$$

With this approximation, one finds for the frequency spacing between neighboring "longitudinal" modes (with $\Delta q = 1$ and $m = n = $ constant)

$$\delta v_q = v_{mn(q+1)} - v_{mn(q)} = \frac{c}{2L}. \tag{3.21}$$

This is the well-known textbook relation found for (one-dimensional) resonators. For neighboring "transverse" frequencies with $\Delta m = 1$ or $\Delta n = 1$ (but $q = $ constant in both cases), one finds for the transverse mode spacing

$$\delta v_m = \frac{1}{\pi} \cdot \sqrt{\frac{2L}{R}} \cdot \frac{c}{2L} \text{ or } \delta v_n = \frac{1}{\pi} \cdot \sqrt{\frac{2L}{R}} \cdot \frac{c}{2L}. \tag{3.22}$$

Equation 3.22 tells us that for a Fabry–Perot type cavity, the spacing of transverse modes is substantially smaller than that for longitudinal modes. These longitudinal and transverse mode spacings are conceptually shown in Figure 3.11.

However, it should be kept in mind that the mode-spacing relations 3.21 and 3.22 are based on the crude approximation given in Equation 3.20. A rigorous treatment for a Fabry–Perot cavity can be found in the groundbreaking publication by Schawlow and Townes (1958), which precedes the first demonstration of the laser itself; many of the narratives and formalism in standard laser textbooks are based on this. It should be noted that some of the simplifying approximations used here to derive the mode frequencies and their spacing are slightly different from those used by Schawlow and Townes, who start with a fully fledged treatment of the full cavity rather than the Gouy phase treatment.

As a further note, we would like to add that the above relations were derived for an "empty" cavity, i.e., no refractive materials between the mirrors. Of course, this is not correct since at least the laser-active medium has to be included in the assessment. In the simplest case, the laser medium fills the complete cavity, like it is found for semiconductor laser diodes. Then Equation 3.21 needs to be modified for the refractive index of the medium and one finds

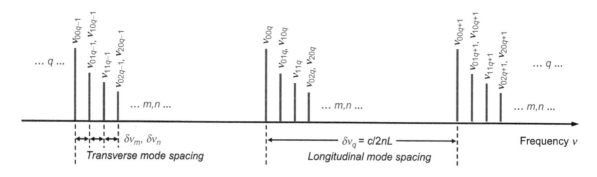

**Figure 3.11**    Longitudinal and transverse (frequency) mode spacing for a Fabry–Perot cavity, indexed $q$ and $m,n$, respectively.

$$\delta v_q = (c/2) \cdot (1/n_L L). \qquad (3.21b)$$

More generally, one finds that lasers comprise more than one optical element in the cavity, with different thicknesses and refractive indices. Then one needs to calculate the mode frequencies and their spacing according to

$$\delta v_q = (c/2) \cdot \left( 1/\sum_i n_i L_i \right), \qquad (3.21c)$$

where $n_i$ is the refractive index of the $i$th element of length $L_i$.

Finally, it is common to rewrite the expression for the mode frequencies, i.e., Equations 3.19a and 3.19b, in a slightly different way. In practical laser systems, the separation distance of the mirrors $L$ is, in general, significantly larger than the wavelength $\lambda$ of the laser transition. Accordingly, the relevant values of the longitudinal (or axial) mode counter $q$ are quite large, very often around $10^5$–$10^6$. However, not all possible values are relevant with reference to the frequency of the laser transition, $v_0$. Therefore, it is convenient to reference the global (approximated) mode frequencies $v_{mnq}$ to this laser frequency in the form

$$v_{mnq} = v_0 + q' \cdot \delta v_q + m \cdot \delta v_m + n \cdot \delta v_n,$$

where $q'$ now counts from $-q'$ to $+q'$, with $q' = 0$ being associated with the line center of the transition.

### 3.3.2 Line profiles and widths

By and large, the conceptual formulation of laser action and resonator behavior, as described in this and the preceding sections of this chapter (and as covered in nearly all laser textbooks), is greatly simplified. Specifically, for ease of use, it is assumed that (1) the energy levels have a singular $\delta$-shape value and (2) as a consequence transitions between those energy levels exhibit a single frequency. However, it is already clear from the use of transition probabilities and level lifetimes in the theoretical quantum framework that this certainly constitutes an oversimplification, and that energy levels and photon transitions possess a finite width associated with the Heisenberg uncertainty relation $2\pi \Delta v \Delta t \geq 1$.

The shape of the related energy level or transition line function is found to be a Lorentzian profile (see Table 3.3); it is derived from the Fourier transformation of the intensity originating from an assembly of radiating particles (spontaneous emission).

It should be noted that in many textbooks, the linewidth parameter is associated with the natural lifetime of the (laser) upper level, i.e.,

$$\Delta v_n = \frac{1}{2\pi} \cdot A_u = \frac{1}{2\pi} \cdot \tau_u^{-1}, \qquad (3.22a)$$

where $A_u$ and $\tau_u$ stand for the spontaneous transition probability and lifetime, respectively. But, like in the case of a four-level laser system, also the lower level decays further, meaning that it has a lifetime and thus an associated width (see Figure 3.12). Fortunately, this is easy to take into account since the convolution of two Lorentzian functions yields a Lorentzian again, and the width parameters are

**Table 3.3**   Linewidths Encountered in Laser Systems

| Mechanism | Causes | Functional Description |
|-----------|--------|------------------------|
| Lorentz (line) width | Spontaneous emission lifetime | $L(v') = \dfrac{1}{\pi^2 \Delta v_n} \cdot \left[1 + \left(\dfrac{v'}{\Delta v_n/2}\right)^2\right]^{-1}$ with $\Delta v_L = (2\pi\tau)^{-1}$ |
| Gaussian (line) width | Doppler broadening (particle movement); pressure/phonon broadening (particle "collisions" in gases/solids) | $G(v') = \dfrac{\sqrt{\pi \ln 2}}{\pi^2 \Delta v_G} \cdot \exp\left[-\ln 2\left(\dfrac{v'}{\Delta v_G/2}\right)^2\right]$ with $\Delta_v G = \sqrt{8RT \cdot \ln 2/Mc^2} \cdot v_0$ |
| Gain bandwidth | (a) Broadened single upper laser level  (b) Multiple upper and/or lower laser levels (or energy bands) | Lorentz/Gauss function molecular/solid-state energy level structure, and level populations |
| Schawlow–Townes width | Cavity-mode "phase" noise | $\Delta v_{mode} = \dfrac{4\pi \cdot hv \cdot \Delta v_c^2}{P_{out}}$  $\sim \dfrac{hv \cdot \alpha_{tot} \cdot (1 - MR_{OC})}{4\pi \cdot t_{RT}^2 \cdot P_{out}}$ |

*Note:*   For details on the meaning of the parameter in the formulae see text. $v' = v - v_0$; $v_0$ = line center.

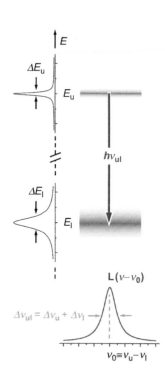

**Figure 3.12**   The transition line profile between two energy levels, $E_u$ and $E_l$, with different lifetimes, leading to a composite-width Lorentz profile.

additive. Thus, for a transition between two energy levels (upper and lower) with associated lifetimes, one finds for the composite Lorentzian (indexed with L)

$$\Delta v_L = \frac{1}{2\pi} \cdot \left(\tau_u^{-1} + \tau_l^{-1}\right). \tag{3.22b}$$

However, this is not necessarily the final adjustment that is required to describe the Lorentzian line shape of the transition from an ensemble of emitters, as encountered in a practical laser system or applying laser radiation to spectroscopic problems.

In the simplest case of a gaseous medium, the particles can move freely (this motion effect will be addressed further below) or may undergo elastic and inelastic collisions, which both affect the Lorentzian line shape function. In inelastic collisions, some small amount of energy is transferred, causing nonradiative quenching of the emitter. Literally, the lifetime of the upper level is shortened, resulting in an "effective" lifetime, and one finds for the linewidth parameter

$$\Delta v_L = \frac{1}{2\pi} \cdot \tau_{eff}^{-1} = \frac{1}{2\pi} \cdot \left(\tau_u^{-1} + \tau_l^{-1} + 2 \cdot \sigma_u^{inel} \cdot \sqrt{1/\mu kT} \cdot p\right), \tag{3.22c}$$

where $\sigma_u^{inel}$ is the inelastic collision cross section; $\mu$ is the reduced mass of the collision system; $T$ is the temperature; and $p$ is the pressure of the gas. Because of the latter dependence, the process is often called pressure broadening: the higher the pressure is, the larger the linewidth becomes.

The treatment of inelastic collisions is slightly more involved. Put simply, during a collision, the partners experience an interaction potential ($E_{pot}$) when coming close to each other. In general, each individual energy level exhibits a differently shaped interaction potential, as a function of particle separation. As a consequence, when a photon is emitted during the collision process, it experiences a shift in frequency, which is given by

$$\delta v_L^{el} = \frac{1}{2\pi} \cdot \sigma_u^{el} \cdot \bar{v} \cdot N, \tag{3.23}$$

where $\sigma_u^{el}$ represents the elastic collision cross section; $\bar{v}$ and $N$ are the average relative velocity and number density of the collision partners, respectively. The shift of the peak frequency can be to lower or higher frequencies, depending on the nature of the interaction potential. The effects of broadening and shift of a Lorentz peak are shown schematically in Figure 3.13.

When the emitting particles (atoms or molecules) are allowed to move freely in space, like in a gas, the spectral transition lines exhibit broadening associated with the well-known Doppler effect. The cumulative effect of the emitting particles, moving with different velocities, results in a (statistical) broadening profile from the Doppler shifts of the individual emitters. Notably, when a particle emitting with frequency $v_0$ moves with velocity $v$, an observer (in the $z$-direction) detects radiation with frequency $v = v_0 \cdot (1 \pm v_z/c)$, where the "plus" and "minus" signs relate to approaching or receding particles, respectively. Assuming that the particle ensemble obeys the thermal Maxwell–Boltzman distribution, the light intensity distribution follows the Gaussian distribution function included in Table 3.3, with the width parameter noted there as well. Three things are noteworthy.

Firstly, the observed Gaussian function constitutes the envelope of emission processes from individual (spontaneous) emitters with Lorentz line profile, distributed according to their thermal statistics for individual frequency components (see Figure 3.14).

Secondly, as shown in Table 3.3, the Doppler width $\delta v_D$ is proportional to the transition frequency $v_0$. This means that Doppler broadening is particularly large for short wavelengths in the visible and the UV region of the spectrum; for example, at room temperature ($T \cong 300K$), one typically finds $\delta v_D \leq 1GHz$, although for very light atoms and molecules, up to about 10–30 GHz may be encountered. However, for spectral lines in the IR, the Doppler broadening is much smaller, and the Doppler width may be of the same order or even smaller than the natural linewidth of the emitter. In that case, the convolution of the Gaussian and Lorentzian functions, with comparable widths, gives rise to a so-called Voigt profile, which is frequently encountered in laser spectroscopy.

Thirdly, the Doppler effect is only really relevant in gases. Atoms or molecules embedded within crystalline or amorphous solids (like solid-state laser-active media) are "locked" into their position and thus cannot move; as a consequence, they do not exhibit Doppler broadening.

### 3.3.3 Laser linewidth, gain bandwidth, and laser spectrum

While the Lorentzian and Gaussian line profiles described above (associated with so-called *homogeneous* and *inhomogeneous* broadening processes, respectively) may be seen as "intrinsic" to the quantum structure of the emitting (or absorbing) particle, in a laser, the "extrinsic" properties of the cavity have to be taken into account.

Assuming for the sake of argument that the laser is oscillating on a single-frequency mode only, it is clear that this laser radiation will nevertheless have a finite linewidth. Loosely speaking, this arises from phase noise (or quantum noise) associated with spontaneous emission in any free-running laser oscillator. It can be shown that the noise of the instantaneous frequency is called *white noise*, i.e., its power spectral density is constant and the spectrum itself is of

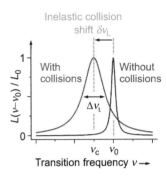

**Figure 3.13** Schematic of the broadening (by inelastic collisions) and shift (by elastic collisions) of a Lorentz line profile; $v_0 =$ line center for spontaneous emission; $v_c =$ collision-induced line center.

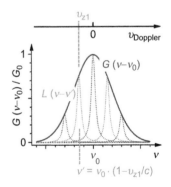

**Figure 3.14** Gaussian envelop of the distribution of line emitters according to their different velocities, resulting in Doppler-shifted line positions (selected Doppler subgroups, with Lorentz profile, are shown as dotted curves).

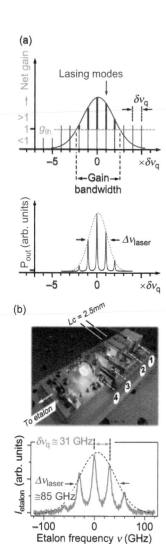

**Figure 3.15**  Gain profile and longitudinal mode structure. (a) Relation between gain bandwidth, laser threshold, and mode spectrum. (b) Measured mode spectrum of a 100 mW green DPSS Nd:YAG laser; the mode spacing calculated using Equation 3.21c matches the measured numerical values. 1 = 808 nm laser diode; 2 = focusing lens; 3–4 = collimating telescope and 1064 nm filter.

Lorentzian shape. Even before the laser was experimentally demonstrated, Schawlow and Townes (1958) calculated the fundamental (quantum) limit for the width of such a laser mode; this limit is widely known as the Schawlow–Townes equation, and it is included for comparison to other linewidth functions and parameters in Table 3.3. It should be noted that it is given here in the form of full width at half-maximum (FWHM) of the resonator cavity bandwidth $\Delta v_c$. For greater transparency of interpretation, the generic laser bandwidth expression can be reasonably well approximated by replacing the cavity bandwidth $\Delta v_c$ with known or calculable parameters, including specifically the transmission of the output coupler, $T_{OC} = 1 - R_{OC}$, and the cavity round-trip time, $t_{CR}$. The formula given in Table 3.3 thus suggests that the laser linewidth becomes narrower when the output coupling is high and/or the cavity round-trip time is large.

In real laser systems, the Schawlow–Townes limit is, in general, rarely reached because, in addition to quantum noise, a variety of "technical" noise sources are encountered, including, e.g., mechanical vibrations and temperature fluctuations, which are difficult to suppress. In any case, certain compromises have to be made in general when designing a laser for optimally narrow linewidth. For example, while a long laser resonator may suggest a small Schawlow–Townes linewidth in principle, it normally becomes more difficult to achieve mechanical stability, and invariably stable single-frequency operation without mode hops is less likely. For a summary on the relation between laser noise and laser linewidth, see, e.g., di Domenico et al. (2010).

For example, typical linewidths of stable, free-running single-frequency semiconductor laser diodes are of the order of a few megahertz. This width can be substantially reduced when using external-cavity diode laser setups (see Chapter 5.1); with a relatively simple but well-designed setup, Saliba and Scholten (2009) measured laser linewidths around <100 kHz, still a factor 2 greater than their calculated Schawlow–Townes limit, showing that "technical" noise still contributed significantly to the linewidth. Note that with extreme instrumental efforts, diode laser linewidths in the subhertz regime have been demonstrated when stabilizing the laser system to an extremely high finesse reference cavity (see, e.g., Hirata et al. 2014).

When comparing the width of the various broadening mechanisms (collisional and Doppler broadening have been discussed above) with the (longitudinal) frequency mode spacing, then it becomes clear that only in rare cases will the laser oscillate on a single frequency. For example, because of the high discharge temperature, argon ion ($Ar^+$) lasers exhibit a Doppler-broadened linewidth of $\Delta v_G \cong 3\text{–}5$ GHz. Typical $Ar^+$ lasers have cavity lengths of 15 to 250 cm for low- and high-power systems, respectively. For the short cavity, one calculates for the longitudinal frequency mode spacing value of $\delta v_q = c/2L = 1$ GHz. This means that a multitude of modes fit under the (homogeneous) laser line profile and may oscillate, contributing to the laser output. The frequency range over which laser oscillation is possible above threshold is often addressed as the laser *gain bandwidth* (for the concept, see Figure 3.15). The modes oscillating above threshold, within the confines of the gain profile, contribute to the laser output spectrum.

Unfortunately, the exact meaning of the term *gain bandwidth* may not always be immediately clear since there are several common descriptions in which gain can be quantified: (1) the FWHM of the logarithmic gain, measured in decibels; (2) the FWHM of the amplification factor; (3) the width of the logarithmic gain or

**Table 3.4** Typical Gain Bandwidths for a Range of Common Laser Sources, in the Units of GHz, cm$^{-1}$, and nm that are Usual for Spectroscopy

| Laser Type | $\Delta\nu$ (GHz) | $\Delta\tilde{\nu}$ (cm$^{-1}$) | $\Delta\lambda$ (nm) | @ $\lambda$ (nm) |
|---|---|---|---|---|
| Gas laser (argon ion) | ~5 | ~0.17 | ~4.4 × 10$^{-3}$ | 514 |
| Solid-state laser (Nd:YAG; $\omega$ & $2\omega$) | 30 | 1 | 0.11 (0.28) | 1064 (532) |
| FP laser diode (InGaN) | 2150 | 71 | 2 | 520 |
| Laser dye (Coumarin 6/Ar$^+$-pumped) | 4.3 × 10$^4$ | 1400 | 40 | 530 ($\lambda_{\text{peak}}$) |
| Solid-state laser (Ti:sapphire) | 1.2 × 10$^5$ | 3900 | 250 | 700–950 |

*Note:* As far as possible, lasers with comparable wavelengths were selected.

the amplification factor, measured at a different level, for example, at the points where the gain has decayed to $1/e$ ($\cong 36.8\%$) or to 10% of the maximum gain; or (4) the width of the range where the gain is at most a certain number of decibels below its maximum value (frequently the 3 or 10 dB points). Any of these can lead to different numerical values; thus, the specification of a gain bandwidth is only really meaningful if the definition used is indicated. Besides these difficulties in definition, additional problems may arise should the shape of the gain spectrum is complicated, for example, being extremely asymmetric or containing multiple peaks.

Regardless, the gain bandwidth of the laser medium is an important parameter for consideration in a number of practical cases, including, for example, the following:

- A narrow gain bandwidth is preferable for stable single-frequency operation.

- A small gain bandwidth is a necessary condition for obtaining high gain efficiency.

- Insufficient gain bandwidth can limit the range for wavelength tuning.

- Very wide gain bandwidth is indispensable for the generation of ultrashort pulses by a mode-locked laser (for further details on short laser-pulse duration, see Section 3.4).

A selection of laser sources with different gain bandwidth, ranging from relatively narrow to very broad, is collated in Table 3.4.

## 3.3.4 Single-mode laser operation

According to the discussion above, a laser may oscillate on more than one frequency mode, depending on the gain bandwidth of the laser medium. In fact, it is more the rule than an exception that multimode emission is encountered. On the other hand, for many applications in spectroscopy and imaging, it is desirable that the laser operates on a single mode. In this context, it is worth noting that the frequently encountered term *single-mode operation* is somewhat ambiguous and is used with different meanings. Quite often, it means that the laser operates on a single transverse (spatial) mode, which is almost invariably the Gaussian mode, TEM$_{00}$; implicitly, it is associated with "good, diffraction-limited laser beam quality." In other cases, the term really indicates that the laser operates on a single (axial) resonator frequency mode, which usually also is a Gaussian mode, TEM$_{00q}$. For example, a novice using laser diodes interprets the term *single mode* in manufacturer's data sheets as single frequency when in fact it means the laser diodes exhibit a Gaussian beam profile.

The excitation of higher-order transverse modes $TEM_{nm}$ is normally avoided by pumping only the volume covered by the axial (00) modes, or by introducing a suitable, restricting aperture into the cavity. Multiple longitudinal (axial) modes $TEM_{00q}$ are still excited in the case when the gain bandwidth is larger than the longitudinal mode spacing. The number of oscillating modes can be reduced in various ways, most frequently by one of the following methods: (1) by decreasing the gain bandwidth above threshold—which is possible, e.g., by carefully reducing the pumping but which therefore restricts the available output power; (2) by increasing the axial mode spacing $\delta v_q$, i.e., by using very short laser cavities—which for many lasers is impracticable; or (3) by inserting a suitable intracavity filter, such as an etalon—which constitutes a coupled-cavity configuration. Note that instead of intracavity filtering, external-cavity feedback is used to enforce oscillation on a particular, single frequency. Note also that gain saturation in inhomogeneous media, caused by so-called spatial hole burning in a standing-wave cavity, can make it difficult to achieve single-frequency operation; for this reason, often unidirectional ring lasers are utilized to achieve reliable and clean single-frequency operation.

## 3.4  LASER CAVITIES: TEMPORAL CHARACTERISTICS

As a last parameter in the discussion, the factor of time will be addressed. From the rate equations alone, it should be clear that one always will encounter time evolution of the level populations and, linked to this, the laser action. For example, in Figure 3.6, it was shown that, in general, one does not encounter a smooth transition to laser action after switch-on of the pumping process, but that one encounters relaxation oscillations. Under the right conditions, the system relaxes to an equilibrium condition of level populations and laser output. The same figure also showed that no equilibrium might be reached in the case that the pumping was switched off prematurely.

Overall, one encounters "continuous" or "pulsed" laser output; accordingly, the classification of lasers is into the classes of *continuous wave* (CW) and pulsed lasers, with the latter further subdivided into specific groups depending on the actual pulse duration, which in general is associated to different "technical" processes in the laser cavity. Very crudely, one often uses the classification of *slow pulses* (derived by modulating a CW laser output); *short pulses* (normally in the nanosecond regime and realized by Q-switching); and *ultrashort pulses* (in the picosecond, femtosecond, or even subfemtosecond regimes, which are by and large generated by mode-locking techniques). All these will be briefly addressed in the paragraphs below.

### 3.4.1  CW operation and laser output modulation

CW operation of a laser means that the laser is continuously pumped sufficiently strongly that population inversion is maintained and that laser radiation is continuously extracted from the cavity. This emission can occur in a single or multiple resonator mode(s). While nowadays a multitude of CW lasers are marketed, it should be noted that CW operation is not necessary to realize that in lasers with low-gain media transitions, CW operation is difficult to achieve and normally high pump power is required, or the attainable laser output power is low; and so-called self-terminating laser transitions are not suitable at all for CW operation. While it is great to have CW laser radiation, for many applications in spectroscopy and imaging, it is advantageous to have amplitude-modulated light available (frequency modulation is also very valuable and common but is not discussed here). For example, the detection techniques of lock-in and boxcar integration (see Chapter 6.6) are frequently used to suppress unwanted contributions from spurious light, and these require suitably modulated laser light; or in some laser imaging applications, "stroboscopic" pulsing is a must. Light modulation can be and is implemented in two main forms, namely "analog" and "digital" modulation. In the former case, the light amplitude is modulated following a certain waveform, like, e.g., a smooth, well-behaved sine function (but even arbitrary waveforms may have their merits); in the latter case, the light intensity is switched between "on" and "off" or between predetermined intensity levels.

Probably the simplest implementation of light amplitude modulation is a mechanical (rotating) light beam chopper, external to the laser cavity. This approach has been and is being used by generations of laser spectroscopists. However, while simple in its approach, the technique has its limitations: repetition rates and pulse durations are by and large limited to the $10^{-3}$–$10^{-5}$ s range, and specifically for the short-duration end of the range, the rise and fall times of the modulation are limited by how quickly the chopper blade passes across the spatially extended laser beam. When higher modulation frequencies are desired, into the hundreds-of-kilohertz or megahertz regime, mechanical choppers can be replaced by so-called refractive modulators, which are based on electro-optic (EO) or acousto-optic (AO) effects (for the concepts, see Figure 3.16).

An *electro-optic modulator (EOM)* is a device that incorporates a nonlinear crystal whose refractive index is modified by an applied electric field (this is also known as the Pockels effect). EOMs can be used for controlling the power, phase, or polarization of a laser beam. One minor drawback is that EOMs, or Pockels cells, usually require hundreds or even thousands of volts, so that a high-voltage amplifier is required, which are not always easy to design for high modulation frequencies. Combined with polarizers, Pockels cells can serve as light amplitude modulators: the Pockels cell modifies the polarization state, and a subsequent polarizer converts this into a change in transmitted laser-light amplitude and power.

An *acousto-optic modulator (AOM)* is a device based on the modification of the refractive index by the oscillating mechanical pressure of a sound wave. As for EOMs, it can be used for controlling the power, frequency, or spatial direction of a laser beam by using an electrical drive signal. The active element of an AOM is a transparent crystal to which a piezoelectric transducer is attached, which excites a sound wave within the crystal. The sound wave generates a traveling periodic refractive index grating at which the light experiences Bragg diffraction (therefore, AOMs are also known as Bragg cells). For sufficiently high acoustic power, more than 50% of the optical power can be diffracted.

Devices based on any one of the two techniques can provide modulation frequencies up to about 100 MHz. If very high contrast ratios between the minimum and maximum transmitted light amplitudes are required, EOMs need to be used.

MECHANICAL CHOPPER

ACOUSTO-OPTIC MODULATOR

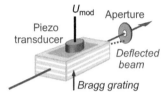

ELECTRO-OPTIC MODULATOR

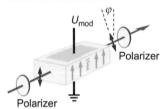

**Figure 3.16** Methods for intensity/amplitude modulation. Top: mechanical chopper. Middle: acousto-optical modulator. Bottom: electro-optical modulator. Note that intensity maximum and minimum can be set to values different to 100% and 0%, respectively.

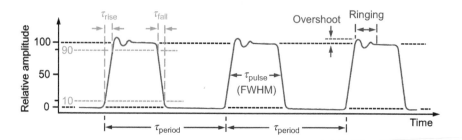

**Figure 3.17** Pulse train from an internally modulated CW laser. The important parameters of pulse repetition period, rise and fall times, pulse width, and over-shoot and ringing are indicated. Note that the pulse baseline may not represent zero light intensity.

Instead of using these modulators externally, i.e., modulating the laser beam, they also can be incorporated into the laser cavity, and thus modulate the laser action itself. It should be noted that cavity-internal EOM/AOM modulation devices are predominantly used for short-pulse and ultrashort-pulse generation (see below) rather than for modulating a CW laser, unless there is no other means to do this. In order to modulate the laser action itself, be it by analog or digital wave forms, modulation is achieved by directly altering the pump rate of the laser medium, pushing the laser to higher or lower efficiency, or above or below the laser threshold. However, because of the laser dynamics discussed earlier, careful design considerations need to be observed when trying to optimize the modu-lation period, with respect to (1) rise and fall time, (2) overshoot and ringing, (3) extinction ratio, and (4) duty cycle. In particular, the first two are often dif-ficult to control easily since they depend critically on the lifetimes of the upper and lower laser levels as well as the pump and laser transition rates. For the overall concept, see Figure 3.17.

By and large, laser modulation is exploited in semiconductor diode (and quantum cascade) lasers, which are directly pumped by the injection current, and therefore lend themselves ideally for modulation by electrical signal waves, or in lasers that are pumped by diode lasers. Some examples are highlighted in Chapter 5.

### 3.4.2 Pulsed laser operation

One may define as a "pulsed" laser any system that does not operate in the "CW" mode: the laser output appears in pulses of some duration, with a given repe-tition rate. In this context, a frequently encountered misconception is that, to some extent, one may treat modulated and pulsed lasers as synonymous. This is because the pulse shape and pulse sequence from a pulsed laser may look very similar to the wave train from a modulated CW laser, as shown in Figure 3.17. The difference between the two is that a pulsed laser emits a variable burst of light energy, depending on the actual pulsing conditions, while a modulated laser alters the laser output between the "low" and "high" (CW steady-state limit) levels, regardless of the speed of the modulation. When tailoring the pulse behavior suitably (mostly respective rise time, duration, and repetition rate), it becomes possible to generate—for a short duration of time—peak pulse power levels, which by far surpass the (CW) steady-state power level.

When generating said laser pulses, one of the most important parameters in their description is the pulse duration or pulse width. Probably the most common characterization of the width of "well-behaved" pulses (i.e., they can be described by a simple pulse-shape function) is to use the FWHM parameter of the laser output power variation versus time; note that this is rather insensitive to any weak background pedestals, as often observed with laser pulses. On the other hand, for complicated temporal pulse profiles, it becomes more appropriate to define the width in relation to the second moment of the temporal intensity profile. Finally, it is sometimes useful to utilize a so-called "effective pulse duration"; this is defined as the pulse energy divided by the peak power, and constitutes a particularly helpful quantity in the context of estimating laser-induced damage.

Broadly speaking, the generation of these energetic pulses of nanosecond duration via Q-switching comprises four phases:

1. Initially, the losses in the cavity are set to a sufficiently high level so that amplification on the laser transition cannot commence. The energy channeled into the laser system by the pumping mechanism accumulates in the upper laser level as population (inversion); the amount of stored energy is mostly limited by the unavoidable spontaneous emission, and it can reach a high multiple of the saturation energy (the steady-state limit without the additional loss insertion).

2. Once the highest possible energy accumulation has been reached, the losses are suddenly reduced to small-enough values so that the laser amplification process is no longer hindered. Normally the process starts from the "noise" (i.e., a rather small amount) of spontaneous emission, which is quickly amplified to high power levels in just a few cavity round trips.

3. The temporally integrated buildup of intracavity power continues until it has reached the order of the saturation energy of the gain medium; consequently, the gain saturates. The large intracavity power now rapidly depletes the remaining stored energy, and the output power decays; once the population in the upper laser level has fallen below the threshold value, the laser pulse ceases. The whole process lasts only a few nanoseconds.

4. After the end of the laser pulse, the cavity losses are increased again to prevent the scenario that any remaining pump energy may repopulate the upper laser level beyond the threshold, and thus secondary, spurious laser action might start.

The overall sequence is sketched schematically in Figure 3.18. Laser pulse energies in the range of a few millijoules are easy to achieve, and peak powers are, in general, orders of magnitude higher than would be achievable from the same gain medium under CW operating conditions.

The switching aforementioned resonator losses between high and low basically can be realized in two ways, namely using *active* or *passive* Q-switching methods.

*Active Q-switches* constitute variable attenuators that are controlled externally. While mechanical devices have been used for this purpose, more commonly AO or EO devices are being used, which are in their operational principle not too different from the AO/EO modulators mentioned further above. These Pockels or Kerr cells are triggered externally, typically by an electrical signal; see Figure 3.19. The energetic laser pulse is formed shortly after the arrival of the electrical trigger signal; the achievable pulse energy and pulse duration strongly depend on the energy stored in the gain medium.

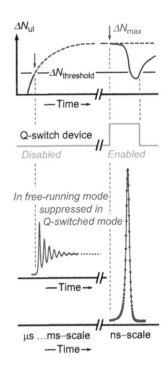

**Figure 3.18** Timing sequence for a Q-switched nanosecond-duration laser pulse. Top: population inversion $\Delta N_{ul}$. Middle: conceptual control signal; the gain is disabled again before the population inversion surpassed the threshold again. Bottom: free-running (suppressed with Q-switch device in the cavity) and Q-switched laser pulses. Note the change in time scale.

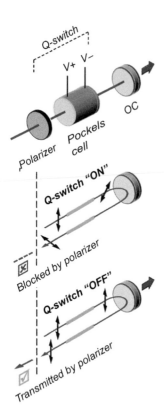

**Figure 3.19**   Concept of an active "off"-Q-switch based on an EO device (here a Pockels cell), in combination with a linear (Glan) polarizer.

The actual evolution of the laser pulse and the associated entities of gain and level populations can be modeled using the coupled rate equation pair (Equations 3.8a and 3.8b), but in which the coupling coefficient between photons and active emitters in the upper laser level also is a function of time, linked to the temporal characteristics of the Q-switch. For the conceptual timing sequence, see Figure 3.18.

It should be noted that the use of active Q-switches has several advantages over their passive counterparts (see below). In particular, certain Q-switch designs allow one to couple an externally generated laser beam into the cavity through the modulator (i.e., no additional optical element needs to be incorporated into the cavity, which would invariably also add additional losses). This "seed" beam may already be tailored for some desired characteristics, such as, e.g., transverse mode profile (such as TEM$_{00}$), narrow bandwidth (such as single longitudinal frequency mode), or wavelength (to preselect certain wavelengths from within a broad gain profile). Then when the Q-value is raised, i.e., the losses are lowered, laser action builds up from the initial seed photons. Consequently, the so-generated Q-switched pulses inherit their characteristics from the seed radiation.

As stated earlier, the pulse duration achievable using Q-switching is in the nanosecond regime. Normally, the pulse duration is at least equal to the resonator round-trip time, and often substantially longer, dictated by the fact that the amplification process requires several passages through the gain medium to deplete its stored energy. In particular, this is true if the laser gain and/or the resonator losses are low; this situation may be encountered when high repetition rates are desired, in which case normally much lower peak energies are stored. Consequently, it can become difficult to generate reasonably short nanosecond pulses.

This problem can be overcome by using a method known as *cavity dumping*. The main difference to ordinary Q-switching is that the standard "passive" output coupler mirror of the cavity is replaced with an "active" mirror whose degree of losses for output coupling is variable from full reflection to full transmission (usually a combination of a total reflector and a fast, optical switch redirecting the beam). Basically, during the Q-switch pulse generation phase, the mirror's reflectivity is set to as close as possible 100%: effectively this constitutes a "closed," very low loss resonator. As soon as all energy stored in the gain medium has been transferred into the circulating laser pulse, the pulse energy is dumped out of the cavity within one resonator round-trip time, fully independent of the actual time required for pulse buildup (see Figure 3.20, where cavity dumping is exemplified for a passive Q-switching/mode-locking configuration).

*Passive Q-switching* devices are based on so-called saturable absorbers, materials whose transmission increases when the impinging light intensity exceeds a certain threshold. The most common materials used for this include bleachable dyes (very common in the early days of Q-switching), ion-doped crystals like Cr: YAG (used, e.g., for Q-switching of Nd:YAG lasers), or passive semiconductor devices (mostly in the form of a semiconductor saturable absorber mirror— SESAM—for the use in high-repetition rate microchip lasers; some details on these devices are given further below). The action of the saturable absorber may be described as follows. Initially, the cavity losses are high because the photons originating in the active medium are largely absorbed. However, the absorptivity is set not too high so that there is still some laser action to develop and subsequently a large amount of energy is stored in the gain medium. From this sort of seeding, the laser power gradually increases, and becoming sufficiently large, it saturates the absorber (no more photons can be absorbed in it). In turn, this rapidly reduces the resonator loss, so that the laser power increase becomes faster and faster, and within a short time all energy is extracted from the gain medium and the pulse ceases. After the pulse, the saturable absorber recovers to its high-loss state, and the sequence repeats itself.

As for active Q-switching, the temporal evolution of the laser pulse, laser gain, and level populations can be modeled using the coupled rate equation pair (Equations 3.8a and 3.8b), which, however, has to be modified insofar as to include an additional population term related to the level structure and level lifetimes of the saturable absorber. A more detailed discussion of these rate equations can be found, e.g., in Siegman (1990), "Chapter 26.3—Passive (Saturable Absorber) Q-Switching."

Overall, passive Q-switching is simple and cost-effective (the normally high costs for AO or EO modulators and their associated electronics are avoided); however, the achievable pulse energies are typically lower, and—detrimental in time-critical experiments, which require mutual synchronization—external triggering of the pulses is not possible.

### 3.4.3 Mode locking: Generation of ultrashort picosecond and femtosecond pulses

Q-switching is no longer applicable if pulses of ultrashort duration, in the picosecond to femtosecond regime ($10^{-12}$–$10^{-15}$ s), are required; instead, one needs to utilize a methodology known as *mode locking*. The technique requires a broadband gain medium, and that within the gain bandwidth, a large number of longitudinal modes resonate in the laser cavity above threshold. If one then induces a fixed-phase relationship between those many oscillating modes, for short intervals (in time), interference between these modes causes the laser light to build up to circulating, short pulses. In the steady-state limit, the parameters influencing the pulse behavior no longer change for individual, complete round trips. Thus, each time the pulse arrives at the output coupler mirror, a usable amount of laser energy is transmitted and a regular pulse train emanates from the laser.

If one assumes that the resonator dimensions are such that only a single pulse circulates, then the repetition period between the emitted pulses corresponds to the resonator round-trip time $\tau_{CRT}$; this is typically of the order of a few nanoseconds, thus corresponding to pulse repetition rates of up to 100 MHz. The duration of individual pulses is substantially lower, typically in the range of a few femtoseconds to a few picoseconds, primarily depending on the number of modes that are locked to each other. Linked to this difference in time scale, one finds that the peak power of an individual pulse can be orders of magnitude higher than the average power in the pulse train.

For example, within the bandwidth of a gas laser-active medium, only a handful of modes is supported, while the ultrawide bandwidth encountered in a Ti: sapphire laser supports many thousands of modes.

Normally, these modes oscillate independently of each other, and the laser emits on a set of independent laser modes at slightly different frequencies. Since the phases of the individual laser mode waves are not fixed to each other but may

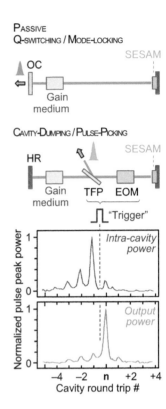

**Figure 3.20** Concept of cavity dumping in comparison to normal (passive) Q-switching or mode locking, exemplified for EOM dumping; EOM = electro-optic modulator; TFP = thin-film polarizer; OC = output coupler; HR = high reflector. Note the much higher dumped-pulse peak power in comparison to the normal output prior to round-trip *n*.

vary randomly, one finds the following: in lasers with only a few oscillating modes, for the above-mentioned interference between the individual modes, one observes a "beating" effect in the laser output, manifesting itself in intensity fluctuations; in lasers oscillating on many thousands of modes, these interference effects tend to average out and one observes near-constant output laser intensity.

However, if all modes operate with a fixed-phase relation between them, a quite different behavior of the laser output is observed. Instead of relatively random or constant output intensity, the laser modes with the given periodicity will constructively interfere resulting in an intense, short pulse of light. If—as surmised further above—only a single such pulse periodicity is encountered in the cavity at any one time, the laser output pulses occur separated in time by $\tau_{CRT} = 2L/c$, i.e., $\tau_{CRT}$ is the time taken for the light to make exactly one round trip in the laser cavity.

The principle and the effect of random and mode-locked laser oscillation can be gathered from the conceptual Figure 3.21.

The duration of each mode-locked laser pulse $\delta\tau_{MLP}$ is linked to the number of modes oscillating in-phase. In the case that $N$ modes (with inherent frequency separation $\delta v_q$) are locked to each other, the overall mode-locked bandwidth is $N \cdot \delta v_q$, meaning that the wider this bandwidth, the shorter the laser pulse

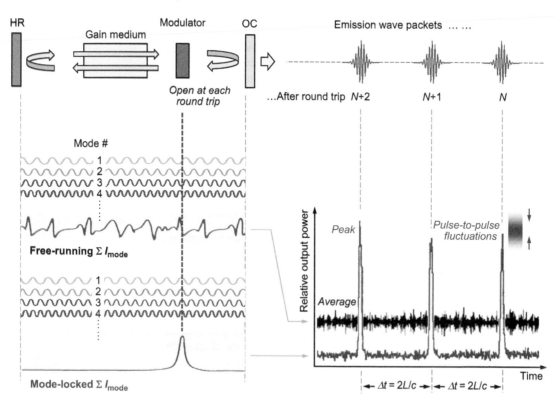

**Figure 3.21**  Conceptual comparison between a free-running and a mode-locked laser system. On the left, a selection of cavity modes (indicated by different color waves) are shown, which add up randomly to a small average power, without locking their phases (top traces) or when in phase add up to "giant" pulses (bottom traces). On the right, the laser output is shown, with a fluctuating ("noisy") average power in the free running case and a sequence of "sharp" peaks in the mode-locked case; note that the relative amplitudes are not to scale.

duration. Note here that in a real laser, not necessarily all possible laser modes will be phase-locked. Note also that the actual pulse duration is influenced by the exact amplitude and phase relationship of each of the contributing longitudinal modes. The minimum possible pulse duration $\delta\tau_{MLP}$ is given by

$$\delta\tau_{MLP} = TBP/(N \cdot \delta v_q), \tag{3.24}$$

where TBP is known as the "time-bandwidth product" of the pulse, which varies depending on the actual pulse shape. For example, for a Gaussian temporal shape, TBP = 0.441, while for the profile often assumed for ultrashort pulses, namely sech$^2$ (hyperbolic secant squared), TBP = 0.315. Based only on the simplified Equation 3.24, and assuming the respective typical laser cavity dimensions, one would find for a Gaussian pulse shape that $\delta\tau_{MLP} \cong 500$ ps for a gas laser medium and $\delta\tau_{MLP} \cong 5$ fs for the Ti:sapphire laser.

Specific methods to achieve mode locking in a laser may be classified as *active* or *passive*, as encountered for the pulse-shortening by Q-switching, i.e., an external signal is used to induce the intracavity modulation in the former case, or a suitable element is inserted into the cavity, which causes self-modulation of the circulating light wave in the latter case.

*Active mode locking* involves (1) the periodic modulation of the resonator losses or (2) periodically changing the round-trip phase. As for Q-switching, such active devices are by and large based on AOMs or EOMs. Provided that the modulation is synchronous with the resonator round-trip time $\tau_{CRT}$, ultrashort pulses with picosecond duration are generated. It should be noted that the actual pulse duration is governed by the interplay of the modulator and cavity effects; in particular, a gain bandwidth adversely dictates how short the pulse may become. There are three main implementations of mode locking, namely *FM mode locking* (FM = frequency modulation); *AM mode locking* (AM = amplitude modulation); and *synchronous mode locking*.

The most common approach to active mode locking, in the *frequency domain*, is *FM mode locking*. Here an AOM in the laser cavity is driven by a suitable electrical signal wave to cause a sinusoidal amplitude modulation of the laser light oscillating in the cavity. For a mode with optical frequency $v_q$, which is amplitude-modulated at a frequency $f_{mod}$, the modulated wave signal has sidebands at the frequencies $v_q \pm f_{mod}$. If the modulator is driven at the same frequency as the cavity-mode spacing $\delta v_q$, these sideband frequencies coincide with the two cavity modes adjacent to the original mode, i.e., $v_{q+1}$ and $v_{q-1}$; and since the two sidebands are driven in-phase, the three modes with $v_{q-1}$, $v_q$, and $v_{q+1}$ are phase-locked. Now the modulation process acts in the same way on those three modes, and phase-locking to the subsequent $v_q \pm 2f_{mod}$ modes ensues, and so on until all modes within the gain bandwidth are locked to each other.

The modulation process can also be considered in the *time domain*, which leads to *AM mode locking*. In this case, the AOM acts like a "shutter" to the circulating laser light wave in the cavity; for this it is driven by an electrical signal wave to set the modulator device to some "open" and "closed" state, changing the amplitude of the transmitted wave. If the modulation frequency $f_{mod}$ is synchronous to the cavity round-trip time $\tau_{CRT}$, then only a single pulse of laser light will travel in the cavity. Note that it suffices for the actual modulation depth, i.e., the amplitude difference between the open and closed states, to be of the order of 1% only, since the same parts of the traveling wave pattern are repeatedly let through or attenuated as it traverses the cavity.

The final method of active mode locking is *synchronous mode locking* (also known as synchronous pumping). For this, the pump source for the laser is itself modulated, which thus effectively generates pulses (normally the pump source is itself another mode-locked laser). Note that this technique requires that the cavity lengths of the pump laser and the driven laser are delicately matched to each other to guarantee that the pulse round-trip times are equal or exact multiples of each other.

In general, when using active mode-locking procedures, the pulse duration is typically in the picosecond range. In simple cases, the pulse duration, which prevails after reaching the steady state, can be calculated using the theory developed by Kuizenga and Siegman (1970). The basic idea is that in active mode locking, two competing mechanisms influence the duration of the circulating pulse: (1) the modulator device slightly attenuates the leading and trailing wings of the pulse, effectively shortening it; and (2) the finite bandwidth of the gain medium tends to reduce the bandwidth of the pulse, and hence leads to an increase in pulse duration. The two effects come into balance for a certain pulse duration, which then constitutes the steady-state value. Subject to some simplifying assumptions, the Kuizenga–Siegman theory yields a relatively simple equation to quantify the steady-state pulse duration:

$$\delta\tau_{\text{MLP}} \cong 0.45 \cdot (G/M)^{1/4} \cdot \left(f_{\text{mod}} \cdot \delta\nu_{\text{gain}}\right)^{-1/2}, \qquad (3.25)$$

where $G$ is the intensity gain, $M$ is the modulation depth, $f_{\text{mod}}$ ($=\tau_{\text{CRT}}$) is the modulation frequency, and $\delta\nu_{\text{gain}}$ is the (FWHM) gain bandwidth.

For *passive mode locking*, saturable absorbers are used—as in passive Q-switching; these include liquid organic dyes, doped crystals, and semiconductor devices. In particular, the latter exhibit very fast response times, of the order of 100fs. Essentially, the saturable absorber can modulate the resonator much faster than any (active) electronic devices, thus allowing for much shorter pulse durations than are achievable by active mode locking.

Inside the laser cavity, the saturable absorber will attenuate low-intensity light waves (namely in the pulse wings), while more intense "spikes" (from initial, nonmode-locked laser action on the many cavity modes under the gain profile) will be transmitted. With continued oscillation of the laser modes in the cavity, this preferential transmission action is repeated, leading to the selective amplification of the high-intensity spikes, and the absorption of the low-intensity light: a train of short pulses develops and mode locking of the laser ensues. Note, however, that the saturable absorber may also lead to passive Q-switching (or Q-switched mode locking) if its properties are not selected appropriately.

When considering the process in the frequency domain, the following is observed. For a mode of optical frequency $\nu_q$, which is amplitude-modulated by the absorber at a frequency $n \cdot f_{\text{mod}}$, the resulting signal has sidebands at optical frequencies $\nu_q \pm n \cdot f_{\text{mod}}$ (recall that $f_{\text{mod}}$ is related to the cavity round-trip time); these then are locked to each other. In the steady state, the mode-locked pulses will have shortened significantly due to the absorber action, which always attenuates the leading wing of the circulating pulse; if in addition the absorber recovery time after saturation is fast enough, then also the trailing pulse wing will be suppressed, further contributing to pulse shortening.

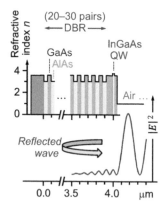

**Figure 3.22** Typical SESAM device used for passive mode locking, consisting of a distributed Bragg reflector (DBR) and the saturable absorber layer (InGaAs quantum well). The refractive index profile and the optical intensity distribution within the SESAM are shown.

The most widely used saturable absorbers for passive mode locking are devices known as semiconductor saturable absorber mirrors, or *SESAM*. These are very compact semiconductor devices, whose operating parameters can be tailored to meet the requirements for a wide range of applications, and which—when appropriately designed—can be used for mode locking as well as Q-switching of many lasers. The use of a SESAM for mode locking was demonstrated by Keller et al. (1992).

In general, a SESAM device comprises a semiconductor Bragg reflector, a single quantum-well absorber layer (near the surface), and a passivation layer on the top surface (for protection, to increase the device lifetime); normally the devices are manufactured by epitaxial growth technology. The alternating material layers making up the Bragg reflector have a bandgap energy that is larger than the design laser wavelength so that essentially no absorption occurs in that region. The thickness of the quantum-well absorber layer determines the modulation depth; thinner layers with small or moderate absorption are utilized in mode-locking applications; thicker layers are required for passive Q-switching. The (saturable) absorption is related to an interband transition in the quantum-well layer: the electrons transferred to the conduction band by photon absorption thermalize rapidly, within less than 100 fs, followed by carrier recombination on a time scale of 10–100 ps. For low laser light intensities, the absorption remains unsaturated, while at high intensities, electrons accumulate in the conduction band, and depleting population in the valence band, thus yielding so-called Pauli blocking (no further absorption is possible prior to relaxation/recombination). The structure of a typical SESAM and its conceptual reflection/absorption behavior are shown in Figure 3.22.

Finally, as an alternative to directly absorbing materials for passive mode locking, it is possible to exploit nonlinear optical effects in intracavity components for the purpose. One of the most successful methods exploits the Kerr effect (see Figure 3.23). The optical Kerr effect is a nonlinear effect, which occurs in most optical materials when intense laser light propagates through them; it is due to nonlinear polarization and refractive index changes generated in the medium, which then modify the propagation properties of the light wave. For example, the refractive index is modified by high laser intensity $I$ according to $\Delta n = n_K \cdot I$, where $n_K$ stands for the nonlinear (Kerr) refractive index (typically for glasses, one finds $n_K = 10^{-16}$–$10^{-14} \text{cm}^2 \cdot \text{W}^{-1}$). Thus, when an intense Gaussian laser beam propagates through the medium, the Kerr effect leads to a phase delay; this is largest on-axis (where the intensity is highest) and smaller off-axis. Assuming that $n_K$ is positive, this is similar to the action of a lens and the beam is focused; the effect is normally known as *self-focusing*. If said Gaussian laser beam (of power $P$ and beam radius $w_0$) propagates through a thin piece (of thickness $d$) of glass, across the beam diameter, a "Kerr lens" is formed with focal length

$$f = \left( \frac{4 n_K d}{\pi w_0^4} P \right)^{-1}. \tag{3.26}$$

Equation 3.26 is derived by calculating the radially dependent (nonlinear) phase changes and then comparing them with those for a lens made from the same optical material. The equation shows that the effect of Kerr lensing increases with laser power. For short, intense laser pulses, the effect gives rise to the so-called

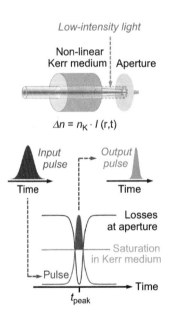

**Figure 3.23** Concept of pulse shortening and mode locking exploiting the Kerr lens effect. The aperture serves as a time-dependent variable loss element, in conjunction with the saturation in the Kerr medium.

*Kerr lens mode locking*, also known as "self-mode locking." By careful arrangement of an aperture in the laser cavity (to curtail the weaker intensity parts of the Kerr-focused beam; see Figure 3.23), the Kerr lens plus aperture constitute the equivalent of an ultrafast response-time saturable absorber. Kerr lensing for femtosecond-laser pulse generation was first realized by Spence et al. (1991).

### 3.4.4 Group delay dispersion: Shortening and lengthening ultrashort (chirped) pulses

When designing actual mode-locked laser systems, a number of practical, physical effects have to be taken into account. By and large, the most important aspects are (1) phase dispersion in the cavity and (2) nonlinear optical effects. The former effect is associated with the fact that waves of different frequency travel with different velocity in optical media, and hence the waves exhibit a different phase shift after passage. If the associated net group velocity dispersion (GDD) over the whole cavity is excessively large, then the phases of all the oscillating modes cannot be locked any longer, thus reducing the participating bandwidth (see Equation 3.25), and consequently lengthening the mode-locked pulses. The result is the generation of "chirped" pulses. To a large extent, this deficiency can be overcome by introducing optical elements in the cavity, which exhibit negative (anomalous) net GDD, and which thus compensate the inherent positive GDD of laser cavities with standard optical components.

There are several methods for negative-GDD shortening of optical pulses, grouped into two main categories. In *linear pulse compression*, the chirp is (at least partially) removed by sending the pulses through an optical element with a suitable amount of *chromatic dispersion*; such elements can comprise grating pairs, prism pairs, lengths of optical fiber, or chirped mirrors (including Bragg gratings). In the ideal case, bandwidth-limited pulses are once again recovered. *Nonlinear pulse compression* comprises a two-step procedure. First, the optical bandwidth is increased, typically using a nonlinear interaction method (such as self-phase modulation), to generate a chirped pulse whose duration is larger than that of the original pulse. This is followed by strong linear (dispersive) compression, which removes the added and original chirps (at least to below the original pulse width). Nonlinear pulse compression is often utilized in the power amplification of ultrashort pulses: here the amplifier stage is set between the stretching and compressing stages to reduce the temporal peak power of the laser pulse, to below the damage threshold, within the amplifier gain medium.

The most common pulse compressors/stretchers are based on prism or grating pairs, both having their distinct advantages and disadvantages. While it is easy to create huge negative or positive GDD with gratings, their drawback is that gratings exhibit rather large losses (diffraction as well as absorption) and are therefore unsuitable for intracavity application. By and large, they are used in ultrashort-pulse amplifier configurations, as just mentioned. Prism compressors, on the other hand, may experience losses of less than about 2% (with appropriate antireflection coatings) and thus are suitable for intracavity use. Lasers based on prism pairs were first characterized by Fork et al. (1984). The principle of a prism-pair compressor/stretcher is shown schematically in Figure 3.24.

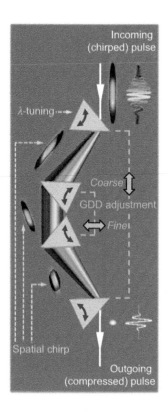

**Figure 3.24** Concept of femtosecond-pulse compression/stretching by using prism pairs, here for the case of compression. The related negative and positive GDD can be set by adjusting the lateral and inline distances of associated prisms, as indicated in the figure. Synchronous rotation of all prisms affords wavelength "shaping" in the output pulse.

While prisms have inherent positive dispersion (nearly all optical materials transparent to visible light have normal dispersion), i.e., the refractive index decreases with wavelength, and consequently longer-wavelength light travels faster through the material, this can be counteracted by the increased distance that the longer wavelength components have to travel through the second prism. With appropriate alignment, the prism pair can become a positive- or negative-GDD element, simply by shifting the second prism $P_2$ up and down. Actual design criteria for sizes, angles, and materials of prisms may be found, e.g., in an application note by Newport Technology and Applications Center (2006).

# 3.5 POLARIZATION AND COHERENCE PROPERTIES OF LASERS AND LASER BEAMS

As the last parameters of laser operation and laser beam properties, polarization and coherence will be briefly discussed in this section.

## 3.5.1 Laser polarization

By and large, most laser systems are polarized, and normally with this one means linear polarization, i.e., the oscillation plane of the electric field of the wave is in a direction perpendicular to its propagation direction, and is stable over time. In a certain way, this is not surprising since a standing wave has to develop in the cavity to be amplified in the gain medium and to generate laser light output. So at least temporarily, (linear) wave polarization has to exist. However, if the gain medium and the laser cavity were both isotropic and homogeneous, then, after switch-on, the laser wave would start to oscillate in an arbitrary polarization plane due to the random nature of the initial spontaneous photons that self-start laser action. But that initial polarization state is in most cases unstable and may switch randomly between different directions, as a consequence of temporary instabilities in the cavity (e.g., thermal drifts and associated localized changes in material birefringence). On the other hand, if there are "permanent," very small differences in gain/loss difference for various polarization directions, these are often sufficient to result in stable linear polarization.

Different phenomena in the gain medium or the laser cavity can be responsible for directional anisotropy and hence for linearly polarized laser emission. These include, for example, the following: (1) the gain medium itself is polarization dependent, as is the case for many laser crystals; or (2) the resonator losses are polarization dependent. For the latter, even a slightly tilted optical component (sometimes merely due to non-optimal alignment) may suffice to result in a preferred linear polarization direction. But more often than not, the polarization dependence of an optical element is inherent, such as wavelength-dependent devices like gratings, prisms, tuning etalons, etc., or Brewster-angled optical surfaces that exhibit near-zero losses for the associated $p$-polarization direction. Recall that the Fresnel equations predict that light with the $p$-polarization (i.e., the light wave oscillates in the same plane as the incident ray and the surface normal) will not be reflected provided that the angle of incidence is the so-called Brewster angle $\theta_B = \tan^{-1}(n_2/n_1)$, where $n_1$ is the refractive index of the "incident" optical medium (often air), and $n_2$ is the one for the other optical medium (see Figure 3.25).

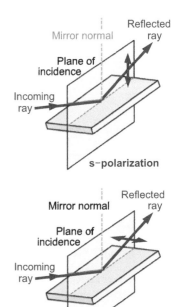

**Figure 3.25** Definition of s- and p-polarized waves, whose electric field vector oscillates parallel and perpendicular to the plane of incidence, respectively. Note that for p-polarization, one obtains vanishing reflectivity from a surface at Brewster's angle.

### 3.5.2 Tailoring the polarization of a laser beam: Linear, circular, and radial polarization

While the aforementioned polarization purity of 1% is sufficient for many applications, others require much higher degrees of polarization. For example, in Raman depolarization measurements, changes in polarization direction of less than $10^{-4}$ are not rare; and in precision ellipsometry, even higher values are required. To achieve such purities, a number of optical components are common, which help in improving the degree of linear polarization. In summary, these include as the most common *absorptive* (or dichroic) *polarizers* (up to $10^3$–$10^4$:1 in the visible but with low transmission and low damage threshold); *thin-film polarizers* based on dielectric coatings (larger $10^4$:1 but exhibiting limited bandwidth and moderate damage threshold); *nanoparticle linear film polarizers* (extinction ratios up to $10^7$:1 combined with high damage threshold); and polarizing beam splitters based on birefringence, including the popular family of *Glan polarizers* (extinction ration $\sim10^5$:1, exhibiting very high damage threshold). For details on the different types of linear polarizers, one may consult standard optics textbooks and specialist application notes from component manufacturers. Note that linear polarizers can also be used for controlled, external laser beam attenuation by adjusting the angle between the original laser beam polarization and the axis of the external polarizer, allowing for varying the laser beam power from its full value (polarizer axis parallel to the laser polarization direction) down to the specified extinction ration for crossed polarization directions.

The polarization orientation of a laser beam emerging from a source is given by the internal orientation of the polarization-determining optical element within the laser cavity. Normally, for ease of use, the polarization is horizontal or vertical to the base of the laser unit. This polarization direction is of key importance for any further beam steering optics in the laser light beam, such as, e.g., dielectric mirrors whose coatings are normally optimized for *p*- or *s*-polarization (the polarization direction is in the plane formed by the incident beam and the mirror surface normal, or perpendicular to it, respectively; see Figure 3.25).

Should the polarization plane be "incorrect" for the particular optics or potential application, this can be adjusted by using a so-called half-wave plate or half-wave retarder. These consist of a double-refracting uniaxial crystal plate whose refractive faces are parallel to the optical propagation axis. The thickness $d$ of the retarder plate is chosen so that the effective path difference $\Delta$ between the ordinary (*o*-)beam ray and the extraordinary (*e*-)beam ray is an odd-integer multiple of $\lambda/2$, i.e., $\Delta = (n_e - n_o) \cdot d = (2m + 1) \cdot (\lambda/2)$, with $n_e$ and $n_o$ being the refractive indices for the extraordinary and ordinary axes, respectively. When this half-wave plate is rotated and the angle between the *e*- and *o*-axes is $\theta$, then the (linear) polarization of the transmitted beam is rotated by twice this angle, i.e., $2\theta$ (see Figure 3.26). This means that, in principle, the polarization direction of a laser beam can be adjusted at will, although in practice, the polarization directions are rotated by normally 90°.

In a slightly different picture of polarization direction, one can envisage a particular direction as a superposition of waves of different amplitude oscillating in the, say, *x*- and *y*-directions. For equal amplitudes, therefore, the resultant superposition would oscillate in a direction at 45° to the *x*- and *y*-axes. If, in

HALF-WAVE PLATE

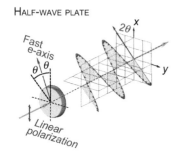

QUARTER-WAVE PLATE

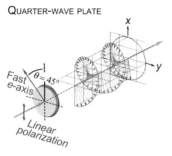

**Figure 3.26**  Change in polarization induced by half- and quarter-wave plates.

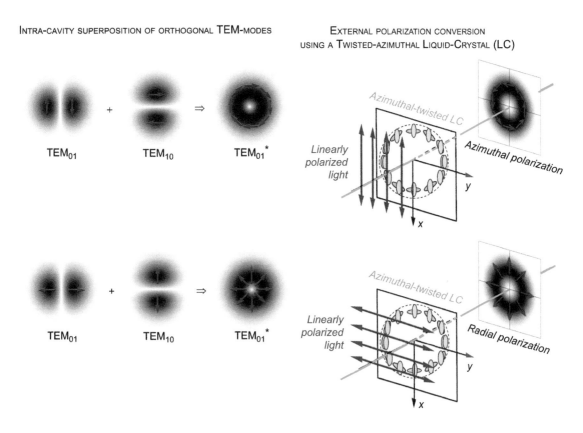

**Figure 3.27** Concepts for the generation of laser beams with radial or azimuthal polarization. For details see, e.g., Oron et al. (2000) and Wang et al. (2013).

addition, if said two waves of equal amplitude experience a phase shift of $\pm\pi/2$, the phenomenon of circular polarization is encountered. During wave propagation, the superposition of the two $E$-fields results in a total $E$-vector, which rotates clockwise/anticlockwise at the angular wave frequency $\omega$. The former case is known as $\sigma^+$ polarization, while the latter one is associated with $\sigma^-$ polarization. Frequently, the linear-polarization case is said to be a superposition of two $\sigma^+/\sigma^-$ polarization waves, which yield the so-called $\pi$ polarization. Linear and circular polarized waves can be transformed between each other by using a quarter-wave plate (or $\lambda/4$-plate). Sending a linear polarized wave through a quarter-wave plate, whose $e$- and $o$-axes are adjusted at 45° to each other, results in a circular polarized wave, and vice versa (see Figure 3.26). The $\sigma^+/\sigma^-$ polarization states can be thought of as related to photon polarization states. When the light intensity is reduced to its smallest possible level, one deals with single photons, i.e., quantized light. At the photon level, the "polarization" is related to its intrinsic angular momentum (spin), $s_{photon} = \pm h$. This means that on photon absorption/emission in an interaction experiment, not only energy conservation but also angular momentum conservation have to be considered.

Finally, an interesting class of polarized laser beams is worth mentioning, depicted conceptually in Figure 3.27. This is the class of cylindrically symmetric vector beams, whose polarization vectors are oriented in the radial and azimuthal directions, respectively, i.e., at every position in the beam, the polarization vector points toward the center of the beam, or is parallel to it. Two great advantages for laser spectroscopy can be derived from laser beams exhibiting these exotic polarization properties. Firstly, when a radially polarized laser beam is focused with a short-focal length lens, the electric field in the focal region possesses a relatively large longitudinal component; this property is exploited in, e.g., $z$-polarization spectroscopy and confocal microscopy (see, e.g., Saito et al. 2008). Secondly, a beam with radial polarization can be focused to a spot size that is substantially smaller than that of a beam with ordinary linear polarization; this property is utilized, e.g., in optical trapping and micromanipulation (see, e.g., Zhan 2004).

Radial and azimuthal polarization can be generated in a number of ways, either by appropriate control components within a laser cavity (see, e.g., Oron et al. 2000; Thirugnanasambandam et al. 2011) or—more widely utilized today—by using a liquid-crystal polarization modulator device to convert the (linear) polarization of a beam into a radial/azimuthal configuration (see, e.g., Wang et al. 2013).

### 3.5.3 Coherence

One of the intrinsic properties of laser radiation is coherence; it is associated with the amplification process due to stimulated emission, which fixes the phase relation between the photons from different emitters in the active medium. Depending on how much in space and how long in time this fixed-phase relation does exist, the related coherence property is imprinted as well on the laser output beam; one therefore distinguishes two different aspects of coherence, namely *spatial* and *temporal* coherence.

*Spatial coherence* means that one encounters a strong, fixed-phase correlation between the laser light field at different locations across the beam profile. For an (ideal $TEM_{00}$) diffraction-limited laser beam, the electric (laser-light) fields at different positions across the beam profile oscillate fully correlated. Note that this condition holds even for a superposition of different frequency components (longitudinal modes).

*Temporal coherence* means that one finds a strong correlation between the laser-light fields at one particular location but at different times. In this context, the output of a single-frequency CW laser exhibits high temporal coherence; the laser field evolves temporally as a clean, sinusoidal oscillation, and does so over extended periods of time. In particular, temporal coherence can be visualized when splitting and then recombining, after some time delay, a particular coherent wave. The duration over which a visible interference pattern is generated is known as the coherence time, $\delta t_c$, which in turn is related to the finite frequency bandwidth of the laser by

$$\delta t_c \propto 1/\delta\nu_{laser}, \tag{3.27}$$

i.e., the narrower the bandwidth, the better the temporal coherence. This coherence time corresponds to a coherence length, $\lambda_c$, through

$$\lambda_c = \delta t_c \cdot c, \tag{3.28}$$

where $c$ is the velocity of light. For example, typical multilongitudinal mode gas lasers have a coherence length of a few centimeters, while the same laser but running in stabilized, single-longitudinal mode exhibits a coherence length of a few 100 m; and for a solitary laser diode with bandwidth $\delta\nu_{laser} \cong 1$ nm, one finds $\lambda_c < 1$ mm, while the same laser diode incorporated into a single-mode external cavity yields a coherence length of tens of meters.

In this context, it is worthwhile to contemplate ultrashort (picosecond and femtosecond) pulses. By definition, their bandwidth is large, so that this would suggest that according to Equation 3.27, their coherence length is short. Indeed, when considering a single, ultrashort pulse, its coherence function quickly decays simply because its wave train (intensity) decays so quickly. However, when considering the (periodic) pulse train from a mode-locked laser, which originates from a single, low-noise pulse circulating in the laser cavity, a strong phase correlation between subsequent pulses is encountered, although pulses are separated by relatively large temporal spacing of normally many nanoseconds. As many as a few 1000 pulses may be well correlated, and thus still exhibit some temporal coherence. Of course, relation 3.27 between coherence time and laser bandwidth constitutes an invalid oversimplification, although there is a relation between spectral width and coherence time. But one actually has to take into account that the Fourier conversion of the pulse train is not a continuum but constitutes an (equidistant) frequency comb; it is the width of a single peak in the comb whose Fourier transform is related to the "real" coherence time, i.e., the long-time phase relation between pulses distant in time.

Finally, it is worth noting that there are a number of different ways of how to quantify the degree of coherence. More or less, all rely on the degree of correlation (correlation functions of different order) as a function of a spatial or temporal interval(s). For example, first-order correlation functions are related to the optical spectrum content; second-order correlation functions describe intensity correlations (such as the effects of photon bunching, or antibunching); and higher-order correlation functions describe further, more subtle details. As an example, in Figure 3.28, the so-called fringe visibility is shown, which links the coherence length to the interference contrast. The fringe visibility is defined as

$$V = (I_{max} - I_{min})/(I_{max} + I_{min}),$$

where $I_{max}$ and $I_{min}$ are the maximum and minimum interference intensities, respectively, recorded by the detector when the Michelson reference arm is scanned. The coherence length $\lambda_c$ is given by

$$V(\lambda_c) = V(0)/e = 0.37 \times V(0).$$

The property of coherence is important for a range of practical, spectroscopic applications that require both high spatial and high temporal coherence, such as various interferometric uses and holography. On the other hand, there are other applications that require high spatial coherence but very low temporal coherence; an example for this type of application is *optical coherence tomography* (OCT) in which interferometric images with very high spatial resolution are created.

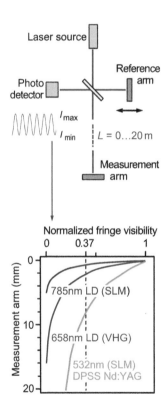

**Figure 3.28** Measurement of the coherence length with the aid of fringe visibility in a long-arm Michelson interferometer. Typical data for some selected laser sources are shown. LD = laser diode; VHG = volume holographic grating; SLM = single longitudinal mode.

## 3.6 BREAKTHROUGHS AND THE CUTTING EDGE

It is next to impossible to identify any singular theoretical development as being the overriding breakthrough for the description of lasers: the number of "ingredients" that have to be contemplated to describe and predict the behavior of any particular laser has multiplied since the demonstration of the first ruby laser. Indeed, laser science predates the invention of the laser itself. In this respect, one may see Einstein's 1917 paper as the cradle for the laser invention nearly half a century later; in the said publication, he rederived Planck's law of radiation, but using a formalism based on probability coefficients (today known as Einstein coefficients) for absorption, spontaneous emission, and the newly postulated stimulated emission of electromagnetic radiation. The early 1960s then saw a flurry of activity to provide semiclassical theoretical descriptions of the dynamic behavior in the rapidly evolving family of laser systems to include phenomena like threshold behavior, gain saturation, hole burning, and short-pulse generation, to name but a few (see the pioneering works of Lamb 1964; Haken 1964/1965).

### 3.6.1 Breakthrough: Theoretical description of modes in a laser cavity

In all the theoretical work, probably one of the key publications is that of Fox and Li (1961), who developed a generalized model to describe the overall EM-field distribution in a laser cavity to include cavity configurations and properties of the gain medium (although the latter only as an effective amplification contributor). Their aim was to calculate—for the first time numerically, using an IBM 704 computer—the field distribution of a wave propagating back and forth between two mirrors. The procedure started with an arbitrary field distribution at the first mirror, giving rise to a (calculated) field distribution at the second mirror after the first cavity transit; this then served as the originating field for the second transit to produce a field distribution at the first mirror again (for the geometry, see Figure 3.29). This was iterated until reaching a steady-state pattern.

Using this approach, Fox and Li could model the spatial amplitude and phase distribution within the cavity (both plane–plane and confocal spherical mirror configurations were addressed in the original publication) for the lowest $TEM_{00}$ and $TEM_{01}$ modes; in addition, the Q-factor of the cavity—in particular related to mirror losses and active-medium gain—could be included in the calculations, meaning that the quantum properties of the gain medium find their way into the modeling. In a certain way, one could classify their work as the first truly generalized laser cavity theory.

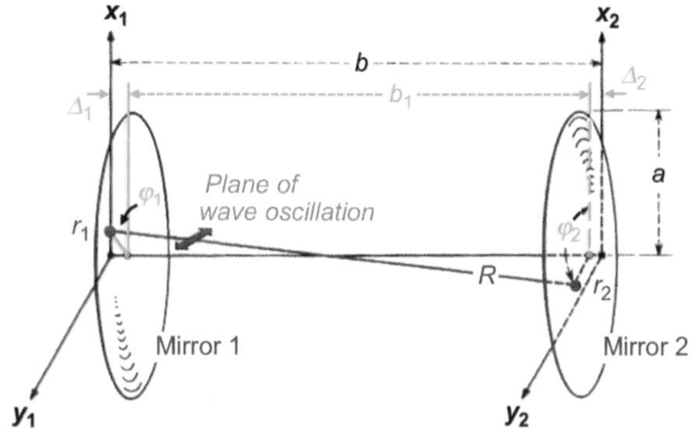

### 3.6.2 At the cutting edge: Steady-state ab initio laser theory for complex gain media

Similarly to the difficulty in identifying original breakthroughs in laser theory, it may be rather contentious as to which aspects of development one is tempted to classify as "cutting-edge"; this may very much depend on

**Figure 3.29** Geometry of the confocal spherical mirror arrangement to calculate the intracavity field distribution. (Adapted from Fox, A.G. and T. Li. *Bell Sys. Tech. J.* 40, no. 2 (1961): 453-488.)

the vista of the interests of a particular laser physicist. So here we arbitrarily stay with the same aspect as in the "breakthrough" segment, namely the modeling of (complex) laser systems and their behavior.

The aforementioned semiclassical laser theory that evolved from the works of Haken and Lamb is sufficient to describe all common mode properties in lasers. What semiclassical theory does not cover are phenomena that arise from quantum fluctuations.

In this context, a few years ago, a new, generalized theoretical framework was introduced by A.D. Stone's group at Yale University, now normally known as *Steady-state Ab initio Laser Theory*—SALT for short. The theory fully incorporates all spatial degrees of freedom of the (EM) fields and any "openness" of the laser system; it is suitable to correctly describe multimode lasing, and it is essentially exact when applied to the case of single-mode lasing, essentially replicating the results from semiclassical theory. The SALT equation systems can be solved efficiently for the steady-state properties of laser cavities of (nearly) arbitrary geometry, in any number of dimensions, and for complex gain media (see, e.g., Türeci et al. 2006; Cerjan et al. 2015).

The theory has been further generalized in a methodology dubbed *complex-SALT* (C-SALT), in which—in addition to the effects covered by SALT—an arbitrary number of levels and lasing transitions can be included, and inhomogeneous gain diffusion can be treated. The ability of C-SALT to treat multimode lasing in the presence of diffusion allows one to model many fundamental aspects of laser physics, including, e.g., the competition between spatial hole-burning and gain saturation.

As an example, the inversion in a four-level laser system, with a single laser transition, is plotted in Figure 3.30 for different (carrier) diffusion coefficients, which are associated with a nonlinear gain coefficient. The actual cavity is of the slab-type cavity—as encountered in, e.g., semiconductor laser diodes and amplifiers, or quantum cascade lasers.

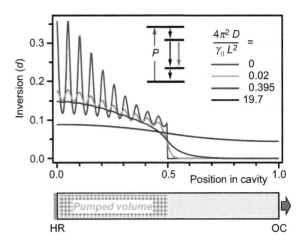

**Figure 3.30** Inversion in a partially pumped, one-sided dielectric slab cavity, for different values of the "carrier" diffusion coefficient (as a function of position in the cavity). The gain medium is a four-level, single-transition gain medium, with $n = 1.5$. (Data adapted from Cerjan, A. et al., *Opt. Express* 23, no. 5 (2015): 6455–6477; see the publication for further details.)

# CHAPTER 4

# Laser Sources Based on Gaseous, Liquid, or Solid-State Active Media

The first functioning laser, based on a flash lamp-pumped synthetic ruby crystal and emitting light at $\lambda = 694$ nm, was demonstrated by Maiman (1960) at Hughes Research Laboratories in California; see Figure 4.1. Since that extraordinary event of the birth of the laser, research groups around the world have been involved in perpetual activities to investigate laser sources themselves and to utilize them for a plethora of applications. It is probably a fair judgment that even now, some 50 years after the heydays of laser research, hardly a month passes during which a spectacular improvement of laser performance, a novel design, or a new commercial product is not publicized.

It cannot be the purpose of this and the next chapters to provide a comprehensive survey of all the different lasers ever demonstrated (literally many hundreds of laser materials and configurations); for details of many of the common or more exotic lasers, the reader may consult, for example, the books by Demtröder (2008) or Silfvast (2008). Instead, we will briefly summarize the key facts about those laser classes and devices that are commonly in use in laser spectroscopy and laser imaging, the topics of this book. Nevertheless, it is worthwhile to give a very brief picture of the historical development of the most important laser families to appreciate the evolutionary cycle in laser technology and their application to scientific and real-world problems.

Crudely, and at times with fuzzy borderlines, one may compartmentalize laser sources into four main classes, namely gas lasers, dye lasers (mostly liquid laser medium but occasionally solid media are used as well), solid-state lasers, and semiconductor lasers. In addition, laser sources based on nonlinear or parametric processes are widely used, and a few "exotic" laser sources have found their way into laser spectroscopy.

*Gas lasers.* When introducing lasers in textbooks, probably one of the most utilized examples is the helium–neon (HeNe) laser, because it constituted the first continuous-wave (CW) laser source [albeit operating on its IR transition rather than on the famous 632.8 nm line; it was constructed by Ali Javan and coworkers in 1960, shortly after the first experimental demonstration of laser action (Javan et al. 1961)]. A real work horse in laser spectroscopy, the argon ion laser (one in the family of rare-gas ion lasers) was invented only a few years later

Cooling enclosure

Ruby rod

Spiral flash lamp

**Figure 4.1** Maiman: the first ruby laser—dissembled into its components. (From Nelson, A. Video: "Maiman's first laser light shines again." *SPIE Newsroom*, May 20, 2010. DOI: 10.1117/2. 3201005.04. http://spie.org/x40717.xml)

(Bridges 1964). Besides these two well-utilized lasers, a number of other gas lasers have been in use, including the nitrogen ($N_2$) laser, the carbon dioxide ($CO_2$) laser, the excimer laser (like XeCl), and metal vapor lasers (like the copper vapor laser). However, other than for a number of niche applications, gas lasers slowly disappeared from the laser scene mainly because of their low efficiency and often high operating costs (see further below).

*Dye lasers.* This class of laser uses an organic dye (usually dissolved in a liquid solvent) as the lasing medium. Because of being complex organic molecules in solution, dyes normally offer a wide, continuous wavelength range of operation, which makes them suitable as tunable laser sources. This is in contrast to most gas lasers that emit light associated with narrow atomic or molecular transitions. Dye lasers were realized in 1966, independently by two groups (see Sorokin and Lankard 1966; Schäfer et al. 1966). While still in wide use, dye lasers are slowly being replaced by much more convenient, other laser devices.

*Solid-state lasers.* As stated above, the first laser ever—the ruby laser—was a solid-state laser; however, it rarely was used for spectroscopic applications. The most known and probably most used solid-state laser is the neodymium laser (with Nd:YAG as the laser-active material), which was first demonstrated at Bell Labs in the United States (Geusic et al. 1964); a typical green Nd:YAG laser is shown in Figure 4.2. Other rare-earth doped host materials have been added over the years (see further below), but the neodymium laser still remains the most popular. It took nearly another 20 years until tunable solid-state lasers were demonstrated, namely titanium-doped sapphire (Moulton 1982); for a long time, it was the laser exhibiting the widest tuning range (about 500 nm in the NIR). Before these two lasers were invented, another type of doped solid-state laser was demonstrated already after the invention of the laser, namely the optical fiber laser (Snitzer 1961b). But only in recent years did this type of laser gain wide penetration into laser spectroscopy and laser imaging.

Scattered light
on cell window

Incoming \ laser beam

**Figure 4.2** Laser beam from a CW DPSS Nd:YAG laser (second harmonic green emission).

## 4.1 PARAMETERS OF IMPORTANCE FOR LASER SPECTROSCOPY AND LASER IMAGING

In order to optimally implement a particular spectroscopic or imaging application, the laser source and its operating parameters have to be selected accordingly. The four key parameters are wavelength, temporal behavior, photon flux, and geometrical beam shape. Here we only briefly summarize important aspects of these linked to the lasers discussed in the sections below. A more detailed description, with close relation to the needs in laser spectroscopy and imaging, is provided in Chapter 3.

The *laser wavelength* is directly related to the stimulated energy transition and thus constitutes an intrinsic parameter linked to the laser-active material. Thus, when a specific wavelength is required for a particular application, a suitable laser will have to be selected (an example for spectral/temporal analysis is shown in Figure 4.3). By and large, commercial "direct" laser sources are available for nearly all wavelengths ranging from the visible, through the near- and mid-IR, up to the terahertz regime. On the other hand, with just a few exceptions, nowadays commercial-device UV wavelengths are generated by "indirect" means. By direct and indirect we mean that the radiation from a (direct) quantum-level laser transition is translated into an (indirect) wave of different frequency/wavelength by utilizing an added converter. Finally, in order to generate wavelengths toward the vacuum-UV (VUV), and even the x-ray regime, more "exotic" setups are required, which rarely are off-the-shelf devices. An overview of wavelength coverage of the most common sources of laser radiation is shown in Figure 4.4.

Note that in addition to the wavelength of a laser transition, one often will have to contemplate the bandwidth of said transition. For example, a narrow bandwidth is of importance if high-resolution spectroscopy is envisaged, while broad bandwidths are essential for the generation of ultrashort laser pulses.

In its most basic meaning, one associates with *temporal behavior* how long the device emits a train of laser photons, often ignoring for simplicity that over said "on" period, the emission may not be constant, or assuming that the "on" shape can be replicated by a Gaussian distribution function. Many of the commercial laser families described further below are available to run in CW mode, i.e., the laser is "always" on, or in pulsed mode. The duration of laser pulses can vary over a wide range, although the time-duration parameter can only in exceptional cases be altered at will. Commonly available laser sources emit nanosecond pulses (normally realized by means of Q-switching)—the $10^{-9}$ s regime; picosecond pulses (normally realized by means of mode locking)—the $10^{-12}$ s regime; or femtosecond pulses (by and large tailored from picosecond pulses)—the $10^{-15}$ s regime. In addition, one should not forget the common time–bandwidth product, associated with the uncertainty relation $\Delta v \cdot \Delta t \geq h$; for more on this aspect, see Chapter 3 or relevant laser textbooks (e.g. Silvast 2008).

When it comes to describing the magnitude of the light output from a laser, it seems that at times even the "experts" tend to mingle the actual definitions associated with the buzz words *laser energy*, *power*, and *intensity*. Laser power $P_L$ has to be understood as the rate at which the laser generates a flux of laser photons $h\nu_L$, i.e., the total output photon energy $E_L$ per time interval $\Delta t$ (in general, normalized as 1 s): $n \cdot h\nu/\Delta t = E_L/\Delta t = P_L$ in units of $(J \cdot s^{-1}) = (W)$. If the laser is working in pulsed mode, its pulse energy $E_{L,p}$ is a more appropriate parameter than pulse power, because the amplitude of the pulse more often than not changes over the duration of the pulse. However, for pulses, one often indicates their peak power $P_{L,peak} = E_{L,p}/\Delta t_p$, i.e., the pulse energy divided by its duration. In common (commercial) laser sources, a wide spread of output powers is encountered, ranging from a few nanowatts to many watts for CW lasers, and a few nanojoules to more than a joule for pulsed lasers. Note here that peak powers in ultrashort pulses can be enormous; for example, a 100 fs laser pulse of just 1 μJ energy exhibits a peak power of already 1 MW (assuming for simplicity that the power does not change during the pulse duration).

Finally, *geometric characteristics* of the laser beam need to be carefully considered; here the beam profile and the beam divergence are the key (extrinsic) parameters, governed predominantly by the physical dimensions and geometry of the laser cavity.

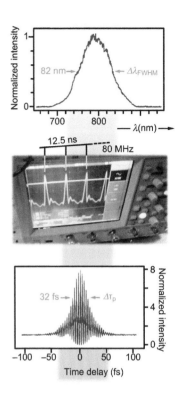

**Figure 4.3** Analysis of a femtosecond pulse from a Ti:sapphire laser. (Top) Spectral data; (middle) pulse repletion data; (bottom) temporal (autocorrelation) data.

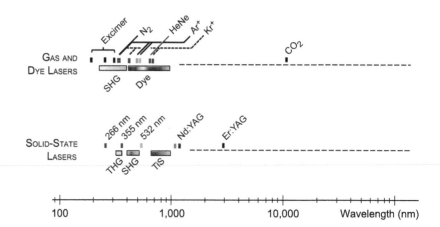

**Figure 4.4** Wavelength coverage by commercially available laser sources. SHG, THG = second and third harmonic generation, respectively; TiS = Ti:sapphire laser.

Most lasers these days strive to achieve a $TEM_{00}$ *spatial beam profile* (see Figure 4.5, Chapter 3.2, or standard laser textbooks). Not only does this simplify the formalistic parameterization of the profile, but also often this constitutes the best utilization of the optical laser medium, and thus the highest efficiency. Higher-order $TEM_{lm}$ (with either or both $l$, $m \neq 0$) profiles are sometimes unavoidable in high-gain media or for high-power laser sources. On the other hand, for specific bespoke laser sources, non-Gaussian profiles are generated on purpose, e.g., for the so-called "flat-top" (pulsed) lasers. The lateral dimension of a laser beam may change substantially the further away from the laser it is measured; the effect is known as *beam divergence*. In general, laser manufacturers and researchers alike prefer low divergence since it avoids that optical components for beam shaping are required even when the laser beam has to travel sizable distances from the source to the location of application.

In this context, two quantities related to laser power and pulse energy are relevant, namely the laser *intensity I* or *power density* (in units of $W \cdot cm^{-2}$) and the *laser fluence* (in units of $J \cdot cm^{-2}$). Both depend on the area through which the photon flux passes, and thus the beam diameter (or spot size) at the interaction location between laser light and sample. For example, for tight focusing of the laser beam by a lens, the spot size can be many orders of magnitude smaller than the original beam diameter; this means that the whole laser light energy is channeled into a minute area, and this density of energy can easily surpass the destruction threshold for the material: then so-called laser-induced damage occurs. It is worth noting that for certain applications, this threshold is surpassed on purpose, e.g., when laser ablation or plasma generation is the objective.

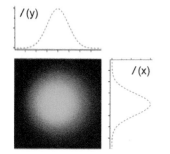

**Figure 4.5** $TEM_{00}$ profile of the beam from a green CW DPSS laser.

## 4.2  GAS LASER SOURCES (MOSTLY FIXED FREQUENCY)

As the name suggests, the laser-active media of this type of laser are either single atoms or molecules; often the gas filling comprises a mixture, with constituents other than the laser-active one present having auxiliary functions. Normally, the population inversion required for gain by stimulated emission is via an electric discharge in the gas (mixture); during operation, the gas comprises a significant concentration of electrically charged particles. The attraction of most gas lasers is that they emit their light with very high beam quality, often close to being diffraction-limited. On the downside, gas lasers usually require high voltages to maintain the discharge, and more often than not a substantial amount of electrical power is required. The latter is associated with the very low conversion efficiency from electric input power to laser light output power, which for the particular devices used in laser spectroscopy and imaging is rarely better than $10^{-3}$ and frequently less than $10^{-4}$. This means that for a target output power of 1 W, already 1 kW of electrical power is required as a minimum; this means that substantial cooling capacity is required to divert (discharge) heat away from the laser medium, often in the form of circulating water cooling. Because of this, powerful gas lasers are normally rather bulky and have often high operating costs, in particular if the discharge tube has to be replaced when the laser is operated perpetually over lengthy periods of time.

Because of their various drawbacks, it is probably fair to say that other than for a few specialist applications, gas lasers are slowly disappearing from applications in routine laser spectroscopy since much more convenient alternatives do exist. Notwithstanding this, we briefly describe one particular type of laser that still is frequently encountered in research laboratories and for some practical applications, namely the argon ion laser.

This laser belongs to the group of powerful rare-gas ion lasers (the other commercial, but less frequently used laser is the krypton ion laser), which can generate multiple watts of optical power. Its core component is an argon-filled discharge tube made from heat-resistant material (e.g., beryllium oxide ceramics for the highest output powers). The discharge is driven by a high-current high-voltage source to generate the high density of excited argon ions required for population inversion and laser gain. Typically, for a laser tube of 1 m length—generating of the order 10 W multiline laser light—the voltage drop across the tube is of the order of a hundred to a few hundred volts, while the current can be up to several tens of amperes. As stated above, the dissipated heat is removed by water flowing around the tube (for better control often in a closed-circle cooling system with a chiller); for lower-power devices yielding up to a couple of hundred milliwatts, air cooling may suffice (see the internal view of a low-power argon laser in Figure 4.6). The typical configuration of a rare-gas ion laser is shown in Figure 4.7.

Air-cooled discharge tube

**Figure 4.6** Inside view of a low-power air-cooled argon ion laser.

While run in multiline operation for numerous applications, like pumping tunable dye or solid-state lasers, individual lines can be selected by using an intracavity tuning prism (see option a in Figure 4.7). The most powerful and commonly used lines are those at 488.0 and 514.5 nm, although two other lower-wavelength lines also have been utilized for specific applications (at 457.9 nm and a narrowly spaced multiple at around 351 nm).

In general, rare-gas ion lasers are run in CW mode, but classically, these lasers were often utilized as well to generate short pulses of picosecond duration by means of mode locking; the principle of mode locking has been outlined in Chapter 3.4. A rather elegant solution to achieve this was demonstrated by Kitahara (1987) who introduced a compact cavity mirror–acousto-optic modulator combination into the laser cavity (see option b in Figure 4.7). In addition to the mode-locking capability, two further modification of a standard rare-gas ion

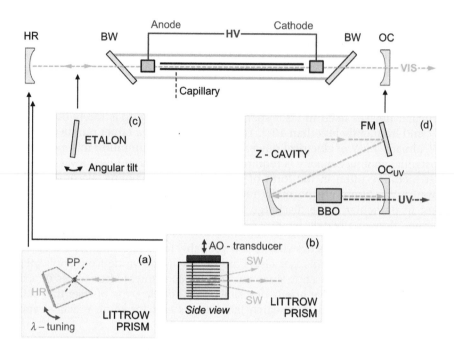

**Figure 4.7** Concepts of high-power rare-gas ion lasers. Cavity options: (a) line tuning with Littrow prism; (b) mode locking based on acousto-optic modulation; (c) single-mode operation with intracavity etalon; and (d) intracavity frequency doubling. BW = Brewster window; HR = high reflector; OC = output coupler; FM = folding mirror; PP = pivot point; BBO = b-barium borate crystal; SW = scattered wave.

laser cavity have frequently been used. One is the use of a (tilted) intracavity etalon to drive the laser into single-mode operation (see option c in Figure 4.7), which goes back to the late 1960s (see Hercher 1969). The other is intracavity frequency doubling to generate, e.g., 244.0 nm from the 488.0 nm line of the argon ion laser (see option d in Figure 4.7); typically, a so-called z-cavity is utilized for this (see, e.g., Asher et al. 1993). All these options were or are available with commercial rare-gas ion lasers.

Traditional applications for multiwatt rare-gas ion lasers include, for example, the pumping of (tunable) dye lasers and titanium–sapphire lasers. However, in recent years, they are rivalled and replaced by frequency-doubled diode-pumped solid-state lasers (see Section 4.4). These solid-state laser sources are far more power efficient (up to $10^{-1}$ rather than $<10^{-3}$), are normally much more reliable, and have substantially longer lifetimes. Nevertheless, rare-gas ion lasers are still in use in a few selected cases where no alternative wavelength (e.g., 488 nm) at the same laser light power was available. However, even here, the rapid progress in solid-state laser technology does suggest that most likely the days of rare-gas ion lasers in routine laser spectroscopy are numbered.

## 4.3 DYE LASERS (TUNABLE FREQUENCY)

The gain medium of dye lasers is—as the name suggests—an (organic) dye, mostly in the form of a liquid solution; typical solvents are methanol or ethanol for pulsed laser excitation, and ethylene glycol for CW laser excitation. While not very common, it is noteworthy that gain elements based on dyes imbedded in solid matrices do exist (see, e.g., Russell et al. 2005). A wide range of emission wavelengths from the UV to the NIR region is covered when utilizing different laser dyes. Typically, the wavelength range covered by an individual dye is of the order of 30–100 nm, and with about the 10 most-common laser dyes (out of the 100 or so commercially available compounds; see, e.g., Brackmann 2000), the complete visible range of the spectrum can be covered, although particular dyes may be selected for optimum tuning and efficiency in certain cases.

The typical characteristics of a dye gain medium may be summarized as follows. Firstly, in general, dyes exhibit a broad gain bandwidth, which permits wide wavelength tunability on the one hand and ultrashort-pulse generation (via mode locking) on the other hand. Secondly, the gain per unit length for most dyes is rather high, of the order $10^3$ cm$^{-1}$, which means that for nanosecond-pulse excitation with high pulse energy, gain saturation is easily encountered and thus oscillator–amplifier chains are required. Thirdly, for common laser pumping, be it CW or pulsed, high conversion efficiency can be reached, typically in the range 10–30% (see the example in Figure 4.8). As the figure demonstrates, the absorption/excitation wavelength is shorter than the emission wavelength; depending on the dye's absorption spectrum, excitation is afforded for short-wavelength laser dyes by pump lasers in the UV and violet (like excimer lasers, third harmonic Nd:YAG or Ar/Kr ion lasers) or the blue-green (like second harmonic Nd:YAG or Ar ion lasers). Finally, it should be stressed that many laser dyes and some of the solvents are poisonous and/or carcinogenic. Hence, one should be careful to avoid exposure of the skin to dye solutions. A particularly hazardous solvent is dimethylsulfoxide (DMSO); this is sometimes used for cyanide dyes (many near-IR dyes), and its particular hazardous action is that it greatly accelerates the transport of dyes into the skin. Therefore, it is not surprising, for these hazard reasons alone, that dye lasers are more and more being replaced by other types of tunable lasers, which in addition offer the advantage of lower maintenance (note that many laser dyes quickly deteriorate due to photochemical reactions, and thus their efficiency decreases rapidly as a consequence).

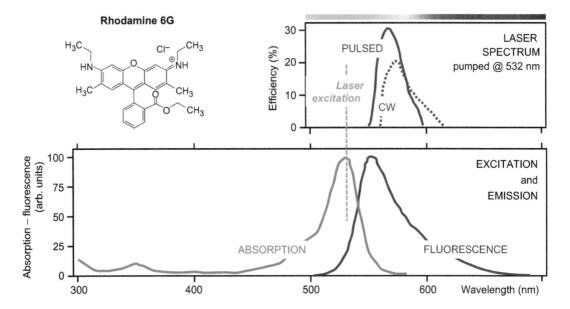

**Figure 4.8** Dye absorption and fluorescence spectra, together with laser emission spectra for CW and pulsed laser excitation, exemplified for the dye Rhodamine 6G.

As stated in the previous paragraph, dye lasers can be operated in both CW and pulse modes (today mostly nanosecond-pulse duration), although the configurations required for the two modes are substantially different. In particular, they differ in the way (1) the active dye medium is pumped; (2) wavelength selection is achieved; and (3) highest conversion efficiency is achieved. Here, we only briefly address the core concepts; for greater details, the reader is referred to relevant textbooks (see, e.g., Schäfer 1990; Duarte and Hillman 1990).

For CW operation, the dye is typically provided by a fast-flowing thin (of the order 100–200 μm) jet of the liquid solution (see Figure 4.9). The solvent has to have a sufficient viscosity to guarantee a smooth laminar flow; ethylene glycol is one of the few solvents meeting this requirement while at the same time providing adequate solubility for the majority of useful dyes. The fast flow is required to avoid the population of triplet states in the dye, which besides photochemical degradation constitutes the major loss mechanism in CW dye lasers (see, e.g., Chapter 7, "Photochemistry of Laser Dyes" in Duarte and Hillman 1990). In order to minimize reflection losses, the jet is oriented at the Brewster angle for the dye laser wavelength. The pump laser beam is tightly focused into the dye jet to provide the required pump photon flux to support population inversion for CW operation; this pumping is "longitudinal," i.e., roughly along the dye laser axis. This is afforded by the fact that the refractive index variation with wavelength provides an angle of incidence for the pump beam, which is different from the dye laser beam direction outside the jet.

**Figure 4.9** (Top) Laminar dye jet. (Bottom) Excitation of the dye, inside a laser cavity, by a green CW DPSS pump laser.

Because of the short length of the active medium, the beam emerging from the dye jet exhibits a rather large divergence so that the cavity mirrors should ideally be in near-concentric configuration to provide efficient feedback and amplification. In general, at least a third (normally plane or long-focal length) mirror is added to complete the cavity and serves as the output coupler. In this way, any frequency selective element encounters the laser beam in its collimated section— provided the focal position of the convex cavity mirrors is adjusted appropriately. The frequency-selective element has to be low loss and normally is a so-called Lyot (or birefringent) filter, oriented at the Brewster angle for the dye laser wavelength. In addition, etalon elements may be introduced into the cavity to reduce the bandwidth to a single longitudinal mode (SLM). The conceptual configuration for this type of CW dye laser is shown in Figure 4.10a. It is noteworthy that the majority of modern, commercial CW dye lasers are based on ring cavities (see Figure 4.10b).

The aforementioned CW dye laser setups in some modification may serve for the generation of ultrashort pulses. In fact, dye lasers dominated the field of ultrashort-pulse generation before the emergence of solid-state lasers like the Ti: sapphire laser. The wide dye gain bandwidth of often around 100 nm allows for pulse durations from a few picoseconds down to the order of 10 fs. Note that mainly due to the disadvantages associated with the handling and lifetime of laser dyes, femtosecond dye lasers are hardly any longer in use.

The configuration for short-pulse dye lasers whose pulse duration mimics that of the nanosecond-pulse pump laser is completely different to that of CW lasers, basically because of the much higher instantaneous photon flux from the pump pulse. Instead of a dye jet, now a flow cell is used; pumping for most commercial products is transverse rather than longitudinal, and the pump laser is focused into the dye cell by a cylindrical lens, to generate a pumped "stripe" volume along

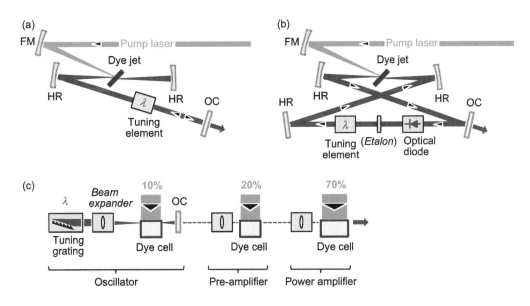

**Figure 4.10**   Conceptual dye laser cavity configurations. (a) CW linear laser cavity; (b) CW ring laser cavity; (c) nanosecond-pulse laser cavity with (optional) amplifier stages.

the dye laser axis (see Figure 4.10c). Because of the nanosecond-pulse duration (typically in the range 3–30 ns), a Lyot filter would not allow for a sufficient number of passages to achieve narrow bandwidth laser operation. Therefore, by and large, gratings are used for wavelength selection, in a variety of configurations, like Littrow and grazing-incidence configurations (for more details, see, e.g., Duarte and Hillman 1990). The high-peak pump-photon flux and the normally high gain of the laser dye quickly lead to gain saturation; as a consequence, the dye laser output does not any longer follow the increasing pump-pulse energy. In order to avoid this problem, normally a dye laser oscillator with its wavelength-selective element is run at suitably low pump-pulse energy, and then the oscillator output is augmented in a (chain of) amplifier stages (see Figure 4.10c). Overall, tunable nanosecond-pulse output energies of up to 100 mJ can easily be generated, depending of course on the available pump-pulse energy. It is worth noting that as a consequence of the transverse pumping geometry of both oscillator and amplifier dye cells in the majority of commercial dye lasers, their output beam profile is rather poor, and rarely does one encounter the ideal $TEM_{00}$ profile common to CW dye lasers. Thus, commercial lasers often use spatial filtering or so-called Bethune cells (see Bethune 1981) to at least provide a reasonably round beam profile. True Gaussian profiles can be realized using sophisticated, but normally noncommercial resonator and amplifier configurations (see, e.g., Corless et al. 1997).

As a final remark on dye lasers, it is likely foreseeable that this type of laser will more and more become a niche product, primarily because of the great disadvantages of laser dyes, i.e., their aforementioned (often rapid) photochemical degradation and thus high maintenance and operating cost, and because of the toxicity and carcinogenicity of dye solutions. Solid-state lasers, in particular titanium sapphire lasers and more recently fiber lasers, started to replace dye lasers (at least in the domain of ultrashort-pulse generation) as soon as they were sufficiently developed. Still, dye lasers are used, for example, in spectroscopic applications utilizing wavelengths that are still difficult to generate with other laser sources.

## 4.4 SOLID-STATE LASER SOURCES (FIXED AND TUNABLE FREQUENCY)

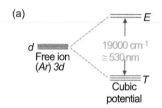

(a)

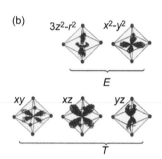

(b)

**Figure 4.11** (a) Splitting of the $Ti^{3+}$ valence $d$-orbital in a TiS crystal (simplified level diagram). (b) $3d$-electron orbital orientation with respect to the octahedral symmetry coordinates of the nearest-neighbor oxygen atoms.

The active medium of solid-state lasers is based on crystalline (or also glassy) "host" materials, which are "doped" with fourth row or sixth row atoms, by and large present in their triply ionized form.

The most common dopants belong to the group of lanthanide rare-earth elements, such as erbium (Er) or neodymium (Nd), because the excited states of their ions couple only weakly with the (thermal) phonons of their crystalline host lattices (phonons), and thus losses are relatively low, which as a consequence also means low pumping thresholds. Since the ruby laser was demonstrated just over 50 years ago, laser action has been demonstrated in literally hundreds of solid-state media; however, only a few are in widespread use. Probably the best-known and widely used solid-state laser material is neodymium-doped yttrium aluminum garnet or Nd:YAG; lasers based on Nd-doped active media will be discussed further below.

It is noteworthy that some solid-state laser materials lend themselves for tuning; these are doped with transition metals, such as chromium (Cr) or titanium (Ti). The tunability of these materials is due to the fact that, in contrast to the lanthanides, transition metal atoms interact with the lattice of the host. In particular, the $3d$ valence electron level of $Ti^{3+}$ splits in the presence of the crystal field, and the displacement potentials of these levels with respect to the nearest-neighbor oxygen atoms of the sapphire host give rise to a wide vibronic band structure (see Figure 4.11). The widest exploited commercially is titanium-doped sapphire, or Ti:sapphire (or TiS); laser devices based on this active medium will also be discussed in this section, further below.

Finally, over the past 10 years or so, lasers have become increasingly popular whose host material is "glassy," in the form of a fiber, which is doped with rare-earth elements. These are related to doped fiber amplifiers, which provide light amplification without lasing and are widely used in telecommunications. Examples of fiber lasers are described separately in Section 4.5.

By and large, solid-state lasing media are typically photon-pumped, using either a flash lamp or laser sources. Laser-pumped devices tend to be much more efficient because the pump wavelength can be or is tailored to the active medium absorption. A rather unique example for such tailored excitation is the chain of a diode-pumped Nd:YAG (solid-state) laser—known as a DPSS laser—which in turn pumps a Ti:sapphire laser.

### 4.4.1 Nd:YAG lasers

Probably the best-known and widely used solid-state laser material is neodymium-doped yttrium aluminum garnet, $Nd:Y_3Al_5O_{12}$ or Nd:YAG. The actual dopant in the crystal matrix is triply ionized neodymium, $Nd^{3+}$, which typically replaces a small fraction (of the order of 1%) of the host's yttrium ions. These Nd ions give rise to the laser activity in the crystal. It should be noted that instead of the YAG crystal, a few other host materials are common nowadays, including yttrium ortho-vanadate ($YVO_4$) and yttrium lithium fluoride ($YLiF_4$), or YLF. Said host materials are often chosen because of superior laser performance in certain configurations. Nd:YAG is a four-level gain medium (except for the 946 nm transition mentioned below). Because of the low interaction of the Nd levels with the host crystal atoms, the energy states are only marginally broadened and thus the gain bandwidth is relatively small. On the other hand, this allows for a high gain efficiency and thus low threshold pump power.

Nd:YAG exhibits the highest absorption in the wavelength band 790–820 nm, corresponding to the transition from the ground state (lowest fine structure level of the $^4I_{9/2}$ state) to the excited state $^4F_{5/2}$. Other, higher-lying pump levels of strong absorption probability are found around 500–600 nm (the $^4G$ manifold). This lends itself to excitation by standard flash lamps (normally for nanosecond-pulse laser operation) or most efficiently by laser diodes operating at 808 nm (normally for CW or short-pulse operation). The pumped (excited) Nd levels undergo rapid relaxation to the upper laser level, $^4F_{3/2}$. From here, the strongest and most commonly used laser transition is that to a fine structure sublevel of the $^4I_{11/2}$ state, at $\lambda = 1064$ nm. A number of other, less strong laser transitions can be utilized; in particular, the transitions to the $^4I_{13/2}$ state, at $\lambda = 1322$ nm, and to the $^4I_{9/2}$ state, at $\lambda = 946$ nm, are exploited in a number of frequency-doubled DPSS laser sources to provide laser light in the blue and the red parts of the spectrum. For a more detailed description of the energy level manifold and dominant transitions, see, e.g., Zagumennyi et al. (2004); here, we only show a simplified diagram scheme (Figure 4.12). Note that the absorption and emission lines of Nd: YLF are slightly shifted to shorter wavelength values, with respect to the other host materials (in particular, Nd:YAG).

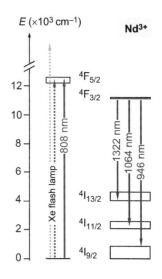

**Figure 4.12** Simplified $Nd^{3+}$ energy level diagram for the main transitions in Nd:YAG lasers.

The cavities for CW and pulsed operation of Nd:YAG lasers are very different, in general, due to the different way efficient gain exploitation has to be implemented. CW lasers these days are predominantly pumped longitudinally using laser diode radiation at around 808 nm, and specifically for high output power devices, efficient ring cavities have become the norm. An example for such a ring-cavity setup is shown conceptually in Figure 4.13, with the addition of an (optional) SLM element and intracavity frequency doubling from inherent 1064 nm wave to 532 nm. Note that in the setup shown here, the intracavity wave is forced to travel in one direction only by an optical diode.

Linear cavities are normally used in devices of lower output power, up to a few hundreds of milliwatts (e.g., in the smallest 532 nm Nd:YAG device shown in Figure 4.14).

Pulsed Nd:YAG lasers are available in two commercial variants, namely lasers with nanosecond-pulse duration, based on Q-switching, and devices for pico-second-pulse operation, based on mode locking. For the latter, both passive and active mode locking can be implemented, the former utilizing a semiconductor saturable absorber mirror (SESAM) element and the latter normally an acousto-optic modulator device; see, e.g., Ling et al. (2010) and Spühler et al. (1999), respectively.

Q-switched lasers exhibit pulse durations of a few nanoseconds and provide output pulse energies of up to a few joules, while mode-locked Nd:YAG lasers

(a)

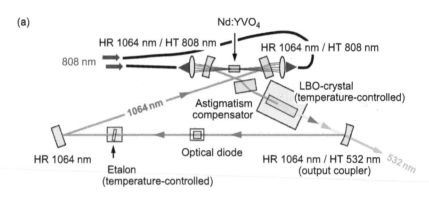

(b)

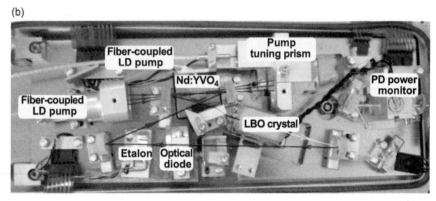

**Figure 4.13** Green ring-cavity DPSS laser. (a) Schematic of the laser cavity, with longitudinal pumping by fiber-coupled 808 nm laser diodes; HR/HT = high-reflector/high-transmitter at the indicated wavelength. (b) Top view of *Verdi* 5 W single-longitudinal mode laser (Coherent plc); with the components labelled according to the schematic.

**Figure 4.14** Smallest DPSS Nd: YAG laser emitting up to 50 mW at 532 nm. (Image courtesy of SnakeCreek Lasers. "MicroGreen™ APC Series." Friendsville (PA), USA: SnakeCreek Lasers, 2013. Available at http://www.snake creeklasers.com/products/laser heads.aspx)

normally run at multimegahertz repetition rates, providing laser pulses of 10–30 ps duration and pulse energies of a few microjoules. Should those pulse energies not suffice for certain applications, regenerative amplifiers may be used, which run at up to a few kilohertz repetition rates and amplify individual pulses up to several millijoules; of course, for this, a so-called pulse picker needs to be used to synchronize specific pulses out of the normal multimegahertz pulse train to the pulses from the pump laser for the amplifier. The typical operating parameters for commercial Nd:YAG lasers are collated in Table 4.1.

Finally, it should be noted that, more often than not, the fundamental wavelength of 1064 nm is shifted by harmonics generation to 532, 355, or 266 nm (second, third, or fourth harmonic, respectively).

Typical applications to laser spectroscopy and laser imaging for Nd:YAG lasers are found in (1) the pumping of other laser devices (like dye or Ti:sapphire lasers); (2) laser-induced breakdown spectroscopy (LIBS; utilizing nanosecond-pulse laser sources); (3) Raman spectroscopy (by and large using CW lasers); and (4) image velocimetry in fluid dynamics (using high-repetition short-pulse lasers), to name but a few. Some of these examples will be discussed later (see Chapters 12 and 15).

### 4.4.2 Ti:sapphire lasers

Titanium-sapphire, also addressed as Ti:sapphire or TiS, is a laser-active medium based on a sapphire ($Al_2O_3$) crystal. Titanium is doped into the crystal, typically at around the 0.1% level, and is encountered in its triply ionized form, $Ti^{3+}$. Its

**Table 4.1**    Common Commercial Solid-State Laser Systems and Their Characteristics

| Laser | Operating Characteristics | Typical Applications |
|---|---|---|
| Nd:YAG (CW) | Wavelength, $\lambda$: 532 nm ($2\omega$ of 1064 nm)<br>Output power, P: 1 mW ... 20 W<br>Linewidth, $\Delta v$: a few megahertz ... a few gigahertz<br>Multi- and single longitudinal mode | Pumping of other (tunable) lasers;<br>Raman spectroscopy |
| Nd:YAG (pulsed) | Wavelength, $\lambda$: 1064 nm (plus 532 nm/355 nm/266 nm harmonics)<br>Output pulse energy, $E_p$: 1 ... 1000 mJ<br>Linewidth, $\Delta v$: a few gigahertz<br>Pulse duration, $\Delta t$: 1 ... 25 ns<br>Pulse repetition rate, PRR: 10 Hz ... 1 kHz<br>*Note*: also available systems with $\Delta t \sim 10$ ps/PRR=1 MHz/$P_{ave}$=1 W | Pumping of other (tunable) lasers;<br>Laser-induced breakdown spectroscopy (LIBS) |
| Ti:sapphire (CW) | Wavelength, $\lambda$: 690 ... 1100 nm<br>Output power, P: up to 5 W<br>Linewidth, $\Delta v$: a few megahertz ... a few gigahertz<br>Multi- and single longitudinal mode | Wavelength-tunable laser for atomic and molecular spectroscopy |
| Ti:sapphire (pulsed) | Wavelength, $\lambda$: 700 ... 1000 nm<br>*Nanosecond-pulse operation*<br>Output pulse energy, $E_p$: 1 ... 500 mJ<br>Linewidth, $\Delta v$: a few gigahertz<br>Pulse duration, $\Delta t$: 10 ... 30 ns<br>Pulse repetition rate, PRR: 10 Hz ... 100 kHz<br>*Femtosecond-pulse operation*<br>Output pulse energy, $E_p$: 1 ... 500 nJ<br>Linewidth, $\Delta v$: Fourier-limited<br>Pulse duration, $\Delta t$: 10 ... 100 fs<br>pulse repetition rate, PRR: 1 ... 100 MHz | Wavelength-tunable laser for atomic and molecular spectroscopy<br><br><br><br>Ultrafast laser spectroscopy and imaging |

free-ion 3d $^2$D ground state is fivefold degenerate; however, this degeneracy is removed by the sapphire crystalline field (cubic field), giving rise to a $^2T_{2g}$ and a $^2E_g$ state, the latter being about 19,000 cm$^{-1}$ higher in energy. The upper state undergoes Jahn–Teller splitting into a doublet, while the lower state splits via trigonal field interaction and spin-orbit effects into a triplet. For a full description of the level structure and splitting mechanisms, see Ma et al. (2006). Because the Ti$^{3+}$ ions do interact with the lattice—other than the Nd$^{3+}$ ions in the YAG crystal—the energy levels broaden into a vibronic structure, within the framework of the next-neighbor oxygen atoms, in O$_h$-symmetry. The level splitting was shown conceptually in Figure 4.11a, while the relevant vibronic potentials are shown (reasonably to scale) in Figure 4.15; note that, for clarity, only the potential associated with the lowest $^2T_{2g}$ state (E$_{3/2}$) is shown.

Because of the vibronic potential structure, a wide range of possible excitation and emission transition energies is available, respectively, into the excited E$_{1/2}$ and E$_{3/2}$ states, and from the vibrational minimum of the upper E$_{3/2}$ potential into high-lying vibrational levels of the ground state E$_{3/2}$ potential. The related wavelength ranges are shown in the panels to the left and right of the potential energy diagram in Figure 4.15. According to these, the absorption for optical pumping peaks at just under 500 nm, while the maximum of the (laser) emission is encountered at around 800 nm.

Traditionally, Ti:sapphire lasers were pumped by argon ion lasers, predominantly using its powerful 514.5 nm radiation. However, with the advent of

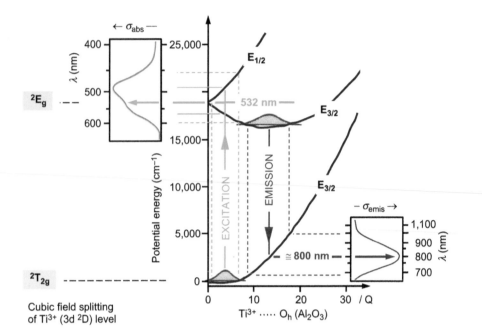

**Figure 4.15** Schematic, simplified energy level diagram for Ti:sapphire. The unperturbed Ti$^{3+}$ (3d $^2$D) level splits up into the $^2E_g$ and $^2T_{2g}$ multiplets in the cubic field of the host crystal, and gives rise to vibronic potentials due to the interaction of the Ti$^{3+}$ ion with nearest-neighbor oxygen atoms, in octahedral O$_h$ symmetry.

affordable high-power DPSS lasers, specifically frequency-doubled Nd-doped lasers (e.g., Nd:YAG(2$\omega$) at 532 nm), the very inefficient argon ion laser is hardly used any longer. With the availability of powerful blue diodes at 455 nm of up to 1 W of output power, whose development was by and large driven by the digital projection market, now even direct laser diode pumping has become feasible, albeit mostly for CW and ultrashort-pulse operation (see Roth et al. 2009; Durfee et al. 2012).

The width of the emission band (from about 650 nm to beyond 1100 nm) distinguishes Ti:sapphire from nearly all other solid-state laser materials. Association with this very impressive tunability has been demonstrated for CW and nanosecond-pulse implementations. But it may be stipulated that this enormous bandwidth has helped to develop the world's highest-performance femtosecond laser systems, and at least at present, Ti:sapphire dominates the commercial market for high-power ultrafast lasers. As is evident from these global statements, Ti:sapphire lasers are available in three main commercial versions, that is, CW, nanosecond-pulse duration, and femtosecond-pulse duration (ultrafast) implementations. The interesting points, advantages, and pitfalls encountered for the three regimes are briefly summarized next.

CW Ti:sapphire lasers come in two basic implementations, namely based on linear (but usually "z-folded") cavities and ring cavities, not very much different in concept to the dye laser cavities shown in Figure 4.10; in Figure 4.16, the typical crystal pumping geometry is displayed. The second one is found in the majority of commercial systems since it normally exhibits higher conversion efficiency from pump to output laser radiation, in well-designed and well-aligned lasers 20–30%, which rivals most dye lasers, and output powers exceeding 2W are not uncommon. The tuning range is normally 700–1000 nm, although this can be extended to 650 and 1100 nm at the lower and upper ends, respectively.

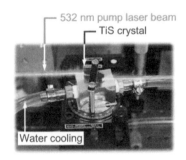

**Figure 4.16** Ti:sapphire crystal pumped by a green CW DPSS laser.

However, for the extended ranges, normally different mirror sets are required, which adds the tedious task of exchanging mirrors and realigning the laser to an otherwise simple-to-use laser. Still, while impressive, the range of available wavelengths suffers from a gap in coverage between the lower end of the fundamental tuning range and the upper end of the second harmonic range, i.e., roughly for the interval 500–650 nm where dye lasers still are competing well. In general, the tuning element is a birefringent filter, as for CW dye lasers, which yields bandwidths of 20–30 GHz. To achieve narrower linewidth values, etalon(s) will have to be added in the cavity; some commercial systems are available with SLM options, narrowing the laser output to a few megahertz or below.

While a number of commercial Ti:sapphire laser systems with nanosecond-pulse duration have been marketed, they have rarely found major penetration, for similar reason as for CW implementations, that is, the wavelength gap for a substantial part of the visible spectrum. Nevertheless, they have proven to be superior when it comes to the coverage of their inherent near-IR wavelength range. Ti:sapphire lasers operating in the nanosecond regime suffer from one caveat: this is related to the relatively long lifetime of the upper laser level of $\tau^*$ $(E_{3/2}) \sim 3.2\ \mu s$, rather than the few nanoseconds that laser dyes exhibit. Thus, the population inversion and the development of the cavity gain are strongly time-dependent when the pump laser is a Q-switched laser of 10–20 ns pulse duration. Depending on the pump pulse energy lasing threshold is reached at different times; as a consequence, the laser output pulse can be time-shifted in a somewhat uncontrollable manner with respect to the well-defined nanosecond-pump pulse (pulse jitter), associated with statistical pulse-to-pulse intensity fluctuations of the pump laser. This can severely affect experiments in which precise synchronization against an independent, stable trigger source is required. Overall, output pulse durations are of the order 10–25 ns, and pulse energies of a few hundreds of millijoules are achievable, depending on the pump pulse energy; for high-power pumping, conversion efficiencies of the order 25–30% are common.

Commercial Ti:sapphire laser systems generating ultrashort pulses are mostly exploiting passive mode-locking configurations, incorporating a saturable Bragg reflector (SBR), or a semiconductor saturable absorber mirror (SESAM)—see its conceptual construction in Figure 4.17; or using Kerr lens mode locking (KLM). Pulse durations of the order of 100 fs are common place, and some commercial devices even offer pulse durations down to around 10 fs. The shortest pulses obtained directly form a Ti:sapphire laser were around 5.5 fs, corresponding to less than two light cycle oscillations (see Morgner et al. 1999); see Figure 4.18. Typically, pulse repetition rates are in the range 70–100 MHz, and their average output powers are of the order of 100–1000 mW.

The common operating parameters for all three types of Ti:sapphire lasers discussed above are collated in Table 4.1.

It is worth noting that besides being the laser-active medium, Ti:sapphire is also frequently used as the medium for femtosecond-pulse amplifiers. Commonly, pulse durations achievable from commercial amplifiers are in the range 20–100 fs, and pulse energies of 1 mJ at 1 kHz repetition rate are easily generated using standard, frequency-doubled Nd:YAG or Nd:YLF nanosecond-pulse lasers; the corresponding pulse peak powers reach several terawatts (TW). Even higher pulse energies and peak powers—into the petawatt (PW) regime—are available at some of the national laser facilities, albeit at hugely reduced repetition rate.

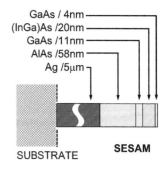

**Figure 4.17** SESAM device (relative layer thickness roughly to scale) utilized in the experiment by Morgner et al. (1999).

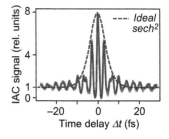

**Figure 4.18** Shortest pulse directly from a TiS laser, with FWHM $\cong$ 5 fs, equivalent to <2 oscillations. (Data adapted from Ell, R. et al., *Opt. Lett.* 26, no. 6 (2001): 373–375.)

Today, virtually all TW- and PW-scale experiments around the world rely on Ti:sapphire laser amplifiers.

Two principle design configurations are used for femtosecond amplifiers: regenerative and multipass amplifiers.

*Regenerative amplifiers* operate by amplifying an individual input pulse in a cavity-type configuration. However, instead of an ordinary laser cavity with an output coupler, the amplifier cavity incorporates a high-speed optical switch through which a pulse is injected and then extracted once the desired intensity has been reached. Note that, in general, "chirped-pulse" configurations (see Chapter 3.4), for intermediate stretching and compression of the pulse duration, are required to prevent the pulse from damaging the optical components in the amplifier.

A *multipass amplifier* does not require an optical switch; instead, using beam steering mirrors, a femtosecond pulse is guided through the Ti:sapphire crystal repeatedly, with slightly different directions. A pulsed pump beam is also guided repeatedly through the crystal, overlapping spatially with the femtosecond-pulse paths. Normally, the first "signal" passes through the center of the first (spot-like) pumped region for maximal amplification; in the following passes, the diameter of the amplified pulse is increased so that the power density stays below the damage threshold, and to suppress amplification of the outer parts of the femtosecond-pulse profile (see Figure 4.19). This increases the beam profile quality but also eliminates some of the unavoidable amplified spontaneous emission.

Typical applications of Ti:sapphire lasers in laser spectroscopy and imaging are found, for example, in multiphoton microscopy and imaging techniques for which the wavelengths and pulse durations can be ideally tailored.

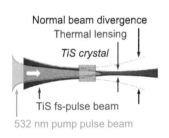

Normal beam divergence
Thermal lensing

*TiS crystal*

TiS fs-pulse beam

532 nm pump pulse beam

**Figure 4.19** Conceptual femto-second-pulse beam shaping in an individual pass through a multipass Ti:sapphire amplifier.

## 4.5 FIBER LASER SOURCES

Nowadays, fiber lasers are addressed as forming their own class of lasers, separate from solid-state lasers, although in principle their optical media exhibit key similarities; that is, by and large, they rely on the same doping ions and their inherent transitions. Here we give only a basic description of the phenomena and devices of importance for laser spectroscopy. More in-depth discussions may be found in Digonnet (2001) and Okhotnikov (2012).

Traditional solid-state lasers incorporate a gain medium consisting of a bulk host crystal, which is doped with metal ions (like Nd:YAG or Ti:sapphire above); the pump radiation and generated laser beams propagate freely through the gain medium, and the laser cavity is made up from standard, free-space optical components. In contrast, in fiber lasers, the gain medium is based on an optical fiber that in most cases is not crystalline in nature, and with the light being confined to a small core, it acts as a limiting waveguide. The fiber is doped with the laser-active (rare-earth) ions and the waveguide itself becomes a gain medium. Most common are the doping ions of erbium ($Er^{3+}$)—with emission in the range 1500–1600 nm and around 2700 nm; ytterbium ($Yb^{3+}$)—with emission in the range 1000–1100 nm; thulium ($Tm^{3+}$)—with emissions in the range 1450–1530 and 1700–2100 nm; and praseodymium ($Pr^{3+}$)—with emission around 1300 nm. Note that these emission bands, as well as the absorption transition wavelengths, may change slightly, depending on in which host glass material the ions are implanted, because of the different Stark-splitting encountered in the various host crystal fields (see Yang et al. 2014).

Although laser gain in a fiber was already demonstrated in the early 1960s (e.g., Snitzer 1961b), historically much of today's success and market penetration for fiber lasers stem from the use of these gain materials in diode-pumped telecommunication boosters, like erbium-doped fiber amplifiers for the C- and L-bands and thulium-doped fiber amplifiers for the S-band. By now, the reliability, compactness, cost, and quality advantages of fiber laser sources have resulted in a wide range of sometimes very novel and versatile laser sources.

Note that fiber laser media are slightly different from most other commonly used laser gain media. Their operating behavior is often categorized as a quasi-three-level scheme. This scheme may be seen as an intermediate between the common three- and four-level laser descriptions: here, the lower laser level is so close to the ground state that at the operating temperature of the laser, it may be thermally populated to a measurable degree. This then means that sizable reabsorption losses can occur at the laser wavelength. Still, fiber lasers can be extremely efficient. For example, when pumping ytterbium-doped silica at 946 nm and monitoring the 1026 nm transition (see Figure 4.20), one encounters a photon conversion efficiency of over 90%. In contrast, for neodymium-doped solid-state laser media, pumped on the standard 808 nm line and emitting at 1064 nm, one finds a conversion efficiency of just around 75%. Of course, not all of the nominal quantum efficiency transgresses into laser efficiency. This is because the spectral shape of the optical gain in such quasi-three-level media depends on the balance between emission and reabsorption, and on resonator losses at the laser wavelength.

Initially, one of the main problems in the development of fiber lasers has been the fact that the fiber core is too small to efficiently focus light of low-quality beam profiles, like that from common laser diodes, into the gain medium. However, with advances in fiber technology now, many of these problems have been overcome, specifically with the development of double-clad fibers (see Figure 4.21). In these, the pump radiation is coupled into the first cladding layer, which is substantially larger than the doped core. This also means that the requirements on the pump source concerning beam profile and mode structure

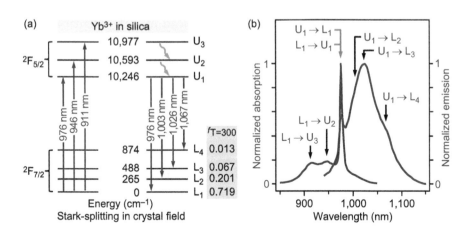

**Figure 4.20** Absorption and emission in ytterbium-doped silica fibers. (a) Energy level diagram of the Stark-split $^2F_{7/2}$ and $^2F_{5/2}$ manifolds, with the transition wavelengths and fractional thermal population in the lower levels. (b) Normalized absorption and emission spectra, with the transition maxima from (a) indicated. (Data adapted from Yang, B. et al., *Opt. Lett.* 39, no. 7 (2014): 1772–1774.)

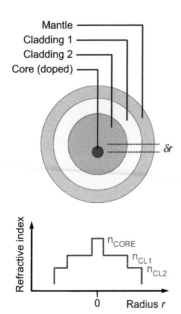

**Figure 4.21** Double-clad step-index fiber whose doped core is offset from the symmetry center; pump laser radiation guided in cladding 1.

are much less stringent. In fact, the first cladding is frequently shaped to match the beam profile of a particular pumping source. Regardless of the cladding shape and the pump beam characteristics, still near-perfect $TEM_{00}$ fiber laser beams can be generated. This so-called "cladding pumping," which was first demonstrated by Po et al. (1989), has clearly revolutionized the development of fiber lasers.

Finally, it should be noted that a rather novel type of fiber gain media, the so-called photonic band gap fibers, offers even larger flexibility in fiber laser parameters.

One caveat of exposing long-path media to laser radiation (from the pump laser) is the occurrence of unwanted nonlinear effects, namely stimulated Brillouin and stimulated Raman scattering. This normally limits the power scaling of fiber lasers. These unwanted effects become less prominent provided the fiber gain medium is short; on the other hand, for efficient operation, the pump radiation needs to be absorbed in the core along this short length. It is clear that these requirements ask for carefully engineered fibers, with respect to cladding profiles, dimensions, and refractive indices.

For various reasons, the design of fiber laser resonators is slightly more complex than for lasers based on bulk gain media. In particular, since nearly all fiber laser gain media exhibit a quasi-three level nature—as pointed out above—one requires strong optical pumping to achieve laser gain. As a consequence, strong saturation effects, associated with the high optical pump intensities in the tiny laser-active fiber core, are also observed.

Most fiber lasers are pumped by one (or several) fiber-coupled laser diode, whose emission wavelength is matched to the doping ion in the laser-active fiber core. In general, dual-clad fibers are used, and the pump light is coupled into the core along the whole length of the active fiber length.

The conceptual setup for a fiber laser resonator and its various mirror options is shown in Figure 4.22.

Laser resonators that incorporate a fiber can be designed as linear systems (with some kind of reflectors) or as fiber ring cavities. Several conceptual "mirror" solutions for linear fiber laser resonators are common. In "quick-to-try" laboratory setups, ordinary dielectric mirrors can be "butted" to the perpendicularly cleaved fiber ends. Clearly, such an approach is not very reliable and stable, and also not very practical for mass fabrication. Instead of such external mirrors, dielectric coatings can be deposited directly on (polished) fiber ends; these coatings can be tailored to the pump and laser wavelengths.

One may not even need any coating at all since in favorable cases the Fresnel reflection from the bare fiber end face may be sufficient to provide cavity gain. For commercial products, it is common to use the so-called fiber Bragg gratings. These either can be written directly in the doped fiber or are incorporated in a separate, nondoped fiber that is spliced to the actual laser-active fiber.

Another option is based on the use of standard fiber couplers (whose splitting ratio may be selected to achieve the desired feedback ratio) and a short piece of not-laser-active fiber. From a theoretical point of view, fiber loop mirrors may be described as sort of a Sagnac interferometer. If one assumes that (1) a single-mode fiber is used; (2) nonlinear effects can be excluded (i.e., the circulating power is not too high); and (3) a fiber coupler with 50:50 power splitting ratio is

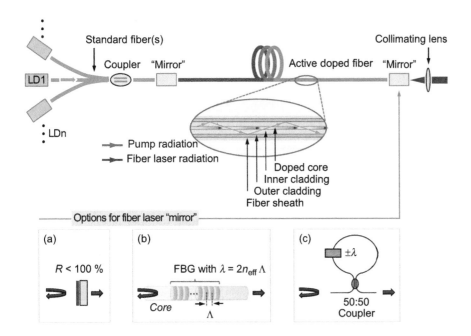

**Figure 4.22** Concept of a fiber laser resonator. The laser-active doped dual-cladding fiber may be straight or coiled (as shown). LD1 ... LDn = pump laser diode(s). Three main types of cavity mirrors are shown: (a) external dielectric mirror; (b) FBG optimized for a selected wavelength; and (c) Sagnac fiber-loop mirror, incorporating a 50:50 fiber coupler. Optionally, mirrors (b) and (c) can be wavelength-tuned.

used, then a fiber-loop mirror can be treated as a perfect reflector, which operates over a wide range of wavelengths and for any input polarization state. This makes them very attractive for fiber lasers and is found in many commercial designs.

Overall, in general, the high efficiency of commercial fiber lasers is due to two facts. Firstly, already the pump lasers, in the form of semiconductor laser diodes, are extremely efficient, often reaching electrical-to-optical power conversion of the order of 50% or more. Matching the laser diode emission wavelength carefully to the absorption line(s) of the fiber laser gain medium, the pump efficiency may reach unity. Secondly, the optical-to-optical conversion efficiency—from pump to laser photons—can be extremely high. For transitions with high quantum yield (see above), and using a carefully designed cavity (optimized for excitation and extraction efficiency), optical-to-optical conversion efficiencies of the order of 60–70% are reported for quite a few of the common fiber lasers. Combining the two, wall-plug efficiencies in the 25–35% range are achievable. Output powers of a few watts are very common, and even values above 1 kW have been reported already in the early 2000s (see Jeong et al. 2004); lasers with above-kilowatt output powers can be found in the portfolio of various commercial fiber laser manufacturers.

## 4.5.1 Wavelength selection and tunability

It is clear from the emission profiles from fiber laser gain media (in Figure 4.20, the example for ytterbium-doped silica fiber was shown) that the output can have, in principle, a rather wide bandwidth. On the other hand, fiber laser cavities can be set up to achieve laser operation on a SLM, whose linewidth can be as narrow as a few kilohertz (see Huang et al. 2005); commercial SLM systems, with kilohertz linewidth and multiwatt power output, are available from a number of manufacturers.

To achieve narrow linewidths, the resonator ends need to incorporate narrow-bandwidth fiber Bragg gratings (FBGs), selecting the actual single resonator mode. Typical SLM output powers are of the order 1–100 mW, although SLM fiber lasers with up to about 1 W output power have been demonstrated. The FBGs, used for wavelength selection in the fiber laser cavity, also provide the possibility to tune the laser over its gain bandwidth. For example, Royon et al. (2007) demonstrated a CW fiber laser, with ytterbium-doped gain medium, which could be tuned over nearly the complete $Yb^{3+}$ bandwidth range 976–1120 nm, as shown in Figure 4.20, yielding output powers of more than 10 W with adjustable linewidth in the range 0.1–1.0 nm.

Besides the very useful NIR wavelengths provided by fiber lasers, their energy level structure often lends itself for the realization of so-called "up-conversion" lasers. While for up-conversion one often has to utilize weak transitions, thus requiring high pump intensities to reach laser gain, sufficient gain efficiency is achievable provided that the fiber length is long enough. However, it has to be noted that, in most cases, doped silica glass media are not suitable because the up-conversion scheme requires relatively long lifetimes of the intermediate levels in the stepwise excitation. In silica fibers, the lifetimes of the relevant metastable states are often shortened significantly because of the relatively large phonon energy of silica glass. Therefore, silica glasses are, by and large, substituted using heavy-metal fluoride fibers instead, for example, one of the fluoro-zirconates, ZBLAN, which normally exhibit low phonon energies. The most popular up-conversion fiber lasers are based on thulium-doped fibers (see the level scheme in Figure 4.23)—for blue laser light generation; praseodymium-doped media—for wavelengths in the red, orange, green, or blue (commercially available are the values around 715 nm and in the bands 635–637, 605–622, 517–540, and 491–493 nm); and erbium-doped fibers—for blue and violet laser light generation. In general, output powers are in the range 1–100 mW, but for selected transitions, outputs above 1 W have been demonstrated. For a short survey on up-conversion fiber lasers, see Zhu and Peyghambarian (2010).

Complementary to up-conversion (which shifts the laser emission to shorter wavelengths in relation to the pump laser), fiber laser down-conversion to longer wavelengths has been utilized extensively, initially in telecommunication applications. The down-shift is realized by exploiting the aforementioned, normally unwanted effect of Raman shifting in the fiber medium. The first successful demonstration of a CW Raman laser with an optical fiber as the gain medium was demonstrated by Hill et al (1976).

Modern, commercially available fiber-based Raman lasers utilize $P_2O_5$-doped silica fibers exhibiting a vibrational (Stokes) shift of about 1320 $cm^{-1}$. Today it is quite common to design the Raman fiber laser resonators as a "cascading" device. This means that the light generated from the first Stokes line in the medium is also allowed to circulate in the laser resonator and is enhanced to such a level that it acts itself as the pump for the generation of the second Stokes line, which is shifted by the same vibrational frequency again. This process can be repeated by adding reflectivity at further Stokes line wavelengths to the resonator. Cascading through several discrete steps, many useful output wavelengths in the range 1200–2100 nm have be generated (for example, see Supradeepa et al. 2013), and even wide, continuous tunability over more than

Tm³⁺ – doped ZBLAN fiber

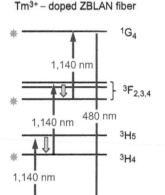

**Figure 4.23** Energy levels and transitions for wavelength up-conversion in a thulium-doped ZBLAN fiber. ✳ = metastable state.

300 nm has been demonstrated (see Krause et al. 2003). Finally, using frequency-doubling add-ons, often-difficult-to-reach visible wavelengths were generated using a primary Raman fiber laser (see Georgiev et al. 2005).

All these principles have been realized in a range of commercial Raman fiber laser systems, providing multiwatt output powers at many wavelengths, ranging from below 600 nm to beyond 2000 nm. Probably one of the most technology-advancing developments in the early 2000s has been the introduction of photonic crystal fibers (PCFs) as a laser-active medium. PCFs' great advances over common fibers are as follows: (1) their increased flexibility in single-mode core sizes; (2) their increased numerical aperture of pump cores; and (3) their high thermal stability of low-loss all-glass structures. All these are of particular importance when scaling fiber lasers to higher powers and, in general, prove to be the limiting factor in standard doped fibers.

As standard fiber laser media, most PCFs guide light based on the principle of total internal reflection between a core with a high refractive index and a cladding with a lower index. In PCFs, the different refractive indices are not provided by two different solid glass materials but are typically provided by air holes/air channels (normally an array) in silica glass. This structural composition may be seen as a hybrid material, with properties that solid materials normally cannot provide, such as a very low effective index or novel dispersion. For example, the air hole–silica matrix may serve as a low-index cladding, provided the holes/channels are of sufficiently small diameter. In the latter case, the matrix structure cannot be resolved by the electro-optic field with the result that the effective refractive index of the structure is determined by the fraction of air (channel volume) to the silica (bulk volume).

Classically, many of the commercial PCFs are based on a triangular-cladding, single-core photonic crystal (see Figure 4.24). Because the refractive index of the cladding region is wavelength-dependent, PCFs can be designed to be "endlessly single-mode," i.e., only the fundamental mode of propagation is supported, regardless of the wavelength or the core diameter. If one defines the spacing of the holes by the parameter $\Lambda$, for a hole diameter $d$, then one obtains the aforementioned endlessly single-mode scenario when $d/\Lambda \leq 0.45$. Note that this approximation only holds for a hexagonal air-hole pattern, with only a single hole missing in the (active) center.

From a practical point of view, one often does not need the bandwidth of endlessly single-mode design; for many laser and amplifier systems, it suffices to design the PCF to support single mode at the operation wavelength; higher-order modes cut off at shorter wavelengths.

It should be noted that the aforementioned (tailored) microstructuring of the PCF allows one to control the refractive index profile, much more so than is possible for any other fiber technology (like step-index or graded-index profiles). One particular bonus is that by carefully designing the PCF microstructure, one can realize extremely small numerical apertures for very large, but still strictly single-mode cores, or very high numerical apertures that are often required for high-brightness pump cores. Typically, a PCF for high-power applications is made up of a single active core, surrounded by an air-silica microstructured inner cladding, followed by a ring of large air holes (often referred to as "air cladding"), which confine light to the pump core, and a final (protective) solid silica overcladding.

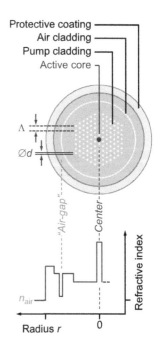

**Figure 4.24** Typical structure of a PCF, with solid single core, a triangular-lattice air-hole cladding (hole diameter $d$ and hole spacing $\Lambda$), and an addition cutoff air cladding. The outermost protective coating is a high-index polymer.

A laser based on a PCF as the active medium was first demonstrated by Wadsworth et al. (2000), yielding just over 10 mW of output power at a conversion efficiency from the pump of about 5%. Since then, PCFs have found their way into numerous commercial fiber laser products, with wall plug efficiencies surpassing 30% and output powers reaching the 100 W level, and also including high-repetition rate ultrashort-pulse lasers and widely tunable (in the range 400–2400 nm), bright supercontinuum sources; see further below. For more details on PCF-based fiber lasers, see Knight (2007) or Raineri et al. (2014).

### 4.5.2 Q-Switched and mode-locked pulse generation

Probably one of the most striking developments in fiber laser technology is the reliable generation of short and ultrashort pulses, based on relatively simple and rugged cavity designs.

When applying passive or active Q-switching methodologies in fiber lasers, pulses of typically 10–100 ns duration can be generated. However, due to the normally high gain encountered in fiber lasers, the temporal details of the Q-switching processes are qualitatively different from those of a bulk laser medium, and also more complicated. In particular, one often is confronted with a temporal substructure, with multiple sharp spikes. Individual pulses are normally of the order 50–500 nJ, although with large mode-area fibers, one may achieve up to 1–2 mJ; the limiting effects normally are gain saturation and fiber damage threshold.

The simplest Q-switched fiber laser setup utilizes passive Q-switching based on a SESAM; a related, basic Q-switch fiber laser resonator is shown in Figure 4.25.

Typically, a SESAM comprises a semiconductor Bragg mirror and a single quantum-well absorber layer. The Bragg reflector is made from a material with larger band-gap energy than the absorber layer, so that essentially no absorption occurs in that reflector region. Note that SESAM devices conceptually are the same whether being used for passive Q-switching or passive mode locking; basically the only difference is the thickness of the absorber layer.

Although passive Q-switching fiber laser setups are attractively simple, their pulse repetition rate and the pulse timing are often uncontrolled. Active Q-switching based on external activation means will provide substantially more control over the characteristics of the output pulses; therefore, actively Q-switched laser designs are the usual choice for commercial devices. There are ways to accomplish active Q-switching, frequently based on a tuning FBG, which is modulated by an acousto-optic modulator and a piezoelectric actuator, or is temperature controlled. For various practical designs, see Ter-Mikirtychev (2014). For the generation of picosecond pulses and ultrashort femtosecond pulses, as usual the principle of (passive or active) mode locking is incorporated into the fiber laser system, normally in the form of a saturable absorber (like a SESAM) or an active modulator. Once again, in the simplest case, one can set up an ultrashort-pulse fiber laser using more or less the same cavity configuration as shown above for the Q-switched fiber laser, incorporating a suitable SESAM. The (active) fiber length determines the pulse repetition rate, while the pulse duration (usually of the order of 100–500 fs) is determined by the interplay among fiber dispersion, fiber nonlinearity, and the resonator gain.

For shorter than aforementioned pulse durations, more sophisticated resonators are required. Particularly elegant is the so-called "figure-of-eight" fiber laser

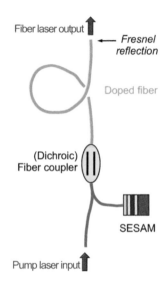

Fiber laser output

← *Fresnel reflection*

Doped fiber

(Dichroic) Fiber coupler

SESAM

Pump laser input

**Figure 4.25** Conceptual Q-switched fiber laser resonator, based on a SESAM. The output coupler comprises the Fresnel reflection from the bare fiber end.

setup, which was first demonstrated by Duling (1991) in a picosecond-pulse mode-locked erbium-doped fiber laser. Here, an "artificial" saturable absorber can be constructed using a nonlinear fiber loop mirror; for the conceptual configuration, see Figure 4.26. In short, one ring of the eight forms the main resonator, acting as the nonlinear (amplifying) loop mirror, while the other ring of the eight is a unidirectional (incorporating a Faraday isolator) passive mirror. The two ring loops are connected via a 50:50 fiber coupler. The main resonator is "asymmetric," i.e., it comprises a short length of laser-active fiber and a longer segment of ordinary fiber. The light circulating in the resonator is split into two counter-propagating components in the fiber coupler of the loop mirror. Because of the asymmetry in the fiber path, the light propagating in, say, the clockwise direction first passes through a long section of ordinary fiber, at low power, and is then amplified in the doped, laser-active fiber section. The anticlockwise propagating light is amplified first.

It can be shown that the nonlinear phase shift for the two components is different; if said difference is $\pi$ (the ideal case), the two components will interfere in such a way that effectively the setup acts like the combination of a laser gain medium with a (artificial) saturable absorber. This artificial saturable absorber generates a single pulse circulating in the fiber resonator, and thus a pulse train can be extracted from this figure-eight resonator by a suitable output coupler (another fiber coupler, normally placed in the non-active part of the eight).

It should be noted that in order to achieve shorter pulse durations, one needs to set up significantly more sophisticated resonators than described in the examples above, in particular including complex pulse shaping. However, with careful optimization, fiber lasers with pulse durations well below the 100 fs mark can be constructed. For further insight into ultrashort-pulse generation from fiber lasers, see the textbook by Binh and Ngo (2010).

In general, commercial femtosecond-pulse laser systems are based on erbium-doped or ytterbium-doped fiber media, emitting laser radiation in the regions 1500–1600 and 1000–1100 nm, respectively, with the actual wavelength selected by an appropriate Bragg reflector. Some of the typical values for various commercial fiber laser types are summarized in Table 4.2.

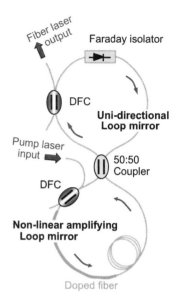

**Figure 4.26** Conceptual unidirectional figure-of-eight fiber laser configuration for ultrashort-pulse generation. DFC = dichroic fiber coupler.

**Table 4.2**    Common Commercial Fiber Laser Systems and Their Characteristics

| Laser | Operating Characteristics | Typical Applications |
|---|---|---|
| Fiber (CW) | Wavelength, $\lambda$: selected values in the NIR<br>Output power, $P$: up to 10 W<br>Linewidth, $\Delta v$: a few kilohertz ... a few gigahertz<br>Multi- and single longitudinal mode | Absorption and cavity ring-down spectroscopy in selected NIR bands;<br>Up-conversion lasers for, e.g., Raman spectroscopy; and ion trapping and spectroscopy |
| Fiber (pulsed) | Wavelength, $\lambda$: selected values in the NIR<br>Output pulse energy, $E_p$: up to a few microjoules<br>Linewidth, $\Delta v$: Fourier-limited<br>Pulse duration, $\Delta t$: 10 ... 1000 fs<br>Pulse repetition rate, PRR: 0.1 ... 100 MHz | Ultrafast laser spectroscopy and imaging;<br>high-repetition rate LIBS |
| Fiber/PCF (continuum) | Wavelength, $\lambda$: 400 ... >2000 nm<br>Output power, $P$: >3 W in the visible<br>    >10 mW/nm<br>Linewidth, $\Delta\lambda$: <5 nm ... >100 nm<br>Pulse duration, $\Delta t$: 2 ... 200 ps<br>Pulse repetition rate, PRR: 1 ... 100 MHz | Broadband spectroscopy;<br>Time-resolved fluorescence imaging;<br>Optical coherence tomography;<br>Terahertz spectroscopy |

### 4.5.3 Supercontinuum sources

Supercontinuum generation is the formation of a continuous spectrum, with large bandwidth, by propagation of short laser pulses of high peak power through any (highly) nonlinear medium. Its origins go back to the early 1970s, but the first supercontinuum generated in a glass fiber was demonstrated by Chernikov et al. (1997).

Today, supercontinuum generation mostly relies on the use of so-called photonic crystal fibers (PCFs), mainly because (1) they allow a strong nonlinear interaction over a significant length of fiber and (2) their chromatic dispersion characteristics can be tailored for particular pump laser sources and desired output spectra. In one of the early demonstrations of using tailored PCF as the nonlinear medium, Ranka et al. (2000) produced a rather flat continuum in the range 400–1450 nm, pumping their PCF with 100 fs pulses of 0.8 μJ pulse energy, at a wavelength of about 800 nm. For a brief review on supercontinua in PCFs, see Dudley et al. (2006) or Paschotta (2010).

The physics underlying supercontinua is rather complex since it cannot simply be described by a single phenomenon but a range of nonlinear effects combined, leading to the extreme spectral pulse broadening. The factors playing the main role in the generation of supercontinua are (1) the pump pulse length; (2) the dispersion of the fiber—relative to the pump wavelength; and (3) the pump peak power. Details on supercontinuum sources, and their physics and design, may be found in the books by Alfano (2006) or Dudley and Taylor (2010). Here we will summarize only a few remarks to highlight the complexity of the problem.

*The dependence on pulse duration.* When pumping the fiber with pulses of nanosecond or picosecond duration, Raman scattering and four-wave mixing are the most dominant contributors. It is worthwhile to note that supercontinuum generation is even possible with multiwatt CW laser excitation in (very) long fibers (see Cumberland et al. 2008). When pulses of femtosecond duration are used, the spectral broadening is normally dominated by self-phase modulation (SPM).

*The dependence on fiber dispersion.* The dispersion, and specifically the sign of the dispersion (associated with normal and anomalous dispersion), determines the type of nonlinear effects contributing to the formation of the continuum, and thus the shape and nature of the spectrum. When pumping with femtosecond pulses in the normal dispersion regime, SPM dominates, as well as Raman scattering broadening toward the wavelengths longer than the pump wavelength. However, it should be noted that other nonlinear effects start to participate if the pump wavelength moves from the normal dispersion regime close to the zero-dispersion wavelength. When pumping in the anomalous dispersion regime, SPM and dispersion combine to give rise to complicated soliton dynamics, including the split-up of higher-order solitons into multiple fundamental solitons, often known as "soliton fission" (see Figure 4.27).

*The dependence on pumping power.* As just stated, pumping the fiber medium near its zero-dispersion wavelength, the spectral broadening extends into the anomalous dispersion region, driven by SPM and Raman scattering. As a consequence, some solitons are generated which move via self-frequency shift to longer wavelengths in line with increasing the pumping power (see Figure 4.27 and Dekker et al. 2011).

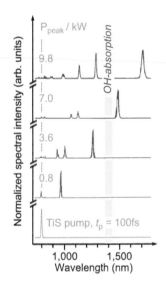

**Figure 4.27**   Soliton self-frequency shift spectrum in high-OH PCF, pumped by 100 fs pulses from an 800 nm Ti:sapphire laser, as a function of peak pulse power. (Data from Dekker et al. 2011, with permission.)

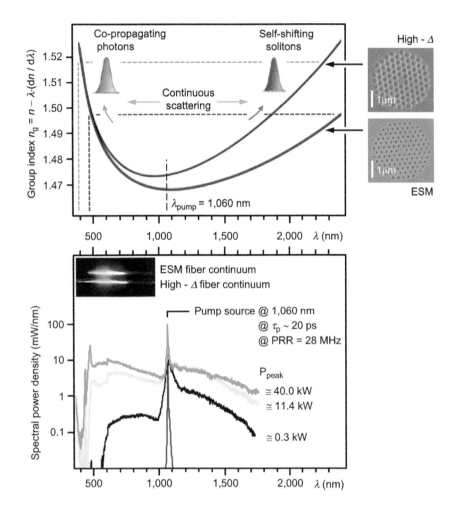

**Figure 4.28** Supercontinuum generation in PCFs. (Top) Modelled group index curves for a high-$\Delta$ fiber (blue) and an ESM fiber (red). The conceptual generating mechanisms are included. (Bottom) Supercontinua generated in an ESM by a picosecond-pulse 1060 nm pump source, for different peak powers. The insert shows spectra for the two fiber types. (Data adapted from Stone, J.M. and J.C. Knight, *Opt. Express* 16, no. 4 (2008): 2670–2675; Chen, K.K. et al., *Opt Express* 18, no. 6 (2010): 5426–5432. With permission.)

In order to highlight the various aspects of supercontinuum generation and its dependence on the various parameters, a very brief "overview case study" is given, based on some data published by Stone and Knight (2008) and Chen et al. (2010), with relevant results collated in Figure 4.28.

As just stated, the observed supercontinuum strongly depends on the system parameters, in particular on the refractive indices of the PCF and the proximity of the pump laser wavelength to the zero-dispersion point, as well as the characteristics of the pump laser in terms of pulse duration and peak power. It is rather easy to tailor PCF dimensions and their refractive index ratio $n_{core}$ to $n_{cladding}$ so that, on the one hand, the minimum of the group-velocity dispersion, $n_g$, is close to the wavelength of the pump laser and, on the other hand, the shape of the group-velocity dispersion curve favors the widest continuum span. This is shown in the top panel of Figure 4.28 for two PCFs with different air-channel sizes and spacing ($n_{cladding}$ reduces significantly with increased air-to-silica ratio). For a constant $n_{core}$, one finds that the lower the value of $n_{cladding}$, the steeper the group-index curve. The long- and short-wavelength limits of the supercontinuum are related to each other, based on the following mechanism: when a self-frequency-shifting soliton propagates in the anomalous-dispersion regime of the fiber (toward the IR—up to about the limit of a strong OH⁻ ion absorption band beyond ~2400 nm), it effectively "traps" short-wavelength photons with the same group index propagating in the normal-dispersion regime (toward the blue); the "blue" photons are scattered to shorter wavelengths, in a cascaded four-wave mixing process. This conceptual process is clearly evident from the dispersed spectra shown in the inset of the lower panel, exemplifying the supercontinuum emission from standard ESM and blue-optimized high-$\Delta$ PCFs, respectively. Under the same excitation conditions (picosecond pulses at 1060 nm), the latter fiber yields the broader emission spectrum.

The influence of the pump peak power on the shape and intensity of the supercontinuum emission is shown in the lower panel of Figure 4.28. With increasing peak power, the spectral coverage becomes wider, both toward the IR and the blue (note that here only the wavelengths up to ~1700 nm are shown, limited by the range of the spectrometer used in the measurements); and the spectral intensity distribution changes in shape, significantly favoring the visible part of the supercontinuum emission. The main concept described in this case study is applied in the design of most commercial supercontinuum sources, which nowadays provide spectral coverage of 400–2400 nm, with super-continuum power densities of up to 10 mW/nm (which is several orders of magnitude higher than an incandescent broadband lamp).

We like to conclude this section on supercontinuum generation and sources with one final remark. By and large, commercial and many self-build supercontinuum sources rely on pump lasers with short pulses (in the few picoseconds to 100fs regime). But because of the high pulse repetition rates of up to 100 MHz, these sources may be seen as "quasi-CW." However, it has been demonstrated by a few research groups that even true CW supercontinuum generation is possible. For example, Cumberland et al. (2008) used a 50 W CW ytterbium fiber laser, operating at 1071 nm to pump a PCF tailored to exhibit a zero-dispersion wavelength of 1068 nm, of length 100 m. They succeeded to generate super-continua in the range 650–1400 nm, with very high spectral power densities up to tens of mW/nm (see Figure 4.29).

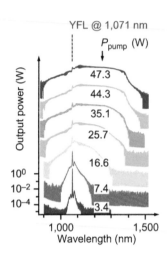

**Figure 4.29** Supercontinuum generated in a PCF, pumped by a CW ytterbium fiber laser (YFL) operating at 1071 nm, as a function of the pump power. (Data adapted from Cumberland, B.A. et al., *Opt. Lett.* 33, no. 18 (2008): 2122–2124. With permission.)

### 4.5.4 Fiber lasers versus bulk solid-state lasers

It is worthwhile to finish off this chapter on fiber lasers with a brief comparison between them and the closely related class of bulk solid-state lasers.

Standard solid-state lasers, in general, consist of a (doped) bulk-crystal medium, pumped by flash lamps, laser diodes, or a suitable secondary pump laser matching the absorption bands in the solid-state laser medium. The laser beam propagates without space restriction through the gain medium, and the cavity is set up with standard free-space optics, including mirrors, lenses, diffraction gratings, temporal pulse-shaping units, etc.

In contrast, in fiber lasers, light propagation is confined to a small (waveguide) core. If the guiding fiber is doped with laser-active rare-earth elements, this waveguide itself becomes the gain medium, which is suitably pumped by another laser, normally through and along a dual-cladding structure. While for temporal pulse-shaping in the early days free-space components have been utilized, modern sources have become all-fiber designs and thus are extremely rugged and often much more user-friendly than their bulk solid-state counterparts.

Table 4.3   Comparison of Bulk Solid-State and Fiber Laser Sources

| Laser Type | Advantages | Drawbacks |
|---|---|---|
| Bulk solid-state | Wide wavelength, narrow bandwidth tuning (e.g., Ti:sapphire laser) <br> High-energy density from large beam cross sections <br> "Clean" ultrashort pulses | Often low stability (due to drift of a free-space cavity) <br> Need for regular service when operating 24/7 <br> High-pulse jitter (for Ti:sapphire laser) <br> Prone to relaxation oscillation noise <br> Sensitive to environmental changes <br> Substantial cooling required for high-power systems <br> Relatively high cost for full systems <br> Often large form factor for full systems (specifically for high power) |
| Fiber | Large gain bandwidth <br> Often diffraction-limited beam <br> High stability <br> High reliability: 24/7 over months and years <br> Low-pulse jitter <br> Low amplitude noise <br> Only moderate cooling required for high-power systems <br> Largely immune to environmental changes <br> Relatively low cost <br> Compact form factor (fibers can be coiled) | Pulse pedestals for high-energy pulses, due to non-ideal spectrum distribution <br> Trade-off whether effects like SPM and Raman scattering are desired or not <br> Alignment for free-space pump into fiber critical <br> Complicated temperature-dependent polarization evolution <br> At high-power risk of fiber damage |

Thus, it is not surprising that fiber lasers are beginning to supplant the traditional solid-state laser in many application areas.

However, as with many other things in instrumentation technology, it is not always possible to only propagate advantages without carrying some inherent drawbacks. A brief summary of the advantages and disadvantages of each of the two laser classes is provided in Table 4.3.

## 4.6  BREAKTHROUGHS AND THE CUTTING EDGE

The demonstration of the ruby laser in 1960 certainly marks the beginning of all laser development. But it is probably the success of two other types of laser that spawned the remarkable march of lasers into nearly all facets of life. These were the semiconductor laser diode (see Chapter 5) and the solid-state Ti:sapphire laser.

### 4.6.1 Breakthrough: Ti:sapphire lasers

A laser based on titanium-doped sapphire (Ti:sapphire) was first demonstrated in the early 1980s at MIT's Lincoln Laboratory (Moulton 1982). The development of lasers using this solid-state active medium, both for CW and pulsed operation, quickly gathered pace. Not the least this was due to its very broad gain bandwidth of nearly 400 nm.

In 1988, the first commercial, broadly tunable CW Ti:sapphire laser was introduced, which complemented and replaced dye laser sources in the near-IR at about 670–1050 nm. Like with so many laser systems, also Ti:sapphire lasers evolved over time toward smaller but at the same time more powerful devices. For example, modern small-footprint (less than the size of a briefcase), monolithic-resonator Ti:sapphire laser units are now available, which operate in

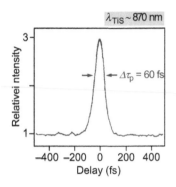

**Figure 4.30** Autocorrelation trace from the first self-mode-locked Ti:sapphire laser with intracavity prisms, demonstrating a pulse width of 60 fs. (Data adapted from Spence, D.E. et al., *Opt. Lett.* 16, no. 1 (1991): 42–44.)

single-longitudinal mode (with linewidth $\Delta v < 5$ MHz), and which provide up to more than 2 W of output power over nearly the full tuning range.

While initially these lasers were pumped by $Ar^+$ lasers (at $\lambda_{pump} = 488$ nm or $\lambda_{pump} = 514$ nm), today most Ti:sapphire laser systems comprise all-solid-state solutions, with Nd:YAG or Nd:YVO$_4$ pump lasers operating at $\lambda_{pump} = 532$ nm. More recently, it was demonstrated that even fiber-laser-based pump sources constitute suitable alternatives (see, e.g., Samanta et al. 2012).

Following the success of CW Ti:sapphire lasers, the first ultrafast Ti:sapphire lasers, both in the picosecond and femtosecond regimes, became commercially available in the late 1990s. The majority of devices were based on research reported by Spence et al. (1991). Today off-the-shelf oscillators are available that offer pulse durations as short as $\tau_{pulse} < 10$ fs (compare this to the first sub-100 fs pulse from a self-mode-locked Ti:sapphire laser; see Figure 4.30). These oscillators typically operate at pulse repetition rates of about PRR = 80 MHz, thus providing pulse energies of the order of $E_{pulse} \sim$ 5–10 nJ. These small individual pulse energies can be amplified significantly—of course at the expense of pulse repetition rates, often up to the millijoule level and beyond.

Beginning with those steps of commercialization, the ultrafast and tunable laser communities quickly replaced many of the cumbersome dye lasers utilized until then. Technological development now makes Ti:sapphire lasers the dominant force in nearly all ultrashort-pulse laser applications, even down to the timescale of attoseconds and into the realm of multiterawatt pulse peak power.

### 4.6.2 At the cutting edge: OFCs for high-resolution spectroscopy

The notion that ultrashort femtosecond pulses with broad spectral bandwidth lend themselves to high-resolution spectroscopy seems to be far-fetched at first glance. However, the so-called optical frequency combs (OFCs), associated with a train of femtosecond pulses, can be used for just that, exploiting its series of evenly spaced optical frequencies. One of the earliest examples of high-resolution direct frequency-comb spectroscopy (DFCS) is the work of Gerginov et al. (2005), who measured the hyperfine features of the Cs D-lines with resolution rivalling that of the best single-mode CW lasers. While an oddity only a few years ago, DFCS is now beginning to penetrate main stream high-resolution spectroscopy.

A femtosecond OFC utilizes a mode-locked femtosecond-laser. In the time domain (see the left-hand part of Figure 4.31), one observes a periodic train of ultrashort pulses, with pulse-to-pulse separation of the round-trip time $T$ for an individual pulse within the laser cavity, which is associated with the pulse repetition rate $v_{REP}$. However, due to unavoidable intracavity dispersion, the individual pulses experience pulse-to-pulse phase shifts $\Delta\phi_{CE}$ with respect to the peak center of the pulse envelope. In the frequency domain—accessible via simple Fourier transform (FT)—the laser output constitutes a "comb" of equally spaced mode frequencies separated by exactly the pulse repetition rate $v_{REP} = 1/T$ (see the right-hand part of Figure 4.31). The FT frequency of the $m$th mode of the comb is given by $v_m = m \cdot v_{REP} + v_0$, where $v_0$ is the displacement of the modes from integer multiples of $v_{RE}$—an effect associated with the aforementioned carrier-envelope phase shift $\Delta\phi_{CE}$ (one finds that $v_0 = v_{REP} \cdot [\Delta\phi_{CE}/2\pi]$).

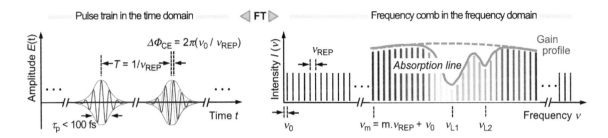

**Figure 4.31**    Principle of OFC generation from a train of femtosecond pulses ($\tau_p$ < 100 fs) at pulse repetition $\nu_{REP} \approx$ 80 MHz, yielding an equidistant frequency comb to span a wavelength range of up to 50–100 nm; conceptually, a DFC molecular absorption spectrum is indicated.

While not trivial, the frequencies $\nu_{REP}$ and $\nu_0$ can be measured with very high accuracy, and thus the mode frequencies $\nu_m$ are also precisely known. In the early days, Ti:sapphire lasers were used to generate OFCs, but increasingly sources based on Er- and Yb-fiber lasers—in conjunction with nonlinear frequency "stretching" methodologies—are being used, yielding comb bandwidths in excess of 1000 cm$^{-1}$ (see Schliesser et al. 2012), ideally suitable for molecular spectroscopy.

# CHAPTER 5

# Laser Sources Based on Semiconductor Media and Nonlinear Optic Phenomena

In this second chapter on laser sources commonly used in spectroscopic or imaging applications, we describe devices that (1) are based on semiconductor gain media, and that (2) exploit frequency/wavelength conversion effects in nonlinear optical (NLO) media.

*Semiconductor lasers.* This is another class of laser materials that has been around since the early days of laser development. The first working laser diode was demonstrated in 1962, a device made from gallium arsenide and emitting laser light at around 850 nm. However, it could be operated only in pulsed mode, and needed to be cooled to liquid nitrogen temperature (Hall et al. 1962). The real breakthrough for semiconductor diode lasers started in 1970 when independent developments in the then USSR and the United States led to room-temperature CW devices (Alferov et al. 1969; Hayashi et al. 1969). Another milestone for the application of semiconductor devices in laser spectroscopy was the discovery of quantum cascade lasers (QCLs) in the mid-1990s (Faist et al. 1994). These lasers—which are based on intraband rather than interband transitions (as in standard laser diodes)—emit light in the mid- to far-IR part of the spectrum. They quickly established themselves as the near-ideal laser source for spectroscopic investigations of numerous molecules of practical interest. Both types of laser diodes will be discussed in more detail below.

*Laser sources based on NLO effects.* Light generation in nonlinear media, at different wavelengths than the stimulating radiation, has a long tradition dating back nearly to the birth of the laser itself. The first demonstration of the related (sum) frequency-mixing phenomenon was the frequency-doubling of light from a ruby laser, at 694 nm, to generate photons at 347 nm (Franken et al. 1961). This process is "passive," i.e., no active laser medium or resonators are required. In contrast, optical parametric oscillators (OPOs) are light sources akin to a laser. However, any optical gain stems from parametric amplification in a nonlinear crystal rather than from amplification of stimulated emission in a resonator. A working OPO was first demonstrated in 1965 (Giordmaine and Miller 1965).

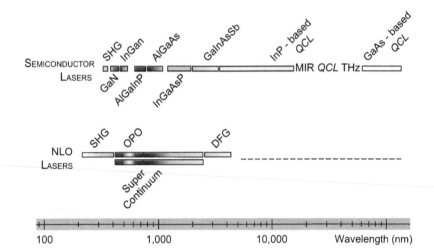

**Figure 5.1** Wavelength coverage by commercially available lasers. NLO = nonlinear optics; SHG, THG = second and third harmonic generation; DFG = difference-frequency generation; OPO = optical parametric oscillator; QCL = quantum cascade laser; MIR = mid-IR.

Devices and their properties for both processes, and their most common applications, are discussed in more detail below.

As for the lasers discussed in Chapter 4, one has to contemplate the key parameters that govern the operation of the above laser classes, namely wavelength, temporal behavior, photon flux, and geometrical beam shape. The latter three parameters will be discussed in conjunction with the individual laser types; for an easy survey as to the usefulness of these laser sources in specific applications, a summary of the wavelength ranges covered by them is shown schematically in Figure 5.1.

## 5.1 SEMICONDUCTOR LASER SOURCES

Since their inception in the early 1960s, semiconductor laser devices have evolved into a class of laser sources that practically tick all the boxes one requests from a "perfect" laser. They are extremely robust and reliable; wall-plug conversion efficiencies for many devices surpass 50%; CW output powers of 1–10 W are not uncommon, and up to several kilowatts has been realized; modulation/pulse rates larger than 10 GHz can routinely be achieved; and now basically all wavelengths from just below 400 nm up to beyond 100 μm can be realized.

Most common mass-produced semiconductor lasers operate as laser diodes, i.e., replicating a p–n junction. The laser is pumped with an electrical current, which injects carriers into the so-called depletion zone at the junction between the n-doped and p-doped material. Then, in this zone, photons are generated by radiative recombination of electrons and holes. The first successful operation of a p–n junction laser diode was reported by Hall et al. (1962) and Nathan et al. (1962).

As an alternative to electrical-current pumping under certain conditions, one also can pump the semiconductor materials optically: carriers (electrons) are generated by absorbing pump photons that lift electrons into the conduction band. This principle was successfully demonstrated by Kuznetsov et al. (1997), with the first commercial device being marketed in 2001.

Finally, a very attractive class of semiconductor lasers for spectroscopy applications are the so-called quantum cascade lasers (QCLs). In contrast to laser diodes that generate photons by *interband* electron–hole recombination (between the conduction and the valence bands), QCLs rely on *intraband* transitions between discrete energy levels in the quantum wells of the conduction band of a semiconductor device structure. The first mid-IR QCL was demonstrated by Faist et al. (1994).

Here we provide a brief, general summary of the underlying physics and some of the device configurations commonly used in semiconductor lasers used in spectroscopy and imaging. Deeper insight into semiconductor laser theory, device fabrication, and properties can be found in the dedicated textbooks by Suhara (2004) or Numai (2014), or in most laser physics textbooks.

## 5.1.1 Principles of laser diodes

Laser diodes belong to the group of so-called "direct-bandgap" semiconductors. Other than single-element diodes well known from electronics, laser diodes are made from compound binary, ternary, or quaternary materials—like GaAs, AlGaAs, or InGaAsP, to name but three examples—which are then p- and n-doped. Electrically forward-biasing a p-n junction laser diode (i.e., the positive terminal of the current source is connected to the p-end of the diode) "injects" the two charge carriers in a diode—the holes and electrons—from opposite sides of the p-n junction into the depletion region of the junction where, as stated above, the two may recombine radiatively across the bandgap from (near) the bottom of the conduction band to (near) the top of the valence band; the photon carries at least the bandgap energy.

Of course, recombination can also be nonradiative, via phonon coupling; in laser diodes, this constitutes an ever-present loss mechanism, which reduces the efficiency of the device. For the concept of the photoemission in a semiconductor laser diode, see Figure 5.2.

Electrically pumped p-n junction lasers are qualitatively always similar—differing only in the choices of the base bulk materials, the dopants, and actual layer thicknesses. The wavelength, efficiency, and output power capabilities of laser diodes are affected by these material properties. The operating voltage is determined by the bandgap of the semiconductor material, augmented by a linear factor caused by the series resistance of the device when injecting a current. In general, laser diodes can be extremely efficient when designed and operated properly, and "wall-plug" approaching 60–70% is not uncommon.

It should be noted that the basic p-n junction laser diode design is rather inefficient and also leads to spread-out laser beam profiles. This is associated with the relatively broad depletion layer, which can extend to more than 2–3 μm. Today's commercial diode laser designs are by and large based on so-called double heterojunction devices. These integrate an additional layer of different bandgap between the two p- and n-doped diode segments. If this layer is very thin, it is known as "quantum well"; instead of a continuum band distribution, the vertical variation of the electron wave function, and thus a component of its energy, is quantized. This is shown schematically in Figure 5.2. It has the effect that now the photon emission is restricted to a narrow channel, normally substantially less than 1 μm in width; this also means that the subsequent laser action is enhanced, leading to more efficient devices.

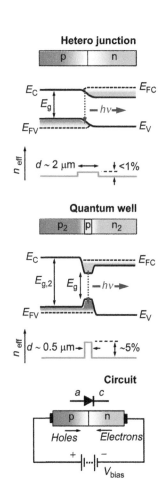

**Figure 5.2** Conceptual energy band structure of a simple GaAs heterojunction laser diode and a $Al_xGa_{1-x}As/GaAs/Al_xGa_{1-x}As$ quantum well laser diode. $E_v$, $E_c$ = valance and conduction band energies; $E_{Fv}$, $E_{Fc}$ = Fermi energies; $E_g$, $E_{g2}$ = bandgap energies; $n_{eff}$ = effective refractive index.

**5.6 mm TO-can**

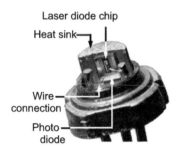

Laser diode chip

Heat sink

Wire connection

Photo diode

**Figure 5.3** Common FP laser diode packaged in a standard ⌀ 5.6 mm TO can. In the "cut-open" view, relevant elements are annotated.

The most common laser diode materials now are based on ternary and quaternary compounds. This is because by altering the relative elemental composition, strained lattices with different bandgaps can be "tailored," thus allowing to influence the base transition wavelengths. It has to be noted that commonly the development of certain compound classes was driven by demand for cheap, mass-produced commercial laser devices used in data handling and telecommunications. Almost as an "afterthought" were these materials further modified to also cater for the normally small laser spectroscopy and imaging market. These materials include InGaN for the 400–450 nm range (originally developed for Blu-ray disks and data projection); AlGaInP for the 630–760 nm range (primarily for DVD disks and, to a lesser degree, as HeNe laser replacement); GaAlAs for the 780–1100 nm range (CD disk, laser printers, solid-state laser pumping); InGaAs for the 900–1000 nm range (predominantly solid-state and fiber laser pumping); InGaAsP for the 1300–1650 nm range (basically telecommunications); and GaInAsSb for the NIR/MIR range of 1850–3330 nm (actually for gas-sensing applications, i.e., real spectroscopy).

One of the major attractions for the use of diode laser in all fields of applications has always been the small size of the devices but still being able to deliver almost up to 1 W of laser light power in quite a few cases, and all that with a laser chip size of normally less than 1 mm$^3$. Evidently, these chips have to be mounted, connectorized, and packaged. Many of the diodes used these days in laser spectroscopy (where often a few milliwatts of optical power is sufficient) are packed in standard TO cans, their form and size well known from standard early days of electronic transistors. An example is shown in Figure 5.3 for a device with TO can of 5.6 mm diameter; in the bottom part of the figure, the can is cut open to reveal the internal structure (notice the photodiode placed under the rear of the laser diode chip, which is used for power monitoring).

## 5.1.2 Laser diode resonators

As in all lasers for proper laser action, the semiconductor gain medium has to be placed within a cavity. The beauty of standard semiconductor lasers is that, in the simplest case, the resonator mirrors are an inherent, integral part of the laser diode chip. The two ends of the crystal are cleaved along the crystal planes and thus form a perfect Fabry–Pérot (FP) resonator. The high refractive index of the semiconductor material (e.g., for GaAs, one has $n_{GaAs} \sim 3.5$) guarantees that the reflectivity is of the order $R \sim 0.3$—one can calculate this from the Fresnel equations for light waves perpendicular to an interface with different refractive indices. This reflectivity is high enough to support laser action, even without an additional (dielectric) coating.

However, while perfectly viable and useful for many applications, these very simple "miniature" resonators carry a number of drawbacks that require additional measures to make them useful for spectroscopy and imaging applications. There are two main deficiencies associated with the short laser diode chip resonator, namely (1) beam divergence and (2) wavelength and mode behavior.

The cross section of the active laser area is very small and, for standard heterojunction laser diodes, is spatially asymmetric. The thickness of the depletion of quantum-well layer (nominally the "vertical" direction) is normally less than 1 μm, while the in-layer width of the laser light emission (nominally the "lateral" direction) is of the order of at least several micrometers, even when confinement structures are incorporated into the device during the epitaxial growth process. This means that due to diffraction, the beam emerging from the laser diode chip cavity diverges rapidly, typically at 20–30° in the vertical and 10–15° in the lateral directions. This is shown schematically in Figure 5.4. This means that, firstly, a lens must be used in order to form a collimated beam; and secondly, the collimated beam ends up being elliptical in shape because of the different vertical and lateral divergence. Despite the fact of being asymmetric, in general, the intensity profile is that of single transverse mode, normally the lowest $TEM_{00}$ mode, because of the waveguiding architecture built into most commercial laser diode devices.

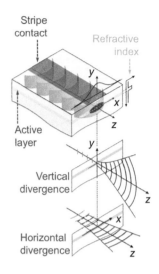

**Figure 5.4** Divergence of the output beam from a single-facet laser diode (index-guided).

Since the length of edge-emitting laser diodes is much larger than the laser wavelength (typically 300–1500 μm as opposed to ~1 μm), the resonator will support multiple longitudinal modes, i.e., multiple wavelengths are emitted simultaneously. Ordinarily, the maximum gain is encountered for photons whose energy is slightly larger than the bandgap energy; the modes nearest the gain curve maximum will be amplified the strongest. The width of the gain curve will then determine how many additional side-mode wavelengths will contribute to the emission spectrum, depending on the operating conditions.

These multiple longitudinal modes from the normal FP resonator are stable. However, normally the wavelength pattern fluctuates as a function of temperature (both variations in the environmental temperature as well as ohmic heating of the diode laser chip as a consequence of the injection current), because of the thermal expansion of the diode laser chip. On the one hand, this behavior is annoying when one desires a specific, predetermined wavelength; on the other hand, the phenomenon can be exploited to the user's advantage, since deliberate tuning of the laser wavelength becomes possible.

The chip expansion is not the only thermal effect: in addition to changes in the chip dimension, the gain profile maximum itself is a function of temperature because of thermal population effects in the depletion/quantum well layer. Unfortunately, these two effects are not in unison but change with different temperature vs wavelength slope. The consequence is the phenomenon of "mode hopping," which is schematically shown in Figure 5.5a. As the temperature increases, the laser mode wavelength in general follows the cavity-mode lines. However, if the actual lasing mode deviates too far from the gain maximum, the laser mode will "jump" to a more favorable one, as indicated in the figure. It should be noted that the whole pattern is made even more complex for controlled tuning because the mode hops experience hysteresis when changing the temperature upward or downward. In Figure 5.5b, an actual example of a tuning map for a quantum-well laser diode is shown (here only for increasing temperature). The tuning map clearly reveals the competition between modes and that only very rarely does the device emit only one single longitudinal mode (SLM). Typically, the shift of the center wavelength of laser diodes is of the order $\Delta\lambda \sim 0.2$–$0.3$ nm/K. Note also that the spectral width of an individual laser diode mode is of the order of a few megahertz.

It is quite clear from this that the use of simple, edge-emitting laser diodes in spectroscopy is daunting. By and large, it is difficult to control the emission wavelength pattern precisely, specifically to achieve narrow-bandwidth (SLM) operation and to keep the system long term on a given center wavelength. For example, in a typical gas-phase atomic or molecular absorption experiments in the visible, one encounters Doppler widths of the order of a few hundreds of

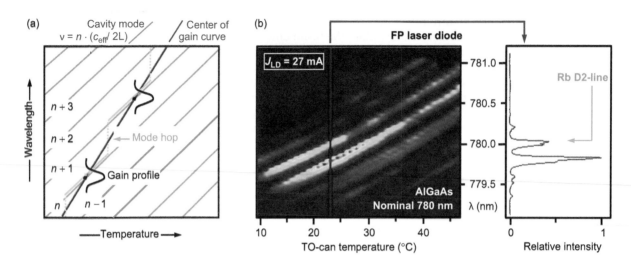

**Figure 5.5**  Wavelength change of the output from an FP laser diode, as a function of its temperature. (a) Conceptual pattern, indicating the different slopes for the cavity modes and the gain profile, leading to mode hops for a severe mismatch between the actual mode wavelength and the gain curve center. (b) Example for an actual AlGaAs laser diode used in an Rb D2-line absorption experiment.

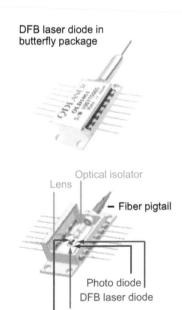

**DFB laser diode in butterfly package**

**Figure 5.6**  DFB laser diode in a butterfly package, with integrated TEC element for tuning. In the cut-open view (bottom), the relevant building blocks are indicated.

megahertz. With the thermal wavelength shift/drift from above, this would mean that the laser diode temperature needs to be adjustable and stable on the millikelvin scale, which is rather challenging.

As a consequence, already early in the use and development of spectroscopy laser sources based on laser diodes, additional wavelength selective elements and external cavity (EC) designs were introduced, namely devices based on monolithic designs (i.e., specific alterations to the laser chip design) and EC setups. However, both come normally at hugely increased cost; compared to the common edge-emitting laser diodes, typically available for 5–50€ (or US$), one easily has to multiply this by a factor of 10–100 for monolithic or EC devices.

### 5.1.3 Monolithic diode laser devices

There are two main approaches that are followed in the realization of monolithic diode laser sources: distributed feedback (DFB) and distributed Bragg reflection (DBR). These monolithic devices are normally packaged in TO cans or butterfly enclosures, and their output is often (single-mode) fiber-coupled; some examples are shown in Figure 5.6.

A DFB laser diode is a single-frequency device in which the lasing wavelength is selected/stabilized by a diffraction grating etched into the laser diode chip close to the p–n junction. This grating causes a single wavelength (related to the grating period) to be fed back to the gain region. Since the grating provides the feedback that is required for laser action, reflection from the laser chip facets is not required. Normally one or both facets are antireflection coated; this makes them also suitable to act as a gain element in EC setups (see further below). The wavelength of a DFB laser diode is set during manufacturing, when the grating with given periodicity is etched into the chip; the emission wavelength is very stable but can only be tuned slightly with temperature. In comparison to solitary edge-emitting laser diodes, the temperature dependence of the laser wavelength is reduced to $\Delta\lambda_{\mathrm{DFB}} < 0.1$ nm/K.

In DBR laser diode implementations, one of the facets of an FP laser cavity is replaced by a grating, which is etched into the laser diode chip (similar to that of the DFB solution but outside the active gain region). The main drawback of a simple DBR design is that mode hops are introduced when changing the diode injection current, since the temperature change of the active diode region and the passive grating segment are different. Often a second current contact to the grating region is introduced, which provides coordination between the gain and grating wavelength shifts, and thus results in wideband tunability.

There are more sophisticated device designs, exploiting a kind of Vernier effect based on a so-called sampled grating (SG). A SG-DBR diode laser consists of four sections: two grating sections, and in between them a gain element and a phase section. Each of the two grating sections is periodically interrupted, but with different periodicity, creating SGs that provide a series of "grating bursts." Lasing only occurs when the reflections from both gratings coincide; this will only be the case at certain widely spaced wavelengths, due to the different mode spacing of the two gratings. Using the phase section control, one can shift the reflection combs relative to each other so that coincidence for different wavelengths is enforced. This can be interpreted as follows: the laser wavelength is tuned by a "Vernier scale" effect. SG-DBR devices offer tuning ranges of 30–70 nm, much wider than the typical 1–2 nm of simple DBR devices (see Yang et al. 2010).

It should be noted that a class of diode lasers known as vertical-cavity surface-emitting lasers (VCSELs) are actually based on a DBR design, but are normally not dubbed as such; the term DBR laser diode is normally associated with edge-emitting devices.

Finally, one should not forget wavelength-stabilized laser diodes, which utilize a volume-holographic grating (VHG) element—which is a transmission grating—in very close proximity to the actual laser chip (see Steckman et al. 2007). While not really a monolithic device any longer—the laser diode chip itself is not affected—the design is small enough to fit into standard TO cans, and thus has the appearance of a monolithic device. The devices provide extremely stable wavelength output, of the order $\Delta\lambda_{\mathrm{VHG}} \sim 0.01$ nm/K.

The principle can even be applied to laser diode stacks, which, because of their odd active region shape and thus odd beam profile, often elude wavelength selection and stabilization; this approach is becoming increasingly popular when high-power (up to watts) but narrow-bandwidth diode laser operation is desired.

An interesting variant of monolithic diode laser sources are the so-called vertical-cavity surface-emitting lasers, or VCSELs. In contrast to the devices described above, in which the light propagates in the plane of the semiconductor wafer and is emitted from the edge, in VCSELs, the light travels vertical to the wafer structure and is emitted from the surface (see Figure 5.7). The second difference is that they have no cleaved facets, which can serve as the cavity mirrors; instead DBR structures on both sides of the (quantum well) gain medium have to be built into the device during the epitaxial growth manufacture.

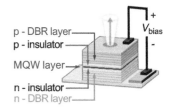

**Figure 5.7** Simplified general structure of a VCSEL laser diode; MQW = multiple-quantum well active laser; DBR = distributed Bragg reflector. The laser output is tailored circular, although not necessarily in TEM$_{00}$ mode.

Because the light emission is vertical to the epitaxial plain structure, one has to build lateral confinement into the chip; this is achieved by either incorporating gain-guiding features or depositing a confinement boundary of oxidized AlAs (usually in the form of a circular ring).

VCSELs have certain advantages, in particular from the manufacturing point of view. Since no cleaved facets are required, they can be fabricated and tested at the wafer level, which makes the whole manufacturing simpler. Furthermore, because of the extremely thin gain area, only low power is required to achieve laser action—laser thresholds of below 1 mA are not uncommon. As a consequence, this type of laser diode has become very popular in a number of spectroscopic sensing applications in which low divergence, nearly circular beam profile, and cost-efficient narrow-bandwidth emission are desirable. On the other hand, on the downside of VCSELs is their normally much lower output power, in comparison to edge-emitting laser diodes, because of the high reflectivity of the output DBR mirror.

### 5.1.4 External cavity diode lasers (ECDL)

The need for having to tailor the production process to fabricate DFB and DBR laser diodes made their commercial availability rather limited; only for wavelengths of interest to the mass laser market (like telecommunications at 1550 nm) are such devices affordable for the ordinary spectroscopist. Thus, free-space ECs using standard reflection gratings, and other optical elements, have been set up successfully by many research groups. By and large, laboratory and commercial configurations use the grating in two alternative configurations, the so-called Littrow or Littman configurations. In the former, the geometry is such that the diffracted beam is back-reflected into the direction of the incident beam, i.e., the grating itself acts as a wavelength-selective mirror. In the latter, the grating is aligned so that the incoming beam is in near-grazing incidence with the grating; however, an additional mirror is required to reflect the diffracted radiation of the desired wavelength back into the cavity. Because of the different geometries, the laser output beam direction differs: in Littrow configuration, the output beam direction changes with wavelength, while in Littman configurations, it does not. The principle is shown schematically in the middle section of Figure 5.8; for more details on the two configurations and their technical variants, one may consult the textbooks by Suhara (2004) or Numai (2014), or tutorial documents usually provided by the main manufacturers of ECDLs.

ECDLs are versatile in their configuration and are compatible with most standard free-space laser diodes. Thus, a wide range of wavelengths is available, naturally depending on the internal laser diode gain element. When designed and operated carefully, SLM operation is easy to achieve, with side mode suppression of larger than a factor 100 (see Figure 5.9). Another advantage of ECDL setups is that because of the relatively long cavity, very narrow linewidths of less than 1 MHz are encountered. They also exhibit much wider tuning ranges than an ordinary FP laser diode, and tuning is often possible across the whole gain bandwidth of normally 30–100 nm. On the other hand, ECDLs are rather prone to mode hops because of the much closer mode spacing introduced by the long EC. Clearly, it depends on the quality of the mechanical design how far one may tune without mode hop. In addition, the competition between laser-chip internal and external-cavity mode frequencies—and thus mode competition—can be reduced by antireflection (AR) coating of the exit facet of the laser diode.

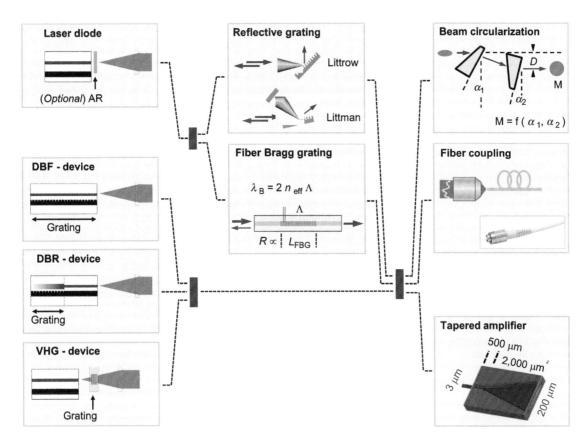

**Figure 5.8** Principles of ECDL design. On the left: device integration within a standard laser diode package (TO can or 14-pin butterfly units); in the center: external wavelength control components (diffraction or fiber Bragg gratings); on the right: "manipulation" of the laser output, including circularization, fiber coupling, and power amplification.

As an alternative to free-space grating configurations, one may choose a setup incorporating a fiber Bragg grating. This is a transmission grating "burned" into a (single-mode) fiber, which acts as the wavelength-selective reflector and output coupler, similar in its action to the DBR or VHG gratings. The fiber can be attached directly to a laser diode enclosure and thus constitutes a very rigid design without the mechanical drawbacks of free-space setups. Tuning of the wavelength can be achieved by heating or piezo-stressing the grating region, thus changing the grating period and with it the wavelength; the principle is also shown in the middle part of Figure 5.8. However, in contrast to free-space cavities with diffraction gratings, the tuning range is reduced. On the other hand, because of their ruggedness, they are often used in molecular detection and other sensor applications (see Gladyshev et al. 2004).

Finally, it should be pointed out that from early on, always laser diodes used in spectroscopy and imaging, whatever their cavity design, have a "coupling" component between the laser and the experimental setup in which the laser radiation is used. For example, it might be necessary to transform the laser beam shape from the normally elliptical to a circular profile. This can be done in two

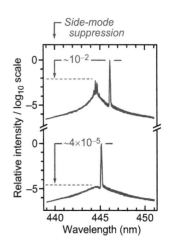

**Figure 5.9** Spectra from an ECDL system tuned to two different wavelengths, revealing that side-mode suppression is superior in the center of the gain curve.

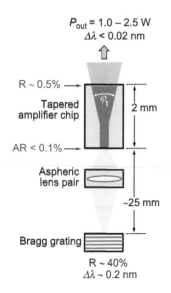

$P_{out}$ = 1.0 – 2.5 W
$\Delta\lambda$ < 0.02 nm

R ~ 0.5% ⟶

Tapered amplifier chip

2 mm

AR < 0.1% ⟶

Aspheric lens pair

~25 mm

Bragg grating

R ~ 40%
$\Delta\lambda$ ~ 0.2 nm

**Figure 5.10**  ECDL based on a tapered semiconductor diode amplifier and a VHG; for details, see Lucas-Leclin et al. (2008).

ways; either one uses an (adjustable) circularizer, made up from a so-called anamorphic prism pair, which expands the beam in only one direction, or one couples the laser radiation into a single-mode fiber at whose output end a nearly perfect Gaussian beam emerges. However, it has to be noted that in the latter case, this goes hand in hand with a severe power loss because only a fraction of the original asymmetric beam profile can be coupled into the fiber. On the other hand, fiber delivery is much more mechanically robust if the radiation has to be delivered over relatively long distances. These circularization/coupling options are shown schematically on the right of Figure 5.8.

In some applications, high laser powers may be required, which are well beyond what is possible to achieve with systems based on standard laser diode chips: rarely does the power extracted from an individual mode surpass the 100 mW level due to gain saturation. If higher narrow-bandwidth power levels are desired, one normally has to use amplifier stages. For this purpose, commercial tapered, semiconductor diode amplifier chips are used—some of them are capable to provide larger than 1 W output power at the seed wavelength. These tapered amplifier chips can also be used as the gain medium in an EC (see Figure 5.10). For example, Lucas-Leclin et al. (2008) in one of their setups, incorporating a VHG as the wavelength-selective grating and feedback element, achieved output powers up to 2.5 W at around $\lambda$ = 810 nm, with linewidth $\Delta\lambda$ < 0.02 nm.

## 5.1.5 Optically pumped ECDLs

The majority of semiconductor laser gain media are used in the form of laser diodes, and they are pumped electrically by an injection current through the diode junction and the depletion layer or quantum well. However, it is also possible to transfer electrons from the valence to the conduction band by optical pumping. Because of this, the gain medium chip requires neither a p–n junction nor electrical contacts; this greatly simplifies epitaxial growth and processing of the laser wafer. The semiconductor gain medium is normally manufactured as a disk with a DBR mirror structure integrated at the back.

Devices using this type of gain chips are commonly known as *optically pumped semiconductor disk lasers* (OP-SDL) or—because of the specific cavity implementation—as *vertical-external-cavity surface emitting lasers* (VECSELs). A survey of OP-SDL sources and development is provided by Guina et al. (2012).

OP-SDLs predominantly make use of two main base materials, namely InGaAs and GaSb, both being III–V semiconductors. For the former, gain chips can be tailored by appropriate doping to provide laser wavelengths in the range 700–1200 nm, while the latter emits strongest at around 2000 nm. Here we exemplify the description for lasers based on InGaAs. The gain region of the chip constitutes a binary layer sequence: GaAs layers that absorb the IR pump light (normally from a suitable laser diode whose wavelength has to exceed the bandgap of the absorbing semiconductor) alternate with InGaAs quantum wells for the laser transition. This means that charge carriers generated in the pump absorption layers drive recombination in the quantum well layers. The stoichiometry and physical dimensions of the quantum wells determine the central gain wavelength of the device. At the bottom of this active layer, a low-loss DBR mirror is

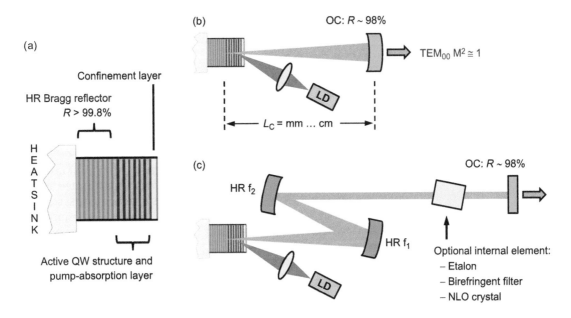

**Figure 5.11**  Schematics of an optically pumped semiconductor disk laser (OP-SDL). (a) Semiconductor gain element, including the gain segment, the wavelength-selective Bragg reflector, and a confinement section, with total thickness of a few micrometers. (b) Basic OP-SDL configuration; LD = pump laser diode; OC = output coupler. (c) Extended OP-SDL configuration with a z-shape cavity, hosting optional cavity-internal elements like an etalon or birefringent filter for wavelength selection and tuning, or an NLO element for wavelength conversion.

incorporated, optimized for the design wavelength of the actual OP-SDL. This conceptual structure is shown in Figure 5.11a.

The pump radiation enters the gain chip at an angle, normally around about 25–30°. While nonlinear, this geometry still introduced relatively little ellipticity and pumps the mode volume of the final OP-DSL external; in addition, the pump beam optics do not impede the laser resonator. Furthermore, such off-axis pump geometry allows one to arrange more than one pump laser azimuthally for power scaling.

In the simplest case, the laser resonator is completed by a single, external mirror whose curvature is matched to the cavity length and the lateral dimension of the pumped mode volume; see Figure 5.11b. However, the EC design of OP-SDLs allows for great flexibility, which is normally only found for nonsemiconductor laser systems. The external resonator can be extended to provide room for the insertion of intracavity elements, such as frequency selection or stabilization elements (etalon, birefringent tuner, VHG, etc.); NLO elements for wavelength conversion; and passive mode-locking devices. In particular, the latter is an interesting option since normally semiconductor (diode) lasers do not lend themselves for mode locking. These options and the related cavity design are shown in Figure 5.11c.

In particular, the combination of power scalability and wavelength conversion from the IR to the visible has proven to be very attractive and lead to a series of very popular commercial laser systems. In particular, some of the "heritage" wavelengths in spectroscopy have benefitted, for example, those previously only available with high power from less convenient laser sources (like the 488 nm argon ion laser line), or powerful radiation in the green/yellow/orange part of the spectrum (selected wavelengths in the range 540–640 nm; see Fallahi et al. 2008).

## 5.2 QUANTUM CASCADE LASERS

As the last type of semiconductor laser, we discuss QCLs. QCLs are so-called unipolar devices whose photon-emission transitions are not from the conduction band down into the valence band, but take place within the conduction band between sublevels (or bands) of a quantum well; in fact, most QCLs consist of a series of multiple quantum wells (MQWs). Normally, they emit laser light in the mid- to far-IR spectral region. The first QCL was demonstrated by Faist et al. (1994), but it took nearly a decade until easy-to-manage systems emerged working CW at room temperature (see Beck et al. 2002). The majority of modern, commercial QCL devices now work at room temperature, supported by Peltier temperature-control elements (see Figure 5.12 for a miniature OEM module). With the knowledge and tools of quantum mechanics, and of course the capabilities of epitaxial growth processes, one can now engineer devices for a vast range of wavelengths, without the constraint of available material bandgaps. It is well beyond the scope of this short section to cover all aspects of QCLs in detail; for wider information on production techniques, device performance, and applications, the reader may consult the textbook by Faist (2013), as well as one of the frequent review articles or short tutorials by several manufacturers. Here, we just present the basic underlying physics—and that only in a simplified way—making QCLs so attractive, and provide a few examples of the key performance relevant for spectroscopy.

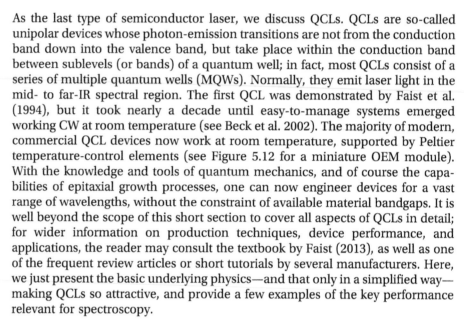

**Figure 5.12** Smallest, widely tunable QCL OEM module *Mini-QCL*™. (Image courtesy of Block Engineering. "Mini-QCL™: Widely-tunable mid-IR OEM laser module." Marlborough, MA, USA: Block Engineering, 2017.)

In brief, the major difference in operation between ordinary laser diodes and quantum cascade devices is the following. In the *bipolar* laser diode, electrons and holes recombine (i.e., "annihilate" each other under generation of a photon) across the bandgap, and thus do not play any further role after the emission of the photon. In the *unipolar* scenario, the electron undergoes an intraband transition between energy sublevels in the first quantum well, emitting a photon in the process. Then the electron is not lost but can tunnel into a subsequent period of the MQW structure and generate a further photon, and so on. This means that a single electron can cause multiple generations of photons while traversing the MQW structure; in other words, the quantum efficiency is greater than unity, which makes QCLs so efficient (often up to 20–50% wall-plug efficiency) and leads to normally much higher output power than ordinary bipolar laser diodes (up to watt-level peak powers at room temperature, for short pulsing).

Two key parameters determine the wavelength behavior of QCLs: (1) the depth of the quantum well—which determines the short-wavelength emission limit; and (2) the width of the quantum well—which can be adapted during manufacture to more or less tailor any center wavelength larger than the lower limit, at will. Still very common—and in fact the material in the first working QCL—is the material combination InGaAs/InAlAs, which provides a well depth of ~0.5 eV. Since then, a range of other material combinations have been developed, which provide deeper quantum well depths, and thus short wavelengths. For example, QCLs have been fabricated based on InGaAs/AlAsSb with a well depth of ~1.6 eV and InAs/AlSb, which has a well depth of ~2.1 eV, and others. Based on the latter material, QCLs have been fabricated, which generated laser emission as short as

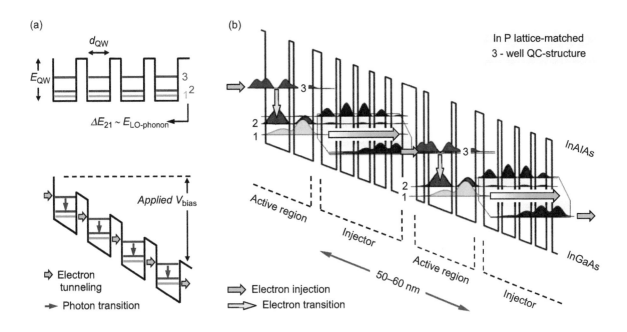

**Figure 5.13**   Conduction band profile and transition/transfer processes in a quantum-cascade gain material. (a) Conceptual multiple-quantum well structure, without (top) and with (bottom) applied bias voltage; $E_{QW}$ = quantum well depth; $d_{QW}$ = width of quantum well. (b) Detailed view of a typical InGaAs/InAlAs quantum cascade gain material, with coupled-well and collector/injector segments; the probabilities $|\psi_i|^2$ for the three quantum well sublevels are provided, together with the electron transfer in the injector miniband region.

2.5–2.7 µm. All these QW materials are lattice-matched to an InP substrate. A different combination is that of GaAs/AlGaAs lattice-matched to GaAs-base; performance-wise, these QCLs do not match InP-based systems; however, they prove to be very useful in the terahertz region of the spectrum (longer than ~60 µm).

For simplicity, treating the quantum well as a one-dimensional potential well of finite depth $V_0$ in the conduction band, the electron carriers are confined to discrete levels (or subbands) of energy $E_n$. Solving the time-independent Schrödinger equation for this system, one finds the straightforward solution for the energy levels within the said well $E_n = -(\pi^2\eta^2/m)\cdot(n/x_q)$, where $x$ is along the direction of layer growth and $x_q$ is the width of the quantum well. Thus, in an individual quantum well, one encounters energy levels with increasing spacing; in general, for QCLs, the lowest three are considered. Such quantum wells and their energy levels are not in isolation, but in QCLs, a sequence of wells—mostly 10–25—are joined, as shown in the top of Figure 5.13a.

When applying a bias voltage across the structure, the potential well becomes (linearly) skewed, and the levels within a specific quantum well become shifted in energy relative to its neighbor. If the barrier width between the wells is not too wide, then quantum tunneling through the barrier is encountered, which becomes "resonant" in case the energy levels match up. As shown in the bottom of Figure 5.13a, this is the basis for a three-level laser system. Electrons tunnel through the barrier to populate the uppermost level 3. From there, radiative decay to level 2 can occur, although other nonradiative processes also need to be considered. The energy spacing between levels 2 and 1 is designed such that it is about equal to the longitudinal optical (LO) phonon energy (a few tens of millielectronvolts). Thus, level 2 depopulates rapidly via resonant LO phonon-electron scattering. From level 1, the electron tunnels into the next well, and the intrawell decay process repeats itself; a cascade of photon decays ensues until the electron has reached the end of the well sequence.

Evidently, this very simplistic model does not reflect the reality of concrete QCLs. For those, the single-quantum well is replaced by coupled-well structure, which allows one—utilizing appropriate variation of the well widths and well barriers—to tailor wave function probabilities for the three energy levels in such a way that in the upper laser level, it favors the feeding of electrons into it; the lower laser level is adjusted to achieve good coupling to the ground state in order to rapidly empty it for sustained laser action; and the ground state is tailored to again favor electron tunneling out of the well. This is shown schematically in Figure 5.13b for a typical InGaAs/InAlAs active material structure. The figure also shows that one active region does not simply follow another one, but that each is separated from the next by a doped superlattice (a sequence of quantum wells) injector/collector segment. This maintains electric field uniformity across the whole structure and assists in effective collection of electrons from level 1 in one well and optimum injection into level 3 in the subsequent one, providing ease of level/band alignment.

The quantum cascade gain medium just described can be used on its own, acting more or less like a superluminescent configuration (incoherent) light source. However, most commonly it is incorporated into a laser cavity to provide better control over all the typical laser operating parameters, like wavelength, linewidth, output power, etc.

In its simplest form, QCLs can be fabricated as simple FP devices, as for the ordinary edge-emitting laser diodes described above. For this, the quantum cascade material is fabricated to form an optical waveguide and ends are then cleaved to form the FP resonator. Such devices are capable of generating high output powers but typically are multimode, like their FP laser diode cousins. The wavelength of the FP QCL can be changed by altering the temperature of the gain medium.

As for laser diodes, monolithic devices can be fabricated, which include DBRs or DFB; their wavelength is tuned mostly by temperature. An interesting tuning variant is sometimes employed for DFB-QCLs, driving it with a pulse whose slope is appropriately "tailored." In this way, the wavelength of the DFB-QCL changes rapidly over the period of the pulse (the process is often called "chirping"). Such devices are used for rapid scanning of a spectral region, in particular for molecular sensing applications (see Wysocki and Weidmann 2010).

Finally, as for laser diode gain media, quantum cascade media can be incorporated into ECs; for such configurations, the facets of the gain medium are AR-coated. And if appropriate wavelength-selective elements are introduced into the cavity, tuning and single-wavelength operation of the device become possible.

The gain profile of a QCL can be very wide, in some favorable media >500 cm$^{-1}$. Using a DFB quantum cascade medium in an external grating cavity, such as the "double-ended Littrow" configuration shown in Figure 5.14, one is able to tune the EC-DFB-QCL over the entire width of the gain profile, and with the careful mechanical setup of the grating tuning mechanism, coupled with appropriate temperature control of the gain medium, some commercial systems today offer mode-hop free tuning over >100 cm$^{-1}$, with linewidths as narrow as a few $10^{-5}$ cm$^{-1}$.

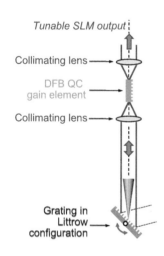

**Figure 5.14** Conceptual setup of a DFB-QC gain element within an external grating cavity (in Littrow configuration).

# 5.3 LASER SOURCES BASED ON NLO: SUM AND DIFFERENCE FREQUENCY CONVERSION

Nonlinear response of media to laser radiation, as well as NLO-generated wavelengths, has been mentioned repeatedly throughout this chapter. Indeed, it is the beneficial exploitation of NLO that has benefitted laser technology and spectroscopy tremendously. In fact, it might be fair to say that without NLO-generated wavelengths—first demonstrated by Bass et al. (1962)—some of the laser sources and specific UV-spectroscopy applications would not be what they are today. Here we provide a short summary of the basic principles behind NLO-wavelength generation, and describe a few examples of common laser sources exploiting them. For any further detail, the reader may consult laser textbooks like Silfvast (2008), Demtröder (2008), or Guha and Gonzalea (2014).

## 5.3.1 Basic principles of frequency conversion in nonlinear media

In principle, nonlinear response to strong laser radiation occurs in nearly all media, and very often is unwanted and uncontrolled. It is nonlinear processes in ordered crystalline media that allow for the real beneficial exploitation of the effects. The most common commercially used frequency conversion crystal materials are $\beta$-barium borate (BBO), monopotassium phosphate (KDP), potassium titanyl phosphate (KTP), lithium triborate (LBO), and lithium niobate (LN). If a medium is subjected to a strong electromagnetic (radiation) field, $E$, its macroscopic polarization, $P(t)$, associated with the nonlinear susceptibility, $\chi$, can be approximated as a power series expansion in terms of the said incident light field:

$$P(t) \propto \underbrace{\chi^{(1)} \cdot E(t)}_{P^{(1)}(t)} + \underbrace{\chi^{(2)} \cdot E^2(t)}_{P^{(2)}(t)} + \underbrace{\chi^{(3)} \cdot E^3(t)}_{P^{(3)}(t)} + \dots \quad (5.1)$$

The coefficients $\chi^{(n)}$ are the $n$th-order susceptibilities of the medium. Most relevant for the NLO exploitation discussed here are three-wave mixing processes, i.e., those in which two incident waves generate a new, third outgoing wave. For any three-wave mixing process, the second-order term in Equation 5.1 is the important one: only this is nonzero in media that exhibit broken inversion symmetry. Two incident waves $E_1(\omega_1,t)$ and $E_2(\omega_2,t)$ can be represented as a superposition wave $E(t) = (E_1 \cdot \exp(i\omega_1 t) + E_2 \cdot \exp(i\omega_2 t)) + c.c.$ Inserting this into Equation 5.1, one obtains for the second-order polarization term

$$P^{(2)}(t) \propto \sum \chi^{(2)} \cdot n_0 \cdot E_1^{n_1} \cdot E_2^{n_2} \cdot \exp(i(m_1\omega_1 + m_2\omega_2)t). \quad (5.2)$$

The summation in Equation 5.2 is over the five variables $n_0$, $n_1$ and $n_2$, and $m_1$ and $m_2$, each of the value 1 or 2. The newly generated wave, $E_3$, has the (angular) frequency

$$\omega_3 = m_1\omega_1 + m_2\omega_2 \quad (5.3)$$

Basically, Equation 5.3 constitutes the *energy conservation* in the three photons involved in the wave-mixing process. Three fundamental mixing processes are observed (conceptionally these are shown in Figure 5.15):

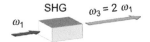

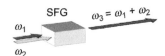

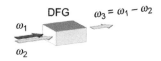

**Figure 5.15** Three-wave mixing processes in an NLO crystal.

- *Second harmonic generation* (SHG), or *frequency doubling*: for this process, one starts with two equal waves (then the five variables are all equal, i.e., $n_0 = n_1 = n_2 = m_1 = m_2 = 1$), or two photons from a single wave contribute (i.e., $n_0 = 1$; $n_1 = m_1 = 2$ and $n_2 = m_2 = 0$, or $n_1 = m_1 = 0$ and $n_2 = m_2 = 2$). Two photons of equal energy are converted into a third one with double the energy/frequency (half the wavelength), meaning that $\omega_3 = 2\omega_1$ or $\omega_3 = 2\omega_2$.

- *Sum frequency generation* (SFG): the two waves interact additively, with the five variables having the values $n_0 = 1$ and $n_1 = n_2 = m_1 = m_2 = 1$, and generate light with a frequency that is the sum of two frequencies of the incoming waves, i.e., $\omega_3 = \omega_1 + \omega_2$. Note that SHG is a special sample of this case.

- *Difference frequency generation* (DFG): the two waves interact subtractively, with the five variables having the values $n_0 = 1$; $n_1 = n_2 = m_1 = 1$; and $m_2 = -1$ (assuming that $\omega_1 > \omega_2$), and generate light with a frequency that is the difference of two frequencies of the incoming waves, i.e., $\omega_3 = \omega_1 - \omega_2$.

It should be noted that two further frequency/wavelength transformation procedures are frequently utilized: *third harmonic generation* (THG) and *fourth harmonic generation* (FHG). In the former case, a third photon of the same energy is mixed with an SHG photon in an additional sum-frequency crystal; in the latter case, the SHG photons pass through a second SHG crystal. One of the most encountered examples for this recipe is utilized for the Nd:YAG laser, with $\lambda_{\text{fundamental}} = 1064$ nm, $\lambda_{\text{SHG}} = 532$ nm, $\lambda_{\text{THG}} = 355$ nm, and $\lambda_{\text{FHG}} = 266$ nm, with all these outputs available from commercial devices.

### 5.3.2 Phase matching

Up to this point, the fact that the light fields exhibit vector and positional properties has been neglected. In order to include these, a travelling light has to be used whose electric field vector is given by $(r,t) = E \cdot \exp(i(\omega t - k \cdot r))$, with $k = n(\omega) \cdot (\omega/c)$; here $k$ is the wave vector—whose absolute value is related to the wavelength by $|k| = 2\pi/\lambda$; $n(\omega)$ is the wavelength-dependent refractive index of the nonlinear medium; and $r$ is the position vector. With this vector-related presentation, the second-order polarization term from Equation 5.2 changes to

$$P^{(2)}(r, t) \propto E_1^{n_1} \cdot E_2^{n_2} \cdot \exp(i(\omega_3 t - (m_1 k_1 + m_2 k_2) \cdot r)) \tag{5.4}$$

Note that also the description of the susceptibilities $\chi^{(n)}$ needs to be altered from the scalar description in Equation 5.1 to a vector representation—in the form of $n$th-order tensors, whose components depend on the combination of frequencies.

As a consequence of the vector properties of $k$, the microscopic contributions from different spatial locations in the nonlinear medium only add up to generate an outgoing wave with useful intensity if the vectors of the phase velocities of the incident and newly generated waves match. Thus, in addition to the energy-conservation condition in Equation 5.3, one has the so-called *phase-matching condition* for the wave vectors:

$$k_3(\omega_1 \pm \omega_2) = k_1(\omega_1) \pm k_2(\omega_2) \tag{5.5}$$

This condition is shown schematically in Figure 5.16.

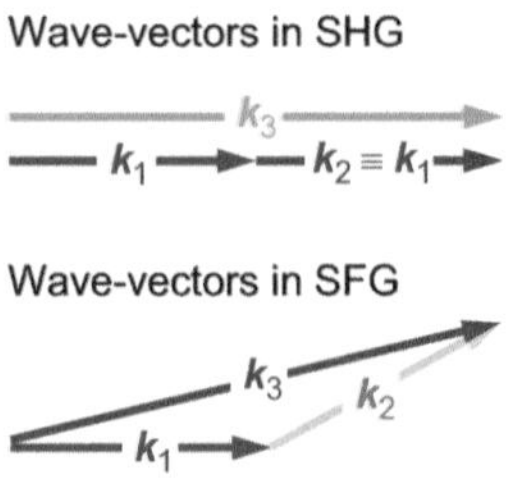

Typically, nonlinear crystal materials exhibit (three) different symmetry-axes, one or two of which have a different refractive index than the other one(s).

Wave-vectors in SHG

Wave-vectors in SFG

**Figure 5.16** Vector diagrams for SHG and SFG.

For example, common uniaxial crystals have one preferred axis, called the extraordinary axis—normally indexed with "e," and the remaining two are ordinary axes—normally indexed "o." Within this framework of symmetry, the direction of propagation of the fundamental wave can be selected so that the index of refraction of the fundamental and second-harmonic are the same. Then, at a mutual angle $\theta$ with the optical axis of the crystal, both waves will remain in phase while propagating through the crystal. The crystal now is *phase-matched* or "index-matched." Of course, when the wavelength (and hence the frequency) of the incoming laser wave(s) changes, the angle $\theta$ needs to be adjusted suitably; this adjustment is termed *angle tuning*.

Two variants of angle tuning are common, which differ in the orientation of the polarization vectors of the incoming $\omega_1/\omega_2$ and outgoing $\omega_3$ waves, with respect to the "o" and "e" directions of the refractive index directions (these are shown schematically in Figure 5.17):

- So-called *Type I phase matching* corresponds to the directional cases $\omega_1\uparrow\uparrow e$, $\omega_2\uparrow\uparrow e$, and $\omega_3\uparrow\uparrow o$ (for positive birefringent uniaxial crystals) and $\omega_1\uparrow\uparrow o$, $\omega_2\uparrow\uparrow o$, and $\omega_3\uparrow\uparrow e$ (for negative birefringent uniaxial crystals)

- So-called *Type II phase matching* is characterized by $\omega_1\uparrow\uparrow o$, $\omega_2\uparrow\uparrow e$, and $\omega_3\uparrow\uparrow o$ (for positive birefringent uniaxial crystals) and $\omega_1\uparrow\uparrow e$, $\omega_2\uparrow\uparrow e$, and $\omega_3\uparrow\uparrow o$ (for negative birefringent uniaxial crystals)

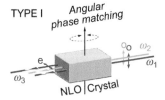

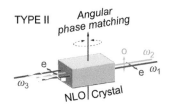

**Figure 5.17** Wave polarization orientation in Type I and Type II phase matching for SHG/SFG.

There is one major caveat one often encounters during operation of NLO-conversion devices. When implementing phase matching by angle tuning, the optical waves involved are not normally collinear with each other. This is because the Poynting vector of the "e"-wave propagating through a birefringent crystal is not parallel with the wave-propagation vector. A consequence of this is a phenomenon known as *beam walk-off*, which limits the NLO conversion efficiency. The problem of walk-off would not occur if one could force the waves to propagate at angle $\theta = 90°$ with respect to the optical axis of the crystal. Since the "e"-index is, in general, more temperature-dependent than the "o"-index, one can adjust the birefringence of the crystal by varying the temperature until perfect phase matching is obtained. However, for any given crystal cut, this condition can only be achieved over a relatively narrow range of frequencies. Further details on these aspects, and additional solutions, can be found in Yao and Wang (2012).

### 5.3.3 Selected nonlinear crystals and their common uses

Many NLO materials are commercially available, serving a wide range of general or specific applications. Here, we only provide a brief summary of the most commonly used crystals and their general fields of application. All crystals need to be cut at particular angles relative to the optical symmetry axis, optimized for any desired application (for a typical example, see Figure 5.18).

*KDP/KD\*P crystals.* For many decades, potassium dihydrogen phosphate ($KH_2PO_4 \equiv KDP$) and potassium dideuterium phosphate ($KD_2PO_4 \equiv KD^*P$, or DKDP) have served as the "work horses" for doubling, tripling, and quadrupling of radiation, predominantly from Nd:YAG but also other laser sources. They exhibit good homogeneity over large volumes and have a high damage threshold. On the other hand, they have the following drawbacks: (1) have relatively low nonlinearity and (2) are hygroscopic—the crystals need to be kept dry (normally

**Figure 5.18** KD\*P crystal cut at the optimum angle for Type II THG at 355 nm.

in sealed, temperature-controlled enclosures) to avoid damage to the polished entrance and exit surfaces.

*KTP crystals.* Potassium titanyl phosphate (KTiOPO$_4$ ≡ KTP) is commonly used for frequency doubling of diode pumped solid-state (DPSS) lasers; it has a relatively high optical damage threshold, significant optical nonlinearity, and superb thermal stability. In practice, however, also KTP crystals—like all other NLO crystals—need to be temperature-stabilized to operate reliably without drift in conversion efficiency. Besides the aforementioned use for frequency doubling to the green, KTP is also used as the NLO element in OPOs for near-IR generation up to about 4000 nm, particularly because of its high damage threshold and large crystal aperture, which allow for high-power applications.

*BBO/LBO crystals.* Of the many borates that have been tested and used for NLO applications, the most popular and widespread are β-barium borate (β-BaB$_2$O$_4$ ≡ BBO) and lithium triborate (LiB$_3$O$_5$ ≡ LBO). Both materials exhibit (1) wide transparency and phase matching ranges (from well below 500 nm to beyond 2600 nm); (2) large nonlinear coefficients (about 5–10 times larger than KDP); (3) high damage threshold; and (4) excellent optical homogeneity. While being expensive, BBO and LBO crystals provide all-round solutions for a wide range of NLO applications in frequency conversion as well as optical parametric generation (OPG) and amplification.

*LiNbO$_3$ crystals.* Lithium niobate (LiNbO$_3$) crystals—particularly in the form of periodically poled stack configuration (short PPLN)—are widely used for frequency doubling of wavelengths > 1000 nm and ultrashort-pulse duration (picosecond- and femtosecond-pulse duration) OPOs. Lithium niobate has a relatively low damage threshold, but due to its high nonlinearity, its normally low pump intensities are sufficient for decent conversion efficiencies.

### 5.3.4 Conversion efficiency and ways to increase it

The probability factor $\eta$ for converting the incoming wave(s) into a sum- or difference-frequency wave in a nonlinear crystal is normally small and depends on quite a few parameters, including, in particular, the susceptibility value for the particular material, $\chi^{(2)}$, the length of the NLO crystal, the laser beam overlap, and the (peak) powers of the laser radiation.

From Equation 5.2, one finds that for low incoming intensities, the conversion wave increases linearly with the intensities of those parent waves. For example, this means that for SHG, the frequency-doubled wave grows with the square of the incoming wave's intensity, i.e.,

$$I_3(2\omega) = \eta[I_1(\omega)]^2 \qquad (5.6)$$

Once the incoming waves become significantly depleted, the slope in the SHG decreases; of course, the converted intensity $I_3$ can never grow to larger values than $I_1$ (see Figure 5.19 for typical experimental data). If the incoming beam has high intensity, is of good beam quality, and has relatively narrow optical bandwidth, then—for good phase matching of the NLO crystal—second harmonic conversion efficiencies can easily exceed the 50% mark; in some favorable cases, values of >80% have been demonstrated (see Ou et al. 1992).

It should be noted that in SFG, of course, both incoming waves experience depletion; however, for efficient conversion, one should make the *photon fluxes*

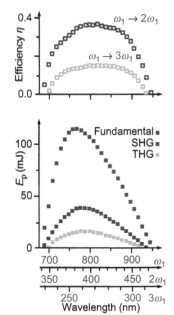

**Figure 5.19** SHG and THG conversion tuning curves and conversion efficiencies for a nanosecond-pulse Ti:sapphire laser (*LX329* plus *LG312* combination; Solar Laser Systems.)

of both input waves similar. For example, if the two waves had frequencies $\omega_1 = \omega$ and $\omega_2 = 2\omega$—for THG to yield a wave $\omega_3 = 3\omega$—then for equal photon fluxes, one should provide $I_2 = I_1/2$. If one input wave had a lower photon flux, its intensity would be depleted faster on traversing the NLO crystal, and hence the overall conversion efficiency would diminish.

### 5.3.5 Outside- and inside-cavity NLO-crystal configurations

When the incoming waves originate from a Q-switched or mode-locked laser system, then the pulse intensities are normally high enough that high conversion efficiency is achieved just for a single pass through a moderately long NLO crystal. Note, however, that for ultrashort femtosecond pulses, effective conversion can adversely be influenced by group-velocity mismatch. This means that the NLO crystal should not be too long, since otherwise one would encounter temporal walk-off. Overall, for frequency conversion of pulsed lasers, the NLO crystal is simply placed in front of the laser and angle-tuned to the appropriate position to meet phase matching.

For CW lasers, the available (peak) intensity is relatively low and would be insufficient to compensate for the small conversion efficiency values of most NLO materials. For example, for SHG, these are by and large in the range $\eta_{SHG} = 10^{-4}$-$10^{-2}$ W·cm. To convert the radiation from low-power CW lasers efficiently, so-called "enhancement" configurations need to be utilized, either by placing the NLO crystal inside the laser cavity itself—as is done very successfully for SHG of DPSS Nd:YAG lasers—or one has to set up an external high-finesse enhancement cavity that builds up the incoming CW wave to a manifold of its inherent intensity—for good coupled cavities, enhancement of >100 is routinely achievable. The latter approach is predominantly used for lasers for which it is difficult or impossible to place the NLO crystal inside the laser cavity itself, e.g., in laser diodes.

There are three common approaches that have been successfully demonstrated, all based on ring cavities that are normally much easier to align and control than linear cavities. These are (1) monolithic ring cavities, in which the sides of the NLO crystal are cut so that at the back the coupled laser beam undergoes perfectly lossless total internal reflection, and the front end is suitably HR/AR-coated; (2) a three-mirror ring cavity in which the entrance and exit surfaces of the NLO crystal are cut as close as possible to the Brewster angle, for minimal loss; and (3) a four-mirror cavity in which the NLO-crystal surfaces are appropriately AR-coated. The latter setup is normally realized in commercial doubling devices, because of the higher flexibility of the arrangement, and the easier handling of the NLO crystal during manufacture; see Figure 5.20 for the conceptual setup of the three implementations.

## 5.4  LASER SOURCES BASED ON NLO: OPTICAL PARAMETRIC AMPLIFICATION (DOWN-CONVERSION)

A widely used NLO process exploiting the $\chi^{(2)}$-nonlinearity is optical parametric down-conversion, normally dubbed *optical parametric amplification* (OPA). Formalistically, OPA may be seen as being akin to DFG, which was discussed

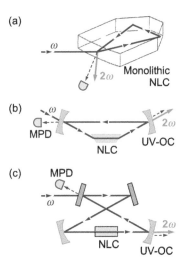

**Figure 5.20**  Resonant cavities for SHG. (a) Monolithic nonlinear crystal (NLC) ring cavity; (b) ring cavity with the NLC surfaces at the Brewster angle; (c) ring cavity with AR-coated NLC. UV-OC = output coupler for UV; MPD = monitor photodiode for feedback to cavity control.

further above. The main difference between the two is that the conditions for the interaction between the incoming waves, with frequencies $\omega_1$ and $\omega_2$, and the NLO crystal are chosen such that in the DFG case, energy from the incoming waves is optimally channeled into the difference-frequency wave $\omega_3$; while in the OPA case, energy from the $\omega_1$-wave is channeled predominantly into the $\omega_2$-wave, thus amplifying it, but also a $\omega_3$-wave is generated. From the point of view of energy content in the individual waves in DFG, both amplitudes of the incoming waves are depleted while the amplitude of the difference-frequency wave increases; in OPA, only one of the incoming waves is depleted, while both the other one and the difference-frequency waves grow in amplitude (see the arrow width in Figure 5.21 for the concept). Traditionally in OPA, the three wave frequencies $\omega_1$, $\omega_2$, and $\omega_3$ are named $\omega_p$, $\omega_s$, and $\omega_i$, which stand for "pump," "signal," and "idler" waves, respectively.

Of course, as all other nonlinear three-wave conversion processes, the three waves in OPA must obey the energy conservation relation, i.e., $\omega_p = \omega_s + \omega_i$, and fulfill the phase-matching condition, i.e., $k_p = k_s + k_i$. For some details on the mathematical framework and experimental design considerations, see Zhang et al. (1995) or Ross (2009). The conceptual comparison between DFG and OPA is shown in Figure 5.21.

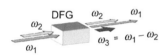
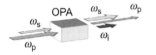

**Figure 5.21** DFG and OPA; the difference in wave intensity is indicated by the thickness of the arrows.

Parametric amplifiers have a number of features that make them very attractive for a range of applications. In most practical systems, OPA is implemented based on pulsed laser sources, ranging from nanosecond- to femtosecond-pulse durations. Since the conversion efficiency of the process scales well with the available pump pulse energy, the parametric gain can be very high, a few orders of magnitude even for NLO crystals of only a few millimeters in length. OPA has two distinct advantages, namely (1) that amplification can be realized over a much wider range than standard laser gain media, basically over the complete wavelength interval over which the NLO crystal is "transparent" to the three wave frequencies (of course, phase matching needs to be maintained as well); and (2) that only minimal heating of the NLO crystal occurs due to the fact that the pump photons are directly converted into signal and idler photons, without absorption, and hence deposition of energy in the crystal.

OPA is now routinely used in chirped-pulse amplification of femtosecond pulses and has also fostered the development of femtosecond OPAs with tunability across the full visible and infrared spectral ranges (see Cerulla and De Silvestri 2003; Cerulla and Manzoni 2007; Liebel et al. 2014).

### 5.4.1 OPG and OPOs

The process of *optical parametric generation* (OPG)—which occasionally is also called "spontaneous parametric down-conversion"—is related to OPA, but there is only one input wave: the pump wave frequency $\omega_p$. Then two output waves of lower frequency are generated in the NLO crystal with $\chi^{(2)}$-nonlinearity, dubbed before as the signal ($\omega_s$) and idler ($\omega_i$) waves, with the requirement $\omega_p = \omega_s + \omega_i$ (see Figure 5.22). The frequencies $\omega_s$ and $\omega_i$ are determined by the phase-matching condition, which can be changed by temperature or angular adjustment; thus, the wavelengths of the signal and the idler photons can be tuned.

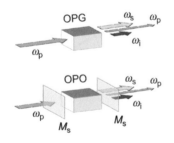

**Figure 5.22** OPG and OPO principles; the signal and idler waves grow at the expense of the pump wave, with the condition $\omega_p = \omega_s + \omega_i$. The OPO cavity mirrors, $M_s$, are optimized for $\omega_s$.

The conversion efficiency of OPG can be substantially increased when dressing the once-through NLO-crystal configuration with a cavity, resulting in an *optical*

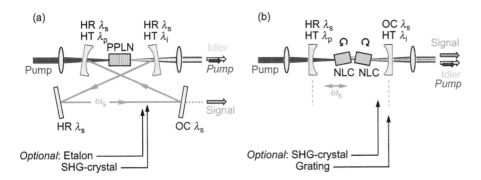

**Figure 5.23** Conceptual setup of OPO resonators (a) for CW operation and (b) for nanosecond-pulse operation. HR = high reflector; HT = high transmitter; OC = output coupler; $\lambda_p$, $\lambda_s$, and $\lambda_i$ are the wavelengths of the pump, signal, and idler waves, respectively.

*parametric oscillator* (OPO). Thus, the OPO is a coherent source of light like a laser, but since no energy levels are involved in the amplification process, it is possible to tune an OPO over a very broad range of wavelengths. Note that, like an ordinary laser, the OPO exhibits a threshold for the pump power, below which the output in the signal and idler waves is negligible. Normally, a pump source with rather high intensity and relatively good spatial coherence is required; this constitutes one of the downsides of OPOs since for decent conversion efficiencies, pump powers often have to be close to the damage threshold of the NLO crystal, and without careful adjustment of the pump power, either the output becomes unusably low or the crystal may be destroyed catastrophically.

Most OPOs are so-called "singly resonant," meaning that the system only resonates at either the signal or the idler wavelength—depending on which wavelength range the user aims at. For the nonresonant wave, optical feedback has to be suppressed, which is normally achieved using dichroic resonator mirrors or a polarizing optical element. A conceptual setup for a typical OPO is shown in Figure 5.23, together with external wavelength separation options. Further details on theoretical aspects and instrumental realization can be found in Erneux and Glorieux (2010) and Thyagarajan and Ghatak (2011). Commercially, both CW and pulsed OPO systems are available, which, in many cases, are pumped by light from standard solid-state lasers, such as the fundamental and harmonics of Nd:YAG lasers.

Optical parametric devices have become very popular over recent years, offering quite a few advantages for spectroscopic applications requiring tunable radiation, such as gas sensing (see Arslanov et al. 2013). Most notable is their extremely wide range of gain from a single material, in contrast to most common laser gain media. The distinct differences of OPOs for CW and pulsed operation are described next, and some of the key aspects comparing ordinary lasers with optical parametric sources are summarized in Table 5.1.

A *continuous-wave OPO* was first demonstrated toward the end of the 1960s (Smith et al. 1968), but it took another 30 years before reliable commercial sources became available. Progress was slow because of limitations in suitable pump sources and nonlinear materials. Conventional NLO materials for bire-fringent phase matching possess only modest nonlinear coefficients, meaning that early demonstrations of CW OPOs required very high pump powers, not easily achievable at the required high beam quality, to generate useable OPO output powers. The advent of so-called quasi-phase-matched nonlinear materials, particularly periodically poled lithium niobate (PPLN), has changed this dramatically, since these materials have reduced the pump-laser power required

**Table 5.1**   Differences between Laser Systems and Optical Parametric Sources

| Source | Potential Advantages | Potential Limitations |
| --- | --- | --- |
| Optical parametric amplifier | Gain throughout NLO-crystal transparency range<br>High gain per unit length<br>No energy storage in gain medium → thermal effects are negligible | High pump intensity required, often close to damage threshold<br>Phase matching required for high efficiency |
| Laser amplifier | Ultrashort-pulse amplifiers, including regenerative amplifiers and chirped-pulse amplifiers | Amplification range limited to gain bandwidth<br>Energy absorbed by gain medium → thermal effects and potential degradation of medium |
| OPO | Tuning over complete NLO-crystal transparence range<br>Amplification is unidirectional<br>No energy stored in gain medium<br>→ thermal effects are negligible<br>Quantum correlation between signal and idler waves | Requires spatially coherent pump source<br>High pump intensity required, often close to damage threshold<br>Phase matching required for high efficiency |
| Laser | Incoherent pump sources (like flash lamps) can be used | Tuning range limited by gain bandwidth<br>Energy absorbed by gain medium<br>→ thermal effects and degradation of medium |

to reach oscillation threshold by an order of magnitude (see Batchko et al. 1998). Now widely tunable OPOs are available; when pumped with a single-mode 1064 nm Nd:YAG laser, wavelength coverage via signal and idler waves extends from about 1400 nm up to nearly 4000 nm, with single-mode output powers of >1 W. This makes CW OPOs very interesting laser sources for numerous applications in molecular spectroscopy.

*OPOs with nanosecond-pulse duration* are pumped by Q-switched (solid-state) laser sources, most commonly at 1064, 532, or 355 nm. The available high pump pulse energies mean that it is easy to overcome the threshold. The output pulses are normally slightly shorter than the pump pulses, since parametric oscillation sets in with some delay, needing the buildup of the signal/idler waves in the (singly resonant) cavity. While nanosecond-pulse OPOs are relatively easy to operate, they often suffer from two drawbacks, namely that (1) their output linewidth is relatively large—normally a few hundreds of gigahertz; and that (2) their pulse-to-pulse amplitude fluctuations are significant. Both are a consequence of the fact that the OPO system often has insufficient time during a nanosecond pulse to settle to the steady state (unlike for CW OPOs). Utilizing both signal and idler waves, commercial nanosecond-pulse OPOs can be tuned in the range 400–2500 nm (which can be extended down to 200 nm by SHG), depending on the pump laser wavelength, with only small gaps around the "degeneracy" points close to double the pump wavelength, and provide pulse energies of often > 50 mJ. Some typical tuning curves are shown in Figure 5.24. The caveat of large output bandwidth can be overcome by two different measures, albeit at the expense of increased system complexity. SLM operation of nanosecond-pulse OPOs can be achieved by injection seeding, i.e., to basically utilize the OPO as an amplifier for an external narrow-bandwidth laser source (like a dye or semiconductor diode laser), or by introducing line-narrowing elements (gratings and/or etalons) into the OPO cavity. A few commercial OPOs exploit the latter principle, reaching even the Fourier limit (see Mes et al. 2002).

In order to realize *OPOs with ultrashort picosecond or femtosecond pulses*, they are synchronously pumped with a mode-locked laser source. The length of the OPO resonator is usually adjusted such that the round-trip time matches the pulse repetition rate of the mode-locked pump, typically a 1064 nm Nd:YAG or

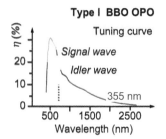

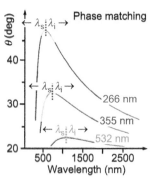

**Figure 5.24**   Tuning and phase-matching curves for a BBO-crystal OPO, pumped by the different Nd:YAG laser wavelengths. Note that the signal and idler waves are generated simultaneously.

800 nm Ti:sapphire laser. Over the duration of many resonator round trips, the OPO can reach steady-state conditions, meaning that noise and peak-power fluctuations can be relatively low. The average pump power required for driving a femtosecond-pulse OPO ranges from a few hundreds of milliwatts to a few watts, at typical pulse repetition rates of 50–100 MHz; the OPO pulse duration is usually comparable to the pump pulse duration. The overall conversion from pump pulse energy to OPO output pulse energy is of the order of 30–50%. Synchronously pumped OPOs are particularly attractive when few-cycle broadband radiation is desired at wavelengths different from the common Ti:sapphire range (see Kumar et al. 2014).

## 5.5 REMARKS ON LASER SAFETY

The often high power available from many common, commercially available lasers can be exploited to "destruct" materials, for example, in the industrial applications of laser cutting and welding, or in the laser (analytical) technique of laser-induced breakdown spectroscopy. Consequently, it is not surprising that laser radiation also affects or destroys human tissue. In particular, our eyes are in danger of being damaged because of their high sensitivity to light. The adverse experience from laser accidents teaches us: even rather moderate laser power is sufficient to cause irreversible damage. Hence, it is prudent to point out to anybody who is working, or intends to work, with lasers in spectroscopy and imaging to follow safety measures to minimize any hazards and to avoid risks to people.

**Figure 5.25** Warning label to be used for lasers of Class 2 and higher.

By and large, working safely with lasers involves, beyond the use of common sense and diligence, three procedural elements:

- *Administrative control* measures involve procedures and information rather than devices or mechanical systems. Some important administrative controls are the posting of warning signs and labels (see the standard laser warning symbol in Figure 5.25), the establishment of standard operating procedures, and safety training.

- *Engineering controls* involve design features or devices applied to the laser, the laser beam, or the laser environment that restrict exposure or reduce irradiance. Such controls include beam shutters, beam attenuators, remote firing and monitoring systems, and—ideally—protective housings placed entirely around the most potent laser systems.

- *Personal protective equipment* is worn by personnel using the laser or in the vicinity of the laser. It includes at least protective eyewear, and potentially gloves, and special clothing.

### 5.5.1 How do laser wavelengths affect our eyes?

The risk of affecting eyesight through laser radiation is due to the optical properties of the human eye. A summary of the risks is given in Figure 5.26.

UV-light below 400 nm is absorbed at the "surface" of the eye (UV-C, UV-B) or advances to the lens (UV-A). A consequence of exposure to high-power light at these wavelengths is an injury to the cornea by ablation or photokeratitis (inflammation of the cornea, similar to "sunburn"), or formation of photochemical cataract (clouding of the eye lens).

**Figure 5.26** Effects of (laser) radiation on the eye, at different wavelengths.

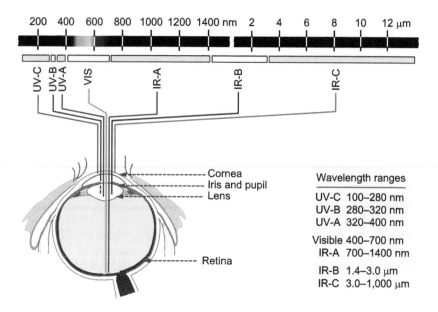

Wavelength ranges

UV-C  100–280 nm
UV-B  280–320 nm
UV-A  320–400 nm

Visible 400–700 nm
IR-A  700–1400 nm

IR-B  1.4–3.0 μm
IR-C  3.0–1,000 μm

Light of visible wavelengths (400–700 nm) advances to the retina, and there it may cause photochemical damage. As a natural protective mechanism, when the light seems too bright—which means the power density exceeds a damage threshold of the eye—we automatically turn away and/or close our eyes (blink reflex). This automatic reaction is effective for radiation up to about 1 mW power. With higher powers, too much energy reaches the eye before the blink reflex can work, with the resulting consequence of possible irreversible damage. The near-infrared wavelengths (700–1400 nm) are a type of radiation that is especially dangerous to the human eye; it still can reach the retina, but no natural protection reaction exists like in the case of visible light, and any harmful exposure may only be noticed when the damage is already done.

Mid-infrared radiation (1400–3000 nm) reaches as far down as into the region of the lens and can cause the so-called aqueous flare—affecting proteins in the aqueous humor, i.e., the layer between the cornea and the lens, or giving rise to corneal burn. Far-infrared radiation (3–1000 μm) is absorbed at the surface of the eye; it leads to overheating of tissue and burning, or ablation of the cornea.

### 5.5.2 Maximum permissible exposure and accessible emission limit

In order to assess whether a laser radiation hazard exists, two limiting quantities are being considered, namely (1) the *maximum permissible exposure* (MPE) and the *accessible emission limit* (AEL). These limits are based on calculations that include the main parameters characterizing laser sources—wavelength, power/energy, and emission duration, as well as the nature of exposure (direct, specular reflection, diffuse reflection).

For the eye, the *MPE* is the maximal radiation level one can be exposed to before immediate or long-term damage occurs; it is derived from the energy density limits, in units of $J \cdot cm^{-2}$, or the power-per-surface-unit (intensity) limits, in units of W·cm$^{-2}$, of exposure (note that similar calculations are performed for hazards related to the skin). The calculation of the MPE for ocular exposure takes into account how light of the various wavelengths interacts with the eye (see Figure 5.26). For example, deep-UV light causes accumulative damage to the cornea, even when the power is very low; mid-IR light with wavelengths longer than ~1400 nm is strongly absorbed by the transparent parts of the eye (the vitreous humor) before reaching the retina, which means that its MPE is higher than that for visible light, which reaches the retina without much attenuation. In addition to the wavelength and exposure time, the MPE takes into account the spatial distribution of the light; in this context, collimated visible and NIR wavelength laser beams are particularly dangerous because the eye lens focuses the light onto a tiny spot on the retina.

The MPE for visible/NIR laser light is based upon the total energy, or power, collected by a fully dilated (night-adapted) human eye; for this situation, it is assumed that this limiting pupil aperture diameter is LA $\cong$ 7 mm, equivalent to a pupil aperture area of ~0.39 mm$^2$. The calculation philosophy errs on the side of caution: normally, the calculated MPE values relate to about 10% of the light radiation dose that has a 50:50 chance of causing damage, under the worst-case scenario conditions (one assumes that the eye lens focuses the light onto the smallest possible spot size on the retina, for a fully open pupil). The principal geometry underlying the MPE estimation is shown in Figure 5.27. While the calculation procedure itself is straightforward, the following should be kept in mind. The most common profile of a laser beam is a Gaussian (TEM$_{00}$) profile. The diameter of a Gaussian beam is defined by either using the $1/e$ or the $1/e^2$ point in the beam profile. Laser manufacturers often base the information in their data sheets on the $1/e^2$ definition for the diameter since the related area encompasses 90% of the total beam power/energy. However, safety calculations use the $1/e$ diameter definition, so care has to be taken that the correct values are used. The two diameter definitions obey a simple relation: if $a$ denotes the diameter, then $(1/e^2) = [2 \cdot a(1/e)]^{1/2}$.

In the figure, another useful parameter is indicated: the nominal hazard zone (NHZ). This zone describes the region within which the level of direct, reflected, or scattered (diffuse) laser radiation surpasses the allowable MPE. The purpose of a NHZ is to define an area in which (physical) control measures are required. It can be estimated from the generic equation

$$\text{NHZ} \cong (1/\theta) \cdot [(4P/\pi \cdot \text{MPE})^{1/2} - a_{\text{laser}}],$$

where $\theta$ is the laser beam divergence; $P$ is the (average) radiant power in watts; and $a_L$ is the beam diameter at the laser aperture. For example, for a laser with beam divergence $\theta = 1$ mrad, one finds the following NHZ distances, for the laser beam to be eye-safe for unintentional exposure: NHZ = 7 m for a 1 mW (Class 2) laser; NHZ = 16 m for a 5 mW (Class 3R) laser; NHZ = 160 m for a 500 mW (Class 3B) laser; and NHZ = 275 m for a 1.5 W (Class 4) laser.

The *AEL* is an entity that is used to classify lasers (see below) according to the related hazard, depending on their characteristics. AELs are based on the laser emission, while MPE limits are based on the radiation received by the eye (or the skin), directly or after reflection. AEL is calculated according to the relation AEL = MPE × $\pi(\text{LA}/2)^2$, where LA is the limiting aperture, based on physical factors such as the (dilated) pupil size (7 mm diameter) and beam "hotspots" (1 mm diameter).

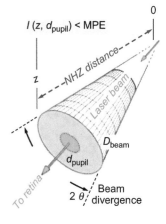

**Figure 5.27** Parameters of importance to determine the light exposure to the eye and assess the NHZ distance.

As shown in this chapter, the range of wavelengths, power levels, pulse energies, and pulse durations that are available from the various laser sources suggests that they are the source of very different hazards. In order to assess MPE, NHZ, and AEL values for certain lasers and specific experimental conditions, complex calculations may have to be performed. Guidelines of how to do this can be found in the publications of the regulatory framework for laser safety, as defined in the EU standard IEC/EN 60825-1 and the US standard ANSI Z136.1 (Ziegelberger 2013; International Electro-technical Commission 2014; Laser Institute of America 2014). Software based on these standards is available to ease the calculations (*The Evaluator* 2014; *LaserSafe* 2015).

### 5.5.3 Laser classification

It is next to impossible to lump all laser sources together as one unique group with common safety limits; however, it is possible to group certain laser sources, and their use, into "classes" according to their MPE and AEL values. Within the regulatory framework for laser safety, lasers are assigned into four broad hazard classes (1, 2, 3, and 4), depending on their potential to cause biological damage, in particular to the eye. Brief descriptions and properties for these, and specific subclasses, are collated in Table 5.2. It should be noted that laser users may still find old (prior to 2007) classification labelling in Roman numerals (classes I to IV), although these should no longer be used; cross-correlation to these classes is included in the aforementioned laser standard documents.

### 5.5.4 Laser safety eyewear

Laser eye protection products should safeguard against accidental exposure to laser radiation; they are not meant to withstand deliberate, prolonged interaction with laser light beyond the timescales laid down in the regulations, i.e., 10 s for CW radiation or for 100 pulses. While the principle guidelines for laser safety glasses are covered in the IEC/EN 60825-1 and ANSI Z136.1 frameworks, more specific norms for eyewear can be found in *EN207:2009* and *EN208:2009* (for general protection and alignment, respectively).

In general, laser eye protection products are labeled with three items of information: (1) for which operating regime (CW or pulsed) they are suitable; (2) the wavelength range covered; and (3) the optical density (OD) by which transmission is reduced. In particular, the working mode is of importance since peak powers for short and ultrashort pulses can be enormous. The four regulatory working modes include "D" (CW, continuous for >0.25 s); "I" (pulsed, duration 1 μs–0.25 s); "R" (giant pulses, duration 1 ns–1 μs); and "M" (mode-locked pulses, duration < 1 ns).

When wearing laser safety glasses, some wavelengths of the spectrum that would usually reach the eyes are filtered out. If light components from the visible spectrum are blocked or attenuated, this inevitably changes the perception of the environment. Firstly, by attenuation of the transmission, the environment appears darker than normal daylight; this effect is defined by the so-called *visible light transmission* (VLT). Secondly, the blocking of certain wavelengths will change the perception of color; this is associated with the phenomenon of *color vision*.

**Table 5.2**  Laser Classification According to IEC/EN 60825-1:2014 and ANSI Z136.1:2014

| Class | Description | Remarks on MPE and AEL |
|---|---|---|
| 1 | Applies to very low power lasers, or encapsulated laser systems; radiation in this class is considered safe.[*] <br> **Hazard**: Safe under all operating conditions. | **MPE**: Never exceeded, even for long exposure duration (>100 s), and even with the use of optical instruments. <br> **Typical AEL**: 40 μW CW in the blue |
| 1M | Applies to very low power lasers, either collimated with large beam diameter or highly divergent. <br> **Hazard**: Safe for viewing directly with the naked eye, but may be hazardous to view with the aid of optical instruments. | **MPE**: MPEs are not exceeded for the naked eye, even for long exposure durations, but may be exceeded with the use of optical instruments.[†] <br> **Typical AEL**: Same as Class 1 |
| 2 | Applies to visible-wavelength (400–700 nm) low-power lasers; radiation in this class is considered low risk. <br> **Hazard**: For accidental exposure, the natural aversion response will be sufficient to prevent damage; prolonged viewing may be dangerous. | **MPE**: Not exceeded for exposure of 0.25 s, even with the use of optical instruments (the blink reflex limits exposure duration to nominally 0.25 s). <br> **Typical AEL**: 1 mW CW; any emissions outside the 400–700 nm range must be below the Class 1 AEL |
| 2M | Applies to visible-wavelength (400–700 nm) low-power lasers, either collimated with large beam diameter or highly divergent; radiation in this class is considered low risk. <br> **Hazard**: The same as Class 2 but may be hazardous to view with the aid of optical instruments. | **MPE**: Not exceeded for exposure of 0.25 s, but may be exceeded with the use of optical instruments. <br> **Typical AEL**: 1 mW CW; emissions outside the 400–700 nm range must be below the Class 1 AEL |
| 3R | Applies to low-power CW lasers; radiation in this class is considered low risk, but potentially hazardous <br> **Hazard**: Safe when handled carefully; but carries a small hazard for accidental exposure. <br> ***Eye protection should be used*** | **MPE**: May be exceeded up to 5×, equally for naked eye and using optical instruments. <br> **Typical AEL**: 5× Class 1 limit in the visible (<1 mW); 5× Class 2 in the UV and IR (<5 mW). |
| 3B | Applies to medium-power lasers; radiation in this class is very likely dangerous. <br> **Hazard**: MPE for eye exposure may be exceeded and direct viewing of the beam is potentially serious. Diffuse reflections from surfaces usually should not be hazardous. Not normally hazardous to the skin. <br> ***Eye protection must be used in the nominal hazard zone.*** | **MPE**: May be exceeded up to 5×, equally for naked eye and using optical instruments. <br> **Typical AEL**: Must not exceed 500 mW for CW lasers with wavelengths >315 nm; must not exceed 125 mJ in less than 0.25 s for pulsed laser systems. |
| 4 | Applies to high-power lasers. <br> **Hazard**: Damaging under both intrabeam as well as diffuse reflection viewing conditions. May also cause skin injuries, pose fire hazards, and generate hazardous fumes. <br> ***Eye protection must always be used.*** | **MPE**: Exceeded for eye and skin, diffuse reflection may exceed MPE for the eye. <br> **Typical AEL**: Exceeds 500 mW for CW lasers; exceeds 125 mJ in less than 0.25 s for pulsed laser systems. <br> **Note**: No upper limits. |

[*]A Class 1 laser **product** is a laser unit or device that may include lasers of a higher class but whose beams are confined within a suitable enclosure so that access to laser radiation is physically prevented.

[†]Optical instruments are magnifying lenses, telescopes, and microscopes; but not prescription glasses.

VLT is determined and evaluated according to the spectral sensitivity of the eye when daylight-adapted (photopic) or night-adapted (scotopic). In case that the VLT value drops below 20% of normal, users should ensure that their working environment receives additional illumination.

If color vision is impaired or restricted, some colors may not be recognized correctly. This effect, of course, may apply to warning lights or displays, or the ability to distinguish between instruments or vessels marked by color. Thus, care

*View through filter*

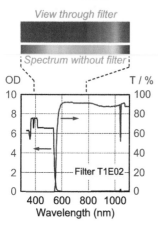

**Figure 5.28** Laser eye safety glasses for the second, third, and fourth harmonics of Nd:YAG lasers. The specific OD and light transmission result in a distorted, "orange" color vision of the environment. (Data provided courtesy of LaserVision. "Laser safety products," 22nd ed. Fürth, Germany: LaserVision GmbH, 2014.)

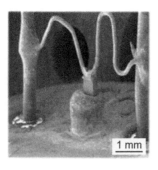

**Figure 5.29** One of the first GaAs laser diode devices. (From Popov, Y.M., *Semicond. Sci. Technol.* 27, no. 9 (2012). Reproduced with permission. © IOP Publishing. All rights reserved.)

should be taken not to introduce secondary hazards by excessively minimizing the laser hazard.

An example of the effects of laser safety glasses on vision is shown in Figure 5.28, for glasses protecting the eye against the harmonics of the Nd:YAG laser (the view of the world is "orange").

Finally, it should be noted that specific precautions should be followed when aligning laser systems. Full protection of the eye may prevent one from actually seeing a laser beam spot (in the visible), which may serve as an alignment aid, and thus accidental good alignment resulting in optimum laser output could present a severe hazard to the operator.

Therefore, dedicated safety glasses are recommended for alignment purposes, which attenuate the actual incident power to the power of a Class 2 laser so that the blink reflex provides adequate protection.

We wish to conclude this section on safety, and the chapter on lasers overall, with two simple messages to those who want to or are using lasers in spectroscopy and imaging.

First, when working with lasers, know your laser system (it always is a good idea to read the manual) and follow as closely as possible the policies and regulations of your workplace, which should be based on the standard safety regulation framework. In addition, use common sense in whatever you do. Second, treat lasers and their radiation with respect; your eyes will thank you for it!

# 5.6 BREAKTHROUGHS AND THE CUTTING EDGE

The demonstration of the ruby laser in 1960 certainly marks the beginning of all laser development. But it is probably the rise of semiconductor lasers that spawned the remarkable march of lasers into nearly all facets of life.

## 5.6.1 Breakthrough: Semiconductor laser diodes

In many ways, it has been the successful demonstration of the GaAs semiconductor laser diode in 1962 (see Figure 5.29), emitting at ~850 nm, that really has to be seen as the beginning of the laser revolution (Hall et al. 1962). From the first demonstration, it took nearly another decade to achieve successful CW operation of laser diodes at room temperature (Hayashi et al. 1969; Alferov et al. 1970). And a few further years went by until the lifetime of laser diodes was extended to levels that made them really usable.

The first such commercial laser diode was marketed in 1975, being specified to emit a few milliwatts for several thousands of hours, at a cost of a few thousands of US dollars. This has to be compared to today's laser diodes, which have become much more powerful, with extended lifetime, and some of which only set one back less than US$1.

Besides many technical applications, laser diodes are now used in a multitude of applications in laser pumping, spectroscopy, and imaging. About half of all the lasers sold today are semiconductor diode devices, with total sales figures exceeding US$4 billion.

## 5.6.2 Breakthrough: Widely tunable QCLs

The QCL—like the diode laser a semiconductor material-based device but operating on intraband quantum well transitions rather than across-bandgap transitions (see Section 5.2)—has certainly revolutionized vibrational molecular spectroscopy and analytical sensing. These lasers can be tailored through their quantum well structure nearly seamlessly to any wavelength in the mid-IR to the terahertz regime, and their broad gain profiles also guarantee wide tuning ranges. For obvious reasons, commercial as well as scientific interests are highest in the mid-IR (range $\lambda_L \sim 3.5\text{-}13 \, \mu m$) and the terahertz spectrum (range $\lambda_L \sim 60\text{-}150 \, \mu m$, equivalent to $\nu_L \sim 2\text{-}5$ THz).

Since their first demonstration by Faist et al. (1994), QCLs have had a similar rapid development as seen with the semiconductor laser diodes two decades earlier. Not only could the devices be manufactured and marketed, which operated at ambient temperature rather than needing cooling to cryogenic temperatures, but also tuning ranges of individual devices were expanded to as much as nearly 1000 $cm^{-1}$ (covering the range of $\lambda_L \sim 5.5\text{-}12.5 \, \mu m$), while costs for such systems have come down significantly. Also, for high-resolution spectroscopy, QCL chips have been incorporated into external-cavity configurations, providing SLM operation, with $\Delta\nu_{SLM} \sim 1\text{-}10$ MHz rather than the free-running linewidth of $\Delta\nu_{QCL} \sim 10\text{-}30$ GHz. Off-the-shelf QCL units are now available, which provide mode hop-free SLM tuning across a range of up to nearly 100 $cm^{-1}$.

## 5.6.3 At the cutting edge: HHG and attosecond pulses

It is probably fair to say that hardly a week goes by during which new research results for laser development are not reported, or that new/improved laser products are not announced by manufacturers, covering all types of lasers from CW to ultrashort-pulse duration. Therefore, it may be somewhat subjective and depend on personal experience and use of lasers which particular developments one categorizes as "cutting edge." We have selected one particular complex of developments, which is exciting for various applications in spectroscopy and imaging, namely high-harmonic generation (HHG) and attosecond pulses.

HHG is an NLO process in which the frequency of laser light is converted into (high) odd-integer multiples. Nowadays, these are generated from atoms or molecules exposed to intense (ultrashort) near-IR laser pulses; for gases, HHG was first demonstrated by McPherson et al. (1987). Typically, the optical intensities required for HHG are of the order of $>10^{14}$ W/cm$^2$; the wavelengths span the extreme UV (EUV or XUV, with wavelength range $\lambda = 124\ldots10$ nm, equivalent to photon energies $10\ldots124$ eV). While only a very small fraction of the incoming laser pulse is converted into higher harmonics, the generated output of photons can still be sufficient for a range of interesting experiments, and HHG has now been established as one of the best methods to produce ultrashort coherent light pulses. For quite a few applications, high harmonics are now used instead of synchrotron radiation.

In most cases, HHG relies on Ti:sapphire laser sources (including amplifiers to generate the required pulse peak-powers), and in general repetition rates are in the range of a few hertz up to a few kilohertz. Now that enhancement resonators have started to replace amplifier stage(s); this (1) allows for higher repetition rates—up to >100 MHz; and (2) provides the means to tailor the pump pulses to aim at specific harmonics (see Pupeza et al. 2014). Even more exciting is that one can generate these EUV wavelengths in pulses in the sub-femtosecond regime, i.e., attosecond regime (see Figure 5.30). Depending on the pump laser's pulse energy, wavelength, and pulse duration, transform-limit pulses shorter than 100 as have been demonstrated (see Chen et al. 2014). Typical applications for such pulses of EUV-photon energy and sub-femtosecond duration are, e.g., in bound-state wave packet dynamics or for imaging of atomic/molecular electron motion (see Chapter 20).

Until a few years ago, it was necessary for researchers to go to (national) laser facilities to gain access to attosecond laser pulses. This has changed to a certain degree with the commercialization of a table-top device that generates highly coherent EUV "laser-like" light, generated in selected noble gases (propagation of the laser beam is controlled in a gas-filled hollow capillary waveguide), whose pulse duration can be on the attosecond scale and whose peak wavelength is user-selectable in the range 47...13 nm, equivalent to 26...95 eV (KMLabs 2009).

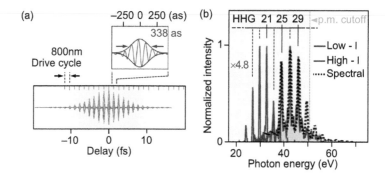

**Figure 5.30** HHG pulses generated in argon by a 10-cycle Ti:sapphire pump laser. (a) Autocorrelation data; (b) HHG spectra, extracted from the FFT of the autocorrelation data (blue/red traces for low/high pump intensity), and recorded spectrally (black dotted trace). (Data adapted from Chen, M.C. et al., *Proc. Natl. Acad. Sci.* 111, no. 23, (2014): E2361–E2367.)

# CHAPTER 6

# Common Spectroscopic and Imaging Detection Techniques

In the most general terms, one may define laser spectroscopy and laser imaging as specific measurements of the effect of the interaction of (exciting) laser photons with matter. Namely, these are measurements in which energy- or wavelength-specific information about the laser photons themselves or the quantum states of the target matter is accessible, or in which the spatial distribution of the interaction process and/or its outcome is probed. Depending on the answers sought from the measurements, a range of detection methods may be utilized, including (1) photon detection methods; (2) charged particle—electron or ion—detection methods; and (3) methods exploiting changes in the (macroscopic) physical properties of the medium with which the laser light interacts.

In this chapter, only the basic principles behind signal retrieval and the instrumentation relevant for this are discussed. For in-depth discussion of construction details and/or operational parameters, the reader will be referred to relevant textbooks and research articles included in the Bibliography.

## 6.1 SPECTRAL AND IMAGE INFORMATION: HOW TO RECOVER THEM FROM EXPERIMENTAL DATA

As pointed out in the introduction to this chapter, analytically probing a sample with laser radiation by and large relies on the selectivity afforded by the transitions between atomic/molecular energy levels, regardless whether absorption, fluorescence, or inelastic scattering is exploited, or if neutral or charged particles are involved. Therefore, methodologies are required, which allow one to access this energy level information—for identification and/or quantification, meaning that dissection into spectroscopic components is required. In some cases, the laser itself may serve as the selective entity if its wavelength can be tuned; in other cases, wavelength-selective devices are required. In this section, a brief overview over the most common wavelength-selective "filtering" methodologies is provided. For a good summary and comparison of the functionality and properties of spectroscopic instrumentation, see Demtröder (2008).

### 6.1.1 Spectral information and its retrieval from photon events

Wavelength-analyzing instrumentation for laser-induced processes yielding optical spectra basically falls into two categories, namely spectrometers and interferometers. In the former case, the (dispersed) spectral information is directly accessible. In the latter case, only in rare cases it is possible to extract the spectral information directly and unambiguously; normally sophisticated mathematical procedures are required to access said information in full (see further below).

*Spectrometers*: Spectrometers rely on the dispersion of the radiation by a wavelength-separating element, in modern instruments by and large a grating. Depending on the actual spectrometer design, wavelength segments of up to the complete visible spectrum can be sampled, with spectral bandwidth $\Delta\lambda$ of a few nanometers down to a hundredth of a nanometer. Of course, the particular wavelength range or segments and the spectral resolution are often optimized to suite a particular experiment. The principle of the most common grating spectrometer design, the so-called Czerny–Turner configuration, is shown in Figure 6.1; here we outline only its main features.

In general, the incoming light is focused onto the entrance slit so that as much of the light from a source as possible enters the spectrometer. Normally, care has to be taken that the solid angle of the "light cone" entering the instrument matches the so-called $F$-number, $F\#$, of the spectrometer. This $F$-number is related to the dimension and focal length of the collimating mirror (or lens):

$$F\# = f/D \cong 1/(2 \cdot \mathrm{NA}) = 1/(2 \cdot n \cdot \sin\theta), \tag{6.1}$$

where $f$ and $D$ are the focal length and size of the mirror (lens), respectively; $\mathrm{NA} = n \cdot \sin\theta$ is the numerical aperture, with $\theta$ being half the acceptance-cone

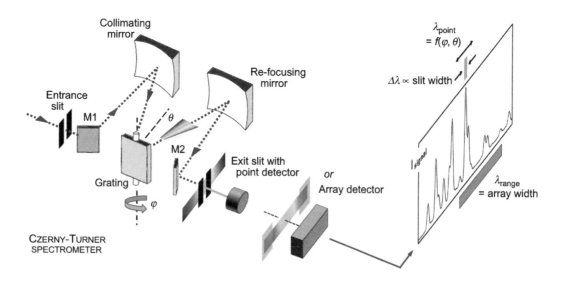

**Figure 6.1** Simplified schematic of a (reflective-element) Czerny–Turner spectrometer, including the options of wavelength "tuning" via rotation of the grating, and of (1) pointwise recording using an exit slit—single-element photodetector combination; or (2) segment-wise spectrum recording using an array photodetector. The two recording options are indicated at the top and bottom of the sample spectrum, respectively.

angle; $n$ is the refractive index (of air). Note that the numerical aperture is quite often used to define the so-called light collection power, meaning the fraction of light that can be collected from a source using any specific spectrometer configuration.

The collimator directs the radiation entering via the entrance slit as "parallel light" beam onto the grating, which then separates light of different wavelengths with an associated, unique diffraction angle $\theta_m$ (with diffraction order $m = 0,1,2,...$).

The so dispersed light is focused by a further (refocusing) mirror, or lens, in the spectral-image plan, to be interrogated pointwise using an exit slit–photosensor combination, or a many-pixel array detector. Note that the wavelength position $x_\lambda$ in the image plane is not linear, but a function of grating-related angles $\varphi_g$ (compensating rotation) and $\theta_g$ (dispersion angle), and the groove spacing $d_g$ of the grating (see Equation 6.2a). The aforementioned spectral bandwidth $\Delta\lambda$ is normally addressed in terms of the spectral resolution $R_\lambda = \lambda/\Delta\lambda$. Primarily, it depends on the spectrometer slit widths $d_s$ (entrance as well as the optional exit slit); the angular dispersion (which is associated with the number of grating grooves per millimeter, or groove spacing $d_g$); and the number of illuminated grating grooves (or width dimension of the grating $d_w$), as given in Equation 6.2b:

$$x_\lambda = f\left(\theta_g; d_g^{-1}\right); \tag{6.2a}$$

$$R_\lambda = f\left(d_s^{-1}; d_g^{-1}; d_w\right). \tag{6.2b}$$

This means that the spectrum becomes wider for decreasing groove spacing and increasing dispersion angle $\theta_g$; and the spectral resolution improves (1) with decreasing slit width, (2) with decreasing groove spacing (or higher groove density), and (3) with increasing grating size (or total number of illuminated grooves). Further details can be found in standard textbooks on optics and spectroscopy, like Demtröder (2008). The key characteristics of typical grating spectrometers are collated in Table 6.1.

Note that the Czerny–Turner spectrometer described here is only one of numerous, possible configurations. Other reasonably popular types of instruments are, e.g., Echelle spectrometers (for the principles, see Harrison 1952) or transmission spectrometers based on volume-holographic gratings (VHGs; see Owen 2007). The former type makes it possible to realize spectrometers with ultrabroad spectral range and high resolution, without any moving part; the latter type has become popular for very compact, high light-collection power systems.

*Interferometers*: The principle of interferometric instruments is based on constructive and destructive interference within a passive resonator. By and large, the spectral segments covered by individual instruments are normally much smaller than those for spectrometers; on the other hand, for high mirror reflectivities, and thus high cavity finesse, much better spectral resolution than for grating spectrometers can be achieved.

The most basic implementation of an interferometric device is that of plane-parallel cavity, the so-called Fabry–Pérot (FP) etalon. They are not frequently used in general spectroscopy due to their normally very small spectral segment, which can be allocated unambiguously to incident wavelengths. However, FPs are frequently used in the characterization of narrowband laser sources (in the form of, e.g., wavemeters or spectrum analyzers; see Table 6.1), or to provide precise calibration markers for the wavelength scale in spectroscopic applications

**Table 6.1** Comparison of Important Parameters for Spectrometric Instruments

| Device Property | Grating Spectrometer | | Wavemeter/Etalon | | FTIR Spectrometer | |
|---|---|---|---|---|---|---|
| Focal length | 100/150...750 | (mm) | 0.2...200[1] | (mm) | 1...25[2] | (mm) |
| Gratings | 150...3600 | (gr·mm$^{-1}$) | – | | – | |
| Linear dispersion | 1...30[3] | (nm·mm$^{-1}$) | – | | – | |
| Optical range | 200...1100 (2500) | (nm) | 200...11,000 | (nm) | 1000...24,000 | (nm) |
| Spectral interval | 10...full range | (nm) | 0.005...5 | (cm$^{-1}$) | 10...8000 | (cm$^{-1}$) |
| Spectral resolution | 0.02...5 (0.7...175) | (nm) (cm$^{-1}$) | 0.0001...0.1 | (cm$^{-1}$) | 0.01...10 | (cm$^{-1}$) |

[1] Mirror separation.

[2] Typical travel of Michelson interferometer mirror.

[3] The linear dispersion is a function of the focal length of the spectrometer, the spacing of grating grooves, and the center wavelength.

that rely on scanning, tunable lasers. The type of interferometer that is most frequently encountered in molecular spectroscopy experiments is the (scanning) Michelson interferometer. By and large, it is used in IR absorption and Raman spectroscopy, and has given rise to—in conjunction with the spectral evaluation by Fourier transform (FT) algorithms—the high-resolution molecular spectroscopy techniques of FTIR absorption spectroscopy (FTIR) and FT Raman spectroscopy (FT-Raman). Details on FT spectroscopy techniques and instrumentation may be found in Griffiths and de Haseth (2007) and Smith (2011). Here, we exemplify the overall underlying principles for FTIR spectroscopy.

In their simplest form, FT spectrometers comprise a basic Michelson interferometer, as shown in Figure 6.2. When scanning the movable mirror (at location $L$) of the setup over some distance $\Delta z$, normally a large multiple of free spectral ranges of the device $\Delta z = N \cdot \delta\lambda$, with $\delta\lambda \cong \lambda_0^2/2n_{air}L$, an interference pattern, or *interferogram*, is produced that "encodes" the absorption spectrum of the sample placed within the interferometer. The light source is a broadband IR emitter, covering all absorption wavelengths for the molecules of interest. The photodetector records this interferogram as the "raw" data of the experiment; a conceptual interferogram is shown in the center part of Figure 6.2. Note that the peak at the center is the so-called ZPD position, or zero-path difference: irrespective of wavelength, all the light passes through the interferometer since its two arms are of equal length.

The basic principle of the Michelson interferometer, which is exposed to a single-wavelength light source, yields a pure, sinusoidal interference pattern. Adding further, different wavelength sources (ultimately, the continuum of an IR radiation source) result in interference patterns of ever increasing complexity. Therefore, in general, the raw data (interferogram) collected from a FT spectrometer will be quite difficult to "read" immediately, in stark contrast to the spectrum from a dispersing spectrometer. The recorded intensity is a function of the path length difference $z$ in the interferometer and the wavelength components of the light source, normally expressed in terms of wavenumber $\tilde{v} = 1/(n_{air}\lambda)$:

$$I_{det}(z, \tilde{v}) = I(\tilde{v}) \cdot [1 + \cos(2\pi\tilde{v}z)], \tag{6.3}$$

where $I(\tilde{v})$ represents the (absorption) spectrum to be determined. Then the total intensity $I(z)$ measured by the detector for any specific path length difference is given by the integral over all spectral contributions, i.e.,

$$I_{\text{det}}(z) = \int_0^\infty I(\tilde{v}) \cdot [1 + \cos(2\pi\tilde{v}z)] \cdot d\tilde{v}. \qquad (6.4)$$

This has just the form of a Fourier cosine transform. Thus, one can calculate the desired spectrum $I(\tilde{v})$ by applying an inverse FT, and one obtains

$$I(\tilde{v}) = 4 \cdot \int_0^\infty \left[I_{\text{det}}(z) - \frac{1}{2} \cdot I_{\text{det}}(z = 0)\right] \cdot \cos(2\pi\tilde{v}z) \cdot dz. \qquad (6.5)$$

An example for such a resolved FT spectrum is also included in Figure 6.2.

As in all spectroscopic methodologies, also FT spectra exhibit linewidths. The recorded interferograms are associated with the *length* domain; since the FT procedure inverts the dimension, the generated spectrum belongs to the *reciprocal-length* domain, i.e., the *wavenumber* domain, of course. As a consequence, the spectral resolution (in cm$^{-1}$) is equal to the reciprocal of the maximum path length difference (in cm). For example, for a maximum path length difference of 0.25 cm, a spectral resolution of 4 cm$^{-1}$ resolution would be obtained. Clearly, the resolution could be much improved simply by increasing the maximum path length difference; however, this is difficult to achieve since the mirror would have to travel on a near-perfect straight line over large distances. Nevertheless, modern high-resolution FTIR instruments routinely achieve sub-1 cm$^{-1}$ resolution, and even instruments with resolutions as low as $10^{-3}$ cm$^{-1}$ are now available commercially. The key characteristics of FT interferometers also are included in Table 6.1.

## 6.1.2 Image information and its retrieval from photon events

Imaging an object is an age-old technique, which in modern times has become the domain of 2D image sensors, like charge coupled device (CCD) cameras. In a certain way, they have even become sort of "spectroscopic" devices recording color pictures, based on the principle of color decomposition/reconstruction of its blue/green/red pixilated sensors. However, this color information is a far cry from the fine spectral details contained in a full dispersed spectrum. Looking at the problem from the side of said dispersed spectrum, it is clear from the above that the wavelength separation of the incoming light in the spectrometer exit/image plane inevitably brings with it the loss of spatial distribution, at least in the direction of the dispersion. Thus, different methods have to be applied to realize full spectral imaging.

Probably the simplest and most widely known approach is that of *multispectral imaging*. For this, several images are recorded using discrete, narrow spectral band-path filters (for the concept of the approach, see Figure 6.3). The approach is akin to that used in a color CCD, with the difference that normally more than three band-path filters are used, covering the spectrum in a suitable manner to provide sufficient detail to answer a particular spectroscopic problem. For example, this principle is encountered in laser fluorescence microscopy.

When finer spectral details are required, then a methodology known as *hyperspectral imaging* (HIS) is used. The underlying principle of this technique is to record full spectrum for each pixel in the image of an object; this allows one to generate images at any of the spectral "pixels," thus making it possible to find and

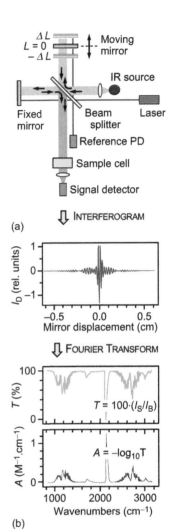

**Figure 6.2** Simplified schematic of a Michelson interferometer (MI) used in FTIR spectroscopy (a), with a reference laser to provide spectral frequency markers. A typical interferogram from a sample within the MI is shown in the center part of the figure; $I_D$ = signal intensity from the detector. FT of the interferogram generates the transmissivity spectrum (b) from the ratio of the signal and "blank" background intensities, $I_S/I_B$; the (molar) absorbance or absorption spectrum is derived from this.

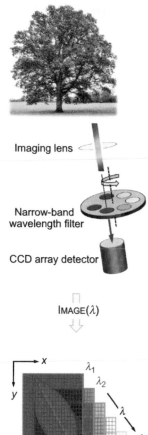

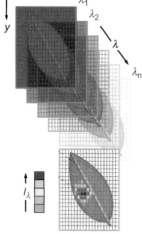

**Figure 6.3** Concept of (global) multispectral imaging, based on wavelength-filtered recording of an object. Wavelength-sensitive features can be identified in the respective image (false-color intensity scale).

identify certain features associated with a "spectral fingerprint"—such as a specific transition in a molecule and its spatial distribution across the sample surface. The method had its origin in earth remote sensing but is now routinely used in such diverse fields as in astronomy, biomedical imaging, geosciences, physics, and security surveillance.

Basically, there are two approaches to collect hyperspectral data: either a global (*x*,*y*) image is recorded, sending the light from the object area through narrow-band filters and acquiring the image at each particular wavelength setting (this approach is commonly known as *spectral scanning*); or one carries out a pixilated (*x*–*y*) scan of the object, recording the full wavelength-dispersed spectrum at each pixel location (this approach is known as *spatial scanning*; note that this may be done in line-scan [1D] or point-scan [pixel-by-pixel] mode). In both cases, a three-dimensional *(x–y–λ) hyperspectral data cube* is generated, which then can be used for processing and analysis. Note that besides spectral and spatial scanning, two further acquisition modes have been developed, namely *nonscanning* (or snapshot) HIS and *spatiospectral scanning*. In the former technique, a 2D CCD sensor output contains all spatial (*x*,*y*) and spectral (*λ*) data, and the complete HIS data cube is obtained simultaneously, with neither spatial nor spectral scanning required. In the latter, recently developed approach (see Bodkin 2009), one records wavelength-coded—also known as "rainbow-colored"—($λ = λ(y)$), spatial (*x*,*y*) image maps of the object. For further details on HIS and its application, see Chang (2003); also see Chapter 17.1.

### 6.1.3 Spectral/image information and its retrieval from charged-particle events

If the laser photon energy is large enough, atoms or molecules may become ionized, generating a charged-particle pair of a negative electron and a positive ion. Mostly now the spectroscopy is different since the photon transition is no longer between certain bound states, which give rise to characteristic spectral lines, but the final state is in the (ionization) continuum. By and large, in charged-particle analysis, one exploits the difference in mass-to-charge ratio (*m/e*) of electrons, or ionized atoms/molecules. This methodology of "mass spectrometry" allows one to distinguish individual species of atoms or molecules, and even to extract chemical and structural information. Charged-particle analyzers exploit electric and/or magnetic fields, which exert a force on charged particles (electrons and positive and negative ions). The basic, simple relationships among force, mass, and the applied fields are given by Newton's second law and Lorentz's force law:

$$\boldsymbol{F} = \underbrace{m \cdot \boldsymbol{a}}_{\text{Newton's 2}^{\text{nd}} \text{ law}} = \underbrace{q \cdot (\boldsymbol{E} + \boldsymbol{v} \times \boldsymbol{B})}_{\text{Lorentz's force law}}, \qquad (6.6)$$

where $\boldsymbol{F}$ and $m$ are the force applied to and the mass of the (charged) particle, respectively; $\boldsymbol{a}$ is the particle acceleration; $q = z \cdot e$ is the electron (with $z = -1$) or ion (with $z = +1, 2,...$) charge; $\boldsymbol{E}$ and $\boldsymbol{B}$ are the applied electric and magnetic fields, respectively; and $\boldsymbol{v}$ is the particle velocity.

In general, charged-particle analyzers comprise an energy- and/or mass-selective filter, and a charged-particle detector. In addition, certain instrument designs utilize extraction and acceleration sections in order to transfer the charged particles from the region where they are generated to the detector. Mass analysis instrumentation typically encountered in laser spectroscopy includes (1) electron energy analyzers, (2) magnetic-sector mass spectrometers, (3) quadrupole mass spectrometers, and (4) time-of-flight (ToF) mass spectrometers. For many laser spectroscopic problems, the latter are the instrumentation of choice since they allow one to maintain much of the spatial information from the location where the charged particles were generated, thus making also imaging possible, in principle. Thus, here we only outline the basic concepts of ToF mass analyzers; for further details on these, as well as other mass-analyzing instrumentation, see Throck Watson and Sparkman (2007), de Hoffmann and Stroobant (2007), or Gross (2011).

Basically, a ToF mass spectrometer measures the mass-dependent time it takes charged particles to travel from the generation region to the detector. For this, the point in time at which the charged particle was generated needs to be well defined: laser sources of suitably short pulse duration (much shorter than the expected particle transit time through the complete analyzer) are required. The simplified concept of ToF mass analysis is shown in Figure 6.4.

In principle, one strives to achieve that the charged particles acquire nearly unikinetic energy in the acceleration region, which is substantially larger than the initial kinetic energy of the charged particles generated by the photon interaction, i.e.,

$$E_{\mathrm{kin},h\nu} \ll E_{\mathrm{kin,acc}} \cong q \cdot V_{\mathrm{acc}} \cong 1/2 \cdot m_i v_{\mathrm{acc}}^2. \tag{6.7}$$

Thus, when entering the field-free ToF region of length $L_{\mathrm{ToF}}$, particles of different mass $m_i$ travel at different velocities and thus arrive at different times $t_{\mathrm{ToF}} = L_{\mathrm{ToF}}/v_{\mathrm{acc}}(m_i)$ at the detector. Relating this to the acceleration voltage $V_{\mathrm{acc}}$, one finds quadratic dependence for the arrival time of particles of different mass at the detector:

$$m_i = \left(2q \cdot V_{\mathrm{acc}}/L_{\mathrm{ToF}}^2\right) \cdot t_{\mathrm{ToF}}^2. \tag{6.8}$$

Note that, in all this, it is the laser that probes the wavelength-dependent quantum transition properties of the ionized atom or molecule; the mass spectrometer only provides the global mass selectivity. For example, this is the underlying philosophy in resonance-enhanced multiphoton ionization (REMPI), in which an initial transition between bound levels is followed by a second photon-excitation step into the ionization continuum, and the so-generated charged particle is detected (for further details, see Chapter 16.1).

In chemical reaction dynamics (associated with, e.g., photofragmentation or inelastic and reactive collision processes), it is often desirable not only to identify any intermediate or final products as such but also to gain insight into their stereodynamic behavior. For this, one will have to measure flux-velocity contour maps for quantum state-selected species; in favorable cases, such contour maps may reveal information about the complete chemical process.

Probably one of the most elegant developments in this respect is the technique of *photo-ion imaging*, which is based on a combination of ToF and laser spectroscopic methods, and which was pioneered by Chandler and Houston (1987) to study photofragmentation in a molecular beam. More recently, crossed-beam experiments have been realized, which allow one to study full chemical reaction processes with well-defined initial conditions (see Wester 2014). The simplified

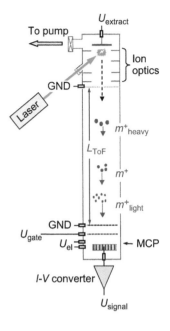

**Figure 6.4** Concept of a (charged-particle) ToF mass analyzer. $L_{\mathrm{ToF}}$ = field-free ToF tube; $m_x^+$ = ion masses; MCP = microchannel plate; GND = ground potential; $U_{\mathrm{extract}}$ = ion extraction potential; $U_{\mathrm{gate}}$ = inhibit (gate) potential; $U_{\mathrm{el}}$ = electron-amplification potential across MCP; $U_{\mathrm{signal}}$ = ion signal output from the I–V (current-to-voltage) converter.

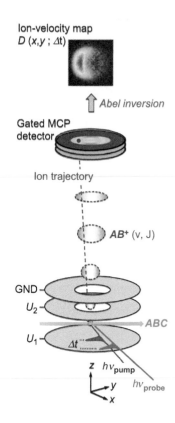

**Figure 6.5** Concept of photo-ion imaging, exemplified for a photo-fragmentation–photo-ionization reaction ABC + $h\nu_{pump}$ → AB + C + $h\nu_{probe}$ → $AB^+$ + C, with time delay $\Delta t$ between fragmentation and ionization. $U_1$, $U_2$ = ion extraction potentials; MCP = microchannel plate.

concept ion imaging, here based on interrogating a molecular beam, is schematically shown in Figure 6.5.

Molecules in the beam are interrogated by laser radiation to generate ions via quantum-state selective steps. In typical molecular reaction or fragmentation experiments, products migrate with velocities of the order $1-10 \times 10^3$ m·s$^{-1}$. This means that charged particles will have evolved measurable distances of a few millimeters in a few microseconds. Note that the ion trajectories can be controlled by applying electric fields, which are poled and tailored (ion optics) and direct (accelerate) the ions toward the field-free ToF tube. At the end of the flight tube, the ions impinge upon a position-sensitive charge detector, most commonly a pair of Chevron-type microchannel plates (MCPs) coupled to a fast-reacting phosphor screen, which allows one to record two-dimensional ToF ($x$-$y$, $t_{ToF}$) profiles for the ions.

At first sight, one might think that the information on the coordinate of the velocity component parallel to the propagation direction of the ion extraction (in Figure 6.5, the $z$-direction) is lost; only the two velocity components perpendicular to the ion propagation ($x$- and $y$-directions) are directly measured in the detector plane, although the full initial three-dimensional distributions of the arriving ions are projected onto the two-dimensional detector. However, as long as the detector plane is parallel or perpendicular to the laser polarization vector, the initial three-dimensional product distributions can be reconstructed from the time-resolved 2D projections. This is afforded by a direct mathematical transformation, the so-called *inverse Abel transformation* (for details, see Heck and Chandler 1995).

# 6.2 PHOTON DETECTION: SINGLE ELEMENT DEVICES

Single-element photon detectors have been, and still are, workhorses in many spectroscopic applications, which do not require that any spatial information from the source region of the spectrum needs to be maintained. The type of detector device very much depends on a range of measurement parameters of interest, such as wavelength, intensity, and temporal requirements. By and large, photodiodes (PDs) and photomultipliers are the most common devices, and an outline of their principles and operation is given next. These photon detector devices are commercially available in a wide range of "packaging," often being tailored specifically to meet any of the aforementioned parameters for a particular application.

Of course, also other photon detection devices are in use, like bolometers or pyroelectric sensors, to name but two; but for spectroscopic applications, they are less frequently used today. On the other hand, new detection techniques and devices are still being developed, such as superconducting nanowire single-photon detectors (SNSPDs), applied to single-photon detection in the near-IR (for a summary, see Natarajan et al. 2012).

## 6.2.1 PDs and their principal modes of operation

PDs are p–n junction semiconductor devices that are based on materials sufficiently "transparent" for the photons to penetrate into the structure but be absorbed and not exiting again without interaction. For the process to work, the bandgap energy, $E_g$, has to be smaller than the photon energy, $E_{h\nu}$, i.e., $E_{h\nu} \geq E_g$. Then incident photons can be absorbed to transfer an electron from the valence band to the conduction band in the process, but only those absorbed in the depletion layer of the p–n junction contribute to the net current in the device. Note that functioning devices appeared in the late 1940s, with the first systematic study of a PD behavior published by Benzer (1947); but it took a further 10 years before PD detectors were mature enough to be used in low-light-level applications. The measured photocurrent linearly follows the incident light intensity, normally over many orders of magnitude, until saturation is reached (typically for an incident light intensity of about 1–2 mW). Note that each absorbed photon transfers only a single electron into the conduction band, but not every incident photon will necessarily create a charge carrier.

PDs are commonly operated in two distinct modes, namely in the *photovoltaic* or *zero bias*, and *photoconductive* or *reverse-bias* modes. Which mode of operation one selects by and large depends on the speed requirements (the former is suitable for slow- and the latter for fast-varying signals) and the amount of tolerable dark current, meaning that for ultralow-light intensities, the photovoltaic mode is the more suitable. The principles of the two modes are shown in Figure 6.6, together with some simplified measurement circuitry. Note that, in general, the PD signal can be measured as a voltage or a current; for the latter, normally a so-called trans-impedance amplifier configuration (also known as a current-to-voltage converter) is used, as shown in the figure.

*Photovoltaic (P-V) mode.* No external voltage is applied (zero-bias) in this mode; therefore, the photocurrent out of the device is restricted, and a voltage builds up across the diode. The measurement circuit is one of high impedance, meaning that the load resistance $R_L$ across which the voltage signal is measured is high (normally larger than several megaohms), e.g., the input resistance of a low-noise JFET OPAMP (an operational amplifier based on junction gate field-effect transistors), with typical input impedance of $10^{13}$ $\Omega$||10 pf. Because of the high impedance, the influence of temperature on the PD responsivity is relatively small, and any device dark current is kept at a minimum. Note that a PD in P-V mode may be treated as a voltage source that generates a current through a load resistor.

*Photoconductive (P-C) mode.* For the detection of fast signals, this is the mode of choice. Here one applies a reverse-bias voltage, $U_{bias}$, to the PD (with the cathode driven positive with respect to the anode). This bias increases the width of the PD's depletion layer, and hence decreases the junction's capacitance; as a consequence, the time (or frequency) response improves. While the bias leaves the photocurrent nearly unaffected, on the downside, the dark current increases, which means that the light detection threshold worsens. In order to maintain the high response speed of the PD in P-C mode, measurement circuit normally utilizes high-speed OPAMP (an operational amplifier based on junction gate field-effect transistors), with typical input impedance of the order $10^6$ $\Omega$||1 pf. Note that a PD in P-C mode may be treated as a current source that generates a voltage drop across a load resistor.

## 6.2.2 Types of PDs

As stated earlier, PDs in their simplest form constitute p–n junction devices, and in the majority of measurement configurations (at least small), reverse-bias

PHOTO-VOLTAIC MODE

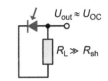

Amplifier circuit

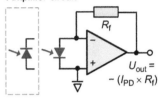

PHOTO-CONDUCTIVE MODE

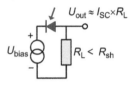

Amplifier circuit

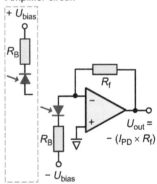

**Figure 6.6** Typical operating modes for PDs, with related (inverting) trans-impedance amplifier; (top) photovoltaic mode; (bottom) photoconductive mode, with reverse-bias voltage $U_{bias}$ applied to the PD. $R_f$ = feedback resistance; $R_L$ = load resistance; $R_{sh}$ = shunt resistance; $R_B$ = bias-limiting resistance; $I_{SC}$ = short-circuit current; $U_{OC}$ = open-circuit voltage.

*p-n type photodiode*

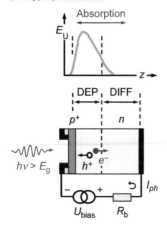

*p-i-n type photodiode*

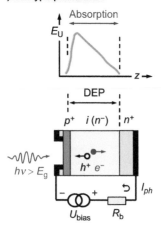

**Figure 6.7** Schematics of basic reverse-biased PD types; the depletion and photon absorption regions are indicated, together with the internal longitudinal $E$-field distributions. (Top) Standard $p$–$n$ junction PD; (bottom) $p$-$i$-$n$ PD.

voltages are applied. Across the p–n junction, a depletion region is generated, which separates the regions with (static) positively charged donor atoms in the $n$-type material from (static) negatively charged acceptor atoms in the $p$-type material. Mobile carriers move to their majority region under the influence of intrinsic or extrinsic (bias) electric fields. As a consequence, the width of the depletion region depends (1) on the doping concentrations and (2) on the magnitude of the applied reverse bias. In general, one finds that the lower the doping levels, the wider the depletion region.

*"Ordinary" p–n junction PDs.* In principle, photons can be absorbed in all regions of the device—the $n$-type, the depletion, and the $p$-type regions—as long as the photon energy is larger than the bandgap energy. The actual location and width of the absorption region primarily depend on the PD materials, but also on the actual energy of the photon energy and its penetration depth into the material (the weaker the absorption probability, the larger the penetration depth). Thus, one may find that charge carriers, due to photon absorption, can be generated and transported by the external field both in the depletion region (the so-called drift region) and the static doped regions (the so-called diffusion region). Normally, the drift process is much faster than the diffusion processes. Therefore, it is important that the photons are absorbed in the depletion layer; otherwise, the response time of the PD deteriorates. In order to have efficient absorption of all available photons while maintaining response speed, the depletion region is made as wide as possible, which is afforded by low doping in the $n$-type material of the PD. By and large, the depletion region has a width of 1–3 μm but is often optimized for the wavelength range over which the device is required to operate. The structure for a standard $p$–$n$ junction PD, its depletion region, and the internal field distribution are shown schematically in the upper part of Figure 6.7. Note that, in general, PD devices are built for the light to enter the $p$-layer, which mostly is antireflection (AR) coated as well, to compensate for the otherwise large losses caused by the high refractive index of semiconductor materials, $n_{SC}$ (for typical values, see Table 6.2).

*PIN PDs.* Whatever the wavelength range, the longer the wavelength, the deeper the light penetrates into the material. This means that for the aforementioned condition, namely that the PD functions best if the photons are absorbed in the depletion layer, that for long wavelengths, the depletion region should become wider. This can be achieved by lighter doping of the $n$-type material and making the device longer. But as a consequence of this, the PD's resistance and capacitance would increase, making it noisier and exhibit slower response. Therefore, an additional "intrinsic" (undoped) layer—normally annotated as $i$-layer—is added between the $p$- and $n$-layers, generating a $p$-$i$-$n$ structure (sometimes also written as $p^+$-$\pi$-$n^+$ if the intrinsic $i$-layer is slightly $p$-doped, as often encountered in commercial devices), commonly known as a PIN PD. Besides aiding the length dimension and therefore the width of the depletion layer, it also adds the benefit of a relatively low resistance and low capacitance, which has a positive effect on the response speed of the device; therefore, nearly all high-speed PDs are PIN devices. Their principle construction is shown in the lower part of Figure 6.7. Note that specific devices are available, which exhibit response times as short as the order $\tau_R \approx 10$ ps.

*Avalanche PDs (APDs).* As stated earlier, in general, each photon absorbed in the PD generates only one individual electron. However, for very low photon fluxes—in the limit down to single-photon events—one would like to have an amplifying

**Table 6.2** Comparison of Important Parameters of Commonly Used PD Detector Devices

| Device Property | Units | Si | In$_x$ Ga$_{1-x}$ As | PbS | Hg$_{1-x}$ Cd$_x$ Te |
|---|---|---|---|---|---|
| Refractive index $n_{SC}$ | | >3.45[1] | >3.34 | >4.2 | >2.9[2] |
| Sensitive area | mm$^2$ | 0.1...100 | 0.05...20 | 1...100 | 0.02...4 |
| Spectral range[3] | nm | 200...1050 | 800...1700[4] | 900...3200 | 2000...16000[5] |
| Responsivity (P-V)[6] | A·W$^{-1}$ | 0.36...0.62 | 0.90...1.20 | – | – |
| Responsivity (P-C)[6] | V·W$^{-1}$ | – | – | 0.3...7.5×10$^5$ | 0.1...2.0×10$^5$ |
| Dark current | nA | 0.02...2 | 0.03...3000 | – | – |
| NEP | W·Hz$^{-1/2}$ | <2×10$^{-13}$ | <1×10$^{-12}$ | <1.5×10$^{-11}$ | <3×10$^{-9}$ |
| Rise time | ns | 0.1...150 | 0.3...25 | 0.5–1.2×10$^3$ | ~100 |
| TE-cooling (integrated) | °C | Optional (–20) | –20...–85 | –20...–85 | Liquid N$_2$ |

[1] Refractive index becomes smaller for $\lambda < 280$ nm.

[2] For $x \cong 0.5$.

[3] 10% points relative to peak responsivity.

[4] Extended-range types up to 2600 nm are available (In$_x$Ga$_{1-x}$As, with $x = 0.53$ to $x = 0.85$).

[5] Common device options: sapphire-window – $\lambda_{max} \cong 5500$ nm; ZnS-window – $\lambda_{max} \cong 22,000$ nm. Note that TE-cooled sapphire-window devices, operating in P-V mode, are also common.

[6] P-V = photovoltaic mode; P-C = photoconductive mode.

mechanism in which the photon generates sufficient charge carriers to become an easily measurable current. By and large, photomultiplier tubes (PMTs) are common for this purpose (see further below); but there are also PD devices around that are suitable for the task, namely APDs.

From the point of view of layer construction, APDs are closely related to PIN PDs. But in addition to the three-layer structure $p$–$i$–$n$, an additional, narrow $p$-doped layer is added the resulting layer structure normally is of the type $p^+$-$\pi$-$p$-$n^+$ (see Figure 6.8). This particular structure entails a very high electric field gradient across the narrow $p$–$n^+$ junction (of the order of $10^5$ V·cm$^{-1}$), as shown in the figure. For a sufficiently high bias voltage, this region then acts like an internal amplification region: as a consequence of the high electric field, any charge carrier entering this region is accelerated and generates secondary charge carriers, which then again can be accelerated—an avalanche of carriers ensues. Note that the bias voltage is, in general, much higher than for ordinary PIN diodes, of the order 20–200 V. The multiplication factor $M$ reached by APDs is typically in the rage 10–1000; $M$ increases as a function to the applied reverse voltage $U_{bias}$, below the diode breakdown voltage $U_{bd}$, according to

$$M \cong \left[1 - \left(\frac{U_{bias} - U_S}{U_{bd}}\right)^n\right]^{-1}, \qquad (6.9)$$

where $U_s$ is the voltage drop across the APD, and the exponent $n$ (<1) depends on the structure and the material of the APD. Care has to be taken when operating an APD close to its breakdown voltage $U_{bd}$ since then large currents flow, which may cause permanent damage to the APD.

APDs can be operated in two distinct modes. In the so-called *linear* or *proportional mode*, the bias voltage is kept suitably below the breakdown voltage; then the recorded signal is directly proportional to the number of incident photons. In the so-called *Geiger mode*, the bias voltage is set to values slightly above the breakdown voltage; then already a (single) incident photon yields an "infinitely"

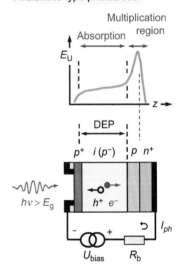

*Avalanche* type photodiode

**Figure 6.8** Schematics of reverse-biased *avalanche* photodiode (APD); the depletion and photon absorption regions are indicated, together with the internal longitudinal $E$-field distributions.

large signal with a multiplication factor of $M \sim 10^6$. However, in order to prevent the aforementioned damage to the device, a "quenching" circuit is required, which after such a photon-triggered, huge avalanche event reduces the current to a safe value again, basically by lowering the bias voltage well below the break-down threshold. As a consequence, the APD becomes insensitive to further photons until the high bias voltage is reapplied again. APDs in Geiger mode are commonly used in photon-counting applications.

### 6.2.3 Important operating parameters of PDs

A number of parameters are of importance when it comes to the selection of a PD for a particular application. These are briefly summarized below, and typical values for a range of commonly used PD types are collated in Table 6.2; these include silicon—Si; indium–gallium–arsenide—$In_xGa_{1-x}As$; lead-sulfide—PbS; and mercury-cadmium-telluride (MCT)—$Hg_{1-x}Cd_xTe$, with $x$ denoting the hetero-epitaxy parameter for the crystal.

*Spectral range.* The usefulness of any photon-detecting device naturally depends on the wavelength range over which it is usable. For PDs, the spectral range depends on the band structure of its particular semiconductor material. This means that a minimum photon energy (or longest wavelength) exists below which the device does not work; this is roughly given by the semiconductor bandgap. At the high-energy photon (short wavelength) end of the spectrum, the limit is normally given by the depth-dependent absorption coefficient and the transmissivity of any window material, which is commonly used in device fabrication.

The *spectral responsivity* $R(\lambda)$ is a quantity defining how much photocurrent $I_{PD}$ is generated by a certain incident light intensity $P_{in}$, at a given wavelength, i.e.,

$$R(\lambda) = I_{PD}/P_{in}; \tag{6.10}$$

it is normally given in units of $A{\cdot}W^{-1}$. But recall that for devices operating in the photoconductive mode, the PD is serving as a current source for a voltage signal across a load resistor; in this case, the spectral sensitivity is normally given in $V{\cdot}W^{-1}$.

*Dark current.* Even without any external light present, a small leakage—or dark—current may flow when a PD is operated in photoconductive mode. Primarily, it is the current flowing when a bias voltage is applied to the PD. It varies directly with temperature; as a rule of thumb, the dark current doubles for every 10 K increase in temperature or, vice versa, halves for every 10 K decrease in temperature. Therefore, if low dark currents are desired, it helps to either (1) reduce the bias voltage or (2) cool the PD to suitably low temperatures. In addition to temper-ature and bias voltage, the dark current also depends on the actual PD semi-conductor material, as well as the size of the device (the smaller the photosensitive area, the lower the dark current).

A signal can only be detected if it is larger than all noise sources. The quantity that characterizes this property is the *noise-equivalent power (NEP)*. It is defined as the minimum incident optical intensity, $P_{in}$, which is necessary to generate a photocurrent that is equal to or larger than the overall RMS noise current (including the aforementioned dark current), within a bandwidth of 1 Hz.

In general, the NEP increases proportionally to the photosensitive area of the device, $A_{PD}$, and is given by

$$\text{NEP} = \frac{P_{in} \cdot A_{PD}}{\text{SNR} \cdot \sqrt{\Delta f}}, \text{ in units of } W \cdot Hz^{-1/2}, \tag{6.11}$$

where SNR stands for the signal-to-noise ratio, and $\Delta f$ is the noise bandwidth.

The *response time* and with it the *signal bandwidth* of PDs depend on a number of different quantities. Overall, the internal and external capacitances and resistances give rise to a combined RC response time, $\tau_{RC}$. The individual contributions stem from the PD's internal junction capacitance ($C_j$), and its shunt and series resistances ($R_{sh}$ and $R_S$, respectively; the latter can normally be ignored), and the external load resistance ($R_L$), as shown in the equivalence circuit for PDs (Figure 6.9):

$$\tau_{RC} = f\left(C_j\left(A_{PD}, U_{bias}^{-1}\right); R_{sh}(\text{material}, T); R_L\right), \tag{6.12}$$

where the device temperature $T$ is measured in units of K; for further details, see the *Opto-semiconductor Handbook* published by Hamamatsu (2014). Note that in many practical electronic circuits, (a small) $R_L$ dominates over (the relatively large) $R_{sh}$.

### 6.2.4 Photomultiplier tubes

In contrast to PDs, PMTs are light-detecting devices that rely on the photoelectric effect rather than interband transitions in a semiconductor material. The photoelectric effect has been known since the late nineteenth century, and usable, quantitative photon detectors exploiting it have been around since the very early twentieth century.

Typically, a PMT consists of four basic elements, namely (1) a photo-emissive cathode; (2) focusing electrodes; (3) a series of electron-multiplying stages, so-called dynodes; and (4) an electron-collecting anode. All are imbedded in a vacuum tube; the overall structure and the electron-multiplication processes are conceptually shown in Figure 6.10.

Basically, photons impinging onto a suitable (cathode) material emit photo-electrons, albeit only beyond a material-dependent energy threshold; electrons leave the photocathode with a low (kinetic) energy of

$$E_{e^-} = E_{h\nu} - W_{C_{ph}}, \tag{6.13}$$

where $W_{C_{ph}}$ is the work function of the photocathode material. The electrons ejected from the cathode are accelerated toward a series of metal plates (so-called dynodes); each of them is kept at a positive potential of about +100 V with respect to the preceding dynode. This means that the so-accelerated photoelectrons (with energy of ~100 eV) release secondary electrons on impact, leading to the number of outgoing electrons being larger than the incoming by a certain factor, i.e., $N_{out}(e^-) = M_D \times N_{in}(e^-)$, for each dynode stage. In most practical devices, this multiplication factor is $M_D = 5\text{-}10$; thus, for typical commercial PMTs with $n = 8\text{-}12$ dynode stages, one can easily achieve a multiplication gain of the order $10^6$.

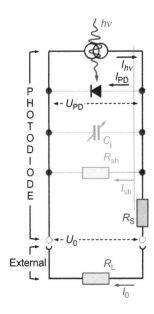

**Figure 6.9**  PD equivalent circuit. $I_{h\nu}$ = current generated by incident photons; $I_{PD}$ = PD current; $U_{PD}$ = voltage across PD; $C_j$ = junction capacitance; $R_{sh}$ = shunt resistance; $I_{sh}$ = shunt resistance current; $R_S$ = series resistance; $R_L$ = external load resistance; $I_0$ = output current through $R_L$; $U_0$ = output voltage across $R_L$.

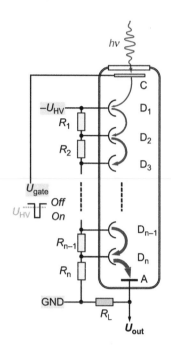

**Figure 6.10**  Principle of a photomultiplier device. $C$ = photosensitive cathode; $D_n$ = dynodes; $A$ = anode; $R_n$ = HV-divider resistors; $R_L$ = load resistor; $U_{HV}$ = high-voltage for divider chain (to accelerate secondary electrons); $U_{out}$ = output voltage signal; $U_{gate}$ = gating voltage ("off" slightly positive, "on" negative with respect to $U_{HV}$).

The overall gain, $G_{PMT}$, in a PMT with $n$ dynodes can be described by

$$G_{PMT} = (SER)^n = \left[ A \cdot \left( \frac{U}{n+1} \right)^\alpha \right]^n = \underbrace{A^n \cdot (n+1)^{-\alpha n} \cdot U^{\alpha n}}_{\text{constant}}. \quad (6.14)$$

Here, SER is the secondary electron emission ratio; $A$ and $\alpha = 0.7$–$0.8$ are dynode-specific coefficients that are determined by its material and geometrical structure; and $U$ is the total voltage applied across the PMT, with $U/(n+1)$ being the interdynode acceleration voltage (provided by the resistive voltage divider chain; see Figure 6.10).

Because of their high multiplication for electrons from incident photons (recall that in a PD only a single electron can be generated per incident photon), photomultipliers have been and are used to detect very low light levels; with careful optimization and good stray-light suppression, single photons can be detected and counted. The first practical PMT based on the aforementioned principles was demonstrated in the late 1930s by Zworykin and Rajchman (1939); their conceptual, basic structure is still the most common one in currently used PMTs.

## 6.2.5  Important operating parameters of photomultipliers

As for PDs, also for photomultipliers a number of parameters are important for their optimal operation in particular applications. These are very much of the same nature since they are meant to detect and measure incident photons, but because of the different underlying physical effects and device structures, they differ in their contribution to the overall performance of a PMT.

*Spectral range and spectral responsivity.* As noted above, the photocathode of the PMT is the primary element that converts the photon energy into free electrons; the conversion efficiency or photocathode sensitivity varies strongly with the energy (wavelength) of the incident photon. Primarily it is determined by the work function of the cathode material, which commonly comprises alkalis. Probably the most versatile and common cathode type are the so-called multialkali cathodes; a very popular combination is Na–K–Cs–Sb, which covers the wavelength range from the deep UV up to about 930 nm. As for PDs, the spectral range is normally defined by the 1% point relative to the maximum sensitivity. In general, the maximum radiant responsivity of the majority of cathode materials is in the range 0.04–0.08 A·W$^{-1}$, with a quantum efficiency of photon-to-electron conversion of QE~0.25.

*Dark current and noise.* Because of the different device structure, the nature of the dark current contributions is different as well, but can be subdivided, as for PDs, into thermal and device-related contributions. Dark currents originate from *thermionic emission* of electrons from the photocathode; this is not surprising since these thermal electrons are successively multiplied by the dynode chain. In addition to the multiplication of the thermal electrons, one has to contemplate the stability of the dynode acceleration voltages: any slight fluctuation in its value(s) will be magnified $n$-fold, according to the number of dynodes, thus affecting both the noise electrons (as well as the signal electrons). Consequently, PMT HV-supplies have to be of low-voltage ripple and drift. Besides this statistical, "continuous" noise, one has to contemplate effects that feed electrons into the amplification chain. Foremost, these are associated with the following:

(1) *ionization of residual gas particles*—these may have undergone ionization as a consequence of a collision with electrons, with the impact of the much more energetic ion on a dynode generating huge noise (after-) pulses; and (2) *field emission electrons*—these can be emitted from the dynodes when the PMT is operated at near its maximum rated voltage, resulting in large, random dark current pulses.

*Time response and signal bandwidth.* By and large, the temporal behavior is dominated by two factors, namely the transit time, $\tau_{TT}$, of electrons travelling through the PMT and the spread of this parameter for arrival times at the anode, $\Delta\tau_{TTS}$. In general, the liberation of an electron from a surface, either by photon or electron impact, is deemed to be instantaneous on the timescale of just a few picoseconds. The overall duration of travel through the PMT strongly depends on its geometry and the number of dynodes, which determine the path length. For most PMTs, one finds $\tau_{TT}$ = 15–50 ns. Since the travel path length depends on the spatial location where electrons are hitting a dynode and secondary electrons are re-emitted, it is clear that the transit time will vary, resulting in the spread of an initial "$\delta$"-pulse arriving at the anode. For many common PMTs, the FWHM of this arrival spread is of the order $\Delta\tau_{TTS}$ = 1–5 ns. These time parameters are usually measured for characterization of a PMT by utilizing a light pulse of duration $\Delta\tau_{hv} \leq 50$ ps. The typical time evolution for electrons generated by such a light pulse and arriving at the PMT anode is shown in Figure 6.11. The extent to which the original temporal characteristics are maintained in a signal measurement strongly depends on the subsequent electronic circuit use for measuring the photocurrent arriving at the PMT anode.

Further information on the design, specifications, and operating parameters of PMTs, as well as electronic circuits for signal measurements, may be found in Hakamata (2007).

Finally, when comparing photomultipliers to PDs, the former certainly has the edge when it comes to light sensitivity because of the built-in amplification factor for photon-to-detectable-current conversion. Also, mostly PMTs exhibit very fast response times, of the order of a few nanoseconds, and thus make them superior to many of the standard PDs, in particular for low-light-level applications. However, there are caveats, too. Firstly, the devices require the provision of high voltage for the dynode-resistor chain. Secondly, the responsivity varies much more dramatically with wavelength in comparison to PDs because of the wavelength-dependent nature of the photocathode materials. Thirdly, the longest detectable wavelength for PMTs is, in general, less than 1000 nm, by and large restricting them to the visible (and UV) spectral ranges, whereas PDs also cover the IR to beyond 20 μm.

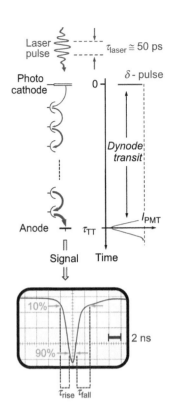

**Figure 6.11** Broadening of a "$\delta$-shaped" electron pulse, from a 50 ps laser pulse, on transit through the PMT dynode chain.

## 6.3 PHOTON DETECTION: MULTIELEMENT ARRAY DEVICES

The electronic provision of spectral and image data has been a major task in the development of spectroscopic techniques. As was stated in the previous section, single-element detectors, which convert photon energies into electron currents, have been around and used since more than a century. While they certainly assisted the development of truly quantitative spectroscopy, the tedious process of scanning spectra wavelength by wavelength, or recording images pixel by pixel, frustrated many researchers. Thus, the drive for array detectors that allowed for simultaneous, spatially resolved spectral data gathering is not surprising. This became a reality with the rapid evolution of semiconductor electronics and miniaturization of devices in the 1960s. It is interesting to note that much of the efforts were driven by observational astronomy where the imaging of faint astronomical objects was hampered by the low dynamic range achievable using photographic plates, and for which the orders-of-magnitude better direct photodetectors promised huge advantages.

By and large, 1D array (line) detectors and 2D array detectors may be divided into three general classes: PD array (PDA) sensors; complementary metal-oxide semiconductor (CMOS)/CCD array sensors; and the so-called intensified devices, which incorporate PD or CMOS/CCD sensors.

## 6.3.1 PDA sensors

As the name suggests, these devices are based on standard $p$–$n$ junction PD elements, which are fabricated on a single substrate of silicon semiconductor material. In the late 1970s/early 1980s, scientists succeeded in their demand of manufacturers of array detectors that devices be developed specifically aimed at scientific, spectroscopic applications, in particular to allow for simultaneous recording of spectra with high resolution (see, e.g., Talmi and Simpson 1980). These devices had up to 1000 PD elements on a pixel raster of 25–50 μm and element heights of up to 2500 μm. Individual elements exhibit the same properties as their single-element equivalents (see Section 6.2), including spectral range, spectral responsivity, noise characteristic, and dynamic range of normally $>10^6$.

While the early 1D line arrays were all based on Si-PDs, with spectral range of $\lambda \cong$ 200–1000 nm, now also NIR ($\lambda \cong$ 1–2.5 μm) and IR ($\lambda \cong$ 2.5–20 μm) capable arrays are available, based on InGaAs and HgCdTe PDs, respectively. In particular, the latter material has been exploited since the 1990s to fabricate 2D focal-plane PD arrays for IR imaging, with array sizes of up 1024×1024 pixels and beyond (see, e.g., Bajaj 1999).

## 6.3.2 CCD and CMOS array sensors

As stated above the development of, in particular, 2D array detectors was driven by the imaging community on the scientific front predominantly by observational astronomers. The rapid development of semiconductor device integration led to the pioneering invention of CCD and CMOS image sensors, both in the late 1960s and early 1970s. The breakthrough came with the development of CCD technology at Bell Labs by, in particular, W.S. Boyle and G.E. Smith (see Boyle and Smith 1970; Amelio et al. 1970)—who in 2009 were awarded the Nobel Prize for this invention. In the early years, CCD technology dominated the field of sensor technology, and only with the huge improvement of lithography capabilities in the 1990s have CMOS devices become competitive. Each of the two technologies has its unique strengths and weaknesses, giving advantages in different applications; these will be briefly outlined in the following paragraphs.

CCD and CMOS imagers both depend on the photoelectric effect to create electrical signal from light. Both types of imagers convert light into electric charge and process it into electronic signals.

Both are most commonly based on $p$-doped MOS capacitors as the light-sensitive element, as shown schematically in Figure 6.12. When a positive voltage $U^+$ is applied to the metal contact (or "gate"), a depletion region is generated below the gate electrode: holes are pushed far into the substrate, with nearly no mobile electrons at or near the surface; the MOS capacitor thus is operational in so-called deep depletion, resembling a potential well for electrons. Then, when a photon is absorbed in this depletion layer, generating an electron–hole pair, the applied electric field separates these charge carriers, and electrons accumulate in the potential well, mostly near the surface. This accumulation of electrons continues, either until the exposure to light is terminated or until thermal equilibrium is reached (then the well is said to be full). The maximum capacity for electron accumulation in the well is typically of the order $10^3$–$10^5$ electrons; this accumulation capacity is often addressed as "well depth" (see Table 6.3).

This photoactive MOS capacitor forms only part of the CCD pixel. Conceptually, any pixel consists of three independent MOS capacitors, each of which can be biased independently, as shown schematically in Figure 6.12. The additional two MOS capacitors serve as a sort of electronic "shift register."

Once the (array) pixel exposure is complete, a control circuit is used to apply—in time sequence—voltages to the individual causes each capacitor to first couple and then transfer the electrons accumulated in the light-sensitive MOS capacitor (s) to its neighboring capacitor well, as shown in Figure 6.13.

The shift operation shown in the figure for an individual CCD pixel is continued through an array "line" (made up of many of these individual structural elements), until the charge has reached the last capacitor in the array. This last capacitor then dumps the charge into a charge amplifier, which converts the

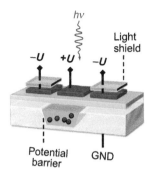

**Figure 6.12** Basic concept of a CCD pixel, comprising a biased metal-oxide–silicon (MOS) micro-sensor structure, and including (light-shielded) coupling/transfer electrodes.

**Table 6.3** Comparison of Important Parameters of Commonly Used Array Detector Devices

| Device Property | Units | PD | CMOS | CCD | EMCCD | ICCD |
|---|---|---|---|---|---|---|
| Sensor material[1] | | Si, IGA, HCT | Si, IGA, HCT | Si, IGA | Si | Si + MCPI |
| Sensor illumination[2] | | BI/FI | BI/FI | BI/FI | BI/FI | BI (FI) |
| Pixel number 1D (max.) | | 2048 | 2048 | 4096 | – | – |
| Pixel number 2D (max.)[3] | | 1024×1024 | 2048×2048 | 2048×2048 | 1024×1024 | 1024×1024 |
| Pixel size (width)[4] | μm | 25...50 | 8...50 | 8...50 | 13...25 | 13...25 |
| Spectral response range[5] | | VIS, NIR, IR | VIS, NIR, IR | VIS, NIR | VIS | VIS[6] |
| Full-well capacity | $\times 10^3\ e^-$ | $5 \times 10^4$ | 35 | 350 | 150 | 350 |
| Dark current (per pixel)[7] | $e^-/s$ | 10...100 | 0.15 | 0.0005 | 0.002 | 0.2 |
| Readout noise | $e^-$ | $10^3$ | 2 | 3 | 1...100 | 5 |
| TE-cooling (1–4 stages)[8] | °C | 0...−20 | −35...−40 | −35...−90 | −35...−90 | −35...−40 |

*Note:* All numerical entries represent typical values.
[1] Si = silicon; IGA = InGaAs; HCT = HgCdTe.
[2] BI = back-illuminated; FI = front-illuminated.
[3] For spectroscopy applications (coupling to spectrometer), the number of pixels in 2D arrays is often asymmetric, e.g., 1024 × 64.
[4] For spectroscopy applications (coupling to spectrometer), the pixel height in 1D arrays often differs from the width; height typically 100... 2500 μm.
[5] VIS = (200) 350...1000 nm; NIR = 900...2500 nm; IR = 2500...20000 nm.
[6] Spectral range depends on the cathode material; for full coverage of VIS, multialkali photocathodes are used.
[7] Typical dark currents for lowest cooling temperature.
[8] Array devices based on HgCdTe sensors are often cooled to liquid-nitrogen temperature of 77 K.

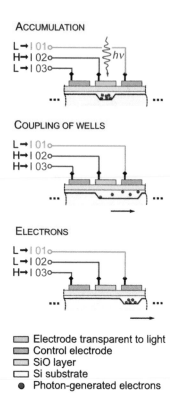

ACCUMULATION

COUPLING OF WELLS

ELECTRONS

☐ Electrode transparent to light
☐ Control electrode
☐ SiO layer
☐ Si substrate
● Photon-generated electrons

**Figure 6.13** Conceptual light accumulation and transfer of the photon-generated charge carriers through a pixel structure. (Top) Light accumulation phase. (Middle) Coupling of neighboring potential wells and sharing of charge carriers. (Bottom) Confinement of charge carriers in the well neighboring the light-collection well; transfer between wells is complete. The associated gate voltages $+U$ ("ON"-status) and $-U$ ("OFF"-status) to the control electrodes I 01/I 02/I 03 are indicated.

charge into a voltage signal (around a few microvolts per electron). By repeating this sequence, the controlling circuit converts the entire contents of the array in the semiconductor to a sequence of voltages. Overall, four main stages are encountered in the CCD reception and transmission of light information:

- *Expose* the detector pixels to light, which generates photoelectrons, and *accumulate/store* these photoelectrons in the potential wells directly beneath the active pixel electrode (with only this electrode in its status "ON")

- A cycle of *couple* (with two neighboring electrodes in their status "ON"), and *transfer* the stored charges (with the trailing electrode returned to its status "OFF"), until the charges have reached the end of the pixel line

- *Read out* the charge-related signal

- *Process* the analog signal (normally utilizing further on-chip manipulation of the voltage signal) and *A-to-D conversion* (signal amplitude resolution is normally in the range 8–16 bit)

CMOS detectors exhibit a different electronic structure to that just described. Rather than having to shift the charge through the complete array structure, a CMOS-sensor pixel has its own charge-to-voltage conversion circuitry, and even amplifiers and digitization are often included on-chip. Of course, such additional functionality increases the design complexity and may reduce the area available for light capture. On the other hand, the full operation of the sensor chip is massively parallel (the readout procedure is often dubbed the "X–Y readout") and hence allows for high total bandwidth and high speed.

As any semiconductor-based photodetector, also CCD and CMOS array detectors face the problem of thermally generated dark charges. This becomes particularly evident when dealing with very low light intensities for which electrons need to be accumulated in the pixel wells over extended periods of time; the photon-generated electrons "compete" against signal-unrelated thermal electrons, with those thermal electrons rapidly filling the well depth. Therefore, in particular, for low-light-level applications in spectroscopy, or imaging of very faint objects, the detector array chip has to be cooled. In modern devices, this is normally achieved using thermoelectric (TE) cooling based on Peltier elements or stacks. As a rule of thumb, one finds that the thermal noise electrons are suppressed by about one order of magnitude for every decrease in operating temperature of about 20–25 K. This is crudely the temperature difference that can be supported by a single Peltier layer; in many of the sensitive CCD array detectors on sale today, up to four Peltier elements are stacked, resulting in operating temperatures down to about –80°C (or about 240 K). Typical values for related operating temperatures and dark electron noise are listed in Table 6.3.

Finally, it is worth noting that scientific CCD and CMOS array detectors, as used in spectroscopic applications, come in two main guises, namely as *front-illuminated* (FI) or *back-illuminated* (BI) devices.

The traditional *FI* sensors are constructed in an orientation that places the sensor pixel matrix and its associated electronics and wiring on the front surface of the detector chip. This greatly simplifies manufacturing, but on the downside, this matrix and wiring layer reflects and/or obscures some of the incident light, and thus lowers the number of photons potentially available to be captured according to the detector pixel size.

A *BI* sensor contains the same structural elements, but now the wiring is integrated behind the photocathode layer. For this, the wafer containing the CCD array structure needs to be flipped over during manufacturing. In addition, the reverse side of the chip has to be thinned so that the photons can still reach the photocathode layer. Since now the photons do not have to pass through the wiring layer, this structure can improve the collection efficiency from the typical 50–60% for an FI device to well over 90% for typical BI devices. However, arranging the active matrix circuitry behind the photocathode layer can result in a number of problems, in particular the effect of so-called "etaloning." This is caused by the constructive/destructive interference of light by the (relatively thick) substrate layer, which results in periodic, wavelength-dependent responsivity fluctuations. In most modern commercial sensors, this detrimental effect has been greatly minimized.

### 6.3.3 On-chip amplified image sensors: EMCCD and e-APD devices

As is clear from the last statements about dark-electron charges in (very) low-light-level applications, CCD/CMOS detectors would benefit from amplification processes, which favor photoelectrons over thermal electrons. In principle, a procedure like secondary-electron generation would be advantageous, as encountered in PMTs. Of course, such a structure is impossible to implement on the miniature scale of pixel sizes of often less than $10{\times}10 \ \mu m^2$. Here, devices known as *electron multiplication CCDs* (EMCCDs) have come to the rescue.

EMCCD technology was introduced into imaging and spectroscopy applications in the early 2000s. In an EMCCD, a solid-state (active) electron multiplying pixel "register" is added at the end of the normal serial CCD register; for the conceptual layout of EMCCDs, see Figure 6.14.

Essentially, this register is nothing more than a chain of MOS pixel capacitors through which the signal charge is transferred. Within each of these gain pixels,

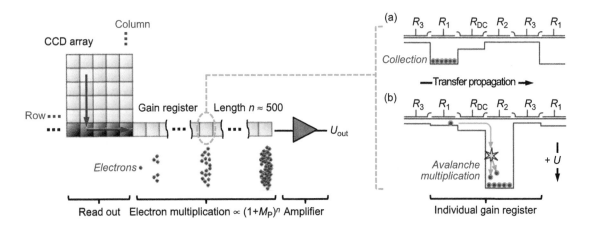

**Figure 6.14**   Concept of electron multiplication in an EMCCD. (Left) CCD array and readout register, followed by a gain register chain (number of elements up to $n \approx 500$). (Right) Individual gain register cell (electron transfer from left to right); (a) potential settings for "electron collection"; (b) potential settings for "avalanche multiplication." The individual register electrodes and related potentials $+U$ are indicated: $R_1$ = collection potential; $R_{DC}$ = constant buffer potential; $R_2$ = avalanche amplification potential; $R_3$ = transfer coupling potential.

one of the normally three phase electrodes is replaced with two electrodes. The first is held at a fixed potential while the second is set to a much higher voltage of 40–60 V when clocked. The large voltage potential difference between the fixed and clocked electrodes is sufficient for some electrons to cause the so-called "impact ionization" while being transferred, and thus generating additional electrons. In a certain way, the process is similar to the electron multiplication in an avalanche diode. Overall, the multiplication factor (i.e., the statistical average increase in electron number) per chain pixel is small—normally of the order of $M_P \cong 0.010$–$0.015$. However, since in common, practical devices, the number of multiplying elements is large (of the order of $n = 500$ or more), one would find for those numbers an overall register gain up to $G_{EM} = (1 + M_P)^n \approx 150$–$1700$. Thus, EMCCDs are useful for very low-light signal applications and can be sensitive down to the single-photon level. It should be noted that the gain function can be switched off by suitably adjusting the voltage of the clocked gain electrode; then the EMCCD operates as a standard CCD. For some further details on EMCCD functionality, see, e.g., "Fundamentals of EMCCD—A Tutorial" (EMCCD Forum 2005).

While the electron multiplication structure in EMCCDs is very easy to implement and is very efficient to multiply electrons post light acquisition, the drawback is a slowdown in temporal response: the travel through the multiplying register chain prolongs the readout time for each pixel, and therefore EMCCDs are not suitable for high-speed applications. An alternative is found in the form of electron-injection APDs (e-APDs) as the active elements in the array. As the name says, the active pixel elements have the device structure of APDs; they are not read out in series, as are CCD array detectors, but are X–Y addressable, i.e., selected parts of the array can be read out directly, at high speed. At the same time, the avalanche process amplifies weak photon signals sufficiently to be easily detectable. Practical devices based on this principle are mainly found in HeCdTe array detectors used for IR imaging applications. While the actual principle of e-APD HeCdTe PDs has its origins in the mid-1990s, viable array detectors only were fabricated more than 10 years later (see Beck et al. 2008).

### 6.3.4 Externally amplified and gated image sensors: ICCD devices

Two specific aspects of amplification of very weak photon signals, as highlighted in the previous segment above, are those of (1) overall amplification/multiplication and (2) fast time response. Both EMCCD and e-APD array detectors foster limitations in both. In general, their maximum multiplication factor for photoelectrons is of the order $10^3$, and their time response is normally governed by the readout speed of the whole or part of the array pixels, meaning that they are unable to follow real-time events on, say, the nanosecond timescale, as for example required in temporally resolved fluorescence spectroscopy.

A device architecture that overcomes both limitations is found in the form of the so-called *intensified CCD* (ICCD) detectors. These devices optically couple an image intensifier to a CCD array detector; they comprise four key functional elements: (1) a photocathode, (2) an MCP electron multiplier (see Section 6.4 for further details on MCPs), (3) a phosphor screen, and (4) the CCD-array chip. These elements are mounted in close sequential proximity to each other. Note that often—because of geometrical constraints dictated by device structure and properties—the phosphor screen and the CCD array are coupled via a short, ordered-fiber bundle of a few millimeters in length, with the fiber core diameter normally matched to the CCD pixel size by tapering; such devices are also known as proximity-focused intensified detectors. The conceptual construction of an ICCD detector device is shown in Figure 6.15, and key operating properties are included in Table 6.3.

The functionality of ICCDs is as follows. Light impinging on the cathode generates photoelectrons; these are accelerated toward the MCP by a control voltage, applied between photocathode and MCP. Because of the short distance, the photoelectron trajectories preserve the spatial information of where the photon did originally strike the photocathode. On their passage through the MCP, electrons are multiplied (up to $10^6$-fold, the actual gain depending on the bias potential across the MCP) and from the exit of the MCP, the amplified electron avalanche is accelerated toward the phosphor screen. These electrons are back-converted into photons, which then can be detected by the CCD-array sensor. By and large, all along the path from the photocathode to the array sensor, the spatial information is maintained, allowing for precise imaging. Some further details about the properties and functionality of image intensifiers may be found in manufacturers' product brochures (see, e.g., Hamamatsu 2009).

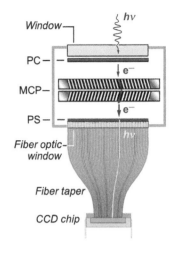

**Figure 6.15** Device structure of gated, intensified CCD (ICCD) detectors; PC = photocathode; MCP = multichannel plate; PS = phosphor layer.

Note that ICCD detectors—based on the principle configuration shown in Figure 6.16—inherently exhibit shutter or gating functionality. If the acceleration voltage between the photocathode and the MCP is reversed, the photoelectrons are no longer accelerated toward the MCP, but their direction is reversed to the photocathode; consequently, no electrons reach the MCP to be amplified and subsequently be visualized by the phosphor screen. The CCD sensor does not receive any light under these circumstances, as if a shutter were closed. The process of reversing the cathode-to-MCP acceleration voltage can be very fast; the voltage of typically around 200 V can be switched with sub-nanosecond rise and fall times, thus allowing for very short light "gating" in the range of just a few nanoseconds. For some types of ICCD cameras, shutter response times as short as 200 ps have been realized. It also should be noted that ICCDs have one particular drawback, and that is the range of spectral sensitivity. The spectral response of an ICCD detector is primarily dictated by the choice of the photocathode material, and thus exhibits similar properties to the photocathodes encountered in photomultipliers. As for those, the widest spectral response is obtained when using multialkali photocathodes.

## 6.4 CHARGED PARTICLE DETECTION

The interaction of an atom or molecule with a photon may not leave it in its neutral, original species form, but the particle can be "destroyed" by breaking it up into fragments through the processes of dissociation or ionization. In the specific case of ionization in general, an ion–electron pair is formed. Being charged particles, both can easily be manipulated by electric and magnetic fields, and be detected by charge-sensitive detectors. The choice of detector for charged particles (electrons or positive/negative ions) depends on the nature of the particular experimental investigation.

Common to all detector implementations is that a current signal is derived from the incident electrons or ions, either by directly measuring the current associated

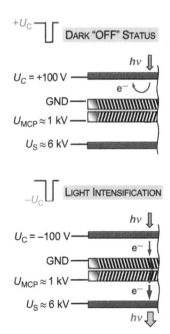

**Figure 6.16** Principle of gated ICCD light amplification. (Top) "Off"-state for light blocking, with positive voltage $+U_C$ applied to the photocathode. (Bottom) "On"-state for light amplification, with negative $-U_C$ applied to the photo cathode.

with the flux of incident charges or by generating secondary electrons, which can be further augmented in a chain of charge amplifiers. The most widely used detector types are briefly described.

### 6.4.1 Direct charge detectors—Faraday cup

The simplest type of charge detector is the so-called *Faraday cup*. It basically constitutes a metal (conductive) "cup" with a flat or appropriately shaped surface, which is placed in the path of a charged particle beam or which captures secondary charged particles. When charged particles impinge on a metal surface, the (electrically isolated) metal will collect the charge; note that ions are normally neutralized in the process. Subsequent to the collection process, the metal charge collector is discharged as a (normally small) current into a measuring circuit; the current flow is equivalent to the number of deposited electrons/ions. Faraday cup detectors are relatively insensitive but are very robust because of their extremely simple construction. To overcome this problem, Faraday cups are often equipped with an additional dynode electrode, which generates an amplified charge current by secondary-particle emission (normally electrons); for the principle operation and functionality of dynodes, see Section 6.2.

### 6.4.2 Single-element amplifying detectors—Channeltron

A *channeltron* is a "horn-shaped" so-called continuous-dynode structure that is coated on the inside with an electron-emissive material. Any charged particle, but also high-energy UV or x-ray photon, striking the channeltron cathode (basically its funnel-shaped entrance opening) creates secondary electrons that repeatedly interact with the walls of the narrow channel itself on their through-travel, creating an avalanche-type, amplified final current at the exit of the device. Other than the simple Faraday cup, the channeltron always provides cascade-current gain; the typical amplification achievable depends on the detector bias voltage and can be as high as $\sim 10^6$. However, it should be noted that, in many practical applications, single-element channeltrons are used less and less, being replaced by MCPs.

### 6.4.3 Multiple-element amplifying detectors—MCP

An MCP is an array of $10^4$–$10^7$ miniature electron multipliers, oriented parallel to one another; these channels replicate sort of continuous dynodes. In most commercial devices, these channels have diameters in the range 10–25 μm; their length-to-diameter ratio is normally of the order 40–50, i.e., the plate thickness is typically about 0.4–1.0 mm. In order to improve the conversion efficiency, most modern MCPs have their channel structure at a bias angle of about 5–10° with respect to the entrance/exit surfaces of the device. The front and back surfaces of the MCP are metal-coated and constitute the input and output electrodes. The principle construction and functionality of an MCP are shown in Figure 6.17.

MCP detectors exhibit a charged-particle amplification factor of $10^3$–$10^4$; but often they consist of cascaded setups (up to three individual plates) and then can reach multiplication factors of larger than $10^6$. Note that, like photomultipliers, they also exhibit very fast response times; but because of the much smaller

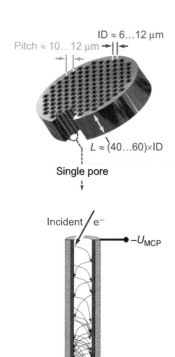

**Figure 6.17** Principles of microchannel plates (MCPs). (Top) Typical MCP, with dimensions indicated; ID = inner diameter of channel (pore); $L$ = length of channel, with typically $L/\mathrm{ID} \sim 40$...60; pitch = separation between channels. (Bottom) Individual channel, with the continuous-dynode electron avalanche indicated; $U_{\mathrm{MCP}}$ = voltage across the MCP.

dimensions between cathode and anode, also their transit times are only of the order of a nanosecond or so.

Particularly popular are dual-plate designs in so-called "Chevron" arrangement; these are found in most modern MCP detectors. In this configuration, the two MCPs, with angled channels, are rotated by 180° with respect to each other, so that the channels are oriented "V"-shaped. By this measure, charged-particle feedback is reduced, and the achievable cascade gain is substantially higher than for a single plate of twice the channel length. The conceptual structure of a Chevron dual-MCP detector is shown in Figure 6.18b.

Finally, with their spatial resolution only limited by the channel dimensions and the spacing between channels (typically both of the order 10–20 μm), they are ideally suited for imaging applications. For this, the amplified charged particle flux exiting from the MCP impinges on a scintillator (phosphor) plate, which, in turn, is coupled to a 2D-array CCD sensor recording the spatial distribution of the light emitted from the phosphor scintillator (see Figure 6.18c). All three structural components are mounted in close proximity to maintain the micrometer resolution; in many cases, the coupling between scintillator screen and CCD detector array is done by a "pixel"-matched, ordered fiber bundle of short length, as shown in Figure 6.15.

It should be noted that these charged-particle imaging devices can also be used for intensified photon detection. Then a photocathode is inserted in front of the MCP so that the electrons generated in the photocathode serve as the charged particles to be amplified by the MCP (see Figure 6.15). The interesting additional feature of such photon-imaging devices (sometimes called image intensifiers) is their capability of "gating." If the voltage applied between photocathode and MCP is reversed, then no electrons reach the detector and thus no signal is recorded. This blocking/unblocking of photoelectron transmission can have a very fast response time, down to less than 5 ns.

## 6.5 DETECTION BY INDIRECT PHENOMENA

In general, the common perception of "spectroscopy" is that one observes the direct effect of the (state-selective) photon–matter interaction, i.e., how photons are absorbed, emitted, or inelastically scattered, or that selectively atoms or molecules are ionized for direct particle detection. Mostly these processes are investigated for minute numbers of individual particles, i.e., microscopically small amounts of matter. Nevertheless, all are associated in one way or another by deposit or extraction of the transition energy from the probed sample. If, however, the number of interacting particles becomes sufficiently large so that they can be treated as a macroscopic ensemble, then also macroscopic effects become observable. Namely, this means that, in particular, the absorption of photons denotes an increase in the sample-internal energy, i.e., heat. Such changes in (local) heating are exploited in the technique of *photothermal* or *photoacoustic spectroscopy* and imaging. For a detailed description of the technique, its experimental realization and application, as well as dedicated data evaluation, see the textbook by Michaelian (2010). Photoacoustic spectroscopy and imaging will also be briefly covered in Chapters 8.6 and 21.2, respectively.

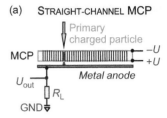

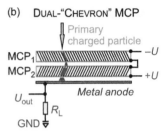

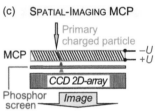

**Figure 6.18** MCP detector configurations: (a) straight-channel MCP, configured for global electron or ion detection; (b) dual "Chevron"-type MCP (channels angled at 5°...12°), configured for global electron or ion detection; (c) single "Chevron"-type MCP, configures for spatially resolved imaging of charged particles.

### 6.5.1 Photothermal/photoacoustic spectroscopy

The underlying phenomenon of the so-called photoacoustic effect was discovered by A.G. Bell in the 1880s. He showed that if a sample was irradiated with modulated light, it emitted "acoustic" waves as a consequence of the local heating and thus thermal expansion: a pressure wave of the same frequency as the interacting light wave (acoustic because the modulation frequencies were in the range of a few hundreds to thousands of hertz, i.e., frequencies of sound, and the detection was via a microphone).

With the development of microphones of very high sensitivity, going hand in hand with advances in lasers and modulation-sensitive electronics, photoacoustic spectroscopy evolved into a sensitive detection technique, initially to probe gaseous samples. The wavelength-sensitive absorption of molecules (ro-vibrational transitions) allowed for state-resolved spectroscopy and species selectivity for gas composition analysis. In conjunction with widely tunable laser sources, specifically diode laser and quantum cascade laser sources whose wavelengths coincide with vibrational band transitions, photoacoustic spectroscopy has evolved into a powerful laser-analytical technique to study concentrations of gases, with sensitivities reaching the part per billion (ppb) and, in favorable cases, even the part per trillion (ppt) levels (see Patimisco et al. 2014).

The conceptual realization of the photoacoustic measurement principle is shown in Figure 6.19, both for gaseous and solid samples. In the simplest case for gas analysis, the (wavelength-tunable) laser light beam—modulated at acoustic frequencies, or utilizing a pulse train of acoustic-frequency repetition rate—passes through a closed cell, and the pressure waves generated by state-selective absorption are measured using a highly sensitive microphone, usually a capacitor microphone (Figure 6.19a). Alternatively, for even higher sensitivity, the laser beam passes through an acousto-optic resonator, most commonly in the form of a piezo tuning fork, which picks up any absorption-related pressure waves between its fork arms (Figure 6.19b); for a comparison between the two types of sensing techniques, see, e.g., Haisch (2012). When solid samples are investigated, two different approaches for signal detection are common. In the first, one measures the secondary pressure waves in the gas adjacent to the solid in which the laser absorption has caused a thermal-expansion wave (Figure 6.19c); in the second, today more common approach, a piezo contact sensor is utilized, which is in close contact with the solid and directly responds to the laser-induced pressure waves in the sample (Figure 6.19d).

The mechanisms underlying photoacoustic spectroscopy are rather complex, in particular in solids, and a complete theory applicable to all possible situations has still to be developed. In a simplified picture, the photoacoustic effect can be treated using the formalism developed by Rosencwaig and Gersho (1976), basically constituting a two-step process: (1) the (repetitive) heat flow—after absorption—from the absorbing solid medium to the (ambient) gas, which is followed by (2) propagation of the acoustic pressure wave through the gas to the (microphone) detector. It is in particular the first step—based on thermal diffusion equations—that is the crucial one; simplistically, it depends on only three main parameters: incorporating the sample thickness, the light absorption coefficient, and the thermal diffusion coefficient.

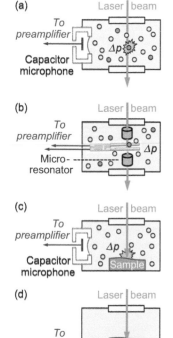

**Figure 6.19** Concepts for photoacoustic spectroscopy, with experimental setups for gaseous samples, with (a) microphone detection and (b) acoustic-resonator detection; and variations for solid samples, exploiting (c) pressure waves in the adjacent gas, using a microphone for detection, or (d) directly sensing sample expansion, using a piezo-transducer.

## 6.5.2 Photoacoustic imaging

One particularly intriguing and frequently exploited aspect of photoacoustic spectroscopy is that, as long as the light can penetrate sufficiently far into the (solid) sample, hidden entities invisible to the eye or molecular compounds can be probed. This has led to a plethora of biomedical applications, exploiting *photoacoustic imaging*; for short surveys, see, e.g., Beard (2011) or Wang et al. (2013).

In photoacoustic imaging, biological tissue is exposed to (non-ionizing) laser radiation, normally from a tunable pulsed laser source. The selectively absorbed radiation causes transient thermoelastic expansion, and the associated (ultrasonic) waves are detected by suitable, common transducers. The optical absorption in the tissue may be associated with *endogenous* molecules, such as hemoglobin or melanin; but it is also quite common to inject *exogenously* delivered contrast agents, which match certain laser wavelengths and which deposit at preferential locations. For example, at certain near-IR wavelengths, both oxygenated ($HbO_2$) and deoxygenated (Hb) hemoglobin exhibit distinctly different and much stronger absorption than ordinary tissue. As a consequence, there is sufficient endogenous absorption contrast that it becomes possible to visualize *in vivo* low-lying blood vessels, using photoacoustic imaging. Similarly, in various studies, it has been shown that photoacoustic imaging is also suitable, in principle, to monitor tumor angiogenesis, to map blood oxygenation, to skin melanoma, and many more (see Michaelian 2010).

In general, photoacoustic imaging is implemented in two distinct configurations, namely *photoacoustic (computed) tomography* (PAT) and *photoacoustic microscopy* (PAM). Typically, in the former, "unfocused" ultrasound detector(s) acquires the signal and image reconstruction is required, normally based on algorithms solving the photoacoustic equation. In the latter configuration, a "spherically focused" ultrasound detector collects the data point-by-point via a 2D scan; now no image reconstruction algorithms are required. Overall, photoacoustic imaging is evolving into an extremely valuable technique for *in vivo* diagnostic purposes, allowing full 3D mapping of subcutaneous tissue; in particular, it comes into its own with respect to the range of depth, as evident from Table 6.4, comparing photoacoustic imaging with other optical alternatives.

**Table 6.4**    Comparison of Photoacoustic Imaging with Other Optical Imaging Techniques

| Technique | Optical Spectroscopy Principle(s) | $\Delta z$ (mm) | $\delta z$ (µm) | $\delta x$, $\delta y$ (µm) |
|---|---|---|---|---|
| Photoacoustic microscopy | Absorption | 3 | 15 | 45 |
| Photoacoustic tomography | Absorption | 50 | 700 | 700 |
| Confocal microscopy | Fluorescence, scattering | 0.2 | 3–20 | 0.3–3.0 |
| Two-photon microscopy | Fluorescence | 0.5–1.0 | 1–10 | 0.3–3.0 |
| Optical coherence tomography | Scattering | 1–2 | 0.5–10.0 | 1–10 |

*Note:*   $\Delta z$ = penetration depth (in mm); $\delta x$, $\delta y$, and $\delta z$ = lateral and depth resolution (all in units of µm).

### 6.5.3 Photoacoustic Raman (stimulated Raman) scattering

Finally, it is noteworthy that besides exploiting absorption in photoacoustic spectroscopy and imaging, one also can make use of scattering in the form of stimulated Raman scattering. Although photoacoustic Raman scattering (PARS) was first demonstrated in the 1980s, only recently has the technique come under the spotlight again, and some developments of the technique for applications in gas sensing (see, e.g., Spencer et al. 2012; Schippers et al. 2014) and biomedical imaging (see, e.g., Yakovleva et al. 2010; Hajireza et al. 2013) are under way, which, however, are still very much in their infancy.

## 6.6 SIGNALS, NOISE, AND SIGNAL RECOVERY METHODOLOGIES

In experimental measurements, the capture and evaluation of signals constitutes one of the central tasks. As should have become clear from the discussion of detection devices in the previous sections, their signal information comes in many guises: one may encounter short, "single" events; one may measure and sample repetitively; information is presented inside a vast matrix of "pixels," each holding part of the overall signal. Also—as is, or should be common knowledge—signals hardly ever occur in their "pure," ideal form but are always accompanied by unwanted, albeit sometimes very small fluctuations or noise components. Here, we briefly describe actual signal and noise contributions to a measured entity, and how one can minimize the influence of noise on the desired result by appropriate sampling methodologies.

### 6.6.1 Signals and noise

When considering signals as a response of an atomic/molecular system to laser radiation, it is clear that the response is directly proportional to (1) the number of particles being exposed to the laser stimulus; (2) the number of photons available to interact; and (3) the probability of an interaction (transition) process to occur. This means that one will not encounter a single, signal amplitude value, but fluctuations due to statistical processes will occur. In addition, also the detecting devices will contribute to signal fluctuations due to the fact that they themselves exhibit statistical probabilities for detecting an event, or they may contribute to the signal as a consequence of some inherent, detector-internal processes. The sum of all fluctuations is normally addressed as "noise"; however, it should be kept in mind that the different noise components may affect a measured signal in different ways. It is therefore worthwhile to briefly look at the various definitions and properties of signal and noise.

*Signals.* The signals recorded in any experiment can be (1) continuous, (2) periodic in time, or (3) random; the most common signal shapes are collected in Figure 6.20. If the amplitude of a signal does not vary over time, such a signal is known as a *DC signal*. However, it should be noted that a continuous DC signal rarely exhibits truly constant amplitude. Thus, in a broader sense, one associates the term "DC" with any signal whose amplitude varies only insignificantly over the measurement time, and which exhibits no sudden changes (except when the source that gives rise to the signal is switched on or off). If the amplitude of a signal varies periodically as a function of time and if the shape of the time-varying

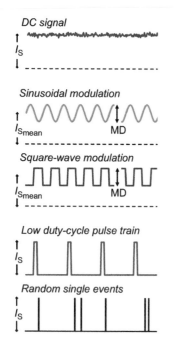

**Figure 6.20** Signal forms encountered in common laser spectroscopy measurements. $I_s$ = signal intensity; $I_{Smean}$ = mean signal intensity; MD = modulation depth.

amplitude is repetitive in time, such signals are classified as *AC signals*. In practice, most commonly sine wave, square-wave, or saw-tooth waveforms are encountered; some examples are shown in Figure 6.20. Note that said waveforms may alter between zero and full amplitude, but also could constitute amplitude modulations around a finite mean value, between a minimum and a maximum. In the case that the signal is still periodic in time, but that the duration of a particular response is (very) short with respect to the repetition of its occurrence, one addressed such responses as *pulsed signals*; their so-called duty cycle—the ratio between "on" and "off" periods—is normally very much shorter than that for AC signals (normally 1:1). For example, in many experiments in which short-pulse (nanosecond, picosecond, or femtosecond duration) laser excitation is utilized, signal duty cycles of the order $10^{-6}$:1 or even less are encountered. When any signal becomes very small at some stage, a point is reached at which individual, *random quantum events* constitute this signal; by definition, random events are not any longer experiencing periodic recurrence in time. But note that partial, accidental periodicity may be observed.

When it comes to signal fluctuation, one can subdivide these into two broad classes, as pointed out further above, namely (1) *intrinsic* signal noise, which is associated with system-internal, statistical quantum processes; and (2) *extrinsic* noise, which may stem from contributions linked to the environment in which the experiment is carried out, or which may originate from the detection device.

*Intrinsic noise.* This type of noise—often referred to as *shot noise*—is associated with the inherent fluctuation in the laser photon–particle interaction. It means that even an "ideal" signal exhibits noise fluctuations because quantum-statistical variations, associated with transition probabilities, by their nature cannot be eliminated. The distribution of this fluctuation in repeat measurements follows a Poisson statistical distribution, and thus it can be described by its width parameter $\sigma_S = \sqrt{I_S}$, where $I_S$ denotes the "ideal" signal intensity, or amplitude.

*Extrinsic noise.* These are contributions to the measured signal that do not originate from laser photon–particle interaction. They can be subdivided into contributions from (1) the environment in which the experiment takes place and (2) detector-internal processes. As such, both are independent of photon-induced signal.

In the former case, one may encounter *environmental background noise* from photons originating, say, from ambient light, or charged particles caused by non-laser ionization processes. By and large, these events contribute an often broadband "socket" to any signal, which carries its own (Poisson statistical) shot noise. Because of this, it can be described again by a statistical width parameter, $\sigma_B = \sqrt{I_B}$, where $I_B$ denotes the noise amplitude.

In the latter case, it depends on which type of detector (photon or charged particle) is utilized. Regardless, any measured signal to be treated by subsequent electronics relies on electron charge carriers, which always will be accompanied by *dark noise* or *thermal noise*, i.e., electrons that are thermally generated within the detector structure. Similar to shot noise, thermal noise follows a Poisson relationship and is equivalent to the square root of the number of thermal electrons generated within the signal-acquisition time interval, with statistical width parameter $\sigma_T$. In addition, one normally encounters the so-called *readout noise*, with width parameter $\sigma_R$.

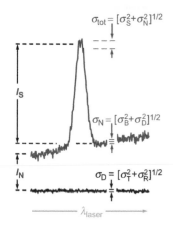

**Figure 6.21** Typical signal encountered in common measurements, identifying the various noise contributions (with their statistical width parameters). Abbreviations: D = detector; T = thermal; R = readout; N = noise; B = background; S = signal.

Assuming that all noise contributions follow stochastic statistical behavior, the total noise distribution can be described via the convolution of the individual contributions, i.e.,

$$\sigma_{\text{total}} = \left[\sigma_S^2 + \underbrace{\left(\sigma_B^2 + \sigma_T^2 + \sigma_R^2\right)}_{background\ noise}\right]^{1/2}. \qquad (6.15)$$

Note that very often the three noise sources of environmental background, thermal, and readout noise are lumped together as *background noise* (with width parameter $\sigma_N$). The noise contributions discussed above are represented schematically in Figure 6.21.

At this point, it is worthwhile to add a few short remarks on signal quantification. It is the background noise alone, i.e., $\sigma_N = [\sigma_B^2 + \sigma_T^2 + \sigma_R^2]^{1/2}$ (with any signal absent), that is commonly used in analytical procedures to determine the limits of detection (LOD) and quantification (LOQ). It is normally linked to the ratio of signal-amplitude to background-noise fluctuation, i.e.,

$$\text{signal-to-background-noise ratio} \equiv \text{SNR} \equiv I_S/\sigma_N, \qquad (6.16)$$

as recommended in the guidelines issued by various accredited organizations (see, e.g., IUPAC's guidelines [Fassel 1976; Voigtman 2008]). Based on said definition (6.16) a factor of three between signal and background noise is recommended to provide sufficient confidence in the calculated LOD value, i.e.,

$$\text{LOD} = 3/\text{SNR} = 3 \cdot \left(I_{\text{noise}}/I_{\text{signal}}\right), \qquad (6.17)$$

while for LOQ, this factor is 10. The fundamental problem in any measurement of the weak signals close to the LOD is to distinguish the actual signal from noise, and the art in the setup of experiments and the design of measurement equipment is to reduce noise contributions to the lowest level possible, i.e., to eliminate as far as feasible any environmental background and to minimize detector noise (e.g., by cooling).

Note that besides experimental/instrumental optimization, one may exploit statistical means to increase the SNR and thus improve the LOD. If a signal can be measured repeatedly (which may not always be possible), one can average the results point by point. This methodology is commonly known as *ensemble averaging*, and—if it can be applied—it is probably the most powerful method for improving the SNR. For a large number of $n$ repeat measurements, each of which exhibits statistical (random) fluctuations, to a good approximation, the SNR improves with $\sqrt{n}$, i.e.,

$$\text{SNR}(n) \equiv I_S/\left(\sqrt{n} \cdot \sigma_N\right). \qquad (6.18)$$

Further improvement to measurement sensitivity can be achieved by applying a range of clever, normally electronic instrument-based procedures whose efficacy naturally depends on (1) the signal intensity itself, (2) the time and frequency distribution of the signal, and (3) the various noise sources, and their time dependence and frequency distribution. As pointed out earlier in this section, one crudely distinguishes between DC and AC measurements. For signal amplitudes that are significantly larger than the noise, both can be measured directly by feeding the signal into a suitable measurement device, in the simplest case merely a volt- or ampere-meter. As soon as signal amplitudes decrease to values similar or below the magnitude of noise components, different instrumental approaches are required to recover signals. Two of the most common approaches are briefly outlined here.

### 6.6.2 Low-intensity "continuous" signals—Lock-in methods

In general, it becomes very difficult, if not impossible, to separate a small continuous DC signal from a pedestal noise background, which in addition may exhibit frequency fluctuations or temporal drifts. In order to overcome this problem, modulation techniques are applied, i.e., the amplitude of the initial DC (laser) excitation is changed periodically, with frequency $f$, in a suitable waveform (see Figure 6.20 for typical modulation functions); as a result, also the signal will be modulated. Now the technique of *phase-sensitive detection* (PSD) can be applied for the recovery of small, periodic signals, which may be obscured by background interference, sometimes being of even larger amplitude than the signal of interest itself.

In essence, a PSD device—often in the form of a so-called *lock-in amplifier*—responds to signals that are of the same frequency $f$ and phase with respect to the reference waveform; all other contributions are rejected. The concept of a lock-in amplifier is shown in Figure 6.22; the functionality of the various subgroups are briefly described.

In the *signal input channel*, the input signal (including its noise components) is amplified by an adjustable-gain, AC-coupled amplifier. Note that the performance of the PSD is usually improved if the bandwidth of the noise voltages reaching it is reduced to the frequency range of the expected signal response frequency, $f$; thus, the signal is passed through an adjustable band-pass filter centered at said reference frequency $f$.

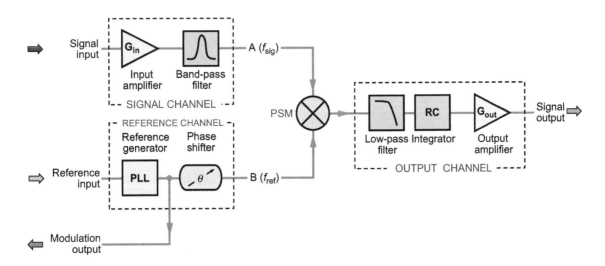

**Figure 6.22**   Concept of a phase-sensitive detector/lock-in amplifier. Abbreviations: $f$ = frequency; G = gain; PLL = phase lock loop; PSM = phase-sensitive mixer; RC = resistor–capacitor integrator; $\theta$ = phase angle.

The *reference input channel* provides a symmetric reference signal of frequency $f$, usually generated using a so-called phase-locked loop (PLL) circuit. It can be driven externally from the source that modulates the laser-driven signal, or it may serve as the actual laser excitation modulator. Note that in most experimental setups, the relationship between the signal waveform from the detector and the reference waveform will not always be exactly in phase. Thus, phase-shifting capabilities in the reference channel are normally included so that the phase relation between signal and reference waveforms can be adjusted, normally to match fixed settings of 0°, 90°, 180°, or 270°.

The waveforms from the signal and reference channels, $A(f_{sig})$ and $B(f_{ref})$, respectively, are mixed within a *phase-sensitive mixer* (PSM); in modern lock-in amplifiers, by and large, digital PSM implementations are encountered.

The final part of a lock-in amplifier is the *signal output channel*. At its heart are a *low-pass filter* and an *integrator/averager*. Their role is to remove AC components from the desired DC output, normally sum and difference frequencies generated in the PSM. If one assumes, for simplicity, that both the signal and reference waveforms are sinusoidal, the output of the PSM will contain components at frequencies of $f_{sig} + f_{ref}$, and $f_{sig} - f_{ref} \cong 0$ Hz. The low-pass filter will eliminate the sum-frequency component; only the difference frequency component near zero hertz passes, effectively generating a DC output. Note that any remaining low-frequency AC noise components are integrated/averaged to a mean value of ~0 $V_{DC}$ using a suitably dimensioned active RC filter.

The DC output signal voltage from a lock-in amplifier is traditionally displayed on a panel meter integral to the lock-in amplifier, or sampled under computer control for further data treatment.

### 6.6.3 Low-intensity pulsed signals—Gating methods

For the measurement of a signal-pulse integral, or an average short-time duration signal, a *gated integrator* or *boxcar averager* is the most suitable instrument. Commercial devices allow one to sample pulses of duration as low as about 100 ps up to pulse lengths of several milliseconds. A gated integrator is typically used in experiments utilizing short-pulse laser radiation (of the order of nanoseconds), in which the pulse repetition rate is often low (in the range 1 Hz to 1 kHz), which means that the duty cycle signal-on to signal-off is quite low.

Conceptually, a gated integrator behaves like an RC filter: its output signal is more or less proportional to the average of the input signal over the integration gate period. This means that frequency components of the input signal, which have an integer number of cycles during the gate, will average to zero. The conceptual realization of a gated integrator is shown in Figure 6.23. Three particular aspects of the device shown in the figure are noteworthy.

First, the input signal only affects the output for the duration of the gating period (i.e., when the gate switch is "closed" to let the signal reach the integrator); at all other times, its level is unimportant. The duration of the gating period and its delay with respect to a (synchronizing) trigger pulse is adjusted to match the conditions of a particular experiment.

Second, the signal is integrated during the gate duration, rather than providing only a "snapshot" measurement of the signal level at one point in time, as

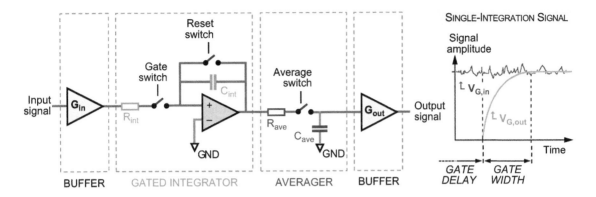

**Figure 6.23**  Concept of a gated integrator/boxcar averager. Abbreviations: C = capacitor; G = gain; GND = signal ground; R = resistor.

common sample and hold circuits do. Over the integration period, the output voltage $V_{out}$ rises exponentially, reaching a value of magnitude $V_{in}$, provided the integration time constant is sufficiently long, as shown in the signal traces in Figure 6.23. Note that for the particular example shown here, the integration gate width is narrow in relation to overall long-term changes in $V_{in}$. Note also that high-frequency (much higher than the reciprocal of the gate-width time) components of the input signal are smoothed out. At the end of the gate period, the signal $V_{out}$ is held sufficiently long by the high-impedance integrator to be available for (slower) post-acquisition treatment.

Third, in an optional add-on (but which is implemented in most commercial instruments), the integrated signal fraction can be averaged over a number of repeat events. This means that low-frequency fluctuations or noise is smoothed, and thus sample-to-sample signal variations are reduced. Note that the signal from the integrator part is allowed to transfer to the averaging circuit only during a short period after completion of any individual signal integration. This is done so that the integrator can be zeroed after each accumulation cycle.

Overall, gated integrators/boxcar averagers are used in two distinct modes, namely in *static-gate sampling* or in *waveform recovery* mode. In *static-gate sampling*, both the integration gate width and delay are fixed with respect to an applied "trigger" from the repetitive laser stimulus: the same relative point in time of the input signal is sampled. This mode is commonly used to follow the time evolution of a laser-related signal. For example, the amplitude of a laser-induced fluorescence signal in a chemical reaction could be studied as a function of varying reagent pressure. In its *waveform recovery* mode, the device mimics time-resolved sampling: the gate delay is swept over a range of values, and the result is a point-by-point record of the input signal waveform. Of course, for this to work, the gate duration has to be sufficiently short to match the temporal variation in the waveform to be studied, for example, the fluorescence decay associated with the upper-level lifetime of a laser-excited atom or molecule.

# 6.7  BREAKTHROUGHS AND THE CUTTING EDGE

It is probably fair to say that the rapid development of semiconductor electronics and photo detection equipment—which happened in parallel to the advances in lasers in the early 1960s—impacted most on laser spectroscopic techniques. In particular, the invention of semiconductor-based photosensor arrays and the transistorizing of signal detection electronics revolutionized spectroscopic and imaging applications. Out of the wealth of inventions of that time, two examples are selected here, which probably had the greatest impact.

### 6.7.1 Breakthrough: First transistorized lock-in amplifier

It is widely accepted that R.H. Dicke of Princeton University, and founder of Princeton Applied Research Corporation (PAR), may be seen as the father of modern lock-in amplifiers. Although the idea and implementation of lock-in techniques can be traced back to the 1940s (see Michels and Curtis 1941), it is the change-over from valve to semiconductor technology that advanced this type of signal processing enormously, starting with the first fully transistorized implementation of the concept by PAR (PAR model HR-8 1965). Without lock-in amplifiers, the retrieval of very weak signals, nearly masked by a noisy environment, would—in many cases—hardly be possible.

### 6.7.2 Breakthrough: First demonstration of CCD imaging

An even more profound impact on laser spectroscopy and imaging has to be attributed to the invention of semiconductor-based array detectors. This starts with the development of CCD technology by G.E. Smith and W.S. Boyle in 1970 for which they received the Nobel Prize in 2009.

Originally, their research was not intentionally aimed at light detection and imaging: the first CCD device, which comprised an eight-element device structure, was operated as a serial memory (shifter). But it was soon tried as a linear scanning imaging device as well (Tompsett et al. 1970); for the first integrated device structure, see Figure 6.24a.

Development was very rapid, and only a year later, the same researchers utilized a 96-element device (made up of two integrated segments of 64 and 32 pixels each), with reasonably fast response to demonstrate usable digital imaging by reproduction of black and white text. Even scanning a picture and recording it in grayscale was demonstrated (Tompsett et al. 1971; see Figure 6.24b).

Today, these originally only linear-array devices have evolved into highly sophisticated 2D-imaging sensors, and beyond scientific applications have made their way into everyday life, e.g., in the form of image sensors in mobile phones.

### 6.7.3 At the cutting edge: Nanoscale light detectors and imaging devices

The advance of technologies and equipment for laser spectroscopic and imaging applications has not shown any sign of abating; on the contrary, every year, new and exciting developments are reported, which promise to revolutionize one or the other field of research and/or practical use. Whether such progress is related to new detector devices, better system integration, or novel methodologies, one often is in awe about what one reads in scientific publications or company publicity.

(a)

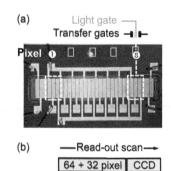

(b)

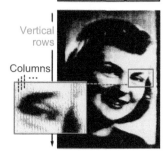

**Figure 6.24** First CCD array devices and their application. (a) First integrated 8-pixel CCD device (adapted from Smith G.E., *Rev. Mod. Phys.* 82, no. 3(2010): 2307–2312.) (b) First grayscale picture, scanned using a 96-pixel (64+32 pixel segments) CCD device; the insert reveals the pixel-column structure. (Adapted from Tompsett, M.F. et al., *IEEE Trans. Electron. Dev.* 18, no. 11 (1971): 992–996.)

Staying with the theme of array detectors, which was touched upon in Section 6.7.2 here we would like to mention just one example where the next revolution of light sensors may come from, namely the use of nanowires as the photosensitive element.

ZnO nanowires turn out to be superb photon detectors; at first sight, this is somewhat surprising, given their very small size (a few hundreds of nanometers in diameter and of the order 10 µm in length) and their limited light absorption. The bandgap of ZnO of $E_g \cong 3.37$ eV has the consequence that nanowires grown from this material respond to UV radiation but not to visible light (the cutoff wavelength corresponding to the bandgap is about $\lambda_{cutoff} \cong 370$–390 nm). The response of ZnO nanowires to photons was first demonstrated by Kind et al. (2002). Loosely speaking, the functionality of ZnO nanowires is related to the change in conductivity induced by the electron-hole generation following photon absorption. A substantial gain in photoconduction, $G_{PC}$, occurs because the holes and electrons produced by an absorbed photon are separated (holes drift to the nanowire surface) and are thus prevented from recombining, allowing current in a biased nanowire to flow more freely (see Figure 6.25a). This, together with the large surface-to-volume ratio and size of the nanowires, could explain the high observed gains of up to $G_{PC} \sim 10^8$, at the very moderate light intensities of $I_{390nm} \sim 10$ µW·cm$^{-2}$ (see Soci et al. 2007).

First attempts on realizing nanowire arrays—albeit using nanowires based on different materials—may be found in Hayden et al. (2006) who demonstrated that their nanoscale *p-n* diodes, consisting of crossed Si–CdS nanowires, could be addressed independently without cross talk. They also showed that these "nano avalanche photodiodes" (nanoAPDs) were extremely sensitive, with detection limits of about 100 photons, and exhibited subwavelength resolution of about 250 nm. Today, the first full imaging arrays based on ZnO nanowire have been demonstrated, e.g., by Han et al. (2015); their exploratory UV-sensitive PD array consisted of 32×40 pixels based on vertically aligned ZnO nanowires, with spatial resolution of ~100 µm (or ~254 dpi). A test image recorded with this device is shown in Figure 6.25b.

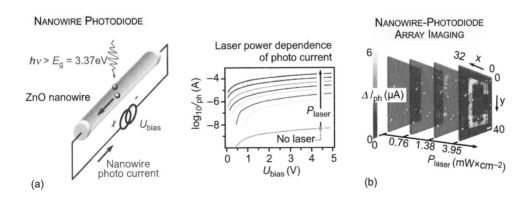

**Figure 6.25**  UV PD detectors based on ZnO nanowires (NW). (a) Concept of a biased nanowire PD, with photocurrent $I_{ph}$ displayed in dependence of incident laser power. (Data adapted from Soci, C. et al., *Nano Lett.* 7, no. 4 (2007): 1003–1009.) (b) Image of an illuminated mask, recorded with an NW-PD array detector. (Adapted from Han, X. et al., *Adv. Mater.* 27, no. 48 (2015): 7963–7969.)

# CHAPTER 7

# Absorption Spectroscopy and Its Implementation

$A$bsorption spectroscopy refers to the absorption of radiation by an atom or molecule when the absorbed photons have the appropriate energy to induce transitions from an initial to an excited state of the atomic or molecular system. The absorption spectrum is obtained by monitoring the absorbed intensity of the incident light after passing through the sample, as a function of its frequency or wavelength. If the transition involves the absorption of a single photon, the spectroscopy is said to be *linear*; otherwise, if several photons are absorbed, the spectroscopy is called *nonlinear* spectroscopy. In what follows, the basic concepts and equations of absorption spectroscopy are described.

## 7.1 CONCEPTS OF LINEAR ABSORPTION SPECTROSCOPY

### 7.1.1 Absorption coefficient and cross section

From a spectroscopic point of view, the key equation to be considered in absorption spectroscopy is the Bohr relation (resonant condition) given by $E_f - E_i = h\nu_{if}$, where $E_i$ and $E_f$ represent the energies of the initial and the final eigenstates, respectively; $h$ is the Planck's constant; and $\nu_{if}$ is the frequency of the radiative, resonant transition between the two energy states. Note that in the most basic form of absorption spectroscopy, $E_i$ is the ground state of the system ($\equiv E_g$), and $E_f$ is any energetically excited state ($\equiv E_e$).

From a quantitative point of view, the basic equation related to this spectroscopy is the Beer–Lambert law that was described in Chapter 2.1; in slightly modified form from Equation 2.6, it can be written as

$$I = I_0 \cdot \exp(-\alpha(\nu) \cdot \ell) = I_0 \cdot \exp(-\sigma(\nu) \cdot N \cdot \ell), \qquad (7.1)$$

in which $I$ and $I_0$ stand for the transmitted and incident light intensities, respectively; $\ell$ is the path length through the sample; and $\alpha(\nu)$ is the so-called *absorption coefficient*, given by the product $\sigma(\nu) \cdot N$, i.e., including the absorption cross section and the (absorbing) particle concentration per unit volume. Note that here for simplicity of notation, $\nu \equiv \nu_{if}$ is used. Note also that according to the rate equation formalisms discussed in Chapter 2, Equation 7.1 constitutes an approximation since more correctly $N$ should be replaced by $\Delta N_{if} = N_i - N_f$, i.e., the population difference between the two states involved in the transition. Only as long as the incident light intensity is very low, the population density in the excited j-state is negligible and hence $\Delta N_{if} \cong N_i \cong N$. Essentially, the main goal of the absorption spectroscopy is to measure the absorption coefficient, either in absolute or relative values, as a function of the frequency or wavelength of the absorbed radiation.

### 7.1.2 Spectral line profiles

For a given transition between two energy levels, the frequency-dependent absorption cross section is given by

$$\sigma(v) = S_T\, g(v - v_0).  \tag{7.2}$$

Here $S_T$ is defined as the transition intensity, also known as the line strength, in units of $cm^{-1} \cdot (molecule\cdot cm^{-2})^{-1}$ when $v$ is given units in $cm^{-1}$ (rather than in standard frequency units $s^{-1}$ or Hz). The function $g(v-v_0)$ describes the line shape, a function centered at $v_0$, which has the same analytical form for all transitions.

Based on the strict, steady-state resonant condition for photon absorption, one would expect that a spectral line constitutes an infinitely sharp feature. However, as a consequence of standard quantum dynamics and the Heisenberg uncertainty condition, in reality, all spectral lines exhibit a linewidth, though it may be extremely narrow. The observed width of a spectral line depends upon several factors like the spectral resolution of the measuring instruments (an extrinsic factor), the lifetime of the excited state produced by the photon absorption (an intrinsic factor), or the velocity distribution of the absorbing particle, to name but a few examples.

In order to characterize the spectral line profile, one often uses the so-called full width at half-maximum (FWHM) value, $\delta v_{1/2}$, which is defined as $\delta v_{1/2} = v_2 - v_1$ with $v_2 > v_0$ and $v_0 > v_1$, such that $I(v_2) = I(v_1) = 0.5 \times I(v_0)$. A conceptual line profile is illustrated in Figure 7.1; the spectral region inside the FWHM is normally known as the *line kernel*, while the regions outside the FWHM are called *line wings* and are indicated.

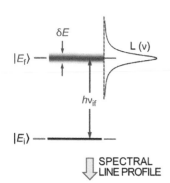

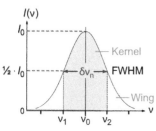

**Figure 7.1** Schematic for the profile function of a spectral transition line, originating from the transition between the ground (initial) state $E_g$ ($E_i$) and an excited (final) state $E_e$ ($E_f$) with finite lifetime; the key parameters are indicated.

## 7.2 LINE BROADENING AND LINE SHAPES IN ABSORPTION SPECTROSCOPY

In general, the "line broadening" factors are classified into two categories: *homogeneous* and *inhomogeneous* line broadening factors. Homogeneous broadening refers to the case in which the contribution to the linewidth is the same for all atoms (or molecules), while inhomogeneous broadening refers to the situation where that contribution is not the same for every atom or molecule. In the latter case, the functional dependence is related to a particle property, like, for example, the particle velocity that differs for individual particles. As key examples for the two broadening mechanisms, natural (homogeneous) and Doppler (inhomogeneous) line broadening will be discussed in more detail.

## 7.2.1 Natural broadening

If one considers an atom or a molecule that can only populate two energy levels, the ground level $E_g$ and an excited level $E_e$, and if the lifetime of the excited state is $\tau_e$, then one may write the related Heisenberg uncertainty principle as $\delta E_e \cdot \tau_e \geq \hbar$, with $\hbar = h/2\pi$ ($h$ is Planck's constant). If one assumes that the ground state lifetime is "infinite" without external intervention, then its related energy uncertainty is "zero." This then can be exploited to approximate the frequency uncertainty in a transition $E_e - E_g = \Delta E_{eg} = h\nu_{eg}$ as $\delta E_{eg} = \hbar/\tau \equiv h\delta\nu_{eg}$. This $\delta\nu_{eg}$ (or often $\delta\nu_n$) is known as the *natural linewidth* of the transition. Note that, for easy distinction, the symbol $\Delta$ is used for the difference between energy levels/frequencies, while $\delta$ is associated with the (uncertainty) width of a particular level/frequency. Recall from Chapter 2 that the lifetime of the upper state in a two-level system can be related to the Einstein $A$ coefficient for spontaneous emission, i.e., $A_{eg} = \tau^{-1} = 2\pi \cdot \delta\nu_n$. If one then substitutes $A_{eg}$ with the expressions provided in Equations 2.3 and 2.5, one finally obtains for the natural linewidth

$$\delta\nu_n = \left(\frac{8\pi^2}{3\varepsilon_0 hc^3}\right) \cdot \nu^3 \cdot |\mu_{eg}|^2, \tag{7.3}$$

where $c$ is the light speed, $\mu_{eg}$ is the transition moment, and $\varepsilon_0$ is the vacuum permittivity. It is important to note that the cubic dependence on transition frequency explains the different order of magnitude of the natural linewidth depending upon which particular spectral region is under consideration. Typically, one finds that in the far-IR (e.g., transitions between rotational energy levels), $\delta\nu_n \sim 10^{-4} - 10^{-5}$ Hz; in the mid-IR (e.g., transitions between vibrational energy levels), $\delta\nu_n \approx 2 - 20$ kHz; and in the visible/UV (in general, transitions between electronic states), $\delta\nu_n \geq 30$ MHz.

The line profile associated with natural broadening is given by the (normalized) Lorentzian line shape function

$$L(\nu-\nu_0) = (\delta\nu_n/2\pi) \cdot \left[(\nu-\nu_0)^2 + (\delta\nu_n/2)^2\right]^{-1}. \tag{7.4}$$

with the *Lorentz width* at half maximum $= (2\pi \cdot \tau)^{-1}$, where $\tau$ is the intrinsic, natural lifetime of the particular (upper) energy level (see Figure 7.2a).

## 7.2.2 Collisional or pressure broadening

In general, particles are not isolated but within an environment of other particles. For example, in a gas, randomly moving particles can come close to each other, undergoing collisions. If such a collision occurs during the time interval of a radiative transition (absorption or emission) within a particle, the transition process will be affected by the collision.

There are two types of collision processes, namely *elastic* and *inelastic* collisions; both affect the photon transition process but in different ways.

In an *inelastic* collision, energy is transferred between kinetic and internal excitation energy. For example, an excited state of particle A can be "quenched" into a different, lower energy state by a collision with a particle B, but without a

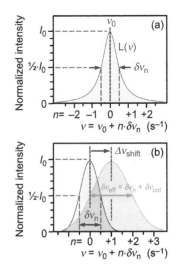

**Figure 7.2** Schematic line profiles for (a) a natural Lorentz-shape function and (b) the shift and broadening of the profile function by collisions.

photon being emitted. One can associate this nonradiative inelastic collision with a quenching probability out of the excited state $E_e$, which in terms of Einstein coefficient can be written as

$$A_e^{inel} = N_B \cdot \bar{v} \cdot \sigma_e^{inel} = N_B \cdot \sqrt{\frac{8kT}{\pi\mu}} \cdot \sigma_e^{inel} = 2\sigma_e^{inel} \cdot \sqrt{\frac{2}{\pi\mu kT}} \cdot p_B, \qquad (7.5)$$

where $N_B$ and $p_B$ are the number density and pressure of the quenching particle B, respectively; $\bar{v}$ is the mean relative velocity of the particles; $\mu$ is the reduced mass of the colliding particles A and B; and $\sigma_e^{inel}$ is the inelastic collision cross section for energy state $E_e$. Equation 7.5 means that the probability for the collision process depends on the pressure of the quenching particle B. With this, one then can write for the total transition probability

$$A_e = A_{eg}^{rad} + A_e^{inel} = \frac{1}{\tau_e^{rad}} + \frac{1}{\tau_e^{inel}} \equiv \frac{1}{\tau_{eff}}, \qquad (7.6)$$

with the superscript "rad" denoting the photon transition process; and $\tau_{eff}$ standing for the apparent collision-affected lifetime of the excited energy state. As a consequence, one needs to replace the purely radiative transition probability in Equation 7.3 with the total transition probability of Equation 7.6. This then leads to an increased—or broadened—pressure-dependent width for the transition line

$$\delta v_{eff} = \delta v_n + \delta v_{col}(p_B). \qquad (7.7)$$

In the case that no internal energy of the collision partners A and B is transferred nonradiatively during the collision, it is termed *elastic*. Such collisions do not alter the amplitude but affect the phase of the emitted wave due to the frequency shift $\Delta v_{shift}(R_{AB})$ during the collision. This shift can be understood from the particle interaction potentials during the collision, which are normally different for the ground and excited state of a collision pair. One finds that the higher the density of particles (or pressure), the larger the shift:

$$\Delta v_{shift} = N_B \cdot \bar{v} \cdot \sigma_e^{el} = 2\sigma_e^{el} \cdot \sqrt{\frac{2}{\pi\mu kT}} \cdot p_B, \qquad (7.8)$$

where $\sigma_e^{el}$ stands for the elastic collision cross section; all other parameters are as in Equation 7.5. It should be noted that this frequency shift can be positive or negative, depending on the actual form of the interaction potentials in the ground and excited states.

In summary, one finds that (1) inelastic collisions broaden the line shape function and (2) elastic collisions shift the peak frequency (higher or lower, depending on the interaction potential); for further details and derivations, see, e.g., Demtröder (2008) or Buldyreva et al. (2011). As a consequence, the natural (Lorentz) line shape function in Equation 7.4 changes to the effective line shape function

$$L(v-v_0) = ((\delta v_n + \delta v_{col})/2\pi) \times \left[((v-v_0) - \Delta v_{shift})^2 + ((\delta v_n + \delta v_{col})/2)^2\right]^{-1}. \quad (7.9)$$

The conceptual effect of line broadening and shift associated with collisions is shown in Figure 7.2b; a typical experimental example is included in Figure 7.3.

As stated earlier, the actual line profile encountered in a collision depends on the interaction potentials between A and B. The probability for a photon transition depends on the energy difference $\Delta E(R_{AB}) = E_e(R_{AB}) - E_g(R_{AB})$, as a function of internuclear separation $R_{AB}$ at the instance of the photon transition. In general, this energy difference does not uniformly rise or fall but is represented by a non-uniform function that may have extrema. As a consequence, the associated line profile is, in most cases, no longer of Lorentzian shape, but exhibits an asymmetric profile.

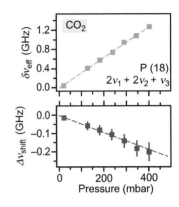

Figure 7.3 Experimental example for the line-shift and line-broadening for a ro-vibrational line of $CO_2$. (Data adapted from de Rosa, M. et al., *J. Quant. Spectrosc. Rad. Trans.* 61, no. 1, (1997): 97–104.)

### 7.2.3 Doppler broadening

Due to the Doppler effect, the observed frequency for the photon absorbed or emitted by an atom or molecule depends on the absorbing/emitting particle's velocity relative to the detector. For example, if the particle is moving toward or away from the detector with a speed $v_z$, the observed frequency of the transition is given by

$$\nu = \nu_0 \cdot (1 \pm (v_z/c))^2, \tag{7.10}$$

where $\nu_0$ is the intrinsic, quantum transition frequency, $c$ is the light speed, and the sign − or + applies for the particle moving away from or toward the detector, respectively. Without going into actual details (for these, see, e.g., Demtröder 2008), the Doppler broadening parameter—based on the Maxwell velocity distribution—may be written as

$$\delta\nu_D = \left(\frac{8RT \cdot \ln 2}{Mc^2}\right)^{1/2} \cdot \nu_0 = 7.16 \times 10^{-7} \cdot \left(\frac{T}{M}\right)^{1/2} \cdot \nu_0 . \tag{7.11}$$

Here $R$ stands for the universal gas constant, $T$ is the temperature (in units of K), $M$ is the particle mass (in units of kg). Then, the ensemble average over all particles participating in this inhomogeneous broadening process gives rise to a Gaussian line shape function, given by

$$G(\nu-\nu_0) = \frac{(\pi \ln 2)^{1/2}}{\pi^2 \delta\nu_D} \cdot \exp\left[-\ln 2 \left(\frac{\nu-\nu_0}{\delta\nu_D/2}\right)^2\right], \tag{7.12}$$

which should be compared with that of natural broadening given by the Lorentzian line shape function (see Equation 7.4).

Note that, in general, one finds in the visible and infrared (IR) spectral ranges that the Doppler width is several orders of magnitude larger than that of the natural linewidth and, therefore, it is not possible to directly estimate the magnitude of the latter. In order to overcome the Doppler broadening limitation, one measures the absorption or emission of radiation from atoms or molecules, which have zero-velocity components along the direction of the light propagation. This can be implemented by using a well-collimated atomic or molecular beam, which is interrogated by the laser radiation perpendicular to the particle propagation direction or, even better, by performing the so-called Lamb dip spectroscopy, as will be discussed further below.

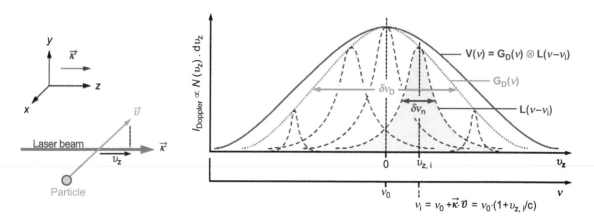

**Figure 7.4** Conceptual (Gaussian) Doppler line profile, assembled from a superposition of Lorentz profiles for individual particle velocity components. Note that for $\delta v_L$ being of similar magnitude to $\delta v_D$, the Gauss profile function morphs, by convolution with the Lorentz line shape function, into a (broadened) Voigt profile function. The coordinate convention for the particle $\leftrightarrow$ photon interaction is shown on the left.

Finally, it should be noted that the Doppler-broadened spectral line is not exactly represented by a pure Gaussian profile function. This is because any full class of molecules, having a definitive velocity component $v_z$, does not emit or absorb radiation at the exact frequency, as deduced from Equation 7.6. Rather, each individual contributing particle carries a finite lifetime, and thus—because of the aforementioned uncertainty principle—the emission or absorption frequencies will not be associated with a unique value, but will be given by all values of the respective Lorentzian distribution. Consequently, the Gaussian line profile of the ensemble may be seen as a superposition of weighted Lorentzian profiles (see Figure 7.4).

### 7.2.4 Combined line profiles—The Voigt convolution profile

When more than one broadening effect contributes to the line profile, the resulting line shape function will be the convolution of the constituent profiles. It is easy to show from functional convolution theory that the convolution of Lorentz functions will yield a Lorentz profile again, but with different width parameter (see the example above for the simple, additive effect of natural and collisional broadening). Similarly, the convolution of Gaussian distribution functions will once more replicate a Gaussian shape.

However, as stated earlier, more often than not, processes associated with both Lorentzian and Gaussian distributions contribute to the overall broadening. Hence, a more correct line profile will be obtained by convoluting the actual Lorentzian and Gaussian line profile functions. In spectroscopy, this type of mixed-convolution profile is commonly known as the *Voigt profile function* $V(v)$, or more informatively $V(v;\delta v_D,\delta v_n)$:

$$V(v) \equiv V(v; \delta v_D, \delta v_n) = \int_{-\infty}^{\infty} G(v'; \delta v_D) \cdot L(v\text{-}v'; \delta v_n) \cdot dv', \qquad (7.13)$$

where $\delta v_D$ and $\delta v_L$ are the Gaussian and Lorentzian linewidth parameters, respectively, as before. The convolution integral is often parameterized in a different form in order to gauge the impact of the two contributors to the Voigt line shape profile; one particular representation is (see, e.g., Amamou et al. 2013)

$$V(v) \equiv \sqrt{\frac{4 \cdot \ln 2}{\pi}} \cdot \frac{1}{\delta v_D} \cdot V(a, u), \tag{7.14a}$$

where

$$V(a, u) = \frac{a}{\pi} \cdot \int_{-\infty}^{\infty} \frac{\exp(-y^2)}{a^2 + (u - y)^2} \cdot dy \tag{7.14b}$$

and

$$a = \sqrt{\ln 2} \cdot \frac{\delta v_n}{\delta v_D}; \; u = \sqrt{4 \cdot \ln 2} \cdot \frac{v - v_0}{\delta v_D}. \tag{7.14c}$$

The parameter $a$ determines the importance of the Lorentzian in the profile: for $a \to 0$, the Lorentzian is negligible; for $a \to \infty$, the Lorentzian is dominant. As an indication regarding the interplay between Lorentz and Gaussian profile functions, an associated Voigt profile is included in Figure 7.4.

## 7.2.5 Other effects impacting on linewidth

There are a number of additional effects that influence the width and shape of an optical transition line. These are only touched on here; for details, the reader is referred to Demröder (2008).

*Power broadening*: this is an effect encountered for laser intensities, which are sufficiently large that the optical pumping rate on an absorbing transition becomes larger than the emitting transition (relaxation) rates, i.e., the apparent lifetime is shortened. As a consequence of the strong pumping, one observes a quite noticeable decrease in the population in the initial (ground state) energy level. This phenomenon of *partial saturation* of the population densities results in additional line broadening. Without going into further detail, one finds that

$$\delta v_{\text{power}} = \delta v_0 \cdot (1 + S)^{1/2} = \delta v_0 \cdot (1 + P/P_0)^{1/2}; \tag{7.15}$$

where $S$ is the so-called saturation parameter; $P$ and $P_0$ stand for the actual and threshold laser power, respectively; and $\delta v_0$ is the linewidth at threshold power $P_0$. Note that the spectral line profiles of the said partially saturated transitions are different for homogenously and inhomogeneously broadened lines. An example is shown in Figure 7.9. Note also that the effect is observed and exploited in nonlinear saturation spectroscopy (see Section 7.3).

*Transit-time broadening*: this effect is observed if the interaction time of the absorbing particle with the radiation field is short in comparison to the (natural) spontaneous lifetime of the excited levels. This situation is encountered in quite a few laser spectroscopy experiments, in particular those that involve transitions between ro-vibrational levels of molecules with spontaneous lifetimes in the range of a few microseconds to milliseconds. For molecules with a mean thermal velocity $\bar{v}$, passing through a (Gaussian profile) laser beam with beam waist $w_0$, their transit time $T_T = w_0/\bar{v}$ may be substantially shorter than the spontaneous lifetime $\tau_e$ (up to several orders of magnitude). In such cases, the linewidth of a Doppler-free transition is no longer limited by the spontaneous emission lifetime $\tau = 1/A_{eg}$, but by the flight time $T_T$ of the particle through the laser beam; this, in turn, determines the interaction time of the particle with the laser radiation field. One finds for the transit-time broadening parameter

$$\delta v_{TT} = (2/2\pi) \cdot (2\ln 2)^{1/2} \cdot \left(\frac{\bar{v}}{w_0}\right) \cong 0.38 \cdot \left(\frac{\bar{v}}{w_0}\right), \qquad (7.16)$$

*Opacity broadening*: in a high-density medium, the radiation emitted by an excited atom or molecule, at a particular point in the medium, can be reabsorbed as it continues to travel through the medium. This reabsorption is wavelength-dependent since photons in the line wings exhibit a lower reabsorption probability than those at the line center. As a consequence, one observes a broadening of the line. For very dense media, the reabsorption near the line center can become so large as to cause the so-called *self-reversal*, meaning that the intensity at the center of the line is lower than at the wings, i.e., the line experiences an intensity dip (see Figure 7.5). Note that the process is also known as *self-absorption*. In general, one wishes to avoid line distortion away from a simple, well-behaved functional representation (like Gaussian or Lorentzian shapes). On the other hand, self-reversed spectral lines provide a powerful diagnostic tool for the determination of temperatures, e.g., in high-density plasmas. For more information on temperature determination and details on self-reversal, see, e.g., Schneidenbach and Franke (2008) or Goto et al. (2010).

## 7.3 NONLINEAR ABSORPTION SPECTROSCOPY

The linear Beer–Lambert law was derived in Chapter 2 (and reiterated earlier in Section 7.1) under the assumption that the population density of the states involved in the absorption did not depend on the incident light intensity, meaning that $\Delta N \cong N_i \cong N$. However, when the light intensity increases significantly, this approximation is no longer valid; now both $N_f$ and $N_i$ will change (increase/decrease, respectively, since $E_f > E_i$). It is precisely the dependence of $\Delta N$ on (high) laser light intensity, $I_L$, that makes the absorption become nonlinear, thus leading to *nonlinear absorption spectroscopy*.

Under nonlinear absorption conditions, the differential light attenuation is given by $dI = -I_L \cdot \sigma_0(v) \cdot \Delta N(I_L) \cdot d\ell = -I_L \cdot \alpha_0(v) \cdot d\ell$, with $d\ell$, $\sigma_0(v)$, and $\alpha_0(v)$ being the differential light path, the absorption cross section, and the absorption coefficient for low light intensities, respectively; note that the latter depends on the light intensity.

In general, one can expand the population difference into a power series

$$dI = -I_L \cdot \sigma_0(v) \cdot \Delta N(I_L) \cdot d\ell$$

$$= -I_L \cdot \sigma_0(v) \cdot \left[\Delta N(0) + \frac{d}{dI_L}(\Delta N(I_L)) \cdot I_L + \dots\right] \cdot d\ell, \qquad (7.17)$$

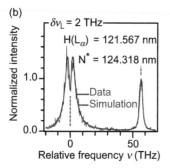

**Figure 7.5** Effect of self-reversal in the absorption in dense (plasma) media. (a) Simulated absorption line profiles for the hydrogen Lyman-$\alpha$ line $H(L_\alpha)$, for a range of particle densities, as a function of $\alpha_0 \cdot L$ ($\alpha_0$ is the absorption coefficient at line center, and $L$ is the absorption depth). (b) Spectrum of a hydrogen/nitrogen discharge plasma, revealing self-reversal in the $H(L_\alpha)$ line but negligible self-reversal for the nitrogen $N^*$-line (between excited energy levels). (Data adapted from Laity, G. et al., *Appl. Phys. Lett.* 102, no. 18 (2013): article 184104.)

**Table 7.1**  Determination of the Saturation Parameter $S$ and the Light-dependent Absorption Coefficient $\alpha(v)$, for a Nondegenerate Two-level System

| Process/Operation | Equation |
|---|---|
| Rate equations for a nondegenerate two-level system with $E_f > E_i$, in the presence of a radiation field with spectral density $\rho(v)$, with Einstein coefficients for induced processes $B_{if} = B_{fi}$ and spontaneous emission $A_{fi}$ (see also Section 2.2) | $\left(\dfrac{dN_i}{dt}\right) = -N_i B_{if}\rho(v) + N_f B_{fi}\rho(v) + N_f A_{fi}$ <br><br> $\left(\dfrac{dN_f}{dt}\right) = +N_i B_{if}\rho(v) - N_f B_{fi}\rho(v) - N_f A_{fi}$ |
| Solving for steady-state field conditions: <br> using $\left(\dfrac{dN_i}{dt}\right) = -\left(\dfrac{dN_f}{dt}\right)$ | $(N_i\text{-}N_f)B_{fi}\rho(v) - N_f A_{fi} = (N_f\text{-}N_i)B_{fi}\rho(v) + N_f A_{fi}$ |
| Conservation of total population: <br> using $N = N_i + N_f$ | $N_f = N \cdot \dfrac{B_{fi}\,\rho(v)}{2B_{fi}\,\rho(v) + A_{fi}}$ <br> with $N_f = 0$ for $\rho(v) = 0$ <br> and $N_f = N_i = N/2$ for $\rho(v) \rightarrow \infty$ |
| State population difference: <br> using $\Delta N = N_i - N_f$ | $\Delta N = N \cdot \dfrac{A_{fi}}{2B_{fi}\,\rho(v) + A_{fi}} \equiv N \cdot \dfrac{1}{1+S}$ <br> with $\Delta N = N$ for $\rho(v) = 0$ <br> and $\Delta N = 0$ for $\rho(v) \rightarrow \infty$, i.e. the transition becomes saturated |
| Saturation parameter $S$: <br> using $B_{fi} = \dfrac{c \cdot \sigma_0(v)}{hv}$ and $\rho(v) = \dfrac{I(v)}{c}$ <br> Dependence of absorption coefficient on laser light intensity: <br> using $\alpha(v) = \sigma(v) \cdot N$ | $S = \dfrac{2B_{fi}\rho(v)}{A_{fi}} = c \cdot \sigma_0(v) \cdot I(v)/hv$ <br><br> $\alpha(v) = \dfrac{\alpha_0(v)}{1+S} = \dfrac{\alpha_0(v)}{1 + I(v)/I_S(v)}$ <br> with the saturation intensity $I_S(v) = hv/2\sigma_0(v)$ |

where $\Delta N(0)$ stands for the population difference at zero intensity, i.e., $I_L = 0$, given by the difference between the unperturbed state populations at zero light intensity $\Delta N(0) = N_i(0) - N_f(0)$. Thus, the first term in Equation 7.17 corresponds to one-photon linear absorption, the second term is associated with two-photon nonlinear absorption, and so on. For further details on the formalisms encountered in the treatment of saturation spectroscopy, see, e.g., Levenson (1982) or Demtröder (2008).

In Table 7.1, the dependence of the state populations on the light intensity is summarized for a two-level system; with increasing laser light intensity, the absorption saturates, giving rise to a regime in which the spectroscopy is called *saturation spectroscopy* whose absorption coefficient $\alpha(v)$ now depends on the intensity $I_L(v)$.

## 7.3.1 Saturation spectroscopy

In order to understand Doppler-free saturation spectroscopy, one may start with the basic experimental setup shown in Figure 7.6. Here, radiation from a (polarized) tunable laser is split in pump and probe beams, which are sent in a counter-propagating way through a vapor cell containing an atomic or molecular sample. For simplicity, it is assumed that in the atom/molecule, only two quantum states have to be considered: the ground state and an excited state. Normally, the pump beam has a much higher intensity than the probe beam (of the order ×10 to ×100; the ratio can be adjusted by rotating the $\lambda/2$ wave plate in

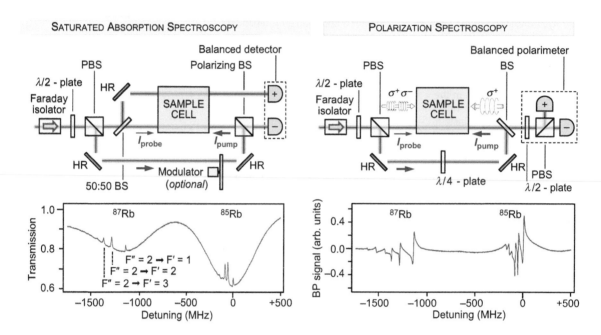

**Figure 7.6**   Conceptual experimental setups for Doppler-free spectroscopy. (Left) Saturated absorption spectroscopy; (right) polarization spectroscopy. Key components: BS = beam splitter; PBS = polarizing beam splitter; HR = high-reflectivity mirror; with the $\lambda/2$-plate adjacent to the Faraday rotator, the ratio $I_{pump}$-to-$I_{probe}$ can be adjusted; with the $\lambda/4$-plate the pump beam in polarization spectroscopy is converted from linear into circular polarization; balanced detection methods allow one to remove nonzero background. (Bottom) Typical signal traces associated with the two techniques, for the fine structure of the $D_1$-line of rubidium. (Data adapted from Pearman, C.P. et al., *J. Phys. B: At. Mol. Opt. Phys.* 35, no. 24 (2002): 5141–5151.)

front of the polarizing beam splitter, altering the linear polarization direction). With the arrangement shown in the figure, one measures the transmitted laser intensity of the probe beam, $I_{probe}$, as its frequency is scanned across the atomic absorption band.

It is worth noting that optionally one can split the probe beam into two beams of equal intensity, one experiencing particles exposed to $I_{pump}$ and the other not. Comparing the two signals in a so-called balanced detector, one can subtract the Doppler profile out of the saturated absorption spectrum. Alternatively, instead of using a second probe-beam path, one may "chop" (modulate) the stronger pump beam, normally at audio frequencies of a few kilohertz. If the chopper blade blocks the pump beam ($I_{pump}$ OFF-condition), the probe beam is ordinarily absorbed on passage through the gas; when the chopper blade does not block the pump beam ($I_{pump}$ ON-condition), it partially saturates the level populations of the atoms/molecules; as a consequence, the probe beam is less attenuated than for the OFF-condition, and thus a stronger probe-beam signal reaches the detector. Note, however, that this modulation only occurs if the two beams interact with the same molecules, i.e., when the laser light interacts with molecules either at rest or at least with zero-velocity component along the direction of the laser beams.

Using a phase-sensitive detector, one can electronically extract only the saturated absorption signal, free from the Doppler background profile, generating the difference signal $I_{ON}-I_{OFF}$. The signal $I_{ON}$ would result in an absorption profile as shown in the bottom panel of Figure 7.6, i.e., a typical Doppler broadened profile with a linewidth determined by the sample temperature superimposed by the saturated absorption peaks natural linewidth.

Those spectral "dip" features with very narrow width at the resonant frequency of the atomic/molecular transition are known as *Lamb dips,* named after the scientist William Lamb who quantitatively described this effect (see McFarlane et al. 1963; Lamb 1964); experimentally, Lamb dips were demonstrated only a few years later (Costain 1969).

*The origin of the Lamb dip.* To understand the Lamb dip, one should consider the Doppler effect for photon absorption by a moving particle inside the irradiated absorption cell. If one denotes a moving particle with velocity (component) $v_{z,i}$ moving from the left to the right, it interacts with two laser beams, namely the pump laser (or saturating) beam (propagating from the right to the left) and the probe laser beam (propagating from the left to the right). Although the two light waves have the same frequency $v_L$, the moving particles will "see" the pump wave as having the frequency $v_{pump} = v_0 \cdot [1 + (v_{z,i}/c)] \equiv v^+$, and the probe wave as having the frequency $v_{probe} = v_0 \cdot [1 - (v_{z,i}/c)] \equiv v^-$. It is clear that particles moving with velocity $v_{z,i}$ toward the saturating beam will absorb at the frequency $v_{pump}$, and thus will deplete the blue-shaded fraction of the velocity distribution marked in Figure 7.7b; note that as a mirror image, a much less pronounced depletion would occur at the frequency $v_{pump}$ (shaded orange in the figure). The situation would invert for particles copropagating with the pump laser beam. Normally, one says that a *hole has been burnt* in the lower state population distribution $\Delta N(v_z)$. Hence, for a given laser frequency $v_L$, any class of atoms having a velocity different from zero, i.e., $v_z \neq 0$, will not absorb the pump and probe laser radiation at the same apparent frequency. Furthermore, the closer the laser frequency is to the resonant transition value ($v_L \approx v_0$), the closer the pump and probe holes move to the center of the absorption profile.

Both holes coincide in the central position when the laser frequency is resonant with that of the atomic transition, i.e., $v_L \equiv v_0$, a condition that will only happen when the particle has a zero-velocity component along the laser-beam propagation direction. In other words, only if $v_z = 0$ will the particle see both laser beams to be at the same frequency, and thus will absorb photons from both. In fact, due to the high intensity of the pump laser beam, the population of the lower state is significantly depleted in favor of that of the excited state, and the remaining particles populating the ground state will also absorb the probe beam; but this absorption would be significantly reduced giving rise to the observed Lamb dip in the transmitted intensity $I_{probe}$ (see Figure 7.7c). For the same reason, when the pump laser is OFF, one would only recover a normal Doppler broadened profile, as shown in Figure 7.7a.

For the experimental technique just described, Smith and Hänsch (1971) coined the name *saturation spectroscopy* or *Lamb dip spectroscopy*. Note that when an absorbing sample cell is located inside a laser cavity, the Lamb dip in the

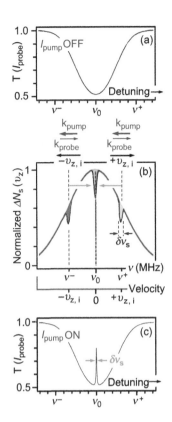

**Figure 7.7** Absorption of (strong) pump and (weak) probe beams by a Doppler-distributed ensemble of particles. (a) Normal absorption profile without saturation ($I_{pump}$ OFF), detuning the probe laser frequency $v_L$ from the resonance transition frequency $v_0$. (b) Hole burning into the level populations when detuning the pump laser frequency to $v_L (\pm) = v_0 \cdot (1 \pm v_{z,i}/c)$; at $v_L = v_0$ a Lamb dip is generated. (c) Absorption profile with saturation ($I_{pump}$ ON), detuning the probe laser frequency $v_L$ from the resonance transition frequency $v_0$, exhibiting a narrow sub-Doppler Lamb dip at line center; for the Lamb-dip width $\delta v_s$, see text. Note that $k_{pump}$ and $k_{probe}$ are the wave vectors of the two counterpropagating laser beams.

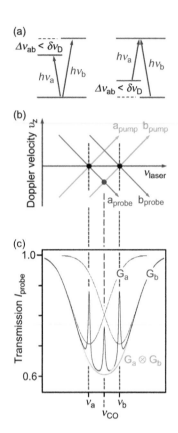

**Figure 7.8** Schematic of crossover resonances. (a) "V"-shape and "Λ"-shape level configurations. (b) Absorption coincidences for velocity groups $v_z$ from counter-propagating pump and probe laser beams, as a function of laser frequency. (c) Saturated absorption dips for the three coincidences in (b), at the transition resonance frequencies $v_a$ and $v_b$, and the crossover frequency $v_{CO}$.

absorption profile results in an increase in laser output power at the very frequency; this maximum in laser output, called an *inverse Lamb dip*, is superimposed on the normal Doppler line profile. Since the saturation of the intracavity absorption cell results in a narrow peak in the laser output, one can use a feedback system to lock the laser output frequency at this peak value. Consequently, one can stabilize the laser frequency; values of frequency stabilization to 1 part in $10^{12}$ or even better are not uncommon (see, e.g., Cerez and Bennett 1979).

Finally, note that the higher the pump laser intensity is, the more populated the excited state will become; in fact, for "infinite" intensity of the pump laser beam, the population of both ground and excited states will tend to be equal, and consequently no absorption of the probe beam can take place; we then say that the atomic/molecular transition is *fully saturated*. Strong saturation will impact the width of the observed Lamb dips; notably they experience power broadening (see further below).

*Multilevel effects and crossover peaks.* As seen in the saturation spectra of Figures 7.6 and 7.7, not only do Lamb-dip peaks occur at the center of the Doppler transition profile, as expected, but also additional peaks are observed whose frequency does not coincide with any of the anticipated Lamb dips. This is due to the fact that in any real atom or molecule, one encounters multiple upper and lower energy levels, which may have to be considered in a transition scheme; then the simple two-level model used to explain the Lamb dips is no longer adequate. If transitions between multiple lower and upper levels are accessible to the laser radiation utilized in a particular experiment, additional *crossover resonances* may be observed. Crossover resonances are additional narrow absorption dips arising because several upper or lower levels are close enough in energy that their Doppler-broadened profiles overlap. They can be understood within the framework of the lowest-number multilevel system, namely a three-level scheme. As shown in Figure 7.8a, two cases are possible, either of the so-called "V"- or "Λ"-configurations with two closely spaced upper or lower levels, respectively. The phenomenon will be illustrated for the "V" configuration.

Having two excited energy levels $|a\rangle$ and $|b\rangle$, the resonance frequencies from the ground state $|g\rangle$ are $v_a$ and $v_b$, which are assumed to be spaced less than a Doppler width apart. Without the pump laser beam, each excited state would absorb the probe laser beam with a Doppler-broadened profile, and the net absorption would be the sum of two Gaussian profiles, centered at $v_a$ and $v_b$, respectively. Note that if the separation $|v_a - v_b|$ were small compared to the Doppler width, they would appear as a single broadened absorption profile; for the issue of line-pair resolution, see Section 7.4. As soon as the pump laser beam is turned on, two holes are burned in the ground-state velocity distribution, at velocities that put the particle transition into resonance with $v_a$ and $v_b$; the respective velocities depend on the laser frequency $v_L$. For example, at $v_L = v_a$, the probe laser absorption to upper state $|a\rangle$ is from particles with velocity $v_z \approx 0$, while the probe laser absorption to upper state $|b\rangle$ arises from some particles with nonzero velocity $v_z$. Accordingly, the pump laser beam burns two holes into the ground-state distribution at the related locations. As with the two-level system, the hole at $v_z \approx 0$ leads to a decreased absorption to upper state $|a\rangle$ and generates a saturated absorption (Lamb) dip at $v_a$. Similarly, one predicts a saturated absorption dip at $v_b$. A third Lamb dip, the crossover resonance, arises at a frequency midway between the two saturated absorption peaks at $v_{CO} = (v_a + v_b)/2$. At $v_{CO}$, the pump and probe laser beams are resonant with the same two, opposite velocity groups $v_z \approx \pm(v_b - v_a)/2v_{CO}$; particles at one of these two velocities will be resonant with the transition to one excited state, and particles at the opposite velocity will be resonant with the transition to the other excited state. The pump laser beam burns a hole in the ground-state populations at both velocities, and these holes affect the absorption of the probe beam, which is simultaneously also arising from atoms with these two velocities (see Figure 7.8b).

*Broadening encountered in Lamb-dip spectra.* Like for all quantum transition phenomena, the Lamb dip exhibits finite width, the most fundamental reason being the uncertainty principle. In the limiting case for which (1) the pump and probe laser beams are monochromatic and weak; (2) dephasing particle collisions are rare; and (3) the particles interact for a sufficiently long time with the laser radiation to achieve steady-state-level populations; the width of the Lamb dip is equal to the natural width of the transition, $\delta v_n$ ($=A_{if}^{-1}$). For example, for allowed dipole transitions in atoms and molecules, one typically finds natural width of the order $\delta v_n \sim 10–100$ MHz. This means that, indeed, one needs to utilize good single-mode laser sources if one wishes to trace the Lamb-dip shape. Certainly in the early days of saturation spectroscopy, when pulsed laser sources with typical linewidth $\delta v_L \geq 1$G Hz were the norm, the width of the Lamb dip was limited by the laser linewidth. But even for the case that a narrow-bandwidth single-mode laser is used, the natural linewidth limit for the Lamb dip may not be achieved. This is the case when any of the broadening mechanisms stated in conditions 1 to 3 cannot be neglected. In particular, that is true for condition 1 requiring weak laser beams: in order to achieve sufficient saturation and population transfer incurred by the pump laser radiation so that the probe beam encounters a measurable change, the pump laser power often needs to be substantial. In that case, power broadening (see Section 7.1) will most likely be encountered: with increasing laser power, the Lamb-dip amplitude increases but so does the width of the feature when the pump laser intensity $I_{pump}$ substantially surpasses the saturation intensity $I_s$. The modeling of the Lamb-dip shape is shown in Figure 7.9, for a range of power ratios $I_{pump}/I_s$.

*Experimental examples for saturation spectroscopy.* Saturation spectroscopy in one form or another has become an indispensable tool when it comes to resolving very narrowly spaced quantum features, like fine- and hyperfine-level structures in atoms and molecules.

Perhaps one of the most impressive, early examples of the capability and resolution achievable with saturation spectroscopy is the study of the fine structure of the Balmer series $H_\alpha$-line (at $\lambda = 656.3$ nm) of the hydrogen atom, illustrated in Figure 7.10a. In the top panel, the allowed transitions between the $n = 2$ and $n = 3$ levels, and their fine-structure sublevels, are shown.

Because of the Doppler limit encountered in discharge (required to generate H-atoms) experiments, one would be hard-pressed to resolve the two groups of transitions (1–3 and 4–7) as two peaks. Exploiting saturated absorption spectroscopy, nearly all transitions are visible and resolved in the spectrum, safe for the low-intensity components 1 and 2, which are below the noise limit (see Hänsch et al. 1972). The composite of the laser linewidth of $\delta v_L \cong 30$ MHz utilized in Hänsch's experiments and the natural width of the individual fine-structure transitions $\delta v_n \leq 100$ MHz, associated with the respective $A_{if}$, would be less than the observed Lamb-dip widths of $\delta v_s \approx 250 - 300$ MHz (power and pressure

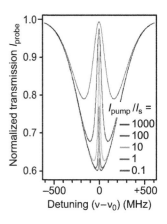

**Figure 7.9** Modeling of a Doppler-free absorption spectrum of a two-level atom, experiencing power broadening of the Lamb dip, as a consequence of $I_{pump} > I_s$ ($I_s$ = saturation intensity).

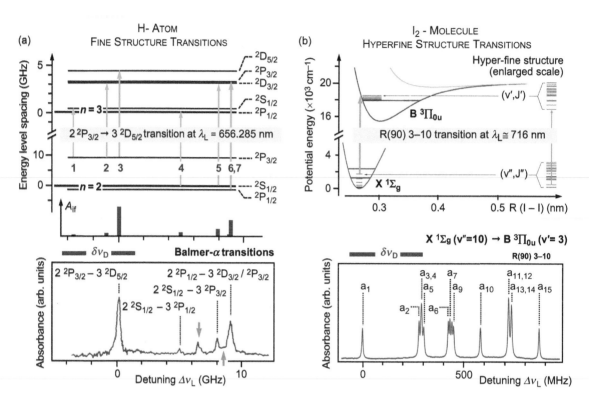

**Figure 7.10**    Examples for high-resolution saturation spectroscopy. (a) Fine structure of the hydrogen Balmer $H_\alpha$-line. (Top) Allowed transitions between the $n = 2$ and $n = 3$ energy levels of atomic hydrogen, together with their relative transition probabilities $A_{if}$. (Bottom) Early example of a Doppler-free absorption spectrum. (Data adapted from Hänsch, T.W. et al., *Nature* 235, no. 5333 (1972): 63–65.) Note the crossover resonance marked by the gray arrows. (b) Hyperfine structure of the R(90)3-10 line of the X→B transition of the $^{127}I_2$ molecule. (Top) Transitions between ro-vibrational levels (hyperfine structure not to scale). (Bottom) Doppler-free spectrum of the 15 hfs-components of the transition. (Data adapted from Huet, N. et al., *J. Opt. Soc. Am. B* 30, no. 5 (2013): 1317–1321.) For further details, see text.

broadening in said experiments were nearly insignificant). A significant additional width contribution stems from the overall short lifetimes of the $^2P$ states (inverse of the sum of all $A_{if}$ commencing in those levels) involved in the transitions; furthermore, the unresolved hyperfine components with $\Delta\nu_{hfs} \leq 60$ MHz increase the overall width. The latter point is an interesting one since it exposes the dilemma: even if the laser linewidth were sufficient to resolve much of the hyperfine structure in the transitions, intrinsic quantum characteristics may cause broadening larger than the hfs-splitting. However, it is worth noting that this is a peculiar feature of the hydrogen atom; resolved hfs-splitting of transitions is routinely demonstrated for the alkaline atoms, which even have become demonstration experiments for undergraduate students (see Jacques et al. 2009).

One of the first examples of Doppler-free saturation spectroscopy, in which complex hyperfine-structure components were resolved, is for molecular iodine. Hänsch et al. (1971) measured and analyzed the P(J = 117)21-1 line of the X→B transition (Kr$^+$-laser excitation at $\lambda_L = 568.2$ nm) of the $^{127}I_2$ molecule; all but one pair of the 21 hfs-components could be spectrally resolved, exhibiting (pressure-broadened) Lamb dips of FWHM $\delta\nu_s \approx 10(\pm 2)$ MHz within the Doppler envelop $\delta\nu_D = 1400$ MHz. Note the large number of hfs-components for the iodine molecule: odd-J levels are split into 21 hyperfine sublevels, while even-J levels are split into 15 sublevels. Overall, the authors demonstrated a spectral resolution of ~$10^8$ at the laser excitation wavelength.

It is also noteworthy that the interest in Doppler-free spectroscopy of transitions in molecular iodine has remained strong throughout the nearly 50 years of those early experiments by Hänsch and coworkers because iodine still serves as a frequency standard, providing thousands of lines with hfs substructure in the visible part of the spectrum; an example from a recent study of the R(90)3-10 line of the X→B transition (Huet et al. 2013) is shown in Figure 7.10b, together with a conceptual transition term scheme (the hfs-splitting is not to scale); laser excitation by a Ti:sapphire laser at $\lambda_L \cong 716$ nm, with linewidth $\delta\nu_L < 1$ MHz.

## 7.3.2 Polarization spectroscopy

Essentially, in this type of spectroscopy, one measures the change of the polarization state of a probe laser photon induced by a polarized pump photon. To understand the basic principle of polarization spectroscopy, one may look at the basic setup illustrated in Figure 7.6b. As in saturation spectroscopy, the radiation from a monochromatic tunable laser is split into (weak) probe and (stronger) pump beams.

The pump laser beam is circularly polarized by means of a $\lambda/4$-plate and pumps molecules from the ground to the excited state. For a given atomic/molecular transition $(J'',m_{J''}) \rightarrow (J',m_{J'})$, the absorption cross section is $m_J$-dependent; consequently, the residual ground-state $m_{J''}$-distribution is different from the initial one, exhibiting some non-uniform character, i.e., some degree of anisotropy. The probe laser beam is linearly polarized and is directed counter-propagating to the pump laser beam through two crossed polarizers, one of those before and the other after the sample cell. When the laser frequency is not resonant with the given $(J'',m_{J''}) \rightarrow (J',m_{J'})$ transition, i.e., $\nu_L \neq \nu_0$, the pump and probe beams interact with different classes of molecules. This is a consequence of the fact that linear polarized light is associated with $\Delta m_J = 0$ transitions, while circularly polarized light induces $\Delta m_J = \pm1$ transitions. Thus, the pump photon absorption will not influence the probe photon absorption, and hence the detector will not register the probe laser beam, as it is blocked by the second (crossed) polarizer.

If the laser frequency is in resonance with the transition, i.e., $\nu_L = \nu_0$, the anisotropic (oriented) $m_J$-distribution of the saturated ground-state population will act like a birefringent medium when exposed to the linearly polarized probe laser beam. As a result, its polarization plane will be slightly changed on passage through the sample, and the detector will register a signal rise because the probe laser will no longer be completely blocked by the second polarizer. Note that using a balanced polarization detector, as shown in the figure, normally the recorded signal has a "dispersive" shape rather than a peak shape, as exemplified in the experimental data included in Figure 7.6.

A good conceptual example of the high resolution of the Doppler-free polarization spectroscopy, with a link to the saturated absorption spectroscopy in hydrogen, is the investigation of the fine and hyperfine structure of the hydrogen Balmer H$_\beta$-line—see, e.g., Wieman and Hänsch (1976) and Hermann et al. (1993). As in the case of saturated absorption spectroscopy, the three fine-structure features are clearly resolved (see the top panel of Figure 7.11), as well as the two crossover resonances. Again, the hyperfine splitting is masked by the

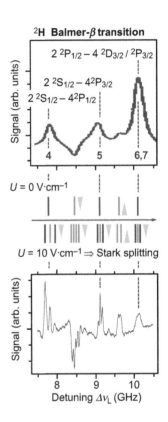

**Figure 7.11** Polarization spectrum of the deuterium ($^2$H) Balmer H$_\beta$-transition, without and with applied Stark field $U$ to reveal hyperfine components. (Top) Field-free polarization spectrum, with $\delta\nu_L \approx 650$ MHz. (Bottom) Polarization spectrum with applied Stark field of $U = 10$ V·cm$^{-1}$. The relevant transitions are indicated; crossover lines are marked with up and down arrows, stemming from transition line pairs with common upper or lower level, respectively. (Stark splitting data are adapted from Wieman C. and T.W. Hänsch, *Phys. Rev. Lett.* 36, no. 20, (1976): 1170–1173.)

resonance peak broadening. However, when an axial electric field of moderate strength is applied (here $U = 10$ V · cm$^{-1}$), the resulting Stark splitting makes the hyperfine components visible (see the bottom panel of the figure). The spectral resolution was sufficient also to reveal the additional crossover features associated with the separate Stark components.

## 7.4 MULTIPHOTON ABSORPTION PROCESSES

An important and widely used type of nonlinear spectroscopy is multiphoton absorption spectroscopy, a technique in which $n \geq 2$ photons are absorbed simultaneously to reach an excited state from the ground state. Note that now an absorption signal can no longer be recorded directly because of the low probability of the process via "virtual" intermediate levels; rather, successful multiphoton absorption into a final state is detected via fluorescence out of that state, or by ionization/dissociation of the atom or molecule. The lowest-order multiphoton absorption processes are shown schematically in Figure 7.12.

### 7.4.1 Two-photon absorption spectroscopy

The simplest case of multiphoton spectroscopy is that of the *two-photon absorption* (TPA), as depicted in Figure 7.12a, where the two photons are of equal energy. The transition moment $\mu_{if}$ for this process $|i\rangle + 2 \cdot h\nu \rightarrow |f\rangle$ via the intermediate, virtual state $|v\rangle$ was derived already in the 1930s (see Goeppert-Mayer 1931), albeit in slightly different notation, as

$$\mu_{if} = \frac{1}{2} \cdot \sum_v \frac{\langle f|\mu \cdot E|v\rangle \langle v|\mu \cdot E|i\rangle}{E_i - E_v - h\nu}, \tag{7.18}$$

where the sum is over all "virtual" intermediate states $|v\rangle$; $\mu$ is the transition dipole moment; $E$ is the radiation field strength; $E_i$ and $E_f$ are the energies of the initial and final states, respectively; $E_v$ ($\equiv E_i + h\nu$) is the energy of the intermediate (virtual) state; and $\nu$ is the frequency of the radiation that induces the transition. Since the TPA transition moment is a sum of products of two electric dipole moments, one may conclude that only those intermediate states for which the $|i\rangle + h\nu \rightarrow |v\rangle$ and $|v\rangle + h\nu \rightarrow |f\rangle$ transitions are allowed by electric dipole selection rules will contribute to the summation of the TPA transition moment. As a result, it should be noted that the overall TPA selection rules are $\Delta L = 0, \pm 2$ and $\Delta J = 0, \pm 2$, basically the same as those for Raman spectroscopy (see Chapter 11).

In analogy to Beer's law, one may investigate the differential absorption of laser light when traversing a medium of length $\ell$, but taking into account that now two photons have to be absorbed simultaneously,

$$dI(x) = -\sigma^{(2)} \cdot N \cdot I^2 dx, \tag{7.19}$$

where, as usual, $I$ is the laser light intensity and $N$ is the density of absorbing particles; $\sigma^{(2)}$ is the TPA cross section (analogous to the one-photon absorption cross section $\sigma \equiv \sigma^{(1)}$). The TPA cross section is traditionally quoted in the units of Goeppert-Mayer, or GM (after its discoverer); in SI unit, this is equivalent to 1 GM = $10^{-50}$ cm$^4$·s·photon$^{-1}$. The reason for this unit is evident, resulting from

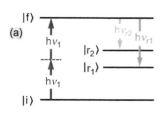

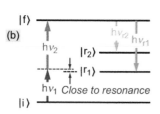

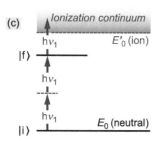

**Figure 7.12**    Examples of multiple-photon absorption processes. (a) One-color TPA. (b) Two-color TPA. (c) One-color (2+1) absorption into the ionization continuum, via the intermediate resonance state $|f\rangle$. Observation of the absorption is indirect by monitoring fluorescence emission from final state $|f\rangle$ (cases a and b) or by recording the ion current (case c).

the product of two cross-section areas (one for each photon, in units of cm$^2$) and an interaction time (within which the two photons must arrive to be able to interact simultaneously with the particle). On integration, Equation 7.19 yields

$$I(\ell) = I_0 \cdot \left[ 1 + \sigma^{(2)} \cdot N \cdot I_0 \cdot \ell \right]^{-1}. \tag{7.20}$$

This is distinctly different from the Beer–Lambert law $I(\ell) = I_0 \cdot \exp(-\sigma^{(1)} \cdot N \cdot \ell)$ and the concepts of absorbance and molar absorption coefficient, as defined in linear absorption spectroscopy (see Section 7.1), are no longer valid.

For example, strong one- and two-photon absorption cross sections may have typical values of $\sigma^{(1)} \cong 2 \times 10^{-17}$ cm$^2$ and $\sigma^{(2)} \cong 2 \times 10^{-48}$ cm$^4$·s·photon$^{-1}$; even for huge laser photon fluxes of the order $2 \times 10^{24}$ photons·cm$^{-2}$·s$^{-1}$ (typically this equates to a laser power of ~$10^6$ W·cm$^{-2}$ for $\lambda_L = 400$ nm), one achieves a possible $\sigma^{(2)} \times I_0 \cong 2 \times 10^{-24}$ cm$^2$, a value almost *seven orders of magnitude smaller than* the one-photon cross section $\sigma^{(1)}$. Thus, in TPA spectroscopy, high laser intensities need to be used, in general, to make the signal detectable above noise and laser intensity fluctuations; in addition, a sensitive detection method has to be utilized, which is also capable of discriminating against one-photon absorption background.

Before the advent of the laser, TPA experiments exploited TPA transitions with photon quanta in the microwave (MW) regime. After lasers became available, in particular IR lasers, TPA experiments used either one MW-photon and one IR-photon, or two laser photons—with the same or different frequencies (as sketched in Figure 7.12a and b). One of the basic and severe limitations of TPA is the very low sensitivity if a direct detection method is used (i.e., the absorbed laser power is measured against the initial, incident laser power). To circumvent this limitation, several indirect methods were developed, as for example the techniques of two-photon induced fluorescence and multiphoton ionization (MPI).

### 7.4.2 Doppler-free TPA

The possibility of recording spectral lines with their Doppler broadening suppressed, by exploiting TPA, was first suggested by Vasilenko et al. (1970). This interesting type of spectroscopy can be better understood with the aid of the sketch shown in the top part of Figure 7.13. Here one considers an arbitrarily moving particle, with velocity component $\upsilon_z$ from left to right, encountering two counter-propagating laser beams with frequency $\nu_{2ph}$.

The molecule will "see" the two photon waves (in co- and counter-propagating directions) having different frequencies $\nu_F$ and $\nu_B$, given by $\nu_F = \nu_{2ph} \cdot (1 - \upsilon_z/c)$ and $\nu_B = \nu_{2ph} \cdot (1 + \upsilon_z/c)$, with $c$ being the speed of light, $\nu_{2ph}$ the laser frequency for the two-photon transition, and $\upsilon_z$ the component of the particle's velocity along the laser beam propagation direction. The following will be observed on tuning the laser frequency $\nu_{2ph}$ (see Figure 7.13 for reference).

As long as $2 \cdot \nu_{2ph} \neq \Delta E_{if}$, the two counter-propagating beams interact with different velocity groups, and hence a two-photon transition is only possible for the absorption of the two photons out of the same laser beam. Thus, when tuning, one would follow the Doppler population, replicating a Doppler profile for the absorption process.

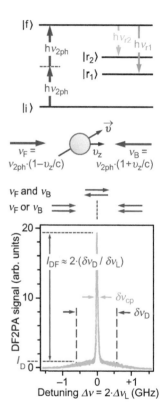

**Figure 7.13** (Top) Concept of Doppler-free TPA spectroscopy in which the absorption of photons from counter-propagating laser beams cancels out the Doppler shift. (Bottom) Line profile of a Doppler-free two-photon signal, exemplified for the transition $5\,^2S_{1/2} \rightarrow 5\,^2D_{5/2}$ in $^{85}$Rb; the observation two-photon transition is via fluorescence decay into $5\,^2P_{3/2}/6\,^2P_{3/2}$. (Data adapted from Ryan, R.E. et al., *J. Opt. Soc. Am. B* 10, no. 9 (1993): 1643–1648.)

As soon as $2 \times v_{2ph} = \Delta E_{if}/h \equiv v_F + v_B$, the two laser beams interact with the same velocity group (within the natural width of the excited, final state $\delta v_{n,f}$). Because of this, now one photon each can be absorbed out of the two counterpropagating beams, since the $-v_z/c$ and $+v_z/c$ terms in $v_F + v_B$ cancel out each other. This then means that the two-photon process becomes velocity independent, and the resulting signal shape follows—in first approximation—the (Doppler-free) Lorentzian line shape associated with $\delta v_{n,f}$. In addition, the on-resonance transition probability is higher than the off-resonance probability, (1) because of the quantum transition-resonance condition, and (2) because of the two-fold increase in selecting the participating photons, i.e., first a photon from the forward beam and then one from the backward beam, or *vice versa*. This means that the intensity of this Doppler-free two-photon signal, $I_{DF}$, can be substantially higher than that for the Doppler-limited signal, $I_D$ (see Figure 7.13). In many cases, it can be approximated by the ratio between the Doppler and natural widths (see, e.g., Biraben et al. 1979; Demtröder 2008)

$$I_{DF} \approx 2 \cdot \delta v_D / \delta v_n. \tag{7.21}$$

For the particular example data shown in Figure 7.13—the transition $5\,{}^2S_{1/2} \rightarrow 5\,{}^2D_{5/2}$ in ${}^{85}$Rb—the observed Doppler-free signal amplitude is $I_{DF} \cong 19$ (see Ryan et al. 1993); this is much less than the value expected from using Equation 7.21. With the Doppler width of the said experiment of $\delta v_D \cong 1.2$ GHz and a natural width for the $5\,{}^2D_{5/2}$ state of $\delta v_n \cong 0.7$ MHz (deduced from the state's lifetime of $\tau = 246$ ns; see Sheng et al. 2008), one would expect an $I_{DF} \cong 3400$. However, the discrepancy is easily explained: it is not the natural width of the upper level, which determines the actual Doppler-free signal width, but in this particular case, the laser linewidth of $\delta v_L \cong 20$ MHz dominates; also, some broadening mechanisms need to be taken into account as well. Thus, Equation 7.21 is somewhat perceptive, and in general, one should utilize the actual Doppler-free signal width $\delta v_s$ (which includes the natural width, the laser linewidth, and additional broadening) instead of the pure natural width $\delta v_n$. On the other hand, for very short-lived states, the natural width may indeed be the dominant contributor.

One additional remark to be made with respect to the aforementioned data is that because of the laser linewidth, the hyperfine splitting of the upper $5\,{}^2D_{5/2}$ level (in the range $\Delta v_{hfs} \sim 3$–$30$ MHz for the two isotopes ${}^{85}$Rb and ${}^{87}$Rb) is not resolved.

Doppler-free TPA spectroscopy effect was first demonstrated in atomic sodium vapor, simultaneously by several groups (Biraben et al. 1974; Levenson and Bloembergen 1974; Hänsch et al. 1974; Pritchard et al. 1974) who measured transitions from the $3\,{}^2S$ ground state to the excited $4\,{}^2D$ or $5\,{}^2S$ level structure. These transitions are only allowed for TPA, but because of the Doppler-free features just discussed, these transitions could be measured with high resolution, revealing fine and hyperfine structure splitting.

### 7.4.3 Multiphoton absorption and molecular dissociation

As stated further above, TPA requires, in general, that the process has to be monitored indirectly because of the low transition probability, and that this is mostly done via recording of the fluorescence emission from the final (upper) level of the transition. As an alternative, an additional photon may be utilized to transfer the atom or molecule from the upper level of the two-photon transition into the ionization continuum, and then measure the resulting ion or electron by a suitable charge-sensitive detection method.

The methodology is widely known as *multiphoton ionization* (MPI), and its most basic implantation is shown in Figure 7.12c, where two photons of the same frequency induce the transition $|i\rangle + 2 \cdot h\nu \rightarrow |f\rangle$ via the virtual intermediate state $|v\rangle$, and a third photon (of the same frequency again) ionizes the excited atom/molecule. This process is denoted *one-color 2+1 MPI*. In this nomenclature, the two digits refer to the number of photons that transfer the particle into the excited eigenstate and the number of those absorbed by the excited eigenstate to result in ionization, respectively. Note that more than two (one) photons may actually be required in the scheme, depending on the actual state energy and the available photon energy. MPI spectroscopy has various advantages with respect to the two-photon fluorescence implementation, specifically because (1) charged particle detection is far more sensitive that photon detection, and (2) the technique does not require that the excited particle exhibit (strong) fluorescence, which would be a particular restriction for tracing metastable states. The first experimental observation of MPI spectra observed was that of benzene (for the excitation laser wavelength range $\lambda_L = 360–450$ nm); the spectra were interpreted as a 2+1 MPI process (see Johnson 1975). MPI studies will be discussed more extensively in Chapter 16.3.

For molecules, a method different from MPI can be utilized, in principle, in which the additional photon results in molecular dissociation rather than in ionization. When investigating organic photochemical mechanisms in the late 1960s and early 1970s, visible luminescence was observed when high-power (IR) laser radiation interacted resonantly with absorption bands of polyatomic molecules, such as ammonia ($NH_3$), boron-trichloride ($BCl_3$), and chloro-trifluoroethane ($C_2F_3Cl$)—to name but a few examples. In these and all other cases, it was found that the luminescence threshold was lower than the optical breakdown (i.e., a plasma with charged particles ensues) threshold; the analysis of the luminescence spectra revealed that they were associated with emission from molecular dissociation products. Indications were that the underlying process could only be due to the simultaneous absorption of (many) IR photons—typically of the order of 20–30, or more—depending on the particular bond dissociation and photon energies. The process is normally called *infrared multiphoton dissociation* (IR-MPD).

The mechanism of IR-MPD is not a simple one; in the multiphoton ladder, one has to distinguish three different excitation regimes, as illustrated in the level scheme in the left-hand part of Figure 7.14.

In the first regime (region I), resonant absorption of a number of IR photons is encountered. Because of the anharmonicity of most molecular potentials, it would be unlikely to maintain photon-resonance conditions when climbing up the vibrational state ladder. The problem can be overcome by exploiting the so-called *rotational compensation*, as shown in the enlarged segment of the level scheme. There, the three photon excitation steps shown involve $\Delta v = +1$ transitions; however, they differ in the rotational energy increment; here $\Delta J = -1$, $\Delta J = 0$, and $\Delta J = +1$ for the first, second, and third photons, respectively.

For higher vibrational states (in the schematic example of the figure, $v \geq 5$) polyatomic molecules exhibit a high density of vibro-rotational states. Together with Doppler broadening, this gives rise to a *quasi-continuum* density of states in region II. For example, the density of states for $SF_6$ is 10 per $cm^{-1}$ at $E_{v,J} \cong 5,000$ $cm^{-1}$ and $10^6$ per $cm^{-1}$ at $E_{v,J} \cong 10,000$ $cm^{-1}$. It is this quasi-continuum density of states that facilitates the simultaneous absorption of photons through the vibro-rotational ladder, ending in the multiphoton dissociation of the excited molecule.

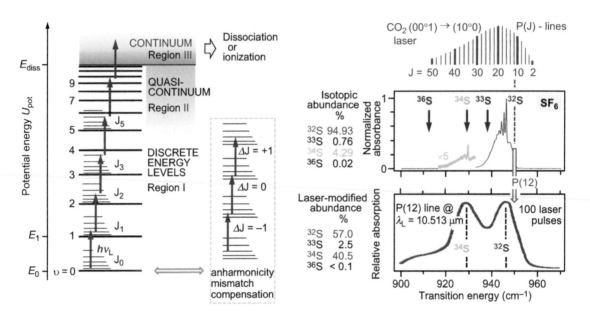

**Figure 7.14** (Left) Energy level diagram illustrating the distinct excitation steps relevant in multiphoton dissociation (ionization) experiments. (Right) Isotopic enrichment of $SF_6$ by multiphoton dissociation, following irradiation with light from individual $CO_2$-laser lines, matching the $3_0^1$ vibrational band of $SF_6$. The displayed absorption spectra were recorded prior to laser irradiation (Top panel: data adapted from Mathi, P. et al., *J. Photochem. Photobiol. A: Chem.* 194, nos. 2–3 (2008): 344–350.) and after irradiation with 100 laser pulses from the P(12)-line of a $CO_2$-laser (bottom panel: data adapted from Ambartsumyan, R.V. et al., *Sov. Phys. JETP* 42, no. 6 (1975): 993–1000. [Russian original version: *Zh. Eksp. Teor. Fiz.* 69, no. 6 (1975): 1956–1970].). The enrichment from the natural-abundance concentrations for the four stable sulfur isotopes is indicated.

Typical applications of IR-MPD are, e.g., the induction of selective chemistry for both fundamental and applied studies in chemical synthesis, or for the separation of isotopes. A pioneering and interesting example of isotope separation by IR-MPD is shown in the right-hand part of Figure 7.14, in which the enrichment of particular sulfur isotopes (in the form of $SF_6$)—following irradiation with light from a pulsed $CO_2$ laser—was demonstrated by Ambartsumyan et al. (1975). In this study, the laser was turned to the rotational $CO_2$-laser line P(12) at $h\nu = 945$ cm$^{-1}$ ($\lambda_L = 10.513$ μm) to induce the resonant absorption of the $SF_6$ ($3_0^1$) vibrational transition (corresponding to the fundamental mode of the $\nu_3$-bending vibration).

The top and bottom spectral traces correspond to the IR-absorption spectrum before and after irradiation, respectively. Note that the before-irradiation spectrum is shown for higher spectral resolution (see Mathi et al. 2008), revealing part of the ro-vibrational band structure; the positions of the Q-branch maxima for the stable isotopes of sulfur are indicated by the arrows; the spectral features for the most abundant isotopes $^{32}$S and $^{34}$S (the natural abundance of all stable isotopes is tabulated next to the spectrum) are shown as blue and green traces, respectively. In addition, the positions of the $CO_2$-laser lines, relevant for this particular $SF_6$ vibrational transition, are included in the figure.

After $CO_2$-laser irradiation (for the data shown here, 100 laser pulses), the two peaks appear with equal intensity; this is due to the selective and significant depletion of the $^{32}SF_6$ isotope by (resonant) IR-MPD, while the peak intensity of the $^{34}SF_6$ remains unchanged. It should be noted that, for the same reason, the previously invisible $^{36}SF_6$-peak can be made out, despite the very small, natural abundance of $^{36}S$ of only 0.017%. The relative isotope enrichments after laser irradiation are tabulated to the left of the bottom spectrum.

## 7.5 KEY PARAMETERS AND EXPERIMENTAL METHODOLOGIES IN ABSORPTION SPECTROSCOPY

When it comes to the practical implementation of absorption spectroscopy, a number of factors need to be considered for optimum performance. In particular, dealing with atoms and molecules and their electronic, vibrational, and rotational transitions, a wide range of wavelengths/energies is encountered, ranging from the deep UV to the MW and even radiofrequency (RF) ranges. Particular ranges require different laser excitation sources, and other bespoke equipment. In addition, since it is desirable to record absorption spectra for very high or ultralow sample concentrations, each application may require a specific methodology to optimize quantitative measurements. Below a short summary for the spectral consideration and key experimental implementations is given.

### 7.5.1 Wavelength regimes

In Figure 7.15, those parts of the electromagnetic spectrum are shown, which are most relevant for atomic and molecular absorption spectroscopy. While fluorescence emission spectroscopy is mostly related to electronic transitions, predominantly in the visible (VIS) part of the spectrum, this is not the case for absorption spectroscopy, which—as shown—extends over a much wider range. Besides the standard scale for laser absorption spectroscopy in wavelength (in units of nanometers), also the most common alternative representations are included, i.e., frequency $v$ (in units of $s^{-1}$, or Hz); photon energy $hv$ (in units of eV); and state energies $\tilde{v}$ (in units of wavenumbers, $cm^{-1}$).

Progressing from the left to the right part of the electromagnetic spectrum, one first encounters part of the RF and MW regions. These spectral segments have been of particular importance to accumulate knowledge about the rotational structure of molecules, aided in particular by absorption spectroscopy. Likewise, the more energetic vibrational structure of molecules has been unraveled by absorption spectroscopy in the far-, mid-, and near-IR spectral regions. In this respect, absorption spectroscopy of vibrational transitions/bands has become the backbone of a plethora of fundamental and applied studies. In absorption spectroscopy, the far- and mid-IR (FIR, MIR) regions are often dubbed the "fingerprint" regions for molecular vibrations, while the near-IR (NIR) is associated with "overtone" (vibrational band) spectroscopy. For a number of selected, but important, small molecules in the gas phase, their ro-vibrational spectra (calculated based on HITRAN 2012 and normalized for each molecule against its most intense peak) have been included in the bottom of Figure 7.15. Such overviews constitute a useful selection tool in choosing the best-suitable absorption laser source. The various IR regions are conveniently accessible by using tunable diode lasers (TDLs; in conjunction with nonlinear conversion techniques) and quantum cascade lasers (QCLs); for specific types and details of these laser sources, see Chapter 5. Some representative surveys of relevant

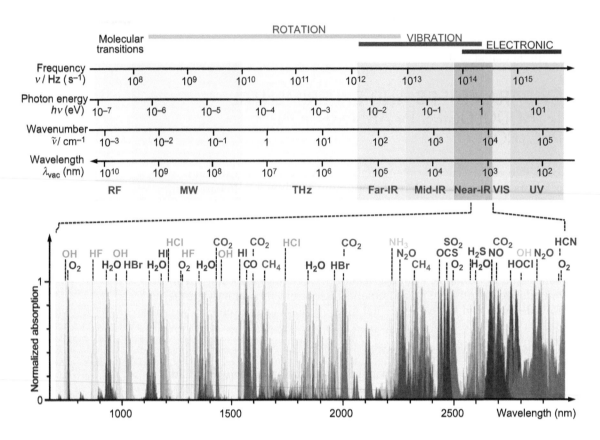

**Figure 7.15**   (Top) Section of the electromagnetic spectrum relevant in laser absorption (and other) spectroscopic techniques. (Bottom) Vibrational-band (fundamental and overtone) absorption features of selected, small molecules in the gas phase. (Band calculations based on HITRAN. "The HITRAN data base." Cambridge, MA, USA: The Harvard-Smithonian Center for Astrophysics, 2012. Available at https://www.cfa.harvard.edu/hitran/.)

molecular vibrations, wavelength ranges, and suitable laser sources are given, e.g., by Kosterev et al. (2008) and Zeller et al. (2010).

The spectroscopy of electronic transitions requires light sources with wavelengths in the visible (VIS) and ultraviolet (UV). In the early days of absorption spectroscopy, high power discharge lamps were used, in particular to provide narrow-bandwidth sources based on atomic emission lines generated in gas discharges. It was only after the advent of tunable laser sources in these wavelength regimes that the structure of excited molecular states could be investigated systematically.

Finally, particular interest is now emerging in a spectral regime, called *terahertz spectroscopy* (1 THz = $10^{12}$ Hz). It refers to a wavelength regime between ~0.03 mm (the long-wavelength edge of the far-IR) and ~3 mm (the EHF band); in various definitions, it is given as the range 0.1–10 THz, although the range is frequently quoted with different values, depending on specific equipment or applications. This type of spectroscopy has only been developed over the last few years but is rapidly gaining importance in a diverse range of important applications (see, e.g., Pawar et al. 2013).

## 7.5.2 Spectral resolving power

In the previous sections, a statement encountered a few times was that closely spaced spectral features could be resolved, or not. In general, spectral resolution is a key concept in any spectroscopy method. Its definition is often somewhat vague; for example, in the IUPAC Gold Book, one reads that spectral resolution is "...the wavenumber, wavelength or frequency difference of two distinguishable lines in a spectrum" (see Znamenáček 2014). However, in general, a more quantitative description would be desired as to when two overlapping lines can be considered as "resolved."

Since the early days of spectroscopy, imaging spectrographs were used (and are still very much so today); quantitative criteria for resolving spectral features are based on the assumption that such instruments are used, i.e., an entrance slit is imaged onto an exit slit via a wavelength-selective element. In this context, it is the so-called *Rayleigh criterion*, which is generally accepted for the minimum resolvable spectral detail. It was introduced by Lord Rayleigh in 1879 as a measure to determine whether two diffraction spots were distinguishable or not: two intensity maxima are separated if the maximum value of the one imaged spot (or spectral line) coincides with the first minimum of the second imaged spot (or spectral line). Quantitatively, the problem can be formulated for two diffraction spectral line profiles $I_{v1}(x)$ and $I_{v2}(x)$ of equal intensity, along an imaging axis $x$. From diffraction theory, one finds for the "dip" between the two pattern maxima (see Figure 7.16) that

$$\min\{I_{v1}(x) + I_{v2}(x)\} = (8/\pi^2) \cdot \max\{I_{v1}(x)\} \cong 0.81 \cdot \max\{I_{v1}(x)\}. \quad (7.22)$$

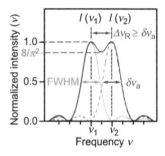

**Figure 7.16** Spectral resolving power, based on the Rayleigh criterion.

As an alternative, the full width at half-maximum (FWHM) criterion is often used; it is also known as the *Taylor criterion*. Originally, it had been devised to determine the minimum resolvable frequency increment of an interferometer; it states that two spectral lines of equal intensity are resolved if the sum of the two individual lines at the midway point is at most equal to the intensity of one of the individual contributing lines.

In practice, both Rayleigh and Taylor criteria give very similar results; some digital simulation procedures provide intuitive and understandable insight into the meaning of spectral resolution (see, e.g., Juvells et al. 2006).

It should be noted that according to condition 7.22, the distance between the two lines has to be larger than their FWHM, i.e., $\Delta v_R \geq \delta v_a$. In principle, $\Delta v_R$ can be calculated from diffraction theory in case of the Rayleigh criterion, and from line convolution in case of the Taylor criterion (if one were using the same 0.81-amplitude value).

## 7.5.3 Experimental methodologies

When it comes to absorption spectroscopy, one encounters a rich variety of experimental methodologies. Two key criteria may be used for their classification:

1. Whether they use *broadband* or *wavelength-tuned* techniques to record the spectrum

2. Whether the method is *direct* (i.e., one measures the depleted photon intensity due to the sample absorption) or *indirect* (i.e., the observable is not

the decrease in photon intensity but a different property, induced by the photon absorption, such as fluorescence emission; or sample-heating leading to photothermal spectroscopy [PTS]).

A selection of the most common absorption spectroscopic techniques is briefly summarized next, and examples for their practical use are given in Chapter 8.

*Methodologies based on broadband sources.* Traditionally, prior to the advent of lasers, classical absorption spectroscopy has been carried out using broadband light sources, like incandescent tungsten lamps. Three different concepts were used: (1) wavelength-swept light sources, i.e., the wavelength was selected prior to passing through the sample; (2) Fourier transform methodologies, in particular in the IR—hence FTIR, i.e., the spectrum was deduced from the broadband interferogram of the light transmitted through the sample; and (3) dispersive spectrometer solutions, i.e., the complete absorption is dispersed after light passage through the sample. In particular, in commercial, analytical absorption spectroscopy, all three techniques are still going strong; the specific methodology used depends on the actual application. Since a few years ago, a particular type of laser broadband source has been challenging broadband lamp sources in the NIR, namely supercontinuum sources generated by picosecond- or femtosecond-laser pulses (see the example in Section 7.5).

*Absorption spectroscopy using frequency combs.* Frequency combs are a phenomenon associated with ultrashort picosecond and femtosecond pulses. In the frequency domain, one can show that coupling the longitudinal modes of the pulsed laser (mode-locking) results in an evenly spaced comb of spectral lines; the separation between two modes, or comb lines, is just equal to the laser repetition frequency, $v_r$. This regular structure may serve as a "precision frequency ruler," and indeed this property has now become a standard tool in precision spectroscopy and optical frequency metrology. Probably the earliest successful experiment in this context is related to the measurement fine- and hyperfine-structure frequency intervals in atomic sodium (see Eckstein et al. 1978). This optical frequency metrology approach culminates in the measurement of the 1S–2S transition frequency of the hydrogen atom, which today is the transition known with the highest precision, approaching $10^{-15}$ (for the latest results and the historic evolution, see Matveev et al. 2013). It should be noted that all modern scientific and commercial realization and exploitation of frequency combs are in one way or another linked to the famous six-page proposal by T.W. Hänsch, dated March 30, 1997, for an "octave-spanning self-referenced universal optical-frequency comb synthesizer." In short, the underlying idea is that the frequency spectrum of femtosecond-laser oscillators can be broadened in a nonlinear optical medium, in the best cases spanning more than an optical octave without destroying the integrity of the comb lines. For details on frequency-comb generation and applications, see, e.g., the textbook by Ye and Cundiff (2005).

In a certain way, the aforementioned frequency comb constitutes a broadband light source, which in contrast to a typical continuous-frequency source exhibits a spectrum with precisely known, equally spaced frequency components. Nowadays, frequency combs can be generated to cover spectral segments ranging from the UV to the far-IR, exploiting broadening by continuum generation and nonlinear frequency conversion (sum and difference frequency generation as well as parametric processes). Individual spectral intervals extend over tens to hundreds of nanometers, with spacing between individual comb frequencies of the order

100 MHz. Thus, a laser frequency comb may be viewed to resemble a multimode laser with narrow-linewidth modes at known optical frequencies; consequently, basically all methods of laser spectroscopy can be applied but with the advantage of automatic, absolute frequency calibration.

The application of such a broadband laser frequency comb to absorption spectroscopy is shown schematically in the upper part of Figure 7.17; the approach is commonly known as *direct frequency-comb spectroscopy* (DFCS). In principle, all spectroscopic analysis techniques may be applied and reveal the absorption of frequency-comb components on passage through a sample (see, e.g., Foltynowicz et al. 2011). Two methodologies dominate: spectral dispersion and Fourier-transform approaches. In the former case, normally a setup incorporating a virtually imaged phased array (VIPA) and a grating is used, in which the comb modes are spatially separated in two dimensions and imaged on a detector array. In the latter case, standard Fourier transform spectrometers based on a Michelson interferometer are the norm; an example for a measurement based on this methodology is shown in the bottom part of the figure. Note that increasingly so-called *dual-DFCS* is exploited in precision molecular spectroscopy (see Coddington et al. 2016). This is analogous to traditional Fourier-transform spectroscopy but utilizing two frequency combs with slightly different repetition rates, which mimics the effect of a fast-scanning delay stage. The two combs are heterodyned against each other on a photodiode; the optical spectrum, including sample absorption, is then obtained by Fourier-transforming the time-domain signal provided by the photodiode.

A selection of examples for the application of frequency combs in high-precision spectroscopy will be discussed in Chapter 8.2.

*Absorption spectroscopy using tunable diode and quantum-cascade laser sources.* For any of the high-resolution absorption spectroscopic techniques described earlier in this chapter, nearly any type of laser has been utilized, depending on specific wavelength required for probing the atom or molecule under investigation. By and large, today, solid-state and semiconductor-based laser sources dominate the field, and in particular in routine analytical applications, (tunable) semiconductor diode laser (DL) and QCL sources are the systems of choice, because of their ease of use and relatively moderate price tags.

Because of the narrow bandwidth and normally wide tenability, external-cavity DLs (ECDLs; see Chapter 5.1) have come to great prominence over recent years. Wavelengths being utilized in experiments range from the deep UV (then including nonlinear conversion techniques as well) to the FIR (e.g., lead-salt DLs); a few examples have been given earlier in this chapter. In particular, for quantitative gas analysis and process control, these laser sources have gained wide acceptance, specifically in the form of TDL absorption spectroscopy (TDLAS); this will be discussed in more detail in Chapter 8.3.

QCLs are now available for the majority of wavelengths in the MIR and FIR, and because they are widely customizable and can be operated at room temperature (see Chapter 5.2), they begin to oust many of the traditional laser sources in those wavelength regimes. Furthermore, certain wide-bandwidth devices (now with gain widths substantially larger than 500 $cm^{-1}$, covering the range $\lambda_L \sim 3$–$20$ μm) can be rapidly wavelength-modulated, and thus are suitable, in principle, to be applied in standard absorption methodologies, which traditionally were the domain of tungsten broadband sources.

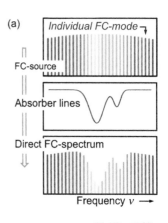

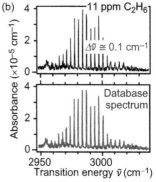

**Figure 7.17** Absorption spectroscopy based on direct frequency combs (DFCS). (a) General concept of DFCS. (b) High-resolution DFC spectrum of trace amounts of ethane (11 ppm in 790 mbar of $N_2$), utilizing an OPO-idler frequency comb centered at $\tilde{v} = 3000$ $cm^{-1}$. (Data adapted from Adler, F. et al., *Opt. Express* 18, no. 21 (2010): 21861–21872.)

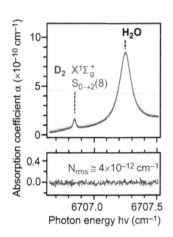

**Figure 7.18** Detection of the weak electric quadrupole transition of the S(J=8) line of the first overtone band $v''=0–v'=2$ of $D_2$, using cw-CRDS. The residuals of the data fit are close to the noise-equivalent absorption (lower panel). (Data adapted from Kassi, S., *J. Chem. Phys.* 136, no. 18 (2012): article 184309.)

*Cavity-enhancement techniques.* When it comes to ultrasensitive detection, i.e., the measurement of very low concentrations of atoms and molecules, standard absorption techniques are often insufficient: the signal might be smaller than the inherent laser intensity fluctuations or background noise. Referring back to Beer's absorption equation (Equation 7.1), lower particle numbers can be compensated for by increasing the absorption length. Conventionally, in absorption spectroscopy, this has been done by the aid of multiple-path cells; a multitude of configurations has been used, including the well-known White and Herriott cells, with effective absorption lengths exceeding 100 m (for a concise review, see Robert 2007).

In absorption spectroscopy using laser radiation, high-finesse cavities have been and are being exploited. A highly collimated beam is reflected between two mirrors (typically with reflectivities exceeding $R \geq 0.999$), with a small amount of radiation leaking through the mirrors at each reflection. One measures the signal decay time, as a laser pulse emerges through one of the mirrors after successive reflections; when a barbing species in the cavity is in resonance with the laser frequency, one observes an increase in the rate of decay as a result of attenuation by absorption. By comparing the cavity decay times without and with absorption, one can extract the concentration of the absorber inside the cavity. The method is known as *cavity ring-down spectroscopy* (CRDS); it was first introduced by O'Keefe and Deacon (1988).

The effective optical path lengths that have been achieved are as high as a few kilometers, while maintaining tabletop dimensions and relatively small sampling volumes. Utilizing CRDS detection limits of gas-phase molecules down to ppb and even ppt levels has been achieved. It also is worth mentioning that not only is it trace concentrations that benefit from the sensitivity of CRDS, but also it is feasible to measure absorption signals with very low small transition probabilities (or cross sections $\sigma_a$), like quadrupole transitions; for an example, see Figure 7.18. Knowing the particle concentration and the cavity properties, one is even able to derive absorption coefficients and cross sections; for the particular example shown in the figure, the S(J = 8) line of the first overtone band $v'' = 0 – v' = 2$ of $D_2$, the line intensity is of the order of $1.8 \times 10^{-31}$ cm · molecule$^{-1}$, one of the weakest transitions ever detected in laboratory experiments at room temperature.

A variant to CRDS was developed in 1998 independently by two different groups (O'Keefe and Deacon 1998; Engeln et al. 1998), known as *integrated cavity output spectroscopy* (ICOS) and *cavity enhanced absorption spectroscopy* (CEAS). In this approach, the absorption spectra are obtained via integration of the total light signal transmitted through the high-finesse cavity, e.g., by oscillating the cavity length. The transmission signal is measured with and without the absorbing species, and from the difference of the signals, the absorption coefficient $\alpha$ is extracted, and—if the absorption cross section is known—the concentration can be deduced, in principle. Note, however, that the cavity transmission depends on several factors other than the absorption itself, mostly related to the laser behavior (including linewidth, stability, and mode structure). Only if the laser effects are eliminated from the transmission signal is it possible to directly convert it to absorption coefficients. For this, a variant of CEAS is used, namely *optical-feedback CEAS* (OF-CEAS); see, e.g., Salter et al. (2012).

*Terahertz spectroscopy.* Most molecular species exhibit strong absorption with frequency-specific (rotational) line patterns in the terahertz regime (1 THz $\equiv 10^{12}$ Hz;

see Figure 7.15). Equally, several of the relevant states in many-body systems (including liquids and solid-state specimen) have transition energy differences, which match with the energy of terahertz photons. However, due to the lack of suitable sources and detectors, the terahertz regime has for a long time been inaccessible to direct measurements and was known as the *terahertz gap*. This changed dramatically with the advancement of ultrafast laser and detector technology since the mid-1980s. The particular technique that has reached great prominence in research and practical applications is based on coherent and time-resolved detection of the electric field of ultrashort radiation bursts in the far-IR; it has become known as *terahertz time-domain spectroscopy* (THz-TDS). The terahertz-pulse generation and detection are sensitive to both amplitude and phase of the sample's response to this terahertz radiation; thus, it can provide additional information (e.g., information on the refractive index of the sample; or on conformational change in the physical structure of the molecular complex) in comparison to conventional Fourier-transform spectroscopy, which is only sensitive to the response amplitude. More detailed information on key aspects of terahertz spectroscopy and its applications may be found, e.g., in Jepsen et al. (2011) or Peiponen et al. (2013).

Terahertz radiation can be generated most efficiently and economically by using optoelectronics. When focusing NIR laser light onto a suitable semiconductor, a photocurrent is induced; this then serves as the source for terahertz waves. Both continuous-wave sources are now available; the former produce narrow-band-width radiation (for high-resolution spectroscopy), while the latter convert femtosecond-laser pulses into picosecond-pulse duration radiation (which allows one to monitor and control ultrafast processes). It should be noted that for frequencies $v \geq 1$ THz, also quantum-cascade laser sources can be used, although these need to be cooled to $T = 40$–$80$ K.

Recall that in optical spectroscopy, one typically measures the light intensity rather than the electromagnetic field, since no detectors exist that can directly respond to it in the optical frequency range (see Chapters 6.2 and 6.3). On the other hand, there are quite a few techniques suitable to directly measure the time evolution of $E_{THz}(t)$. Common detection schemes for pulsed THz-TDS include photoconductive antenna sampling and electro-optical sampling, both in pump-probe configurations. Continuous-wave terahertz detection utilizes thermal detectors (like bolometers); semiconductor devices; electro-optic setups; and heterodyning with a quantum-cascade laser as the local oscillator. Of course, because of the low frequencies, cooling of the devices is indispensable.

Over recent years, terahertz spectroscopy has become an invaluable tool in numerous fields, equally in scientific and practical applications, as well as imaging. For example, many molecules provide fingerprint spectroscopic lines in the terahertz regime and spectroscopy in gaseous samples (e.g., of importance for trace gas analysis or specimen of astrophysical interest—for an example, see Figure 7.19); in liquid phase environments (e.g., libration modes of water and other liquids, or hydration shells around amino acids, etc.); and in solids (e.g., charge carrier dynamics in semiconductors and nanomaterials). Examples for some of these applications are given in Chapter 8.5, together with key aspects of terahertz-radiation generation and detection.

Finally, terahertz radiation is now also exploited in a range of imaging applications including medicine, security screening, nondestructive industrial testing,

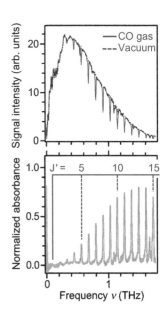

**Figure 7.19** Terahertz spectroscopy of CO (X $^1\Sigma^+$;$v$=0). (Top) Transmission signal with and without CO gas in the absorption cell. (Bottom) Normalized absorbance; the rotational R-lines are annotated with their quantum number $J'$. (Data adapted from Deng, Y. et al., *Metrologia* 51, no. 1 (2014): 18–24.)

and scientific imaging in chemistry and biochemistry. Some representative examples are included in Chapter 21.

*Photoacoustic and photothermal spectroscopies.* All methodologies described thus far all base the detection of the absorption process on measuring photons, either recording the "missing-photon" signal or the photons that are emitted subsequent to the absorption step. However, it is also possible to exploit indirect effects. The absorption of a photon means the deposit of energy in the sample. Therefore, if the sample is irradiated intermittently, it undergoes periodic heating (with light "on") and cooling (with light "off"), which in turn generates "pressure fluctuations," commonly in the acoustic frequency range. The phenomenon is widely known as the photoacoustic effect, which is known since the 1880s. In principle, any type of sample material can be probed in this way—be it gaseous, liquid, or solid in nature, and irrespective of whether it is translucent, opaque, or nontransparent to the probe light. When changing the wavelength of the exciting (laser) light, different compounds in the sample will absorb the incident photons, and hence the resulting wavelength-dependent response constitutes *photoacoustic spectroscopy* (PAS). One also encounters the description *photothermal spectroscopy* (PTS); conceptually, it is no different from PAS, but most commonly, today one associates PAS with gas analysis and PTS with the probing of solid samples and interfaces (in particular, when it comes to imaging applications, both in lateral and depth dimensions). Note that the detected signals depend on two key parameters; they are proportional to the thermal diffusion length, $L_{th}$, and inversely proportional to the absorption coefficient, $1/\alpha$. Details on the methodologies and common applications can be found in the textbook by Michaelian (2010).

Any sample for PAS/PTS measurements is normally contained in a small, well-sealed cell, which should be isolated as much as possible from potentially interfering sources of acoustic waves within the range of the modulation frequency of the probe laser beam. In general, by making the cell volume as small as possible, the measurement sensitivity increases. The PAS/PTS response signal is recorded using a high-sensitivity microphone or piezoelectric transducer devices. In principle, PAS/PTS is based on the same basic concepts as conventional absorption spectroscopy; however, there are some important differences. Notably, the absorption (which is proportional to the concentration) is measured directly and not relative to a background (the laser intensity itself), and thus measurements based on this technique are in general highly accurate and very sensitive.

It is worth noting that while PAS/PTS are not necessarily seen as a mainstream technology in absorption spectroscopy, specific applications have greatly benefitted from the advancement and innovation in cell configurations and detection methodologies. Two specific fields are (1) trace-gas analysis exploiting quartz-enhanced cavity configurations—the approach is often addressed as *quartz-enhances PAS*, QE-PAS (see, e.g., the review by Patimisco et al. 2014); and (2) biomedical studies and specifically the analysis of single cells in the form of *photoacoustic microscopy* (PAM), providing insight into some of their anatomical, biomechanical, and functional properties (see, e.g., the recent review by Strohm et al. 2016).

Selected examples related to the spectroscopy and imaging exploiting photoacoustic principles are included in Chapters 8.6 and 21.1/21.2, respectively.

# 7.6 BREAKTHROUGHS AND THE CUTTING EDGE

Absorption spectroscopy is the most "fundamental" type of spectroscopy, often being the first stepping stone for other techniques, like fluorescence spectroscopy. The advent of narrow-bandwidth, tunable lasers in particular gave rise to key developments beyond basic absorption spectroscopy. For example, Doppler-free saturated absorption introduced the capability of unprecedented resolution to reveal spectral details such as fine and hyperfine structure. Many more milestones stand out over the past decades, along the development of new laser sources or novel experimental techniques.

As one of the more recent breakthroughs, one might see the introduction of picosecond/femtosecond-laser-based supercontinuum (SC) sources (see Chapter 4.5) to "traditional" (nonlaser) absorption spectroscopy.

## 7.6.1 Breakthrough: Absorption spectroscopy utilizing SC sources

Broadband femtosecond (or picosecond) SC sources are commercially available since about 2005/2006, with spectral widths well exceeding 1000nm in the near-IR (up to about 2400nm), and also including the visible part of the spectrum. One of the major advantages of SC sources is their far superior spectral power density. Admittedly, the vibrational-fingerprint regions for molecules in the MIR and FIR are not covered, but many gaseous species of interest for chemical processes or atmospheric applications have vibrational overtone transitions in the NIR (see Figure 7.15). All typical spectroscopic setups for classical absorption spectroscopy have been utilized, including wavelength-swept, FTIR, and dispersive spectrometer setups (see Hult et al. 2007; Mandon et al. 2008; Werblinski et al. 2016).

Particularly intriguing is the approach by Hult et al. who exploited a so-called dispersed SC source. For this, they launched the SC (pulse) radiation into a dispersion compensating module with large negative dispersion, which temporally dispersed the light at a rate of 1–2 ns · nm$^{-1}$. Using an ultrafast photodiode and a high-speed signal digitization system, with a combined bandwidth of 8–10 GHz, the full spectrum of each SC pulse is swept within a few microseconds, with spectral resolution of about $\Delta\lambda \approx 0.04$ nm (this is comparable to the width of molecular spectral lines at atmospheric pressure). Absorption spectra could be recorded and accumulated at repetition rates exceeding 1 MHz, allowing for real-time monitoring on the millisecond timescale. As an example, part of a broadband spectrum for the $2\nu_3$ overtone band of $CH_4$ is shown in Figure 7.20; also included are spectra of the same region, recorded using a standard FTIR spectrometer, and a TDLAS setup. Clearly, the absorption spectra based on the dispersed SC source compare quite favorably, in particular when taking into account the combined spectral coverage, spectral resolution, and recording speed.

## 7.6.2 At the cutting edge: Precision laser spectroscopy of hydrogen: Challenging QED?

The spectroscopy of hydrogen and related hydrogenic species has played a key role in the understanding of fundamental physics. In particular, the refinement of Doppler-free (two-photon) spectroscopy in conjunction with ultra-narrow and stable laser sources has led to absolute frequency measurement accuracy for

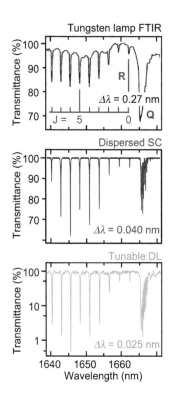

**Figure 7.20** Segment of the absorption spectrum of the $2\nu_3$-band of $CH_4$. (Top) Standard FTIR spectrum using a tungsten lamp source; absorption cell length $L = 10$ cm; 1000 mbar (pure) $CH_4$. (Middle) Absorption spectrum using an SC source; absorption cell length $L = 10.5$ cm; 197 mbar $CH_4$ in air. (Data adapted from Hult, J. et al., *Opt. Express* 15, no. 18 (2007): 11385–11395.) (Bottom) TDLAS spectrum (for comparison); absorption capillary length $L = 5$ m; ~19 mbar $CH_4$ in air. (Data adapted from Cubillas, A.M. et al., *Sensors* 2009, no. 9 (2009): 6261–6272.)

hydrogen transitions to become benchmarks for quantum electrodynamics (QEDs) or the standard model. For example, spectroscopy of the H(1S-2S) transition has reached uncertainties of just a few $10^{-15}$ (see, e.g., Matveev et al. 2013). It is worth noting that high-precision spectroscopy of the 1S-2S transition in anti-hydrogen has been realized as well, with a relative precision of about $2 \times 10^{-10}$ (see Ahmadi et al. 2017); experiments aiming at even better precision are already under way. These studies may be seen in the light of the CPT (charge conjugation, parity reversal and time reversal) theorem, a cornerstone of the Standard Model, which requires that hydrogen and anti-hydrogen have the same spectrum. Together with other hydrogen transition measurements, precise values for fundamental constants like the Rydberg constant and the proton charge radius can be derived (see the most recent CODATA results; Mohr et al. 2012).

Currently, one of the persisting discrepancies related to QED is the determination of the proton charge radius. Comparing spectroscopic data from muonic hydrogen ($\mu$-p) and normal hydrogen (e-p), the analysis reveals a $4\sigma$ discrepancy between the $\mu$-p results and the hydrogen mean value from 11 different transitions (see Antognini et al. 2013; Beyer et al. 2013). This discrepancy is commonly addressed as "the proton radius puzzle." Proton charge radius values, derived from a range of one- and two-photon transition data, are shown in Figure 7.21. Because the muon has a mass about 200 times heavier than the electron, the $\mu$-p uncertainties are substantially smaller than for the e-p (hydrogen) system. However, it is only the average of all hydrogen laser spectroscopy measurements, together with data from hydrogen Lamb shift and electron–proton scattering, which shows the aforementioned discrepancy: individual hydrogen transition data were not precise enough to determine whether the discrepancy is real or not (note the error bars in the figure, overlapping with $\mu$-p). For the transitions marked with $\leftrightarrow$ in the figure, efforts are well under way to improve the statistical errors sufficiently to alleviate this. While successful in principle (see Matveev et al. 2013; Galtier et al. 2015), at the time of writing, no conclusive, final results had been published yet.

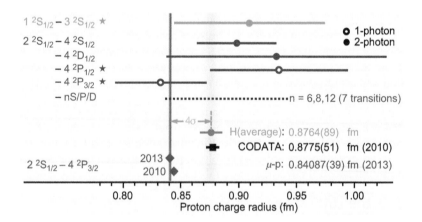

**Figure 7.21**  Determination of the rms charge radius of the proton, from laser-spectroscopic data of hydrogen (data points in green and blue) and myonic hydrogen, $\mu$-p (data points in red); the blue-shaded bar indicates the average of all laser-spectroscopic hydrogen measurements. For comparison, the CODATA-2010 value is included, which includes hydrogen Lamb shift and electron–proton scattering data. (Adapted from Beyer, A. et al., *J. Phys. Conf. Ser.* 467 (2013): article 012003.)

# CHAPTER 8

# Selected Applications of Absorption Spectroscopy

With the maturing of the laser into commercial devices in the 1970s—both fixed-frequency and tunable laser sources—absorption spectroscopy experienced an explosion of new methodologies, and a wealth of applications to fundamental and applied analytical problems evolved. Absorption spectroscopy benefitted equally from the enhanced sensitivity and the increased spectral resolution laser sources offered. Now nearly 50 years have passed since the first implementation of laser absorption spectroscopy, and one might think that such a mature field of spectroscopy would therefore have well passed its cusp of novelty. However, the opposite seems to be the case. A number of new developments in laser technology and experimental implementation have begun to make inroads, even into well-established and commercially exploited pre-laser methodologies, and exciting new applications are emerging that promise to expand the boundaries of fundamental physical and chemical understanding.

In Section 7.5, a brief summary of now common but still evolving methodologies of laser absorption spectroscopy was given. In this chapter, more extensive details on those methodologies and their implementation are given, and a range of representative examples are described. These are meant to highlight the versatility and maturity of laser absorption spectroscopy on the one hand, and provide a flavor of vibrant, cutting-edge developments on the other.

As already pointed out in Chapter 7, (laser) absorption spectroscopy offers a rich variety of experimental implementations, which utilize either *broadband* or *wavelength-tuned* laser sources in the recording of spectra. By and large, three basic concepts are used:

1. *Wavelength-swept light sources*, i.e., the wavelength is selected prior to passing through the sample, either by wavelength tuning or wavelength filtering, depending on the specific laser source

2. *Fourier transform methodologies*, i.e., the spectrum is deduced from the transform of an interferogram of the light transmitted through the sample

3. *Dispersive spectrometer solutions*, i.e., the overall absorption spectrum is dispersed after light passage through the sample

The key points of the various techniques discussed here are summarized in Table 8.1, together with information on spectral coverage and resolution, and detection sensitivity.

**Table 8.1**    Main Implementations of Laser Absorption Spectroscopy

| Methodology | Remarks | Spectral range $\Delta\tilde{v}$ Resolution $\delta v$ Sensitivity $\Delta N$ |
|---|---|---|
| Broadband source ("static") | SC laser sources—mainly based on fiber lasers; wavelength coverage $\lambda \sim 0.4 \ldots 2.3$ μm<br>Requires a spectrometer for recording | 500 ... 10,000 cm$^{-1}$<br>1 ... 100 GHz<br>10 ... 1000 ppm |
| Broadband source (wavelength swept) | SC laser sources—mainly based on fiber lasers; wavelength coverage $\lambda \sim 0.4 \ldots 2.3$ μm<br>Selection/tuning by acousto-optic filter | 500 ... 10,000 cm$^{-1}$<br>1000 ... 2000 GHz<br>10 ... 1000 ppm |
| Tunable laser source | Diode laser and QCL sources; wavelength coverage $\lambda \sim 0.2 \ldots 16$ μm<br>Resolution depends on laser cavity type | 1 ... 500 cm$^{-1}$<br>0.01 ... 3000 MHz<br>$10^{-5}$ ... 1000 ppm |
| Fourier transform setups | Wavelength coverage $\lambda \sim 1 \ldots 16$ μm<br>Requires broadband (laser) source, static or rapidly wavelength swept<br>Resolution depends of Michelson travel | 100 ... 10,000 cm$^{-1}$<br>3 ... 300 GHz<br>1 ... 1000 ppb |
| Dispersive spectrometer setups | Scanning or static spectrographs; wavelength coverage $\lambda \sim 0.2 \ldots 16$ μm<br>Resolution depends on spectrometer length and grating dispersion | 1000 ... 10,000 cm$^{-1}$<br>3 ... 300 GHz<br>1 ... 1000 ppb |

*Note:*    Key features are indicated in the middle column. Typical values for wavelength coverage $\lambda$, spectral range $\Delta\tilde{v}$, spectral resolution $\delta v$, and particle density detection sensitivities $\Delta N$ are included. Note that laser frequency-comb sources are not included (for details, see Section 8.2).

# 8.1 BASIC METHODOLOGIES BASED ON BROADBAND SOURCES

The key ingredients of any basic absorption spectrometer are a light source, the sample, a dispersive element (e.g., a monochromator that can be located before or after the sample), and a detector (single-element or array). Before the development of lasers, light sources were typically continuum sources, and the measured spectra were categorized as falling into the class of *broadband absorption spectroscopy* (BB-AS).

In general, BB-AS has been proven to be a useful technique for the determination of absolute particle densities of species encountered in gas samples, plasmas, open-field atmospheres, liquids, and transparent solid media. This spectroscopic technique provides absorption rates down to $10^{-4}$, which makes possible the absolute measurement of absorption coefficients in the range of $10^{-5}$ to $10^{-6}$. Typical light sources used in conventional BB-AS are high-pressure Xe-arc lamps (dominant spectral content in the UV), light-emitting diodes, or incandescent tungsten filaments (dominant spectral content in the VIS, near-IR, and mid-IR), sources that all require a high degree of stability to guarantee an acceptable signal-to-noise ratio (SNR).

In recent years, a very significant improvement has taken place with the incorporation of so-called supercontinuum (SC) radiation sources, realized by coupling intense laser pulses into a photonic-crystal fiber (PCF), as outlined in Chapter 4.5.

In many fields of application, a spectroscopic technique is required that offers broad spectral coverage in order to perform multicomponent analysis. Traditionally, this basic requirement has been achieved using thermal light sources in conjunction with *Fourier transform infrared spectroscopy* (FTIR) and *nondispersive infrared spectroscopy* (NDIR). These techniques, in particular FTIR, have been—and still are—the workhorses of chemical analysis in multicomponent mixtures and molecular identification through vibrational-mode analysis; however, they suffer from important drawbacks, in particular when minute absorption is encountered, or if *in situ* absorption diagnostic is needed.

## 8.1.1 BB-AS utilizing SC sources

In recent years, a very significant improvement has taken place with the development of a new type of (laser) light source, called supercontinuum (SC) sources, sometimes also known as white-light laser sources. These sources have demonstrated to offer most of the advantages of the traditional broadband sources on the one hand (with the exception of wavelength coverage beyond $\lambda \geq 2500$ nm), and those of the narrow-linewidth laser sources on the other hand. As already mentioned earlier in this chapter, light from a short (picosecond- or femtosecond-) pulse duration, high-repetition rate, and high-power laser is focused into a PCF, in which new wavelengths are generated via nonlinear processes, resulting in broadening of the incident pulse spectrum; this broadened output can span wavelength ranges that are commonly broader than one octave (see Chapter 4.5) As a consequence, the repeatability and stability of the pulses guarantees, in conjunction with pulse averaging, that absorption spectra can be acquired over very wide wavelength intervals, ranging from the UV to the mid-IR (see, e.g., Werblinski et al. 2016); an example for an absorption spectrum using an SC source was shown earlier in Chapter 7.6.

Since the SC-source output is intense and highly coherent, the use of *SC absorption spectroscopy* (SC-AS) basically combines BB-AS with high spectral resolution and good limits of detection (LODs). In Figure 8.1, these three features of broadband SC-AS are illustrated.

In the top panel of Figure 8.1a, the complete transmission profile of the SC source utilized by Yoo et al. (2016) is shown, after passing through a cell containing water vapor—over the wavelength segment $\Delta\lambda \sim 1000$–1400 nm. Note that the vibrational band positions for $CH_4$, $C_2H_2$, and $C_2H_4$ are indicated, which were later used by the authors in their study to measure the content of gas mixtures, at various pressures and relative concentrations. The normalized integrated absorbance values for these three hydrocarbon molecules, measured at 1 bar and 295 K and for different gas mixtures, are depicted in the bottom panel of the figure. From these data, the authors estimated LODs for $CH_4$, $C_2H_2$, and $C_2H_4$, at room temperature and atmospheric pressure, of 0.1%, 0.09%, and 0.17%, respectively.

In Figure 8.1b, an example for ultrasensitive SC-AS is shown. Here, the first use of SC-radiation for broadband trace gas detection is demonstrated, in which cavity-enhanced absorption was realized (Langridge et al. 2008), a technique whose acronym is SC-CEAS. The shown spectrum was measured for a gas sample containing $NO_3$ at a concentration of $c = 38$ pptv; the spectrum was acquired at a resolution of $\delta\lambda = 0.1$ nm FWHM, for a signal integration time of 2 s. From this, the authors estimated a LOD = 3 pptv for said data acquisition time.

These results should be compared with an earlier measurement of the same molecule by Venables et al. (2006); they utilized an incoherent broadband source in conjunction with cavity enhanced absorption spectroscopy (IBB-CEAS) for *in situ* detection of $NO_3$ in an atmospheric gas mixture. The authors deduced

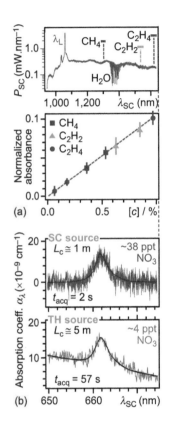

**Figure 8.1** Three aspects of BB-AS using a SC source. (a) Wavelength coverage interval $\Delta\nu$ and multispecies analysis. (Data adapted from Werblinski, T. et al., *Opt. Lett.* 41, no. 10 (2016): 2322–2325.) (b) Spectral resolution $\delta\nu$ and molecule detection sensitivity $\delta N$, exemplified for the radical $NO_3$ in ambient air. (Data adapted from Langridge, J.M. et al., *Opt. Express* 16, no. 14 (2008): 10178–10187.)

from comparison with other, quantitative spectral data that their measured signal corresponded to a concentration of $c$ = 4.1 pptv, and they estimate from this a LOD = 1.8 pptv for a data acquisition time of ~1 min. Compared to the laser-based method of SC-CEAS, IBB-CEAS suffers from the drawback of lower spectral resolution, in addition to the fact that much longer acquisition times are required. In addition, it has to be noted that Venables et al. used a much longer absorption pass ($L_c$ = 5 m instead of $L_c$ = 1 m employed in Langridge's measurements). While the said IBB-CEAS is impressive, it is clear that on relative normalization to each other, SC-CEAS is superior. Nevertheless, both implementations of broadband cavity-enhanced absorption spectroscopy demonstrated the great potential for *in situ* detection of atmospheric trace gases and radicals.

It should be noted that, at times, spectral resolution poses the limiting factor in BB-AS, in particular when it comes to the identification of species with closely spaced spectral-line patterns and vastly different concentrations. This limitation can be overcome to a certain degree by incorporating a narrow-bandwidth tunable diode laser as the light source into the setup. In fact, tunable laser diode spectroscopy (TDLAS) is currently used for high sensitive and high resolution analysis in gases, flames, combustion processes, etc. (see Section 8.3). The combination of narrow-bandwidth and well-collimated output beams has made possible to perform extremely sensitive measurements and consequently probe minute concentration for a variety of gases. However, it has to be acknowledged that the wavelength range accessible in TDLAS is still somewhat limited, even if one exploits the wider tuning ranges provided by external cavity diode lasers (ECDLs).

The main advantage of BB-AS over other absorption techniques, like the just-mentioned TDLAS, stems from its capability to cover extremely large spectral ranges, and hence to perform simultaneous multispecies analysis of samples whose constituent molecules absorb at very different wavelengths. On the downside, one finds that BB-AS exhibits certain limitations when it comes to spectral and spatial resolution.

## 8.1.2 Minimum detectable concentrations and LODs

In the discussion above, low concentration values and LODs occurred repeatedly. It seems therefore prudent in this context to provide a tighter definition of *minimum detectable molecular concentrations*, or *limits of detection*; this has to be seen in the light that in the literature, the usage and definition of LOD in data analysis are not always consistent.

Following on from the formulation given in Chapter 7 for the Beer–Lambert law (see Equation 7.1), for small absorption—nearly always encountered for very low concentrations—it can be approximated as

$$I = I_0 \cdot \exp(-\sigma(v) \cdot N \cdot \ell) \cong I_0 \cdot (1 - \sigma(v) \cdot N \cdot \ell), \qquad (8.1)$$

where the various parameters have the usual meaning, i.e., $I_0$ and $I$ are the (laser) intensities before and after passing through the absorption volume, respectively; $\sigma(v)$ is the wavelength-dependent absorption cross section (in units of $cm^{-2}$); $N$ is the number of absorbing molecules (in units of $cm^{-3}$); and $\ell$ is the absorption path length (in units of cm). This means that the fractional absorbed laser intensity $\Delta I$ can be written as

$$\Delta I = I_0 - I = I_0 \cdot \sigma(v) \cdot N \cdot \ell, \tag{8.2}$$

yielding by rearrangement the number of absorbed molecules $N = \Delta I/(I_0 \cdot \sigma(v) \cdot \ell)$. Then for the minimum detectable laser intensity $\Delta I_{min}$, which by and large is dictated by technical characteristics and instruments sensitivity, the minimum detectable molecular concentration $N_{min}$ depends inversely on the initial laser intensity and the absorption path length. From Equation 8.2, it follows that

$$N_{min} = \Delta I_{min}/(I_0 \cdot \sigma(v) \cdot \ell). \tag{8.3}$$

In other words, for low-absorption conditions, one finds that the higher the incident laser intensity or the longer the absorption light path, the lower the minimum concentration that could be detected of a given analyte, whose absorption cross section is $\sigma(v)$. Therefore, it is not surprising that tremendous efforts have been made to develop absorption spectroscopic techniques, which could substantially lower $N_{min}$. Relevant key techniques encountered in this context are described in more detail further below.

As mentioned earlier, the minimum detectable molecular concentration and the LOD are closely linked. In spectroscopy, said quantities are extracted from the actual spectra. Therefore, at this point, it is probably worthwhile to recall Section 6.6 where the signal fluctuations were discussed in the context of *intrinsic* signal shot-noise and *extrinsic* noise contributions (from external background and detector noise). Based on said discussion and the procedures recommended by IUPAC (1976) in their guidelines, one can calculate the minimum concentration $N_{min}$, which corresponds to a signal-to-background-noise ratio with value SNR = 3:

$$N_{min} \equiv \text{LOD} = 3/\text{SNR}(N) = 3 \cdot \left(I_{noise}/I_{signal}\right) \cdot N_{signal}; \tag{8.4}$$

here $N_{signal}$ stands for the number density for which the actual, measured signal $I_{signal}$ was generated; and $I_{noise}$ constitutes the average fluctuations of the background (signal baseline). This means that one can estimate detection limits already from a single spectrum, provided that the concentration of the analyte ($N_{signal}$) associated with the particular measurement is known. For a description of the extraction of $I_{signal}$ and $I_{noise}$ from the (absorption) spectrum, see, e.g., Shrivastava and Gupta (2011).

Note that in direct absorption spectroscopy, i.e., the transmitted light intensity is measured, the laser light fluctuations normally dominate the background noise. Assuming stochastic shot-noise behavior for the laser light, this means that while the signal increases with laser intensity $I_0$, as noted in Equation 8.2, at the same time, the baseline/background (shot) noise increased by a factor $\sqrt{I_0}$, hence adversely affecting the SNR, and hence the LOD and $N_{min}$.

# 8.2 ABSORPTION SPECTROSCOPY USING FREQUENCY COMBS

The notion that (mode-locked) ultrashort picosecond- and femtosecond-pulse laser sources can be used for ultrahigh resolution and precision spectroscopy seems somewhat odd at first sight, recalling that the spectral bandwidth of such laser light pulses is rather broad. This is a consequence of the paradigm that mode locking actually is a *frequency domain* concept, while mode-locked lasers themselves and their applications are typically discussed in the *time domain*. The central concept of understanding the underlying phenomena is that the pulse train generated by a mode-locked laser exhibits a frequency spectrum that consists of a discrete, regularly spaced series of sharp lines across the broad gain spectrum; this frequency spectrum is now normally addressed as an *optical frequency comb*. This idea that a regularly spaced train of pulses in the time domain corresponds to a comb of narrow frequency spikes in the frequency domain, and might thus be used to excite narrow atomic or molecular transition resonances, was first realized and exploited in the late 1970s (see, e.g., Eckstein et al. 1978). With the advent of femtosecond-laser optical frequency-comb synthesizers in the late 1990s (see, e.g., Udem et al. 2002), measurements in the field of frequency-comb metrology and precision spectroscopy were enormously simplified, which led to an avalanche of uses in fundamental science and practical applications.

In this section, a brief introduction into the basic frequency-comb generation and control is provided (for more detailed descriptions, the reader is referred to, e.g., Ye and Cundiff 2005), and a number of key applications are described.

## 8.2.1 Basic concepts of frequency combs

The key features of importance for the understanding of optical frequency combs, generated by a mode-locked femtosecond laser, are summarized in Figure 8.2; all relevant parameters are indicated, both in the time and frequency regimes.

In the time domain, one encounters a periodic train of ultrashort pulses, with pulse width of typically $\tau_p \approx 100$ fs (e.g., from a Ti:sapphire laser). The associated pulse-to-pulse separation is related to the round-trip time for an individual pulse within the laser cavity, $\tau_R$, which is linked to the *pulse repetition rate*, $f_R$. Typical mode-locked Ti:sapphire lasers with cavity length $L_c \sim 1$–2 m operate at repetition rates of $f_R \sim 80$–100 MHz, with the associated round-trip time of about $\tau_R \approx 10$ ns.

While the temporal separation between pulses is constant, one observes that the individual pulses experience pulse-to-pulse phase shifts $\Delta\phi_{CE}$ with respect to the peak center of the pulse envelope, due to unavoidable intracavity dispersion; after $n$ pulses, the total phase shift accumulates to $n \cdot \Delta\phi_{CE}$. This can be expressed in terms of the average phase and group velocities, $v_P$ and $v_G$, respectively, inside the laser cavity:

$$\Delta\phi_{CE} = \left( v_G^{-1} - v_P^{-1} \right) \cdot (2L_c) \cdot 2\pi v_c, \tag{8.5}$$

where $L_c$ is the cavity length, and $2\pi v_c$ is the so-called "carrier" frequency.

For a train of identical, fixed time-interval pulses, Fourier transform analysis provides the spectral description in the frequency domain, yielding a comb of regularly spaced frequencies. The spacing between neighboring comb frequencies is inversely proportional to the time between pulses, i.e., the laser's pulse repetition rate $f_R$. However, because of the aforementioned (regular) pulse-to-pulse phase shift $\Delta\phi_{CE}$ in the time domain, then in the frequency domain, a rigid shift $f_0$ will be encountered for those frequencies, for which the pulses add constructively. This carrier-envelope offset-frequency $f_0$ is determined as

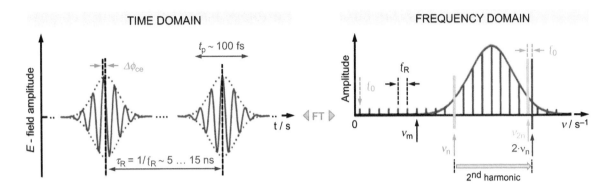

**Figure 8.2**   Summary of the time-frequency correspondence for a pulse train with evolving carrier envelope. (Left) Time domain; $\tau_R$ and $f_R$ = pulse repetition time and frequency, respectively; $t_p$ = pulse duration; $\Delta\varphi_{ce}$ = pulse-to-pulse phase shift. (Right) Frequency domain; $f_0$ = comb offset; $\nu_m$, $\nu_n$, and $\nu_{2n}$ = specific mode frequencies. Note that the spectral width should include both $\nu_n$ and $\nu_{2n}$ (equivalent to an octave). For further details, see text.

$$f_0 = (1/2\pi) \cdot f_R \cdot \Delta\varphi_{CE}. \tag{8.6}$$

Thus, the optical frequencies $\nu_m$ of the comb lines are given by

$$\nu_m = m \cdot f_R + f_0. \tag{8.7}$$

## 8.2.2 Measuring and controlling frequency-comb parameters

The two frequencies $f_R$ and $f_0$ are in the radio-frequency (RF) regime and thus can be measured experimentally with very high accuracy, although determining the carrier offset frequency $f_0$ is not trivial. If both are known, then also the mode frequencies $\nu_m$ become precisely known.

Measurement of the carrier-envelope offset frequency is usually done by using self-referencing techniques, in which the phase of one particular mode $\nu_n$ in the spectrum is compared to its harmonic $\nu_{2n}$; note that for this the frequency bandwidth of the pulse has to span at least a decade. One frequently used approach to do this is the *frequency minus two-times frequency* methodology, as indicated conceptually in Figure 8.2. For this, light from mode $\nu_n$ (at the low-energy side of the comb) is frequency-doubled to $2\nu_n \equiv 2 \times (n \cdot f_R + f_0)$ using second harmonic generation in a nonlinear crystal. This second harmonic light is then mixed with light from comb mode $\nu_{2n} \equiv 2n \cdot f_R + f_0$ (at the upper energy side of the spectrum), generating a heterodyne beat signal. This beat signal, which is easy to detect using a photodiode, includes a difference-frequency component, which is equal to $f_0$:

$$|2\nu_n - \nu_{2n}| \equiv f_0. \tag{8.8}$$

In the absence of active stabilization, the repetition rate and carrier-envelope offset frequency would be free to drift. The repetition rate can be stabilized by using, e.g., a piezoelectric transducer, which moves one of the cavity mirrors to change the cavity length, and thus adjusts the round-trip time. In common femtosecond lasers with double-chirped mirrors, the phase shift depends strongly on the Kerr effect; then, by modulating the pump power, one changes the peak intensity of the laser pulse, and as a result the Kerr phase shift can be controlled.

### 8.2.3 Spectroscopic metrology based on frequency combs

With the availability of optical frequency combs, it became relatively "easy" to precisely measure the frequency of light from any stable laser. With a few exceptions, all mode-locked lasers operate at repetition rates $f_R < 1$ GHz; therefore, it is straightforward and easy to measure this. The offset frequency $f_0$ is also accessible, utilizing the methodologies just described in the previous segment. With the knowledge of these two, the frequencies of all (evenly spaced) comb are also known, and thus it becomes possible to access the frequency $\nu_{laser}$ of an "unknown" laser, as long as its emission is at a wavelength covered by the frequency comb. The frequency difference between an unknown laser line and the closest comb line will be a frequency less than the repetition-rate frequency, i.e. $f_{beat} \leq f_R$. This heterodyne beat can finally be measured with high precision relative to a frequency standard, such as, e.g., a cesium-fountain atomic clock.

With these three RF-frequency measurements—of $f_R$, $f_0$, and $f_{beat}$—the optical frequency of a mode-locked laser can be evaluated from

$$\nu_{laser} = n \cdot f_R + f_0 + f_{beat} \text{ or } \nu_{laser} = n \cdot f_R + f_0 - f_{beat}. \quad (8.9)$$

The integer $n$ (associated with the particular comb spike) and the correct sign for $f_{beat}$ can be determined with a little bit of "prior knowledge" of the laser frequency $\nu_{laser}$. For example, today's commercial wave meters exhibit sufficient accuracy for such an estimate; but even without that, they can be determined if $f_R$ and $f_0$ are varied systematically while $f_{beat}$ is being monitored. Thus, frequency combs can serve as high-precision "yardsticks" for the exact measurement of laser line frequencies, and hence optical transitions in atoms and molecules, which were induced by that laser radiation.

One of the earliest and most clear-cut examples for using comb patterns in high-resolution spectroscopy is the measurement of fine-structure and hyperfine-structure splitting in sodium by Eckstein et al. (1978). In the introduction to their publication, they actually state "In this paper we report on experiments which establish the feasibility of high-resolution spectroscopy with a synchronously-pumped actively mode-locked cw dye laser." The researchers used a cw synchronously pumped, mode-locked dye laser, which had a pulse length of $\tau_p \approx$ 500 ps. This is rather moderate when compared to modern femtosecond-laser systems used to generate frequency combs, which meant that also the comb structure extends only over a moderate frequency range. The (comb) mode spectrum could be translated continuously in frequency over several gigahertz by simultaneously scanning the cavity length and the air-spaced intracavity etalon of the dye laser. The laser was tuned to the two-photon transition 3s $^2S_{1/2} \rightarrow$ 3d $^2D_{3/2,5/2}$ at $\lambda = 578.732$ nm, and then scanned across the transition region to provide spectra with com-frequency related markers. Exact frequency analysis of these spectra then yielded values for the fine-structure splitting of the upper 3d-level $\Delta E_{FS}(J = 3/2 \leftrightarrow J = 5/2) = 1028.4 \pm 0.4$ MHz and for the hyperfine-splitting of the lower 3s-level $\Delta E_{HFS}(F = 1 \leftrightarrow F = 2) = 1771.6 \pm 0.4$ MHz. These were in excellent agreement with values determined by alternative spectroscopic methods, and thus demonstrated the viability of the overall approach for precise, high-resolution spectroscopy.

Since then, quite a number of experiments have been undertaken to determine absolute values for atomic and molecular transitions, and—in addition—to provide an easy means to link spectroscopy and precise laser frequency measurements to frequency standards such as atomic clocks.

Probably one of the most amazing series of experiments, which have driven the precision of spectroscopic measurements close to the boundaries of theoretical limits, is the investigation of the 1s–2s transition in hydrogen. This spectroscopic system is seen as one of the most important test beds for accurate comparison between precision experiments and state-of-the-art quantum theory. In three major measurement campaigns, commencing in the late 1990s and spanning about a decade, the group of Hänsch and co-workers conducted Doppler-free two-photon spectroscopy of the 1s $^2S_{1/2} \rightarrow$ 2s $^2S_{1/2}$ transition of hydrogen, using frequency-comb-linked absolute frequency measurement configurations (see Niering et al. 2000; Fischer et al. 2004; Parthey et al. 2011; Matveev et al. 2013).

The experimental setup concept was nearly identical throughout (see the schematic at the top of Figure 8.3), although improvements in all technical aspects and in certain methodologies evolved over time.

In short, the hydrogen atoms were generated in an RF-discharge, in conjunction with cooling-gas expansion ($T < 10$ K), and interacted with the excitation laser radiation (at $\lambda = 243$ nm) inside an enhancement cavity. In the 1999 and 2003 campaigns, a frequency-doubled, ultrastable and ultranarrow linewidth dye laser (at $\lambda = 486$ nm) was used, which was referenced to the frequency comb from a Ti: sapphire laser, locked to a Cs-fountain clock. For the 2011 measurements, the laser system was upgraded to an external-cavity diode laser system (at $\lambda = 972$ nm), which was referenced to the frequency comb from an Er-doped fiber laser, locked to a hydrogen maser. Key results on the determination of the 1S–2S transition frequency are shown in the three data panels of Figure 8.3. A clear improvement from campaign to campaign is evident.

Summing up all measurement uncertainties and taking into account a range of frequency corrections, the latest, revisited data evaluation yielded a frequency for the 1S–2S transition in hydrogen of $\nu_{1S-2S} = 2466061413187018(11)$Hz, with relative uncertainty of less than $5 \times 10^{-15}$. This is by far the most precise, direct measurement of an atomic transition frequency. Such measurements are extremely valuable for the determination of fundamental entities like, e.g., the 1S-Lamb shift and the Rydberg constant; both can be derived by comparison of two transition (precisely known) frequencies in the hydrogen atom.

## 8.2.4 Direct frequency comb spectroscopy—DFCS

Since their inception, frequency combs from femtosecond-laser sources have been demonstrated to be "ideal" broadband light sources for absorption spectroscopy: they combine broad spectral coverage (in some cases up to 1000 cm$^{-1}$ or more), dense spectral sampling (of the order of ~100 MHz), high accuracy (down to a few kilohertz), and fast measurement (submillisecond sampling is feasible). This means that frequency combs allow for simultaneous interrogation at many thousands of precisely known optical frequencies. In addition, resonant-cavity enhancement for high sensitivity (see Section 8.4) can efficiently be implemented since, by their very nature of generation, frequency combs are coherent light sources.

There are a number of different approaches of how to utilize the individual modes of a frequency comb to generate and record (line-resolved) broadband

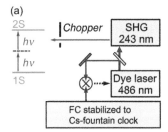

(a)

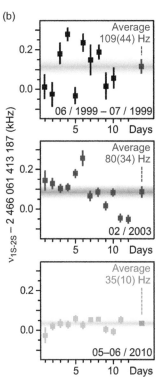

(b)

**Figure 8.3** Determination of the absolute transition frequency for hydrogen 1S-2S. (a) Basic experimental concept used to measure the atomic transition frequency (see text for details). (b) Three measurement campaigns from 1999, 2003, and 2010 are compared. The data points represent an average per day, with the error bars being the total 1σ-uncertainty; the final average values are labeled with the numerical result, in units of Hertz. FC, frequency comb. (Data adapted from Matveev, A. et al., *Phys. Rev. Lett.* 110, no. 23 (2013): article 230801.)

spectra, e.g., complete ro-vibrational bands of molecules. By and large, these fall into two main categories, namely implementations based on (1) *dispersive* and (2) *Fourier transform* spectral analysis.

*DFCS utilizing dispersive spectral analysis.* In principle, absorption features will diminish the intensity of the transmitted light from a broadband source at the frequency/wavelength position of the absorber transition, as described in Chapter 7. The only difference for FC sources is that they exhibit discrete, although tightly spaced frequency components rather than the continuous spectrum of conventional broadband sources. As long as the linewidth of transitions in the absorber atom/molecule is markedly wider than the spacing between neighboring comb frequencies, the observed spectra very much resemble the traditional line (envelope) appearance, as seen, e.g., in Figure 4.31 in Chapter 4 (where frequency comb absorption was first highlighted) or in Figure 8.6.

The overall absorption spectrum might then be recorded and analyzed using a standard, dispersive (grating) spectrometer. Note here that—as stated above—the mode frequencies for combs from most mode-locked laser systems are far too dense to be resolved with a common grating spectrometer. Thus, without resolving the comb lines, the frequency comb resembles nothing more than a "broadband lamp" in terms of resolution. In a certain way, this would defeat the object of using a frequency comb with exactly known frequency components of narrow spacing and narrow width.

This problem can be overcome by an approach that is now commonly addressed as *Vernier-type comb spectroscopy*; the principle of this technique is described in Diddams et al. (2007) and Gohle et al. (2007). In short, the frequency-comb light is sent through a high-finesse resonator, like a Fabry–Perót (FP) cavity, whose length is adjusted so that the spacing of its mode frequencies is slightly different to that of the frequency comb. As a consequence, now only every $m$th mode from the frequency comb will be resonant with every $k$th mode of the cavity; therefore, only these will be transmitted, and all others are suppressed. Thus, the two mode sets resemble a Vernier, with the Vernier ratio given by the ratio of the free spectral range of the FP and the frequency-comb mode spacing: $\mathrm{FSR_{FP}}/f_R = m/k$.

As a consequence of this FP filtering, the submitted comb now exhibits a frequency spacing of $m \cdot f_R$, so that for sufficiently large $m$, it can easily be resolved with a small grating spectrograph. However, while the spectrograph may be able to resolve the wider-spaced comb frequencies, one may have lost the absorption line, which may fall between the comb frequencies. This deficiency can be reversed by stepwise-tuning the FP resonator; with a minute change in resonator length, primarily the Vernier shifts, and the next set of $m \cdot f_R$-spaced comb frequencies comes into resonance with the filter and is thus transmitted. After scanning the resonator round-trip length by one FSR, $m$ groups of the $m \cdot f_R$-comb have passed and the initial group is transmitted again. Out of such a set of reduced-comb spectra, one can assemble the high-resolution comb spectrum. Rather than recording a sequential set of 1D dispersed spectra, one exploits modern 2D array CCD detectors, namely to "stack" the FP-tuned set of individual spectra synchronous with the FP stepping. This two-dimensional array of mode-frequency light spots can be uniquely ordered into a single frequency axis (see Diddams et al. 2007).

It should be noted that in modern implementation of high-resolution DFCS, the generation of the 2D spectral arrays is done by combining a so-called virtually imaged phased array (VIPA) spectral disperser with a conventional grating in an orthogonal arrangement; for the operating principle of a VIPA, see Shirasaki (1999). Such a VIPA-plus-grating spectrometer was used by Hébert et al. (2015) in their study of isotopologues of HCN. They utilized a fully automatic control system, which dynamically locked the FP-filter cavity to the frequency comb and which sequentially measured the decimated, interleaved comb subsets. The results were complete spectra with spectral sampling resolution of $\delta\nu_s \equiv \Delta\nu_{FC} =$ 250 MHz over a spectral interval of $\Delta\nu > 4$ THz, within a total measurement time of less than 10s. Examples for complete band spectra and individual rotational lines are shown in Figure 8.4.

It is worth noting that combining a VIPA-DFCS setup with a simple Michelson interferometer, which simultaneously measures the complex phase, both absorption and dispersion spectra of the sample could be measured (see Scholten et al. 2016).

*DFCS utilizing Fourier transform spectral analysis.* When it comes to spectral analysis, Fourier transform plays a major role in absorption spectroscopy; many commercial instruments aimed at quantitative molecular analysis are based on FTIR spectroscopy. In general, it is desirable that advanced spectroscopic instruments combine broad detection bandwidth (to access a wide variety of molecular species), high resolution, high detection sensitivity, and short acquisition times. However, combining the latter two requirements poses a tremendous challenge for standard FTIR spectroscopy due to its typically incoherent light source. Fourier transform spectrometers based on optical frequency combs—in the literature often abbreviated as FC-FTS—have been shown to address these problems.

Frequency combs exhibit high spectral brightness and spatial coherence, and operate at high repetition rates of $f_R \approx 100$ MHz, or above. The latter not only provides the high-resolution requirement associated with the comb spacing but also means that the $1/f$ technical noise is reduced by many orders of magnitude when compared with the very best classical interferometer measurements at an upper modulation frequency of ~50 kHz. In addition, the coherent nature of the FC source provides absorption and dispersion parameters in a single measurement, based on in-phase and in-quadrature detection utilizing a lock-in amplifier.

Basic descriptions of FC-FTS and impressive proof-of-principle measurements of quite a range of molecules—including many smaller hydrocarbon molecules and nitric oxides—can be found in Mandon et al. (2009) or Adler et al. (2010). In their experiments, both groups used mode-locked laser sources with repetition rates of $f_R \approx 140$ MHz, which generated frequency combs with comb spacing $\Delta\nu_{FC} \approx$ 140 MHz, but based on different laser media and thus different wavelength/ wavenumber ranges. The former used a $Cr^{4+}$:YAG source operating in the wavelength range $\lambda \approx 1460$–1645 nm (equivalent to $\tilde{\nu} \approx 6100$–6800 $cm^{-1}$); the latter was based on an idler-OPO source operating in the wavelength range $\lambda \approx 2800$–4700 nm (equivalent to $\tilde{\nu} \approx 2100$–3600 $cm^{-1}$); both used similar Michelson interferometers as their Fourier transform instrument. Detection sensitivities in the low-ppm to ppb range were achieved, depending on the specific molecule, over total acquisition times of just a few seconds.

While the two spectroscopic approaches discussed above—VIPA-DFCS and FC-FTS—have been very successful implementations for high-resolution

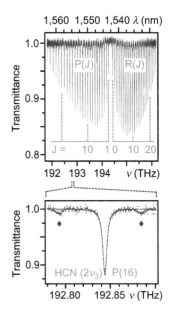

**Figure 8.4** High-resolution frequency-comb absorption spectrum of the $2\nu_3$ overtone band of the hydrogen cyanide isotopologue $H^{13}C^{14}N$; the P- and R-branch lines are annotated. At the bottom, the P (16) line is shown on an enlarged scale, revealing some additional, weak absorption features (marked by the red diamonds), attributed to hot-band lines. (Data adapted from Hébert, N.B. et al., *Opt. Express* 23, no. 11 (2015): 13991–14001.)

spectroscopy using broadband FC sources, the most widely pursued form of direct frequency-comb spectroscopy is dual frequency-comb spectroscopy, or dual-DFCS, which eliminates external (dispersive or Fourier transform) spectrometer altogether. The state-of-the-art of dual-DCFS is reviewed in Coddington et al. (2016); the basic concepts of the methodology are summarized in Figure 8.5.

Briefly, the light from two frequency combs with slightly different repetition rates, $f_1$ and $f_2 = f_1 + \Delta f_r$, is heterodyned on a single photodiode detector. This generates an RF comb composed of distinguishable heterodyne beats between pairs of optical comb (frequency) teeth. The frequency spacing between teeth of this RF comb is $\Delta v_{\text{RFC}} \equiv \Delta f_r$. corresponding to a $(\Delta f_r / f_1)$-scaling between the optical- and RF-comb teeth. As a consequence of this scaling, the RF comb—which still carries all relevant spectral information contained in the original optical comb—is easily accessible with standard RF electronics.

Absorption spectroscopy can be performed in two ways, namely to pass only the light of one of the combs through the sample, or from both; sample-response information is carried by the beam(s) interrogating the sample. The former approach is often addressed as asymmetric (dispersive) dual-comb spectroscopy—simultaneous full phase and amplitude response are measured, thus yielding information of both absorptive and dispersive properties of the sample; while the latter is known as symmetric (collinear) dual-comb spectroscopy—this only gives access to the absorption properties of the sample and is analogous to the typical FTIR measurement. It should be noted that if signals are accumulated for long periods (of some seconds), then individual comb lines become resolved; this means that spectra are only sampled at discrete frequencies, but frequencies that are known with the ultrahigh accuracy. In the case that more complete coverage of the frequency space is required, the comb spectrum can be scanned (by altering the repetition rate of one of the FC sources), and the set of spectra are interleaved.

Dual-DFCS has several inherent advantages over common Fourier transform spectroscopy. First, as noted above, it uses coherent laser light rather than incoherent light. Second, no moving parts are required. Third, one can set the spectrometer speed to values mimicking interferometer-mirror scanning speeds of up to $10^4 \text{m·s}^{-1}$; as a consequence, spectrum acquisition speed can be increased by orders of magnitude, and $1/f$ noise in the measurements is greatly reduced. Fourth, FC sources are now available for many spectral regions, ranging from the UV to the far-IR (and even terahertz), although some frequency ranges do still come with experimental challenges. However, as FC sources continue to evolve, so most likely will frequency-comb spectroscopy.

Based on the above, it is not surprising that since the mid-2000s applications of dual-DFCS to high-resolution spectroscopy of molecules are becoming ever more popular. As a key representative example, here the study of acetylene is presented (see Ideguchi et al. 2014).

In their experiment, the authors used two free-running, commercial femtosecond Er-doped fiber lasers, emitting in the $\lambda \sim 1550$ nm region (actual spectral coverage $\lambda \approx 1500–1590$ nm, corresponding to $v \approx 188–200$ THz, or $\tilde{v} \approx 6270–6670$ cm$^{-1}$). The lasers operated at repetition rates of $f_1, f_2 \approx 100$ MHz, with a difference of $\Delta f_r = 350$ Hz. For adaptive signal referencing, each of the femtosecond-laser frequency combs was independently beat with two free-running cw Er-doped fiber lasers (emitting at $\lambda_1 = 1557$ nm and $\lambda_2 = 1532$ nm, respectively). The two FC-source beams were combined and the molecular gas

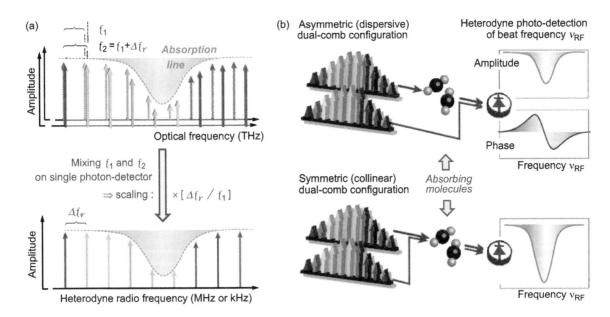

**Figure 8.5** Principles of dual direct frequency-comb spectroscopy. (Left) Simplified concept of dual-DFCS; the two combs with repetition rates $f_1$ and $f_2 = f_1 + \Delta f_r$ are mixed and detected by a single photoreceiver. As a result, the heterodyne signal generates a unique RF comb with spacing $\Delta f_r$; typically, $f_1 \sim 100$ MHz and $\Delta f_r = 0.1...1$ kHz are utilized. (Right) For dual-DFCS, either one or both combs are passed through the sample. The actual absorption (or phase shifts) on individual comb teeth is encoded onto the corresponding amplitude (or phase) of the measured RF-comb teeth.

sample was interrogated (acetylene at pressure $p = 2.13$ mbar, in a cell of length $\ell = 70$ cm). The transmitted comb radiation was beat on a fast InGaAs photodiode detector, with a measurement time of 2.7 s.

The resulting (overview) spectrum for the $v_1 + v_3$ combination band of $C_2H_2$ is shown in the upper part of Figure 8.6, covering the entire frequency domain $v \approx 188$–$200$ THz of femtosecond fiber-laser frequency combs, with more than $10^5$ well-resolved individual comb lines. In the data panel, the frequencies of the two cw reference lasers are indicated by the green lines.

In the lower part of the figure, zoom-ins in three different spectral regions are shown, not only revealing that the coherence between the two FCs is maintained over the full measurement time, but also highlighting the high-resolution features encountered in dual-DFCS. On the left, one can see the very narrow width of an individual comb tooth, revealing the potential spectral resolution step-width of $\delta v \approx 200$ kHz when interlacing spectra. In the center, the spacing between individual teeth of the frequency comb(s) $\Delta v_{FC} \equiv f_1(f_2) = 100$ MHz is shown. On the right, an individual Doppler-limited rotational line of $C_2H_2$ is shown, with $\delta v_s \approx 1.2$ GHz.

## 8.3 ABSORPTION SPECTROSCOPY USING TUNABLE DIODE AND QUANTUM-CASCADE LASER (QCL) SOURCES

Tunable lasers have played an important role in absorption spectroscopy, right from the beginning. However, it is probably fair to say that the progress in

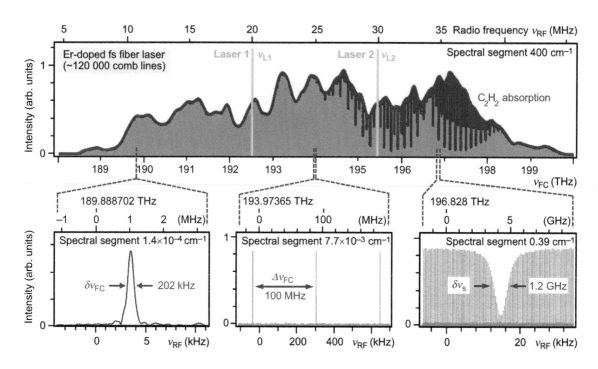

**Figure 8.6** NIR dual-comb absorption spectrum of acetylene ($C_2H_2$), with resolved comb lines, utilizing the emission from an Er-doped femtosecond fiber laser. The optical instrumental linewidth (of an individual comb line) is $\delta\nu_{FC} \approx 200$ kHz (lower-left panel), and the comb-frequency spacing is $\Delta\nu_{FC} = 100$ MHz (lower-middle panel). An individual molecular absorption line is shown in the lower-right panel, exhibiting a signal (Doppler) width of $\delta\nu_s \approx 1.2$ GHz. (Adapted from Ideguchi, T. et al., *Nat. Commun.* 5, (2014): article 3375.)

semiconductor laser technology has to be seen as key in the development of absorption techniques, which opened the door to a plethora of now-common applications. The field has become known as *tunable diode laser absorption spectroscopy* (TDLAS), which now encompasses as well QCLs (although they are not diode lasers in the common sense).

### 8.3.1 Tunable diode laser absorption spectroscopy

The most widely used light source used in infrared (IR) absorption spectroscopy is the (tunable) diode laser (TDL). These lasers offer many advantages based on their tunability, small size, and modulation capability. Probably one of their most relevant features is that the laser output and wavelength can be controlled by changing the injection current, as for example in the VCSELs, which can be tuned as much as 30–150 GHz·mA$^{-1}$ (see Lytkine et al. 2006). More details on diode laser systems can be found in Chapter 5.1.

TDLs have been successfully applied in many areas of science and technology, including—to cite just a few examples—high-resolution molecular spectroscopy, atmospheric trace gas analysis, biomedical applications, or plasma diagnostics. In particular, its high spectral resolution and remote nondestructive character have made *TDL absorption spectroscopy* or TDLAS one of the most widely used methods in harsh environmental conditions, encountered, for example, in combustion or industrial studies (see Bolshov et al. 2015).

The principle of measurement in this type of spectroscopy relies on the assumption that the frequency-dependent light transmitted through an absorbing sample can be completely modeled by the Beer–Lambert law. In this view, the transmitted intensity $I(v)$ is given by

$$I(v) = I_0(v) \cdot \exp(-S_T \cdot g(v - v_0) \cdot N \cdot \ell), \tag{8.10}$$

where $I_0(v)$ is the incident (laser) light intensity, $S_T$ is the temperature-dependent line strength, $g(v - v_0)$ is the normalized (area = 1) line-shape function, and $N$ and $\ell$ have the usual meaning of number density of particles (atom, molecules, aggregates) and absorption path length, respectively. Using the expression for spectral absorbance $A(v) = -\ln[I(v)/I_0(v)]$, one derives from Equation 8.10 that the spectral absorbance can be written in differential or integral form as

$$A(v) = S_T \cdot g(v - v_0) \cdot N \cdot \ell \tag{8.11a}$$

$$A_{line} = \int A(v)dv = S_T \cdot N \cdot \ell \tag{8.11b}$$

Here $A_{line}$ is the area obtained by integration of the absorbance over the full spectral band. From Equation 8.11b, one finds for the number density $N = A_{line}/S_T \cdot \ell$, meaning that it can be deduced from the measured spectrum, provided that the absorption path length and the line strength are known. Although the latter quantity can be measured, it is often obtained from spectroscopic databases like HITRAN (see Rothman et al. 2013).

In many analytical applications in which the analyte is not the only species present in the probed volume, one uses the concept of the amount-of-substance fraction of the absorbing species, $x_{species}$, given by

$$x_{species} = \frac{p_{species}}{p_{total}} = \frac{k_B T}{S_T \cdot \ell \cdot p_{total}}; \tag{8.12}$$

underlying this equation is the well-known ideal gas law $p = N \cdot k_B \cdot T$, with $k_B$ and $T$ being the Boltzmann constant and the gas temperature, respectively. In line with the discussion in Section 8.1, the detection limit $\Delta x_{species}$ can be defined as

$$\Delta x_{species} = \frac{\Delta A_{residual}}{A_{peak}(v_0)}; \tag{8.13}$$

where $A_{peak}(v_0)$ and $\Delta A_{residual}$ correspond to the peak absorption at the line center frequency and to the standard deviation ($1\sigma$) of the fit residual over the entire line profile, respectively.

A basic TLDAS setup is illustrated in Figure 8.7. The laser light is split into two channels to simultaneously record $I(v)$ and $I_0(v)$; the two signals are usually processed in a balanced detector unit. Depending on the specific application, often a multipass cell is utilized to increase the absorption path length $\ell$ and hence the detection sensitivity. A gas manifold system controls the gas mixture to the preselected pressure, and both a temperature monitor and a pressure gauge are incorporated to measure the gas pressure and the sample cell wall temperature, respectively. For spectral scaling, the laser wavelength is monitored in parallel.

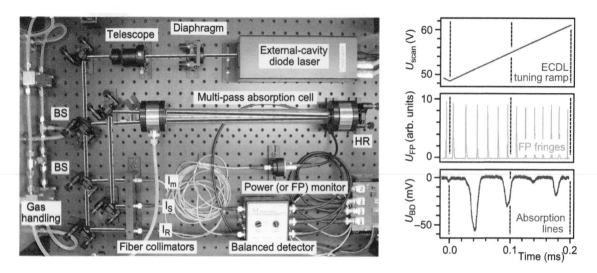

**Figure 8.7** Implementation of a simple TDLAS system, with a tunable, external-cavity diode laser system, a simple multipass absorption cell, and balanced detection (absorption signal references against non-absorption laser intensity). On the right, a single tuning (ramp) cycle is indicated, in which the absorption spectrum is wavelength-referenced against a monitor FP wavelength analyzer.

As seen in the panels on the right-hand side of Figure 8.7, the diode laser wavelength is altered by a tuning voltage (here a saw-tooth waveform), and sending part of the radiation through a monitor FP interferometer, precise frequency markers are available for wavelength calibration. The absorption signal (here four line features) can easily be averaged—to improve the SNR—by repeating the tuning cycle as often as required to achieve a particular SNR.

Among the many experimental realizations of the TDLAS, two of the most frequently used are the so-called *direct absorption* (DA-TDLAS) and *waveform modulation* (WM-TDLAS) techniques.

Certainly, the simplest one is that of direct detection (either based on fixed-wavelength or scanned-wavelength implementations): the light absorption is detected directly after the absorption cell, without any further electronically induced variation of the laser light source or the detector signal. It was traditionally used and deemed acceptable in such cases in which SNR was not critical, i.e., for strong absorption signals with little noise. However, as soon as a signal becomes rather small, reaching the noise level, or if instabilities in the gas flow, turbulences, or signal fluctuation (due to mechanical vibrations in the experimental apparatus) occur, then DA-TDLAS becomes less suitable, or even unusable; for accurate data acquisition, techniques that are able to retrieve signals from noise or compensate for irregular signal fluctuations are required: here WM-TDLAS comes to the rescue.

The basic principle of the waveform-modulation spectroscopy (WMS; not restricted to absorption spectroscopy alone) is well documented in the literature (see, e.g., the review on TDLAS for combustion monitoring by Bolshov et al. 2015); thus, here only a brief description of the most common "calibration-free" modality (WMS-$2f/1f$) is given, which incorporates first harmonic normalized waveform (laser light) modulation with second harmonic detection. Essentially, the method utilizes the modulation of the diode laser injection current with angular frequency $\omega = 2\pi v$, which results in a laser optical frequency of the form $v(t) = <v> + A \cdot \cos(\omega t)$, where $\langle v \rangle$ is the time-averaged frequency value. When the TDL injection current is varied, not only does the laser output wavelength vary but with it also a linked intensity modulation ensues. Then the transmitted signal (laser light minus absorbed light) is measured using phase-sensitive detection, using a lock-in amplifier capable of extracting the first and second harmonic output signals $v(t)$ and $2v(t)$, respectively. As a consequence of the frequency modulation, the TDL intensity $I_0(t)$ is synchronously modulated, and in general its temporal evolution can be well described retaining only the first and second harmonic components (see Bolshov et al. 2015). Typically, while most of the 2f-harmonic signal depends on the light absorption, the 1f-signal contribution is mainly associated with the laser-intrinsic linear modulation with little, if any, contribution from the light absorption. It can then be demonstrated that in an optically thin line-center measurement of an isolated absorption transition, the $(2f/1f)$-signal ratio is given by

$$(2f/1f) = S_T \cdot N \cdot \ell \cdot F(g(v)), \qquad (8.14)$$

where $F(g(v))$ is the integral term that depends on the line-shape function.

Normally, WMS measurements require *in situ* calibration in order to account for the actual line shape function. On the other hand, if the line shape is known *a priori*, this calibration measurement can be avoided since the absorption line shape and strength are intrinsic properties of the molecule(s) under examination. This "calibration-free" feature (see Rieker et al. 2009) constitutes an important advantage of TDLAS with respect to other spectroscopic methods, and it is exploited in applications requiring the monitoring in harsh environments where *in situ* calibration may not be possible (see Sur et al. 2015).

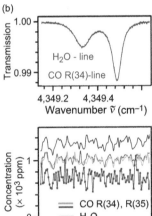

Nowadays, pollutant emissions as well as depletion of fossil fuel resources are subjected to continuous environmental regulations, which force the development of alternative fuel technologies, e.g., those based on high-temperature energy conversion and combustion, or gasification processes.

A good example of those new methodologies is that of entrained-flow gasification of biomass, which currently plays a decisive role in the syngas production of motor fuel synthesis. In practice, the complexity associated to the combustion or gasification of many new biomass fuels requires a precise monitoring of physical parameters during processing, like temperature or gas pressure, and chemical information on gas and soot concentrations. Within this scenario, Sepman et al. (2016) have developed and applied the first TDLAS sensor capable of accurately determining the key parameters in the gasification process. These authors used a single diode laser for determining more than one species and soot concentration from the measured spectra.

As shown in Figure 8.8, the authors applied the sensor for diagnostics of CO, $H_2O$, and soot concentrations in the reactor core of a pilot-scale, atmospheric air-blown, entrained-flow gasifier located at SP Energy Technology Center in Piteå (Sweden). They selected the spectral region near 4350 cm$^{-1}$ ($\lambda_{laser} \sim 2.3$ μm) containing strong transition lines for CO but also includes a considerable number of $H_2O$ and $CH_4$ transitions.

The exemplary spectrum shown in the figure was recorded for a $CH_4$/air flame at a flow rate of $\sim 10$l·min$^{-1}$; the two dominant spectral lines for $H_2O$ and CO are

**Figure 8.8** Diagnostics of CO, $H_2O$, and soot concentrations in the reactor core of an entrained-flow gasifier setup (photograph show in [a]), using a TDLAS sensor; a typical line absorption spectrum is shown in the upper data panel (b). The traces in the lower data panel (b) are indicative for the sensitivity (at the few ppm level) of the *in situ* measurements, which demonstrate the viability of TDLAS for real-time monitoring and process control. (Adapted from Sepman, A. et al., *Appl. Phys. B* 122, no. 2 (2016): article 29.)

annotated. Spectral scans were of duration ~11 ms each, and 100 such scans were averaged for diagnostics evaluation, equating to a monitor response of the order 1 s. Note that the variation in the CO mole fraction and the soot concentration are rather well correlated with each other.

### 8.3.2 QCL in absorption spectroscopy

As discussed in Chapter 5, in the traditional laser diode (LD), the radiation emission is produced by the recombination of electrons and holes, moving in their conduction and valence bands, within the depletion layer of a semiconductor diode device. Therefore, the wavelength of laser light emission is determined by this interband transition, i.e., the laser radiation frequency $\nu_{laser} \geq E_g/h$, where $E_g$ is the energy band gap, and $h$ is Planck's constant. In contrast, in the QCLs, the electrons are the only charge carriers that perform quantum jumps between the "artificially" created levels occurring in multiquantum well layer structure of certain semiconducting materials. Now the laser radiation is generated by electron transitions between said quantum well levels, and the photon energy is determined by the thickness of the wells and barriers, and thus is decoupled from the material band gap. This provides tremendous flexibility in the tailoring of laser transitions, covering the wavelength region from the high-energy MIR end at ~4 μm way into the terahertz regime at ~300 μm (see Chapter 5.2).

Significant impact on the use of QCLs in many practical applications was afforded by device technology, which (1) with DFB-QCLs provides highly tunable laser sources, and which (2) is now able to operate at room temperature. For example, DFB-QCLs with wavelengths in the atmospheric IR windows now allow for the detection of atmospheric trace gas molecules with sensitivities reaching a few pptv (see Curl et al. 2010).

Note also that "broadband" multiple-DFB laser spectrometers have been realized using arrays of DFB-QCLs, which are able to cover a very wide range of MIR frequencies. This type of BB-DFB-QCL spectrometer compares favorably with conventional FTIR spectrometers in terms of wavelength coverage but wins hands down when it comes to spectral resolution. Typically, in the QCL spectrometers, the frequency resolution is $\delta\nu_{laser} \approx 0.01$ cm$^{-1}$ in pulsed-mode operation and $\delta\nu_{laser} \approx 0.001$ cm$^{-1}$ in CW operation. This spectral resolution should be compared to that of a typical "bench-top" FTIR instrument with $\delta\nu_{FTIR} \approx 0.1$ cm$^{-1}$ (see Table 8.1).

Finally, as also noted in the introduction, the utilization of high-brightness broadband laser sources gives BB-DFB-QCL spectroscopy a crucial edge in, e.g., the remote sensing of trace gases, which often require ultrahigh sensitivities. Below the most common QCL configurations and typical applications are briefly outlined, grouped into implementations using (continuous wave) cw-QCL, (external-cavity) EC-QCL, and (pulsed) p-QCL sources.

### 8.3.3 cw-QCL absorption spectroscopy

In principle, setups for cw-QCL IR spectrometers are similar to the basic TDLAS implementation shown in Figure 8.7. They basically consist of three elements: a QCL device (instead of the diode laser); a sample cell that can be used in single-pass or multipass configuration; and an IR photodetector, today most commonly a Hg–Cd–Te detector.

As an example of the use of a cw-QCL source in absorption spectroscopy, an investigation on silane by Bartlome et al. (2009) has been selected. It is well known that the consumption and dissociation of silane ($SiH_4$) are crucial monitoring parameters, e.g., in the manufacture of solar cells. As a consequence, high-resolution spectra of the mother compound and dissociation radicals like $SiH_3$ are essential for the understanding of the deposition mechanism of silicon layers in plasma-enhanced chemical vapor deposition (CVD). In this context, Figure 8.9 illustrates a comparison between high-resolution FTIR and cw-QCL measurements of the $v_3$-band of $SiH_4$.

In the top panel, an overview spectrum recorded with the FTIR instrument is shown; the three ro-vibrational P-, Q-, and R-sequences are indicated, with rotational quantum number assignment for the resolved lines. It is clear from the given spectral resolution information that in the FTIR spectrum, the known multiplet structure of individual transition lines remains unresolved. This structure becomes visible when using the narrow bandwidth cw-QCL source, which exhibits a 100-fold increased spectral resolution. The R(9) multiplet of the $v_3$-vibrational band is shown as an example; note that some of the weaker features most likely represent unassigned transitions in $SiH_4$.

One of the goals of the study was to gain insight into the Si-film deposition rates by comparing selected peak intensities before and after ignition of the plasma, as a function of silane flow rates. An example of the (linear) relation between flow rates and the absorbance of silane vapor is given in the lower panel of the figure. From a comparison of their spectral data of silane absorption lines under plasma deposition and nonplasma conditions, the authors could rather accurately deduce the growth rates of Si-film deposition: based on their deposition model equations and the *in situ* depletion monitoring data, they calculated a growth rate of 0.48 ± 0.02 nm·s$^{-1}$. This value compared rather well with an *ex situ* profilometric measurement on the film thickness that yielded a growth rate of 0.437 ± 0.017 nm·s$^{-1}$. The authors ascribe the small discrepancy between the values to slightly nonuniform deposition regimes.

### 8.3.4 EC-QCL absorption spectroscopy

An important development in the QCL tunability was achieved by setting up systems in external-cavity (EC) configuration (EC-QCLs) that allow for arbitrary wavelength selection across the laser spectral gain curve, without changing the chip temperature; this is analogous to ECDLs. While the (temperature) tunability of commercial cw-DFB-QCLs is only a few wavenumbers, at room temperature, cw-EC-QCLs achieve tuning ranges of 200 cm$^{-1}$ and more. These wide tuning ranges, combined with the high-power and narrow linewidth of the lasers, have facilitated their use in high-resolution spectroscopy, which had to span huge spectral segments in the mid-IR (MIR).

Staying with the above example of silane, recent high-resolution spectroscopy studies of its $v_3$-fundamental band showed the strength of using a cw-EC-QCL for the task (see van Helden et al. 2015). In their experimental setup, the authors employed a water-cooled cw-EC-QCL system, which was tunable in the range 1985–2250 cm$^{-1}$, covering nearly the complete interval of the FTIR spectrum shown in Figure 8.9, and not only one or two individual rotational lines. Utilizing a White-type multipass cell to record absorption transitions of the stretching dyad within the P-branch of the silane $v_3$-fundamental band, the authors

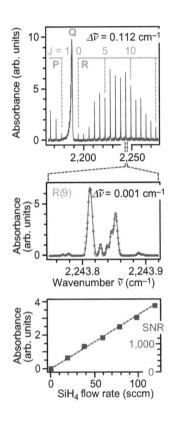

**Figure 8.9** The $v_3$-band of $SiH_4$ acquired (1) by a high-resolution FTIR (top panel) and (2) by cw-QCL absorption spectroscopy (center panel), exemplified for the Doppler-resolved R(9) multiplet. The linear relation between $SiH_4$ flow rates and absorbance signals is shown in the bottom panel. (Data adapted from Bartlome, R. et al., *Appl. Phys. Lett.* 94, no. 20 (2009): article 201501.)

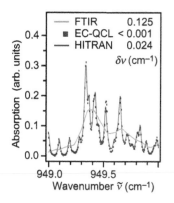

**Figure 8.10** Absorption spectrum of ethylene ($C_2H_4$) at low concentration (0.159% in 130 mbar of $N_2$) in a gas cell of $L_c$ = 15 cm, using fast, mode-hop-free tuning of an external-cavity QCL (EC-QCL). For comparison, the data from an FTIR measurement of the same sample are included, together with a synthetic HITRAN spectrum. (Data adapted from Tsai, T. and G. Wysocki, *Appl. Phys. B* 100, no. 2 (2010): 243–251.)

recorded the whole of said band with a resolution of $2 \times 10^{-3}$ cm$^{-1}$, and determined—for the first time—absolute line strength values for many P-branch transitions and their subcomponents.

In order to demonstrate the versatility of (fast-scanning) cw-EC-QCL spectrometers in absorption spectroscopy, results from a study of a mixture of 0.1% $C_2H_4$ diluted in a buffer gas ($N_2$), in a gas cell with absorption path length $\lambda$ = 15 cm at room temperature, are shown in Figure 8.10 (see Tsai and Wysocki 2010). The data not only reveal the superiority in resolution when compared to FTIR measurements but also show good correlation between the collected data and a HITRAN spectral simulation.

### 8.3.5 p-QCL absorption spectroscopy

In early experimental realizations of spectroscopic measurements using p-QCLs, named *inter-* or *short-pulse mode*, the traditional TDL-scanning approach was utilized, i.e., injecting a burst of a DC current. In QCLs, this particular mode results in the generation of a wavelength-chirped laser pulse of a few nanoseconds in duration; the wavelength tuning is due to refractive index changes, associated with the temperature rise in the active region during the duration of the current injection and laser pulse; a *frequency down-chirp* ensues. Progress in device technology later allowed one to use much longer current injection pulses, up to some hundreds of nanoseconds, and complete absorption spectra of certain species could be measured; this experimental procedure is now normally addressed as the *intrapulse mode*.

This mode of operation offers some great advantages for dynamic chemical trace analysis: an entire absorption structure of up to $\tilde{v}$ = 1 cm$^{-1}$ ($v$ = 30 GHz) can be recorded within a few hundred nanoseconds. A downside of this technique is the inherent increase in the laser linewidth, up to values of $\delta v_{\text{laser}} \approx 1.2$ GHz. Regardless, definitive advantages of the frequency down-chirp method are undisputed when comparing it with the short-pulse method: (1) all spectral elements are recorded in a single laser pulse; (2) a wide, fast tuning range; and (3) when a "top-hat" current pulse is applied, the laser frequency tuning (from high-to-low frequency) is an almost linear function of the scan time.

One of the most interesting applications of this frequency down-chirp p-QCLs methodology is the trace detection of atmospheric gases. The associated, typical p-QCL spectrometer consists of pulsed, linear-frequency down-chirp QCL (excited using a top hat-shaped current pulses of a few tens of nanoseconds up to 300 ns—at repetition rates of up to 100 kHz—and with peak currents of up to a few amperes). To increase the detection sensitivity, a multipass cell is normally incorporated into the setup that can provide effective absorption path lengths up to 100 m. In their study, Nwaboh et al. (2011) investigated the content of CO in well-prepared gas mixtures, in a gas cell with effective absorption path length $\ell$ = 21 m, filled to a total pressure of the order $p_{\text{total}} \approx 100$–300 mbar. Besides $N_2$, $O_2$, and $CO_2$, carbon monoxide (CO) was incorporated into gravimetrically prepared gas mixtures, at a fractional content of 100 μmol·mol$^{-1}$. A QCL operating in the intrapulse mode was used to probe the P(1) line of CO at $\tilde{v}$ = 2139.4 cm$^{-1}$. Spectral measurement data of the cell filling—with and without CO—are shown in the top panel of Figure 8.11, for individual p-QCL laser sweeps. Fringes from a reference FP interferometer (FPI) provide the frequency scale markers for the final absorbance spectrum (lower panel in the figure).

The authors found that their spectroscopically determined CO amount fractions agreed perfectly with gravimetric reference values, and based on the uncertainty analysis of their spectrometry-based data retrieval and the respective traceability of input parameters, they arrived at a measurement reproducibility of better than 1%. Since CO is an important molecule for environmental monitoring, industrial process control, and a biomarker in exhaled human breath, obtaining such reliable and traceable data is indispensable.

It should be noted that in the same publication, the authors also utilize the same type of measurement setup, sample preparation, and evaluation methodology to measure their gas mixtures—for comparison of sensitivity and reproducibility— by TDLAS and cavity ring-down spectroscopy (CDRS; see Section 8.4)

## 8.4 CAVITY-ENHANCEMENT TECHNIQUES

The determination of very small gas concentrations requires detection sensitivities at ppbv or sub-ppbv levels that can only be achieved (1) with long effective optical path lengths or (2) by efficient suppression of optical and laser noise. With respect to noise reduction, the techniques highlighted in Chapter 6.6 may be applied, besides careful system optimization to suppress any extrinsic light noise contributions—basically frequency modulation and electronic signal filtering provide the tools for successful SNR enhancement. Thus, in this subsection, the emphasis will be placed on *long absorption path-length spectroscopy*.

In order not to make this approach excessively cumbersome in the laboratory, so-called multipass absorption cells are utilized to achieve the desired increase in overall absorption path length $\ell$ on a small footprint, and thus to reach the necessary sensitivity. The most commonly used designs are *Herriott* and *White* cells (named after their inventors); these are commercially available or can be custom-tailored to meet specific requirements. The said two cell types and others were recently reviewed for their benefits in TDLAS (see Li et al. 2014). Just to give a flavor of what is feasible in terms of sensitivity, in their work on nitric oxides, Nelson et al. (2006) utilized a Herriott cell with effective path length $\ell = 210$ m, in conjunction with a tunable QCL, and were able to measure NO in atmospheric samples down to a few parts in $10^{12}$, with LOD ≈ 30 pptv within measurement times of less than a minute.

Another way to realize long path lengths is found in the techniques related to so-called *cavity-enhanced (absorption) spectroscopy*. The main idea is to "recycle" the photons by tuning the cavity resonance frequency to the photon frequency of the probe beam, or *vice versa*, as a method to effectively increase the photon path length and thus the effective absorption path length. This can be done either within an active (laser) cavity [then the method is known as *intracavity laser absorption spectroscopy* (ICLAS)] or within a passive (external) cavity [a methodology known as *cavity ring-down spectroscopy* (CRDS)]. These two approaches are outlined next, together with a few representative examples.

### 8.4.1 Intracavity laser absorption spectroscopy

ICLAS is a highly sensitive technique, which was introduced by Pakhomycheva et al. (1970). Here the absorbing sample is placed inside the cavity of a sufficiently broadband laser. The positive feedback mechanism responsible for the laser

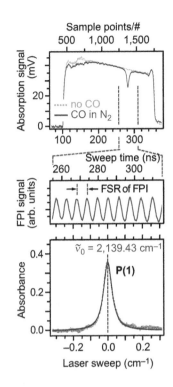

**Figure 8.11** Example for intrapulse p-QCL absorption spectroscopy, for ppm traces of CO in atmospheric gas mixtures. (Top) Frequency-chirp data, with and without CO present in the cell; (center) frequency marker data from a reference FP interferometer; (bottom) absolute absorbance data (open circles) and Voigt-profile fit (red trace). (Data adapted from Nwaboh, J.A. et al., *Appl. Phys. B* 103, no. 4 (2011): 947–957.)

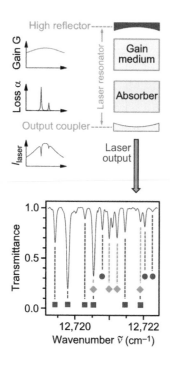

**Figure 8.12** Concept of laser intracavity absorption spectroscopy (ICLAS). The example spectrum is that of a $H_2O$:HDO:$D_2O$ mixture within a Ti:sapphire laser cavity; lines associated with the three isotopologues are indicated by the symbols $\lambda$, $\upsilon$, and $\nu$, respectively. (Spectral data adapted from Naumenko, O.V. et al., *J. Mol. Spectrosc.* 248, no. 2 (2008): 122–133.)

action will produce holes in the laser output spectrum at the frequency positions where small quantities of a narrow-line absorber are present. The principle of the laser gain/loss action is schematically illustrated in the top part of Figure 8.12; the laser cavity is made up of two mirrors M1 and M2, and the medium of broadband gain $G(v)$ and an absorbing sample with loss spectrum $\alpha(v)$ are depicted. The laser output then is depleted at those loss frequencies due to absorption.

The intracavity absorption follows the Beer–Lambert law in the form $I(v) = I_0(v) \cdot \exp(-\alpha(v) \cdot \ell_{eq})$, where as usual $I_0$ and $I$ are the initial and transmitted intensity, respectively; $\alpha(v)$ is the absorption coefficient; and $\ell_{eq} = c \cdot t_g$ is the "equivalent-path" length [with $t_g$ being the generation time, measured from the onset (buildup) of the laser radiation until the moment of observation; $c$ is the speed of light]. One of the key advantages of ICLAS is that the cavity losses are well compensated for by the laser gain, making possible much longer effective light paths than those achievable when using a passive cavity, as in CRDS. As a result, absorption coefficients as low as $10^{-10}$–$10^{-11}$ cm$^{-1}$ have been measured (see Kachanov et al. 1989). However, in order to guarantee an adequate spectral sensitivity, multimode ICLAS applications require that the absorber linewidth is smaller than the gain medium's homogeneous spectral broadening. An illustrative example for the capabilities of ICLAS is shown in the bottom part of Figure 8.12. Here, the effect of the absorption of small amounts of the three water isotopologues on the laser output is shown for part of the $3v_1 + v_2 + v_3$ vibrational combination band (Naumenko et al. 2008). The authors recorded their spectra within the cavity of a Ti:sapphire laser, with an intracavity absorption cell of length $L_{cell}$ = 65 cm that constituted about 40% of the optical laser cavity. Typically, the spectra were recorded using a generation time of $t_g$ = 190 $\mu$s, which equates to an equivalent absorption path length of $\ell_{eq} \approx 24$ km. In their measurements, Naumenko et al. achieved sensitivities (noise-equivalent absorption) of the order of $\alpha_{min} \approx 10^{-9}$ cm$^{-1}$, which allowed them to detect very weak transitions with line strengths as low as ~$5 \times 10^{-28}$ cm·molecule$^{-1}$.

## 8.4.2 Cavity ring-down spectroscopy

From a conceptual point of view, CRDS is a simple technique. As illustrated in Figure 8.13, the absorption medium is situated inside two highly reflective concave mirrors, into which a laser pulse is injected through one of the mirrors and the cavity output pulse detected after exciting the cavity through the second mirror. The number of "rings" between the two mirrors depends on the experimental conditions, i.e., the geometrical configuration, but mostly the mirror reflectivities.

The "decay" of the light after injection into the cavity can be traced as follows (for simplicity, a short injected laser light pulse is considered, and the reflectivity $R$ of the two mirrors is assumed to be equal). If a laser pulse of initial intensity $I_{in}$ is injected into the cavity, for each subsequent round trip, the intracavity intensity will decrease by a factor $R^2 \cdot \exp(-2\alpha(v) \cdot L_c)$; thus, after $n$ round trips, the intensity measured by a detector, placed after mirror M2, would be

$$I_n = I_0 \cdot [R \cdot \exp(-\alpha(v) \cdot L_c)]^{2n} = I_0 \cdot \exp\{2n \cdot (\ln R - \alpha(v) \cdot L_c)\}, \qquad (8.15)$$

where $I_0$ is the transmitted intensity when $I_{in}$ "hits" mirror M2 for the first time, i.e., $I_0 = T \cdot \exp(-\alpha(v) \cdot L_c) \cdot I_{in}$. Here, $T = 1 - R$ is the mirror transmissivity, $\alpha(v)$ is the (frequency-dependent) absorption coefficient, and $I_{in}$ is the intensity of the initially injected laser pulse. For high-quality dielectric mirrors, reflectivities tend to nearly $R \to 1$ ($R \geq 0.999$ is not uncommon); one finds that $\ln R \approx -(1 - R)$, and one can write

$$I_n \approx I_0 \cdot \exp\{-2n \cdot (1 - R) - 2n \cdot \alpha(v) \cdot L_c\}. \qquad (8.16)$$

Normally, the light detector output signal is filtered with a time constant that records only the time envelope of the pulse train—as shown in the bottom part of Figure 8.13a—and, consequently, it is convenient to change from the discrete counting variable $n$ to the continuous time variable $t = 2n \cdot (L_c/c)$, where $c$ is the speed of light. With this change of variables, one obtains for the time-dependent light intensity emerging from the ring-down cavity

$$I(t) = I_0 \cdot \exp\{-t/\tau_0 - c \cdot \alpha(v) \cdot t\} \equiv I_0 \cdot \exp\{-t/\tau\}, \qquad (8.17)$$

where the "empty cavity decay time"—also called *cavity ring-down time*—is defined as $\tau_0 = L_c/[c \cdot (1 - R)]$, and one exploits the key equation in CRDS

$$\tau^{-1} = \tau_0^{-1} + c \cdot \alpha(v), \qquad (8.18)$$

which indicates that the light in a cavity filled with an absorbing sample decays with a specific ring-down time $\tau$. Note that for high-Q cavities, the ring-down time $\tau_0$ (for an empty cavity) can be of the order of many microseconds, or in other words, thousands of round trips take place before the final, initially injected photons have seeped out. This is associated with a very long effective photon absorption path, which can well exceed 50 km or so. Thus, by comparing the ring-down times $\tau$ and $\tau_0$, one can immediately extract quantitative absorption spectra when measuring the ring-down time as a function of laser light frequency, the essence of CDRS; sensitivities of the order $\alpha{\sim}10^{-9}{-}10^{-10}$ cm$^{-1}$are easily achievable.

In Figure 8.13b, an example for a CRDS measurement is shown. In that experiment, the cavity was not completely filled with an absorbing gas, but rather a molecular beam generated in slit-jet expansion, to produce cold sample molecules (without or with a discharge, to obtain free radicals like the OH-molecules shown here); see Wu et al. (2006). The study demonstrated the potential of CRDS for obtaining high-resolution spectra of both reactive and nonreactive species (throughout the entire IR region); for example, OH could be detected with a sensitivity of a few ppm per pass, translating into a minimum number of detectable molecules of $N_{min} = 2.1 \times 10^9$ cm$^{-1}$.

In a second example, results from CRDS measurements of the $(v_2 + v_3)$ vibrational combination band of acetylene ($C_2H_2$) are shown. In their experiment, Santamaria et al. (2016) used a cavity ring-down spectrometer to probe a buffer-gas-cooled molecular beam, at different distances from the expansion nozzle; the experimental scheme is sketched in Figure 8.14a, with two typical results for scanning the laser frequency across a single rotational line—$R(0)$—of the band spectrum. The narrow-linewidth diode laser, stabilized against a rubidium atomic clock via an optical frequency comb synthesizer, allowed for extracting linewidth parameters from the spectra, demonstrating not only high-resolution spectroscopy of molecular transitions but also temperatures for a cooled

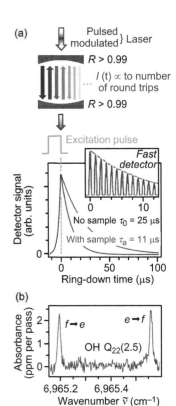

**Figure 8.13** Principle of CRDS. (a) Injection of a laser pulse into a high-finesse cavity and the associated ring-down oscillations associated with the number of cavity round trips. (b) Example spectrum for jet-cooled OH radical within a ring-down cavity, probed by narrow-bandwidth ($\delta v_{laser} \sim 50$ MHz) Ti: sapphire laser pulses of nanosecond duration. (Data adapted from Wu, S. et al., *Phys. Chem. Chem. Phys.* 8, no. 14 (2006): 1682–1689.)

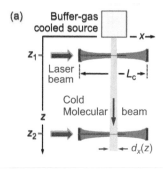

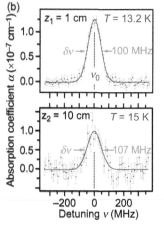

**Figure 8.14**   CRDS of a buffer-gas cooled molecular beam. (a) Schematic of the setup to probe molecular beam density and divergence; relevant parameters are indicated. (b) Example spectra for the $v_1 + v_3$ R(0) line of $C_2H_2$ (with acetylene in helium at flow rates of 5 and 10 sccm, respectively), measured for the indicated distances $z_i$ from the beam nozzle. (Data adapted from Santamaria, L. et al., *Phys. Chem. Chem. Phys.* 18, no. 25 (2016): 16715–16720.)

molecular beam that could be determined from a fit to the experimental line profiles (see Figure 8.14b).

In a final example, the "broadband" character of CRDS is demonstrated, meaning that the laser source is scanned over a large spectral region, either to cover widely spaced features of an absorbing molecule or to probe a multitude of different molecular species. The data shown in Figure 8.15 are for $NO_2$; the laser tuning range is $\Delta\tilde{v} \approx 7550\ldots7900\ \text{cm}^{-1}$ (see Lukashevskaya et al. 2016 and references therein); the CRDS setup of the authors incorporated a fiber-connected cw-DFB diode laser system. The ring-down time of the CRDS cavity was on the order of $\tau_0 \approx 100\ \mu s$; typically, about 30 ring-down events were averaged for each data point, with a resolution of $\delta\tilde{v} \approx 2.5 \times 10^{-3}\ \text{cm}^{-1}$. The noise-equivalent absorption that the authors achieved was $\alpha_{min} \sim 10^{-10}\ \text{cm}^{-1}$.

In the upper panel of the figure, a CRDS overview spectrum for $NO_2$ is shown; the relevant vibrational band transitions are indicated by horizontal bars and annotated with the key quantum numbers of the upper level (the lower level is always 000). To illustrate the high spectral resolution and the high sensitivity of the method, the bottom panels in the figure show some particular details of the weakest band 331–000. On the left, a segment with several vibro-rotational lines is shown, which highlights the spectral resolution (indicated within the data panel). On the right, a different part of the same band is displayed for which the (theoretical) positions of doublets due to spin-rotation splitting are marked by short, red lines; the spectrum reveals indeed that all these spectral transitions are resolved.

## 8.5  TERAHERTZ SPECTROSCOPY

Terahertz radiation broadly comprises frequencies in the range $v \sim 0.1\text{–}10$ THz (1 THz = $10^{12}$ Hz) of the electromagnetic spectrum, and is located between microwave (MW) and IR radiation. Terahertz radiation is also known as T-rays or as millimeter radiation because of the associated wavelength range $\lambda \sim 0.03\text{–}3$ mm. The molecular modes and physical activities in the terahertz region of the electromagnetic spectrum are illustrated in Figure 8.16; it can be seen that, while the rotational modes active in this region are qualitatively similar to those observed in the MW region (except for some higher energies), a significant qualitative difference occurs with respect to the vibrational modes active in the terahertz region compared with those traditionally observed in IR spectroscopy. In particular, in the terahertz region, we deal with vibrational motion associated to weak bonds, weak interactions, and large molecules having large reduced masses. Thus, the type of vibrational spectroscopy pursued in the terahertz region involves biomolecule (DNA, proteins, lipids, etc.) interactions, crystalline phonon vibrations, and other large-molecular scale processes.

### 8.5.1  Basic features and experimental methodologies

Terahertz radiation has some remarkable properties since many materials, including living tissues, are semitransparent and show "terahertz fingerprints" allowing them to be identified, analyzed, and imaged. In addition, their low-energy and non-ionizing, nondestructive properties make them intrinsically safe and very suitable for medical, industrial, and environmental applications.

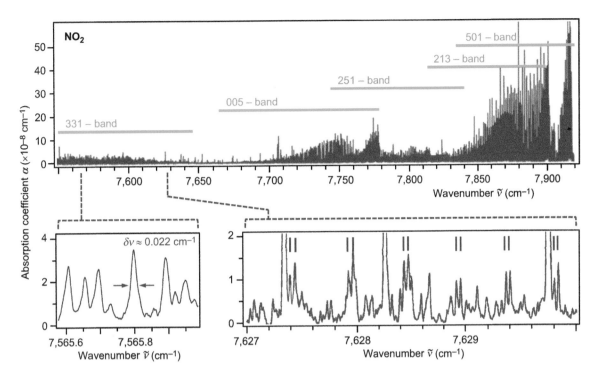

**Figure 8.15** High-resolution CRDS spectroscopy of $NO_2$, at partial pressure $p \approx 11.8$ mbar for the monomer, using a cw-DFB diode laser system. (Top) Overview spectrum; the dominant vibrational bands are indicated. The expanded-scale spectral segments demonstrate the Doppler-limited spectral resolution (left) and the identification of spin-rotation splitting of the $K_a = 5/N = 13$–$18$ transitions of the $3\nu_1 + 3\nu_2 + \nu_3$ band of $NO_2$. (Data adapted from Lukashevskaya, A.A. et al., *J. Quant. Spectrosc. Radiat. Transfer* 177 (2016): 225–233.)

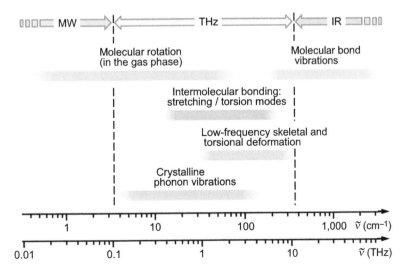

**Figure 8.16** Molecular modes and activity in the terahertz region of the electromagnetic spectrum. Scales are provided in wave number and frequency domain.

For quite some time, terahertz radiation was difficult to access from an experimental point of view. On one hand, the low-frequency terahertz region based on the radiation emission from high-speed transistors is limited at about $\nu \leq$ 300 GHz ($\lambda \geq$ 1 mm), although, in practice, the technology is already rather inefficient for frequencies about $\nu \geq$ 50 GHz. On the other hand, at the high-frequency end of the region, the radiation emitted by semiconductor laser devices, in general, only stretches down to about $\nu \geq$ 30 THz. As a result, from a technological point of view, the terahertz electromagnetic region remained largely out of reach, and it explains why the terahertz region has remained underdeveloped until rather recently, despite its remarkable properties and the identification of various possible applications (such as, e.g., for chemical compound detection, in astronomy, and in medical imaging). Only when Köhler et al. (2002) developed a new semiconductor laser device, capable of delivering intense radiation at $\nu = 4.4$ THz ($\lambda = 67$ μm), was the deadlock broken. Notably, their (monolithic) terahertz injection laser was based on interminiband transitions in the conduction band of a semiconductor (GaAs/AlGaAs) heterostructure.

Today, terahertz radiation can be produced and detected by several methods. The terahertz sources include gyrotrons, backward-wave oscillators, resonant-tunneling diodes, synchrotrons, free electron lasers, narrowband QCLs, and broadband generation utilizing femtosecond-laser pulses.

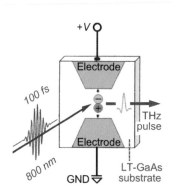

**Figure 8.17** Terahertz wave generation in photoconductive emitters. A voltage bias is applied to the electrodes and accelerates free carriers in the substrate. The movement of these carriers radiates a terahertz electric field according to Maxwell's equations.

In particular, the latter technique involving *photoconductive emitters* has proven to be one of the most efficient approaches for the generation of terahertz waves by converting ultrashort visible/NIR laser pulses to terahertz radiation (Pawar et al. 2013). Essentially, this type of terahertz wave generator goes back over 40 years to a device developed by Auston (1975). The basic principle of the methodology is illustrated in Figure 8.17. A femtosecond-laser pulse generates electron–hole pairs in a semiconductor crystal, above the band gap. Subsequently, the photoexcited carriers are accelerated by an applied electric field, producing a transient current pulse that finally emits terahertz radiation. The process can easily be described by the Maxwell equations.

*Photoconductive receivers* operate in a reverse manner to photoconductive emitters, meaning that the terahertz radiation field incident on the receiver accelerates the excited carriers, generate a photocurrent that is subsequently detected. Thus, the photoconductive detector does not need to support a high voltage (like in the emitters), but only the generated electric field.

In principle, a terahertz device can work using two modalities, namely (1) like a radar system by sending out non-ionizing pulses and recording the echo signal—from which one may reconstruct "images," e.g., to detect explosives, weapons, or drugs hidden under clothes or other "transparent" materials; or (2) like a broadband frequency spectrometer to obtain absorption spectra—revealing, e.g., the chemical composition of an investigated sample. In other words, terahertz scanners, which now start to appear as commercial instruments, can provide not only 3D images of an object but also information on its chemical composition. Three forms of terahertz spectroscopy are typically used (in addition to the common terahertz imaging, as discussed in Chapters 21.1 and 21.2):

- Terahertz time-domain spectroscopy (THz-TDS)

- Time-resolved terahertz spectroscopy (TRTS)

- Terahertz emission spectroscopy (TES)

Of these three methodologies, THz-TDS is probably the most common one, and a more detailed discussion of terahertz spectroscopy thus focuses on this particular type. The simplified, schematic experimental arrangement for THz-TDS is illustrated in Figure 8.18, both for refection and transmission geometries.

The terahertz pulses are created and detected using short NIR laser pulses ($\tau_p = $ 10–100 fs), by and large from Ti:sapphire lasers operating at a center wavelength of $\lambda \approx 800$ nm. The terahertz pulses of duration of $\tau_{THz} \approx 1$ ps are generated and detected by photoconductive antenna devices, conceptionally similar to those shown in Figure 8.17. This type of THz-TDS has the advantages of using coherent radiation and, in addition, of providing picosecond or subpicosecond time resolution.

Typically, in a terahertz spectrometer—for example in transmission geometry—one measures the electrical field strengths $E_i$ and $E_t$ before and after transmission, respectively. The ratio of these two field quantities—often known as the transmission factor $T(v)$—is given by (see Walther et al. 2000)

$$T(v) = E_t/E_i = RL_{n(v)} \cdot \exp\{-\alpha \cdot d - (i2\pi/c) \cdot v \cdot d\}; \qquad (8.19)$$

here $\alpha$ and $d$ are the absorption coefficient and thickness of the sample, respectively; $v$ is the frequency of the terahertz radiation; $c$ is the speed of light (in the vacuum); and $RL_{n(v)}$ is a factor that accounts for reflection losses at the sample surface (as a function of the frequency-dependent refractive index $n(v)$). Thus, by measuring $T(v)$, both the absorption coefficient and refractive index of the probed material can be determined.

As indicated in Figure 8.18, normally the electric field of the terahertz pulse is measured coherently in the time domain using a gated detection scheme; an optical delay unit changes the time delay between the terahertz generation and terahertz detection arms of the spectrometer in order to retrieve the terahertz waveform.

A typical example is shown in Figure 8.19a, where the terahertz pulses before and after passing through a 9-cis retinal sample are depicted (Walther et al. 2000). Note also how the terahertz radiation passing through a sample is *attenuated* and *delayed*, according to the sample absorbance and refractive index, respectively, as described earlier in the transmission equation 8.19. The corresponding frequency spectrum obtained by Fourier transform of the time-domain reference and sample waveforms is shown in the lower panel of Figure 8.19a. Appropriate analysis of these types of data allows determination of the sample absorptivity (see Figure 8.19b) and refractive index, which is at the core of THz-TDS.

Clearly, the development of coherent terahertz radiation generators and detectors has been the driving force for tremendous activity in terahertz spectroscopy, both for fundamental and applied research applications. Several key examples are given below, to illustrate the tremendous progress experienced in this relatively new field of spectroscopy.

## 8.5.2 Applications of terahertz spectroscopy in molecular structure and chemical analysis

The first demonstration of a THz-TDS system was carried out by Grischkowsky et al. (1990) who reported the absorption spectrum of water vapor; due to its

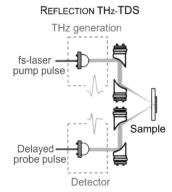

REFLECTION THz-TDS

THz generation

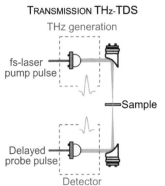

TRANSMISSION THz-TDS

THz generation

**Figure 8.18** Basic experimental arrangements for THz-TDS, both for reflection and transmission geometries. The terahertz pulses are generated by femtosecond-laser pulses in a photoconductive switch (antenna) and are detected by a second, similar photoconductive antenna, stimulated by a femtosecond-laser pulse, which is delayed relative to the generating pulse.

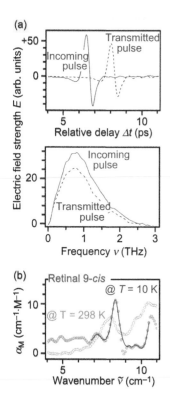

**Figure 8.19** Typical THz-TDS experiment, exemplified for probing a 9-*cis* retinal sample. (a) Terahertz pulses are shown (top) prior to and after transmission through the sample; Fourier transform of the pulses yields the corresponding frequency spectrum (middle). (b) Deduced absorption spectra, revealing structural information about the probed molecule; note that at room temperature this information is lost. (Data adapted from Walther, M. et al., *Chem. Phys. Lett.* 332, nos. 3–4 (2000): 389–395.)

crucial role in the Earth's atmosphere balance, a detailed knowledge of this compound spectroscopy is essential for climate modeling, and terahertz spectroscopy data have greatly contributed to a better understanding of the underlying processes, as continued studies and results have shown. For example, Hoshina et al. (2008) carried out detailed measurements of the pressure broadening of pure rotational transitions in water (utilizing low-frequency range using THz-TDS), without and with the inclusion of gaseous nitrogen and oxygen, aiming at understanding exchange reactions and inelastic collisional processes.

The new analytical capabilities provided by terahertz spectroscopy have also been applied to the investigation of more complex molecular entities, such as, e.g., explosives—because of their great importance in security applications.

In Figure 8.20, some terahertz absorption spectra of a range of common explosives are shown, once again recorded using the technique of THz-TDs (see Fan et al. 2007). The range of explosive materials investigated in said work included 1,3,5-trinitroperhydro-1,3,5-triazine, known as RDX; 1,3-dinitrato-2,2-bis (nitratomethyl)propane 6, known as PETN (a "plastic" explosive); *SX2*, trade name of an RDX-based plastic explosive; and *Semtex-H*, trade name of a plastic explosive containing both RDX and PETN. In support of these measurements, *Density Functional Theory* (DFT) calculations of the terahertz frequency vibrational modes of many of these materials have also been carried out to help in the identification and in the tracing of manufacturing origin (see Chen et al. 2007).

### 8.5.3 Applications of terahertz spectroscopy in biology and medicine

It is well known that the collective modes and population of intramolecular hydrogen-bond modes of a protein depend on its tertiary structure: low-frequency vibrational modes are related to hydrogen bonding and van der Waals interactions. Hence, terahertz spectroscopy provides an access route to investigate protein dynamics by probing conformational changes—they directly affect the absorption and dielectric response in the terahertz range. Protein processes/interactions to which terahertz spectroscopy would be sensitive are protein-ligand binding, hydration, photoactive processes, oxidation, denaturation, and fibrillation.

There are now far too many examples to do justice to the field of terahertz spectroscopy in biology and medicine within the constraints of this chapter; only a few highlights can be provided, and for further insight into the wealth of applications, the reader is referred to the relevant literature, or—as a starting point—to the book by Son (2014).

*Photoactive processes.* In the late 2000s, Castro and Johnston (2008) observed how the partial unfolding of photoactive yellow protein leads to an increase in absorption at terahertz frequencies; this was explained to be due to an increase in the density of delocalized vibrational modes in the more flexible partially unfolded state.

*Intermolecular vibrations.* Terahertz radiation has been used to probe intermolecular vibrations due to van der Waals forces and hydrogen bonding, i.e., weak interactions that determine the conformation and dynamics of biomolecules, like

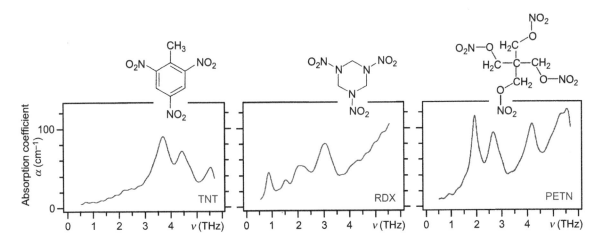

**Figure 8.20** Terahertz absorption spectra of a range of common explosives; the pure explosive samples (TNT, RDX, and PETN) were imbedded in a PTFE matrix. (Data adapted from Fan, W.H. et al., *Appl. Spectrosc.* 61, no. 6 (2007): 638–643.)

the aforementioned proteins. An example of such sensitivity is given in Figure 8.21 for an *in vivo* study of DNA nucleobases in human skin (see Pickwell-MacPherson and Wallace 2009). Terahertz spectra of three of the four nucleobases of DNA—cytosine (C), guanine (G), and thymine (T)—are displayed, showing their distinct spectroscopic signatures. For each nucleobase, spectra were recorded at $T = 300$ K and $T = 10$ K; it should be noted that, at room temperature, the spectral features are basically broad resonances, while at cryogenic temperatures, they split up into several narrowbands at slightly higher frequencies, as is evident from the displayed data. This observation was attributed to the shortening of the molecular bond lengths at lower temperatures (see Fischer et al. 2002).

*Cancer diagnosis.* As mentioned earlier, the photon energy of terahertz radiation is, unlike that for x-rays, low enough not to damage tissue and DNA. Certain frequencies of terahertz radiation not only can penetrate several millimeters of tissue with low water contrast and be back-reflected, but also can sense differences in water content and tissue densities; these key features are currently exploited for medical diagnosis of cancer. The discrimination between normal and tumorous tissues is mainly based on the higher cell density and higher water content in the tumor region. Consequently, the refractive index and absorption coefficient of tumor tissues are higher than those of normal tissues, as was recently demonstrated in fresh rat brain tissue using THz-TDS (see Yamaguchi et al. 2016).

*Crystalline polymorphism.* Terahertz spectroscopy can provide information on collective vibrational modes, namely—as already stated further above—in biomolecules; for intermolecular vibrations between biomolecules and water; and for different types of biological tissues. The high sensitivity of terahertz spectroscopy to both intermolecular and intramolecular vibrations in chemical species makes it an ideal tool to investigate crystalline states, i.e., its polymorphism. It can identify the different crystalline forms of drug molecules, which may have different solubilities and bioavailabilities. Therefore, terahertz spectroscopy may be utilized to assess drug efficiency on and after administration (see Ikeda et al. 2010).

## 8.6 PHOTOACOUSTIC AND PHOTOTHERMAL SPECTROSCOPY WITH LASERS

(a)

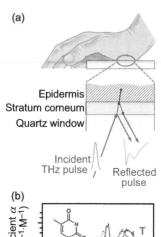

(b)

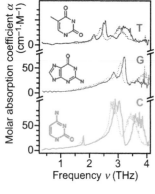

**Figure 8.21** *In vivo* terahertz spectroscopy to differentiate hydrate forms of lactose. (a) Conceptual experimental procedure for *in vivo* skin measurements. (b) Molar absorption coefficients of the nucleobases C, G, and T, recorded at 10 K (solid curves) and 300 K (dashed curves). The molecular structure for each nucleobase is drawn next to the corresponding spectrum. (Data adapted from Pickwell-MacPherson, E. and V.P. Wallace, *Photodiagn. Photodyn. Ther.* 6, no. 2 (2009): 128–134.)

Photoacoustic spectroscopy (PAS) is a calorimetric spectroscopy technique based on the photoacoustic effect: an acoustic wave is generated by the absorption of (modulated) light on passage through an absorbing medium. The effect was first described by Bell (1880). Essentially, the absorption of the modulated (laser) radiation by molecules generates excited energy states; via collisional relaxation processes, the excitation energy can be converted into local heating synchronous with the excitation light modulation frequency.

In gaseous media, the periodic heating/cooling is associated with the generation of pressure waves—when also adjusting the photon energy, one normally now speaks of *photoacoustic spectroscopy (PAS)*, while in the liquid or solid phase, it means that periodic spatial expansion is observed—this is usually addressed as *photothermal spectroscopy* (PTS), which in turn are converted to pressure waves again at the interface with a gaseous medium, e.g., the ambient air. These pressure waves are detected with a microphone, and in general, the modulation frequency is filtered by using a lock-in amplifier in order to suppress ambient noise contributions.

The conceptual experimental arrangement for PAS is shown in Figure 8.22a. The laser beam is aligned with the cylindrical axis of the photoacoustic cell, and a microphone is incorporated into the enclosure, which isolates the probed sample from the laboratory environment. Note that, in most cases, a gas-handling system is added to the cell; this allows for static filling or to follow the changes in dynamic flows.

Since in general PAS signals are very small—specifically if the absorption in trace molecules inside the cell is to be detected—quite often, an acoustic enhancement cavity is incorporated into the setup, in which the acoustic resonator frequency matches the (laser) light modulation frequency; an acoustic standing wave develops inside the resonator. The microphone is located at the center of the resonator where the acoustic standing wave intensity exhibits its maximum (see Figure 8.22b). In addition, the signal can be enhanced even further when the effective optical path length is increased by adding a cavity for local laser light enhancement (the principles of laser cavity enhancements have been discussed in Section 8.4). This variant is also indicated in Figure 8.22b; setups like these are now frequently used in trace gas analysis. The methodology is known as cavity-enhanced (resonant) PAS (CEPAS or CERPAS); as a further improvement for detection sensitivity and signal stability, part of the transmitted radiation is utilized in a feedback loop to control the probe laser—both its emission and modulation frequencies; the resulting modality is known under the acronym OF-CERPAS. Using the various photoacoustic enhancement techniques, absorbing gas-phase species can be detected with extremely high sensitivity; they have been demonstrated to reach a normalized noise-equivalent absorption coefficients of $\alpha_{min} = 2.6 \times 10^{-11}$ $cm^{-1} \cdot s^{1/2} \cdot W$, for laser powers of $P_{laser} \sim 1$ W and data acquisition times of $t_{acq} \sim 1s$ (see Hippler et al. 2010).

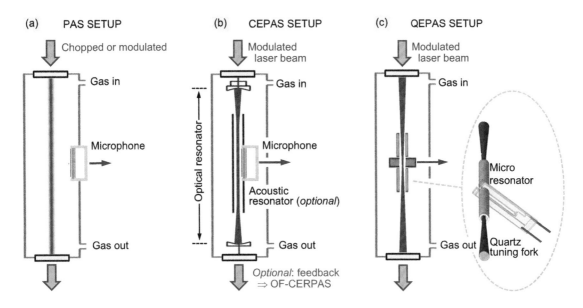

**Figure 8.22**  Typical experimental arrangements for photoacoustic measurements, utilizing modulated laser light; (a) basic configuration for PAS; (b) cavity-enhanced configuration (CEPAS), with the optional variant of laser-feedback stabilization (OF-CEPAS); (c) configuration in which the microphone is replaced by a quartz tuning-fork unit (QEPAS).

As stated in Section 8.1, the signals for small laser light absorption follow the Beer–Lambert law along the approximations given in Equations 8.1 and 8.2. Since the photoacoustic response signal recorded by the microphone is a direct consequence of the conversion of the absorbed laser intensity, $\Delta I$, into an acoustic wave, the generated acoustic pressure change, $\Delta p_{ac}$, should be directly proportional, i.e., $\Delta p_{ac} \propto \Delta I$. One finds

$$\Delta p_{ac} = k_{cell} \cdot \Delta I = k_{cell} \cdot I_0 \cdot \sigma(v) \cdot N \cdot \ell, \qquad (8.20)$$

where $k_{cell}$ is a cell-response constant, which is by and large related to the geometry of the cell. For example, for cylindrical layouts, the cell-constant value depends on whether one considers a resonant or nonresonant cylindrical configuration (Harren et al. 2012); in the case of a cylindrical resonant cell, the associated cell constant is

$$k_{cell} = \frac{G \cdot \leq (\gamma - 1) \cdot L_{res} \cdot Q}{2\pi v \cdot V_{res}}; \qquad (8.21)$$

here $G$ is a geometric factor; $\gamma$ is the specific heat constant, $v$ is the modulation frequency; $L_{res}$ and $V_{res}$ are the length and volume of the resonator, respectively; and $Q$ is the so-called quality factor of the generated acoustic resonance, which is equal to the ratio of the energy stored in the acoustical standing wave over the energy losses per cycle.

From the spectroscopy point of view, the MIR wavelength region is one of the most relevant in PAS since many molecules of analytical interest have strong absorption lines in the range $\lambda = 2.5$–$25$ μm. This is the main reason why in the 1970s and 1980s, the (fixed-wavelength) output from $CO_2$ lasers (range $\lambda_{CO2} \approx 9...$ 11 μm) and CO lasers (range $\lambda_{CO} \approx 4.6...8.2$ μm) have been so popular.

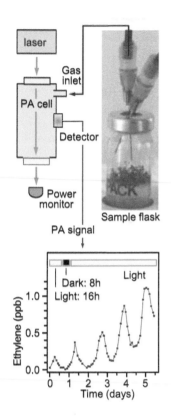

**Figure 8.23** Influence of day length on ethylene production in *Arabidopsis*. (Data adapted from Thain, S.C. et al., *Plant Physiol.* 136, no. 3 (2004): 3751–3761.)

One particular example is that of the plant hormone ethylene ($C_2H_4$) that—although primarily associated with fruit ripening—has a diversity of effects throughout the plant life cycle (see de Martinis et al. 2015). In addition, ethylene has been extensively investigated by using PAS in a multitude of scientific areas, including analytical chemistry, environmental gas analysis, and plant physiology (see Harren et al. 2012). All these investigations were facilitated by the fact that the Q-branch of the $v_3$-band of ethylene is in near resonance with the $10P_{14}$-line of the $CO_2$ laser; detection limits of LOD=6pptv for $C_2H_4$ in $N_2$ have been achieved (see Harren et al. 1990). Overall, PAS is particularly popular for trace-gas monitoring in a wide range of applications in atmospheric, chemical, environmental, biological, agricultural, and human health sciences; a wealth of detailed examples is given in the reviews by Johnson and Ecker (1998) or Harren et al. (2012).

Here, just one illustrative PAS example related to plant physiology is given. In their studies, Thain et al. (2004) investigated the release of ethylene by *Arabidopsis thaliana*, a well-known model plant system. *Arabidopsis* seedlings were grown for 6 days, with periods of 16 h light followed by 8 h of darkness. Subsequently, they used PAS on gas samples from the growth vessel to test for ethylene emission during one further light–dark cycle; this was followed by measurements during which the light was permanently on. As can be seen in the data trace in Figure 8.23, $C_2H_4$ released by the plant displays a circadian rhythm, manifested by high emission during the light and low emission during dark, with a peak in the mid-subjective day.

### 8.6.1 Quartz-enhanced PAS

In (gas-phase) PAS, the common approach today is to use a resonant photoacoustic cell and therefore "store" the absorbed energy in the gas volume of the acoustic resonator—the CEPAS principle outlined above. An alternative method is to accumulate this energy in a sensitive response element, such as a quartz turning fork (QTF), which allows for the detection of weak photoacoustic signals generated in an extremely small volume; this approach is known as *quartz-enhanced PAS* (QEPAS). The high-Q quartz crystals used in these devices are the same as those found as frequency standards in clocks and smart phones, with resonant frequencies around $v_Q \equiv v_{QTF} \sim 32.768$ kHz (see Patimisco 2014). Typically, said QTFs exhibit $Q \approx 20,000$, or higher, in vacuum and around $Q \approx 10,000$ at normal atmospheric pressure (the lowering of the Q-value is associated with the viscous properties of air).

It should be noted that the only effective way to cause the QTF to resonate via the photoacoustic effect is to produce the sound waves from a source located between the two QT-prongs (see Figure 8.22c); the standard way to realize this is for the excitation laser beam to pass through the gap between the prongs, without touching them. For instantaneous V–T (vibrational-to-translational) relaxation, i.e., the generation of the pressure wave, the detected photoacoustic signal can be expressed as

$$S_{QEPAS}(v) \propto Q \cdot \alpha(v) \cdot (I_0/v_{QTF}),  \qquad (8.22)$$

where $\alpha(v)$ is the absorption coefficient and $I_0$ is the incident laser intensity. With $Q$ typically in the range $10^4$–$10^5$, depending on the carrier gas and the gas pressure, QEPAS signals can be substantially larger than ordinary PAS signals. An additional, advantageous feature of QTF elements is that their resonance widths

$\delta\nu_{QTF} \approx 4$ Hz are extremely narrow at normal pressures, which provides high spectral resolution since only frequency components within this window can efficiently excite the QTF vibration.

An interesting recent study based on QEPAS was carried out by Cremer et al. (2016). They demonstrated photoacoustic measurements on optically trapped single nanodroplets, thus providing a direct, broadly applicable method to measure absorption with attoliter sensitivity. The authors used the QTF-configuration illustrated in Figure 8.24a. The droplets of nano- to micrometer size were VIS441/TAG solution droplets (a molecule photosensitive near $\lambda \sim 445$ nm dissolved in tetraethylene glycol).

The droplets were trapped by gradient forces, employing a counterpropagating optical tweezer generated from a cw laser beam at $\lambda = 660$ nm (an image of one single submicron droplet between the tines of the fork is shown in the upper part of the figure). Note that, in principle, the droplet size can be determined from laser light elastically scattered by the droplet. Probing of the droplets is done by a diode laser ($\lambda = 445$ nm) whose power could be varied in the range $P_{445} = 0.3...$ 40 mW; laser modulation was at the resonant frequency $\nu_{QTF}$ of the fork device. The amplified QEPAS-signal was averaged for a few hundreds of milliseconds (roughly equal to the QTF resonance buildup time).

The authors were not only able to measure the presence of the microdroplet within the QTF but also successfully followed the decay in absorption resulting from the population decay of the photoactive substance in the droplet, which is associated with photolysis of said photoactive substance on exposure to the probe laser radiation. An example for this phenomenon is shown in Figure 8.24b, where it is seen that after switch-on of the probe laser, the photoactive substance is nearly completely destroyed after only a few seconds. Based on measurements like the one shown in the figure and quantitative photolysis and diffusion models, the authors estimated an absorption coefficient of $\alpha_{min} = 7.4 \times 10^{-11}$ cm$^{-1}$ for the VIS441-analyte under investigation.

## 8.7  BREAKTHROUGHS AND THE CUTTING EDGE

Absorption spectroscopy may be seen as a fundamental cornerstone in spectroscopy, and thus it is probably not surprising that despite its maturity, efforts have always continued to improve on the overall methodology, or to push it to even greater performance, along with the emergence of new or improved hardware (including laser sources, spectrometers, or detectors). Thus, one encounters "breakthroughs" on a regular basis. In this context SC sources—which provide single-mode, spatially coherent light covering the wavelength range from the UV to the near-IR (NIR), at power levels of several mW·nm$^{-1}$—in combination with enhancement techniques have started to push BB-AS into realms rarely seen before.

### 8.7.1  Breakthrough: Cavity-enhanced absorption spectroscopy utilizing SC sources

Recently, Werblinski et al. (2016) coupled a coherent, broadband SC-source with a cavity-enhancement configuration (CEAS) and a high-speed NIR spectrometer unit. The aim of their study was to provide a proof-of-principle, namely that main

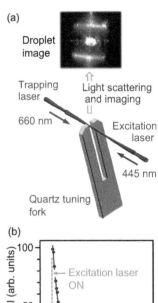

(a)

Droplet image

Trapping laser

660 nm

Light scattering and imaging

Excitation laser

445 nm

Quartz tuning fork

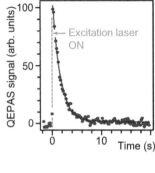

(b)

Excitation laser ON

**Figure 8.24** QEPAS measurement of light-trapped nano-micro-droplets. (a) Conceptual tuning fork setup, with a snapshot photo of a single droplet trapped between the tines of the tuning fork (view from top). (b) Decay of the QEPAS signal, caused by photolysis inside the solute droplet (of radius $a = 1.3$ μm), after the excitation laser ($\lambda = 445$ nm; $P_\lambda = 4$ mW) is switched on. Data points (in blue) are fitted with an exponential decay curve (in red). (Adapted from Cremer, J.W. et al., *Nat. Commun.* 7 (2016): article 10941.)

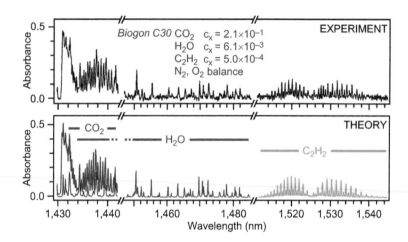

**Figure 8.25** Cavity-enhanced absorption spectrum, measured with an SC source; (top) experimental spectrum at $T = 298$ K (species concentrations are indicated); (bottom) best-matching theory spectrum (vibrational band regions for the three minor species $CO_2$, $H_2O$, and $C_2H_2$ are indicated by the horizontal bars. (Data adapted from Werblinski, T. et al., *Opt. Lett.* 41, no. 10 (2016): 2322–2325; reproduced with the permission of The Optical Society of America.)

and trace gas components encountered in combustion processes (including $H_2O$, $CO_2$, $C_2H_2$, CO, OH, $CH_4$, …) could be measured rapidly (sufficiently fast for later process control) and with high sensitivity (to trace minor chemical reaction products).

An example for an absorption spectrum, recorded with the author's setup for a mixture of acetylene and *Biogon C30* (a premix gas with certified concentrations), is shown in Figure 8.25; only the spectral segments relevant to three selected trace gases are shown here. The experimental data are mimicked nearly to perfection by the theoretical model spectra. Spectra could be recorded with rates of up to 50 kHz, at which the authors achieved minimal absorption coefficients of $\alpha_{min} = 1.56 \times 10^{-5}: 1.01 \times 10^{-5}: 1.13 \times 10^{-5}$ cm$^{-1}$ for $H_2O$: $CO_2$: $C_2H_2$, respectively. Besides the simultaneous detection of multiple species, the determination of gas properties—such as temperature (from a fit to the rotational line spectra) and partial pressures—was possible as well.

### 8.7.2 At the cutting edge: CRDS of optically trapped aerosol particles

As stated quite a few times in this and the previous chapter, absorption spectroscopy is an established and quite mature technique. Besides the steps toward improvements to specific configurations and modalities, the possibility of novel uses and applications has been a motivation for researchers. One particular challenging field for laser analysis is the study of aerosols of nanometer to micrometer size. This requires (1) (single) aerosol particle manipulation through trapping and transport; and (2) a measurement technique that exhibits sufficient spatial and time resolution to be able to pinpoint and track the (single) particle. At the same time, the technique should have chemical (spectroscopic) sensitivity as well, so that differences or changes in the aerosol properties can be detected.

Here an example is described, in which some physical parameter of single aerosol particles was measured, rather than their wavelength-dependent

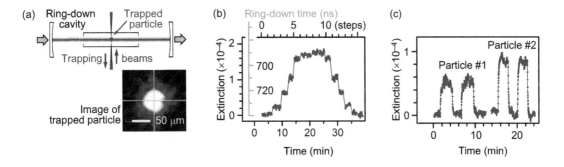

**Figure 8.26**   Optical-trapping CRDS of single-particle aerosols. (a) Conceptual experimental setup, with a scattered-light image of a trapped aerosol particle (perpendicular to the plane of the laser beams). (b) OT-CRDS response when walking a trapped particle through the ring-down beam, step by step ($x_{step}$ = 25.4 μm). (c) OT-CRDS response for single particles of different size. (Adapted from Wang, C. et al., *Appl. Phys. Lett.* 107, no. 24 (2015): article 241903.)

chemical properties. In their experiments, Wang et al. (2015) characterized the physical parameters (size, motion, and restoring-force constant) of single aerosol particles using a technique they call *optical trapping CRDS* (OT-CRDS).

As shown at the top of Figure 8.26, the counterpropagating trapping laser beams, whose position can be adjusted in the $x$-$y$ plane, intersect the laser light buildup volume of a ring-down cavity (in the $z$-direction). A pulsed single-mode Gaussian beam (TEM00) laser source is utilized for trapping (operating at $\lambda$ = 405 nm, with the repetition rate of 1 kHz and pulse duration of $\tau_{p,OT}$ = 0.9 ms). The ring-down laser source comprised a pulsed tunable dye laser (operating at $\lambda$ = 315...320 nm, with the repetition rate of 20cHz and pulse duration of $\tau_{p,CRDS}$ = 10 ns).

When the position of the optical trapping region was laterally adjusted, any trapped particle could be "stepped through" the CRDS probe laser beam. Due to the different shadowing of the probe beam, its attenuation in the ring-down cavity changes proportionally to the particle position; in the cited experiments, the particle-walking step was $x_{step}(y_{step})$ = 25.4 μm. Such a stepped attenuation response of the ring-down cavity is shown in the central panel of the figure, for an aerosol particle of size $d \approx 54$ μm; the particle is walked in about five steps through the center of the ring-down beam. From this and some other, known parameters, the authors deduced single-particle extinction coefficients of the order $\varepsilon_{ext} \equiv \alpha\ell) = 1.7 \times 10^{-4}$ (at $\lambda_{CRDS}$ = 315 nm); this extinction is different for particles of different size (see bottom panel of the figure).

Finally, the authors were able to directly measure an oscillation restoring force constants as small as $k \sim 10^{-10}$N·m$^{-1}$, based on the change in the extinction with $v_{osc}$, as a function of time.

# CHAPTER 9

# Fluorescence Spectroscopy and Its Implementation

The first documented report related to "fluorescence emission" is that by Nicolas Monardos, a Spanish physician and botanic, in 1565. He describes the bluish opalescence of a water infusion from the wood of a tree used by the Aztecs in treating kidney and urinary diseases (see Valeur and Berberan-Santos 2011). Since then, other similar observations were described (for a summary, see, e.g., Harvey 1957), including the one reported in 1845 by the British scientist John Herschel, who investigated the blue emission produced when illuminating a quinine sulfate solution (Herschel 1845). But it was only in 1852, when George Stokes (Stokes 1852) carried out a detailed experimental investigation of the phenomenon (see Section 9.7), that it was discovered that the wavelength of the dispersed reemitted light was always longer than that of the incident light, an effect now known as Stoke's shift.

From the conceptual point of view, a major breakthrough in the understanding of light absorption and emission by a molecule and, therefore, of the fluorescence emission dates to the early 1930s, nearly a century after the phenomenon of fluorescence was initially formulated. In those years, the Polish scientist Aleksander Jablonski developed a type of diagram—in its refined form nowadays widely known as a Jablonski diagram—to illustrate the various molecular processes that can take place after a molecule has been excited by light (Jablonski 1933, 1935), and distinguishing fast processes from resonant states and slow processes from metastable states. Incidentally, in those days, fluorescence times could already be measured quite accurately. For example, a measurement of fluorescence lifetimes of nanosecond duration was reported as early as the mid-1920s (Gaviola 1926).

A further breakthrough for fluorescence spectroscopy is linked to the invention of the laser. In particular, its narrow bandwidth made it possible to study state-selective, interlinked absorption and emission processes with high precision; normally the methodology is known as laser-induced fluorescence (LIF). Nowadays, *fluorescence spectroscopy* and *laser-induced fluorescence spectroscopy* (LIFS) constitute well-developed fields of spectroscopy, and are used in a wide range of disciplines, in particular in biochemistry and biophysics (see, e.g., Lakowicz 2006; Jameson 2014).

**Table 9.1**  Key Fluorescent Aminoacids (Dissolved in Water)

| Aminoacid | Maximum of Absorption $\lambda_{abs}$ (nm) | Maximum of Emission $\lambda_{emis}$ (nm) |
|---|---|---|
| Tyrosine | 274 | 303 |
| Phenylalanine | 257 | 282 |
| Tryptophan | 280 | 348 |

In this chapter, the basic concepts of (laser-induced) fluorescence will be discussed; the description of specific applications and their experimental implementation can be found in Chapters 10 and 18 for spectroscopy and imaging, respectively.

# 9.1  FUNDAMENTAL ASPECTS OF FLUORESCENCE EMISSION

## 9.1.1  The concept of fluorophores

A key concept in fluorescence spectroscopy is that of *fluorophore* (or sometimes called *fluorochrome*), as the often used *chromophore* in color studies. It refers to a fluorescent compound, typically an organic molecule made up of 20 to 100 atoms. Common fluorophores are organic dyes derived from rhodamine, coumarin, fluorescein, etc. At present, fluorescence probes include not only fluorescent dyes and proteins but also quantum dots, or other nanoparticles (Sreenivasan et al. 2013). Since their fluorescence emission is significantly affected by the environment in which they are embedded, this property is exploited to investigate the nature of the fluorophore interaction and bonding with its surrounding medium.

Fluorophores can be classified into two classes: intrinsic and extrinsic, depending upon whether they occur naturally or are added to the studied sample, respectively. One of the most important intrinsic fluorophores in protein study is the indole group of tryptophan. Table 9.1 lists the three major amino acids that show significant fluorescence, with their absorption and emission maxima, when dissolved in water.

Since DNA by itself shows only weak fluorescence emission, one normally uses dyes bound to DNA as extrinsic fluorophores. A good example is the 4′-6-diamidino-2-phenolindole (DAPI) used in DNA probing.

A relevant application of fluorophores consists in the use of so-called fluorescent indicators whose fluorescence emission properties are very sensitive to a specific species (atom, molecule, or ion). Nowadays, there is a rich variety of fluorescence indicators for $Ca^{2+}$, $Mg^{2+}$, $Na^+$, and $Cl^-$, as well as for pH determination.

There are a large number of dyes, nowadays commercially available, which are widely used as noncovalent protein-labeling probes. Typically, these dyes display no or little fluorescence in water, but they do fluoresce strongly upon binding to proteins or membranes. Perhaps one of the most important fluorescence probes

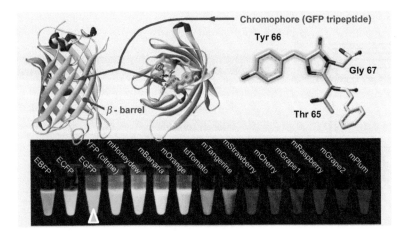

**Figure 9.1** Example of a chromophoric moiety used in fluorescent labeling—*A. victoria* wildtype green-fluorescent protein (GFP). (Left) Molecular structure (front and side view). (Right) Structure of the GFP 238 amino acid triplet. (Bottom) Common chromophoric fluorescent-label compounds, ranging from the blue to the red part of the spectrum; the GFP is indicated by the arrow. (Adapted from Ishikawa-Ankerhold, H.C. et al., *Molecules* 17 (2012): 4047–4132.)

in biology is the green fluorescence protein (GFP) naturally present in the jellyfish *Aequorea victoria*, which was discovered in 1962 by Shimomura et al. (1962) and whose structure is shown in Figure 9.1. This protein has the relevant feature that its chromophore (the oxidized form of the tripeptide sequence serine–tyrosine–glycine), shown in the right of the figure, forms spontaneously upon folding of the polypeptide chain; the specific chromophoric tripeptide is localized within the protective $\beta$-barrel skeleton. The particular chromophore of the wild-type GFP (from *A. victoria*) shown here exhibits a major absorption peak at $\lambda_{abs} \sim 395$ nm and fluoresces strongly at $\lambda_F \sim 509$ nm (indicated by the green "glow" in the GFP structure in the figure), with a quantum yield of QE = 0.77.

## 9.1.2 Principal processes in excited-state fluorescence

Luminescence is nowadays defined as "spontaneous emission of radiation from an electronically- or vibrationally excited species not in thermal equilibrium with its environment" (Braslavsky 2007).

One may distinguish various types of *luminescence* depending on the excitation mode of the species emitting the radiation. In particular, photoluminescence refers to the case in which the species was excited by light, and it may manifest in three distinct forms: fluorescence, phosphorescence, and delayed fluorescence. This chapter is dedicated to the fundamental aspects of *fluorescence*, a term that was first introduced by G.G. Stokes in 1852 (Stokes 1852). Fluorescence is the emission of radiation from an excited state of the same spin multiplicity to the lower state involved in the transition, usually the ground state. The fluorescence transition between the first excited to the ground singlet states, $S_1 \rightarrow S_0$, is usually a quick process with lifetimes of 1ps to 1µs ($10^{-12}$–$10^{-6}$ s).

Before the development of molecular spectroscopy and photochemistry, the empirical distinction between these radiative processes was often based on the duration of the emission. Shorter (longer) lifetimes were typically considered as a distinctive feature associated with fluorescence (phosphorescence). Since the lifetime criterion is not a rigorous one, due to the existence of long-lived

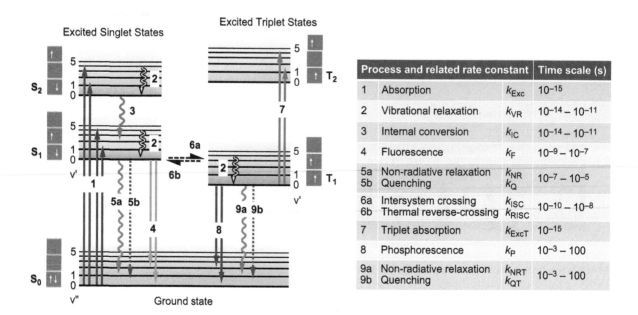

**Figure 9.2**   Jablonski diagram illustrating radiative and nonradiative processes encountered in fluorescing macromolecules. Optical transitions are indicated by full straight arrows, internal nonradiative processes are indicated by wiggly arrows, and quenching processes are marked by dotted arrows. The summary on the right includes the relevant rate coefficients, as well as the timescales of the various processes.

fluorescence (as for example the so-called delayed fluorescence) and short-lived phosphorescence, nowadays a more rigorous distinction is based on conservation or not of the molecular spin multiplicity. In this context, if the multiplicity of the excited state is the same as that of the lower state, the emission is called fluorescence. On the other hand, if the spin multiplicity changes, the emission is called phosphorescence. This is illustrated in the Jablonski diagram of Figure 9.2.

After the absorption of radiation, from the ground state to an excited state, vibrational relaxation induced by molecular collisions brings the excited state down to the lowest vibrational level $v' = 0$ of the excited state. Since the fluorescence emission may end up in higher vibrational levels $v'' > 0$ of the electronic ground state results in the fluorescence spectrum appear to be shifted to lower frequencies (higher wavelengths) than the absorption spectrum, as shown in Figure 9.3.

Note that the transitions between electronic states of molecules involve a change in the electronic structure of the molecule, and the spacing between the constituent atoms of the molecule (associated with a particular electronic state and described by potential energy curves). The photon emission (as well as absorption) processes are related to fast electronic motion with respect to the nuclear (rearrangement) motion; this gives rise to the so-called Franck–Condon principle—the transitions are represented by "vertical" transitions in the molecular potential energy diagram, i.e., without a change in internuclear spacing.

Finally, strictly speaking, phosphorescence would be prohibited under the electric dipole selection rules. However, due to weak spin-orbital coupling, the $T_1 \rightarrow S_0$ transition becomes partially allowed, and (normally) very weak emission is observed. Thus, the two basic characteristics of phosphorescence can be taken as follows: (1) very low light intensity and (2) a lifetime longer than that of fluorescence, usually in the range $10^{-6}$ s up to 1 s.

## 9.2 STRUCTURE OF FLUORESCENCE SPECTRA

If the vibrational spacing is similar for the two electronic states involved in the fluorescence, this spectrum will be the mirror of the absorption spectrum image, as shown in Figure 9.3. In fact, if there were no interactions with the environment, the $(v'' = 0) \leftrightarrow (v' = 0)$ emission and absorption bands would be identical. However, normally this does not occur due to the losses of energy between the molecule and the (embedding) environment.

For molecules in solution, the interactions between the solvent and the electronic excited and ground states of the molecule are different, since the structure of the "cells" that surround these distinctly excited molecules is different. Given that an electronic transition occurs at a higher rate than the reorganization of the respective cells, the energy changes involved in the absorption and emission of photons are therefore different. Because of this, the maxima of the absorption and fluorescence spectra are separated by some energy value; this energy difference is known as the *Stokes shift*.

The Stokes shift is given by $\tilde{v}_{abs} - \tilde{v}_{emis}$ and can be estimated using the so-called Lippert equation (Lippert 1957):

$$\tilde{v}_{abs} - \tilde{v}_{emis} = \frac{2\left(\mu_e - \mu_g\right)^2}{c \cdot h \cdot a^3} \cdot \left[\frac{2(\varepsilon_s - 1)}{2(\varepsilon_s + 1)} - \frac{2(n^2 - 1)}{2(n^2 + 1)}\right], \tag{9.1}$$

where $n$ and $\varepsilon_s$ stand for the refractive index and the dielectric constant of the solvent, respectively; $\mu_g$ and $\mu_e$ are the dipole moments of the fluorophore in the ground and excited states, respectively; $c$ is the speed of light; $h$ is Planck's constant; and $a$ is the Onsager radius of the fluorophore in the associated solvent. Note that in Equation 9.1, the dependence on the dipole moment difference between the excited and ground states of the molecule is highly relevant, as are the solvent properties. Of particular importance is the solvent polarity, giving rise to an effect known as *solvatochronism*.

The occurrence of the Stokes shift explains the high sensitivity of the fluorescence detection. Since elastically scattered light from the excitation radiation can be filtered without difficulty, the fluorescence emission, which is shifted to longer wavelengths, exhibits very low or negligible background.

In many organic molecules in condensed phase, fluorescence usually originates from the $v' = 0$ level of the lowest excited singlet state $S_1$, and likewise phosphorescence from the $v' = 0$ level of the lowest excited triplet state $T_1$. This rule is known as *Kasha's rule* (Kasha 1950), which states that "... the emitting level of a given multiplicity is the lowest excited level of that multiplicity...." The basis of this rule is that internal conversion and vibrational relaxation from upper electronically excited states $S_n$ toward the first excited electronic state $S_1$ are faster than the direct radiation process to the ground state $S_0$. However, there are certain organic compounds whose fluorescence emission does not follow Kasha's rule; an example of this is the non-alternant hydrocarbon azulene, whose fluorescence

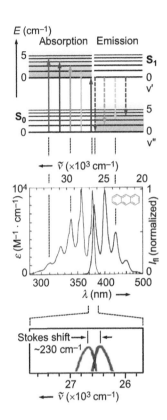

**Figure 9.3** Absorption and fluorescence spectra of anthracene in cyclohexane, for its $^1A_{1g}$-$^1B_{2u}$ transition. The correspondence of spectral features to vibrational transitions is indicated by the arrows in the Jablonski diagram (top). The Stokes shift region of the $(v'' = 0) \leftrightarrow (v' = 0)$ band is enlarged in the bottom panel. (Data adapted from Byron, C.M. and T.C. Werner, *J. Chem. Educ.* 68, no. 5 (1991): 433–436.)

emission could only be interpreted as an $S_2 \rightarrow S_0$ transition (Beer and Longuet-Higgins 1955). Normally, this upper singlet emission is due to a large energy gap between the $S_2$ and $S_1$ states (Turro 1978).

## 9.3 RADIATIVE LIFETIMES AND QUANTUM YIELDS

Let us consider the spontaneous emission process

$$M^* \rightarrow M + h\nu', \tag{9.2}$$

where $h\nu'$ is the energy of the photon emitted, in which the prime stands to denote a lower value than that of the absorbed one (i.e., $h\nu' < h\nu$). Spontaneous emission is a random process that follows first-order kinetics described by the rate equation:

$$-d[M^*]/dt = k_F \cdot [M^*], \tag{9.3a}$$

where the subscript "F" means the (radiative) fluorescence transition $S_n \rightarrow S_0$. In analogy, the subscript "P" refers to the (radiative) phosphorescence transition $T_1 \rightarrow S_0$. The origin of the superscript "0" will be explained below. If we integrate the above equation, one gets

$$[M^*] = [M^*]_0 \cdot \exp(-k_F \cdot t), \tag{9.3b}$$

where $[M^*]_0$ denotes the concentration of $M^*$ for $t = 0$. In photochemistry, one normally defines the natural lifetime $\tau_0$ of a quantum state (solely via radiative decay) through the relation

$$\tau_0 = (k_F)^{-1}. \tag{9.3c}$$

By replacing the time $t \equiv \tau_0$ in the integrated equation 9.3b, one obtains the concentration value $[M^*]_{t=\tau_0} = [M^*]_0/e$; this is associated with the notion that $\tau_0$ is often called the $(1/e)$ lifetime.

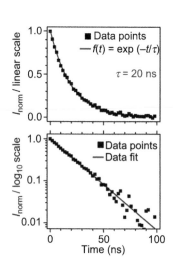

**Figure 9.4** Time dependence of the (normalized) fluorescence emission of an excited species with lifetime of $\tau_0 = 20$ ns. (Top) Linear intensity scale; (bottom) logarithmic scale. Note the linear slope of the fit.

As an example, the fluorescence emission decay of an excited state is shown in Figure 9.4 for the model case of a radiative lifetime value of $\tau_0 = 20$ ns; in the lower part of the figure, the same fluorescence decay is presented in the form of a semilog plot, i.e., $\ln(M^*/M)$ vs. $t$. Note that such a representation exhibits a linear behavior whose slope is precisely the inverse of the lifetime, i.e., $-\tau_0^{-1}$. It is also important to note that, although the lifetime of the excited species $M^*$ in the example is $\tau_0 = 20$ ns, fluorescence radiation is emitted for much longer times, albeit with rapidly decreasing amplitude. One should thus emphasize that the lifetime is a statistical measure associated not to a single molecule but to a molecular ensemble.

It should be noted that the spontaneous emission rate coefficient is a magnitude of relevance in the theory of photophysical processes because it is equivalent to the *Einstein A coefficient*, which in turn is directly proportional to the natural lifetime $\tau_0$ (see Chapter 2). With this knowledge, one can refine Equation 9.3a as

$$-d[M^*]/dt = k_F \cdot [M^*] = (\tau_0)^{-1} \cdot [M^*] = A_{eg} \cdot [M^*], \tag{9.3d}$$

where in the Einstein $A$ coefficient $A_{eg}$, the subscripts e and g denote the excited and ground states, respectively.

It was shown by Strickler and Berg (1962) that the fluorescence rate coefficient and therefore its associated lifetime are related to properties in the absorption and fluorescence spectra, yielding the relation

$$k_F = \left(\tau^0\right)^{-1} = 2.881 \times 10^{-9} \cdot \tilde{v}_{max}^2 \cdot n^2 \cdot \mathcal{A}. \tag{9.4a}$$

Here, $n$ is the refractive index of the fluorophore, assumed to be the same as that of the medium (in particular, over the maximum of the absorption band); $\tilde{v}_{max}$ is the wavenumber at the maximum of the absorption band; and $\mathcal{A}$ is *the integrated absorption coefficient.* If one approximates the absorption by a Lorentzian profile, this integrated absorption coefficient is given by $\mathcal{A} = \pi \cdot \varepsilon_{max} \cdot (\Gamma/2)$, where $\varepsilon_{max}$ is the (decadic) molar absorption coefficient at the maximum of absorption band; and $\Gamma$ represents the full width at half-maximum (FWHM) of the distribution. Using this approximation, the natural life can be calculated from the absorption spectrum using the relation

$$\tau_0 \approx \frac{2.2 \times 10^8}{\tilde{v}_{max}^2 \cdot n^2 \cdot \varepsilon_{max} \cdot \Gamma}. \tag{9.4b}$$

Nowadays, fluorescence lifetime measurements utilize different experimental approaches, as will be described further below.

If one considers the Jablonski diagram shown in Figure 9.5, in which in addition to the fluorescence other processes like internal conversion and intersystem crossing compete with the fluorescence, the temporal evolution of the excited state concentration is more complex, and instead of the simple Equation 9.3 one finds

$$-d[M^*]/dt = k_F \cdot [M^*] + k_{ISC} \cdot [M^*] + k_{IC} \cdot [M^*], \tag{9.5a}$$

which can be integrated to give

$$[M^*] = [M^*]_0 \cdot \exp(-k_{eff} \cdot t). \tag{9.5b}$$

Here the total "effective" rate coefficient is $k_{eff} = k_F + k_{ISC} + k_{IC}$, i.e., $k_{eff} = \sum_i k_i$.

Therefore, the related, composite fluorescence lifetime can be written as $\tau_{eff} = (k_{eff})^{-1} = \left( \sum_i k_i \right)^{-1}$; note that this is shorter than the lifetime in the absence of the competing depletion processes, i.e., $\tau_{eff} < \tau_F$.

This leads naturally to the concept of "quantum yield" that is often represented by the letter $\Phi$. For a given process $i$, $\Phi_i$ is defined as the fraction of excited molecules that undergo the said process. In other words,

$$\Phi_i = \frac{\text{number of photons experiencing process } i}{\text{number of absorbed photons}}$$
$$\equiv \frac{\text{rate of process } i}{\text{number of absorbed photons per unit time per unit volume}} \tag{9.6}$$

**Table 9.2** Example of How to Determine the Quantum Yield (According to Equation 9.6) for a Particular Process, Here Exemplified for the Case of Fluorescence ($\Phi_F$), When in Competition with Intersystem Crossing from $S_1$ into $T_1$. Absorption is Assumed to be a Single-Photon Process

| Process | | Rate Equation |
|---|---|---|
| Formation of $S_1$: | $S_0 + h\nu \xrightarrow{I_{abs}} S_1$ | $d[S_1]/dt = I_{abs}$ |
| Fluorescence from $S_1$: | $S_1 \xrightarrow{k_F} S_0 + h\nu$ | $d[S_1]/dt = -k_F \cdot [S_1]$ |
| Loss out of $S_1$ (here intersystem crossing): | $S_1 \xrightarrow{k_{ISC}} T_1$ | $d[S_1]/dt = -k_{ISC} \cdot [S_1]$ |
| Steady-state solution for population in $S_1$: | $[S_1]$ | $d[S_1]/dt = I_{abs} - k_F \cdot [S_1] - k_{ISC} \cdot [S_1]$ $\equiv 0$ |
| Quantum yield for fluorescence: | $\Phi_F$ | $\Phi_F = \dfrac{k_F \cdot [S_1]}{I_{abs}} = \dfrac{k_F}{k_F + k_{ISC}}$ |

Information and relevant mathematical expressions for an example of how to calculate such quantum yields are collated in Table 9.2, namely, for the fluorescence quantum yield, $\Phi_F$, for the simple case that the fluorescence competes with a single, nonradiative (intersystem crossing) process.

The concept of quantum yield, in combination with well-known expressions for photon absorption, can be used to introduce a frequently used, fundamental equation for fluorescence, which, on the one hand, allows one to determine (absolute) fluorescence intensities and, on the other hand, provides a means to actually measure quantum yields. One finds that the fluorescence intensity, $I_F$, emerging from a sample is proportional to the amount of light that had been absorbed, and the fluorescence quantum yield, $\Phi_F$

$$I_F(c) = C_p \cdot I_0 \cdot \Phi_F \cdot (1 - 10^{-\varepsilon c \ell}) \tag{9.7a}$$

where $C_p$ is a proportionality constant, $I_0$ is the incident light intensity, $\varepsilon$ is the molar absorptivity, $c$ is the concentration of the (absorbing) sample particles, and $\ell$ is the path length over which absorption takes place. Note that the fluorescence intensity is a function of the concentration. Recall also that the exponent expression in the equation is the molar absorptivity, which was already defined in Equation 2.7.

Equation 9.7a has two limits, namely (1) that of an infinite dilute sample and (2) that of the infinite absorption case.

The former limit corresponds to the case that entails $\varepsilon c \ell \rightarrow 0$; after Taylor expansion of Equation 9.7a, this leads to

$$I_F(c) = C_p \cdot I_0 \cdot \Phi_F \cdot 2.3 \cdot \varepsilon c \ell, \tag{9.7b}$$

which constitutes a simple relation of great analytical value, indicating linear dependence of the fluorescence intensity on the sample concentration.

In the latter limit, one has $\varepsilon c \ell \rightarrow \infty$ and thus Equation 9.7a reduces to

$$I_F(c) = C_p \cdot I_0 \cdot \Phi_F, \tag{9.7c}$$

which allows one to determine the absolute fluorescence quantum yield of the sample under investigation.

# 9.4 QUENCHING, TRANSFER, AND DELAY OF FLUORESCENCE

Rarely can molecules be treated in isolation so that only their internal quantum properties are responsible for the observed fluorescence intensity. Normally, they are "imbedded" in an environment of other particles, with which they can interact. If it is probable that an excited molecule (say, in its $S_1$ state) interacts with its environment within its lifetime, then losses in $S_1$ population are expected, and as a consequence, a change in fluorescence intensity is observed. The processes that are commonly encountered are (1) quenching, i.e., radiationless deactivation from $S_1$ to $S_0$; (2) transfer of the photon energy to a close-proximity molecule that can absorb the emitted photon, thus decreasing the parent fluorescence; and (3) thermal retransfer from $T_1$ back into $S_1$, after initial intersystem crossing, which increases the observed fluorescence intensity. These three processes are detailed next.

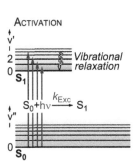

## 9.4.1 Fluorescence quenching and the Stern–Volmer law

Let us assume that a new Q substance is added to the above photophysical reaction scheme, as illustrated in Figure 9.5.

Obviously the effect of the last collisional process is to reduce the fluorescence yield. This is why this process is called collisional deactivation or in this case fluorescence quenching. Applying the steady-state approximation again for $[S_1]$, as in Table 9.2, but now also including a quench term, one can show that fluorescence intensity $I_F = k_F \cdot [S_1]$, is now equal to

$$I_F = \frac{k_F \cdot I_{abs}}{k_F + k_{ISC} + k_Q \cdot [Q]} \qquad (9.8a)$$

where $k_Q$ is the rate coefficient of the quenching process, and $[Q]$ is the concentration of the quenching particles. Note that when $[Q] = 0$, Equation 9.8a reduces to the quenching-free fluorescence intensity

$$I_0 \equiv \frac{k_F \cdot I_{abs}}{k_F + k_{ISC}} \cdot \qquad (9.8b)$$

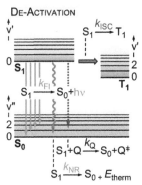

**Figure 9.5** Jablonski diagram depicting the main excitation and deactivation processes, including the relevant rate constants, highlighting the "competition" to the fluorescence emission.

The ratio of the fluorescence intensities without and with quenching yields the so-called *Stern–Volmer equation* (Stern and Volmer 1919):

$$\frac{I_0}{I_F} = \frac{k_F + k_{ISC} + k_Q \cdot [Q]}{k_F + k_{ISC}} = 1 + \frac{k_Q \cdot [Q]}{k_F + k_{ISC}} \cdot \qquad (9.9)$$

This equation predicts that a plot of $I_0/I_F$ versus $[Q]$ should be linear, with a slope of $k_Q/(k_F + k_{ISC})$. From this, $k_Q$ can be estimated provided that the value for $k_F + k_{ISC}$ is known. It should be noted that the Stern–Volmer equation is only valid for processes in the gas phase. In solution, the situation is more complex, because the collisional deactivation processes are faster than those in the gas phase and, in addition, may also be substantially influenced by diffusion.

The quantum yield in the presence of quenching is now given by

$$\Phi_F = \frac{k_F}{k_F + k_{ISC} + k_Q \cdot [Q]} \; . \tag{9.10a}$$

In fact, the quantum yield expression can be then generalized to any set of nonradiative processes and, therefore, can be written in the form

$$\Phi_F = \frac{k_F}{k_F + k_{NR}} \equiv \frac{\tau_{eff}}{\tau_0} \; . \tag{9.10b}$$

Here, $k_{NR}$ represents the sum of all nonradiative rate coefficients, i.e., $k_{NR} = \sum_i k_{NR}(i)$ with $k_{NR}(i)$ being the rate coefficient of the $i$th nonradiative process. It should be noted that the right part of Equation 9.10b denotes the ratio of the "effective" lifetime value in presence of all (radiative and nonradiative) processes, $\tau_{eff}$, and the "natural" lifetime value, $\tau_0$, that refers to the ideal situation that only fluorescence emission contributes to the depletion of $S_1$.

## 9.4.2 Förster resonance energy transfer

One of the most interesting fluorescence quenching cases is the so-called Förster resonance energy transfer (FRET). Essentially, it is a bimolecular fluorescence quenching process in which the excited state energy of a (donor) fluorophore is nonradiatively transferred to an (acceptor) molecule initially in its ground state; the related Jablonski diagram is shown in Figure 9.6.

The phenomenon was correctly described by Förster (1946) using a dipole–dipole coupling interaction; it has been widely exploited to investigate protein interactions or conformational changes, with selected examples described in Chapter 20. A short summary of the key concepts underpinning Förster theory is outlined below.

When a molecule is excited to an upper electronic state, e.g., $S_1$, energy transfer may occur out of the (donor) parent molecule to another (acceptor) molecule in close proximity; this transfer process will manifest itself by the onset of fluorescence of the acceptor molecule. For the FRET mechanism to work, several key conditions need to be fulfilled:

i. The emission and absorption spectra of the (donor) fluorophore transferring energy and the molecule receiving energy (acceptor), respectively, must overlap (see Figure 9.7).

ii. For the energy transfer to be efficient, the two fluorophores involved in the transfer must be in reasonably close proximity, of the order 1–10 nm of each other (this distance is comparable to the diameter of proteins).

Förster derived an equation for the rate of energy transfer between a donor and an acceptor, $k_{ET}(R_{DA})$, in dependence of the distance $R_{DA}$ between the two (donor and acceptor) molecules, namely

$$k_{ET}(R_{DA}) = \frac{1}{\tau_0} \cdot \left( \frac{R_{FD}}{R_{DA}} \right)^6 , \tag{9.11}$$

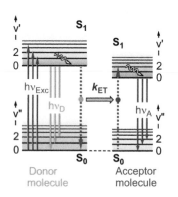

**Figure 9.6** Jablonski diagram for the FRET between a donor and an acceptor molecule. Note that the rate constant for the transfer process, $k_{ET}$, depends on the distance between the two molecules, $R_{DA}$, as $(R_{DA})^{-6}$.

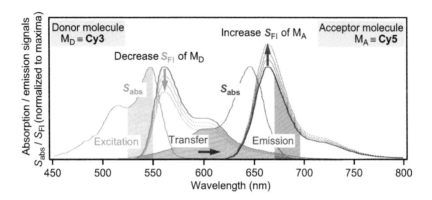

**Figure 9.7**   Example of the overlapping absorption/emission spectra for a FRET cyanine dye pair, with the fluorescence of the donor molecule (here Cy3) reduced and that of the acceptor molecule (here Cy5) increased. (Data adapted from Ishikawa-Ankerhold, H.C. et al., *Molecules* 17 (2012): 4047–4132.)

where $\tau_0$ is the fluorescence lifetime of the donor in the absence of an acceptor, and $R_{FD}$ is the so-called Förster distance. This distance represents the separation distance between a donor molecule (D) and an acceptor molecule (A) under which the rate of energy transfer equals that of donor fluorescence decay, i.e., $k_{ET} = \tau_0^{-1}$ at $R_{DA} \equiv R_{FD}$.

The efficiency $E(R_{DA})$ of the FRET process for a single donor–acceptor pair, as a function of their separation, is normally given by

$$E(R_{DA}) = \frac{R_{FD}^6}{R_{FD}^6 + R_{DA}^6} \ . \tag{9.12}$$

Thus, from the measurements of the donor fluorescence lifetime in the presence and absence of an acceptor (i.e., the fluorescence quenching by the presence of the acceptor), one can estimate the distance between the acceptor and the donor. An example for a (photophysical) FRET process and its efficiency—here for the cyanine dye pair Cy3/Cy5—is shown in Figure 9.8. In close proximity of less than 10 nm, a substantial transfer of energy is observed, leading to fluorescence of the acceptor at the expense of the donor's, while for large distances between the two,

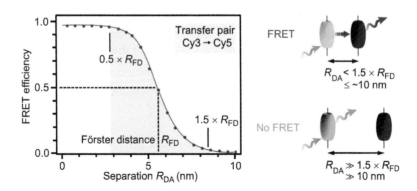

**Figure 9.8**   Dependence of the FRET efficiency on the distance $R_{DA}$ between the donor and acceptor molecules $M_D$ and $M_A$, respectively, exemplified for the cyanine dye pair Cy3 and Cy5; $R_{FD}$ = Förster distance. (Data adapted from Ishikawa-Ankerhold, H.C. et al., *Molecules* 17 (2012): 4047–4132.)

only donor fluorescence will be detected. Of course—as mentioned before—for the FRET process to occur, the Cy3-fluorescence and Cy5-absoprtion spectra need to overlap (see Figure 9.7).

### 9.4.3 Delayed fluorescence

In some specific cases, a molecule can display two types of fluorescence, known as "fast" and "delayed" fluorescence. The former is that of "direct" fluorescence, as discussed in the previous subsections; the latter refers to a slowly evolving fluorescence, which exhibits nearly identical spectral features as normal fluorescence, but whose lifetime is of similar length as observed for phosphorescence. Delayed fluorescence occurs via either of the two types of mechanisms, namely (1) P-type or (2) E-type delayed fluorescence.

The name "P-type" indicates that it was observed for the first time for pyrene. Its formation involves the collision of two triplet states to form the singlet state (the so-called triplet–triplet annihilation); the actual mechanism includes the following steps:

a. *Absorption*: $M + h\nu_1 \rightarrow M^* (\rightarrow M + h\nu_{2,\text{prompt}})$

b. *Intersystem crossing*: $M^* \rightarrow M^{**}$

c. *Triplet–triplet annihilation*: $M^{**} + M^{**} \rightarrow M^* + M$

d. *Delayed fluorescence*: $M^* \rightarrow M + h\nu_{2,\text{delayed}}$

Here, one asterisk represents the excited singlet state, while two asterisks denote the triplet state. Hence, the fluorescence intensity depends on both the triplet state concentration and the rate coefficient of the forming reaction.

E-type delayed fluorescence was observed for the first time in eosine. Again, its fluorescence is spectrally near-identical to that of normal, prompt fluorescence, but its apparent lifetime is that of phosphorescence. The latter observation was interpreted that the mechanism-limiting step implies the formation of the triplet state. This type of delayed fluorescence is illustrated in Figure 9.9, representing the simplified Jablonski diagram for the two steps involved in the E-type delayed fluorescence (including the triplet state formation and the thermal activation of the relaxed $T_1$ state to provide sufficient energy for the re-crossing into $S_1$), and the fluorescence decay observed for the overall process. Because of the need for (thermal) activation energy of the triplet state, in order to overcome the $S_1$ barrier, the process is also known as thermally activated delayed fluorescence (TADF). An experimental example for the TADF process is shown in Figure 9.10.

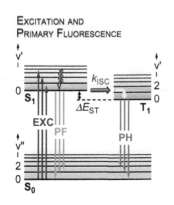

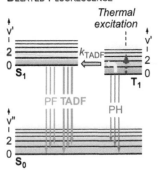

**Figure 9.9** Jablonski diagram showing the different steps involved in the process of thermally activated fluorescence. EXC = photon excitation; PF = primary fluorescence; PH = phosphorescence; TADF = thermally activated delayed fluorescence. For further details, see text.

## 9.5 FLUORESCENCE POLARIZATION AND ANISOTROPY

The absorption of a photon by a fluorophore preferentially occurs when the photon electric vector is aligned parallel to the transition moment of the fluorophore. In fact, from quantum mechanics, we know that the probability, $P_{\text{abs}}$, that a molecule whose transition moment vector is denoted by $\boldsymbol{\mu}$ absorbs a photon whose electric field vector is $\boldsymbol{E}$ is given by

$$P_{\text{abs}} = |\boldsymbol{\mu} \cdot \boldsymbol{E}|^2 \propto \cos^2 \theta, \tag{9.13}$$

with $\theta$ in the angle between $\mu$ and $E$. In view of the anisotropy associated to the photon absorption, the selective absorption of polarized light results in a partially oriented population of the excited fluorophores, even if the fluorophore distribution were initially oriented randomly, as occurs for, example, in an isotropic solution.

Can these photoselected excited fluorophores emit polarized fluorescence? In principle, the answer would be yes in the case that the fluorophore orientation does not change during the entire fluorescence emission. Is this aforementioned possibility very likely? The answer to this question is not straightforward, as it depends on several factors. Since the transition moment has a unique orientation with respect to the fluorophore molecular axis, it is clear that molecular rotational diffusion can significantly decrease the anisotropy of the fluorescence emission.

In solution, the fluorophore rotational period is typically of a few tens of picoseconds, and therefore it can rotate many times during the fluorescence lifetime, which is typically much longer (1–20 ns) than the rotational period of the excited fluorophore. Then, the initial polarization of the excitation is lost, and the fluorescence emission would be completely randomized.

The degree of polarized fluorescence emission is measured by the two parameters called *fluorescence anisotropy* ($r$) and *polarization ratio* ($p$), which are defined by

$$r = \frac{I_{\parallel} - I_{\perp}}{I_{\parallel} + 2 \cdot I_{\perp}} \quad \text{and} \quad p = \frac{I_{\parallel} - I_{\perp}}{I_{\parallel} + I_{\perp}}. \tag{9.14a}$$

Here $I_{\parallel}$ and $I_{\perp}$ denote the fluorescence intensities of the parallel ($\parallel$) and vertical ($\perp$) polarized emission components (with respect to the polarization of the exciting light); it is also assumed that the sample is excited by vertically polarized light, as shown schematically in Figure 9.11. Using the definitions given in Equation 9.14a, one can easily show that

$$r = 2p/(3\text{-}p) \quad \text{and} \quad p = 3r/(2 + r). \tag{9.14b}$$

Note that in many biological applications, use is made of the emission anisotropy $r$ since it offers certain advantages in the physical interpretation of measured data. On the other hand, one advantage of using the polarization ratio is that the measurements are ratiometric and thus do not depend on the actual light intensity values or the fluorophore concentration.

Fluorescence polarization studies are based on the main assumption that the dominant cause of fluorescence depolarization is rotational diffusion of fluorophores. Thus, if it were possible to prevent fluorophore rotation, the initial fluorescence polarization would be maintained. In this context, binding the fluorophore to another, much larger and heavier entity—such as a protein—the effect of rotational diffusion would significantly decrease.

A basic mathematical framework for fluorescence polarization was first given by Perrin (1926); in the case of free rotational diffusion (i.e., no other external actions affect the fluorophore rotation), the fluorescence anisotropy can be described by the so-called Perrin equation:

$$r_0/r = 1 + (\tau/\theta). \tag{9.15}$$

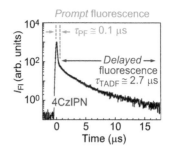

**Figure 9.10** Fluorescence decay exhibiting "prompt," primary fluorescence (PF) and TADF components, exemplified here for the 4CzIPN, a phthalonitrile compound utilized in organic light-emitting diodes (OLEDs). (Data adapted from Sandanayaka, A.S.D. et al., *J. Phys. Chem. C* 119, no. 14 (2015): 7631–7636.)

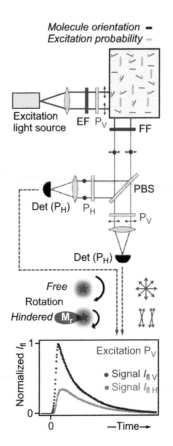

**Figure 9.11** Schematic illustration of a setup for measuring fluorescence (polarization) anisotropy. The sample is illuminated with vertically ($P_V$) polarized light; the fluorescence emission is measured for both vertical ($P_V$) and parallel ($P_H$) polarization components. EF = excitation light filter; FF = fluorescence light filter; PBS = polarizing beam splitter. Note that for unhindered (not linked to a receptor molecule $M_r$) fluorophores, the fluorescence polarization is random.

Here, $r$ is the measured anisotropy; $r_0$ is the fundamental anisotropy (i.e., in the absence of any rotational diffusion); $\tau$ is the fluorescence lifetime (i.e., the average time a fluorophore stays in the excited $S_1$ state before emitting a photon); and $\theta$ is the so-called rotational correlation time. For a brief summary of fluorescence anisotropy, polarization, and rotational correlation time, see, e.g., the appendix in Owicki (2000). Finally, it is worth noting that the said rotational correlation time $\theta$ exhibits the proportionality $\theta \propto (\eta V/RT)$, where $\eta$ is the viscosity of the sample environment in which the fluorophore or the fluorophore-labeled molecule is imbedded; $V$ is the volume of the rotating unit; and $R$ and $T$ are the gas constant and sample temperature, respectively.

Looking at Equation 9.15, it is clear that the smaller the rotational correlation time, i.e., lesser rotational hindrance through binding to a larger molecule, the lower the fluorophore's emission polarization (see also the bottom part of Figure 9.11). Thus, the measurement of the fluorescence polarization constitutes nowadays a sensitive probe to study cell membrane composition or interaction of proteins, to name but two examples.

## 9.6 SINGLE-MOLECULE FLUORESCENCE

Single-molecule spectroscopy (SMS) permits observation of exactly one molecule within a condensed-phase environment by using tunable optical radiation (Moerner 2002). To achieve SMS at any temperature, at least two important requirements must be satisfied: (1) that only one molecule is in resonance in the volume interrogated by the laser radiation; and (2) that the signal-to-noise ratio (SNR) of the single-molecule signal is greater than unity for the required average time.

The first condition is usually achieved by dilution. Then, achieving the necessary SNR requires—among other conditions—(1) a small focal volume for the laser radiation, (2) large absorption cross sections; and (3) a high fluorescence quantum yield (see Moerner 1994, 2002 for further details).

To gauge the second condition for the SNR, it needs to be remembered that the single molecule whose fluorescence is to be measured is imbedded in a vastly higher number of "matrix" (solvent) molecules. Even if these do not fluoresce when exposed to the radiation, which is used for exciting the molecule under investigation, one may still experience background photons, e.g., generated by Raman scattering in the solvent (although Raman cross sections are significantly smaller than absorption and fluorescence cross sections).

In this context, it is worthwhile to estimate how much Raman light contribution can be tolerated so that one still achieves SNR ≥ 1 for the fluorescence signal. One may contemplate the following example.

For a fluorescein molecule, dissolved in water and excited by laser light at $\lambda =$ 565 nm, one finds for the associated absorption cross section a value of about $\sigma_{abs,565} \cong 0.6 \times 10^{-16}$ cm$^2$ (Kastrup and Hell 2004). At the same laser wavelength, water exhibits a Raman cross section of about $\sigma_{Raman,565} \cong 1 \times 10^{-28}$ cm$^2$. If one assumes for simplicity that the fluorescein quantum yield is unity, and that both the fluorescence and Raman emission appear in the same spectral region, then the SNR for a single fluorescein molecule, solvated in a volume containing $N$ water molecules, will be SNR = $0.6 \times 10^{-16}/(10^{-28} \cdot N) = 0.6 \times 10^{12}/N$. Thus, the condition SNR $\geq$ 1 can only be satisfied if the number of solvent molecules is $N \leq$ $0.6 \times 10^{12}$. Taking the density $\rho_{H_2O} \cong 1$ g/cm$^3$ and molar weight $M_{H_2O} \cong 18$ g for water, one would calculate for the above number $N$ a critical (maximum) volume of

$$V_{max} \leq \left(M_{H_2O}/\rho_{H_2O}\right) \cdot \left(N/N_A\right)$$
$$\cong 18 \cdot (0.6 \times 10^{12}/6 \times 10^{23}) = 18 \times 10^{-12} \text{ cm}^3 = 18 \times 10^{-15} \ell. \qquad (9.16)$$

In other words, to guarantee the said SNR, the probed volume (of water with a single fluorescent molecule) needs to be smaller than about 18 f$\ell$ (femtoliter). This clearly demonstrates the necessity to reduce the overall sample size to the nanoscale in order to measure the signal from a single fluorophore with a reasonable SNR. In practice, it is not easy to prepare nanosamples containing a single molecule, but even if it were not overly difficult, it would be inconvenient, as one often wishes to observe a region from a heterogeneous biological sample or homogeneous solution. To overcome this difficulty, several optical techniques have been developed, e.g., confocal fluorescence microscopy, which is particularly suited to limit the observed volume; this will be discussed and demonstrated in Chapter 18.2.

For SMS, typically an electronic transition of the molecule is resonantly excited, and the induced transition is detected either directly, by measuring the absorption signal, or via the associated fluorescence emission (see Orrit and Bernard 1990). Both absorption and fluorescence methodologies opened the way for detecting and identifying biomolecules at the single-molecule level of sensitivity. The first experimental demonstrations of efficient fluorescence detection of single fluorophores were carried out in the late 1980s and early 1990s. In order to minimize the probed/observed volume, either microflow conditions or imbedding in host crystals at low temperature was utilized; for both methodologies, typical examples are given below.

In Figure 9.12, experimental results for the detection of single rhodamine 6G molecules in a 10fM aqueous solution are shown (Castro and Shera 1995).

In the top (red) trace, the clear large-amplitude bursts correspond to individual molecules. High-amplitude bursts are due to molecules passing through the center of the laser beam, undergoing as many as 150–200 photon absorption-emission cycles during the transit through the laser beam, while smaller bursts correspond to molecules that pass through the edges of the beam. For comparison, the lower (blue) trace shows the data set obtained for pure water; these were recorded under identical experimental conditions as with fluorophores in aqueous solution.

Optical absorption-cum-fluorescence spectra from single-dopant molecules were pursued since the late 1980s by Moerner and coworkers (see, e.g., Moerner and Kador 1989). As an example, the fluorescence excitation spectra for single pentacene molecules, isolated in a p-terphenyl matrix at 1.5 K, are shown in Figure 9.13. The displayed data were gathered using a tunable dye laser source with linewidth $\Delta\lambda_L \cong 3$ MHz, at a center wavelength of $\lambda = 592.321$ nm. The "spikes" in the top (green) and middle (blue) traces represent reproducible fluorescence signals from individual molecules. In the bottom (red) trace, an individual spike, i.e., single-molecule emission, is shown for the excitation

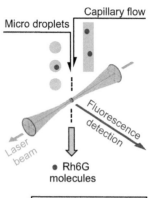

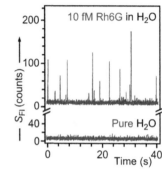

**Figure 9.12** Single-molecule fluorescence detection in dilute solutions (in capillary flows or microdroplets), exemplified for a capillary flow of a 10 fM rhodamine 6G (Rh6G) aqueous solution. (Data adapted from Castro, A. and E.B. Shera, *Appl. Opt.* 34, no. 18 (1995): 3218–3222.)

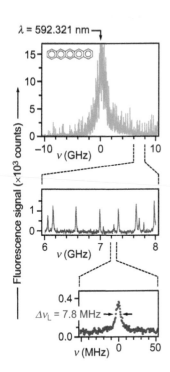

**Figure 9.13** Fluorescence excitation spectrum for pentacene molecules, isolated in a p-terphenyl matrix at 1.5 K; the laser detuning frequency is referenced to the line center. (Top) Scan of the (inhomogenously broadened) transition profile. (Middle) Expanded part of the tuning range, showing emission from individual, single molecules. (Bottom) Scan of a single molecule exhibiting a lifetime-limited Lorentzian profile. (Data extracted from Moerner, W.E., *Science* 265, no. 5168 (1994): 46–53.)

wavelength $\lambda$ = 592.407 nm; the (Lorentzian) line profile exhibits a lifetime-limited width of $\Delta v$ = 7.8 MHz (Moerner 1994).

One may ask, what new information is provided by SMS that is not available by standard (ensemble averaged) measurements? Clearly, by observing a single molecule, the usual ensemble average is removed, which allows one to investigate properties of individual molecules. Thus, for a given experimental parameter, one can access the actual distribution of values rather than the ensemble-averaged value from a large number of molecules. The knowledge of such parameter distributions provides crucial information to explore, e.g., the heterogeneity of the (matrix) system where the single molecule is located. Thus, this converts the single molecule into a "... local reporter of its nanoenvironment ..." (Moerner 2002).

## 9.7 BREAKTHROUGHS AND THE CUTTING EDGE

That matter can absorb or emit light had been known for centuries. But it was only during the second half of the nineteenth century that the link between those two phenomena was realized, and that light emission subsequent to absorption of photons became known as fluorescence. Fluorescence spectroscopy then evolved into a cornerstone for many fundamental scientific and practical applications. However, fluorescence spectroscopy did not always rely on the link absorption $\rightarrow$ emission, but other excitation mechanisms to fluorescent quantum states were quite common. The said link then became crucial with the emergence of the laser, when fluorescence transformed into LIF.

### 9.7.1 Breakthrough: Coining the term "fluorescence"

In the early 1850s, G.G. Stokes published a paper entitled "On the Refrangibility of Light" (Stokes 1852), in which he described a remarkable set of experiments on illuminating a solution of quinine inside a test tube with light filtered from the solar spectrum. One of the most original aspects of the experiment was to disperse the solar light by a prism and then to expose the test sample to different spectral segments of light. He noticed that for the visible part of the spectrum, the solution remained transparent and no scattering was observed. However, when the tube was moved beyond the violet portion of the spectrum to (invisible) ultraviolet radiation, the solution evidently absorbed part of the radiation and reemitted light at blue wavelengths. He repeated this experiment for a range of substances and named the observed phenomenon *dispersive reflection*. One of his additional findings was that the reemitted light was always shifted to longer wavelengths with respect to that of the incident light; this feature is nowadays known as the *Stokes shift*. In his second paper on the very topic, he dropped the term *dispersive reflection* in favor of the word *fluorescence* for the light reemitted by the substance subsequent to absorption (Stokes 1853).

### 9.7.2 Breakthrough: First LIF spectroscopy

It was realized nearly immediately with the invention of the laser that its narrowband radiation could be used to excite individual quantum states of atoms and molecules; the first rotational state-resolved LIF spectroscopy experiments were reported in 1967 and 1968 by the research groups of R.N. Zare at Joint Institute for Laboratory Astrophysics (JILA, Boulder, CO) and S. Ezekiel at Massachusetts Institute of Technology (MIT, Cambridge, MA). They used HeNe and $Ar^+$ gas laser lines to excite individual quantum levels, and recorded the associated fluorescence line sequences for alkali dimers (Tango et al. 1967, 1968) and iodine (Ezekiel and Weiss 1968). The same groups from then on used LIF spectroscopy to derive very precise spectroscopic constants, often far superior to what was achievable before (e.g., Demtröder et al. 1969); a partial LIF recording of their ro-vibrational line sequence for $Na_2$ is shown in Figure 9.14. While the aforementioned experiments all relied on the accidental coincidence of available fixed-frequency laser lines, they laid the foundation for today's sophisticated LIF spectroscopy exploiting tunable laser sources.

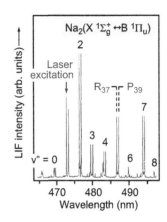

Figure 9.14 Quantum state-resolved LIF spectrum of $Na_2$, excited by the blue $Ar^+$ laser line with $\lambda = 472.7$ nm on the ($v'' = 1|$ $J'' = 37$) → ($v' = 9|J' = 38$) transition; the vibrational levels $v''$ for the P (39)-R(37) fluorescence doublet sequence are indicated. (Data adapted from Demtröder, W. et al., *J. Chem. Phys.* 51, no. 12 (1969): 5495–5508.)

### 9.7.3 At the cutting edge: Laser-stimulated fluorescence on the macroscopic level—Fluorescing fossils

In general, LIF excitation is applied to investigate samples at the microscope or molecular level utilizing well-collimated or confocal microscope configurations and sensitivity detectors. On the other hand, recent developments in commercial laser technology have allowed researchers to apply LIF even at the macroscopic level.

The list of nondestructive techniques that are both affordable and accessible for use in paleontology is rather short. By and large, photographic techniques are used to reveal structural, topographic features, and illuminating the sample with UV-light causes some minerals like hydroxyapatite (a component of bone) to fluoresce, thus providing limited compositional information. But in order to gain access to "hidden" specimens in fossil rock matrices, expensive and bulky imaging equipment is normally required, such as x-ray imaging or CT scanning.

In this context, Kaye et al. (2015) proposed and demonstrated that LIF, or laser-stimulated fluorescence (LSF) as they call it, can provide an instantaneous, noninvasive, geochemical fingerprint of the relatively large paleontological specimen. To a certain degree, LSF is akin to confocal laser-scanning microscopy, but is much simpler in its scanning concept and also lacks depth resolution. Furthermore, LSF offers a significant improvement in the SNR (up to an order of magnitude) when compared to widely used (standard) UV-illumination techniques.

Using LSF at carefully selected laser wavelengths, Kaye and coworkers showed that one is able to distinguish between components like fossilized bone; soft tissue; protective layers like skin; and, of course, the surrounding rock matrix material. An example for this—the skull of a *Microraptor* specimen—is shown in Figure 9.15.

Under white-light illumination conditions Figure 9.15a, nothing unusual stands out on first inspection, other than a few subtle color differences. However, utilizing LSF did reveal dramatic differences in the fluorescence different parts of the skull, like those of the skeletal ring or the tooth roots. Differences like those seen could be extremely useful for meaningful interpretation of properties of fossil

(a)    White light image    (b)    Laser-stimulated
fluorescence image

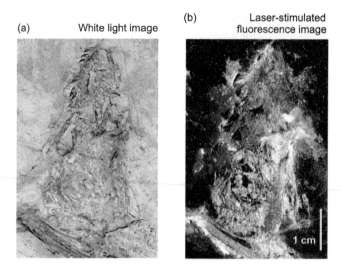

**Figure 9.15**   Skull of a *Microraptor* specimen (IVPP V13320). (a) Photograph under white light. (b) Photograph under LSF conditions. (Adapted from Kaye, T.G. et al., *PLoS One* 10, no. 5 (2015): e0125923. With permission under CCAL.)

specimen. In conclusion, the method of LSF provides a sensitive, fast, and portable fossil imaging methodology with partial speciation, at the macroscopic level, although it lacks much of the macromolecular information provided, e.g., by Raman spectroscopy.

# CHAPTER 10

# Selected Applications of Laser-Induced Fluorescence Spectroscopy

In Chapter 9, the fundamental aspects of the phenomenon of fluorescence and key features of fluorescence excitation and emission spectroscopy were discussed, including fluorescence intensity, spectral structure, polarization, lifetime, and so on. In particular, the analysis followed the line of the fluorophore structure, its interaction with the surrounding medium, and the main variables responsible for matter–light interaction.

The present chapter is essentially dedicated to exploring the main experimental methods and techniques currently available to generate and collect (laser-induced) fluorescence—LIF—emission, both in the steady state and time-resolved. Illustrative examples are given to provide an overview of the wealth of applications in distinct scientific fields.

## 10.1 LIF MEASUREMENT INSTRUMENTATION IN SPECTROFLUORIMETRY

A widely used and common approach for measuring the fluorescence of a substance is to use a general type of apparatus, known as fluorimeter or fluorometer (following the nomenclature for static, steady-state, or time-resolved measurements, respectively—see, e.g., Jameson 2014).

Fluorimeter/fluorometer instruments typically work under two different operation modes, depending on whether the excitation or emission wavelength is kept at a preselected value. When the former mode of operation is adopted, the fluorescence emission is resolved at a fixed excitation wavelength—typically near or at the absorption maximum; this type of spectrum is normally known as the *fluorescence emission spectrum*. In the latter case, one collects the fluorescence emission at a given wavelength—typically near or at the maximum of the emission spectrum, while the excitation wavelength is scanned across the absorption band; this type of spectrum is called the *fluorescence excitation spectrum*. The key difference between absorption and an excitation spectrum is that while the former measures the wavelengths at which a molecule absorbs light, the latter describes the light wavelengths that excite the molecular fluorescence.

By and large, the instrumental building blocks are similar in whatever guise one encounters in any LIF implementation, be it a commercial fluorimeter or a bespoke, specialized configuration. The particular building blocks are (1) the light excitation source; (2) the interaction region of the excitation light with a sample, including light-guiding components; (3) the light analysis components and photon detectors; and (4) the electronic signal acquisition and data analysis system. The overall conceptual outline is depicted in Figure 10.1.

Note that many of the components required in setting up a workable LIF experiment are common with those found in other laser-spectroscopic methodologies; the most common devices were already covered in Chapter 6. Thus, here only a brief summary is proved, highlighting in particular aspects that become important for specific LIF applications.

*Light sources.* Traditionally, xenon arc lamps have been the common light source in commercial fluorimeter instruments. In particular, its wide wavelength coverage, ranging from the UV to the NIR, was unbeatable for a large variety of fluorescence studies, including specifically those involving biological samples. However, in the past years, these sometimes cumbersome discharge lamps are more and more substituted by dedicated-wavelength diode lasers; or the so-called "white-light lasers" (or supercontinuum sources) serve as a direct replacement. The latter offer the same spectral coverage as do xenon arc lamps, but with much higher output power over the complete wavelength range, and the additional advantage low spatial divergence. Recall that typical (commercial) supercontinuum laser sources are based on a laser-pumped (normally at $\lambda_p \cong 1064$ nm) photonic-crystal fiber (PCF), with spectral coverage of typically 400–1600 nm (see Chapter 4.5). Besides the aforementioned lasers, nearly any type of CW and pulsed laser source can be found in one or the other application; these will be addressed where appropriate when describing particular LIF spectroscopy applications.

*Probing of the sample.* In commercial fluorimeters, the sample is usually provided in the form of a sample cuvette. However, many other ways can be envisaged and have been realized, including direct, *in situ* probing of, e.g., biological tissue, or the provision of analytes within a molecular beam, to name but two examples. The optical components used to guide or select the light from the source to the sample or the fluorescence light from the sample to the detector, respectively, comprise optical filters (including narrow-bandwidth filters, wavelength-selective modulators, or fully fledged monochromators); polarizers; light-beam steering components (like mirrors or optical fibers); and imaging elements (like focusing mirrors and lenses). Frequently, the optical filter and polarization elements are already incorporated in the light source or the light analysis system. Typically, the fluorescence is measured at 90° with respect to the excitation in order to avoid or minimize the interference of the (strong) incident radiation with the (normally weak) fluorescence emission.

*Light detectors.* In principle, any of the light detectors discussed in Chapter 6 are encountered in LIF implementation, ranging from photomultipliers (PMTs) through avalanche photodiodes (APDs) to high-sensitivity 2D array detectors (although the latter are normally only used in imaging applications). Three main operating modes for photon detection are encountered, namely continuous DC, modulated AC, and photon-counting modes. The DC modality is widely applied for high light signal levels. The AC mode of operation is suitable when the light source is modulated and the fluorescence signal is of only moderate strength—frequency filtering allows one to separate the signal from noise background of

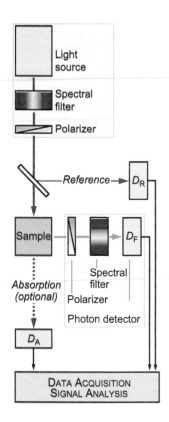

**Figure 10.1** Basic scheme of a fluorimeter; main block on the excitation side—light source, with wavelength selection and polarization control; main block on the detection side—photon detector ($D_F$), with polarization and wavelength selection of the fluorescence light; optional—recording of excitation reference ($D_R$) and absorption ($D_A$) spectra.

similar amplitude. For very low fluorescence signals, the most adequate method is the photon-counting mode [photon counting will be revisited further in Section 10.3 when discussing the technique of time-correlated single-photon counting (TCSPC)]. In this modality, the photon detector generates a pulse output associated with a single photon; only pulses with amplitude higher than a preset amplitude threshold are counted and accumulated as photon signals (to eliminate sporadic pulses from detector dark-current events).

*Signal acquisition and data analysis systems.* The detected signals associated with the fluorescence event—whether in DC, AC, or individual photon-pulse form—are processed further by appropriate electronic instruments. These include signal amplifiers, lock-in amplifiers, gated integrators, and counting equipment [comprising as the most important parts constant-fraction discriminators (CFD) and time-to-amplitude converters (TACs)]. Most of these components have been described already in Chapter 6.6. Furthermore, most modern data acquisition and analysis systems also incorporate computer hardware and software, to control the instrumentation and to analyze and process the recorded data.

With respect to the aforementioned components and building blocks, further information can be found in Chapters 4 through 6. In addition, there are numerous specialized texts where the reader can find precise information about optical components, light sources, and detectors normally used in fluorescence spectroscopy (see, e.g., Moore et al. 1983; Lakowicz 2006; Telle et al. 2007; Jameson 2014).

One of the most serious limitations of fluorimetric techniques is that many species do not fluoresce. In such cases, a widely used alternative is the use of methods of derivatization: a substance that, in principle, does not exhibit fluorescence undergoes a chemical reaction and is converted into another compound that does fluoresce. For example, in molecular biology and biotechnology, it is quite common to use the so-called fluorescent tags, also known as a label. In general, fluorescence labeling implies the chemical attachment of a given fluorophore to the investigated biomolecule (protein, antibody, or amino acid). Information about a wide range of fluorophores is available through web sites of dedicated organizations (e.g., fluorophores.org).

Clear advantages of the fluorescence-labeling methodology are as follows: (1) it provides high fluorescence sensitivity even at low concentration; (2) it is non-destructive thus preserving the target structure and function; and (3) it affords high chemical selectivity with respect to distinct substructures or chemical environments of the target. A rich variety of fluorescence labels is commercially available, ranging from synthetic fluorescence probes to natural compounds, such as the green fluorescent protein (GFP) mentioned earlier in the introduction to Chapter 9.

## 10.2 STEADY-STATE LASER-INDUCED FLUORESCENCE SPECTROSCOPY

The first applications of laser-induced fluorescence (LIF) in the late 1960s and early 1970s were primarily directed at fundamental atomic and molecular spectroscopy; for a personal perspective of those years of LIF, see, e.g., Zare (2012). Because of limitations in photon detector and signal electronics, by and large, LIF observations were primarily made in the "steady-state" regime, averaging the signal over reasonably long times to measure signals of sufficient amplitude and signal-to-noise ratio for sensible analysis. Time resolution down to the picosecond and femtosecond regime—as discussed in Section 10.3—was still out of reach. With the evolution of laser, photon detector, and signal processing technologies, LIF rapidly evolved into one of the workhorses of spectroscopy, finding its way into a vast range of applications beyond basic spectroscopy. Here we address a handful of selected examples to demonstrate the versatility of LIFS, ranging from fundamental spectroscopy and reaction dynamics through analytical chemistry applications to practical uses in medical diagnostics.

### 10.2.1 LIF in gas-phase molecular spectroscopy

An elegant method to remove pressure broadening as well as the Doppler broadening to arrive at high-resolution spectral lines consists in using the laser beam–molecular beam methodologies (see, e.g., Campargue 2001; Telle et al. 2007). As is well known, when a gas is expanded from a source at high stagnation pressures into vacuum, through a small-diameter pinhole (nozzle), the huge number of collisions taking place in the vicinity of the nozzle converts the random motion of the particles inside the source into a hydrodynamic flow. As a consequence of this isentropic expansion, these "jets" exhibit little spread in their translational velocities (often to translational temperatures below 1 K) on the one hand, and significant cooling in their internal degrees of freedom (i.e., vibration and, specifically, rotational energy) on the other hand.

The cooling in vibrational energies leads to the formation of complexes and van der Waals molecules, which survive due to the collision-free character of the molecular beam. The rotational cooling leads to the simplification of the molecular spectra due to the rotational congestion in the low J-states. These conditions, combined with the highly directional character of the molecular beam, allow one to overcome Doppler broadening when the laser beam and molecular (or atomic) beam cross each other perpendicularly, a configuration usually employed to obtain high-resolution rotationally resolved LIF spectroscopy. Specifically for polyatomic molecules, rotational resolution would rarely be achievable without this approach.

An example of rotationally resolved LIF of a polyatomic molecule is shown in Figure 10.2 for the fluorescence excitation spectra of aniline and the aniline···Ar complex (Sinclair and Pratt 1966). The specific spectral regions shown here with rotational resolution are the $0_0^0$-bands of $S_0 \rightarrow S_1$ transition (or more specifically $X^1A_1 \rightarrow A^1B_2$).

The spectra were obtained using a ring dye laser with a linewidth of $\Delta v_L = 0.5$ MHz. The expanded regions of the recorded spectra (green traces in the figure) show well-resolved rotational line transitions, exhibiting linewidths of about $\Delta v_F = 30$ MHz. From the spectra, precise rotational spectroscopic constants could be extracted; the spectra modeled on these (blue traces in the figure) replicate the experiment rather well (Sinclair and Pratt 1996).

### 10.2.2 LIF applied to reaction dynamics

Molecular beam techniques not only allow for fundamental spectroscopy of a particular molecule traveling in it, but also to apply LIF to molecules that are generated as a result of a chemical reaction within the well-defined molecular

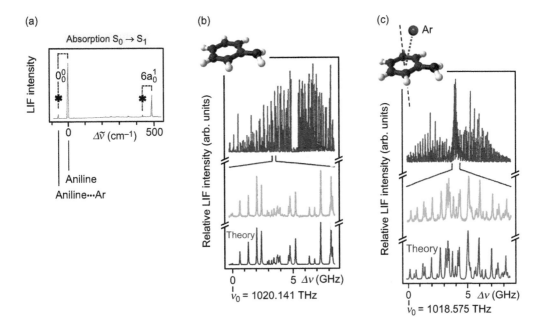

**Figure 10.2** LIF spectroscopy of aniline seeded in an argon beam. (a) Overview spectrum; (b) and (c) rotationally-resolved bands for aniline and the van der Waals complex aniline⋯Ar, respectively. (Data adapted from Sinclair W.E. and D.W. Pratt, *J. Chem. Phys.* 105, no. 18 (1996): 7942–7956.)

beam geometry. Specifically, in crossed molecular beam configurations, LIF may reveal the product center-of-mass (CM) angular scattering distributions of elementary reactions. Essentially, the basic idea of the method is to determine the orientation of the recoil velocity of products, from the Doppler shift of the laser radiation absorbed by the nascent product in a given ro-vibrational state. This type of experiment was first carried out by Kinsey and coworkers (Murphy et al. 1979) who investigated the angular distribution of the product OH, generated in the reaction $H + NO_2 \rightarrow NO + OH$.

In crossed-beam experiments, it is very useful to consider the kinematics of the collision using the so-called Newton diagram, which represents the laboratory velocities of reagents and products, as well as the desired CM velocities (see, e.g., Lee 1988). A general Newton diagram for the elementary reaction $A + BC \rightarrow AB + C$ is shown in Figure 10.3.

Traditionally, a detector (e.g., a mass spectrometer) is rotated around the interaction region, to measure the laboratory angular distribution $P(\psi)$ of the product AB. Then, the respective CM distribution $P(\theta)$ is obtained using a (complex) deconvolution procedure, taking into account the initial distributions and dispersions of velocities (angle and modulus).

The originality of the new LIF approach relies on the fact that the CM-distribution $P(\theta)$ can be extracted from the shape of fluorescence profiles, which are primarily determined by the experimental Doppler profiles, associated with the product recoil velocity in the CM frame. This can be understood if one bears in mind the Newton diagram and, specifically, the orientation of the product CM velocities. Indeed, since the scattering angles are usually referenced to the direction of the incoming atom, forward scattering will correspond to AB molecules, which

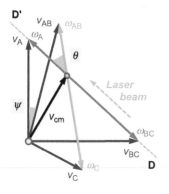

**Figure 10.3** Newton diagram for the prototype reaction A + BC → AB + C. AB-products scatter at the angle ($\theta$) with respect to the velocity $\omega_A$ of the incoming atom in the center-of-mass (CM) frame; $v$ and $\omega$ stand for laboratory and CM velocities, respectively.

scatter in the direction of A-atom and, likewise, C-atoms, which scatter in the direction of BC molecules, i.e., to deflection angles $\theta$ smaller than 90° (see Figure 10.3). In the absence of external fields, the probability $P(\theta)$ is symmetric with respect to the collision axis; it is given by

$$P(\theta) = \left(\frac{dF}{d\Omega}\right)/F, \text{ in units of steradian}^{-1}, \tag{10.1}$$

where $dF$ represents the flux of AB molecules, which scatter at $\theta$ in the solid angle $d\Omega$, and $F$ is the total flux of scattering molecules; note that $P(\theta)$ is proportional to the differential cross section $d\sigma/d\Omega$, with the total cross section $\sigma$ and the scattering solid angle $\Omega$. The Doppler frequency shift—associated with a given AB CM velocity, $\omega_{AB}$—is $\Delta v = v - v_0 = (v_0 \cdot \omega_{AB}/c) \cdot \cos\theta$, where $\theta$ is the angle of $\omega_{AB}$, with respect to the laser propagation axis; $v_0$ is the resonance frequency; and $c$ is the velocity of light.

In the case of a parallel arrangement, as the one shown in Figure 10.3, the laser beam propagates along the DD′-axis, i.e., parallel to the relative velocity of the reagents, or the collision axis. Therefore, each spectral element of the resulting Doppler profile, $D(v - v_0)$, is due to AB products that scatter with velocity $\omega_{AB}$ at angle $\theta$. The flux of AB-product molecules, which scatter between $\theta$ and $\theta + d\theta$, is then given by $dN = 2\pi \cdot P(\theta) \cdot \sin\theta \cdot d\theta$. These product molecules will be excited in the frequency interval $dv = v_0 \cdot \omega_{AB} \cdot \sin\theta \cdot d\theta/c$; hence the Doppler frequency profile is found to be

$$(v - v_0) \propto dF/dv = 2\pi \cdot P(\theta) \cdot N \cdot c/(v_0 \cdot \omega_{AB}). \tag{10.2}$$

This equation illustrates that the shape of the LIF profile directly reflects the distribution in the velocity component parallel to the direction of propagation of the incident laser beam. In other words, it demonstrates how the product CM angular distribution can be extracted from the LIF Doppler profile.

Here the study of the reaction Cs*(7P) + $H_2$ → CsH($X^1\Sigma^+|v'' = 0, J''$) + H, carried out by Vetter and coworkers (L'Hermite et al. 1990), is an example to illustrate the LIF Doppler technique described above. In their crossed molecular-beam experiment, the reagent Cs-atom was electronically excited to its $7p^2P_{1/2,3/2}$ levels; the product molecule CsH was probed by LIF via the ro-vibronic excitation transitions ($X^1\Sigma^+|v'' = 0, J''$)→ ($A^1\Sigma^+|v' = 5, J = J''+1$). The CM angular scattering probabilities $P(\theta)$ of the CsH($v''= 0, J''$) product were determined for all rotational levels in the range $J'' = 1$ to $J'' = 14$, and for various values of the collision energy between $E_C = 0.03$ eV and $E_C = 0.12$ eV. A CsH-product Doppler profile measured at $E_C = 0.09$ eV is shown in Figure 10.4a—for both cases of the laser beam being parallel or perpendicular to the collision axis. The respective $P(\theta) \cdot \sin\theta \cdot d\theta$ angular distribution is depicted in Figure 10.4b—together with the LIF-derived distributions, three different collision energies, but for the same $J''$-level.

It is clear from the figure that the molecular product CsH scatters preferentially at small angles, in the forward direction with respect to the initial Cs-atom direction. As a matter of fact, the best fit, shown in Figure 10.4a by dots, corresponds to a forward $P(\theta)$ distribution, peaking close to $\theta \sim 0°$ degrees (for a more detailed analysis and modeling, see L'Hermite et al. 1990). The forward scattering is proportional to the area of the probability curve between $\theta = 0°$ and $\theta = 90°$; note the dipping of $P(\theta)$ with increasing collision energy $E_C$. Overall, these results were interpreted by the authors that a "harpooning mechanism" is responsible of the dynamics of this elementary reaction.

This type of harpooning reaction dynamics is well characterized, among other features, by a very large cross section, generally attributed to large distance electron jumps; a marked forward scattering of products; and a high excitation of the product vibration (see Levine and Bernstein 1987).

### 10.2.3 LIF in analytical chemistry

As is well established, a major route in atmospheric ozone cycling is linked to $NO_2$ photolysis, according to the following reaction scheme:

$$NO_2 + hv(\lambda_L < 420\,nm) \rightarrow NO + O(^3P)$$

$$O(^3P) + O_2 + M \rightarrow O_3 + M \qquad (10.3)$$

$$NO + O_3 \rightarrow NO_2 + O_2,$$

in which no net conversion of ozone is present. However, the ozone balance can be changed by the presence of additional reactions in the troposphere. In this context, the great majority of volatile organic compounds (VOCs) emitted into the atmosphere are reactive with the radical OH (see, e.g., Cox 1987). The reaction mechanism can be summarized as follows (see, e.g., Telle et al. 2007):

$$RH + OH \rightarrow R + H_2O$$

$$R + O_2 \rightarrow RO_2$$

$$RO_2 + NO \rightarrow NO_2 + RO \qquad (10.4)$$

$$RO + O_2 \rightarrow HO_2 + R*CO$$

$$HO_2 + NO \rightarrow OH + NO_2$$

whose net reaction is $RH + 2O_2 + 2NO \rightarrow 2NO_2 + R*CO + H_2O$. Hence, the sequence produces the oxidation of two NO molecules into two $NO_2$ molecules, which in turn produce two molecules of ozone. It is then obvious that the rate of $RO_2$ production is directly dependent on the rate of OH attack on hydrocarbon molecules.

In view that the OH plays a central role in tropospheric chemistry, the *in situ* measurement of its concentration has long been a goal and considerable progress has been made over the last decades. In particular, the development of LIF techniques contributed greatly to the understanding of the complex mechanisms and the ability of monitoring this radical. Despite the challenges associated with the low concentration of OH and its short lifetime, great achievements were made (see, e.g., Crosley 1995; Heard and Pilling 2003). One of the most common techniques used to measure the tropospheric OH by LIF is the methodology of the so-called *fluorescence-assay by gas expansion* (FAGE), which was pioneered in the late 1970s (Hard et al. 1979) and which was later improved to overcome the interference $O_3$, as encountered in the early measurements (for a detailed description of distinct FAGE applications in the measurement of OH and $HO_2$ in the troposphere, see Heard and Pilling 2003).

The basic equipment utilized in FAGE of OH is depicted in Figure 10.5. Ambient air is pumped into a detection cell at low pressure through a pinhole nozzle. Radiation from the excitation laser, tuned to $\lambda_L = 308$ nm, passes perpendicularly across the gas beam, causing on-resonance fluorescence of the OH radicals. Since the sampled gas expands into a low-pressure cell, collisional quenching of the laser-excited OH radical is significantly reduced; this significantly increases the fluorescence lifetime, facilitating temporal discrimination between the laser

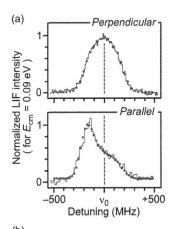

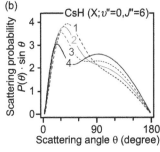

**Figure 10.4** LIF of the molecular product $CsH(X|v''=0,J'')$ generated in the reaction $Cs* + H_2 \rightarrow CsH + H$. (a) Recorded Doppler profiles with the laser beam perpendicular and parallel to the collision axis; dots represent the best fit calculated profile. (b) Evolution of the product scattering probability $P(\theta) \cdot \sin\theta$, for $J'' = 6$, with different values for collision energy $E_c$; 1: $E_c = 0.03$ eV; 2: $E_c = 0.06$ eV; 3: $E_c = 0.09$ eV; 4: $E_c = 0.12$ eV.

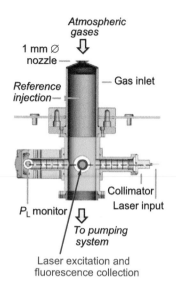

**Figure 10.5** Implementation of a ground-based FAGE OH detection cell (for resonance LIF of OH at $\lambda_L = 308$ nm). (Adapted from Heard, D. "Fluorescence assay by gas expansion (FAGE)." Atmospheric Chemistry and Astro-chemistry, Department of Chemistry, University of Leeds, UK, 2011.)

pulse and OH fluorescence. The fluorescence signal is sampled by photon counting, and finally the recorded signal is converted into an OH concentration, based on suitable off-line instrument calibration. At present, the limit of detection of the OH radical is of the order $LOD_{OH} < 5 \times 10^5$ molecules·cm$^{-3}$. The main advantages of the FAGE technique are (1) the direct excitation of OH following *in situ* sampling and (2) the excellent sensitivity of LIF spectroscopy to monitor OH at ultra-trace concentrations, in near-real time.

*In situ* FAGE-assaying of OH-radical concentrations, and those of the associated three-atomic radical HO$_2$ (see process equations 10.4), have been made at many locations around the globe, ranging from urban environments to remote locations, like in Antarctica. An example of such a measurement cycle is shown in Figure 10.6.

The particular data shown here are based on measurements that were performed at the British Antarctic Survey's Halley Research Station, related to a 6-week period during the austral summer of 2005 (see Bloss et al. 2007). The displayed FAGE data were obtained by on-resonance LIF to detect OH [i.e., the laser was tuned to $\lambda_L = 308$ nm to excite the radical on its transition $X^1\Pi_1(v''= 0) \rightarrow A^1\Sigma^+(v'= 0)$, followed by fluorescence back to the ground state], with HO$_2$ measured in a subsequent chemical conversion process, through addition of NO. The data displayed in the figure clearly show the fluctuations over the day, with typical maximum levels, at local noontime, of $C_{OH} \sim 7.9 \times 10^5$ molecules·cm$^{-3}$ and $C_{HO_2} \sim 1.5$ pptv, respectively. The authors identified the photolysis of O$_3$ and HCHO as the main sources of HO$_x$ generation; of the measured OH sinks, the reactions with CO and CH$_4$ were found to dominate, although comparison of the observed OH concentrations with those calculated via the steady-state approximation indicated the likely presence of further coreactants. Note that the displayed data are complemented by flux data for electronically excited atomic oxygen, $j(O^1D)$, as determined by a spectral radiometer (black data points in the upper panel of Figure 10.6). This rate of photolytic production of these metastable oxygen atoms depends upon the chemical environment and ultimately influences the concentration of OH.

## 10.2.4 LIF for medical diagnosis

One of the most interesting applications of the high species sensitivity associated with LIF is in biomedicine and specifically has been applied in cancer diagnosis. Here this application is illustrated for the case of esophageal cancer (related to tobacco use and excessive alcohol consumption), which is the most fatal cancer worldwide (see, e.g., Parkin et al. 1990; Cancer Statistics Center 2016). A typical, schematic instrumental setup of the technique, in which the endogenous fluorescence of normal and malignant tissues (MTs) is measured *in vivo* using a fiber-optic probe inserted through an endoscope—without requiring biopsy—is shown in Figure 10.7 (right-hand part).

The procedure typically employed in this cancer diagnosis application is *differential normalized fluorescence* (DNF); the procedure consists of normalization of the spectral intensities and baseline fitting, followed by a comparison with a baseline curve deduced from a set of normal tissues (NTs).

An example for the application of DNF in *esophageal adenocarcinoma* diagnosis is shown in the left-hand part of Figure 10.7. From the information provided in the figure, it has to be noted that the (maximum) fluorescence intensity of the MT is much weaker than that of the NT; in the particular example, $I_{max,NT} = 10.8$ versus $I_{max,MT} = 1.2$. One consequence of this significant difference is that it would be rather difficult to extract conclusive information directly from the comparison of such spectra. Thus, according to the DNF protocol, one plots the data on the same intensity scale, i.e., dividing the intensity at each wavelength by the integrated area under the total spectrum. This has been done for the spectra shown in the figure; based on the intensity scaling procedure, a depleted area in the range $\lambda_F \sim 470$–$490$ nm can be observed in the spectrum of the MT sample (highlighted by the shaded area). In their analysis, the authors attributed this spectral depletion to a deficiency of certain components in MTs, which normally would fluoresce in the said range. Thus, such spectral features could provide important criteria in MT diagnosis, as argued by Vo-Dinh et al. (1997).

## 10.3 TIME-RESOLVED LIF SPECTROSCOPY

As was discussed in Chapter 9, fluorescence happens subsequent to population of an excited quantum level. Thus, once the excitation process has terminated and the system is left to evolve unperturbed, the excited population gradually decays to return to the ground state—and hence the fluorescence intensity diminishes as a function of time. If the quantum system obeys first-order kinetics, then the so-called *fluorescence decay law* can be used to describe the fluorescence decay:

$$I_F(t) = I_0 \cdot \exp(-t/\tau), \tag{10.5}$$

where $I_0$ is the intensity at time $t = 0$, $t$ is the time after the absorption process has terminated, and $\tau$ is the lifetime of the excited level.

In order to measure fluorescence lifetimes, two complementary techniques may be used, namely measurements in the *time domain* (TD) and in the *frequency domain* (FD). In the time domain, a short light pulse is used to excite the

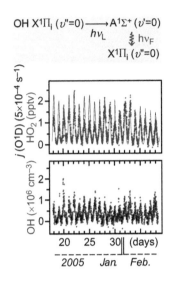

**Figure 10.6** Atmospheric OH and $HO_2$ concentrations, measured with FAGE instrumentation, during a 2005 Antarctica campaign (15 min averaging for each data point). (Data adapted from Bloss, W.J. et al., *Atmos. Chem. Phys.* 7, no. 16 (2007): 4171–4185.)

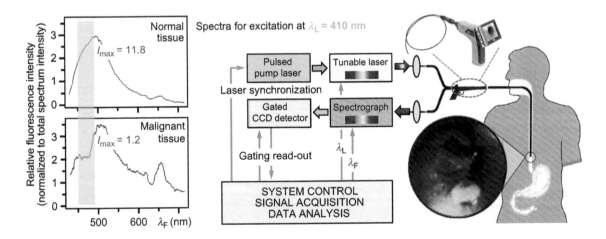

**Figure 10.7** (Right) Schematic instrumental setup employed for laser-induced fluorescence diagnosis of esophagus cancer. (Left) Normalized fluorescence spectra of a normal esophageal tissue (top trace) and of an esophageal adenocarcinoma (bottom trace). The spectra were normalized with respect to the total intensity integrated over the full spectrum. (Data adapted from Vo-Dinh, T. et al., *Appl. Spectrosc.* 51, no. 1 (1997): 58–63.)

sample, and the subsequent fluorescence emission (normally on the nano-second timescale) is recorded as a function of time. In the FD, amplitude-modulated light is used to excite the sample. The fluorescence will follow the excitation waveform, but since the fluorescence emission is a random process, the wave becomes demodulated and is phase-shifted from the excitation curve. Regardless, the lifetime can be calculated from the observed modulation parameter $M$ and phase shift $\phi$.

It should be noted that both methodologies yield equivalent results; it depends on the actual quantum system and under which conditions measurements can be conducted that determine which of the two approaches may be more advantageous. The respective key features are highlighted in the examples outlined below.

### 10.3.1 Measurements of lifetimes in the FD

The concept of the methodology for measurements in the FD, in which the sample is excited with intensity-modulated light, is illustrated in Figure 10.8.

Note that, typically, the frequency of the light modulation is of the order of a few hundreds of megahertz, so that the reciprocal of the frequency is of the same order of magnitude as the fluorescence decay time.

When a fluorophore is excited by an intensity-modulated light, the emitted fluorescence will be modulated at the same frequency, but the lifetime of the excited fluorophore will introduce a delay with respect to the excitation, which is manifested by a phase shift. This finite response of the excited sample also causes a demodulation $\Delta M$, as depicted in the middle part of the figure. Normally the modulation of a sine-wave function can be described by the ratio

$$M = \frac{1}{2} \cdot \frac{S_{\text{p-p}}}{\langle S \rangle}, \tag{10.6}$$

where $S_{\text{p-p}}$ is the peak-to-peak amplitude of the modulated wave, and $S$ is the signal average value. For the case of single-exponential decay of the fluorescence

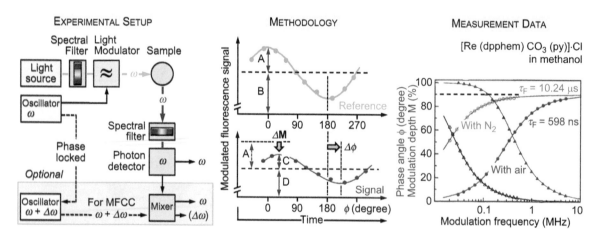

**Figure 10.8** Concept of time-resolved LIF in the FD. (Left) Schematic experimental setup; (middle) excitation and fluorescence sinusoidal waveforms and quantities used in FD fluorometry (for details, see text); (right) FD fluorescence data for Re(dpphem (CO)$_3$ (py) in methanol, in the presence (air) and absence of oxygen (N$_2$). (Data adapted from Szmacinsky, H.K. and Q. Chang, *Appl. Spectrosc.* 54, no. 1 (2000): 106–109.)

emission, one can show that the phase angle $\phi$ and modulation $M$ are given by

$$\tan \phi = \omega \cdot \tau_F \quad \text{and} \quad M = (1 + \omega \cdot \tau_F)^{-1/2}. \tag{10.7}$$

According to Figure 10.8, the modulation of the exciting and emitted light waves, $M_{exc}$ and $M_{emis}$, respectively, are given by

$$M_{exc} = A/B \quad \text{and} \quad M_{emis} = C/D. \tag{10.8}$$

In FD experiments, the emission modulation is always measured relative to that of the excitation, i.e., one uses a modulation parameter (often addressed as demodulation factor), which is given by

$$\Delta M = M_{em}/M_{exc} = (C/D)/(A/B). \tag{10.9}$$

Based on Equations 10.7 through 10.9, the fluorescence lifetime $\tau_F$ can be extracted from the measured demodulation data.

An example for such FD lifetime measurements is shown in the right-hand part of Figure 10.8, exemplified for a long-lifetime Re(I)-fluorophore complex—[Re (dpphem) (CO)$_3$(py)]Cl in methanol. Measurements were carried out for modulation frequencies in the range $\omega = 0.02–5$ MHz, with and without oxygen present (Szmacinsky and Chang 2000). Data analysis resulted in a single-exponential intensity decay, with lifetimes of $\tau_F = 10.23$ µs in the absence of oxygen (a pure-$N_2$ environment) and $\tau_F = 0.6$ µs with oxygen present (sample air-equilibrated).

Finally, it should be noted that FD measurements are conveniently performed using the so-called multifrequency cross-correlation (MFCC) phase-and-modulation methodology; this is included in the schematic of Figure 10.8 as "optional." In this approach, the fluorescence photon detection is also modulated, albeit at a shifted frequency, $\omega + \Delta\omega$, slightly different to that of the laser excitation modulation, $\omega$. This can be done by either directly modulating the gain of the photodetector or mixing the photon signal wave with the shifted reference wave (the latter is the implementation depicted in the figure). By appropriate filtering of the wave-mixed signal, the low-frequency difference wave at $\Delta\omega$ can be extracted. This $\Delta\omega$ signal contains the same phase-angle shift ($\phi$) and demodulation ratio ($\Delta M$) as the original fluorescence signal at frequency $\omega$. This methodology has been known since the 1980s and has been successfully used for lifetime measurements with picosecond resolution (see, e.g., Gratton and Limkeman 1983).

## 10.3.2 Measurements of lifetimes in the time domain: TCSPC

In the conventional steady-state laser fluorescence measurements described in Section 10.2, a single light excitation pulse leads to the detection of the full decay profile. In order to increase the signal-to-noise ratio, one often uses signal averaging of successive decay profiles. Regardless, it is not easy to fully eliminate accompanying noise embedded in the recorded signal (e.g., laser or electronic noise). An elegant method to overcome much of this problem is to use time-resolved techniques coupled to photon counting, a method now widely known as *time-correlated single-photon counting* (TCSPC).

Basically, TCSPC is a "start–stop" technique that measures the time each molecule takes to emit a photon, after excitation, rather than measuring the photon emission from many molecules as a function of time. The concept of the technique is depicted in Figure 10.9.

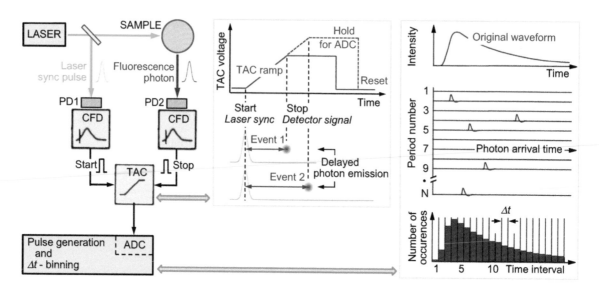

**Figure 10.9** Concept of the TCSPC. (Left) Schematic experimental setup; (middle) principle of the start–stop, or "stopwatch" triggering sequence; (right) accumulation of photon events into time-delay bins. ADC = analog-to-digital converter; CFD = constant-fraction discriminator; PD1, PD2 = photon detectors; TAC = time-to-amplitude converter.

A laser pulse is sent to a beam splitter, which directs a fraction of the light intensity to a photon detector, generating—after amplitude discrimination—the so-called START-PULSE that activates then the TAC unit; on arrival of this start-trigger, the TAC voltage ramp starts (see the middle part of the figure). The other fraction of the laser pulse intensity is directed to the sample, using appropriate optics and suitable attenuation of the light level, to maintain single-photon emission conditions in the fluorescence. Upon excitation, the sample will emit a (single) photon, sometimes none, as displayed in time-period sequence on the right part of the figure. The detection of this single fluorescence photon produces the so-called STOP-PULSE for the TAC. The action of this stop-trigger is to halt the increase in linear voltage ramp; the voltage is held long enough so that its amplitude value can be converted for digital processing. The actual final voltage is directly proportional to the time constant of the ramping, and hence time bins can be generated into which to "count" the single-photon fluorescence events related to the particular time delay.

Note that, in order to increase the time resolution, CFDs are normally used for both START and STOP channels. They generate the trigger pulses for the TAC in the steepest ascent of the photon-detector signal, hence providing the best time resolution.

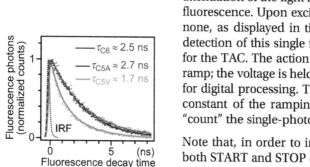

**Figure 10.10** Fluorescence decay curves measured by TCSPC. The data represent the influence of matching (C5V) or nonmatching (C5A) protein–protein in FRET, affecting the fluorescence lifetime of the donor (*Cerulean*) in the presence of a matching/nonmatching acceptor (*Venus/Amber*). C6 = reference fluorophore *Coumarin* 6; IRF = instrument response function. (Data adapted from Sun, Y. et al., *Nat. Protoc.* 6, no. 9 (2011): 1324–1340.)

Of course, the experiment must be repeated many times to collect photons (at most one associated with each excitation pulse) throughout the complete fluorescence emission decay curve, with sufficient counting statistics. The overall acquisition procedure is akin to the counting of events in nuclear decay experiments. Through time-calibrated computer conversion, counting events are sent to the time bin associated with the related TAC voltage, and the fluorescence decay curve is generated in the form of a histogram, in which the number of photons emitted during each time-bin interval is represented.

An example for a TCSPC lifetime measurement is shown in Figure 10.10, exemplified for an experiment on protein–protein fluorescence transfer, demonstrating how the choice of a particular acceptor influences the fluorescence lifetime of the donor. The particular case shown here is that of a FRET experiment in which a *Cerulean* donor protein (color "cyan") is coupled to a *Venus* acceptor protein (color "yellow")—C5V, or an *Amber* acceptor protein (color "orange")—C5A; see Sun et al. (2011). In the latter case, the FRET transfer is hampered because of the mismatch between the fluorescence emission of the *Cerulean* protein and the absorption of the *Amber* protein. This manifests itself in a significantly longer *Cerulean* fluorescence lifetime of the C5A compound in relation to the C5V compound. For comparison, the fluorescence decay curve for the *Coumarin 6* is included—a dye that is often used as a lifetime reference standard. In addition, the instrument response function (IRF), represented by the black data points in the figure, provides a guide for the potential time resolution of the measurement system (in the experiment shown here, $FWHM_{IRF} \cong 300$ ps is encountered).

### 10.3.3 LIF applied to femtosecond transition-state spectroscopy

Before the development of ultrashort laser pulses, provided by femtosecond lasers, it was only possible to investigate the dynamics of a chemical reaction by preparing reactants in a given state or by probing the reaction products in their nascent states. This limitation was imposed by the inherently short time duration of elementary reactions, which typically range from a few tens of femtoseconds (direct reactions) to a few picoseconds (complex reactions). Therefore, with the advent of femtosecond-pulse laser sources, it became possible to investigate bond-breaking or bond-forming in real time.

In order to illustrate the method, the general "model" case of the photodissociation of a triatomic molecule is discussed here, i.e., the unimolecular reaction

$$ABC + h\nu_1 \rightarrow [ABC]^{\ddagger} \rightarrow AB + C, \tag{10.10}$$

via the transition state [ABC $]^{\ddagger}$. The evolution of this dissociating transition state can be traced by LIF in which time-delayed excitation (with respect to the dissociation laser pulse) allows one to probe the transition state potential at different locations, linked to the time it takes for the molecular constituents to retract from each other. Conceptually, this probing process may be written as

$$[ABC]^{\ddagger} \underset{h\nu_2}{\rightarrow} [ABC]^{\ddagger\,*} \rightarrow AB^* + C \rightarrow AB + C + h\nu_F; \tag{10.11}$$

here $h\nu_2$ stands for the probe photon (time-delayed with respect to the pump photon $h\nu_1$); and $h\nu_F$ stands for the fluorescence decay photon of the excited dissociation product AB*.

A key example for a pump-probe experiment along the lines of Equations 10.10 and 10.11 is the photodissociation of the linear molecule ICN, carried out by Zewail and coworkers (Dantus et al. 1988), using lasers of femtosecond-pulse duration for both excitation steps. Research prior to the said femtosecond experiment, i.e., work using nanosecond-pulse duration lasers and quantum chemical calculations, had concluded that—to a first approximation—the CN radical could be treated as a single (atom-like) unit in the dissociation process and, hence, that ICN could be treated as a quasi-diatomic molecule, with a single internuclear coordinate, $R$, representing the separation between the iodine atom, I, and the center of mass of the cyan-radical, CN (see the left-hand part of Figure 10.11).

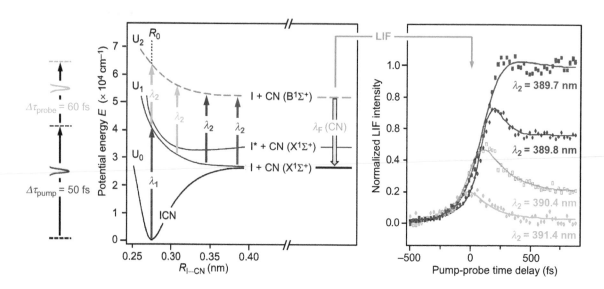

**Figure 10.11** Transition state spectroscopy applied to the I⋯CN photodissociation process, probed with femtosecond-laser pulses. (Data adapted with premission from Rosker, M.J. et al. *J. Chem. Phys.* 89, no. 10 (1988): 6113–6127. Copyright 1988, American Institute of Physics.)

The experimental pump-probe LIF data (right-hand part in the figure) reveal the formation of the free CN radical, after excitation to the transition state [I⋯CN]˙ at $\lambda_1 \cong 306$ nm. All LIF traces shown in the figure, for different probe wavelengths $\lambda_2$, reveal the same rise time in the increase of fluorescence intensity. In their study, the authors tuned their probe laser through the range $\lambda_2 = 387$–396 nm, which includes the wavelength $\lambda_2 = 388.5$ nm for the on-resonance transition of the free radical CN (not included in the figure, but this specific excitation served as a timing marker for all other measurements, yielding a value of $\tau_{1/2} = 205 \pm 30$ fs for the decay of the transition state into the final products). The selected LIF data traces shown here demonstrate the different stages in the evolution of the transition state, from its formation through to the full separation into final products.

## 10.4 LIF SPECTROSCOPY AT THE SMALL SCALE

The applications of LIF discussed so far were, by and large, addressing the analysis of "bulk" samples. But like all other laser spectroscopic techniques, which benefited from better lasers and detection instrumentation with vastly improved sensitivity, LIF also followed the same trend. In particular, higher sensitivity meant that smaller samples could be probed. In LIF spectroscopy, the push to smaller samples and interaction volumes benefited, in addition to the aforementioned instrumental improvements, from the integration of microscopy technology into experimental setups. Here we will briefly outline the evolution of LIF spectroscopy from microscopic volumes through nanoscale resolution measurements to the probing of individual molecules.

## 10.4.1 LIF microscopy

In a conventional light microscope, visible light ($\lambda = 400$–$700$ nm) is used to illuminate the sample under study, and a magnified image—with resolution on the micrometer scale—is basically produced by light reflection and absorption. In a (laser-induced) fluorescence microscope, the image is generated by collecting the fluorescence emitted by illuminated fluorophores.

Fluorescence microscopy is widely used in biology and biomedical science, and is routinely employed in the investigation of small samples, down to the scale of individual microbes. The associated instruments are commonly known by the name of *epifluorescence microscopes*, or wide-field microscopes.

A simplified scheme of an epifluorescence microscope is shown in Figure 10.12a. Here, the excitation (laser) light is focused onto the sample through a microscope-objective lens; and likewise, the fluorescence emission is collimated toward the detector through the same objective. Thanks to the dichroic (long-path) filter, installed between the detector and the objective, only longer-wavelength radiation reaches the detector. In excitation schemes based on one-photon laser transitions, the illuminating light excites the complete thickness of the sample in a conical pattern, as shown in the enlarged segment related

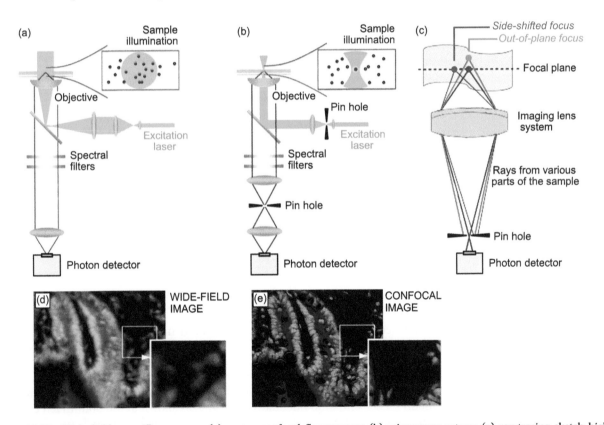

**Figure 10.12** Wide-field or epifluorescence (a) versus confocal fluorescence (b) microscopy setups; (c) ray-tracing sketch highlighting the improved spatial resolution encountered in confocal microscopes. (d) and (e) Images of a triple-labeled cell aggregates (mouse intestine section), recorded with the respective epifluorescence and confocal setups. (Reproduced with permission; © Carl Zeiss Micro-Imaging GmbH, 2011.)

to this configuration. This means that all specimens inside this excitation volume will contribute to the observed signal, since all light from this volume including the desired fluorescence emission but also unwanted Raman scattering and emission from impurities is collected though the microscope objective. In addition, spatial information is somewhat limited, simply because not only particles from the exact focal point contribute but also a larger ensemble—both in sideways and depth dimensions.

Great progress in fluorescence microscopy was made with the development of *confocal laser microscopy*. This type of configuration allows one to realize the so-called optical sectioning (for an explanation, see further below), and therefore one obtains much improved resolution in the image from the fluorescent sample. The basic setup scheme for a confocal microscope is depicted in Figure 10.12b. Rather than illuminating a wider region of the sample, now the radiation from the exciting laser is focused into/onto the sample (the focal spot size on target aided by a limiting pinhole in the excitation laser beam path). The high background signal and other nondesired signal can be significantly reduced by confocal imaging, i.e., the focal spot in the focal plane is imaged onto a small pinhole, installed in the image plane of the light-collection path. With this restriction, light originating from below and above the focal plane will not reach the detector; neither does light that is associated with emission from regions outside the laser-spot area (for a ray-tracing visualization of this concept, see Figure 10.12c). Thus, a key feature of this technique is its ability to acquire focused images from various depths within the sample, a procedure called "optical sectioning." As a result, the observed volume can be reduced and the spatial resolution significantly enhanced with respect to that of an instrument not being configured for confocal imaging.

An example in which a wide field (epifluorescence) and a confocal microscope image are compared is shown in the bottom panels of the figure. Here, the gain in resolution is illustrated for a triple-labeled (thus the three colors blue, red, and green in the picture) cell aggregate of a mouse intestine section.

Clearly, the confocal image reveals extraordinary detail—specifically comparing the enlarged sections, while in the epifluorescence mode, the image is distinctly blurred. In the wide-field image (Figure 10.12d), specimen planes outside the focal plane degrade the information of interest from the focal plane, and differently stained specimen details appear in mixed color. In the confocal image (Figure 10.12e), specimen details become distinctly visible and the image contrast is greatly improved.

## 10.4.2 Fluorescence-correlation spectroscopy

Fluorescence-correlation spectroscopy (FCS) has been around for just over a decade. It is a method in which one measures the fluorescence intensity—as a function of time—from extremely small, optically well-defined volumes, typically of the order 1 fL. Since the fluorescence intensity is proportional to the number of fluorescent molecules present in the investigated volume, the spontaneous fluorescence emission fluctuates with time, as shown in the bottom left corner of Figure 10.13. These fluctuations are due to (1) molecules diffusing (driven by Brownian motion) in and out the laser–interaction volume; and (2) chemical reaction or photophysical processes that change the concentration of fluorescent molecules inside the said volume. It can be shown that the temporal autocorrelation of the fluorescence intensity can provide information about the average

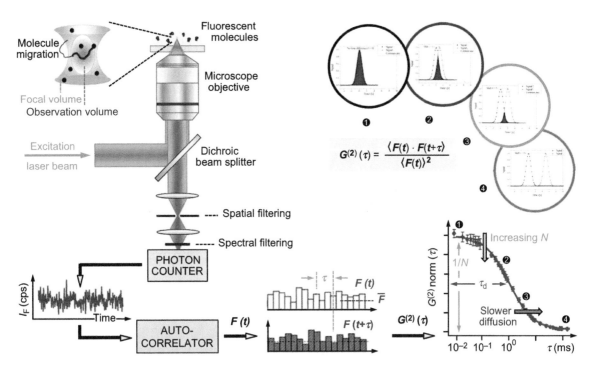

**Figure 10.13**   Schematic description of a typical setup for FCS (right panel) and basic data treatment (left panel) to obtain the fluorescence autocorrelation function.

number of fluorophores $N$ in the excited volume, as well as on the temporal dynamics of the physical processes underlying the observed fluctuations.

From a mathematical point of view, the normalized autocorrelation function $G(\tau)$ can be calculated using the equation

$$G(\tau) = \delta F(t) \cdot \delta F(t + \tau)/F(t)^2, \qquad (10.12)$$

where $\delta F(t)$ is the fluctuation of the fluorescence, at time $t$, and $\delta F(t + \tau)$ stands for the fluctuation at the delayed time $t + \tau$. In Equation 10.12, the product of the two fluctuations is normalized by the squared time average of the fluorescence emission.

As an illustrative example of the capability of the FCS, it can be shown (Maiti et al. 1997) that for the case of only one species, and with no chemical reaction present in the confocal volume, $G(\tau)$ is given by

$$G(\tau) = (1/N) \cdot \left(1 + \frac{4D\tau z}{r^2}\right)^{-1} \cdot \left(1 + \frac{4D\tau}{l^2}\right)^{-1/2}, \qquad (10.13)$$

where $D$ is the diffusion coefficient of the probed species, and $r$ and $l$ stand for the half axes of the three-dimensional Gaussian function used to describe the so-called detectivity profile, normally determined by the optical system employed for light excitation and collection. In Figure 10.14, an illustration of such a fluorescence autocorrelation function is given for different concentrations of the dye *rhodamine 6G* (R6G), in sucrose aqueous solutions, in the range $C = 0.62\text{--}10$ nM.

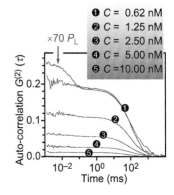

**Figure 10.14**   Fluorescence autocorrelation functions, as a function of time, for several concentrations of *rhodamine 6G* (R6G) in 70% sucrose aqueous solution. (Data adapted from Maiti, S. et al., *Proc. Natl. Acad. Sci. USA* 94, no. 22 (1997): 11753–11757.)

It can be seen how the autocorrelation decays on the timescale of about 100 ms due to diffusion of molecules in and out of the excitation volume. Note that the top-most curve (in red) is obtained for the same $C = 0.62$ nM R6G solution as in the trace (1), but at 70 times higher laser excitation power. The authors ascribed the higher $G(\tau \cong 0)$ observed in this latter scenario to the fact that the molecular ground-state population was lowered by long-lived triplet excited-state formation due to the strong laser-light pumping. The then rapid decay of the excess autocorrelation, within much less than 1 ms, reflects the triplet-state lifetime (see Maiti et al. 1997).

## 10.5 BREAKTHROUGHS AND THE CUTTING EDGE

During the 1970s, one of the major breakthroughs in the field of molecular reaction dynamics was the advent of the so-called "universal machines." The "universal machines" essentially consisted of a molecular beam apparatus, which incorporated a "detector" whose combined role was to first ionize the product(s) emerging from the elementary reaction and then detect them by mass analysis of the ionized species. Despite the great advance, these apparatuses were not able to probe the internal (ro-vibrational) energy distribution of the reaction products.

### 10.5.1 Breakthrough: First LIF measurements to resolve the internal state distribution of reaction products

The group of Zare and coworkers was the first to combine molecular beam techniques, as encountered in the universal machines, with LIF, in order to probe the product state distribution of elementary reactions with $v$- and $J$-selectivity. The particular molecule they investigated was BaO, formed in the reaction Ba + $O_2 \rightarrow$ BaO$(X^1\Sigma;v'',J'')$ + O$(^3P)$; see Schultz et al. (1972). The product molecule was excited by a tunable, pulsed dye laser (with $\tau_{pulse} \sim$ 2–10 ns), thus probing the product molecule's ground-state distribution via transitions to the first excited electronic state, namely BaO$(X^1\Sigma;v'',J'')$ + $hv \rightarrow$ BaO*$(A^1\Pi;v',J')$. The fluorescence light emitted by the excited state was then detected at right angles to the laser beam using a fast-response PMT.

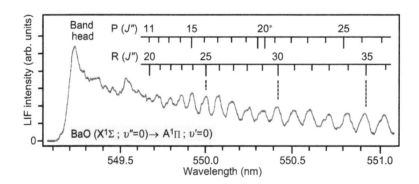

**Figure 10.15**   BaO $(X^1\Sigma – {}^1\Pi)$ LIF spectrum, as a function of the excitation laser wavelength, showing the (partially resolved) P- and R-branch line sequence of the $(v'' = 0,J'') \rightarrow (v' = 3,J' = J'' \pm 1)$ band. The star * marks a rotational-level perturbation caused by the interaction between the $A^1\Pi$ and $a^3\Pi$ states. (Data adapted from Schultz, A. et al., *J. Chem. Phys.* 57, no. 3 (1972): 1354–1355.)

As an example, in Figure 10.15, a BaO fluorescence spectrum is shown, as a function of the laser wavelength, for the ro-vibrational band $(v''= 0, J'') \rightarrow (v'= 3, J'=J'' \pm 1)$; the spectrum reveals the (partially resolved) P- and R-branch lines.

## 10.5.2 At the cutting edge: FRET measurement of gaseous ionized proteins

Our knowledge on the structure of proteins in the gas phase has been significantly increased, thanks to the combined use of soft-ionization methods—such as electrospray ionization (ESI) and trapping mass spectrometry (TMS). However, any correspondence between the protein's structure in its native environment and in the gas phase remained rather unclear. The comparison between the structure of a gas phase (i.e., desolvated) protein and that found in nature (i.e., in solution) can shed light onto the protein ↔ solvent interactions (those interactions are removed upon transfer to the gas phase). This lack of information could mainly be ascribed to the basic limitations of available tools for gas-phase structural characterization.

In this respect, Czar et al. (2015) reported the first FRET measurements of a gaseous ionized protein, a 59-residue variant of the immunoglobulin G-binding domain of protein G (GB1). This protein has a common structural motif, composed of a four-stranded ß-sheet, which is spanned by a single $\alpha$-helix, thus forming a densely packed hydrophobic core (as shown schematically at the top of Figure 10.16). Although both the structural and dynamic properties of GB1 had been studied in the condensed phase, there was no information about its structure in the gas phase. In their study, the authors investigated the FRET efficiency ($E_T$) of the gaseous protein, as a function of charge state, by measuring both the dispersed and time-resolved fluorescence from the donor dye. The derived gas-phase FRET efficiencies and distance constraints were compared with the results from solution-phase single-molecule FRET experiments to obtain insight into the structural desolvated protein.

Representative gas-phase fluorescence emission spectra for GB1-DA generated by positive nano-ESI are shown in the bottom part of the figure for distinct charge states (here $5^+$ through to $7^+$ out of the full measured range $4^+$ through to $8^+$). Two bands clearly manifest themselves in the spectra, with maxima at around 539 and 640 nm, corresponding to donor and acceptor emission, respectively. Those band spectra indicate that the higher the charge state, the lower the FRET efficiency; from the spectra (and the trend in lifetimes), the authors estimated FRET efficiencies in the range $E_T = 0.77$ for the $4^+$ charge state to $E_T = 0.06$ for the $8^+$ charge state. This charge-state dependence of $E_T$ was attributed to the protein structure in which the termini of the protein are progressively farther apart with increasing charge, which is in agreement with the notion of an expansion of the protein's conformation, driven by electrostatic repulsion. This was expected, in principle, to be magnified in the gas phase due to the absence of charge screening, which is normally provided by the solvent in the condensed phase. The measured efficiency values $E_T$ confirmed this, since for all charge states of the gaseous protein, they were lower than those in the solution phase. Studies of gas-phase FRET of mass-selected proteins thus may provide a good method to gain insight into protein folding in the complete (or partial) absence of any (native) solvent.

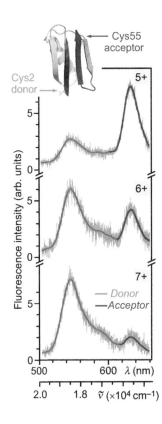

**Figure 10.16** Charge-state ($n^+$) resolved emission spectra of fluorescently labeled, gaseous GB1-DA conjugates (structural schematic at the top), produced by nano-ESI and excited with light at $\lambda = 485$ nm. The colored lines in spectra highlight the donor (green) and acceptor (red) emission bands. See text for comments. (Data adapted from Czar, M.F. et al., *Anal. Chem.* 87, no. 15 (2015): 7559–7565.)

# CHAPTER 11

# Raman Spectroscopy and Its Implementation

The third fundamental type of spectroscopy relies on the scattering of photons rather than absorption or emission between (discrete) energy levels. As was already pointed out in Chapter 2, the scattering of photons from molecules is predominantly elastic. This means that the scattered photons have the same energy (frequency)—and therefore wavelength—as the incident photons. However, a very small fraction of photons (of the order of one in a million) are scattered inelastically, meaning that their optical frequencies, or wavelengths, differ from that of the incident photons. This change in photon energy is linked to a change in the internal energy of the scattering molecule, associated with vibrational and rotational energy levels, and in rare cases electronic energy levels, too. This inelastic scattering process is known as the Raman effect. Its existence was predicted by Smekal (1923) and experimentally confirmed by Raman and Krishnan (1928), after whom the phenomenon is named. The years after its discovery, Raman spectroscopic investigations with a flurry of activities ensued and a myriad of Raman spectra for many different molecules were recorded and interpreted in terms of the (vibrational and rotational) energy level structure of the individual molecules. From the theoretical point of view, by and large, based on the interpretation of the observed spectra, the most comprehensive treatment is probably still that of Placzek and Teller (1933) and Placzek (1934), which is still very much in use today. A further key framework publication is the textbook on infrared (IR) and Raman spectra by Herzberg (1945). It is noteworthy that much of the detailed theory of the Raman effect and structure of Raman spectra pre-dates the advent of the laser. And it is certainly clear when studying the textbook by Long (2002), which presents a unified theory of Raman scattering including modern quantum theory and specific laser properties, that it still builds on those early works by Placzek and Herzberg.

In the first two decades or so after its discovery, the data from Raman spectroscopy provided the first comprehensive catalog of molecular vibrational (and rotational) frequencies. But the experimental efforts can only be described as "heroic," keeping in mind that relatively weak light sources (mostly wavelength-filtered gas discharge lamps, like high-pressure mercury lamps) and relatively tedious detection methods (by and large, photographic plates were used to visualize and record spectra) made the task of measuring the low-intensity Raman signals even more daunting. Nevertheless, the conceptual realization of how to measure Raman spectra is still the same as in the early days—only the instrumental components have significantly matured (see Figure 11.1).

In light of the aforementioned difficulties, it is not surprising that the use of Raman spectroscopy regressed, favoring IR (absorption) spectrophotometers, which were commercialized in the 1940s. However, Raman spectroscopy was revived by the advent of lasers and by the parallel, rapid evolution of spectroscopic instruments, which enormously boosted the sensitivity of the technique.

Today, Raman spectroscopy has become a common, mainstream analytical technique and characterization tool, which, for many applications, has become the method of choice. By and large, the great popularity of Raman spectroscopy relates to the fact that samples of any form and shape may be used, whether they are gaseous, liquid (including gels and pastes as well), or solid (small-scale particles, powders, pellets, thin films, bio-matter, etc.). It is therefore fair to say that the scope and emphasis of Raman spectroscopy, and its multitude of applications, are very much in flux. Consequently, it is often difficult to keep track of new developments, and in particular, specialist books on different aspects of Raman spectroscopy may become outdated rather quickly.

In particular, some enhancement and nonlinear techniques have developed more or less into branches of spectroscopy in their own right. For one, enhancement of the otherwise weak Raman response has been afforded, on the one hand, by resonant enhancement associated with excitation into an excited molecular state manifold in general and, on the other hand, by wave-guiding both the laser excitation and Raman scattering light waves; in particular, the latter methodology has become paramount for many microfluidic applications in biochemistry and medicine (see, e.g., Ashok and Dholakia 2012 or Zhou and Kim 2016).

Even more so, it can be justified to treat some key nonlinear Raman techniques as independent laser-spectroscopic modalities. Among these are specifically the methods of surface-enhanced Raman spectroscopy (SERS) and coherent anti-Stokes Raman scattering (CARS), which have now evolved into well-established, nearly stand-alone techniques (see, e.g., the relatively recent textbooks by Le Ru and Etchegoin 2008, Eesley 2013, and Prochazka 2016).

Since both aforementioned implementations require rather different theoretical treatment and exploit different experimental approaches, they will be covered later on in their own, dedicated chapters (13 and 14).

In the context of linear Raman spectroscopy (whose key aspects are outlined below and for which practical applications are given in Chapter 12), excellent up-to-date annual surveys have been published over the past few years (see Nafie 2007, 2008, 2009, 2010, 2011, 2012, 2013, 2014), which are likely to continue for the foreseeable future; they are certainly not fully comprehensive but constitute a good guide as to the actual trends in the field.

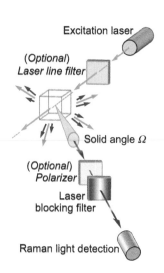

**Figure 11.1** Conceptual arrangement for Raman spectroscopy, observing Raman/Rayleigh light scattered into a solid angle $\Omega$.

## 11.1 FUNDAMENTALS OF THE RAMAN PROCESS: EXCITATION AND DETECTION

Phenomenologically, the scattering of photons by molecular matter can be interpreted as "collisions" between the incident photons and vibrating (and/or rotating) molecules, or a lattice comprising molecular constituents. These collisions can be elastic or inelastic.

If the collision is *elastic*, the energies of neither the photon nor the molecule change but remain the same as before the collision. As already briefly discussed in Chapter 2, the elastic photon scattering process is known as Rayleigh scattering.

If the collision is *inelastic*, both the photon and the molecular energies change. Two cases can be distinguished (see Figure 11.2):

1. The internal energy of the molecule is increased after the collision, while the energy of the scattered photons decreases by the same amount (for energy conservation). The scattered photons exhibit longer wavelengths than the incident photons; the spectral displacement is known as the *Stokes shift*.

2. The internal energy of the molecule is decreased after the collision, while the energy of the scattered photons increases by the same amount (for energy conservation). The scattered photons exhibit shorter wavelengths than the incident photons; the spectral displacement is known as the *anti-Stokes shift*.

In its simplest approach, Raman scattering and the related spectral features can be treated in the framework of classical, electromagnetic field theory. Namely, one finds that the dipole moment, $P$, induced in a molecule by an externally applied electric field (the exciting light), $E$, is given by

$$P = (\alpha)E, \tag{11.1}$$

where $(\alpha)$ is the polarizability (tensor) of the molecule. Raman scattering occurs in the case that the molecular motion (vibration or rotation) changes said polarizability. The actual change in polarizability is described by the (nonvanishing) derivative with respect to the normal coordinates, $Q$, of the vibration:

$$\partial\alpha/\partial Q \neq 0. \tag{11.2}$$

Note that the scattering intensity is proportional to the square of the induced dipole moment and thus to the square of the polarizability derivative, i.e., $I_{\text{Raman}} \propto (\partial\alpha/\partial Q)^2$. Note also that the appearance of the polarizability tensor and the use of specific normal coordinates require one to treat symmetries in the molecule in the framework of group theory.

While the classical polarizability approach provides an intuitive picture and also reveals the difference between Raman-active and vibrational–absorption active species, in modern analytical science, a quantum–mechanical approach to Raman scattering theory is preferred, which relates scattered-light frequencies and intensities to internal energy states of the molecule. Note that in the standard perturbation theory treatment, it is assumed that the frequency (energy) of the incident photon is small in comparison to the frequency (energy) of the first electronically excited energy state of the molecule. Conceptually, one can write the expression for quantitative Raman scattering as

$$I_S = k_{\tilde{\nu}} \cdot \tilde{\nu}_L \cdot \tilde{\nu}_S^3 \cdot N_j \cdot \Phi(\varphi, \theta, p_i, p_s) \cdot I_L \tag{11.3}$$

In this equation, $I_L$ and $I_s$ stand for the incident laser photon and the Raman scattered photon fluxes, respectively; $\tilde{\nu}_L$ and $\tilde{\nu}_s$ are the laser and Raman scattered photon energies, in units of wavenumbers; $k_{\tilde{\nu}}$ is an entity that comprises unit conversion factors as well as experimental factors (including Raman excitation/collection geometries and response functions); $N_j$ is the population factor in the

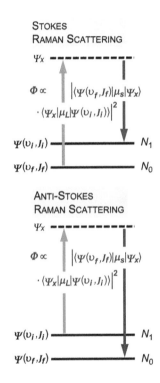

**Figure 11.2** Conceptual Raman processes: Stokes Raman scattering (red-shifted) and anti-Stokes Raman scattering (blue-shifted); the state wave functions $\psi$, population numbers $N$, and transition probabilities $\Phi$ are indicated.

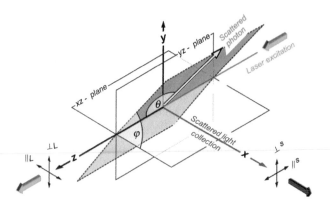

**Figure 11.3** Definition of the Raman scattering angles. The notation of the variables is according to Long (2002).

initial energy state(s) of the molecule, in units of cm$^{-3}$, which depends on the sample temperature and which includes as well the statistical weight factors for each individual (vibrational–rotational) state; $\Phi$ is the Raman scattering line strength function, which depends on the transition moments and which is parameterized in terms of the angles $\varphi$ and $\theta$, describing the direction of the scattered beam relative to the scattering plane; and parameters $p_i$ and $p_s$ represent the polarization state of the incident and scattered light, respectively (see Figure 11.3).

Note that the line scattering function $\Phi$ incorporates a two-step dipole transition moment (polarizability) function that mediates between the initial, intermediate ("virtual"), and final quantum states of the molecule, whose respective wave functions $\psi$ are indexed $i$, $v$, and $f$:

$$\Phi \propto |\langle\psi_f|\mu_s|\psi_v\rangle\langle\psi_v|\mu_L|\psi_i\rangle|^2; \tag{11.4}$$

here, $\mu_L$ and $\mu_s$ are excitation and scattering transition operators.

With reference to Figure 11.3, the scattering plane is determined by the light excitation $z$-axis and its azimuth angle $\varphi$, with respect to the light collection $x$-axis. The zenith angle $\theta$ is defined as the angle between the $z$-axis and the scattered ray in the scattering plane. In the direction of the light collection, imaging lenses and polarization optics are found. For more details on the theory behind Raman scattering and the description of specific molecular cases see, e.g., Long (2002). It is worth noting that Long's formulation is predominantly given in terms of the polarizability tensor description. A few additional remarks on quantum state selection rules associated with Equation 11.4 will be given further below in Section 11.2.

When inspecting Equation 11.3, one finds that any observed Raman signal exhibits the following dependencies:

- It is directly proportional to the intensity of the excitation light source (nowadays predominantly a laser) and thus can be increased by raising the excitation laser power.

- It is directly proportional to the number of participating molecules (in a particular state) and thus can be increased by higher particle densities or larger interaction volumes.

- Since, in many cases, $\tilde{v}_L$ and $\tilde{v}_s$ do not differ too dramatically, one observes a photon energy dependence of approximately $\tilde{v}^4$, meaning that the Raman signals become significantly stronger for short-wavelength excitation.

- It experiences a molecule-specific scattering transition probability.

Note that in "normal" Raman scattering, it is assumed that the energy of the photons initiating Raman scattering is far away from any electronic (or vibrational) absorption of the probed molecule. And because $\psi_v$ does not relate a real energy state, the transition probability described by Equation 11.4 is in general extremely small, normally down by at least six orders of magnitude in comparison to allowed transitions between discrete energy levels. However, in certain cases, the frequency of the incident radiation may nearly coincide with the frequency of an electronic transition of the interrogated molecule. This then means that because of the proximity to a "resonance," the Raman scattering probability increases dramatically, thus compensating the normally low intensities encountered in normal Raman scattering. This so-called technique of *resonance Raman spectroscopy* (RRS) will be discussed further in Chapter 13.3.

## 11.2 THE STRUCTURE OF RAMAN SPECTRA

In Figure 11.2, the energy difference between the incident (laser) and scattered (Raman) photons is indicated by the arrows of different color and length. Numerically, the energy difference between the initial and final ro-vibrational levels $\Delta E = E(v'', J'') - E(v', J') \equiv \tilde{v}_{v,J}$ is associated with the wavelength shift between the excitation and Raman lines (photons) $\Delta\lambda = \lambda_L - \lambda_s$. Normally, the former is provided in units of wavenumbers $(cm^{-1})$, while the latter is measured, in air, in units of wavelengths (nm); the two are linked together via

$$\tilde{v}_{v,J} = \left( n_L \lambda_L^{-1} - n_s \lambda_s^{-1} \right) \cdot 10^7. \tag{11.5}$$

Note that the values in the visible and IR ranges are given in terms of air wavelengths and have to be converted to vacuum wavelengths by multiplication with the refractive index of air at the particular wavelength, $n_\lambda$. This is normally done by using the current procedure accepted by the National Institute of Standards and Technology described in Ciddor (1996).

It is far beyond the scope of this chapter to describe and discuss all possible spectral features that may be encountered in the recording and study of Raman spectra, which also are affected by the aggregate state of the probed sample (gaseous, liquid, or solid). Here, we provide only a short generic summary for the simplest (diatomic) "textbook" molecules, in the gas phase, which exhibit the majority of observable Raman features.

As stated already, Raman lines associated with rotational and vibrational molecular levels are encountered and internal energy can be gained (Stokes transition) or lost (anti-Stokes transition). The line strengths of the various spectral lines reflect the (thermal) population of the initial energy levels of the probed molecule; in addition, the statistical weight (partition function), which incorporates the nuclear spin constraints as well, plays a significant role. The actually observed transitions obey certain selection rules. These are

$$\Delta J = 0, \pm2 \text{ and} + \leftrightarrow +, - \leftrightarrow - \quad \text{for rotation, and} \quad (11.6a)$$

$$\Delta v = 0, \pm1, \pm2, \dots \quad \text{for vibration} \quad (11.6b)$$

The + and − symbols in the rules in Equation 11.6a are related to the even/odd character of the rotational quantum state. As in absorption and fluorescence spectroscopy, the various rotational "bands" traditionally carry capital-letter denotations, namely, "Q" for $\Delta J = 0$, "S" for $\Delta J = +2$ and "O" for $\Delta J = -2$. Note that for the spacing between subsequent lines of the S- and O-branches, one roughly finds $\Delta E_{J \leftrightarrow J+1} \cong 4B$, with the rotational constant $B$; because of the normally low anharmonicity in the rotational constants of the majority of molecules, the lines of the Q-branch "pile up" about the same wavelength location in the spectrum. Also, as in absorption spectroscopy, transitions with $\Delta v \geq 2$ are called "overtones," and transitions commencing in excited vibrational levels are dubbed "hot bands." All features are schematically summarized in Figure 11.4.

Note that for polyatomic molecules, slightly different selection rules for rotational transitions hold, since, now, in addition to the rotational quantum number $J$, the quantum number $K$ for the angular momentum component about a symmetry axis has to be considered. Furthermore, additional vibrational (fundamental) modes exist, which are governed by symmetry group theory and which accordingly are not all Raman active; because of more than one vibrational motion, now "combination" bands—made up from a combination of fundamental vibrations—also supplement overtone bands, associated with $\Delta v \geq 2 \ (= v_1 + v_2)$.

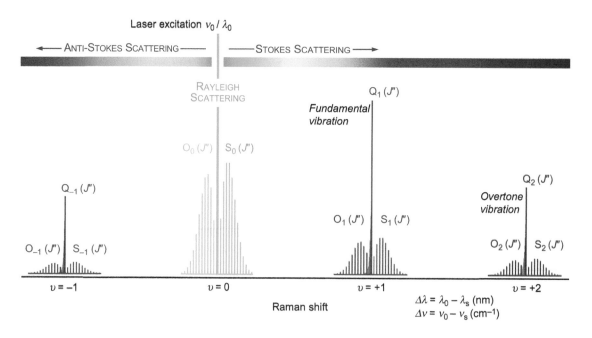

**Figure 11.4**   Conceptual Raman spectrum, featuring Rayleigh, Stokes, and anti-Stokes scattering; the ro-vibrational bands are annotated according to the selection rules $\Delta J = 0, \pm2$ and $\Delta v = 0, \pm1, \dots$.

### 11.2.1 Stokes and anti-Stokes Raman scattering

It is clear from the transition schematic in Figure 11.2 that Raman bands will be observed "symmetrically" on either side of the excitation line (Rayleigh scattering peak) but with different absolute intensities. The Stokes bands are shifted to lower wavenumbers (or red-shifted in wavelength) and exhibit higher intensities than the anti-Stokes bands, which are shifted to higher wavenumbers (or blue-shifted in wavelength). This is a consequence of the anti-Stokes lines originating from higher energy levels, which, in thermal equilibrium, exhibit a lower Boltzmann population. In particular, for molecules with large vibrational constants, the anti-Stokes ro-vibrational bands are significantly lower in amplitude than their Stokes counterparts (see the generic Figure 11.4). Therefore, in quantitative analytical Raman spectroscopy, nearly exclusively the Stokes part of the spectra is utilized, which also brings certain advantages in the experimental realization (see Section 11.3).

### 11.2.2 "Pure" rotational Raman spectra

When contemplating "pure" rotational Raman spectra, one has to keep in mind that the majority of molecules carry rotational constants of just a few wavenumbers in magnitude; only for the very lightest molecules (such as, e.g., $H_2$) are larger rotational constants of tens of wavenumbers encountered. Thus, the related Stokes ($\Delta J = +2$) and anti-Stokes ($\Delta J = -2$) lines are very close to the laser excitation line. This constitutes a severe experimental problem since scattered light from the exciting laser and the elastic-scattering Rayleigh contribution are substantially more intense than the Raman signals, which may become masked by the Lorentz wings of the former (also see the discussion in Section 11.3). Regardless, exploiting some suitable experimental and spectrum evaluation procedures, rotational Raman spectra can be recorded and analyzed. An example for the molecule $N_2$ is shown in Figure 11.5.

Clearly, the aforementioned lesser intensity of the anti-Stokes spectrum is revealed, together with the overall intensity envelope of a typical, thermal Boltzmann distribution for rotational level population. In addition, one observes an intensity alteration between neighboring rotational lines. This is due to the statistical weight alteration associated with the nuclear spin $I$ in molecules with nuclei axis-symmetry in their point group ($D_{\infty h}$ for the hetero-nuclear molecule $^{14}N_2$ shown here); this is known as the *ortho–para* (odd-$J$/even-$J$) asymmetry between rotational levels, whose statistical weight alters as $(I + 1)/I$. In the example shown here, the nuclear spin of $^{14}N$ is $I = 1$, and thus the intensity alteration between odd and even rotational levels is 1:2.

It has to be noted that, for polyatomic molecules, the rotational level structure and the associated selection rules become slightly more complex. As already pointed out further above, in addition to the rotational quantum number $J$, now the quantum number $K$ for the angular momentum component about a symmetry axis has to be considered as well. This leads to a change in the selection rules: in addition to $\Delta J$, now the change in $\Delta K$ also has to be taken into account.

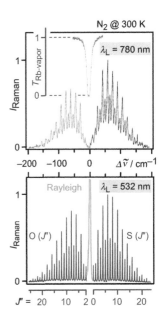

**Figure 11.5** Pure-rotation Raman spectrum of $N_2$ at atmospheric pressure and room temperature; the red-shifted Stokes $S_0(J'')$- and blue-shifted anti-Stokes $O_0(J'')$-branches are indicated. Top panel: Ti:sapphire laser excitation ($\lambda_L = 780$ nm), with the Rayleigh scattering filtered by an Rb-vapor cell; for details, see Lee and Lempert (2003). Bottom panel: Nd:YAG DPSS laser excitation ($\lambda_L = 532$ nm), with the Rayleigh scattering filtered by a notch filter; for details, see Newton et al. (2014). Note the slightly different spectral resolution in the two experiments.

For $\Delta K = 0$, the selection rule for $\Delta J$ given in Equation 11.6a is maintained; however, for $\Delta K \neq 0$, one now finds

$$\Delta J = 0, \pm 1, \pm 2 \quad \text{and} \quad + \leftrightarrow +, - \leftrightarrow -, \qquad (11.6c)$$

meaning that now P- and R-branches will also appear in the Raman spectra, in addition to the standard S- and O-branches.

For a full discussion of the rotational energy level structure and the expected spectral branches and their line strength, the reader should consult standard textbooks covering the theory of Raman spectra (see, e.g., Herzberg 1945 or Larkin 2011).

### 11.2.3 Ro-vibrational Raman bands

When the Raman transition ends in a different vibrational level compared to when it started, the complete spectrum is shifted a substantial distance away from the excitation line, usually of the order of hundreds to a few thousands of wavenumbers. Clearly, this makes the spectroscopy of the Raman features much easier since no interference from the elastic Rayleigh scattering and stray light occurs. Staying initially with the simple diatomic molecule of $N_2$, as discussed for the pure rotational case, now in addition to the S- and O-branches, the Q-branch with $\Delta J = 0$ occurs (see Figure 11.6). Note that the odd–even line strength alternation in the S- and O-branches mimics that for the pure rotational Raman spectrum shown in Figure 11.5.

In fact, the intensity of this Q-branch is the strongest of the three, even taking into account the pileup of rotational lines. This is due to the fact that the (diagonal) polarization tensor elements associated with the Q-branch transitions are, for the majority of molecules, much larger than those for the S- and O-branch transitions. This is relatively easy to deduce from perturbation theory (see Long 2002) but was already introduced and formulated by Placzek and Teller (1933) and Placzek (1934). Note also the additional feature observed near 2290 cm$^{-1}$; this is the Q-branch of the minor isotopologue $^{14}N^{15}N$ with a natural abundance of about 0.73%.

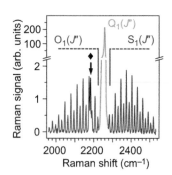

**Figure 11.6** Fundamental vibrational Raman spectrum of $^{14}N_2$; the rotational branches $O_1(J'')$, $Q_1(J'')$, and $S_1(J'')$ are indicated. ♦ = $Q_1(J'')$-branch of the minor isotopologue $^{14}N^{15}N$.

Each fundamental vibrational mode of a (polyatomic) molecule, i.e., the energy difference between the $v = 0$ and $v = 1$ levels of said mode, relates to a different Raman shift; this is because the frequencies of vibrational transitions depend on the atomic masses and the bond strengths: the heavier the atoms, the lower is the corresponding vibrational frequency, and the stronger the bond, the higher is the associated vibrational frequency. Therefore, Raman spectroscopy can be used to identify different chemical compounds and specific bonds; a schematic example is shown in Figure 11.7. The number of fundamental vibrational modes scales with the number of atoms in the molecule; thus, Raman spectra from large molecules can be rather complicated (some examples for this will be shown in Chapters 12 and 19).

Note that not all vibrational transitions are *Raman active*; i.e., some vibrational transitions do not appear in the Raman spectrum. This constitutes a similar restriction as was encountered for IR absorption spectroscopy (see Chapter 7) although the restriction is slightly less severe, due to the fact that the change in polarizability required for Raman activity is induced by the exciting radiation field. Overall, one finds that (1) in a centro-symmetric molecule, Raman and IR activity are mutually exclusive, i.e., asymmetrical stretching and bending modes are IR active but Raman inactive, while symmetrical stretching and bending is Raman active and IR inactive; and that (2) for molecules without a center of symmetry, each vibrational mode may be Raman active or not, but that (3) in particular symmetrical stretch and bending modes of said asymmetric molecules tend to be Raman active.

## 11.2.4 Hot bands, overtones, and combination bands

Besides the "fundamental" rotational and ro-vibrational Raman bands that refer to the quantum transitions $(v'' = 0|J'') \rightarrow (v' = v''|J' = J'' \pm 2)$ and $(v'' = 0|J'') \rightarrow (v' = v'' \pm 1|J' = J'', J'' \pm 2)$, respectively, transitions with different vibrational quantum numbers often need to be considered. Namely, these are (1) the so-called hot bands, (2) overtone bands, and (3) combination bands.

*Hot bands* originate in excited vibrational levels, i.e., levels with $v'' \geq 1$. However, as stated already further above, the energy spacing for the majority of molecules is much larger than that for the rotational levels; normally, the vibrational spectroscopic constants describing these levels are in the range of a few hundred to a few thousand wavenumbers. This means that their thermal population at room temperature (~300 K) is normally rather small. For example, for values of the vibrational constant $\omega_e = 500$, 1000, and 2000 cm$^{-1}$, one finds for the relative thermal population of the first excited vibrational level $N_{v''=1} = 9.1 \times 10^{-2}, 8.3 \times 10^{-3}$, and $6.9 \times 10^{-5}$, respectively.

The characteristic vibrations for a multitude of organic molecules—for whose identification Raman spectroscopy is routinely used—are, by and large, in the range 1000–3000 cm$^{-1}$ so that, in general, the relative population in excited vibrational states is well below 1% and thus are difficult to observe. The difficulty in their detection is even further exacerbated by the fact that, in general, the vibrational anharmonicity is small, and therefore the related energy shift between hot band transitions $(v'' \geq 1) \rightarrow (v' = v'' \pm 1)$ is small as well—often in the range of just a few wavenumbers. Hence, hot bands are normally very difficult to separate from the much stronger fundamental bands (see Figure 11.8 for an example).

*Overtone bands* are transitions for which the change in vibrational quantum number is $\Delta v > 1$. While these bands are much easier to observe than hot bands—because the thermal population and spectral shift problems do not play a role—they experience problems of a different nature. First, the probability for transitions between vibrational levels favors transitions with $\Delta v = \pm 1$, associated with the anharmonic oscillator transition selection rules. On the other hand, in quite a few polyatomic molecules, the symmetry and substate configuration selection rules may determine that the fundamental transition is "forbidden" but that the overtone transition is "allowed," and thus the overtone may be of similar intensity as the fundamental band (see the schematic example for $CH_4$ in Figure 11.7, and for a general summary of the related transition bands and their strengths, see, e.g., Albert et al. 2009). Second, if the spectral shift for the fundamental band is already large, then overtones quickly fall outside the observable spectral window.

Finally, in most Raman spectra of polyatomic molecules, one commonly observes the so-called *combination bands*; these excitations are due to a combination of more than one of the fundamental vibrations in a specific molecule.

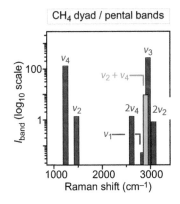

**Figure 11.7** Bar graph vibrational mode spectrum of $CH_4$ (dyad and pentad subsystems), displaying the relative intensities of the $Q_1$-branch peaks for the fundamental vibrations (in red), overtones (in blue), and combination bands (in green).

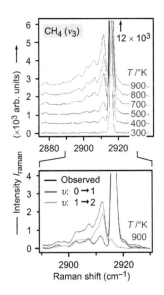

**Figure 11.8** Evolution of the hot band of $CH_4(v_3)$, as a function of temperature. In the lower panel, theoretical modeling for the components 0→1 and 1→2 is shown, representative for $T = 900$ K.

Like in the case of overtones, the transition probabilities for combination bands are weaker than those for fundamental vibrations, but again, one finds that the combination of selection rules for the various vibrational sublevels results in transition probabilities that "enhance" links to otherwise forbidden vibrations. For example, the $v_1$-vibration in $CH_4$ is Raman inactive, with $I(v_1) \cong 5 \times 10^{-2}$, while the $v_4$-vibration is Raman active, with $I(v_1) \cong 128$; their combination vibration $v_1 + v_4$ exhibits an intensity of $I(v_1 + v_4) \cong 8$ and is therefore reasonably strong to be observable (see Figure 11.7 and Albert et al. 2009). The related spectral information can then be used to deduce information about the nonactive vibrational level structure.

## 11.2.5 Peculiarities in the Raman spectra from liquids and solid samples

When moving from the gas phase to the liquid and solid phases of matter, the significantly higher particle density and spatial structure (often ordered into "lattices") also affect the observed Raman spectra. In particular, two effects will affect the shape and appearance of Raman spectra, namely, the interaction with neighboring molecules, and the partial restriction of free motion (specifically free rotation is largely hindered or absent).

While, in principle, all freedoms of motion (translation, rotation, and vibration) are still allowed in liquids, it is in particular the rotational motion that tends to restrict bond bridges to neighboring molecules. The restricted rotations now become rocking motions since the bonding forces drive the molecule back to its original orientation when it tries to commence rotation. This rocking motion is normally known as *libration*; their frequencies depend on the moment of inertia of the molecule and the bond network's nature. Overall, the Raman spectra of liquids look distinctly different from their gas phase brethren. The interaction of a molecule with its neighbors introduces broadening and shift of spectral features, and the rich structure of individual rotational lines observed in the gas phase disappears. The conceptual differences in the spectra in the gas and liquid phases are shown in Figure 11.9, exemplary for the molecule $H_2O$.

In addition to the normal vibrations and the vibration-like librations, one often observes "cluster vibrations" or translational vibrations. For example, in liquid water translational vibrations involving the combination hydrogen bond $O–H\cdots\cdots O$, stretching and bending give rise to a band structure in the 180–200 $cm^{-1}$ range. Also, combination bands between normal vibrations and librations are observed; taking liquid water as an example again, a weak but significant feature at ~2125 $cm^{-1}$ is due to the combination of the bending ($v_2$) and libration modes. Note that all features that involve interactions with neighboring molecules strongly depend on temperature. It also should be noted that it is often feasible to not only identify compound molecules in the liquid, but by recording Raman spectra as a function of temperature, it may become possible to extract information about local (and global) structural changes (see, e.g., Lin et al. 2013).

In the solid phase, the constraints of rotational movement become even more stringent, particularly if the solid is ordered in a crystalline lattice structure. On the other hand, the majority of features encountered in liquids are also observed for solids. Very often, the peaks become narrower but not necessarily so;

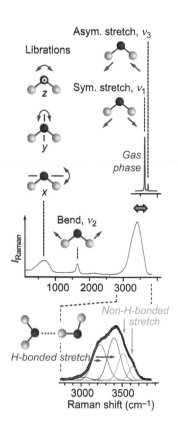

**Figure 11.9** Raman spectrum of liquid water; the (shifted) gas-phase symmetric/asymmetric ($v_1/$ $v_3$) stretch region is shown; the various fundamental $H_2O$ vibrations are indicated as cartoons. A component simulation for the stretch bands is shown at the bottom, including hydrogen bonding (for details, see, e.g., Li et al. 2010).

much depends on the nature and strength of interaction with neighboring molecules. In addition to fundamental, overtone, and combination vibrations, and the librations associated with the hindered rotation of the molecules, one also will encounter interaction of the scattering photons with (global) lattice vibrations; the related low-frequency phonon interactions give rise to Raman features in the terahertz regime. In fact, Raman spectroscopy is increasingly used to rapidly measure the phonon spectra, without excessive sample preparation.

### 11.2.6 Polarization effects in Raman spectra

As already indicated in Section 11.1 and Figure 11.3 therein, Raman scattering exhibits at least partial polarization, even for molecules in the gas (or liquid) phase, where the molecules are randomly oriented. In isotropic media, polarization arises simply because the induced electric dipole exhibits spatially varying components with respect to the coordinates of the molecule (see Figure 11.10).

Of course, polarization effects in the Raman spectra can only be sensibly analyzed if the exciting (laser) source is polarized, normally plane polarized. Spectra acquired with a polarization analyzer, set at both perpendicular and parallel to the excitation plane (for the concept, see Figure 11.3), can be utilized to calculate the so-called *depolarization ratio*. This spectral analysis approach is extremely useful to establish links to group theory, molecular symmetries and orientation, and Raman activity; examples will be briefly addressed in later chapters. In particular, Raman depolarization measurements are applied to shed light into the orientation of macromolecules in crystal lattices or polymers, to name but two examples.

## 11.3 BASIC EXPERIMENTAL IMPLEMENTATIONS: KEY ISSUES ON EXCITATION AND DETECTION

Conceptually, a basic Raman spectroscopy system is relatively simple and includes as main building blocks the following: (1) a laser excitation source, (2) delivery optics for the excitation laser light, (3) the actual sample, (4) Raman light collection optics, (5) a wavelength separation device, (6) a photon detector and its associated electronics, and (7) signal acquisition and data analysis equipment. This overall concept is visualized in Figure 11.11.

Because the Raman scattering process itself is weak, and thus Raman spectral signals are very faint, optimal performance of each of the individual components into the setup is critical. Only a fully optimized system will be capable of realizing the full potential of Raman spectroscopy, i.e., to (simultaneously) capture the widest range of molecular species and to measure the lowest concentrations within the shortest amount of time possible. In addition, the experimental system has to be capable of maximizing the detection of the Raman signals, while at the same time minimizing adverse contributions from Rayleigh-scattered and fluorescence light, which are often much more intense than the Raman signal itself. The individual building blocks and measures for their optimization are briefly summarized below, with further details included in the later specialist chapters on individual Raman measurement methodologies.

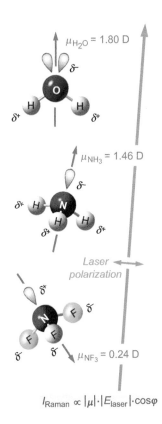

$$I_{\text{Raman}} \propto |\mu| \cdot |E_{\text{laser}}| \cdot \cos\varphi$$

**Figure 11.10** Conceptual presentation for the polarization dependence of the Raman signal, from the interaction of a linearly polarized laser field $E_{\text{laser}}$ with the molecular dipole moment $\mu$, exemplified for (randomly oriented) molecules of $H_2O$, $NH_3$, and $NF_3$ (bent and trigonal-pyramidal symmetry, respectively).

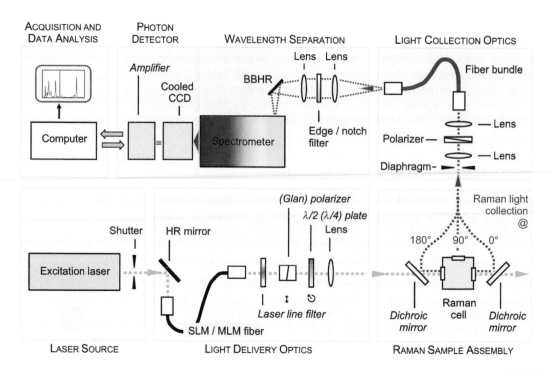

**Figure 11.11**   Conceptual implementation of a Raman scattering experiment for spectroscopy or imaging. HR, high reflector for laser wavelength; BBHR, broadband high reflector for Raman light; components indicated in *italics* are optional.

## 11.3.1 Laser excitation sources

In principle, there is no restriction on the laser sources to initiate Raman scattering, since (1) the process is nonresonant and therefore doesn't require a particular wavelength; and (2) it is instantaneous, meaning that any time scale from femtosecond pulses to continuous wave (CW) can be utilized. It really depends on the particular application what type of laser may be the most suitable for the task. However, certain "restrictions" may be encountered; these are instrumental, or sample related.

For example, not every laser wavelength is available from commercial laser sources at the power level to generate Raman intensities that are sufficiently above the noise level for detection. In addition, one normally requires the rejection of laser stray light and Rayleigh-scattered light, and not for all wavelengths are suitable rejection filters available off-the-shelf (although many more wavelengths are now covered than used to be the case only less than 10 years ago). On the other hand, standard systems today rely very much on common solid state or semiconductor diode laser sources from the UV (e.g., 244 and 355 nm) through the visible (e.g., 488, 532, and 633 nm) to the near-infrared (NIR) (e.g., 785, 830, and 1064 nm).

But not all of these wavelengths are equally suitable for all types of samples. In particular, when investigating biomedical specimens, excitation in the visible part of the spectrum also generates a substantial amount of fluorescence that may swamp the Raman signal completely. In this case, excitation wavelengths in the NIR or, more recently, in the UV are preferred or indispensable. On the other hand, prolonged exposure to NIR laser radiation may "cook" the sample and cause irreversible damage; thus, one may have to compromise whether one can tolerate a significant fluorescence background or one needs to preserve the integrity of the sample.

Finally, from the point of view of the duration of laser radiation, most common Raman systems utilize CW laser sources. On the other hand, increasingly ultrashort femtosecond pulse lasers are becoming popular, particularly in applications of stimulated Raman technologies.

## 11.3.2 Delivery of excitation laser light

The schematic experimental setup shown in Figure 11.11 indicates that, in general, reflecting and transmitting optical elements are encountered by the laser radiation to be used for Raman excitation. These essentially include (1) mirrors—often with dielectric coating optimized for the specific wavelength; (2) lenses—used for focusing the laser radiation into the Raman scattering sample volume, but in addition, setup-specific elements are frequently required as well. The latter comprise, for example, (3) dichroic beam splitters, to separate laser and Raman light in backward scattering configurations, and/or (4) optical fibers, to deliver the laser light to the sample in a convenient fashion—avoiding complicated optical guidance paths—or to reach otherwise inaccessible locations.

Any optical component in the path of a powerful laser will need to be considered carefully in order to minimize any potentially adverse effects leading to a deterioration of the already weak Raman signal. For example, dielectric coatings—be they high-reflective coatings of beam-steering mirrors or anti-reflection coatings for transmitting optics—should be as optimal as possible for the excitation laser wavelength, in order to avoid excessive laser power losses.

In particular, critical aspects are the choice of material for transmitting optics and the number of optical components in the beam path (which should be minimal). This is because the powerful laser radiation, required for the Raman process, may cause the generation of detrimental (broadband) fluorescence in said optical components or even stimulate (narrowband) line features associated with material–internal conversion processes. The latter is specifically encountered in optical fibers when exposed to high-power transmission of laser radiation, and one may encounter both major power losses and the generation of secondary spectral lines of nonnegligible intensity. The most relevant of these limiting effects in fused silica fibers are stimulated Brillouin scattering (SBS) and stimulated Raman scattering (SRS); these are due to the inelastic scattering interaction with either acoustical or optical phonons. Both effects are correlated to the $\chi(3)$ nonlinearity of a medium.

In fused silica, the frequency shift induced by these effects is in the order of 11 GHz for SBS and about 12–13 THz for SRS. In the former case, light contributions shifted by the equivalent 3–4 $cm^{-1}$ can result—if the radiation is strong enough—in broadening of the Raman lines, potentially affecting the spectral resolution. In the latter case, where the lines exhibit an equivalent shift to about 400 $cm^{-1}$, said light may generate a secondary Raman spectrum with lines at completely different wavelength positions; this would make the interpretation of already complex spectra even more difficult. The typical frequency shift spectrum for SBS and SRS are shown in the top panel of Figure 11.12.

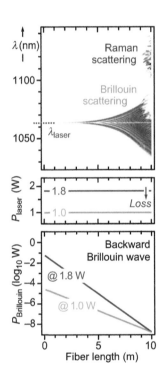

**Figure 11.12** Effects observed on transmission of high-power laser radiation through optical fibers. Top panel: Spectral broadening of 1064-nm laser radiation by Brillouin and Raman scattering, as a function of fiber length. Bottom panel: power in the back-scattered Brillouin wave, as a function of fiber length. For higher incident laser power, the Brillouin wave reaches significant power levels, causing reduced laser power exiting the fiber. Calculated using the *RP FiberPower*. (From Paschotta 2015.)

The losses encountered in the fiber transmission of powerful laser radiation are associated with the gain experienced by the stimulated wave For SBS, strong, nonlinear optical gain for the back-reflected wave results in an originally weak counterpropagation being amplified, at the expense of the original (pump) wave amplitude; for SRS, it is the forward-scattered wave that undergoes amplification. If the initial light power is suitably strong and the length of the fiber is substantial, then the scattered waves can rise to power levels comparable to those of the pump wave; this is exemplified for SBS in the bottom part of Figure 11.12.

Note that while here the effects of stimulated scattering losses are detrimental to power transmission, it is also exploited beneficially in some types of fiber lasers (for example, to shift the light emission to wavelengths not normally provided or to narrow the bandwidth of a particular laser wave).

Note that the effect of SRS-generated secondary light can be eliminated by using ultranarrow laser line filters (elements with full width at half maximum [FWHM] of ~50–60 $cm^{-1}$ are available for many of the standard Raman laser wavelengths); their transmission at the original wavelength can be as high as >95% while the suppression at 200 $cm^{-1}$ away from line center is of the order $10^5$, increasing even further with larger shifts (for an example of such a filter curve, see Figure 11.13.

### 11.3.3 Samples and their incorporation into the overall setup

As pointed out frequently in this chapter, Raman spectroscopy is being used for the analysis of gaseous, liquid, and solid specimens.

The former two generally need some sort of container or cell to confine the sample within the laser interaction region. Such cells allow for a variety of Raman configurations, normally set up (1) for light collection at 0°, i.e., for forward scattering or transmission observation of the Raman light, in the same direction as the laser beam; (2) for light collection at 180°, i.e., for backward scattering in reverse direction of the laser beam; and (3) for light collection at 90°, i.e., avoiding any direct sight of the laser beam. On the other hand, standoff solutions have been devised for the two in which open gas volumes or liquids can be probed, normally in backward-scattering arrangements. Of course, the same standoff instrumentation would also be suitable for solid samples, in general.

Solid samples are normally provided "as is" and are placed in the focal regions of the exciting laser beam and the Raman light collection optics. By and large, setups are configured for 180° Raman light collection. However, in some variants of Raman spectroscopy, such as transmission Raman spectroscopy and offset-Raman spectroscopy, the geometrical arrangement for excitation laser radiation delivery, sample positioning, and Raman collection optics may significantly differ; some selected examples will be shown in Chapters 12 through 14.

### 11.3.4 Raman light collection

Like in the majority of spectroscopic techniques, the Raman light emerging from the laser interaction region has to be collected for efficient coupling to the spectrometer and its detector. In principle, a wide range of imaging arrangements—mainly based on lenses and mirrors—can be implemented. For best efficiency, the imaging optics should transpose the origin dimensions to the spectrometer

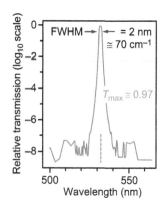

**Figure 11.13** Transmission curve for a 532-nm laser line clean-up filter, with FWHM $\cong$ 2 nm and maximum transmission $T_{max} \geq$ 0.97. Suppression beyond ±5 nm from laser line center >$10^{-6}$.

slit dimensions, and emission and acceptance solid angles lossless light propagation should match.

In quite a number of cases optical fibers or fiber bundles are also used for geometrical light path flexibility or for matching the (sometimes odd-shaped) interaction volume to the entrance slit of a spectrometer. For example, in forward- and backward-scattering Raman light collection, the round ($TEM_{00}$) laser beam profile has to be transformed into a "slit-type" profile if the losses are to be minimized—a "dot-to-slit" profile fiber bundle is required; on the other hand, for 90°-Raman setups, the viewed interaction volume resembles a very elongated cylinder, and accordingly, a "slit-to-slit" profile fiber bundle achieves minimized losses. Of course, low-loss imaging into and out of the fiber (bundle) has to follow the same geometrical optics requirements for focusing and acceptance solid angles as for the direct Raman volume–to–spectrometer slit light transfer.

Because the intensity of the Raman light is in general very weak, one of the main difficulties in Raman spectroscopy is its separation from the intense Rayleigh scattering and any residual stray laser light. The problem here is not so much to spectrally separate Raman and Rayleigh components from each other; rather, it is the intensity ratio between the two: Raman features close to the Rayleigh peak may drown in the Rayleigh peak if the dynamic range of the detector is insufficient. In addition, if a fiber (bundle) is used for Raman light transfer, the Rayleigh peak and laser stray light may be strong enough to generate a secondary signal in the fiber itself. Then, detection and interpretation of the actual Raman spectra might become very difficult indeed. Therefore, with nearly no exception, laser line "blockers" are used in standard, spontaneous Raman spectroscopy.

Two main filter types are normally used today, namely, notch and edge filters; with the former type, both Stokes and anti-Stokes spectra can be recorded (but normally the laser line suppression is not optimal), while with the latter type, only Stokes spectra can be measured (but the laser line suppression may be two to three orders of magnitude better than that for notch filters). Today, probably the best-known and best-performing varieties are the so-called volume-holographic grating (VHG) notch filters and razor-edge ultra-steep long-pass filters. Some typical response curves are shown in Figure 11.14. Some filters allow one to observe Raman features as close as a few cm$^{-1}$ from the Raman excitation line, suppressing this laser line by up to eight to nine orders of magnitude while at the same time still transmitting more than 95% of the Raman light.

### 11.3.5 Wavelength separation/selection devices

Most modern Raman detection systems incorporate standard dispersive spectrometers, comprising, by and large, gratings (both reflective and transmission gratings). These devices range from the size of fitting into a hand to common desktop dimensions, depending on the application (field use or laboratory based) and the required spectral resolution (from low-resolution survey spectroscopy to high-resolution precision spectroscopy). As a rule of thumb, common instruments provide spectral resolutions in the range 1–10 cm$^{-1}$. Today, most spectrometers are of the "imaging" type; i.e., a complete dispersed spectrum element is recorded using an array photon detector (see below).

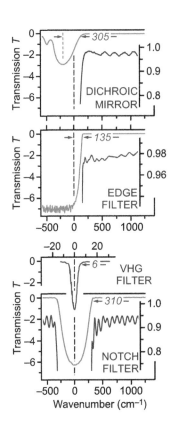

**Figure 11.14** Transmission curves for 532-nm Raman laser filters; top panel: dichroic *RazorEdge* beam splitter; middle panel: *RazorEdge* long-pass filter; bottom panel: *StopLine* notch filter and *SureBlock* VHG notch filter (red trace). The data are displayed in logarithmic (green trace) and linear (blue trace) scale. The 50% transition widths are indicated. (Adapted from Semrock Inc. and Ondax Inc. data sheets.)

In addition to spectrometers, which constitute the most widely used family of wavelength separation and selection devices, narrow bandwidth (interference) filters are also used. Their wavelength separation power is normally much lower than that for gratings, of the order of 100 cm$^{-1}$ only. However, this lower resolution can be sufficient in applications where the spectra are sparse and the spectral features exhibit large spacing (such as, e.g., in global spatial imaging applications of pharmaceutical samples) or in cases where the frequency selection is provided by other means (such as, e.g., by the stimulating laser in CARS).

Finally, the class of Fourier transform Raman spectrometers exploits interferometric separation of the Raman features; spectral resolution to well below 1 cm$^{-1}$ is feasible.

### 11.3.6 Photon detectors

In the very early days of dispersive spectroscopy, photographic plates were used to record the spectrum as a whole. But the evaluation of these spectra was tedious and not necessarily fully quantitative, due to nonlinear saturation of the light-sensitive layer; in addition, photographic emulsions were not very sensitive to extremely weak light, such as Raman emissions.

With the advent of photo-multiplier tubes (PMTs), which provide direct conversion from the detected photon into an electrical signal, much of the sensitivity, nonlinearity, and signal evaluation problems could be overcome. But this came at the expense that the spectrum had to be recorded step by step, which meant that a Raman spectrum with decent resolution required extremely long recording times. For a long time, this hampered the progress of Raman spectroscopy as an analytical technique, in both research and industry; well into the 1980s, commercial systems made use of PMT detectors.

The advance of light-sensitive semiconductor diode technology then made it possible to return to the full-spectrum recording principle, with the emergence of charge-coupled device (CCD) technology and array detectors in the late 1970s, but it took more than another 10 years before device technology and diminishing cost let CCD array detectors become viable for spectroscopy instrumentation. Today, these sensors come both in 1D (linear array) and in 2D (image sensors) varieties, and their use is near exclusive now in commercial Raman instrumentation, for both spectroscopy and imaging applications; some examples are shown in Figure 11.15, together with a few single-element detectors.

With respect to array detectors, two particular device classes are noteworthy. First, in addition to CCD sensors, the so-called complementary metal oxide semiconductor (CMOS) sensors are gaining popularity. Each sensor type has advantages and disadvantages, and determining which is the most appropriate will depend on the particular application. CCD-type detectors have the edge when it comes to dynamic range and uniformity, while CMOS-type devices are in the lead with respect to responsivity and speed. Second, driven by demanding applications in astronomy and medical imaging, now affordable 1D and 2D array detectors in the NIR have also become available for applications in laser spectroscopy and imaging in general and Raman spectroscopy in particular. These detectors are based on InGaAs diodes and typically cover the spectral range 900–1700 nm (or even wider for enhanced types).

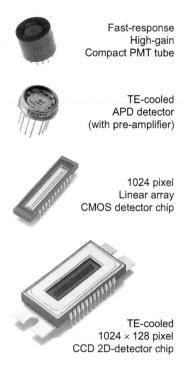

Fast-response
High-gain
Compact PMT tube

TE-cooled
APD detector
(with pre-amplifier)

1024 pixel
Linear array
CMOS detector chip

TE-cooled
1024 × 128 pixel
CCD 2D-detector chip

**Figure 11.15** Typical photon-detector elements used in Raman spectroscopy, including a compact PMT, a silicon avalanche photodiode device with integrated pre-amplifier and TE-cooling element, a linear array CMOS detector chip, and a 2D-array CCD detector device with integrated TE-cooling element. All devices are shown to scale.

A general survey of a wide range of array detectors and their properties and performance can be found, e.g., in Rogalski (2012).

### 11.3.7 Signal acquisition and data analysis equipment

Today's recording of Raman spectra and images is, by and large, based on array detectors whose operation and signal readout is computer based. In general, manufacturers provide software for their detectors with built-in electronics, or integrated Raman systems, which (1) controls the detector, (2) reads out the accumulated light signal pixel by pixel or averaged on-chip, and (3) displays a 1D spectrum or spectrally filtered 2D image. Only in the rare cases, in which single-element detectors (photo-multipliers or avalanche diodes) are utilized, is additional acquisition and/or analysis equipment required; this is normally in the form of lock-in amplifiers (for CW applications) or boxcar integrators (for pulsed applications).

# 11.4 RAMAN SPECTROSCOPY AND ITS VARIANTS

In all the discussions thus far, the Raman process and the observation of Raman scattered radiation were treated as "spontaneous"; i.e., Raman scattering molecule(s) and the laser photons were treated as isolated entities, and the three common observation methods in forward, right-angle, and backward geometry were addressed. It also should now have become clear from the previous chapter segments that, even with modern laser sources and photon detection equipment, the recording of Raman spectra remains challenging. This is because of the generally very low efficiency of the Raman process itself.

Complementing the basic methodology of spontaneous Raman light detection, a number of variants were developed over the past decades since the inception of laser-based Raman spectroscopy. Common goals have been to increase the sensitivity of the process, to improve on the spatial resolution of probing a sample, or to acquire specific information about molecular and material properties. Here, we give a short summary of the most common and widely used methodologies; many of the techniques highlighted below will be discussed in more depth in Chapter 12 ("conventional" linear Raman spectroscopy), Chapter 13 (enhancement techniques), Chapter 14 (nonlinear Raman spectroscopy), and Chapter 19 (Raman imaging).

### 11.4.1 Spontaneous Raman spectroscopy variants

Despite the fact that Raman signals are in general very weak, researchers succeeded—aided by the availability of powerful laser sources and highly sensitive CCD array detectors—to overcome some of the problems related to, e.g., (1) the distance between a Raman scatterer and a detection system, (2) the swamping of the Raman signal by background fluorescence, or (3) very small sample sizes.

Nowadays, *standoff/remote Raman spectroscopy* is a rather popular method to analyze samples over distances ranging from fractions of a meter to many kilometers. Systems are normally configured for backward scattering observation. For example, quite a few handheld Raman spectroscopic devices are marketed, which are used in the detection of traces of explosives or gas leaks, to name but two common applications. Raman LIDAR systems—utilizing pulsed laser sources—are used to monitor the atmosphere for common and trace molecule components or to deduce temperature information from the $J$-level intensity distribution, and that even up into the stratosphere.

A methodology that is applied to the analysis of solid samples is the so-called *spatially offset Raman spectroscopy—SORS*. In particular, this technique is used to obtain Raman spectra from a specimen beneath an obscuring (but still light-transmitting) surface. Multiple measurements—but usually just two—are made to determine the spectrum of the "hidden" specimen. The Raman measurement at the location of laser illumination is normally dominated by Raman light (and fluorescence) from the surface layer—e.g., skin or a coating. The Raman signal measured at a location offset from the excitation location, however, will contain larger Raman contributions from the subsurface specimen (see Figure 11.16). A suitably scaled subtraction of the two measurements results in a (nearly) clean Raman spectrum free from surface-layer contributions; the difference spectrum can now be used to analyze/identify the hidden probe specimen.

When small specimens (down to the nanoscale) are to be investigated in suspension, it is paramount that the particle is "kept in place." This is necessary to allow for Raman light acquisition over extended periods of time, in order that spectra exhibit sufficient signal-to-noise ratio for analysis. Here, *optical tweezers Raman spectroscopy—OTRS* has become a method of choice, since individual particles can be characterized and it is even possible to follow, e.g., biochemical processes in a single cell. This is afforded by trapping the particle in the focal point of optical tweezers, and the same laser field providing the trapping force also induces Raman interaction with the particle.

### 11.4.2 "Enhanced" Raman techniques

Referring back to Equation 11.3, the Raman signal depends on a number of extrinsic experimental and intrinsic quantum physics parameters; the former include (1) the particle density in the light–matter interaction volume—i.e., how many particles participate in the Raman process; (2) the (local) excitation laser power—i.e., how many photons are available to initiate the Raman process; (3) the excitation laser wavelength—this is associated with the approximate $\tilde{v}^4$-dependence; and (4) properties of the detection system—solid angle of light collection, transmission of the spectrometer, and detector sensitivity. For the conceptual influence on the various parameters on the Raman signal strength, see Figure 11.17.

The first two parameters—particle number density and laser power—are often interlinked. For example, the local photon flux can be substantially enhanced by focusing the laser beam, but as a consequence, the interaction volume becomes smaller, and thus for unchanged particle number density, a lesser number of particles is examined. Hence, the product $N_j \cdot I_L$ is largely unaffected and the Raman signal remains unchanged. In order to increase the Raman signal to useful levels for analysis for low-$N_j$ samples, one of them needs to be enhanced without influencing the other.

For gaseous samples, which often suffer from very low particle densities, two methodologies are becoming rather popular.

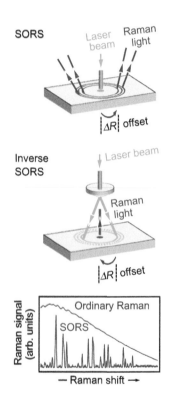

**Figure 11.16** Top panel: principle of spatially offset Raman spectroscopy (SORS); the Raman signal is picked up at a location offset by $\Delta R$ from the laser excitation position. Middle panel: principle of *inverse* SORS; ring-type laser excitation is employed, using an axion lens; the Raman scattered light is collected at the center of the ring. Bottom panel: conceptual spectra for probing a turbid sample by ordinary Raman spectroscopy and SORS.

If the gas is enclosed in a reflective wave guide (hollow-core fiber) and the laser light is coupled (focused) into this wave guide, then the laser light interacts over a much longer path and hence with many more molecules. Equally, the Raman photons are guided to a single (wave guide) aperture, thus making the light capture much more efficient than without it. Signal enhancements in the range 10–100 have been reported (see, e.g., James et al. 2015).

Often, the laser power is limited by the available commercial laser sources. One method to enhance the local laser power density available for the Raman process even for low-power lasers is to couple the light into a resonant cavity, which enhances the intracavity circulating power substantially beyond the external input power. In a good, high-finesse optical cavity, power buildup by factors up to $10^4$ can easily be achieved. If the cavity mirrors sufficiently transmit wavelengths different from the resonance one, then Raman spectroscopy becomes possible at extremely high local power levels; the method is known as *cavity-enhanced Raman spectroscopy—CERS* (see, e.g., Salter et al. 2012).

The latter two experimental parameters are, by and large, linked to the available equipment where it is not always feasible to change or improve it at will, not least due to possible financial constraints. Assuming that one has state-of-the-art equipment, not very much can be done to improve system performance other than careful alignment and clever data acquisition procedures. The only really "free" choice would be that of excitation wavelength to optimize the $\tilde{v}^4$ factor. However, limitations are normally imposed by the availability of suitably powerful laser sources, by the fact that the optics and detection equipment may not cover the Raman spectral range adequately, or by the occurrence of laser-dependent effects that may mask the Raman signal (such as, e.g., fluorescence).

Finally, enhancement of the quantum-optical parameter $\Phi$ is possible, although this is not always easy to achieve—due to being intrinsic to the molecule, and thus might not be possible universally. Two particular approaches are frequently used.

In the so-called method of *resonance Raman spectroscopy—RRS*, the excitation wavelength is carefully chosen to overlap with (or be very close to) an electronic transition in the molecule. This (near-)coincidence can result in Raman scattering intensities that are larger by factors of $10^2$ to $10^6$, due to the transition matrix elements now involving an (allowed) electronic quantum state rather than a "virtual" energy state. Thus, detection limits and measurement times can be reduced by the same margin. However, on the downside, one finds that fluorescence backgrounds can become significant, potentially reaching levels that nullify the gains from involving the resonance transition in the Raman process. In addition, by varying the excitation wavelength, vibration mode-specific resonance Raman spectra of the molecule can be obtained, meaning that the spectral contribution from a specific part of the molecule is selectively enhanced and hence can be separated from the rest of the molecule. Similarly, the same holds for trace molecules in a dominant "matrix," provided that the molecular resonances are different for the two.

While in RRS, the enhancement of the line strength function $\Phi$ is influenced by the proximity of the photon energy to an allowed electronic transition, in *surface-enhanced Raman spectroscopy—SERS*, it is aided by placing the molecule under study in close proximity to a (most often colloidal) metal surface. When exposing this metal substrate to laser light, surface plasmons are excited

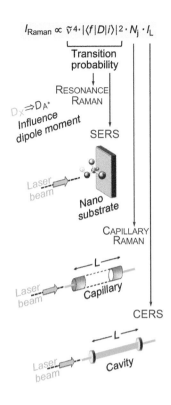

**Figure 11.17** Enhancement strategies affecting the Raman signal strength $I_{\text{Raman}}$; SERS, surface-enhanced Raman spectroscopy; CERS, cavity-enhanced Raman spectroscopy; $D$, transition dipole moment.

and the electric fields close to the metal surface are greatly enhanced. Since the Raman process is proportional to the electric field, a large increase in the measured signal can be observed, which can be as large as up to 10 orders of magnitude. In a variant of SERS, known as *tip-enhanced Raman spectroscopy—TERS*, a metallic ultratip (usually etched from silver or gold wires) is used to push the spatial resolution into the nanoscale, reaching approximately the size of the tip apex of typically 20–30 nm. This is well beyond what can be resolved by "normal" Raman scattering from a diffraction-limited laser spot, which is of the order of 1 μm. Thus, TERS has particular potential for biomedical analysis applications (a few examples will be discussed in Chapter 19).

### 11.4.3 Nonlinear Raman techniques

Like all nonlinear laser processes, nonlinear Raman spectroscopy also requires the involvement of two photon (fields) to initiate the observed response signal, which will be proportional to $I_1 \cdot I_2$. The two photon energies can be equal or different; in the former case, a technique known as *hyper-Raman spectroscopy* ensues, while in the latter case, various *stimulated Raman spectroscopy* variants can be realized.

In *hyper-Raman spectroscopy—HRS*, the vibrational modes (and the associated rotational sublevels) of the molecules interact with two photons of the excitation laser beam. The two pump photons are converted into one photon of Raman scattered light and an excited ro-vibrational state of the molecule. As for all "second-harmonic" processes, the effect is rather weak, unless the laser power is very high; thus, hyper-Raman spectroscopy is, by and large, realized using pulsed laser sources. In addition, SERS-type enhancement techniques are frequently added to boost the sensitivity. Therefore, one could argue that this technique of "doubly weak" nature would not be attractive: in a sense, a spontaneous Raman spectrum is observed, but with two-photon rather than one-photon excitation. However, associated with the different selection rules, hyper-Raman spectroscopy can provide information on molecular vibrations for which ordinary Raman scattering is suppressed, due to symmetry issues (so-called "silent" modes).

*Stimulated Raman spectroscopy—SRS* constitutes a two-color two-photon process in which population is transferred from a ground to an excited (ro-vibrational) state of the molecule. One can think of the process as an amplifying or gain process for the Raman Stokes photon. When the difference frequency (or Raman shift) $\Delta\omega = \omega_\mathrm{p} - \omega_\mathrm{S}$ matches a particular molecular ro-vibrational frequency, amplification of the Raman signal is proportional to the stimulated rate of the transition into the Stokes level of the molecule:

$$r_\mathrm{SRS} \propto \sigma_\mathrm{Raman} \cdot n_\mathrm{p} \cdot (n_\mathrm{S} + 1), \tag{11.7}$$

where $\sigma_\mathrm{Raman}$ is the Raman scattering cross section, related to the scattering matrix relation (Equation 11.4), and $n_\mathrm{p}$ and $n_\mathrm{S}$ are the number of photons in the pump and Stokes laser fields, respectively. Of course, in the absence of the stimulating Stokes beam (i.e., $n_\mathrm{S} = 0$), Equation 11.7 reverts back to that for spontaneous Raman scattering.

Using typical (tunable) lasers for downward stimulation, $n_\mathrm{S}$ is typically six to seven orders of magnitude larger than the number of spontaneous Raman photons at the same wavelength; thus, SRS provides amplification at the vibrational transition rate. As a consequence, the intensity of the Stokes beam, $I_S$, experiences stimulated Raman gain (SRG = $\Delta I_S$) while the intensity of the pump beam, $I_p$, suffers a loss (SRL = $\Delta I_p$). Either change in gain or loss can be used for detection, and one finds

$$\Delta I_S \propto N \cdot \sigma_{Raman} \cdot I_p \cdot I_S \text{ and } \Delta I_p \propto -N \cdot \sigma_{Raman} \cdot I_p \cdot I_S, \qquad (11.8)$$

where $N$ is the number of molecules in the probe volume generated by the two overlapping laser beams. Note that both SRG and SRL signals are measured unidirectional with the respective laser beams. Note also that no spectrometer is required for detection but that by tuning the second laser, the Raman line spectrum is built up sequentially.

For sufficiently powerful laser pulses, up to about half of energy can be channeled coherently from the pump to the Stokes laser beams. In comparison, in spontaneous Raman spectroscopy, only $10^{-5}$ to $10^{-6}$ of the pump photons are converted to Stokes photons, meaning that SRS enhancement of up to four to five orders of magnitude is feasible. Because of the huge difference between the spontaneous and stimulated Raman signals, SRS is virtually background free, which is, e.g., exploited in high-contrast SRS chemical imaging (see Chapter 19).

An interesting variant to SRS is *photo-acoustic Raman spectroscopy—PARS*, which has proven to be one of the most sensitive stimulated Raman techniques. In PARS, the pump and stimulating laser beams are focused into a photo-acoustic cell. If the energy difference between the laser photons corresponds to an allowed transition, the molecules are promoted to excited ro-vibrational (V-R) states. Subsequent to the Raman process, the V-R excitation of the molecules is converted by collisions into local heating (V-R→T), which results in a pressure wave, which is detected by a microphone (thus the name photo-acoustic) or a sensitive piezo tuning fork (see, e.g., Schippers et al. 2011). This very sensitive, indirect detection of Raman signals was predominantly applied for gaseous samples, with detection limits down to a few parts per million (see, e.g., Siebert et al. 1980), but in recent years, it has been adapted for use in position-resolved PARS measurements in solid samples (see, e.g., Yakovlev et al. 2010).

Another widely used nonlinear Raman spectroscopy technique is that of *coherent anti-Stokes Raman spectroscopy—CARS*. The method of CARS is very much akin to the process of SRS, just discussed; the major difference is that it is a three-color four-photon process, and the observation is not on the (stimulated) Stokes transition but on an anti-Stokes line. Again, no spectrometer is required and any spectrum is built up sequentially by tuning the stimulating laser. The two (pump and stimulating) laser beams with photon energies $\tilde{\nu}_p$ and $\tilde{\nu}_S$, with ($\tilde{\nu}_p > \tilde{\nu}_S$), measured in $cm^{-1}$, interact coherently with the molecule to generate scattered light whose photon energy is

$$\tilde{\nu}_{AS} = 2\tilde{\nu}_p - \tilde{\nu}_S = \tilde{\nu}_p + \tilde{\nu}_m, \qquad (11.9)$$

where $\tilde{\nu}_{AS}$ is energy of the anti-Stokes photon and $\tilde{\nu}_m$ is the energy of the matching ro-vibrational molecular state; for the transitions and connected energy levels, see the top of Figure 11.18.

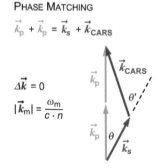

FREQUENCY MATCHING

$2\,\omega_p - \omega_s = \omega_{CARS}$

PHASE MATCHING

$\vec{k}_p + \vec{k}_p = \vec{k}_s + \vec{k}_{CARS}$

$\Delta\vec{k} = 0$

$|\vec{k}_m| = \dfrac{\omega_m}{c \cdot n}$

**Figure 11.18** Top panel: energy levels $E$, transitions, and transition frequencies $\omega$ in coherent anti-Stokes Raman scattering (CARS). Bottom panel: phase-matching diagram for the wave vectors $\boldsymbol{k}$ in CARS.

Note that, normally, the anti-Stokes Raman light would be rather weak because of the low thermal population of excited vibrational states. However, because of the transfer of substantial population into the excited vibrational level in the first (SRS) step, now in CARS, the anti-Stokes emission can be orders of magnitude more intense than that for normal Raman scattering, although the second phase in the CARS process is a spontaneous process.

Note that CARS emission is highly directional; this makes the collection of the anti-Stokes Raman signal substantially more efficient than in ordinary, spontaneous anti-Stokes Raman scattering. This is a direct consequence of the fact that CARS constitutes a third-order nonlinear (four-wave mixing) process in which not only energy conservation according to Equation 11.9 is required, but momentum conservation (phase-matching) as well:

$$2 \cdot \boldsymbol{k}_\text{p} - \boldsymbol{k}_\text{S} - \boldsymbol{k}_\text{AS} = 0; \qquad (11.10)$$

see the bottom of Figure 11.18 and, e.g., Chandra et al. (1981). This phase matching is required since waves of different frequency have different propagation velocities in dispersive media, which may result in a mismatch between their respective phases. As a result, a destructively interfering phase mismatch can take place, and the intensity of the final anti-Stokes wave, and hence the CARS signal intensity, would be reduced. To compensate for the phase mismatch, normally, noncollinear overlap configurations between the pump and Stokes-stimulating incident beams are chosen, matching the vector directions outlined in Figure 11.18.

As a final remark, we like to add that because the CARS signal is blue-shifted, with respect to the incident beams, spectral overlap with any red-shifted fluorescence background is, by and large, avoided. This dramatically improves signal-to-noise levels because of the absence of (incoherent) underlying fluorescence background.

A range of applications of CARS in chemical analysis and species-specific 3D spectral imaging will be touched upon in Chapters 14 and 19, respectively.

## 11.5 ADVANTAGES AND DRAWBACKS, AND COMPARISON TO OTHER "VIBRATIONAL" ANALYSIS TECHNIQUES

In the early days of Raman spectroscopy, it was a challenge even to simply measure a useable Raman spectrum. With the great advances of instrumentation development, it has become possible not only to qualitatively assess the chemical composition of the sample but also to delve into quantitative analysis, allowing for chemical species concentrations to be measured, monitored, and controlled, often even in real time and in situ.

Raman spectroscopy is now commonly used in a wide range of different areas and disciplines, covering problems in chemical analysis, providing insight into fundamental physicochemical processes, and aiding in biomedical, diagnostic, and other fields. This is because Raman spectroscopy is specific to the ro-vibrational information associated with chemical bonds and the symmetry of molecules. Therefore, it provides a spectroscopic "fingerprint" by which the molecule can be identified. Here, we list just three typical examples:

- The analysis of gaseous samples by Raman spectroscopy has found many practical applications, ranging from, e.g., industrial gas leak testing to real-time monitoring of anesthetic and respiratory gas mixtures during surgery.

- In the biopharmaceutical industry, Raman spectroscopy can be used to not only identify certain active pharmaceutical ingredients but also identify the polymorphic form of a specific compound, as well as reasonably quantify concentrations.

- In solid-state physics, Raman spectroscopy is used—among other things—to characterize the composition of bulk and surface-layer materials, to measure localized temperature profiles, or to find the crystallographic orientation of a sample.

Despite all the successes of Raman spectroscopy and imaging, being applied to all facets of scientific disciplines and industrial-scale problems, not everything is plain sailing. The main adverse issue is that of fluorescence emission "competing" with the weak Raman spectra.

## 11.5.1 The problem of fluorescence

Fluorescence is a problem that has bedeviled and is bedeviling Raman spectroscopy, particularly when using laser excitation sources in the visible spectral range. As stated repeatedly, spontaneous Raman signals are weak, in general, and fluorescence initiated by the excitation laser can easily overwhelm the Raman signals. Even if the analyte molecule does not fluoresce itself, on interaction with the Raman excitation laser light, fluorescence may easily originate from molecules of the sample "matrix" (for example, from a solvent or contaminant in the sample). This fluorescence problem is also and in particular encountered in RRS because the excitation photons have energies near that of a molecule's electronic transition: the excitation radiation may become more likely to be absorbed rather than Raman scattered, resulting in fluorescence as a possible de-excitation mechanism for the electron's return to the ground state. An example of the adverse presence of fluorescence masking Raman signals is shown in Figure 11.19a.

While it is not always possible to avoid fluorescence completely, there are a number of ways of overcoming or, at least, substantially improving the interfering contribution from fluorescence.

One possible approach is to simply use an excitation laser source whose photon energy is low enough to only weakly generate fluorescence or none at all, as demonstrated in Figure 11.19 where the fluorescence background is nearly nonexistent when using a NIR laser excitation source (nowadays, predominantly semiconductor diode or solid-state lasers at 780, 830, or 1064 nm). However, this may be at the expense of Raman sensitivity due to the approximate $\tilde{\nu}^4$ dependence, and such solutions may be much costlier than visible-laser solutions, due to the more expensive IR-sensitive photo-detection equipment.

If the change of excitation laser wavelength to the NIR is not feasible or desirable, a number of solutions exist, which rely on background subtraction (see, e.g.,

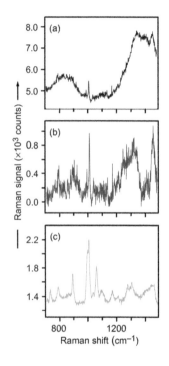

**Figure 11.19** Raman spectra of a bacteria sample. (a) Spectrum of a bacteria sample on a BK7 glass substrate (raw data); the spectrum is dominated by the fluorescence from the glass. (b) Background-corrected spectrum (using rolling-circle removal algorithm). (c) Spectrum of a bacteria sample on a gold-coated BK7 glass substrate (raw data); the fluorescence is substantially reduced, by a factor >5. (Data adapted from Al-Marashi, J.F.M. and H.H. Telle, "Raman spectra of *S. epidermidis*." Unpublished results, 2013.)

James et al. 2013 and references therein). However, for complex spectra, this may not always be straightforward since the shape of the background is not necessarily easy to determine, and thus subtraction may lead to substantial errors; see, e.g., Figure 11.19b. In another approach, the differences in excitation lifetimes for Raman and fluorescence excitation/decay can be exploited. The latter is substantially longer and thus—when using ultrashort picosecond or femtosecond laser pulses—by time gating the detection, the fluorescence and Raman components can be separated (see, e.g., De Luca et al. 2015 and references therein).

It may also be possible to reduce the fluorescence contribution to the Raman spectrum by using nonfluorescing sample substrates; for an example, see Figure 11.19c.

Finally, a different, rather elaborate but quite successful approach was proposed and used by Matousek et al. (2002); it constitutes a combination of the aforementioned time gating (in the form of a Kerr gate) and shifted-excitation Raman difference spectroscopy (SERDS). SERDS utilizes two slightly different excitation wavelengths—shifted by about the bandwidth of the Raman signal—to generate two Raman spectra. These two spectra are subtracted from each other, and from the difference spectrum, a nearly background-free Raman spectrum is reconstructed.

### 11.5.2 Advantages and drawbacks of Raman spectroscopy, and comparison to (IR) absorption spectroscopy

To a certain degree, Raman spectroscopy combines the analytical advantages of the techniques of IR ($100$–$4000$ $cm^{-1}$) and NIR ($4000$–$12500$ $cm^{-1}$). Although Raman and IR spectroscopy are based on different physical processes, namely, scattering and absorption, both excite fundamental molecular vibrations (and/or rotations), which can easily be analyzed or interpreted as constituting sort of the molecule's "fingerprint." The different physical processes give rise to distinctly different pros and cons for using either of the techniques. As mentioned a few times already in this chapter, one of the greatest obstacles for Raman spectroscopy as an analytical method has been that inelastic Raman scattering is a very weak phenomenon when compared to Rayleigh scattering (elastically scattered light), but even more so when compared to the resonant transition processes of absorption and fluorescence. Nevertheless, Raman spectroscopy is now a widespread technique in qualitative and quantitative analysis, as well as spectroscopically resolved imaging. The most important advantages and drawbacks for Raman spectroscopy are provided in Table 11.1.

While it is beyond the scope of this short summary to fully compare Raman spectroscopy, and its multitude of variants, with the absorption-spectroscopic techniques that probe molecular vibrations, it is worthwhile to put all of them into context. This is done in Table 11.2 where the various spectroscopic and instrumental features are summarized. Note that for the NIR, the variant of tunable diode laser spectroscopy—TDLAS—forms part of the comparison, and for the mid-IR range, its Fourier transform variant—FTIR—is included (although FTIR has also been applied to the NIR). Further details of the aforementioned absorption techniques can be found in Chapters 7 and 8.

**Table 11.1**   General Advantages and Drawbacks Encountered in Raman Spectroscopy and Imaging

| Experimental Issues | Remarks |
|---|---|
| **Advantages** | |
| Laser sources (wavelengths) | In principle, lasers at any wavelength can be used, but certain wavelengths dominate applications due to optical filter availability. Depending on the actual Raman system implementation and application, either CW or pulsed lasers are used. |
| Raman signal (ease of light delivery) | Both the laser excitation and Raman scattered light can be transmitted through optical fibers, even over long distances (remote analysis). |
| Samples (accessible types) | Samples in any aggregate form (gaseous, liquid, or solid) can be analyzed; nearly all organic and inorganic compounds exhibit distinct Raman signatures. In particular, analysis in aqueous systems is superior to IR spectroscopy, because of the noninterference with water absorption bands. |
| Sample preparation (mostly none) | In general, Raman does not require special sample preparation, but samples mostly can be used "as is." |
| Analytical capabilities | *Specificity*: Raman signals are (mostly) related to fundamental vibrations, which often are nonoverlapping; thus, "fingerprinting" of samples is feasible.<br>*Quantitative analysis over a wide range of concentrations*: the intensity of a Raman band is directly proportional to the number of molecules associated with that band. This provides a measure of the concentration of a molecular species. With sufficiently high SNR fractional species concentrations in the range 100% down to parts per million can be measured. |
| Measurement speed (measurement times) | Typically, Raman spectra can be acquired on a timescale of subseconds to several minutes; thus, in principle, Raman spectroscopy can be used to monitor samples in "real time." |
| Spatial resolution | Raman spectra can be collected from very small areas or volumes (dimensions of less than 1 μm can be probed). |
| **Drawbacks** | |
| Laser sources (high power) | The normally intense laser radiation required for strong Raman signals can significantly heat some types of sample and evaporate or burn certain or all components. Thus, the normally stated nondestructive nature of Raman spectroscopy may be lost. |
| Raman signal (low intensity) | The Raman effect is very weak; thus, the detection instrumentation needs to be highly sensitive. Impurities or matrix compounds in/of the sample may cause fluorescence that can be strong enough to completely mask the Raman signal itself. |
| Samples (inaccessible types) | Unsuitable for metals and the majority of alloys. |

**Table 11.2**   Comparison of the Most Common Spectroscopic Techniques Probing Molecular Vibrational Modes

| Spectroscopic Technique | Raman | NIR (incl. TDLAS)[a] | Mid-IR (incl. FTIR)[b] |
|---|---|---|---|
| Molecular interaction | Scattering | Absorption | Absorption |
| Observed ro-vibrational bands | Fundamental | Overtone/combination | Fundamental |
| Sample molecules | Organic/inorganic | Organic | Organic/inorganic |
| Sample preparation | None | Seldom | Normally required |
| Sample aggregate state | Gaseous/liquid/solid | Mainly solid | Gaseous/liquid/solid |
| Remote sampling | Yes | Yes | No |
| Spectral range ($cm^{-1}$) | 100–5000 | 100–1000 | 10–1000 |
| Spectral resolution ($cm^{-1}$) | 1–20 | 0.001–1.0 | 0.05–5.0 |
| Spectral accuracy ($cm^{-1}$) | ± 0.1–1.0 | ± 0.01–1.0 | ± 0.01–0.1 |
| Signal-to-noise ratio ($cm^{-1}$) | $10^2$ to $5 \times 10^4$ | $10^2$ to $5 \times 10^3$ | $10^2$ to $10^5$ |
| Acquisition time (s) | 0.1–100 | 0.1–10 | ~1 |

[a]   TDLAS, tunable diode laser absorption spectroscopy.
[b]   FTIR, Fourier transform IR absorption spectroscopy.

# 11.6 BREAKTHROUGHS AND THE CUTTING EDGE

In the rise of Raman spectroscopy to prominence, there are many developments that may be classified as breakthroughs, and to highlight them all would be beyond the scope of this chapter. Here, we have chosen just one particular approach, which has proven to be invaluable for applications of Raman spectroscopy to complex organic and biomedical specimen.

## 11.6.1 Breakthrough: UV Raman spectroscopy

In general, Raman spectroscopy is faced with two common limitations, namely, (1) that the cross section for Raman scattering is small, meaning small and noisy signals; and (2) that an already poor signal-to-noise ratio is often lowered further by fundamental, intrinsic noise sources, in particular fluorescence. Moving from the most commonly used excitation wavelengths in the visible and NIR to UV wavelengths, the above limitations can be substantially alleviated. For a more detailed survey, see, e.g., Asher (2002).

First, one gains substantially in signal strength due to the fact that the Raman cross section is roughly proportional to $v_{\text{laser}}^4$, thus gaining more than an order of magnitude, say, when using the fourth harmonic of a Nd:YAG laser ($\lambda = 266$ nm) instead of its second harmonic ($\lambda = 532$ nm).

Second, UV-B/UV-C photon energies below 300 nm often already reach electronically excited states of the molecule; thus, the intensity of Raman-active vibrations can increase by many orders of magnitude, due to near-resonance conditions in the transition matrix elements—one encounters the above mentioned effect of resonance Raman scattering. For an early example of UV RRS, see, e.g., Fodor et al. (1985).

Third, while UV-B/UV-C photons tend to be sufficiently energetic to give rise to fluorescence, its wavelengths are typically above 300 nm, regardless of the UV laser photon energy. Thus, when, e.g., using a laser wavelength of $\lambda = 266$ nm, even Raman features Stokes-shifted by up to nearly 4000 cm$^{-1}$ are still below 300 nm—fluorescence simply doesn't interfere with the Raman signal, in contrast to laser excitation wavelengths in the visible. This is shown schematically in Figure 11.20.

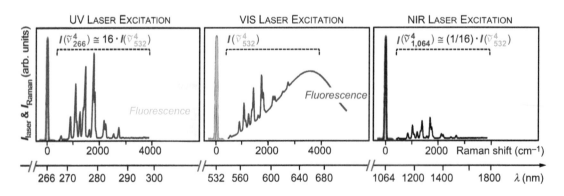

**Figure 11.20** Conceptual UV-excited ($\lambda_{\text{laser}} = 266$ nm) Raman spectrum and noninterfering fluorescence contribution, in comparison to laser excitation at $\lambda_{\text{laser}} = 532$ nm and $\lambda_{\text{laser}} = 1064$ nm.

## 11.6.2 At the cutting edge: Atomic properties probed by Raman spectroscopy

With the rapid advances in laser and detector technologies, Raman spectroscopy has very much entered the mainstream of laser (analytical) spectroscopy and imaging. Thus, it is not surprising to frequently read about new and exciting developments and applications. Choosing any of these at random probably wouldn't do justice to the field.

In our view, probably one of the most exciting aspects at present seems to be the success in pushing the detection sensitivity and resolution to levels that make it possible to study nanostructures and individual molecules. In particular, fundamental structural and chemical bonding information can be extracted from the related spectra. Even atomic properties have become accessible—which sounds mindboggling, keeping in mind that, ordinarily, Raman spectroscopy deals with ro-vibrational states and transitions in molecules.

Measuring the physical dimensions of atoms with radii in the order of a few $10^{-10}$ m (or angstrom, Å) is ordinarily beyond the limit of optical, visible-light techniques. One needs other approaches capable of subatomic resolution, such as, e.g., scanning tunneling microscopy; however, certain limitations apply. However, quite recently, Wang et al. (2013) showed that indeed atomic sizes can be accessed optically using Raman spectroscopy.

In their specific example, they determined the so-called van der Waals radius of iodine atoms. For this, polarized Raman spectroscopy was applied to iodine diatomic molecules, $I_2$, confined in nanoscale channels of zeolite single crystals of aluminophosphate, $AlPO_4$, in their so-called AFI and AEL framework structures. AFI and AEL nanochannels are well-established scale references, particularly AFI with a round-profile channel dimension of $2R = 7.3 \pm 0.1$Å.

The (vibrational) model of $I_2$ imbedded in the AFI nanochannel is shown in Figure 11.21. Here, it is assumed that one atom "touches" the left-hand wall of the channel, and the molecule vibrates freely around the equilibrium internuclear separation, $r_e$, between the inner and outer potential turning points, $r_{ti}$ and $r_{ta}$. The vibration amplitude increases for the higher vibrational quantum states, $\upsilon$. As the outer turning positions approach the wall, the vibrational motion becomes restricted, and vibrational excitation beyond this limit is suppressed because of the boundary condition $2r_a + r_{ta} \leq 2R$. This is reflected in the overtone Raman spectrum of the molecule, meaning that transitions to those levels do not occur. For $I_2$ confined in the AFI nanochannels, Wang et al. found $\upsilon = 14 \pm 1$ (see the spectrum in the lower part of Figure 11.21); normally, in the free gas phase, one observes overtones up to $\upsilon > 30$.

Based on the value for $r_{ta,\ \upsilon=14} = 3.02$ Å (from the potential energy curve for $I_2$), the authors estimate the radius of the iodine atom to be $r_a = (2R - r_{ta,\ \upsilon=14})/2 = 2.10 \pm 0.05$Å.

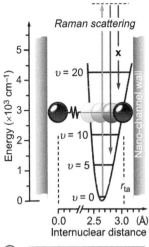

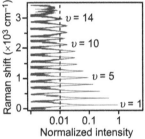

**Figure 11.21** Schematics of diatomic (here $I_2$) molecules, confined in a cylindrical-cage nanochannel of diameter $2R$. The internuclear potential, vibrational energy levels (equilibrium distance $r_e$ and inner/outer turning points $r_{ti}$ and $r_{ta}$) and Raman transitions between vibration levels at the electronic ground state are sketched. Data panel: polarized Raman spectrum of $I_2$ inside an AFI nanochannel, for laser excitation at $\lambda = 511.5$ nm; the "vibrational-constraint" limit is defined for the transition at which the Raman signal amplitude has decreased to about 1% of the maximum. (Adapted from Wang D. et al. *Nature Sci. Rep.* 3 (2013): article 1486.)

# CHAPTER 12

# Linear Raman Spectroscopy

Raman spectroscopy has become one of the most popular and widespread analytical laser spectroscopy methods. In principle, this is because only one single, fixed-frequency laser source is needed to tackle basically any molecule, even in complex mixtures—with the sample being in gaseous, liquid, or solid form, and mostly needing little sample preparation—yielding both qualitative and quantitative information about the analyte.

However, in practice, laser spectroscopic methods are rarely as straightforward as introductory texts might imply, and often the promises of instrument manufacturers regarding "simplicity and universality" do not stand up to scrutiny; frequently, both add to the frustration of novices and experienced users of a particular method. This general notion is not different for Raman spectroscopy. Consequently, careful considerations have to go into a planned application to guarantee that meaningful results become available after a particular Raman measurement.

When planning an analytic Raman spectroscopy campaign, it is important that a few central questions are posed and answered before commencing a measurement; the selection of equipment and the likely outcomes from a measurement heavily depend on this question–answer interplay. In general, the following key issues need to be addressed:

1. Is the molecule to be analyzed present in pure form or within a mixture of chemical compounds?

2. In which aggregate form does the sample exist (gas, liquid, condensed matter)?

3. Is sample preparation potentially required?

4. Will any type of background light interfere detrimentally with the Raman signals?

5. Does the user wish to obtain qualitative or quantitative results?

It may already become clear from this short, "simple" list of questions—which for specific cases may not even be sufficient—that a Raman measurement may not be so simple after all. With reference to Figure 12.1, a few general answers to the

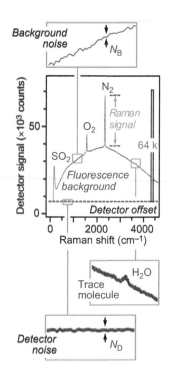

**Figure 12.1** Concept of qualitative and quantitative Raman spectroscopy, exemplified for a spectrum of ambient air, contained in a cell with (fluorescent) glass windows. The bar indexed with 64k stands for the maximum dynamic range in 16 bit A-to-D signal conversion.

above questions are summarized here, before the more contentious issues are addressed in more detail in Section 12.1.

A selected number of representative examples for the use of Raman spectroscopy are described in the subsequent four sections, highlighting strengths and weaknesses of the technique.

With respect to question (1), in the case that only a single molecular compound is present in the sample, which may be true when conducting pure spectroscopy investigations, no problem should occur—although sometimes surprises are in store should the sample not be as "pure" as claimed. In any case, each molecule exhibits spectral "fingerprints" associated with its vibrational (and rotational) motion; thus, one will expect one, a few, or many peaked features in the spectrum, which are characteristic of a particular molecule, and which may become ever more complex the larger the molecule and its number of vibrational modes. If there are several compounds in a mixture, the resulting Raman spectrum constitutes a superposition of the spectra of each of the individual components. This will be complicated by the fact that the vibrational modes of certain molecular subgroups in a bigger entity are the same (such as, e.g., the CH-stretch mode in nearly all organic molecules), and thus such a peak would not be suitable to identify a particular molecule. Thus, often a prior knowledge of the constituents of the sample and their general spectral contributions need to be available to interpret the Raman spectrum of the sample. In any case, depending on the actual sample and desirable compound distinction, the choice of spectrometer equipment and the respective spectral resolution might be a key issue.

As already stated in Chapter 11, in which the principles of Raman spectroscopy have been covered, it is possible to obtain Raman spectra from nearly any molecular sample, i.e., from gases, liquids, solids, gels, slurries, powders, films, and so on. However, in the context of question (2) above, it is important to be aware of any particularities of a sample "matrix" when searching for a minor constituent in a mixture of many contributors. Specifically, it is important to have appropriate knowledge about the physical–chemical environment; this may play a substantial role in the position of certain spectral peaks (the interaction with the matrix may affect the vibrational modes and thus their spectral position). Also, the analyte may not be distributed homogenously in the matrix, and thus one may need to contemplate how to access spatially distinct parts of the sample; this latter aspect may be highly significant when investigating biological tissue samples.

Usually, Raman spectra can be recorded using samples "as is," and samples do not require pretreatment before a measurement. However, samples may be "coated" with layers of material that are not transparent to both laser and Raman light or detrimentally contribute their own Raman signature, which would interfere with the Raman spectrum of the actual sample; in such cases, these layers need to be removed. Also, if one wishes to detect molecules at very low number densities, preconcentration of a sample might be required.

The answer to question (4) is intimately linked with issues (2) and (3); the nature of the matrix of a sample or any cover layer may heavily contribute to structured background in the form of overlaying Raman peaks or broadband background due to fluorescence, a process that should, by and large, be avoided since the transition probabilities for fluorescence are orders of magnitude larger than those for Raman scattering. This problem can be minimized by various means that are briefly discussed further below.

Finally, in answer to question (5), Raman spectra may be interpreted both qualitatively and quantitatively. As stated repeatedly, Raman spectra constitute unique molecular "fingerprints," which means that a search of, e.g., spectral libraries can be used to identify (qualitatively) participating constituent molecules of the sample. It becomes substantially trickier when quantitative information on molecular concentrations is sought. Typically, quantitative analysis of a mixture is performed by measuring the relative intensities of bands that are directly proportional to the relative concentrations of the compounds. But for this to work correctly, substantial prior knowledge has to be incorporated into the evaluation of the spectra, including, as key issues, relative quantum transition probabilities, molecular concentration calibration, and spectral sensitivity of the Raman instrumentation.

## 12.1 THE FRAMEWORK FOR QUALITATIVE AND QUANTITATIVE RAMAN SPECTROSCOPY

It is probably clear from the introductory remarks that for the generation and evaluation of Raman spectra, the correct treatment of extrinsic experimental parameters and intrinsic molecular and quantum properties is indispensable. And while qualitative analysis requires certain knowledge about all the above contributing factors, it is quantification that places more stringent boundary conditions on proceedings. Whatever the evaluation approach—qualitative or quantitative—it is the evaluation of the respective wavelength/wavenumber versus intensity of the peaks in the Raman spectrum that yields the desired information about the probed molecule(s). In this context, one has to return to Equation 11.3 from the previous chapter that describes the observed Raman signal in dependence of all relevant intrinsic and extrinsic parameters encountered in a measurement. This equation for the observed Raman signal $I_S$ is repeated here, together with an abridged annotation of terms:

Stokes/anti-Stokes Raman line $\tilde{v}_s = \tilde{v}_L \pm \tilde{v}_{if}$

Raman scattering line strength function

$$I_S(\tilde{v}_s) = k_{\tilde{v}} \cdot \tilde{v}_L \cdot \tilde{v}_s^3 \cdot N_j \cdot \Phi(\varphi, \theta, pi, ps) \cdot I_L(\tilde{v}_L) \cdot \tag{12.1}$$

Laser intensity

Particle density

Laser line position

Response and geometry functions

All factors and parameters have the same meaning and units as defined in Chapter 11.1, with the intrinsic and extrinsic factors noted above and below the equation, respectively. Note that the intrinsic factor of the Raman line position $\tilde{v}_s$ is parametrically dependent on the laser line position $\tilde{v}_L$, but the observed Raman shift relative to the laser excitation line remains constant and is equal to intrinsic vibrational mode frequency $\tilde{v}_{if}$. Note also that the Raman excitation/observation geometry reference frame is that depicted earlier in Figure 11.3.

This means that, in a Raman spectrum, in which the intensity $I_{sig}(\tilde{\nu}_s)$ is recorded in dependence of wavelength $\lambda$ or wavenumber $\tilde{\nu}$, respectively, for quantification, one requires both accurate wavelength (or wavenumber, or frequency) and Raman photon response calibration. The relevant issues associated with calibration are addressed below.

### 12.1.1 Determining and calibrating the Raman excitation laser wavelength

As is evident from Equation 12.1, the exact knowledge of the line emission wavelength of the laser used in the Raman experiment is paramount since it is necessary to extract the Raman shift and, with it, the frequency of the molecular vibration mode, from the spectra.

Although, in principle, any laser source could be used in Raman spectroscopy experiments, today, only a handful of popular laser sources/laser wavelengths dominate in commercial Raman spectrometer equipment. These (still) include (atomic) gas lasers, like the $Ar^+$ laser, but, by and large, solid-state and semiconductor diode lasers are the devices of choice.

For gas lasers, no wavelength calibration is required in general, since their emission wavelength is determined by atomic transitions whose energy difference and, hence, wavelength are known with high accuracy, and their emission is stable over long periods of time; typical values for laser line position accuracy are of the order $\Delta\lambda_L \sim 0.001$ nm, which at $\lambda_L = 500$ nm equates to an uncertainty in frequency of $\Delta\nu \sim 1.2$ GHz or in wavenumber of $\Delta\tilde{\nu}_L = 0.04$ cm$^{-1}$. This is normally well below the resolution of standard spectrometers.

The output wavelengths of solid-state and semiconductor diode lasers are not necessarily exactly known because of their intrinsically relatively broad gain profile, and they are much less stable over time, often depending critically on how well their operating temperature can be held constant. Manufacturers of Raman instrumentation recommend to calibrate (i.e., check) the laser wavelength on a regular basis. Depending on the resolution required in a particular experiment, regular could mean daily, or even hourly. Such a wavelength calibration can be done in three ways, in general.

First, one can use an external wavemeter. Precision wavemeters are based on reference-stabilized single-mode HeNe lasers and thus exhibit measurement accuracies significantly better than the aforementioned gas laser sources. This is the most precise approach, but in most cases, this may be seen as overkill.

Second, the laser wavelength can be measured using a wavelength-calibrated spectrometer; in general, one uses, for convenience, the one with which the Raman spectra are measured later.

Third, the laser line can be referenced against a Raman standard. This latter method is frequently incorporated as a routine check and correction method in commercial Raman microscopes, utilizing a crystalline silicon sample that exhibits a narrow Raman peak whose shift is precisely known. In turn, with these linked data, the spectrometer software calibrates the Raman shift scale.

## 12.1.2 Calibrating the spectrometer wavelength and Raman shift scales

Raman spectra are in general displayed in two scale representations, either directly in measured wavelengths (or alternatively in absolute wavenumber scale) or in converted Raman shift scale, i.e., relative to the Raman laser line position. The latter provides a direct measure of the molecule's vibrational modes, which represent the desired result of any Raman measurement. The relation between the two scales is given by (here for the Stokes wavelength and Raman shift, $\lambda_S$ and $\tilde{v}_S$)

$$\lambda_S = \tilde{v}_S^{-1} = (\tilde{v}_L - \tilde{v}_{if})^{-1} \quad \text{(wavelength/wavenumber scale)} \qquad (12.2a)$$

$$\tilde{v}_{if} = \lambda_S^{-1} - \lambda_L^{-1} \quad \text{(Raman shift scale)} \qquad (12.2b)$$

It should be noted that these direct relations between measured wavelength and the quantum state information, noted usually in an energy scale, are only valid in a vacuum. Unfortunately, because experimental measurements are normally conducted in air, the light wave propagation in the medium of air is associated with a wavelength scaling, due to the refractive index of the medium

$$\lambda_{air} = \lambda_{vac}/n_\lambda \quad \text{(wavelength scaling in a medium)}, \qquad (12.2c)$$

where $n_\lambda$ is the wavelength-dependent refractive index of the medium (here air). Because of Equation 12.2c, the wavelength ↔ wavenumber conversion process is nonlinear; usually, one applies the so-called Edlén formula (see, e.g., Birch and Downs 1993) for this conversion process, which incorporates the refractive index in parameterized form. In commercial spectroscopy software, this conversion process is normally automated, without the need for user intervention.

The wavelength calibration of modern spectrometer systems, which, by and large, comprise a grating spectrometer and CCD array detector, is further complicated because the dispersed image on the linear pixel-distance detector is a function of the angular dispersion of the grating, i.e., a nonlinear transformation. Thus, one has to establish a scaling of recorded spectral lines at a particular detector pixel with the actual wavelength of the line; normally, this is done with the aid of atomic transition standards, whose wavelengths are well known, usually in the form of discharge lamps; the concept is summarized in Figure 12.2.

Based on the (nonlinear) pixel–wavelength correlation pairs for individual spectral lines, the appropriate wavelength scale can be generated by curve fitting of the data with the suitable sine function (based on grating dispersion equations), and with the aid of Equation 12.2, the desired conversion into wavenumber or Raman shift scales can be done.

It should be noted that spectrometers are normally calibrated once globally on installation, using the strong emission lines from a mercury lamp (which span the near-ultraviolet [UV] and visible spectral range) and grating dispersion formulas.

For low-resolution application (of the order $\Delta\tilde{v} \geq 1 \ cm^{-1}$), this is sufficient in most cases. However, if high-precision measurements are conducted, or closely spaced lines are to be resolved, such a global calibration is in general unreliable,

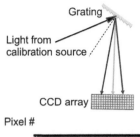

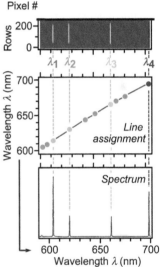

**Figure 12.2** Concept of wavelength calibration for a CCD array detector spectrometer.

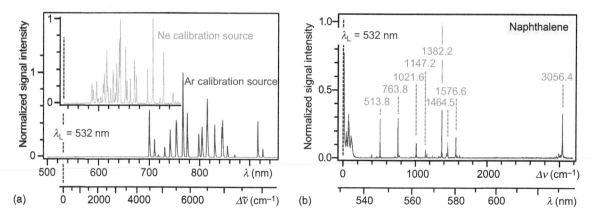

**Figure 12.3**  Emission spectra from spectral calibration sources. (a) Wavelength calibration line spectra of Ne and Ar discharge calibration devices. (b) Raman calibration spectrum (vibrational bands) of naphthalene, a common Raman spectroscopy calibration standard. (Data adapted from ASTM E1840. "Standard guide for Raman shift standards for spectrometer calibration." West Conshohocken (PA), USA: ASTM International. Active Standard ASTM E1840 - 96 (2014).)

because it does not take into account any potential localized deviations from the formula framework (such as mechanical or geometric nonlinearities). Then, a dedicated wavelength calibration within the spectral region of interest is required. Since the majority of commercial Raman systems utilize laser excitation sources in the visible and near-infrared, one can make use of rare-gas discharge calibration lamps (e.g., spectra, see Figure 12.3a).

Their many spectral lines are tabulated (see, e.g., National Institute of Standards and Technology [NIST] Atomic Transition Database 2016); note, however, that sometimes care has to be taken when trying to identify and associate closely spaced, weak lines since not all transitions between high-energy levels are necessarily tabulated. After recording the lamp emission spectrum in the range of interest, one fits the recorded lines with standard line shape functions to determine their exact line center, which will have most likely noninteger values associated with pixel numbers. The tabulated correlation pairs can then be used to generate a very precise wavelength scale. Note that, more often than not, the recorded line profiles are not necessarily replicated by standard line profiles, like, e.g., a Lorentz function, but are asymmetric because of imperfect imaging in the spectrometer–detector assembly. This deficiency can be overcome by utilizing line-fit procedures, which incorporate procedures to take into account spectrometer system imperfections (see, e.g., James et al. 2013b and SpecTools 2012).

It should be noted that for commercial Raman spectroscopy instruments whose spectral scale is software prepared with a Raman shift scale, molecular standards whose narrow vibrational band peaks are precisely known are utilized (see, e.g., ASTM E1840 2014); one example spectrum (for naphthalene) is shown in Figure 12.3b.

### 12.1.3 Intensity calibration for quantitative Raman spectra

Referring back to Equation 12.1, there are numerous factors to consider that all affect the absolute signal intensity $I_{sig}(\tilde{v}_s)$ or $I_{sig}(\lambda_s)$, depending on whether wavenumber or wavelength calibration of the spectrometer is utilized. This would not pose a real problem but a few of the terms in the equation depend on wavelength, and hence, the total change in intensity across a whole spectrum is in general a nonlinear function, linking the "fixed" excitation laser intensity to the wavelength-variable Raman spectral intensity. As a consequence, unless anything is ever changed in an experimental setup, which is rather unlikely, one requires intensity calibration of the whole system. In particular, accurate calibration is indispensable if users wish to compare spectra, which were recorded (1) with different spectral detection equipment or (2) using different Raman excitation lasers. An example to demonstrate the need for calibration is shown in Figure 12.4: although the same molecule is probed, the measured spectral response amplitudes of its various vibrations appear to be very different. And finally, the calibration has to be even rather precise (3) if absolute Raman response functions $\Phi$ need to be measured.

When inspecting all parameters in Equation 12.1, it is clear that, except the response and geometry term $k_{\tilde{\nu}}$ (or $k_\lambda$ in wavelength notation), all others are "easy" to handle since they exhibit simple functional dependences. The parametric calibration function $k_{\tilde{\nu}}$ ($k_\lambda$) is affected by a wide range of responses, including the transmission properties of optical components, filters, and the grating; the polarization configuration and dependence in the light path; and the response sensitivity of the (CCD array) detector. And all these can be affected differently if the geometrical configuration of the setup is altered. Thus, it is not surprising that the measurement of Raman spectra with absolute intensities is in general a daunting task, although on paper the procedure appears to be straightforward, even simple.

In order to generate said calibration functions $k_\lambda$, one normally proceeds as follows. First, one should configure the spectrometer system, including light delivery passes, as close as possible to the final setup of how the sample is probed. Then, instead of the sample (illuminated by the Raman excitation laser), one places a calibrated, standard light source at the location of the sample. Then, from the ratio between the recorded intensity spectrum and the nominal (certified) emission from the source, the response calibration curve can be constructed. The (conceptual) response calibration functions $k_\lambda$, necessary to arrive at the "true" Raman spectra, are also included in Figure 12.4.

Traditionally, such standards are based on blackbody radiation sources (often approximated by tungsten filament lamps), whose emission spectrum is well described by Planck's formula for the blackbody intensity distribution as a function of frequency or wavelength: $I_\lambda = (8\pi hc/\lambda^5)\cdot[\exp(hc/\lambda kT)-1]^{-1}$; a typical spectrum is shown in Figure 12.5a. The main problem with these sources is their critical temperature dependence; because of the exponential behavior, only a few degrees of deviation from a nominal value can distort the derived response function beyond acceptable tolerances. In general, the nominal temperature value is linked to an exactly known current through a calibrated load resistor, and guaranteeing this temperature during operation is far from easy.

Because of the aforementioned difficulties with blackbody radiation standards, when used in ordinary laboratory environments, dedicated, passive intensity calibration standards have been developed at NIST for the most common Raman laser excitation sources (Choquette et al. 2007). These standards consist of an optical glass element (in the shape of a thick slide) that emits a broadband luminescence spectrum when irradiated with light from the Raman excitation laser (e.g., luminescence curves related to common Raman excitation lasers are

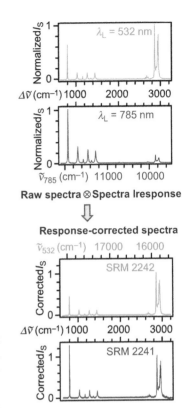

**Figure 12.4** Raman spectra of cyclohexane, measured excitation lasers at $\lambda_L = 532$ nm (top raw-data panel) and $\lambda_L = 785$ nm (bottom raw-data panel). Intensity calibration functions for the respective wavelength intervals were utilized to generate the "intensity-corrected" spectra in the lower panels; the remaining differences in the calibrated spectra are mainly due to the $\tilde{\nu}_L^3 \cdot \tilde{\nu}_S^3$.

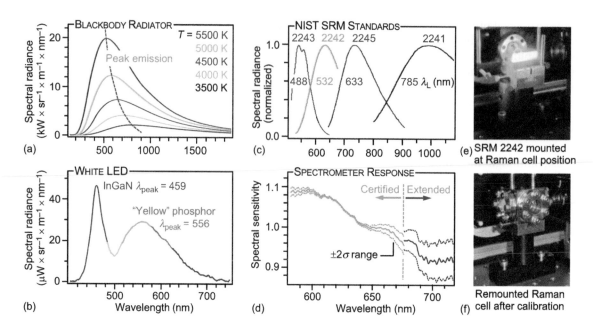

**Figure 12.5** Spectral distribution curves of active and passive light sources, used for intensity calibration of Raman spectrometer systems. (a) Emission spectra from a blackbody radiation source for different temperatures. (b) Emission spectrum from a broadband LED source. (Data adapted from Muray, A.J. et al. "A stabilized HBLED suitable as calibration standard." Presentation at CIE Conference '*Light and Lighting Conference with Special Emphasis on LEDs and Solid State Lighting*'. Budapest, Hungary (May 27–29, 2009).) (c) Spectra of NIST luminescence standards, excited by the appropriate Raman laser source; emission data adapted from NIST certification sheets, together with the spectral response calibration of a 90° Raman cell–spectrometer combination (d). (Data adapted from Schlösser, M. et al. *J. Mol. Struct.* 1044 (2013): 61–66.) Images of the mounting of the standard and the Raman cell are shown in (e) and (f), respectively.

shown in Figure 12.5c). The actual shape of the luminescence spectrum is approximated a polynomial expression, based on NIST calibration, which replicates the shape to better than ~2% over the complete spectral range for which it is certified. Currently, standards exist for the Raman laser wavelengths $\lambda_L = 488/514$, 532, 633, 785, and 1064 nm. Utilizing this mathematical description of the luminescence spectrum in conjunction with the measured luminescence spectrum of the standard, one can determine the correction function for the intensity response of the complete spectrometer system.

The great advantage of these passive luminescence intensity standards is that because of their small and convenient size, they can be used exactly in situ, more or less replacing the sample at its normal measurement position. In this way, any potential geometrical mismatch that might influence the spectral intensity distribution can be controlled rather well. It should be noted that, originally, these NIST standard sources were developed for Raman microscopy applications in which the Raman signal is collected in backward-scattering configuration. However, it was shown in a recent study that with carefully optimized mounting conditions, these standards can also be used in 90° Raman configurations, which are common in Raman spectroscopy of gases (see Schlösser et al. 2015). By replacing the gas cell in the said experiments with the appropriately oriented luminescence standard, the setup does not only address the problem of spectral response but also eliminates, to a large extent, any geometrical problems, replicating exactly the Raman interaction volume; this principle is demonstrated in Figure 14.5d.

It is noteworthy that another type of spectral intensity standard has been introduced toward the end of the 2000s (Muray et al. 2009) and has now matured into commercial products supplied by a number of manufacturers; these sources gain popularity because of their ease of use and high reliability, being based on high-brightness LEDs with traceable emission properties. They rely on light emission, like blackbody radiation sources, and span the complete visible part of the spectrum; they may be seen as complementary to blackbody emitters, providing much higher photon fluxes. On the other hand, their spectral distribution cannot be described with a simple formula (like for blackbody emitters) or well-defined polynomials (like for the NIST fluorescent glasses), but the certificate comprises tabulated data from which the emission spectrum can interpolated for each wavelength. A typical example spectrum is included in Figure 12.5b.

### 12.1.4 Quantification of molecular constituents in a sample

In any analytical measurement, the goal is to relate a measured signal directly to the number of particles contributing to the process, in particular in mixtures of species. This is no different for Raman spectroscopy; the key equation for this task is Equation 12.1, which links the observed Raman light intensity $I_{sig}(\tilde{v}_s)$, or $I_{sig}(\lambda_s)$, to the particle density $N_j$. Thus, in principle, the amount of probed molecules in the sample could be directly deduced from the observed signal amplitude, were the intensity calibration factors $k_{\tilde{v}}$ (or $k_\lambda$) and the transition moment functions $\Phi$ known. However, while it has become quite feasible to accurately calibrate the wavelength-dependent intensity response—as just discussed above—the transition moments for molecules are, by and large, theoretically determined quantities and have not always been measured experimentally; both theoretical and experimental values for $\Phi$ are in many cases not very accurate. As a consequence, the particle density $N_j$ of molecular species $j$, derived from the Raman signal, would carry a similar, often large uncertainty.

An additional complication arises from the fact that the total number of molecules is distributed over a range of vibrational and rotational states, described by the Boltzmann distribution, which, in turn, is dependent on temperature, $T$; thus, one needs to consider the summation of particles over all thermally populated vibrational/rotational quantum states, i.e.,

$$N_j = \sum_{v,J} N_{v,J}(T). \qquad (12.3)$$

Since the line positions of particular ro-vibrational bands may spread over a nonnegligible spectral range, for a quantitative answer on $N_j$, one would need to integrate the Raman signal intensity over the related wavelength range.

Because of all these complications in the direct, quantitative link between Raman peak intensity and the density of participating particles, one normally measures molecular particle densities relative to a Raman intensity versus known concentration calibration relation; an example for this is shown in Figure 12.6. The concentration of an unknown amount of a molecular species in a sample can then be read off the calibration graph, related to the measured Raman peak intensity.

With reference to Figure 12.1, in general, one will encounter two limitations in the quantification of molecular concentrations, extracted from a single Raman

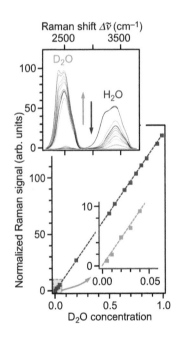

**Figure 12.6** Calibration of the concentrations of $D_2O$ in $H_2O$, derived from the intensity integrated over the respective vibrational Raman bands of $D_2O$ and $H_2O$ (inset panel at top); note that the green and red arrows indicate that the $D_2O$ signal increases synchronously with the decrease in the $H_2O$ signal. (Data from the author's laboratory.)

spectrum. These are the largest span of relative concentration values, which, by and large, is determined by the dynamic range of the photon detector system, and the limit of detection, which is associated with the noise level in the measurement (as pointed out repeatedly, normally, the $3\sigma$ noise value is used). Modern CCD array detectors are equipped with 16-bit analog-to-digital signal conversion, which corresponds to a range of ~$64 \times 10^3$ count levels. The dynamic measurement range is defined as the ratio between the largest and the smallest detectable signal. Since noise fluctuations are in general larger than a single count step, particularly in Raman measurement for which the shot noise of background fluorescence may significantly contribute to the noise fluctuations, rarely will one reach the upper theoretical limit of the $64 \times 10^3 / 1 = 64 \times 10^3$. Typically, dynamic ranges for single Raman spectrum measurements are in the range $1 \times 10^3$ to $5 \times 10^3$, rarely above. However, measurement ranges for concentrations may be extended beyond the dynamic range limit by "stitching" together concentration intervals that were recorded by suitable "signal attenuation or amplification," which can be achieved by light collection/averaging over shorter or longer time intervals. In the former case, the highest peak is lowered, thus allowing for higher concentrations to be probed; in the latter case, the noise level is lowered, thus potentially accessing lower concentration values.

## 12.2  MEASURING MOLECULAR PROPERTIES USING LINEAR RAMAN SPECTROSCOPY

Referring back to Chapter 11 and the previous Section 12.1, measured Raman spectra contain *intrinsic* and *extrinsic* parameter information (see Equation 12.1), specifically about molecular quantum structure and properties, and about number densities of the species under investigation. In this context, one encounters two general types of studies, namely,

- Those addressing qualitative or quantitative analysis

- Those that probe for intrinsic properties of molecules

In the former type of studies, the goals normally are threefold. First, one may wish to identify individual molecular species, even in complex mixtures. This is afforded by the Raman shift term $\tilde{v}_s = \tilde{v}_L \pm \tilde{v}_{if}$, in which $\tilde{v}_{if}$ is in general quite different for the vibrational modes of individual molecular species and thus allow one to identify it from the actual wavelength/wavenumber position in the recorded spectrum. Second, on a more fundamental level, one can deduce vibrational and rotational energy level structures (by and large from the Raman shift of the recorded lines and their substructure; see Chapter 11.2). Third, since $I_{sig}(\tilde{v}_s) \propto N_j$, one can derive actual number densities of a detected species in an unknown sample, comparing it to the Raman response intensity from a known, calibrated sample. And fourth, from the spectral shape of the Raman bands, one may be able to derive extrinsic entities, like sample temperature, from the relative line intensity distribution in a ro-vibrational band, which is associated with $N_j(v,J)$ via the Boltzmann distribution function. Examples for all these are discussed in Sections 12.3 through 12.5, individually for gaseous, liquid, and solid samples (since experimental setups and spectral behavior can be distinctly different).

In the latter type of studies, one exploits from Equation 12.1 that $I_{sig}(\tilde{v}_s) \propto \Phi(\varphi, \theta, p_i, p_s)$, meaning that the recorded Raman signal intensity directly reflects the magnitude of the transition moment matrix element, provided that all other parameters in the equation are known with precision. By and large, for this purpose, one draws on the fact that there are distinct relations between the polarized excitation laser and Raman light waves and the inherent and induced dipole moments of the molecule. In addition to this measurement of intrinsic quantum functions, it is also possible to determine stereo-structural information about large (organic) molecules, like, e.g., proteins, via Raman-related measurements based on circularly polarized light waves, a methodology known as *Raman optical activity* (ROA). The relation between polarized light waves, polarizabilities of molecules, and associated Raman signal strengths will be discussed in the remainder of this section, together with a few key examples.

## 12.2.1 Raman scattering of polarized light waves

The Raman scattered light is induced by the electric field of the incident, normally polarized laser light. Therefore, the direction of the electric field from the induced molecular dipole, giving rise to the Raman scattered light wave, is directly related to that of the incident light wave and provides a dominant polarization direction. In the framework of polarizability tensors $\overline{\overline{\alpha}}$, the Raman scattered light intensity $I_s$ is associated with the induced dipole polarization $p = \overline{\overline{\alpha}} \cdot E$, i.e., $I_s \propto (\overline{\overline{\alpha}} \cdot E)^2$. The tensor $\overline{\overline{\alpha}}$ contains all orientation aspects of a space-fixed molecule (in Cartesian coordinate system $x$, $y$, $z$—see Figure 11.4 in the previous chapter) and its interaction with the incident and scattered radiation fields. In component form, this tensor is normally written as

$$\overline{\overline{\alpha}} = \begin{pmatrix} \alpha_{xx} & \alpha_{xy} & \alpha_{xz} \\ \alpha_{yx} & \alpha_{yy} & \alpha_{yz} \\ \alpha_{zx} & \alpha_{zy} & \alpha_{zz} \end{pmatrix}. \tag{12.4}$$

For mathematical convenience, and in order to phenomenologically describe the induced polarization processes, commonly the polarizability is divided into isotropic and anisotropic parts, often with further subdivision of the anisotropy into matrix-symmetric and matrix-antisymmetric contributions, i.e.,

$$\overline{\overline{\alpha}} = \overline{\overline{\alpha}}_{\text{iso}} + \overline{\overline{\alpha}}_{\text{aniso}}^{\text{s}} + \overline{\overline{\alpha}}_{\text{aniso}}^{\text{as}}. \tag{12.5}$$

Note that the elements of the transition polarizability tensor relate to the molecular coordinate frame ($x$, $y$, $z$). However, in any actual Raman experiment, one measures the scattering in the laboratory coordinate frame ($X$, $Y$, $Z$). Thus, the Raman scattering signal is the average of all (random) molecular orientations in this laboratory frame. Traditionally, such orientation averaging has been carried out using direction cosines (with coordinate and angle conventions as shown in Figure 11.3). The averaging process leads to rotational invariants, i.e., entities that, for a Raman measurement, are independent of molecular orientation.

The very first derivation of the three rotational invariants relevant to Raman scattering was carried out by Placzek (1934); a concise summary can be found in Long (2002). The so-called Placzek-invariants $\mathcal{G}^{(0)}$, $\mathcal{G}^{(1)}$, and $\mathcal{G}^{(2)}$ are expressed in terms of squares of the components $\alpha_{\rho\sigma}$ of the polarizability tensor (associated with the Raman light intensity) and correspond to the isotropic, antisymmetric anisotropic, and symmetric anisotropic parts of the polarizability tensor, respectively; they are collated as Equations 12.6a through 12.6c in Table 12.1.

**Table 12.1**   Summary of Raman Scattering Light Properties, in Terms of the Elements of the Polarizability Tensor $\overline{\overline{\alpha}}$

| Entity | Equation | |
|---|---|---|
| ***Placzek polarizability invariants*** | | |
| Isotropic invariant | $\mathcal{G}^{(0)} = 1/3 \cdot \left[ \alpha_{xx} + \alpha_{yy} + \alpha_{zz} \right]^2$ | (12.6a) |
| Antisymmetric anisotropy invariant | $\mathcal{G}^{(1)} = 1/2 \cdot \left[ \left( \alpha_{xy} - \alpha_{yx} \right)^2 + \left( \alpha_{xz} - \alpha_{zx} \right)^2 + \left( \alpha_{zy} - \alpha_{yz} \right)^2 \right]$ | (12.6b) |
| Symmetric anisotropy invariant | $\mathcal{G}^{(2)} = 1/2 \cdot \left[ \left( \alpha_{xy} + \alpha_{yx} \right)^2 + \left( \alpha_{xz} + \alpha_{zx} \right)^2 + \left( \alpha_{zy} + \alpha_{yz} \right)^2 \right]$ | |
| | $+ 1/3 \cdot \left[ \left( \alpha_{xx} - \alpha_{yy} \right)^2 + \left( \alpha_{yy} - \alpha_{zz} \right)^2 + \left( \alpha_{zz} - \alpha_{xx} \right)^2 \right]$ | (12.6c) |
| ***Raman scattering light intensities*** | | |
| Intensity for $E_s$ and $E_i$ perpendicular to scattering plane | $I_{\perp^s, \perp^i}(\varphi, \theta) \propto \bar{\alpha}^2 \cos (\varphi)^2$ | |
| | $+ b^{(2)} \left( \gamma^2 / 45 \right) \left( 4 - \sin (\varphi)^2 \right)$ | (12.8a) |
| Intensity for $E_s$ perpendicular and $E_i$ parallel to scattering plane | $I_{\perp^s, \|^i}(\varphi, \theta) \propto \bar{\alpha}^2 \sin (\varphi)^2$ | |
| | $+ b^{(2)} \left( \gamma^2 / 45 \right) \left( 3 + \sin (\varphi)^2 \right)$ | (12.8b) |
| Intensity for $E_s$ and $E_i$ parallel to scattering plane | $I_{\|^s, \|^i}(\varphi, \theta) \propto \bar{\alpha}^2 \cos (\theta)^2 \cos (\varphi)^2$ | |
| | $+ b^{(2)} \left( \gamma^2 / 45 \right) \left( 3 + \cos \theta^2 \cos (\varphi)^2 \right)$ | (12.8c) |
| Intensity for $E_s$ parallel and $E_i$ perpendicular to scattering plane | $I_{\|^s, \perp^i}(\varphi, \theta) \propto \bar{\alpha}^2 \cos (\theta)^2 \sin (\varphi)^2$ | |
| | $+ b^{(2)} \left( \gamma^2 / 45 \right) \left( 3 + \cos \theta^2 \sin (\varphi)^2 \right)$ | (12.8d) |
| Placzek–Teller factor for rotational quantum number dependence | $b_{J,J}^{(2)} = J(J+1) \, / \, ((2J - 1)(2J + 3))$     example for $\Delta J = 0$[a] | (12.8e) |
| ***Measurement of polarizability invariants*** | | |
| Isotropic invariant | $\bar{\alpha}^2 = (1/90K) \cdot \left[ I_{\|^s, \|^i} (\theta = 90°) - (2/3) \cdot I_{R^s, R^i} (\theta = 180°) \right]$[b] | (12.9a) |
| Antisymmetric anisotropic invariant | $\delta^2 = (1/12K) \cdot I_{R^s, R^i} (\theta = 180°)$ | (12.9b) |
| Symmetric anisotropic invariant | $\gamma^2 = (1/20K) \cdot \left[ I_{R^s, R^i} (\theta = 180°) - 2 \cdot I_{\perp^s, \|^i} (\theta = 90°) \right]$ | (12.9c) |

*Note:*   Coordinate and angle convention as in Long 2002. Copyright Wiley-VCH Verlag GmbH & Co. KGaA. Reproduced with permission.

[a]   For other $b_{\Delta J}^{(2)}$-factors, see Long (2002).

[b]   $K$ is a proportionality factor common in the determination of $\bar{\alpha}^2$, $\delta^2$, and $\gamma^2$. The index R indicates circularly polarized light.

Note that, according to Equation 12.6a, the isotropic part $\mathcal{G}^{(0)}$ is the square of the sum of the diagonal tensor elements (or the trace). Therefore, the other two terms represent the deviation of the polarizability from spherical symmetry. Note also that, in nonresonant Raman scattering (i.e., the Raman excitation laser wavelength is far from any allowed transition), the polarizability tensor is symmetric, i.e., $\mathcal{G}^{(1)} = 0$. The symmetric anisotropic part $\mathcal{G}^{(2)} \neq 0$ requires that nonzero off-diagonal terms exist and/or that the diagonal tensor elements are different. This will be addressed in Section 12.2.2. Finally, in many texts and publications, the mean polarizability $\bar{\alpha}^2$ and the antisymmetric/symmetric polarizabilities $\delta/\gamma$ are used to describe the Raman signal intensities. Their relation to the Placzek rotational invariants is given by (see, e.g., Long 2002)

$$\bar{\alpha}^2 = 1/3 \cdot \mathcal{G}^{(0)}, \, \delta^2 = 3/2 \cdot \mathcal{G}^{(1)}, \, \text{ and } \gamma^2 = 3/2 \cdot \mathcal{G}^{(0)}. \tag{12.7}$$

While, in principle, it is possible to select any relative laser excitation–Raman light observation geometry (with the coordinate and angle convention of Figure 11.3), in common Raman scattering experiments, only three specific setup configurations are normally utilized, namely, those in which the observation of the Raman light is conducted (1) perpendicular (at 90°, i.e., in $X$-direction), (2) in forward direction (at 0°, i.e., in $Z$-direction), or (3) in backward direction (at 180°, i.e., in $-Z$-direction) with respect to the propagation direction of the laser beam ($Z$-direction), i.e., $\varphi = 0$ and, respectively, (1) $\theta = \pi/2$; (2) $\theta = 0$; or (3) $\theta = \pi$. The perpendicular 90° scenario is depicted schematically in the top part of Figure 12.7.

In addition to these scattering-configuration criteria, in general, one selects the laser polarization direction conveniently with respect to the experimental scattering plane (here, the $X$–$Z$ plane), normally parallel or perpendicular to this plane, i.e., in $X$- or $Y$-direction, respectively. Furthermore, for ease of signal intensity evaluation, the Raman scattered light is "filtered" (analyzed) using a polarizer oriented parallel or perpendicular to the laser polarization direction. The four related Raman light intensity expressions are summarized in Table 12.1 (Equations 12.8a through 12.8d).

Of course, instead of linear polarization configurations, one may opt for circular polarized laser excitation and/or Raman light observation. This type of experiment is much less common but is exploited in ROA measurements (see further below). A related schematic setup configuration is shown in the bottom part of Figure 12.7. The intensity expressions equivalent to Equation 12.8 re not listed here but can be found tabulated in Long (2002).

It is also worth noting that utilizing a combination of selected measurement configurations, one is able to isolate the aforementioned Raman tensor invariants—$\bar{\alpha}^2$, $\delta^2$, and $\gamma^2$—by combining the results from Raman intensity measurements and using at least two different scattering angle geometries (normally from right angle and back-scattering angle geometries). The relations of these invariants to the measured Raman scattering intensities are included in Table 12.1 as well (Equations 12.9a through 12.9c).

Finally, it has to be pointed out that the discussion thus far has centered on the aspect associated with molecular vibration and its relation with molecular symmetry. However, at least in the gas phase, molecules are free to rotate and hence it has to be clarified which role they may play in the determination of Raman scattering intensities (recall that rotation is clearly noticeable in most ro-vibrational Raman bands when recorded with sufficient spectral resolution). Indeed, since an increase in rotational energy results in a change in vibrational bond energies, associated with the effective molecular potentials, this will essentially affect the anisotropic contribution to the molecular polarizability and hence the intensity of the Raman scattered transition lines. This is taken into account by the so-called Placzek–Teller factors $b_{J,J}^{(2)}$, which were introduced by Placzek and Teller (1933) in their groundbreaking work on pure rotational and ro-vibrational Raman scattering. One simple example for the Q($J$)-branches, with quantum number $K = 0$, is given by Equation 12.8e in Table 12.1. A concise description and tabulations of Placzek–Teller expressions can be found in Long (2002).

## 12.2.2 Depolarization ratios

As stated already, the Raman scattered light is linked to the electric field of the stimulating, incident laser light. Intuitively, one might expect that the direction of the Raman light polarization direction might be the same as that of the incident

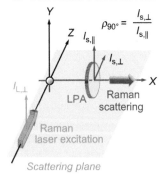

90° Scattering Configuration

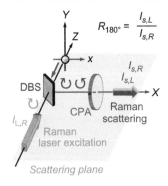

180° Scattering Configuration

**Figure 12.7** Common scenarios for depolarization measurements; top panel: 90°-scattering configuration, based on linear polarized laser radiation; bottom panel: 180°-backscattering configuration, based on circular polarized laser radiation. DBS, dichroic beam splitter. Note that the capital letters refer to the laboratory-frame coordinate system.

light (i.e., the polarization vectors are parallel to each other). However, in reality, some fraction of the Raman scattered light has a polarization component differing from that of the incident light. It is then convenient to split the total polarization vector into a parallel and a perpendicular component; the associated Raman light intensities are usually addressed as the polarized and depolarized Raman intensities, $I_{pol}$ and $I_{depol}$, respectively. The ratio between these two components is known as the *depolarization ratio* $\rho$ and is given by

$$\rho = I_{s\perp}/I_{s\parallel} \equiv I_{depol}/I_{pol} \qquad (12.10)$$

where $I_{s\perp}$ and $I_{s\parallel}$ stand for the (measured) Raman light intensities with the excitation and detection polarization directions perpendicular or parallel to each other (utilizing, e.g., a pair of cross- or parallel-oriented polarizers in the light paths). Note that $I_{s\perp}$ and $I_{s\parallel}$ are associated with the Raman light intensity expression given by Equation 12.8 in Table 12.1; e.g., $\rho$ = (Equation 12.8b)/(Equation 12.8a) = (Equation 12.8d)/(Equation 12.8a) for the case that one changes the polarization direction of the incoming laser beam, keeping the analyzer orientation for the recording of the Raman scattered light equal for both (first equality), or maintaining the polarization direction of the incoming laser beam, and changing the analyzer orientation for the recording of the Raman scattered light (second equality).

When measuring the polarized and depolarized components of the Raman scattered light, one basically probes the "projection" of the polarization vector $\boldsymbol{p}$ in the laboratory frame (which of course incorporates the polarizability tensor $\overline{\overline{\alpha}}$) on the polarization axes, also described in the laboratory frame. Conceptually, it is straightforward to calculate these components, and thus the expected depolarization values. For the experimental scheme depicted in Figure 12.7 and using Equations 12.6 in the transformation from the $(x, y, z)$ to the $(X, Y, Z)$ reference frames, one finds for the polarized and depolarized components (see Placzek 1934 or Long 2002)

$$I_{s\parallel} \propto \left|p_{Y,Y}\right|^2 = (1/3) \cdot \mathcal{G}^{(0)} + (2/15) \cdot \mathcal{G}^{(2)} \qquad (12.11a)$$

and

$$I_{s\perp} \propto \left|p_{Z,Y}\right|^2 = (1/6) \cdot \mathcal{G}^{(1)} + (1/10) \cdot \mathcal{G}^{(2)} \qquad (12.11b)$$

Hence, the expression for the depolarization ratio from Equation 12.10 becomes

$$\rho = (5 \cdot \mathcal{G}^{(1)} + 3 \cdot \mathcal{G}^{(2)}/(10 \cdot \mathcal{G}^{(0)} + 4 \cdot \mathcal{G}^{(2)}) \qquad (12.11c)$$

The measurement of depolarization ratios has become a quite common approach to identify the nature of specific molecular vibrational modes. This is because the polarizability tensor reflects in its components the symmetry of the molecule and its vibrational modes. Hence, utilizing Equations 12.6 and 12.11, one can often associate distinct depolarization ratios with specific vibrational modes. In particular, one can easily distinguish between *totally symmetric* and *non-totally symmetric* vibrational modes.

### Totally symmetric vibrational modes

The polarizability tensor for these modes preserves symmetry and has only diagonal tensor components $\alpha_{xx}$, $\alpha_{yy}$, and $\alpha_{zz}$; all other components $\alpha_{\rho\sigma}$ with

**Table 12.2** Depolarization Ratios $\rho$ for Selected Molecular Vibrational Modes, Calculated According to the Polarizability Invariants $\mathcal{G}^{(k)}$, Using Equation 12.11c

| Molecular Mode Symmetry | $\mathcal{G}^{(0)}$ | $\mathcal{G}^{(1)}$ | $\mathcal{G}^{(2)}$ | $\rho$ |
|---|---|---|---|---|
| ***Totally symmetric modes*** | | | | |
| Spherically symmetric molecules, with $\bar{\alpha}^2 = (\alpha_{xx} + \alpha_{yy} + \alpha_{zz})^2 \equiv 9\alpha^2$ | $3 \cdot \alpha^2$ | $0$ | $0$ | $0$ |
| Symmetric-top molecules with $\bar{\alpha}^2 = (\alpha_{xx} + \alpha_{yy})^2 \equiv 4\alpha^2$ and $\alpha_{zz} = 0$ | $(4/3) \cdot \alpha^2$ | $0$ | $(2/3) \cdot \alpha^2$ | $1/8$ |
| Asymmetric-top molecules with $\bar{\alpha}^2 = (\alpha_{xx})^2 \equiv \alpha^2$ and $\alpha_{yy} = \alpha_{zz} = 0$ | $(1/3) \cdot \alpha^2$ | $0$ | $(2/3) \cdot \alpha^2$ | $1/3$ |
| ***Non-totally symmetric modes*** | | | | |
| Molecules, or mode coupling with $\bar{\alpha}^2 = (\alpha_{xx} + \alpha_{yy} + \alpha_{zz})^2 = 0$ | $0$ | $0$ | $2 \cdot [(\alpha_{xy})^2 + (\alpha_{xz})^2 + (\alpha_{zy})^2]$ | $3/4$ |

*Note:* See text for remarks on the particular values for the diagonal polarizability tensor elements.

$\rho \neq \sigma$ are zero. One can distinguish three particular cases of totally symmetric modes for

- Spherically symmetry molecules, with $\alpha_{xx} = \alpha_{yy} = \alpha_{zz}$ (e.g., $CH_4$ or $SF_6$)

- Symmetric-top molecules, with $\alpha_{xx} = \alpha_{yy} \neq \alpha_{zz}$ (often with $\alpha_{zz} = 0$, e.g., in planar molecules like $C_6H_6$)

- Asymmetric-top molecules with $\alpha_{xx} \neq \alpha_{yy} \neq \alpha_{zz}$ (e.g., $NH_3$)

Depolarization ratios for some selected combinations of tensor-element values, which were selected for the ease of calculation (assuming specifically that the nonzero terms have equal values), are collated in Table 12.2.

From the tabulated values for the depolarization ratio, one can see (1) that for non-totally symmetric (or asymmetric) Raman modes, one encounters a constant value $\rho \equiv \rho_s$ irrespective of their exact molecular or vibrational symmetry, and (2) that for totally-symmetric Raman modes, $\rho \equiv \rho_{as}$ can vary over a distinct range:

$$0 \leq \rho_{as} = 1/3 \quad \text{for symmetric Raman modes;} \quad (12.12a)$$
$$\rho_{as} = 3/4 \quad \text{for asymmetric Raman modes} \quad (12.12b)$$

### Non-totally symmetric vibrational modes

For these modes, $\text{Tr}(\bar{\bar{\alpha}})$ vanishes and only off-diagonal elements from the polarizability tensor contributes to the Raman scattering cross section. Because of the vanishing $\text{Tr}(\bar{\bar{\alpha}})$, one also has that $\mathcal{G}^{(0)} = 0$. Note that for nonresonant Raman scattering, one normally also has $\alpha_{\sigma\rho} = \alpha_{\rho\sigma}$.

Note that the values for $\rho_{as}$ are not restricted to the distinct values included in Table 12.2; depending on the actual values for the diagonal elements $\alpha_{xx}$, $\alpha_{yy}$ and $\alpha_{zz}$ of the polarizability tensor, any value within the range given in Equation 12.12a may be encountered. Note also that the same depolarization ratios can be derived utilizing the representation of the rotational polarization invariants in terms of $\bar{\alpha}^2$, $\delta^2$, and $\gamma^2$ instead of $\mathcal{G}^{(0)}$, $\mathcal{G}^{(1)}$, and $\mathcal{G}^{(2)}$.

### 12.2.3 Measuring depolarization ratio

Conceptually, in the theoretical framework for the depolarization ratio, one usually utilizes a simplified geometrical configuration, in which it is assumed that (1) the scattered light only originates from a single point and that (2) scattering is observed for vanishing solid angle, i.e., in the limit of a single ray line. This also has been the case in the discussion in this section so far. It means that the depolarization ratio defined in this way is not influenced by the measurement apparatus and therefore should appropriately to be unambiguous. For example, it may be indexed by "SP0SA" (standing for "single point, zero solid angle"), as suggested in James et al. (2013a). From an experimental point of view, this situation is hardly ever realistic. First, the scattered light originates along the extended volume in which the laser radiation interacts with the probed molecules, and second, usually the Raman scattered light is collected over fairly large solid angles so that nonzero solid angles need to be taken into account. Both have significant effects on the measured depolarization ratio and thus have to be included in the theoretical description by integrating over the scattering geometry.

As a consequence of actual experimental conditions, any measurement will normally result in an *effective depolarization ratio*, based on the measured intensity ratios according to Equation 12.10, whose value can be significantly different from that predicted by theory, according to Equation 11.12. Methods of how to take account of finite collection angles in the measurement of depolarization ratios have been described, e.g., by Teboul et al. (1992). However, while taking into account the solid-angle problem, the methodology was only accurate for point-like Raman emission and for circular apertures. Furthermore, because of computational limitations at the time, the calculation was based on a fourth-order series expansion, which becomes inaccurate for larger (realistic) solid angles. A more inclusive treatment in which polarization-affecting properties of optical components were also taken into account has been described by James et al. (2013a).

A representative example of a Raman depolarization study, in which corrections to the experimentally measured values for $\rho$ were incorporated, and which also reveal the influence of the Placzek–Teller factors for rotational transition lines, is shown in Figures 12.8 and 12.9 for the simplest and lightest of all molecules, namely, the isotopologues of hydrogen (see James et al. 2013a).

In the said experiments, mixtures of hydrogen isotopologues were probed, and polarized and depolarized ro-vibrational Raman lines were measured, to confirm, in addition, that for lines with a polarized character, indeed the theoretical value of $\rho = 0.75$ is obtained. The experimental setup was for 90° observation, as shown schematically in Figure 12.7 further above. Rather than measuring the Raman scattered light only for the two positions required by the definition of $\rho$, i.e., $I_{s\perp}$ and $I_{s\parallel}$, the laser polarization direction was altered continuously over a full rotation of $2\pi$, in small incremental steps (a segment of the laser polarization scan for two $Q_1(J)$-branch lines is shown in Figure 12.8). While such a scan is not necessary, in general, it nevertheless helps in identifying the influences of experimental "aberrations," i.e., the aforementioned solid-angle and Raman scattering volume dependence (reflected in values deviating from the theoretical $\rho_{SPOSA}$), and the imperfection of the optical components' respective polarization behavior (caused by dielectric coatings and curvature of optical surfaces, reflected in small angular shifts of the polarization curves).

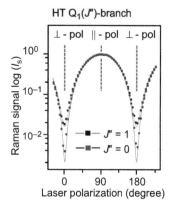

HT $Q_1(J'')$-branch

$\perp$ - pol    $\parallel$ - pol    $\perp$ - pol

Raman signal log $(I_s)$

$J'' = 1$
$J'' = 0$

Laser polarization (degree)

Depolarization Values

| | $Q_1(0)$ | $Q_1(1)$ |
|---|---|---|
| $\rho_{obs}$ | 0.0060(1) | 0.0233(1) |
| $\rho_{SPOSA}$ | 0.0000(6) | 0.0173(6) |
| $\rho_{theory}$ | 0 | 0.0179 |

**Figure 12.8** Depolarization measurement as a function of (linear) laser polarization angle, with respect to the orientation of the linear polarization analyzer (orientation according to Figure 12.7), exemplified for the two lowest-$J$ line in the $Q_1(J)$ branch of HT. The corrected experimental values are compared to theoretical values. (Data adapted from James, T.M. et al. *J. Raman Spectrosc.* 44, no. 6 (2013): 857–865.)

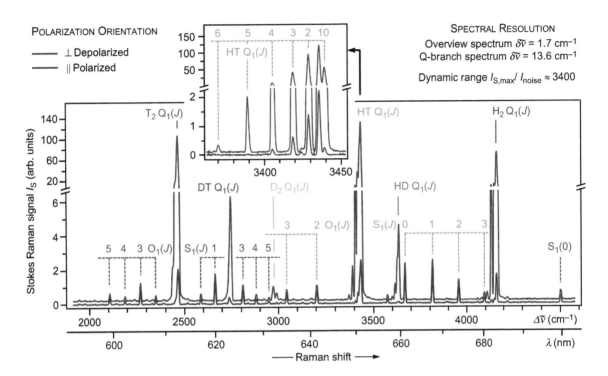

**Figure 12.9**  Depolarization spectrum of $H_2$ isotopologues, revealing the polarized and depolarized character of the $S_1(J)$, $O_1(J)$, and $Q_1(J)$, respectively. (Data adapted from James, T.M. et al. *J. Raman Spectrosc.* 44, no. 6 (2013): 857–865.)

According to the Placzek–Teller factors, one should find $\rho(Q_1(0)) = 0$, while for all other rotational $J$-transitions, the depolarization ratio should be small but finite. However, a cos-square fit—according to the angular dependence in Equation 12.8—to the data reveals a deviation from the zero value outside the measurement accuracy, $\rho_{exp}(Q_1(0)) = 0.0060(1)$. When correcting for the identified aberrations, the authors obtained $\rho_{SPOSA\text{-}corr}(Q_1(0)) = 0.0000(6)$, which now reflects the expected $\rho_{theory} = 0$ value.

In Figure 12.9, the complete $I_{s\perp}$ and $I_{s\parallel}$ spectra for all six hydrogen isotopologues are shown, from which, in principle, polarization ratios for all rotational branch lines $Q_1(J)$ and $O_1(J)$, $S_1(J)$ can be extracted. Of course, for best accuracy, the intensities of the relevant specific isotopologue under investigation should exhibit large amplitudes, which isn't the case for all in the example shown here. One would need to alter the gas mixture appropriately. Nevertheless, what is quite evident from the annotated spectra is that all $Q_1$-branches are depolarized in nature while the $O_1$- and $S_1$-branches are polarized, exhibiting the expected $\rho = 3/4$ value within the error bars. The experimental values were found to be $\rho_{exp}(S_1(J)) = 0.753(9)$ and $\rho_{exp}(O_1(J)) = 0.759(11)$, which yield the corrected value $\rho_{SPOSA\text{-}corr}(S_1(J);O_1(J)) = 0.7503(8)$. Note that the numbers in brackets denote the uncertainties in the least-significant digit of the measurement value.

The example for the (diatomic) hydrogen molecule was discussed here to demonstrate various fine details, which need to be considered when measuring depolarization ratios (such as instrumental configuration and imperfection issues, as well as the dependence on rotational quantum numbers). The actual

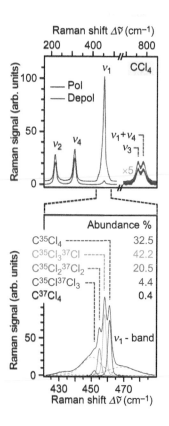

**Figure 12.10** Depolarization spectra of CCl$_4$, revealing the polarized and depolarized character of the vibrational bands $v_2/v_3/v_4/v_1 + v_4$, and $v_1$, respectively. The expanded-scale display in the lower panel reveals the isotopologue shifts in the $v_1$ band, associated with $^{35}$Cl ↔ $^{37}$Cl isotope substitution. (Data adapted from Chakraborty, T. and S.N. Rai. *Spectrochim. Acta A* 62, no. 1-3 (2005): 438–445.)

strength and general usage of depolarization measurements is the characterization of polyatomic molecules and how their structural point symmetry influence and vibrational mode character can be deduced from Raman spectra measured under suitable polarization conditions. An example of this is shown in Figure 12.10 in which $I_{s\perp}$ and $I_{s\parallel}$ spectra for (liquid) CCl$_4$ are shown. In their experiments, Chakraborty and Rai (2005) demonstrated that polarized and depolarized Raman features could easily be distinguished and be associated to the appropriate vibrational modes. The carbon tetrachloride molecule belongs to the class of tetrahedral XY$_4$ molecules of the cubic point group $T_d$, exhibiting the fundamental vibrations $v_1$ (symmetric stretch, nondegenerate), $v_2$ (symmetric deformation, twofold degenerate), $v_3$ (asymmetric stretch, threefold degenerate), and $v_4$ (asymmetric deformation, threefold degenerate); for further details, see, e.g., Herzberg (1945, 1990). These modes are clearly visible in the experimental spectra, and in addition, the $v_1 + v_4$ combination band forms a Fermi resonance dyad together with $v_3$-vibration. Except for the $v_1$-vibration—which is spherically (totally) symmetric—all other vibrations in the observed spectral range are non-totally symmetric; according to Table 12.2, one thus expects the associated Raman bands to exhibit $\rho \approx 0$ in the former case and $\rho = 0.75$ in the latter case. Indeed, the authors find $\rho(v_1) = 0.0039(4)$, $\rho(v_2) = \rho(v_4) = 0.72$, and $\rho(v_3) = \rho(v_1 + v_4) = 0.69$, which are close to exactly as expected. The small discrepancies are most likely due to the values not being corrected for SPOSA-representation.

In addition, the expanded part of the spectrum in the lower panel of Figure 12.10 shows the "splitting" of the $v_1$ Raman band into isotopologue components, as a consequence of isotopic substitution $^{35}$Cl ↔ $^{37}$Cl; the observed intensities closely mimic the natural abundance of the individual isotopologues. The isotope substitution is minutely breaking the complete spherical symmetry and thus should be reflected in the depolarization ratio. Unfortunately, the partial overlap of the bands and the relatively low signal-to-noise ratio did not allow the authors to precisely determine the depolarization ratios fully independently for each individual isotopologue, in order to provide an unambiguous answer to the theoretical controversy whether isotopic substitution alters the depolarization ratio of the CCl$_4$ isotopologues, or not. However, the overall trend of the wavelength-dependent depolarization spectrum suggests some sort of change, although making a unique association to individual isotopologues is not possible because of the spectral overlap.

### 12.2.4 Raman optical activity

A powerful measurement technique, which exploits the polarization dependence of Raman excitation and scattered light observation, is *Raman optical activity* (ROA). In ROA, one measures the (normally) small difference in the intensity of Raman scattered light from chiral molecules when examining them with right and left circularly polarized laser light or, equivalently, the intensity of a small, circularly polarized contribution in the Raman scattered light when probing the molecule with linearly polarized laser light. The two related measurement strategies are known as *incident circular polarization* (ICP) and *scattered circular polarization* (SCP) ROA, respectively. Since ROA is sensitive to chirality, it adds an extra dimension of sensitivity to Raman spectroscopy, namely, to measure details of the three-dimensional structure.

Recall that chirality is a geometric property of (some) molecules, meaning that such a molecule is nonsuperposable on its mirror image. For example, the presence of an asymmetric carbon center is one of several structural features that are responsible for chirality in molecules. Note that left- and right-hand configurations of chiral molecules are often dubbed *enantiomers*. The left- and right-handed variants carry the labels S and R, respectively; they indicate the absolute configuration of the molecule; i.e., they refer to the actual orientation in space of the substituents around the stereo-center. This nomenclature allows one to describe the steric configuration of molecules, without the need for a 3D picture representation.

ROA was pioneered in the early 1970 by Barron and colleagues; the first experimental observation was made for the enantiomers of neat 1-phenylethanol and 1-phenylethylamine (see Barron et al. 1973). Further key milestones in the development and application of ROA were the extension of the methodology to biomolecules in the late 1980s and early 1990s, with the earliest experiments of vibrational ROA of peptides and proteins (see, e.g., Barron et al. 1990), and then the introduction of dedicated, commercial ROA spectroscopy instrumentation in the mid-2000s. Since then, ROA has evolved into a powerful chiroptical spectroscopy tool for the study of a multitude of biomolecules in their natural environment, i.e., in "aqueous solution" (see, e.g., the dedicated reviews by Barron et al. 2000 and Barron 2015). Among other things, ROA provides information about

- Motif and fold of proteins, as well as their secondary structure

- Solution structure of carbohydrates

- Polypeptide and carbohydrate structure of intact glycoproteins

- Structural elements present in unfolded protein sequences

- Protein and nucleic acid structure of intact viruses

Quantum-chemical (first-principle) calculations and simulations of observed ROA spectra now can provide the complete three-dimensional structure of selected biomolecules, together with information about conformational dynamics (see, e.g., Kessler et al. 2015).

The basic principle of ROA is the interplay of light waves scattered in association with the polarizability tensor and optical activity tensors of a chiral molecule. This leads to a difference between the intensities of the right- and left-handed circularly polarized Raman scattered light. Practically, ROA is measured using the conceptual experimental setup depicted in the bottom part of Figure 12.7. The measured entities of importance for ROA are the difference and the sum of the two circularly polarized light components in the Raman scattering process, in either ICP or SCP configuration, i.e., $I_R - I_L$ and $I_R + I_L$; the former constitutes the ROA signal while the latter is simply the full Raman signal. In general, ROA is treated in the dimensionless entity of the so-called circular intensity difference (CID)

$$\Delta = (I_R - I_L)/(I_R + I_L) \tag{12.13}$$

In a certain way, the concept may be seen as akin to the depolarization ratio measured for linear polarized light.

Theoretically, this expression can be described in terms of the electric dipole–electric dipole molecular polarizability tensor $(\overline{\overline{\alpha}})_{\rho\sigma}$ and the electric dipole–magnetic dipole and electric dipole–electric quadrupole optical activity tensors $(\overline{\overline{G'}})_{\rho\sigma}$ and $(\overline{\overline{A}})_{\rho\sigma\tau}$, respectively (note that $(\overline{\overline{A}})_{\rho\sigma\tau}$ is a third-rank tensor). For example, for ICP-ROA measurements in backward-scattering configuration, for

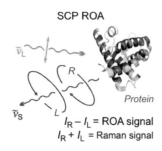

SCP ROA

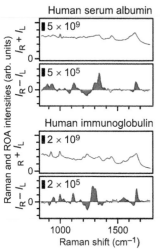

$I_R - I_L$ = ROA signal
$I_R + I_L$ = Raman signal

Human serum albumin

Human immunoglobulin

Raman and ROA intensities (arb. units)

Raman shift (cm$^{-1}$)

**Figure 12.11** ROA spectroscopy of biomolecules, measured in backward-scattering SCP configuration (the schematic principle is shown at the top). The example spectra are for the proteins human serum *albumin* and human *immunoglobulin*. For comparison, both Raman spectra (measurement of $I_R + I_L$) and ROA spectra (measurement of $I_R - I_L$) are shown. (Data adapted from Zhu, F. et al. *Structure* 13, no. 10 (2005): 1409–1419.)

an isotropic ensemble of chiral molecules with dimensions much smaller than the wavelength of the incident laser light, the associated CID is given by (see, e.g., Barron 2015)

$$\Delta(180°) = \frac{24 \cdot \left[\beta(G')^2 + 1/3 \cdot \beta(A)^2\right]}{c \cdot \left[45 \cdot \alpha^2 + 7 \cdot \beta(\alpha)^2\right]}, \tag{12.14}$$

where $\alpha$ and G′ are the isotropic rotational tensor invariants, and $\beta(G')^2$, $\beta(A)^2$, and $\beta(\alpha)^2$ are the anisotropic rotational tensor invariants. Note here that the notation is slightly different to that used in the (linear) depolarization treatment above, but also common, namely, that the averaged anisotropic component is expressed as $\beta(x)^2$—with $x = \alpha$, G′, A—rather than $\gamma^2$. For example, the total Raman signal in the denominator of Equation 12.14 is exactly the same as that found by adding Equations 12.8a and 12.8b; only the anisotropy parameter has a different notation. Note that the tensor components are normally referenced to molecule-fixed axes, but that all invariants are independent of the choice of origin, and thus each of them is accessible to experimental measurements.

An example of a backward-scattering SCP-ROA measurement is shown in Figure 12.11. The data shown here are for two proteins, human serum *albumin* and human *immunoglobulin* (see Zhu et al. 2005). In the figure, one can glean only scarcely noticeable differences in the Raman spectra (measurement of $I_R + I_L$); on the other hand, the ROA spectra (measurement of $I_R - I_L$) exhibit distinct differences, which allow one to easily differentiate between the two; this is not shown here. Further analysis of the spectra allows one to deduce information on the structure of the said proteins, e.g., about secondary $\alpha$-helical or $\beta$-sheet structures. For example, without going into detail here, one can confirm from the inspection and quantification of certain peaks the all-$\alpha$ and all-$\beta$ character of human serum *albumin* and human *immunoglobulin*, respectively.

## 12.3 RAMAN SPECTROSCOPY OF GASEOUS SAMPLES

Raman spectroscopy of gaseous samples always has been a difficult task because of the normally quite low number density $N_j$ in a molecular quantum state $j$ in the probed volume, since according to Equation 12.1 above, this is directly proportional to the Raman signal amplitude. In particular, high-resolution spectroscopy to resolve the rotational structure of in vibrational bands has been tedious.

It has to be said that high-resolution Raman spectroscopy of gases had its heyday in the 1950s and 1960s and did not utilize lasers at all, but rather strong emission lines of high-power mercury lamps (specifically, the UV-line at $\lambda_{Hg} = 253.6517$ nm). Also, at the time, no array light detectors were available yet, which today are conveniently used; instead, long-time exposure of photographic plates was the order of the day.

In this context, the most pioneering measurements with long-lasting impact were those of Boris Stoicheff at Toronto, Canada. During the 1950s, he and his colleagues studied a wide range of linear and symmetric-top molecules, providing an extensive rotational transition database that, e.g., served to supplement bond-length determinations from other methods. The high resolution was afforded by spectrometers of very long focal length, in particular a high-dispersion spectrograph with focal length $f = 6.4$ m exhibiting spectral resolution matching that of the Hg excitation source, $\delta\tilde{v} \cong \delta\tilde{v}_{Hg,243nm} = 0.25$ cm$^{-1}$. However, this came at the expense of very long spectral accumulation times, often requiring plate exposure of up to 48 h. Probably one of his major successes at the time was the recording of the first high-resolution Raman spectrum of the threefold degenerate stretching mode $v_3$ of methane (a tetrahedral XY$_4$ molecule); see Stoicheff et al. (1952). In those experiments, the researchers found evidence for 14 of the 15 theoretically predicted rotational subbranches. They revisited the molecule about a decade later, determining spectroscopic constants of both the upper and the lower vibrational states, including—for the first time in a tetrahedral molecule—the fine-structure splitting of the levels of the ground state (Herranz and Stoicheff 1963).

It is interesting to note that with the advent of lasers and their narrow linewidth in many spectroscopic techniques, resolution could be driven to very extreme levels. In a certain way, the opposite happened to Raman spectroscopy. Because of the rather low sensitivity and thus the long acquisition times required, Raman spectroscopy ceded its role in high-resolution spectroscopy to absorption and laser-induced fluorescence spectroscopy. Instead, the drive was toward small-size instrumentation and speed of spectral recording. In this regard, it gained recognition and acceptance as an indispensable analytical technique, mainly because of the aforementioned reasons, and because of its universality of being able to record spectra for any Raman-active molecule, requiring only a single, fixed-wavelength laser excitation source. Only in the mid-1980s did high-resolution Raman spectroscopy of molecular gases reappear in the form of Fourier transform (FT) Raman spectroscopy. For example, ro-vibrational band structures of molecules like acetylene were recorded with a resolution of $\delta\tilde{v} = 0.17$ cm$^{-1}$ (see Jennings et al. 1986). However, FT Raman spectroscopy never gained the scientific importance as a high-resolution spectroscopic technique, in comparison to the prelaser Raman era. The development of high-resolution Raman spectroscopy of gases, spanning the period from the pioneering studies of Stoicheff until the end of the 1990s, is reviewed in Jones (2000).

Below, a few selected examples are presented, to highlight the role of Raman spectroscopy of gases in spectroscopic and analytical measurements.

## 12.3.1 Spectroscopy of rotational and vibrational features

As an example for the spectroscopic versatility of Raman spectroscopy of gases, even at medium spectral resolution, part of the vibrational pentad system of CH$_4$ ($v_1$, $v_3$, $2v_2$, $v_2 + v_4$, and $2v_4$) is shown in Figure 12.12; this particular spectrum may also serve as a link to the prelaser measurements of the same molecule and spectral range. The important spectral aspects are briefly discussed, and conclusions are drawn about molecular symmetries from the polarization-dependent recordings of the spectra.

The data shown in Figure 12.12 have been recorded for a 90° setup configuration; i.e., the Raman scattered light is collected at right angles to the laser excitation beam, with a small gas cell at about atmospheric pressure; for excitation radiation at $\lambda_L = 532$ nm, a continuous-wave frequency-doubled Nd:YVO$_4$ laser was used. The top (black) spectral trace has been recorded under "polarized" conditions; i.e., the polarization of the laser excitation light was parallel to the direction of the analyzing polarizer in the Raman collection light pass. The clearly

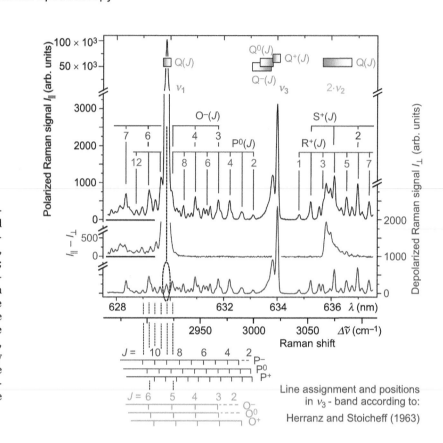

**Figure 12.12** Depolarization spectra of methane ($CH_4$) gas, recorded at atmospheric pressure, comprising the vibrational bands $v_1$, $v_3$, and $2v_2$ from the pentad-system; black trace: polarized Raman signal $I_\parallel$; red trace: depolarized Raman signal $I_\perp$; blue trace: difference signal $I_\parallel - (4/3) \cdot I_\perp$, revealing the partially masked $2v_2$ band. The rotational structure of the O-, P-, R-, and S-branches is not fully resolved; its triply-degenerate nature is indicated for selected O- and P-lines. (Spectral data from the author's laboratory.)

dominant feature is the $Q_1$-branch ($\Delta J = 0$) of the symmetric-stretch vibrational band $v_1$, which is—as mentioned before—one member of the $CH_4$ pentad system. Accompanying this major peak, which incidentally is normally used in analytical Raman spectroscopy, is a multitude of weaker features, smaller in amplitude by a factor of around 100. These will comprise a mixture of the other $\Delta J$-branches of the $v_1$-band and contributions from the other two vibrational bands indicated in the figure.

In order to unravel at least some of this multitude of intermingled Raman transition lines, one can utilize Raman depolarization measurements, as described in Section 12.2. A depolarized spectrum (the laser polarization was turned by 90° for this, with the Raman analyzer remaining unchanged) is included in the lower (red) trace of the Figure 12.12 (note that the spectra are offset with respect to each other for clarity).

Recall that $CH_4$ belongs to the point group $T_d$, and the symmetries of the fundamental vibrational modes $v_1$ (symmetric stretch) and $v_3$ (asymmetric stretch) are therefore $A_1$ and $T_2$, respectively. For linearly polarized laser excitation, their theoretical values of the Raman depolarization ratio are 0 and 3/4, respectively. For the individual rotational lines within a Raman band, the depolarization ratio is expected to be dependent on the rotational transition and quantum number (see, e.g., Long 2002): for transitions with $\Delta J = -2, -1, 1,$ and 2 (i.e., the O-, P-, R- and S-branches), the depolarization ratio is $\rho = 3/4$, while for transition $\Delta J = 0$ (i.e., the Q-branch), the value can take values in the range $0 \le \rho \le 3/4$, depending on the symmetry of vibrational mode.

When inspecting the two spectra (polarized and depolarized), one sees immediately that the features associated with the $Q_1$-branches of the $v_1$- and overtone $2v_2$-bands have vanished; this is consistent with the theoretical expectations. Note that from the data presented in the figure, one finds $\rho(v_1) \leq 0.002$, in agreement with other experimental results (see, e.g., Yu et al. 2007).

Equally, the $\rho = 3/4$ expectation for all branches of the $v_3$-band can easily be confirmed when generating the difference spectrum with $I_{\parallel}-(4/3)\cdot I_{\perp}$; this is shown in the middle (blue) trace of Figure 12.12. Indeed, within the measurement noise, in this spectrum, the $v_3$-band is completely eliminated, confirming the theoretical expectation of its $\rho = 3/4$ value. This also means that the depolarized spectrum, by and large, represents the $v_3$-band alone. Thus, one could attempt to assign the rotational transition lines. In the lower part of the figure, the line assignment according to the high-resolution Raman measurements by Herranz and Stoicheff (1963) is included. This shows that crudely the $J$ dependence is resolved but not the subbranch fine structure.

As a final remark, it is noteworthy that by theoretically modeling the experimental spectrum of the $v_3$-band, one should be able to remove its overlapping contribution from the $v_1$-band, hence enabling one to derive an even better (lower) value for $\rho(v_1)$.

## 12.3.2 Analytical Raman spectroscopy and process monitoring

As has been pointed out earlier, Raman spectroscopy of gases has been exploited successfully as an analytical tool, particularly in the monitoring and control of combustion processes (for some of the more recent trends, see, e.g., Kojima and Fischer 2013 or Fuest et al. 2015). However, it is worth pointing out that more and more combustion diagnostics has become the domain of nonlinear Raman techniques, like CARS.

In order to demonstrate the strength of Raman spectroscopy as a means for monitoring processes and possible chemical exchange processes, here we show an example in which the radioactive molecular hydrogen isotopologue, $T_2$, undergoes exchange reactions with surface and bulk contaminants during extended circulation through steel tubes. The example shown in Figure 12.13 is related to measurements undertaken during tests for the tritium injection and circulation system of the KATRIN (Karlsruhe Tritium Neutrino) experiment; see Fischer et al. (2011).

The Raman spectrum reflecting the molecular composition of the circulating gas filling is shown on the left-hand side of the figure, showing a spectrum at the very beginning and close to the end of the test run; note that during the said measurement campaign, Raman spectra were recorded in 3-min intervals. At the beginning, one observes Raman peak intensities in the gas mixture that closely represent the concentrations provided in the gas chromatographic (GC) filling certificate, revealing the predominance of $T_2$ ($c_{T2} \geq 97\%$) with respect to the other, minor isotopologues ($c_{DT} = 2.25\%$, $c_{HT} = 0.5\%$, and all others were below the sensitivity threshold of ~0.15% for the Raman system used in this study). Note that the apparent discrepancy of the DT content between the GC analysis and the relative Raman intensities can be attributed to the known overlap of the $Q_1$-branch of DT and the $S_1(J'' = 2)$ line of $T_2$. Note also that the spectra shown here were not corrected for fluorescence background.

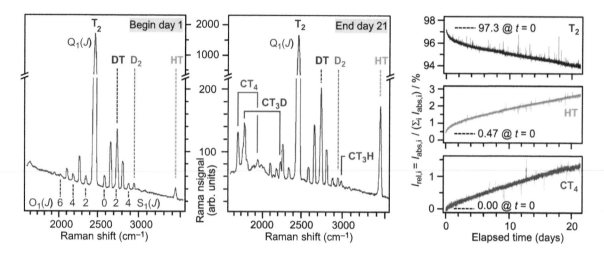

**Figure 12.13** Dynamic Raman measurement of a mixture of hydrogen isotopologues, circulating in the LOOPINO test setup for the Karlsruhe Tritium Neutrino experiment (KATRIN) for 21 days (total pressure $p = 200$ mbar; initial $T_2$-content, 97%). The representative spectra recorded at the beginning and the end of the measurement series reveal the generation of tritiated hydrogen isotopologues and tritiated methane, due to exchange reactions with the stainless-steel tubing. The right panel shows the time evolution of the process over the measurement time. (Data adapted from Fischer, S. et al. *Fusion Sci. Technol.* 60, no. 3 (2011): 925–930.)

After 3 weeks of operation, the spectra look markedly different. First, the relative amount of HT has increased dramatically; second, in particular at the low-wavenumber end of the spectrum, substantial additional Raman bands have evolved, which could be attributed to tritiated methane isotopologues.

The evolution of these features over time is shown in the right-hand panels of the figure, revealing a gradual decrease in the concentration of $T_2$, accompanied by an increase in concentration of HT and $CT_4$ (for the change of other compounds in the mixture, see Fischer et al. 2011). The observed compositional changes can be linked to hydrogen isotope exchange reactions and gas–wall interactions with the stainless steel tube walls, which contain traces of hydrogen and carbon; both formation processes are initiated by the energy released in the tritium $\beta$-decay. Therefore, the increase in the occurrence of HT (and DT), as well as tritiated methanes, is not unexpected. For example, the formation of the latter in the interaction between tritium gas and stainless steel has been reported by Morris (1977). The time evolution of the hydrogen isotopologue concentrations can be fitted by a double-exponential, with the shorter time constant related to gas–gas molecular interactions, while the longer one is likely to reflect the more gradual, slower extraction from the steel in wall–gas interactions.

These results quite nicely describe and confirm that Raman spectroscopy is suitable, in principle, for process monitoring and control, detecting molecular species and reaction products over a wide range of concentrations, differing by as much as a factor of $10^3$.

### 12.3.3 Remote sensing using Raman spectroscopy—The Raman LIDAR

All principal laser spectroscopic techniques—absorption, fluorescence, and scattering—lend themselves for contactless, remote applications. One particular measurement modality is that of **light-detection and ranging**, or LIDAR.

The first Raman LIDAR measurements go back to the late 1960s, with the measurement of molecules with the highest concentrations in the atmosphere (first $N_2$ and $O_2$, and later $H_2O$); lesser-abundant trace gases became more a playing field for differential absorption LIDAR—DIAL—because of the much larger interaction cross sections associated with laser absorption spectroscopy. A brief history of using Raman LIDAR systems in the profiling of (trace) gases and aerosols in the atmosphere can be found, e.g., in Turner and Whiteman (2002).

The basic principle underlying LIDAR in general is to send laser pulses into the atmosphere, where they are scattered and absorbed by atmospheric molecules and aerosols. The backscattered light is collected using a telescope for high light-collection efficiency. The atmospheric response to the laser radiation consists of several spectral components. These are mainly represented by elastically scattered light (Rayleigh and Mie scattering, see Section 2.3) and pure rotational as well as rotational-vibrational Raman spectra of atmospheric molecules. Therefore, isolating individual spectral features from each other, one can measure, e.g., atmospheric minor constituents such as ozone, carbon dioxide, or water vapor, as well as additional information on derived properties such as wind speed and direction, temperature, and cloud structure. The vertical and temporal resolution of LIDAR measurements are typically on the order of 100 m (up to a height of 10 km, or so) and 2–3 min, respectively. Modern LIDAR systems are multichannel systems with several detection channels; the most common yield data of water vapor mixing ratio, temperature, aerosol backscatter coefficient, extinction, and depolarization ratio. Time gating provides the necessary height information, linked to the short but finite travel time of the laser pulse up to a specific height and the return travel of the scattered wave. The LIDAR concept is shown schematically in the left-hand part of Figure 12.14.

In general, modern LIDAR systems operate with three laser wavelengths, normally the fundamental emission wavelength and its harmonics of a nanosecond pulse Nd:YAG laser (at $\lambda_L$ = 1064, 532, and 355 nm). A wealth of atmospheric parameters can be extracted from a single, multiwavelength laser excitation, multichannel light detection measurement. These include the following:

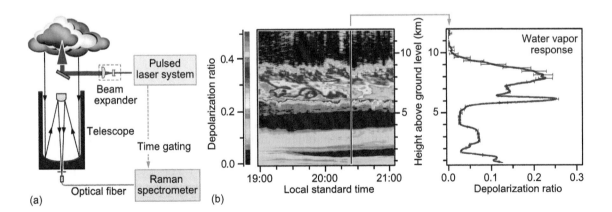

**Figure 12.14**   Raman LIDAR measurement of water vapor in the atmosphere. Left: conceptual measurement setup with telescope. Right: height profile of depolarization ratio of the Raman signal associated with the Raman signal from $H_2O$, revealing the temporal variation of water concentration (on the very right a time-slice trace is shown from which the water vapor concentration can be deduced. (Data adapted from Wu, S., X. Song, B. Liu, G. Dai, J. Liu, K. Zhang, S. Qin, D. Hua, F. Gao and L. Liu. *Opt. Express* 23, no. 26 (2015): 33870–33892. With permission of Optical Society of America.)

- The **particle backscatter**, which is derived directly as the ratio of atmospheric LIDAR returns in the elastic (Rayleigh/Mie) scattering and the pure rotational Raman scattering channels.

- The **particle extinction**, which is calculated directly from the atmospheric attenuation of the pure rotational Raman signal, incorporates information about particle densities.

- The **(linear) depolarization ratio**, which is normally measured at $\lambda_L = 532$ nm, gives information about the spherical shape of particles.

- The **water vapor mixing ratio**, which is derived as a ratio of backscatter signals in vibrational Raman branch of water vapor and nitrogen molecules. Note that the *water vapor mixing ratio*, *w*, is the absolute measure of the amount of water vapor that is in the air and is measured in units of grams of vapor per kilogram of dry air. For example, in meteorology, the mixing ratio is very important for tracing the properties of vast air masses as they rise or fall in the atmosphere, as well as for the understanding of cloud formation and precipitation (see, e.g., Sakai et al. 2013).

- The **air temperature**, which can be derived from the pure rotational Raman spectrum of $N_2$.

- The **aerosol** type and concentration, which can be obtained from a combination of (1) the particle backscatter measured at three wavelengths, (2) the particle extinction measured at two wavelengths, and (3) the particle depolarization ratio.

Examples for a combined measurement of the majority of all the above parameters can be found, e.g., in Wu et al. (2015). Representative for these, the depolarization ratio data $\rho$ for a 2-h measurement period are shown in Figure 12.14. In the left data panel, the variation (color coded) of $\rho$ with height (with spatial resolution $\Delta h \approx 4$ m) in the atmosphere is shown, as a function of time (with time resolution $\Delta t = 16$ s); on the left, a single time slice is presented, which reveals increased values for $\rho$ at about $h = 6$ km and $h \approx 8$ km, and which can be associated with cloud layers.

## 12.4 RAMAN SPECTROSCOPY OF LIQUID SAMPLES

From a conceptual experimental point of view, Raman spectroscopy of liquids should not be fundamentally different to that of gaseous samples, but because of the much higher molecular particle densities $N_j$, all scattering process, including Raman scattering, should be much more intense according to Equation 12.1. This can be already noticed by the naked eye when a laser beam passes through a liquid sample: normally, the beam is easily visible as a consequence of (elastic) Rayleigh scattering, as opposed to the beam passing through, e.g., air.

However, in general, substantial differences in the spectral shape are observed, which can be attributed to interactions between neighboring molecules. First, one normally observes that the vibrational bands shift, as a consequence of the influence the molecule–molecule interaction has on the bond lengths within the molecules and thus the vibrational frequencies. A second consequence of the closer particle distances is an increase in the probability of collisions, which in turn results in substantial line broadening. And finally, again as a consequence of (polar) interactions, free rotation is, by and large, suppressed. Rather, one encounters a type of reciprocating motion, in which the molecule rotates slightly back and forth around a nearly fixed orientation; this phenomenon is known as *libration* (the molecule's free motion is affected by external [interaction] forces that constrain its orientation).

As before, separate examples are given, which either highlight spectroscopic aspects or are related to practical, analytical applications.

## 12.4.1 Spectroscopic aspects of Raman spectroscopy in liquids

From the spectroscopic point of view, one can address two different aspects, namely, Raman line or band positions, or the shape and broadening of features. Other than simply identifying individual molecular species, both shift in position from and broadening of the gas-phase Raman spectral features allow one to draw conclusions about the nature of the molecular interaction in the liquid phase. For example, one can identify ionic interaction, hydrogen bonding, and aggregate formation, to name a few of the most important ones.

In general, the first call of port in Raman spectroscopy of liquids is to record spectra of a "pure" sample, in order to provide a "fingerprint" spectrum for specific molecules. Such fingerprint spectra have been accumulated in a number of Raman spectral libraries (e.g., those available from Fiveash Management or Thermo Fisher Scientific). Fingerprint spectra are highly relevant when it comes to the identification of many of the organic compounds, which all exhibit characteristic subvibrations of, e.g., their C = C or C–H bonds (being common to many compounds, these would not help in the identification process). Thus, vibrations specific to a particular molecular compound need to be identified.

An example of the importance for such a comparison to library Raman spectra is shown in Figure 12.15, where a comparison between unleaded regular (ULR) and premium (ULP) petrol samples is made. While the same crude oil might have been used in the petrol production process, which thus would not lead to different spectra (note that crude oils from different sources may exhibit distinct Raman spectra; see, e.g., Orange et al. 1996), the different (minor) additives will show up as markers. In the upper panel, representative spectra for ULP and ULR are displayed, with the locations of a selection of xylene additives marked. It is quite clear from the comparison of pure-compound and composite-compound Raman spectra that a careful selection of spectral features is required for unique identification; often, visual inspection might not be enough and multivariate analysis techniques are required, which, in addition to feature positions, take their relative amplitudes into account. Overall, with the help of spectral libraries, Raman spectroscopy has the capability to distinguish whether petrol was produced by different manufacturers, or if it was adulterated (blended); see, e.g., Tan et al. (2013).

As stated earlier, on the second spectroscopic aspect in the evaluation of Raman spectra is the shift, broadening, and shape of the observed features in comparison to the gas-phase spectra, since these are bound to provide substantial insight into molecular interaction processes as the environment around individual molecules changes.

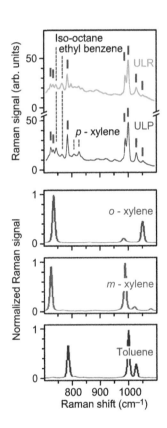

**Figure 12.15** Raman spectra of unleaded regular (ULR) and premium (ULP) gasoline samples (top panel). Pure spectra of xylene additives are given in the lower panels; the positions of the most relevant line positions are marked in the gasoline spectra (color coded). (Gasoline spectral data courtesy of Sacher Lasertechnik GmbH, Marburg, Germany (2016).)

One particular case study is that of water, not only because $H_2O$ is a strongly polar molecule, and therefore is extremely perceptible for long- and short-range molecular interactions, but also because of its relevance in a great number of biological processes. In water, the dominant vibrational motions are those of the O···H bond, in the form of the symmetric and asymmetric stretch vibrational modes $v_1$ and $v_3$, respectively. For the free (gas-phase) water molecule, these are found at the related Raman bands found at $\tilde{v}(H_2O_{(vap)}, v_1) \approx 3659$ cm$^{-1}$ and $\tilde{v}(H_2O_{(vap)}, v_3) \approx 3756$ cm$^{-1}$.

In the liquid phase, these main stretching bands are shifted to lower frequencies, and now one observes the related Raman bands at $\tilde{v}(H_2O_{(liq)}, v_1) \approx 3280$ cm$^{-1}$ and $\tilde{v}(H_2O_{(liq)}, v_3) \approx 3490$ cm$^{-1}$. This is due to the increase in intermolecular H bonding; the stronger this effect is, the larger is the observed shift. These shifts are affected by the macroscopic parameters temperature and pressure, since they affect the particle density. For example, one observes that, in liquid water, the OH stretch vibrations shift to higher frequencies when raising the temperature; this is because hydrogen bonding weakens, thus strengthening the covalent O···H bonds and meaning that their vibrational frequencies increase again. On the other hand, when increasing the pressure, the O···O distances decrease, which, in turn, increases the O···H distances and thus lowers their stretch frequencies. In general, a raise in pressure causes a reduction in long, weak or broken bonds but an increase in bent and short, strong hydrogen bonds.

Overall, one finds that the Raman spectra of liquid water are far more complex than those of water vapor, due to vibrational overtones and combinations with librations $L_1$ and $L_2$ (the aforementioned restricted rotations, or rocking motions); they are observed at $\tilde{v}(H_2O_{(liq)}, L_1) \approx 396$ cm$^{-1}$ and $\tilde{v}(H_2O_{(liq)}, L_2) \approx 686$ cm$^{-1}$. Note that librations depend on the moments of inertia. For example, the moment of inertia for $D_2O$ is nearly double that for $H_2O$; as a consequence, the libration frequencies reduce by about a factor of $\sqrt{2}$. Note also that one will observe (translational) cluster vibrations that involve combinations of hydrogen bond O-H···O stretching and bending. It is clear from this discussion that unambiguous molecular-level interpretation of experimental spectral features in many organic liquids in general and water in particular remains a challenge, due to the complexity of the underlying hydrogen-bonding network. Utilizing "first-principles" representations with centroid molecular dynamics, successful attempts are being made to simulate Raman spectra of liquids under ambient conditions, without relying on any ad hoc parameters; such simulations are now in reasonably good agreement with corresponding experimental data (see, e.g., Medders and Paesani 2015 and Lin et al. 2010).

### 12.4.2 Analytical aspects of Raman spectroscopy in liquids

Addressing analytical aspects of liquids by Raman spectroscopy has a long tradition; for both relative (comparative) aspects for identification purposes and absolute concentration measurements, this laser spectroscopic method has demonstrated its merits. Examples were already briefly touched upon earlier, namely, (1) the absolute concentration measurements of constituents of mixtures, exemplified for $H_2O/D_2O$ mixtures (as described in the conceptual discussion on concentration calibration in Section 12.1), and (2) for the identification of gasoline additives by comparison to reference Raman spectra (as presented further above in this section).

Here, a further example is added, in which both types of qualitative and quantitative aspects are encountered, namely, the analysis of edible oils, aiming at distinction of different species and quality control. In their study, Vaskova and Buckova (2014) acquired Raman spectra for a wide range of common edible oils, and on the basis of characteristic line ratios of unsaturated fatty acids contained in oils, they were able to distinguish different species. A selection of spectra is shown in Figure 12.16 (upper panel), together with a characteristic spectral region that was exploited for unique species identification (lower panel).

The Raman spectra of different oils all look all very much alike; this is not surprising because they contain similar components, such as, e.g., triacylglycerols, and saturated and unsaturated fatty acids (with straight aliphatic chains, predominantly with 16 or 18 C-atoms in the chain). However, each oil species has different concentrations and ratios of these components; this is reflected in the relative intensities of the associated Raman features and thus can be exploited for their identification.

The spectra were normalized to a band that did not exhibit any changes; this particular band at $\tilde{v} \approx 1747$ cm$^{-1}$ could be associated with the C = O ester-carbonyl stretch mode. Then, a spectral region in which clear relative changes occur, clearly noticeable by the naked eye, is around $\tilde{v} \sim 1250$–$1350$ cm$^{-1}$. In the expanded-scale spectrum in the lower part of the figure, it is quite evident that the ratios between the two peaks at $\tilde{v} \approx 1267$ cm$^{-1}$ and $\tilde{v} \approx 1303$ cm$^{-1}$ (the =C–H symmetric rocking and CH$_2$ in-phase twist modes, respectively) are distinctly different for the various oils. This ratio (of unsaturated-to-saturated carbon bonds) reflects the share of linoleic acids, with, e.g., sunflower and grape oils evidently having larger amounts than rice or almond oils; this is in line with common knowledge.

The authors were also able to determine the degradation of oils in response to heating, based on the monitoring changes in the characteristic vibration Raman peaks and the appearance of additional features, which could be linked to decomposition products. For example, as a result of oil heating, the activity of CH$_2$ bonds increases while that of C = C bonds downgrades. This was reflected in changes of the two associated Raman peak intensities and therefore indicated the deterioration of linoleic acid content.

It is worth noting that similar studies using Raman spectroscopy for the quantitative determination of lipids and oils in bio-organisms have been reported by Samek et al. (2011), in which the ratio of unsaturated-to-saturated carbon bonds, derived from the Raman spectra, was quantified and cross-calibrated to provide the so-called iodine value that constitutes part of the current food industry and biofuel standards.

The final short example for analytical applications of Raman spectroscopy concerns "operational safety" issues. Here, the particular modality of *spatially offset Raman spectroscopy* (SORS) comes to prominence, a technique that was first introduced by Matousek et al. (2005). SORS allows for highly accurate chemical analysis of objects beneath obscuring surfaces, such as, e.g., biological tissue, coatings, or bottle walls. In this context, SORS instrumentation has been developed, as a response to aviation security requirements (adopted in 2014) that at most international airports, liquids, aerosols, and gels need to be screened for the presence of liquid explosives (like, e.g., hydrogen peroxide), with extremely low false-alarm rates. It is suitable to probe a wide range of containers, including

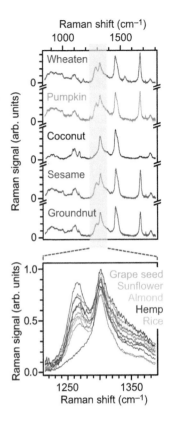

**Figure 12.16** Raman spectra of a range of edible oils. Top: overview spectra; individual spectra offset for clarity. Bottom: expanded spectral region including the =C–H symmetric rocking and CH$_2$ in-phase twist modes, whose ratio is used to determine the linoleic acid content. (Data adapted from Vaskova, H. and M. Buckova 2014.)

(1) colored, opaque, or clear plastics; (2) glass and paper; and (3) duty-free bottles in secure tamper-evident bags. Instrumentation based on SORS is currently deployed in some 500 systems at airports worldwide; false-alarm rates of the order <0.5% are being reported (see Loeffen et al. 2016 and references therein).

In brief, SORS operates by making more than a single Raman measurement on a particular container. First, a so-called "zero-offset" measurement is made. For this, the Raman light detector is positioned axially to the laser spot on the container. Then, a second (or more) measurement is made in which the laser spot location and the detector are offset by a small distance $\Delta d$. With increasing offset, the "surface" component (i.e., the Raman signal generated in the container material) will, in general, diminish more rapidly than the "content" component.

In diffusely scattering media (either container or contents), the "content" component of the signal is excited by photons that have migrated over the distance $\Delta s$ by elastic scattering, it less sensitive to the offset. Making multiple measurements at different offset locations $\Delta d_i$, with the associated varying signal contributions from the "container" and "contents," it is possible to separate the signal into its respective components. Thus, with no prior knowledge of the bottle, or its contents, one can obtain clean spectra of the contents, without contributions or interference from the container. Note that the methodology works even with diffuse scattering media, such as, e.g., white, high-density polyethylene, or with highly fluorescent materials, such as, e.g., colored glass (Loeffen et al. 2016).

### 12.4.3 "Super-resolution" Raman spectroscopy

In Section 12.3, it was mentioned that with modern Raman spectroscopy instrumentation, the analytical aspects were more important than spectral resolution per se, as long as the spectral features allowed for the extraction of the desired qualitative or quantitative information about the sample molecule(s). Basically, the spectral resolution of a dispersive Raman detection system is limited by the (diffraction-limited) image of the entrance slit of the spectrometer on the pixel array of a CCD detector and the spectral dispersion of the instrument. Fitting the observed line shapes, their center position can be determined to be better than a pixel width; however, what cannot normally be achieved is to separate lines whose line centers are spaced by less than the Rayleigh criterion. As a consequence, one may not be able to reveal, e.g., fine-structure information for different molecular substances. The challenge therefore is to come up with an approach that preserves chemically specific Raman scattered photons in such a way that allows one to retrieve overlapping/hidden/buried Raman signatures; this means than one needs "super-resolution" (SR).

The concept of SR is known from 2D image processing, in which one retrieves unresolved information, caused, e.g., by detector decimation, by combining a sequence of low-resolution frames. Using appropriate deconvolution algorithms, one can then reconstruct images, which reveal higher details than any of the single, individually grabbed frames. In the context of laser imaging, the SR concept is encountered in some of the laser-induced fluorescence imaging modalities that achieve below-diffraction-limited resolution (like, e.g., the technique of stochastic optical reconstruction microscopy—STORM; see Section 18.4).

In the context of spectral super-resolution, a solution proposed and demonstrated by Chen et al. (2015) is a multichannel acquisition framework that is based on (1) shift excitation and (2) slit modulation, followed by (3) mathematical postprocessing. The authors demonstrated a significant improvement in the spectral specificity of Raman characterization when compared to standard dispersive Raman measurements. In short, the technique—dubbed *shift-excitation blind super-resolution Raman spectroscopy* (SEBSR)—uses multiple degraded spectra to extract high-resolution information from dispersion-limited raw data. Note that part of the methodology—namely, shifted-wavelength laser excitation—utilized in this approach has been adopted from a widely used approach to remove strong fluorescence background from Raman spectra (see, e.g., Cooper et al. 2013). SEBSR promises to overcome a fundamental problem normally plaguing Raman spectroscopy, namely, that high spectral resolution requires large dispersion (which is accompanied by extreme optical loss due to a very narrow solid-angle light collection cone and associated with this unwieldy large equipment). The principle underlying SEBSR is shown schematically on the left-hand side of Figure 12.17.

To obtain shifted-excitation laser wavelengths, the authors used a single-mode, external-cavity diode laser, with a linewidth of about $\delta\nu_L \approx 200$ kHz; this laser was locked to an optical frequency comb that resulted in stabilized-frequency uncertainties of the order $\Delta\nu_L \sim 1$ kHz. This allowed for extremely well-defined Raman excitation wavelengths, which could be used in Raman spectra with precisely known subpixel shift differences. The key point here is that the subpixel shift is necessary to implement SR recovery algorithms to provide new spectral

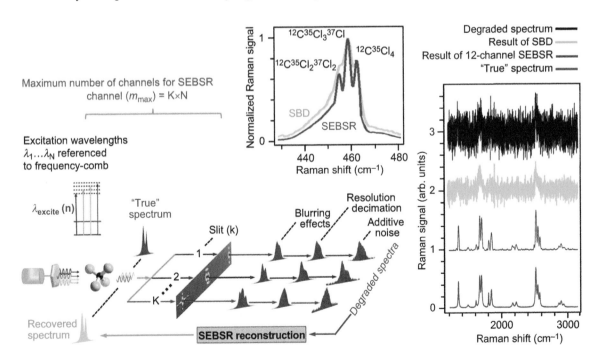

**Figure 12.17**  Implementation and application of SEBSR. Left: concept of the experimental realization, with multiple laser excitation wavelengths and multiple spectrometer-slit settings. Right: concept of the spectrum recovery, exemplified for a synthetic spectrum. Center: experimental results for CCl$_4$, resolving individual isotopologue contributions. (Adapted from Chen, K. et al. *Sci. Rep.* 5 (2015): article 13952.)

information. In addition to the shifted excitation, the authors added a multiple-blind deconvolution methodology, requiring each of the "blind" channels to carry spectroscopic differences in addition to the shift. In order to generate different detecting channels, the authors utilized slit modulation of the spectrograph; the slits could have a different shape (e.g., single slit, double slit, cross-shaped slit, or triangular slit) and could also be modulated in width. Overall, in the work described here, up to 12 independent channels were used in the demonstration of SEBSR.

In order to demonstrate the potential of their approach, the authors tested their method on synthetically blurred spectral data (also introducing different, challenging signal-to-noise scenarios). An example is shown in the right data panel in Figure 12.17, for a very challenging SNR = 3. As is evident, all the spectral features in the original spectrum (bottom, red trace) are recovered using the maximum number of 12 channels—four laser excitation wavelengths and three slit variations—utilized in this study (blue trace), while the individual channel raw-data information (top, black trace) hardly let any Raman signal be recognized. For comparison, the single-channel raw data were treated with a semiblind spectral deconvolution methodology to improve on the spectral data information (green trace); this shows that Raman features start to become visible above the noise, but because of the single-channel nature of the approach, no gain in spectral resolution ensues.

The method was also demonstrated for actually recorded Raman spectra, resented here in the top panel of Figure 12.17 for the C–Cl bending mode of $CCl_4$; clearly, using the SEBSR technique, one is able to recover the isotopologue structure, which was unresolved in ordinary single-channel spectra.

It is noteworthy that the super-resolution approach describe here is not the only one attempted for linear Raman spectroscopy. Malka et al. (2013) proposed an approach that also incorporates shifted excitation. But other than in the SR approach described above, involving "filtering" of the signal by different slit widths, here, the Raman light is sent through a stepwise tunable Fabry–Perot filter, which selects slightly different Raman features for different laser excitation wavelengths. Encoding the Raman spectra in this way, before analysis by the spectrometer, provides a set of time-multiplexed data channels over the spectral domain from which super-resolved spectra can be reconstructed. The given resolution of a spectrometer can be improved, in principle, up to the narrowest, resolvable feature from the FP-filter encoder, i.e., $\delta v_{FP}$ (which is determined by the free spectral range and finesse of the device). In initial tests, the authors demonstrated an increase in spectral resolution up to a factor ×5 over the inherent spectrometer resolution.

## 12.5 RAMAN SPECTROSCOPY OF SOLID SAMPLES

By the very nature of solid samples, thye spectra of molecules within them are expected to be even more different from those of free, gas-phase molecules than was experienced for molecules in the liquid phase. Now, rotational movement is completely hindered and librations also become rather subdued. On the other hand, additional vibrational features tend to appear; in particular, in ordered, crystalline structures, one expects contributions from global-oscillation motions associated with acoustic and optical phonons. Also, morphological changes in the solid's structure should manifest themselves in the shift of Raman bands, or even the appearance of new ones. Condensed-phase sample specimens, without periodic structure, are often even more challenging for the interpretation of Raman spectra. For example, soft matter like biological tissue contains a multitude of organic compounds, which give rise to very complex spectra on the one hand, and without certain prior knowledge or sophisticated mathematical evaluation methodologies, it might be next to impossible to distinguish between similar species. In addition, in general, most (organic molecule) soft matter samples suffer from severe fluorescence background.

Finally, it is worth noting that most practical applications for solid samples are based on commercial Raman microscope instrumentation. It is obvious that the use of a microscope as the optical interface immediately opens up Raman spectroscopy to imaging; indeed, the imaging aspect has become a rather important one when investigating soft tissue specimen (see Chapter 18 for representative examples). Below, a few selected (nonimaging) examples are presented, highlighting the distinct aspects of Raman spectroscopy of ordered and soft matter samples.

## 12.5.1 Spectroscopic and structural information for "ordered" materials

Out of the wealth of Raman spectroscopic investigations of ordered, crystalline materials, a family of examples has been selected, which demonstrates the versatility of Raman spectroscopy for the understanding of "engineered" materials of small size and low dimensionality, in comparison to "bulk" materials. For example, graphene, nanotubes, and other carbon nanostructures have shown their potential as candidates for advanced technological applications, due to the different coordination of carbon atoms in the structure. It is now possible to easily fabricate 2D structures, like graphene sheets, or even 1D systems in the form of carbon nanowires (or carbon atom wires [CAWs]), which potentially would allow one to downscale devices to the atomic level.

CAWs can be arranged in two possible structures, namely, (1) a sequence of double bonds, exhibiting "metal-like" conduction behavior, or (2) an alternating sequence of single and triple bonds (commonly known as *polyynes*), which are expected to possess "semiconducting" properties. Polyyne structures were investigated, e.g., by Milani et al. (2015); an example spectrum for a biphenyl-capped CAW sample is shown in the lower data panel of Figure 12.18, together with a comparison of the CAW spectrum to other carbon structure Raman spectra in the upper data panels of the figure. Clearly, using Raman spectroscopy, one is able to distinguish the individual structures unambiguously.

Density functional theory (DFT) calculations predicted that the vibrational features are strongly dependent on the CAW (see Milani et al. 2011, who computed the active Raman modes for single carbon atom wires with 6 to 18 carbon atoms). They found that the wavenumber $\tilde{\nu}_n$ of the Raman vibrational modes decreased for increasing wire length, while in parallel, the Raman intensity increased. It should be noted that DFT calculations are not very accurate in estimating the Raman intensity as a function of chain length. However, using the DFT results in summing up the contributions from the individual chain-length cases and properly weighting their quantity in the mixture, the authors were able to generate

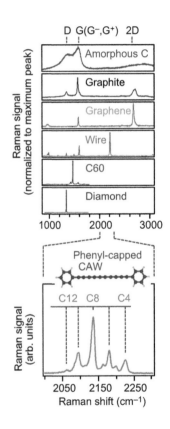

**Figure 12.18** Raman spectroscopy of carbon structures. Top: experimental Raman spectra of carbon solids and nanostructures. Bottom: expanded-scale spectrum of diphenyl-capped carbon-atom wire (CAW); the peaks are annotated respective to the number of carbon atoms $C_n$. (Data adapted from Milani, A. et al. *Beilstein J. Nanotechnol.* 6 (2015): 480–491.)

theoretical spectra that were reasonably fair representations of the experimental data.

A quite different problem in which the understanding of evolving carbon structures might be essential is that of carbon accumulation at interfaces, e.g., at the metal contact–semiconductor interfaces of SiC high-temperature devices. During the fabrication process, such devices are normally annealed at high temperatures, up to the order of $T_{anneal} \sim 1000°C$. However, it was found that, frequently, the metal contact was peeling off or had high contact resistance (see, e.g., Dunstan et al. 2004); this was later ascribed to the accumulation of carbon islands at the metal–semiconductor interface. A deficiency of the measurement methodology in the said investigation was that the method of scanning tunneling microscopy (STM) only provides structural and conduction information, but lacks chemical specificity. This problem could be overcome in part by using Raman spectroscopy on the same samples; provided that the Ni-contact layer was not too thick (less than about $d_{Ni} < 100$ nm), laser excitation and Raman scattered light could still penetrate the layer to yield meaningful Raman signals. In Figure 12.19, example spectra are shown for Ni/4H-SiC interfaces; clearly, the spectra reveal a substantial accumulation of carbon in the form of unstructured graphite.

It should be noted that the spectro-chemical characterization of hidden structures can be driven even further to deeper locations provided that the sample is reasonably transparent for both laser excitation and Raman scattering radiation, even for strongly elastic-scattering materials. For example, Sil and Umapathy (2014) demonstrated that they could obtain molecular structural information of materials buried well below the sample surface (up to a few centimeters) by using Raman scattering in the modality of *universal multiple-angle Raman spectroscopy*.

## 12.5.2 Analytical and diagnostic applications for "soft tissue" samples

Traditionally, the assessment and analysis of soft, biological tissue are done by optical microscopy, in which an "image" of the sample is inspected by eye or recorded digitally using (often color-filtered) CCD image sensors. The advent of laser light sources and laser spectroscopic techniques did not significantly change the imaging microscopy approach, with imaging based on laser-induced fluorescence and Raman scattering being the dominant conceptual approaches (for relevant key examples, see Chapters 18 and 19, respectively). However, there are a number of applications, specifically in diagnostics, in which the imaging aspect itself does not play the central role. The microscope image is only used to guide the probing laser to a particular sample location, and it is then the average laser-induced response from the interaction area that is used for diagnosis, based on the recorded spectra.

Over the past few years, Raman spectroscopy is gaining ever-increasing importance in fields as diverse as environmental science, food hygiene and safety, and medical diagnosis, all of which benefit from fast and reliable detection and identification of microorganisms; Stöckel et al. (2016) provide a concise updated survey of these developments. The growing popularity that Raman spectroscopy enjoys is due to a number of factors that set it apart from other analytical diagnosis techniques, namely, that (1) it constitutes an easy application methodology at

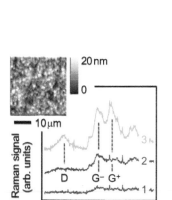

**Figure 12.19** Raman spectrum of graphite accumulation below the nickel-contact layer on a SiC device substrate; an STM image of the topological structure of part of the device, with a 50-nm Ni-contact layer; the Raman response are recorded through the Ni-layer: 1, 100-nm Ni-layer; 2, 50-nm Ni-layer; 3, 50-nm Ni-layer, after annealing the sample for 300s at $T = 900°C$. The well-known graphite Raman features are annotated. (Data from the author's laboratory.)

relatively modest cost; (2) it provides qualitative and quantitative results, requiring only short times for measurement and data analysis; and (3) it carries high information content on both the chemical composition and the structure of biomolecules within the microorganisms. With respect to the latter, it has become possible, e.g., to follow slight changes in the chemical composition of microorganisms over time, and the Raman spectra carry sufficient molecular compound information so that one may be able to differentiate genera, species, or even strains and clones.

Here, only one particular example of the capabilities of Raman spectroscopy will be discussed, namely, its application in the diagnosis of bacterial infections and diseases; this particular aspect has been followed by numerous research groups around the world. In Figure 12.20, some representative results from the authors' own laboratories are shown. The spectral traces in the top data panel demonstrate one of the common problems in microbiology laboratories, which is associated with the difficulty of distinguishing/recognizing different strains of a particular microorganism, here for two strains (9142 and 1457) and an additional clone (1457M10) of the genus *Staphylococcus epidermidis* (see Samek et al. 2010). The spectra themselves are very similar insofar that nearly all molecular band peaks seem to be the same, only differing marginally in their amplitude: distinction by naked eye is next to impossible.

Only after applying complex but well-established data analysis, which take into account the complete spectral content—e.g., in the form principal component analysis—does one start to see statistically relevant differences. This is shown in the bottom panel of the figure for a correlation (cluster) plot of the first two principal components PC1 and PC2 from the analysis of a set of spectra from different measurements (of the order of 100 for statistical relevance). For clarity, only the weighted center of the distribution and its spread are shown by a symbol and a surrounding boundary, respectively. The data show that using only two principal components, one can easily type different genera (*S. epidermidis*, *Staphylococcus aureus*, and *Escherichia coli*) as well as *S. epidermidis* strains (9142 and 1457), but genetic mutants (of type M10) cannot be separated from their parent (note that the strain 1457 is biofilm-positive while its mutant 1457-M10 is biofilm-negative). To achieve this, all principal components have to be carried into further analysis steps, like, e.g., cluster analysis and dendrogram generation, to achieve this final distinction. While seemingly complex, the complete process of data taking and Raman spectral analysis is still fast (normally just a few minutes) in comparison to other techniques like the currently most prominent method for bacterial typing—pulsed-field gel electrophoresis.

It should be noted that, in principle, it is possible—and has been shown experimentally—that one can even identify specific intracellular pathogens and the interaction of the bacteria with antibiotics (for a brief overview, see, e.g., Stöckel et al. 2016). This has been afforded by a range of studies to identify important, individual biomolecules and accumulate reference Raman spectra of these in libraries (see, e.g., de Gelder et al. 2007).

Finally, the analysis of individual, single bacteria may be seen as the ultimate "gold standard" in analytical Raman spectroscopy, which would open up in situ diagnostics without the need for cultivation into colonies. In this context, several research groups succeeded in the quest for single-bacteria analysis and typing, although in said studies, original cultivation was still the point of sample origin. Strola et al. (2014) utilized confocal Raman microscopy; with a time span

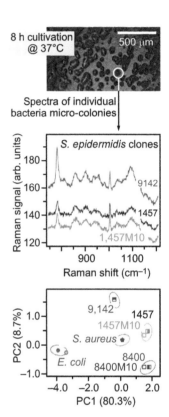

**Figure 12.20** Raman spectra of (individual) bacterial micro-colonies of *S. epidermidis* strains and clones. After multivariate analysis, the classification into well-separated strain groups and distinction against other bacteria is possible, while genetic clones of particular strains are difficult to separate. (Data adapted from Samek, O. et al. *Laser Phys. Lett.* 7, no. 5 (2010): 378–383.)

of just 60 s, they were able to first localize small bacteria aggregates (typically 10–20 individuals), then align on a single individual, and finally collect the Raman scattering signal with sufficient signal-to-noise ratio for data analysis and bacterial typing. Recently, the authors extended their efforts toward identifying single bacteria in environmental samples; i.e., measurements had to be performed in nonideal conditions and introducing matrix diversity; they showed that it was still possible to type bacterial strains by comparison to well-defined reference training sets (Baritaux et al. 2016). However, the authors point out that because of the huge matrix diversity of environmental samples, there is a trade-off between sample preparation and direct in situ measurements.

## 12.6 BREAKTHROUGHS AND THE CUTTING EDGE

Linear Raman spectroscopy is an established spectroscopic technique, which has gradually evolved—after beginning to use lasers for excitation—with the availability of new laser, spectrometer, and detector equipment, but, by and large, has remained unchanged. However, Raman spectroscopy with lasers close to the excitation line has always been difficult, due to the need for filters to reject the laser light from the Raman light detection channel, with transmission cutoffs in general more than 100–200 cm$^{-1}$ away from the laser line, thus very much excluding the important terahertz spectral region (i.e., $\lesssim$ 200 cm$^{-1}$).

This deficiency has been overcome elegantly with the introduction of so-called volume-holographic grating (VHG) filters, which now allow for Raman spectroscopy as close as <5 cm$^{-1}$ to the laser line (see, e.g., Glebov et al. 2012).

### 12.6.1 Breakthrough: Raman spectroscopy in the terahertz range

Because of the ability to record Raman spectra in the terahertz regime, many important Raman features in this ultralow optical-frequency regime have become accessible, which were closed to investigation before. Thus, new successful applications evolved in both fundamental spectroscopy and analytics.

For example, it has become possible to distinguish polymorphic forms of molecules, with differences scarcely evident in Raman spectra recorded in the common "fingerprint" region (containing the predominant *chemical* information), but which exhibit distinct differences at terahertz frequencies (the range is associated with *structural* information).

One such example is shown in Figure 12.21, clearly showing distinct differences in the polymorphic forms of *carbamazepine* (CBZ) in the terahertz regime (top panel of the figure); see Carriere et al. (2013). CBZ is an important anticonvulsant and mood-stabilizing drug commonly prescribed for the treatment of epilepsy and bipolar disorder. It has four different polymorphic forms, but only one—form 3 (blue trace in the spectra)—is approved for medical use; using terahertz Raman spectroscopy, it now can easily be identified in tablets. Besides this pharmaceutical application, the authors also demonstrated the potential usefulness of terahertz Raman spectroscopy for the detection and identification of, e.g., explosives and illicit drugs.

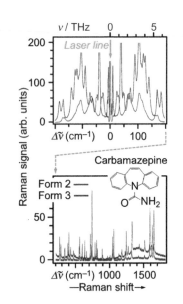

**Figure 12.21** Raman spectra of two polymorphic forms of *carbamazepine* (CBZ). Top: terahertz region; bottom: "fingerprint" region (note the difference in Raman shift scale). (Data adapted from Carriere, J.T.A. et al. *Proc. SPIE* 8710 (2013): article 87100M.)

Another interesting application for terahertz Raman spectroscopy that is gaining increasing prominence is the probing of lattice vibrations in crystals and low-dimensional structures, whose (optical and acoustic) phonon frequencies and interlayer vibrational modes reside in the terahertz regime. For example, in their review on the topic, Zhang et al. (2015) describe lattice vibrations of two-dimensional structures from monolayer to multilayer to bulk material, including the various types of phonons, and Raman features revealing interlayer coupling, spin-orbit splitting, and external perturbations.

## 12.6.2 At the cutting edge: Raman spectroscopy in the search for life on Mars

By their very nature of light excitation/detection, nearly all laser (analytical) techniques are "contactless" and therefore lend themselves to standoff and remote applications, in principle. In this context, Raman spectroscopy has struggled somewhat in the past because of the very low transition probabilities of the Raman scattering process, which severely hampered sensitivity in cases of less-than-optimal light collection situations. However, because of its universality and chemical specificity, in conjunction with requiring only a fixed-wavelength laser source, standoff Raman spectroscopy has been high on the wish list of researchers for out-of-the-laboratory applications under harsh environmental conditions. A major drive toward its practical implementation has been under way since the mid-2000s, with a view to incorporating Raman spectroscopy on future space (Mars) missions, in the quest (1) to identify organic compounds and search for signatures of life, (2) to identify the mineral products and indicators of biological activities, and (3) to characterize mineral phases produced by water-related processes (see, e.g., Jehlička et al. 2009).

In this context, two (landed) Mars missions are planned for the next suitable launch window in July 2020; both will incorporate the capability for Raman spectroscopy. For the NASA "Mars 2020 Rover" mission, it is a combination instrument, the *Scanning Habitable Environments with Raman & Luminescence for Organics and Chemicals* (SHERLOC) unit; in the ESA/Roscosmos "Exomars 2020" mission, a dedicated Raman spectrometer will be used. A flurry of activities over a 10-year period has come to fruition, in time for the planned launches, having (1) provided viable instrumentation (see, e.g., Blacksberg et al. 2016), (2) resulted in novel data evaluation and interpretation strategies (see, e.g., Ferralis et al. 2016), and (3) verified the required performance in the field (see, e.g., Mittelholz et al. 2016). Specifically, for the latter, a Raman spectrometer was included in the rover prototype payload, within the framework of the Exploration Surface Mobility project. During this "CanMars" mission, a time-resolved Raman instrument (incorporating a pulsed Nd:YAG laser) was used in mast-mounted, standoff configuration; the device proved usable over standoff distances of up to 13 m.

Most important in all Mars mission-related Raman spectroscopy studies was to prove that materials associated with signs of life and organic chemistry could reliably be identified. Thus, the aforementioned CanMars and other analytical tests were carried out on samples from regions with Martian-analog geology (see, e.g., Hutchinson et al. 2014). For example, *gypsum*—a sulfate mineral found on

Mars—is closely associated with water and is known to harbor fossil life on Earth; *apatite* incorporates calcium phosphates (phosphates may be evidence of life because they are part of the backbone of DNA); and calcite is important to life because L- and D-amino acids are known to adsorb to it. An example Raman spectrum for a Martian-relevant mineral, *jarosite* (a water-containing sulfate mineral), is shown in Figure 12.22.

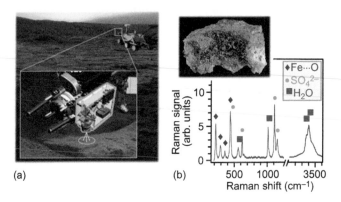

(a)  (b)

**Figure 12.22** Raman spectroscopy goes to Mars. (a) Artist's impression of the "SHERLOC" spectroscopy unit of the NASA's *Mars 2020 rover*. (Courtesy of NASA/JPL-Caltech 2014). (b) Raman spectrum of a *jarosite* sample from the Jarosco Ravine, Spain, serving as a model mineral from a Mars-like geological environment. (Data adapted from Pérez, F.R. and J. Martinez-Frias. *Spectrosc. Europe* 18, no. 1 (2006): 18–21; assignment of Raman bands according to Sasaki, K. et al. *Can. Mineral.* 36, no. 5 (1998): 1225–1235.)

# CHAPTER 13

# Enhancement Techniques in Raman Spectroscopy

In Chapter 11, expressions for (quantitative) Raman scattered light were introduced (see Equations 11.3 and 11.4) in which the observed Raman light was related to a number of parameters and functions, including

(i) The number of particles participating to the Raman scattering process;

(ii) The incident laser photon flux stimulating the Raman transition; and

(iii) The Raman transition energies and the Raman transition dipole moment function.

Amalgamating those two equations, and rearranging terms, one arrives at the proportionality expression

$$I_S \propto \underbrace{k_{\tilde{\nu}}}_{F1} \cdot \underbrace{\tilde{\nu}_S^3 \cdot \tilde{\nu}_L}_{F2} \cdot \underbrace{I_L}_{F3} \cdot \underbrace{N_i}_{F4} \cdot \underbrace{|\langle \Psi_f | \mu_S | \Psi_r \rangle \langle \Psi_r | \mu_L | \Psi_i \rangle|^2}_{F5}. \tag{13.1}$$

This may be subdivided into five contributing functional groups (as indicated), each of which influences the scattered Raman light intensity in one way or another:

F1: $k_{\tilde{\nu}}$ is an entity which, by and large, incorporates experimental factors, including Raman excitation/collection geometries and equipment response functions.

F2: transition frequency/energy terms for the laser radiation ($\tilde{\nu}_L$) and scattered Raman light ($\tilde{\nu}_S = \tilde{\nu}_L - \tilde{\nu}_{if}$), both in units of wavenumbers. Note that, often, the two frequencies are crudely approximated as being close-to-equal in magnitude, leading to the widely utilized $\tilde{\nu}_L^4$-law, meaning that the Raman light intensity increases roughly with the fourth power of the laser transition energy.

F3: $I_L$ and $I_S$ stand for the incident laser photon and scattered Raman photon fluxes.

F4: $N_i$ represents the population number density in the initial energy state(s) of the molecule, in units of $cm^{-3}$; it depends on the sample temperature and also includes the statistical weight factors for each individual (vibrational-rotational) state. Note that this has to be multiplied with the interaction/

observation volume to provide the total number of particles associated with the actually observed Raman signal.

F5: the term is related to the Raman scattering line strength function. Here, only the all-important transition matrix element for the two-step Raman process is shown, which mediates between the initial, intermediate ("virtual"), and final quantum states of the molecule and whose respective wave functions $\psi$ are indexed $i$, $r$, and $f$; $\boldsymbol{\mu}_L$ and $\boldsymbol{\mu}_S$ are the excitation and scattering dipole transition moments.

When inspecting Expression 13.1, it may seem straightforward to approach enhancement along the lines of optimizing (increasing) an individual or up to all of the three main parameters responsible for the Raman scattering intensity, i.e., the particle number density involved in the scattering process, the incident laser photon flux, and the transition dipole moment function. However, on second glance, one will recognize that not everything is necessarily as simple as initially imagined, because the five functional groups indicated in the expression are not fully independent of each other but are sometimes (strongly) interlinked or limited by operational constraints.

The aforementioned three key parameters that potentially yield enhancement of the Raman signal and common implementation methodologies are addressed in the following subsections. In this context, we like to stress that signal enhancement by general setup optimization or efficiency evolution of equipment components (all included in $k_{\tilde{\nu}}$, i.e., the functional group F1) is not discussed separately here. However, where appropriate and relevant, signal enhancement linked to specific instrumental choices, which indeed may yield rather significant increases in Raman signal, will be exemplified in conjunction with specific enhancement techniques.

## 13.1  WAVEGUIDE-ENHANCED RAMAN SPECTROSCOPY

When it comes to increasing the number of particles participating in the Raman process, two approaches are possible, namely, (i) increasing the particle density $N_i$ while maintaining the interaction volume or (ii) keeping the particle density $N_i$ constant but enlarging the interaction volume. Note that the volume factor is included in the factor $k_{\tilde{\nu}}$; i.e., the functional groups F1 and F4 are interlinked. Thus, keeping everything constant other than the number density and the volume, one would have

$$I_S = Q_{NV} \cdot N_i \cdot V, \tag{13.2}$$

where $Q_{NV}$ is the associated proportionality factor containing all the other (kept-constant) parameters in Equation 13.1.

Since merely in the gas phase is it possible to substantially alter the particle density, only the second approach is viable, in general. But simply increasing the potential interaction volume in itself does not necessarily yield an increase in the observed Raman signal, as becomes clear from Figure 13.1.

Two aspects need to be considered when increasing the particle volume, since it impacts on other factors as well.

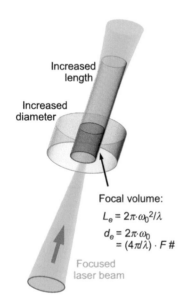

Increased length

Increased diameter

Focal volume:

$L_e = 2\pi \cdot \omega_0^2 / \lambda$

$d_e = 2\pi \cdot \omega_0$
$= (4\pi/\lambda) \cdot F\#$

Focused laser beam

**Figure 13.1**  Increase in particle interaction volume by an increase in the diameter or length of volume: requires adjustment to laser excitation focusing or Raman signal collection configuration.

In the first case, when increasing the diameter of volume (assuming for simplicity a circular cross section, as shown in the figure), nothing would be gained unless the diameter of the laser excitation beam were increased as well: the increase in volume cross section would require an increase in laser beam diameter by the same amount, which in turn would mean a decrease in laser photon flux if the laser power were not altered. Recall here that the photon flux is given by the number of photons per cross-section area per time interval, being associated with the laser intensity (in units of W·cm$^{-2}$). Thus, in this case, the factor $Q_{NV}$ would decrease by the same amount as $V$ increases, and $I_S$ would not be affected if all Raman photons could be collected. However, the scenario could be even worse, since the image area of the increased volume diameter might be larger than the acceptance aperture for the Raman photon flux, and thus, rather than an increase in signal, a reduction would be observed. This means that also the light collection geometry would need to be adapted to the altered excitation volume dimension.

The second option would be to increase the length of the interaction volume while keeping its diameter constant. This would mean that that the photon flux deficiency encountered in the first case would not apply. However, still, the collection geometry would most likely need to be altered, in order to capture all of the Raman light from the extended part of the interaction volume; this might not always be possible, and the increase in $I_S$ would not necessarily be proportional to the increase in interaction volume. A second caveat is that, in general, the laser excitation beam is not collimated, but focused. This is nearly always the case, to overcome the low Raman efficiency by increasing the photon flux within the nearly constant beam diameter of the focal depth regime; this means that outside this regime, the photon flux density decreases further away from the focal volume.

Nevertheless, a simple increase in the length of the interaction volume provides already a means to increase the observed Raman signal. As it turns out, with a "trick," the increase in length of interaction volume becomes the solution for the problem, albeit normally only for samples in the gas and liquid aggregate states. This solution presents itself in form of a waveguide for both laser excitation and Raman light, as shown schematically in Figure 13.2.

As the figure reveals, now the laser photon flux restrictions for a focused laser beam are removed since all laser light is confined to the inner diameter of the waveguide and the light intensity does not alter along the guide (provided the reflection losses at the waveguide surface can be neglected in first approximation). Equally, all Raman light is guided to the waveguide exit regardless of which direction the Raman photon was emitted. Thus, indeed, an increase in the recorded Raman light intensity would be directly proportional to the length in waveguide. For example, increasing the typical Raman interaction length of about 1 cm in a free-space gaseous sample to a waveguide length of about 1 m would mean that the observable signal would be larger by a factor of ~100. Note that for waveguide setups, the observation of the Raman light is in forward or backward direction; configurations at right angles are not suitable (for general Raman spectroscopy configurations, see Chapter 11.3).

The concept of using waveguide techniques for increasing the Raman signal by serious amounts predates the laser age, going back to nearly the beginning of Raman spectroscopy when lamps were used as excitation sources. Designs were based on capillaries which promoted total internal reflection (TIR) within a

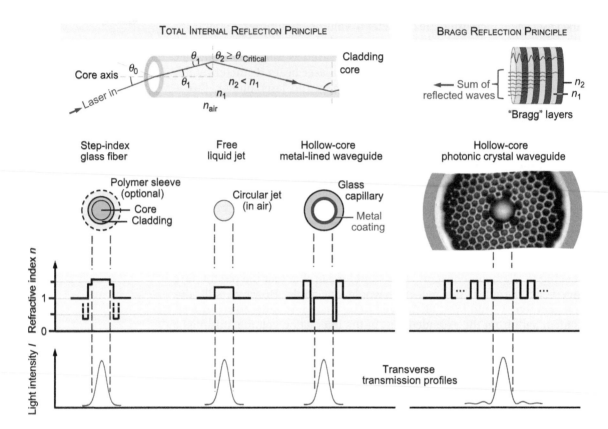

**Figure 13.2** Concept of the waveguide structure to increase the number of particles participating in the Raman process. From left to right: standard optical fiber (for comparison of principle of a total-internal reflection waveguide); free liquid jet (in air); hollow-core metal-lined (glass) waveguide; hollow-core photonic-crystal waveguide.

sample liquid; thus, they were regularly addressed as liquid-core optical fibers (LC-OFs). Besides these conventional waveguides, some modern alternatives have been developed, which exhibit wider applicability; these include predominantly waveguides based on hollow-core metal-lined capillaries and hollow-core photonic-crystal fibers (HC-PCFs). These three waveguide structures and their application are discussed in more detail in the Sections 13.1.1–13.1.3.

## 13.1.1 Raman spectroscopy using liquid-core waveguides (LC-OF)

While LC-OFs were already used in one way or another in the early days of laser Raman spectroscopy, probably the most significant technical advance in waveguide capillary sampling dates to the early 1970s. Then, LC-OF cells of up to 25 m were developed, and intensity enhancements of as much as ×3000 were demonstrated, relative to conventional Raman sampling arrangements (see, e.g., Walrafen and Stone 1972). Their cells were based on fused-silica LC-OFs in which they measured some high-refractive index liquids. However, the drawback of their particular system was that it could only be used for liquids having refractive indices higher than that of fused silica could be utilized ($n_{SiO_2} \sim 1.46$ for the most common Raman laser excitation wavelengths in the visible), to maintain

TIR necessary for light wave-guiding, i.e., $n_{liquid} - n_{SiO_2} > 0$. Recognizing this limitation, they also suggested the use of fibers having lower refractive index, such as Teflon $n_{Teflon} \sim 1.35$.

If one considers only spontaneous Raman scattering (i.e., any stimulated, nonlinear effects are neglected), one can calculate the intensifying of the Raman scattered light $I_S$ within a length segment of the fiber:

$$dI_S = G_S \cdot I_L \cdot dx - \alpha \cdot I_S \cdot dx, \tag{13.3a}$$

which—under the assumption that the conversion laser light $I_L$ into Raman light $I_S$ is small—on integration over the total length $L$ of the waveguide yields

$$I_S(L) = G_S \cdot I_L \cdot L \cdot \exp(-\alpha L). \tag{13.3b}$$

In these equations, $G_S$ represents the Raman gain factor $G_S = g_S \cdot \pi \cdot \theta_c^2$, with the Raman scattering cross section ($g_S$) and solid angle ($\pi \cdot \theta_c^2$) of light guided in the waveguide core as parameters, and $\alpha$ stands for the effective overall loss coefficient for the light travelling in the waveguide (assuming this to be equal for all wavelengths, for simplicity). It is clear from Equation 13.3b that there must be an optimum length for the waveguide after which no increase in Raman light signal is expected but rather that the signal amplitude diminishes again. This optimum length can be calculated from the derivative of the said equation, leading to

$$\frac{dI_S}{dL} \equiv 0 = -G_S \cdot I_L \cdot \alpha \cdot L \cdot \exp(-\alpha L) + G_S \cdot I_L \cdot \exp(-\alpha L), \tag{13.4a}$$

and consequently giving the optimum (effective) $L$ as

$$L = 1/\alpha. \tag{13.4b}$$

A range of excitation/collection geometries incorporating LC-OF Raman cells (based either on fused silica or Teflon hollow-core fibers) have been utilized over the years, with quite some success; a summary—with figures of merit for the individual designs—may be found, e.g., in Altkorn et al. (2001).

Over the years, a plethora of Raman spectroscopy applications using liquid-core waveguide setups have been utilized, and it is beyond the scope of this chapter to give credit to them all. Here, only a small handful of examples are outlined, to demonstrate the versatility and sensitivity of the methodology.

In their implementation using an LC-OF setup for Raman spectroscopy of chemical concentrations in clinical blood serum and urine samples—i.e., a biomedical application—Qi and Berger (2007) followed the "classical" concept of Walrafen and Stone (1972); see Figure 13.3 for the conceptual setup.

Briefly, in their setup, the bio-fluid sample is confined within an LC-OF of length $L = 30$ cm and inner diameter $d = 600$ μm, made of Teflon; the total excitation volume thus is about 85 mm$^3$, i.e., less than 0.1 mL. An 830-nm laser source is used for Raman excitation, with 160 mW of power coupled into the LC-OF. The Raman signal is recorded with a charge-coupled device (CCD) spectrograph in backward scattering geometry and is decoupled from the laser excitation radiation in the usual manner via a dichroic beam splitter. Since the authors were aiming at quantitative results to enable clinically relevant conclusions to be drawn, signals needed to be corrected for the various light loss mechanisms in the sample. In general, this is always the case if quantitative information is sought from a (nonideal) waveguide setup: the loss coefficient $\alpha$ in Equation 13.3b needs to be known or measured. The said global loss coefficient comprises contributions

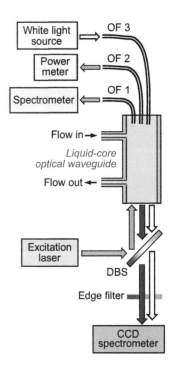

**Figure 13.3** Conceptual liquid-core optical waveguide setup for Raman spectroscopy, with the capability to simultaneously or sequentially measure the laser and Raman light attenuation coefficients of the LC-OF.

from the attenuation of the laser light, $\alpha_L$, and the Raman light, $\alpha_L(\lambda)$, as well as the effective scattering losses, $\alpha_s$. The coefficients $\alpha_L$ and $\alpha_L(\lambda)$ could be measured frequently during a measurement campaign, by measuring the laser power exiting the LC-OF (in forward direction) and the wavelength-dependent attenuation of a white-light source (in backward direction, using the Raman spectrometer for recording). This type of correction procedure has been described, e.g., in Qi and Berger (2006); see also Figure 13.3 for its experimental implementation.

The LC-OF Raman setup was used to measure a statistically relevant number of clinical urine and blood serum samples, and it was demonstrated that multiple key analytes (13 in all) could be determined at physiological concentrations. Including the aforementioned absorption-based correction for variations in LC-OF efficiency, the concentrations for most constituents could be predicted well within the accuracy of a clinical reference analyzer. For example, for glucose (blood sugar), the authors determine a concentration prediction error of about 12 mg/dL, within a measurement time of 20 s; this is better than the accuracy requirements for blood sugar measurement instruments, as recommended by the Food and Drug Administration or by European ISO norms. The authors conclude that their approach based on LC-OF Raman spectroscopy may well be suitable for high-volume (fast), automated sample analysis.

Over the last decade or so, miniaturization of analytical techniques has been high on the agenda of many researchers, trying to integrate optical spectroscopic methodologies into lab-on-a-chip devices. In particular, approaches based on opto-fluidic waveguides have been at the center-stage of these efforts. Spectroscopic opto-fluidic devices are now finding significant applications in the field of flow cytometry for cell counting, analysis, and sorting.

While integrated opto-fluidic waveguide devices have been demonstrated to be suitable for laser absorption and laser-induced fluorescence applications, Raman spectroscopy has been more elusive. This is not really surprising because of the very small dimensions of the excitation volumes encountered in on-the-chip optical waveguides. In conjunction with further signal enhancement, associated with surface-enhanced Raman scattering/spectroscopy (SERS) (see Chapter 14.1), it was demonstrated that such devices indeed possess the capability to detect minimal molecular concentrations in interaction volumes of only a few picoliters (pL), or $10^{-15}$ m$^3$ (see Measor et al. 2007).

In their experiment, the researchers interconnected a solid-core waveguide (to provide optimized coupling of the laser beam) and a liquid-core antiresonant reflecting optical waveguide (normally addressed now as ARROW) on a single chip device, forming a planar beam geometry that allowed for high laser light mode intensity along the microfluidic channel. The planar liquid-core ARROW structure was integrated in an Si-substrate. Its hollow core was 12 μm wide and 5 μm high, with a length of approximately 750 μm; this equates to an excitation volume of about $45 \times 10^{-15}$ m$^3$ (or 45 pL). The radiation from a HeNe laser ($\lambda_{HeNe} = 632.8$ nm) was coupled into the ARROW structure via a single-mode optical fiber; the laser light could be controlled for intensity and polarization. The schematic of the experimental setup is shown in the bottom part of Figure 13.4.

The device was successfully demonstrated to function well with a solution of rhodamine 6G (Rh6G), of varying concentrations in the range 0.3–30 μM; as SERS-enhancers, the solution also contained sodium chloride-induced aggregated silver nanoparticles (1 nM). This system is well characterized and is frequently used as a good proof-of-concept system for SERS detection experiments (see, e.g., Hillebrandt and Stockburger 1984). Some representative results are shown in the top part of Figure 13.4. The limit of detection (LOD) for the solution was found as $\cong$ 30 nM (corresponding to $\sim 4 \times 10^5$ molecules in the ARROW excitation volume); this result was comparable with other SERS-type measurements in LC-OFs.

A rather different approach to liquid-core waveguides is to exploit the fact that a (round) jet stream of a liquid may serve as the waveguide because light traveling inside the liquid experiences TIR at the interface with the surrounding air, provided that its angle of incidence on the boundary is less than the critical angle. Thus, in principle, with a liquid jet, one overcomes the problem that one has with capillary envelopes, namely, that their refractive index is very often larger than that of the liquid and that therefore no TIR condition is encountered. This concept has been applied relatively recently, and the potential of such a liquid-core jet Raman spectroscopy system to serve as a sensor in liquid analysis has been successfully demonstrated (Persichetti et al. 2015).

In their setup, the authors used a collinearly aligned configuration, which allowed them to achieve direct coupling between the liquid jet waveguide and (multimode) optical fibers used to deliver the laser excitation light and to collect the Raman scattered light for spectroscopic analysis. This approach avoids any necessity to contain the solution; i.e., no solid enclosure wall is required, and it minimizes any background signal coming from solid-substrate materials that commonly and adversely affect Raman signals. Obviously, the generation of the jet and its performance to maintain a smooth, continuous waveguide are rather critical, and strongly depend on a range of parameters, such as gravitational and inertial forces, surface tension, viscous drag, and liquid velocity. If the jet nozzle diameter and the liquid velocity are chosen suitably, a continuous regular-cylinder shape develops up to the breakup length. Using a nozzle with inner diameter $d \cong 0.5$ mm, the authors achieved usable jet lengths of about 25 mm, corresponding to an excitation volume of roughly 5 mm$^3$ (or 5 μL). It should be noted that in their particular configuration, separate fibers were used for laser excitation light delivery and Raman light collection; in principle, one could also envisage to use a single fiber for delivery and collection and separate wavelengths externally by means of a dichroic beam splitter, as is customary in backward Raman spectroscopy configurations (see Chapter 11.3).

In test measurements using ethanol diluted in water, an LOD = $2.2(5) \times 10^{-4}$ was achieved; this was an order of magnitude higher than a measurement on bulk material of the same excitation length, demonstrating once more the superiority of wave-guiding configurations. The potential of the system as a water quality sensor was demonstrated with a KNO$_3$ water solution, to monitor nitrate concentrations (based on evaluating the symmetric stretch Raman mode of the NO$_3$ radical; Raman shift = 1049 cm$^{-1}$). An LOD = 0.08 g/L was achieved within measurement periods of just a few seconds. This value is already rather close to the legal concentration limit (0.05 g/L) in drinking water, as per guidelines of the World Health Organization.

## 13.1.2 Hollow-core metal-lined waveguides

The total internal refraction condition exploited in LC-OFs for liquid sample Raman spectroscopy no longer applies if one wants to utilize hollow-core

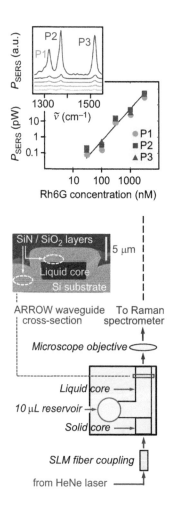

**Figure 13.4** The on-chip optofluidic ARROW concept used for SERS measurements of low-concentration solutions. (Bottom) typical experimental setup, including an ARROW waveguide cross section. (Top) spectral measurement data and a concentration plot for Rh6G/Ag-nanoparticle solutions. (Data adapted from Measor, P. et al., *Appl. Phys. Lett.*, 90 (2007): art. 211107.)

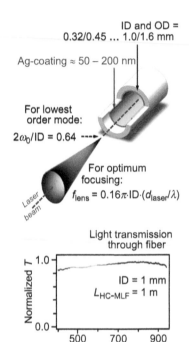

ID and OD =
0.32/0.45 ... 1.0/1.6 mm

Ag-coating ≈ 50 – 200 nm

For lowest
order mode:

$2\omega_0/ID = 0.64$

For optimum
focusing:

$f_{lens} = 0.16\pi \cdot ID \cdot (d_{laser}/\lambda)$

Laser
beam

Light transmission
through fiber

Normalized $T$

1.0

ID = 1 mm
$L_{HC\text{-}MLF}$ = 1 m

0.0

500    700    900
Wavelength (nm)

**Figure 13.5**  Transmission through an HC-MLF with internal silver coating. The optimum laser coupling conditions for focusing and laser beam waist to achieve minimum-loss transmission are indicated. A typical transmission curve for a waveguide of ID = 1 mm is shown in the lower panel. (Data adapted from Opto-Knowledge Inc, *VNIR Hollow Fiber Optics*, 2015.)

waveguides for the analysis of gaseous samples. Then, of course, the refractive index of the gas core would be much lower than that of the glass capillary wall, and no TIR would ensue. As a promising approach to overcome this deficiency, internally metal-coated hollow glass fibers spring to mind as an alternative. Hollow-core glass fiber waveguides with an internal metallic layer (HC-MLFs) were first identified in the 1960s as a useful tool to transmit and guide (high-power) laser and other light (see, e.g., Eaglesfield 1962).

The first mention of the principle for use in sensitivity-enhanced Raman spectroscopy goes back to a US patent (Mitchell 1996), although it took quite a number of years until suitable HC-MLFs could be reliably produced, although those devices were initially intended not for spectroscopy but to guide high-power pulsed laser radiation (see Osawa et al. 1995; Matsuura et al. 2002). These HC-MLFs have become available commercially; the available waveguides come in lengths of up to 2 m, with core diameters of typically 1 mm, and a silver (Ag) metal lining (optionally protected with a polymer coating); some key parameters and wavelength transmission data are shown in Figure 13.5. It is thus not surprising that the idea to use them as Raman enhancement cells was taken up in the mid- to late-2000s.

A handful of research groups demonstrated that enhancement factor of a few orders of magnitude were feasible, in comparison to standard unguided Raman configurations. This enhanced sensitivity is thought to be vital for applications in dynamic gas analysis and process control for which enhanced Raman signals are essential to achieve detection of compounds present only at trace-level amounts and to realize real-time response for feedback control of system operating parameters. A few representative examples of these uses of HC-MLFs for Raman spectroscopy are outlined here.

One of the earliest demonstrations of near-real-time Raman monitoring using HC-MLFs is that of composition analysis of (combustion) fuel gases (see Buric et al. 2009, 2012). Primarily, the authors analyzed flows of natural gas to monitor its composition and the speed with which changes to the composition could be measured if gas mixing valves were switched on or off. Their experimental setup closely resembled that shown schematically in Figure 13.1, with the difference that the HC-MLF is embedded in an external tube enclosure so that controlled gas circulation through the hollow waveguide could be achieved. Natural gas in general is mainly composed of methane ($CH_4$) and the lesser constituents ethane ($C_2H_6$) and propane ($C_3H_8$); further trace components like CO, $CO_2$, and $H_2$ may also be detected or be found in the combustion flue gases, besides—obviously— $N_2$ and $O_2$. Using a low-power excitation laser (a Nd:$YVO_4$ laser, emitting 30 mW of 532 nm radiation) and recording the Raman spectra for accumulation times of only 1 s, the authors demonstrated in their earlier experiments (Buric et al. 2009) that all the aforementioned compounds could be detected down to or below the 1% level in atmospheric pressure gas mixtures. For example, analyzing a sample from a liquid natural gas supply, they found relative concentrations of 0.9566(22), 0.0357(23), and 0.0077(27) for the main constituents methane, ethane, and propane, respectively; numbers in brackets are the uncertainties in the last digit. In their later work (Buric et al. 2012), the aim was to explore the suitability of Raman monitoring based on an HC-MLF setup for actual process control, to preadjust the burner parameters typical of gas turbine and boiler, based on the incoming fuel composition. In these experiments, an HC-MLF waveguide of length $L =$ 100 cm and inner diameter of $d = 0.3$ mm was utilized; this allowed for a full sample exchange in about 0.4 s, or less; with a Raman signal acquisition time of only 0.1 s, this meant that reference data on the fuel composition could be returned back to the controller within less than 1 s.

When analytical answers down to trace-level concentrations (well below the 1% level) are sought, but speedy response may still be an issue, a rather careful approach in system design is required when using HC-MLFs for Raman spectroscopy. In this context, James et al. (2015) provide a systematic investigation into the sources that often limit the sensitivity (in particular, detrimental for low-amplitude Raman signals is fluorescence in the glass walls of the waveguide) and implemented a few improvements pushing the detection limits for some gas-phase molecules into the few-ppm regime. In particular, they performed Raman trace gas measurements in (static) ambient air—aiming at environmental air monitoring—and incorporated their HC-MLF cell into a bespoke gas flow system, including a catalytic converter element, to monitor the mixing of gases and to follow the catalytic conversion into product molecules—all that in as close as possible to real-time (with spectrum accumulation and evaluation within a few seconds, or less). Here, we briefly outline the catalysis-related measurements.

When mixing hydrogen ($H_2$) with deuterium ($D_2$), the rate for the chemical isotope exchange transformation $H_2 + D_2 \rightleftharpoons 2\,HD$ is extremely low under normal gas phase conditions. So, unless one wishes to wait "forever" for the build-up of easily measurable amounts of the isotopologue HD, assistance by catalytic conversion is required. In the case of hydrogen, the catalyst for the surface-mediated exchange reaction often comprises $Al_2O_3$ pellets coated with 0.5% Pt. In the experiment described by the authors, the HC-MLF circulation system is filled sequentially with $D_2$ and $H_2$ at similar quantities (up to about atmospheric pressure), and the spectra show no discernible amount of HD, even after lengthy circulation. As soon as the mixture is allowed to pass through the catalyst, the Raman branches (predominantly $Q_1(J)$) for HD appeared rapidly, reaching the chemical equilibrium for the three isotopologues after about only 6.5–7.0 s, corresponding to just three complete circulations of the gas volume through the catalytic converter. Raman spectra were acquired for 0.5 s each and evaluated in parallel to the next spectrum accumulation, thus maintaining real-time monitoring. Typical snapshot spectra and the HD generation, as a function of time, are shown in Figure 13.6.

Overall, the authors showed that their HC-MLF waveguide Raman system was capable of following rapid changes in gas composition and pressure, on the time scale of less than 1 s or even lower (limited by the spectral data processing time of about 0.25 s in real-time). Thus, the principal suitability of the setup for process control or the monitoring of rapidly changing composition samples was demonstrated. Note that applying a rolling-average procedure to the data, smaller concentration variations can be identified, at the expense of a slightly slower response time (see the bottom part of Figure 13.6).

Finally, the spectral intensity data were converted into partial pressure information. Based on the reservoir/system, filling volumes and filling pressures for $H_2$ and $D_2$ used in these test measurements, yielding a total system pressure 838 ±5 mbar, a $3\sigma$-detection limit of LOD = 2.3 ± 0.4 mbar was derived for the three hydrogen isotopologues, for spectrum acquisition times of just 1 s. This corresponds to a relative concentration of $2.7 \times 10^{-3}$, well below the 1% mark one can normally hope to achieve in standard Raman spectroscopy setups without enhancement, at least at such short acquisition times. According to the authors, these values are subject to substantial further improvement, if certain measures

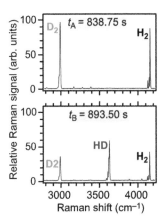

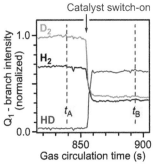

**Figure 13.6** Raman spectral data for a mixture of hydrogen isotopologues, measured in a catalytic circulation system incorporating an HC-MLF. (Top) Spectra of circulating $D_2/H_2$ mixture before (A) and after (B) switch-in of the catalyst. (Bottom) HD concentration data extracted from the spectra, with step-size 1 s, demonstrating that chemical equilibrium is reached within 6–7 s. (Data adapted from James, T.M. et al., *Anal. Methods*, 7 (2015): 2568–2576.)

are incorporated in the experimental setup to minimize or nearly eliminate the background in the Raman spectra (primarily fluorescence from the glass body of the HC-MLF); the issue of background minimization will be briefly addressed further in this section.

### 13.1.3 Hollow-core photonic-crystal fibers

In Chapter 4.5, HC-PCFs were introduced for use in laser absorption spectroscopy; as it turns out, HC-PCFs are also suitable for applications in Raman spectroscopy. In fact, similar to the hollow-core waveguides discussed earlier in this section, they provide enhancement in Raman signal strength in comparison to ordinary Raman cell setups. The operating principle and the properties of HC-PCFs were outlined in Chapter 7 and thus will not be repeated here. However, for Raman spectroscopy, slightly different considerations need to be taken into account since waves of different wavelengths propagate through the waveguide, and the propagation and attenuation properties will most likely differ substantially. While the actual HC-PCF waveguide confinement mechanisms are different from that of the LC-OFs and HC-MLFs, more or less the same formalism can be applied to the attenuation of light waves over distance travelled within the waveguide. But rather than assuming the oversimplification introduced for Equations 13.3 and 13.4, for HC-PCFs with much smaller core diameters (normally $d_{HC-PCF} < 10$ μm in contrast to $d_{LCOF} > 100$ μm and $d_{HC-MLF} > 500$ μm) in absorption coefficients $\alpha_L$ and $\alpha_R$ differ substantially, in general. According to the formalism outlined in Altkorn et al. (2001), one needs to needs to distinguish between the allowable physical length, $L_p$, of the fiber and the effective length, $L_e$, associated with the maximum recordable Raman signal; these lengths also differ whether the Raman signal is observed in forward or backward scattering configuration. One finds the following:

Forward Raman scattering (FRS):

$$L_p^{\mathrm{FRS}} = \frac{1}{\alpha_R - \alpha_L} \cdot \ln\left(\frac{\alpha_R}{\alpha_L}\right) \tag{13.5a}$$

$$L_e^{\mathrm{FRS}} = \frac{\left(\dfrac{\alpha_R}{\alpha_L}\right)^{-\alpha_L/(\alpha_R - \alpha_L)} - \left(\dfrac{\alpha_R}{\alpha_L}\right)^{-\alpha_R/(\alpha_R - \alpha_L)}}{\alpha_R - \alpha_L} \tag{13.5b}$$

Backward Raman scattering (BRS):

$$L_p^{\mathrm{BRS}} \to \infty \tag{13.6a}$$

$$L_e^{\mathrm{BRS}} = \frac{1}{\alpha_R + \alpha_L} \tag{13.6b}$$

From these equations, it is clear that the setup of BRS configurations is less critical than that for FRS, and thus, this is why most implementations reported in the literature utilize backward observation geometries. As long as the length of the hollow-core waveguide is equal to, or longer than, that given by the combined attenuation coefficients (Equation 13.6b), the maximum Raman signal is achieved—but it doesn't pay to make the length substantially longer (see Figure 13.7). Note that the attenuation coefficient $\alpha_R$ is wavelength dependent and normally increases substantially the further away the (Stokes-shifted) Raman band is from the wavelength of the excitation laser.

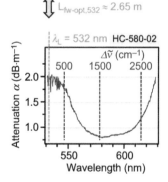

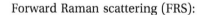

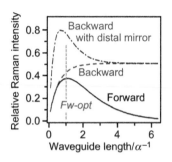

**Figure 13.7** Simulated relative Raman intensity vs. effective waveguide length $L_e$ (in units of $\alpha^{-1}$), exemplified for an HC-PCF for laser excitation at $\lambda_L = 532$ nm; the wavelength-dependent attenuation data used in the simulation are those for the HC-PCF "HC-580-02" (NKT Photonics 2015). Note: fw-opt indicates the optimal waveguide length for forward scattering configurations.

HC-PCFs suitable for laser spectroscopy have been commercially available since about the mid-2000s, and their first application in Raman spectroscopy was reported by Buric et al. (2008). They used a HC-PCF-based setup to demonstrate that Raman signals could be measured as a function of the concentrations for major species in natural gas. Detection limits of at least two orders of magnitude higher than for standard free-space configurations were achieved. Since then, HC-PCF Raman spectroscopy has found its way into a range of different analytical applications where high sensitivity coupled with low excitation volumes proved beneficial. It is worth noting that HC-PCFs may not be suitable when high flow rates are required, due to the extremely narrow waveguide channel of only a few µm. In particular, (real-time) environmental gas and human breath analysis have substantially benefited from HC-PCF Raman spectroscopy (see, e.g., Hanf et al. 2014, 2015).

Besides the main constituents of $N_2$ and $O_2$, low concentration constituents like $CO_2$ and $H_2O$ as well as trace molecules are of high relevance (like $H_2$, $CH_4$, CO, $NH_3$, and some volatile organic compounds as biomedical marker gases; or $CH_4$, $CO_2$, and $N_2O$ as climate-relevant atmospheric trace gases). In both cases, the simultaneous quantification of as many of the gaseous components as possible, in combination with chemometric analysis, is paramount to realize the highest analytical potential. The minor gases appear in the lower ppm range, as do various natural isotopes like $^{14}N^{15}N$ or $^{13}CO_2$, which can be used as artificial or natural markers of metabolic or geological processes. Therefore, high sensitivity and dynamic range are required, and small sample volumes are required; in their experiments, Hanf et al. (2014) realized all these. In their Raman spectroscopic setup, they used an HC-PCF waveguide cell of length $L = 1$ m and hollow-core diameter of $d_{HC} \cong 7$ µm; this equates to a total sample volume probed by the Raman excitation laser of less than 40 µL.

In order to counteract the extremely low number of molecules participating in the Raman process under normal atmospheric conditions, for many of their measurements, the cell-internal pressure could be raised to about 20 bar. The particular HC-PCF (HC-580-02, NKT Photonics) exhibits a central transmission wavelength of $\lambda = 580$ nm and a low-loss, guided-wave spectral width of about 3000 cm$^{-1}$. For laser excitation at $\lambda_L = 532$ nm (at the low wavelength edge of the HC-PCFs transmission range), the Raman bands of all the aforementioned molecular species are within the fiber's transmission range, and thus could be detected simultaneously with high sensitivity. With a typical excitation laser power of about $P_L \sim 2$ W and a high-resolution spectrometer with cooled-CCD detector, a dynamic range of $\sim 10^6$ was achieved, with all relevant molecular species spectrally resolved. When filling the cell with biogenic multigas mixtures or human breath samples, at pressure $p_{HC} \cong 2$ bar, lower LODs of just a few ppm in relative concentration were achieved, simultaneously for a wide range of molecular isotopologues. A list of the observed and analyzed species is provided in Table 13.1, together with the most relevant, quantitative analytical results.

HC-PCF waveguide cells have also been proposed for applications of Raman spectroscopy of liquid samples. In principle, this should not pose a conceptual problem; however, in practice, one is faced with two problems, which adversely affect reproducibility and sensitivity in (quantitative) Raman spectroscopy.

First, light coupling and wave-guidance are severely affected by the frequent formation of air gaps or discontinuities in sample distribution within the hollow fiber core (as well as in the surrounding photonic gap channel structure). Mainly,

**Table 13.1** Isotopologues Observed in Multiple Biogenic Gases and Human Breath Samples, Using HC-PCF Enhanced Raman Spectroscopy

| Observed Molecule | Relative $\sigma_{Raman}$ @ 532 nm[a] | Atmospheric Concentration[b] | Abundance (Literature)[c] | Abundance (Measured)[d] | LOD, ppm[d] |
|---|---|---|---|---|---|
| $^{14}N_2$ | 1.0 | 0.77524 | 0.9926 | – | 9 |
| $^{14}N^{15}N$ | 1.0 | 0.00576 | 0.0074 | 0.0080 (5) | |
| $^{15}N_2$ | 1.0 | $0.78 \times 10^{-6}$ | $1 \times 10^{-6}$ | Below LOD | |
| $^{16}O_2$ | 1.02 | 0.20839 | 0.9952 | – | 8 |
| $^{16}O^{18}O$ | 1.02 | 0.00084 | 0.0040 | NE | |
| $^{12}C^{16}O_2$ | 1.1 | 0.00037 | 0.9851 | – | 4 |
| $^{13}C^{16}O_2$ | 1.13 | $4.13 \times 10^{-6}$ | 0.0109 | 0.0109 (3) | |
| $^{12}C^{16}O^{18}O$ | 1.1 | $1.48 \times 10^{-6}$ | 0.0039 | NE | |
| $^{12}C^1H_4$ | 8.6 | $1.77 \times 10^{-6}$ | 0.9899 | – | 0.2 |
| $^{14}N_2{}^{16}O$ | 0.51 | $0.32 \times 10^{-6}$ | 0.9906 | – | 19 |
| $^1H_2$ | 3.9 | $0.55 \times 10^{-6}$ | 0.9997 | – | 4.7[e] |

*Note:* The table includes relative abundance measurement data and LOD data. The abundances were measured in pure nitrogen and carbon dioxide samples. NE = not evaluated. For comparison, information is provided (1) for the normalized Raman cross-sections (at $\lambda_{laser} = 532$ nm), (2) for the atmospheric concentrations of individual species, and (3) literature values for relative isotopologue abundances. Numbers in brackets refer to the uncertainty in the least-significant digit of the measured value.

[a] The relative Raman cross sections are normalized to $^{14}N_2$. (Data adapted from Fenner et al. 1973.)

[b] Concentrations of molecular gases, in dry air at standard conditions.(Data adapted from Kaye & Laby Online 2012.)

[c] Relative isotopologue abundances (Data adapted from Eiler 2007; and Coursey et al. 2013.)

[d] Experimental data, recorded for gas pressure $p_{HC\text{-}PCF} = 20$ bar and laser power $P_{532nm} = 2$ W.

[e] Experimental data, recorded for gas pressure $p_{HC\text{-}PCF} \cong 1$ bar; laser excitation at $\lambda = 607/660$ nm (Hanf et al. 2015).

this can be associated to the surface tension between the liquid and the fiber channel walls, which in themselves are not evenly smooth along the full length of the HC-PCF. As a consequence, in repeat measurements, poor-quality Raman spectra with wildly varying intensity are encountered, which thus rarely correlate directly with chemical concentration.

Second, because of the significantly higher density of any liquid, in comparison to gases, fluorescence contributions from the liquid can be significant, in particular if bio-fluidic samples are probed.

The latter problem is one that is rather difficult to overcome. In general, in Raman spectroscopy of biochemical samples (which are prone to fluorescence background in their Raman spectra), near infrared (NIR) laser excitation sources are utilized, which are as much as possible outside any molecular electronic transition, thus minimizing the generation of fluorescence. But on the down side—as noted a few times earlier (e.g., in Chapter 12)—spectroscopic instrument limitations may arise, and—with reference to Equation 13.1—the $\tilde{\nu}^4$-factor severely diminishes the actual Raman signal strength.

For the former problem of uneven filling, it has been demonstrated that carefully designed filling configurations allows for reliable and repeatable provision of liquid sample in the HC-PCF cell.

In experiments aimed at the inclusion of HC-PCFs in point-of-care analysis of bio-fluidic samples, Kethani et al. (2013) integrated an HC-PCF waveguide cell into a differential pressure system that allowed effective (repetitive) filling, draining, and refilling of samples into an HC-PCF under identical optical conditions. In the study, the authors used a common NIR laser excitation source at wavelength $\lambda_L = 785$ nm; accordingly, the HC-PCF (HC-1550-04, NKT Photonics) was chosen appropriately to cover the Raman spectral range of interest. The hollow-core diameter of this particular HC-PCF was $d_{HC} \cong 11$ μm, and the cell channel length was $L = 10$ cm. The pressure difference across the HC-PCF could be adjusted approximately in the range $p_{diff} \sim 1–4$ bar; at the lower end of the range, the cell filling time was about 20 minutes, while this decreased to only about 4 minutes at the higher end.

The system was applied to the analysis of clinically important molecules, namely, *heparin* and *adenosine* (diluted in blood serum). The former compound is a blood anticoagulant, which is commonly administered during heart surgeries; the latter is a regulative compound influencing extracellular physiological activity. From the Raman spectra, the concentration of both molecules in blood serum could be extracted, over the clinical important range, with confidence limits $\chi^2 \sim 0.995$ and root mean-square error of ~1.5%. This demonstrated the immense potential of their HC-PCF cell scheme, with differential pressure filling capabilities, for rapid monitoring of multiple chemical species in solution, using Raman spectroscopy.

### 13.1.4 Measures to reduce fluorescence contributions in backward Raman setups

It might have become clear from the discussion of the various types of waveguide, all based on hollow-core silica- or Teflon-walled fibers, that fluorescence background from the wall material to the Raman spectra constitutes a major nuisance—in many cases, merely the Raman signal amplification is noted but not the accompanying increase in background. This is due to the fact that for certain diagnostic applications, it might not be overly relevant. However, as soon as one aims at quantitative and precise measurements of trace-level concentrations, when the Raman signal amplitude barely surpasses the background level, the fluorescence signal amplitude often becomes the dominant and limiting factor affecting the signal-to-noise ratio (SNR), and therefore, the LOD $\cong 3 \cdot \sigma_{noise}$. For example, if one follows one of the widely used definitions of SNR in Raman spectroscopy, i.e., the SNR is equal to the Raman signal amplitude divided by the root mean square (RMS) fluctuations of the background signal, one is dealing with the fluorescence shot-noise $(S_{fl})^{1/2}$ of the fluorescence signal $S_{fl}$. Thus, if one were able to lower the fluorescence by a factor of 100, its shot-noise would be smaller by a factor of 10, and consequently, the LOD would improve by the same margin for unaltered Raman signal amplitude.

Over the past few years, systematic efforts have been undertaken (1) to trace the actual causes for fluorescence contributions from the hollow-core fibers and (2) to devise methods of how to minimize their generation and collection.

For HC-MLFs, James et al. (2015) carried out a systematic study of all potential contributors to laser-induced fluorescence from system components to the background of Raman spectra. In particular, they showed that the coupling of the laser radiation into the fiber was of critical importance, meaning that any exposure of the silica fiber front face should be avoided and that the laser beam should undergo as few reflections at the fiber walls as possible, meaning a nearly collimated beam (or focusing with a long-focal length lens) traveling along the (straight) waveguide core. Both measures minimize fluorescence (as well as silica Raman signals) generated in the glass wall material. This can be achieved by "pinhole" aperturing, or placing a suitable metal cap over the hollow-core waveguide, whose diameter is smaller than the hollow-core diameter, i.e., $d_a <$ $d_{HC} \approx 1$ mm. Thus, fiber face exposure is eliminated and light traveling through the fiber wall is largely reduced. Some typical results for fluorescence reduction are shown in Figure 13.8a for an HC-MLF of length $L = 60$ cm. The data show clearly that the reduction in fluorescence background improves the SNR by a factor of 8 to 9; of course, it decreases again as soon as the aperture also limits the Raman light collection and not only suppresses fluorescence light contributions, as the data show.

Similarly, albeit with very different aperture diameters (micrometer rather than millimeter), one finds for fluorescence elimination in HC-PCF setups (Hanf et al. 2015) that, using a limiting pinhole, the SNR changes with an equivalent trend; i.e., it increases with the degree of fluorescence suppression but tends to diminish again when the Raman light collection efficiency reduces. The relevant data are shown in Figure 13.8b, for an HC-PCF of length $L = 10$ cm. As in the case for the HC-MLF waveguide, the SNR increases when cutting off the fluorescence in the silica bandgap layer, but as soon as the hollow-core transmission profile is affected, the trend is reversed.

Finally, very recently, the potential advantages of all-metal hollow waveguides for Raman spectroscopy have been explored in a study comparing various types of hollow-core waveguides and implementing additional means for further reduction in fluorescence generation in optical components (see Rupp et al. 2015, 2016).

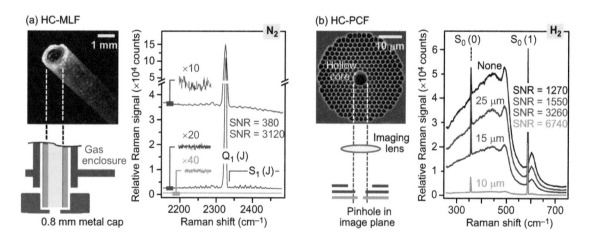

**Figure 13.8** Fluorescence background reduction in hollow-core waveguide-based Raman spectroscopy, for (a) a hollow-core metal-lined fiber (HC-MLF) with $d_{HC} = 1$ mm. (Data from James, T.M. et al., *Anal. Methods*, 7 (2015): 2568–2576.) (b) A hollow-core photonic crystal fiber (HC-PCF) with $d_{HC} = 7$ μm. (Data from Hanf, S. et al., *Anal. Chem.* 87, no. 2 (2015): 982–988.) Raman excitation for (a) was at $\lambda_L = 532$ nm and for (b) was at $\lambda_L = 660$ nm.

In those measurements, no optical elements (like lenses, windows, or beam splitters) are exposed to laser radiation, which could cause fluorescence in these components. The experimental Raman spectral data were recorded for a so-called "light-pipe" of length $L = 20$ cm and hollow-core diameter of $d_{HC} = 2$ mm. This light-pipe comprised a nickel tube internally lined with a hard gold coating, which is produced by galvano-forming and guarantees high reflectivity even for small diameters (custom-manufactured by Epner Technology Inc.). Raman spectra recorded for ambient air with this setup were compared with configurations incorporating a silver-lined HC-MLF waveguide, for consistency also of length $L = 20$ cm but with narrower hollow-core diameter of $d_{HC} = 1$ mm, either using the same parabolic mirror light delivery/collection geometry, or a standard lens/dichroic mirror optical assembly. The data impressively show that despite the much lower reflectivity of gold in comparison to silver, the light-pipe wins in terms of Raman background light by a large margin, due to the nearly zero fluorescence contribution (very close to the detector noise floor).

## 13.2 CAVITY-ENHANCED RAMAN SPECTROSCOPY

As highlighted in the introduction to this chapter, the second approach by which the Raman signal amplitude can be augmented is the laser photon flux associated with the Raman process: the larger the number of photons that participate in the scattering process, the higher the amplitude of the Raman signal if no other parameter is changed (factor $F_3$ in Equation 13.1). This means that by simply jacking up the laser power, it should be easy to enhance the observable Raman signal, in principle.

However, while ingenious in its conceptual simplicity, in practical terms, this approach often has its limitation.

For once, the overall power output from the typical, commercially available lasers used in Raman spectroscopy is not infinite, and it strongly depends on the wavelength of a particular laser type. For example, 532-nm solid-state laser sources suitable for Raman spectroscopy are available with output powers of up to $P_L \sim 100$ W; on the other hand, at other wavelengths that are becoming increasingly popular, available output powers may reach only $P_L \sim 0.1$ W. Typically, many of the Raman spectroscopy experiments reported in the literature involve lasers with output power in the range $P_L = 1–5$ W. Therefore, if it were possible—both from the laser type and free of budgetary constraints—to utilize a 100-W source instead of a 1-W device, the expected Raman signal would be higher by a factor of ~100, according to Equation 13.1.

A limiting factor in how high one could push the laser power evidently is the damage threshold of optical component materials, such as, e.g., cell windows (for a focused laser beam, the damage threshold for soft antireflection coatings may already be exceeded) or of the probed sample itself, like, e.g., biological tissue (tissue may already be "cooked" and destroyed at rather moderate laser power levels).

If an increase in laser power say—for the sake of argument—of the order of $P_L \sim 0.1$ W is not possible, two lines of action may be followed, namely, (1) changing the laser wavelength or (2) increasing the localized power by use of a (resonant) enhancement cavity.

In the former case, one exploits the fact that for constant laser output power, the Raman signal increases dramatically, if the laser emission were shifted to shorter wavelengths; this is because of the approximate $\tilde{v}_L^4$-dependence of factor $F_2$ in Equation 13.1. For example, replacing a 532-nm solid state laser source with a 405-nm diode laser source or a 266-nm source (frequency doubled from 532nm), specified for the same output power, one would gain a factor of approximately $532^4/405^4 \cong 3$ or $532^4/266^4 \cong 16$, respectively. However, while clearly workable in principle, this approach of change in laser excitation wavelength may not always be feasible or advisable.

In the latter case, one exploits the fact that in low-loss cavities, radiation can be built up for those light wavelengths that are resonant with cavity mode frequencies. This is the principle of "active" laser resonators and "passive" etalon cavities (often used for frequency content analysis of laser radiation).

The enhancement (gain) factor G of a cavity, into which laser radiation is injected, is—in first approximation—defined as the ratio of the intracavity power $P_{IC}$ to the incident power $P_{in}$, i.e.,

$$G = \frac{P_{IC}}{P_{in}} \cong \frac{T_M}{1 - 2 \cdot [(1 - T_M) \cdot (1 - L_C)]^{1/2} + [(1 - T_M) \cdot (1 - L_C)]}, \quad (13.7)$$

where $T_M$ stands for the transmissivity of the mirror through which the laser radiation is injected (at the fundamental laser wavelength) and $L_C$ combines all losses in the cavity, including absorption and scattering in optical components of and in the cavity, and the transmissivity of the second (end) mirror of the cavity. From Equation 13.7, one can derive that the gain G is maximum if the transmissivity of the input coupling mirror equals the losses, i.e., $T_M = L_C$, a condition known as impedance matching. For example, if the cavity losses were $L_C = 0.01$, being matched by a $T_M = 0.01$ of the cavity input mirror, one would find an enhancement factor of $G = 100 \equiv 1/L_C$. For this to work correctly, the incoming radiation must be mode-matched to the cavity configuration, using, e.g., appropriate optics for focusing or collimation and alignment; note that, ideally, the laser beam quality should be diffraction limited.

The first realization of cavity-enhanced power buildup for Raman spectroscopy is found in the inverse situation to the one just described, i.e., rather than coupling a laser beam into a (external) cavity, the Raman sample was placed inside the cavity of the laser itself (see, e.g., Hickman and Liang 1973). For their experiments, the authors modified the cavity configuration of a commercial continuous wave (CW) $Ar^+$-laser, operating at $\lambda_L = 488.0$ nm (its optics optimized to suppress the parasitic laser line at $\lambda_L = 488.9$ nm); they achieved an intracavity laser power of about $P_{IC,488} \sim 120$ W. Placing a Raman cell within the resonator, they were able to record rotationally resolved Raman spectra with very high SNR.

A simple passive, external cavity setup, along the concepts outlined earlier, was used by Miller et al. (1993). In their measurements, radiation of power $P_{in} \sim 250$ mW from a CW $Ar^+$-laser was coupled into a confocal enhancement cavity, which contained a sample cell with solid HD at its focal position. Because of the high losses associated with reflection from windows and other optical components, and imperfect sample transparency, only very moderate power enhancement factors of $G = 4$–$10$ were achieved. Nevertheless, the then available photon flux $I_L$ was sufficiently high to record rotationally resolved Raman features, with a dynamic range of the signal up to about $10^6$. Even very weak features

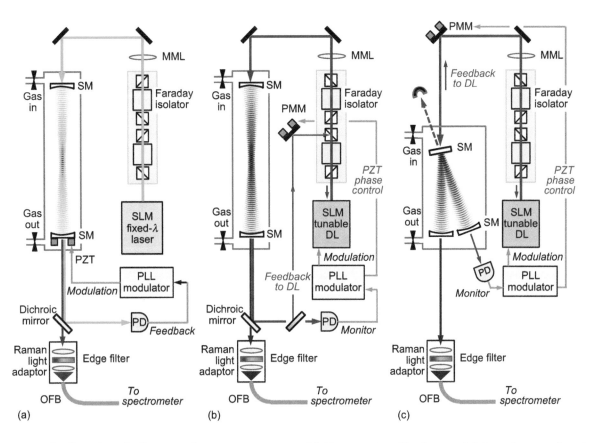

**Figure 13.9** Configurations for Raman enhancement resonators. (a) Linear cavity with piezo-modulation of one cavity mirror. (b) linear cavity with frequency modulation of the laser source and phase-coupling compensation. (c) V-shape cavity with frequency modulation of the laser source and phase-coupling compensation. DL, diode laser; MML, mode-matching lens; OFB, optical fiber bundle; PD, photo diode; PLL, phase-locked loop; PMM, phase-control mirror; PZT, piezo transducer; SLM, single-longitudinal mode; SM, mirror with high-reflection at laser wavelength, transmitting at Raman wavelengths.

associated with Raman phonon branches could be detected. While the work by Miller et al. (1993) demonstrated that external-cavity enhancement does work for applications in Raman spectroscopy, their experimental implementation has the distinct drawback that the enhancement resonance condition is easily lost if the laser wavelength drifts or the effective cavity length changes through thermal variations.

The aforementioned deficiency can be alleviated by actively keeping the cavity in resonance, e.g., via piezo-controlled mirror movements linked to an optical feedback signal. This type of cavity-resonance control is used as a standard in external-cavity frequency-doubling applications for low-power CW lasers (see Chapter 5.3). The concept for applications in Raman spectroscopy is shown in Figure 13.9a.

The concept was first demonstrated by Taylor et al. (2001) with a system intended for the sensitive measurement of hydrogen isotopologues. The laser used in their experiments was a frequency-doubled, laser-diode pumped $Nd:YVO_4$ laser, emitting up to $P_L = 5$ W at $\lambda_L = 532$ nm, on a single longitudinal mode (SLM). The (nearly confocal) enhancement cavity had mirrors with reflectivities of $R_{in,532nm} \cong 0.997$ and $R_{out,532nm} \cong 0.9998$. Assuming no other losses in the cavity than those incurred by the mirror transmissions, i.e., $L_C \sim 0$, according to Equation 13.7, one would expect a cavity enhancement factor of the order $G \cong$ 250–300; indeed, this was found by the authors when comparing their Raman signals with and without the enhancement cavity in place. The cavity was actively stabilized by modulating the cavity length, moving the output cavity mirror by a piezo transducer, which kept the cavity in frequency-mode resonance via electronic proportional–integral–derivative (PID) servo control. When measuring ambient air in this servo-controlled enhancement cavity setup, they were able to easily detect minor components like $CO_2$ and determined from the observed signal strength that the enhanced sensitivity yielded an LOD of about 4.5 ppm. As stated, the authors intended to measure hydrogen isotopologue mixtures in a gas-circulation environment, over a wide range of component ratios. For this, they had to introduce a cell with optical windows, which introduced significant losses into the cavity even though the cell windows were anti-reflection (AR) coated ($R_{AR,532nm} \cong 0.0012$). Accordingly, the gain factor dropped from $G \cong 250$ to $G \cong 50$, but this was still sufficient to easily measure the minor residual isotopologues $H_2$ and HD in an ultrapure $D_2$-supply, deducing a detection limit of the order of LOD $\sim 10$ ppm for these two "contaminants."

The same principle of enhancement based on a servo-controlled cavity can be applied even if the laser used in the setup is not operating in SLM but emits on multiple longitudinal mode (MLM) frequencies. This has been shown, e.g., by Thorstensen et al. (2014), who aimed at setting up a Raman-based multigas sensing system using low-cost components but which nevertheless provided detection limits for individual gas components of the order of 0.5% at atmospheric pressure, for measurement times of less than a minute. A simple frequency-doubled Nd:YAG laser pointer with MLM-output power $P_L = 50$ mW was used, whose mode field was matched to that of a Fabry-Perot cavity, comprising low-cost mirrors with reflectivity $R_{in,532nm} \cong 0.99$ and $R_{out,532nm} \cong 0.9996$. Again, the simple calculation according to Equation 13.7 would result in a gain factor of $G \cong 150$ for a lossless cavity, which compares favorably with the value of $G_{measured} \cong 50$. From the SNR measured for $O_2$ and $N_2$ in atmospheric air, the authors deduced that they have met their Raman sensitivity goal with such a simple enhancement setup using a low-power laser source.

A slightly different cavity-enhancement configuration was utilized by Salter et al. (2012), who modulated the mode-frequency of their laser source rather than the length of the enhancement cavity; for the conceptual setup, see Figure 13.9b. Note that the cavity setup and the laser control principles are very similar to that used in cavity-enhanced absorption spectroscopy (CEAS), as described in Chapter 8.4.

The authors used a single-mode CW laser diode, operating at around $\lambda_L \approx$ 635 nm, providing typically $P_L \leq 10$ mW of linearly polarized laser radiation. The laser output was collimated and mode-matched to the external optical cavity (with length $L = 35$ cm), consisting of two curved mirrors (with $f = 2$ m) in linear configuration. The mirror reflectivities were nominally $R_{635nm} \geq 0.9999$. The diode laser current offset and modulation (saw-tooth function of typically 1.5 kHz) depth were set so that the laser's emission mode-frequency matched the cavity resonance frequencies in the middle of each modulation period, with a duty cycle of ~50%. With this configuration, cavity gains of $G \sim 800$–1000 were achieved, meaning that for a laser power of $P_{in} \approx 3$ mW coupled into the cavity, the intracavity circulating power was of the order $P_{IC} \approx 2.5$ W. By optical feedback (including mode and phase matching as described previously) from the cavity output into the diode laser, its offset current could be adjusted to keep the diode laser in resonance with the enhancement cavity. The enhancement cavity input-side was decoupled from the laser source by a set of two Faraday isolators, but note that the actual modulation feedback from the cavity output was injected back into the laser diode via the first of these isolators. Finally, note that the authors added an additional resonance control mechanism in later measurements, namely, a piezo-modulation of the output mirror of the enhancement cavity (see Hippler et al. 2015).

With this cavity-enhanced Raman spectroscopy (CERS) setup just described, a multitude of gaseous samples was measured, including pure gases, specific gas mixtures, ambient air, and natural fuel gases. In the various mixtures, the authors were able to detect, and quantify, numerous of the trace components, including, e.g., $CO_2$, $CH_4$, ethane, and propane. Sensitivities in relation to relative abundances were in the range of LOD $\sim$ 50–1000 ppm, at a sample pressure of 1 bar and spectrum recoding times of only 30 s. A few representative results are shown in Figure 13.10.

An interesting variant to the previous CERS implementations is the use of a folded, "V"-shape resonator, a configuration that has been proposed and used by a number of research groups for applications in cavity-enhanced absorption spectroscopy (see, e.g., Courtillot et al. 2006 or Manfred et al. 2015). The concept of this CERS configuration is shown schematically in Figure 13.9c. However, this type of setup has rarely been used for CERS measurements, although it has the great advantage that no feedback-prevention is required to suppress unwanted coupling from light back-reflected from the cavity entrance mirror into the laser source. On the downside, it becomes more difficult to achieve perfect mode matching between the incoming laser beam and the intracavity modes.

In a range of explorative tests, Schlösser et al. (2015) utilized such a V-shaped resonance cavity, in conjunction with an MLM laser diode, emitting up to $P_L \leq$ 100 mW at $\lambda_L \cong 660$ nm, with 10–20 mW in individual frequency modes. The reflectivities of the bespoke cavity mirrors were optimized for high reflection at the laser wavelength (a variety of flat input mirrors with $R_{in,660nm} \geq 0.995...0.9995$, curved end mirrors with $R_{out,660nm} \geq 0.9999$) and high transmission at the Raman scattering wavelengths (with $R_{out,>750nm} \leq 0.05$). With these mirrors, theoretically, one should have been able to obtain gain factor in the range $G \cong 200$–2000; in reality, normally, only $G_{measured} \cong 60$–100 was realized; this much lower value was associated predominantly with the incomplete mode matching.

Note also that other than in the linear-cavity configuration depicted in Figure 13.9b, and discussed earlier, control-feedback to the laser diode is from the input mirror, not the output mirror, which in the experiments described here could not yet be adjusted for optimum mode coupling.

Two experimental scenarios for the Raman analysis of ambient (intracavity) air were carried out, namely, (i) without any active feedback adjustment by piezo-modulation of the phase-control mirror and (ii) with the modulation switched on under PID control. In the former case, one expects only "accidental" but adjustable frequency-mode match and, thus, related power buildup inside the cavity, similar to the situation described in Miller et al. (1993). However, as the

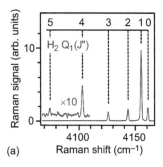

(a)

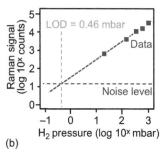

(b)

**Figure 13.10** (a) CERS spectrum of $H_2$ using a diode laser at $\lambda_L =$ 635 nm, generating an intracavity power of $P_{IC} \geq 2.5$ W; acquisition time = 50 s. (b) $H_2$-concentration calibration, with indication of the LOD. (Data adapted from Salter, R. et al., *Analyst*, 137 (2012): 4669–4676.)

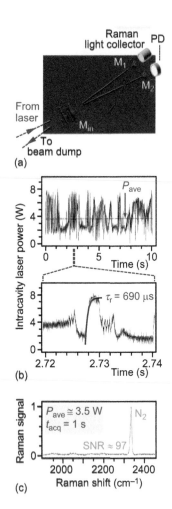

(a)

(b)

(c)

**Figure 13.11** CERS utilizing a V-shape enhancement cavity. (a) Visualization of the intracavity enhanced beam from a laser diode at $\lambda_L =$ 661.816 nm; $M_1$, $M_2$ = concave high-reflectors; $M_{in}$ = input coupler; PD = photo diode. (b) Temporal fluctuation of the intracavity laser power, without feedback stabilization of the laser diode; lower trace = expanded segment, highlighting the time scale for full power build-up. (c) Typical CERS Raman spectrum of $N_2$ in ambient air, for intracavity power of 3.5 W and accumulation time of 1 s. (Data adapted from Schlösser et al., unpublished data, Universidad Complutense de Madrid, 2014.)

data in Figure 13.11 show, even without any control applied and the MLM nature of the diode laser ($P_{MLM} \cong 100$ mW), the average enhancement-cavity gain was $G \sim 40$, sufficient to record Raman spectra of air with a decent SNR of >70 (implying a LOD $\cong 10$ mbar), for spectrum accumulation times of only 1 s. With feedback control switched on, the achievable Raman sensitivity improved significantly.

# 13.3 RESONANCE RAMAN SPECTROSCOPY

As was stated at the beginning of Section 13.2, the Raman signal increases substantially according to the $\tilde{\nu}_L^4$-dependence in Equation 13.1 when the laser excitation is shifted to shorter wavelengths. This trend seemingly ends when the wavelength becomes short enough that an electronically excited state of the molecule can be reached. Then, one begins to deal with the actual absorption of the photon to an allowed quantum state, followed by subsequent emission of (laser-induced) fluorescence. The large dipole-transition matrix elements for these two processes thus give rise to process probabilities that are orders of magnitude higher than those encountered in Raman scattering. Thus, the fluorescence signal would seemingly swamp the Raman signal to an extent that it might not be observable at all. However, "not everything is lost"; if the laser wavelength is shifted to even lower values, so that the transition is well above the $\nu' = 0$ level in the ro-vibrational manifold of the electronically excited state, then favorable "competition" between the fluorescence and Raman transitions ensues. This is shown schematically in Figure 13.12.

## 13.3.1 Basic concepts of resonance Raman scattering

As a reminder from Chapter 12, the time scale of the internal (vibrational) relaxation into $S_1(\nu' = 0)$ is of the order $\tau_{VR} = 10^{-10}$–$10^{-12}$ s for the majority of polyatomic molecules, i.e., normally just a few ps; this has to be compared to fluorescence lifetimes of a few ns or longer and the Raman process, which is considered to be instantaneous. Other facts to remember are (1) that the shortest fluorescence energy is $\tilde{\nu}_{min, F} \cong E^*(S_1(\nu' = 0)) - E(S_0(\nu'' = 0))$ and that the energy difference between laser excitation and fluorescence is $\Delta\tilde{\nu}_F \cong \tilde{\nu}_L - \tilde{\nu}_{min, F}$ and (2) that for all molecules, except for $H_2$, one finds for the Raman shift $\Delta\tilde{\nu}_{Raman} \lesssim 3500$ cm$^{-1}$. One also finds that for many (polyatomic) molecules, $E^*(S_1) \gtrsim 4$ eV $\simeq 33,000$ cm$^{-1} \,\hat{=}\, 310$ nm.

As a consequence, one finds the following for the Raman and fluorescence spectral contributions, as depicted schematically on the right-hand side of Figure 13.12. If the laser excitation wavelength is just surpassing the first electronic state energy $E^*(S_1)$, then the Raman and fluorescence spectra overlap, and most likely some (or all) Raman features may be difficult to measure above the fluorescence noise level. However, if the laser excitation wavelength is very short so that one reaches $E^{**}(S_1) \gtrsim E^*(S_1(\nu' = 0)) + \Delta\tilde{\nu}_{Raman}$, then the two spectral contributions become well separated. Thus, one finds that the Raman spectrum is nearly free of any background. Note that the background shown here is not associated with fluorescence from the Raman-active molecule itself but from the measurement environment.

The particular type of Raman spectroscopy just described, i.e., the excitation laser-photon energy is close to or larger than the electronic transition energy

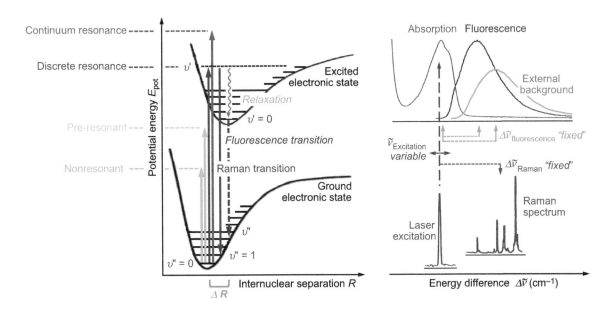

**Figure 13.12** Concept of RRS, with the distinct regimes of resonance enhancement indicated on the molecular potential scheme on the left. The photon energy relations between laser excitation, absorption and fluorescence, and Raman scattering are indicated schematically on the right.

$E^*(S_1)$–$E(S_0)$ of the molecule—as shown in the left-hand part of Figure 13.12—is known as *resonance Raman spectroscopy/scattering* (RRS).

The observed Raman scattering intensities, benefitting from the presence of an electronic state as the transition intermediate, are enhanced by a few orders of magnitude. The reason for this enhancement is that the transition dipole moments $\mu_L$ and $\mu_S$ tend to values of (allowed) dipole transitions between real energy states, with cross-sections of the order $\sigma_{RR} \cong 10^{-26}$ cm$^2 \cdot$sr$^{-1}$, rather than involving a low-probability "virtual" energy level, with cross-sections of the order $\sigma_R \cong 10^{-29}$ cm$^2 \cdot$sr$^{-1}$. In general, one distinguishes two main cases in RRS, namely, that of pre-resonance and resonance; in the former case, the laser excitation energy is just below the $S_1$-state minimum, by a few 100 cm$^{-1}$, while in the latter case, the laser excitation reaches into the vibrational manifold of $S_1$. A short summary of the properties of RR scattering, in comparison to ordinary off-RRS, is provided in Table 13.2.

It should be noted that the Raman band intensities are generally derived from the Kramers–Heisenberg–Dirac (KHD) dispersion theory; for this, two distinct approximations are commonly used. In the case of off-resonance (or non-resonant) Raman scattering, the so-called *Placzek polarizability theory* is used for its simplicity; it is based on approximating the Raman scattering tensor by a Taylor expansion of the polarizability tensor (see, e.g., Section 4.7 in Long 2002).

A more general approach for describing all types of Raman scattering intensities is to use the framework of Herzberg-Teller approximation in determining the Raman scattering tensor elements $a_{pq}$. This is based on a first-order perturbation expansion of the complete set of zero-order Born-Oppenheimer functions for the molecular state wave functions and transitions, yielding a four-term

**Table 13.2** Distinction in the Vibrational Raman Process, Associated with the Proximity of an Electronically Excited State

| | Off-resonant Raman | Pre-resonant Raman | Resonant Raman |
|---|---|---|---|
| **Laser excitation** | | | |
| Photon energy $\tilde{v}_L$ | $\tilde{v}_L \ll E^*(S_1(v' = 0))$ | $\tilde{v}_L = E^*(S_1(v' = 0)) - \Delta E$ with $\Delta E \approx 100{-}500$ cm$^{-1}$ | $\tilde{v}_L > E^*(S_1(v' = 0))$ |
| **Raman transitions** | | | |
| Approximation for the Raman intensity tensor | Placzek polarizability theory: $a_{pq} \cong \langle i|\alpha_{\rho\sigma}|f\rangle$ | Albrecht vibronic theory: $a_{pq} \sim A_{FCF} + B_{VC}$ | Albrecht vibronic theory: $a_{pq} \sim A_{FCF} + B_{VC}$ |
| $\Delta v$-selection rule | $\Delta v = 1$ (overtones weak) also $(\partial\alpha_{pq}/\partial Q_k)_0 \neq 0$ | | |
| "Polarization rule" | TSM: $0 \leq \rho = 0.75$ Non-TSM: $\rho = 0.75$ | TSM: $\Delta v \geq 1$ Non-TSM: $\Delta v = 1,2$ $0 \leq \rho = \infty$ | TSM: $\Delta v \geq 1$ Non-TSM: $\Delta v = 1,2$ $0 \leq \rho = \infty$ |
| **Raman signal** | | | |
| Relative signal strength | 1 | 5–10 | $10^2$–$10^6$ |
| Dependence on laser wavelength change $\Delta\lambda_L$ | Smooth | Smooth | Sequence of resonances |
| Signal amplitude change associated with $\lambda_L$ | $\tilde{v}_L^4$ | $\tilde{v}_L^4 \times f(\Delta E^{-1})$ | Irregular, depending on resonance with given $v'$ |

*Note:*  For comparison, the off-resonance Raman process has been included. $a_{pq}$ = Raman scattering tensor elements; $\alpha_{pq}$ = polarizability tensor elements; $Q_k$ = normal coordinates; $A_{FCF}$, $B_{VC}$ = Franck-Condon and vibronic coupling term in the KHD perturbation theory (for more details see text); $\rho$ = Raman polarization ratio; TSM = totally symmetric (vibration) mode.

representation of the Raman scattering tensor in the form (see, e.g., Section 4.8 in Long 2002):

$$a_{pq} = A_{FCF} + B_{VC} + C + D; \tag{13.8}$$

here, the A- and B-terms are indexed for mnemonic purposes only. The A-term is associated with the Franck-Condon approximation, determined by the electronic transition dipole moment function and the vibrational wave function overlap integral; in the B-term, one encounters the coupling between the resonantly excited vibronic state $|v_e^r\rangle$ and one other excited state $|v_e^s\rangle$ the C-term represents the vibronic coupling of the ground electronic state to an excited electronic state; and the D-term describes the vibronic coupling of the excited electronic state to two further excited (electronic) states.

Note that the C- and D-terms by and large are associated with higher-order Raman processes and are therefore often neglected. This leaves the A- and B-terms (or here $A_{FCF}$ and $B_{VC}$) as the relevant ones for resonant Raman spectroscopy; this description is normally associated with the often-used *Albert's vibronic theory* (see Albrecht 1961), although it is a general theory for all Raman processes. Note also (1) that for strong dipole-allowed transitions, the resonant Raman spectra are dominated by contributions from the A-term; (2) that the A-term does not contribute to nontotally symmetric vibrational modes (non-TSMs); and (3) that for weak dipole-allowed transitions in general, both the A- and B-terms contribute to the Raman scattering spectra of totally symmetric vibrational modes (TSMs).

Overall, one finds that in RRS, one not only obtains the vibrational energy level structure of the electronic ground state $S_0$, but because of the quantum-state related nature of the resonant excitation, one also gleans information about the vibrational structure of the excited electronic state $S_1$ (or any higher one if the excitation laser wavelength is short enough). In the context of resonance with an electronically excited energy state, one has to distinguish two scenarios.

If the laser excitation wavelength is tuned across the sequence of excited ro-vibronic quantum level, i.e., $h\nu_L \cong E^*(S_1(v',J'))$, the observed Raman signal will change in an irregular fashion, reflecting the presence of the various vibrational modes and their transition strengths. Also, because of the transition selection rules, resonance Raman spectra will tend to exhibit much fewer spectral features that are encountered in normal off-RRS. This property is particularly useful when probing mixtures of molecular compounds, as, e.g., encountered in the analysis of biological or medical samples: by selecting the laser excitation wavelength appropriately, one may be able to selectively probe an individual molecular species, due to the fact that its $S_1$-state is different in energy to that of others. A couple of examples for this are described in the following two sections.

If the laser excitation wavelength is short enough to access the excited state continuum, i.e., $h\nu_L > E^*(S_1(v'=0, J'=0)) + D_e(S_1)$, one accesses—in principle—dissociative and fragmenting channels for the molecule.

## 13.3.2 Applications of RRS to probing of excited electronic state quantum levels

Following up from statements made so far on RRS, from quantum-physics and quantum-chemistry viewpoints, the technique opens the door to quite a number of aspects and properties of electronically excited molecules, whether they are as small as diatomic entities or as large as complex biomolecules. As already stressed a few times, this is due to the resonant nature of the laser excitation when coinciding with a particular quantum level in the electronically excited molecule; and that the Raman scattering response can reach enhancements to make them of similar amplitude as dipole-allowed fluorescence transitions.

The application of RRS to molecular quantum-resolved probing of electronically excited molecular states began in earnest in the early 1970s, with the availability of suitable (tunable) laser sources, specifically in the deep ultraviolet (UV), which—as shown in the schematics of Figure 13.12—by and large freed Raman spectra from interfering fluorescence and augmented the observable Raman signal substantially through quantum resonances. But it should also be noted that even for (tunable) laser excitation wavelengths, which do give rise to fluorescence close to the laser line, the resonance enhancement of the Raman scattering probability is normally high enough to generate Raman signals with very good signal-to-background noise ratio. Here, we give a few representative examples for the observation and assignment of quantum levels of the electronic states probed by RRS, as well as addressing localized, steric structure of compositional groups within a macromolecule, and how it is influenced by its environment. While in the identification and state analysis of polyatomic molecules by Raman scattering, rotational information and resolution were considered only to be of secondary interest, it was increasingly recognized that the coupling of rotational and vibrational degrees of freedom could play a significant role in determining intramolecular energy relaxation and subsequent dynamical behavior in polyatomic molecules. It was realized already very early on in the development of RRS that this technique had the potential for disentangling some of the phenomena observed in the distributions of vibrational and rotational energy in unimolecular reaction rates.

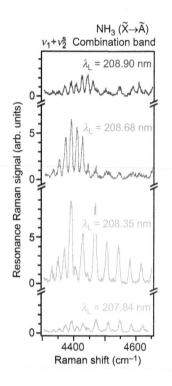

**Figure 13.13** Resonance Raman spectra (R(J)/S(J) rotational sequence of the $v_1 + v_2^S$ combination band), excited to the NH$_3$ ($\tilde{X} \to \tilde{A}$, $v_2' = 2$) resonance for different laser excitation wavelengths. (Data adapted from Ziegler, L.D., *J. Chem. Phys.*, 86 (1987): 1703–1704.)

In this context, evidence for rotational-dependent reaction mechanisms was inferred from rotational-level specific lifetimes. For example, Ziegler (1987) showed that the rates of photo-dissociation on the Ã-state surface of the tetrahedral molecule NH$_3$ exhibited significant rotational quantum dependence, specifically when the resonance Raman excitation was into its $v_2' = 2$ vibronic state (laser wavelengths $\lambda_L = 207$–$210$ nm, coinciding with the NH$_3$($\tilde{X} \to \tilde{A}$) absorption band). From the analysis of the rotational branch sequences, as a function of excitation wavelength, and the fitting of the rotational Raman excitation profiles, *J*-specific rovibronic lifetimes could be inferred, decreasing with higher rotational quantum number, namely, from $\tau_{J=2} = 140$ fs to $\tau_{J=8} = 70$ fs. This rotational-level dependence was interpreted as evidence of coupling rotational- and nuclear-coordinate degrees of freedom associated with particular shape of the photo-dissociative potential energy surface. Representative rotation-specific data from Ziegler's measurements are shown in Figure 13.13.

In a second example, vibrational structure and transitions associated with RRS will be highlighted. As was pointed repeatedly, in RRS, one encounters the competition with (resonant) fluorescence, with the Raman signal reaching similar amplitudes at or very close to a transition resonance; this has been shown to hold in numerous studies since the inception of RRS. For demonstrating the principle, here, one investigation will be discussed, which in particular addressed the differentiation between RRS and single-vibrational level fluorescence, namely, the study of benzene excited into its $^1B_{2u}$ electronic state (Harmon and Asher 1988).

In addition, said study also demonstrated that in interpreting resonant Raman intensity profiles, additional electronic states were involved, other than the one involved in the actual resonant excitation process, is important. In their experiment, the authors excited benzene in the vapor phase, utilizing a pulsed dye laser source, tunable around about $\lambda_L \sim 230$ nm, with spectral width of $\Delta \tilde{v}_L \cong 0.25$ cm$^{-1}$, pulse duration of $\tau_p = 6$ ns and pulse repetition rate of PPR = 20 Hz. The scattered light was analyzed using a high-resolution spectrograph, coupled with a gated, intensified multichannel optical detector. The noted excitation wavelength specifically probed the $6_0^1 1_0^5$ vibronic transition to the $^1B_{2u}$ excited electronic state of benzene; in particular, the involvement of the $a_{1g}$ ring breathing mode was highlighted (for a sequence of resonance Raman spectra, for different laser excitation wavelengths, see Figure 13.14).

Briefly, the spectra sequence shown in the figure commences with an excitation that is not in resonance with any structured (quantum level) absorption feature, in the particular case below the resonant energy level; see trace A for $\lambda_L = 231.642$ nm (or $\tilde{v}_L = 43,170$ cm$^{-1}$). As a consequence, only the two totally symmetric ring-breathing mode, $v_1$, and the C-H stretching mode, $v_2$, are expected to exhibit measureable intensity. The excitation at $\lambda_L = 231.321$ nm (or $\tilde{v}_L = 43,230$ cm$^{-1}$) is in resonance with $6_0^1 1_0^5$ absorption directly into a $^1B_{2u}$ vibronic level. The resulting spectrum shown in trace b is significantly different from that in trace a. First, the intensity of the $v_1$-mode itself increases substantially; second, in addition, a combination/overtone band progression in $2v_6 + n \cdot v_1$ appears, as annotated in the figure. Note that this kind of overtone and combination scattering is typical for resonances with forbidden electronic transitions. Finally, for $\lambda_L = 231.046$ nm (or $\tilde{v}_L = 43,280$ cm$^{-1}$), the excitation has passed to energies above the resonance, and hence, the spectrum has returned to the usual nonresonance appearance in the $^1B_{2u}$ region with Raman activity (see trace c).

By means of an excitation profile analysis of the $v_1$ fundamental mode, the authors were able to distinguish between RRS and the so-called hot luminescence (HL) components of the resonant light scattering process. Note that HL originates from different molecules, which have populated iso-energetic $^1B_{2u}$ molecular eigenstates, via dynamic intramolecular vibrational redistribution (IVR), prior to the emission of a photon. IVR is symptomatic for statistical effects of a high density of background states within the molecule, which constitute a bath for dephasing any state, which is coupled by the incident laser radiation field. This means that prior to IVR, basically RRS will occur; while after IVR, only HL from the populated molecular eigenstates will be observed.

### 13.3.3 Applications of RRS to obtain structural information for large molecules

It is well established that any change in electron density can be confined to a particular segment of the molecule, namely, the chromophore, which participates in the electronic transition, which in turn leads to a change in bond length in the excited state of the said chromophore. Then resonance Raman transition enhancements will also be affected by this locality of confined electron density changes. For example, resonance with a $\pi \rightarrow \pi^*$ transition will enhance stretching modes associated with the $\pi$-bonds involved with the transition; other modes by and large will remain unaffected. Consequently, this selectivity will help in identifying observed spectral bands and associating them to distinct vibrational modes of specific parts of the molecule.

This aspect of (resonance) Raman spectroscopy is particularly useful when probing large biomolecules that have one or more chromophores embedded in their structure. It means that in a molecule with a multitude of vibrational modes, RRS allows one to probe just a few vibrational modes at any one time; this reduces the complexity of the spectrum and allows for easier identification of, e.g., an unknown protein or chromophores therein. In particular, deep-UV ($\lambda_L \approx$ 200 nm) RRS can be used to determine the secondary structure of proteins by measuring the vibrational modes and their intensities of, e.g., amide-backbone modes. Thus, the ability to track both backbone and aromatic side chain modes makes RRS a very powerful technique to trace protein folding and amyloid formation.

One of the earliest studies of this kind was carried out for *hemoglobin* and *cytochrome c* in dilute solution in the early 1970s (see, e.g., Spiro and Strekas 1972). In their study, the authors used laser excitation at wavelengths within the visible absorption envelopes of the *heme-proteins*. Their Raman scattering detection system incorporated polarization-selectivity, providing the ability to record *depolarization ratios* (see Chapter 14.2 for details). Linked to changes in oxidation state of hemoglobin, they observed selective changes in the relative intensities of relevant vibrational modes. The authors' main findings were that they could assign resonance Raman bands with inverse-polarization trend to $A_{2g}$ symmetry (such transitions are inactive in ordinary Raman spectroscopy) and depolarized bands to $B_{1g}$ or $B_{2g}$ symmetry of porphyrins. This selective behavior provides some insight into the structure of heme-proteins inasmuch as their intensities correlate inversely with the degree to which the iron atoms are out of plane.

This type of RRS studies for "structural" analysis of biological molecules has gained popularity over the years. This is not surprising since the RRS signal is

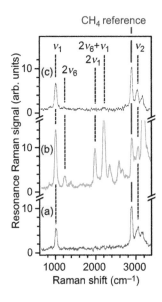

**Figure 13.14** Resonance Raman spectra for gaseous benzene. (a) Pre-resonance excitation at $\lambda_L =$ 231.642 nm. (b) On-resonance excitation at $\lambda_L =$ 231.321 nm. (c) Beyond-resonance excitation at $\lambda_L =$ 231.046 nm. Dominant vibrational modes are indicated. (Data adapted from Harmon, P.A., and Asher, S.A., *J. Chem. Phys.*, **88** (1988): 2925–2938.)

sensitive to any type of intermolecular interactions or processes that affect the electronic structure and that consequently also impact on the vibrational modes in given electronic states. For example, RRS signals from aromatic amino acids, such as tyrosine, have been used to gauge the influence of the local environment within a protein. In this context, UV–RRS was used to shed light into hydrophilic interactions, specifically investigating the H-bond strength between $H_2O$ solvent molecules and polar molecular groups on aromatic side chains; their impact is normally reflected in vibrational frequency shifts and changes in the Raman band intensities (see, e.g., Hildebrandt et al. 1988).

However, it should be noted that, often, it is experimentally impossible to directly deduce some of the subtleties in electronic structure changes from the resonance Raman spectra; modeling of the electronic structure and electronic transitions is required, which—with the aid of modern computing methodologies—has become common practice. And in the context of electronic structure, the environment in which the molecule under RRS probing is found also has to be noted, e.g., a chromophore embedded in a protein skeleton or a biomolecule in a specific solvent or serum. For example, Cabalo et al. (2014) showed in their RRS work on tyrosine (in its neutral form, solvated in pure water) and tyrosinate (in its anionic form, solvated in a high-pH environment) that the environment noticeably affected the vibrational frequency and their relative transition strengths. For example, the removal of the phenoxy-hydrogen in the high-pH environment results in a significant change in molecular symmetry, and associated with this, very noticeable red-shifts of some vibration peaks and changes in band strength were observed (see the example spectra in Figure 13.15). The authors underpinned their experimental data with extensive model calculations (based on time-dependent density functional theory) and found very satisfactory agreement when including more than the actual vibronic state in the RRS excitation. They showed for the specific case of the tyrosine amino acid that its $B_{a,b}$ electronic states significantly influenced the resonance Raman transition to the $L_b$ electronic, even though those former states are nearly 2 eV higher in energy than the one undergoing resonant excitation.

Finally, time-resolved RRS—spanning the range from picoseconds to seconds—provides a wealth of dynamic information in the selective study of chromophores, protein backbones, and side chains.

### 13.3.4 Applications of RRS to analytical problems

As has been pointed out already frequently, the intensities of RR spectra may be enhanced by up to about six orders of magnitude. Thus, components at rather low concentrations can now be detected and analyzed, which would have been below the detection limit of nonresonant Raman spectroscopy. Also, by changing the laser excitation wavelength, different RR spectra of the same molecule can be obtained. This does not mean only the various vibrational modes; even more importantly, specific parts of the molecule are probed selectively, if the excitation wavelength matches the absorption of a particular chromophore. The thus said chromophore is spectrally separated from the rest of the molecule, and in general, RR spectra exhibit relatively few lines. This feature makes the technique particularly useful for the analysis of larger, organic molecules, e.g., biomolecules. But also other chemical compound groups like hydrocarbons (including carbon-chain molecules and polycyclic aromatic hydrocarbons

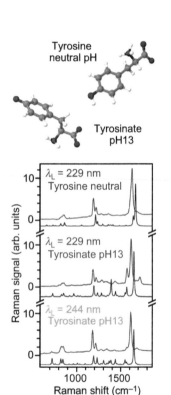

**Figure 13.15** Resonance Raman spectra of tyrosine in a neutral and pH = 13 environments, excited at $\lambda_L$ = 229 nm (top and middle traces, respectively) and excited at $\lambda_L$ = 244 nm (bottom traces). Experimental data in red, theoretical modeling in black. (Data adapted from Cabalo, J.B. et al., *J. Phys. Chem. A*, 118 (2014): 9675–9686.)

[PAHs]) have been shown to benefit from the selectivity and the enhancement properties of RRS, in particular when present at low concentrations in complex mixtures.

Out of the wealth of proven applications, only a few selected examples can be discussed here to demonstrate the analytical power of RRS.

PAHs for long have been identified as potential health hazards because quite a few of these species are known to be carcinogenic. Thus, significant efforts have been made to devise techniques for their identification and quantification in complex host matrices. UV–RRS has become a competitive methodology for these analytical tasks, specifically since Raman shift and intensity differences allow for differentiation even of structurally similar derivatives of a particular compound. The application of UV–RRS to the analytical measurement of PAHs goes back to the mid-1980s to the work of Asher (1984) and Jones et al. (1985). In their work, they demonstrated the selectivity of RRS for the speciation of naphthalene, anthracene (and various of its derivatives), phenanthrene, and pyrene; in addition, they showed that RR enhancement was sufficient to study these compounds at concentration levels as low as 20 ppb and that they could be identified and quantified in complex matrices. In these measurements pulsed, tunable dye laser sources were used providing wavelengths down to $\lambda_L \geq 217$ nm, operating at a pulse repetition rate of PRR = 20 Hz and yielding an average laser power of $P_L = 3$–12 mW. For recoding the Raman spectra, a standard grating spectrometer was used, coupled to a gated, intensified multichannel *Reticon* detector.

When tuning the Raman excitation laser to the absorption maximum of individual PAH compounds, the resulting Raman spectra showed distinctly different vibronic band structure, which made it easy to clearly distinguish between all of them and quantifying them in mixtures based on the relative intensities of a range of vibration-mode features. When diluting various PAH solutions, the RR signal followed the molar concentrations linearly over at least four to five orders of magnitude, with a detection limit of below the 1 ppm level.

Also, the RR spectra showed a strong sensibility to structural and peripheral substitution changes in a particular PAH. For example, for a series of anthracene derivatives (namely, 2-methyl, 9-methyl, 9-phenyl, and 9,10-diphenyl), the RR spectra excited at $\lambda_L = 254$ nm allowed one to easily reveal distinct differences (see Asher 1984). It would be next to impossible to observe any differences for these derivatives by using either absorption or fluorescence spectroscopy techniques, at least not without major postacquisition data treatment and modeling calculations. A few representative spectra from these measurement campaigns are shown in Figure 13.16.

The second example selected here addresses the detection of explosives. The need for methods to rapidly (and ideally remotely) detect explosives has increased enormously in recent years, by and large due to the use of improvised explosive devices (IEDs) by terrorists and insurgents. Raman spectroscopy was identified as one of the most promising techniques for rapidly detecting and analyzing explosive materials, and in particular, UV–RRS emerged from the fold because of its high sensitivity and compound selectivity. In their work, Asher and coworkers systematically investigated a range of explosives—including, among others, trinitrotoluene, used widely in civilian applications; pentaerythritol tetranitrate, used predominantly by the military; and ammonium nitrate

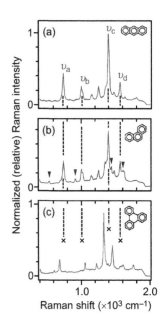

**Figure 13.16** UV–RR spectra of (a) anthracene, (b) phenanthracene, and (c) triphenylene. (Data adapted from Jones, C.M. et al., *Trends Anal. Chem.*, 4 (1985): 75–80.) Selected vibrational features for anthracene are annotated: $\upsilon_a = \gamma_{CC}$ out-of-plane; $\upsilon_b = \gamma_{CH}$ out-of-plane; $\upsilon_c = $ ring-CCC-stretch; and $\upsilon_d = $ ring-CC-stretch (see, e.g., Abdullah, H.H. et al. *Z. Naturforsch* 58a, no. 11 (2003): 645–655). These are missing for triphenylene (indicated by the crosses), and additional features appear for phenanthracene (indicated by the arrow-type symbols).

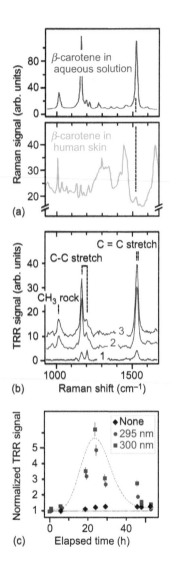

**Figure 13.17** Raman spectroscopy of β-carotene in biological tissue. (a) Reference spectrum of pure β-carotene in aqueous solution and in human skin. (Data adapted from Fluhr, J.W. et al. *J. Biomed. Opt.*, 16 (2011):article035002.)(b)Transition resonance Raman (TRR) spectra of β-carotene in grapes, illuminated by LED radiation; spectra are offset for clarity. (c) Evolution of the TRR signal of β-carotene in grapes, after illumination with UV–LED radiation.(DataadaptedfromGonzálvez, A.G. et al., *Chem. Phys. Lett.*, 559 (2013):26–29.)

($NH_4NO_3$), used in IEDs (see Ghosh et al. 2012). For proof-of-principle that UV–RRS was sufficiently selective and sensitive, solutions of these explosives were excited at selected wavelengths in the range $\lambda_L$ = 200–260 nm, either using CW laser sources (at $\lambda_L$ = 229, 244, 257 nm) or high-PRR pulsed laser source (at $\lambda_L$ = 204 nm); the average laser power on the sample was in the range $P_L$ = 0.3–2 mW. The recorded RR spectra revealed very distinct vibronic bands and vibrational modes, which made it extremely easy to identify individual explosive formulations. But not only could they identify compounds, they were also able to quantify the amounts found in their solutions, down to very low concentrations. In their preliminary study, the authors achieved detection limits of less than 1 ppm for these explosives, dissolved in acrylonitrile/water solutions and excited at $\lambda_L$ = 229 nm.

Finally, in the third example, the selectivity of RRS in detecting and quantifying individual compounds in complex matrices is addressed. The specific example is that of detecting, *in vivo*, carotenoids (here β-carotene) in biological tissue. Bio-organisms have developed a protection system against the destructive action of free radicals in the form of so-called antioxidants. The said antioxidants consist of substances such as vitamins, enzymes, carotenoids, and so on, which also might act in synergy.

For example, the human skin acts as a multifunctional and self-regulating barrier to exogenous influences and chemical compounds; this includes exposure from UV radiation (from the sun) or contact to hazardous substances, both of which can produce free radicals. The consequences of the reaction chains triggered by these radicals are damage on a cellular and molecular level, giving rise to skin irritation, collagen and elastin destruction, or skin cancer, to name but the most common damages. However, most of the antioxidants cannot be produced by the human body itself and have to be supplied by nutrition (e.g., in the form of fruit and vegetables containing a high amount of antioxidants). The presence or absence of a particular antioxidant—like the aforementioned carotenoids—can be used as marker for the functioning of the antioxidative network in general. In this context, it has been shown that carotenoids can be detected noninvasively in the human skin by exploiting RRS (see, e.g., Fluhr et al. 2011).

Because of the link between human health and disease prevention and the wide use of plant secondary metabolites, it is not surprising that ways of detecting a series of phytochemicals (including the carotenoids) in the food chain have become highly relevant. The specific interest in carotenoids is that all those that contain a so-called β-ring can be converted to retinol, which is the precursor of vitamin A. It is well known that plant abiotic stress, like UV–irradiation, triggers a plant defense mechanism enhancing the contents of some secondary metabolites, such as the carotenoids. In a recent study, Gonzálvez et al. (2013) showed in experiments on grapes that a low dose of UV-B irradiation significantly enhanced their carotenoid content; the complete experiment—UV irradiation and the quantitative analysis of the carotenoid content (specifically β-carotene)—was carried out in real-time and *in vivo*.

In the experiments, the authors used grapes (of the muscatel variety), divided into three subgroups, which were (1) untreated or irradiated for about 30 minutes by narrow-band UV–LEDs of low power ($P_{LED} \cong 20$ μW), with peak emission at (2) 295 nm and (3) 300 nm, respectively. The effect of the irradiation on the concentration of β-carotene was monitored continuously over the duration of 2½ days subsequent to the UV-irradiation, transmission RRS. Here, transmission means that the Raman signal is collected in forward direction, through the whole of the grape sample. The Raman excitation wavelength was $\lambda_L = 513.5$ nm (from a CW Ar$^+$ laser); this laser wavelength satisfied the resonance-enhancement conditions for the $\beta$-carotene (in the tail of its absorption band), while at the same time contributions from fluorescence were minimal. As usual, in modern Raman spectroscopy, the spectrum was recorded using a standard spectrometer coupled to a high-sensitivity, cooled CCD array detector. Some of the key results from the experiments are shown in Figure 13.17.

The observations are the following. First, the RR spectra from the grape samples (traces c) more or less only comprise the vibrational bands of $\beta$-carotene, in their appearance, rather like Raman spectra from pure $\beta$-carotene (trace b), with no interference from other, more abundant compounds in the bio-matrix. This can be associated to the resonance of the laser excitation with a vibronic state of $\beta$-carotene. In contrast, when the excitation is off-resonance, stark interference from Raman signals of the matrix nearly masks the $\beta$-carotene bands, as shown in trace a (recorded for a human skin sample; see Fluhr et al. 2011). Second, with some time delay, metabolic production of $\beta$-carotene was observed, increasing and decreasing over the duration of a couple of days. These results clearly show that RRS is quite capable of following the rather low, time-varying concentrations of an antioxidant molecule, present within a complex bio-matrix, which in itself might have varied as well over time in its composition. In this context, it is noteworthy that quite recently RRS (in backward-scattering geometry) of $\beta$-carotene was exploited for quantitative monitoring of postharvest time evolution and ripening of tomatoes (see Martin et al. 2017).

## 13.4 BREAKTHROUGHS AND THE CUTTING EDGE

Probably the highest impact on the early efforts in Raman measurements is associated with RRS. While some of the inherent universality in general Raman spectroscopy was lost (including the use of only one single excitation wavelength), the gain in sensitivity and selectivity far outweighed the need for adaptable laser excitation wavelengths and often custom-made, expensive detection filters.

### 13.4.1 Breakthrough: First RRS of heme-proteins

In the context of RRS, the experiments carried out by Spiro and Strekas (1972), and additional references therein, may be seen as the key stepping stones toward the success and popularity of the technique, specifically applied to the identification, characterization, and monitoring of biomolecules (e.g., proteins, among many others). In their work, the authors investigated a range of important heme-proteins, in the cited publication specifically oxy-hemoglobin (HbO$_2$) and ferro-cytochrome c (*cyt-c*), using laser excitation wavelengths of $\lambda_{L,HbO2} = 568$ nm and $\lambda_{L,cyt-c} = 514$ nm, respectively. The scattered Raman intensities were recorded 90°-configuration for the components $I$ and $I_\perp$, parallel and perpendicular to the incident laser polarization; example spectra for *cyt-c* are shown in Figure 13.18. From the band amplitudes, the authors were able to extract depolarization ratios $\rho_i = I_\perp/I$, which in turn served as indicators for the symmetry of individual molecular vibrations (for further details on the principles of depolarization ratios, see Section 12.2). The most prominent bands are assigned in the figure (following

**Figure 13.18** Resonance Raman spectra of cytochrome c, excited at $\lambda_L = 514$ nm into the vibronic Q-state. The Raman scattered light is analyzed for the components $I_\parallel$ and $I_\perp$, parallel and perpendicular to the incident laser polarization, respectively. The polarization behavior of selected vibrational bands is annotated by the symbols at the top of the figure (for further details, see text). (Data adapted from Spiro, T.G., and T.C. Strekas, *Proc. Natl. Acad. Sci. USA*, 69 (1972): 2622–2626.)

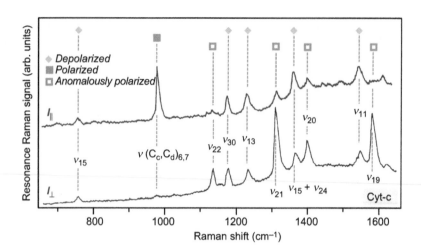

the nomenclature given in Hu et al. 1993). Clearly, one can identify fully "depolarized" features (i.e., $\rho_i = 3/4$), as well as "polarized" (i.e., $\rho_i < 3/4$) and "anomalously polarized" (i.e., $\rho_i > 3/4$) bands. The band assignments, together with the dependence of Raman intensity on the laser exciting wavelength $\lambda_L$, led to the conclusion the experimenters were dealing with vibrations that were vibronically active through the electronic absorption in the heme-proteins.

### 13.4.2 At the cutting edge: Low-concentration gas sensors based on HC-PCFs

While extremely attractive due to its multispecies capabilities, Raman scattering measurements of gaseous samples have always struggled to reach trace-level sensitivities (i.e., to detect and quantify multispecies molecules at ppm concentrations), due to the inherently low particle densities. Waveguide enhancement techniques now show a way out of this dilemma, specifically those that involve HC-PCFs (see Section 13.1).

Reliable and sensitive measurement devices of tabletop or handheld dimensions, lending themselves for fast (near real-time) *in situ* monitoring of multimolecular atmospheres, are crucial in a wide range of applications, including, e.g., human breath analysis or the monitoring of environments requiring a well-controlled gas composition.

Very recently, sensing devices for human breath analysis have been successfully tested, which are based on Raman signal enhancement in HC-PCFs, serving simultaneously as the enhancement waveguide and as the flowing-gas cell (see Chow et al. 2014 or Hanf et al. 2015). Both showed that gas mixtures consisting of $H_2$, $CH_4$, $N_2$, $O_2$, and $CO_2$ could easily be quantified within a few seconds, reaching sensitivities of the order 10–20 ppm, with a dynamic range of four to five orders of magnitude, averaging over just a few exhalations. The authors indicate that their HC-PCF-based Raman sensors may pave the way for fast, noninvasive point-of-care diagnosis of exhaled breath, thus allowing for rapid metabolic disease diagnosis, and that their systems offer great potential for miniaturization, but still being of affordable cost.

Another, very promising prototype development of an HC-PCF-based Raman sensor system is that reported by Jochum et al. (2016), who targeted the precise monitoring of controlled atmospheric environments. They specifically addressed the monitoring of critical process gases encountered in fruit conservation rooms for the ripening of harvested fruit. For example, ripening is delayed by precise management of the interior $O_2$ and $CO_2$ concentration levels, while $C_2H_4$—a natural plant "hormone"—is commonly used to trigger fruit ripening shortly before being shipped to markets. In laboratory measurements, carried out with typical fruit conservation gas mixtures, the sensor showed to be capable of quantifying $O_2$ and $CO_2$ concentrations with an accuracy of 3% or less (with respect to reference concentrations), exhibiting sensitivities to well below the 100 ppm level and a dynamic range of about four orders of magnitude. In addition to $O_2$ and $CO_2$ ethylene could be quantified simultaneously in said multicomponent mixtures. Some representative Raman spectra are shown in Figure 13.19. The authors conclude that their HC-PCF enhanced Raman sensor has the potential to become universally usable as an on-site, real-time gas sensor for controlled atmosphere applications, such as the one demonstrated in post-harvest fruit management.

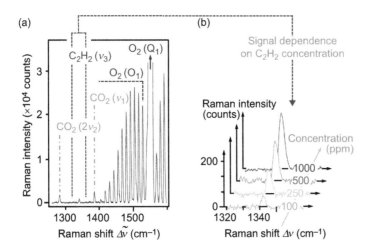

**Figure 13.19** (a) Background-corrected Raman spectrum of a simulated fruit-chamber atmosphere (containing 20% $O_2$, 500 ppm $CO_2$, and 250 ppm $C_2H_4$, balance $N_2$), measured within an HC-PCF of length $L = 30$ cm. The relevant vibrational bands of carbon dioxide and ethylene are well separated from the dominant ro-vibrational bands of $O_2$. (b) Enlarged spectral segment for $C_2H_4(v_3)$, exhibiting signal linearity with increasing concentration of ethylene. (Data adapted from Jochum, T. et al., *Analyst*, 141 (2016): 2023–2029.)

# CHAPTER 14

# Nonlinear Raman Spectroscopy

$\mathbf{A}$s emphasized repeatedly in the previous chapters on absorption, fluores-cence, and Raman spectroscopies, the interaction with *one* incident (laser) radiation field, which is weak, can be treated as a *linear response* between light and matter. If matter interacts with *two or more* independent incident fields, or if the linear-response theory is inadequate to describe the material behavior (as, e.g., in the case of very intense incident laser radiation), then one often addresses these as *nonlinear response* cases of spectroscopy.

In the classical light–matter interaction picture, the total *dipole moment* $\boldsymbol{p}$ of the molecule in an oscillating electric (laser light) field $\boldsymbol{E}(t) = \boldsymbol{E}_\mathrm{A} \cdot cos(2\pi v \cdot t)$ with amplitude $\boldsymbol{E}_\mathrm{A}$ is given as

$$\boldsymbol{p}(t) = \boldsymbol{\mu}_0 + \overset{=}{\alpha} \cdot \boldsymbol{E}(t) = \boldsymbol{\mu}_0 + \boldsymbol{\mu}(t), \tag{14.1}$$

where $\boldsymbol{\mu}_0$ and $\boldsymbol{\mu}$ are the *permanent* and *field-induced* dipole moments, respec-tively, and $\overset{=}{\alpha}$ is the *polarizability* tensor.

By adding up all $N$ (molecular) electric dipoles per unit volume, and assuming weak (in comparison to the electron binding) electric fields, one obtains a macroscopic polarization of the sample as

$$\boldsymbol{P}(t) \propto N \cdot \boldsymbol{\mu}(t) \equiv \varepsilon_0 \cdot \chi \cdot \boldsymbol{E}(t), \tag{14.2}$$

where $\varepsilon_0$ is the electric permittivity in vacuum and $\chi$ is the susceptibility of the sample material. Note that Equations 14.1 and 14.2 constitute the microscopic and macroscopic description of molecular polarization, respectively.

When the intensity of the incident laser light is sufficiently large, the induced oscillation of the dipole moment becomes *nonlinear* and has to be expressed as an expansion with terms of higher than linear order in $\boldsymbol{E}$, i.e.,

$$\boldsymbol{p} = \boldsymbol{\mu}_0 + \overset{=}{\alpha} \cdot \boldsymbol{E} + \overset{=}{\beta} \cdot \boldsymbol{E} \cdot \boldsymbol{E} + \overset{=}{\gamma} \cdot \boldsymbol{E} \cdot \boldsymbol{E} \cdot \boldsymbol{E} + ..., \tag{14.3}$$

where $\overset{=}{\beta}$ and $\overset{=}{\gamma}$ are the first-order and second-order *hyperpolarizability tensors*, respectively. Note that the higher-order $\boldsymbol{E}$ terms may result from the same or from two (or more) different sources.

Analogous to Equation 14.2, the nonlinear macroscopic polarization of the sample can once again be written as a power series in the field strength $E$, yielding

$$P_{NL}(t) = \varepsilon_0 \cdot \{\chi^{(1)} \cdot E(t) + \chi^{(2)} \cdot E(t) \cdot E(t) + \chi^{(3)} \cdot E(t) \cdot E(t) \cdot E(t) + ...\} \tag{14.4}$$

where $\chi^{(n)}$ is the $n$th order susceptibility of the sample material. For a more detailed description of nonlinear Raman phenomena, in both classical and quantum formulations, see, e.g., Potma and Mukamel (2013).

Note that in this chapter, the description will be restricted to a description of effects associated with the $\chi^{(2)}$- and $\chi^{(3)}$-term. The most prominent of these are

- The *incoherent* (spontaneous) processes of hyper-Raman scattering/ spectroscopy (HRS) and surface-enhanced Raman scattering/spectroscopy (SERS) (note that the nonlinearity is associated with laser-induced localized surface plasmons);

- The *coherent* (stimulated) processes of stimulated (gain and loss) Raman scattering/spectroscopy (SRS, SRG, SRL); and

- Coherent anti-Stokes Raman scattering/spectroscopy (CARS) (to a much lesser extent, coherent Stokes Raman spectroscopy [CSRS] is used because of interference with spontaneous Raman and fluorescence background).

These processes are schematically sketched in Figure 14.1, showing the interaction of the laser light waves with the molecular energy level structure. All processes will be briefly described in the following sections, including representative examples.

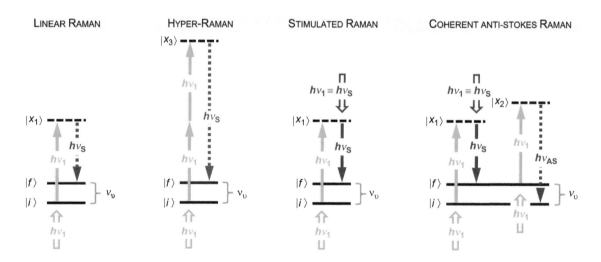

**Figure 14.1**   Schematic level schemes for ordinary linear RS and for the nonlinear processes of hyper-Raman stimulated Raman and coherent anti-Stokes Raman scattering. The incident laser fields are at $v_1$ and $v_2$; the Stokes and anti-Stokes frequencies are $v_S$ and $v_{AS}$, respectively; and $v_v$ is the difference in frequencies between molecular levels $|i\rangle$ and $|f\rangle$; the levels $|x_n\rangle$ represent the intermediate (virtual) levels in the Raman process. Spontaneous and stimulated transitions are indicated by the dotted and solid line arrows, respectively; the external laser fields are represented by double-line arrows.

# 14.1 BASIC CONCEPTS AND CLASSIFICATION OF NONLINEAR RAMAN RESPONSES

Before discussing the aforementioned Raman effects, it is probably useful to define and briefly discuss a number of properties encountered in nonlinear Raman scattering (RS) and the related spectra and to put these processes into context with each other and with ordinary, linear RS. Whether an observed (Raman) signal is linear or not is defined through its dependence on the intensity $I_{sig}$ of the incident radiation $I_{in}$; if $I_{sig}$ exhibits a quadratic or higher-order dependence on $I_{in}$, it is classified as nonlinear.

For further classification of Raman signals, one has to consider at least (1) whether the signal is *incoherent* or *coherent*; (2) whether the Raman process is *spontaneous* or *stimulated*; and (3) whether the signal field is at a frequency that is identical to any of the incident field frequencies (*heterodyne* detection) or not (*homodyne* detection). The association of these classification criteria with the various nonlinear Raman techniques is summarized in Table 14.1.

## 14.1.1 Incoherent vs. coherent signal character

The signal wave observed in RS can be classified as *incoherent* or *coherent*, depending on whether the optical waves radiated from molecular dipole emitters exhibit a well-defined phase relationship or not. When averaging over all dipole emitters, in the incoherent case, the wave phases are random and the total field averaged to zero, i.e., $\langle |E| \rangle = 0$; note that the signal intensity defined by $\langle |E \cdot E^*| \rangle$ can be finite. In the coherent case, the field average is nonvanishing, i.e., $\langle |E| \rangle \neq 0$.

Thus, in incoherent RS, the observed signal intensity $I_{sig\text{-}incoh}$ constitutes an incoherent superposition of waves scattered by individual molecules, i.e., the phases and directions are uncorrelated; one finds

$$I_{sig\text{-}incoh} \propto \sum_i |\mu_i|^2, \text{ with } E_{sig} \text{ varying as a function of } \sin\theta, \qquad (14.5a)$$

where $\theta$ is the relative angle between the incoming laser radiation and the spontaneous (random-direction) Raman signal wave. This is true for not only conventional RS but also HRS and SERS.

Table 14.1    Classification of Nonlinear Raman Techniques

| | Hyper-Raman HRS | Coherent Homodyne CARS | Coherent Heterodyne CARS | Pump–Probe SRS |
|---|---|---|---|---|
| Raman process | Spontaneous | Stimulated | Stimulated | Stimulated |
| Detection mode | Spontaneous | Spontaneous | Stimulated | Stimulated |
| Scaling with $N$ | $N$ | $N^2$ | $N$ | $N$ |
| Scaling with $I$ | $I$ | $I^3$ | $I^2$ | $I^2$ |

$N$ = Number density of scattering molecules. $I$ = Intensity of the incident laser radiation.

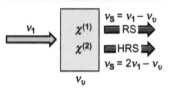

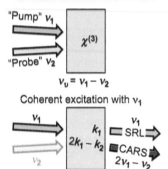

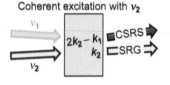

**Figure 14.2** Nonlinear Raman processes. (Top) Incoherent HRS (associated with $\chi^{(2)}$), compared to common linear RS (associated with $\chi^{(1)}$). (Bottom) Coherent RS processes, related to the resonant interaction of $v_v \equiv v_1 - v_2$ with an incoming waves $v_1$ or $v_2$, in a medium with nonzero $\chi^{(3)}$. CSRS/CARS = coherent Stokes/anti-Stokes RS (note that for these two processes, phase-matching conditions apply; for details, see Section 14.6).

If one or more of the input fields act coherently on the dipoles of the sample molecules, a macroscopic (oscillating) polarization ensues, which may be seen as a signal source radiating into a well-defined direction. Then the observed signal intensity $I_{\text{sig-coh}}$ is

$$I_{\text{sig-coh}} \propto \left| \sum_i \boldsymbol{\mu}_i \right|^2, \quad \text{with } \boldsymbol{E}_{\text{sig}} \parallel \boldsymbol{k}_{\text{sig}}, \quad (14.5b)$$

where $\boldsymbol{k}_{\text{sig}}$ is the wave vector resulting from the addition of the wave vectors of the incoming waves. This is the case for SRS and CARS. As a consequence of Equation 14.5 in incoherent RS, the signal is conveniently observed *perpendicular* to the incident laser beam, to avoid adverse contributions of the excitation light to the signal (although forward and backward configurations are also often utilized); in coherent RS, the signal is observed *in same direction* as incident laser beam(s), or small angles to it. Finally, it should be noted that one might view the coherent RS processes as an interaction of an incident wave with the (coherently) generated vibrational excitation $v_v \equiv v_1 - v_2$, where $v_1$ and $v_2$ are the frequencies of the excitation and probe laser fields, respectively, in a medium with nonzero susceptibility $\chi^{(3)}$. The association of the observed coherent Raman process with respect to the interaction of $v_v$ with either of the incident waves is shown schematically in Figure 14.2.

## 14.1.2 Spontaneous vs. stimulated scattering processes

The stimulated character of a Raman techniques can either be associated with (1) the Raman (resonance) process itself or (2) the mode of Raman signal detection.

### *Stimulated Raman resonance*

The Raman resonance is said to be stimulated if the molecule is driven into resonance by two incident laser fields. Thus, the initial phase of the Raman oscillation $v_v$ is determined by the relative phase difference of the two driving laser fields. In all nonlinear Raman techniques, the Raman resonance is driven in a stimulated fashion. In contrast, the Raman resonance is created in a spontaneous way if the molecule interacts with a single laser input field; the phase of the Raman oscillation is random at equilibrium. Spontaneous Raman resonances are relevant to all incoherent Raman techniques, including linear RS, HRS, and basic SERS.

### *Stimulated detection mode*

The nature of the detection mode depends on how the observed Raman wave was actually generated, i.e., whether, initially, the Raman field was "empty" or "occupied," or, from another point of view, whether the detected Raman light field is at an optical frequency different from the frequencies carried by the input laser fields or not. The former is the case not only for all incoherent Raman techniques but also for homodyne-detected coherent RS. Here, it should be kept in mind that a spontaneous signal is not necessarily incoherent; as just stated, homodyne CARS is spontaneous in the detection mode, but the detected signal is nevertheless coherent. The latter is the case where the signal frequency is equal to the frequency of one of the input laser fields, which holds for SRS and heterodyne CARS; then the signal field and, thus, the detection mode are stimulated.

### 14.1.3 Homodyne vs. heterodyne detection

The difference between homodyne and heterodyne detection may be defined in terms of classical field theory. The signal is classified as *homodyne* if the Raman light is detected at an optical frequency different from the incident laser light; the Raman signal intensity is proportional to $|E_{sig}|^2$. The signal is classified as *heterodyne* if the Raman light is detected at the optical frequency that is identical to any of the frequencies contained in the incident laser light fields $E_{in}$, then the Raman signal intensity is proportional to $|E_{sig} + E_{in}|^2$. This means that the detected Raman intensity contains a "mixing" term $(E_{sig}^* \cdot E_{in} + E_{sig} \cdot E_{in}^*)$, which is defined as the heterodyne contribution to the Raman signal.

For example, with these definitions, it is clear that CARS has to be classified as a *coherent homodyne* technique: it is coherent because the waves emitted by the molecules in the sample volume exhibit a distinct phase relation, while the Raman light detection is homodyne since the anti-Stokes frequency is different from those of the input laser radiation fields (see also Table 14.1).

Overall, the nonlinear Raman processes—in particular the stimulated techniques SRS and CARS—offer certain advantages with respect to common linear RS; these include, among others,

- Higher signal accumulation efficiency because of the strong directionality of the Raman radiation in the stimulated techniques; and

- Reduction in adverse fluorescence background for HRS and CARS, for which the Raman radiation is at a lower frequency than the excitation laser frequencies, and for stimulated Raman, when heterodyning in the excitation field (stimulated Raman loss spectroscopy).

## 14.2 NONLINEAR INTERACTION WITH SURFACES: SERS

The amplification of the Raman signal—with respect to that of free, isolated molecules—in SERS is by and large associated with the electromagnetic (EM) interaction of laser light with a nanostructured metal (surface); this interaction produces large amplifications of the laser field through excitation of metal-intrinsic electron resonances, generally known as *plasmon* resonances. The underlying cornerstone effects and procedures may be summarized from the composite name SERS as following: *S (surface)*—SERS is a surface-spectroscopy technique, namely, that the molecules under investigation must be adsorbed on or close to the metal surface; *E (enhanced)*—the signal enhancement is coupled to the aforementioned plasmon-resonances in the metal substrate (note that "plasmon resonances" constitute a family of effects associated with the interaction of EM radiation with metals); *R (Raman)*—as in ordinary RS, this technique provides Raman signals of the probed molecules and thus provides an insight into their chemical (vibrational) structure; *S (Scattering or Spectroscopy)*—is used depending on whether the emphasis lies one the optical effect (scattering) itself or on the technique and its applications (spectroscopy). The conceptual implementation of this principle is shown in Figure 14.3.

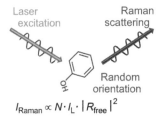

(a) GAS-PHASE RAMAN

$$I_{Raman} \propto N \cdot I_L \cdot |R_{free}|^2$$

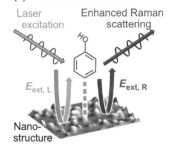

(b) PHYSISORPTION SERS

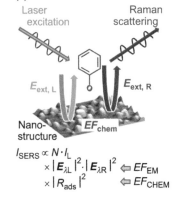

(c) CHEMISORPTION SERS

$$I_{SERS} \propto N \cdot I_L$$
$$\times |E_{\lambda L}|^2 \cdot |E_{\lambda R}|^2 \Leftarrow EF_{EM}$$
$$\times |R_{ads}|^2 \quad\quad \Leftarrow EF_{CHEM}$$

**Figure 14.3** Conceptual illustration of SERS, in comparison to ordinary RS. $E_{\lambda,L}$ and $E_{\lambda,R}$ are the enhancement fields mediated by a nanostructure substrate; $R_{free}$ and $R_{ads}$ are the transition dipole functions for free and adsorbed molecules, respectively.

A SERS signal was observed for the first time by Fleischmann et al. (1974) from pyridine molecules adsorbed on electrochemically roughened silver electrode, although term SERS was coined only a few years later. Early theories explaining said SERS effect were based on rather simple models and were introduced by Jeanmaire and Van Duyne (1977), who proposed an EM effect, and by Albrecht and Creighton (1977), who based their treatment on a charge-transfer effect; both worked with the same or similar dye molecule—silver electrode systems. Note that, although from today's point of view somewhat oversimplified, these two approaches are still accepted as valuable, first-approach theories. Since those early days, SERS has matured beyond recognition and is the most utilized Raman enhancement technique, manifested in nearly 10,000 publications since its inception just over 40 years ago and in the publication of several dedicated textbooks (see, e.g., Aroca 2007; Le Ru and Etchegoin 2009) and monographs (see, e.g., Kneipp et al. 2006; Schlucker 2010; Osaki et al. 2014; Kumar 2015; Procházka 2016). Because of this in-depth coverage of SERS, here, only its key aspects will be summarized, highlighting its possible pros and cons, and the evident application breadth will be highlighted by a handful of selected examples.

### 14.2.1 Trying to understand SERS spectra

While the observation of SERS signals of molecules attached to metallic nanostructures conceptually seems easy, the interpretation of "quickly recorded" SERS spectra has been—and still is—an often frustrating task. This is not really surprising since any observed SERS spectrum is associated with a multivariate function of factors, and in most cases, these elude proper or full control. Thus, for a complete understanding of the SERS process and the interpretation of particular spectra, it is indispensable to examine and closely analyze the set of variables that may affect the observed SERS spectra.

In this context, it is rather straightforward to interpret Raman vibrational spectra of molecules in the gas phase, since the basic components involved in the RS process are just the molecule itself and the incident laser excitation radiation. In the analytical interpretation of spectra, it is thus normally sufficient to consider (1) the (stationary) vibrational energy levels of the molecule; (2) the monochromaticity, polarization, and intensity of the incident laser radiation; and (3) the dynamics of the interaction between the molecule and the incoming radiation field, i.e., the strength of the interaction and the Raman transition selection rules (determined by and large by the molecular point symmetry group).

In contrast, the basic components involved in SERS are the molecule and the (EM) laser radiation—as before—but in addition, now a metal nanostructured substrate needs to be included. Evidently, this introduces a much greater complexity of a SERS measurement relative to a simple Raman measurement in the gas phase. Thus, the often-encountered controversy in the interpretation of SERS spectra and the underlying physics is entirely due to the complexity brought about by the multitude of factors contributing to SERS, now including at least a superposition of the interaction light ↔ molecule, light ↔ substrate, and light ↔ molecule–substrate complex. A brief summary of these processes and other aspects of SERS are given in Table 14.2. For further details, the reader is referred

**Table 14.2** Summary of Processes and Effects Encountered in SERS, and Their Comparison to Interaction-Free (Gas Phase) RS

| Process | Relevance to SERS | Relevance to "Ordinary" RS |
|---|---|---|
| Interaction of molecules with nanostructured metal | Physisorption or chemisorption to metal structure<br>Absorption and scattering of light by the molecule depend on the shape and size of the metal nanostructure<br>Shifts in vibrational modes are possible. | Not applicable in the gas phase<br>For molecules on/in (nonmetal) substrates/matrices shifts in vibrational modes are common |
| Interaction of incident photons with molecules | Generation of Raman signal<br>Photo-dissociation, photon-mediated reactions, or photo-desorption processes are possible. | Generation of Raman signal |
| Dynamics of interaction with molecules and vibrational selection rules | "Surface selection rules" for a "fixed," spatially oriented molecule at the surface of the enhancing nanostructure apply in addition to "molecule-only" selection rules.<br>Symmetry properties of the dipole transition and the modification of the intensities due to the components of the local electric field vector at the surface have to be considered. | Selection rules in the gas phase determined by symmetry point group<br>For molecules on/in substrates, in general, symmetry properties are influenced and bond interactions may be altered. |
| Interaction of incident photons with (nanostructured) substrate | Substrate excitations include electron–hole pairs, surface plasmons and/or surface phonons.<br>In nanostructures, strong local electric fields can be generated. | Not applicable in the gas phase<br>For molecules on/in substrates, in general, fluorescence and Raman features from the substrate are observed. |
| Excitations in the molecule-nanostructure complex | Charge-transfer transitions from states above the Fermi level to LUMO level of the molecule may become possible, leading to excitation in resonance with the electronic transition of the adsorbed molecule–metal complex.<br>Associated signals may be due to surface-enhanced RRS (SERRS). | Not applicable |

*Note:* LUMO = lowest unoccupied molecular orbital.

to the textbooks by Aroca (2007) and Le Ru and Etchegoin (2009); also see the review on SERS and various derivatives by Ding et al. (2016).

It should be noted here that, when a molecule is "embedded" in a liquid or solid environment matrix, i.e., it is not any longer in the interaction-free gas phase, interactions with "neighbors" has to be taken into account. However, in most cases, the interaction affects the molecular vibrations rather marginally, and the "only" effects noticed in the spectra are some broadening and shift of the vibrational transitions and the presence of spectral contributions from the substrate matrix, in the form of additional Raman features and fluorescence.

Despite all the difficulties addressed previously, accumulated experience and extensive, systematic SERS studies as well as sophisticated theoretical model calculations, which incorporate all aspects of SERS accurately, are beginning to bear fruit. Now, the interpretation of observed phenomena has become much more quantitative in nature, and predictions for novel applications are more than wild guesses. Having said all this, on the positive side of arguments stands the fact that—even with the multiple-variable nature of SERS—relatively simple models do exist that allow one to tackle how to design specific experiments or how to interpret observed SERS spectra. By and large, those models were developed around EM enhancement mechanisms; an in-depth discussion of many aspects related to SERS modeling can be found in Le Ru and Etchegoin (2009). Probably the easiest-to-assess and widely used model case is that of a single spherical metallic nanoparticle—it is applicable to colloid- or nanoparticle-based SERS; its key features are briefly outlined in the following section.

### 14.2.2 Single spherical nanoparticle model for SERS

As discussed in Chapter 11, RS is directly proportional to the square of the induced dipole moment $\mu_{ind} = \alpha \cdot E$, where $\alpha$ is the Raman polarizability and $E$ is the EM field of the incident laser radiation. In the case that the laser radiation excites a localized surface-plasmon resonance (LSPR) of a nanostructured or nanoparticle precious-metal surface (gold or silver), the local EM field amplitude $E$ is enhanced. Consequently, the induced dipole moment encountered for the Raman process of a molecule near the said nanoparticle is also enhanced. LSPR occurs when the collective oscillation of valence electrons in the precious-metal nanoparticle is in resonance with the frequency of incident laser light. In general, the exact modeling of the effects is complex; relevant details may be found, e.g., in the aforementioned texts by Aroca (2007) and Le Ru and Etchegoin (2009).

For simplicity, here, we describe the framework of SERS EM by a simplified quasi-static approach, based on a spherical nanoparticle of radius $a$, which is irradiated by $z$-polarized light of wavelength $\lambda$. In the model, the so-called long-wavelength limit is assumed, meaning that the dimension of the nanoparticle is substantially smaller than the laser light wavelength; in this limit, one has that $a/\lambda < 0.1$. In this approximation, one can assume that the electric field around the nanosphere is uniform. A concise summary of this model approach can be found, e.g., in Stiles et al. (2008); its concept is shown in Figure 14.4.

In brief, the analytical solution of the Laplace equation of electrostatics for the plasmon-enhanced EM field outside the nanoparticle, $E_{ext}$, is given by

$$E_{ext}(x,y,z) = E_0 \cdot \hat{z} - \alpha E_0 \cdot \left[ r^{-3} \cdot \hat{z} - \left(3z \cdot r^{-5}\right) \cdot (x\hat{x} + y\hat{y} + z\hat{z}) \right], \qquad (14.6)$$

where $x$, $y$, $z$ and $\hat{x}$, $\hat{y}$, $\hat{z}$ are the usual Cartesian coordinates and unit vectors and $r$ is the radial distance. The parameter $\alpha = g \cdot a^3$ is the polarizability of the metal, where $a$ is the radius of the nanosphere and

$$g = (\varepsilon_{int} - \varepsilon_{ext})/(\varepsilon_{int} + 2\varepsilon_{ext}) \qquad (14.7)$$

incorporates the internal and external (complex) dielectric constants of the metal nanoparticle and the environment, respectively. Note that Equation 14.6 is wavelength dependent because the real part of the nanoparticle's dielectric constant is a function of wavelength, $\lambda$; hence, also $E_{ext}$ exhibits wavelength dependence.

Two observations can be made when inspecting Equations 14.6 and 14.7 further. First, the polarizability of the metal nanosphere reaches a maximum value when the denominator of g approaches zero, i.e., $\varepsilon_{int} \approx -2\varepsilon_{ext}$, and thus also the external field $E_{ext}$ is maximized for the said dielectric resonance condition. It should be noted that nanoparticles (or structures) other than spheres can be used, which requires a generalized polarizability parameter, $\alpha$. This generalization is primarily associated with replacing the factor 2 in Equation 14.7 with the nanoparticle shape factor $\chi$, which describes the deviation from spherical geometry into higher aspect ratio structures; values as large as $\chi \cong 20$ can be encountered. The generalized resonance condition $\varepsilon_{int} \approx -\chi \cdot \varepsilon_{ext}$ is met by gold and silver metal nanoparticles of large aspect ratio, $\chi$, for visible-light wavelengths. The second observation one can make, from Equation 14.6, is that the

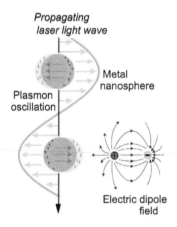

*Propagating laser light wave*

Metal nanosphere

Plasmon oscillation

Electric dipole field

Extinction efficiency for metal nanosphere $\dfrac{|E_{ext}|^2}{|E_0|^2}$

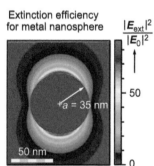

$a = 35$ nm

50 nm

50

0

**Figure 14.4** Illustration of the localized surface plasmon resonance (LSPR) effect. (Top) Induction of surface plasmons in a metal nanosphere by a laser light wave. (Bottom) Extinction ratio for a spherical silver nanosphere with radius $a = 35$ nm, for a wavelength near its plasmon extinction maximum. (Data adapted from Stiles, P.L. et al., *Annu. Rev. Anal. Chem.* 1 (2008): 601–626.)

external field $E_{\text{ext}}$ decays with $r^{-3}$ away from the nanosphere; this means that the SERS sensing volume around the nanoparticle is finite and that, indeed, SERS is a surface effect.

## 14.2.3 $E^4$-enhancement in the Raman response

Recall that the intensity of Raman scattered light is linear with the incident light intensity, $I \equiv |E_0|^2$. At and near the surface of the nanoparticle ($r \cong a$), the electric field is amplified according to Equation 14.6. Consequently, also the Raman intensity is amplified and is now related to $|E_{\text{ext}}|^2$. Manipulating Equation 14.6, one finds for a small metal sphere

$$|E_{\text{ext}}|^2 = E_0^2 \cdot \left[ |1 - \text{g}|^2 + 3 \cdot \cos^2\theta \cdot \left( 2\text{Re}(\text{g}) + |\text{g}|^2 \right) \right], \qquad (14.8)$$

where $\theta$ is the angle between the incident field vector and the vector to the position of the molecule at/near the surface. According to the equation, the maximum enhancement is obtained for $\theta = 0°$ or $\theta = 180°$, meaning that the scattering molecule is on-axis with the light propagation. In the case that g is large, the maximum enhancement approaches $|E_{\text{ext}}|^2 \cong 4 \cdot E_0^2 \cdot |\text{g}|^2$. This treatment for the enhancement of the incident field in the interaction with the nanosphere is simple and straightforward. Arriving at the enhancement for the observed Raman intensity is more complex; this was first carried out rigorously by Kerker and coworkers (see, e.g., Kerker 1984). In a simplified approach, often, an expression similar to that in Equation 14.8 is used, which is evaluated at the Raman-shifted wavelength and not at the incident laser light wavelength. In this approximation, one obtains the so-called SERS EM enhancement factor $\text{EF}_{\text{EM}}$ as

$$\text{EF}_{\text{EM}} = \left( |E_{\text{ext}}|^2 \cdot \left| E_{\text{ext}}' \right|^2 \right) / |E_0|^4 = 4 \cdot |\text{g}|^2 \cdot |\text{g}'|^2. \qquad (14.9)$$

Here, the primed entities refer to the fields evaluated at the RS wavelength. Note that for small Raman shifts, both g and g′ are at approximately the same wavelength, and hence, the enhancement factor scales as $\text{EF}_{\text{EM}} \approx \text{g}^4$; this is commonly referred to as $E^4$-enhancement. For example, assuming a value of $\text{g} \cong 10$ for a small nanosphere, the SERS EM enhancement factor is of the order $\text{EF}_{\text{EM}} \cong 10^4$. Note that in high-order silver nanostructures, the value for $|\text{g}|$ can be significantly larger and, consequently, also the enhancement factor $\text{EF}_{\text{EM}}$.

While the theoretical determination of the enhancement factor $\text{EF}_{\text{EM}}$ based on the earlier approximations is straightforward, it is often preferred, or simpler, to experimentally measure it by using the expression

$$\text{EF}_{\text{EM}} = [I_{\text{SERS}}/N_{\text{surf}}] / [I_{\text{RS}}/N_{\text{vol}}]. \qquad (14.10)$$

This expression, evaluated at the laser excitation wavelength, describes the average Raman enhancement and encompasses the enhancement of both the incident laser excitation and the Stokes-shifted Raman fields. Here, $I_{\text{SERS}}$ corresponds to the surface-enhanced Raman intensity; $N_{\text{surf}}$ is the number of molecules at or in close proximity to the enhancing metallic nanosubstrate; $I_{\text{RS}}$ is the Raman intensity without the presence of the nanoparticles; and $N_{\text{vol}}$ is the number of molecules in the laser excitation volume. Note that $I_{\text{SERS}}$ and $I_{\text{RS}}$ must be measured independently.

When analyzing the expected enhancement associated with the increased electric field, it becomes clear from the previous equations that two parameters will play a significant role in optimizing the SERS process, namely, the dependence of the exciting laser radiation and the distance of the molecule experiencing Raman shift from the enhancing nanostructure.

### 14.2.4 Wavelength dependence of the $E^4$-enhancement

It was pointed out in the previous section, in conjunction with Equation 14.10, that the dielectric constants encountered in the problem are wavelength dependent and hence the field strength $E_{ext}$. As a consequence of the subsequent (simplified) $E^4$-enhancement approximation, one would expect the optimum wavelength for the plasmon resonance, i.e., maximum enhancement of the incident field intensity at the nanoparticle surface, to be that of the laser excitation. However, for the SERS process to be successful, one requires that enhancement is present both in the incident field and in the Raman-scattering field, and those are naturally at different wavelengths due to the Stokes shift associated with the vibrational energy spacing. For the majority of molecules, this means substantial wavelength differences associated with the typical Raman Stokes shifts in the range $\Delta \tilde{\nu} \sim 500\text{–}2500 \text{ cm}^{-1}$. Therefore, the small Stokes-shift approximation becomes invalid, and so does the simplified $E^4$-approximation.

From all this, it is clear that any optimization procedure to match laser excitation wavelength, individual Raman shift, and nanostructure properties (specifically its LSPR) is a complex and delicate one. In principle, one would need a wavelength-tunable laser excitation source, and associated with this is a tunable Raman light detection channel as well. While this is feasible in principle, in practice, common off-the-shelf Raman spectroscopic equipment (laser sources as well as detection optics) do not allow for this. In fact, with advanced nanostructure fabrication techniques, it becomes more appropriate to "tune" the maximum of the LSPR extinction spectrum, to match the other experimental conditions. A good compromise for the enhancement of both fields seems to be the situation when the wavelengths of the incident and Raman scattered fields "straddle" the LSPR extinction spectrum; i.e., they are below and above the LSPR maximum (see Figure 14.5), as shown in relevant, systematic studies—for example, those by McFarland et al. (2005).

Consequently, to achieve maximum EM enhancement, one should ideally prepare the sample such that the LSPR extinction of the nanostructure is at the appropriate location for fixed-wavelength laser Raman systems.

### 14.2.5 Distance dependence of the $E^4$-enhancement

SERS EM predicts that a Raman-probed molecule does not have to be in direct contact with the metal nanostructure surface, but it suffices if it is within a certain sensing volume; this is evident in conjunction with Equation 14.6. According to the said equation, the field enhancement around a metal nanosphere decays proportionally to $r^{-3}$; consequently—using the common $E^4$-approximation—the overall field enhancement should scale with $r^{-12}$. If one then considers, for simplicity, that the molecules one wishes to influence reside on a sphere a certain distance away from the nanoparticle, the surface area of this sphere increases

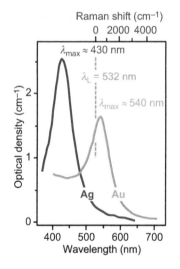

**Figure 14.5** Optical extinction spectra for silver (Ag) and gold (Au) nanospheres with radius $a = 30$ nm, in aqueous buffer solution. For the laser excitation wavelength $\lambda_L = 532$ nm, the relevant Raman shift range is indicated at the top of the figure. Data extracted from manufacturer data sheets.

proportionally to $r^2$. Overall, one therefore should observe a SERS signal enhancement, which scales with $r^{-12} \times r^2 = r^{-10}$ and is given by

$$I_{\text{SERS}} = [1 + (a/d)]^{-10}, \qquad (14.11)$$

where $I_{\text{SERS}}$ is the measured intensity of the Raman scattered light; $a$ is the (average) size of the nanosphere (or features on a nanostructured surface); and $d$ is the distance from the surface to the (adsorbed) molecule(s). That this relation holds rather well has been shown experimentally. For example, in a range of systematic experiments carried out in the van Duyne group, Ag-film over nanosphere targets (AgFON) were coated with $Al_2O_3$ layers of controlled thickness, thus providing well-scaled shells on which to adsorb the molecules of interest (see Dieringer et al. 2006). An example for SERS measurements of pyridine on such coated nanospheres is shown in Figure 14.6, clearly revealing the rapid decrease in the Raman signal amplitude. In the context of a diminishing SERS intensity with distance from the metal nanosphere, often, a parameter dubbed $d_{10}$ is used; this is defined as the surface-to-molecule distance for which the SERS intensity has decreased by a factor of 10. For the particular data shown in the figure, fitting to Expression 14.11 yielded an average size of the enhancing nanoparticle of $a \cong 12.0$ nm, and for the metal-surface to molecule distance, a value of $d_{10} \cong 2.8$ nm. This finding is very much in line with expectations, as highlighted in the conceptual Figure 14.4.

## 14.2.6 Chemical enhancement in the Raman response

In addition to the properties (by and large material properties) and dimensions (geometry and orientation) of the nanoparticle, and its interaction with the excitation light field (predominantly wavelength and polarization), one has to consider properties of the (adsorbed) probed molecule itself. The latter include concentration (or surface coverage), adsorption efficiency, adsorption orientation (whether ordered or random), and the possible modification of the intrinsic Raman polarizability induced by any adsorption "bond" being formed. The very last aspect is associated with the so-called chemical enhancement factor, $EF_{\text{Chem}}$, which is multiplicative to $EF_{\text{EM}}$. In a certain way, one may view the chemical enhancement effect as a modification to the electronic polarizability of the probed molecule, which may be able to induce a sort-of resonant RS at wavelengths at which the nonadsorbed molecule would not be in resonance; but this explanation is rather controversial. Probably, the most widely used interpretation is that of a charge-transfer mechanism, which requires the molecule actually to be chemically adsorbed on the surface. However, its contribution to the overall enhancement is believed to be much smaller than that related to the EM effect and is thus often ignored, but the reader may consult, e.g., Aroca (2007) and Le Ru and Etchegoin (2009) for additional details.

As just mentioned, the two enhancement factors are multiplicative, and although one finds in general that $EF_{\text{Chem}} \ll EF_{\text{EM}}$, the combined effects still account for huge enhancements over standard RS, and—depending on to what extent full optimization could be achieved—one typically finds $EF_{\text{Chem}} \times EF_{\text{EM}} \approx 10^3 \ldots 10^8$. In the best-case scenario, which includes chemical enhancement aided by resonance

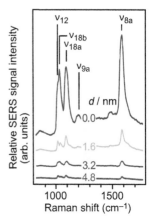

**Figure 14.6** SERS spectra of pyridine adsorbed to AgFON samples coated with alumina layers of with various thicknesses (annotated to the data traces). Excitation laser source: $\lambda_L = 532$ nm, $P_L = 1$ mW. (Data adapted from Dieringer, J.A. et al., *Faraday Discuss.* 132 (2016): 9–26.)

Raman conditions, it is now possible to carry out RS on single molecules (see, e.g., Wang and Irudayaraj 2013).

### 14.2.7 SERS substrates

It probably has become clear from the previous discussion that the actual SERS substrate is of utmost importance, namely, that it has to exhibit sort of a nanostructure and that particular shapes may be advantageous. Therefore, a brief conceptual summary of common SERS substrates is provided here. One may distinguish three main classes, although boundaries between them are somewhat fluid (see Figure 14.7).

Historically, *metallic electrodes* with roughened surface, to provide a type of "nanostructure," played a very important role; however, their importance has decreased substantially, mostly because of normally low enhancement factors and poor reproducibility.

Solutions of *metallic colloids*—by and large made of silver (Ag) or gold (Au)—had and still have preeminence in SERS. Not the least, this is associated with the fact that some of the important applications of SERS are associated with tracing particular molecules in an aqueous environment; by their nature, Ag- and Au-colloids are easily supported in liquids, including water.

In addition to being supported in liquids, metallic colloids can be dried and attached to suitable flat, secondary substrates, thus generating *"planar" metallic structures*.

Note that for the latter two types of SERS substrates, irregular colloids are increasingly substituted by controlled-manufactured nanoparticles or periodic structures generated through nanofabrication techniques (for a recent review, see Wang and Kong 2015). In particular, the aforementioned "tailoring" of the

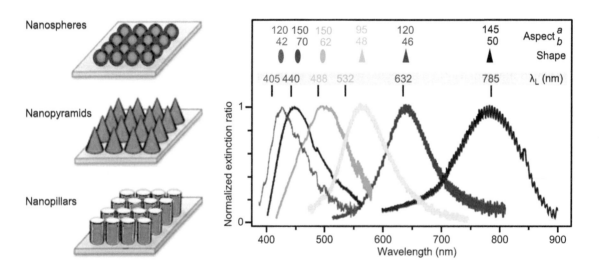

**Figure 14.7**   Matching the resonance EM-enhancement wavelengths of periodic, nanostructured SERS substrates to common Raman laser excitation sources by optimizing the size, shape, and aspect ratio of the nanostructures (the most common—spheres, pyramids, and pillars—are shown on the left). (Data adapted from Dieringer, J.A. et al., *Faraday Discuss.* 132 (2016): 9–26.)

enhancement factors, i.e., the matching of the wavelengths for laser excitation, RS, and plasmon resonance maxima, has become feasible. By adjusting the size, form, and aspect ratios of the nanostructures, the surface-plasmon resonances can be optimally adapted to particular Raman measurement equipment and specific molecules; see, e.g., the right-hand part of Figure 14.7 for the LSPR-matching to common Raman laser sources, using particular shapes, sizes, and aspect ratios.

## 14.3 VARIANTS OF SERS—TOWARD ULTRALOW CONCENTRATION AND ULTRAHIGH SPATIAL RESOLUTION RS

As shown in the earlier sections of this chapter, enhancing the normally weak Raman signal by various means—with the largest signal amplification found for SERS—one is able to detect ever smaller quantities of a sample molecule. This, however, goes hand in hand with the fact that concentrations within a given volume become smaller, too. Ultimately, the concentrations will become so low that only a single molecule may be present in the laser-probed volume. That indeed it is possible to record Raman spectra of single molecules utilizing SERS was demonstrated by Kneipp et al. (1997). Now, it has even become possible to resolve substructures, associated with particular Raman bands, in individual molecules (see, e.g., Zhang et al. 2013). Some of the aspects related to ultralow concentration samples, single molecule detection, and intramolecule resolution bases on SERS are discussed in the following.

### 14.3.1 Preconcentration of ultralow concentration samples—SLIPSERS

As a consequence of ever lower concentrations, it becomes increasingly problematic to "find" the actual analyte molecule, i.e., to achieve an overlap between the laser-probed volume and a sufficiently large number of molecules to result in a meaningful Raman signal. In analytical chemistry, this problem is often alleviated by preconcentrating the sample, for example, by evaporation of any solvent from an originally dilute solution of the analyte molecule. While in many cases this procedure is rather successful, it carries the disadvantages (1) that any intermediate sample handling may cause contamination with unwanted molecular species and (2) that the overall process might be rather time-consuming. On the other hand, preconcentration might actually be necessary. For example, even a tiny μL droplet administered to a substrate often expands over an area up to a few $mm^2$, while in standard Raman instrumentation for SERS, laser beam areas are typically of the order of a few $\mu m^2$. Thus, for very low analyte concentrations of femto-moles, one may not even encounter a molecule in the probed volume.

In this context, a very elegant solution to overcome the aforementioned problem was recently proposed and successfully demonstrated by Yang et al. (2016). The researchers developed a SERS methodology that they dubbed "slippery liquid-infused porous SERS," abbreviated as SLIPSERS. In short, the technique is based on a slippery, omni-phobic substrate that enables one to completely concentrate the analyte and associated SERS substrates (e.g., gold nanoparticles) within an evaporating liquid droplet. A SLIP-surface consists of a film of lubricating fluid locked in place by a microporous/nanoporous substrate, which creates a smooth and stable interface that nearly eliminates "pinning" of the liquid droplet contact line; i.e., its contact line at the solid–liquid–air interface experiences small resistance to movement on the substrate. This means that, as the droplet evaporates, the analyte and the SERS nanoparticles are enriched and delivered into a specific SERS detection region. The so-enriched, small sample volume can now easily be probed by the Raman excitation laser. This is shown schematically in the left-hand part of Figure 14.8.

The authors also demonstrated that SLIPSERS is capable of detecting molecules that were initially in the gas phase (schematics included in Figure 14.8). Inherently, SERS normally struggles in applications for gaseous samples. When exposing a suitable liquid solvent (with high solubility for the analyte), which contains (gold) nanoparticles as well, for a few minutes to the analyte gas, then sufficient molecules become attached to the nanoparticles to be detectable by a SERS measurement.

To demonstrate the potential of SLIPSERS as a practical, general-purpose sensing methodology, the researchers investigated a range of real-life samples suspended in a liquid, dispersed in gaseous media, adsorbed on solid substrates (e.g., soil).

As examples for analytes in liquid media, the authors investigated biomolecules like DNA-bases and proteins, as well as organic contaminants (e.g., the plasticizer di[2-ethylhexyl]phthalate [DEHP], which commonly diffuses into food and

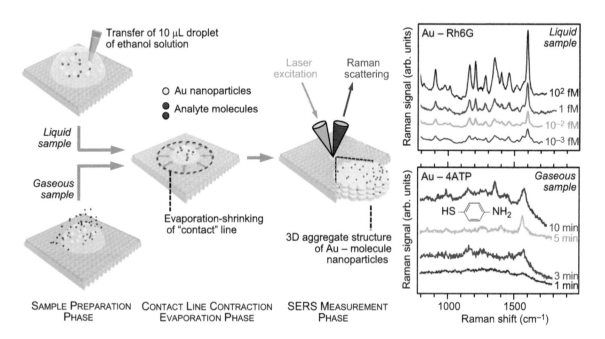

**Figure 14.8**   (Left) Schematic illustration of the concept of SLIPSERS; analyte molecules in solution or in the gas phase can be enriched by droplet-evaporation, to form small nanoparticle aggregates facilitating subsequent SERS detection. (Right) Examples of SLIPSERS spectra for different ultralow concentrations of Rh6G in ethanol solution (top) and of 4-ATP captured from the gas-phase on Au nanoparticles in an ethanol droplet, for different exposure times (bottom). (Data adapted from Yang, S. et al., *Proc. Natl. Acad. Sci. USA* 113 (2016): 268–273.)

water). In all cases, they could detect these compounds to well below the 1 fM concentration levels. In order to show that their SLIPSERS can also return quantitative results, systematic measurements were conducted for solutions of rhodamine 6G (Rh6G) in the range of concentrations of 10 orders of magnitude ($7.5 \times 10^{-7}$ M to $7.5 \times 10^{-17}$ M); selected data traces are shown in Figure 14.8. Quantitative linearity was found over about three orders of magnitude, from about 75 pM down to about 75 fM (the related plot is not shown here). For statistical purposes, areas of about $20 \times 20$ $\mu m^2$ (probe laser beam diameter $d_L \sim$ 1 $\mu m$) are averaged over at least 20 recorded spectra. Note that the previously quoted upper concentration limit saturation occurred (all SERS hotspots were occupied by molecules) and that for concentrations below fM, SERS signals could be detected only at random sites over the scan area.

For the demonstration that airborne molecules can be tackled using SLIPSERS, the researchers captured the aromatic compound 4-aminothiophenol (4-ATP) from air into 60 $\mu L$ droplets of gold nanoparticles in ethanol, for different lengths of time. The experimental data in Figure 14.8 reveal that over a few minutes, a sufficient amount of 4-ATP has accumulated to allow for quantitative measurements, when comparing the spectra with those of SERS spectra of control samples with known concentrations. In the particular case shown here, the number density of 4-ATP molecules on the Au nanoparticle aggregate was estimated to be >240 $\mu m^{-2}$.

## 14.3.2 Single-molecule SERS

As just stated, for very low concentrations of analyte molecule (on metallic nanoparticles), one may encounter just one of these aggregates with the Raman laser probe volume. Then, a SERS signal associated with analyte molecule vibrations would be observable if the product out of nanoparticle SERS enhancement factor $EF_{EM}$ and the number of participating, adsorbed analyte molecules $N$ were sufficiently large to lift the SERS signal intensity above the noise level, i.e.,

$$I_{SERS} \propto EF_{EM} \cdot N > I_{noise}, \tag{14.12}$$

for a sufficiently long signal integration time.

This spatial-sparsity feature has spurred the prospect to use precious-metal nanoparticles for spatially resolved SERS measurements of analytes within individual (living) cells. For example, Kneipp et al. (2002) deposited gold nanoparticles inside cells and succeeded to carry out SERS spectroscopy, (1) achieving sensitive and structurally selective detection of native chemicals inside the cell (e.g., DNA and phenylalanine) and (2) monitoring the intracellular distribution of said chemicals. Since this early work, numerous studies have been directed toward intracellular SERS, with increasing emphasis on single-molecule SERS (smSERS) of/in biomolecules and chemical imaging of intracellular distribution of (single) molecules (for a recent survey of the methodology, see, e.g., Radziuk et al. 2015).

The dye molecule Rh6G has played a major role in smSERS for the understanding of physicochemical processes and in the benchmarking of different experimental approaches, aided by the facts (1) that it is soluble in numerous liquid solvents and (2) that quite a few investigations have been published on single-molecule fluorescence of Rh6G. The first measurements of smSERS of Rh6G were reported, more or less simultaneously, by two groups in the late 1990s (Nie and Emory 1997 and Kneipp et al. 1997). Note that Nie and Emory compared their smSERS results directly with single-molecule fluorescence data, measured in the same confocal excitation/observation geometry, which included the capability of polarization control in both channels.

Realizing a smSERS signal means that in Equation 14.12, the number of molecules is $N = 1$; this means that in order to achieve $I_{SERS} > I_{noise}$, the amplification factor $EF_{EM}$ (as well as $EF_{Chem}$) had to be extremely large. In the early days of smSERS studies, the seemingly random alteration between success and failure in repeat measurements was often frustrating. It was realized that the parameters contributing to $EF_{EM} \times EF_{Chem}$ had be carefully optimized to achieve the required process cross-sections of the order $\sigma_{smSERS} \sim 10^{-21} cm^2$. In a range of systematic studies, it was shown that the success of smSERS hinged one highly local effects of plasmonic nanostructures, occurring at sharp edges, interparticle junctions, particle crevices, or other geometries with a sharp nanofeatures—now commonly addressed as "hotspots." The emission of an individual molecule thus depends primarily on not only the local enhancement field of the hotspot location but also on the binding affinity and positioning at a hotspot region. The "design" of plasmonic hotspots for smSERS and a summary of underlying enhancement mechanisms are provided, e.g., in Radziuk et al. (2015).

The method of smSERS has become a reasonably well-established methodology, but it was not always plain sailing, and the claims for observing a SERS signal to be due to a single molecule were frequently met with skepticism. This is not overly surprising since even under the best Raman probing conditions, the interrogated volume is of the order 1 $\mu m^3$, while nanoparticles and colloidal aggregates normally have dimensions of the order $d \cong 30$–100 nm, and many of the investigated molecules, adsorbed to a nanoparticle, are often even smaller than that. In addition, in principle, more than one molecule can be adsorbed to the surface of the nanoparticle, and thus, a SERS signal according to Equation 14.12 does not necessarily originate from a single analyte molecule, although it might be easily possible to ascertain that only a single nanoparticle was probed. The observed signal $I_{SERS}$ could actually be a superposition of $N > 1$ molecules.

It is now accepted that for the verification of single-molecule detection based on SERS, the use of two independent analytes is required, utilizing either the bianalyte or two-isotopologue approach. In both cases, the preferential detection of a selected, individual analyte—recorded for a range of mixture combinations of both analytes—is used to deduce that only a single molecule contributed to the signal. While both approaches do work, the two-isotopologue methodology (introduced by Dieringer et al. 2007) is now the preferred one because by and large isotopologues exhibit identical surface binding affinities and Raman cross-sections; this is not normally the case for bianalyte pairs. However, while the isotopologue method certainly is the preferred one, it may be subject to practical limitations. For example, synthesis of particular isotopologues could be complex, and the associated costs may be high. Also, the various isotopologues must have spectral peak shifts for the vibrational bands affected by the isotopic substitution, which are sufficiently large to allow for unequivocal identification of the two isotopologues. Thus, the use of (additional) bi-analyte implementation might be unavoidable. Recently, the issues of proving smSERS via bianalyte and/or two-isotopologue methodologies have been critically reviewed (Zrimsek et al. 2016).

The concept underlying both approaches is summarized in Figure 14.9. Here, we briefly describe the principles of the two-isotopologue approach. Instead of merely using the strength of the Raman signal, the spectral information of the two competing analyte molecules (here isotopologues) is exploited. Part of the results of the study shown here are for the rhodamine isotopologue pair Rh6G-$d_0$ and Rh6G-$d_4$; these were also used in a range of earlier studies and thus serve well for comparison to results published by other researchers. In addition, a second set of isotopologues, CV-$d_0$ and CV-$d_{12}$ (crystal violet), was used as a cross check for the isotopologue validity for a molecule of substantially different enhancement factor (caused by different quantum and adhesion properties). Spectral traces of all four specimens are shown in the top part of the figure; single-molecule behavior can be deduced from the spectral signature of the individually tracked nanoparticles to which the dye molecules have adsorbed. A few relevant isotopologue features of the substitution-affected vibrational band are indicated in the figure, which have been used in the statistical data analysis.

Of course, for this isotopologue approach to work correctly, the concentrations for both analyte molecules have to be low enough so that, on average, only one type of molecule and only one single molecule are adsorbed to each nanoparticle. Thus, the SERS spectra recorded from individual nanoparticles should reveal spectral features of only a single isotopologue, based on a 1:1 mixture of the two isotopologues. As soon as one increases the overall concentration to a level such that both isotopically substituted analytes should be present on individual nanoparticles, one should observe the vibrational characteristics of both isotopologues. A spectrum associated with this scenario is included as well (the middle, gray trace in Figure 14.9). Thus, one can distinguish single-molecule and multiple-molecule SERS events simply by "counting" the number of peaks in the SERS spectrum, specifically those that are substantially affected by the isotopic substitution.

The count rates for the occurrence of isotopologue 1 or 2 only or both from repeat measurements are then statistically evaluated (SERS events or the number of molecules detected per spectrum should follow a Poisson distribution). The results from typical analysis are shown Figure 14.10. In brief, the statistics for the Rh6G isotopologue pair reveal that, indeed, those can be associated with smSERS, while the situation for the CV isotopologue pair is less clear-cut. The latter is due to the fact that the concentrations of CV were about a factor of 10 higher than for Rh6G; the higher concentration was selected to obtain similar SERS signal count rates for both species. In addition, the statistics part of the figure includes, in the form of the shaded areas, the total count rates for the isotopologue pairs and the cross-term in which both specimen CV and Rh6G are detected. For a more detailed discussion of the statistical evaluation procedure and the verification criteria for smSERS spectra, see Zrimsek et al. (2016). The overall strategy presented by the authors constitutes a promising approach for the verification of smSERS, based on using multiple analytes with isotopologue-pairs. Not only is this useful for the actual verification, but it also may be an essential step toward future experiments, e.g., for the measurement of chemical reactions at the single molecule.

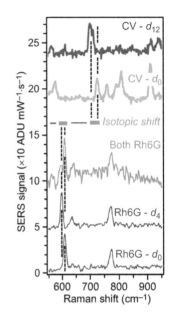

**Figure 14.9** Examples for smSERS spectra of isotope-substituted Rh6G (Rh6G-$d_0$ and Rh6G-$d_4$) and CV (CV-$d_0$, and CV-$d_{12}$) in dilute Ag-colloid solutions; characteristic isotope-shifted vibrational bands for the identification and evaluation of each substituted analyte are indicated by the dashed vertical lines. The spectra are offset with respect to each other for clarity. (Data adapted from Zrimsek, A.B. et al., *J. Phys. Chem. C* 120 (2016): 5133–5142.)

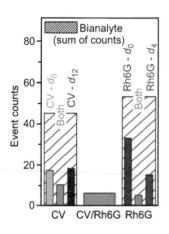

**Figure 14.10** Histogram of counts for a multianalyte experiment, conducted with Ag-colloids dosed with 4 nM of CV and 0.1 nM of Rh6G, respectively (the particular concentrations were selected to observe similar count rates for both CV and R6G). (Data adapted from Zrimsek, A.B. et al., *J. Phys. Chem. C* 120 (2016): 5133–5142.)

### 14.3.3 Principles of tip-enhanced RS

The spatial (lateral) resolution that one can obtain in focused-beam laser spectroscopy—such as in confocal RS (see Chapter 12.5)—is ultimately determined by the Abbe limit of diffraction of light, which is given by $d_{Abbe} = (\lambda/2)\cdot(n_{\lambda}\cdot\sin\theta)^{-1} = (\lambda/2)\cdot(NA)^{-1}$. Here, $\lambda$ is the wavelength of the light; $n_{\lambda}$ is the wavelength-dependent refractive index; and $\theta$ is the convergence angle of the focused light cone. Modern focusing optics can reach the limit numerical apertures of NA $\geq$ 1. For visible light of wavelength $\lambda_L = 500$ nm, this would mean a laser spot size of the order $d_{Abbe,500} \approx 250$ nm. This is small in comparison to many microscopic entities, like, e.g., most biological cells; however, it is large in comparison to simple molecules (~1 nm in size) and even complex organic macromolecules like proteins (~10 nm in size).

In this context, tip-enhanced RS (TERS) is a technique that combines the spatial resolution of atomic force microscopy (AFM) with the chemical/structural capabilities of (enhanced) RS. In effect, it is based on the principle of the SERS effect, in which the electric field is greatly enhanced near the surfaces of a nanostructured metal surface (see Section 14.2). However, instead of the nanoparticle structure, now, the AFM tip—coated with gold or silver—serves as quasi-metal nanoparticle to generate a very localized electric field and, hence, Raman signal enhancement. The Raman signal enhancement is of the order ~$(E_{sig}/E_L)^4$, where $E_{sig}$ is the enhanced electric field of the Raman signal wave and $E_L$ is electric field associated with the excitation laser. Utilizing such AFM/TERS nanotips, one can now reach spatial resolutions in RS of $d_{TERS}$ < 10 nm; lateral displacement can be achieved in the same manner as in ordinary AFM systems, thus allowing to build up two-dimensional maps that come close to resolving individual molecules. Since TERS is predominantly used in nanoimaging applications, it will be discussed in more depth in Chapter 19.2; here, only the basic principles are summarized.

The first reports of TERS go back to the early 2000s (e.g., Anderson 2000 or Stöckle et al. 2000); for some recent overviews of the history and advances of TERS since those early days, see, e.g., Sonntag et al. (2014) and Langelüddecke et al. (2015).

In Figure 14.11, the basic concept behind high-resolution TERS is shown. The two key elements in a TERS setup comprise a (confocal) microscope and a scanning probe (today mostly AFM) microscope, coupled together. The high-NA objective optical microscope is used to focus the laser radiation onto the metal-coated AFM tip, which serves as the SERS-active nanostructure. By and large, three experimental configurations are common: (1) *bottom* illumination, (2) *side* illumination, and (3) *top* illumination; they indicate the direction in which the incident laser light propagates toward the sample substrate. In the figure, case (1) is exemplified. In general, rather than moving the tip to scan to a feature of interest, as is the case for many traditional AFM systems, the sample is moved laterally, in order not to disturb the laser beam—tip alignment (for a description of these configurations and practical TERS more generally, see, e.g., Stadler et al. 2012 and Kumar et al. 2015).

As shown in the top part of the figure, the electric field is predominantly enhanced at the tip end, and thus, also the observed Raman signal originates predominantly from that region. The overall spatial resolution depends on the key dimensions encountered in the setup, including as the most important the tip diameter $d_T$ and the lateral displacement capabilities $d_x$ of the x–y moving stage on the equipment side and the dimensions of the sample structure $d_M$ (individual molecules or nanostructures).

While the ultimate aim of TERS is to record only the tip-enhanced signal, one has to be aware of the fact that ordinary RS will also contribute; i.e., the SERS-type signal from the tip $d_T \approx$ 10–100 nm competes with the "normal" RS from the $d_{Abbe} \approx$ 250–1000 nm focal spot of the laser radiation. Since SERS can provide signal enhancement of up to factors $\sim 10^{10}$, actual nm-scale Raman analysis depends on the TERS signal exhibiting similar or greater intensity than the "normal" Raman signal. This is not necessarily guaranteed, because, in line with the diminished spatial dimension, orders of magnitude of fewer molecules are sampled in TERS when comparing it to the linear Raman signals normally observed without a tip present.

In the near-field of the TERS tip, the intensity of Raman signal from the analyzed molecules is proportional to the fourth power of the local electric field (see Section 14.2 for the $E^4$-factor). As a consequence, based on the near-field and far-field electric-field amplitudes, $E_{NF}$ and $E_{FF}$, respectively, the enhancement of the Raman signal with a TERS probe is proportional to

$$EF_{EM}(\text{TERS}) \propto (E_{NF}/E_{FF})^4 . \tag{14.13}$$

The actual Raman enhancement factor $EF_{TERS}$ reached in a TERS experiment then is

$$EF_{TERS} = \underbrace{\left( \frac{I_{\text{Tip-in}}}{I_{\text{Tip-out}}} - 1 \right)}_{\text{"contrast"}} \cdot \frac{A_{FF}}{A_{NF}} , \tag{14.14}$$

where $I_{\text{Tip-in}}$ and $I_{\text{Tip-out}}$ are the Raman signal intensities measured with the tip close to and retracted from the sample, respectively; $A_{FF}$ and $A_{NF}$ are the area of the far-field laser probe (focal spot of the laser) and the effective area of TERS tip (usually estimated from the geometric diameter of the tip-apex); for further details, see, e.g., Kumar et al. (2015).

Successful TERS measurements have been made for a wide range of sample types, and commercial instruments are now available from various manufacturers; still, achieving meaningful TERS signals with good contrast and being able to interpret them quantitatively is far from trivial. As stated earlier, in general, TERS is applied to imaging applications to gain chemical and structural spectral information on the nanoscale, which is covered in Chapter 19.2.

Here, just one representative example is given to demonstrate the strengths of TERS for chemical and structural analysis, namely, the study of meso-tetrakis(3,5-di-tertiarybutylphenyl)-porphyrin (H2TBPP) molecules on a Ag(111) monocrystal surface, in vacuum and at cryogenic temperatures (Zhang et al. 2013). As such, porphyrins are of great interest in the fields of chemistry, physics, biology, and medicine due to their natural abundance and distinct electronic, photonic, and catalytic properties, which may form ordered ad-layers via self-assembly. In their study, the authors utilized a TERS configuration different from that shown in Figure 14.11, namely, a tunneling-controlled TERS tip in a confocal-type side-illumination configuration. They were able to record plasmon-enhanced Raman spectra from single molecules, located at the scanning

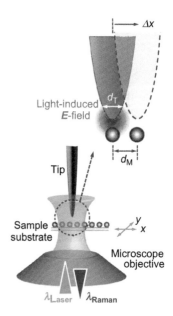

**Figure 14.11** Schematic concept of TERS: laser radiation is tightly focused (using a microscope objective) onto the sample surface and a "sharp" metal tip is positioned within the laser focus; the laser-induced locally enhanced $E$-field at the tip interacts with the sample (surface), giving rise to a huge increase in Raman response. The spatial resolution $d_M$ depends on the tip diameter $d_T$ and the lateral displacement resolution $\Delta x$ (see top part of the figure).

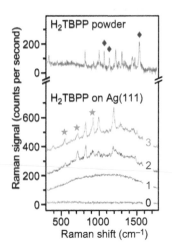

**Figure 14.12** Example TERS spectra of porphyrin-derivative H2TBPP molecules on an Ag(111) monocrystal substrate, for different conditions: 0—spectrum at the same position as trace (3) but with the tip retracted by 5 nm from the surface; 1—substrate-only spectrum, recorded from a spot bare of H2TBPP molecules; 2—spectrum taken with the tip on top of a single H2TBPP molecule; 3—spectrum taken with the tip on top of a molecular island (multiple molecules). For comparison, a regular Raman spectrum recorded from a H2TBPP powder sample is shown (top trace). The "diamond" and "star" symbols indicate spectral bands whose activity strongly depend on the laser polarization conditions. (Data adapted from Zhang, R. et al., *Nature* 498 (2013): 82–86.)

tunneling microscopy nanocavity. Some representative TERS data are shown in Figure 14.12.

TERS spectra (2) and (3) were recorded with the tip on top of an individual, single molecule and with the tip on top of a molecular island (i.e., a small cluster of molecules), respectively. The data clearly exhibit spectral similarity, confirming that, indeed, individual H2TBPP molecules could be identified from their spectral fingerprints. One can also gauge the excellent signal-to-noise ratio associated with the tip enhancement when comparing the molecular data with those when no molecule is close to the tip (trace 1) and when the tip is retracted from the surface and thus removing the enhancement (trace 0).

For comparison, a Raman spectrum recorded from a H2TBPP-powder sample is shown (top trace in the figure), which includes all key Raman-active vibrational modes; the spectrum constitutes an average over an ensemble of randomly oriented molecules. Quite a few of the Raman peaks of the powder spectrum are replicated in the TERS spectra for the molecular island and single molecules, providing clear chemical identification of the H2TBPP molecules on the surface. However, the actual number and relative intensity of the various Raman peaks differ substantially; the authors attribute this to (1) the ordering of the H2TBPP molecules on Ag(111) and to (2) the preferred axial polarization of the laser-induced nanocavity plasmons, which are selective for specific Raman modes. Some of the initial deductions on orientation and Raman activity of specific modes were later confirmed in a theoretical and experimental study of a monolayer of H2TBPP on Ag(111); see Chiang et al. (2015).

## 14.4 HYPER-RAMAN SPECTROSCOPY: HRS

HRS may be seen as a modified version of RS; the process is shown schematically in Figure 14.1. As that conceptual transition diagram shows, HRS exhibits the following characteristics. Firstly, the Raman scattered light is at frequencies well above the excitation laser frequency, $v_{HRS} = 2 \cdot v_L \pm v_{if}$; here, $v_L$ and $v_{if}$ stand for the laser frequency and the difference between the initial and final energy state connected in the process, respectively; and the +/− stand for the anti-Stokes and Stokes transitions, respectively. Second, two photons are necessary to contribute simultaneously, thus making the process nonlinear, with probability proportional to $I_L^2$ and involving the second-order susceptibility tensor $\chi^{(2)}$ (see Equation 14.4). Therefore, the intensity of the observed hyper-Raman light is normally extremely weak. Third, counteracting the low-intensity argument, HRS exhibits certain aspects that make it still interesting for RS. These include (1) that HRS can provide vibrational information on molecules for which ordinary, linear RS is suppressed (associated with molecular symmetry issues (these are so-called "silent" modes) and (2) that the HRS frequencies are normally much higher than the laser excitation frequency, thus avoiding interference with fluorescence which is always at lower frequencies in comparison to the laser excitation. Fourth, the Raman shift $v_{if}$ observed in RS and HRS does not change, in principle, although in many cases, different ro-vibrational states are accessed due to the transition selection rules. For a theoretical description of HRS, see, e.g., Chung and Ziegler (1988), Ziegler (2001), or Simmons et al. (2015).

The first observation of HRS was made in water, in the very early days of lasers being available for spectroscopy (see Terhune et al. 1965). Since then, HRS has been applied regularly but infrequently due to its very low probability and, hence, weak signals. But—as stated in the previous paragraph—it has found its niches when Raman- and/or IR-forbidden transitions were to be probed. An example highlighting the low intensity of HRS on the one hand but the provision of vibrational band information on the other hand is shown in Figure 14.13a for a sample of pyridine (see Nedderson et al. 1989). Indeed, the HRS signals are more than two orders of magnitude smaller than those for many of the RS features; however, clearly, the additional vibrational transitions revealed in HRS do justify using this less-than-perfect spectroscopy solution (the relevant vibrational bands are annotated in the figure).

The HRS probability rate can be substantially increased by exploiting any of the enhancement techniques discussed in the previous chapters and this chapter, namely, to combine HRS with resonance excitation conditions, yielding *resonant HRS* (RHRS), or exploiting the enhancement near surfaces, which gives rise to *surface-enhanced HRS* (SEHRS). Using (predominantly) these two enhanced hyper-Raman approaches, HRS has enjoyed increased popularity in the recent years since it provides access to otherwise inaccessible vibrational energy levels and electronic states; representative examples for the two approaches are included here. A brief illustrative review of the two techniques is given by Meyers Kelly (2010).

In their work on characterization of multipolar variants of the chromophores of TPB(PV)$_1$NO$_2$, Shoute et al. (2004) compared resonant Raman with resonant hyper-Raman data to gain insight into the excited electronic-vibrational structure of the selected chromophores. The said conjugated "push–pull" chromophores with large hyperpolarizability $\beta$ should exhibit strong RHRS when the laser excitation wavelength is a two-photon resonant with any of the transitions into the chromophore's lowest-lying electronic states. As can be seen in Figure 14.13b, the information provided by the resonance RS (RRS) and RHRS data is complementary, giving detailed insight into the energy level structure of the chromophore. In addition, in RHRS, the peaks from the solvent (here acetone) are suppressed, providing much "cleaner" and unambiguous information about the studied chromophore.

Probably the most exploited enhancement technique for HRS is SEHRS; this is very much analogous to SERS, only the transition probabilities and selection rules differ for the associated one- and two-photon processes. Over the past 10 years, a multitude of problems has been investigated utilizing the technique of SEHRS. Probably one of the most intriguing uses is its application to single-molecule spectroscopy; this challenging but rewarding approach was discussed for smSERS in Section 14.3.

In analogy to smSERS for substituted CV dyes (see Zrimsek et al. 2016), the same dye variants were used by Milojevitch et al. (2013) in their single-molecule studies but now based on smSEHRS. Some representative SEHRS data are shown in Figure 14.14a; the spectra show clearly whether either or both of the substituted dyes CV-$d_{30}$ and CV-$d_0$ are present. Note that both research groups used in their tests for single-molecule detection low concentration mixtures of CV-$d_{12}$ or CV-$d_{30}$/CV-$d_0$ in a silver colloid suspension so that, on average, one dye molecule

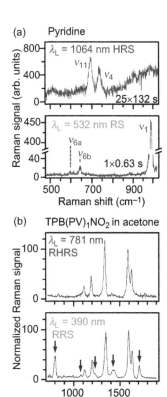

**Figure 14.13** Comparison standard RS vs. HRS, and their resonance implementations RRS vs. RHRS. (a) The RS and HRS spectra for pyridine reveal that different vibrational modes are accessed and that the intensities are substantially different with $I_{RS}/I_{HRS} > 10^3$. (Data adapted from Neddersen, J.P. et al., *J. Chem. Phys.*, 90 (1989): 4719–4726.) (b) The RRS and RHRS spectra for TPB(PV)$_1$NO$_2$ in acetone, demonstrating that spectral contributions from the solvent acetone (indicated by the arrows) are suppressed in RHRS. Spectra are normalized to the peak at $\tilde{v} =$ 1594 cm$^{-1}$. (Data adapted from Shoute, L.C.T. et al., *J. Chem. Phys.* 121 (2004): 7045–7048.)

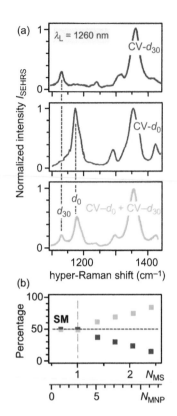

**Figure 14.14** Single-molecule detection of crystal-violet CV-$d_0$ and its per-deuterated isotopologue CV-$d_{30}$, adsorbed on nanoparticles, by using SEHRS. (a) Spectral hyper-Raman signals from CV-$d_{30}$ (top panel) CV-$d_0$ (center panel) and a 1:1 mixture of CV-$d_0$ + CV-$d_{30}$ (bottom panel); the spectral features used for identification are marked by the vertical lines. (b) Experimentally measured distribution for single- and multiple-molecule events, as a function of the average number of dye molecules adsorbed on the nanoparticles. Red data points—CV-$d_{30}$ events only; green data points—combined events. For further details, see text. (Data adapted from Milojevich, C.B. et al., *J. Phys. Chem. Lett.* 4 (2013): 3420–3423.)

per silver nanoparticle is adsorbed. The hyper-Raman excitation wavelength was set to $\lambda_L = 1260$ nm.

The statistical evaluation from a large set of individual spectra showed (i) that they could be sorted according to the number of molecules contributing to individual spectra, $N_{MS}$, assigning them with high fidelity to CV-$d_{30}$ and CV-$d_0$, and (ii) that, indeed, the majority of SEHRS spectra were due to single molecule probing, i.e., $N_{MS} = 1$ (see Figure 14.14b). It should be noted that for solution concentrations of less than three dye molecules per nanoparticle, the statistics of the ensemble data set followed those expected for response from single molecules. With increasing dye concentration, multiple-molecule events became more frequent, i.e., $N_{MS} > 1$; as a consequence, the distribution in CV-$d_{30}$/CV-$d_0$ changes. Note also that the authors found that the number of adsorbed dye *molecules per nanoparticle*, $N_{MNP}$, was higher than that predicted by the statistical distribution; the authors attributed this to the fact that not all adsorbed dye molecules were at sites with sufficient enhancement to be experimentally detectable. The results also confirmed that plasmonic nanostructures are indeed suitable in enhancing higher-order nonlinear scattering processes.

## 14.5 STIMULATED RAMAN SCATTERING AND SPECTROSCOPY: SRS

As stated previously in this and the other chapters on RS, in ordinary (linear) RS, one (pump) photon at frequency $v_p = v_{L1}$ is converted into one lower-energy signal (Stokes) photon with frequency $v_s$. The difference in photon energies is channeled into molecular vibrations of frequency $v_{if}$ (see also the upper part of Figure 14.15). If the RS process commences in an excited molecular level, then also higher-energy signal (anti-Stokes) photons at frequency $v_{as}$ are observed, albeit at much smaller intensity due to normally lesser population in excited molecular levels.

In SRS, two laser sources interrogate the sample molecule, at frequencies $v_p = v_{L1}$ and $v_s = v_{L2}$. In the case that the difference frequency matches a particular molecular vibrational frequency $v_{if}$, i.e., $\Delta v = v_p - v_s \equiv v_{if}$, amplification of the Raman signal is observed; in the framework of photon rate equations, the SRS-signal rate $\mathfrak{R}_{SRS}$ is proportional to the number of photons per mode in the pump and Stokes fields, $n_p$ and $n_s$, respectively:

$$\mathfrak{R}_{SRS} \propto \sigma_{Raman} \cdot n_p \cdot (n_s + 1), \qquad (14.15)$$

where $\sigma_{Raman}$ is the ($v_{if}$-dependent) scattering cross-section of the molecule.

Note that in the absence of the (probe) Stokes laser beam, i.e., $n_s = 0$, Equation 14.15 reduces to the case of ordinary, spontaneous RS. However, for common SRS scenarios, one has that $n_s > 10^7$–$10^8$, and thus, SRS results in amplification at the rate given in Equation 14.15. As a consequence, the intensity of the Stokes (probe) laser beam, $I_s$, experiences a gain, $\Delta I_s \equiv \Delta I_{L2}$, and the intensity of the pump beam, $I_p$, experiences a loss, $\Delta I_p \equiv \Delta I_{L1}$; these are commonly addressed as stimulated Raman gain and loss (SRG and SRL), respectively. This is shown schematically in Figure 14.15. Either SRG or SRL can be exploited to monitor the effect of SRS.

It should be stressed that SRS does not occur if $\Delta v \neq v_{\text{if}}$; i.e., it does not match any vibrational transition resonance in the molecule; as a consequence, no nonresonant background signal is encountered. Following the previous rate equation framework, the differential SRG- and SRL-coefficients are described by

$$G \equiv \Delta I_s / I_s \propto N \cdot \sigma_{\text{Raman}} \cdot I_p \quad \text{and} \quad L \equiv \Delta I_p / I_p \propto -N \cdot \sigma_{\text{Raman}} \cdot I_s, \quad (14.16a)$$

where $N$ is the number of molecules in the volume interrogated by the lasers (see the bottom part of Figure 14.15). The magnitudes of G and L are usually of the order of $<10^{-3}$ (sometimes even less). On the other hand, over longer interaction lengths, they can be as high as 20–30%; this can be and is exploited in higher-order Raman laser light converters.

Note that in the macroscopic, classical picture, SRS is a scattering effect associated with the nonlinear susceptibility term $\chi^{(3)}$, as described in Equation 14.4; utilizing the said macroscopic formalism would also lead to the result given in Equation 14.16a. For example, for collinear laser beams with parallel, linear polarization, for which the susceptibility tensor $\chi^{(3)}$ reduces to the isotropic portion of the polarizability, one finds (see, e.g., Tolles et al. 1977)

$$G \equiv \Delta I_s / I_s \propto \text{Im}\left[\chi^{(3)}(-v_s, v_s, v_p, -v_p)\right] \cdot I_p. \quad (14.16b)$$

As hinted at already further previously, SRS has certain advantages in comparison to the other nonlinear Raman techniques, in particular CARS (which will be discussed in Section 14.6). First, since it is directional, along the axis of either of the pump or probe laser beams, it is nearly free from the nonresonant, spontaneous background, which scatters into the full solid angle $\Omega$ and would be suppressed by the ratio $d\Omega/\Omega$, where $d\Omega$ stands for the solid angle of the narrow light-collection cone. Second, it exhibits spectra that are identical in line positions and amplitude to those obtained in linear, spontaneous RS. And third, SRS is directly (linearly) proportional to the number density of the analyte molecule (see Equation 14.16) and therefore allows for straightforward quantification.

## 14.5.1 SRS using tunable probe laser sources

The first observation of SRS dates back to the early 1970s in glass-fiber optical waveguides (see Stolen et al. 1972). It should be noted that in those experiments, no second laser at frequency $v_s$ was utilized, but the $v_s$ wave built up from initial spontaneous Raman photons that were amplified on passage through the long optical waveguide structure. In subsequent years, SRG and SRL experiments were successfully conducted for a wide range of solid, liquid, and gaseous sample materials, mostly utilizing pulsed, high-power laser sources. That it was possible to also succeed to surpass the measurable SRS threshold (i.e., $G,L > I_{\text{laser-noise}}$) when using low-power CW lasers for gaseous media was, e.g., demonstrated by Owyoung (1978). In his experiments, he obtained high-resolution SRS signals of the $Q_{01}(J=1)$ transition in molecular hydrogen, at a pressure of $p(H_2) \sim 260$ mbar The spectra fully resolved both the real and the imaginary parts of the Raman susceptibility $\chi^{(3)}$, with a resolution of $\delta v < 25$ MHz.

While the experiments just mentioned demonstrate the versatility and selectivity of SRS analysis, in order to drive the methodology to higher sensitivity, in particular for low-density gaseous samples, further enhancements along the lines

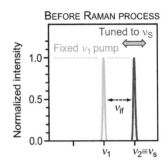

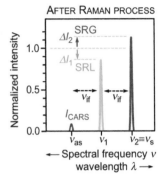

**Figure 14.15** The concept of SRS. Initially equal (normalized) laser intensities $I_1(v_1)$ and $I_2(v_2 \equiv v_s)$ exhibit gain $\Delta I_2$ and loss $\Delta I_1$ if the difference frequency is equal to a molecular transition frequency, $\Delta v \equiv v_{\text{if}}$. Small contributions from anti-Stokes Raman processes (shown in blue) do not contribute to the SRS signals.

discussed in Chapter 13 are required. In particular, waveguide configurations have been used to increase the interaction length between the laser radiation and the molecular sample. One such example is the SRS of gases within a hollow-core photonic crystal fiber (HC-PCF), in which the authors showed that high resolution and high sensitivity were achievable even with low-power cw laser sources (Doménech and Cueto 2013). In their experiments, the researchers used a single-mode Ar+ laser as the pump laser ($\lambda_p$ = 514.5 nm, $P_p$ = 40 mW) and a single-mode tunable dye laser as the Stokes-wave probe laser ($\lambda_s$ ≈ 540–595 nm, $P_s$ = 4 mW); the combined linewidth of the lasers is dominated by that of the probe (dye) laser and is $\delta\nu_{L1\otimes L2}$ ~ 3 MHz (equivalent to $\delta\tilde{\nu}_{L1\otimes L2}$ ~ $10^{-4}$ cm$^{-1}$). The radiation from these two laser sources, with parallel polarization orientation, was coupled collinearly into an HC-PCF of length $L_{HC\text{-}PCF}$ = 1 m. The hollow core of the fiber ($d_{HC\text{-}PCF}$ ~ 5 μm) was filled with the sample gas $CO_2$, at room temperature, and pressure values ranging from a few mbar to atmospheric pressure.

In Figure 14.16, an example spectrum for the Q-branch of the $2\nu_2$ overtone band of the Fermi dyad of $CO_2$ is shown (at $p_{CO2}$ = 29 mbar), demonstrating that all rotational transition lines can be resolved, exhibiting the expected Doppler width of $\delta\nu_D$ ~ 75 MHz (equivalent to $\delta\tilde{\nu}_D$ ~ $2.4 \times 10^{-3}$ cm$^{-1}$). Note that the SNR > 100 encountered for the majority of lines in the spectrum suggests that molecular gas concentrations below 1 mbar would be easy to detect. Overall, the authors achieved more than four orders-of-magnitude enhancement of sensitivity in their HC-PCF setup, in comparison to a single-focus configuration using the same laser sources.

## 14.5.2 SRS using ps- and fs-laser sources (fs-SRS)

One small drawback of SRS is that one loses the universality of common, linear RS; i.e., rather than obtaining a complete (dispersed) Raman spectrum from the interaction of the excitation (pump) laser radiation, normally, at a fixed wavelength $\lambda_p \equiv \lambda_{L1}$, one now needs a second, tunable (probe) laser source with variable wavelength $\lambda_s \equiv \lambda_{L2}$. This second laser has to be scanned in wavelength across the ro-vibrational structure of the molecular Raman transition spectrum. While the scanning process is time-consuming, on the upside, one finds that the use of a laser source, tuned through Raman resonances, does not require the use of a spectrometer to resolve the Raman lines. Indeed, tunable lasers can offer much higher spectral resolution than is normal for dispersive of Fourier-transform spectrometers.

This "deficiency" of needing a tunable laser for the stimulating probe process can be partially overcome by utilizing a broadband fs-laser source as the probe laser, whose spectral components are automatically in resonance with a normally many molecular transition lines. Of course, now, one needs to add a wavelength-analyzing spectrometer again. As a further bonus of fs-time resolution, one finds that SRS incorporating fs-laser sources (this approach is also known as fs-SRS) allows one to follow vibrational structural evolution with a temporal resolution, which is comparable to or faster than the vibrational periods of the probed molecular motion (see, e.g., McCamant et al. 2003).

The concept of fs-SRS is simple and straightforward. A narrow-bandwidth Raman pump pulse of ps duration is overlapped in space and time with a broad-bandwidth Raman (Stokes) probe pulse of fs duration. This will lead to the appearance of SRG features on top of the probe laser pulse envelope whenever

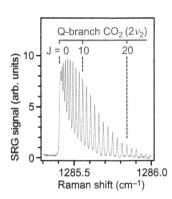

**Figure 14.16** SRS spectrum of the Q-branch of the $2\nu_2$ overtone band of the Fermi dyad of $CO_2$, for a gas sample pressure of $p_{CO2}$ = 29 mbar in an HC-PCF, at room temperature. (Data adapted from Doménech, J.L. et al., *Opt. Lett.* 38 (2013): 4074–4077.)

a spectral component is in resonance with the difference frequency $v_p - v_s = v_{if}$; the width of these SRG lines is by and large determined by the spectral width of the pump laser radiation, i.e., $\delta v_{SRS} \approx \delta v_p$. This is shown schematically in the lower part of Figure 14.17.

The figure reveals that if one uses the SRG signal for analysis, a series of narrow spectral lines sits on top of the fs-probe pulse profile. The line transition gain signal is normally relatively small (the scale of the lines is exaggerated for clarity)—as stated earlier—and therefore one might think of the fs-pulse profile as being detrimental to sensitive measurements. However, the SRG lines are present only as long as the Raman pump radiation is present. Therefore, if one takes the ratio of the spectra recorded with and without the pump pulse being present, the residue will be the SRG lines and one obtains a nearly background-free Raman spectrum $I_{Raman}(v) \equiv I_{SRG}(v) = [(I_p + I_s(v))/Is(v)]$, as shown by the (black) SRG-spectrum trace included above the SRS-signal curve. Indeed, the SRG-spectrum is exactly identical in line position but different in relative amplitude to the standard spontaneous Raman spectrum, which is included in the top of the figure for comparison (the signals in both cases are proportional to the molecular number density $N$).

From the figure, it is also evident that one will normally observe anti-Stokes features in the spectrum as well. With respect to the anti-Stokes Raman signal, it should be noted that in the case of spontaneous, linear RS, the anti-Stokes

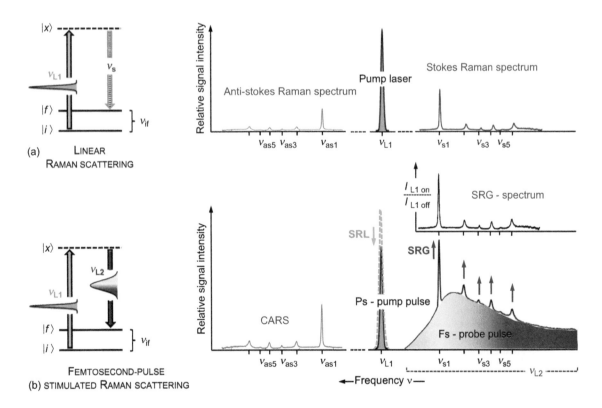

**Figure 14.17** Conceptual comparison of spontaneous, linear RS (a) and stimulated, nonlinear RS—SRS (b). When using a broadband fs-pulse for probing frequency-selective gain is observed for the probe pulse, while the loss measured in the pump pulse is the integral (non-frequency-selective) contribution from all gain lines.

Raman intensity exhibits a fixed ratio, given by the thermal energy level populations.

In contrast, in SRS, this intensity stems from two contributions, namely, the spontaneous Raman signal from the pump laser pulse alone (this contribution is normally rather small) and the stimulated Raman contribution, which depends on the probe-laser intensity resonant with the various spectral lines (this can be seen as sort of a CARS signal; see Section 14.6).

It should be noted that this advantageous concept of fs-SRS stems from the same disentanglement of time and energy resolution exploited by fs-dynamic absorption spectroscopy (see Chapter 8.2). In brief, in the stimulated Raman process, the fs-pulse (stimulating) probe laser field generates—in the presence of the ps-pulse Raman pump field—a macroscopic polarization with high time resolution, which will allow for the observation of time-dependent features, while the dispersed, non-time-resolved detection of the (dispersed) spectrum provides high energy resolution, which allows for the observation of individual rotational transition features. As a consequence, fs-SRS studies increasingly find their way into the analysis of chemical and biochemical reaction dynamics, giving previously unattainable insight into the structural dynamics of reactively evolving molecular systems, with spatial resolution on the atomic scale, high transition energy resolution, and temporal resolution on the fs-scale (see, e.g., Kukura et al. 2007).

Finally, with careful relative timing adjustment between the Raman pump and fs-probe pulses (gating), it is possible to trace the link between near-resonant Raman excitation—to enhance the Raman signal further—and subsequent, stimulated fluorescence from the minutely populated (resonant) electronic level, which will interfere with the Raman signal. This is the drawback in "spontaneous" RRS, in which the intense fluorescence background can mask the largely enhanced, although still weak, Raman response (see Chapter 13.3); unless extreme care is taken, practical applications of RRS are limited.

The aforementioned timing or gating capabilities in fs-SRS was exploited by Kim et al. (2014) in their experiments with the NIR dye 3,3′-diethylthiatricarbocyanine (DTTC) iodide, in which they utilized fs-SRS under pre-resonance conditions, i.e., excitation to energies just below or near the first electronic-state potential minimum of the dye (see Figure 14.18a).

The laser system they used was based on a fs-Ti:sapphire laser ($\lambda_p = 800$ nm, $\delta\tau_p = 130$ fs, average power $P_{p,\,ave} = 1$ W), from which they derived the pump laser radiation (narrowing it down by a pass-filter to guarantee a decent spectral resolution of $\delta\tilde{\nu} \sim 27$ cm$^{-1}$, sufficient to resolve most of the vibrational structure of DTTC), and the probe laser radiation by super-continuum generation in an optical fiber. The precise, relative timing between the two laser pulses was achieved by a delay-line controlled via a motorized translation stage. A sequence of fs-SRS spectra for different time delays between the pump- and probe-pulses is shown in Figure 14.18b; the spectra were recorded in steps of $\Delta\tau_{pump-probe} = 0.2$ ps, with the traces in the figure offset to each other for clarity. Clearly, the onset of stimulated (fluorescence) emission (SFE) due to resonant excitation of the electronically excited state of DTTC is evident, onward from the temporal coincidence between the two laser pulses, until for large delays only SFE remains.

The temporal behaviors of the SRG signal (the SRG contribution is evaluated at $\tilde{\nu} = 1240$ cm$^{-1}$, the position of the CH-bending mode) and the SFE signal are com-

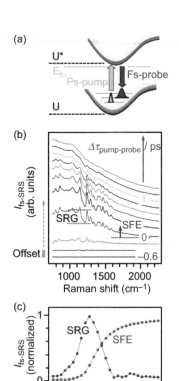

**Figure 14.18**  Near-resonance fs-SRS spectroscopy of DTTC iodine. (a) Schematic potential energy level diagram highlighting the sub-resonance condition in the SRS scheme. (b) Sequence of fs-SRS spectra, for increments of $\Delta\tau = 0.2$ ps delay between the pump and probe Raman laser pulses, revealing SRG only for near temporal overly between the pulses. (c) Evaluation of SRG and SFE contributions, as a function of pump–probe pulse time delay. (Kim, H.M. et al., *Phys. Chem. Chem. Phys.* 16 (2014): 5312–5318. Data adapted by permission of The Royal Society of Chemistry.)

pared to each other in Figure 14.18c. The data show that the best background-free fs-SRS spectra were obtained when the probe laser pulse was delivered to the molecules slightly in advance of the pump laser pulse, a feature that seems to be prevalent in resonant or near-resonant fs-SRS, in general. Note from the figure that no SRG is observed past a delay of ~1 ps; i.e., the stimulated Raman signal is indeed "instantaneous," while the stimulated fluorescence continues for much longer, in line with the knowledge that fluorescence lifetimes of an electronically excited level are of normally a few ns in duration (see Chapter 10.3).

## 14.6 COHERENT ANTI-STOKES RAMAN SCATTERING AND SPECTROSCOPY: CARS

CARS is a nonlinear scattering process, which—in a sense—might be seen as evolving from SRS. This can easily be understood in reference to Figure 14.1 at the beginning of the chapter. In SRS, the Stokes transition is stimulated (coherently) downward by a second laser at frequency $v_{L2} \equiv v_S$, with the condition $v_p - v_S = v_{if}$, i.e., the difference between the initial and the final ro-vibrational energy states in the process. Provided that the intensity $I_{L2}$ is sufficiently high, the final energy level will become substantially populated, usually orders of magnitudes higher than observed for normal, thermal population at room temperature. Then, "secondary" RS occurs for the pump laser radiation $v_{L1} \equiv v_p$, but starting now from an excited ro-vibrational level, giving rise—as usual—to Stokes and anti-Stokes Raman lines. In standard linear RS, the former coincides with so-called hot-band Raman transitions (and is not included in the transition diagram in Figure 14.1); however, it is the anti-Stokes transition, which provides the signal wave of interest.

It should be noted that—looking at the figure—from the spectroscopic point of view, CARS wave provides the same information on $v_{if}$ as does the earlier Stokes wave in the sequence, which mirrors standard, linear Raman information. But it has the advantage of occurring on the high-frequency (short wavelength) side of the pump and Stokes laser lines, hence being free of spontaneous Raman and possible fluorescence light background.

As SRS, also CARS is a third-order nonlinear optical process, being associated with the third-order susceptibility $\chi^{(3)}$, but now involving three laser waves: a pump-laser wave of frequency $v_p$, a laser wave at Stokes frequency $v_S$ (resonant with $v_{if}$), and a probe-laser wave at frequency $v_{probe} \equiv v_p$. These laser waves interact with the molecular sample and generate a coherent optical signal at the anti-Stokes frequency

$$v_{AS} \equiv v_{CARS} = 2v_p - v_S . \quad \text{(energy conservation)} \quad (14.17)$$

One has to treat CARS as sort of a four-wave mixing process, in which the three incoming light waves (two at the same frequency $v_p$) generate a fourth outgoing wave. Since all media possess a (frequency/wavelength dependent) refractive index $n_v$, analogous to Equation 14.17, one finds that for the associated wave vectors,

$$k_{CARS} = 2 \cdot k_p - k_S \quad \text{(momentum conservation)} \quad (14.18)$$

phase matching conditions have to be fulfilled (see Chapter 5.3 and Figure 14.19). This means that for efficient CARS implementation, collinear laser beams

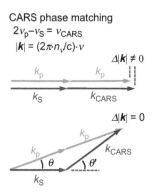

CARS phase matching
$2v_p - v_S = v_{CARS}$
$|\mathbf{k}| = (2\pi \cdot n_v/c) \cdot v$

$\Delta|\mathbf{k}| \neq 0$

$k_p$          $k_p$

$k_S$          $k_{CARS}$

$\Delta|\mathbf{k}| = 0$

$k_p$

$k_p$          $k_{CARS}$

$\theta$     $\theta'$

$k_S$

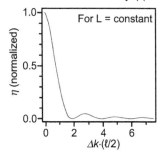

CARS conversion efficiency $\eta (\Delta k)$

For L = constant

$\eta$ (normalized)

$\Delta k \cdot (\ell/2)$

**Figure 14.19** CARS-signal conversion efficiency in dependence of the wave vector phase-mismatch. L = interaction length within the probed medium; $n_v$ = frequency-dependent refractive index of the medium.

(radiation fields) may not suffice to efficiently generate the signal wave at $\nu_{CARS}$ (see the schematic vector diagrams in the figure for the situations $\Delta|\mathbf{k}|=0$ and $\Delta|\mathbf{k}|\neq0$). The CARS conversion efficiency as a function of phase mismatch $\Delta k$ is shown in the lower part of the figure; a numerical expression is given in Section 14.6.1.

As a consequence of the energy conservation and phase matching condition of the three input waves, CARS offers a range of advantages over standard, linear RS. The most important ones are:

- *Spectroscopic information:* Because of the dependence of the CARS signal on the (complex) susceptibility $\chi^{(3)}$, additional information with respect to spontaneous RS (related to the absorptive part of the refractive index) can be obtained, namely, the dispersive part of the refractive index (which is reflected in the CARS line shapes).

- *Spectral selectivity:* the CARS signal generation is stringently conditional upon the two-photon Raman resonant condition for the pump and Stokes laser fields. While this may be deemed a drawback as to the simplicity of ordinary RS, it has the flexibility of selecting the pump–probe frequency pairs to match specific $\nu_{if}$. Hence, it is easier to avoid spectral interference with molecules other than the one of interest. Furthermore, the CARS signal is by and large insensitive to fluorescence interference because of being "blue-shifted" with respect to the pump- and probe-laser waves.

- *Spectral resolution:* This depends on the laser sources being used, but ultimately—when using single-mode laser sources—resolutions of the order $\delta\nu \leq 300$ MHz (or $\delta\tilde{\nu} \leq 10^{-3}$ cm$^{-1}$) have been reached, which corresponds to about the Doppler-limited line width of gas-phase molecules. Normal spontaneous RS is in general much worse.

- *Directionality of the signal wave:* Since the CARS wave is generated in a coherent process, it exhibits rather high directionality. This means the associated collimated signal wave requires only a small solid angle for light collection; in addition, because of the small beam divergence of the CARS signal, any photon detector can be placed some distance away from the sample, to avoid luminous background such as that from flames. Furthermore, because of the phase-matching related different propagation direction for the stimulating waves, they can be spatially separated out, and dumped, thus avoiding any background from these waves.

- *Spatial resolution:* Due to the crossing of the laser beams at different angles (because of the phase-matching conditions), one can achieve high spatial resolution. When using focusing lenses, as is common, it has been demonstrated that spatial resolution of the order $\delta x \leq 25$ μm can be reached.

At the same time, one should not underestimate some of the (depending on the actual application probably minor) drawbacks of CARS spectroscopy. Like all nonlinear methods, CARS signals are significantly affected by fluctuations in laser intensity and are subject to saturation at the higher laser power density levels (these aspects were already discussed for absorption spectroscopy in Chapter 7.3). And in cases where detection limits are an issue, the ever-present background of nonresonant contributions in the Raman process limit detection levels to $10^{-5}$ to $10^{-2}$, depending on thermodynamic conditions and molecular species studied.

Finally, it should be mentioned that CSRS is very similar to the predominant, common CARS. However, it is based on (pump) excitation radiation starting from an excited vibrational level of the molecule and anti-Stokes frequency (probe) stimulation radiation, into a lower quantum state than the initial state. Associated with the second pump step, out of that state, one observes Stokes emission as the final signal. The overall process is rather disadvantageous since anti-Stokes processes start in a less populated excited state, and the CSRS signal may be swamped by contributions from linear Stokes Raman processes associated with the two laser radiation fields. Thus, here, CSRS will not be discussed any further.

## 14.6.1 Basic framework for CARS

The description of the CARS process can be based either on a *classical* oscillator model involving EM waves and polarizability tensors or on *quantum* models, which incorporate the energy levels of the molecule, transition probabilities, and laser photons. For simplicity, here, we will only summarize the classical approach, highlighting the aspects most important in CARS spectroscopy; for further details, see, e.g., the concise review articles by Tolles et al. (1977), Roy et al. (2010), or El-Diasty (2011).

When describing the linear Raman process, many authors treat the (Raman-active) vibrator of the molecule as a (damped) harmonic oscillator, with characteristic frequency of $v_{if}$. In CARS, this oscillator is not driven by a single optical wave, but by the difference frequency $(v_p - v_S)$ wave between the pump and the Stokes laser waves instead. Thus, one experiences a resonance-response function, with maximum at $v_p - v_S = v_{if}$. On a macroscopic scale, this means that a periodic modulation of the refractive index ensues, which—simplistically—the couples pump-, Stokes-, and probe- (the same as the pump) waves to generate the anti-Stokes wave at frequency $v_{CARS}$ (see Equation 14.17). Note that while intuitive, this classical picture does not properly take into account the quantum nature of the energy levels of the molecule.

The classical description of the propagating $E$-field wave (in $z$-direction) is that the three fields from the incoming pump- and Stokes-waves produce a resulting $E$-field

$$E(z, t) = \frac{1}{2} \cdot \left\{ \sum_{j=1}^{3} E_j(z, v_j) \cdot \exp\left[i(k_j \cdot z - v \cdot t)\right] + c.c. \right\}, \tag{14.19}$$

where the indices $j = 1,2$ and $j = 3$ are associated with the two pump-fields $E_p$ and the Stokes-fields $E_S$, respectively. In this simplified presentation, it is assumed that all $k$-vectors have only components in $z$-direction and that all $E$-fields are polarized in the $x$-direction. Recall that the nonlinear response of the medium to strong (incident) $E$-fields gives rise to nonlinear polarization $P_{NL}$, whose $\chi^{(3)}$-term is responsible for the generation of the CARS wave (see Equation 14.4). Then, the wave equation of propagation of the CARS-field through the medium can be written as

$$\frac{\partial^2}{\partial z^2} E_{CARS} + k \cdot E_{CARS} = \frac{4\pi^2}{\varepsilon_0 c^2} \cdot v^2 \cdot P_{CARS}^{(3)}, \tag{14.20}$$

where the third-order polarization component of $P_{NL}$ is given by

$$P_{CARS}^{(3)} = \chi^{(3)} \cdot N_f \cdot E_p^2 \cdot E_S^* . \tag{14.21}$$

Here, $N_{CARS}$ is the number density of molecules contributing to the CARS-wave. The CARS signal intensity generated in a molecular sample by the incident laser fields can be determined from the set of Equations 14.19 to 14.21. The time-averaged CARS signal intensity $I_{CARS}$, integrated over the interaction length $L$, is then found to be, for the resonance condition $v_p - v_S \equiv v_{if}$ (see, e.g., El-Diastry 2011),

$$I_{CARS} \propto v_{CARS}^2 \cdot N \cdot N_f \cdot \left[ \chi^{(3)}(v_{if}) \right]^2 \cdot I_p^2 \cdot I_S \cdot L^2 \cdot \frac{\sin^2(\Delta k \cdot (L/2))}{(\Delta k \cdot (L/2))^2} . \tag{14.22}$$

As before, $I_p$ and $I_s$ stand for the pump and Stokes wave intensities; $N$ and $N_f$ are the total and excited-level molecular number densities, respectively; and $\Delta k$ is the phase-mismatch between the wave vectors (see collinear-wave part of Figure 14.19), which can be expressed as

$$\Delta k = \left| k_p \right| + \left| k_p \right| - \left| k_S \right| - \left| k_{CARS} \right|. \tag{14.23}$$

Equation 14.22 makes clear that the CARS signal depends critically on some experimental parameters. Evidently, it depends on the product of the laser intensities; then it depends on the square of the molecular number density, $N^2$ (note that $N_f =$ factor·$N$ is functionally dependent on the total number $N$); and the CARS-conversion efficiency (the last term in Equation 14.22) is a nonlinear ($\sin^2 x / x^2$) function of the phase mismatch. The latter is shown in the bottom part of Figure 14.19. All this means that, in general, CARS-signal intensities are difficult to predict *a priory*, without carefully considering all individual terms in Equation 14.22.

While one can easily play around with laser light intensities and the associated molecular number densities, phase-matching is the crucial part of any optimization procedure, the more so the higher the mismatch, which, according to Figure 14.19, can severely reduce the signal $I_{CARS}$. As was already discussed in Chapter 5.3, proper phase-matching makes it paramount to adjust. The angles between the various propagating waves are adjusted so that the propagation vectors meet at the same point in the three-dimensional vector space. Two particular cases (so-called BOXCARS configurations) for this are shown in Figure 14.20, one with all light waves in the same plane and the other in which the Stokes and anti-Stokes (CARS) waves are in a plane perpendicular to the one in which the Raman pump laser waves propagate.

The BOXCARS configurations shown in the figure highlight three distinctive issues. First, with careful adjustment of the positions of the incoming laser beams on the focusing lens, laterally from the center propagation axis for all laser light beams, one can achieve perfect phase-matching if the refraction angles of a beam through the lens matches the angle determined for the particular beam from the vector phase diagram. Second, the angles between the focused beams mean that very careful laser beam positioning is required so that they actually overlap in the molecular sample region. A further caveat of angled overlap is that the interaction length $L$ in the sample can be severely reduced in comparison to the collinear beam configuration, which will affect the overall signal, which is dependent on $L^2$, according to Equation 14.22. In fact, a largely reduced $L^2$ may

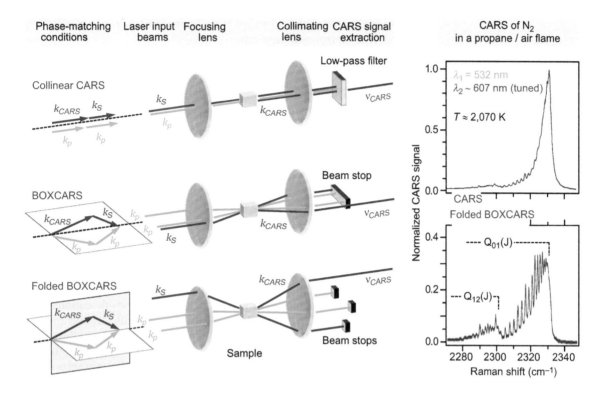

**Figure 14.20**    BOXCARS configurations (in-plane and out-of-plane cases) of incident laser beam propagation directions to overcome phase-mismatch. The parallel input laser beams are positioned appropriately so that their angle toward the focal point of the CARS interaction volume matches the angles from the vector diagrams. On the left, CARS spectra for collinear and BOXCARS configurations are compared; for details, see text. (Data adapted from Alden, A. et al., *Phys. Scripta* 27 (1983): 29–38.)

even be detrimental to the increase in CARS efficiency, $\eta$. Third, the angled laser beam arrangement is rather advantageous since the beams are spatially separated after the collimating light collection lens, so that they can easily be eliminated from reaching the CARS photodetector using beam dumps.

An example for the advantages/disadvantages of the various CARS configurations is included in Figure 14.20, here for the $Q_{01}$-branch spectrum of molecular nitrogen, $N_2$ (see, e.g., Alden et al. 1983). While the BOXCARS result is lower in intensity, on the one hand, on the other hand, it certainly has gained in spectral resolution when compared to the collinear, non-phase-matched configuration. It should be noted here that for a multitude of studies, $N_2$ serves as a test molecule for comparative calibration purposes, in particular for CARS experiments of gaseous samples.

What Figure 14.20 (partially unresolved line structure) and, further, Figure 14.22 (fully resolved line structure) do not reveal is that the line shape of the CARS signal may be rather complicated. Besides the resonant parts, as indicated in the normal, simplified transition level diagrams—like the one shown in Figure 14.1 or 14.23—the total CARS signal also contains an inherent, nonresonant background. This becomes evident when looking closer at the third-order nonlinear susceptibility tensor $\chi^{(3)}$; its form will depend on the level structure of the materials (molecules) to which CARS is applied. The complex susceptibility is often approximated in the so-called Blombergen notation, $\chi = \chi' + i\chi'' + \chi_{\mathrm{NR}}$; for the third-order term, one then has

$$\chi^{(3)} = \underbrace{\chi^{(3)}_{\mathrm{dis}} + i \cdot \chi^{(3)}_{\mathrm{abs}}}_{\text{resonant}} + \underbrace{\chi^{(3)}_{\mathrm{NR}}}_{\text{nonresonant}} , \qquad (14.24)$$

where the real and imaginary terms of the resonant susceptibility are associated with dispersion (indexed "dis") and absorption (indexed "abs") in the medium. The nonresonant contribution can be linked to one or several far-off-resonance transitions, which, however, add coherently, too. In the context of Equations 14.22 and 14.23,) one then finds that the resonant CARS signal exhibits a phase shift of $\Delta\varphi = \pi$ across the actual resonance profile, whereas the nonresonant part does not. Hence, the spectroscopic line shape of the CARS intensity may be asymmetric, as indicated in Figure 14.21: the line center is slightly shifted, and the shape of the line exhibits a "Fano-type" profile. For further details on CARS line shapes, see, e.g., El-Diastry (2011).

Finally, it is worthwhile to briefly address the quantum theory modeling of the CARS process. Simplistically, the simultaneous presence of fluxes of photons $h\nu_{\mathrm{p}}$ and $h\nu_{\mathrm{S}}$ couples the initial and final states, i.e., $|i\rangle$ and $|f\rangle$, coherently via the (virtual) intermediate energy level $|x\rangle$. Thus, the molecule resides in a so-called coherent superposition of states. This coherent superposition of the two states is then interrogated by probe laser photons (conveniently, again, $h\nu_{\mathrm{p}}$ but not mandatory). This probing generates Stokes and anti-Stokes photons in the usual way, via another (virtual) intermediate energy level $|x'\rangle$, and the molecular system ends up in the ground state $|i\rangle$ again or another final state $|f'\rangle$. As a consequence of this probing, the molecule loses its coherent state superposition since it resides once again in a single energy state only. Note that coherent state superposition will be addressed again in Chapter 14.6.3 when discussing time-resolved tr-CARS, using fs-laser sources; there, coherent state superposition plays a central role.

A concise description of the CARS quantum model, based on the density matrix approach, can be found, e.g., in Potma and Mukamel (2013). It may be worth noting that in the quantum model of CARS, no energy is deposited in the molecule during the scattering process. Rather, the molecule acts like a nonlinear medium, converting the frequencies of the three incoming laser light waves into a CARS signal; the process is akin to a parametric conversion process, as discussed in Chapter 5.4.

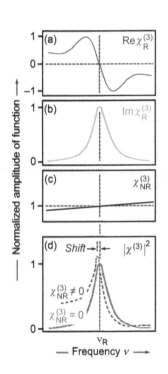

**Figure 14.21** Frequency dependence of the two resonant and the nonresonant terms of the third-order susceptibility $\chi^{(3)}$, as given in Equation 14.24. In the lower panel (d), the line shape function relevant for a CARS transition is given, without and with a nonlinear susceptibility contribution.

## 14.6.2 Tuned single-mode and ns-pulse CARS

Because of the high laser intensities to realize the CARS process, via population of an (real) excited vibrational state, in general, pulsed laser sources were initially utilized. However, already by the mid-1970s, laser and detector technology had advanced sufficiently to also realize CARS using low- to medium-power cw lasers.

The first observation of the third-order effects leading to nonlinear Raman effects, including CARS, was reported by Maker and Terhune (1965), who studied nonlinear polarizability effects in crystalline and isotropic media, stimulated by a giant-pulse ruby laser. At the time, the process was not yet dubbed CARS; that happened only about a decade later. During the subsequent years, and even up until now, CARS has been one of the most used spectroscopic, analytical methods to study combustion processes and plasma samples.

Rather than providing an exhaustive survey—which is beyond the scope of this chapter—here, the aspect of spectral resolution will be addressed through a few key examples. As already highlighted frequently in earlier chapters, the observed width in a spectral line is, broadly speaking, determined by the convolution of the exciting (pump) laser line width and the inherent molecular system line width (e.g., the Doppler width is the limiting factor in gas phase samples). This means that in order to observe molecular-system inherent line widths, one requires narrow bandwidth laser sources, substantially narrower than the molecular line width (ideally single-mode laser sources). As it turns out, the line width of the laser stimulating the downward Stokes Raman transition is only of secondary importance. If it is narrow, it only probes the shape of what the pump laser provides; if it is (much) wider, it comes into resonance with a multitude of final Stokes states $|f\rangle$. This latter property is the key aspect in fs-CARS, discussed in Section 14.6.3.

The example shown in Figure 14.20 is a good one to highlight the advantages/disadvantages when comparing CARS using scanning narrow-bandwidth Stokes laser and broadband laser sources. In their studies of a multitude of molecules ($CH_4$, $CO$, $CO_2$, $H_2$, $H_2O$, and $N_2$) in burner flames, Alden et al. (1983) used ns-pulse laser sources; their CARS pump laser was a frequency-doubled Nd:YAG laser ($\lambda_p = 532$ nm, with estimated line width $\delta\nu_p \approx 30$ GHz equivalent to $\delta\tilde{\nu}_p \approx 1$ cm$^{-1}$) and the operation of the dye laser for the Stokes transition could be changed between scanning configuration (with $\Delta\tilde{\nu}_S = 0.3$ cm$^{-1}$) or broadband configuration (with $\Delta\tilde{\nu}_{S,BB} > 100$ cm$^{-1}$). Depending on the burner conditions, the Doppler width is of the order $\delta\tilde{\nu}_D \lesssim 0.1$ cm$^{-1}$; i.e., the pump laser line width is expected to dominate the observed CARS lines. This was indeed observed by the authors who determined a CARS line width of $\delta\tilde{\nu}_{CARS} = 0.8$ cm$^{-1}$. When inspecting the experimental traces in Figure 14.20, one notices that only for the higher J values are the $Q_{01}$-branch of $N_2$ lines fully resolved. In this context, it should be noted that the displayed data were recorded with the broadband dye laser source, meaning that the spectrometer needed for spectral dispersion of the overall CARS response had to have a better resolution than dictated by the pump laser line width (the spectrometers used by the authors did fulfill this criterion). Not shown are the narrow-bandwidth scans that the authors made for comparison, which revealed exactly the same resolved spectral structure and demonstrated the need for a spectrometer of very high resolution.

Interestingly, more than 30 years on, still the same approach is in use, as demonstrated in Figure 14.22 for the $Q_{01}$-branch of $H_2$. In their experiments on $H_2$ in microwave plasmas, Tuesta et al. (2015) similarly used a pulsed green Nd:YAG laser radiation and a broadband dye laser whose center wavelength of the Stokes laser was adjusted to match the $Q_{01}(1)$ transition in $H_2$. Other than the $Q_{01}(0)$ line, all others are fully resolved, again with a linewidth that was dominated by the line width of the pump laser radiation.

In order to achieve spectral resolutions, which may reveal the intrinsic properties of the molecular system, like natural width, pressure broadening, or Doppler profiles, a (normally single-mode) cw laser source is required for pumping the Raman transition in the CARS scheme. In addition, the laser stimulating the Stokes transition needs to be scanned across the resonance. A broadband system is, in general, unsuitable since then a spectrometer would be required, whose resolution is almost always worse than the molecular-intrinsic line widths. Measurements using cw lasers have been performed since the mid-1970s, again predominantly in applications to combustion processes (see, e.g., Fabelinski et al. 1977).

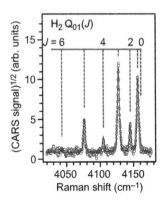

**Figure 14.22** CARS spectrum of the $Q_{01}$-branch of $H_2$ in a high-temperature microwave plasma. The data points (blue circles) are fitted (red line) to a temperature of $T = 1033$ K. (Data adapted from Tuesta, A.D. et al., *J. Micro Nano-Manuf.* 4 (2015).)

In their experiment, the authors used pump and Stokes laser sources with line widths of $\delta v \approx 30\text{--}40$ MHz, so that the width of the CARS spectral lines was not any longer dominated by the lasers.

In another experiment, this time not in a combustion environment but in a supersonic molecular beam arrangement, Gustafson et al. (1983) not only succeeded in performing high-resolution CARS spectroscopy but also were able to trace the influence of line broadening mechanisms. They also used single-mode cw lasers (an $Ar^+$ laser with $\delta v \approx 30$ MHz and a tunable dye laser with $\delta v = 3$ MHz). Evaluating transition line shapes, they derived that in their experiments, transit-time broadening was the dominant contribution to the line width. It is interesting to note that in supersonic beam experiments, the molecular number densities being probed are very small. In the particular case of Gustafson's experiments, the laser–molecular beam interaction volume was less than $10^{-8}$ cm$^3$, and based on the expansion conditions, the net number of molecules contributing to the CARS signal was $N_J < 3 \times 10^{10}$ for a single rotational-$J$ component.

### 14.6.3 Broadband fs-pulse CARS and time-resolved CARS

From the description of the various experimental examples, it probably has become clear that, in general, the trend in CARS has been to use broadband Stokes laser sources in combination with a spectrometer rather than tunable laser sources and scanning their wavelength through Raman resonances. This is because the majority of Raman techniques have been aimed at addressing analytical problems in which ultrahigh spectral resolution is not really an issue. Also, tuning procedures involving processes of low transition probability, in general, proves to be rather time-consuming. Also, the need for careful wavelength selection and wavelength scanning lets CARS be less universal than spontaneous RS. This deficiency has become much less of an issue with the advent of broadband ps- and fs-pulse laser sources; the latter, with bandwidths of several hundreds of wavenumbers, have provided a means to simultaneously access a wide range of vibrational frequencies of a multitude of molecular species, thus replicating spontaneous RS in its universality. This is shown schematically in Figure 14.23, in which the conceptual generation of full Raman spectra is compared for scanning and broadband CARS.

In the broadband case, the Stokes laser radiation stimulates transitions into a manifold of final states $|f\rangle$, and the probing by these states, utilizing the primary pump laser radiation again, yields the complete CARS spectrum in one single shot, in principle (single-shot CARS will be briefly discussed towards the end of this section). In principle, the anti-Stokes spectra will be exactly the same, although they need to be normalized to the wavelength-dependent intensity distribution of the Stokes laser source, of course.

But another property in SRS, and broadband processes in particular, is its coherent nature. As a consequence, the final states $|f\rangle$ are prepared in a so-called coherent superposition, as already stated a few times earlier, which means that the states are coupled in phase. If left alone, this coherent superposition (wave packet in the quantum picture) will lose its phase relation, and the link between the states will disappear over time. The duration of this loss will be determined primarily by the lifetime of the excited vibrational states $|f\rangle$, which is at least of the orders of a few ns, or much longer (associated with the sum of the pure radiative lifetime of the vibrational level and dephasing collisions). This time is rather long

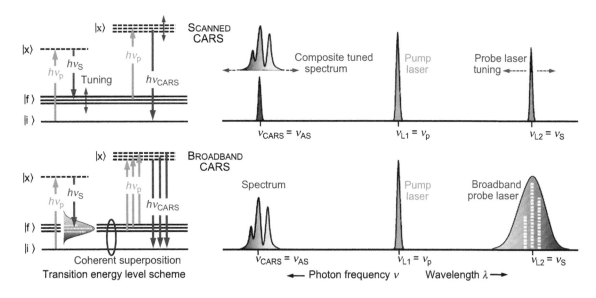

**Figure 14.23** Comparison between scanned, narrow-band and broadband ps-/fs-pulse CARS. (Left) Energy level diagram with pump-, Stokes-, and CARS-transitions; the coherent superposition of stated is indicated by the violet oval. (Right) Conceptual CARS spectra for the two cases.

in comparison to the pump and stimulating laser pulses of ps or fs duration. Thus, it becomes possible to probe the coherent superposition of excited vibrational states $|f\rangle$ with a secondary pump pulse, delayed with respect to the initial pump pulse, which generates a time-delayed CARS signal. The conceptual implementation is shown in Figure 14.26 in Section 14.7.

This particular approach is now known as *time-resolved CARS*, or in short, tr-CARS. Since ps-/fs-pulses are utilized for the excitation, stimulation, and probe steps, it therefore becomes possible to probe molecular motions on the ultrashort time scale—like molecule vibration—during the time interval over which coherent state superposition would normally survive. In principle, this can be viewed as being akin to the femtosecond pump–probe experiments of vibrational motion, performed by Zewail's group, who were the first to follow the vibrational motion by time-delayed probing of the wave packet. In those early experiments, they used photo-dissociation out of the coherently prepared wave packets, in which the product carried the information on the oscillating wave packet (see, e.g., Dantus et al. 1990).

However, while those photo-dissociation experiments probed the evolving vibrational wave packet locally, within the molecular potential, and those could follow the molecular vibration directly, this is not the case in standard CARS: no localized probing within the internuclear distance in the potential is possible when utilizing excitation schemes via "virtual" levels far away from any resonance. On the other hand, when exploiting resonance Raman routes via vibrational levels in an electronically excited state, localized probing of vibrational wave packets within the potentials becomes indeed possible utilizing tr-CARS. This has been demonstrated, for example, in experiments investigating molecular dynamics in iodine, by changing the delay between the primary and secondary pump laser pulses in a resonant tr-CARS scheme involving the electronic ground X and excited B states of $I_2$. The coherently generated wave packet motion on both the electronically excited and ground states could be detected as oscillations in the coherent anti-Stokes signal (see Schmitt et al. 1997).

But in nonresonant CARS, one can visualize actual vibrational motion in a slightly indirect way. Since the Stokes radiation that stimulates the Raman transition is broadband, it normally populates more than one final, vibrational state. Thus, one generates a multitude of coherently superimposed ground–excited state pairs, which are all mutually coherent as well. Thus, one should be able to observe beats between such pairs, with the beat frequency equal to the spacing of the (two) final states, $\Delta\nu(v_1,v_2)$. An example of this is shown in Figure 14.24.

The data shown here correspond to time-resolved fs-CARS measurements of $N_2$ in a hot, atmospheric-pressure $C_2H_4$/air flame (Roy et al. 2010). An oscillation frequency of $\Delta\nu_{beat} = 834$ GHz can be deduced from the time-spacing $\tau_{beat} \approx 1.2$ ps between consecutive beat maxima recorded in the time-delayed CARS probing. This corresponds to an energy difference of $\Delta\tilde{\nu}_{beat} = 27.8$ cm$^{-1}$; the authors associate this to the beating between the fundamental and first overtone vibration of $N_2$. Also evident from the figure is the dephasing of the coherent superposition of the vibrational states within a few picoseconds. The authors also showed that the dephasing duration decreased with higher flame temperature, which is in line with expectation that molecular collision rates increase.

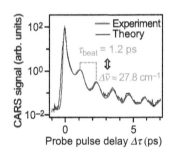

**Figure 14.24** Comparison of the time-resolved fs-CARS signal for $N_2$ and the results from the theoretical model of the authors, based on a best-fit flame temperature value of $T = 2020 \pm 50$ K. (Reprinted from *Prog. Energy Combust. Sci.*, 36, Roy, S. et al., 280–306, Copyright 2010, with permission from Elsevier.)

It should be noted that all CARS studies addressed thus far in this section were gas phase studies. But equally, it is possible to investigate molecules in liquid environments; see, e.g., Fickenscher and Laubereau (1990) and Knutsen et al. (2006) to name but two examples over two decades in time. Equally, CARS has been frequently applied to samples in the condensed phase, in particular to the study of biological materials. In general, those applications were related to spectral imaging of specific macromolecules embedded in the tissue; some key examples will be discussed in Chapter 19.4.

In the context of number density, it is worth noting that the sensitivity of CARS in general and time-resolved CARS in particular has improved significantly over the years, but as a rule of thumb, still, densities of about $N_{v,J} \approx 10^{10}$ cm$^{-3}$ are required to result in meaningful CARS signals. Now, this limit has been broken spectacularly. In conjunction with additional SERS enhancement, Yampolsky et al. (2014) demonstrated that the beating between coherently super-positioned, vibrational modes in a single molecule could be followed, as well as the dephasing of said coherent superposition. The authors investigated single molecules of bipyridylethylene (BPE), which were confined in the gap between two gold nanoparticles. The complete entity was encapsulated in a silica envelope for structural stabilization. The CARS laser excitation system consisted of an fs-laser setup, which provided the pump pulses at $\lambda_p = 714$ nm, the Stokes pulses at $\lambda_S \approx 809$ nm, and the (time-delayed) probe pulses at $\lambda_p = 714$ nm again. The pump–Stokes laser pulse pair prepared the molecule in a coherent superposition of a number of the over 60 vibrational modes of BPE, in particular the pair of the pyridine ring-stretching mode $\nu_{8a}$ at $\tilde{\nu} \cong 161$ cm$^{-1}$ and the stretching mode $\nu_{C=C}$ at $\tilde{\nu} \cong 1647$ cm$^{-1}$, with the highest transition probability. Their energy spacing of $\Delta\tilde{\nu} = 35$ cm$^{-1}$ matches nearly exactly the value deduced from the beat frequency observed in the tr-CARS measurements (see Figure 14.26 in Section 14.7).

Finally, a very interesting variation of tr-CARS will be briefly discussed, the so-called *single-shot tr-CARS*. In this particular approach, chirped probe pulses are used to map the time evolution of the molecular dynamics, stimulated by the pump and Stokes pulse pairs, onto the spectrum of the CARS signal pulses. Specifically, for this, the temporal width of the probe pulse is linearly expanded by sending it through a glass rod of appropriate refractive index and length; in this way, an original fs pulse of about 100 fs duration can be stretched to about 2–3 ps (FWHM) and in which the red part of the spectrum arrives earlier than the blue part. This time-domain single-shot detection technique, based on nonresonant fs-CARS, was developed and applied to tracing molecular processes in turbulent media, like flames and combustion processes, and to deduce temperatures from the spectra on a very short time scale (based on a single tr-CARS event from a laser system operating at kHz repetition rates); see Lang and Motzkus (2002).

In their publication, the authors provide an analytical description of this mapping process, which specifies ranges of linear and nonlinear time-to-frequency mapping, and an experimental realization is presented for single-shot thermometry. Also, the concept of time-to-frequency mapping by chirped probe pulses is intuitively demonstrated for data from a tr-CARS experiment on $H_2$. In principle, the dynamical process detection is as fast as the time that is required to acquire and evaluate the CARS spectrum (acquisition on and readout of the array detector of the spectrometer and the computer analysis); with modern instrumentation, this may be achievable within a few milliseconds.

### 14.6.4 Spontaneous, stimulated, and coherent anti-Stokes Raman spectroscopies in comparison

In the literature, the coherent technique of CARS is often compared to spontaneous RS, since both probe the same Raman active modes. However, beyond this same-active-mode property, there are quite a few differences that one encounters in CARS; in particular, the interpretation of observed spectral line shapes and intensities is not always easy and straightforward. In Table 14.3, a summary of the similarities and differences between spontaneous RS and CARS is provided; for comparison, the same properties are listed in parallel for SRS (including the SRG and SRL variants).

# 14.7 BREAKTHROUGHS AND THE CUTTING EDGE

Of all the nonlinear/enhancement approaches in RS, for sure, SERS has had the biggest impact. While in the early days, the technique struggled with reproducibility issues and researchers had difficulties in interpreting results, since the early 2000s, SERS has seen an explosion of publications, with more than 15,000 papers written since then, addressing substrate manufacture, interpretation of phenomena, and usage in a wealth of applications. By and large, this may be attributed to the ability to fabricate reproducible nanoparticles/nanostructures, the key ingredient in any SERS measurement.

### 14.7.1 Breakthrough: SERS using silver films over nanospheres (AgFON)

Although randomly roughened surfaces and metal–particle colloids exhibited sufficient enhancement to generate measurable SERS signals, many of the details

**Table 14.3** Comparison of Key Features of Spontaneous, Stimulated, and Coherent Anti-Stokes Raman Spectroscopy

| Process | Spontaneous Raman Spectroscopy | Stimulated Raman Spectroscopy | Coherent Anti-Stokes Raman Spectroscopy |
|---|---|---|---|
| Scattering process | Results in energy transfer between light and matter | Results in energy transfer between light and matter | Parametric process (molecules are left unchanged after the interaction) |
| Experimental realization | A single laser source (cw or pulsed) is used | Two laser sources are needed (in general pulsed, less common cw) | Two laser sources are needed (in general pulsed, less common cw) |
| Nature of transition | Spontaneous transition | Coherently driven transition | Coherently driven plus spontaneous transition |
| Spectral line intensity and shape | Proportional to $\mathrm{Im}\{\chi^{(1)}\}$ | Identical spectral positions to spontaneous Raman but shapes $\propto \mathrm{Im}\{\chi^{(3)}\}$ | Distorted complex "Fano" spectral line positions and shapes $\propto |\chi^{(3)}|^2$ |
| Signal property | Incoherent addition from individual molecules | Coherent addition from individual molecules. | Coherent addition from individual molecules; for full quantification, phase-matching is required |
| Signal directionality | Emitted in all directions | Directional along pump or Stokes laser beams | Collimated beam in phase-matching direction |
| Signal amplitude | Linear with interaction length; linear in molecule concentration | Quadratic with interaction length; linear with molecule concentration | Quadratic with interaction length; quadratic with molecule concentration |
| Signal background | Sensitive to background fluorescence | Immune to background fluorescence | Inherent nonresonant background from far off-resonance transitions that add coherently |
| Detection limits | Laser shot noise limited sensitivity | Laser shot noise limited sensitivity | Limited by distinction between resonant and nonresonant parts of CARS signal |

of the EM enhancement mechanism, discussed in Section 14.2, were difficult to access due to the broad distribution of surface feature sizes. In this context, the work reported by Litorja et al. (2001) provided a breakthrough that may be seen as one of the most important cornerstones for much of the subsequent development of nanostructured SERS substrates. In the reported study, the researchers' primary goal was to create SERS-active surfaces that were reproducible, stable, and predictable. Their general approach is nowadays known as AgFON—silver film over nanospheres. In their studies, they deposited silica ($SiO_2$) nanospheres of radius $a = 200$ nm on a polished metal substrate; this layer of nanospheres was then coated under ultra-high vacuum (UHV) conditions with a thin layer of silver (Ag). Rather, large areas of a uniform nanostructured surface could be generated in this way; in measurements with a test analytes (benzene, pyridine, and $C_{60}$), they showed that substantial enhancement in the Raman signals could be achieved, that the spectra allowed for easier interpretation of the molecular adsorption behavior on the surface, and that their AgFON nanostructured substrates could even be regenerated after use.

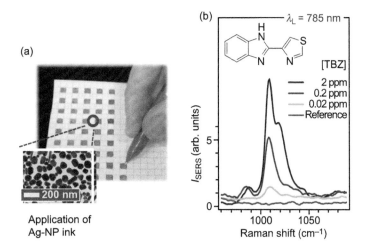

**Figure 14.25** Pen-on-paper approach SERS. (a) Generating SERS substrate areas on a sheet of paper, using a pen loaded with plasmonic nanoparticle ink. (b) SERS spectra of TBZ on AgNP substrate, for different concentrations of TBZ. (La Porta, A., et al.: *Small*. 2014. 10. 3065–3071. Copyright Wiley-VCH Verlag GmbH & Co. KGaA. Reproduced with permission.)

## 14.7.2 Breakthrough: Toward "pen-on-paper" SERS substrates

The fabrication of nanostructured SERS substrates is, in general, relatively complex and expensive; thus, SERS has rarely made it into routine analysis applications. This may well change with a very ingenious demonstration of such substrates, which are easily and cheaply to produce and are as usable as other, commercially produced devices. Polavarapu et al. (2014) utilized commercially available conductive nanoparticle ink and applied this to ordinary photocopier paper, in homogeneous areas of up to $3 \times 3 \text{ mm}^2$ (see Figure 14.25a). With such simple and cheap but nevertheless effective SERS substrates, they were able to quantitatively measure minute amounts of thiabendazole (TBZ)—a common, widely used fungicide—with detection sensitivity of 20 ppb in a 10 μL sample volume (see Figure 14.25b). Thus, this sort of SERS substrate would certainly have the potential to simplify *in situ* field tests for the presence of this pesticide on harvested fruit (see, e.g., the studies by He et al. 2014 and Müller et al. 2014).

## 14.7.3 At the cutting edge: Seeing a single molecule vibrate utilizing tr-CARS

It always has been the "holy grail" of chemists (and biologists) to be able to peek into individual, large molecules and follow the movement of any of its subentities, in real time. However, existing single-molecule methods require >1 ms for spectroscopic (energy) finger-printing, which is insufficient to access molecular vibrational dynamics on the time scale of ≤1 ps.

Researchers at the Center for Chemistry at the Space-Time Limit (UC Irvine) succeeded for the first time to optically measure, with fs-time resolution, vibrational dephasing within a single molecule. They probed single molecules of BPE, using time-resolved CARS in conjunction with plasmonic enhancement (see Yampolsky et al. 2014). The molecule was confined in the gap of a gold nanoparticle dumbbell, where it was prepared by a ps/fs-broadband pump–Stokes pulse-pair (the "pump") in a coherent superposition of vibrational states. The associated wave packet evolves in time, and the time-delayed "probe" pulse generates a "monitor" CARS signal; its amplitude is modulated with a frequency corresponding to the beat between the excited vibrational modes. This overall concept is shown in Figure 14.26a. Only when the stimulating Stokes wavelength $\lambda_2$ was tuned to actually prepare the vibrational wave packet superposition were signal beats observed whose time period corresponded to the energy spacing of the excited vibrational states, as shown in Figure 14.26b. Note that the nonresonant background signal could be attributed to coherent RS at plasmons of the metal nanojunction (see Crampton et al. 2016).

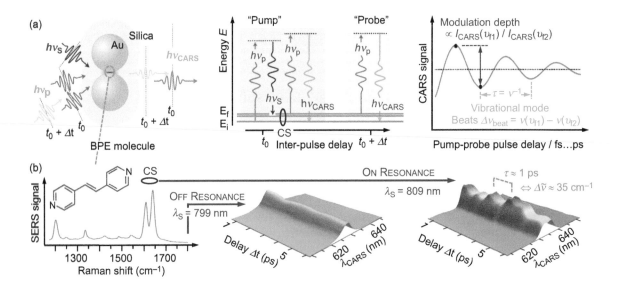

**Figure 14.26**  Probing the dynamics of single molecules of BPE. (a) The measurement concept. (b) Experimental results for on-/off-resonance tr-SE-CARS, revealing the time evolution of wave-packet quantum beats from the coherent superposition (CS) of vibrational levels. (Data adapted from Yampolsky, S. et al., *Nature Photon.* 8 (2014): 650–656.)

# CHAPTER 15

# Laser-Induced Breakdown Spectroscopy

In these final two chapters on laser spectroscopic techniques, two topics with a "twist" will be addressed, namely, spectroscopic methods that are based on laser ablation and laser ionization.

In the first one, *laser ablation* is addressed, which has been and is widely used to globally or selectively remove particles (atoms or molecules) from solid surfaces and transfer them in a controlled manner into the gas phase. Then, laser spectroscopy or other types of analytical spectroscopy are carried out, either directly on the evaporated, low-density species or after their interaction/reaction with other particles (atoms, molecules, surfaces). A plethora of examples of both can be found in the scientific literature, more or less since the time that powerful, pulsed laser sources have been available. A survey and discussion of the many examples of breathtaking beauty would be well beyond the scope of this book, the more so that its emphasis lies on direct laser-induced (optical) spectroscopy. Therefore, here only one specific laser ablation modality will be discussed, namely, *laser-induced breakdown spectroscopy* (*LIBS*), in which a high-energy laser pulse generates luminous plasma, whose light emission is spectrochemically analyzed.

In *laser ionization* spectroscopy (the topic of Chapter 16), the approach is that the resonance of photon energies with atomic/molecular transitions is still exploited (like in absorption and laser-induced fluorescence spectroscopy), but that the interaction with a second laser photon ionizes the (selected) species. The charged particles generated in this two-step process—ions and electrons—can easily be detected with efficiencies reaching 100%, which is in general far superior to photon detection efficiencies.

Of the laser ablation methodologies, only LIBS will be outlined in its concepts, together with key representative examples.

## 15.1 METHOD OF LIBS

LIBS, often also called *laser microanalysis*, is an optical emission technique for analyzing the elemental composition of materials (solids, liquids, and gas samples); it is used in a wide range of scientific and technical applications. This spectroscopy technique is based on the analysis of the spectral lines emitted by microplasma, which is created by the interaction of a laser pulse and a sample. The vaporization and excitation of the sample up to very high energy levels generates atomic and ionic species. Subsequently, as the plasma relaxes and the excited elements/atoms emit light of specific wavelengths; this is collected by a spectrometer and spectrally analyzed to determine—qualitatively or quantitatively—the presence of elements and the chemical composition of the sample.

The first spectral analysis using LIBS is ascribed to Runge et al. (1964); within only a few years thereafter, the technique developed from a qualitative into a quantitative method, used to determine the elemental composition of distinct samples. Nowadays LIBS has become a well-established methodology in many scientific and technical areas, such as, e.g., (1) for basic understanding of plasma physics and spectroscopy; (2) in analytical applications in industry, forensic science, cultural heritage, geochemistry, and biomedicine; and (3) for remote sensing of hazardous materials such as explosives, nuclear waste management, and space exploration—to name but a few.

The main advantages of LIBS analysis are the following (see, e.g., Miziolek et al. 2006; or Noll 2012):

- No initial sample preparation is required.

- It allows simultaneous multielemental analysis.

- It is very fast and nondestructive (i.e., the analysis consumes only very little material of the sample).

- It can work on-line and *in situ*.

### 15.1.1 Basic concepts: Plasma generation and characterization

The first step of a LIBS experiment concerns the interaction of the laser with the sample, which in principle can be a gas, liquid, or solid, although for the present discussion, we will assume it is a solid sample. When the laser power density is higher than a material-dependent value known as the *ablation threshold* (on the order of $10^9$–$10^{14}$ W·m$^{-2}$), the material is ablated and the energy delivered to the material does not only vaporize the sample but also breaks chemical bonds and ionizes the elements, resulting in a plasma plume. The material breakdown takes place immediately after the laser radiation strikes the surface of the sample; typically, this occurs when the electron density has reached about $n_e \approx 10^{19}$ cm$^{-3}$. The transient plasma contains neutral particles, ionic species, and free electrons that expand away from the sample surface while also emitting radiation due to the high plasma temperature, typically on the order of $T_p \sim 10^3$–$10^5$ K.

An important phenomenon that can take place in LIBS is the so-called *plasma shielding*, in which a significant part of the laser pulse is absorbed during the plasma expansion by electrons (inverse *bremsstrahlung*) and/or multiphoton ionization. Consequently, if plasma shielding were to occur too early, a significant fraction of the laser pulse does not reach the sample; thus, the resulting plasma may not be enough to extract an analytical signal. Although an increase in the electron density enhances the laser absorption, favoring the occurrence of plasma shielding, it should be noted also that the enhanced absorption gives rise to a higher plasma temperature; in turn, this increases the plasma emission

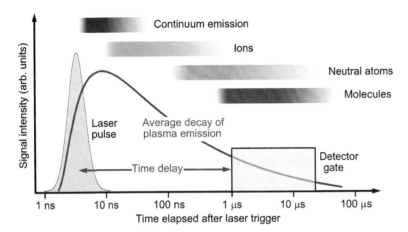

**Figure 15.1**    Schematic representation of the relevant periods of plasma evolution in LIBS.

intensity. In practice, the experimentalist must optimize the working parameters to minimize or even avoid plasma shielding.

One of the most relevant experimental parameters in LIBS performance is the laser pulse duration. In fact, for laser light with a femtosecond-pulse duration, plasma shielding is negligible due to the significantly reduced interaction time with/between electrons and neutral and ionized species. In optically dense plasma, one often refers to the so-called *plasma resonant frequency* $v_p$, which is given by

$$v_p = \left(4\pi \cdot n_e \cdot e^2 / m_e\right)^{1/2} = \left(8.9 \times 10^3\right) \cdot n_e^{1/2}. \tag{15.1}$$

Here, $e$, $m_e$, and $n_e$ are the electron charge, the electron mass, and the electron density (in units of $cm^{-3}$) in the plasma, respectively. Note that a critical electron density does exist, at which the plasma frequency equals the laser frequency $v_L$ (i.e., $v_L \equiv v_p$); for this condition, the plasma strongly absorbs the laser radiation. On the other hand, for $v_L < v_p$, the laser radiation is reflected by the plasma, and, thus, a reduction of the mass ablated from the sample ensues.

The relevant periods of plasma formation and subsequent relaxation are illustrated in Figure 15.1. Essentially, for the first hundred nanoseconds or so following the laser ablation, the emission spectrum is dominated by a continuum-radiation background. During this time period, only the strongest atomic lines can be seen above the background. As the plasma relaxes, the temperature decreases, the electron density diminishes, and the background significantly fades such that the atomic spectral lines can be observed. Thus, in a LIBS experiment, the detector needs to be gated and delayed for typically ~1 µs after the laser pulse, to avoid the continuum background and collect the characteristic atomic line emissions. Once the emission spectrum is measured, qualitative and quantitative information, such as, e.g., elemental composition, can be extracted from the spectra. In addition, emission line properties—such as frequency shifts, widths, and shapes—can be used to deduce information about the plasma temperature and the electron density.

Analytical LIBS applications are normally based on assuming that the plasma is in the state of so-called *local thermodynamic equilibrium* (*LTE*); for its definition and discussion, see, e.g., Hahn and Omenetto (2010). This state is different from that of *complete thermodynamic equilibrium,* for which all processes are balanced and characterized by a single temperature (i.e., all "temperatures" are assumed to be equal: $T_e = T_{ion} = T_{plasma}$). This also means that the radiation absorbed is equal to the radiation emitted. In the hot LIBS plasma, radiative processes are more pronounced than absorption processes, giving rise to a radiative disequilibrium. However, if collisions inside the plasma still dominate, then the same laws describing the full thermodynamics hold, despite the radiative disequilibrium, and one can describe the plasma state by the LTE approximation. In other words, if the collision rate is much higher than the radiative rate, LTE conditions are warranted (see Hahn and Omenetto 2010).

Under the LTE condition, the theoretical Boltzmann–Maxwell and Saha–Eggert expressions (see further below) can be applied to extract basic plasma parameters and the concentrations of analytic species. The most widely accepted criterion for plasma diagnostics, and normally invoked as a proof of the existence of LTE in the plasma, is the so-called McWhirter criterion (McWhirter 1965). It states that to reach the LTE state, the plasma electron density $n_e$ (in units of cm$^{-3}$) should be higher than a threshold value, which is given by

$$n_e \geq 1.6 \times 10^{12} \cdot T^{1/2} \cdot (\Delta E)^2. \tag{15.2}$$

Here, $\Delta E$ (in units of electron volts) is the most energetic electronic transition energy considered and $T$ (in units of kelvin) is the electron temperature. For example, for a typical LIBS experiment in which the most energetic observed transitions correspond to a spectral wavelength of about $\lambda \approx 266$ nm (equivalent to $\Delta E \approx 4.66$ eV) and for $T = 500$ K, the minimum density required to fulfill the LTE condition would be $n_e \approx 1.86 \times 10^{18}$ cm$^{-3}$.

The most relevant parameters for plasma characterization are the linewidth of the atomic emission, the plasma electron density, and the plasma temperature. Line profiles in a LIBS experiment are basically due to Doppler broadening, resonance broadening, and Stark broadening; the latter last is the dominant process. In general, only emission lines of the hydrogen (H) atom and H-like ions exhibit broadening associated with the linear Stark effect. For the H atom, the electron density and the linewidth are connected (Pasquini et al. 2007) via the expression

$$n_e = 8.02 \times 10^{12} \cdot \left(\Delta\lambda_{1/2}/a_{1/2}\right)^{3/2}, \tag{15.3}$$

where $\Delta\lambda_{1/2}$ is the measured full width at half maximum and $a_{1/2}$ is the so-called reduced wavelength that can be obtained from the tabulated values (see Griem 1974). Normally, the H$_\beta$ linewidth is the most used in extracting the plasma electron density. For plasma diagnosis using non-H-like atoms, for which the quadratic Stark effect is predominant, the electron-density dependence on the linewidth can be approximated by (Griem 1964)

$$\Delta\lambda_{1/2} = 2 \times 10^{-16} \cdot w \cdot n_e. \tag{15.4}$$

Here $w$ is the electron impact parameter at an electron density of $2 \times 10^{16}$ cm$^{-3}$ and its values can be found in literature tables (see, e.g., Griem 1964).

From an experimental point of view, the LTE is a good approximation beyond a few hundred nanoseconds after plasma formation and when high power densities are used for laser ablation. Then, the Boltzmann law can be used to describe the energy-level population of the species and the plasma temperature.

The latter can be deduced using the emission intensities of transitions of the atomized species; this is given by (see, e.g., Cremers and Radziemski 2006)

$$I_{ij} = \left(A_{ij} \cdot g_i / \lambda_{ij}\right) \cdot (N/Z) \cdot \exp\left(-E_{ij}/kT\right). \tag{15.5}$$

In the preceding expression, the subscripts $i$ and $j$ refer to the upper and lower levels of the transition, respectively; $I_{ij}$, $A_{ij}$, $\lambda_{ij}$, $g_i$, and $E_{ij}$ are the emission intensity, the Einstein coefficient, the wavelength, the statistical weight of the upper level, and the energy difference of the transition, respectively; $Z$ and $N$ are the partition function and total number density of the species under study; $k$ is the Boltzmann constant; and $T$ is the plasma temperature. It should be noted that Equation 15.5 describes a linear dependence between the line intensity of a given element and its concentration. In other words, the equation could be rewritten as

$$I_{ij} = B \times N, \tag{15.6}$$

where $B$ is a concentration-independent proportionality. Then, the equation can be rearranged into the logarithmic form

$$\ln\left(\frac{I_{ij} \cdot \lambda_{ij}}{A_{ij} \cdot g_i}\right) = -\frac{E_{ij}}{kT} + \ln\left(\frac{N}{Z}\right). \tag{15.7}$$

Thus, plotting the left-side term against $E_{ij}$ will result in a straight line, whose slope is $-1/kT$, and an axis intercept from which $N$ could be extracted (provided the partition function of the element under study is known). It should be noted that without careful calibration, the measurement of relative intensities is not always straightforward, at least not with high precision. Therefore, one often uses multiple measurements of line intensities and extracts the plasma temperature from a plot comparative based on Equation 15.7.

The plasma temperature $T$ can be determined using the intensity ratios line pairs, either from neutral versus neutral lines or from ion versus neutral lines, normally from the same element. In the case of a neutral line pair, one can use Equation 15.7 to obtain the line intensity ratio as

$$I_1/I_2 = (A_1 \cdot g_1 \cdot \lambda_2 / A_2 \cdot g_2 \cdot \lambda_1) \cdot \exp(-(E_1 - E_2)/kT), \tag{15.8}$$

where the indices 1 and 2 refer to the individual lines of the selected pair; the statistical weights refer to those of the excited states of the transitions; and $E_1 - E_2$ is the energy difference between the upper terms of the two transition lines. The plasma temperature can be extracted also by using the *Saha equation*, which relates the line intensity ratio of the ion to the neutral chemical element given by

$$\frac{I_{\text{ion}}}{I_{\text{neutral}}} = \frac{2 \cdot (2m_e kT)^{3/2}}{n_e \cdot h^3} \cdot \left(\frac{g \cdot A}{\lambda}\right)_{\text{ion}} \cdot \left(\frac{\lambda}{g \cdot A}\right)_{\text{neutral}}$$
$$\times \exp\left(-\frac{U^+ + E_{\text{ion}} + E_{\text{neutral}}}{kT_{\text{ion}}}\right). \tag{15.9}$$

Here $I_{\text{ion}}$ and $I_{\text{neutral}}$ are the integrated emission intensities of the transitions in the ion and neutral states; $E_{\text{ion}}$ and $E_{\text{neutral}}$ are the excitation energies of the transitions; and $U^+$ is the ionization potential of the atom (all in units of joules); $T_{\text{ion}}$ is the ionization temperature (in units of kelvin); $n_e$ is the electron density (in units of $cm^{-3}$); $g \cdot A$ is the product of the Einstein coefficient for spontaneous emission of the upper level (in units of $s^{-1}$) and the statistical weight; and $\lambda$ is the transition wavelength (in units of nanometers).

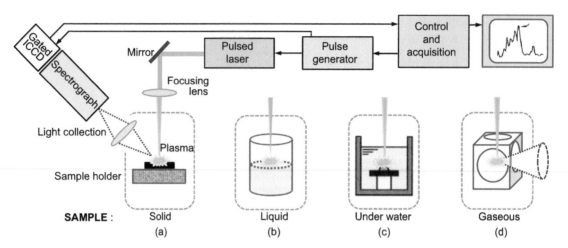

**Figure 15.2** Building blocks for a LIBS system; the basic experimental setup of LIBS is illustrated for (a) solid, (b) liquid, (c) solid underwater, and (d) gaseous samples.

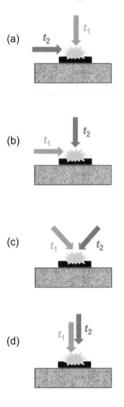

## 15.1.2 Basic experimental setups and ranging approaches

The main building blocks of a LIBS system are (1) a laser system, (2) light delivery and collecting optics, and (3) an optical detection unit with spectral resolution. These distinct blocks are selected and arranged according to any specific LIBS spectroscopic application. Here only the conceptual setup configurations are outlined, as encountered in the usual analytical chemistry applications of LIBS. In addition, some comments are devoted to two-laser/double-pulse implementations, as well as remote- and standoff-LIBS (ST-LIBS) configurations. In Figure 15.2 the basic LIBS instrumentation is shown, comprising the aforementioned building blocks; in principle, the setups are the same for all types of samples—solid, liquid, or gaseous—with only the light excitation/collection geometry needing minor adjustment.

Most commonly, pulsed neodymium-doped yttrium aluminum garnet (Nd:YAG) lasers are used (pulse duration $\tau_p \approx$ 1–20 ns, pulse energy $E_p \approx$ 10–150 mJ); with reference to Figure 15.1, the LIBS signal is typically integrated over a time window of $\Delta\tau_{\text{LIBS}} \approx$ 0.1–2.0 μs and delay times in the range $\Delta\tau_{\text{delay}} \approx$ 0.01–2.5 μs after the laser pulse. Nowadays, with the use of detectors based on intensified charge-coupled device (ICCD) technology, the plasma emission signals can be monitored as a function of time, a methodology which is known as *time-resolved laser-induced breakdown spectroscopy* (TRELIBS); it allows for optimal time resolution, to select the most adequate and reproducible spectral response for analytical applications of LIBS.

**Figure 15.3** Schematic illustration of different double-pulse configurations: (a) orthogonal reheating, (b) orthogonal preablation, (c) crossed-beam reheating, and (d) collinear dual-pulse reheating. The arrows mark the direction of propagation; $t_1 < t_2$ refer to the temporal sequence of the two laser pulses.

## 15.1.3 Double-pulse excitation

An interesting experimental modality in LIBS is the use of double-pulse excitation; this normally comprises two separate laser sources, in different configurations with respect to their propagation directions and relative timing between the pulses. As sketched in Figure 15.3, the configurations are mostly orthogonal or collinear.

For the former, two distinct scenarios are displayed concerning the sequential order of the pulses. While in the *orthogonal reheating* configuration, the pulse from the first (plasma-generating) laser is focused perpendicular to the surface and the second (reheating) is parallel to it. In the *orthogonal preablation* configuration, it is the other way around; i.e., the first laser pulse propagates horizontally to the surface and the second one is directed perpendicular to it, just below where the first pulse was applied.

Evidently, the collinear configuration is easier to align and is therefore the most widely used.

In general, double-pulse excitation in LIBS produces emission enhancement and, therefore, a significant analytical improvement of the technique due to several factors. These include, e.g., a higher ablated mass of the material and the reexcitation of the sample material ablated by the first pulse during the second laser pulse (this increases both the plasma temperature and electron density). An application for which double-pulse LIBS (DP-LIBS) is very appropriate is underwater analysis; this is due to the confinement of the plasma formation as well as the weaker emission intensities when only a single pulse is utilized. In fact, in the double-pulse modality, the first pulse originates a gaseous cavity in the water, just above the surface of the sample being ablated; this cavity is excited in the second pulse. Despite the inherent complexity of underwater plasma generation, detection levels have been achieved that are comparable with those routinely obtained for single-pulse LIBS (SP-LIBS) in air (see, e.g., Sallé et al. 2004).

### 15.1.4 Portable, remote, and standoff LIBS

The "ranging" approach of LIBS has been classified (Fortes and Laserna 2010; Fortes et al. 2013) as portable, remote, and standoff configurations. In a portable system, both the sensor and the operator who manipulates it are close to the target. In contrast, in remote and standoff systems, the target and the operator with its detection equipment are physically separated by some distance. The difference between the two is that in remote configurations, the laser radiation and the LIBS signal are transmitted using a fiber-optic cable, while in a standoff system, the light transmission takes place telescopically via open space.

The first portable LIBS instrument was developed by Yamamoto et al. (1996) for the detection of metal contaminants on surfaces; it could be fitted into a small carry-on case, as depicted in Figure 15.4 (top). It used a small Q-switched Nd: YAG laser source ($\lambda_L = 1064$ nm, pulse duration $\tau_p \approx 4$–8 ns, pulse energy $E_p \approx 15$–20 mJ), a grating spectrograph, a nongateable CCD and data recording on a laptop computer. Among other applications, the instrument was evaluated with very satisfactory results in the analysis of soils and paints, for their content of Ba, Be, Pb, and Sr. The ever-increasing number of applications triggered the development of a series of commercial LIBS instruments; today, quite a few handheld LIBS spectroscopic devices are marketed (see the center part of Figure 15.4), which are used in the detection of chemical elements in a wide variety of samples, including explosives, minerals, gas leaks, paintings, liquids, and food, to name but a few of the most common applications (see, e.g., Pasquini et al. 2007; or Fortes et al. 2013). In the bottom part of the figure, a prototype of a man-portable LIBS (MP-LIBS) system is shown, including an operator backpack system. Equipment such as this is designed for potentially hazardous material

**Figure 15.4** (Top) Portable LIBS surface analyzer developed by Yamamoto et al. (1996). (Reproduced with permission of The Society for Applied Spectroscopy.) (Center) Commercial, handheld portable LIBS system utilized, e.g., by Miziolek (2012). (Image courtesy of Applied Photonics Ltd.). (Bottom) Man-portable/backpack sensor system; see DeLucia et al. (2005). (From Gottfried, J.L. and F.C. De Lucia, Jr. "Laser-induced breakdown spectroscopy: Capabilities and applications." Report ARL-TR-5238. Aberdeen Proving Ground (MD), USA: Army Research Laboratory, 2010. Reproduced with permission of US Army Research Laboratory.)

detection in homeland security and force protection activities. The backpack contains a small-board computer, a lithium ion battery, electronics for the laser, and a broadband high-resolution spectrometer; the laser; a very compact, actively Q-switched Nd:YAG laser, is incorporated into the handle.

Schematic layouts of the remote and standoff LIBS configurations are illustrated in Figure 15.5. One of the main differences of remote LIBS with respect to laboratory-standard equipment is the use of long optical fiber. Commonly, the same type of fiber is used for light delivery and collection. Since laser pulses of high energy (up to $E_p \sim 50$ mJ) have to be transmitted, on the one hand, the core diameter of the fiber needs to be large enough to not exceed the laser radiation damage threshold; on the other hand, it has to be small enough to guarantee a reasonably tight focal spot required for material ablation and subsequent LIBS plasma formation. A good compromise is achieved by choosing a core diameter of a few hundred micrometers for remote systems in which the target is located a few tens of meters. Of further importance for the selection of the fiber type are two properties. First is the transmission in particular in the ultraviolet (UV) spectral region, where many elemental emission lines relevant for quantitative analysis are found; and, second, special care needs to be taken in hazardous environments experiencing high doses of ionizing radiation (see Davies et al. 1995).

With reference to the ST-LIBS configuration, it should be recognized that the detection of hazardous materials, such as chemical/biological warfare agents and explosives, is nowadays an important demand not only for military purposes but also for civilian security. This necessity has stimulated the development of laboratory and commercial ST-LIBS instruments, to give an appropriate response to the problem. The technical and scientific challenges were considerable for various reasons. First, the long distance between the instrument and target imposed stringent requirements in signal optimization and data acquisition. Second, the open-field configuration gave rise to a variable measurement background, as well as changing transmission through the atmosphere; this proved to be challenging for reliable detection, with unambiguous chemical identification, of the considerable number of explosives in common use today. However, recent years have witnessed the development of several ST-LIBS

**Figure 15.5** Remote (top) and standoff (bottom) LIBS configuration systems; all relevant components are annotated.

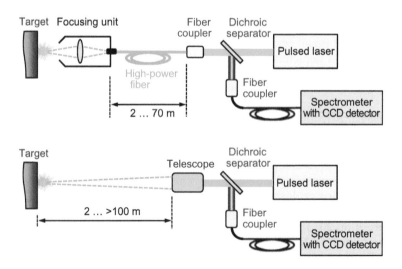

implementations that are capable of reliably detecting and distinguishing explosives and hazardous materials (see, e.g., Gottfried et al. 2007; Fortes and Laserna 2010; and the discussion in Section 15.2).

### 15.1.5 Femtosecond LIBS

To conclude this section, a few remarks about the use of femtosecond-pulse laser sources in LIBS should be made; while femtosecond-pulse lasers are much less widespread than the standard nanosecond-pulse varieties, they nevertheless can be useful for certain applications.

For typical pulses of femtosecond duration, the energy fluence is in the order of $I_p \approx 10^{10}$–$10^{11}$ W·cm$^2$. Thus, the optical components within femtosecond LIBS (fs-LIBS) instruments should be designed to resist such high power densities; these could even be higher than a few $10^{12}$ W·cm$^2$ at the laser focus. In fact, these values are greater than the threshold for optical breakdown of glass and air and, therefore, demand close experimental control. In particular, the focal point position should be adjusted precisely on the sample surface, to avoid undesired optical air breakdown, which could perturb the analytical aspect of the investigation. From the plasma physics point of view, the use of femtosecond-duration laser pulses yields a different plasma evolution behavior. In particular, the laser pulse duration is shorter than the time constant required for thermal coupling in matter (which is in the order of ~1 ps); this favors multiphoton ionization mechanisms over thermal decomposition (the latter is prevalent in nanosecond-pulse laser excitation).

In comparison with conventional nanosecond LIBS (ns-LIBS), fs-LIBS exhibits some rather interesting aspects; these include among others (1) lower ablation thresholds due to higher peak power and lower energy dissipation; (2) higher ablation efficiency due to energy confinement; (3) higher laser pulse-to-pulse stability; and (4) better spatial resolution (see, e.g., Harilal et al. 2014, or Gurevich et al. 2007).

## 15.2 QUALITATIVE AND QUANTITATIVE LIBS ANALYSES

Kicking off the discussion, issues of importance to the analytical aspect of LIBS will be highlighted with reference to some typical LIBS spectra, shown in Figure 15.6. In the related experiment (Pasquini et al. 2007), a pulsed Nd:YAG laser was used ($\lambda_L$ = 1064 nm, pulse duration $\tau_p$ = 5.5 ns, pulse energy $E_p$ = 120 mJ), generating power flux densities of $I_p \approx 1.7 \times 10^{11}$ W·cm$^2$ at the focal location on the surface of a copper-foil sample. The spectra were obtained by integrating the LIBS emission for a period of $\Delta\tau_{LIBS}$ = 1 μs, for a set of delay times $\Delta\tau_{delay}$ whose values are indicated in the figure.

One can clearly see how at the times ($\Delta\tau_{delay}$ = 10 ns) when plasma formation commences, an intense continuous background, due to the bremsstrahlung emission mechanism, dominates; also—less evident on the scale of the display—the elemental emission lines are broad, due to the Stark effect created by the high density of free electrons. After longer delays, elemental lines (both ionic and neutral) can be clearly noticed; they appear as a consequence to the ion–electron recombination, leading to highly excited neutral atoms that decay to lower energy levels. In addition, the line broadening is significantly reduced due to the

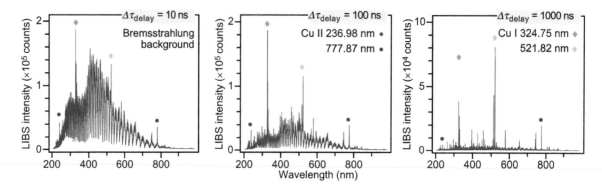

**Figure 15.6** Set of spectra obtained from a TRELIBS experiment on fresh surface locations on a copper foil with $d_{Cu}$ = 1 mm. Each spectrum was obtained by integrating the emitted radiation for $\Delta\tau_{LIBS}$ = 1 μs at the indicated delay time $\Delta\tau_{delay}$ that had elapsed after the laser pulse was applied. (Data adapted from Pasquini, C. et al., *J. Braz. Chem. Soc.* 18, no. 3 (2007): 463–512.)

reduction in the Stark effect and cooling of the (expanding) plasma. After $\Delta\tau_{delay}$ ~ 1 μs, nearly optimal conditions are reached for which the LIBS spectrum is representative only of the most intense lines of the chemical elements present at higher concentration in the sample. Note that for the longest delay time shown, the bremsstrahlung background has all but ceased. Note also that the apparent intensity fluctuations in the bremsstrahlung background are due to the spectral transmission function of the optical collection and recording equipment.

Qualitative (as well as quantitative) LIBS analysis requires the identification of the emission lines of the spectrum. For this, a variety of databases for elemental emission lines are available that provide line wavelengths and relative intensities; probably the most used and complete are the National Institute of Standards and Technology (NIST 2016) tables. The identification of a chemical element from LIBS spectra normally requires the presence of two or more high-intensity lines. From the relative ratios of particular line groups, major and minor elemental constituents of the sample can be allotted, whether it is a "pure" sample, or an "alloy." For organic materials, the identification is not always so straightforward; typically, it has to be done by "fingerprinting," based on multiple lines originating from the "matrix" elements C, N, O, and H, plus relevant "trace" elements. This pattern is then utilized in the analysis and identification of the analyte. An example of this approach is shown in Figure 15.7, in which the LIBS spectra of anthrax and ricin surrogates are compared; while many emission lines are observed in both spectra, it is their relative intensity pattern that allows for unequivocal distinction.

Quantitative LIBS analysis is performed using either of two methods, namely, (1) the *calibration curve* method or (2) the *calibration-free LIBS* (CF-LIBS) approach. In the calibration curve method, one requires certified reference samples comprising the element(s) to be analyzed with different known concentrations. Then, the calibration curve is obtained by plotting the peak area (intensity) of the emission line of the element under investigation as a function of the added concentration. This method is known as the *standard addition method* in analytical chemistry (see, e.g., Harris 2015); it encompasses that an unknown concentration of the elements present in a pure sample is obtained from the

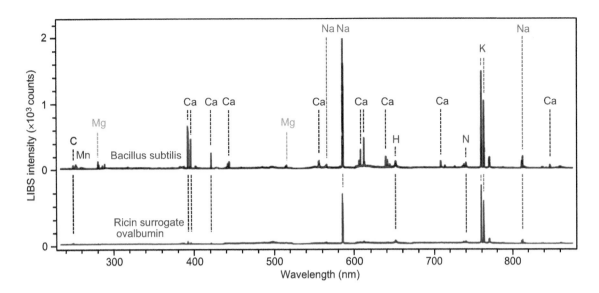

**Figure 15.7**  Single-shot spectra of the anthrax surrogate *Bacillus subtilis* (top—blue spectral trace) and the ricin surrogate ovalbumin (bottom—red spectral trace); spectra were acquired using a field-portable MP-LIBS system. (Data courtesy of US Army Research Laboratory 2010.)

extrapolation of the linear fit of the measurement response data for "added" concentrations, down to the intercept with the concentration axis.

To understand the basic concept underlying this method one may rewrite Equation 15.6 in the form

$$I_{ij} = B \cdot N_{\text{total}} = B \cdot \left(N_{\text{pure}} + N_{\text{added}}\right), \tag{15.10a}$$

where $N_{\text{pure}}$ and $N_{\text{added}}$ stand for the (unknown) pure and added concentrations in the sample, respectively. It is then clear that the intercept of the linear fit to the concentration response data is precisely $B \cdot N_{\text{pure}}$ when $N_{\text{added}} = 0$. Therefore, the concentration of the element in the pure sample is equal to the intercept on the negative concentration axis because only in this case

$$I_{ij} = B \cdot N_{\text{total}} = B \cdot \left(N_{\text{pure}} - N_{\text{added}}\right) \equiv 0. \tag{15.10b}$$

Aside from the aforementioned SAM calibration, another frequently used method is based on calibrating the relative ratio between the intensities of line pairs, one of them belonging to the trace element whose concentration one wishes to measure and the other of the element utilized as a (constant) reference; this reference might be internal, or added at a known concentration. For this procedure, one may start with Equation 15.8, but rather than using two lines from the same element, now one belongs to the trace element and the other to the reference, indexed "t" and "r," respectively:

$$I_t/I_r = (N_t \cdot A_t \cdot g_t \cdot \lambda_r/N_r \cdot A_r \cdot g_r \cdot \lambda_t) \cdot \exp(-(E_t - E_r)/kT). \tag{15.11}$$

The meaning of the variables is the same as in Equation 15.8. In the case that the upper-state energies of the two transitions used in the ratio are rather similar, then $\Delta E = |E_t - E_r| \cong 0$ and the ratio will experience a weak temperature dependence only. In that case, taking the logarithm of Equation 15.11, one obtains

$$\ln\left(\frac{I_t}{I_r}\right) = \ln\left(\frac{N_t}{N_r}\right) + \ln\left(\frac{A_t \cdot g_t \cdot \lambda_r}{A_r \cdot g_r \cdot \lambda_t}\right) - \frac{\Delta E}{kT}. \qquad (15.12)$$

Obviously, since the latter two terms in Equation 15.12 are constant (in the case of $\Delta E \cong 0$), an ln–ln plot of the intensity ratio against the fractional concentration of the trace element will result in a straight line; this is commonly known as the *calibration* or sometimes *working curve*. This can be used to find an initially unknown concentration of a particular element in the sample. To emphasize, to minimize the errors in the application of Equation 15.12 to quantitative analysis, certain conditions should be satisfied. First, as already hinted earlier, the spectral separation between upper levels for the transition of the trace and the reference element should be as small as possible. Second, the higher the intensities of the selected lines, the smaller the error of the calibration, because of reduced shot noise errors. Third, care needs to be taken when measuring lines with high transition probabilities. For example, for resonance transitions, saturation and line-profile reversal in the emission intensity may be encountered due to self-absorption; as a consequence, deviations from a linear slope could not be ruled out.

As a final remark on respective practical calibration, to minimize so-called matrix effects (which may adversely affect the linear behavior of the calibration curve), the emission intensities are normalized, taking the emission from the most abundant element in the sample as the internal standard. Alternately, one can also normalize the analyte signal to the intensity of the spectral background signal (see Bescos et al. 1995).

As an example of the application of the above calibration procedures we refer to a series of experiments from our own laboratories, namely, a study carried out by Samek et al. (2001) for the quantitative analysis of elements potentially toxic in high concentrations—such as, e.g., Al, Pb, and Sr—in calcified tissues, including, for example, teeth and bones. In the study, artificial reference samples, in the form of pressed pellets with a $CaCO_3$ matrix, were prepared to mimic the general physical properties of hydroxyapatite as the main bone compound.

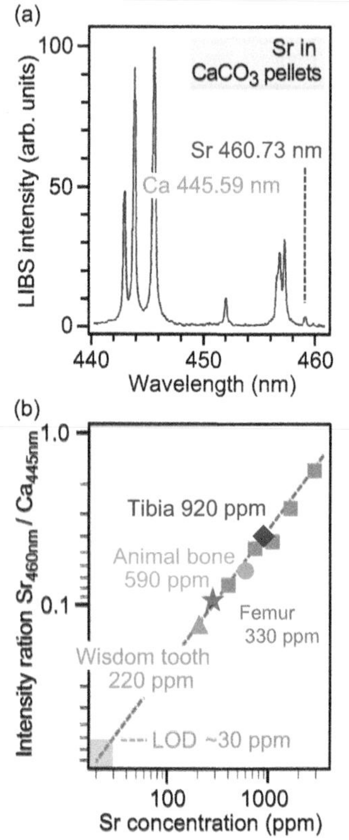

A typical calibration curve for Sr is shown in Figure 15.8; here the results for the ratio between the Sr line at $\lambda_{Sr} = 460.73$ nm and the Ca line at $\lambda_{Ca} = 445.59$ nm are shown (the other line pairs yielded very similar results). Using the common $3\sigma$ rule, an LOD of about 30 ppm was estimated. This calibration curve was used to estimate the Sr concentration in various tooth and bone samples, as indicated in the colored data symbols overlaid with the calibration. One of the main conclusions of this investigation was that cross-calibrated LIBS measurements, in addition to being faster and almost noninvasive, provided results consistent with those obtained by atomic absorption spectroscopy reported in the same study.

Finally, quantitative LIBS analysis can also be achieved without using certified reference materials of the specific element to be analyzed, namely, utilizing the so-called CF-LIBS; thi method was introduced by Ciucci et al. (1999) and is based on the main hypothesis that the LIBS spectrum contains all the necessary information to deduce the composition of the matrix. The mathematical model applied in the CF-LIBS procedure assumes (1) that the plasma composition is that of the matrix; (2) that the plasma temperature and electron density can be obtained via the LTE approximation; and (3) that the plasma is homogeneous and thin. A key to the applicability of the CF-LIBS methodology is that all $N_i$, i.e., the total number densities for all species present in the sample, can be represented by the

**Figure 15.8** Calibration curve for Sr content in various calcified tissue samples, obtained by applying univariate analysis to line ratios between Sr (trace element) and Ca(matrix element). (a) Representative LIBS spectrum. (b) Sr calibration curve. The gray squares represent calibration measurements, recorded from reference pellets of Sr within the $CaCO_3$ matrix. The limit of detection (LOD) is indicated by the gray-shaded area. (Data adapted from Samek, O. et al., *Spectrochim. Acta B* 56, no. 6 (2001): 865–875.)

simple proportionality relation between line intensity and particle number density, as given by Equation 15.6. Also, since $N_i$ is proportional to the weight percentage of the species $i$ in the sample, denoted by $w_i(\%)$, this can be retrieved as described by Ciucci et al. (1999) by using the normalization procedure

$$\sum_i w_i(\%) = Q \cdot \sum_i (N_i \cdot AW_i) = 1. \tag{15.13a}$$

Here $AW_i$ stands for the atomic weight of the species $i$ and $Q$ is an overall normalization constant. Once $Q$ has been deduced, then each species percentage is given by

$$w_i(\%) = Q \cdot N_i \cdot AW_i. \tag{15.13b}$$

## 15.3 SELECTED LIBS APPLICATIONS

A large and ever-increasing number of LIBS application papers appear every year in the scientific literature; with the limited space available, only a handful of representative examples of LIBS in a variety of "exotic" fields are presented here. For further insight into the technique and its wide-ranging applications, the reader is referred to a number of recent books (see Miziolek et al. 2006; Noll 2012; or Musazzi and Perini 2014) and reviews (see, e.g., Hahn and Omenetto 2010; or Fortes et al. 2013).

### 15.3.1 Application of LIBS to liquids and samples submerged in liquids

Quantitative analysis of liquid samples by LIBS has proven to be rather more difficult than that of solid and gaseous samples. Various reasons contribute to this, including among others the following:

- The plasma radiation is weak because of quenching by water or solvent molecules.

- The spectral lines are weakened and broadened due to collisions and the Stark effect.

- The "splashing" from liquid surfaces and the formation of bubbles contribute to the extinction of plasma emission and shortening its lifetime.

All these factors make the LIBS signal an extremely unstable signal from one laser pulse to the next. To overcome these difficulties and drawbacks, different approaches have been proposed, such as (1) initiating the plasma within the bulk liquid; (2) using laminar flows or jets; (3) freezing the liquid samples into icy, solid matrices; or (4) double-pulse plasma generation, to name the most common ones (see, e.g., Cahoon et al. 2012; Rifai et al. 2012; Lazic and Jovićević 2014; or Ohba et al. 2014). Examples of two of these problem-solving approaches are given in the following.

The introduction of DP-LIBS has been extremely beneficial, in particular for bulk liquid analysis. While the first laser pulse produces a cavitation (gas-phase) bubble and plasma, the second one—after an optimal delay, to let the bubble fully evolve—reexcites the thinned sample within the bubble, i.e., doing LIBS excitation under gaseous-like conditions. Thus, the optimized combination of

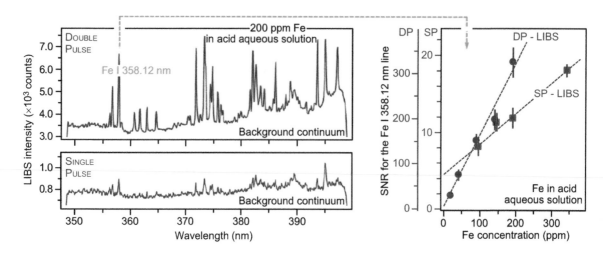

**Figure 15.9** DP-LIBS and SP-LIBS in an acidic aqueous solution containing 200 ppm of iron (Fe). (Left) Spectra for SP-LIBS (blue data trace) and DP-LIBS (red data trace). (Right) Signal-to-noise ratio (SNR) for the Fe reference line $\lambda_{Fe}$ = 358.12 nm for a range of concentration-calibration solutions. (Data adapted from Rifai, K. et al., *J. Anal. At. Spectrom.* 27, no. 2 (2012): 276–283.)

the two laser pulses enhances the emission intensity, reduces the duration of the continuous spectra, and, consequently, increases the analytical performance of LIBS. Rifai et al. (2012) investigated a solution containing about 200 ppm of iron (Fe), comparing LIBS signals gathered using the SP-LIBS and DP-LIBS measurement modalities. Example spectra for this type of comparison are shown in Figure 15.9; each spectrum corresponds to an average of over 25 laser pulses. These two spectra clearly demonstrate that in DP-LIBS, the spectral line emissions are strongly enhanced. For example, the particular line annotated in the spectrum (Fe I 358.12 nm) has increased by a factor of about ×30. Using this particular line for analysis, the authors generated calibration curves for a range of Fe concentrations (about $c_{Fe} \approx$ 10–500 ppm in water); see the right-hand side of the figure. From the SNR plot, it is clear that DP-LIBS is able to reach significantly lower detection limits, based on the common $3\sigma$ rule.

The most sensitive and reproducible method for the analysis of liquid sample seems to be a technique exploiting laminar flows, with the sheath or carrier gas. This liquid-jet technique requires only very simple equipment (see Samek et al. 2000; or Ohba et al. 2014); it constitutes one of the most cost-effective techniques for remote and on-line analysis of liquids. One of the most important advantages of using liquid jets instead of bulk liquid is the significant reduction of sample splashing during the laser ablation.

While in the laser ablation of bulk liquids the energy is mainly consumed by droplet formation instead of plasma generation, the nature of only a small interaction depth within the thin jet leads to more or less immediate vaporization of the very small liquid volume: splashing is almost completely suppressed. In addition, assuming equal laser pulse-to-pulse energy and a constant-thickness laminar flow, the vaporized amount of material is extremely reproducible. Thus, a highly accurate LIBS analysis is possible. Note also that this approach for an exactly defined amount of evaporated liquid is conceptually much simpler than the approach of using liquid-droplet or the aerosol particles (see, e.g., Cahoon and Almiral 2012).

Ohba et al. (2014) investigated the SNR as a function of liquid-sheet thickness for distilled water and an aqueous solution containing 1 ppm of sodium (Na). They found a similar behavior for both samples: the SNR in the LIBS signal increased in line with the jet thickness $d_{jet}$, reaching an optimal value $d_{jet} \sim 20$ μm; beyond this value, the SNR declined again. In addition, it was found that for increasing jet thickness, the probability for splashing/droplet formation did rise as well. Thus, from an experimental point of view, one has to reach compromise conditions for which splashing is minimal without sacrificing the signal intensity of the plasma emission.

Another method for overcoming the difficulties encountered in LIBS of liquid samples—such as splashing, sloshing, and focal length changing—was introduced by Cáceres et al. (2001) in the authors' laboratories. In this method, the water samples were flash-frozen in a bath of liquid nitrogen before the LIBS analysis was carried out, using a Q-switched $CO_2$ laser or a Nd:YAG (both with pulse energies $E_p \approx 20$–200 mJ). Using this then novel approach, well-characterized linear analytical curves were generated for aqueous solutions of aluminum (Al) and sodium (Na) over the concentration range $c_{Al,Na} \approx 0.01$–1%, with estimated detection limits of LOD $\approx 1$–2 ppm.

The capability of analyzing liquids in their solidified aggregate state has made its way into the analysis of ice cores, under arctic conditions (see, e.g., Clausen and Courville 2013).

The authors explored the suitability of LIBS for the analysis of ice cores, with the view to detect, e.g., paleoclimate proxy indicators relating to atmospheric circulation. Their results showed that LIBS analysis was sufficiently sensitive to reveal (1) C and N spectral peaks (consistent with the presence of organic material) and (2) key alkali and alkaline-earth metals (including Ca, K, Mg, and Na), as well as a number of trace metals (such as Al, Cu, Fe, Mn, and Ti). Example spectra of three different ice cores (sliced-core samples of Greenland ice) are overlaid in the bottom data panel of Figure 15.10b; the spectra clearly reveal differences from sample to sample (indexed as 33-9 and 33-18 in the panel), as well as sample inhomogeneities (indexed as 13-18a and 13-18b in the panel). For the measurements, a three-dimensional (3D)-movable, thermoelectrically cooled platform was used, to bring the sample into the laser focus, at preselected locations, while at the same time preserving ice integrity. The authors showed that their LIBS measurements of ice-core samples could be quantitative (when calibrating against reference-concentration standards, see the spectra in Figure 15.10a), and—most importantly—were nearly nondestructive (see the images in the figure that highlight that the ice surface is only minimally affected by the LIBS ablation process). This latter point is of utmost importance for the preservation of archived ice-core specimens.

When analyzing samples submerged in water (underwater LIBS), the aforementioned double-pulse irradiation methodology is rather appropriate; however, there is an alternative to this approach, which is quite simple and also less costly (only a single laser source rather than two is required). The concept is simple: application of a (pressurized) gas flow locally removes the liquid above the solid target, generating a gaseous environment that replicates the creation of the first-pulse vapor bubble, as in a typical DP-LIBS experiment. The use of the gas flow has the additional advantage that the laser pulse energy is neither lost in liquid evaporation nor scattered by the liquid and floating particles. One of the first arrangements for underwater LIBS measurements by applying a gas flow was

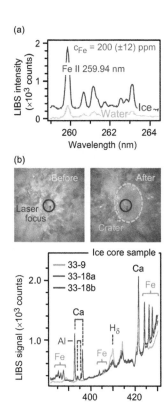

**Figure 15.10**  LIBS spectroscopy of frozen samples. (a) Comparison of a water sample spiked with $Fe_2O_3$; the LIBS intensity of the flash-frozen sample (in liquid nitrogen bath) has increased by a factor of about ×8 (blue data trace) with respect to the liquid sample (green data trace). Data from the author's laboratory. (b) LIBS spectra from a selection of ice cores, demonstrating the ability to detect low-concentration trace elements with only negligible damage to the sample (see images); lines of key elements are annotated. (Data adapted from Clausen, J. and Z. Courville. "Ice-core analysis in a polar environment using laser-induced breakdown spectroscopy (LIBS)." In *13th Polar Technology Conference*, US Naval Academy, Annapolis, MD, April 2–4, 2013.)

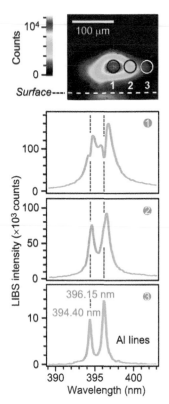

**Figure 15.11** Light emission from a laser ablation plasma, produced at the surface of an Al-plate submerged in ultrapure water, observed with a delay time of $\Delta\tau_{delay}$ = 600 ns after the laser pulse. (Top) False-color intensity image of the plasma; the three measurement locations across the plasma plume are indicated. (Bottom) LIBS spectra recorded at the three plasma locations annotated in the image; for a detailed interpretation, see main text. (Data adapted from Matsumoto, A. et al., *Anal. Chem.* 85, no. 8 (2013): 3807–3811.)

reported by Beddows et al. (2002). The authors employed SP-LIBS analysis of steel samples, using flexible tubing for pressurized gas delivery. The tube also carried the optical fiber (with a core diameter of 550 μm), to deliver the laser radiation and collect the LIBS signal, the assembly had a length of 20 m. The researchers carried out a variety of tests with flows of nitrogen ($N_2$), argon (Ar), and air, finding that the LIBS signal was slightly higher for argon than for the other two gases and that optimal gas pressure was about 2 bar, with a pressure difference between the gas and the water of $\Delta p$ = 1 bar. Detection limits and reproducibility were comparable with those achieved when the same samples were placed in standard ambient air, with all other experimental conditions unchanged. In standard steel samples, detection limits for the trace elements Cr, Mn, and Si of LOD(Cr) = 310 ± 45 ppm, LOD(Mn) = 325 ± 48 ppm, and LOD(Si) = 455 ± 55 ppm were obtained.

Spatially resolved plasma emission of aluminum (Al) samples submerged in water was measured by Matsumoto et al. (2013) by using SP-LIBS. The authors used a Q-switched Nd:YAG laser ($\lambda_L$ = 1064 nm, pulse duration $\tau_p$ = 100 ns, pulse energy $E_p$ = 2 mJ) at a repetition rate of $f_{PRR}$ = 0.3 Hz. The LIBS signal was integrated over a gate period of $\Delta\tau_{LIBS}$ = 1000 ns. LIBS spectra for the Al emission line doublet at $\lambda$ = 394.40 nm/$\lambda$ = 396.15 nm are shown in Figure 15.11, recorded after a delay time with respect to the exciting laser pulse of $\Delta\tau_{LIBS}$ = 600 ns.

The three spectral traces are related to the three spatial positions indicated in the (false-color) intensity image at the top of the figure. They clearly reveal the cooling of the plasma toward the edges of the plasma bubble: the spectral features change significantly with the measurement position, changing from self-reversed to Lorentz-type line shapes. The authors concluded from this finding that the analyte density at the plasma periphery was low enough to rule out self-absorption effects, even when atomic resonance lines with high transition probability were used in the measurements.

## 15.3.2 Detection of hazardous substances by ST-LIBS

Standoff detection and identification of hazardous substances via LIBS analysis has become increasingly popular. A variety of experimental implementations for ST-LIBS have been successfully tested over the past 10 years or so. Two of these are discussed in the following examples.

The first example of ST-LIBS detection is based on double-pulse laser excitation (Gottfried et al. 2007), to enhance the selectivity and sensitivity of the instrument (as noted already earlier, double-pulse excitation increases the ablation rate, the plasma density and temperature, and therefore the intensity of the emission lines). This ST-LIBS system was used to demonstrate that explosive residues could be detected and identified over substantial distances (up to about 20 m). In Figure 15.12, representative single- and double-pulse spectra (averaged over 20 laser pulses) of RDX residue on an aluminum target are compared.

Although in the double-pulse modality the pulse energy of the two lasers was reduced to give the same total energy as that for the single-pulse scheme, the spectra from the former still exhibited a dramatic improvement with respect to the spectral intensity encountered in the single-pulse scheme. For example, many of the Al lines are enhanced by a factor of at least ×20; in fact, the strongest transition lines saturate the detector. It is also evident that the carbon line at $\lambda$ = 247.86 nm has increased substantially as well (the carbon is indicative of the

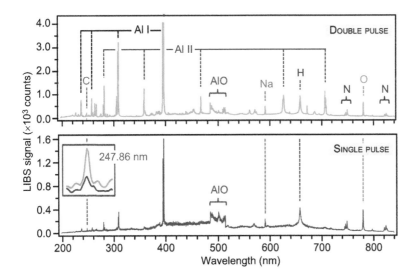

**Figure 15.12** Comparison of single-pulse versus double-pulse ST-LIBS spectra of RDX explosive residue on aluminum; for details, see text. (Data adapted from Gottfried, L.J. et al., *Spectrochim. Acta B* 62, no. 12 (2007): 1405–1411.)

presence of organic materials), meaning that together with other spectral fin-gerprint indicators, the presence of an explosive compound can be deduced.

Such indicators are advantageously accessible in (telescopic) double-pulse ST-LIBS for explosive residue analysis; in particular, the reduction in the depen-dence on atmospheric oxygen and nitrogen entrained during the plasma evolution associated with the time-delayed second laser pulse is beneficial when trying to link these elements to an organic origin. The authors also demonstrated the suitability of the double-pulse ST-LIBS system for detecting biological samples at similar distances of up to 20 m.

In the second example of ST-LIBS, a Raman–LIBS hybrid system was used; the aim was to acquire, in real time, molecular and multielement spectral signatures from the same target cross section at standoff distances of up to 20 m (see Moros et al. 2010). The instrumentation for their spectral analysis was built around a Cassegrain telescope that simultaneously collected both Raman and LIBS signals, produced by the same laser pulse. In Figure 15.13, some representative (intensity-normalized) Raman and LIBS spectra of pure anthracene and 2,6-dinitrotoluene (DNT) are shown, together with their respective cross-correlation patterns used for identification.

Two things should be noted. First, the spectral appearances of the Raman and LIBS spectra are very different; this is not surprising since in Raman spectroscopy, the molecular vibrational bands are recorded, while LIBS spectra comprise electronic transition lines/bands in atoms and molecules. Second, a small gap at the wave-length of the excitation laser is encountered in the LIBS spectra, due to the use of a laser-line rejection (notch) filter. In their investigation, the authors demonstrated that if the experimental conditions allow one to record Raman spectra in parallel with the LIBS spectra, then the capability for quick, clear identification of a molecular compound—from a single, dual-data set—increases significantly.

## 15.3.3 Space applications

One of the most exciting and interesting ST-LIBS applications is encountered in extraterrestrial planetary space exploration. In 2004, the National Aeronautics

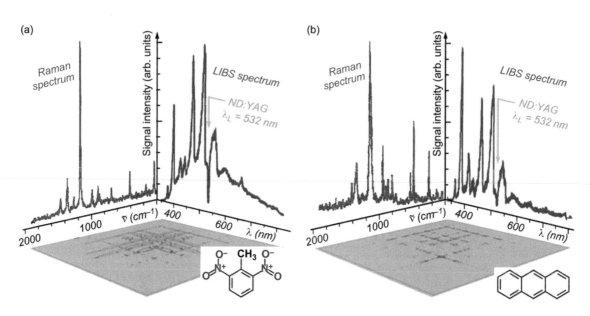

**Figure 15.13** Standoff Raman (blue data traces) and LIBS (red data traces) spectra of (a) pure anthracene and (b) DNT, together with the correspondence patterns (gray area map) for identification. Note that the spectral intensities are normalized to the respective most intense peaks. (Data adapted from Moros, J. et al., "LIBS detection of explosives in traces." In S. Musazzi and U. Perini (eds.), *Laser-Induced Breakdown Spectroscopy: Theory and Application*, Springer Verlag, Berlin, 2014, pp. 349–376.)

and Space Administration (NASA) selected a LIBS instrument to be included in its mobile Mars Science Laboratory rover *Curiosity*. The Chemistry and Camera instrument, known as ChemCam, was one of the scientific instruments onboard of *Curiosity*, which was launched on November 6, 2011, landing on the surface of Mars on August 6, 2012. The main objective of ChemCam was to assess the Martian environment and to explore whether it is capable of supporting microbial life. *Curiosity* landed in a region rich in minerals. As illustrated in Figure 15.14a, the ChemCam unit comprised two standoff sensing instruments: the actual LIBS spectroscopic unit and a remote microimager (RMI), dedicated to elemental analysis. ChemCam was set up to record, without any moving parts in its three subspectrometers, light in the UV, visible, and near-infrared regions of the spectrum, covering the range $\lambda = 240$–850 nm. While the LIBS unit recorded spectra for the determination of elemental composition, the RMI device provided the geomorphological context to assign the LIBS analysis results to particular objects on the surface of Mars.

One of the first LIBS spectra recorded by the ChemCam instrument on NASA's *Curiosity* rover and sent back from Mars on August 19, 2012, is shown in Figure 15.14c (see Wiens et al. 2012); the spectral data shown here were averaged over 30 laser pulses, originating from a single position (laser spot diameter 0.4 mm) on a rook close to the Coronation landing site of the rover (see the photograph in Figure 15.14b). In its first 2 years on Mars, ChemCam has returned more than 160,000 spectra from about 45,000 locations! A closer look at the displayed spectrum reveals the presence of minor elements such as titanium (Ti) and manganese (Mn); also, temporal analysis of the spectra showed that magnesium (Mg) was slightly enriched at the surface. Overall, the spectral analysis of the said sample was consistent with basalt (a volcanic rock material), which was known to

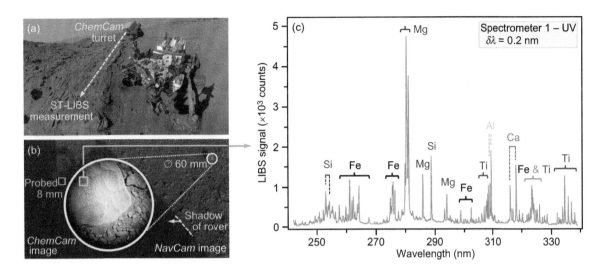

**Figure 15.14**    LIBS measurement data from the Mars mission of NASA's *Curiosity* rover. (a) Self-portrait of the *Curiosity* rover and its ST-LIBS turret. (b) Images of the geological sample area probed by the ChemCam LIBS spectrometer system. (c) First LIBS data sent from Mars by NASA's *Curiosity* ChemCam instrument on August 19, 2012. (Courtesy of NASA/JPL-Caltech, 2012; data adapted from Wiens, R.C. et al., *Space Sci. Rev.* 170, no. 1 (2012): 167–227.)

be abundant on Mars. It should be noted that the spectral peak heights cannot be directly related to the relative abundances of the elements, because of the very different transition probabilities of the observed emission lines; for quantification, comparison with reference samples with known concentrations are required.

Both NASA and the European Space Agency are currently preparing Mars rover missions for the next launch window in the summer of 2020, which will include ST-LIBS and Raman spectroscopy systems, to search for potential traces of life.

## 15.3.4 Industrial applications

In industrial applications of LIBS, three types of process analyses are usually considered, namely, the following:

- *Off-line analysis*, which refers to the sequence of taking samples at the workplace and transferring them to a measurement location (in general a dedicated laboratory space); this is a slow procedure and does not allow for rapid feedback into the industrial scale process.

- *On-line analysis*, which refers to automated measurements in conjunction with the process itself, based on a bypass configuration; from the direct analysis, continuous feedback can be given to control the industrial process.

- *In-line analysis*, which refers to measurements that are taken *in situ*, involving no physical sampling but only optimal monitoring by LIBS; again, the direct analysis allows for real-time control feedback to the industrial process.

The last analytical methodology is the most desirable, as it requires neither mechanical preparation nor does it produce a potentially unrepresentative bias associated with bypassing the main stream. As a consequence, its important advantages are shorter response times and better, representative information on how to regulate a process in real time.

An interesting example of the industrial application of LIBS is the in-line analysis of processed coal. The so-called ash content is a crucial parameter for the use of highly purified coal in the chemical industry. In general, the ash composition of unprocessed, raw coal may differ substantially, depending on where it was extracted and on naturally occurring variations within the deposit. During the combustion process, a variety of oxides associated with trace elements in the coal contribute to the ash content (including $Al_2O_3$, $Fe_2O_3$ and $SiO_2$). In general, in unprocessed coal, this oxide concentration in ash is $c_{ash} > 10$ m.-%. However, for coal to be suitable in the chemical industry, this value has to be below $c_{ash} < 2$ m.-%; thus, it has to be preprocessed and, naturally, be monitored for its purity. The cited notation is common in coal combustion, referring to the percentage of (dry basis) ash = $(M_{ash}/M_{dry}) \times 100$. Routinely, monitoring is based on off-line laboratory measurements, for which random coal samples are taken and combusted, and the resulting ash is weighted.

As pointed out earlier, LIBS analysis can be undertaken in-line, without the need to remove any sample material. In the left panel of Figure 15.15, a schematic view of an in-line coal analyzer based on LIBS is shown, as used by Noll et al. (2014). Typically, the conveyor belt transports 70–90 t·h$^{-1}$, at a speed of about ~1 m·s$^{-1}$. The LIBS measurement head could be moved perpendicularly to the motion of the conveyor, to sample the material at distinct lateral positions. In routine analysis campaigns, the instrument developed by the authors can analyze all relevant elements present in the coal sample every 100 ms.

The LIBS data in Figure 15.15 show the in-line variations of the potential coal-ash content (prior to combustion); these LIBS data (gray and blue data points in the graph) are compared with routine laboratory combustion test results (red data points in the graph). Clearly, the absolute values and the trend of the LIBS and the laboratory measurements are in good agreement, in particular when considering the rather low absolute values of the ash content. Essentially, the authors demonstrated with these results that in-line LIBS monitoring may become one of the most important industrial applications of LIBS, as it provides continuous 24/7 chemical analysis and operational control of industrial scale processing.

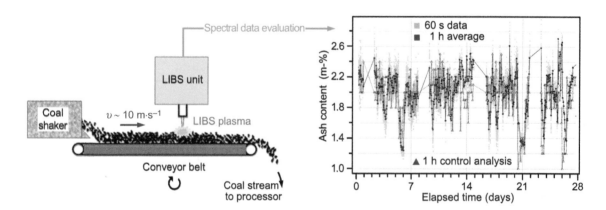

**Figure 15.15** In-line coal analyzer based on LIBS spectroscopy. (Left) Schematic experimental setup. (Right) Ash content variation measured by LIBS (gray and blue data points) compared with the routine laboratory control analysis (red data points); the lines are provided to guide the eye. (Reprinted from Noll, R. et al., 93, *Spectrochim. Acta B* 41–51, Copyright 2014, with permission from Elsevier.)

# 15.4 BREAKTHROUGHS AND THE CUTTING EDGE

Much of the early LIBS research, but still up to the present day, has been directed at the understanding of laser-generated plasmas and the process of laser ablation itself. But it is the elemental (to a much lesser degree molecular) analytical aspect of LIBS that has given it its deserved place among laser spectroscopic methodologies, and it is in particular the capability to provide quantitative analysis of any imaginable sample (requiring suitable calibration, of course).

### 15.4.1 Breakthrough: Quantitative LIBS analysis using nanosecond- and femtosecond-pulse lasers

The first study of luminous laser-induced plasmas, in which light emission from purely laser-supported plasmas was sufficiently strong to allow for reasonably quantitative analytical interpretation, was published in an article by Runge et al. (1964). The authors used a Q-switched ruby laser ($\lambda_L = 694.3$ nm, $\tau_p \approx 15$–$20$ ns, $E_p \approx 50$–$75$ mJ), which was focused by a short-focal length lens ($f = 2$ cm, generating flux densities on target of $I_p \sim 10^{13}$ W·cm$^{-2}$) onto certified stainless-steel standard samples; the dispersed laser plasma light emission was recorded by photographic plate. From the evaluation of the spectra, the authors generated calibration curves for the minor elements nickel and chromium in the samples, based on selected Ni/Fe and Cr/Fe line pairs. Note that the authors state also that the craters from the laser sample interaction were rather substantial (diameter ~ 400 μm, depth ~ 50 μm), much larger than those generated in the then widely used method of microspectral analysis using electrical sparks. The presence of large, irregular craters, in general with substantial redeposition of ejected material (see the top image in Figure 15.16), has remained somewhat problematic throughout the development and practical use of nanosecond-pulse LIBS.

With the availability of femtosecond-pulse lasers, this crater problem could be "solved," at least in part; the craters became much more defined in shape, in all respects (see the center panel in the figure). Also, because femtosecond-laser pulses were shorter than nanosecond-laser pulses by a factor >1000, the pulse energy could be lowered by the same margin to reach the flux densities required for plasma breakdown; rather than multimillijoule pulse energies, now a few microjoules of energy suffice, in principle. This also means much smaller crater dimensions, making LIBS viable for elemental analysis with micrometer spatial resolution (see, e.g., Zorba et al. 2011). The first quantitative application of LIBS using femtosecond-laser excitation was demonstrated by Margetic et al. (2000); in their study, the authors used an amplified femtosecond-pulse Ti:sapphire laser ($\lambda_L = 775$ nm, $\tau_p \approx 170$–$200$ ns, $E_p \approx 5$–$500$ μJ). From the evaluation of the spectral data, the authors generated calibration curves for a range of certified brass samples, based on selected Zn/Cu line pairs.

When comparing spectra from the same target, obtained by ns- and fs-LIBS, it is noticeable that normally the early evolution of the plasma is dominated by ionic lines in the former case, but that in the latter case, hardly any ionic emission is observed. An example of this is given in Figure 15.16, for a silicon monocrystal target, excited by radiation from Ti:sapphire laser sources, at $\lambda_L = 780$ nm.

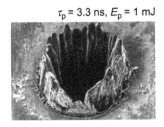

$\tau_p = 3.3$ ns, $E_p = 1$ mJ

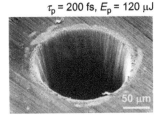

$\tau_p = 200$ fs, $E_p = 120$ μJ

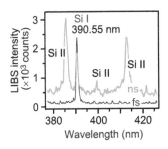

**Figure 15.16** Comparison of LIBS measurements using nanosecond- or femtosecond-laser pulses (Ti:sapphire laser at $\lambda_L = 780$ nm). LIBS craters in Fe foil of thickness 100 μm, for nanosecond-pulse (top) and femtosecond-pulse (center) ablation. (With kind permission from Springer Science+Business: *Appl. Phys. A*, Femto-second, picosecond and nanosecond laser ablation of solids, 63, no. 2 (1996): 109–115, Chichkov, B.N. C. Momma, S. Nolte, F. von Alvensleben and A. Tunnermann.) (Bottom) LIBS spectra of a Si target, for nanosecond-pulse (green trace) and femtosecond-pulse (blue trace) excitation; data from the authors' laboratory.

### 15.4.2 At the cutting edge: Elemental chemical mapping of biological samples using LIBS

Other than Raman and laser-induced fluorescence spectroscopy, LIBS has long remained dormant like a "Sleeping Beauty" when it came to spectrochemical imaging. This was mainly due to the low sampling rates (because of the low-repetition-rate pulsed lasers ($f_{PRR} \sim$ 10–20 Hz, in rare cases up to 1 kHz) and because LIBS is an elemental analysis tool, which for biological samples is perceived to be somewhat restrictive. Furthermore, LIBS is "destructive," leading to at least surface damage, albeit in most cases only minute. Over the past few years, the view on LIBS imaging has been changing; quite a few 2D/3D mapping applications have materialized (see, e.g., the review by Piñon et al. 2013).

Here, a recent proof-of-concept study is highlighted in which LIBS was used to study the migratory behavior of nanoobjects within biological tissues (Gimenez et al. 2016). The authors provided, for the first time, 3D label-free nanoparticle imaging at the entire organ scale. This was done in two different but complementary approaches, namely, by volume reconstruction of a sliced organ and by depth-profiling analysis. The images in Figure 15.17 reveal elemental biodistributions in a coronal murine kidney, 3 hours after Gd-based nanoparticles were administered. In preparation, the kidney was embedded in epoxy resin (a typical protocol used, e.g., for electron microscopy); the sample was then thinly sliced (thickness ~200 μm) to provide access to the axial direction. SP-LIBS spectra were recorded for each raster point, with a step size $\Delta d_{x,y} = 35$ μm (the LIBS ablation craters had diameters of $d_{crater} \approx$ 3–5 μm; see the image in the top left corner); the total lateral scan area was of size 120 × 250 pixels. With their laser system of $f_{PR} = 10$ Hz, the acquisition of the full hyperspectral LIBS data cube was complete within ~1 h. Comparing the elemental distribution images for (exogenous) Gd and (endogenous) Ca reveals clearly that the nanoparticles had accumulated in the cortex of the kidney.

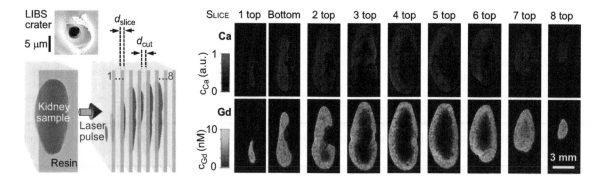

**Figure 15.17**   3D distribution and quantification of Gd-based nanoparticles in a kidney sample, probed by spatially resolved LIBS. (Left) Sample embedded in epoxy resin, with subsequent slicing; slice thickness and cut waste are equal, $d_{slice} = d_{cut} = 200$ μm. (Right) Series of elemental images for Gd and Ca, on the top side of the adjacent coronal sections (the bottom-side image is shown only for the first slice). (Data adapted from Gimenez, Y. et al., *Sci. Rep.* 6 (2016): article 29936.)

# CHAPTER 16

# Laser Ionization Techniques

I n *laser ionization* spectroscopy, the approach is that the resonance of photon energies with atomic/molecular transitions is still exploited (such as in absorption and laser-induced fluorescence [LIF] spectroscopy), but the interaction with a second laser photon ionizes the (selected) species. The charged particles generated in this two-step process—ions and electrons—can easily be detected with efficiencies reaching 100%, which is in general far superior to photon detection efficiencies.

In general, laser ionization spectroscopy is divided into two modalities, depending on whether the ions or the electrons are used for detection. The first of these is commonly known as *resonance ionization spectroscopy* or *resonance-enhanced multiphoton ionization* (*REMPI*). When electrons rather than ions are measured and analyzed, this is normally done in the form of *zero electron kinetic energy* (*ZEKE*); i.e., the ionizing photon step is to energy levels (or the continuum) very close to the threshold, so that the photoelectrons have extremely low kinetic energy. It should be noted that besides these main modalities, a number of experimental derivative techniques have been developed, such as, e.g., mass-analyzed threshold ionization (MATI).

In the following, the two main laser ionization spectroscopy methodologies, REMPI and ZEKE, will be outlined in their concepts, together with key representative examples.

## 16.1 BASIC CONCEPTS OF REMPI

REMPI (also termed *resonance ionization spectroscopy*) is a technique that primarily provides structural and spectroscopic information on the excited states of neutral species. This type of spectroscopy refers to an ionization process that takes place by *resonant m-photon* excitation from the ground electronic state of a molecule to one of its excited electronic states as a first step, followed by a second step in which *additional n-photons* are absorbed to ionize said resonantly excited electronic state (see Boesl 2000). The overall process is abbreviated to *(m+n) REMPI* for one-color REMPI or to *(m+n') REMPI* for two-color REMPI; occasionally, these are also referred to as R($m$+$n$)PI or R($m$+$n'$)PI, respectively.

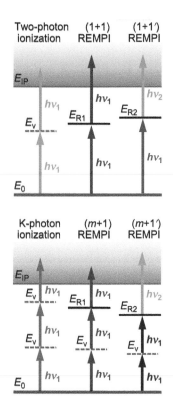

**Figure 16.1** Photoionization schemes, involving two-photon (top, $m = n \equiv 1$) and $k$-photon (bottom, $k = m + n \geq 3$) processes, exemplified for $m = 2$ and $n = 1$ and, thus, $k = 3$. (Left) Nonresonant photoionization. (Center) One-color REMPI. (Right) Two-color REMPI. For further details, see text.

In Figure 16.1, different schemes of (nonresonant and resonant) ionization are summarized; for conceptual simplicity, only the cases for $m = 1, 2$ for the resonant step and $n = 1$ for the ionization step are shown, corresponding to two-photon ($k = 1 + 1$, top part of the figure) and $k$-photon ($k = m + 1 \equiv 3$, bottom part of the figure) ionizations. Three basic scenarios are common, namely, (1) no intermediate state is in resonance with the photon energy, i.e., excitation is via a "virtual state" associated with the energy $E_v$ (left process in the panels); (2) only photons with one color participate in the process and resonant excitation takes place to an intermediate energy state $E_{R1}$ (center part of the panels); or (3) photons with different color participate in the ionization process, again via a resonant intermediate state $E_{R2}$ (right part of the panels). In both resonant scenarios, the intermediate state corresponds to a real excited molecular state, whose lifetime is several orders of magnitude longer than that of the virtual state created by the multiphoton coincidence; this is the key feature that is responsible for the resonant enhancement of the measured ion signal. In the most common case of $m = 1$ and $n = 1$, the one-color and two-color processes are commonly abbreviated as (1+1) REMPI and (1+1′) REMPI, respectively; for $m, n \neq 1$ the "ones" in the brackets are replaced by the actual numbers (e.g., $m = 2$ and $n = 1$ yield (2+1) REMPI; see the bottom part of the figure).

Two great advantages of REMPI are the mass and species selectivity of the technique. The former arises because the final ions, which are produced through the resonant intermediate, are typically detected by time-of-flight (ToF) mass spectrometry, or ToFMS. The species selectivity is due to the spectroscopic selectivity of the initial resonant absorption, which permits distinguishing two ions with the same mass via their REMPI spectra.

From a molecular structure point of view, the main steps involved in, for example, a (1+1′) REMPI scheme are visualized in Figure 16.2 and may be written as follows:

Step 1: excitation to the resonant intermediate state

$$AB\left(v'', J''\right) \xrightarrow{h\nu_1} AB^*\left(v', J'\right) \tag{16.1a}$$

Step 2: photoionization out of the excited intermediate state

$$AB^* \xrightarrow{h\nu_2} AB^+ + e^- \tag{16.1b}$$

$$AB^* \xrightarrow{h\nu_2} A + B^+ + e^- \tag{16.1c}$$

$$AB^* \xrightarrow{h\nu_2} AB^{**}\left(E_{int} > IP\right) \rightarrow AB^+ + e^- \tag{16.1d}$$

in which the three alternative ionization routes include direct photoionization (Equation 16.1b), dissociative photoionization (Equation 16.1c), and autoionization (Equation 16.1d). Here the double stars (**) stand for a highly excited electronic state (including Rydberg states); IP is the adiabatic ionization energy (i.e., the energy necessary to produce an ion with no internal energy and an electron with zero kinetic energy); and $E_{int}$ is the total internal molecular energy (electronic, vibrational, and rotational).

The energy balance of the two steps allows us to write the following relations:

$$h\nu_1 = AB^*\left(v', J'\right) - AB\left(v'', J''\right), \tag{16.2a}$$

$$h\nu_1 + h\nu_2 = IP + E_{ion} + E_{kin}(e^-) + E_{kin}(AB^+) \approx IP + E_{ion} + E_{kin}(e^-), \tag{16.2b}$$

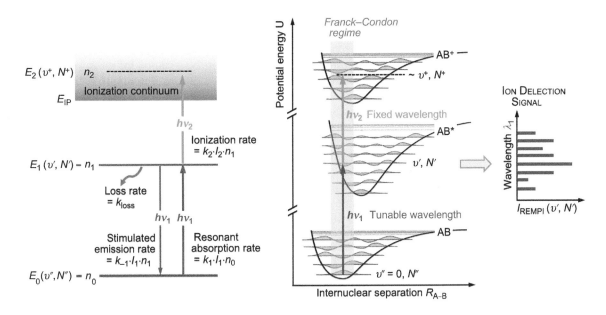

**Figure 16.2**    Concept of $(1+1')$ REMPI spectroscopy. (Left) Basic three-level REMPI scheme, including the key transition rates for rate equation calculations. (Right) Two-step two-color state-selective $(v, N)$ molecular transition scheme; the Franck–Condon regime relevant in the first step is indicated by the blue-shaded bar. A conceptual $(v, N)$-dependent REMPI spectrum, as a function of tuning wavelength $\lambda_1$, is shown on the very right.

where $E_{\mathrm{ion}}$ is the internal energy of the cation (electronic, vibrational, rotational) and $E_{\mathrm{kin}}(e^-)$ and $E_{\mathrm{kin}}(AB^+)$ are the kinetic energies of the free electron and the ion, respectively (the latter is typically assumed to be negligible). As indicated in the $(1+1')$ REMPI scheme in Figures 16.1 and 16.2, one typically uses a tunable laser radiation providing the photons $h\nu_1$ for the resonant excitation step and a fixed wavelength laser providing the photons $h\nu_2$ for the ionization step. As long as $h\nu_1 + h\nu_2 \gg$ IP, the produced ions $AB^+$ will carry a distribution of internal energies (no resonant condition). Thus, the monitoring of the total yield of $AB^+$ ions, as a function of the laser excitation energy $h\nu_1$, will probe the structure and properties of the intermediate state $AB^*$; this constitutes the REMPI spectrum.

## 16.1.1 Quantitative description of REMPI in the framework of rate equations

In general, multiphoton ionization processes are described in the framework of quantum mechanics. However, one often uses a *rate equation model*, which is valid provided (1) that the light intensity is not too high and (2) that coherent effects, such as, e.g., *Rabi oscillations* can be neglected. For conceptual simplicity, only the rate model equation of the $(1+1)$ REMPI scheme is illustrated in Figure 16.2. Here we use a three-level system of the rate equations

$$\mathrm{d}n_0/\mathrm{d}t = -k_1 \cdot I \cdot n_0 + k_{-1} \cdot I \cdot n_1, \tag{16.3a}$$

$$\mathrm{d}n_1/\mathrm{d}t = k_1 \cdot I \cdot n_0 - k_{-1} \cdot I \cdot n_1 - k_2 \cdot I \cdot n_1 - k_{\mathrm{loss}} \cdot n_1, \tag{16.3b}$$

$$\mathrm{d}n_2/\mathrm{d}t = k_2 \cdot I \cdot n_1. \tag{16.3c}$$

In this scheme, $n_i$ ($i$ = 0, 1, 2) are the time-dependent number densities in the energy levels $i$ and $k_i$ ($i$ = 1, –1, 2) are the rate coefficients of the absorption, stimulated emission, and ionization processes, respectively. Note that $k_1 \equiv k_{-1}$ is proportional to the Einstein $B$ coefficient; $k_{\text{loss}}$ stands for the sum of all the rate coefficients contributing to the decay of the intermediate resonant state (i.e., internal conversion, intersystem crossing, etc.); and $I$ is the light intensity. Here the spontaneous emission from the excited state to the initial state is omitted; this is normally admissible because of the large number of final-state pathways accessible from the excited state, which results in only negligible repopulation of the initial ro-vibronic energy state. The rate equation system (Equation 16.3) of linear differential equations can be solved exactly for $n_2$, with the initial conditions that $n_0(t = 0) \equiv n$ and $n_1 = n_2 \equiv 0$ ($n$ is the total number density) and assuming a rectangular laser pulse of amplitude $I$ and duration $\Delta t$. The result is (see, e.g., Swain 1979; or Jacobs et al. 1986)

$$n_2(\Delta t) = n \cdot G(k_1, k_2, I \cdot \Delta t), \tag{16.4a}$$

where

$$G(k_1, k_2, I \cdot \Delta t) = \left( I - \frac{F_1 - F_2}{2B} \right), \tag{16.4b}$$

with

$$F_1 = (A + B) \cdot \exp[-0.5 \cdot (A - B) \cdot I \cdot \Delta t],$$

$$F_2 = (A - B) \cdot \exp[-0.5 \cdot (A + B) \cdot I \cdot \Delta t],$$

$$A = 2k_1 + k_2,$$

$$B = \left( 4(k_1)^2 + (k_2)^2 \right)^{1/2}.$$

In the long-time limit, i.e., for $\Delta t \to \infty$, it can be demonstrated that $F_1 = F_2 \equiv 0$, and, therefore, $n_2(\Delta t) \to n_0$.

## 16.1.2 REMPI signal intensity

By and large, for many molecules the ionization rate is much smaller than that of the resonant absorption rate; therefore, one has that $k_2 \cdot I \cdot \Delta t \ll 1$. In this limit, it can be shown that

$$n_2(\Delta t) \to n_0 \cdot k_1 \cdot k_2 \cdot (\Delta t/2). \tag{16.5}$$

Furthermore, $k_2$ is not expected to change significantly, due to the nature of the nonresonant continuum transition; it thus can be considered as a wavelength-independent rate constant. Hence, in the absence of saturation effects in the first resonant excitation step, or photophysical processes perturbing the lifetime of the intermediate resonant state, REMPI intensities will be proportional to $n_2(\Delta t)$. They can be approximated as

$$I_{\text{REMPI}} = C \cdot \frac{N(v'', J'')}{2J'' + 1} \cdot q_{v', v''} \cdot S(J', J''). \tag{16.6}$$

Here, the quantum numbers $v''$ and $J''$ stand for the vibrational level and total angular momentum in the ground state, while $v'$ and $J'$ are the corresponding quantities in the resonantly excited state; $q_{v', v''}$ and $S(J', J'')$ are the

Franck–Condon (FC) and the rotational line strength factors, respectively, for the transitions between the ground and intermediate electronic states; $N(v'', J'')$ is the actual population in the ground-state $(v'', J'')$ levels of the molecule, with degeneracy $(2J''+1)$; the factor $C$ is a constant that is proportional to the laser intensity, the ionization rate, the quantum yield for ionization, and experimental factors related to the detection efficiency. Note that Equation 16.6 can also be applied to $(1+1')$ REMPI schemes, in which the wavelength of the laser providing the photons for ionization is different (but normally fixed) to that which is used to resonantly transfer the ground quantum-state population to the intermediate excited electronic state.

An example to illustrate the similarity between REMPI and absorption and LIF spectra is shown in Figure 16.3, here for the $(1+1')$ REMPI study of formaldehyde ($H_2CO$), a molecule whose spectroscopy and photochemistry have been thoroughly studied over many years. In particular, it was investigated in the gas phase, to understand the so-called state-selected photodissociation dynamics. REMPI spectroscopy of $H_2CO$ has been implemented using a variety of schemes, to access a range of electronic states, mainly employing three or more photons (see Park et al. 2016, and references therein), which all required rather high laser intensities and offered limited sensitivity.

Recently, Park et al. (2016) utilized a $(1+1')$ REMPI scheme to study, with good rotational line resolution, the vibro-rotationally allowed transition $\tilde{X}\,^1A_1 \rightarrow \tilde{A}\,^1A_2$ $(4_0^1)$, associated with the out-of-plane wagging vibration $v_4$, with $v'' = 0$ and $v' = 1$. They used a tunable photon $hv_1$ (at around $\lambda_1 \sim 353$ nm) for the first (resonant) step and a second photon $hv_2$ (with fixed $\lambda_2 = 157$ nm) that initiated direct ionization out of the $\tilde{A}$-state; the conceptual term scheme is shown on the left panel of Figure 16.3. The figure includes three spectra, namely, LIF data

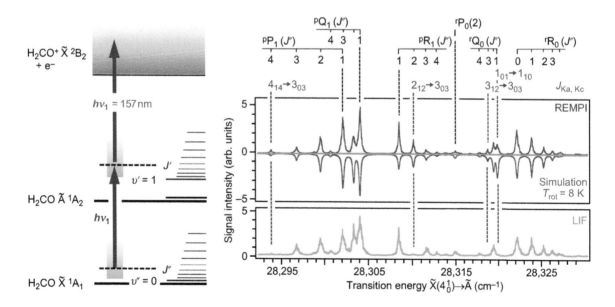

**Figure 16.3** $(1+1')$ REMPI spectroscopy of formaldehyde ($H_2CO$), for the resonant transition $\tilde{X}\,^1A_1 \rightarrow \tilde{A}\,^1A_2$ $(4_0^1)$. (Left) REMPI transition scheme, with the relevant levels indicated. (Right) REMPI spectrum (blue data trace), compared with a simulation of an absorption spectrum, at $T_{rot} = 8$ K (magenta data trace), and a LIF spectrum (green data trace). For further details, see text. (Data adapted from Park, G. et al., *Phys. Chem. Chem. Phys.* 18, no. 32 (2016): article 22355.)

(green spectral trace), REMPI data (blue spectral trace), and simulated absorption data (magenta spectral trace). Note that the experimental data were recorded in a rotationally cold, supersonic-expansion molecular beam ($T_{rot} \approx 8$ K); the theoretical absorption spectrum was calculated for the same rotational temperature.

Quite evident is the great similarity between the REMPI spectrum and the simulated absorption spectrum, in both line position and relative intensities; this was interpreted as a clear indication that the REMPI signal is proportional to the absorption cross section of the transition $\tilde{X}\ ^1A_1 \leftrightarrow \tilde{A}\ ^1A_2\ (4_0^1)$ and, therefore, that the ionization cross section out of the $\tilde{A}$-state showed no dependence on the rotational levels of the intermediate resonant state, at least under the particular experimental conditions. This result led the authors to conclude that the REMPI intensities could be used for quantitative determination of the rotational population distribution in the $\tilde{X}$-state.

The comparison between the REMPI and LIF spectra revealed that a series of peaks (marked by the red annotation in the REMPI spectrum) are not observed in the LIF spectrum. These particular lines correspond to transitions from the ground state to some short-lived rotational levels of the $\tilde{A}$ ($4^1$) state, whose lifetimes are shorter than (or comparable with) the detection delay time used in the LIF measurements. Suitable delay times were required to avoid including light from the laser pulse tail in the integral LIF signal window. Thus, in cases such as this, REMPI may be considered as an advantage over LIF spectroscopy.

More rigorously, $n_2(\Delta t)$ needs to be calculated numerically in cases where the preceding simplifying approximations do not hold. Normally, this is not an easy task; it requires the knowledge of the rate constants $k_1$ and $k_2$, which depend on the specific transition moments of the individual steps, i.e., the resonant absorption and the ionization transitions involved in the (1+1) REMPI, as well as on the laser pulse intensity $I$ and duration $\Delta t$.

While for linear molecules the resonant-transition dipole selection rules $\Delta J = 0$ (Q-branches) and $\Delta J = \pm 1$ (P- and R-branches) are applicable, this is not the case for the (second) ionization step, even for linear molecules. In general, in a REMPI process, one has information neither on the total angular momentum $J^+$ of the final (ion) state nor on its partition into individual parts (the molecular ion rotational momentum, the electronic angular momentum, the free-electron angular momentum, etc.), that is, unless the photoelectron energy spectrum is known. With this in mind, Jacobs et al. (1986) calculated the REMPI rate constants for the case of linear molecules, in the classical limit. The final expressions obtained for the rate constants, based on dipole selection rules, are summarized in Table 16.1.

**Table 16.1**    Rate Constant Values for Linear Molecules in (1+1) REMPI

| Step | Q-branch | P- and R-branches |
|---|---|---|
| Resonant absorption; rate $k_1 =$ | $3 \cdot Z \cdot \cos^2 \theta$ | $(3/2) \cdot Z \cdot \sin^2 \theta$ |
| Ionization; rate $k_2 =$ | $D \cdot [1 + \cos^2 \theta]$ | $D \cdot [(2/3) + \sin^2 \theta]$ |

*Source:*   Jacobs, D.C. et al., *J. Chem. Phys.* 85, no. 10 (1986): 5469–5479.

*Note:*   $Z = H \cdot S(J_0, J_1)/(2J_0 + 1)$, with $S(J_0, J_1)$ = rotational line strength; $H \propto (c^2/8\pi h\nu^3) \cdot \Delta\nu \cdot \tau$, $D = 3 \cdot (\sigma/4h\nu)$, with $\sigma$ = ionization cross section; $\Delta\nu$ = laser bandwidth (if larger than the homogeneous line width of the transition; otherwise, transition width); $\tau$ = vibrational-specific lifetime for the resonant step; $\theta$ = angle between laser field polarization $\boldsymbol{E}$ and molecular angular momentum $\boldsymbol{J}$.

It should be noted that the evaluation of the total ion yield does not need to decouple the total angular momentum into its components; it suffices to sum over all final angular momentum states and to merely specify the fraction of the parallel character (see Jacobs et al. 1986), a basic requirement that is already incorporated in the expressions for $k_2$ given in Table 16.1. Thus, knowing the specific values of the rate constants, integration of Equation 16.4 can be performed; then, one obtains for $n_2(\Delta t)$ of a specific ro-vibronic states

$$n_2(\Delta t) = (1/2) \cdot n_0(J; t = 0) \cdot \int_0^\pi G[k_1(\theta) \cdot k_2(\theta) \cdot I \cdot \Delta t] \cdot \sin \theta \, d\theta; \qquad (16.7)$$

this can be evaluated numerically.

### 16.1.3 Selection rules for the ionization step in REMPI

To complete the selection rules for the general $(m+1)$ REMPI process, the selection rules in the (final) ionization step will be considered; the treatment closely follows that given by Signorell and Merkt (1997).

The starting point is to recall that for dipole matrix elements to be nonzero, the direct product of the initial state, the final state, and the transition dipole operator must contain the totally symmetric representation $\Gamma^s$ of the molecule symmetry group. This condition can be written as

$$\Gamma^f(\text{rev}) \otimes \Gamma_{\text{dipole}} \otimes \Gamma^i(\text{rev}) \supset \Gamma^s, \qquad (16.8a)$$

where $\Gamma_{\text{dipole}}$ is the symmetry of the dipole operator and $\Gamma^f(\text{rev})$ and $\Gamma^i(\text{rev})$ are the total ro-vibronic symmetry (excluding spin) of the final and initial states, respectively. Since the dipole operator is antisymmetric with respect to inversion, one can write $\Gamma_{\text{dipole}} = \Gamma^*$ ($\Gamma^*$ is the antisymmetric representation in the molecular symmetry group). Introducing this condition in Equation 16.8a, one obtains $\Gamma^f(\text{rev}) \otimes \Gamma^i(\text{rev}) \supset \Gamma^*$. If one now splits the final state symmetry as $\Gamma^f(\text{rev}) = \Gamma^e \otimes \Gamma^+(\text{rev})$, where $\Gamma^e$ and $\Gamma^+$ stand for the symmetry of the photoelectron wave function and the ion, respectively, then one obtains the general relation in the form

$$\Gamma^e \otimes \Gamma^+(\text{rev}) \otimes \Gamma^i(\text{rev}) \supset \Gamma^s. \qquad (16.8b)$$

In general, to solve this latter relation, one would need to consider $\Gamma^e$ in the appropriate molecular symmetry group. However, at long range, one typically considers the free electron being subjected to a central potential created by a point charge (namely, the ion core, situated at the coordinate origin). In this scenario, the photoelectron symmetry can be assigned with respect to the inversion coordinates, and one finds $\Gamma^e = \Gamma^*$ (for odd $\ell$) or $\Gamma^e = \Gamma^s$ (for even $\ell$). Finally, introducing these last conditions into the general relations in Equation 16.8, one ends up with the photoionization selection rules:

$$\Gamma^+(\text{rev}) \otimes \Gamma^i(\text{rev}) \supset \Gamma^* \quad \text{(for even } \ell) \quad \text{and}$$
$$\Gamma^+(\text{rev}) \otimes \Gamma^i(\text{rev}) \supset \Gamma^s \quad \text{(for odd } \ell). \qquad (16.9)$$

In general, the application of the selection rules requires the knowledge of the *even* or *odd* character of the photoemitted electron. This can be obtained, for instance, from experimental photoelectron spectroscopy. However, a special case occurs when both the intermediate resonant state and the final ion state are $\Sigma$-states; this particular case is considered in the following.

For $\Sigma–\Sigma$ transitions, it has been demonstrated that the photoionization must obey the selection rule $\Delta N + \ell =$ odd, where $\Delta N = N_+ - N_0$; here $N_+$ and $N_0$ are the quantum numbers for the electronic plus rotational angular momentum of the ionic and initial states, respectively (Dixit et al. 1985). The authors also demonstrated that the selection rule depends neither on how the initial $\Sigma$-state is created nor on the anisotropy of this state; consequently, it should also be applicable to the $(n+1)$ REMPI case via a $\Sigma$-state route. Hence, in $\Sigma$-state-mediated $(n+1)$ REMPI processes, one finds

$$\Delta N = 0, \pm 2, \ldots$$

$$\text{for photoelectron orbital with odd symmetry (s, p,\ldots) and} \tag{16.10a}$$

$$\Delta N = \pm 1, \pm 3, \ldots$$

$$\text{for photoelectron orbital with even symmetry (s, d,\ldots)} \tag{16.10b}$$

It should be noted that for singlet states, $\Delta N \equiv \Delta J$, where $J$ stands for the total angular momentum quantum number.

In general, parity selection rules depend on the Hund case coupling (see Dixit et al. 1985; or Xie and Zare 1990). For example, in Hund's case b limit, the parity selection rule is $\Delta N + \ell + \Delta p =$ odd, where $\Delta p = p^+ - p^i$ (here $p^+$ and $p^i$ are the parities of the ionic and intermediate states, respectively). For $\Sigma–\Sigma$ transitions, $\Delta p = 0$, and, thus, the selection rule in Equation 16.10a applies. For the case of an intermediate coupling scheme, between Hund's cases a and b, the parity selection rule is given by $\Delta J + \Delta S + \Delta p + \Delta q + \ell =$ even (see, e.g., Xie and Zare 1990; or Wang et al. 1995); here $\Delta J = J^+ - J^i$ (with $J^+$ and $J^i$ being the total angular momentum of the ion and intermediate states, respectively, but excluding the nuclear spin; $\Delta S$ is the same but referring to the total spin; and similarly $\Delta q = q^+ - q^i$ (with $q = 1$ for $\Sigma^-$ states and $q = 0$ for all others).

The preceding discourse on rotational-transition selection rules may be exemplified for REMPI experiments of one of the most studied diatomic molecules, nitric oxide (NO). For example, Park and Zare (1993) utilized rotationally resolved $(1+1')$ REMPI to study the process NO $X^2\Pi$ ($v = 0, J = 11.5$) $\rightarrow$ NO $A^2\Sigma^+$ ($v = 0, N^i = 13$) $\rightarrow$ NO$^+$ $X^1\Sigma^+$ ($v^+ = 0, N^+$) + e$^-$ (see the transition term scheme in Figure 16.4), measuring angle- and energy-resolved photoelectron spectra. A typical photoelectron ToF spectrum from the said study is shown in the bottom part of the figure; the fully resolved rotational peaks correspond to the ion signal associated with excitation into different rotational levels of NO$^+$.

In the $\Sigma–\Sigma$ case, the selection rule to be applied is $\Delta N + \ell =$ odd; consequently, $\Delta N = 0, \pm 2, \pm 4, \ldots$ should be observed for odd $\ell$ values, and $\Delta N = \pm 1, \pm 3, \ldots$ for even $\ell$ values. A closer look at the results reveals that both even and odd $\Delta N$ values are observed, with the $\Delta N = 0$ transition being the most intense; this is a sort of a propensity rule, which is encountered in REMPI spectra of many diatomic molecules.

Note that to obtain global REMPI selection rules, one has to combine those of the resonant excitation transition with those of the ionization step. As already mentioned, in many diatomic molecules, the $\Delta N = 0$ transition is the most intense. Therefore, one expects that the global REMPI selection rules are dictated

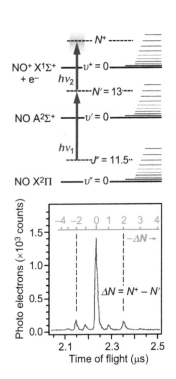

**Figure 16.4** ToF photoelectron spectrum of the $(1+1')$ REMPI excitation NO(X) $\rightarrow$ NO*(A) $\rightarrow$ NO$^+$(X) + e$^-$. The ro-vibrational quantum levels are indicated in the term scheme (top); the 1' laser wavelength for the ionization step is tuned through the rotational transition sequence with $\Delta N = N^+ - N'$. (Data adapted from Park H. and R. Zare, *J. Chem. Phys.* 99, no. 9 (1993): 6537–6544.)

by the one-photon or two-photon dipole matrix transition rules of the resonant excitation to the intermediate state.

## 16.1.4 Conceptual experimental REMPI setups

In Figure 16.5, the concepts of typical experimental REMPI implementations are summarized. In general, these comprise a combination of a molecular beam source, an ion extraction region (where the laser and molecular beam cross), and a ToF mass spectrometer (ToFMS); the two most common modalities are shown, based on linear (Figure 16.5a) or reflectron-type (Figure 16.5b) ToFMS analyzers. Normally, one uses experimental designs with very high collection efficiency for the ions or electrons produced in the REMPI process. This is typically achieved by accelerating the charged particles to a high kinetic energy and detecting them with electron multipliers (e.g., in the form of microchannel plates [MCPs]).

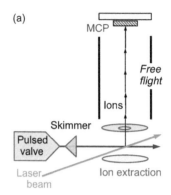

The clear advantage of using ToFMS is related to the use of repetitive-pulse techniques, typically in the nanosecond-duration regime; the use of short-duration laser pulses guarantees that the ions are formed within a very short time window. Subsequently, all ions are accelerated toward the field-free drift region, obtaining the same initial kinetic energy but not the same velocity. The latter condition enables one to exploit the relation

$$E_{\text{pot}} = E_{\text{kin}} = (1/2) \cdot m \cdot v^2 = \text{constant}. \tag{16.11}$$

Here $E_{\text{pot}}$, $E_{\text{kin}}$, $m$, and $v$ stand for the potential energy, kinetic energy, mass, and velocity of the ion, respectively. Hence, one finds that ion mass and flight time obey the simple relation

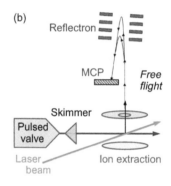

$$t_{\text{ToF}} \approx z \cdot m^{1/2}, \tag{16.12a}$$

where $z$ is a mass-independent proportionality factor, although in practice one will have to account for a small offset due to experimental factors (such as, e.g., time delays by electronic equipment), which is typically taken care of using a zero-order correction, i.e.,

$$t_{\text{ToF}} = z_0 + z_1 \cdot m^{1/2}. \tag{16.12b}$$

From an experimental point of view, one collects the ToF spectra of ions with known masses to calibrate the spectrometer, by determining accurate values of both parameters $z_0$ and $z_1$. The mass resolution of a linear ToFMS depends on several factors; one of them is its length, typically in the range $L_{\text{ToF}} = 20$–$100$ cm.

**Figure 16.5** Basic REMPI configurations, based on a pulsed molecular beam, crossed by a laser beam, and a ToF mass analyzer: (a) linear-type ToF; (b) reflectron-type ToF. MCP = microchannel plate detector. For further details, see text.

Although the voltage applied inside the acceleration region is, in principle, experienced equally by all ions, they will not have acquired exactly the same kinetic energy at the exit of this initiation region. This is due to various factors, including, for example, the presence of different initial velocities associated with the formation process, i.e., before the same potential energy was applied to any initial, distinct spatial ion location inside the acceleration regions (see, e.g., Boest 2000). One particular method for increasing the mass resolution of ToFMS is to replace the linear ToF tube with a so-called *reflectron-type* configuration (this is shown in the bottom part of Figure 16.5b). Essentially, in the reflectron region of the device, a constant electrostatic field reflects the ion beam toward the detector such that the more energetic ions, with the same mass-to-charge ratio, penetrate deeper into the reflectron, thus traveling a longer path until they reach the detector. In contrast, less energetic ions travel a shorter distance into the reflectron and, hence, have a shorter path to the detector. For best mass/time resolution, the ion detector (usually a microchannel plate) is located at the so-called *time-of-flight focus*, which is a point where all ions with the same mass but different energies arrive simultaneously. A reflectron ToFMS can achieve mass resolution from a few thousands up to 20,000, which constitutes a significant improvement over linear ToFMS. It has been shown that even rotational structure (associated with minute energy differences) in large molecules can be resolved. It should be noted that LIF spectroscopy (see Chapter 9) can provide the same spectral resolution as REMPI but without mass resolution; also, LIF requires that the molecule under investigation possesses excited states that fluoresce.

Although REMPI spectroscopy has been, and still is, used in setups based on gas cells, high-resolution REMPI normally aims at resolving the molecular vibro-rotational structure; this is significantly facilitated by the use of supersonic molecular beams. Indeed, the narrow velocity distribution in the transversal direction of the beam is exploited to reduce the Doppler width and, therefore, to increase the spectral resolution. In addition, the inherent cooling associated with supersonic expansion leads to rotational temperatures of a just few degrees kelvin, thus eliminating the spectral band congestion encountered in room-temperature measurements. This makes it possible to even resolve rotational lines in large polyatomic molecules.

In the last decade or so, REMPI has experienced a revolution, particularly in the study of chemical reactions; this is due to the incorporation of the technique of ion velocity mapping. This technique allows one to collect charged particles with a two-dimensional (2D) position-sensitive detector (typically two microchannel plates coupled with a phosphor screen). Thus, the spatial distribution of a photodissociation fragment or reaction product can be mapped. This velocity-imaging REMPI modality will be addressed, with a few key examples, in Section 21.4.

## 16.2 APPLICATIONS OF REMPI IN MOLECULAR SPECTROSCOPY AND TO CHEMICAL INTERACTION PROCESSES

Areas in which REMPI spectroscopy has shown certain advantages over other techniques are the spectroscopy of molecular radicals and the study of chemical reaction dynamics, to reveal in particular molecular mechanism in reactive and nonreactive collisions. To illustrate the versatility of REMPI, a selection of gas phase and surface reactions will be discussed, including bimolecular and unimolecular chemical processes, as well as nonreactive collisions. One should note that one of the most important and impressive applications of REMPI is in the study of reaction stereodynamics, utilizing the technique of so-called *ion imaging*; this topic will be presented in Section 21.4.

### 16.2.1 Molecular spectroscopy utilizing REMPI

*Spectroscopy of the molecule nitric oxide (NO)*

One of the most extensively studied diatomic molecules, using REMPI, is the radical nitric oxide (NO). There are several factors that contribute to this popularity. First, the spectroscopy of NO is well known. Second, NO has a rather low ionization potential IP = 9.2 eV, which could be reached by two photons of moderately short wavelength of $\lambda_L < 268$ nm, in principle. Third, the molecule exhibits a low-energy excited state, the A $^2\Sigma^+$ state, whose lifetime is $\tau = 216 \pm 4$ ns; it is accessible from the ground state X $^2\Pi_{1/2}$ via an allowed electric dipole transition, with photons of $\lambda_L = 226.25$ nm. Note that the long lifetime of the intermediate enhances the probability of the global resonant ionization process.

In Figure 16.6, several REMPI schemes are illustrated (left panel) that have been used in the detection and spectroscopy of NO by REMPI-ToFMS; these include the following:

- One-color (1+1) REMPI via the intermediate state A $^2\Sigma^+$ (or alternatively B $^2\Pi$)

- Two-color (1+1′) REMPI via the intermediate state B $^2\Pi$ (or alternatively A $^2\Sigma^+$)

- One-color (2+1) REMPI via the intermediate state B $^2\Pi$, using laser light of wavelength $\lambda_L = 382$ nm

- One-color (2+1) REMPI via the intermediate state D $^2\Sigma^+$, using laser light of wavelength $\lambda_L = 375$ nm

Clearly, this variety of possible REMPI schemes demonstrates that, on the one hand, convenient routes can easily be adopted when merely the detection of NO is an issue, but on the other hand, dedicated spectroscopy of electronically excited stated can be carried out when selecting the excitation laser wavelength for the resonant step appropriately.

For example, Streibel et al. (2006) recorded an overview REMPI-ToFMS spectrum, tuning their laser source over the range $\lambda_L = 220$–340 nm. The authors noted that the most intense signals occurred around wavelengths of $\lambda_L \approx 226$ nm, corresponding to the well-known X $^2\Pi_\Omega \rightarrow$ A $^2\Sigma^+$ transition (also known as the $\gamma$-band). They also found that most spectral features occurring at larger laser wavelengths were smaller by a factor of about ~500. This therefore explains that many studies, in particular those of analytical relevance for NO trace detection, use the (1+1) or (1+1′) REMPI schemes via the A-state intermediate; an example is given further in the following.

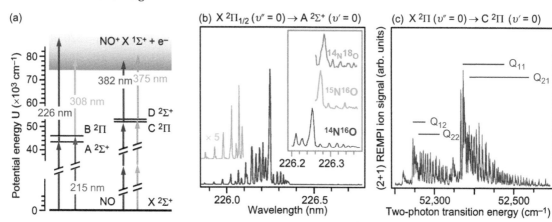

**Figure 16.6** (Left) Schematic view of the most used ($m$+1) REMPI processes for NO detection. (Middle) (1+1) REMPI spectrum revealing ro-vibrational and isotopologue structure in the X($v'' = 0$) → A($v' = 0$) transition. (Data adapted from Schmidt, S. et al., *Anal. Chem.* 71, no. 17 (1999): 3721–3729.) (Right) (2+1) REMPI spectrum revealing the rich rotational branch structure in the X($v'' = 0$) → C($v' = 0$) transition (not all branches are annotated). (Data adapted from Hippler, M. and J. Pfab, *Chem. Phys. Lett.* 243, nos. 5–6 (1998): 500–505.)

To demonstrate the versatility of the various REMPI schemes for the detection and spectroscopy of NO by REMPI, example spectra from two different studies are included.

In the middle panel of Figure 16.6, a partial (1+1) REMPI spectrum for NO is shown, near the Q-branch band head (Schmidt et al. 1999). This particular set of data was recorded about 2 mm in front of the nozzle in a supersonic expansion of NO, diluted in argon; the supersonic expansion resulted in substantial rotational cooling so that the spectrum exhibits a well-resolved, sparse rotational line structure (the authors could fit the data assuming a Boltzmann rotational distribution with a rotational temperature of $T_{rot} \approx 40$ K). Furthermore, in spectra recorded at room temperature, both X $^2\Pi_{1/2}$, $^2\Pi_{3/2} \rightarrow$ A $^2\Sigma^+$ subbands exhibit rather similar intensities; in the rotationally cold supersonic-expansion spectrum, rotational lines originating in the excited X $^2\Pi_{3/2}$ substate are hardy observable at all (on the scale of the figure they are "hidden" in the nearly flat part of the data trace in the range $\lambda_L = 226.5$–226.8 nm). In the inset panel of the figure, an additional, advantageous feature of the REMPI methodology is revealed, namely, that of the spectroscopy of (minor) isotopologues. Besides the main isotopologue $^{14}N^{16}O$, (shifted) line spectra of the trace species $^{15}N^{16}O$ and $^{14}N^{18}O$ (with relative abundances of ~0.004 and ~0.002, respectively) can also be observed. This is afforded by the fact that the three species have different masses; hence, by setting the mass filter for data analysis appropriately, the isotopologue response to the REMPI laser excitation can be extracted; under normal medium-resolution spectroscopy conditions, these features would most likely be unresolved or swamped by the major constituent $^{14}N^{16}O$.

An interesting question concerns the advantages or drawbacks of using one-color or two-color two-photon resonant ionization processes through the same resonant intermediate of the same molecule, such as, for example, using (1+1) and (1+1') REMPI of NO via its intermediate state A $^2\Sigma^+$. Hippler and Pfab (1995) measured an almost 20-fold increase in the signals in the two-color (1+1') REMPI laser photoionization route compared with that of the one-color (1+1) REMPI scheme. The authors attributed this increase in sensitivity to the very different ionization cross sections of NO out of the excited A $^2\Sigma^+$ state, for $\lambda_1 = 226$ nm and $\lambda_{1'} = 308$ nm, respectively. While for the latter, the cross section for ionization is on the order of $\sigma_{ion} \approx 1.2 \times 10^{-16}$ cm$^{-2}$ for a typical $\gamma(0,0)$ resonant ro-vibronic transition, the cross section for the one-photon ionization at $\lambda_1 = 226$ nm out of A-state levels has been measured as $\sigma_{ion} = 7.0 \ (\pm 0.9) \times 10^{-19}$ cm$^{-2}$ (see Zacharias et al. 1980). Thus, the use of a second laser, with a different wavelength, for the ionization step has several advantages with respect to one-color REMPI schemes, despite the inherent complication of needing an additional laser source. The first one is that in many cases, as in the one presented here, the energy of the ionizing photon can be lowered to diminish the probability of ion fragmentation. The second one is that the laser intensity of the resonant excitation step could be reduced, thus eliminating saturation effect, as well as the onset of stimulated emission from the intermediate state, which otherwise would reduce the REMPI signals.

### Spectroscopy of the radicals calcium hydride/calcium deuteride (CaH/CaD)

The combination of the technique of REMPI-ToFMS with the collision-free environment provided in low-density molecular beams has made it possible to investigate the spectroscopy and structure of unstable species or chemical radicals, an objective that otherwise would not have been possible. As an example of this type of application, here a measurement of REMPI spectra of the radicals CaH/CaD is described. The radical was first produced by a chemical reaction of laser-sputtered calcium atoms seeded into a pulsed beam of $H_2/D_2$ (in a carrier gas of helium); subsequently, the rotational levels of its electronic ground-state product were probed by $(1+1')$ REMPI-ToFMS (see Gasmi et al. 2003), as shown in Figure 16.7 according to the overall scheme

$$\text{CaR}\left(X\,^2\Sigma^+, v'' = 0\right) \xrightarrow{hv_1} \text{CaR}^*\left(A\,^2\Pi, v' \middle| B\,^2\Sigma^+, v'\right)$$

$$\xrightarrow{hv_2} \text{CaR}^+\left(X\,^1\Sigma^+, v^+\right) + e^-,$$

(16.13)

where R = H, D indicates the two isotopologues CaH and CaD, respectively. The REMPI spectrum shown in the same figure is that for the overlapping bands of calcium hydride CaH $(X;v'' = 0 \rightarrow B;v' = 0)$ and $(X;v'' = 0 \rightarrow A;v' = 1)$; the rotational bands of these two transitions are annotated in green and red, respectively. The spectral transitions can be described by Hund's case b approximation, at least for the low-rotational quantum levels observed in the investigation whose results are discussed here (the estimated rotation temperature was $T_{rot} \approx 40$ K). Thus, one expects to observe six line branches, namely, $P_1$, $P_2$, $^PQ_{12}$, $^RQ_{21}$, $R_1$, and $R_2$, for each of the transition bands. Note that the indices "1" and "2" refer to the two spin substates $F_1$ and $F_2$ with $J = N + 1/2$ and $J = N - 1/2$; here $J$ and $N$ are the total (excluding nuclear spin) and rotational angular momentum quantum numbers, respectively. Note also that because of the line width of the excitation laser, $\delta\lambda_1$ was of the same order as the spin-splitting constant of the ground vibrational state so that some of the branches could not be resolved from each other. Still, the analysis of the rotational structure of the REMPI spectra allowed the authors to derive the rotational constants for the CaR $(B^2\Sigma^+;v' = 0, 1)$ states, including the spin-splitting parameters, for the first time.

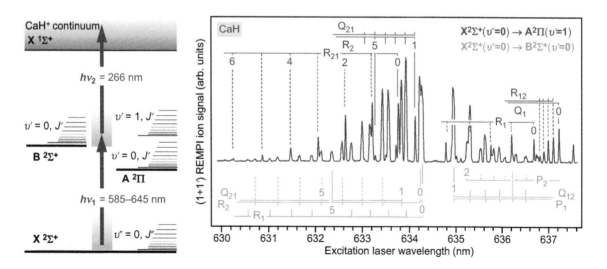

**Figure 16.7** $(1+1')$ REMPI spectroscopy of CaH X $^2\Sigma^+$. (Left) REMPI excitation scheme, via A $^2\Pi$ and B $^2\Sigma^+$ intermediate resonant states. (Right) Overlapping bands $(X;v'' = 0 \rightarrow B;v' = 0)$ and $(X;v'' = 0 \rightarrow A;v' = 1)$; the rotational branches are annotated, together with the relevant rotational quantum numbers $N$. (Data adapted from Gasmi, K. et al., *J. Phys. Chem. A* 107, no. 50 (2003): 10960–10968.)

## 16.2.2 Investigation of chemical reactions utilizing REMPI

### Hydrogen exchange reaction $H + D_2 \rightarrow HD + D$

Thanks to the high selectivity and sensitivity offered by REMPI, it has become possible to measure the quantum-state distribution of reaction products in elementary chemical reactions. The reaction $H + D_2 \rightarrow HD + D$ may be seen as one of the key example for this. Using a (2+1) REMPI scheme, Zare and coworkers measured the internal state distributions for the reaction product $HD(v;J)$ (see, e.g., Marinero et al. 1984, and Rinnen et al. 1989). In their experiments, the authors photodissociated hydrogen iodide (HI) by laser radiation ($\lambda_{pd} = 266$ nm), constituting the source of H atoms. Following the reaction of these atoms with $D_2$, the authors probed the product HD, using tunable laser radiation in the range $\lambda_1 = 200–226$ nm to excite the two-photon transition $X^1\Sigma_g^+(v'' = 0\text{-}3; J'') \rightarrow$ E, $F^1\Sigma_g^+(v' = 0; J')$. Subsequently, the resonantly excited states were ionized by a third photon (also at wavelength $\lambda_1$); the resulting $HD^+$ ion was mass-selected by a ToF mass spectrometer. The authors measured complete rotational and vibrational distributions of the product HD, for a variety of collision energies, specifically to determine the nascent populations of HD ($v' = 0; J = 0\text{-}15$), HD ($v' = 1; J = 0\text{-}12$) and HD ($v' = 2; J = 0\text{-}8$) formed in the reaction.

In REMPI studies of molecular hydrogen and its isotopes, the quantum-state population of the ground state cannot be deduced directly from the measured REMPI ion signals; this is because the ionization cross section of the intermediate excited state (any one of the B-, E-, and F-states) depends on the photon energy in the ionization step, as well as on the particular ro-vibrational state. Therefore, the recorded line intensities need to be corrected with appropriate *correction factors*, which can be determined by calibration against a thermal effusive source of the molecular sample under study, whose vibrational and rotational state populations are known to follow a thermal Boltzmann distribution (see, e.g., Rinnen et al. 1991, for further details). After applying the corresponding corrections factors, the authors concluded from the analysis of their data that of the available energy, 73% was partitioned into product translation, 9% into HD vibration, and 18% into HD rotation. They also found that both rotational and vibrational distributions were in rather good agreement with quasi-classical trajectory (QCT) calculations. To illustrate the comparison between experiment and theoretical calculations, the HD ($v'' = 0; J'$) product rotational state distribution is shown in Figure 16.8, originating from the reaction $H + D_2(v = 0; J = 0\text{-}4)$ $\rightarrow HD(v''; J'') + D$, for a collision energy near $E_c \approx 1.6$ eV (see Bean et al. 2001). The experimental data were compared with QCT calculations carried out by Blais and Truhlar (1989); as seen in the bottom part of the figure, excellent agreement between the two sets of data was found.

More recently, the same research group measured the differential cross section (DCS) for the reaction $H + D_2 \rightarrow D + HD(v'' = 1; J'' = 1, 5, 8)$, at a collision energy of $E_c \approx 1.7$ eV, using a novel two-color three-photon Doppler-free REMPI scheme; see Goldberg et al. (2007). In this particular REMPI scheme, the two-photon resonant transition $X^1\Sigma_g^+(v'' = 0, 1; J'' = 1, 5, 8) \rightarrow$ E, $F^1\Sigma_g^+(v'; J')$ was realized by means of two counterpropagating laser beams whose wavelengths were slightly detuned with respect to each other (one by a few wavenumbers to the blue, the other by the same amount to the red from line center). As a result, the resonant transition to the intermediate state in the REMPI scheme can be produced only by the absorption of one photon from each laser beam. One further photon

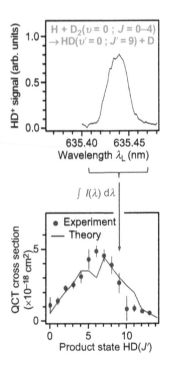

**Figure 16.8** Experimentally measured rotational state distribution of the HD product in the reaction $H + D_2(v = 0; J = 0\text{-}4) \rightarrow HD(v''; J'') + D$ and comparison with QCT calculation data. Each data point is derived from the integration of the REMPI ion signal over the full line profile. (Experimental data adapted from Bean, B.D. et al., *J. Phys. Chem. A* 105, no. 11 (2001): 2228–2233 QCT calculations according to Blais, N.C. and D.G. Truhlar, *Chem. Phys. Lett.* 162, no. 6 (1989): 503–510.)

(from either laser beam) then generates the REMPI ion; the scheme may thus be described as ([1+1′]+[1 or 1′]) REMPI. With this elegant method, the background due to absorption of the two photons from the same laser is eliminated (for the general description of two-photon absorption, see Section 7.4). Therefore, higher detection sensitivity for nascent products from bimolecular reactions can be achieved, which by and large exhibit small cross-section values.

Finally, the fundamental hydrogen exchange reactions have been investigated extensively by several groups, using REMPI coupled with ion imaging; some key results will be presented in Section 21.4.

### Excited-state chemical reaction $O*({}^1D) + N_2O$

The reaction between excited oxygen atoms $O*({}^1D)$ and nitrous oxide ($N_2O$) into products predominantly occurs via two pathways, namely (see, e.g., Tokel et al. 2010),

$$O*({}^1D) + N_2O \rightarrow N_2 + O_2, \quad k = 4.9 \times 10^{-11} cm^3 \cdot molecule^{-1} \cdot s^{-1}, \quad (16.14a)$$

$$\rightarrow NO + NO, \quad k = 6.7 \times 10^{-11} cm^3 \cdot molecule^{-1} \cdot s^{-1}. \quad (16.14b)$$

The pathway in Equation 16.14b is of great relevance to atmospheric chemistry since the production of two NO molecules lower the ozone concentration, according to the "catalytic" reaction

$$2O_3 + NO \rightarrow 3O_2 + NO. \quad (16.15)$$

The relevance of the said reaction of excited atoms with molecules has resulted in increased interest in the study of its detailed reaction mechanism, even at the molecular level, using molecular beams and REMPI spectroscopy (see, e.g., Tokel et al. 2010). These authors investigated the reaction in Equation 16.14b in a molecular beam experiment, in which $O_3$ and $N_2O$ were coexpanded. The atomic reagent $O*({}^1D)$ was produced by photodissociation of ozone at $\lambda_{pd} = 266$ nm; the product molecules NO ($X^2\Pi$) were detected by (1+1) REMPI, with the laser tuned over the range $\lambda_1 = 220$–246 nm. A segment of a representative (1+1) REMPI spectrum from this study, involving the resonant transition NO ($X^2\Pi$) → ($A^2\Sigma^+$), is shown in the bottom panel of Figure 16.9, in comparison with a similar REMPI spectrum from a thermal beam of NO (top panel of the figure, for $v'' = 1 \rightarrow v' = 0$, or $\gamma_{10}$).

Clearly, a much richer line structure is observed in the REMPI spectrum of NO generated by the (rather exothermic) chemical reaction in Equation 16.14b, now also including contributions from $\gamma_{53}$ and $\gamma_{64}$ (note that the indices follow the convention $\gamma_{v''v'}$). The richer line structure is particularly due to the very high rotational temperature, which according to theoretical simulations was estimated to be on the order of $T_{rot} \approx 4500$ K (see the red simulation trace in the figure).

### Doppler-selected REMPI-ToF

Laser spectroscopic techniques have revolutionized the study of unimolecular and bimolecular chemical reactions, providing detailed information about the molecular mechanism of the chemical event (see, e.g., the textbook by Telle et al. 2007). One of the techniques that has significantly contributed to this progress is the so-called *Doppler-selected REMPI-ToF* (Lai et al. 1996). Here the techniques of Doppler-shift selected excitation and ToF ion (mass) analysis are combined in

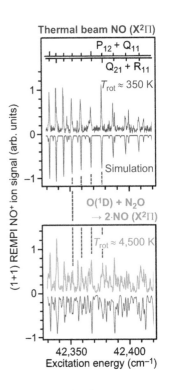

**Figure 16.9** (1+1) REMPI spectrum of NO ($X^2\Pi$), generated in the reaction $O*({}^1D) + N_2O \rightarrow 2NO$ ($X^2\Pi$). (Top) Reference spectrum of thermal NO; selected rotational line branches in the $\gamma_{10}$ band are indicated. (Bottom) Spectrum of NO generated reactively. Both experimental data are compared with simulation data (red traces), for $T_{rot} = 300$ K and $T_{rot} = 4500$ K, respectively. (Data adapted from Tokel, O. et al., *J. Phys. Chem. A* 114, no. 42 (2010): 11292–11297.)

an "orthogonal" manner, such that the three-dimensional (3D) velocity distribution of the reaction or photofragmentation products can be measured directly in the center-of-mass (CM) frame. The approach has been implemented to study a range of photodissociation processes involving small molecules, such as, e.g., $C_2H_2$ or $SH_2$, and to explore chemical kinetics, such as the reaction of excited O* ($^1D$) or S*($^1D$) atoms with $H_2$, $D_2$, and HD molecules, to cite but a few key examples (see Lee et al. 2006).

The basic idea of this relatively young experimental method can be understood with the aid of Figure 16.10. The assumed scenario is to fully probe any one of the two products of the reaction AB + C → A + BC, i.e., either the atom A or the molecule BC. The task then would be to measure not only the internal energy state distribution but also the differential 3D product velocity distribution $S(v_x, v_y, v_z)$. For this, the technique of Doppler-shift selection is employed i.e., one selectively ionizes a subgroup of the (atomic or molecular) product, which has a given value of $v_z$ in the CM frame (assuming that the laser radiation propagates along the relative axis $z$). In the related (1+1) REMPI scheme, the photon energy of the excitation laser would be tuned to match the $v_z$-shifted resonant-transition energy.

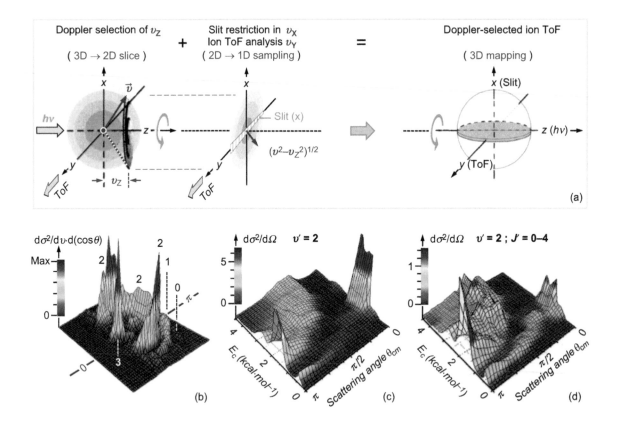

**Figure 16.10** (a) Illustration of the basic idea of the Doppler-selected REMPI-ToF method. (b–d) Measurements of molecular product distributions in the exchange reaction F + HD → HF + D: (b) 3D representation of the product flux–velocity contour map, at a collision energy of $E_c$ = 2.88 kcal·mol$^{-1}$, with vibration-resolved features labeled (note that $\theta$ = 0 ≡ forward scattering, $\theta$ = $\pi$ ≡ backward scattering); (c) 3D representation of the vibrationally resolved angular distribution for the HF($v'$ = 2) product, evolving with the change in collision energy; (d) 3D representation of the rotationally resolved (in part) angular distribution for the HF($v'$ = 2;$J'$ = 0–4) product, evolving with the change in collision energy. (Data adapted from Lee, S.H. et al., *J. Chem. Phys.* 125, no. 13 (2006): article 133106.)

In a traditional Doppler-profile measurement, this REMPI signal would lead to a single data point. However, in this method, all Doppler-selected ions are dispersed (see the left panel in Figure 16.10a), both spatially (a $v_x$-component in the $x$-direction) and temporally (a $v_y$-component in the $y$-direction), assuming that the ToF mass analysis is along the $y$-axis. When now a slit is placed in front of the microchannel-plate ion detector, only ions with $v_z \approx 0$ will be detected, resulting in one-dimensional sampling of the product velocity distribution. Repeating this type of ToF measurement, successively tuning the probe laser wavelength over the Doppler profile of the selected reaction product, the entire 3D distribution can be mapped out (see the right panel of Figure 16.10a). One of the clear advantages of this approach is that both the Doppler slicing and the ToF ion measurement are essentially in the CM frame and thus—because the $v_x$-component associated with the $\mathbf{v}_{CM}$ velocity vector is small—the measurement directly maps out the desired 3D $\mathbf{v}_{CM}$ distribution.

One of the most interesting studies carried out by using this high-resolution technique is the reaction $F + HD \rightarrow HF + D$ (DF + H), in which the existence of a dynamical resonance was shown (see Lee et al. 2006, and references therein). The term *resonance* refers to a transient metastable species produced as the reaction takes place: transient intermediates can live for a few vibrational periods to thousands of vibrational periods before they dissociate and, in general, play a significant role in many kinds of atomic and molecular processes. To illustrate the degree of detail in the investigation of this fundamental reaction, some representative results are summarized in the data part of Figure 16.10.

In brief, the authors carried out several series of Doppler scans of the D atom product, as a function of collision energies; these were acquired under a parallel configuration; i.e., the probe laser propagated along the relative velocity vector of the collision system. The analysis of these Doppler profiles allowed the authors to extract angle–velocity distributions of the HF product, with vibrational and rotational state resolution.

For this, the raw ToF spectral data were converted into speed distributions, each of those being weighted by the solid-angle factor $v^2$ and by applying the appropriate density-to-flux transformation correction. Then, all $v_y$ distributions of a full data set (i.e., measurements over the entire Doppler profile) were combined, yielding the desired 3D contour map of the product flux–velocity distribution, $d\sigma^2/(dv \cdot d(\cos\theta))$; an example is shown in Figure 16.10b, for a collision energy of $E_c = 2.88$ kcal·mol$^{-1}$. Based on energetic arguments, the observed ringlike features can readily be assigned to vibrational energy states of the reaction product HF($v' = 0, 1, 2, 3$). In particular, one notices that the angular distribution for the weakest feature ($v' = 0$) is confined in the backward-scattering hemisphere; a similar but slightly broader and more intense distribution is found for $v' = 1$. The most prominent feature is recorded for the product vibrational state HF($v' = 2$), exhibiting a predominantly sideways-scattering character. Finally, for HF($v' = 3$), one observes a sharp forward peak.

It is clear from this single collision-energy data set that the information content in measurement series such as the one underlying this example is enormous—including the vibrational and rotational (in part) quantum state variables $v'$ and $J'$, as well as the collision energy $E_c$ between the reactants. To address this information overflow and to gain better insight into the quantum-state-dependent dynamics of the reaction, the authors presented their data in a state-resolved fashion, plotting the angular-dependent ($\theta$) differential scattering cross section ($d\sigma^2/d\Omega$) for each individual, as a function of collision energy $E_c$. Representative examples for a vibrational quantum state—HF($v' = 2$)—and a partially resolved range of rotational quantum states—HF($v' = 2; J' = 0\text{-}4$)—are shown in Figure 16.10c and d. The data reveal the aforementioned "resonance" character for certain values of collision energies, resulting in clear directional preferences of reaction product, dependent on the quantum state.

### 16.2.3 Photodissociation studies utilizing REMPI

The bond breaking and making of typical bimolecular reactions becomes simplified in so-called half-collision processes—the molecular photodissociation process is often referred to like this, since it involves the breaking of a molecular bond only. In this context, REMPI probing has proven to be a versatile tool for investigating the quantum-state distribution of photodissociation fragments.

#### Photodissociation of N₂O

One of the molecules whose photodissociation has been extensively investigated by REMPI is nitrous oxide ($N_2O$). This is probably not surprising since $N_2O$ is one of the most potent greenhouse gases. It photodissociates when irradiated by ultraviolet C (UV-C) light in the wavelength window $\lambda = 190\text{-}200$ nm according the photochemical reaction

$$N_2O \xrightarrow{h\nu_{UV-C}} N_2 \left(^1\Sigma_g^+\right) + O\left(^3P_2\right), O^*\left(^1D_2\right) \tag{16.16}$$

thus constituting an important photodepletion mechanism in the global $N_2O$ budget.

For example, Nishide and Susuki (2004) measured (2+1) REMPI spectra of the nascent $N_2$ $\left(^1\Sigma_g^+\right)$ produced. The product molecule $N_2$ was generated by photodissociation of the parent molecule $N_2O$ (prepared in a supersonic beam), by laser light at wavelength $\lambda = 203$ nm. Photons from the same (tunable) laser source were then used in their (2+1) REMPI scheme, probing the molecular fragment

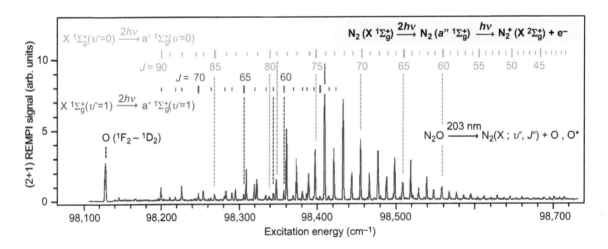

**Figure 16.11** (2+1) REMPI spectrum of nascent $N_2$ in the one-color photodissociation/REMPI detection experiment of $N_2O$ in a supersonic molecular beam. The spectrum shows the Q-branch of the two-photon resonant transition $N_2(X; v'') \rightarrow N_2(v''; v')$, for $v'' = v' = 0, 1$. The rotational quantum assignments are indicated above the spectrum. (Data adapted from Nishide, T. and J. Suzuki, *Phys. Chem. A.*108, no. 39 (2004): 786–7870.)

with rotational quantum-state resolution, on the resonant two-photon transition $N_2$ $(X^1\Sigma_g^+) \rightarrow N_2$ $(a''^1\Sigma_g^+)$; an example spectrum is shown in Figure 16.11. The spectrum is dominated by the rotational Q-branch (with $\Delta N = 0$); while O- and S-branches (with $\Delta N = \pm 2$) should also be observable in two-photon transition schemes, in principle, the authors found that here they were much weaker than the Q-branch, by about two orders of magnitude. Also, the rotational state distribution evidently is nonthermal, peaking at high rotational quantum numbers (at around $J = 75$); the clear intensity alternation of 2:1 is due to the nuclear spin statistics.

One should be noted that the investigation carried out by Nishide and coworkers was also extended to include velocity-map ion imaging of the $O^*(^1D_2)$ fragment, which provided deeper insight into the photofragmentation dynamics of this relevant triatomic molecule; for a description and examples of velocity-map ion imaging, see Section 21.4.

## 16.2.4 REMPI spectroscopy of catalytic reactions

The simplest but most fundamental catalytic surface reaction is the recombination of two hydrogen atoms at a metal surface. Indeed, this associative desorption phenomenon has been extensively investigated by numerous research groups, searching for the influence of the substrate structure and temperature, on the one hand, and measuring the angular, velocity, and quantum state distribution of the desorbed molecules, on the other hand, to fully understand the catalytic mechanism at the molecular level.

### Recombination of $D_2$ at the surface of Pd(100)

In general, ro-vibrational population distributions of desorbing molecules are measured by REMPI. Here some results from the study by Wetzig et al. (2001) are presented, who measured the vibrational and rotational population distributions in the associative desorption of $D_2$ from Pd(100). In their experiments, deuterium was supplied from a permeation source, whose temperature could be adjusted in the range $T = 400$–$850$ K; the desorbing molecules were detected using a $(1+1')$ REMPI scheme, as illustrated in the potential energy diagram in Figure 16.12. Essentially, tunable laser radiation at $\lambda_1 = 106$–$110$ nm was used for the resonant excitation step $D_2$ $(X_1\Sigma_g^+) \rightarrow D_2$ $(B_1\Sigma_u^+)$, followed by ionization out of the excited intermediate state into the ionic ground state $D_2^+(X_2\Sigma_g^+)$, at wavelength $\lambda_{1'} = 266$ nm (see the blue and violet arrows in the figure, respectively).

A typical $(1+1')$ REMPI spectrum of $D_2$ desorbing from a clean Pd(100) surface, at temperature $T_{surf} = 800$ K is shown in Figure 16.12, with full line identification provided. Note that in addition to the Lyman bands $(v'' = 0 \rightarrow v' = 0)$ to $(v'' = 0 \rightarrow v' = 3)$, rotational lines from the "hot" bands $(v'' = 0 \rightarrow v' = 0)$ to $(v'' = 0 \rightarrow v' = 0)$ are also observed. In their study, the authors found that the rotational temperature of the desorbed $D_2$ was always significantly lower than that of the surface, indicating that the rotation is not in thermal equilibrium with the surface.

So the question arose of what mechanism was responsible for the rotational cooling observed in the molecular desorption. The authors concluded that, based on microscopic reversibility, the mechanism responsible for the *rotational cooling in desorption* is the same as that responsible for the *observed rotational hindering in adsorption*. Detailed calculations based on potential energy surface modeling show a large anisotropic interaction with respect to the barrier for hydrogen dissociation (see, e.g., Groß et al. 1996; or Wilke and Scheffler 1996).

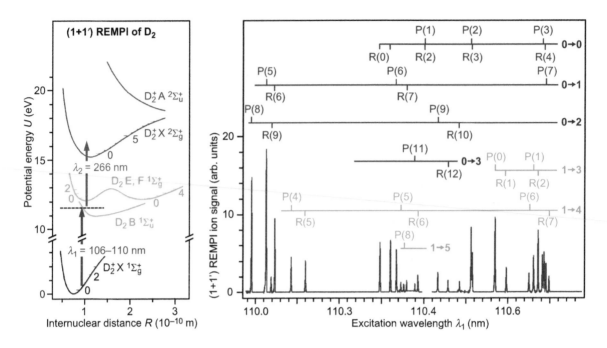

**Figure 16.12** Measurement of the ro-vibrational distribution of $D_2$ desorbing from Pd(100). (Left) (Partial) Potential energy curve diagram, relevant in the (1+1') REMPI probing of $D_2$. (Right) Rotationally resolved REMPI spectrum of $D_2$, desorbing from Pd(100) at surface temperature $T_{surf}$ = 800 K; the relevant ro-vibrational line assignment is included. (Data adapted from Wetzig, D. et al., *Phys. Rev. B* 63, no. 20 (2001): article 205412.)

Thus, while molecules in high-rotational states will rotate out of a preferred orientation toward dissociative adsorption (during the time required to break the molecular bond), those in low-rotational states will be driven by the potential energy surface more easily into a favorable orientation for dissociative adsorption. Hence, microscopic reversibility ensures that molecules in low-rotational states will be predominant in the desorption flux, as opposed to thermal equilibrium with the surface.

## 16.3 REMPI AND ANALYTICAL CHEMISTRY

As just outlined in Section 16.2, REMPI has become a powerful tool in fundamental molecular spectroscopy and in the understanding of chemical reaction pathways and dynamics. Beyond this, its extraordinary selectivity (due to the high species resolution afforded by the combination of resonant photon excitation and molecular-mass-specific ion detection) and its superb sensitivity (afforded by the extremely efficient detection of the REMPI excitation products) have also made the methodology interesting in analytical chemistry, for both qualitative and quantitative applications. A few key examples are discussed in the following, including the distinction of isotopically substituted molecules (isotopologues) and the identification of steric variants (isomers), as well as trace detection of primary and secondary chemical reaction species.

## 16.3.1 REMPI spectroscopy with isotopologue and isomeric selectivity

The mass- and wavelength-selective spectroscopy inherent to REMPI-ToFMS is particularly relevant for the measurement of molecular isotopologues and/or (structural) isomers, which otherwise would exhibit a significant degree of spectral overlap.

As demonstrated by, e.g., Keil et al. (2000), isotopically resolved spectra can give important information on molecular composition and structure and, therefore, facilitate the correct spectral assignment necessary to test theoretical calculations. A decent proof-of-principle example for this is isotope-selective (1+1) or (1+1′) REMPI of the NO radical; the spectroscopic framework underlying REMPI studies of NO has been addressed already in Section 16.2. To recap, in these schemes, one utilizes resonant photon absorption at $\lambda_1 \sim 226$ nm to excite NO from its ground state $X^2\Pi$ into the intermediate state $A^2\Sigma^+$. A second photon of the same (or different) wavelength produces $NO^+$ ions (provided the joint excitation energy of the two photons is greater than the NO ionization potential $IP_{NO} = 9.2$ eV).

In reference to the above, Schmidt et al. (1999) demonstrated that there is a slight shift to longer wavelengths in the UV absorption of NO, with increasing isotopic mass of the molecule. Using the most abundant isotopologue $^{14}N^{16}O$ as the reference, these shifts are $\Delta\lambda = -0.02$ nm and $\Delta\lambda = -0.07$ nm for minor isotopologues $^{15}N^{16}O$ and $^{14}N^{18}O$, respectively. The authors clearly resolved these three isotopologues in their mass-selected (1+1) REMPI spectra, measured in a mixture of 20 ppbV of natural NO in argon. More recently, Mitscherling et al. (2009) reported on specific mass-selective (1+1) REMPI spectroscopic windows for monitoring said three isotopologues, nearly without spectroscopic interference with each other, as illustrated in Figure 16.13 for the minor species $^{14}N^{18}O$ (natural abundance about $1.993 \times 10^{-3}$). The rotational line sequence $^OP_{12}(J'')$ of the transition $(X;v'' = 0) \rightarrow (A;v' = 1)$ is the dominant feature in the spectrum (exhibiting a turning point near $J'' \approx 11.5$); only a few, mostly rather weak lines of

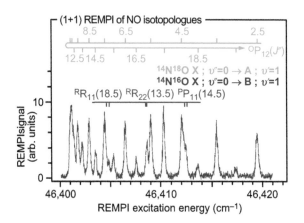
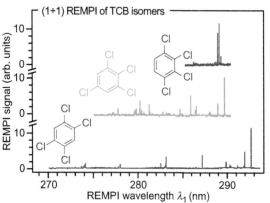

**Figure 16.13**  Isotopologue- and isomer-selective REMPI measurements. (Left) (1+1) REMPI spectrum of $^{14}N^{18}O$, resonantly excited by photons in the range $\lambda_1 = 215.42$–215.51 nm; note the slight interference from a few lines of the major isotopologue $^{14}N^{16}O$ (annotated in red). (Data adapted from Mitscherling, C. et al., *Phys. Scr.* 80, no. 4 (2009): article 048122.) (Right) (1+1) REMPI spectra of the three isomers of tetrachlorobenzene (TCB), rotationally cooled in a supersonic jet expansion. (Data adapted from Weickhardt, C. et al., *Rapid Commun. Mass Spectrom.* 8, no. 5 (1994): 381–384.)

the $(X; v'' = 0) \rightarrow (B; v' = 1)$ transition of the major isotopologue $^{14}N^{16}O$ are observed within the same spectral window (these are annotated in red in the figure). From their data, the authors estimate a limit of detection for $^{14}N^{18}O$ of LOD = 0.8 pptV.

As pointed out earlier, besides the isotopic-mass selectivity, REMPI also allows for the analysis of the conformational isomer and tautomer structure of molecules; this has found wide-ranging applications in analytical chemistry of a multitude of organic compounds. One example is shown in the right panel of Figure 16.13, for (1+1) REMPI of the three TCB isomers, excited via their $S_0 \rightarrow S_1$ transition band. The spectra shown in the figure were recorded in a supersonic beam expansion environment (see Weickhardt et al. 1994). The cooling associated with supersonic expansion greatly simplifies the structure of the REMPI spectrum, due to the reduction in rotational line number and the lines being narrow. As a consequence, the well-separated rotational-transition lines ensue from the selective excitation of the sparse, lowest ro-vibronic level population. From these spectra, one can then find the appropriate laser wavelength(s), to realize selective ionization of a single isomer of the compound under investigation, which can be of high relevance in environmental analysis.

## 16.3.2 REMPI spectroscopy in trace and environmental analyses

### REMPI of NO in exhaled breath

One frequently used application of REMPI spectroscopic analysis in medical investigation addresses trace analysis of nitric oxide (NO) in human exhaled breath. This is linked to the fact that an increased concentration of NO in alveolar ventilation is indicative of inflammatory stress within the lung. It turns out that the sensitivity of the (1+1) REMPI, using laser radiation at $\lambda_1 = 266.121$ nm, is sufficiently high to measure the (temporal) NO concentration profile in exhaled breath. Specifically, the protocol implies that the patient breathes NO-free air for several minutes. Subsequently, the exhaled breath as well as its flow is monitored; the latter is required to calculate the sampled mixing ratio for the NO being exhaled.

A typical result is illustrated in Figure 16.14 in which the concentration-versus-time profile of NO sampled in oral and nasal exhalation is shown (see Short et al. 2006). In the exhalation time profile, peak concentrations of NO of about $c_{NO}$ ~ 20 ppbV early on during oral exhalation are observed, followed by a nearly constant level indicating a steady-state production of NO ($c_{NO}$ ~ 5 ppbV); the measurements confirm the typical "rise–peak–plateau" behavior for NO content in exhaled breath. The data also replicate the common breathing cycle, comprising the inhalation and pause phases, during which the concentration of NO builds up, until it is released during exhalation.

When sampling nasally exhaled breath, the authors measured peak concentrations of up to $c_{NO} \approx 80$–110 ppbV, i.e., three to four times higher than that of the NO measured in association with exhalation via the oral cavity alone (see the bottom panel in Figure 16.14). The authors therefore stress that it is necessary to provide some means of separating measurements taken from the mouth versus that taken from the nose; this might be done, e.g., by providing a slight overpressure of inhaled synthetic air for the former.

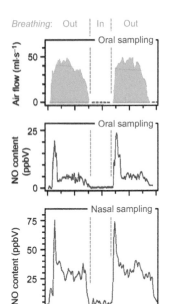

**Figure 16.14** Temporal profile of NO measured by (1+1) REMPI-MS, for a laser wavelength of $\lambda_1 = 266.121$ nm, from orally and nasally sampled exhaled breath (center and bottom, respectively). For quantitative referencing, total airflow profiles are added (top). (Data adapted from Short et al. *Appl. Spectrose* 60, no. 2 (2006): 217–222.)

As pointed out earlier, increased levels of NO in exhaled breath are associated with inflammatory stress to the lungs. For example, in their study of asthma sufferers, Shinkai et al. (2002) found NO concentration profiles with a shape similar to those shown in Figure 16.14, with peak and plateau values of $c_{NO,peak}$ ~ 50 ppbV and $c_{NO,plateau}$ ~ 20 ppbV, respectively. This clearly indicates the relative increase in NO concentration in the breath of individual test patients with pulmonary stress conditions.

## REMPI of polyaromatic hydrocarbons

For environmental gas analysis, REMPI-ToFMS offers a "2D" analytical method (optical transition selectivity and ToF mass-resolved ion analysis), with both high specificity and high sensitivity. This is of particular advantage in situations where purely optical spectroscopy might struggle to disentangle a larger number of organic molecular species in a mixture, such as those found, e.g., in the exhaust gases from combustion processes. An additional advantage of REMPI-ToFMS is that it can be exploited for on-line measurements in gas exhausts (e.g., by using sampling in a flow bypass).

A well-developed REMPI analytical application is the monitoring of waste incineration plants (see, e.g., Heger et al. 1999; or Zimmermann 2005). In the left panel of Figure 16.15a, a representative REMPI-ToF mass spectrum of the flue gas emission from hazardous waste incineration plant is shown (measured for the REMPI laser wavelength of $\lambda_1 = 269.82$ nm). A large number of aromatic organic compounds can be identified in the mass spectrum (for clarity, only a few of those are annotated in the figure).

What is particularly evident in the sample spectrum is the presence of the mass peaks at $m/z = 112$ amu and $m/z = 114$ amu. These can be assigned to the MCBz isotopologues $^{35}C$- and $^{37}Cl$-$C_6H_5$, respectively. Note that MCBz has been identified as a surrogate for the emission of the highly toxic compounds of polychlorinated dibenzo-$p$-dioxins/furans. Note also that the insert spectrum nicely demonstrates the laser selectivity in the REMPI excitation path: for a laser wavelength of $\lambda_1 = 269.82$ nm, the MCBz mass peaks are clearly observed (red data trace), while for a laser wavelength of $\lambda_1 = 266.0$ nm, they are absent (green data trace).

In the right panel of Figure 16.15a, a 3D plot of an on-line measurement of MCBz (quantified by external standardization) is shown. The time evolution segment shown here is over a period of 1000 s (with arbitrary starting point); note that only the mass interval containing MCBz peaks is shown for clarity. Emission spikes such as the one observed at $t$ ~ 375 s are indicative of nonstationary combustion conditions.

The second example given here for the monitoring of incineration/combustion exhaust gases by REMPI-ToFMS is that of a marine medium-speed diesel engine that is operated, in sequence, first with diesel fuel and thereafter with residual HFO; see, e.g., the study by Radischat et al. (2015). It is well established that in aerosol organic combustion, a vast number of aromatic compounds, mostly polyaromatic hydrocarbons (PAHs), is generated; the dominant range of these PAHs was monitored by the authors (mass range about $m/z = 75$–300 amu), utilizing (1+1) REMPI-ToFMS, with a REMPI laser wavelength of $\lambda_1 = 266.0$ nm.

In Figure 16.15b, time-resolved mass data are collated in a 3D plot; for the first 40 minutes of the combustion cycle, the engine was running on diesel fuel, while thereafter it was switched (over a transition time of about 5 minutes) to the consumption of HFO. The following typical behavior was observed. After starting

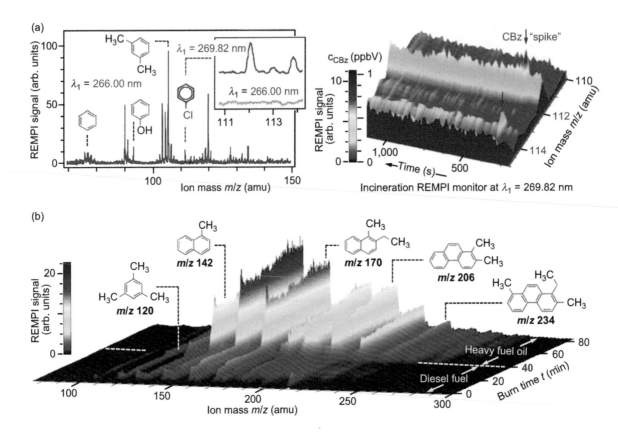

**Figure 16.15**  Monitoring of organic molecules in incineration and combustion flue gases, based on REMPI-ToFMS. (a) On-line measurement of monochlorobenzene (MCBz) and other aromatic organic molecules in the flue gas from a hazardous waste incineration plant, for a REMPI laser wavelength of $\lambda_1 = 269.82$ nm; the MCBz mass peaks are absent for $\lambda_1 = 266.0$ nm (see spectral insert). The temporal monitoring on the right reveals occasional spiking of the MCBz emission (quantified by external standardization). (Data adapted from Zimmermann, R., *Anal. Bioanal. Chem.* 381, no. 1 (2005): 57–60.) (b) Time-resolved REMPI-ToFMS monitoring of combustion engine exhaust gases, running on diesel fuel for about 40 minutes of operation, and thereafter burning heavy fuel oil (HFO); selected mass signals for the observed range of mono- and polyaromatic compounds are annotated. (Data adapted from Radischat, C et al., *Anal. Bioanal. Chem.* 407, no. 20 (2015): 5939–5951, reproduced with permission of Springer.)

the engine, the concentration of the main PAHs (mass range $m/z \approx 120$–250 amu) rises visibly, but remains at relatively low values. As soon as the fuel is switched to HFO, a marked increase in PAH emission is observed, which for some particular compounds is up to a factor of ×10 higher than that for operation with ordinary diesel fuel.

This example clearly demonstrates that REMPI mass spectrometry is capable of on-line monitoring of organic combustion compounds (PAHs) during operation of (diesel) combustion engines; its time resolution is even good enough to follow rapid changes in the combustion conditions.

### 16.3.3 Following biological processes by using REMPI spectroscopy

The analysis of nonvolatile compounds in fruit and vegetable samples faces several difficulties; this is because their concentration levels are low, or they decompose upon heating during volatilization, prior to injection into the analyzer.

To overcome these problems, a range of techniques have been applied for their analysis, including, e.g., fast atom bombardment, laser desorption (LD), field desorption, plasma-desorption mass spectrometry, or secondary-ion mass spectrometry. One of the most powerful methods, specifically designed to perform fast and direct analysis of nonvolatile compounds in fruit and vegetables, is based on the combination of LD followed by REMPI and ToFMS detection. The method is sometimes referred by its sum of acronyms, i.e., LD + REMPI + ToFMS (see, e.g., Orea et al. 2001). In addition to the already mentioned high selectivity and sensitivity of REMPI-ToFMS, the technique offers some additional advantages. In particular, the volatilization and ionization steps are physically separated from each other and occur at different times such that they can be independently optimized; thus, this technique provides additional sampling sensitivity.

Representative of the application of LD-REMPI-ToFMS to the analysis of nonvolatile organic compounds, here the monitoring of *trans*-resveratrol in fruit has been selected. *Trans*-resveratrol (3,5,4′-trihydroxystilbene—for its structural formula see the spectral data panel in Figure 16.16) is an antioxidant and fungitoxic compound naturally produced in many plants, including grapes. This compound is accumulated, e.g., in vine leaves and grape skin, in response to biotic (fungal organisms) and abiotic (UV radiation or chemicals) stress, playing a crucial role in the fruit defense response (see, e.g., Montero et al. 2003, and references therein). In fact, the production of *trans*-resveratrol grape vines has been shown to reach physiological concentrations against *B. cinerea*, the causal agent for gray mold and one of the main pathogens in grapes.

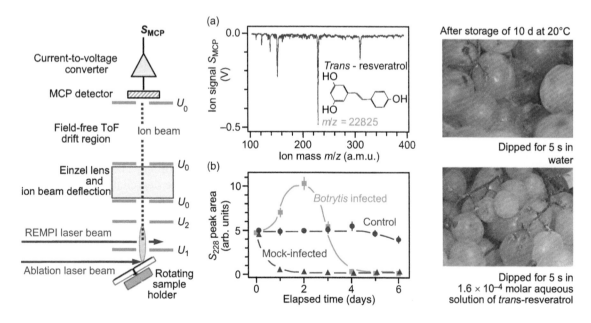

**Figure 16.16** Monitoring of *trans*-resveratrol by LD-REMPI-ToFMS (Left) Schematic experimental setup. (Center, top) Mass spectrum of a grape skin sample, with the REMPI laser wavelength optimized for *trans*-resveratrol detection ($\lambda_1$ = 302.18 nm). (Center, bottom) Comparison of LD-REMPI-ToFMS analysis data for control (uninfected), mock-infected, and *Botrytis cinerea*-infected grape samples, for grapes stored at room temperature of up to 10 days. (Right) Images of untreated (top) and *trans*-resveratrol treated (bottom) grape, after 10 days of storage at room temperature. (Data adapted from Montero, C. et al., *Plant Physiol.* 131, no. 1 (2003): 129–138.)

The conceptual LD-REMPI-ToFMS setup used by the authors is shown in the left panel of Figure 16.16. Essentially, it consists of two independent regions that correspond to two separate high-vacuum chamber segments. The first one is used for both LD and REMPI ionization; the REMPI ions are accelerated (and "beam shaped" by electrostatic deflectors and an Einzel lens) toward the second chamber, which contains the ToF unit with a double-microchannel plate detector. Pulses from a Nd:YAG laser ($\lambda_{LD} = 1064$ nm, $\tau_p \approx 5$ ns) were used for ("soft", i.e., low pulse energy) sample LD. Subsequent to desorption, and spatially separated from it, selective (1+1) REMPI spectroscopy was implemented, covering the wavelength range $\lambda_1 = 300.9–307.5$ nm, with the optimal wavelength for *trans*-resveratrol analysis being $\lambda_1 = 302.18$ nm.

In the center top panel of Figure 16.16, a representative (room-temperature) LD-REMPI-ToFMS spectrum is displayed; it was recorded from a processed grape skin sample (administered to the sample holder, shown in the experimental schematic), which was rotated to a fresh sample position after a few laser ablation pulses. In the spectrum, the main mass peak (associated with *trans*-resveratrol and annotated accordingly) suggests both (1) negligible fragmentation of the analyte molecules under soft desorption conditions and (2) high selectivity in the REMPI process optimized in wavelength for the *trans*-resveratrol resonant transition (only a few smaller peaks from other organic compounds appear, albeit their concentration dominance in the probed wine leaf sample). Indeed, one can perfectly identify and quantify the desired analyte, here *trans*-resveratrol, nearly without any interference from the huge number of organic compounds (in general more than a few hundreds) present in common biological samples.

In their studies, the authors also investigated the postharvest elicitation of *trans*-resveratrol in grapes upon fungal infection, specifically by *B. cinerea*. For this, three batches of samples were monitored, over periods of up to 10 days (and stored at room temperature), for their *trans*-resveratrol content: noninfected, mock-infected, and *B. cinerea*-infected grapes. The results are shown in the center bottom panel of Figure 16.16. Notably, the content in the control samples remains more or less constant over the shown measurement period of 6 days. In contrast, for the mock-infected grapes, a rapid decrease in *trans*-resveratrol content is observed, with hardly any left after only 2 days. For the *B. cinerea*-infected group, a significant increase in the *trans*-resveratrol content is observed for the first 2 days after infection; thereafter, a rapid decrease sets in, and after 5 days, hardly any *trans*-resveratrol signal is recorded. The authors speculate that this decrease can probably be attributed to the degradation of *trans*-resveratrol by a laccase-like stilbene oxidase produced by *B. cinerea*.

Based on these results, the authors came to the conclusion that *trans*-resveratrol may be a good candidate to serve as a natural pesticide against pathogen attack and therefore to improve the natural resistance of grapes to fungal infection. Indeed, the difference in appearance of untreated and *trans*-resveratrol treated (dipped for only 5 s in a $1.6 \times 10^{-4}$ molar aqueous solution of *trans*-resveratrol) is striking: in the former case "moldy" spots are clearly visible on the grapes, while in the latter case the appearance is nearly "as fresh" (see the images in the right panel of Figure 16.16).

## 16.4 ZEKE SPECTROSCOPY

ZEKE spectroscopy is a technique that detects electrons within a small energy range around a selected molecular ionization threshold, combining supersonic free-jet expansion with laser multiphoton ionization and delayed pulsed electric field extraction (Müller-Dethlefs et al. 1984). Essentially, it constitutes a form of high-resolution photoelectron spectroscopy (PES), which determines ro-vibronic states of molecular ions with unprecedented accuracy.

In the conventional description of the photoionization process inherent to traditional PES, as sketched in Figure 16.17, one visualizes the entire process as an electronic transition between the ground state of the neutral molecule and an electronic state of the cation. The energy balance can be written as

$$hv = \text{IP} + E_n^+ + E_{\text{kin}}(\text{M}^+) + E_{\text{kin}}(e^-) \cong \text{IP} + E_n^+ + E_{\text{kin}}(e^-), \qquad (16.17)$$

where IP stands for the ionization potential of the molecule M; $E_n^+$ represents internal energy states of the molecular cation $\text{M}^+$ (with respect to the cation ground state); and $E_{\text{kin}}(\text{M}^+)$ and $E_{\text{kin}}(e^-)$ are the kinetic energies of the cation and electron, respectively. Note that, in general, $E_{\text{kin}}(\text{M}^+)$ is very small compared with $E_{\text{kin}}(e^-)$ and thus can be neglected, as indicated in Equation 16.17. The photoelectron spectrum, which consists of a representation of the electron signal intensity as a function of its kinetic energy (see an ideal simple line spectrum at the right part of the figure) can reveal the ro-vibronic structure of the distinct molecular cation eigenstates produced in the photoionization process. In principle, the resonant condition given by the energy balance equation (Equation 16.17) will obviously be mirrored in the electron spectrum, with the intensities determined by the associated transition probabilities.

Unfortunately, the measurement of the kinetic energy of the electrons has limitations; in particular, experimental constraints normally limit the resolution to about $\delta E \approx 80\text{–}100 \text{ cm}^{-1}$ (or 10–12 meV). To overcome this limitation, a modified technique called *threshold photoelectron spectroscopy* (TPES) was developed, in which the photon energy is selected such that the electrons are emitted only at the threshold of a specific molecular ion eigenstate, as illustrated in the bottom part of Figure 16.17. In TPES, the spectrum is obtained by scanning the laser radiation across each ionization threshold such that one collects threshold photoelectrons with little or no kinetic energy. In other words, the TPES spectrum is a detailed map of the ionization thresholds of the molecular ion states $E_n^+$ as a function of the wavelength of the ionization laser source. Note that in modern applications of TPES, synchrotron radiation sources are often used, which do not exhibit the ultranarrow bandwidth of lasers but still provide adequate resolution, as high as $\delta E \approx 1$ meV.

### 16.4.1 Methodology of ZEKE spectroscopy

The advance of TPES led in a natural way to the development of ZEKE spectroscopy, a technique developed by Müller-Dethlefs et al. (1984), which is capable of resolving rotational energy level structure of ionic states, through the detection of threshold electrons. The basic principle of the original ZEKE method can be understood with the aid of Figure 16.18, where the basic ingredients and key steps of the ionization process are illustrated. In short, in the technique, one combines supersonic molecular beam expansion with laser multiphoton ionization. In the particular example shown in the figure, a (1+1′) REMPI scheme is used, followed by pulsed electric field discrimination. As depicted by the "cloud" in the molecular beam–laser beam interaction region, the ionization laser pulse—tuned to some ionization threshold—produces (at $t = 0$) a mixture of threshold

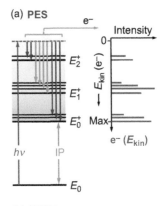

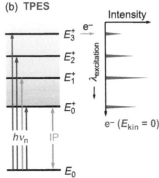

**Figure 16.17** (a) Concept of PES showing the ground and various electronic excited states of the neutral molecule M and cation $\text{M}^+$, respectively; a conceptual PES line spectrum is sketched, for ionization with a fixed-wavelength laser light source. (b) Concept of threshold photoelectron spectroscopy (TPES); the wavelength of the excitation laser is scanned across thresholds of the molecular cation states.

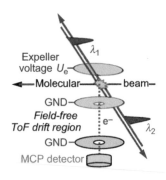

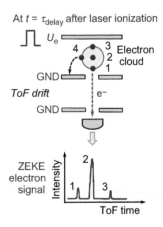

**Figure 16.18** Concept of ZEKE spectroscopy. (Top) Schematic experimental setup, with supersonic molecular beam; (1+1′) REMPI excitation (with counter-propagating laser beams); and electron-extraction electrodes. (Bottom) Principle of delayed pulsed electric field extraction and a schematic ToF spectrum distinguishing fast electrons ($e_1$, $e_3$) from ZEKE electrons ($e_2$). GND = ground potential ($U_{GND} = 0$ V).

electrons, nominally with zero kinetic energy, and fast electrons with a nonnegligible kinetic energy.

The key experimental feature of the technique is the introduction of a delay between the ionization laser pulse and the application of the pulsed electrical field for electron extraction. After a time delay $\tau_{delay}$, the "fast" electrons $e_1$, $e_3$, and $e_4$ would have moved away from the ionization volume, specifically with $e_1$ toward the detector and the electron $e_3$ away from it; the ZEKE electron $e_2$ remains close to the center. Consequently, on extraction to the detector, the spatially separated fast and ZEKE electrons could be discriminated via their respective ToFs to the detector. Applying the electron-extraction pulse after the said delay, conceptually one would observe the following. The fast electrons $e_1$ and $e_3$ would arrive at the detector before and after the ZEKE electron $e_2$, as shown in the schematic ToF spectrum at the bottom of the figure. Note that sideways-moving electrons $e_4$ would be suppressed altogether, being directed toward the ToF-drift tube plate (on ground potential). Since the fast electrons will have dispersed in space, while the ZEKE electrons have not, the electron signal intensity related to the latter should be much larger, as indicated in the figure.

To demonstrate the extraordinary improvement in resolution (about three orders of magnitude), of ZEKE spectroscopy over PES, a comparative example is shown in Figure 16.19 for state-resolved ionization of nitric oxide $NO^+(v^+;N^+)$. In the top panel of the figure, the vibrationally resolved photoelectron spectrum of $NO^+$ is shown, excited by a HeI light source, with $E_{1\alpha} = 21.22$ eV (see Turner et al. 1970).

The spectrum reveals five well-resolved peaks ($v^+ = 0$, ..., 4) of the cation ground state $NO^+$ ($X^1\Sigma^+$), with a further one just recognizable above the noise floor; these vibrational peaks exhibit a width of $\delta E_v \approx 70$ meV (see the middle data panel in the figure). In the bottom panel, the very first ZEKE spectrum of $NO^+$ is shown, as measured by Müller-Dethlefs et al. (1984) by using a (3+1′) REMPI scheme:

$$NO \; X^2\Pi_{1/2} \, (v'' = 0; N'' = 0) \overset{3h\nu_1}{\to} \; NO \; C^2\Pi_{1/2}(v' = 0; N' = 0)$$

$$\overset{3h\nu_2}{\to} NO^+ \, X^1\Sigma^+(v^+ = 0; N^+ = 0, 1, 2, (3)) + e^-. \tag{16.18}$$

A resolution of 1.2 cm$^{-1}$ (equivalent to ~0.15 meV) was achieved; this allowed the authors to fully resolve the lowest rotational energy levels of the cation $NO^+$, namely, $N^+ = 0$, 1, 2. From their spectra, the authors were able to extract, with high resolution, a directly measured value for the ionization potential $IP_{NO} = 74717.2$ cm$^{-1}$; at the time, this constituted the first precision IP value for a molecule, except for $H_2$. Shortly after these experiments, Sander et al. (1987) repeated the ZEKE spectroscopy (in its pulsed-field ionization ZEKE measurement modality; see the following section) of nitric oxide, with improved laser calibration, and obtained a (more accurate) value for its ionization potential of $IP_{NO} = 74719.3(5)$ cm$^{-1}$.

### 16.4.2 Measurement modality of pulsed-field ionization: PFI-ZEKE

It should be noted that the ZEKE signal is made up of two categories of electrons. The first contribution is from so-called *free ZEKE electrons*; these are generated at the threshold (i.e., at the ro-vibronic eigenstate of the molecular cations) and can be separated from any (other) kinetic electrons by the steradiancy principle and their distinct ToF behavior, as illustrated in Figure 16.18. The second contribution originates from *long-lived Rydberg levels* with principal quantum numbers $n$ very close to the ion threshold ($n > 100$, with lifetimes $\tau_n \sim$ 1–100 µs). The phenomenon was first described by Reiser et al. (1988) for a region dubbed the "magic region" of about ~1 meV converging to a certain ion threshold. The authors showed that a very significant portion of the ZEKE signal can be produced by field ionization out of the said high-lying Rydberg states; this experimental innovation—today the most widely used ZEKE measurement modality—is commonly known as *pulsed-field ionization zero kinetic energy* (PFI-ZEKE; the acronym ZEKE-PFI is used as well). The concept of the technique is illustrated in Figure 16.20; essentially, three well-defined steps are implemented in the following sequence:

- First, either one- or two-photon resonant laser excitation is applied (the latter case is shown in the figure) to excite the molecule from its electronic ground state (via an intermediate state) into a high-$n$ excited (neutral) Rydberg state.

- Second, one waits for a few microseconds, so that any energetic electrons escape from the ionization region, without being detected.

- Third, a small pulsed electric potential is applied to field-ionize the remaining long-lived high-$n$ Rydberg states. Note that the delay between the laser excitation and the pulsed electric field ionization is crucial to ascertain that neither background hot electrons and ions nor short-lived (autoionizing) low-$n$ Rydberg states contribute to the electron signal.

A schematic view of the field ionization process is shown in Figure 16.20b. On the left, the Rydberg states allowed within the Coulomb potential, below the ionization limit, are shown; on the right, the potential is modified by an external, linear electric field. It is precisely the energy range marked $\Delta$IP, in which the originally bound high-$n$ (metastable) Rydberg states are found; the delayed electric field pulse ionizes these. In a sense, the very long lifetimes $\tau_n \gg 1$ µs of high-$n$ molecular Rydberg states makes it possible to "temporally store" electrons in *ZEKE Rydberg states*; this was first noted by Chupka (1993). He proposed that the long-lived ZEKE Rydberg states result in part from Stark mixing of high-$n$ and high-$\ell$ Rydberg states, which therefore experience reduced interaction with the *ionic core* and, consequently, lead to increased lifetimes. For a more extensive discussion on the long lifetimes of the high-$n$ Rydberg states see, e.g., Held and Schlag (1998).

In summary, in PFI-ZEKE spectroscopy—like in TPES—one detects zero kinetic energy electrons; however, while in the latter technique electrons are generated above the ionic threshold, in the former method they originate from a very narrow energy band just below the ionization threshold. It is this key feature that ultimately leads to the improvement in resolution of about two orders of magnitude in determining the adiabatic ionization energy of a neutral molecule. Based on PFI-ZEKE spectroscopic schemes, one finds for the energy balance, in close analogy to Equation 16.17,

$$hv = \text{IP} + \left[ E_{\text{int}}^{\text{ion}}(n^+; S^+; v^+; N^+) - E_{\text{int}}^{\text{neutral}}(n, S, v, N) \right] + E_{\text{kin}}(e^-). \qquad (16.19)$$

As before, IP is the adiabatic ionization energy of the neutral molecule, either from the initial ground state for one-photon processes or from the resonant intermediate state for multiphoton absorption processes; $hv$ is the laser photon

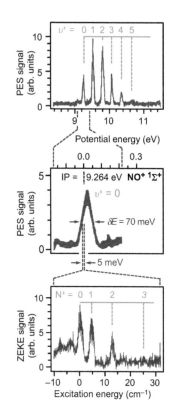

**Figure 16.19** Comparison of the PES and ZEKE spectra for nitric oxide cation NO⁺. (Top) Vibrationally resolved vacuum ultraviolet (VUV) photoelectron spectrum (PES), measured by Turner et al. (1970). (Center) Spectral width of the $v^+ = 0$ band. (Bottom) First rotationally resolved ZEKE spectrum of NO⁺, based on a (3+1′) REMPI scheme; relevant vibrational and rotational levels are annotated. (Data adapted from Müller-Dethlefs K. and E.W. Schlag, *Angew. Chem. Int. Ed.* 37, no. 10 (1998): 1346–1374.)

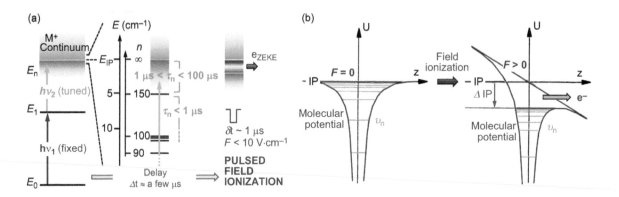

**Figure 16.20** Concept of the *modus operandi* of PFI-ZEKE spectroscopy. (a) Term-level diagram relevant to PFI-ZEKE, exemplified for (1+1′) excitation into molecular Rydberg states, followed by time-delayed pulsed electric field ionization; the important regions for $n < 150$ and $n > 150$ (the magic region) are indicated. (b) Schematic representation of the field ionization process.

energy of the ionization step; $n$, $S$, $v$, and $N$ are the electronic, spin, vibrational, and rotational quantum numbers, respectively (for the ion with superscript "+"); and $E_{kin}(e^-)$ is the kinetic energy of the (ZEKE) electron. Note that if one uses Equation 16.19 to determine IP, care should be taken to correct the value for the electric field shift associated with the extraction pulse. As indicated in Figure 16.20b, the ionization potential is lowered by $\Delta IP \equiv c_F \cdot F^{1/2}$, where $F$ stands for the applied electric field (in units of $V \cdot cm^{-1}$) and $c_F$ is a constant that has to be determined experimentally (see, e.g., Ford et al. 2003).

The technique of PFI-ZEKE can be implemented with distinct combinations of electric field extraction pulses to reach spectral resolutions down to a few tenths of wavenumbers, as necessary to resolve the rotational structure of polyatomic molecules (see, e.g., Cockett 2005). One of the most demonstrative examples of the capability of PFI-ZEKE spectroscopy, with rotational resolution, is the determination of the molecular structure and symmetry of the benzene cation, a subject of extensive investigation, in particular by Müller-Dethlefs and coworkers (see, e.g., Ford et al. 2003, and references therein). These authors demonstrated that the benzene rotational structure can be described consistently only if both the neutral intermediate state and the electronic ground state of the cation, with the $\nu_6$ vibration excited, have $D_{6h}$ symmetry.

Due to the length limitation of this chapter, here only a few selected examples of ZEKE spectroscopy studies can be outlined.

### 16.4.3 Examples of high-resolution ZEKE spectroscopy

#### *Quasi-bound rotational levels of $H_2^+$*

The cation of the hydrogen molecule, $H_2^+$, is the very simplest molecule and thus has played—and is still playing—a crucial role in the development of quantum chemistry. The level structure of $H_2^+$ can be calculated with high precision and accuracy by *ab initio* quantum-chemical methods. In addition, this molecular cation is one of the first species to have been formed in the universe and is therefore also of relevance in astrophysics.

The ground electronic state $H_2^+ X^2\Sigma_g^+$ has 481 vibrational levels (see, e.g., Beyer et al. 2016a), and as many as 58 of these are quasi-bound tunneling (shape) resonances located above the H(1s)+$H^+$ dissociation limit, but below the maxima of the relevant centrifugal barriers (see, e.g., Moss 1993). Furthermore, a significant fraction of these quasi-bound levels, located close to the top of the respective centrifugal potential barriers, have not been yet observed experimentally (Beyer et al. 2016). These shape resonances are interesting from a theoretical point of view insofar as they represent a channel for the formation of $H_2^+$ in the collision process $H^+$+H(1s), via radiative or three-body recombination.

Beyer and Merkt (2016a, 2016b) studied the quasi-bound levels of $H_2^+$ by PFI-ZEKE PES. The spectra were recorded by monitoring the electron signal produced by field ionization of very high Rydberg states ($n > 100$) as a function of the laser excitation energy $h\nu_{VUV} + h\nu_1 + h\nu_2$ (the last tunable). Since the bound and quasi-bound rotational levels of the highest vibrational states of the electronic ground state of the anion $H_2^+$ exhibit a large average internuclear separation and are therefore not directly accessible from the electronic ground state of the neutral $H_2$, a resonant stepwise three-photon excitation sequence via the B- and H-states as intermediates was employed to gradually enlarge the internuclear separation; the scheme is noted in the left-hand side of Figure 16.21. The associated, level-annotated PFI-ZEKE photoelectron spectra of *ortho*- and *para*-$H_2$, in the region of the dissociation limit to $H_2^+ X^2\Sigma_g^+$ are depicted in the lower and upper data panels, respectively (note that the *para*-$H_2$ spectrum is flipped vertically for clarity).

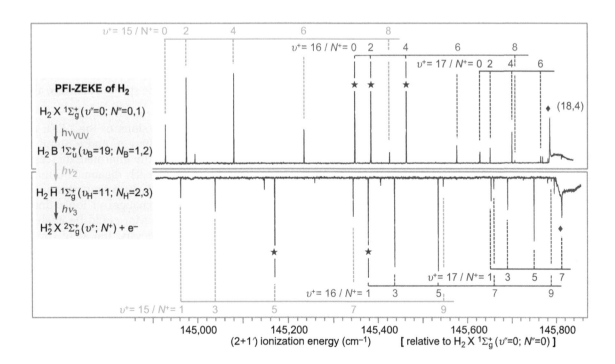

**Figure 16.21**   PFI-ZEKE photoelectron spectra of $H_2$, recorded for levels near the dissociative-ionization threshold out of the electronic intermediate levels [H ($v_H$ = 11;$N_H$ = 2)] (top) and [H ($v_H$ = 11;$N_H$ = 3)] (bottom); note that the *para*-$H_2$ spectrum is flipped vertically for clarity. The intensity scale is linear and in arbitrary units. The lines annotated with the symbol ☆ are out of scale; the lines annotated with the symbol ◇ are shape resonances. (Data adapted from Beyer, M. and F. Merkt, *J. Mol. Spectrosc.* 330 (2016): 147–157.)

Note also that as a further measure of clarity, only the most evident sequences for $H_2^+$ $X_2\Sigma_g^+$ are annotated, namely, $(v^+ = 15, ..., 17; N^+ = 0, ..., 9)$. Other (in general low-intensity) lines from the rotational bands of $v^+ = 2, 14, 18$ were also observed, as well as a few transitions to $H_2^+$ $A^2\Sigma_u^+(v_A^+ = 0)$, near the onset of the ionization continuum (see Beyer and Merkt 2016b). Furthermore, as marked in the spectrum, two shape resonances for $(v^+ = 18; N^+ = 4)$ and $(v^+ = 17; N^+ = 7)$ are observed.

### Line intensities in the vibrational progressions of the ZEKE spectra: The $I_2$ molecule

The energy level structure of diatomic halogens has been extensively studied utilizing (2+1′) REMPI and ZEKE spectroscopy (see, e.g., Cockett et al. 1995, 2002); as examples, here only a few results and findings for molecular iodine $I_2$ are commented on. For example, in said referenced studies, the factors governing the line intensities in the vibrational progressions of the ZEKE spectra were investigated; this constitutes an important issue to fully understand the ZEKE electron dynamics. Due to the inherent complexity of the heavy halogen spectra, these studies were carried out in a molecular beam apparatus, taking advantage of the strong rotational cooling in the supersonic expansion. Molecular iodine $I_2$ was evaporated at thermal temperatures and seeded in helium carrier gas at pressure $p_{He} \approx 1$–2 bar, in preparation for supersonic expansion. The molecular beam was crossed with a (pulsed) pump laser beam to excite the $I_2$ Rydberg states $[^2\Pi_{3/2}]_c$;5d $2_g$ and $[^2\Pi_{1/2}]_c$;5d $2_g$. Subsequently, the wavelength of the probe laser, used for the ionization step, was adjusted to match the desired ionization scheme, i.e., (1) to excite Rydberg states just below the thresholds of the two spin-orbit states of $I_2$, to measure ZEKE spectra, or (2) to excite autoionizing Rydberg states just above the thresholds of the two spin-orbit states of $I_2$, to measure autoionization (AI) spectra. Note that both types of spectra, ZEKE and AI, were implemented using the same resonant intermediate states, rather than an excited valence state (as utilized by the authors in some of their earlier studies). This particular approach simplifies the spectral analysis, allowing for an easier identification of the Rydberg series.

Since the line intensities in the vibrational progressions of the ZEKE spectra should be governed by the FC factors $|\langle v^+| v'\rangle|^2$, where $v^+$ and $v'$ stand for the vibrational quantum number of the final (ion) and the intermediate (resonant) electronic states, respectively, one should expect a nearly diagonal FC matrix for the probe step since $R_c$ and $\omega_c$ (the internuclear distance and vibrational frequency of the Rydberg core, respectively) hardly change on ionization. Furthermore, any deviation from the expected FC behavior could be analyzed by comparing the two types of spectra; from this analysis, the state (or states) responsible for such departure could be identified.

In Figure 16.22, a selection of example spectra is shown, both for (2+1′) PFI-ZEKE ionizing into $I_2^+$ (via the $[^2\Pi_{3/2}]_c$;5d $2_g$ [Figure 16.22a] and $[^2\Pi_{1/2}]_c$;5d $2_g$ intermediates [Figure 16.22b]) and for (2+1′) delayed extraction AI ( via the $[^2\Pi_{3/2}]_c$;5d $2_g$ intermediate only [Figure 16.22c]). The "vertical" FC transition lines in the sequences, with $\Delta v = 0$, are annotated (with the quantum number $v^+$). Note that for the $[^2\Pi_{1/2}]_c$; 5d $2_g$ intermediate, near-perfect FC behavior is observed, with the transition $v^+ = v'$ carrying almost the complete intensity. In contrast, off-diagonal contributions (sometimes nearly as intense as the diagonal FC transition) are observed for the $[^2\Pi_{3/2}]_c$;5d $2_g$ intermediate. This immediately poses the question as to what could be the origin of this FC deviation in the case of the latter ZEKE vibrational progressions.

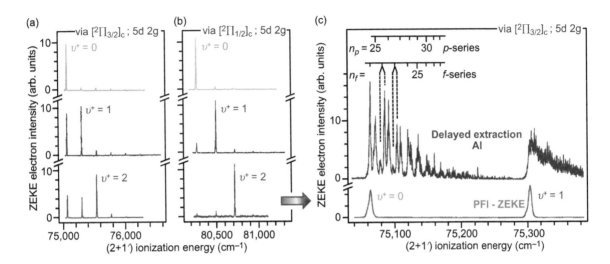

**Figure 16.22** High-resolution ionization electron spectroscopy of $I_2^+$. (a,b) (2+1′) PFI-ZEKE spectra, with ionization via the intermediate electronic states indicated in the panels. The FC transitions $\Delta v = 0$ are annotated with the quantum number $v^+ = v'$. (c) (2+1′) delayed extraction AI spectrum via the intermediate electronic state indicated in the panel, with the $v^+ = 0$ feature exhibiting strong Rydberg structure. For comparison, the $v^+ = 1$ PFI-ZEKE spectrum from panel (a) is included in the lower data trace. (Data adapted from Cockett, M.C.R. et al., *J. Chem. Phys.* 102, no. 13 (1995): 5226–5234; Cockett, M.C.R. et al., *Phys. Chem. Chem. Phys.* 4, no. 8 (2002): 1419–1424.)

As part of the quest to answer said question, Cockett et al. (2002) recorded (2+1′) delayed-extraction AI spectra in the region of the $I_2^+$ $X^+$ $[^2\Pi_{3/2}]_g$ ($v^+ = 0$) ionization threshold (as in the PFI-ZEKE case via the intermediate state $[^2\Pi_{3/2}]_c$;5d $2_g$, with $v' = 1$). A representative AI spectrum is shown in Figure 16.22c; one observes a sort of unstructured photoelectron signal on the high-energy side of the $v^+ = 1$ ionization threshold but a clearly structured spectrum on the high-energy side of the $v^+ = 0$ ionization threshold.

For the interpretation of the spectra, it should be recalled that, while in ZEKE spectroscopy the excited Rydberg states lay a few cm$^{-1}$ below the molecular ion threshold, in AI spectroscopy the autoionizing Rydberg states are a few cm$^{-1}$ above the ion threshold. Essentially, the only difference between the two types of experiments is in the timing of the pulsed fields: in the AI spectroscopy, the field-pulse timing sequence is altered, specifically relaxing the discrimination between free PFI-ZEKE electrons and trapped electrons produced by AI, both detected after a time delay. With this in mind, the authors concluded that slow electrons produced by direct ionization to $v^+ = 1$ are responsible for the observed, near-structureless continuum, revealing that electrons with energies of up to $\sim$100 cm$^{-1}$ can be trapped, although with rapidly decreasing efficiency. In contrast, the region between the $v^+ = 0$ and $v^+ = 1$ thresholds exhibit a peak structure associated with the Rydberg series; individual peak positions could be recorded with decent accuracy and could be assigned to specific $n\ell$ Rydberg series (specifically the $p$ and $f$ series, with $n$ in the range $n \approx 20, \ldots, 30$). This assignment then allows one to identify the particular state responsible for the off-diagonal behavior seen in the PFI-ZEKE spectrum: the 21$f$ Rydberg state, based on $v^+ = 1$, is in near coincidence with the ionization energy of $v^+ = 0$ (by a difference of just $\Delta E \leq 4$ cm$^{-1}$).

### Low-frequency modes: van der Waals complexes and internal rotation of molecular cations

One of the areas where ZEKE spectroscopy has proved to be extremely useful is the study of intermolecular forces (e.g., van der Waals [vdW] interaction and/or hydrogen bonding), as they are key to the interpretation of many physical, chemical, and biological phenomena. These include, e.g., the formation of molecular crystals, solvation dynamics, and protein folding, to name but three, in which relatively simple aromatic compounds are often used as models for large biological molecules.

Representative study cases of such prototype systems are vdW complexes between noble gas atoms and aromatic molecules. For examples, the phenol$\cdots$Ar complex can exist in at least two isomeric forms, namely, one in which the noble gas atom is positioned at the center of the aromatic ring (as found, e.g., in the benzene$\cdots$Ar complex) and the other in which the hydrogen-bonding OH group may induce the argon atom to move from the aromatic ring plane and form a hydrogen-bonding isomer. The aforementioned phenol$\cdots$Ar complex has been investigated using REMPI and (ZEKE) PES, under supersonic jet conditions via four intermediate levels of the excited state $S_1$ of the neutral complex (see Haines et al. 2000). The spectral analysis allowed the authors to identify several in-plane bending modes (at $\tilde{v} = 21$ cm$^{-1}$ and $\tilde{v} = 47$ cm$^{-1}$) and the intermolecular stretch of the cation at $\tilde{v} = 66$ cm$^{-1}$. All these modes are consistent with a vdW structure in which the Ar atom binds above the aromatic ring.

The high resolution associated with ZEKE spectroscopy has been very useful in the study of internal rotation or torsional motions of molecular cations, particularly in molecules containing methyl groups. As an illustrative example, one may take the investigation of the internal rotation of the acetone $n$-radical cation; this was studied, e.g., by Shea et al. (2000) using (2+1′) high-resolution ZEKE spectroscopy. The authors were able to determine the cation transition origin at $\tilde{v}_{\text{acetone-}d_3} = 78{,}299.6$ cm$^{-1}$, as well as the frequencies of the fundamental torsional modes (such as, e.g., the antigearing (b1) mode with $\tilde{v}_{17} = 119.1$ cm$^{-1}$). From the ZEKE spectral analysis, the authors developed an experimentally based potential function for internal rotation in the $^2B_2$ cation ground state; they also concluded that the barrier to synchronous rotation is higher at the ground state of the cation than it is at the ground state of the neutral molecule.

### 16.4.4 MATI spectroscopy

Despite the high resolution achieved in ZEKE spectroscopy, the technique is not free of certain drawbacks because electron detection, in the absence of mass information, can potentially give ambiguous or uninterpretable results. In particular, such ambiguity can occur in situations when more than one molecular entity may be ionized simultaneously, such as, e.g., in cluster, radical, and vdW complex spectroscopy, or when the investigated sample is a product of a complex reaction chain. Different techniques have been developed to overcome this limitation; the two most common ones are (1) *photoelectron–photoion coincidence measurements*, which simultaneously record both charged particles, albeit with very low acquisition rates, and (2) MATI spectroscopy, in which only ions produced at ionization thresholds are measured (MATI was originally developed by Zhu and Johnson 1991).

As in ZEKE, MATI spectroscopy takes advantage of field-ionizing long-lived, high Rydberg states to signal an ionization threshold. However, instead of measuring the ZEKE electrons, one monitors the produced ions at distinct thresholds as the laser wavelength is scanned through the ionization continuum of the molecule. As mentioned further earlier, in a PFI-ZEKE experiment. threshold ionization events are eliminated by waiting long enough until all electrons produced by direct laser ionization have moved away of the ionization region, before the high Rydberg states are field-ionized by a small electric pulse field (for further details, see Zhu and Johnson 1991). It should be noted that the latter scheme does not work for ions because their relatively low mobility does not make it easy to differentiate prompt photoions from ions, which were generated by the pulsed-field ionization of the high-$n$ Rydberg states. In many cases, this difficulty was overcome by employing higher field strengths in MATI than in ZEKE experiments (see, e.g., Cockett 2005).

As an example of the strength of MATI spectroscopy, the investigation of a vdW complex is discussed, namely, that of DABCO$\cdots$N$_2$ (DABCO = 1,4-diazabicyclo [2,2,2]octane). Utilizing a two-color $(1+1')$ excitation scheme, Cockett and Watkins (2004) recorded MATI spectra via a range of intermediate states (for example, the $S_1(0^0)$ band origin mode or the $S_1(1\tau)$ vdW mode). A representative spectrum involving the former intermediate is shown in Figure 16.23 (upper data panel); for comparison, an analogous ZEKE spectrum is also included (lower data panel).

Both spectra show two band features that correspond to the cation origin $(v^+ = 0^0)$ and the fundamental vdW mode, labeled $(1\tau')^+$. The higher field strength necessary for recording MATI spectra clearly manifests itself in the substantially larger full width at half maximum (FWHM) bandwidths, when compared with the narrow ZEKE response. While in the ZEKE experiments the authors measure FWHMs on the order of $\delta E_{ZEKE} \sim 1$ cm$^{-1}$ (which would allow for rotational resolution, in principle), the corresponding FWHM in the MATI spectra was on the order of $\delta E_{MATI} \sim 15$ cm$^{-1}$. A closer look at the MATI spectrum also reveals a well-resolved substructure of Rydberg series that converge to specific ion thresholds, associated with its $(0^0)^+$ and $(1\tau')^+$ vibrations of the cation.

Perhaps one of the most important applications of MATI spectroscopy is the accurate measurement of the dissociation energies of weakly bound complexes, as described, e.g., by Braun and Neusser (2002).

Despite its great success, the initial MATI experimental configuration carried the drawback of a much lower spectral resolution compared with that of ZEKE spectroscopy. This drawback is linked to the large amplitude of the field ionization pulses and separation fields necessary to separate PFI ions from "spontaneous" ions produced by the initial laser pulse. This limitation was resolved by Dessent et al. (1999), who developed an ingenious experimental configuration, which—based on the careful choice of amplitude and timing of the field-ionization pulse—allowed one to record MATI spectra with resolutions similar to those encountered in standard ZEKE spectra.

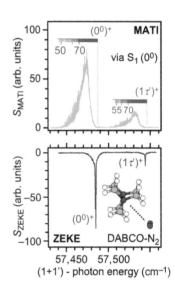

**Figure 16.23** Two-color $(1+1')$ MATI spectrum of the vdW complex DABCO$\cdots$N$_2$, recorded via the $S_1(0^0)$ band origin. For comparison, an analogous ZEKE spectrum in included in the lower panel. (Data adapted from Cockett, M.C.R. and M.J. Watkins, *Phys. Rev. Lett.* 92, no. 4 (2004): article 43001.)

A final remark is reserved for the aforementioned photoelectron–photoion coincidence technique. This elegant but not-easy-to-implement technique has proved to be very valuable for gaining insight into the detailed molecular mechanisms underlying ultrafast chemical reactions. One particular example of this type of experiment is the investigation of the photoinduced intracluster reactions. Stert et al. (2001) implemented time-resolved photoion–photoelectron spectroscopy, using femtosecond-laser excitation, to follow intracluster reaction $Ba{\cdots}FCH_3 + h\nu \rightarrow BaF + CH_3$ in real time (note that this reaction had previously been studied by Skowronek et al. 1997, but using nanosecond-laser excitation). These authors were able to identify the rate-limiting step of the intracluster reaction as the *internal conversion* from the cluster vibrationless excited electronic state $\tilde{A}(^1E)$ to the vibrationally excited low-lying electronic state $\tilde{A}'(^1A_1)$, before the C–F bond breaks and the BaF and $CH_3$ products separate.

## 16.5 TECHNIQUE OF H ATOM RYDBERG TAGGING

To detect hydrogen (H) atoms with high sensitivity and with high energy resolution, Welge and coworkers developed a technique now commonly known as the *H atom Rydberg tagging time-of-flight* methodology (see Schnieder et al. 1990). The central idea behind this technique is a two-step excitation procedure for the H atom, from its ground state $n = 1$ via the resonant intermediate state $n = 2$, into high-lying Rydberg levels ($n_H \approx 40, ..., 90$), without ionizing it:

$$H(n = 1) \overset{h\nu_{121.6 \text{ nm}}}{\longrightarrow} H^*(n = 2) \overset{h\nu_{365 \text{ nm}}}{\longrightarrow} H^{**}(n \sim 50). \qquad (16.20)$$

The overall process is highly efficient, due to several factors, but specifically (1) it exhibits a very high excitation cross section in the first step, on the order of $\sigma_{1\rightarrow2} \sim 10^{-13} \text{ cm}^2$, and (2) the lifetimes of said high H atom Rydberg levels are on the order of $\tau_n^{**} \sim 1\text{–}10$ ms, in small electric fields.

The neutral Rydberg H atoms travel a significant ToF distance toward the MCP detector; note that in most H reaction studies, the ToF range is limited to $t_{ToF} \leq 100$ μs. Just in front of the detector, the Rydberg H atoms pass through a fine metal mesh. Here they are field-ionized by applying an electric field between the said mesh and the front plate of the Z-stack MCP detector. These technical features ensure the absence of space charge and stray field effects, which are typically the limiting factors in ion ToF measurements. It also eliminates any spread in velocity due to the ionic repulsion during the ToF. Additional advantages of the method are (1) that no mass analysis is needed, since the light H atom is selected for detection spectroscopically, and (2) it is "background-free" since the detection is sensitive only to the atoms that initially were moving toward the detector. Due to its high translational energy resolution (nowadays as high as $\Delta E/E \sim 6 \times 10^{-4}$; see, e.g., Yang 2011), the H tagging methodology has been applied to investigate a vast number of unimolecular photodissociation and crossed beam bimolecular reactions, in which H or D fragments or products are formed. To illustrate the high resolution and sensitivity of this type of ionization spectroscopy, a few selected examples are highlighted specifically for H atom exchange reactions, due to their fundamental relevance and because they exhibit rather small reaction cross sections.

### 16.5.1 Reaction H + D$_2$ → HD + D

Schnieder et al. (1995, 1997) carried out systematic studies of the fundamental exchange reaction

$$H(E_c = 0.53; 1.28 \text{ eV}) + D_2(v'' = 0; J'' = 0) \rightarrow HD(v'; J') + D, \qquad (16.21)$$

in a crossed particle beam arrangement (for the concept, see the left panel of Figure 16.24). As is common practice, the H atom beam was generated locally by pulsed-laser photolysis of HI (in a pulsed-nozzle expansion beam), according to $HI + h\nu_{266} \rightarrow H + I(^2P_{3/2}); I^*(^2P_{1/2})$. This provides two well-defined collision energies for the H atom/D$_2$ molecule reaction, namely, $E_c = 1.28$ eV and $E_c = 0.53$ eV (associated with the H atom recoil energies of 1.194 and 0.659 eV, respectively); note that as a consequence of the underlying photodissociation dynamics, the H atom beam exhibits a very narrow velocity spread ($\Delta v/v \sim 10^{-3}$). It then crosses (perpendicularly) a supersonic beam of ortho-D$_2$ ($v'' = 0; J'' = 0$). The D atom product from reaction in Equation 16.21 is "tagged" by using the same type of excitation scheme as highlighted in Equation 16.20 for the H atom, i.e., with the end product of a long-lived high-$n$ Rydberg state ($n_D \sim 70$).

The aforementioned very high energy resolution allows for energetically (spectrally) resolving all ro-vibrational levels of the HD counterpart product provided that the kinetic energy spectrum of the D atom is measured and that both energy and momentum conservation are invoked. From such spectra and from the total laboratory angular distributions, state-to-state (vibrationally and rotationally resolved) DSCs and state distributions could be derived; a typical example is shown in Figure 16.24. These experimental results were compared with theoretical predictions from both quasi-classical and quantum mechanical calculations. Note that, while, in general, quite decent agreement was found for the

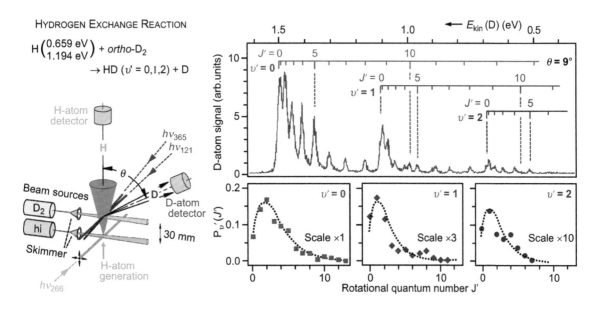

**Figure 16.24**  State-resolved reaction products in H + D$_2$ → HD + D. (Left) Schematic experimental realization. (Right) $E_{kin,LAB}$ spectrum for the D product, for a detection angle of $\theta_{LAB} = 9°$ and collision energy $E_c = 1.28$ eV; the ro-vibrational distributions extracted from the spectrum are displayed below. (Data adapted from Schnieder, L. et al., *J. Chem. Phys.* 107, no. 16 (1997): 6175–6195.)

converged quantum mechanical calculations, for QCT calculations such good agreement was reached only at the higher collision energies.

### 16.5.2 Reaction of F atoms with H₂ molecules: Dynamical resonances

Besides the fundamental hydrogen exchange reaction (see Equation 16.21), probably one of the most extensively studied chemical reactions, theoretically and experimentally, is the halogen exchange reaction

$$F + H_2(v'';J'') \rightarrow HF(v';J') + H \tag{16.22a}$$

and its isotopic variants. It also serves as a benchmark system for so-called *reaction resonances*, i.e., quasi-bound quantum states, transiently trapped in the transition state region (see, e.g., Lee et al. 2006; and Yang 2011).

An example of the search of such reactive resonances is the full quantum-state-resolved scattering study on the isotope-substituted reaction (Equation 16.22a), i.e.,

$$F(^2P_{3/2}) + HD(v'' = 0; J'' = 0) \rightarrow HF(v';J') + D, \tag{16.22b}$$

using the D atom Rydberg tagging of its atomic product (see Yang 2011, and references therein). Yang and coworkers measured the ToF spectra of the D atom product at a wide range of scattering angles, for the range of collision energies $E_c = 0.3$–$1.2$ kcal·mol$^{-1}$. In this study, for the first time, *individual partial-wave resolved resonances* in the reaction in Equation 16.22b were resolved, by measuring the collision-energy-dependent, angle-and-state-resolved DCS with very high energy resolution. Selected results from said study are shown in Figure 16.25, specifically for the product channel HF($v' = 2;J' = 6$) + D, in backward scattering direction.

The three peaks marked in the figure are assigned to partial-wave reactive (Feshbach) resonances, with $J = 12$, 13, and 14; these occur when the energetic transient is temporally trapped (for a few hundred femtoseconds) in an internal coordinate of the transition state, i.e., different from the actual reaction coordinate. Note that the agreement of the data with the high-level theoretical dynamics results (red data line in the figure) confirms the extraordinary sensitivity of the D tagging approach. Note also that the 3D DCS plot shown in the figure (measured for a collision energy of $E_c = 1.285$ kcal·mol$^{-1}$) nicely reveals the forward–backward asymmetry.

### 16.5.3 Four-atom reaction OH + D₂ → HOD + D

In three-atom systems, internal excitation is limited to two rotational and one vibrational degrees of freedom in the reactant or product diatom molecules. In four-atom reactions, the triatomic reactant or product has three vibrational degrees of freedom (provided that it is nonlinear). Thus, when a triatomic molecule is formed in a four-atom reaction, it becomes possible to investigate not only how much energy is deposited into internal (vibrational) energy but also how the energy is distributed among the individual vibrational modes. This means one can, in principle, study potential mode-specific energy disposal, with preferential vibrational excitation of the newly formed bond.

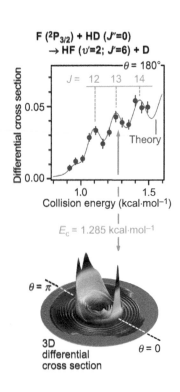

**Figure 16.25** DCS measurement in the reaction F($^2P_{3/2}$) + HD($v'' = 0$; $J'' = 0$) → HF($v';J'$) + D. (Top) Data plot for the product HF($v' = 2;J' = 6$), revealing rotational Feshbach resonances. (Bottom) 3D DCS plot for $E_c = 1.285$ kcal·mol$^{-1}$, revealing forward–backward scattering asymmetry. (Data adapted from Dong, W. et al., *Science* 327, no. 5972 (2010): 1501–1502.)

The first investigation of this kind, i.e., mode-specific energy disposal in a four-atom reaction system, was carried out by Strazisar et al. (2000). The authors measured D atom product flux contours in the reaction system

$$OH(v'' = 0; J'') + D_2(v'' = 0; J'') \rightarrow HOD(m, n) + D, \qquad (16.23)$$

where $m$ and $n$ correspond to the quanta-count numbers of the HOD bending and local OD stretching modes, respectively. Representative results from the authors' work are presented in Figure 16.26. In the top panel of the figure, a ToF spectrum for the D atom product is shown, for a laboratory scattering angle of $\theta_{LAB} = 11°$ (relative to the OH initial velocity vector). The two strongest peaks can be associated with energy predominantly being channeled into the OD stretch mode, either its fundamental (with one quantum, i.e., $n = 1$) or the first overtone (with two quanta, i.e., $n = 2$), in total about 85% of the vibrational state distribution. Also, some weaker contributions from the vibration-free situation ($m = 0$; $n = 0$) or the lowest combination mode ($m = 1; n = 1$) can be identified.

The D atom product flux contour map (in the CM velocity space) in the bottom panel of the figure reveals that the atoms recoils from the triatomic HOD molecule in forward direction, meaning that the HOD products themselves are strongly backward scattered, as is common in direct rebound reactions, with a large fraction of the available energy (>30%) deposited into HOD internal energy (largely vibrational).

Overall, the high energy resolution of the D tagging method allowed the authors to specifically resolve, e.g., contributions from one- and two-quanta excitation in the local OD stretching mode. This observation underpins the mode-specific behavior in the four-atom reaction $OH + D_2 \rightarrow HOD + D$, a dynamical feature that was also forecast in the quantum scattering calculations by Pogrebnya et al. (2000).

# 16.6 BREAKTHROUGHS AND THE CUTTING EDGE

In view of the large extent and diversity of laser ionization techniques that have been covered in this chapter (REMPI, ZEKE, H tagging), it is next to impossible to do justice to the many breakthrough moments for all of them. Here a moment in the history of multiphoton ionization spectroscopy is rekindled, i.e., the recording of the first realization of state-resolved REMPI spectroscopy of a molecule, using a tunable laser source.

## 16.6.1 Breakthrough: First state-resolved REMPI spectrum of a molecule

Although multiphoton ionization and REMPI of molecules date back to the early 1970s, it was probably the work of Johnson (1975, 1976) on state-resolved REMPI spectroscopy of the aromatic molecule benzene that laid the foundation for the rapid development of REMPI as a very sensitive spectroscopic tool for the fundamental and analytical investigation of molecules. He utilized a pulsed dye laser source, tunable in the range $\lambda_1 = 360{-}400$ nm, to measure the ionization of gas-phase benzene; in the spectra shown in his publications, a clear, strong resonance at $\lambda \approx 391.4$ nm occurs, accompanied by a few further but less pronounced peaks (see the spectrum in Figure 16.27).

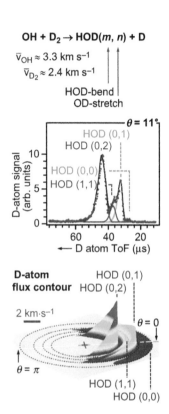

**Figure 16.26** Mode-selective behavior in the four-atom reaction $OH + D_2 \rightarrow HOD + D$. (Top) D atom product ToF spectrum, for a laboratory scattering angle $\theta = 11°$ (relative to the OH initial velocity vector). (Bottom) D atom product flux contour map (in the CM velocity space); the dominant observed vibrational modes are annotated. (Data adapted from Strazisar, B.R. et al., *Science* 290, no. 5493 (2000): 958–961.)

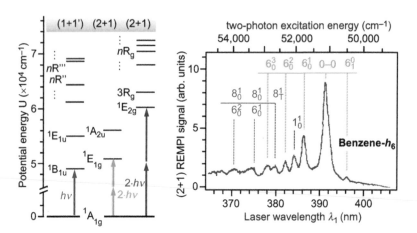

**Figure 16.27** (2+1) REMPI spectroscopy of benzene-$h_6$. (Left) Partial energy level diagram with relevant two-photon resonances for the dominant valence and Rydberg series. (Right) (2+1) REMPI spectrum via the intermediate state $^1E_{1g}$, with vibrational band annotation. (Data adapted from Johnson, P.M., *J. Chem. Phys.* 64, no. 10 (1976): 4143–4148.)

Said peak feature in the benzene spectrum was identified as a two-photon resonance with a bound state at $\tilde{v} \approx 51{,}085$ cm$^{-1}$, whose symmetry was assigned to belong to either $^1E_{1g}$ or $^1E_{2g}$ (see the term-level diagram in Figure 16.27). Based on the findings and assignment from other earlier studies and on isotopic substitution shifts, the minor peaks could be associated with the vibrational structure in the benzene spectrum, specifically the overtones and combination bands with the benzene $v_6$ vibration.

The early uncertainty about the actual electronic state assignment was solved just a few years later in subsequent REMPI measurements, exploiting a wider laser tuning range than what was available to Johnson. Now the state observed by him is associated with the $^1E_{1g}$ valence state (see, e.g., Whetten et al. 1983).

### 16.6.2 At the cutting edge: Ultrahigh sensitivity PAH analysis using GC-APLI-MS

It is well known that several PAHs, such as, e.g., the dibenzopyrenes, are highly carcinogenic, even at very low environmental concentrations in the microgram-per-kilogram range. These circumstances demand highly sensitive chemical analysis to detect and quantify PAHs at ultralow concentrations in a variety of host matrices.

Recently Große-Brinkhaus et al. (2107) developed a method combining gas chromatography (GC) and, for the first time, *atmospheric pressure* laser ionization mass spectrometry (APLI-MS). The combination with laser ionization makes the method much more sensitive in comparison with GC coupled with common electron-impact mass spectrometry (GC-EI-MS). In particular, for the analysis of PAHs, the authors utilized a KrF excimer laser ($\lambda_1 = 248$ nm) in their selective and sensitive (1+1) REMPI implementation to ionize PAHs and selected isomers of alkyl PAHs, reaching LODs of $LOD_{PAH} \approx 5$–$50$ fg·$\mu$L$^{-1}$.

In the left data panels of Figure 16.28, a segment from the total ion chromatograms of a pine needle extract is shown for the range $t_{retention} = 40$–$50$ min, which is centered near $m/z = 228$ amu (associated with triphenylene). The data were recorded using GC-EI-MS; the chromatogram reveals the presence of many nonaromatic substances from the bulk sample (a broad, near-featureless mass peak). In the GC-APLI-MS measurement, the laser ionization selectivity yields

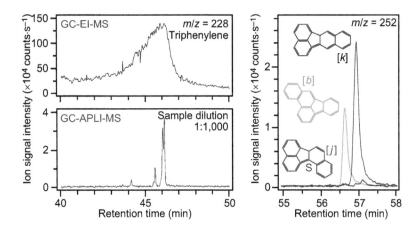

**Figure 16.28** Ultrasensitive detection of PAH compounds by GC-APLI-MS. (Left) Partial chromatograms of pine needle extract, measured by standard GC-EI-MS and GC-APLI-MS. (Right) Partial chromatograms for three structural forms of benzofluoranthene (200 fg·μL⁻¹). (Data adapted from Große-Brinkhaus, S.G. et al., *Anal. Bioanal. Chem.* 409, no. 11 (2017): 2801–2812.)

only the PAH peak (here triphenylene). Note that in GC-APLI-MS, samples were diluted by a factor of 1:1000; thus, the load of matrix material on the capillary column is reduced, yielding better separation and improved mass peak shape. In the right data panel of the figure, GC-APLI-MS partial chromatograms for benzofluoranthene (mass $m/z = 252$ amu) samples are shown (concentration 200 fg·μL⁻¹), revealing the slight mass shifts for the structurally different forms (the low signal for benzo[*j*]fluoranthene is thought to be due to the poor planarity of the compound).

# CHAPTER 17

# Basic Concepts of Laser Imaging

Imaging has been and is the most important methodology to capture the reality of nature for immediate or later interpretation, and to provide records for the future. Whether this has been in the form of "primitive" painting on the walls of caves in prehistoric times, or in the form of magnificent canvases through the last thousand years or so, and whether photographic films were used, or today's state-of-the-art digital image sensors, the key methodologies and overall goals of imaging have remained unchanged. Broadly speaking, in imaging, one exploits the interaction of light with a medium, followed by a response of the said medium (absorption/reflection, fluorescence, scattering) suitable for image projection transfer onto a receiver (canvas, photographic film, photo sensor). For many scientific imaging applications today, traditional light sources have been replaced by laser light sources, which offer tremendous versatility in the control and selectivity of light wavelength and which provide wide flexibility in the choice of spatial illumination, be it over a large area/volume or in the form of tight focusing to spot sizes well below 1 μm.

Whatever the choice of light source—traditional "lamps" or lasers—the basic principles of the imaging process do not change: the specimen to be imaged is illuminated, the response from the light–matter interaction is collected and guided using optical transfer components (like lenses, mirrors, or fiber bundles), and near exclusively today the transmitted light is recorded by single-element or array photo detectors; finally, the image is, in most cases, stored in digital form, ready for display on data displays and for scientific interpretation. This chain of actions is summarized in the conceptual Figure 17.1.

First, the sample is interrogated by light from a laser source whose wavelength can be selected (either device-inherent or by utilizing a tuning element). By and large, using laser light eliminates the need for spectral filtering of a wavelength-continuum lamp, as has been and still is common in many imaging applications, like in standard wavelength-selective microscopy. As already stated, in addition to its wavelength selectivity, laser sources allow for easy manipulation of the illumination modus, in the form of wide-field exposure (then global area imaging can be utilized), or in the form of a tightly focused beam (then the image is built up from a point-by-point scan of the spatial region of interest).

The response light may, or may not, be filtered further (using band-path filters or spectrometers), depending on whether one only requires the overall

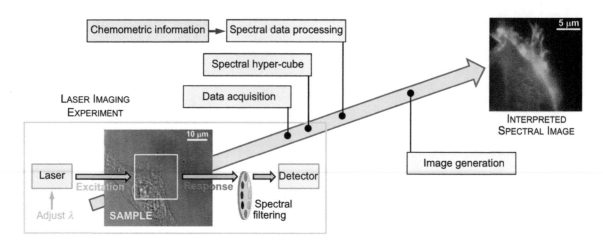

**Figure 17.1** Concept of laser imaging—from probing a sample with laser radiation to a reconstructed image.

summary response associated with the light–sample interaction, or if additional wavelength-dependent molecular information is desired (e.g., in some application of laser-induced fluorescence [LIF] imaging or Raman imaging).

The light received by the (point or array) detector in the image plane is converted into an electrical signal, which is digitized according to the particular light intensity and stored as a digital "pixel" pattern, both for spatial and spectral (if the response is recorded for different wavelength channels) information; the result is a so-called "spectral hypercube." This hypercube consists of a spatial map of the sample for each individually recorded wavelength, and these are then stacked; a closer look at hypercubes and how to utilize them for image construction and interpretation is provided in Section 17.1.

Finally, the data from the hypercube will be processed, including spectrally distinct properties if they were provided and are required for later interpretation (like in spatially resolved chemometric analysis). After processing, an appropriate image is generated and visualized, ready for interpretation (see Section 17.2).

In the remainder of this chapter, the basic concept of imaging is briefly introduced in a general manner, in principle valid for all types of imaging, but also particularities associated with laser light imaging techniques will be addressed. Foremost in this respect are superresolution techniques (see Section 17.3), which have revolutionized in particular LIF imaging.

## 17.1 CONCEPTS OF IMAGING WITH LASER LIGHT

As stated already in the introduction, conceptually the "ballistic photon" (= light propagation generally follows linear-ray propagation rules) imaging process is independent of the light source used for sample interrogation. However, laser light beams are normally much easier to handle and manipulate, particularly because of their inherent properties of low divergence and monochromaticity.

As in ordinary imaging for scientific and application-oriented purposes, the experimental implementation is, in general, slightly different for gaseous and liquid samples, and for condensed-phase specimen. This is because the former are, in general transparent to light, while the latter are, in most cases, nontransparent, or severely scatter diffusely, due to very high molecular particle densities. Accordingly, imaging can be done in nearly all directions in the former case (in forward, backward, or sideways directions—with imaging at 90° the preferred orientation), while for solid samples, imaging is nearly exclusively done in backward direction (occasionally in forward direction provided the sample is sufficiently thin to transmit light without diffuse scattering). The differences for the two sample classes are summarized below.

In the context of the aforementioned scattering processes affecting the path of the exciting and response photons, the theoretical framework for image reconstruction, taking into account absorption, fluorescence as well as elastic and inelastic scattering in the underlying process modeling, is entirely different. Thus, these will not be covered here but are subject to a detailed description in Chapter 20, which is dedicated to diffuse optical imaging.

### 17.1.1 Laser illumination concepts: Point, line, and sheet patterns in transparent gas and liquid samples

The overall concept for the imaging of gaseous and (transparent) liquids is shown in Figure 17.2; in particular, the three most common light-excitation scenarios are highlighted, exemplified for interrogation of a burner flame by LIF and imaged at 90° (here the X-axis) with respect to the laser propagation axis (Z-axis). Note that similar arguments as used below would also apply for Rayleigh and Raman scattering; for the imaging of localized absorption, one would need to image in forward or backward direction.

When focusing the laser beam (tightly) into the flame, the strongest fluorescence will emerge from the small focal volume. Note that for exact point-imaging, one would need to spatially filter the light collection region using a pinhole in the light collection path. A complete cross-sectional *point-by-point* (pixelated) image of the flame would be built up by scanning the position of the laser focus location in *Z- and Y-directions*. However, scanning the laser focus position is somewhat cumbersome, and also ray propagation in the light imaging path may introduce distorting aberrations. Thus, in general, the illumination–light collection geometry is kept unchanged; instead, it is the probed object that is moved laterally and vertically (this may not always be possible, so that one has to revert to the former inconvenience of adjusting the light passes).

Sending a thin, collimated laser beam through the flame will result in "line" illumination. A full cross-sectional image can now be built up by scanning this *line* in the *Z-direction*. Note that this type of line-scan imaging is encountered in many practical imaging implementations, like in machine vision or satellite-based imaging, to name but two common applications.

Finally, the laser beam can be expanded to form a two-dimensional "sheet." This can be achieved by a combination of a collimating beam expander and a cylindrical lens; in the configuration shown in Figure 17.2, the result is a sheet of constant height (in the *Y-direction*), focused in the *X-direction*. To generate a cross-section image now, no scanning is required. Note that the illuminated area in the flame is limited by the size of the optical components; should one wish to enlarge this further, one can utilize a "fan"-like sheet, which is divergent in the Z-direction.

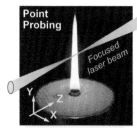

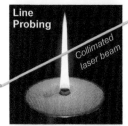

**Figure 17.2** Concept of light illumination modalities, resulting in (a) point-scan imaging; (b) line-scan imaging; and (c) area imaging.

It is worth noting that three-dimensional (3D) images can be generated as well. This is possible by scanning the illumination pattern (point, line, or sheet) additionally in the *X*-direction. For each *X*-position, a *Y–Z* plane image would be recorded, and from this "stack," a 3D representation of the object (here the flame) could be reconstructed; an example for this is given in Chapter 18.1.

### 17.1.2 Laser illumination concepts: Point, line, and sheet patterns in condensed-phase samples

As stated already in the introduction to this section, the imaging conditions for solid, condensed-phase samples are somewhat different than for gases and liquids. In particular, since most solid samples are nontransparent to light, or at best opaque, only 2D imaging of the surface is in general realized; the third, vertical dimension is only accessible for a few micrometers, as far as exciting light beams can penetrate and product photons can emerge again without further interaction. Free-space laser imaging of solid samples has been around for a long time (since the mid-1970s) in the form of airborne, remote, or standoff active LIDAR imaging (its wide versatility may be gleaned from Sitter and Gelbart 2001; Kim et al. 2003; Omasa et al. 2007). But it is its application in microscopy that "steals the show"; a representative selection of examples for LIF and Raman imaging, down to the nanometer scale, can be found in Chapters 18 and 19.

The conceptual setup of laser-imaging microscopy is shown in Figure 17.3; note that as in the case for the imaging of gaseous samples, the *Z*-direction is that of the propagation of the laser beam. But because the observation of the product light is in backward (*–Z*) direction, the image-scan plan is now *X–Y* rather than *Z–Y*. As before, there are three imaging modalities.

In the point-scan mode, the laser light is focused by the microscope objective to a tight spot size, of the order of a few micrometers in diameter (ultimately as small as allowed by the diffraction limit); this focal spot is then scanned point-by-point in the *X*- and *Y*-directions, often in the form of a zigzag pattern, as shown at the bottom of the figure (other patterns are also used, depending on the particular application).

A number of commercial laser imaging microscopes offer the capability to expand the circular laser beam in one direction, resulting in a stripe-like illumination of the sample. Then this "line" is scanned in the *Y*-direction, generating the full data set to construct a 2D image of the sample. Note, however, that the "line" illumination is not homogeneous, but rather constitutes an elongated ellipse resulting from the stretching of the original circular laser beam profile in one direction; hence, the spatial resolution changes with the actual position within this profile (see the bottom right of Figure 17.3).

Finally, moving the laser focus away from the sample surface, a larger area can be illuminated allowing for immediate 2D imaging, without the need for spatial scanning—as in the case of sheet illumination shown in Figure 17.2. Once again, the illumination intensity is non-uniform, following the Gaussian intensity distribution of the laser beam cross section.

Note that, even more than in free-space imaging of gaseous (liquid) samples, spatial scanning is realized mostly by moving the target rather than the laser beam; specifically, this is indispensable in confocal microscopy for which the passage of light through aligned pinholes is paramount. Also, modern motor-driven sample

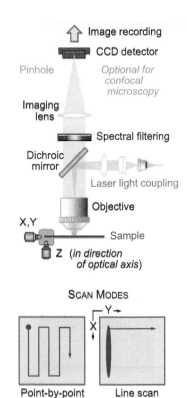

**Figure 17.3** Concept of solid-sample laser microscopy using point-scan and line-scan imaging.

tables for microscope systems exhibit movement precisions down to the nanometer scale, which, in general, is much less than the laser spot size diameter.

As in the case of the gas-sample imaging, it is possible to generate 3D images; for this, the sample is moved up- or downward in the $Z$-direction, as indicated in the microscope schematic of the figure. However, the range of depth is normally limited to just a few micrometers, as opposed to the much larger range in the lateral directions (only limited by the total scan range of the sample table).

### 17.1.3 Image sensing and recording concepts

Starting from the general philosophy of how to probe an object with laser light, a number of steps are involved to arrive at a faithful reproduction of the probed object in the image; these are summarized conceptually in Figure 17.4. The basic idea behind the laser-imaging process is to convert the (analog) information about the object into a (digital) recorded signal from which an image can be generated. Since the imaged object exhibits spatial dimensions but may also respond with different strength yielding different light amplitudes, one is faced with a dual digitizing problem, namely one in the spatial dimensions (normally addressed as *sampling*) and on the intensity scale (in general, known as *quantization*).

Broadly speaking, the chain encompasses the following steps: (1) the laser probing (sampling) of the object, in the figure exemplified by raster-scanning a focused laser beam across the surface of the specimen (bacteria); (2) the light response from this interrogation is collected and (optionally) filtered for spectral content; (3) the collected light is recorded with spatial resolution (most commonly) by a CCD array detector, which then is converted into a digital signal, and is read out for further data treatment and storage; and finally (4) the digital data are used to reconstruct an image of the probed object, for visualization and interpretation. These four individual steps are specified in more detail hereafter and will be discussed more quantitatively in Section 17.2.

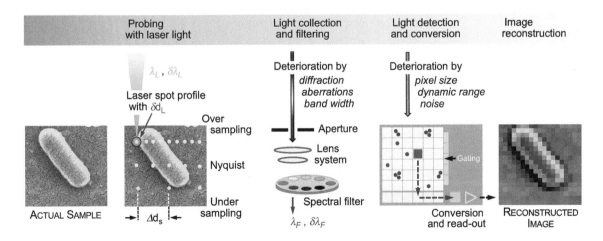

**Figure 17.4** Conceptual sequence of laser imaging of an object, involving the steps of raster probing, light transfer to the detector, signal accumulation and readout; and conversion into a final image. Note that the sampling raster point patterns in the laser-excitation panel are demonstrative only to highlight the loss of resolution once the Nyquist criterion no longer applies.

*The sampling step.* In signal processing, sampling is the reduction of a continuous signal function $f(x)$ to a discrete signal representation $f(x_n)$. A sufficient condition for a sampling rate that permits a discrete sequence of samples to capture all the information from a continuous signal (of finite bandwidth) is the so-called *Nyquist–Shannon sampling theorem*. Originally, it was formulated for time-varying signals, stating that if a function $x(t)$ contains only frequencies not higher than $B$ (in units of hertz), it is completely determined by representing it at a series of ordinate points spaced $1/(2B)$ apart; a sufficient (minimum) sample rate would therefore be $2B$. It was shown that it could easily be transposed to other variables, like the spatial probing $\Delta d_{x,y}$ encountered in the imaging of objects. Therefore, if $\Delta d_{min}$ were the smallest dimensional variation of the object one would like to resolve during imaging, then the sampling step size $\Delta d_s$ would need to obey the condition

$$\Delta d_s \leq \Delta d_{min}/2. \tag{17.1}$$

In the laser-probing sketch of Figure 17.4, two additional scenarios are depicted besides the Nyquist criterion, namely, "undersampling" and "oversampling." In the former case, the sampling distance $\Delta d_s$ is larger than the widths of the bacteria; then not only does one lose information but also one might get a wrong piece of information about the object (note that undersampling is also often addressed as "aliasing"). In the latter case, the signal is sampled at many more raster points than necessary to recover the shape of the bacteria; obviously, the more samples are taken, the higher the quality of the image might become. Note also that certain types of oversampling can be exploited beneficially in image processing to drive the resolution beyond the diffraction limit (see Section 17.2).

*Light collection, transfer, and distortion.* Evidently, the light emerging from the probed sample needs to be collected and guided to the photon detector. This involves, in general, a selection of optical components like lenses, mirrors, and (spectral) filters. All of these will introduce certain types of distortions, like diffraction at the apertures of said components; aberrations in the imaging path associated with curved-surface optics; or influences on the spectral compositions and/or bandwidth. All of these are bound to detrimentally affect the positioning and resolution of the recorded signal in the detector plane, and therefore deteriorate the possible image resolution.

*CCD array detector and its pixels.* In many of the commercial imaging microscopes, and in other spectroscopic imaging applications, the detector is based on 2D pixel-sensor CCD arrays. In the case that "one-to-one" imaging optics is utilized, each probe raster point can be associated with an individual pixel, i.e., the number of samples is directly equal to the number of sensor pixels of the CCD array detector. Thus, the smallest element in a sampled image corresponds to an individual pixel; each of them collects a known fraction (related to the overall system transmission efficiency) of the photons emitted from the associated, sampled location on the object area.

It has to be noted that in repeat measurements, the detected signal is not always equal in magnitude; in particular, the signal varies statistically in amplitude according to the shot-noise (or Poisson) criterion, meaning that an average signal of $n$ photons carries a variance of $\sqrt{n}$ for an individual measurement. In addition, each pixel element exhibits statistical thermal (or dark current) noise. And finally, a further contribution to the noise balance stems from the readout process (readout noise). The effect of noise on the quality of reconstructed images is discussed further in Section 17.2. Overall, the light amplitude variation, which can be recorded, is limited by the so-called dynamic range of the detector device, which is defined as (also see Figure 17.5)

$$\text{dynamic range} = (\text{pixel well-depth})/(\text{pixel noise}), \tag{17.2}$$

where the pixel well-depth is associated with the maximum number of photogenerated electrons, which can be held in an individual CCD pixel element.

It should be noted that the one-to-one imaging mentioned here does not necessarily need to apply; one also could choose an optics configuration, which generates an enlarged- or reduced-size image, i.e., if one were to treat the complete point raster as a "continuous" emission area, then the pattern of detector pixels would over- or undersample the information content from the object. In this context, one may refer to Figure 17.5 and the Nyquist criterion of Equation 17.1. If the probing (circular) laser spot were to be imaged properly, to provide spatial information on it, an enlarging imaging optic would be required, generating a projection of the Gaussian distribution footprint of the sample over the region of pixels, each of them collecting (shot-noise modulated) light according to the number of photons associated with the Airy-disk distribution function. Recall that the Airy disk is characterized by the position of the first minimum of the diffraction pattern, which is given by

$$d_{\text{Airy}} = 2.44 \times (\lambda \cdot f / d_{\text{aperture}}), \tag{17.3}$$

where $\lambda$ is the wavelength of the laser light, and $f$ and $d_{\text{aperture}}$ are the focal distance and aperture diameter of the imaging system, respectively.

According to the Nyquist criterion, the width dimension would need to be sampled at half its value, i.e., a sensing area of $2 \times 2 = 4$ pixels would be required to properly localize the laser spot. Again, this resolution issue will be further explored in Section 17.2. A general convention in microscopy is to choose the size of the pixel to be equal to one half of the Abbé criterion resolution of the optical system. It is also worth noting that a distortion of the expected Gaussian distribution might be expected because detector pixel elements are not ideally isolated from each other. Rather, light-generated electrons "leak" between pixels, constituting a net flow from pixels with a large charge to those with lesser charge (see the bottom part of Figure 17.5). When reading out this exposure information, reconstruction from these data then one generates a slightly distorted image.

A final remark concerning the detector pixel structure has to be made in relation to spectral information. In contrast to digital color-photographic cameras, which record the spectral content via three-element RGB clusters (RGB = red/green/blue), for laser imaging, nearly exclusively monochrome devices are used; these provide denser pixel structures with the associated higher spatial resolution. The spectral information needs to be indexed by a filter analyzer or spectrometer in the light collection path (see the central panel of Figure 17.4).

*Image reconstruction.* As a final step in the imaging path, the image reconstruction step is briefly considered. In its simplest form, the image is assembled pixel-by-pixel from the data readout and stored from the detector sensor; the intensity information for this 2D spatial image would be provided in a gray-scale format or in false-color representation (see Figure 17.9). More often than not, these "raw" images are treated further using pixel-iterative and pixel-interpolating algorithms, often also accompanied by resampling to provide a finer image raster

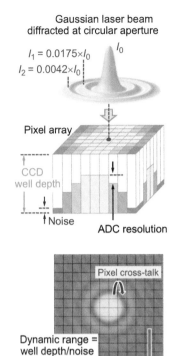

**Figure 17.5** Sampling of the Airy-disk pattern of a Gaussian laser beam. ADC = analog-to-digital conversion.

and thus a "smoother" image representation. Of course, this cannot increase the overall resolution, but it may allow for easier recognition of narrow features. This is a well-known procedure in ordinary, software-supported image processing.

### 17.1.4 Multispectral and hyperspectral recording

The last parameter that needs consideration is the "color content" of the image. As mentioned in the previous subsection, in contemporary photographic technology—encountered in cameras and smartphones—the image sensors are made up of RGB arrays that directly capture, store, and display the wavelength-dependent light components received from the object. For various reasons in laser imaging, as in many scientific and analytical imaging applications, the approach is different, and by and large "monochrome" detector arrays are utilized. This does not mean they record only a single color; far from it. Indeed, they are sensitive to all wavelengths from the UV to the near IR, but they integrate over the complete wavelength range. Thus, in order to provide spectral information, the excitation laser light source has to be monochrome, or a wavelength-selective filter has to be introduced into the light collection path. In fact, filtering of the collected light is nearly always indispensable. This is the consequence of the response of the sample to monochrome excitation, which is rarely at a single wavelength. For example, many dyes and tags that are used in fluorescence imaging can be stimulated by the same laser wavelength but emit in distinctly different fluorescence bands; if one wishes to distinguish them, wavelength filtering is required. This is not only true for fluorescence imaging but equally, or even more so, for Raman scatter imaging.

When using CCD array detectors for laser-imaging purposes, the three modalities—point raster scanning, line scanning, and direct 2D sheet imaging—require slightly different approaches for image recording and reconstruction. This is visualized in Figure 17.6.

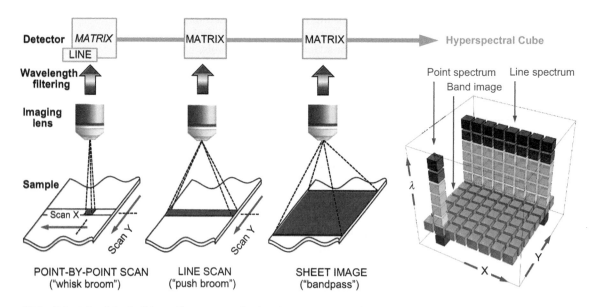

**Figure 17.6**  Principle of the buildup of hyperspectral cubes from point-scan, line-scan, and sheet imaging; for the choice of wavelength filtering, see text.

In order to build up the hyperspectral cube of spatial and wavelength information of the imaged sample, each of the spatial pixels needs to be associated with color information, which can be a single wavelength or a complete spectrum; in the former case, a sparse-element three-dimensional matrix would be encountered; in the latter case, a full cube would be generated. In point-by-point and line-scan applications, which are the most common modalities in laser spectroscopic imaging, the spectral selectivity can be provided by both band-path filters or spectrometers; in direct 2D sheet imaging, only filters can be used. Note that in point-by-point imaging, a CCD array detector would not be required; a single-element detector would suffice; this is, for example, often the case in confocal microscopy (although not compulsory).

With reference to Figure 17.6, in point-by-point raster scan applications, individual spectral columns in the hyperspectral cube are generated; the column is filled with information from the dispersed spectrum. The most efficient way to obtain detailed spectral data is to utilize a spectrograph coupled to a CCD array detector, which records the complete spectrum in one go. With filters, the process could be more cumbersome, since different filters would need to be introduced sequentially into the light collection path. On the other hand, it might constitute a cost-effective approach if only few spectral channels were required for a particular laser imaging application. In line-scan applications, a full spatial stripe with complete spectral information can be added to the cube; when using a spectrograph, the 2D array detector would record the spatial line information in one direction and the spectrally dispersed information in the other one. Note that line-scan hyperspectral implementations are most frequently found in Raman scatter imaging. Finally, in 2D sheet imaging, areal filters need to be utilized, which generate individual color pixel planes; the full cube is built up by sequentially changing or tuning the spectral filter.

As stated already, it is not always necessary to record a full hyperspectral cube, but information at a few selected (filtered) wavelengths suffices to analyze and interpret the images. On the other hand, with insufficient spectral information, it may become difficult to draw full conclusions on the image content, or one even may completely lose the ability for certain types of identification. A comparison between the properties of the most common (laser) spectral imaging modalities, according to the number of individual wavelength channels, is given in Table 17.1.

**Table 17.1** Comparison of the Various Spectral Imaging Modes, with Respect to Spatial and Spectral Information Content, and Analytical Capabilities

| | Monochrome | RGB | Multispectral | Hyperspectral | Spectroscopy |
|---|---|---|---|---|---|
| Spatial information | Yes | Yes | Yes | Yes | No |
| Spectral information | No | No | Limited | Yes | Yes |
| Number of spectral bands | 1 | 3 | 3 … 10 | Up to hundreds | Up to thousands |
| Multiconstituents information | No | Limited | Limited | Yes | Yes |
| Sensitivity to minor constituents | No | No | Limited | Yes | Limited |

## 17.2 IMAGE GENERATION, IMAGE SAMPLING, AND IMAGE RECONSTRUCTION

Images are now almost exclusively recorded by digital means. For this, the total flux of photons originating from an object area is divided into small geometrical pixels of normally a CCD camera device to form the final image. The light intensity in each pixel is then converted into a recordable electric signal by an analog-to-digital converter (ADC) unit and stored as a single number.

The overall digital imaging process may be defined as a sampling and quantization sequence. The spatial and amplitude digitization of an image function $f(x,y)$ is called image *sampling* when it refers to spatial coordinates $(x,y)$ and (gray-level) *quantization* when it refers to the light amplitude. Before taking a closer view at these two processes and associating them with a descriptive, mathematical framework, it is worthwhile to highlight a few fundamental points that will affect the quality of any image.

One of the key parameters in all imaging processes is resolution. If one defines an image as a 2D rectilinear array of sample points, then the sampling raster limits the spatial resolution, and the electronic quantization device limits the light intensity resolution. In general, sampling limitations are governed by the photon detector. For a single-element detector (often used, e.g., in confocal microscopy), the pixilation of the image is limited, on the one hand, by the mechanical motion with which the sample can be scanned in the $(x,y)$ plane and, on the other hand, by the light collection optics (closely linked to the Abbé limit). For 1D line sensors, the resolution is linked to the size and number of photosensitive pixel elements in the strip; and for 2D array detectors, the same holds for its pixel elements. Some of the instrumental detector properties were already outlined in Section 17.1.

Errors in the imaging process are introduced if insufficient spatial or intensity resolution is encountered. In particular, in the former case, the so-called *aliasing* may occur, i.e., image artifacts basically due to undersampling. Thus, in general, one needs to consider the basics of sampling theory, namely (1) how many samples are required to represent a given image feature without loss of information and (2) what type of features can be reconstructed without loss for a given sampling raster. As pointed out already in Section 17.1, a particular signal feature can be reconstructed from its samples, if—according to the Nyquist rate (see Equation 17.1)—the original signal incorporates no features larger than half the sampling rate (in the spatial manifold less than 2 pixels).

Digital image recording may be seen as a dual, interlinked digitization process. As shown in the top part of Figure 17.7, the analog signal function has to be quantized (1) in its space coordinate, the quantization step size being equal to the CCD pixel dimension (note that in point-by-point laser scanning, it also could be directly related to the laser scan step size); and (2) in its intensity coordinate, the quantization step size being determined by the so-called bit resolution of the ADC device (see further below). Note that while the spatial and intensity discretization processes are displayed independently in the figure, for clarity, in real digital CCD imaging, they are not: both exhibit fixed step sizes rather than variable quantities. It is clear that this can mean a severe distortion of the original object when reconstructing its image from the pixel-intensity data. The issues of sampling, quantization, and resolution are detailed in the subsections following

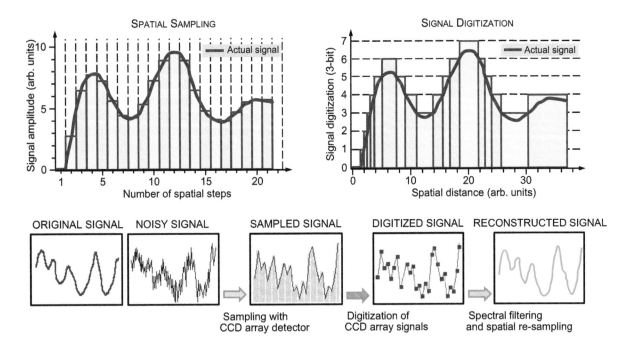

**Figure 17.7**   Quantization of an object signal function. (Top) Spatial sampling related to the pixel-dimension step size, and intensity digitization, related to the ADC step size (shown here for a 3-bit device). Note that in both digitization presentations, the distance between consecutive dashed lines corresponds to the least-significant bit (LSB). (Bottom) Image sampling and reconstruction chain, including the addition of noise and post-acquisition data treatment.

below; details of many of the associated problem points can be found, e.g., in Pawley (2006).

## 17.2.1 Sampling and its relation to signal digitization

As the emission and detection of photons are quantum events, inherently the detected signal incorporates an intrinsic uncertainty in the number of recorded photons, which will vary in repeat measurements. This uncertainty is referred to as *Poisson* (or statistical) noise and is equal to the square root of the number of detected photons. For example, an individual signal reading of 100 photons means that in a repeat measurement, the probability of detection is rather $100 \pm 10$ photons. A noisy signal is shown in the second panel of the image process chain shown in the bottom part of Figure 17.7. Note that, like diffraction, Poisson noise constitutes a rigid physical limitation; the *only* way to reduce the relative uncertainty is to count a larger number of photons or to average over a number of repeat measurements.

In all signal recording—of which imaging may be seen as a distinct subset—the most obvious request to the object-to-image chain is that the intensity value detected in a given pixel should be linearly related to the numerical light signal value stored in the image memory but also to the display brightness of the very image pixel. This request should be easy to fulfil from an instrumental point of view since most electronic photon detectors (like CCDs and single-element photodiodes) and ADCs are linear devices. It also seems a logical request, because how else would one be able to compare repeat measurements.

Indeed, linearity has its undisputed advantages; however, some practical complications ensue in certain digital imaging applications. In particular, in LIF microscopy, fluorescent dyes are frequently used (see Chapter 18), and these often bleach with time and thus may result in nonlinear responses. In addition, in LIF imaging of biosamples, the response signals are normally very weak. Then, because of Poisson statistics, intensity values representing only a small number of photons are inherently imprecise. As a consequence, intensity steps in the digitization process may be smaller than this imprecision—displaying them with full ADC resolution is rather pointless and could even become misleading.

For example, in confocal fluorescence images, the signal in the brightest pixel is of the order 10–30 photons; for the sake of argument, a number of 16 photons is considered here. Thus, in the quantization process, more than 16 digital levels (equal to 4 bit) would not be meaningful. But because of the Poisson (shot) noise limit, the number of "meaningful" intensity steps in this 16-photon signal is even less. The statistically expected, brightest signal is $16 \pm 4$; in order to really distinguish intensities, one needs to surpass the $1\sigma$ threshold (standard deviation); this means the next lower signal levels would be $9 \pm 3$, then $4 \pm 2$, and finally $1 \pm 1$. This means that for a 16-photon peak signal, one can discriminate only four "real" signal levels. Of course, in the digitization, one would still utilize the linearly spaced 16 steps of a 4-bit ADC (note that standard ADCs, utilized in laser imaging and microscopy, commonly have resolutions of 8 or 12 bit).

## 17.2.2 Sampling and its relation to spatial resolution

As stated already, according to the Shannon–Nyquist theorem, the sampling interval between intensity measurements has to be less than half the period of the highest frequency in the signal, if one aims at a faithful reconstruction of the original signal from the recorded, digital values. Translated into optical terms, one usually tends to classify that an object feature is imaged as being useful (for reconstruction and interpretation) if the intensity between features drops to about 25% of its peak contrast (this is associated with the Abbé or Rayleigh criterion noted above), although this is a somewhat arbitrary choice.

If one thinks of an analog signal function as a 1D image (see Figure 17.7), it is rather straightforward to extrapolate to the 2D (and 3D) imaging scenario: the image data now vary in 2D or 3D, but the rest of the imaging and analysis procedures still apply. From the point of view of image resolution, this means that the "bandwidth" of an image must be somehow associated with its "sharpness," and this is related to the highest spatial frequencies it contains. While in consumer photography, it is often problematic to apply an image analysis along the aforementioned criteria (even an out-of-focus or blurred image may still carry memorable impressions), in scientific imaging and microscopy, access to the proper imaging parameters is in most cases feasible. For example, at the very least, one knows (1) that the diffraction limits the maximum sharpness of the data that can be recorded and (2) that the "spatial frequency response" of a microscope can be defined by a suitable calibration sample [e.g., the "1951 USAF resolution test chart"—see Messina (2006); or the so-called "Siemens Star" target—see Thurman (2011)]. As a consequence of the above considerations, the convention is to adapt the relation between the imaging optics properties (ultimately governed by diffraction limit $d_{Abbé}$) and the pixel size $d_{pixel}$ of the imaging array detector to obey

$$d_{pixel} \leq d_{Abbé}/2. \tag{17.4}$$

However, there is one particular caveat to contemplate. While the structural features of a 1D image can only vary in that single dimension (e.g. the $X$-direction), those of a 2D image can vary in more than the $X$- and $Y$-axis directions. This is shown schematically in Figure 17.8 for an arbitrarily shaped object.

For example, for orientations deviating from the "vertical" axes, sampling points are much further apart, with a maximum factor ×1.41 for the 45° orientation. However, in this simplistic analysis, one has neglected that all image "features" extend in 2D. As a result, features running at 45° will also be sampled by pixels other than the diagonal in the array. In order to account for this problem in matching the detector pixel size to the diffraction limit of objects with arbitrary orientation, the denominator in Equation 17.4 should be changed from 2 to at least 2.8. But—as stated repeatedly—it is not only the shape of the object that is of importance but also the light intensity from a particular object location to the associated image pixel (see the matrix representation at the bottom of the figure).

Clearly, the related "third" information axis in an image is difficult to visualize in a paper or computer screen projection. Therefore, this is normally done in a grayscale (or brightness) representation, in which the actual light intensity values are mapped to different grayscale values, normally with "black" representing no signal and "white" representing the maximal possible signal (see the scale example below the sampled image in Figure 17.8). The full grayscale range is, in general, limited to 256 values (or 8 bit); this is due to the fact that imaging (ADC) and display (digital-to-analog—DAC) converters are commonly based on 8-bit electronic devices. Note that higher bit-content converters (12 or even 16 bit) are occasionally encountered for specific applications, but this goes at the expense of conversion speed and may therefore not be suitable for "real-time" applications. It also should be noted that instead of grayscales, which are sometimes difficult to distinguish by eye, "false-color" scales are utilized since, in general, the eye is more perceptible to changes in color across the visible spectrum.

It is straightforward to associate the narrative above with a mathematical representation framework. In short, an image is a function $f$ of space ($\mathfrak{R}^m$; $m = 2, 3$) and light intensity codomain ($\mathfrak{I}^k$; $k$ = number of "colors"); typically, 2D projections of the 3D space are used, dictated by the common utilization of 2D array CCD detectors for recording in the image plane. The said function $f$ can be single-valued or multivalued:

$$f: \mathfrak{R}^m \leftrightarrow \mathfrak{I}; \; m = 2, 3 \text{ (function of total light intensity)}; \tag{17.5a}$$

$$f: \mathfrak{R}^m \leftrightarrow \mathfrak{I}^k; \; m = 2, 3 \text{ (for } k \text{ different color intensities)}. \tag{17.5b}$$

The image functions in Equation 17.5 are representations of the analog world, which for digital signal processing need to be digitized involving pixel-wise spatial discretization and intensity quantization, resulting in the transformation $Q(f)$. Space discretization is associated with the pixels in the 2D detector (with $N$ columns and $M$ rows), and which is normally represented in matrix form with

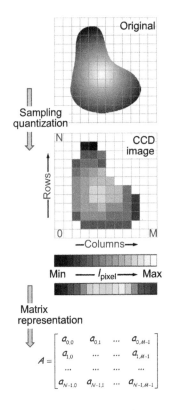

**Figure 17.8** Spatial pixel quantization, related to the resolution of an $M \times N$ CCD array detector, and generation of a data matrix in which the intensity of each pixel element $p(x,y)$ is represented by the numerical value of the related matrix element $a_{n,m}$. Note that the display of intensities in images is either by grayscale or "false-color" representation.

elements $a_{nm}$ $(n, m = 1,..., N, M$ are discrete integer numbers, with $N$ and $M$ usually binary multiples, like $2^8 \equiv 256$), yielding

$$Q(f): \mathfrak{R}^m \rightarrow [1, ..., N] \times [1, ..., M]; \; m = 2, 3 . \tag{17.6}$$

This is followed by the discretization of the intensity values, i.e., the codomain of the function $f$. Thus, after both sampling and quantization, one obtains

$$Q(f): [1, ..., N] \times [1, ..., M] \leftrightarrow [1, ..., L], \tag{17.7}$$

where $L$ is the maximum number of discrete (integer) intensity values; as stated above, 256 levels, or 8 bits, normally suffice. For color image representations, usually 256 levels are used for each of the $k$ colors.

As an example of the impact of discretization on resolution and shape recognition, the imaging of a ring structure onto a detector array with different pixel size is shown in Figure 17.9. Note that the shown sequence with varying pixel size is equivalent to the imaging of a ring with different size onto a constant-size pixel array but zoomed for clarity. Note also that only a 1-bit quantization for the intensity is used ("0" = no light, "1" = light), for simplicity.

In the case of the pixel size being larger than the determining ring dimension $w_{\text{ring}}/2$, one loses all information about the shape of the object; one only can state the "there is something" (see the 2×2 pixel panel of the figure). As soon as the pixel size reaches the Abbé/Rayleigh criterion (i.e., $d_{\text{pixel}} \cong w_{\text{ring}}/2$, where $w_{\text{rin}}$ constitutes the full diameter of the ring object), the image allows one to conclude that "there is an object with a hole in the middle" (see the 4×4 pixel panel). But only when the pixel step size become significantly smaller than $w_{\text{ring}}/2$ can one state with increasing confidence "a ring-shaped object is being imaged."

The very same procedure has been applied to a real, biological tissue sample that exhibits distinct cell structure (see Figure 17.10). Here the pixel resolution is decreased stepwise in multiples of two (i.e., sampling distances ×2, ×4, and ×8). Clearly, some of the finer details become already lost in the ×4 reduction, and in the case of the ×8 reduction in sampling distance one more or less only distinguishes that in the image there is a "green" part and a "red" part (the latter with some apparent structure).

But small pixel dimension is not necessarily sufficient to recognize features in an image. In order for an image to show perceivable structure, the array of pixel

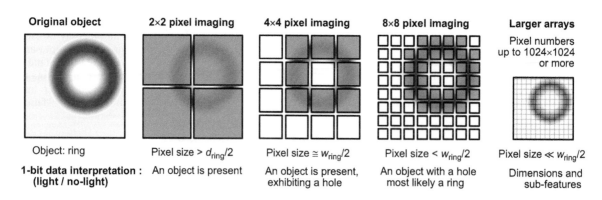

| Original object | 2×2 pixel imaging | 4×4 pixel imaging | 8×8 pixel imaging | Larger arrays |
|---|---|---|---|---|
| | | | | Pixel numbers up to 1024×1024 or more |
| Object: ring | Pixel size > $d_{\text{ring}}/2$ | Pixel size $\cong w_{\text{ring}}/2$ | Pixel size < $w_{\text{ring}}/2$ | Pixel size $\ll w_{\text{ring}}/2$ |
| **1-bit data interpretation :** (light / no-light) | An object is present | An object is present, exhibiting a hole | An object with a hole most likely a ring | Dimensions and sub-features |

**Figure 17.9** Object recognition with increasing pixel resolution, here represented for 1-bit intensity digitization (= light/no-light distinction). For further details, see text.

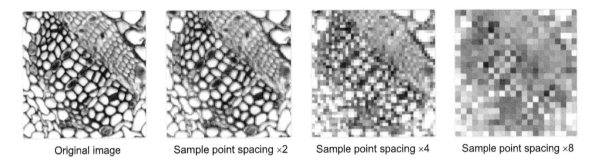

Original image    Sample point spacing ×2    Sample point spacing ×4    Sample point spacing ×8

**Figure 17.10**  Image resolution encountered for a biological specimen with distinct cell structure, as a function of increasing sample-point spacing; with larger spacing, finer details are lost.

points must exhibit *contrast*, i.e., something about the specimen must produce changes in the intensity recorded at different image points. For example, in transmission microscopy, the contrast may be due to structures that partially or fully block the light; or in fluorescence microscopy of biological specimen, chromophores may become luminous or are bleached under laser light excitation. So in addition to the pixel resolution, the amount of contrast (also addressed as visibility) present in the image determines the accuracy with which one may be able to identify and distinguish (small) features. In its most basic form, contrast $C$ is often defined in terms of the so-called Michelson contrast (MC), namely as

$$C = 2 \times \text{MC} = 2 \times [(I_{max} - I_{min})/(I_{max} + I_{min})], \qquad (17.8)$$

where $I_{max}$ and $I_{min}$ represent the highest and lowest luminance, respectively, in a periodically spaced pattern, e.g., that of an aforementioned "Siemens star." It is expressed normally in terms of a spatial frequency in units of periods·mm$^{-1}$. The contrast itself is fundamentally linked to the contrast transfer function (CTF), which constitutes a useful measure for characterizing the information (feature) transmission capability of any optical imaging system.

Although in most laser imaging applications one does not deal with periodic structures, in general, an image can be thought of to constitute a collection of "spacings" and orientations: the criterion of contrast and CTF is then applied as to which smallest feature can still be identified under random orientation. Note that the CTF always drops at higher spatial frequencies, i.e., small features that have low contrast become even less apparent as they pass through successive stages, from structures in the object to an image generated for the viewer. One can take two messages away from this discussion:

- Irrespective of the contrast provided by the optical process defining a feature in the object, in the final image, smaller features always appear with lesser contrast than larger features.

- Features that exhibit low intrinsic contrast in the object will show up with even lower contrast in the image.

In the context of spatial resolution and contrast, it is worth noting that Poisson noise also may play a nonnegligible role. If the feature signals from a structured object (obeying the spatial Rayleigh criterion) are composed of many photons and exhibit little noise, and if it is Nyquist-sampled, then everything is well for optimum image resolution. However, if the signal photon number drops significantly, the signal-to-noise ratio (SNR) drops as well by a large amount (scaling with the inverse square root of the total photon count). Then random variations in the signal can play havoc with the data, and single-pixel noise may mimic "artifact features," which are not present in the original object.

### 17.2.3 Sampling and its relation to spectral resolution

In the discussion so far, it was implicitly assumed, for simplicity, that only light intensity at a single wavelength or the total, integral light intensity needed to be considered in the formalism. But in many imaging applications, this is not sufficient, and the complexity of multicolor imaging was already hinted at in Equation 17.5b and in Table 17.1. In modern photography and printing, the common approach is that the complete color spectrum is realized through the so-called RGB mixing (red, green, blue), each of them—like in the grayscale intensity differentiation—with 256 individual brightness levels. This leads to the well-known "true-color" representation with about $17.8 \times 10^6$ individual spectral values.

While consumer photographic cameras include CCD sensors with RGB pixel-cluster structure, light detector arrays for scientific applications are normally only "monochrome," which records only the total light intensity falling onto an individual pixel but which allows for higher intrinsic spatial resolution. Therefore, spectral recording has to be realized via "color filtering" in the input or output channel (see Figure 17.1).

In the input (illumination) channel, this can be done either by tuning the laser light source to a specific wavelength or—if a broadband source is utilized (which may be, e.g., a common white-light source or a short-pulse continuum laser source)—a wavelength-selective filter needs to be inserted into the beam path. This scenario is commonly encountered in standard and LIF microscopy (see Chapter 18). In the output (light collection) channel, similar wavelength-selective (band-path) filters are utilized, or full spectrometers; the latter is usually required if spectrochemical information is desired as an outcome in the imaging process, like in Raman imaging (see Chapter 19). It also should be noted that spectral filtering in both channels may be required. For example, this is necessary in multichromophore tagging of biological specimen for selective LIF imaging. The tags one wishes to distinguish may fluoresce simultaneously under excitation but at different wavelengths; or one encounters fluorophore-excitation transfer for which one might want to discern between the donor and acceptor fluorescence, like in fluorescence/Förster resonance energy transfer (FRET) microscopy.

There are quite a number of filtering scenarios, which all require recording the full spatial image at each of the selected color or wavelength channels. It is immediately clear that every additional filter channel adds to the amount of digital storage of 2D image matrices with the associated light-intensity values (see the bottom of Figure 17.8 for a representative matrix). The most common scenarios are those collated in Table 17.1, which are repeated here in narrative form:

- At certain times, it may be sufficient to record the image at a *single wavelength*, if all that is required is, e.g., the total response of a chromophore to laser excitation at a given wavelength; the result is a *single* image matrix.

- For "truthful" color reproduction of an object in the image, one will need *RGB filtering*, which results in *three* image matrices.

- For many chromophore-tagging biological applications or, e.g., for component analysis in combustion, *multispectral filtering* is required; this results in commonly *5 to 10* image matrices.

- In Raman imaging of complex specimens, such as biological samples, hyperspectral imaging is necessary (the complete spectrum with small spectral increments is needed for full interpretation); this results in a hyperspectral data cube with sometimes up to a *thousand* image matrices.

The hyperspectral case (to which all the other scenarios may be seen as subsets) is shown schematically in Figure 17.11; note that for the sake of clarity, only a small number of individual-wavelength matrices are shown. For two pixels, their related hyperspectral data are shown at the top of the figure, conceptually demonstrating that for individual image pixels, one may encounter different intensities for certain wavelengths. In the wavelength-selective reconstruction of images, one "pulls" the relevant image matrix and "dresses" the spatial pixel location in the image with the associated light intensity information; the intensity scale is mostly represented in "false-color" scale, for clarity and ease of recognition. Note that for interpretation of the observed spectral images, ordinary photographic images are recorded for comparison.

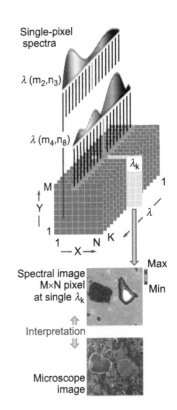

**Figure 17.11** Concept of hyperspectral cube generation. Representative spectra for two selected pixels are shown, revealing different spectral intensities. Then, for one selected wavelength, a pixel image can be extracted whose intensity distribution is normally represented in "false-color" coding. For comparison and feature interpretation, a 1:1-scale microscope image is often added.

## 17.2.4 Image reconstruction

In general, image reconstruction can be a rather straightforward process, at least if one were confronted with only a single-valued image matrix. Then one simply converts the pixel number indices into (scaled) image visualization coordinates and processes the intensity values into an appropriate grayscale or other scale representation. This process is shown schematically in Figure 17.12: the analog photon signal from the object is digitized and stored in a single-valued image matrix, and then retrieved from the said matrix for visualization on a computer screen or for printing. The "raw" result is a "pixilated" image, as shown in the pixel-image panel of the figure.

However, in general, one finds that both computer screens and printers exhibit a higher pixel density than that of the CCD camera detector used to record the image. Thus, commonly one applies image processing, e.g., by the so-called "edge-smoothing" procedure. For this, the pixel data values are interpolated with a suitable function and then resampled using a finer pixel mesh. Quite a few algorithms can be used for this; here only two of the most common ones are summarized:

- *Bicubic algorithms*: Bicubic interpolation is an extension of one-dimensional cubic interpolation, for interpolating data values from a two-dimensional regular pixel grid. The resulting, interpolated surface is much smoother than corresponding surfaces obtained by bilinear interpolation or nearest-neighbor interpolation (both may yield "ragged" edges). Bicubic interpolation is based on algorithms incorporating Lagrange polynomials, cubic splines, or cubic convolution.

- *Fourier transform methods*: Simple Fourier transform-based interpolation based on "padding" of the frequency domain data with zero components

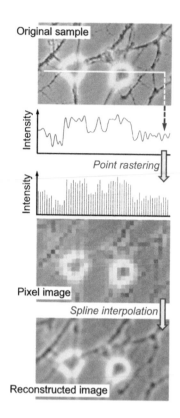

**Original sample**

Intensity

*Point rastering*

Intensity

**Pixel image**

*Spline interpolation*

**Reconstructed image**

**Figure 17.12** Image reconstruction of a monochrome recording of a cross section through neuron tissue. The point-rastering digitization step from the "original" image to the CCD pixel image is indicated schematically. In addition to the direct pixel image, post-acquisition spline interpolation to the data points has been applied, resulting in a smoother display resembling the original but with some "blurring."

preserves or recovers details well; but on the downside, one may encounter noticeable "ringing" and the circular "bleeding" of content from the left border to the right image border (for the prediction of the magnitude of these errors, see, e.g., Su et al. 2015).

An example for data interpolation using cubic spline interpolation is shown in the bottom panel of Figure 17.12; while clearly not all of the original features are reproduced because of reduced pixel resolution of the raw image, the recognition of certain features and their comparison to the original has become much clearer because of the smoothing of the ragged edges. The underlying mathematical framework for this and other image processing methods and algorithms may be found, e.g., in Cristobal et al. (2011) or Pratt (2014).

The reconstruction of images based on more than one image data matrix, as encountered in the RGB or multispectral cases, is evidently slightly more complex. Simplistically one might think that straightforward superposition of, say, the three matrix components from an RGB recording yields a "true" reproduction of object properties, as in photographs with color CCD cameras; however, this may not necessarily be correct for laser-spectroscopic imaging. For example, in LIF imaging, an RGB composition might not recover the actual spectral band component because of the problem of interpolation between the three widely spaced RGB components. However, in recent years, methodologies have been developed in which actual spectral properties could be recovered from few component color-filtered subsets, like a single RGB image. The concept behind this approach is one of "training and learning" by which one generates a mapping function between spectral responses and their corresponding RGB values for a given filter set (see Nguyen et al. 2014).

As the final and ultimate case, hyperspectral image reconstruction will be addressed. Intuitively, the reconstruction process should be much more complex than the "simple" RGB case just discussed. While all the spectral information is available for each pixel, it becomes sometimes problematic which components to choose for image presentation to highlight a specific feature and/or its properties. This is illustrated in Figure 17.13 for a skin sample in which damage was induced by ionizing radiation: biomedical knowledge predicts that it should become "visible" via oxygenation and perfusion.

While in the white-light microscope image (left column in the figure) the damaged area is just visible, when using oxyhemoglobin-selective (Hb-oxy) or deoxyhemoglobin-selective (Hb-deoxy) filter windows (right two columns in the figure), one can clearly identify the damaged areas and follow their evolution over time. But one has to note that the provided scale represents a 0% to 100% "false-color" scale for the individual components, which therefore does not reflect the relative concentration of the two. Only when presenting the image data in appropriate hyperspectral form does the real, biological process become visible. From such representations, the authors were able to extract actual oxygen-saturation data as well as relative and total hemoglobin concentrations from the hyperspectral images. It is evident from the dual-concentration intensity scale for the displayed hyperspectral image how difficult it is already to visualize a "true" (quantitative) spectral image for two-component system. Many of the issues related to hyperspectral image generation and reconstruction are discussed in depth in Chang (2013).

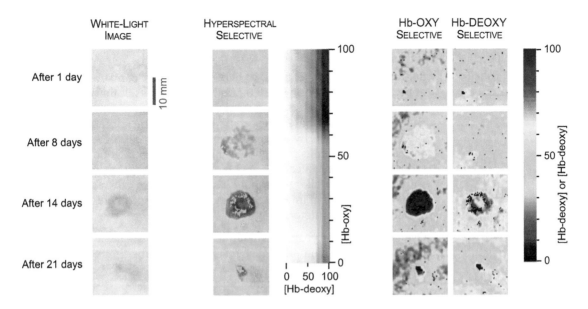

**Figure 17.13** Example of hyperspectral image reconstruction of a skin specimen, exhibiting acute reaction to ionizing radiation. From left to right: visible-light microscope image; hyperspectral selective image (with dual-axis spectral intensity scale); Hb-oxy and Hb-deoxy selective image representation (with false-color intensity scale). (Data adapted from Chin, M.S. et al., *J. Biomed. Opt.* 17, no. 2 (2012): article 026010.)

In order to alleviate some of the aforementioned complications, a novel visualization approach for hyperspectral imaging has been developed within the framework of the open-source "Gerbil" project (see Jordan et al. 2016). The underlying concept involves visualizing the spectral distribution of an image via parallel coordinates, but maintaining a strong link to traditional visualization techniques.

## 17.3 SUPERRESOLUTION IMAGING

In many of the imaging applications, spatial resolution is of utmost importance. In particular, this is true for laser-spectroscopic imaging modalities applied to biomedical problems, which in most cases are conducted with the aid of microscopes. Much of the discussion below will address the problems encountered in laser-spectral microscopy. On the other hand, many of the issues related to spatial resolution apply to any of the laser-imaging implementations covered in the following chapters, involving the laser light–matter interaction processes of absorption, emission, and scattering.

The most critical parameter in all (microscope) imaging applications is—as pointed out already a few times in this chapter—the diffraction limit: one may be able to *detect* an object, but one cannot *resolve* features below a certain threshold. In this context, detection means to determine whether a structure or substance is present in a sample or not; note that the capability of detection is a function of total brightness within the diffraction volume. Resolving means to determine the

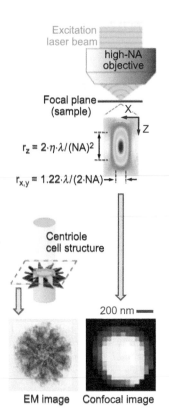

**Figure 17.14** Schematic of the lateral and vertical diffraction limits encountered in laser (spectroscopic) microscopy. The associated limitation in resolution in confocal laser imaging is shown for a centriole cell structure. (EM data adapted from Hagan, I.M. and R.E. Palazzo, *Eur. Mol. Biol. Org. Rep.* 7, no. 4 (2006): 364–371; confocal image data adapted from Lau, L. et al., *Biophys. J.* 102, no. 12 (2012): 2926–2935.)

number, or size, of objects in a sample, or their relative position with respect to each other; as defined earlier in Section 17.2, resolution is a function of Airy disk separation.

Subresolution object features, or light sources, are "convolved" by a microscope appearing as "diffraction-limited" Airy disks. The resolution can be increased by adapting the parameters in the Abbé limit definition, i.e., (1) one may utilize shorter wavelengths or (2) one can try to increase the numerical aperture (NA) of the imaging system. By and large, with standard microscopes, one encounters certain practical limits, namely that imaging optics may not be suitable for UV radiation, and that typically their $NA \geq 1.5$. This means that, in most cases, objects with dimension smaller than about 0.3–0.5 μm cannot be resolved. However, the relevant structures of interest in biological specimen very often are of dimension of only a few nanometers.

Over the past few years, tremendous, ultimately successful efforts have been made to develop methodologies to surpass the Abbé-limited resolution, associated with the diffraction point spread function (PSF). Two distinct approaches have been followed to break this limit:

- Structural superresolution "PSF engineering," like that encountered in, e.g., structured illumination microscopy (SIM) or stimulated emission depletion (STED)

- Single-molecule localization "PSF mapping," like that encountered, e.g., in stochastic optical reconstruction microscopy (STORM) or photoactivated light microscopy (PALM).

All these techniques have predominantly been developed for LIF applications; they will be discussed, together with application examples, particularly in Chapter 18. For further details and technique comparison, the reader may consult Milanfar (2011) or Diaspro and van Zandvoort (2017).

The general limitations in ordinary laser-spectral imaging are highlighted in Figure 17.14, including the diffraction-limited dimensions both in lateral (imaging plane) and depth dimensions. As an example, which will be followed through for all superresolution techniques addressed below, a LIF image of a typical centriole cell is shown in the bottom part of the figure. As stated previously, the confocal microscopy image (on the right) only confirms that "there is an object," but does not reveal the nanometer-scale details present in the electron-microscope image.

### 17.3.1 Sub-Abbé limit localization and "classical" superresolution strategies

From a "classical" point of view, i.e., imaging in general, very often the desire arose to "beat the Abbé or diffraction limit" in image applications, ranging from microscopy to standoff, air-borne, and satellite-based observations. Some ideas of how to tackle this problem—beyond the few, limited-workable approaches pointed out above—surfaced as early as the late 1970s with the emergence of viable scientific and consumer-oriented digital photoarray sensors and cameras.

Those ideas all involved the realization that while one could not really influence the Airy-disk (or PSF) dimensions of a particular imaging system itself, i.e., make them smaller, one could determine its actual center to a much higher precision than its diameter expansion. This meant that one should, in principle, be able to reconstruct the origin of a two- or multiple-center origin distribution, which is completely "washed out" in a diffraction-limited image if only one were able to image the individual contributors to the image independently. This idea is developed schematically in Figure 17.15, for a sensor with pixel dimension $20 \times 20\ \mu m^2$; in the particular example, the positions of two laser light spots with FWHM of 4 pixels, spaced by only 2 pixels (i.e., less than the Abbé limit), were to be determined.

It is clear from the left-hand part of the figure that a direct deconvolution of the two source centers from the overlapping image is next to impossible. However, when the two sources are "switched on" individually, e.g., by moving a single laser beam mechanically by a small amount from one position to the other, then their centers can be extracted with high accuracy via fitting the light distribution profile with a Gaussian function. Superposition of the two center positions then provides the necessary information to reconstruct the original dual-center scenario. Of course, the precision with which this process can be executed quite strongly depends on the strength of the photon signal (or the visibility) of the imaged pattern of a Gaussian-shaped laser spot. The larger the total number of detected photons is, the better becomes the fit of the Gaussian profile, and with it the determination of the maximum-intensity center. In the right-hand part of Figure 17.15, this is depicted for the cases of 200, 800, and 2400 photons, for which the uncertainty for the center position decreases from 40 nm via 20 nm to 10 nm, respectively (recall that for a Gaussian intensity function, $\delta_{pos} \propto d_{PSF} \cdot N^{-1/2}$; see also Chapter 18.3).

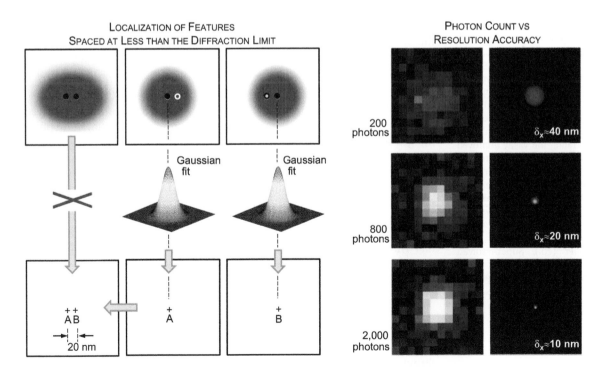

**Figure 17.15** Concept of image reconstruction from PSF images. (Left) Schematic of reconstruction for two overlapping PSFs from individual PSF centers. (Right) Demonstration of the increase in PSF precision with total photon number (with $d_{pixel}$ = 20 nm).

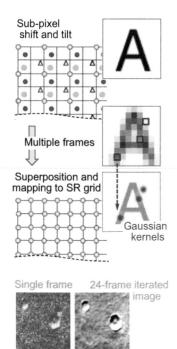

**Figure 17.16** (Top) Concept of superresolution (SR) image reconstruction from a set of subsampled, low-resolution (LR) images, with SR-grid = 1/2·LR-grid. The image positions are shifted/tilted within a pixel area; then their (Gaussian) kernels are localized for the HR-grid, and summed up. (Bottom) Example from the Viking Mars mission, demonstrating the vastly improved spatial resolution by reconstruction from a 24-image set of low-resolution images. (With kind permission from Springer Science+Business Media: *Fundamental theories of Physics, vol. 62, Maximum Entropy and Bayesian Methods*, G.R. Heidbreder (ed.), Super-resolved surface reconstruction from multiple images, 1996, pp. 293–308, Cheeseman, P., B. Kanefsky, R. Kraft, J. Stutz and R. Hanson.)

This concept of "stochastic" recognition of normally unresolved origins within a recorded low-resolution image constitutes the underlying strategy in the majority of "superresolution" imaging modalities. A flurry of activities to develop recording and image reconstruction algorithms occurred in the late 1980s and early 1990s, with the increased popularity and use of airborne and satellite-based observation. Of course, other than in the conceptual case just described, no "switch-on" of individual point of origin within an imaged object is possible in said applications. However, it was realized that the center of gravity for a point source could be determined by utilizing a series of low-resolution, undersampled images (for the meaning of undersampling, refer to Figure 17.4), which were shifted or rotated with respect to each other by subpixel amounts; for a description of the conceptual ideas, related algorithms, and examples, see, e.g., Cheeseman et al. (1996) or Hardie et al. (1998).

Basically, the objective of superresolution imaging was, and still is, to reconstruct a single higher-resolution image from a set of lower-resolution images, recorded with a detector array whose pixel density did not meet the Nyquist criterion during image acquisition. The set of images were acquired from the same object but under slightly different spatial imaging conditions, i.e., shifted and/or rotated with respect to each other. From the pixel-by-pixel PSF determination and mapping onto a denser (subdivided) pixel grid, a sharpened, higher-resolution image could be constructed. This is shown schematically in Figure 17.16, together with a space-exploration example from the Viking Mars mission (Cheeseman et al. 1996). In the said example, which involved the superposition reconstruction based on 24 individual low-resolution images, improvement in resolution by a factor of about ×4 was achieved.

One should note that, for most imaging systems, one encounters two main contributors to the overall system PSF. The first, and normally dominant factor, is the finite detector pixel size (this effect is spatially invariant as long as the detector array can be assumed to be uniform). The second contributor is the optical system itself, which in general contributes some spatial (or chromatic) aberrations to the final image.

As a final remark on superresolution, reconstruction from multiple low-resolution images is related to hyperspectral imaging. Often, the resolution in hyperspectral images is rather coarse because of some limitations in the imaging hardware. However, some signal postprocessing algorithms have been developed, e.g., based on maximum *a posteriori* (MAP) multiframe analysis, which allows one to improve on spatial resolution (see, e.g., Zhang et al. 2012). The approach relies on the fact that spectral composition of light from an object normally varies somewhat over certain spectral bands, mimicking spatially shifted images as discussed above. Thus, combining a selection of images associated with particular wavelength channels of narrow bandwidth, it becomes possible to achieve spatial "superresolution"; the authors demonstrated improvements by a factor of up to ×2.

### 17.3.2 Imaging and reconstruction strategies for structured illumination methods

Structured illumination is a wide-field microscopy technique in which a light-grid pattern is generated through interference of diffraction orders and super-imposed for illumination of the specimen while capturing the image. The interference grid (very frequently in the form of a "bar-code" stripe pattern) is generated by sending the excitation laser light through a transmission grating; of the various diffraction orders, in general, only the $m = \pm1$ orders are allowed to proceed (all others are blocked). The recombination of these two grating orders on target results in the aforementioned interference pattern. If one assumes, for the sake of argument, that the target object exhibits a stripe pattern as well (although this is extremely rare in real samples), and if the illuminating interference pattern is rotated with respect to the target pattern, an effect known as *Moiré fringes* is observed, a pattern akin to standard interference fringes with periodic dark and light zones; this is shown schematically in the top part of Figure 17.17.

The key property of this Moiré pattern is that the observed distance between its fringes $d_{\text{Moiré}}$ is larger than the fringe distance $d_{\text{pattern}}$ of either the target or the illumination structure; also, by adjusting the angle between the two stripe patterns, the Moiré-fringe distances can be altered. This is a quite valuable feature because—when applied properly—the Moiré fringes can be visible even if the object and illumination patterns are spaced by less than the diffraction limit and therefore would not be spatially resolved in an image on their own. Essentially, the reason for this is that the Moiré fringes have a lower spatial frequency than the original structures within the sample.

Since the Moiré pattern is related to the (unknown) object and (known) illumination patterns, one should be able to extract information about the object structure from an image with Moiré-pattern information. From a mathematical point of view, one encounters a transform operation in which the low spatial frequency content in the Moiré image is linked to the product out of the high spatial frequencies of the object and illumination patterns, which constitutes a "modulation." When applying Fourier transform to said modulation product, one finds

$$\mathcal{F}\{h\} = \mathcal{F}\{f \cdot g\} = \mathcal{F}\{f\} \otimes \mathcal{F}\{g\}; \qquad (17.9)$$

here $f$, $g$, and $h$ stand for the functional representations of the object, illumination, and Moiré patterns, respectively. Note that according to the convolution theorem of Fourier transform theory, the multiplication of two functions in the original-variable space is related to a convolution operation in the covariable space (see the right-hand part of Equation 17.9). In other words, by using Fourier transformations, the rather complex information contained in the spatial brightness variations in an image can be transformed into mathematically simpler oscillatory functions.

In this context, a microscopy image in real space might be seen as information with spatially varying light intensities; then in the Fourier-transformed image, the same information is expressed by the amplitudes and phases of a continuous set of frequencies. Deconvolving the image information in the Fourier space, one then can make the image information visible in real space again by applying inverse Fourier transform. A concise tutorial-type introduction into SIM can be found in Saxena et al. (2015).

Note that for full exploitation of this concept, normally a series of Moiré images is recorded, utilizing three independent angular rotations of the object and illumination patterns relative to each other, and three phase settings for the illumination (introduced by rotating the polarization direction of the excitation

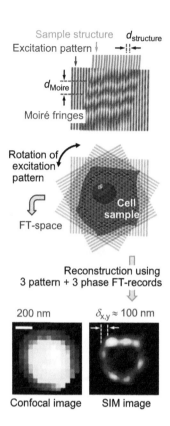

**Figure 17.17** Concept of SIM imaging. (Top) Moiré pattern generated from stripe-type patterns in the sample and the illumination. (Center) Implementation of SIM for reconstruction from frames with different angular and phase orientation. (Bottom) SIM image example for a centriole cell structure. (Data adapted from Lawo, S. et al., *Nat. Cell Biol.* 14, no. 11 (2012): 1148–1158.)

light). This procedure is summarized in the middle part of Figure 17.17. Nowadays, SIM is routinely applied in LIF microscopy; one practical example is shown in the bottom right of the figure, for the imaging of a centriole cell structure (see Lawo et al. 2012). In comparison to the standard confocal fluorescence image of a similar structure, an improvement in resolution of about ×2 to ×3 has been achieved, which is typical for SIM applications. Further details and examples for SIM applied to LIF imaging are described in Chapter 18.3. It is worth noting that while predominantly SIM has been the domain of LIF imaging, the method has been demonstrated to work also for other spectroscopic imaging techniques, such as CARS imaging (see Park et al. 2014).

Dedicated SIM instrumentation is now available commercially, including the necessary imaging control and analysis software. In the context of software, it is worth noting that some open-source plug-in packages have become available, with user-friendly graphical interface, which allow for the processing of two- and three-dimensional data acquired by SIM (see Křížek et al. 2016; Müller et al. 2016).

### 17.3.3 Imaging and reconstruction strategies for local-saturation methods

As summarized in the introduction, superresolution beyond the diffraction limit can be achieved by tailoring or engineering the PSF. While in the SIM approach just described, this is done indirectly by periodic illumination patterning, local saturation methods do this directly by reducing the ordinary PSF to a smaller-diameter "effective" PSF. Probably the most used modality of this is STED imaging; this LIF imaging technique was first proposed by Hell and Wichmann (1994). Further details and selected applications of STED, and related approaches, are discussed in Chapter 18.3; for additional concise summaries, the reader is referred to a recent review by Blom and Widengren (2017).

Briefly, STED is based on standard confocal laser-scanning microscopy in which fluorophores are excited for observation of an image. The tailoring of the PSF is realized by the use of pairs of synchronized, short (picosecond or subpicosecond) laser pulses. The (first) excitation laser pulse is focused onto the sample, resulting in an ordinary diffraction-limited fluorescence spot of excited molecules. This excitation laser pulse is immediately followed by a second (STED) laser pulse, which spatially overlaps the excitation laser spot and whose wavelength $\lambda_{STED}$ is red-shifted with respect to the excitation pulse and the fluorescence spectrum. The "trick" in this arrangement, which is used to affect the PSF, is that the STED laser pulse is toroidal in shape (doughnut profile). This is shown schematically in the top part of Figure 17.18.

The effect is the following. If the stimulated emission rate is high enough, the fluorophores in the (outer) overlapping region will emit (stimulated) photons of wavelength $\lambda_{STED}$ rather than (spontaneous) fluorescence photons of wavelength $\lambda_F$. The stimulated emission process effectively depletes the fluorophores in the overlapping region, leaving only the fluorophores in the (center) non-overlapping region with the ability to fluoresce at $\lambda_F$. Filtering the two fluorescence contributions appropriately leads to an apparent size reduction of the effective fluorescence spot of $PSF_{STEAD} < PSF_{LIF}$. Lateral resolution of typically 30–100 nm can be achieved, constituting a gain in resolution by about ×3 (see Chapter 18.3 for a slightly more elaborate mathematical description, and parameters for optimal generation of the effective PSF). Note that with the decrease in fluorescence spot

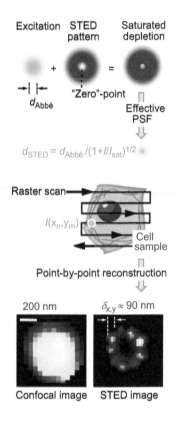

**Figure 17.18** Concept of STED imaging. (Top) Generation of the effective PSF with diameter $d_{STED}$, reduced in size with respect to the excitation PSF with diameter $d_{Abbé}$. The saturation intensity $I_{sat}$ is defined as the value for which the ON and OFF states of the fluorophore are equally populated. (Center) Schematic implementation of STED raster scanning. (Bottom) STED image example for a centriole cell structure. (Data adapted from Lau, L. et al., *Biophys. J.* 102, no. 12 (2012): 2926–2935.)

size, also the scanning process (shown schematically in the center part of the figure) becomes more time consuming, increasing on average roughly 10-fold. An example of a STED image, for a centriole cell structure, is shown in the bottom part of the figure; clearly, the observed lateral resolution is substantially improved with respect to the confocal LIF image.

In summary, STED microscopy allows for superresolution imaging in the range of around 50 nm. However, this increase in optical resolution comes at a price. Because the majority of initially excited fluorophores are quenched by the STED-depletion laser, the result is a much lower fluorescence signal. As a consequence of the Poisson photon statistics, therefore the SNR in the recorded image will also be much lower than in normal confocal imaging. This means that also the normally straightforward deconvolution procedure for converting the PSF-scan pattern data into a proper image is more involved. Modern commercial and open-source (see, e.g., Waithe et al. 2016) software incorporates efficient algorithms, which apply, e.g., time-correlated photon counting and theoretical PSF-construction methods to reduce noise and thus increase the image contrast for better spatial resolution.

### 17.3.4 Imaging and reconstruction strategies for single-molecule response methods

As just pointed out, the reduction in PSF diameter to achieve increased resolution comes with the penalty of greatly reduced signal strength; ultimately, even with sophisticated photon-counting methodologies, the SNR will become insufficient for signal identification. This means that structure information for biomolecular entities, which are often on a few nanometers scale, would still be inaccessible. Here, the very nature of chemical complexity of molecular fluorophores comes to the rescue. Said chemical complexity, in general, goes hand in hand with a complex temporal behavior. This then can be exploited to force closely spaced fluorophores to emit at separate times; therefore, their detection becomes resolvable in time, and associated with these events one can independently determine the localization center of their (stochastically distributed) PSFs. This approach in biomedical imaging is widely addressed as *single-molecule localization microscopy* (SMLM). By definition, it includes all microscopic techniques that achieve superresolution by the following two-step procedure: (1) isolating emitters and (2) fitting their images with the PSF. A multitude of SMLM methods have been developed since the first inception in 2006; basically, they mostly differ in the type of fluorophore used and the way the timing process is implemented. The most recognized approaches are *stochastic optical reconstruction microscopy* (STORM) and *photoactivated localization microscopy* (PALM). A concise review of the SMLM superresolution methodology can be found in Laine et al. (2016), albeit with a slight bias toward neuroscience in the application part. Some further details of STORM and PALM, together with a representative selection of examples, will be discussed in Chapter 18.4.

In brief, during imaging, at any given time only a small subset of fluorophores are activated into their excited, fluorescent energy state. The spacing between these individual fluorophores has to be sufficiently large so that their position can be pinpointed with high precision, by fitting the PSF position of the single-molecule images associated with an individual fluorophore. After imaging, the fluorophores are deactivated, and the next subset is activated and imaged. After complete iteration (in general, several thousand steps) over all available fluorophores, a

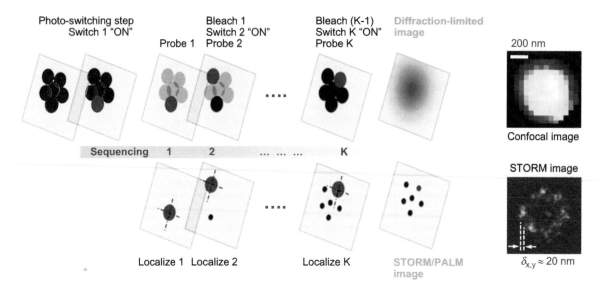

**Figure 17.19** Concept of SMLM, here exemplified for the method of STORM. After the first switch-ON procedure, a sequence of readout, switch-OFF, and switch-ON-again steps follows, until full depletion of fluorophores. (Right) STORM image example for a centriole cell structure. (Data adapted from Mennella, V. et al., *Nat. Cell Biol.* 14, no. 11 (2012): 1159–1168.)

superresolution image can be constructed from the full series of positional image data. The procedure is shown schematically in Figure 17.19.

In normal LIF spectroscopy, the width of the laser-spot PSF would limit the resolution (typically $d_{PSF} \sim 250$ nm). However, for an isolated fluorescence emitter within that PSF, one can precisely determine its location. The localization precision $\delta_{x,y}$ is determined by the number of photons $N$ collected during the multiple excitation–emission cycles of the fluorophore during laser irradiation, according to

$$\delta_{x,y} \cong d_{PSF} \cdot N^{-1/2}. \tag{17.10}$$

Lateral position accuracies of the order of $\delta_{x,y} \approx 20\text{--}30$ nm and axial resolutions of the order of $\delta_{x,y} \approx 50\text{--}60$ nm are routinely achieved. Note that the PSF-fitting process can only be carried out reliably for fully isolated emitters.

The (large) series of images of (sparsely distributed) fluorophores need to be evaluated for the PSFs of the individual light-emitting positions, and then globally have to be reconstructed into the actual image. For this, specific algorithms have been developed, which in particular address the precise localization issues; for some conceptual ideas, see, e.g., Rieger et al. (2015) or Hugelier et al. (2016). Typically, the algorithms attempt to fit a PSF model to the pixel intensities in the raw-data image; basically, for 2D $(x,y)$-localization microscopy, such models consist of a circularly symmetric Gaussian function:

$$PSF(x,y) = \left(1/2\pi \cdot d_{PSF}^2\right) \cdot \exp\left[-\frac{(x-x_c)^2 + (y-y_c)^2}{d_{PSF}^2}\right], \tag{17.11}$$

Note that the Gaussian PSF model is not derived from optical theory but is instead chosen for its conceptual simplicity and computational efficiency; this approximation is normally sufficient for accurate and precise localization. Based on the PSF-model locations, the expected intensities $I_k$ for the $k$th pixel, which carry light information from an individual fluorophore, are fitted to the data

$$I_k(x, y) = I_0 \cdot \int_{A_k} \text{PSF}(u, v) \mathrm{d}u \mathrm{d}v + I_b, \qquad (17.12)$$

where $I_0$ is the total light intensity collected from the fluorophore, and $I_b$ is the (expected) background photon-count intensity; the integration is over the full area $A_k$ of an individual pixel $k$. Note that an important issue in localization microscopy is the precision with which single fluorophores can be localized; because of the relatively low photon count number in individual pixels, in general, Poisson-noise models are adopted in the evaluation.

Because of the wide acceptance of SMLM in biomedical imaging, and its large variety of modalities, it is not surprising that many different approaches have been followed to come to grips with the elaborate image acquisition and reconstruction processes. Some of the oldest and still popular evaluation and reconstruction software packages are "QuickPALM" (Henriques et al. 2010)—a set of programs to aid in the acquisition and image analysis of STORM/PALM data; "GraspJ" (Brede and Lakadamyali 2012)—a set of real-time analysis and image rendering tools for STORM/PALM data; and "rapidSTORM" (Wolter et al. 2012)—a 2D and 3D, multicolor analysis and image generation suite for STORM data. With this wealth of potential analysis algorithms, the choice is often not an easy one; in this respect, the SMLM Software Benchmarking Initiative (2016) provides a comprehensive review of SMLM software packages, in which common reference data sets—simulating biological structures and image formation processes—are used for benchmarking and cross-referencing.

## 17.4  BREAKTHROUGHS AND THE CUTTING EDGE

Throughout the evolution of microscopy, researchers and developers have always thrived to improve on the spatial resolution of images. One key breakthrough came with the introduction of confocal (laser) microscopy, which eliminated some of the "blurring" effects of lacking depth resolution; however, lateral resolution stubbornly remained at the Abbé limit. A recent instrumental development has broken this barrier, exploiting a subpixel shift superresolution strategy without having to introduce actual mechanical movement.

### 17.4.1 Breakthrough: Airy-scan detection in confocal laser microscopy

The image in confocal (laser) microscopy is an Airy disk, generated by the confocal aperture; a smaller aperture means better resolution, but the caveat is reduced transmitted light intensity. Both light throughput and spatial resolution can be improved by using a novel detector geometry, the so-called *Airy-scan* detector. It consists of a multielement (typically 32 elements) device recording the confocal Airy disk (its size is defined in Airy units, AU). Each detector element of size 0.2 AU acts like a pinhole, with a full detector diameter for light capture of 1.25 AU. This means that this detector array will produce 32 images with different, small displacements. As the sample is scanned, so is the Airy disk of the PSF. Then, mathematically one shifts all individual pixel images, with well-known

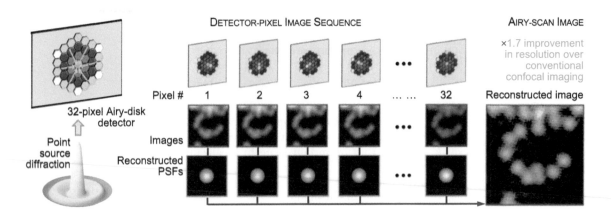

**Figure 17.20** Concept of Airy-scanning detection in confocal laser microscopy. The light data recorded by the individual pixel elements of a detector, covering the confocal Airy disk, are used to reconstruct an image with improved resolution of ×1.7. (Image data adapted from Woese, C.R., "Confocal, Airy-scan and structured illumination super-resolution microscopy," *Lunch with the Core* series. Urbana, IL, USA: Inst. Genomic Biology, University of Illinois, 2015.)

displacement, back to the center position. As in the subpixel-shift approach for superresolution (discussed in Section 17.3), the composite image exhibits improved resolution in comparison to a classical confocal imaging. Together with image deconvolution algorithms, the spatial resolution can be improved by as much as a factor ×1.7. The overall concept of Airy scanning and the framework for image reconstruction are discussed in detail in Weisshart (2014), Huff (2015), and Huff et al. (2015). The process of Airy-scan imaging is shown schematically in Figure 17.20.

### 17.4.2 At the cutting edge: Single-pixel detector multispectral imaging

Multi- and hyperspectral imaging techniques normally require a compromise among spatial, spectral, and temporal resolutions. Traditionally, in multispectral image acquisition, one segregates either spatial (pixel-by-pixel) or spectral (band-path filtering) information, and records the image information by sequential (temporal) scanning. Thus, the full spatiospectral data collection for an image is time-consuming, and huge amounts of data need to be transferred and stored.

Over the years, some efforts have been made to simplify and speed up the recording of multi-/hyperspectral images (see, e.g., Hagen et al. 2012; Cao et al. 2016), and some commercial multispectral snapshot cameras have now become available, aiming at applications as diverse as aerial surveillance mapping or biomedical imaging. All these concepts and devices are based on array CCD sensors.

Very recently, a different technical approach has been proposed and demonstrated for multispectral imaging in which a single-element ("single-pixel") detector is utilized rather than a multipixel array device (see Li et al. 2017). The authors implemented their multispectral imaging based on a single-pixel "compressive-sensing" architecture, in which an all-optical multiplexing technique was utilized. Specifically, a spatial light modulator (SLM) was utilized to split spatially resolved light into structured pattern, whose resolution was the same as that of the later, reconstructed image. This no-moving-part modulation method guarantees stable and fast operation on the one hand, and the spatial multiplexing ensures efficient image acquisition on the other hand. The concept is shown schematically in Figure 17.21.

With a proof-of-principle setup, the authors validated their idea for multispectral imaging. For this, three Wollaston prisms were oriented with their splitting directions perpendicular to each other; such a layout splits light alternately along vertical and horizontal directions. In conjunction with these prisms, quartz phase retarders were introduced into the beam path. Their thicknesses were chosen to assure the required, sequential spectral polarization, so that eight spatially separated spectral band passes were generated (see the optical chain in the top part of the figure). The band passes were (approximately) equally spaced within the wavelength range $\lambda = 420$–$720$ nm. For multiplexing of the spectrospatial pattern (from selected, fixed target areas of nominal $128 \times 128$ pixels), a common, commercial SLM unit with $1920 \times 1080$ pixels was used to direct the light sequentially to high-speed response photodiode.

In their study, the authors used both macro- and micro-imaging system implementations, thus demonstrating the universality of their approach for efficient multispectral sampling, e.g., for the imaging of macroscopic targets (exemplified for a color-checker sheet sample), or in biomedical microscopy (exemplified for an assembly of multicolor microbeads).

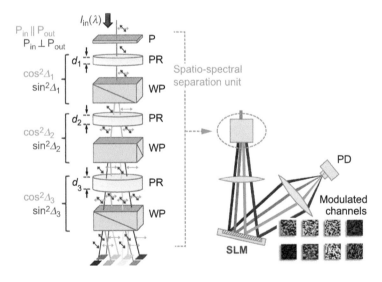

**Figure 17.21** Implementation of multispectral imaging, based on spatiospectral separation of the polarized (P) light from the object by a stack of phase retarder (PR)–Wollaston prism (WP) pairs, and multiplexing of this image information onto a single-element photon detector (PD) by a SLM. (Adapted from Li, Z. et al., *Sci. Rep.* 7 (2017): article 41435.)

# CHAPTER 18

# Laser-Induced Fluorescence Imaging

Fluorescence imaging is the visualization of fluorescence radiation emitted at different wavelengths by molecules present in a given sample that has been illuminated by incident photons. Thus, the color image of the fluorescence radiation obtained with a sufficient resolution provides a map of fluorophore distribution within the investigated object. There are two main criteria for classifying fluorescence imaging methods: the first one is the *size of the object* to be imaged and the second one is based on the *contrast mechanism* employed to obtain the image.

With respect to the first criterion, the fluorescence-detection methodology can be further subclassified into *microscopic* or *macroscopic* imaging. Optical microscopy is the most widely used methodology in biochemical research and clinical applications, but is normally restricted to thin tissue samples; in general, the spatial resolution is diffraction limited and depends on the used wavelength. For larger or deeper tissue structures, the macroscopic imaging approach is utilized; now the resolution is limited by the depth and optical properties of the object.

In relation to the second criterion, fluorescence imaging can again be subdivided, depending on whether an *endogenous* or *exogenous contrast* mechanism is used. Note that these two approaches refer to the use of *natural* or *injected* fluorochromes, respectively, to obtain the image. Two of the most common natural compounds employed in biomedical tissue imaging and spectroscopy are *oxy-* and *deoxyhemoglobin*, whose distinct absorption coefficients in specific spectral regions are utilized for tissue functional imaging. With respect to the exogenous contrast methodology, the most common technique in life sciences is tissue staining. In particular, fluorescence probes are utilized—such as dies, proteins, or quantum dots—which, in addition, contain an active component capable of interacting with the target. Recently the discovery of the *green fluorescent protein* (GFP) and its different color protein variants (see Section 18.2) has significantly expanded the fluorescence imaging field by the development of so-called indirect fluorescence imaging methods that are based on genetic manipulation to induce fluorescence contrast, i.e., the introduction of a gene into a cell that encodes it for a fluorescent protein (FP).

# 18.1 TWO- AND THREE-DIMENSIONAL PLANAR LASER-INDUCED FLUORESCENCE IMAGING

Macroscopic imaging methods can be divided into planar and tomographic imaging techniques. In principle, laser-induced fluorescence (LIF) can be utilized to measure scalar concentration at a point (zero-dimensional), along a line (one-dimensional), on two-dimensional (2D) planes, and in three-dimensional (3D) volumes. The last two approaches correspond to the classes of so-called *planar laser-induced fluorescence* (or PLIF) and *fluorescence tomography*, respectively, constituting the most relevant macroscopic fluorescence imaging techniques; only these will be addressed in this section. For their implementation and application, selected examples are discussed both for gaseous samples and for dense media, specifically tissues.

## 18.1.1 PLIF imaging in gaseous samples

### Basic theory and experimental setup

The LIF signal $I_{LIF}$—under weak excitation conditions, i.e., without any saturation, and recorded by an individual pixel of the image detector—is given by Blomberg et al. (2016):

$$I_{LIF} = \eta \cdot g \cdot S_{if} \cdot \phi \cdot N_{abs} \cdot I_L, \tag{18.1}$$

where

- $I_{LIF}$ is the LIF intensity (often in units of photons per pixel)

- $I_L$ is the laser intensity reaching the pixel-equivalent volume $V_{pe}$, which is imaged onto the related detector pixel ($V_{pe} = a \cdot d_{LS}$, with $a$ the area imaged onto an individual pixel and the laser sheet thickness $d_{LS}$)

- $\eta$ is the instrumental collection efficiency (incorporating optics transmission and detector camera sensitivity)

- $g$ is the overlap function between the laser and absorption line profiles

- $S_{if}$ is the absorption line strength taking into account the Boltzmann fraction $f$ ($T$) of the absorbing species in the lower state of the transition

- $N_{abs} = (V_{pe} \cdot P_{abs}/k_B \cdot T)$ is the number of detected species molecules, with the related partial pressure $P_{abs}$

- $\phi$ is the fluorescence quantum yield

It should be noted that except for $\phi$, all other parameters in Equation 18.1 can be estimated from laser, optics, and camera properties and from spectroscopic databases such as HITRAN (see, e.g., Chapter 7, for further details); fluorescence quantum yield is further described in Chapter 9.3, including the associated basic theory and the experimental method for its determination.

A typical experimental setup for PLIF measurements is illustrated in the left part of Figure 18.1; on the right, both side and top views are shown on how laser light sheets are generated, using a beam expander and a cylindrical lens. The key optical feature of the technique is the incorporation of a cylindrical lens to obtain the planar (2D) illumination focus in only one direction; telescopic laser beam expansion or compression is added, which determines the width of the so-called

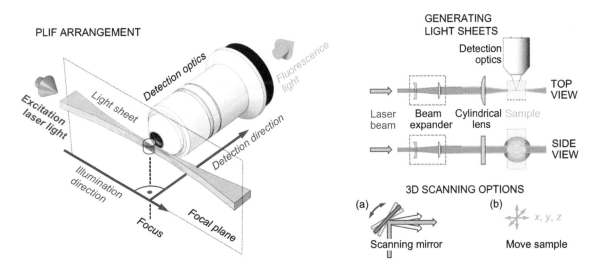

**Figure 18.1**    Basic scheme of a PLIF setup. On the right, laser light-sheet generation based on a beam expander–cylindrical lens combination. For 3D PLIF, either (a) the light sheet is moved using a scanning mirror or (b) the sample position is adjusted to bring the desired region into the focal volume.

*laser light sheet* (see the right-hand part of Figure 18.1). The fluorescence emission from the laser light sheet interaction with the sample gas is normally filtered, to suppress residual laser radiation, and imaged with the aid of appropriate optics onto the detector chip of a charge-coupled device (CCD) pixel camera.

The main advantages of PLIF are the following:

- It is species and quantum-state specific, and, consequently, it is sensitive to species, temperature, composition, number density, and velocity.

- It provides instantaneous information across the full area of the imaged laser light sheet.

- LIF signals are much higher than those of other laser-based spectroscopic (imaging) techniques, such as Rayleigh and Raman scattering.

PLIF has been and still is widely used in fluid dynamics research, since it offers the capability to gain deeper understanding of the underlying physics and chemistry in—for example—gas flow and combustion processes (see, e.g., Orain et al. 2009). In general, the technique requires the use of "molecular tracers," i.e., molecules that fluoresce, to achieve clear diagnostics of the gas flow regime under investigation, such as, for example, flow generation, mixing, or combustion. In those application fields, two types of fluorescent tracers are most widely used, namely, (1) flame-front tracers and (2) fuel tracers.

The flame front is the limiting region of the flame in which most of the chemical reactions responsible for the combustion process take place, and, therefore, it is this region where most of the energy is released. Hence, experimental data on its properties and structure are crucial for testing combustion models.

One of the most important radical intermediates in combustion is the OH radical; thus, it is not surprising that it is frequently used to investigate flame-front location via PLIF. To this end, an electronic transition of the OH is excited by ultraviolet (UV) photons at a wavelength of approximately $\lambda_L \approx 283$ nm obtained by doubling a dye laser operating at around 566 nm. Subsequent to excitation, the OH fluorescence is collected near $\lambda_{fl} \approx 310$ nm. Although the OH radical is the most frequently used flame-front indicator, there are other species utilized as well, such as CH, CO, or CHO.

When investigating combustion processes within the flue, one normally uses a single-component reference with respect to fuel visualization and concentration distribution. In this context, the most common fuel tracers are aliphatic molecules such as, for example, acetone or kerosene (Schulz and Sick 2005).

An important limitation of PLIF systems is that the related 2D fluorescence images reflect events only within the plane of the laser sheet and therefore may not necessarily be representative of the whole 3D combustion region. This limitation can be overcome by using an electro-optic or acousto-optic device that moves a reflecting mirror to sweep the laser sheet with high speed (see, e.g., Cho et al. 2014); this approach is shown schematically in the 3D scanning option (a) in Figure 18.1. The corresponding sequence of 2D images taken for each focal plane position and rotating mirror orientation allows one to build complete 3D fluorescence images (see the example discussed in Section 18.1.2).

## 18.1.2 Selected examples for PLIF of gaseous samples

### *OH imaging in a turbulent nonpremixed flame*

As already stated, OH is one of the most relevant intermediates occurring in combustion chemical reactions and, consequently, PLIF of OH has become an important tool for investigating flame structures. Kaminski et al. (1999) reported on the first measurement of time-sequence PLIF OH images in a turbulent nonpremixed methane–air flame. Representative sequences are illustrated in Figure 18.2, where different phenomena are revealed that are characteristic of

**Figure 18.2** Temporal PLIF sequences of OH, recorded at different height regions I–IV in a turbulent, nonpremixed CH₄-air flame; the time sequence is in steps of $\Delta t = 125$ μs. Each PLIF image represents the response to a single laser pulse. A direct-emission photograph of the flame is included to the left. (With kind permission from Springer Science +Business Media: *Appl. Phys. B* 68, no. 4, 1999, 757–760, Kaminski, C.F. et al.)

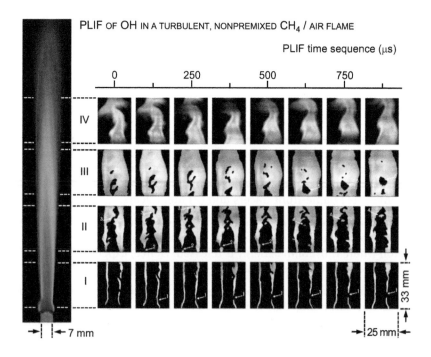

PLIF OF OH IN A TURBULENT, NONPREMIXED CH₄ / AIR FLAME

PLIF time sequence (μs)

turbulent flames—such as extinction, entrainment, or flame detachment. Time-sequence PLIF imaging allows one to follow their evolution in real time.

With reference to the figure, in region I, close to the nozzle exit, the flux is still laminar, as one can see from the typical "braids" of OH; this signals zones in which fuel and oxidizer mix, i.e., the reaction zone. Then, in region II, one can clearly observe the formation of vortices and their subsequent progression, which provides information about the development of the flame structure. The very broad distribution of OH in region III can be interpreted to be a consequence of the long lifetime of the species, basically indicating the presence of burnt gases. Thereafter, in the evolution of the flame, the absence of change in the overall features in the top section (IV) (over the time sequence) may be seen as indicative that all main instabilities have been dissipated and only residual OH in the burnt gas undergoes diffusive mixing with the surrounding air.

### Kerosene combustion in multipoint injectors

Radial instantaneous-fluorescence images of kerosene and OH radicals, at the outlet of a multipoint combustion injector, are shown in Figure 18.3; this study was undertaken to test aircraft and helicopter injectors under real operating conditions (see Orain et al. 2009).

Radial instantaneous image
of kerosene

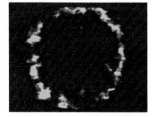

In the experiments, the injection configuration comprised (1) a single pilot injector that was utilized (located on the combustion axis) to stabilize the flame and (2) a multipoint injector located at a larger radial distance. The gas flow $F_G$ could be split into different ratios between the two injectors; for the data displayed in Figure 18.3, the equivalence ratio was set, i.e., $F_{pilot}/F_{main} = 1$. Both kerosene and OH fluorescence images were taken with the pilot and the main injector operating at a pressure of $P = 0.95$ MPa and an air inlet temperature of $T = 590$ K. Two laser beams were shaped into two superimposed, collimated sheets by a unique set of cylindrical and spherical lenses; this dual-wavelength approach allowed the researchers to simultaneously perform PLIF measurements on OH and kerosene vapor components. The OH fluorescence is produced by exciting the rotational $Q_1(J = 5)$ line of the $(1,0)$ vibrational band of the electronic transition OH $(X^2\Pi – A^2\Sigma^+)$. The kerosene fluorescence is performed by single-photon excitation at $\lambda_L = 266$ nm (fourth harmonic from a Nd-doped yttrium aluminum garnet laser). A delay time of $\Delta t = 200$ ns (much shorter than the timescale of the flow conditions in the experiment) was set between the two laser pulses, to avoid "cross talk" between the fluorescence signals from the OH radical and the kerosene.

Radial instantaneous image
of OH radical

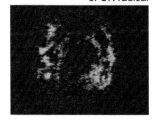

**Figure 18.3** Radial instantaneous-fluorescence images of kerosene top and OH radicals (bottom) at the outlet of a multipoint injector; operating conditions are noted in the main text. The size of the imaged "ring" is about 50 mm in diameter. (Adapted from Orain, M. et al., *C. R. Mécanique* 337, no. 6–7 (2009): 373–384.)

Looking at the flame-front image, one can notice a sort of a double structure: one near the combustor axis that originates from the single pilot injector and one (peripheral) flame-front that can be attributed to the main multipoint injector. A closer look at the two images reveals that kerosene is concentrated at larger axial distances in comparison with the OH radical. This feature is interpreted as a clear indication that combustion takes place toward the inner side of the kerosene cone. Furthermore, the axial symmetry of both kerosene and OH radicals demonstrates the excellent performance of the multipoint injector, mirrored by the apparent homogeneous distribution of kerosene inside the combustor.

### Gelled fuel droplet combustion

Gelled propellants are currently of great interest in combustion/propulsion, due to their potential safety and benefits compared with solid propellants. This is

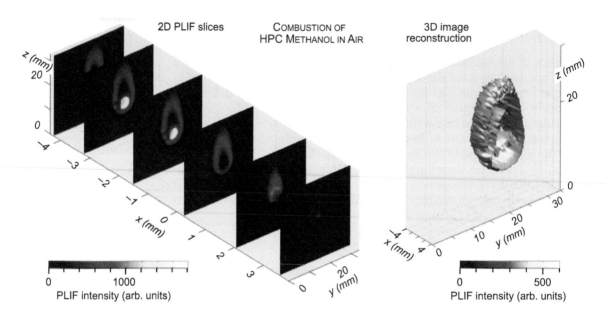

**Figure 18.4** Steady combustion flame of hydroxypropyl cellulose (HPC) methanol gel (3 wt.%) in air (at atmospheric pressure). (Left) Six positional slices through the flame. (Right) Isosurface construction for a PLIF intensity surface value of 330 (arbitrary units). (Data adapted from Cho, K.Y. et al., *Appl. Opt.* 53, no. 3 (2014): 316–326.)

because gels behave like a solid under low stress and like a liquid under high stress. For this reason, they are receiving increased attention in combustion studies; in such studies, PLIF techniques are specifically invaluable in providing contact-free 3D spatial and temporal information about the combustion process.

An interesting example of such a 3D PLIF study is shown in Figure 18.4; it stems from a combustion experiment in which methanol had been gelled into fuel droplets with HPC, at 3 wt.% (see Cho et al. 2014). In the figure, the OH radical distribution, produced during combustion, is imaged in three dimensions by sweeping a 2D laser light sheet through the combustion region.

In the left-hand part of the figure, a subset of six image slices (from a single sheet sweep), at different locations, is shown. Here, the *x*-axis coincides with the direction of the laser sheet movement; note that because of constraints due to sheet forming/scanning optics, the spatial extent covered in the scan direction is much less than the width (*y*-direction) and height (*z*-direction) of the image. From the full set of scanned PLIF sheet images, one generates 3D surfaces (by using, e.g., the isosurface function in MATLAB™; this connects all "pixels" in a data set exhibiting the same pixel value). In general, for this process, one selects signal amplitude values that best describe the flame. In the 3D image construct of the OH fluorescence shown on the right in Figure 18.4 one can easily identify inner and outer layers from droplet combustion. From such 3D data, the flame standoff distance and flame thickness can be calculated.

Studies such as the one described here are of great importance in understanding the combustion behavior of gelled propellants and specifically in elucidating the choice of gellant. This is because, due to the difference in evaporation temperatures of the fuel and gellant, a layer of less volatile components (i.e., the unburned gellant) can remain on the outside after the fuel is vaporized and burned. The nature of the outer shell is different for different gellant types, which, as a consequence, may result in drastically different microexplosion behaviors.

Evidently, 3D PLIF has an advantage over 2D PLIF when viewing steady flames (droplet combustion) because a static laser sheet probes an arbitrary location (not necessarily coinciding with the center) of the droplet, and, therefore, one may miss potential symmetries or asymmetries in the combustion process.

## PLIF imaging in catalysis

Another interesting and current PLIF application is found in research addressing catalysis. In recent years, a significant number of experiments have been carried out related to conditions encountered in industrial scale catalysis (see, e.g., the topical review by Blomberg et al. 2016). These studies have provided new information on the dynamics at and close to the catalyst surface while the actual reactions take place, i.e., in real time. Because the catalyst structure during the chemical reaction is sensitive to the surrounding gas phase, the knowledge of the gas phase composition and spatial distribution in the vicinity of the catalyst seems to be crucial for a complete understanding of the catalytic process. The application of PLIF to the investigation of model catalysis has several advantages with respect to other more conventional techniques, such as mass spectrometry (MS), near-ambient X-ray photoemission spectroscopy, and surface X-ray diffraction, to name but a few. The most notable among those advantages are (1) the capability to investigate the actual gas composition in the vicinity of the catalyst surface; (2) the spatial resolution of PLIF, which provides an opportunity to compare distinct catalysts within the same reactant environment; and (3) its high spatial information content and its excellent time resolution (on the order of milliseconds or even microseconds).

One well-studied catalytic reaction, utilizing PLIF, is the production of $CO_2$ when mixing CO and $O_2$ over a Pd(100) single crystal (Blomberg et al. 2016); Pd(100) single crystals are used because they mimic well the surfaces of the catalyst nanoparticle found in commercial devices. In related studies, the $CO_2$ molecule was excited using laser radiation at $\lambda_L = 2.7$ μm laser, on the ro-vibrational transition of the band $(00^00) \rightarrow (10^01)$; subsequently, the $CO_2$ molecule was visualized using the fluorescence emission from the band $(10^01) \rightarrow (10^00)$, at $\lambda_{PLIF} = 4.3$ μm.

At the beginning of the cited experiments, the crystal temperature was below $T_{cryst} = 100°C$, and no carbon dioxide was produced. On increasing the temperature gradually, the $CO_2$ signal followed suit up to about $T_{cryst} = 270°C$, but increasing only very moderately. Beyond this point, the $CO_2$ signal became significantly larger, indicating that a sort of threshold is surpassed. The extent of the $CO_2$ cloud above the catalyst surface is shown in the vertical-slice PLIF images in Figure 18.5, for the two temperatures $T_{cryst} = 270°C$ and $T_{cryst} = 290°C$, close to and well above the threshold for the catalytic reaction. Clearly, for the latter temperature, the $CO_2$ cloud, often referred to as the $CO_2$ boundary layer, is much more extended. Both image slices demonstrate that the carbon dioxide concentration inside the sphere, limited by the boundary layer, is significantly higher than that in regions of the chamber farther away from the catalyst.

Note that gas partial pressures in catalytic test chambers are normally monitored (e.g., by MS techniques) relatively far away from the catalysis location, which thus would not mirror the actual species partial pressure profile; this confirms

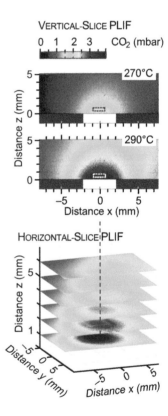

**Figure 18.5** CO oxidation using a Pd(100) surface as a catalyst. (Top) Vertical-slice PLIF images of the $CO_2$ distribution above the surface, at temperatures $T_{cryst} = 270°C$ and $T_{cryst} = 290°C$. (Bottom) Sequence of horizontal-slice PLIF images of the catalysis volume, at $T_{cryst} = 270°C$. (Data adapted from Blomberg, S. et al., *Phys. Condens. Matter* 28, no. 45 (2016): article 453002.)

the importance of localized PLIF measurements. Note also that with the addition of a series of horizontal-slice PLIF images of $CO_2$, its 3D concentration distribution can be accurately mapped.

### 18.1.3 PLIF imaging of biological tissues

PLIF is widely applied in recording fluorescence from tissues, with distinct experimental configurations, depending on the contrast mechanism and illumination geometry. The two most common geometries are referred to as (1) *epi-illumination (reflectance) imaging* and (2) *transillumination imaging* (see also Section 18.2.1).

In the reflectance configuration, the illumination source and imaging detector (normally a CCD camera) are located on the same side of the object and, typically, the technique is known as *fluorescence reflectance imaging* (FRI). This configuration is suitable when the sample fluorochromes are located near the surface, since otherwise the fluorescence radiation may become significantly attenuated. Thus, FRI exhibits a few important limitations: (1) The technique permits only small penetration depths (of a few millimeters); the signal from deeper lesions is substantially attenuated and suffers from background noise—mainly fluorescence—produced by superficial layers. (2) It cannot quantify the concentration of the fluorochromes since, for example, different fluorescent lesions associated with distinct tissue penetrations can result in the same appearance on the surface. Note that these limitations are resolved by the fluorescence tomography method discussed in the following section.

For the case of substantial fluorescence attenuation within the tissue sample, the transillumination geometry is preferred; here the light source and detector are located on opposite sides of the sample to be imaged. The technique is known as *fluorescence transillumination imaging* (FTI). With respect to FRI, it has some advantages, since the sum of the light excitation and fluorescence emission paths is the same, regardless of the specific depth location of the fluorochromes.

Regardless of the actual imaging geometry, FRI or FTI, in general, one encounters a common cause of imaging quality deterioration in PLIF, namely, the presence of tissue inhomogeneities; these produce absorption differences and, as a consequence, distinct fluorescence attenuation, which can severely affect the quality of the image. To circumvent this limitation, Ntziachristos et al. (2005) developed a normalization method that consists of normalizing the fluorescence image to an image recorded at the laser excitation wavelength.

An example of FRI is shown in the upper part of Figure 18.6. Normally, FRI images are accompanied by another image without fluorescence, i.e., just a photograph of the tissue sample, taken at the excitation wavelength for calibration, to record the spatial information about the object. The mouse shown here has developed a tumor in the upper posterior thorax.

The sequence shown in the figure begins with an intrinsic high-resolution image of the animal surface structure, taken at the excitation wavelength (for the results shown here $\lambda_L = 672$ nm). The second image is an FRI image of the same nude mouse, with a developed cathepsin B-rich HT1080 fibrosarcoma in the posterior upper thorax (the fluorescence is filter for the wavelength $\lambda_{FRI} = 705$ nm). In the final image, the superposition of the two images is depicted; this yields an improved visual effect while retaining the high resolution.

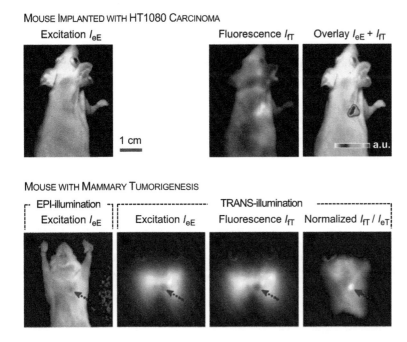

Mouse Implanted with HT1080 Carcinoma

Excitation $I_{eE}$          Fluorescence $I_{fT}$     Overlay $I_{eE} + I_{fT}$

1 cm

a.u.

Mouse with Mammary Tumorigenesis

EPI-illumination        TRANS-illumination

Excitation $I_{eE}$     Excitation $I_{eE}$     Fluorescence $I_{fT}$     Normalized $I_{fT} / I_{eT}$

**Figure 18.6**  FRI and FTI of mice, after tumor development. (Top row) FRI measurement; from left to right: intrinsic light image obtained without the fluorescent band-pass filter; FRI image; overlay image of the two. (Adapted from Ntziachristos, V. et al., *Eur. Radiol.* 13, no. 1: 195–208 2003.) (Bottom row) FTI measurement; from left to right: epi-illumination image; transillumination images at the excitation and fluorescence wavelengths; normalized FTI image. (Adapted from Ntziachristos, V., *Annu. Rev. Biomed. Eng.* 8 (2006): 1–33.)

Optical transillumination, also known as *diaphanography* or *shadowgram optical imaging* was first employed by Cutler, who investigated pathology issues of the human breast (see Cohn 1929). The technique was initially superseded by the higher resolution and robustness of the X-ray technology. However, this development is being reversed today with the evolution of planar laser imaging, in which both the relative attenuation of the excitation light (shadowgrams) and the emitted fluorescence are recorded. This is mainly due to the fact that one now can achieve molecular tissue specificity.

As mentioned earlier, when transillumination FTI measurement results at the emission wavelength are divided by geometrically identical transillumination data at the excitation wavelength, image quantification and contrast can be improved, as shown, e.g., by Ntziachristos et al. (2005).

In the lower part of Figure 18.6, such an example of transillumination and normalization images is shown, associated with the superficial and deep-seated fluorescence activity from a whole transgenic mouse that exhibits multifocal spontaneous mammary tumorigenesis (to record the FTI images, it was injected with a cathepsin-sensitive probe). Clearly one can see how the aforementioned correction operation improves the contrast and suppresses nonspecific signals.

Finally, it is worth noting that rather recently PLIF imaging for applications in biomedicine started moving toward miniaturization, particularly under the umbrella of "lab-on-a-chip" initiatives. In this context, Takehara et al. (2017) merged contact-fluorescence microscopic imaging with microfluidic chip technology, applying it successfully in pilot studies for fluorescence detection and imaging of cellular activity. Their layered devices consisted of a lithography-tailored microfluidic structure, on an ultrathin glass substrate, which was directly sandwiched to a complementary metal–oxide–semiconductor array detector (120 × 268 pixels of size 7.5 × 7.5 $\mu m^2$) via a fiber-optic plate. The performance of these devices was proven in spatial resolution tests with fluorescent microspheres of diameter $d_{sphere} \approx 10$ μm, which yielded images of said spheres with a full width at half maximum (FWHM) of $d_{fluor} = 19.0 \pm 0.7$ μm. To actually demonstrate the suitability of their device for the evaluation of cellular activity changes, induced by extracellular agents, the authors cultured HeLa cells inside the microchannels of the device, for comparison with or without extracellular signal-regulated kinase. These samples were prepared with FRET fluorescence probes; only cells cultured with endothelial growth factor (EGF) exhibited an increase in fluorescence in response to laser sheet irradiation, revealing well-resolved individual HeLa cells.

## 18.2 FLUORESCENCE MOLECULAR TOMOGRAPHY

Fluorescence tomography is a type of optical tomography in which the fluorescence emission from the fluorochromes present in the sample is collected using multi-illumination configurations (e.g., light slicing) to obtain 3D images of optical contrast.

### 18.2.1 Basic concepts

The basic concept of the tomographic technique resembles that of X-ray computer tomography since the tissue is illuminated under different projections and the 3D images are reconstructed from the ensemble of projections. The collected light, and its spatial distribution, is treated with mathematical models describing the photon propagation in tissues. One of the most important differences, however, between these two methodologies is that in optical tomography, the near-infrared (NIR) or visible light photons are highly scattered inside the tissues, other than X-rays. In fact, in the presence of the highly scattering medium, the detected photon field $\Phi$ exhibits a nonlinear dependence that can be described, to a first approximation, by

$$\Phi = \frac{\exp(-ikr)}{r},\tag{18.2}$$

where $r$ is the distance between the source and detector and $k = |\boldsymbol{k}|$ is the photon propagation wavenumber that depends on the optical and diffusing-tissue properties, as well as on the medium-dependent speed of light. In general, fluorescence tomography uses the theoretical framework employed in *diffuse optical tomography* for image reconstruction (see Chapters 17 and 20 for details), where the propagation of NIR light through tissue is reasonably well described by the diffusion equation, as an approximation of the more general radiative transport equation (see, e.g., Arridge 1999; Gibson et al. 2005; or Jacques and Pogue 2008).

A particular widespread fluorescence tomographic technique is the so-called *fluorescence molecular tomography* (FMT); its aim is to resolve the distribution or activation of fluorescence probes with molecular specificity and thus to provide information about their deep penetration and quantification, two of the major limitations of planar fluorescence imaging. Macroscopic fluorescence imaging, based on tomographic systems, has been developed with great success and diagnostic potential. Mainly fluorescence probes emitting in the NIR spectral region were used, because macroscopic fluorescence in the visible spectum suffers from a limited penetration depth of $d_T \approx 1\text{-}2$ mm.

On the downside, in the NIR, biological tissues exhibit only low absorption and limited detection sensitivity. Initially, fluorescence tomography (FMT) methods involved fiber-based systems, to couple light to and from tissue, as well as matching fluids to eliminate optical discontinuities and heterogeneities at the sample (animal) tissue surfaces.

In modern FMT implementations, the said systems have been replaced by noncontact technologies, in conjunction with multiview imaging and CCD camera detectors. In the noncontact methods, neither the light source nor the detector comes into contact with the tissue. Note that one can set up those systems in a free-space arrangement, in which no matching fluids are required. In general, such FMT imaging systems provide higher image quality compared with fiber-based systems.

In Figure 18.7, the experimental setup configurations for free-space, noncontact FMT of small animals (in general mice) is depicted schematically. Illumination is made using continuous-wave laser radiation, and recording is for 360° of angular coverage; the test animal is confined to the (rotatable) cylinder. As illustrated in the figure, the animal can be scanned by a laser beam, focused on its surface, which can be translated on the $x$–$y$ plane ($z$ is the direction of the "rotation" axis) by means of two orthogonally arranged galvanometric mirrors. The animal is placed in a vertical rotation stage and imaged in free space over multiple projections. The transmitted and fluorescence photon fields are collected in the transillumination configuration (configuration as in the center of the figure), using a high-sensitivity low-noise CCD camera.

## 18.2.2 Examples of FMT

Here a small selection of representative examples is given to demonstrate the capabilities and the potential of FMT for *in vivo* applications. The first examples are based on the exogenous administration of fluorescence probes, with emission in the NIR spectral region; i.e., they represent direct-imaging modalities. The subsequent examples correspond to FMT in the visible region; for these FPs are used, and the methodology may be classified as indirect imaging. Their *modus operandi* involves a transgene—the reporter gene—that encodes for an FP. After the transgene is introduced into the cell, its transcription leads to FP production, whose emission is imaged.

The first example, shown in the left part of Figure 18.8, describes an investigation, in which a cypate polypeptide fluorescent probe was injected; this breast-specific protein was found to localize in human MDA-MB-361 breast cancer xenografts and in the kidneys of nude mice (see Patwardhan et al. 2008).

FMT imaging can be also applied to other diseases, as illustrated in the second example, shown in the right part of Figure 18.8; here it is demonstrated that FMT can be applied to the imaging of lung inflammation (see Ntziachristos 2009). The FMT and MRI images shown in the figure are for a disease-affected mouse (top row) and a healthy control mouse (bottom row). While the fluorescence probe

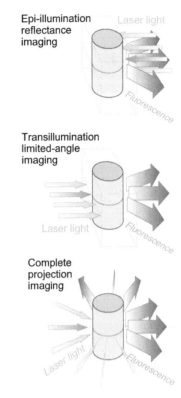

**Figure 18.7** Schematic of free-space projection FMT imaging. (Top) Reflectance imaging. (Center) Transillumination imaging. (Bottom) Full 360° transillumination imaging.

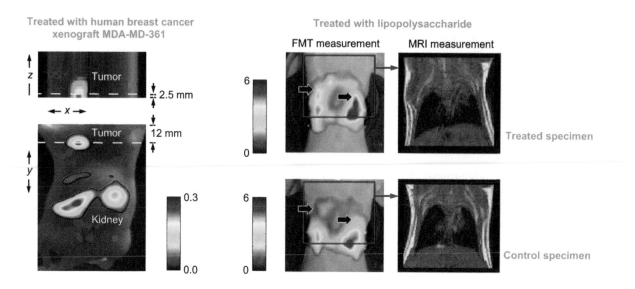

**Figure 18.8** FMT of tumor growth in mice. (Left) Representative *x–y* and *x–z* FMT slices from 3D tomographic reconstruction of a nude mouse with a subcutaneous breast-specific cancer (xenograft MDA-MD-361). (Adapted from Patwardhan, S.P. et al., "Fluorescence molecular imaging: Microscopic to macroscopic." In A.P. Dhawan et al.(eds.), *Principles and Advanced Methods in Medical Imaging and Image Analysis*, World Scientific, 2008, pp. 311–336.) (Right) FMT slices for mice specimen, with lung inflammation (top row) and for control animals (bottom row); the (false-color) FMT images are superimposed onto (grayscale) mouse photographs. Fluorescence probes accumulate in the liver and the affected lung (the lung region is indicated by the black arrows). Note that the comparative MRI images do not reveal these details. (Adapted from Ntziachristos, V., *Proc. Am. Thorac. Soc.* 6, no. 5 (2009): 416–418.)

can be seen to accumulate in the liver, no significant activity is visualized in the lung, as marked by the black arrows. Clearly, the FMT images showcase a marked difference in fluorescence signals when comparing the diseased and control specimen. The authors interpreted this as a sign of the higher activation of the probe, associated with protease upregulation in the inflamed lungs.

Certainly, FPs have revolutionized the twenty-first-century status of the life sciences and, in particular, of *in vivo* imaging of tissues and whole-body animals (see the short reviews by Hoffman 2005, 2015). This revolution has become possible due to several factors, including (1) FP emissions are so bright that only very simple equipment is required to implement *in vivo* imaging; (2) the GFP could be genetically linked with many proteins, thus providing a stable and efficient labeling mechanism for investigating protein function and location in living cells; (3) in addition to the GFP from the jellyfish *Aequorea victoria*, a rich variety of FPs of different color emission have been found in nature or were engineered in the laboratory; and (4) the distinct FPs currently available are nontoxic to living organisms, allowing long-term tumor growth and progression to be investigated.

This "FP family" of proteins allows for the multiple labeling of live cells and visualization of different processes simultaneously, enabling researchers to visualize—even at subdiffraction limit spatial resolution—phenomena such as gene expression, cell division, protein localization and interactions, intracellular transport pathways, and tumor evolution.

It should be noted that, because the used FPs emit in the visible part of the spectrum—a spectral region in which living tissues and animals exhibit high absorption coefficients—the tomographic imaging of FPs faced the initial reconstruction problem, associated with the lack of a model for photon propagation in tissue in the visible. The visible light problem was not included in the basic assumption for the development of the mathematical treatment for NIR tomographic imaging, because in the NIR, light is predominantly scattered by tissues, rather than absorbed.

A method for performing tomographic imaging of FPs in whole animals was first developed by Zacharakis et al. (2005a, 2005b). The method is based on a modified solution to the diffusion equation, taking into account both the high absorption and high scattering of tissue in the visible spectral region. These authors demonstrated that they could properly reconstruct the superficial and deep-seated FP activity *in vivo*, as traced by noninvasive 3D fluorescence imaging.

One area in which FP imaging is particularly useful is the investigation of tumor progression, using multiple-colored proteins—as, for example, employing dual-color fluorescence imaging— to visualize the human tumor–host interaction. An illustrative example of this type of FP application is shown in Figure 18.9 from a study carried out by Yang et al. (2009). In their work, the authors exploited results from an earlier investigation by Vintersten et al. (2004), who developed and characterized a transgenic RFP. Specifically, in the nude mouse experiments shown, the RFP was used as a host for GFP, constituting GFP-RFP-labeling human cancer cells. Subsequently, they used transgenic animals to visualize the growth, metastases, and tumor–host interaction of tumor cell lines, expressing the cited FPs. From the authors' study, a whole-body image of the RFP-labeled nude mouse is shown in the top panel of Figure 18.9, revealing a homogeneous appearance indicative of a healthy animal. The lower panels show the imaging of the cancer cells labeled with GFP, orthotopically transplanted in RFP nude mice. The GFP-expressing human cancer cell lines included MDA-MB-435-GFP human breast (center panel) and HCT116-GFP human colon (lower panel) cancers. The authors found that the actual human tumors had similar growth rate when compared with the tumors growing in the nontransgenic nude mouse specimen.

In conclusion to this section, we may point out that nowadays there is little doubt that the use of FPs is the method of choice for whole-body *in vivo* imaging.

## 18.3  SUPERRESOLUTION MICROSCOPY

Traditional and LIF microscopy has been and is of utmost relevance in life sciences. However, for many years, users lamented the fact that the methods were hampered by the diffraction limit encountered in optical microscopy. This discontent was associated with the fact that researchers knew from the use of other techniques with higher spatial resolution that subdiffraction limit features in biological tissue were highly relevant for the understanding, e.g., of cell metabolism.

In optical microscopy, the diffraction of light converts a sharp point on the object into a finite-sized spot in the image whose 3D intensity distribution is known as the *point spread function* (PSF). Obviously, the PSF size determines the resolution of the microscope, and it is well known that the FWHM of the PSF in the lateral directions $x$ and $y$, i.e., perpendicular to the optical axis, is given by

Whole-body image
transgenic RFP nude mouse

GFP-expressing MDA-MB-435
human mammary cancer

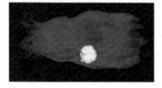

GFP-expressing HCT116-RFP
human colon cancer

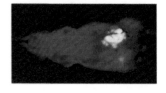

**Figure 18.9**  Whole-body dual-color FMT imaging of orthotopically growing GFP-expressing human tumors in transgenic red fluorescent protein (RFP) nude mice; note that all major organs and tissues fluoresce red, under excitation with blue light. (Top) Unaffected specimen. (Middle) Specimen exhibiting GFP-expressing MDA-MB-435 human mammary cancer. (Bottom) Specimen exhibiting GFP-expressing HCT116-RFP human colon cancer. (Adapted from Yang, M. et al., *J. Cell Biochem.* 106, no. 2 (2009): 279–284.)

$$\Delta x_{\mathrm{PSF}}(\Delta y_{\mathrm{PSF}}) \approx 0.61\lambda/\mathrm{NA}. \tag{18.3a}$$

Here, $\lambda$ is the wavelength of the light and NA is the numerical aperture of the microscope objective, defined as $\mathrm{NA} = n \cdot \sin\alpha$, with $n$ being the medium refractive index and $\alpha$ the half-cone angle of the light focused by the objective. For high-NA objectives, one finds that the axial width of the PSF is significantly larger than the lateral width, with

$$\Delta z_{\mathrm{PSF}} \sim (2-3) \times \Delta x_{\mathrm{PSF}}(\Delta y_{\mathrm{PSF}}). \tag{18.3b}$$

This resolution limit of optical microscopy, a condition known as the *diffraction* or *Abbe limit*, was theoretically demonstrated in the late nineteenth century by Ernst Abbe, who stated that "... the smallest resolvable distance between two points using a conventional light microscope cannot be smaller than half the wavelength of the imaging light ..." (Abbe 1873). The shortest visible light wavelength is at about $\lambda = 400$ nm (blue-violet light); thus, according to the Abbe limit, the smallest lateral resolution that can be achieved in a conventional microscope is $d \approx 200$ nm. While a microorganism—such as a bacterium whose size is about that of a diffraction limit—can be imaged by an optical instrument, there is little chance of imaging objects inside the cells, such as, for example, individual proteins. Images with much better resolution than that of optical microscopes can be obtained by using electron microscopes (images are generated by scattered electron beams rather than photon beams), but this technique has significant limitations such as, for example, that the sample has to be installed in vacuum, a requirement that immediately excludes its application to living specimens.

Therefore—as stated at the beginning of this section—in general, biologists suffered from the "unavoidable" limitation that no optical microscope was capable of visualizing the interior of the nanostructures in living cells. Therefore, it was clear that to fully understand how a cell functions, the Abbe diffraction limit needed to be broken to perform nanostructure and single-molecule spectroscopy.

A concept that revolutionized optical microscopy and helped in breaking the Abbe limit was developed around the turn of the twenty-first century, namely, the exploitation of the property of fluorophores in attaching them spatially specific to parts in larger molecular entities. The related technique is now known as superresolution microscopy and lead to the (commercial) development of the so-called *nanoscope*. For this development, the 2014 Nobel Prize in Chemistry was awarded to Eric Betzig, Stefan W. Hell, and William E. Moerner "e.g. the development of super-resolved fluorescence microscopy ..." (see, for example, Dickson et al. 1997; Hell 2003; Betzig et al. 2006). In particular, the laureates contributed—among other applications—(1) to monitoring of the interplay between individual molecules inside cells; (2) to observing disease-related proteins aggregate; and (3) to tracking cell division at the nanoscale.

Normally, superresolution fluorescence techniques are classified according to the approach by which they introduce or achieve the subdiffraction limit. By using this criterion, they can be grouped into two main categories: techniques that introduce the subdiffraction limit by *spatially patterned excitation* and techniques that go beyond the diffraction limit implementing *single-molecule imaging*. One often also addresses these groups as *deterministic* and *stochastic* superresolution techniques. In the former, one exploits the fact that most fluorophores used in biological microscopy exhibit a nonlinear response to excitation, which can be exploited to enhance resolution. In the latter, one exploits that many molecular fluorescence emitters follow a complex temporal behavior; closely spaced fluorophores can be made to emit light at different times, linking spatial to time resolution.

To the first group belong techniques such as stimulated emission depletion (STED) microscopy, reversible saturable optically fluorescence transition (RESOLFT) microscopy, and structured-illumination microscopy (SIM)/saturated structured-illumination microscopy (SSIM). Examples of superresolution fluorescence microscopies belonging to the second group are stochastic optical reconstruction microscopy (STORM), photoactivated localization microscopy (PALM), and fluorescence photoactivation localization microscopy (FPALM). The main concepts underlying these techniques together with selected examples are discussed in the following sections.

## 18.3.1 STED microscopy

One of the most widely used superresolution scanning fluorescence techniques is STED microscopy (see, e.g., Hell and Wichmann 1994). The first instrument developed by these authors was capable of resolving dimensions of $d = 35$ nm in the far field. The diffraction resolution limit was overcome by employing stimulated emission to inhibit the fluorescence process in the outer regions of the excitation point. The basic principle of the STED methodology can be understood with the aid of Figure 18.10.

In the first place, a laser excites the fluorescence of the fluorophore (the PSF 1 spot in the figure). A second laser with a doughnut-shaped beam profile, generated by an optical spatial modulator (the PSF 2 spot in the figure), is then imaged concentric to the first one and depletes—by stimulated emission—the fluorescence of the excited region, except for the central zone (the PSF 1 $\otimes$ PSF 2 convolution spot in the figure). The net fluorescence is a very small spot of dimension below the diffraction limit, which is represented by the PSF spot in the figure.

The key feature of the reduction in resolvable spot size relies on the use of the second laser, which inhibits the fluorescence emission by the stimulated emission from the excited fluorophore state. Hence, while the first laser excites the molecule, the second one brings it back to its ground state, thus appearing "dark" to the detector: only fluorescent molecules inside the area of subdiffraction dimension contribute. Note that, as shown in the bottom part of the figure, the diameter of the contributing central region can be influenced by the strength of the stimulating doughnut-shaped laser beam. The stronger the stimulating radiation, the smaller the remaining central part becomes, which contributes to the observed fluorescence signal.

The potential of STED microscopy is demonstrated here in three examples.

In the first example, the authors aimed at revealing molecular substructures in living cells; the specific system in their study were immune-labeled nuclear pore complexes (NPCs) of cultured *Xenopus* cells (Göttfert et al. 2013). Using STED microscopy—with a stimulating laser at $\lambda_{\text{depletion}} = 775$ nm, which simultaneously could deplete the green and red fluorophores used in the study—they were able to resolve particular subunits of this protein complex, specifically the

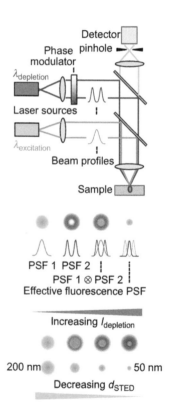

**Figure 18.10** Concepts of a STED microscope. (Top) Experimental realization, in confocal configuration. (Bottom) Concept of the interaction between the fluorescence-generating laser beam 1 and the stimulating laser beam 2 (with doughnut-shaped profile), together with the effect that the stimulating laser power has on the fluorescence image spot size.

typical eightfold symmetry of the peripheral transmembrane protein *gp210* homodimers (antibody labeled red), along with a set of other proteins (labeled green) in the central pore channel (see Figure 18.11a). As the scales in the right-hand part of the image and its inserts show, spatial resolution down to $d \sim 20$ nm could be achieved; in the equivalent confocal laser spectroscopy image, the structures are blurred and unresolved, also resulting in the "mixing" of the colors in the fluorescence response; i.e., yellowish areas appear.

The second example, shown in Figure 18.11b, represents one of the first images using the STED microscope methodology (see Klar et al. 2000). In their study, the authors imaged an *E. coli* bacterium, labeled with the dye pyridine 4. As is clearly evident, in the STED image, the bacterial membrane is substantially better outlined than in the confocal spectroscopy counterpart (on the left), recorded at the same time. The authors demonstrated, even at the infancy stage of STED, that significant improvements in spatial resolution were possible, in the particular case shown here a threefold improvement over conventional confocal imaging.

The third example, shown in Figure 18.11c, illustrates results from a recent study, in which STED nanoscopy has moved into the realm of living systems. This is of the highest physiological interest and is exemplified here for visualizing neurons in a living mouse (see Berning et al. 2012). For this, the authors used heterozygous TgN(Thy1-EYFP) mice, expressing EYFP as a nonfusion protein in neuronal cytoplasm. The STED image is of the molecular layer of the somatosensory cortex, located 10–15 μm below the surface; from the insert, it is clear that structural elements of the order ≤70 nm can be resolved (constituting about 1/4 of the diffraction limit at the operating wavelength $\lambda_{STED} = 592$ nm in the

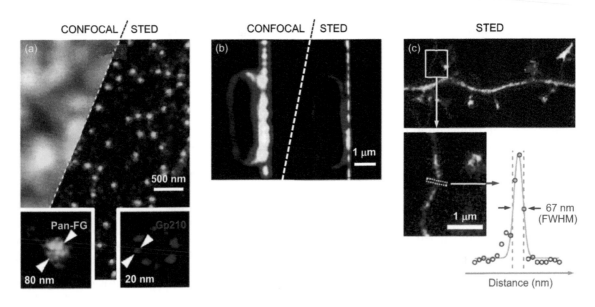

**Figure 18.11**    Examples of STED microscopy, in comparison with conventional confocal microscopy. (a) Images of immune-labeled NPCs of cultured *Xenopus* cells, revealing protein substructures down to $d \sim 20$ nm. (Adapted from Göttfert, F. et al., *Biophys. J.* 105, no. 1 (2013): L01–L03.) (b) One of the first images (of an *Escherichia coli* bacterium), utilizing a STED microscope. (Reproduced from Klar, T.A. et al., *PNAS* 97, no. 15 (2000): 8206–8210. With permission.) (c) STED microscopy in the molecular layer of the somatosensory cortex of a mouse with enhanced yellow fluorescent protein (EYFP)-labeled neurons, showing dendritic and axonal structures; the insert reveals that structural elements could be resolved down to about a quarter of the diffraction limit. (Adapted from Berning, S. et al., *Science* 335, no. 6068 (2012): 551.)

experiment). In addition, the authors observed that, when recording STED images of the same structural area over extended periods, dendritic spines underwent morphological changes and movements on the timescale of minutes.

## 18.3.2 RESOLFT microscopy

There are other mechanisms for suppressing the undesired fluorescence, in addition to stimulated emission exploited in STED. One of them is the use of saturable depletion; this is the basic concept underlying another superresolution fluorescence technique, named *RESOLFT* microscopy. In this technique, one uses fluorescent probes, which can be reversibly photoswitched between its fluorescent state (on) and a dark state (off); such dark states can be triplet states, any dark state of a photoswitchable fluorophore, or even the ground state of a fluorophore (as in the case of STED). The concept is shown schematically in the left-hand part of Figure 18.12. Essentially in this case of saturated depletion, the dimension of the "effective" part of the PSF, $\Delta d_{\mathrm{RESOLFT}}$, is given by

$$\Delta d_{\mathrm{RESOLFT}} \cong \Delta d_{\mathrm{PSF}}/\sqrt{(1 + (I/I_{\mathrm{sat}}))}, \qquad (18.4)$$

where $\Delta d_{\mathrm{PSF}}$ is the FWHM of the PSF (as defined in Equation 18.3); $I$ is the peak intensity of the depletion laser radiation; and $I_{\mathrm{sat}}$ is the saturation intensity for the fluorophore (the saturation intensity is defined as the value for which the on and off states of the fluorophore are equally populated). It becomes clear from Equation 18.4 that the larger the ratio $(I/I_{\mathrm{sat}})$, the smaller becomes the effective size $\Delta d_{\mathrm{RESOLFT}}$ of the PSF; therefore, the spatial resolution increases, with the condition of superresolution reached if the intensity ratio is substantially higher than unity.

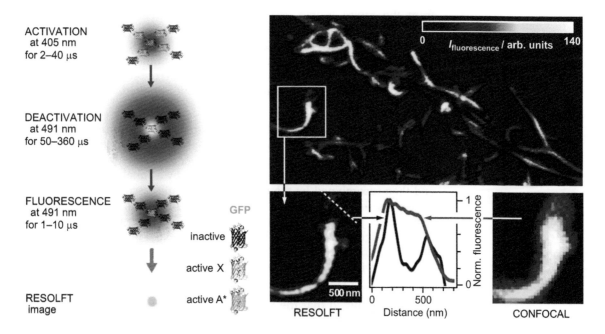

**Figure 18.12**  (Left) Concept of RESOLFT nanoscopy; GFP = green fluorescent protein. (Right) RESOLFT image of brain tissue, revealing actin fine structures in dendrites and dendritic spines; these details are unresolved in common confocal fluorescence imaging (see the enlarged presentation in the bottom part of the figure). (Data adapted from Testa, I. et al., *Neuron* 75, no. 6 (2012): 992–1000.)

In RESOLFT microscopy, one utilizes optical transitions that require low $I_{sat}$ values, meaning that superresolution imaging can be achieved with lower depletion laser intensities than in STED, for which typical values are on the order of $I_{sat} \approx 10^7$ W·cm$^{-2}$. This value can be lowered significantly by reverting to long-lived (microsecond to millisecond) molecular states, which can be populated or depopulated with light intensities in the range of only $I_{sat} \approx 1$–$1000$ W·cm$^{-2}$. Typically, *reversible switchable FPs*—such as Drompa, EGF promoter, or yellow fluorescent protein—are used in RESOLFT microscopy, reaching lateral resolution of $d_{RESOLFT} < 100$ nm. An example is shown in Figure 18.12, in which neuronal proteins in a brain slice are well resolved, unraveling different dendritic spine morphologies (Testa et al. 2012). This is particularly highlighted by the enlarged insert: the cross-sectional intensity plot confirms a ring-shaped feature, which is not recognizable in the comparative confocal fluorescence image.

### 18.3.3 SIM and SSIM

To understand the advantages of SIM/SSIM for microscopy, one first should recall the basic concept behind structured illumination. In short, structured illumination (microscopy) utilizes illumination of a (structured) sample with patterned light, which generates a moiré pattern. The latter is an interference pattern that is created when two grids with different angles or mesh sizes are overlaid.

An example of a moiré pattern is shown in the center row of Figure 18.13, where two fine patterns are superposed multiplicatively. It should be noted how the moiré fringes are much coarser than either of the original patterns and appear with a spatial frequency much lower than that of the sample. Hence, the sample structure—which is unobservable under uniform illumination—can be retrieved from the moiré fringes, using suitable mathematical algorithms for image reconstruction (general recipes and relevant references for such algorithms may be found, e.g., in Saxena et al. 2015; or Lal et al. 2016). It is probably worth noting that under structured illumination, the original sample structure would be easily observable in the microscope image, even if both (or one) of the original patterns were too fine to be resolved by a diffraction-limited, uniform illumination instrument.

In SIM applications, the sample pattern with (unknown) structure is represented by the spatial distribution of the fluorophores; the excitation light pattern structure is well known if it is generated, e.g., as an interference pattern form a laser light beam that is initially split into two by a suitable transmission grating and then recombined by the collimation plus microscope objective lenses; this concept is shown in the left-hand part of Figure 18.13. Since the fluorescence emission is proportional to the fluorophore concentration as well as to the excitation light intensity, the observed light collected in the image, and therefore the moiré fringes, will contain information on both patterns. Consequently, information from the unknown sample structure can be retrieved if the illumination patter is known.

It should be noted that the moiré-patterned image maintains the information about the sample structure, even after "blurring" associated with diffraction-limited optical imaging; this is because—as stated earlier—the moiré pattern exhibits much lower spatial variation than the original sample, which becomes larger than the diffraction limit for a suitable angular rotation between the two

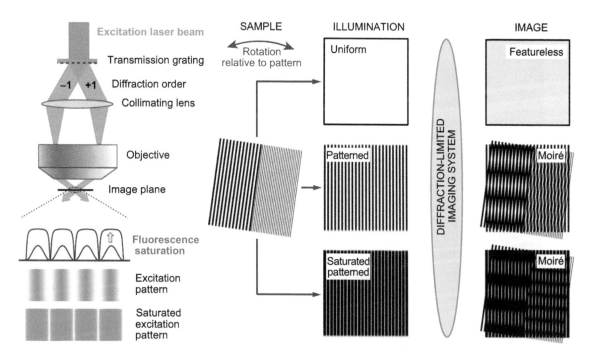

**Figure 18.13** (Left) Principle of the generation of light excitation patterns in SIM and SSIM. (Right) Comparison of image formation using conventional wide-field illumination (top row), structured illumination (middle row), and saturated structured illumination (bottom row); only the SIM and SSIM images carry the structural information in their moiré pattern.

patterns. That information is lost for uniform illumination, as indicated in the top row of the figure. It is also worth noting that the principle behind SIM is somewhat similar to that of heterodyne detection, where the signal of interest—typically a time-varying signal—is convolved with a (time-varying) reference signal. The mathematical product (the moiré fringes in SIM) contains a new signal at a lower frequency (called beat frequency) that is easier to detect, given the bandwidth of the detector. By analogy, in SIM imaging, the high spatial frequency information of the fluorescent sample is shifted to a lower-frequency range and can be recovered as beat patterns of lower spatial frequencies of the moiré fringes and, consequently, captured by the microscope.

The improvement on the spatial resolution can be demonstrated mathematically using the space- and frequency-domain description and inverse Fourier transform to reconstruct fluorescence images with high spatial frequency information (see, e.g., Yamanaka et al. 2014). However, since the spatial frequencies that can be created optically are also limited by diffraction, the resolution improvement of SIM is only a factor of 2 (see Gustafsson 2000). An example of SIM is shown in Figure 18.14, in which images of the actin cytoskeleton of a HeLa cell are compared, recorded with conventional microcsopy and SIM. The FWHMs of the observable structures seen in the two images are on the order of 280–300 nm for conventional microscopy and 110–120 nm for SIM.

Much higher resolution than that obtainable in SIM can be achieved by using *nonlinear* SIM, also known as saturated structured-illumination microscopy or SSIM. This technique was developed by Gustafsson and is based on the saturation of the fluorescence emission in the SIM scheme. As illustrated in the

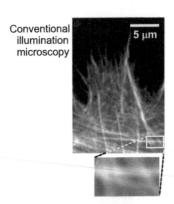

Conventional illumination microscopy

5 μm

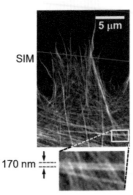

SIM

5 μm

170 nm

**Figure 18.14** Image slices of the actin cytoskeleton of a cell, using conventional wide-field micros-copy (top panel) and SIM (bottom panel). (Adapted from Gustafsson, M.G.L., *J. Microsc.* 198, no. 2 (2002): 82–87. Reproduced with permission, © The Royal Microscopic Society.)

left-hand part and the lower row in Figure 18.14 above, the excitation pattern becomes flat topped, and thus broader, due to saturation, while still no fluorescence is emitted from the zero points in the light interference pattern. The flat-topped pattern means—from the point of view of Fourier analysis of an asymmetric, repetitive pattern—that "higher frequencies" are added to the excitation pattern. As pointed out earlier, it is precisely the mixing of the excitation pattern with the high-frequency spatial features in the sample that allows reconstruction of the images with subdiffraction-limited resolution. Therefore, what limits the spatial resolution of SSIM is not the light diffraction but rather the fluorescence saturation that can be reached experimentally. The SSIM resolution can be approximated by the simple relation

$$R_{\text{SSIM}} \approx R/(2 + h), \tag{18.5}$$

where $R$ is the conventional resolution limit and $h$ is the number of higher (Fourier) harmonics achieved when the saturated flat-topped pattern is used.

It should be noted that as the fluorescence intensity approaches its saturation level, the emitted fluorescence is no longer proportional to the excitation light intensity; a nonlinear regime ensues in which ultimately the fluorescence lifetime becomes the limiting factor of the rate of emission (see Gustafsson 2005).

An example of the improvement of resolution by SSIM is shown in Figure 18.15. For the generation of the displayed test measurement data (Gustafsson 2005), a thin-(mono)layer sample was prepared from a dilute (in water) suspension of beads with a diameter of $d_{\text{bead}} = 50$ nm. The images show a nonuniform distribution of these beads across the displayed area. On sequential inspection of the images from top to bottom, the following emerges. The small bead cluster in the center appears as an almost shapeless blob in the conventional (diffraction-limited) microscopy image (first image); even with so-called linear filtering, the reconstructed image does not resolve said structure (see the second image). Once linear structured illumination is utilized, a distinct shape can be recognized, revealing already that it seems to consist of individual particles. Finally, when applying the new nonlinear saturation technique—SSIM—the structure is clearly resolved into individual beads (19 to be exact). To quantify the resolution of the SSIM method in the presented example, the author measured the fluorescence intensity profiles for a large number of isolated beads; a value of the FWHM average of $d_{\text{bead,SSIM}} = 58.6 \pm 0.5$ nm was derived, which constituted an improvement of a factor >5 in comparison with the lateral FWHM obtained in conventional microscopy, $d_{\text{bead,CM}} \cong 265$ nm.

As a final comment, it should be pointed out that the increased resolution of SSIM does not come completely free but is at the expense of a limited choice of fluorophores, which need to be either highly photostable or photoswitchable.

## 18.4 SUPERRESOLUTION FLUORESCENCE MICROSCOPY BASED ON SINGLE-MOLECULE IMAGING

The fundamental principle of superresolution fluorescence microscopy, based on single-molecule imaging, requires that the position of a spatially isolated fluorophore can be resolved with a precision better than the width of the PSF.

From this point of view, the first step therefore is to locate the fluorophore with lateral accuracy below the diffraction limit. Once this has been accomplished, the image of the molecule/fluorophore can be reconstructed, as briefly outlined later.

The problem of single-molecule localization (SML) is best addressed using the example of imaging a single or two closely spaced nanoparticles (in the example squares with dimension smaller than the diffraction limit, representing molecules), using a diffraction-limited detection system (see Figure 18.16). The position of a single nanoparticle can be determined by fitting the lateral intensity distribution across the CCD image with a Gaussian distribution; the center of the fitted curve then represents the center of the PSF. Thus, the uncertainty in determining the position of the nanoparticle is linked to the uncertainty in fitting the center of the PSF.

It is well known that the fit of a signal distribution function becomes better the larger the signal amplitude is; in the case of a photon signal, this means a larger number of detected photons $N_{ph}$. Theoretically one finds that the uncertainty $\Delta_{pos}$ in localizing the center position in the (Gaussian) distribution function scales with the square root of the number of photons contributing to the detected signal, i.e.,

$$\Delta_{pos} \sim \Delta_{PSF}/\sqrt{N_{ph}}, \tag{18.6}$$

where $\Delta_{PSF}$ is the FWHM of the PSF. Once $\Delta_{pos}$ is obtained, the position image of the nanoparticle can be reconstructed, as shown in the last panel of the sequence. Note that for a finite number of photons the reconstructed particle position is blurred by $\Delta_{pos}$ and that the shape of the particle is in general not recovered.

In the bottom row of Figure 18.16, the same image reconstruction procedure is being applied for two nanoparticles whose lateral spacing is less than the diffraction limit. Because of this, the superposition of the emission from the two particles would result in a single, spatially unresolved CCD image; in the very best case, it might exhibit some recognizable asymmetry. Fitting this CCD image intensity distribution function automatically, without prior knowledge and certain fit restrictions, will nearly invariably result in a single-center solution, with its $\Delta_{pos}^{(2)}$ likely somewhat larger than that for a single particle. Thus, the reconstruction would result in an apparent single-particle position, which is blurred by $\Delta_{pos}^{(2)}$, constituting a false image and suggesting a single particle rather than two.

Only if the separation of the particles were sufficiently large, so that the convolution of the emissions resulted in a "dip" of intensity in the diffraction-limited imaging process, would one be able to derive positional separation in the image reconstruction. This is the well-known Rayleigh criterion in spectroscopy, used to classify whether overlapping lines are separated or not.

In general, biological samples—such as, e.g., cells—have such a high density of proteins that, when labeled, they do not fulfill the requirement of spatially distinguishable particles (molecules); thus, under diffraction-limited conditions, their imaging would always result in blurred, unresolved structures. This limitation was overcome by using fluorescence-labels that can be cycled between activated and deactivated states (see, e.g., Patterson et al. 2002, or Bates et al.

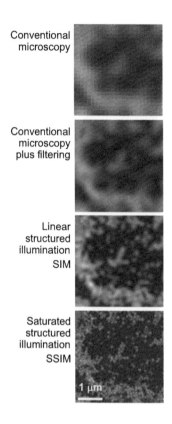

Conventional microscopy

Conventional microscopy plus filtering

Linear structured illumination SIM

Saturated structured illumination SSIM

1 µm

**Figure 18.15** Comparison of feature resolution, exemplified for a field of fluorescent beads, with diameter $d_{bead} = 50$ nm. Images from top to bottom: conventional microscopy, conventional microscopy plus linear filtering; linear structured illumination (SIM); and saturated structured illumination (SSIM). (Adapted from Gustafsson, M.G.L., *PNAS* 102, no. 37 (2005): 13081–13086.)

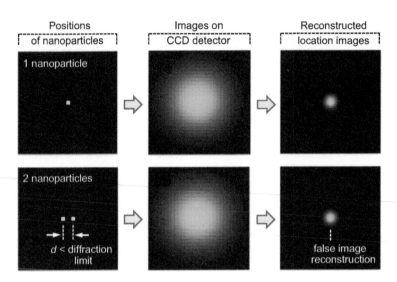

**Figure 18.16** Limitation of resolution closely spaced nanoparticles in diffraction-limited imaging. Image reconstruction from observed CCD images, for a single nanoparticle (top row) and two nanoparticles spaced by less than the diffraction limit (bottom row). In the latter case, the reconstruction result is "false image" reconstruction, suggesting only a single-particle origin.

2005), i.e., fluorescence probes which can switch between a fluorescent and a dark state. In brief, the basic trick in their use is that the number of activated fluorophores is adjusted to be small and, therefore, the probability of having two or more of them within the diffraction-limited volume is negligible. These then are activated at different times; thus, subsets of well-separated molecules are imaged, one at a time, so that their individual positions can be determined accurately. Subsequently, the complete image is assembled by adding all images of the distinct subsets together (see, e.g. the review by Huang et al. 2009, in which the concepts of this and other superresolution techniques are compared).

This new concept for overcoming the diffraction limit barrier "by separating in the time domain the spatially overlapping fluorescence images" was independently developed by three research groups, albeit with slightly different approaches in experimental realization and image reconstruction philosophies. The names associated with these three developments are STORM (see Rust et al. 2006), PALM (see Betzig et al. 2006), and FPALM (see Hess et al. 2006).

### 18.4.1 Basic principles of STORM/PALM

The concomitant development of STORM and PALM/FPALM as a high-resolution biophysical imaging method is based on the discovery of fluorescent probes, which could individually be activated and deactivated. The main differences lay primarily in the nature of the fluorescent probes. For PALM, endogenously expressed fluorophores are utilized, conceptually in the form of genetic-fusion constructs to an FP (e.g., GFP); their photochromism is controllable. In STORM, one uses immune labeling of the biostructure with antibodies and tagged with organic fluorophores. Originally, these constituted paired cyanine dyes: when excited near its absorption maximum, the first molecule of the pair (called activator) activates the second molecule (called reporter) to the fluorescent state. The choice of a technical approach—STORM or PALM—and the use of a specific fluorophore ultimately depend on the particular application. For a review of the biochemical and photophysical properties of photocontrollable FPs and dyes, which are used today in superresolution microscopy, see, e.g., Shcherbakova et al. (2014).

Common to all aforementioned single-molecule superresolution methodologies is that the activation of states of a photoswitchable fluorophore relies on two fundamental criteria. Firstly, excitation of the fluorophore must lead to the consecutive emission of a sufficient number of photons to generate an image-light distribution, which allows one to extract precise localization, before it enters a dark state or becomes deactivated by photobleaching. Secondly, the actual number of activated fluorescent molecules must be spare, so that their spacing exceeds the Abbe diffraction limit; under excitation with visible light, this is on the order of $d_{Abbe} \approx 250$ nm. Only then can a multitude of individual emitters be recorded in parallel and a distinct set of lateral coordinates can be determined.

The basic steps involved in the creation of a STORM/PALM superresolution image are summarized schematically in Figure 18.17.

The actual target structure is illustrated in the top panel of the figure, showing a structure with a dense label spacing below the diffraction limit.

In the following panels of the figure, a sparse set of the fluorescent probes is activated, here represented by single (pink) entities. Single-molecule images that do not overlap are generated for each cycle, in the sequence "activation → readout → localization → deactivation." Note that normally low-power readout laser radiation is used, which is applied until the fluorophores spontaneously photobleach or return to a dark state. The "localization" step means that after an individual fluorophore image has been captured by the CCD camera, the PSFs of the individual molecules are determined with high precision. That an image has been generated for particular fluorophores is indicated by the cross in the sub-sequent panel. The process is repeated (stochastically) until all fluorophores are exhausted.

The final superresolution image shown in the bottom panel of the figure is constructed by superimposing all measured fluorophore positions. Of course, the resolution of this image is limited by the number of localizations and the precision to which each was determined; however, normally this resolution exceeds that of the diffraction limit by quite a margin. By the nature of the reconstruction process, such superresolution images are "pointillistic"; i.e., they constitute representations of the lateral coordinates of all the localized molecules, rendered by a 2D Gaussian amplitude distribution, which is proportional to the number of photons recorded for each fluorophore. For a concise, interactive tutorial-style description of the principles of STORM/PALM and their comparison with other superresolution methodologies, see, e.g., Huang et al. (2010) or Leung and Chou (2011).

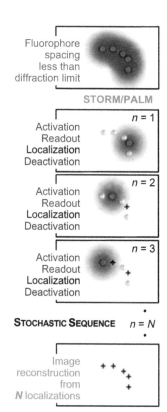

**Figure 18.17**  Concept of the build-up of a superresolution STORM/PALM image based on consecutive recording of a sparse distribution of fluorophores, involving activation → readout → localization → deactivation sequences.

## 18.4.2 Fluorophore localization

The concept of localizing individual molecules with high precision is closely linked to Heisenberg's work on uncertainty in the 1930s, but its mathematical (and computational) treatment was put into a formalistic framework only more than 50 years later.

As already pointed out in the introduction to this section, one finds that the CCD image of a single fluorescent emitter has a finite size; this is limited by the diffraction of the optical detection system. On the other hand, the exact position of the imaged molecule can be determined to a much better precision than the size of the diffraction-limited image provided that a statistically significant number of photons contributed to the image.

Basically, most methods for determining said molecular position coordinates are based on statistical curve-fitting algorithms of the measured photon distribution to a Gaussian function. The results of this fitting procedure are the mean value of the fitted position and its positional uncertainty $\Delta_{pos}$, which is given by (see, e.g., Shtengel et al. 2009)

$$\Delta_{pos}^2 = [\underbrace{(\Delta_{PSF}^2/N_{ph})}_{\substack{\text{photon} \\ \text{short noise}}} + \underbrace{(a_p^2/12N_{ph})}_{\substack{\text{pixel size} \\ \text{of detector}}} + \underbrace{(8\pi \cdot \Delta_B^2 \cdot \Delta_{PSF}^4)/(a_p^2 \cdot N_{ph}^2)}_{\text{background noise}}]^{1/2} \qquad (18.7)$$

Here, as earlier, $\Delta_{PSF}$ represents the width of the Gaussian function, fitted to the (diffraction-limited) PSF of the fluorescent molecule; $N_{ph}$ is the number of photons recorded; $a_p$ is the pixel size of the CCD detector; and $\Delta_B$ is the standard deviation of the background fluorescence and detector noise. Note that for a large number of photons from a single fluorophore, the background noise term becomes negligible, and the expression is approximated by Equation 18.6. In conclusion, Equations 18.6 and 18.7 mean that the critical element in achieving the high-precision molecular localization in STORM/PALM imaging is (1) minimizing the background noise while (2) maximizing photon emission from the fluorescent probe.

For example, assuming that 1000 photons can be accumulated from an individual fluorophore molecule (in the absence of background contributions), position accuracies on the order of $\Delta_{pos} \approx 10$ nm would be achievable for diffraction-limited imaging onto a CCD detector, using fluorophore excitation with visible-wavelength laser light. Of course, this number would increase/decrease in line with the change in the number of recorded photons $N_{ph}$.

An example of the quality and resolution achievable by single-molecule super-resolution fluorescence imaging is exemplified for a STORM measurement in Figure 18.18. The images shown here (in false-color representation, used to label the positional light intensities) revealed a new membrane–skeleton structure in neurons (see Xu et al. 2013). It should be recalled that dendritic spines are cellular structures of small dimensions (with dimension $d \sim 500$ nm); these are of great importance not only because they compartmentalize the excitation sites of neurotransmission in neurons, but also because their fast morphological changes seem to play an essential role in the plasticity of synaptic transmission.

Overall, superresolution laser fluorescence methods such as STORM/PALM have proven to be rather fast (complete images such as the one shown can be accumulated within a few seconds) and are well suited for imaging the neuron-spine morphology with greater detail and resolution than that achieved by conventional fluorescence microscopy (for dimensions, see the inserts in the lower image panel).

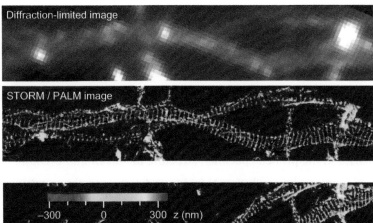

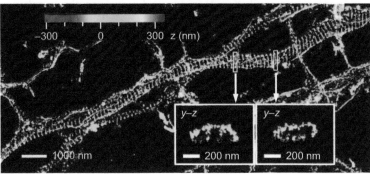

**Figure 18.18** Three-dimensional STORM imaging of actin in the axons of neurons; the actin is labeled with phalloidin, conjugated to photoswitchable dyes. (Top panel) Comparison of conventional (upper) and 3D STORM (lower) images. (Courtesy of X. Zhuang, 2016.) (Bottom panel) x-y map of a 3D STORM image; the z-positions are color coded, with violet and red corresponding to positions closest to and farthest from the substratum, respectively. The y-$z$ cross-sectional insets correspond to the white-boxed regions. The 3D STORM images reveal a periodic, actin-spectrin-based membrane skeleton in axons. (Adapted from Xu, K. et al., *Science* 339, no. 6118 (2013): 452–456.)

## 18.4.3 Factors affecting the resolution in STORM/PALM imaging

A number of factors affect the actual outcome of any particular STORM/PALM measurement; among those, the most critical are the following:

- The accuracy of each individual localization measurement

- The molecular density of fluorescent probes that have been localized in the final image

- The temporal behavior of the on/off switching cycles for the measurement sequence

Firstly, the relationship between resolution and the localization precision of a single molecule is relatively easy to determine. Put simply, the ability to separate two individual fluorophores as different entities is limited by the precision to which they can be localized. In turn, the localization precision primarily depends on the number of photons that are collected from the fluorophore during a single activation–deactivation cycle (under the assumption of negligible background noise).

Secondly, the relationship between molecular density and final image resolution is best described in terms of the so-called Nyquist sampling theory. This states that at least two data points per resolution unit are required. Should this not be the case (e.g., insufficient density of fluorophore labeling), artifacts—such as discontinuity in fine structural details—can appear in the superresolution images. To resolve (in the two lateral dimensions) spatial features of size $d_{\text{feature}}$, the minimum molecular density of localized fluorophores needs to meet the Nyquist criterion, i.e.,

$$\text{Nyquist molecular density} \approx (2/d_{\text{feature}})^2. \tag{18.8}$$

For example, to achieve a resolution of $d_{\text{feature}} = 20$ nm in two dimensions, one fluorophore has to be positioned at least every 10 nm in the probed structure. This translates into a rather high molecular density of around $10^4$ molecules·$\mu m^{-2}$. In practice, much lower molecular densities are often sufficient if geometries in the probed specimen structure are taken into account.

Overall, to increase the resolution $N$-fold in $D$ dimensions, the number of pixels that need to be acquired has to be increased $N^D$-fold. If such a resolution gain is desired (and compromising neither the imaging speed nor the signal-to-noise ratio in the process), the signal collection rate must increase by at least the same factor $N^D$. As a consequence, an $N^D$-fold increase in exposure to the excitation laser for each image is required.

Thirdly, in the single-molecule superresolution techniques STORM/PALM, which temporally separate the fluorescence emission by utilizing photo-switchable fluorophores, the ratio of the on- and off-state switching kinetics is critical and has to be appropriately adjusted to the actual experimental conditions (in particular the molecular density).

Finally, it should be noted that in view of the single molecule nature, STORM/PALM sets the current record in spatial resolution for fluorescence microscopy in biological samples, reaching a lateral spatial resolution of $\Delta_{\text{pos}} \approx 10$–30 nm. In Figure 18.19, these resolutions are compared with those of other superresolution and "conventional" methods.

### 18.4.4 Toward 3D superresolution imaging: Interferometric PALM

It is clear from the inspection of Figure 18.19 and with reference to Equation 18.3b that regardless of the technique—whether "ordinary" or superresolution spectroscopy—the depth-achievable resolution is always less than the lateral resolution.

This conundrum has been addressed recently with the development of a technique called *interferometric photoactivation and localization microscopy* (*i*PALM); in this novel approach, the researchers combined SML with multiphase interferometry (see Shtengel et al. 2009).

The first step of SML (in lateral $x$–$y$-directions) is exactly the same as in all single-molecule superresolution microscopy modalities.

To achieve the desired improvement in the axial $z$-direction in *i*PALM, one utilizes additional interferometric principles. In classical interferometry, a coherent laser light beam is split into two paths, which then are overlaid again—after some path difference—resulting in an interference signal. From the pattern of this interference signal, one can derive the path difference to within a few nanometers, or less. In the adaptation of this interference principle to work with *i*PALM, the researchers exploited the fact that a single fluorophore can be treated as a

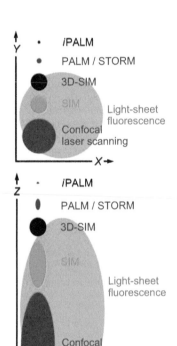

**Figure 18.19** Three-dimensional spatial resolution comparison for the common superresolution techniques PALM (*i*PALM), (interferometric) photoactivated localization microscopy; STORM, stochastic optical reconstruction microscopy; and SIM (3D-SIM), (3D) structured-illumination microscopy. For comparison, conventional light-sheet fluorescence microscopy and confocal laser scanning microscopy are included as well.

quantum source; i.e., the photons emitted by a single fluorescent molecule can interfere with themselves. To provide two light paths, as required for interferometry, for the fluorescent signal to traverse, use is made of two objectives to image the sample, one at the bottom and one at the top. This dual-objective arrangement has an added efficiency benefit, namely, that nearly the complete $4\pi$ emission from the fluorophores is collected. For a detailed description of the methodology and some applications of $i$PALM, see the original publication by Shtengel et al. (2009), or for an instructive, video-based demonstration, see Wang and Kanchanawong (2016).

Using this technique up to then unparalleled isotropic resolution, in both lateral and axial directions, was achieved down to values of $\Delta_{x,y,z}$ <25 nm. Thus, $i$PALM is suitable for studies that require resolution approaching that of electron microscopy, but it can be performed on intact, fluorescently labeled samples. It should be noted that $i$PALM is limited to relatively thin biological samples, such as, e.g., cell culture monolayers, to maintain the interference capability of the system; maximum thicknesses on the order of $d_z$ ~ 750 μm have been reported in the literature (see, e.g., Brown et al. 2011).

## 18.5 BREAKTHROUGHS AND THE CUTTING EDGE

For many of the traditional (laser-induced) fluorescence microscopy imaging applications in biology and biomedicine, samples were "stained" with a dye to enhance the contrast or to target specific sites in the specimen under investigation. By and large, such studies were *in vitro* because of the strong phototoxicity of most small fluorescent molecules (such as, e.g. fluorescein isothiocyanate). This very much changed with the discovery of FPs; it was found that such proteins were usually much less harmful when illuminated in living cells. This has triggered the development of the field of live-cell fluorescence microscopy, which today has to be seen as one of the most powerful tools in the imaging of bioorganisms.

### 18.5.1 Breakthrough: GFP as a marker for gene expression

GFP is a protein produced by a jellyfish, *A. victoria*; it produces green glowing light points at the edges of its umbrella. While initially studied as a curiosity of nature, it was found that this protein could be used for labeling and that it could act as a reporter molecule for gene expression. The latter was first demonstrated by Chalfie et al. (1994), specifically producing *E. coli* with a GFP expression construct. For the discovery and development of GFP, Chalfie, Tsien, and Shimomura shared the 2008 Nobel Prize in Chemistry.

The key result from those early studies is shown in Figure 18.20. The cultured bacteria with GFP expression exhibited significant fluorescence following illumination with light from a long-wave UV source; no such fluorescence was observed in control bacteria colonies. GFP absorbs blue light (with a maximum at $\lambda_{\text{excite}}$ ≈ 395 nm) and emits green light, with a peak at $\lambda_{\text{fl}}$ ≈ 509 nm (see the spectral traces in the lower part of the figure), under stable conditions and in the absence of photobleaching. The authors also observed that the GFP had no toxic effect on the cells. Interestingly, the biological GFP expressions exhibited fluorescence excitation and emission spectra that were indistinguishable from those of purified *A. victoria*-form GFP.

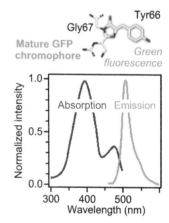

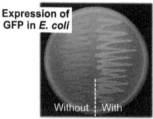

**Figure 18.20** (Top) Expression of GFP in *E. coli*; the bacteria on the right side of the figure incorporate the GFP expression plasmid; those on the left do not. For the photograph, the petri dish was irradiated with a long-wave UV source. (Adapted from Chalfie, M. et al., *Science* 263, no. 5148 (1994): 802–805.) (Bottom) Normalized absorption and fluorescence spectra for the GFP chromophore.

Since exogenous substrates and cofactors were not required to enable GFP fluorescence and because GFP can be introduced into many species through transgenic techniques (and maintained in their genome and that of their offspring), this opened the way to investigation of many processes, in real time, in living cells, including gene expression, *in vivo* fluorescence tagging, protein–protein interactions, protein trafficking; FP-expressing tumors and metastasis, and so on.

In addition to GFP, a range of different FPs have been developed over the years (see, e.g., Shcherbakova et al. 2014), all serving specific expression purposes, in particular in superresolution LIF microscopy applications.

## 18.5.2  At the cutting edge: Nanometer resolution imaging

The holy grail in the microscopy of biomolecules always has been to increase the resolution to such levels that individual, closely spaced entities (molecules) are distinguishable. The superresolution techniques discussed in this chapter have come a long way, and in particular, STED and STORM/PALM have paved the way to routine spatial resolutions in the 20–30 nm range. While really impressive, some biomolecular problems require even better resolution. A new type of superresolution microscopy technique has just emerged, in which resolutions well below 10 nm have been achieved (see Balzarotti et al. 2017). In addition, it is much faster and it requires collecting far fewer fluorescence photons than any of the other currently utilized superresolution microscopy techniques.

The new technique, which the authors named *minimal emission fluxes* (MINFLUX), combines the main concepts employed in the superresolution microscopy modalities STED and PALM. As usual in superresolution methodologies, one follows the sequence excitation → readout → determination of position.

Like STORM/PALM, MINFLUX switches individual molecules randomly on and off. At the same time, their exact positions are determined using a doughnut-shaped readout laser beam, as in STED. But, in contrast to STED, this doughnut-shaped beam initiates the fluorescence, rather than quenching it. This means that the molecule will fluoresce when it is on the ring, but will remain dark if it is at the dark center of the doughnut beam. Using their novel algorithm, which exploits the properties of the doughnut-shaped beam, the authors were able to achieve a best positional precision of $\Delta_x \approx 1$ nm for molecules that were spaced by just $d_x \approx 6$ nm.

In Figure 18.21, results from test measurements of an ordered matrix of molecules are shown, using the new MINFLUX technique; the superiority in spatial precision and resolution of MINFLUX over STORM/PALM is evident in these proof-of-principle measurements.

In addition to improved spatial resolution, the MINFLUX measurements were also fast enough and required rather low laser intensities so that the researchers were able to use the method to track individual molecules, within living cells, at various short timescales.

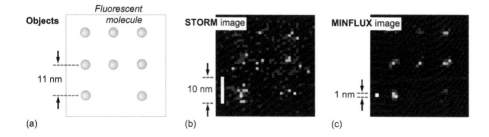

**Figure 18.21**   Comparison of the spatial resolution capabilities for an ordered array of fluorescent molecules (a), utilizing PALM/STORM (b) and the novel technique of MINFLUX (c); with MINFLUX a resolution of ~1 nm is achievable. (Adapted from Balzarotti, F. et al., *Science* 355, no. 6325 (2017): 606–612.)

The authors point out that in those early experiments, the conceptual limits of the methodology had not yet been reached. Thus, it is quite feasible that this new localization modality MINFLUX might open up new horizons for observing the dynamics, distribution, and structure of macromolecules in living cells.

# CHAPTER 19

# Raman Imaging and Microscopy

Besides fluorescence imaging, the various incarnations of Raman imaging have become the most used (analytical) laser-spectroscopic methods for the chemical analysis of samples. The popularity and wide penetration of Raman imaging is reflected in the publication of dedicated, general books; some of the most recent ones are those of Dieing et al. (2011), Zoubir (2012), and Salzer and Siesler (2014).

In addition to basic, spontaneous Raman spectroscopy, also nearly all the enhancement methodologies described in Chapters 13 and 14 have found their way into Raman imaging applications, including surface-enhanced and tip-enhanced techniques—surface-enhanced Raman spectroscopy/scattering (SERS) and tip-enhanced Raman spectroscopy/scattering (TERS)—as well as the nonlinear techniques of stimulated Raman spectroscopy/scattering (SRS) and coherent anti-Stokes Raman scattering (CARS). These will be described in Sections 19.2 and 19.3, subsequent to a more general introduction of Raman imaging principles and examples of spontaneous Raman imaging.

## 19.1 RAMAN MICROSCOPIC IMAGING

Raman microscopy (RM) and spectrochemical imaging has evolved into one of the most important analytical techniques, which at the same time is nondestructive and universal, meaning that not only selected molecular compounds in a sample can be probed, but all chemical components in the sample are accessible simultaneously—admittedly with, in general, complex numerical and statistical data evaluation procedures. Over the years, numerous reviews have been written about the subject, and several dedicated books have been published. To name but a few of the more recent ones, the reader is referred to the review by Opilik et al. (2013) on modern Raman imaging on the micrometer and nanometer scales or to the books by Zoubir (2012) on Raman imaging techniques and applications and Salzer and Siesler (2014) on infrared and Raman spectroscopic imaging. It is beyond the scope of this short chapter to comprehensively cover all aspects of spontaneous Raman spectrochemical imaging; only a flavor can be given. In particular, the key issues of importance to its implementation are described, giving a handful of representative examples, which highlight the versatility of the methodology.

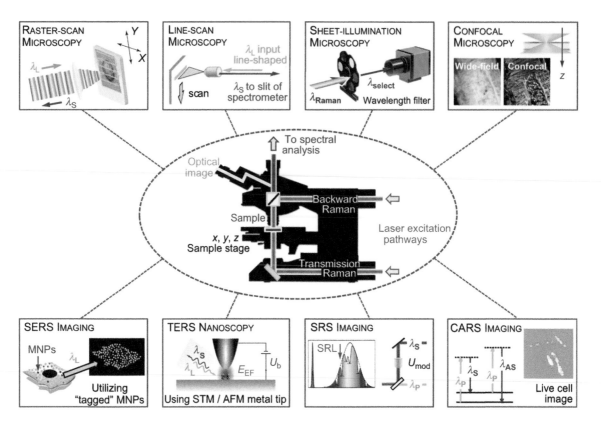

**Figure 19.1**   Experimental realizations of Raman microscopic imaging. (Top) Standard imaging modalities. From left to right: (point) raster scanning, line scanning, two-dimensional (2D) sheet illumination, and confocal microscopy. (Bottom) Enhanced Raman configurations. From left to right: surface-enhanced Raman imaging (SERS); tip-enhanced Raman imaging (TERS); stimulated Raman imaging (SRS), normally in the form of stimulated Raman loss (SRL); and CARS imaging. MNPs, metal nanoparticles; AFM, atomic force microscopy; STM, scanning tunneling microscopy.

### 19.1.1  Concepts of Raman imaging and microscopy

As stated in the introduction, by an overwhelming margin, Raman imaging is dominated by microscopic implementations. To begin with, a Raman microscope is based on a standard optical microscope, to which excitation laser source(s) and spectroscopic analysis components are added (see the schematic setup in the center part of Figure 19.1).

In general, spectral analysis is exercised by using a spectrometer to which a sensitive, optical detector is coupled, either in the form of a single-element detector—such as a photodiode (PD) or a photomultiplier tube—or in the form of an array detector—such as a charge-coupled device (CCD). Note that the laser and detection equipment components are not shown in the figure, but only the conceptual light passes.

As in standard Raman spectroscopy, any type of laser source is, in principle, suitable for the task (see Chapter 11). Note, however, that for the RM of biomedical specimens, one uses, in general, laser sources in the near-infrared (NIR) (commonly diode lasers operating at $\lambda_L = 785$ nm or $\lambda_L = 830$ nm or a neodymium-doped yttrium aluminum garnet [Nd:YAG] laser source at $\lambda_L = 1064$ nm).

This reduces the risk of photochemical damage of the specimen and minimizes tissue fluorescence; however, this comes at the expense of much reduced Raman scattering efficiency (due to the approximate $\lambda^{-4}$ dependence), and, consequently, data acquisition times steeply increase.

In addition to the mentioned components, a laser-line rejection filter is incorporated into the Raman light detection path, as is common for all Raman spectroscopic setups (see Chapter 11).

In the "traditional" implementation of RM, one measures the Raman spectrum at a single location on the probed sample and then moves to the next location, building up a point-by-point map. This imaging approach, i.e., point-by-point rastering, is known as *hyperspectral* or *chemical imaging*; here a large number of Raman spectra are acquired, one for each raster point over the scanned area (their number can run into the thousands). The data can then be used to construct wavenumber-selective images, exposing the location and amount of different chemical components.

More recently, some extended configurations have been introduced to implement Raman spectroscopy for direct chemical imaging over a complete 2D area. This latter imaging mode is commonly addressed as *direct imaging*; i.e., the whole field of view is examined for a narrow spectral range of Raman shift wavenumbers (using a band-path filter), adjusted to the characteristic spectral feature associated with a specific chemical compound. Because the whole image is recorded in "real time," rather than by (time-consuming) point-wise buildup, the methodology lends itself favorably for *in vivo* time- (dynamic) and space-resolved Raman imaging. Thus, the evolution of the selected, chemical compound across the sampled region can be followed.

The various imaging modes—from zero-dimensional point via one-dimensional line to 2D sheet laser illumination—are shown schematically in the top-row panels of Figure 19.1; in addition, the methodology of confocal imaging is included, which has become one of the most popular incarnations in RM and which is outlined in the figure. Note that for completeness, in the bottom row of Raman imaging methodologies, popular enhancement and nonlinear techniques are shown as well, including surface-enhanced and tip-enhanced Raman imaging (SERS and TERS) and the nonlinear processes of stimulated and coherent Raman imaging (SRS/SRL and CARS). These will be discussed in more detail in Sections 19.2 and 19.3.

## 19.1.2 Confocal Raman imaging

Just combining a microscope with a Raman spectrometer does not necessarily result in high-quality imaging. Two specific problems are encountered. Firstly, the actual sampling volume is not necessarily very well defined, specifically when it comes to axial depth restriction. Secondly, associated with the primary "Raman-active" probe volume is a larger "fluorescence" volume since the fluorescence yield is orders of magnitude higher than the Raman scattering probability; hence, the Raman signal is often completely swamped by fluorescence. To a large extent, these problems can be overcome by appropriate spatial filtering—in the so-called *confocal microscopy* configuration—which provides the ability to spatially filter the analysis volume of the sample, both in the lateral $X$- and $Y$-directions and in the axial (depth) $Z$-direction (see, e.g., the discussion on diffraction limits in Sections 16.3 and 17.3).

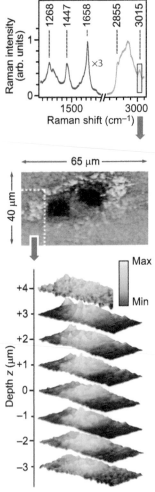

**Figure 19.2** Confocal Raman imaging of HAoECs with lipid droplet (LD) accumulation. (Top) Characteristic Raman spectrum. The bands at $\tilde{v} \approx 1660\ cm^{-1}$ and $\tilde{v} \approx 3015\ cm^{-1}$ are indicative of the lipid molecule. (Middle) Raman map of a single HAoEC, at $\tilde{v} \approx 3015\ cm^{-1}$; mapped area $d_X \times d_Y = 65 \times 40\ \mu m^2$. (Bottom) imaging stack (at $\tilde{v} \approx 1660\ cm^{-1}$), for different depth locations, of LD accumulation in the cell (imaged cell segment indicated by the white dotted area). (Data adapted from Majzner, K. et al., *Anal. Chem.* 86, no. 13 (2014): 6666–6674.)

To briefly summarize again, confocality is implemented by a set of two correlated pinhole apertures. The first one is placed in front of the laser light source. This "point source" is refocused by the microscope objective and illuminates only a "magnification-scaled" single point on a sample. As a consequence, scattered light and fluorescence light from any other part of the sample are eliminated, and the background contribution to the Raman signal is reduced by a few orders of magnitude, without affecting the focal laser illumination intensity. It effectively constitutes a spatial filter on the *X–Y* plane. The second pinhole aperture is placed between the microscope objective and the image plane; it rejects any residual light originated from any out-of-focus points in the sample and functions as a spatial filter in the *Z*-direction.

The mapping can be achieved in two ways, either (1) by scanning the laser light beam with a system-internal set of oscillating mirrors or (2) by moving the sample itself by using an *X–Y–Z* scanning stage. This ability to control the Raman probe position (in three dimensions) opens up the capability to implement a number of distinct sample mapping modi:

- *X–Y optical sectioning (surface scanning):* Mapping of a (flat, thin) sample layer in two dimensions, generating a single optical *X–Y* section.

- *X–Z cross sectioning (depth profiling):* Generation of a 2D optical cross section of the sample in the *X–Z* direction by moving the focus position (thus probing the internal structure of the sample).

- *Z-series scanning (full three-dimensional [3D] X–Y–Z scanning):* A combination of *X–Y* and *X–Z* scanning. This accumulates multiple *X–Y* optical cross sections along the *Z*-axis.

- *Surface profiling (utilizing autofocusing):* While scanning across the sample surface, the focus of the microscope objective is adjusted, so that the focus lies exactly at the sample surface. This can be seen to resemble a profilometer that provides not only information about the surface morphology but also detailed chemical compound information.

Of course, the spatial resolution commonly achievable in Raman microscopic imaging depends on the laser wavelength and the magnification of the imaging objective; typically, one encounters lateral and axial resolutions of $\Delta d_{X,Y} = 0.25$–$0.50\ \mu m$ and $\Delta d_Z = 2$–$3\ \mu m$, respectively. An example of a *Z*-series Raman image scan is shown in Figure 19.2.

In their work, Majzner et al. (2014) studied liquid droplets (LD) in single endothelial cells—specifically human aortic endothelial cells (HAoEC)—by using 3D linear Raman spectroscopy imaging. The goal was to determine the composition of LDs and how the uptake of exogenous unsaturated free fatty acids resulted in detrimental effects in the cells, known to be responsible for increased morbidity due to obesity. The results demonstrate the usefulness of Raman spectroscopy for characterizing intracellular lipid distribution, in both two dimensions and three dimensions, and for determining the degree of saturation.

Note that LDs, as well as unsaturated fatty acids, can be identified in Raman spectra by the vibrational bands at $\tilde{v} \approx 1660\ cm^{-1}$ and $\tilde{v} \approx 3015\ cm^{-1}$, which are linked to the stretching mode of =C–H; their intensity is approximately proportional to the number of C=C double bonds in the lipid molecule. Note that the Raman bands in the region $\tilde{v} = 1000 - 2000$ cm$^{-1}$ are multiplied by a factor of 3 relative to the higher-wavenumber bands. To obtain multidimensional information about the 3D distribution and size of LDs, confocal Raman microimaging was used to illustrate the distribution of LDs in a single HAoEC (see the bottom part of the figure); the change in lipid signal strength, as a function of depth position (in steps of $\Delta d_Z = 1$ µm), reveals the size and distribution of LDs in the cell.

This example demonstrates that, indeed, confocal Raman imaging can be a powerful tool in the investigation of biological tissue; in fact, confocal RM has emerged as a field of research in its own right (for a comprehensive summary, see, e.g., the textbook by Dieing et al. 2011).

### 19.1.3 Hyperspectral Raman imaging in two dimensions and three dimensions

As noted earlier, in the most common Raman imaging implementations, for each position in the probed sample a complete Raman spectrum is recorded; this results in the generation of a hyperspectral data cube (for hyperspectral image information, see Section 16.1). To reiterate, the collection of spectra taken from a 2D $X$–$Y$ area scan is represented by a 3D data cube, whose $X$- and $Y$-dimensions hold spatial $X$- and $Y$-coordinate information and whose $Z$-dimension is associated with the spectral information.

Expanding the sample scan to three dimensions, i.e., including depth profiling in the $Z$-direction, then the 3D spectral scan is represented by a four-dimensional (4D) data cube, which comprises the three spatial dimensions plus an additional spectral dimension.

However, in general, the human mind struggles to easily comprehend dense 3D and 4D array structures of data. Therefore, image reconstruction software handling the interpretation of such data cubes normally includes potent data mining functionality; i.e., with intelligent questioning and/or prior knowledge, useful information from the multidimensional data structures is extracted by slicing them along any of spatial or spectral coordinates.

Indications of the necessary mathematical framework applied to the decomposition of hyperspectral chemical image data for biological cells (but which is equally suitable for other scenarios) can be found, e.g., in Klein et al. (2012) or Li et al. (2017).

In the quest to unravel molecular compound information from the hyperspectral data cube and to construct spectrochemical maps of the sample, one of the most prominent methods is *spectral unmixing*. This refers to procedures for first decomposing mixed-component spectra into a series of constituent spectra (often called endmembers) and then deducing the spatial abundances of the estimated spectral signals, which represent the ratio of each endmember in any given pixel. The potential success of spectral unmixing relies heavily on the expected type of mixing; by and large, the procedure encompasses a simple yet normally rather representative model, the so-called *linear mixing model*

(LMM)—see, e.g., Keshava and Mustard (2002). The key advantage of the LMM is that each pixel in hyperspectral image cube may be seen as a linear mixture of each endmember. The mathematical description of the LMM is given by

$$x[n] = \sum_{i=1}^{P} a_i \cdot s_i[n] + v[n], \qquad (19.1a)$$

or in abbreviated form

$$X = \overline{\overline{A}} \cdot S + V. \qquad (19.1b)$$

Here $x[n] \in \mathbb{R}^L$ for each $n = 1, ..., N$, where $N$ is the total number of pixels and $L$ is the number of spectral channels; $a_i \in \mathbb{R}^L$ for each $i = 1, ..., P$, where $P$ is the number of endmembers—$a_i$ is called often the *endmember signature vector*, which indicates the constituent spectra of pure components in the sample; $s_i[n]$ is a fractional measure of the abundance of endmember $i$ in a given pixel $n$, often called the *coefficient vector*; and $v[n]$ represents the *noise components*. Note that $\mathbb{R}^m$ represents the respective $m$-dimensional image data space. In the abbreviated notation, one has $X \in \mathbb{R}^{L \times N}$; $A = [a_1, ..., a_P] \in \mathbb{R}^{L \times P}$; $S = [s_1[n], ..., s_P[n]]^T \in \mathbb{R}^{P \times N}$; and $V \in \mathbb{R}^L$. These are called the *mixture matrix*, the *endmember matrix*, the *abundance matrix*, and the *noise*, respectively.

Since the linear mixture of components has to represent physical significance, the coefficient vector underlies two constraints, namely,

$$s_1[n] \geq 0 \quad \text{for } i = 1, ..., P \text{ and } \sum_{i=1}^{P} s_1[n] = 1. \qquad (19.2)$$

It is beyond the scope of this chapter to discuss hyperspectral unmixing and image reconstruction in detail; it is noted only that several methods can be used to identify the endmembers, based on geometry, statistics, and other theories utilized in hyperspectral analysis. Anyway, the estimation of the component abundances can be done by solving the constrained least-squares problem of Equations 19.1, utilizing, e.g., quadratic programming algorithms; for a few relevant methodic approaches, see, e.g., Li et al. (2017).

In the following, some aspects of data slicing and multivariate analysis (MVA) approaches are outlined with reference to two examples.

The first one addresses the "simple" slicing procedure and is shown in Figure 19.3; the Raman imaging data presented here are associated with a study of single, living T-lymphocyte cells, obtained in line-scanning mode (see Schie and Huser 2013).

As indicated already, Raman spectra recorded from a cell contain a wealth of information about its chemical composition, for reliably identifying major constituents, such as nucleic acids, proteins, carbohydrates, and lipids. This ability makes Raman spectroscopy an invaluable tool for the investigation of a range of biochemical aspects of living cells and—potentially—their dynamic changes.

However, the normally complex spectra are, in general, difficult to analyze, due to the fact that the majority of biological compounds share overlapping molecular vibrational modes and are thus often hard to distinguish. This means that more likely than not, sophisticated, multivariate statistical analysis tools are necessary to reduce the data and extract components of interest. The issue is highlighted by the data displayed in the figure. In the spectrum, a series of relevant Raman peaks

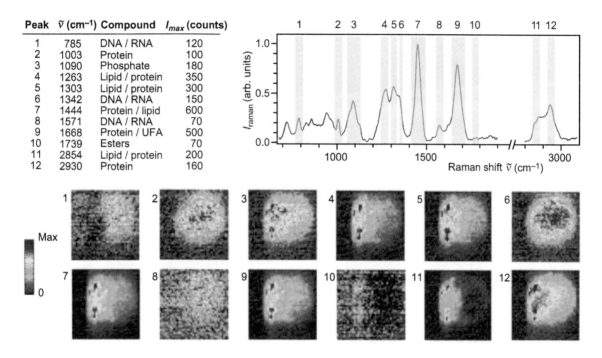

| Peak | $\tilde{\nu}$ (cm⁻¹) | Compound | $I_{max}$ (counts) |
|------|------|----------|-------------|
| 1 | 785 | DNA / RNA | 120 |
| 2 | 1003 | Protein | 100 |
| 3 | 1090 | Phosphate | 180 |
| 4 | 1263 | Lipid / protein | 350 |
| 5 | 1303 | Lipid / protein | 300 |
| 6 | 1342 | DNA / RNA | 150 |
| 7 | 1444 | Protein / lipid | 600 |
| 8 | 1571 | DNA / RNA | 70 |
| 9 | 1668 | Protein / UFA | 500 |
| 10 | 1739 | Esters | 70 |
| 11 | 2854 | Lipid / protein | 200 |
| 12 | 2930 | Protein | 160 |

**Figure 19.3**   Raman microspectroscopy for single-cell analysis. (Top) Raman spectrum of a living T lymphocyte obtained in line-scanning mode; the numbered spectral features are associated with the tabulated, vibrational modes biomolecules (the blue-hatched areas indicate the band-integration ranges). DNA, deoxyribonucleic acid; RNA, ribonucleic acid; UFA, unsaturated fatty acid. (Bottom) Scaled Raman images of the cell, for each of the indicated vibrational bands; the range of the (false-color) intensity scale is included in the tabulated information. (Data adapted from Schie, I.W. and T. Huser, *Appl. Spectrosc.* 67, no. 8 (2013): 813–828.)

is indicated, whose assignment to associated molecular bonds, and the corresponding biomolecules, is included in the summary table.

In the image panels in the bottom part of Figure 19.3, the intensity distribution for each of the highlighted Raman peaks is shown. From these spectrochemical images, it is possible, in principle, to deduce the presence and the location of particular molecular vibrations and their relative abundance. For example, one might use image 1 (associated with the ring-breathing mode at $\tilde{\nu} = 785$ cm⁻¹) to determine the DNA distribution throughout the cell. This distribution looks very similar to that of image 8 (at $\tilde{\nu} = 1571$ cm⁻¹), but rather dissimilar to that in image 6 (at $\tilde{\nu} = 1342$ cm⁻¹). This shows that the method of Raman peak slicing seems to work reasonably as long as the vibrational peak feature is isolated. But as soon as partial or full overlap is encountered, like, e.g., peak 6 sitting on the shoulder of peak 5, identification may indeed become very challenging to trace minute spectral changes in a particular measurement. Thus, as mentioned earlier, multivariate statistical analysis becomes of utmost importance in the analysis of Raman spectra and images of complex biological samples.

Note that for multivariate statistics the Raman band data cube information is represented in the form of a vector that is pointing to a specific location within an $n$-dimensional vector space, where $n$ is the number of data points (these count in the thousands). However, despite this complexity, a set of Raman spectra has a low-rank structure; in other words, the whole spectrum can be described by a relatively small number of (uncorrelated) variables. This is because (1) peaks

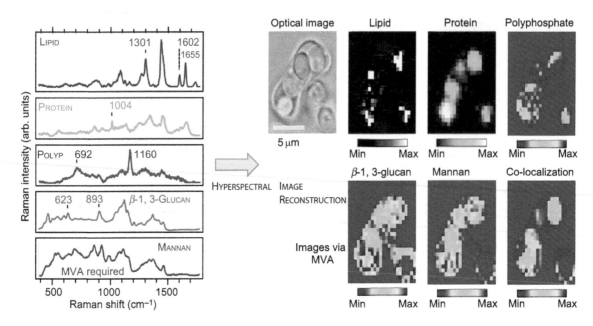

**Figure 19.4**    Label-free chemical imaging of fungal ascospores. (Left) Raman spectra of selected constituents encountered in the analysis of fungal ascospores (note that the composite fungal spectrum is not shown here); relevant Raman peaks are indicated. (Right) Component distributions images; top row: based on single- or few-peak slicing; bottom row: based on MVA. For shape comparison, an optical image is included. (Data adapted from Noothalapati, H. et al., *Sci. Rep.* 6 (2016): article 27789.)

normally extend over several data points and (2) groups of peaks may correlate with the same molecular vibration (which, thus, will scale jointly). This transformation from a complete data cube to a reduced set of variables is the basis for all statistical MVA methods, also known as chemometrics; commonly used techniques utilized in the analysis in Raman spectral imaging include least-squares fitting procedures, principal component analysis, cluster analysis, and blind source-separation ratios.

An example of the utilization of MVA analysis in the construction of Raman images, and the comparison with sliced-data cube extraction, is shown in Figure 19.4.

The displayed example is from a study of fungal spores, with the aim of distinguishing and directly visualizing structurally similar cell wall components, such as various polysaccharides (see Noothalapati et al. 2016). The pure-component spectra shown in the left-hand part of the figure suggest that in the analysis of the complete Raman spectrum (not shown here) from cells with a multitude of components, for some, single- or few-band analysis is sufficient—see the top row images for lipids, proteins, and polyphosphates—but that for others, multivariate data analyses are clearly needed—see the bottom row images for the naturally occurring polysaccharides $\beta$-(1,3)-glucan and $\beta$-(1,4)-D-mannan. The authors interpreted these images, together with colocation maps and optical images of the cell structure; their conclusion was that—specifically based on colocation results—that the ascus walls and spores both contained glucans but mannans were primarily distributed in the spores.

## 19.1.4 Examples of Raman imaging in biology and medicine

As already stated in the preamble to this chapter, and as might have become evident from the examples used earlier in the explanation of common principles, by and large, RM is dominated by applications in biology and medicine. This is for the obvious reason that Raman spectroscopy is one of the few laser-spectroscopic analytical methods that provide universal chemical specificity, although this cannot always be easily achieved. Over the years, quite a number of reviews have been written on the subject; here—without any claim of completeness—only a few recent ones will be mentioned. Downes and Elfick (2010) address Raman spectroscopy and related techniques and their use in biomedicine in general (note that the authors also included outlines of SERS and TERS, which will be covered in Section 19.2). Klein et al. (2012) summarize the quest for label-free live-cell imaging based on confocal RM and provide sorts of recipes for appropriate sample preparation and data treatment. Since such recipes are crucial for successful experiment and reproducibility in the wider research community, Palonpon et al. (2013a) provide some explicit protocols for molecular imaging experiments of live cells by RM. And finally, a rather up-to-date state of play of RM for cellular investigations is found in Kann et al. (2015).

Out of the wealth of studies two examples have been chosen, which both address Raman imaging of live cells and the prospects and ability to follow chemical dynamics within the cell. Here it has to be noted that the temporal resolution of spontaneous Raman imaging is still not competitive with that of fluorescence imaging. This is due to the fact that the Raman scattering probability is orders of magnitude smaller than that for fluorescence, so that in general, Raman imaging requires long measurement time (typically a few seconds per spectrum), and as a consequence, the buildup of a full 2D image utilizing confocal Raman imaging (i.e., point rastering) may take up to some hours. This approach is therefore unrealistic for imaging a living sample, which changes its condition over such long time intervals.

Nevertheless, slow dynamics of cells—such as, e.g., cell division—can still be tackled, utilizing "parallel" detection techniques, namely, the so-called line-scan mode mentioned earlier in the introduction. Instead of a single laser focus used in a conventional confocal Raman imaging, the sample is illuminated by a line-shaped focus (for details, see Section 16.1); thus, the equivalent of up to a few hundred points can be measured simultaneously, and, consequently, the image acquisition time decreases proportionally to the number of simultaneous detection points. An example of the application of this approach is shown in Figure 19.5, the Raman images obtained from a HeLa cell during the final stages of cell division (see Palonpon et al. 2013a).

The authors monitored HeLa cells over periods of up to an hour, recording Raman images about every 5 minutes. Such a time-lapse sequence for a single cell is shown in the lower part of the figure. One clearly can follow the division of the cell body from the mapping of selected chemical constituents of the cell, e.g., proteins (the associated key Raman peak in the spectrum—shown in the upper part of the figure—is at $\tilde{\nu} = 1686$ cm$^{-1}$). The movements of mitochondria and LDs, associated with cell division, are also visualized—colored green and red in the images. Not shown here is that the authors also imaged HeLa cells undergoing division of chromosomes.

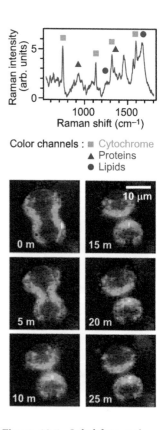

**Figure 19.5** Label-free molecular imaging of live HeLa cells by RM. (Top) Representative Raman spectrum from a single HeLa cell; spectral features for the biocompounds cytochrome (green symbols), proteins (blue symbols), and lipids (red symbols) are annotated. (Bottom) Time-lapse Raman images of an apoptotic HeLa cell, following the final stages of cell division and the evolution of the spatial distribution of certain biochemical compounds (color coded in the images). (Reprinted from *Nat. Protoc.* 8, no. 4, Palonpon, A.F. et al., 677–692, Copyright 2013, with permission with Elsevier.)

The second example shown here follows the dynamics of fatty acid uptake and its storage in the form of triglycerides in monocyte-derived macrophages (Matthäus et al. 2012); macrophages play a key role in atherogenesis since their transformation into so-called foam cells is responsible for the deposit of lipids in plaques within arterial walls. The authors tried to image the metabolism of such lipids and to trace their subsequent storage patterns within the cells. To distinguish the different individual lipid species of interest, such as fatty acids or cholesterol, from other naturally occurring lipids, deuterium labels were introduced by incubation in various deuterated compounds. A sample Raman image sequence for the intracellular lipid metabolism in THP-1 macrophages, incubated with palmitic acid-$d_{31}$, is shown in Figure 19.6.

The authors recorded Raman spectral images for incubation times, ranging from half an hour up to one-and-a-half days. To easily follow the (quantitative) uptake of the deuterated species, the spectra were normalized to the CH stretch vibrational mode, at about $\tilde{v} \approx 2850$ cm$^{-1}$; the progressive increase in the CD stretch intensities, at about $\tilde{v} \approx 2100$ cm$^{-1}$, is easy to notice for the observed period (see the spectra in Figure 19.6a).

Based on MVA of the spectra, the images show a gradual increase and spatial distribution of the deuterated fatty acid in comparison with the inherent cellular proteins. Not shown in the spectra and the images are other important compounds and their ratios to inherent, nonincubated cell contents. Specifically, the authors followed the evolution of the Raman peak at $\tilde{v} \approx 2850$ cm$^{-1}$—associated with the C=O stretch mode of the carboxy groups; the increase in its intensity implied that the deuterated fatty acid molecules were predominantly stored in the form of triglycerides, as expected.

The spectral information could be translated into quantitative values for the uptake of fatty acids and the storage as (transformed) triglycerides; this is shown

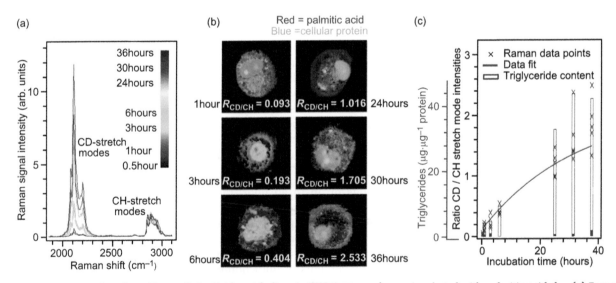

**Figure 19.6**   Raman imaging of intracellular lipid metabolism in THP-1 macrophages, incubated with palmitic acid-$d_{31}$. (a) Raman spectra obtained from LDs in THP-1 macrophages at different times of incubation. (b) Raman images of THP-1 macrophages, in dependence on incubation time; the CD/CH band intensity ratios reflect the amounts of deuterated fatty acid molecules normalized to the protein composition of the cells. (c) Uptake dynamics for palmitic acid-$d_{31}$ and quantification of intracellular storage of the fatty acids as triglycerides. (Data adapted from Matthäus, C. et al., *Anal. Chem.* 84, no. 20 (2012): 8549–8556.)

in Figure 19.6c, in which the CD/CH spectral ratios are correlated with actual triglyceride concentrations (in units of micrograms per microgram of protein), which exhibit an exponential dependence and leveling off after about 36 hours.

## 19.1.5 Nonbiological applications of Raman imaging

While certainly biomedical cell and tissue imaging dominate RM, there is a range of other applications that have evolved into routine analytical tasks. For example, it is quite common today to check pharmaceutical samples in high-throughput screening for, e.g., drug content and its homogeneous distribution throughout individual tablets. To name but a few further key examples, RM is also utilized in the following:

- Geology and mineralogy (e.g., to trace fluid inclusions or mineral and phase distribution in rock sections)

- Analysis of carbon materials (e.g., single-walled carbon nanotubes [CNTs], defect/disorder analysis in carbon materials)

- Semiconductor analysis (e.g., to characterize intrinsic stress/strain, doping effects, or superlattice structures)

Here two particular examples were selected to demonstrate the versatility of Raman imaging, namely, (1) an application in forensics and (2) an example of mineralogical studies of extraterrestrial objects.

It is well established that fingerprints are one of the best and speediest forms of personal identification for criminal investigation purposes. While in general, forensic fingerprinting it is only the actual print pattern that is utilized, for identification, RM of prints would, in addition, open the prospect of detailing, from the spectrochemical composition of the residue deposited by the finger contact to whether the perpetrator had, e.g., contact with explosives or drugs. In their study, Deng et al. (2012) explored this possibility. In particular, the authors studied latent fingerprints, the most common type used for criminal evidence. It is well established that fingerprints mainly comprise sweat (sweat gland secretions contain $\beta$-carotene [betaC]) and fat. In their proof-of-principle study, the authors used $\beta$-carotene as the "trace" molecule, mixed into a fatty petroleum solution. Some representative results are shown in Figure 19.7.

The Raman map image shown in the figure was recorded for a relatively large concentration of $\beta$-carotene in the sweaty print; the image was constructed from the spectral intensity information in the characteristic peak of $\beta$-carotene at $\tilde{v} = 1545$ cm$^{-1}$. It should be noted that a similar plot utilizing the other characteristic Raman peak gave similar clear results. The authors also demonstrated that they could still map fingerprint images with very low $\beta$-carotene concentrations in their test solution, with the limit of detection on the order of $c_{LOD}$(betaC) = 3.4 × $10^{-9}$ mol·L$^{-1}$. These results suggest that Raman imaging has the potential to simultaneously map different materials in fingerprints, down to trace-level concentrations, on different substrates, and with high speed and accuracy.

In the second example, an application from mineralogy, in the widest sense, is being described. In mineralogy, not only it is often indispensable to identify the mineral and perform chemical composition analysis, but also the knowledge of individual mineral phases in, e.g., fine-grained clusters is often of equal importance. Here Raman imaging has a clear advantage over other analytical methods,

Picture of fingerprint

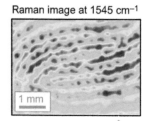

Raman image at 1545 cm$^{-1}$

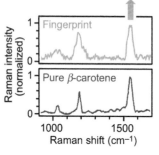

**Figure 19.7** Line-scan Raman spectrochemical imaging of fingerprints. (Top) Camera picture of the (thumb)print sample. (Middle) 2D distribution map of the $\beta$-carotene Raman signal (for the characteristic peak at $\tilde{v} = 1545$ cm$^{-1}$). (Bottom) Comparison of Raman spectra from the print and a pure $\beta$-carotene sample. (Data adapted from Deng, S. et al., *Appl. Opt.* 51, no. 17 (2012): 3701–3706. With permission of Optical Society of America.)

providing all this information in a single spectrum. A further bonus is that, besides the chemical analysis capability, Raman imaging is "nondestructive" by nature. This latter fact is particularly important in cases where the samples are only very small or are only one of a kind, so that taking away even a small amount for destructive analysis would soon completely consume said sample.

A striking example of this is the analysis and characterization of extraterrestrial specimens, which one would not be able to replace once they were destroyed. The investigation of such unique and irreplaceable samples, such as (Martian) meteorites therefore requires close control, and any proposal for using a destructive analysis method is usually frowned upon (although it cannot always be avoided). For demonstration of all aforementioned points, in Figure 19.8 a Raman spectrochemical image of a polished slice from the Martian meteorite MIL 03346 is shown (see Fries and Steele 2011); the Raman image is complemented by a reflected-light photograph, for identification of structural features. This nakhlite meteorite belongs to the so-called Miller Range (MIL); these have been studied extensively, utilizing a wide range of analytical techniques (including Raman spectroscopy), because of their unique, complex secondary mineral phases and their potential implications for the hydrologic history of Mars. The practical result from the image shown here is that RM can identify individual minerals and their phases (here jarosite, goethite, and clay minerals), even in fine-grained clusters of micrometer to submicrometer dimension. For a comprehensive Raman imaging study of this very meteorite, see Ling and Wang (2015).

*MIL 03346* Martian Meteorite
Reflected light image

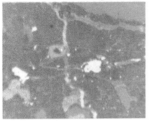

Raman image

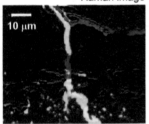

Red : jarosite
Green : geothite
Blue : clay minerals

**Figure 19.8** Confocal Raman imaging of an alteration vein in the Martian meteorite MIL 03346. (Top) Reflected-light image of a polished thin section. (Bottom) Raman spectrochemical image, with color-coded mineral phases highlighted. (With kind permission from Springer Science+Business Media: *Confocal Raman Microscopy*, T. Dieing et al. (eds.), Raman spectroscopy and confocal Raman imaging in mineralogy and petrography, 2011, pp. 111–135, Fries, M. and A. Steele.)

## 19.2 SURFACE- AND TIP-ENHANCED (SERS AND TERS) RAMAN IMAGING

With the constant stream of publications related to imaging and exploiting Raman enhancement techniques, specifically surface-enhanced Raman imaging (here dubbed i-SERS) and tip-enhanced Raman imaging (here dubbed i-TERS), it is next to impossible to decide which one to declare the state of the art. For example, in the field of the application of i-SERS in biomedicine—probably the most active at the moment—dozens of new research results are published every month: one may literally take a random pick to have found something novel and exciting.

Therefore, only key fundamental aspects of importance for imaging applications of the two techniques are outlined in this section (for the basics of SERS and TERS, the reader is referred back to Chapter 14); the versatility of i-SERS and i-TERS will be discussed with the help of a handful of representative examples.

### 19.2.1 Biomedical imaging based on SERS

In the basic configurations of SERS spectroscopy, the predominant aim is to enhance the signal associated with minute amounts of sample materials (i.e., molecules at trace-level concentrations) sufficiently to be detectable. In imaging (of, e.g., live biological tissue), this basic approach is not necessarily sufficient, since the size of a cell might be too large so that particularly interesting segments may lie outside the enhancement distance or the molecular compound signals may not be sufficiently selective. In this context, i-SERS microscopy is a novel method of vibrational imaging, for the selective, localized detection of biomolecules; conceptually, it combines biofunctionalized metal nanoparticles (NPs) and Raman microspectroscopy.

Based on localized surface-plasmon resonances (LSPRs) of metallic NPs, gold nanoparticles (AuNPs) or silver nanoparticles (AgNPs) were developed to serve as sensitized, plasmonic nanosensors that could be introduced into the cell itself and act as localized enhancement intermediators. Now i-SERS microscopy has enabled one to both visualize and quantify the distribution of a range of biologically important target molecules. The method combines several advantages, specifically (1) rather simple sample preparation, (2) good reproducibility, (3) ultrasensitivity, and (4) compatibility with many biomolecules. In this respect, NPs have proven to be a sort of a holy grail in nanomedicine, now serving as biomarkers, drug-delivery vehicles, spectroscopic fingerprint enhancers, or cell visualizers or ideally all of these at the same time.

SERS-active AuNPs and AgNPs have been conjugated with a large variety of targeting ligands, such as, e.g., peptides, proteins, antibodies, antibody fragments, DNA, and affibodies, for molecular imaging applications (see Figure 19.9).

Besides the desired biochemical functionality, a main driving force in the engineering of LSPR-based plasmonic biosensors has been the improvement of their sensitivity; strategies for addressing this may be categorized based on their different sensing mechanisms (for a concise review, see, e.g., Guo et al. 2015; and Li et al. 2017a).

Key examples of i-SERS based on sensitized metal NPs are the imaging of epidermal growth factor receptors (EGFRs), including members of the HER family of receptors that controls a complex network of ligand–receptor interactions and cellular responses; the imaging of prostate-specific antigen; the imaging of folate receptor; the imaging of nuclear targeting probes; or the imaging of apoptosis, to name but a few. A brief summary of now popular applications, some of them even demonstrated in live animals, was given, e.g., by Zhang et al. (2011). Out of the overwhelming wealth of applications, here only two are highlighted.

The topic discussed in the first example of enhanced Raman imaging of cells is that of cell metabolism and the generation of reaction products in the presence of molecular radicals. It is known since the late nineteenth century that nitric oxide (NO) is a physiological messenger and effector molecule in biological systems. For example, excessive and unregulated NO synthesis has been implicated in various studies as a contributing factor to pathophysiological conditions, such as, e.g., endothelial dysfunction and neurodegenerative diseases. However, the accurate determination of intracellular concentrations of NO has remained challenging. In their work, Xu et al. (2017) designed specific gold (Au) nanostructured probes with designed messenger molecules, which enabled sensitive, quantitative imaging and biosensing of NO in live cells.

For this, the authors synthesized a novel organic compound, 3,4-diaminobenzene-thiol (DABT), which comprises two distinct parts, namely, an *o*-phenylenediamine group for specifically recognizing NO to generate benzotriazole, and a thiol group for immobilizing the DABT molecule onto trisoctahedral Au nanostructures (hereafter abbreviated as Au/DABT).

When the *o*-phenylenediamine group in DABT reacted with NO, over time a new Raman peak appeared in the spectrum at $\tilde{\nu} = 698$ cm$^{-1}$ (see Figure 19.10a); this

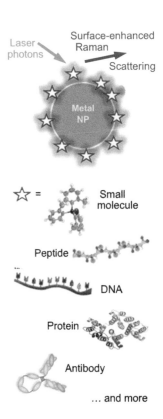

**Figure 19.9** Conceptual process of SERS from sensitized metal NPs; a selection of possible sensing receptors are indicated.

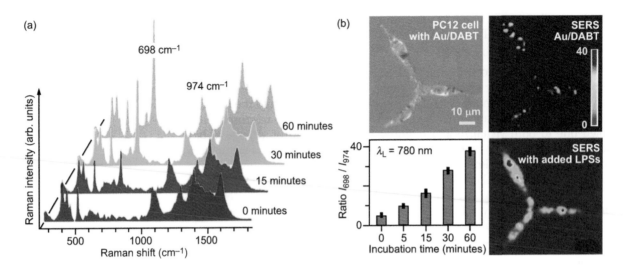

**Figure 19.10** SERS imaging and biosensing of NO-triggered reactions in live PC12 cells, utilizing Au/DABT nanoprobes. (a) SERS spectra from Au/DABT-laced cells, revealing the reaction between DABT and NO to generate $1H$-benzo[$d$][1,2,3]-triazole-6-thiol (BTAH), with the enhancement over time of its characteristic peak at $\tilde{v}$ = 698 cm$^{-1}$ in evidence (the ratio values are plotted on the right). (b) Optical image of cell treated with Au/DABT nanoprobes (top left), as well as ratiometric SERS images of $I_{698}/I_{974}$ obtained after 4 hours of incubation with Au/DABT nanoprobes (top right) and after 1-hour stimulation with additional lipopolysaccharides (LPSs) (bottom right). (Data adapted from Xu, Q. et al., *Chem. Commun.* 53, no. 11 (2017): 1880–1883.)

is attributed to the triazole ring scissor mode of the reaction product BTAH. On the other hand, the signal from the out-of-plane bending mode of the C–S bond at $\tilde{v}$ = 974 cm$^{-1}$ remained unchanged. Notably, the observed peak intensity ratio $I_{698}/I_{974}$ exhibited good linear correlation with the concentration of NO over a wide range of about 50–1000 nM; thus, this ratiometric measure can serve as an accurate, quantitative indicator for NO (see the Raman line ratio plot in Figure 19.10b).

The remarkable analytical performance of the Au/DABT SERS probe, together with its excellent biocompatibility and stability over long periods, enabled the authors to conduct real-time monitoring and imaging of the NO content in live cells.

When comparing the optical images of the PC12 cells with the Raman ratiometric images, it is evident (1) that the Au/DABT probes easily enter live cells and (2) that these nanoprobes mainly accumulate in the cytoplasm. Note that after additional stimulation by lipopolysaccharides, which can produce NO in cytoplasm, the signal amplitude in the ratiometric SERS map increases dramatically (see the top right and bottom right of Figure 19.10b).

The second example of i-SERS is from a theme of rapidly growing interest and progress, namely, that of theranostics based on NPs. Inspired by the ability of SERS "nanoantennas" to serve as a platform for enhanced *in vivo* disease targeting, Conde et al. (2014) aimed at developing a highly sensitive probe for *in vivo* tumor recognition, with the additional capacity to target specific cancer biomarkers. In their study of (engineered) antibody–drug–AuNPs, the authors quite convincingly showed how their engineered vehicles readily lend themselves for tumor theranostics, specifically to monitor the effectiveness and selectivity of the drug treatment respective cancer growth, comparing SERS fingerprinting and tomographic epifluorescence imaging from the Raman reporters. A concept summary of their approach, including relevant data, is shown in Figure 19.11.

The authors used AuNPs of radius $r_{NP} \approx 90$ nm as their vehicle, covered with 3,3′-diethylthiatricarbocyanine iodide (DTTC) dye molecules as Raman reporters and encapsulated by a common polyethylene glycol (PEG) coat. These AuNP-DTTC-PEGs served as the substrate for binding cetuximab (Cab), a monoclonal (Food and Drug Administration-approved) antibody–drug conjugate that specifically targets EGFRs that often appear in high amounts on the surface of cancer cells. In their studies, the authors used tagged and untagged specimens—AuNP-DTTC-PEG-Cab and AuNP-DTTC-PEG, the latter as a nontherapeutic reference. The *in vivo* epifluorescence images show (1) that AuNPs accumulate preferentially in the region of the tumor and (2) that drug–Raman NPs have led to an extensive tumor size reduction, evidenced by the much reduced luminescence signal. In the lower part of the figure, representative SERS spectra from the tumor site are shown (measured through the tumor tissues); the characteristic SERS peak at 508 cm$^{-1}$ is indicated. The notion of the tumor cells as attractors for attack by AuNPs, whether drug tagged or not, is evidenced by the near-absence of a SERS signal recorded off the location of the tumor (green data trace in the figure). Overall, these findings highlight the value of drug–Raman AuNPs as SERS reporters for *in vivo* theranostics.

In a similar fashion, other research groups have exploited SERS nanotags to study, *in vivo*, other theranostic problems, such as, e.g., multiplexing of triple bioconjugate SERS nanotags (to detect intrinsic cancer biomarkers) and time monitoring their presence in the system (Dinish et al. 2014).

## 19.2.2 Raman imaging at the nanoscale: TERS imaging

Nowadays the imaging of surface morphology generally is carried out by utilizing atomic force microscopy (AFM) and scanning tunneling microscopy (STM); spatial resolutions down to subnanometer precision are routinely achieved. However, it still remains a major challenge to simultaneously investigate the surface chemistry and composition with the same spatial resolution. In particular, the visualization of individual molecules, with chemical recognition, has been on the wish list of researchers in, e.g., catalysis, molecular nanotechnology, or biochemistry.

In this context, vibrational spectroscopy in the form of TERS comes into its own, allowing one to access the spectromolecular signals of even subnanometer features at the sample surface (note that TERS constitutes a surface-analysis tool). Molecular species very efficiently probed by enhanced Raman scattering, via the strongly localized plasmonic fields produced by the Raman laser light at the tip apex. The basic concepts of TERS probing were already outlined in Section 14.3; here a brief summary is given, with reference to Figure 19.12.

The principles of TERS and SERS are closely related: for both, metallic NPs or nanostructures are responsible for the enhancement of the normally weak Raman signals from the probed molecules, often by several orders of magnitude. The key difference is that the enhancement in TERS originates from only a single, individual metallic nanostructure, namely, the end tip of the SPM/AFM probe; in contrast, in SERS, a multitude of enhancing nanostructure "hot spots" (locations at which analyte molecules are adsorbed) contribute to an "averaged" spectrum.

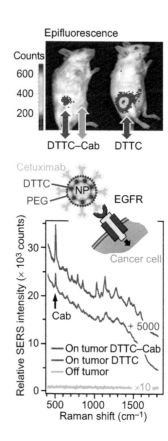

**Figure 19.11** Tumor assessment after AuNP administration. (Center) Conceptual interaction between tagged NP and EGFR. PEG, polyethylene glycol. (Top) Whole-body mice tomographic *in vivo* epifluorescence images, related to the uptake of AuNP-DTTC-Cab or AuNP-DTTC (i.e., with or without antibody–drug conjugate). (Bottom) SERS spectra of the tumor site, measured through the tumor tissue (left) and an off-tumor site (right). The spectral data are color coded to the positions in the tomographic images; the characteristic SERS peak is at $\tilde{\nu} = 508$ cm$^{-1}$. (Data adapted from Conde, J. et al., *J. Control. Release* 183 (2014): 87–93.)

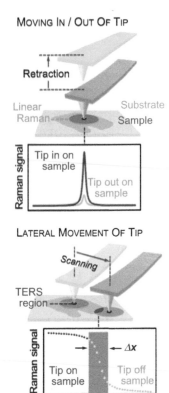

MOVING IN / OUT OF TIP

Retraction

Linear
Raman                    Substrate
                         Sample

Raman signal

Tip in on
sample
          Tip out on
          sample

LATERAL MOVEMENT OF TIP

Scanning

TERS
region

Raman signal

← Δ*x*

Tip on        Tip off
sample        sample

**Figure 19.12** Schematic illustration of the TERS principle. (Top) An effective contrast enhancement is provided by the comparison of Raman signal strengths for the tip in close proximity and being retracted. (Bottom) A lateral scan across the edge of a Raman-active sample exhibits an intensity slope; Δ*x* is a direct measure of the lateral dimension of the near-field contribution and, thus, the TERS image resolution. Note that green and red symbolize the size of the laser focus spot and the effective near-field enhancement area, respectively.

As in common SPM/AFM measurements, in TERS the sample can also be probed with spatial resolution on the nanometer scale, placing the tip at a specific initial position and then raster scanning the sample for building up a point-by-point (pixel) image, with the full Raman spectroscopic information recorded at each individual pixel. Note that the SPM/AFM resolution is maintained in TERS because the original (large) "far-field" laser illumination spot with area $A_{FF}$ is reduced to a much smaller, effective "near-field" area $A_{NF}$ for which the $E$ field enhancement, originating from the proximity of the tip, is sufficient to enhance the Raman signal as well (see the top part of Figure 19.12, where the far-field laser light spot is shown in green and the effective near-field enhancement spot is marked in red).

Note that when considering TERS for imaging applications, the spectroscopic contrast is of key importance, i.e., the enhancement over the far-field Raman signal, associated with the field confinement in the TERS area/volume. Unless the tip follows the surface topography during a scan (see the second example below), the TERS signal may experience different enhancement along the way. Commonly, the enhancement factor $EF_{TERS}$ is given by Equation 14.17 in Section 14.3. In Raman imaging, in general only the "contrast" part $C_{TERS}$ of that equation is contemplated, since during a scan it is not normally easy to keep track of the actual spot sizes $A_{FF}$ and $A_{NF}$:

$$C_{TERS} = \frac{I_{NF}}{I_{FF}} = \frac{I_{Tip\text{-}in} - I_{Tip\text{-}out}}{I_{Tip\text{-}out}} = \frac{I_{Tip\text{-}in}}{I_{Tip\text{-}out}} - 1, \tag{19.3}$$

where $I_{NF}$ and $I_{FF}$ are the pure near-field and far-field Raman intensities and $I_{Tip\text{-}in}$ and $I_{Tip\text{-}out}$ are the Raman signal intensity when the TERS tip is in close proximity to or retracted from the sample, respectively. For further general details and a wide selection of representative examples, one may consult some recent review articles, e.g., Schmid et al. (2013), Kumar et al. (2015) or Langelüddecke et al. (2015), Shi et al. (2017), and Verma et al. (2017).

Here three examples of TERS imaging will be presented that highlight the key intricacies of the method; these are spectrochemical imaging (1) of nanostructures—like e.g. carbon nanotubes (CNT)—and (2) individual molecules adsorbed on surfaces and (3) revealing of localized physical effects (here in the form of spin waves) during TERS scanning.

As pointed out earlier in the section, the surface morphology of materials and nanostructures is routinely imaged by using STM and AFM instrumentation. Adding laser illumination to the metal tip—thus realizing a TERS probe—enables one to add the spectrochemical dimension to the morphologic data set.

This principle is nicely exemplified for the simultaneous chemical and structural analysis of (individual) CNTs by STM-based TERS (see, e.g., Chen et al. 2014). In Chen et al.'s (2014) experiment, the authors deposited CNTs on a gold substrate and then illuminated the gap between the STM tip and the substrate with radiation from a HeNe laser ($\lambda_L = 632.8$ nm). During the scan of the tip, both the STM and TERS signals were recorded; this provides collocated spectrochemical and morphological images of the sample, as shown in Figure 19.13.

In the scan area, three individual CNTs can clearly be identified in the STM images; they are annotated 1, 2, and 3 (the rightmost image in the figure). When illuminating the gap between the tip and the gold substrate, strong (enhanced) Raman spectra are recorded when the tip hovers above a CNT; in contrast, the

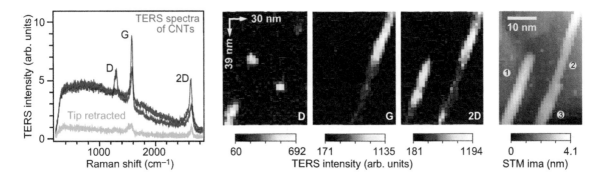

**Figure 19.13**   STM and TERS imaging of CNTs. (Left) Comparison of a far-field Raman spectrum (green, tip retracted from sample) with TERS spectra (red and blue traces, recorded at different locations of the CNTs); the characteristic D, G, and 2D bands are indicated. (Right) Spectrally resolved TERS images and a collocation STM image. Scan step size $\Delta d_{x,y} = 1$ nm; the intensity scales are offset by the averaged far-field background. (Data adapted from Chen, C. et al., *Nat. Commun.* 5 (2014): article 3312.)

spontaneous far-field Raman spectra recorded when the tip is retracted from the surface are much weaker. As expected, in the spectra, one can easily identify the signature Raman peaks of $sp^2$ carbon materials, namely, the D, G, and 2D bands (see the spectral traces in the left part of the figure). Note that the relative intensity of individual Raman peaks varies substantially between locations, exemplified by the red and blue spectral data curves; the authors speculate that this might be due to the structure of individual CNTs or could be associated with the actual, local environment.

When translating the TERS spectral intensities associated with the three signature Raman bands into spatiochemical images, a few facts are immediately obvious. First, the spatial resolution of the TERS images closely mimics that of the STM image, meaning that spectral analysis of the CNTs is possible with nanometer resolution (for the measurement results shown here, $\delta d_{x,y} \approx 1.7$ nm). Second, in general, the D band is related to forbidden phonon oscillations and for it to be observable requires either structural deformation or defects to be activated. This is evident from the associated image (leftmost in Figure 19.13) in which hardly any Raman intensity is observed, except for the ends of the CNTs: there "deformations"/"defects" occur due to the CNT end structure. Third, the G and 2D bands are observable in the images for all three CNTs; however, strong variations in intensity are also present, suggesting that localized effects are important for these as well, although they have totally different origins. Overall, the authors demonstrated that they were able to simultaneously carry out chemical and structural analyses of individual CNTs, with spatial resolution on the nanometer scale.

In STM/AFM surface scanning, it is now routine that atomic and molecular structures can be revealed. However, visualizing individual molecules with chemical recognition—a long-standing dream of researchers studying, e.g., catalytic or biochemical processes or nanostructure problems—has been rather elusive. In principle, TERS allows one to access the spectral fingerprints of molecular species, exploiting the strong nanometer-localized Raman signal enhancement. However, TERS imaging resolution is still mostly limited to about $\Delta d \geq 3$ nm; this is still insufficient to resolve the structure of a single molecule with chemical attributes. In their experiment, Zhang et al. (2013) demonstrated that they could drive the TERS spatial resolution below $\Delta d \leq 1$ nm. They achieved this by spectrally matching the resonance of the nanocavity plasmon (at the TERS tip) to specific molecular vibronic transitions. Some of their key results from TERS imaging of individual H2TBPP molecules (meso-tetrakis(3,5-di-tetrarybutyl-phenyl)porphyrin) are shown in Figure 19.14.

From electron microscopy and STM imaging, it is well established that the H2TBPP molecule exhibits a characteristic four-lobe pattern. This pattern is replicated in the TERS images of an isolated, single H2TBPP molecule on a Ag(111) monocrystal shown in the figure. Clearly, the intensities from the locations of the molecular lobes are stronger than those from the center; note that the TERS signal from the nearby Ag surface exhibits only a broad continuum.

This highlights the strongly localized nature of the TERS response. It is also clear from the two images that the contrast becomes less pronounced if the Raman band is weak; this is in line with the aforementioned contrast arguments earlier in this section. On the other hand, the observed Raman image contrast and frequency dependence can also be used to interpret the relative localization of the various vibrational modes of the H2TBPP molecule, with regard to the axial polarization of the highly confined local plasmonic field at the TERS tip. The authors conclude that the low-wavenumber vibrational modes have their origin predominantly in the lobes, but that the high-wavenumber modes contain substantial contributions from the porphyrin core. The authors support this interpretation with the (modeled) vibrational motion information provided in the supplementary video accompanying their publication: when simulating TERS images based on said localized vibrational mode, they closely resembled the experimental results.

It is clear that the ability to optically (i.e., nondestructively) access the conformal structure of a single (macro)molecule, providing both subnanometer spatial resolution and associated chemical (vibrational band) recognition, opens up a vast potential for exploring the nanometer-scale world.

In the last example, an issue related to the fundamental understanding of nanostructures will be addressed, namely, that nanomaterials may exhibit characteristic physical properties that are different from those of their bulk counterparts. Furthermore, additional complexity normally arises when certain NPs interact with metallic (substrate) nanostructures. One such possible interaction is the collective excitation of the electrons' spin structure in a crystal lattice, giving rise to so-called *spin-wave magnons*.

The occurrence of this phenomenon was seen, e.g., by Rodriguez et al. (2015), who studied the interaction of iron oxide NPs deposited on metallic nanostructure surfaces, on semiconducting Si(111) crystals, and on insulating glass substrates. When probing the interface regions of these samples by TERS, they observed the localized appearance of spin-wave magnons, but only when the tip was in close proximity to a nanocrystal edge. The authors concluded from this that the coupling of a localized plasmon with spin waves arises due to broken symmetry at the NP interface and the additional electric-field confinement associated with the TERS tip–laser light interaction. The measurement results shown in Figure 19.15 were conducted for $\alpha$-Fe$_2$O$_3$ NPs (of rhombohedral shape, with edge dimension of $D_{NP} \approx 40$ nm) adsorbed on a glass substrate; the TERS spectra were acquired using a Ag tip, under excitation with green light from a Nd: YAG laser at $\lambda_L = 532$ nm. Note that the experiment was performed in tip-scan mode and that the scan was only over a region sufficiently small to guarantee

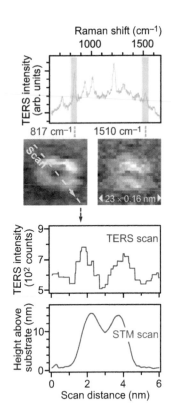

**Figure 19.14** TERS imaging of single H2TBPP molecules absorbed on Ag(111). (Top) Representative single-molecule TERS spectrum of a flat-lying molecule on Ag(111). (Middle) TERS maps (23 × 23 pixels, with step size $\Delta d_{pixel} = 0.16$ nm) of a single molecule for two different Raman peaks. (Bottom) TERS intensity profile scan along the line indicated in the TERS map, in comparison with the STM topographical profile scan along the same line. (Data adapted from Zhang, R. et al., *Nature* 498, no. 7452 (2013): 82–86.)

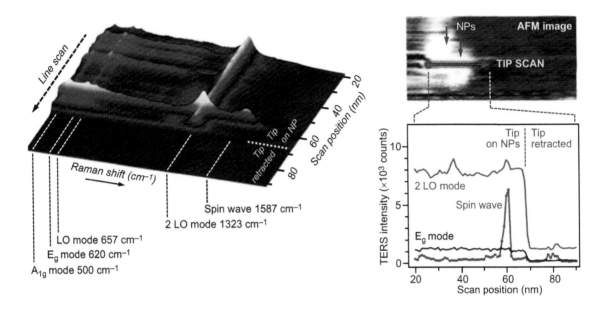

**Figure 19.15** TERS line scan of an iron oxide NP/glass interface, revealing spin waves. (Left) Full spectral TERS intensity profile; prominent Raman modes are indicated, together with key scan information. (Top right) AFM image of iron oxide NPs deposited on a glass substrate; the location and direction for the TERS line scan is indicated by the blue arrow. (Bottom right) TERS intensity profiles for three distinct Raman wavenumbers. (Data adapted from Rodriguez, R.D. et al., *Nanoscale* 7, no. 21 (2015): 9545–9551.)

constant far-field contribution to the signal. Thus, any observed spectral change could be attributed to the near-field response generated at the apex of the Ag tip.

As the spatiospectral TERS scan map on the left-hand side of the figure shows, one observes the characteristic Raman vibrational modes commonly expected from crystalline iron oxide (nano) samples, namely, the modes $A_{1g}$ (at $\tilde{v} = 500 \text{ cm}^{-1}$), $E_g$ (at $\tilde{v} = 620 \text{ cm}^{-1}$), LO (at $\tilde{v} = 657 \text{ cm}^{-1}$), and 2LO (at $\tilde{v} = 1323 \text{ cm}^{-1}$) (LO: longitudinal optical phonon mode). A further strong spectral feature is observed at $\tilde{v} = 1587 \text{ cm}^{-1}$, but only in close proximity to the interface edge between the NP and the glass substrate. This is associated with the localized spin-wave magnon surmised earlier.

The authors attribute this phenomenon to the specific rhombohedral geometry of the $\alpha$-$Fe_2O_3$ NPs (for a detailed discussion and further experimental evidence, see Rodriguez et al. 2015). Note also that for all other modes, a sudden increase in TERS intensity is observed at the same localization as the spin-wave magnon; again, the explanation for this is given within the framework of NP geometry.

All three TERS examples described here and the vast majority of investigations reported in the literature are for solid samples or for molecules adsorbed on solid surfaces. Therefore, for completeness, a final note is appropriate. As already mentioned in the introduction to this section, it is, in particular, the imaging of living cells and their spatiochemical composition that intrigue many researchers in bioscience and biochemistry. But TERS in a liquid medium environment has eluded efforts for a long time and is still difficult to achieve. For example, Ag tips are very popular in TERS experiments, but they suffer detrimental tip oxidation when coming in contact with many liquids. Another difficulty comes with the delicate optical alignment in refractive-index media, which is necessary to achieve accurate focusing of the laser beam on the TERS tip apex on the one hand and to ensure efficient collection of the scattered signal through a layer of liquid on the other hand. While an increasing number of examples for TERS imaging of biological cells have been published (see, e.g., Rusciano et al. 2014), work on TERS in the presence of liquid media is still extremely sparse. Probably one of the most promising approaches is that of Touzalin et al. (2016), who tried to develop an approach to imaging opaque samples in organic liquids.

## 19.3 SRL (STIMULATED RAMAN LOSS) IMAGING

In Chapter 17, a range of fluorescence-based imaging techniques have been discussed, including, e.g., confocal laser scanning, single-molecule microscopy, and superresolution imaging. However, many molecular species fluoresce only weakly or are intrinsically nonfluorescent, so that fluorescent labels, natural or artificial, need to be introduced; frequently, these are often perturbative. While spontaneous RM provides a label-free imaging solution (see Section 19.1), which in addition adds spectrochemical analysis capabilities, Raman signals are notoriously weak and, thus, image acquisition times are in general rather long. Even some of the Raman signal enhancement techniques outlined in Section 19.2 are not free of caveats; either they also require that external tags are administered, in the form of NPs, or—such as in the case of tip-enhanced Raman imaging (TERS)—extensive sample preparation is normally required, and live cells are mostly out of reach for TERS.

Here (coherent) nonlinear molecular spectroscopy techniques come to the rescue: a plethora of optical signals can be generated that do not rely on fluorescence emission and which lend themselves for high-contrast, label-free chemical imaging. Two key methodologies will be addressed in this section, namely, SRS—a nonlinear, dissipative optical spectroscopy technique in which the molecules exchange energy with two incident laser radiation fields (at different frequencies/wavelengths)—and CARS—a parametric generation spectroscopy technique in which incident and resulting light fields exchange energy with each other while the interacting molecules will have returned to their ground state at the end of the process.

Some of the key advantages encountered in nonlinear Raman imaging are, e.g., (1) nonfluorescent molecules can be imaged; (2) the techniques utilize high-frequency modulation (because of the use of high-repetition-rate, short-pulse laser sources), resulting in often shot-noise-limited sensitivity; or (3) overall nonlinear intensity dependence, which allows for intrinsic 3D optical section in Raman microspectroscopic applications. For in-depth studies, a comprehensive description of SRS and CARS, and its applications, can be found, e.g., in the book on coherent Raman scattering microscopy by Cheng and Xie (2012) or in the relatively recent reviews by Min et al. (2011), Winterhalder and Zumbusch (2015), and Kawata et al. (2017), to name but a few.

### 19.3.1 Concepts of SRL imaging

The principles of SRS have been outlined in Section 14.5; further general aspects related to SRS imaging may be found in some of the numerous concise surveys that have been published over the past few years (see, e.g., Nandakumar et al.

2009; Freudiger and Xie 2013; Tipping et al. 2016; or Prince et al. 2017). Here only a brief summary is provided, specifically highlighting the differences and advantages of SRS imaging over common spontaneous Raman (micro) imaging.

Recall that in SRS the sample is coherently excited by two lasers, namely, the pump laser radiation with frequency $\tilde{\nu}_P$ and the (stimulating) Stokes laser radiation with frequency $\tilde{\nu}_S$. When the frequency difference $\tilde{\nu}_{vm} \equiv \Delta\tilde{\nu} = \tilde{\nu}_P - \tilde{\nu}_S$ matches the frequency of a particular Raman-active vibration mode of molecules in the sample, $\tilde{\nu}_{vm}$, a nonlinear SRS response ensues, which manifests itself as a linked decrease in pump laser intensity and increase in Stokes laser intensity. Thus, one can measure the effect either by monitoring the pump laser intensity— this is known as stimulated Raman loss or SRL; or by monitoring the Stokes laser intensity—this is known as stimulated Raman gain (SRG). This is shown schematically in the top part of Figure 19.16.

It should be noted that in deviation from the representation in Section 14.5, the pump laser radiation is altered in frequency/wavelength, for both tuned and broadband realizations, rather than the Stokes laser radiation; both are equally feasible although for practical, experimental reasons. The scenario shown here is the common one encountered in SRS microscopy. Also, for historical and practical reasons alike, it is the SRL channel that is normally used to monitor the stimulated Raman response.

The typical experimental setup for SRS microscopy is shown in the bottom part of the figure. The SRS interaction is afforded by collinearly combining the pump and the Stokes laser beams and then focusing them through the microscope objective onto the sample. The SRL signal is detected in the forward (transmission) direction; note that the Stokes radiation is eliminated by the use of a suitable laser-line rejection filter. Scanning of the image area can be realized in two ways, either by lateral movement of the sample (as is common in spontaneous RM) or by piezo-driven mirror scanning of the laser beams. The latter approach is the preferred one in SRL imaging, since it is much faster than the slow translation of a heavy sample stage.

By intensity modulation (using acousto-optic or electro-optic modulators) of one of the laser beams, typically the Stokes laser beam, one can use lock-in filtering of the signal to suppress low-frequency noise. Recall that only if the two laser spectral frequencies obey the equality $\tilde{\nu}_{vm} \equiv \Delta\tilde{\nu} = \tilde{\nu}_P - \tilde{\nu}_S$ that a SRL response signal is observed; this means that no nonresonant background is encountered. This renders quantitative image analysis rather simple, since, in addition, the SRL signal is linearly proportional to the concentration of the analyte molecules.

As just stated, in general, in SRL ultrahigh-repetition-rate (mode-locked) lasers with picosecond- or femtosecond-pulse durations are utilized; the majority of these commercial laser sources operate at pulse repetition frequencies of $f_{PRF} \approx 80$ MHz. As shown in the experimental setup section of Figure 19.16, the stimulating Stokes laser beam is modulated; because of the very high basic pulse frequency, the modulation frequency can also be high and very often is in the order of $f_{mod} \approx 1$ MHz. In this way, signal fluctuations during a measurement are normally small since laser amplitude noise occurs primarily at low frequencies (associated with acoustic-frequency thermal and mechanical fluctuations following the common $1/f$ noise law). Exploiting this approach of combining the ultrahigh laser repetition rates with high Stokes wave modulation, it has been

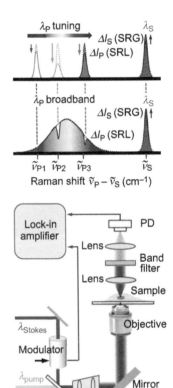

**Figure 19.16** Concepts of SRS microscopy. (Top) Result of resonant interaction with vibrational modes of the sample molecules, exhibiting loss and gain in the pump and Stokes laser intensities, respectively, for narrow-bandwidth tunable (upper panel) and broadband (lower panel) pump laser scenarios. (Bottom) Experimental realization of SRS microscopy; the Stokes laser source is modulated and the pump intensity loss (SRL) signal is monitored in the forward (transmission) direction. DBS, dichroic beam splitter.

demonstrated that sensitivities in the order of $\Delta I_P/I_P \leq 10^{-6}$ can be achieved, even for pixel dwell times of less than 100 µs. Note that for typical, biocompatible laser power levels, this comes close to the theoretical shot-noise limit.

One interesting attribute of SRL is that its sensitivity limits are nearly independent of the focal spot-size; i.e., the actual numerical aperture (NA) of the microscope objective does not play a role; some simple estimates confirm this. The focal volume can be approximated by

$$V_{foc} \propto A_{foc} \cdot Z_{foc} \propto r_{foc}^2 \cdot Z_{foc}, \tag{19.4}$$

where $r_{foc}$ and $A_{foc}$ are the radius and the area of the laser beam focus and $Z_{foc}$ is the focal depth for a Gaussian laser beam. According to geometrical optics $r_{foc} \propto 1/NA$ and $Z_{foc} \propto 1/(NA)^2$, which means that $V_{foc} \propto 1/NA^4$.

Furthermore, the excitation laser power densities in the focal spot plane can be approximated as $I_P, I_S \propto 1/A_{foc} \propto NA^2$. Therefore, the SRL signal is proportional to

$$S_{SRL} \propto I_L^2 \cdot V_{foc} \propto \left(NA^2 \cdot NA^2\right)/NA^4 \equiv const, \tag{19.5}$$

confirming the opening statement presented earlier. Of course, this crude estimate holds only as long as the laser spot size is smaller than the probed sample structure containing the Raman-active molecules and as long as the focal depth is small enough so that elastic scattering can be neglected. Note that in SRL imaging, as in all other basic microscopy implementations, the spatial resolution is diffraction limited; i.e., for laser wavelengths in the visible and NIR regions of the spectrum, $\Delta d_{x,y} \approx 0.5$ µm and $\Delta d_z \approx 1.5$ µm for the lateral and axial dimensions.

When comparing conventional spontaneous RM with stimulated Raman imaging (e.g., in the form of SRL), one finds some key advantages for the stimulated imaging process:

- The full image frame rate for RM is rather slow (usually on the order of many minutes per frame), while with SRL, frame rates of up to video rates have been demonstrated (up to ~25 frames·s$^{-1}$).

- RM has no inherent axial depth resolution, while SRL offers intrinsic optical z-sectioning.

- RM normally suffers from background fluorescence, while SRL is nearly free from it.

- On the downside, SRL provides only selective spectral information, while in RM, full Raman spectra are collected for each image pixel point.

This brief SRS imaging summary concludes with a few remarks on chemical specificity in the recorded SRL images. The simplest way to generate chemically specific SRL images is to tune the laser sources into resonance with a desired vibrational mode, record the image, and then sequentially continue the exercise for all vibrational bands of interest. While this can still be reasonably fast, keeping in mind the high laser pulse repetition frequencies, one may lose the colocalization information for different Raman modes, in particular if the analytes are in a dynamically changing environment. To overcome this, two multicolor, multiplex methodologies have been utilized in SRS microscopy.

The first approach is based on double modulation and double demodulation, in which, in addition to the amplitude modulation of the Stokes laser beam, the broadband pump laser radiation is modulated by a multichannel acousto-optical tunable filter device. This provides a sequence of different wavelengths for realizing multicolor imaging with multiple chemical contrasts for the quantitative imaging of different chemical species. Unfortunately, the scanning speed is limited, due to the occurrence of secondary frequency modulation and the need for demodulation procedures.

The second approach also utilizes broadband pump laser radiation. But rather than modulating the spectral content to match vibrational mode resonances, the full spectrum of the pump laser beam, including also the SRL components, is dispersed in the recording beam path and individual detectors are placed at the spectral SRL wavelength positions. This allows for parallel detection of all spectrochemical channels, and hardly any speed penalty has to be paid in comparison to single-color, single-molecular-mode SRL imaging.

In the following section, two representative examples of SRL microscopy of biological tissue samples are provided, highlighting key elements of importance in stimulated Raman imaging.

### 19.3.2 Selected applications of SRL imaging

In the first example, a topic is addressed that is paramount for the understanding of living-cell evolution, namely, the conversion of an external "stimulating" molecule into a new internal compound via metabolic reactions. For example, the synthesis of new proteins is a key biological process by which cells respond rapidly to environmental stimuli. However, selective visualization of such newly synthesized proteins in living cells, with subcellular spatial resolution, has been rather challenging. That indeed such distinction is possible and that the metabolic conversion process can be followed has been demonstrated, e.g., by Wei et al. (2013), who utilized SRS microscopy to visualize nascent and newly synthesized proteins in living cells, exemplified for the metabolic incorporation of deuterium-labeled amino acids.

In their studies, the authors used picosecond-pulse ($\tau_p \sim 5$–7 ps) duration lasers with pulse repetition rates of 80 MHz; the pump pulse wavelength was tunable, while the wavelength of the modulated ($f_{mod} = 10$ MHz) Stokes radiation was fixed. The SRL signals were recorded via a lock-in amplifier; the dwell time per pixel was set to $\Delta t_{SRL} = 100$ μs, which translated into a full-frame time of about $t_{frame} \approx 25$ s for the chosen image area of $512 \times 512$ pixels. These settings were used to image the newly synthesized proteins in live HeLa and HEK293T cells, with high spatial–temporal resolution. Some representative results for HEK293T cells are shown in Figure 19.17.

In the top part of the figure, a spontaneous Raman spectrum of an HEK293T cell is shown; the cells were incubated with a deuterium-labeled set of all amino acids for duration of 12 hours. Clearly, the signal at $\tilde{v} = 2133$ cm$^{-1}$ (associated with the C–D vibrational mode of deuterated proteins) is substantial, nearly as high as native amide I response at $\tilde{v} = 1655$ cm$^{-1}$. This is well in line with other results for the incorporation of deuterium-labeled amino acids into living cells (for a similar process, see, e.g., the study of THP-1 macrophages in Section 19.1 and the Raman spectral data in Figure 19.6).

The recorded SRL-image at $\tilde{v} = 2133$ cm$^{-1}$ reveals bright responses from the synthesized, deuterated proteins, with an intense pattern residing in nucleoli

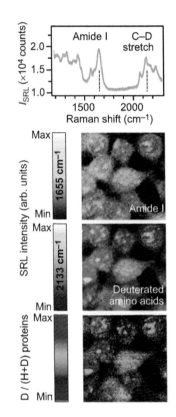

**Figure 19.17** SRS imaging of metabolic protein synthesis following exposure to (deuterium-labeled) amino acids in live human embryonic kidney (HEK293T) cells. (Top) Spontaneous Raman spectrum of HEK293T cells incubated with deuterium-labeled amino acids; vibrational mode relevant to the image analysis are labeled. (Bottom) Live-cell SRS images, targeting the C–D vibrational mode indicative of new protein synthesis (top image) and intrinsic cell proteins such as amide I (center image), as well as a ratiometric image illustrating the spatial distribution for nascent proteins (bottom image). (Data adapted from Wei, L. et al., *PNAS* 110, no. 28 (2013): 11226–11231.)

(top image). In contrast, the amide I image resembles a raster consistent overall protein distribution (center image). The ratio image between the two really highlights the nucleoli for active protein turnover in HEK293T cells (bottom image).

It should be noted that incorporating deuterium-labeled amino acids into live cells can be seen as a minimally perturbative process. Furthermore, SRS imaging of such exogenous carbon–deuterium bonds (C–D) exploits the normally cell-silent Raman spectral region; therefore, the method is very specific and sensitive and moreover quantitative. Thus, it looks quite likely that this technique of nonlinear vibrational SRL imaging of stable isotope incorporation may become a valuable analysis tool for *in vivo* studies of complex spatial and temporal dynamics of newly synthesized compounds in cells.

As in the example just discussed, in the majority of stimulated Raman imaging applications, probing of the sample by SRL spectroscopy has been implemented by narrowband laser excitation, accessing a particular, isolated Raman band. While being fast with respect to hyperspectral Raman imaging, single-color SRL lacks the ability to resolve overlapping Raman bands, which stem from a combination of (several) target molecules and background tissue components. Thus, single-color SRL is not able to provide compositional information; it is by and large applicable only to studying known species by using their respective, isolated Raman bands.

On the other hand, (label-free) spectrochemical imaging on the scale of milliseconds to a few seconds is desirable to monitor a number of "slow" dynamic cellular states and processes. For this, Liao et al. (2015) proposed and successfully demonstrated a microsecond-scale vibrational spectroscopic imaging approach that utilized lock-in-free, parallel detection of the spectrally dispersed SRL signal. The authors used a synchronized, dual-output femtosecond-pulse laser system, providing the (tunable) broadband pump pulses with a spectral width of about $\delta\tilde{v} \sim 200-300$ cm$^{-1}$ and pulse duration $\tau_{p,pump} \approx 120$ fs and a fixed-value Stokes pulse (stretched to $\tau_{p,Stokes} \approx 2.3$ ps for narrow bandwidth), both running at pulse repetition rates of 80 MHz; the Stokes radiation was modulated at $f_{mod} = 2.1$ MHz. For hyperspectral imaging, the SRL spectrum was dispersed onto a fast PD array, providing 16 spectral channels (see the spectral traces in Figure 19.18). Their signals were processed by a 16-channel tuned-amplifier array and an associated parallel analog-to-digital converter unit, allowing the researchers to record spectrally dispersed SRL signals with a modulation depth as small as $dI/I \approx 10^{-6}$, with a pixel dwell time of $\Delta t_{SRL} = 32$ μs. With the time required for signal data treatment and pixel scanning, this meant frame rates of $t_{frame} \approx 3$ s for the image areas of 200 × 200 pixels, as shown in the figure.

Incorporating MCR analysis, the authors mapped a range of dynamic processes in cells, with multicomponent spatial distribution capability. The particular example shown here is that of molecular diffusion *in vivo* through (mouse) skin tissue. Such diffusion processes are of particular interest, e.g., for drug delivery into a cell. Ordinary drug delivery probes are invasive, and the technique is time consuming. Therefore, (near) real-time optical visualization of drug diffusion into biological tissues *in vivo* would provide a noninvasive tool for assessing drug delivery efficiency.

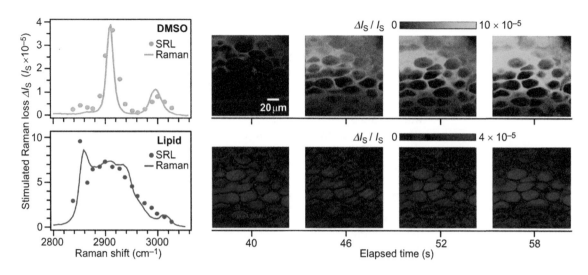

**Figure 19.18**   *In vivo* imaging of molecular diffusion of dimethyl sulfoxide (DMSO) molecules through skin tissue. (Right) Time-lapse multivariate curve resolution (MCR)-analyzed SRS images of DMSO molecule concentration (top row) and cell lipid content (bottom row). (Left) Comparison of multiplexed SRL and spontaneous Raman spectra for DMSO and lipids. (With kind permission from Springer Science+Business Media: *Light Sci. Appl.* 4, 2015 article e265, Liao, C.S. et al.)

As a proof-of-principle demonstration of the feasibility of mapping *in vivo* the migration of molecules whose spectral features overlap with the tissue signal itself, the authors administered a DMSO solution to an intact mouse ear *in vivo* and monitored DMSO penetration into the adipose tissue, located about $d_z \sim$ 50 μm beneath the skin surface. The selection of time-lapse image frames shown in Figure 19.18 clearly reveal that only very small amounts of DMSO penetrated into the adipose tissue within the first 40 s, but that over the following 20 s or so, substantial diffusive redistribution of DMSO over the skin cell structure was observed. At the same time, the images of cell material itself, here for the spectrally overlapping lipids, do hardly change at all, as expected.

While both examples shown here demonstrate that *in vivo* cell processes such as slow metabolic conversion or migratory transport can be imaged using SRL, the image frame rate of a few seconds is still too long to follow fast intratissue variations, which in cell biology often are well on the subsecond timescale. Therefore, it is noteworthy that some efforts have been made, and successfully so, to realize video-frame imaging of such fast events *in vivo*. Saar et al. (2010) developed a bespoke, single-color SRL setup. As utilized by most other research groups, their laser system was based on high-repetition-rate (~80 MHz) pico-second-pulse ($\tau_p$ = 7 ps) sources, namely, a tunable optical parametric oscillator providing the Raman pump laser radiation, and the fundamental output of the associated Nd:YAG driving laser as the stimulating Stokes laser radiation. To aid high-speed imaging, the Stokes laser radiation was intensity modulated $f_{\mathrm{mod}}$ = 20 MHz; as a key addition to their signal treatment, the authors used a custom-built, all-analog lock-in amplifier, with a response time of ~100 ns. At the time of the study, available commercial lock-in amplifiers offered response times of only 10–100 μs, a factor of 10–100 slower; of course, today faster lock-in amplifiers have become available commercially, even surpassing the aforementioned response time of 100 ns.

For their video-rate SRL measurements, the authors utilized collinear laser beams, wavelength tuned to match the molecular vibrational mode of interest. The laser beams were raster scanned (512 × 512 pixel) across the sample by a resonant, galvanometer-driven mirror, with a line-scan rate of 8 kHz. Together with a pixel dwell time of about 100 ns, this equated to an achievable SRL image rate of up to 25 frames·s$^{-1}$. Note that this high-frequency detection at >1 MHz also provides efficient noise suppression since most laser intensity fluctuation processes occur at much lower frequency and thus are by and large eliminated by the lock-in filtering; the authors achieved sensitivities of $dI/I < 10^{-8}$ when integrating the signal for 1 s.

The authors utilized their setup for a number of different *in vivo* processes, including the real-time *in vivo* imaging, e.g., (1) of blood flow, pinpointing the motion of individual red blood cells, and (2) of the delivery of a topical drug administered through the skin, along the shaft of a hair. In both cases, the frame acquisition times were 37 ms; the authors included phenomenal, instructive videos of these events in the supplementary material to their publication (Saar et al. 2010).

## 19.4 CARS IMAGING

### 19.4.1 Concepts of CARS imaging

The origins of CARS microscopy date back to the early 1980s, but it was not until the turn of the new millennium that it became a viable microimaging tool, with the work of Zumbusch et al. (1999), who used collinearly propagating pump and Stokes (NIR) laser beams, tightly focused by the microscope objective, and detected the CARS signal in the forward direction through a second microscope objective. This type of configuration has become one of the most commonly used CARS imaging implementations. A series of technical advances, in particular for laser, scanning, and detector equipment, has transformed CARS microscopy into a powerful technique that allows for vibrational Raman band imaging with high sensitivity, high spectral resolution, and 3D sectioning capability.

It should be noted, however, that unlike SRL and fluorescence imaging, CARS detection is not background-free. This is because of electronic contributions to the third-order susceptibility $\chi^{(3)}$—which constitutes the key quantity in describing CARS—from the sampled molecule and the embedding-matrix molecules result in a nonresonant background signal, which therefore reduces the vibrational-band contrast in the image.

Substantial efforts have therefore been directed toward efficient suppression of nonresonant background signals in CARS microscopy; most prominently, these include configuration changes in the form of epidetection and counter-propagating CARS implementations (see Figure 19.19), polarization-sensitive detection, pulse-sequenced detection, and the use of coherent control techniques.

As in SRL microscopy, there are two complementary approaches to CARS microscopy, namely, *single-frequency* CARS and *multiplex* (hyperspectral) CARS.

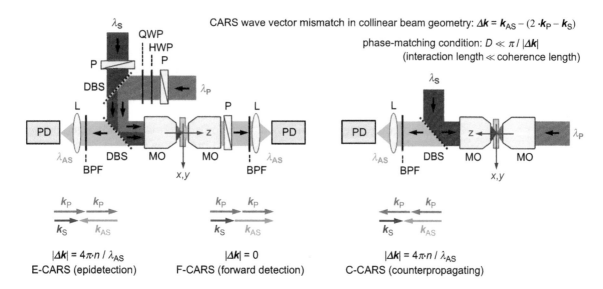

**Figure 19.19** Schematic configurations for the most common collinear CARS microscopy modalities. (Left) E-CARS (backward detection or epidetection of $I_{CARS}$) and F-CARS (forward detection of $I_{CARS}$). (Right) C-CARS (with counterpropagating pump and Stokes laser beams). The corresponding wave vector diagrams along the laser light propagation direction ($z$-axis), together with the phase-matching conditions, are indicated at the bottom of the figure. BPF, band-path filter; DBS, dichroic beam splitter/combiner; HWP, QWP, half-wave and quarter-wave plates, respectively; L, light collection lens; MO, microscope objective; P, polarizer; PD, photodetector (single-element or array).

In single-frequency CARS, only a single, individual vibrational mode is targeted across the image area. It is most suitable when images are sufficient that only provide qualitative image contrast for certain vibrational-mode differences in the sample molecules; however, the lack of spectral information by and large is the limiting factor of single-frequency CARS for quantitative imaging applications.

In multiplex CARS, one can address a range of molecular vibrational modes simultaneously, as in spontaneous Raman imaging and broadband SRL imaging: full vibrational Raman spectra are acquired at each spatial location in the image (hyperspectral imaging). However, not merely does the nonlinear and coherent nature of CARS result in an enhanced response signal but this amplified signal also depends nonlinearly on the incident laser intensities, which makes quantification less straightforward.

It is beyond the scope of this chapter to delve into all theoretical details associated with CARS. For this, the reader is referred, e.g., to Cheng and Xie (2012), who extensively cover theoretical, practical, and application aspects of coherent Raman scattering microscopy. Also, two relatively recent reviews may be useful: Volkmer (2005) addresses CARS microscopy with particular emphasis on collinear CARS in the forward direction, or F-CARS, including a set of useful equations, and Day et al. (2011) provide a concise description of the quantitative aspect in CARS microscopy, again together with the relevant mathematical framework. Here only a brief summary is provided of aspects that are important to CARS imaging and that differ from other Raman imaging techniques.

The rationale behind the beam geometry in conventional, collinear CARS spectroscopy, as described in Chapter 14.6, is that one should meet the wave vector phase-matching condition $\Delta|\mathbf{k}| \cdot L \ll \pi$ as closely as possible; for this, it is necessary to minimize the wave vector mismatch $\Delta k = k_{AS} - (2k_P - k_S)$ while maximizing the interaction length $L$. As shown in Section 14.6, in general this requires careful adjustment of (spatially separated) laser beam pass positions, to provide the angular direction conditions for phase matching after passage through a focusing lens. In collinear CARS microscopy, the latter problem is largely eliminated. Because of the small dimension of the excitation volume and the large cone of angled wave vectors, resulting from the high-magnification microscope objective, the wave vector mismatch is by and large compensated for; this is due to the spectral dispersion properties of the refractive index of the optical elements of the objective and the probed sample. Therefore, it is not surprising that collinear CARS beam geometry has become the configuration of choice in CARS microscopy, exhibiting superior spatial resolution and image quality. The three most common modalities of F-CARS and E-CARS (for which the recording of images is in the forward and backward directions, respectively) and C-CARS (for which the pump and Stokes laser beams counterpropagate) are shown in Figure 19.19.

Note that to theoretically describe the CARS signal generation for a collinear CARS microscope configuration, one needs to keep two distinct features in mind. First, caused by the *tight focusing* by a microscope objective, the laser excitation field distribution deviates from that of a Gaussian beam profile, because of the breakdown of the paraxial approximation. Second, the amount of *wave vector mismatch* is controlled by the *distribution in propagation directions* for both the incident laser beams and the generated CARS radiation; both are influenced by the actual geometry of the microscope setup.

Some final thoughts are directed at the quantization aspect in CARS microscopy. CARS is a coherent process; therefore, the resonant and nonresonant Raman responses of the sample molecules interfere according to

$$I_{AS} \propto \left| \chi^{(3)} \right|^2 = \left| \chi_R^{(3)} + \chi_{NR}^{(3)} \right|^2 = \left| \chi_R^{(3)} \right|^2 + 2 \cdot \chi_{NR}^{(3)} \cdot \mathrm{Re}\left[ \chi_R^{(3)} \right] + \left| \chi_{NR}^{(3)} \right|^2, \qquad (19.6)$$

where $\chi_R^{(3)}$ and $\chi_{NR}^{(3)}$ are the resonant and nonresonant parts of the complex susceptibility tensor, respectively. This means that the two parts cannot be simply separated from each other to access the quantitative information contained within $\chi_R^{(3)}$. As a consequence, there is no simple, linear correlation between $I_{AS}$ and the concentration of molecules. To extract the quantitative information from $I_{AS}$, one has to take a closer look at $\chi_R^{(3)}$. When probing only a single vibrational (mode) resonance, the electrodynamic theory of the four-wave mixing process underlying CARS gives a dispersion-like behavior $\chi_R^{(3)}$ when tuning across the resonance,

$$\chi_R^{(3)} = A_n \cdot \{ v_n - (v_P - v_S) - i\Gamma_n \}^{-1} \cdot N, \qquad (19.7)$$

where $A_n$, $v_n$, and $\Gamma_n$ stand for the amplitude, frequency, and line width of the vibrational mode $n$ and $N$ is the number of molecules per unit volume participating in the scattering process. From Equation 19.7, one can extract the absorptive (imaginary) part of the susceptibility:

$$\mathrm{Im}\left[ \chi_R^{(3)} \right] = A_n \cdot \{ (v_n - (v_P - v_S))^2 + \Gamma^2 \}^{-1} \cdot N. \qquad (19.8)$$

This expression may be compared with the vibrational-mode intensity profile for spontaneous Raman scattering:

$$I_{\text{Raman}} = A_n \cdot \left\{ (v_n - v_{\text{L}})^2 + \Gamma^2 \right\}^{-1} \cdot N, \tag{19.9}$$

where $v_{\text{L}}$ is the frequency of the (tunable) excitation laser radiation; the two expressions (Equations 19.8 and 19.9) look very much analogous. Thus, since spontaneous Raman signals are linearly dependent on concentrations, it should be possible to also derive quantitative information from the CARS-intensities $I_{\text{AS}}$ if one were able to extract $\text{Im}[\chi_R^{(3)}]$ from it. Besides the quantitative nature of the result, this would have the additional advantage that a spectrum constructed in terms of $\text{Im}[\chi_R^{(3)}]$ would be directly comparable with a spontaneous Raman spectrum of the sample, and, thus, one could correlate observed features to tabulated charts of known Raman bands. For these obvious reasons, in most CARS (imaging) applications, spectra are represented in this form (see the following examples). Of course, in general this requires some computational efforts; in particular, when the spectra comprise a range of overlapping features, deconvolution methods are required, as usual. However, in some cases, the extraction of $\text{Im}[\chi_R^{(3)}]$ might be straightforward. Because the nonresonant part of the susceptibility tensor, $\chi_{\text{NR}}^{(3)}$, is instantaneous and thus comprises a real component only, it turns out to be easy to retrieve $\text{Im}[\chi_R^{(3)}]$. For this, it is sufficient to know the phase $\varphi$ of the CARS field; one obtains

$$\text{Im}\left[\chi_R^{(3)}\right] = \left|\chi^{(3)}\right| \cdot \sin \varphi \propto (I_{\text{AS}})^{1/2} \cdot \sin \varphi \tag{19.10}$$

The preceding data evaluation strategies and various image analysis approaches are discussed in more detail, e.g., in Volkmer (2005) and Day et al. (2011).

Finally, a few properties inherent to CARS imaging are noteworthy:

- In addition to its higher sensitivity, CARS microscopic imaging also exhibits superior spatial resolution, in comparison with spontaneous RM. This is due to the facts (1) that the CARS response exhibits a cubic dependence on the incident laser intensities and (2) that CARS microscopy it is intrinsically confocal and directional.

- The CARS signal $I_{\text{AS}}$ is blueshifted with respect to the incident pump and Stokes laser beams; as a consequence, (long-wavelength) single-photon fluorescence does not interfere with the CARS signal. This results in better image contrast.

- In general, much shorter acquisition times are required in multiplex CARS applications than in hyperspectral spontaneous Raman experiments to achieve a comparable signal-to-noise ratio. Because of the shorter acquisitions, it also becomes feasible to visualize dynamical processes.

In the following, two representative examples are presented, which highlight the key aspects of CARS imaging outlined earlier.

## 19.4.2 Selected applications of CARS microscopic imaging

In CARS microscopy, the weak Raman signal is boosted sufficiently to allow one to perform label-free visualization of molecular distributions within a sample and their dynamical behavior in living cells and tissues with much higher speed than what is possible for spontaneous Raman imaging.

As with nearly all Raman spectroscopic variants, CARS also constitutes, in principle, a hyperspectral imaging method; it opens up the ability to simultaneously acquire both spectral and spatial image information. This is of particular importance in the field of biomedical imaging. It was already pointed out in Section 19.1 that hyperspectral representation (i.e., the superposition of a set of spectral subsets) aids the better visualization of certain molecular component distributions, which might become easier to recognize. Here such an example is provided for a CARS study of living L929 cells (see Kano et al., 2016).

In this particular study, a passively Q-switched Nd:YAG laser (with pulses of $\tau_p <$ 1 ns, at a repetition rate of $f_p = 33$ kHz) formed the basis of a dual-output CARS laser source. Part of its split output served as the CARS pump radiation ($\lambda_P = $ 1064 nm). The other part was coupled into an air–silica photonic crystal fiber, generating "white" supercontinuum laser radiation in the range $\lambda_S = 1100-$1700 nm; this ultrabroadband laser source served as stimulus for the Stokes transition of the CARS process. Recall that for efficient generation of the CARS signal, the phase-matching condition $k_{CARS} = 2 \cdot k_P - k_S$ should be satisfied as closely as possible. This is normally not straightforward when high-NA imaging objective lenses are in use, which exhibit substantial angular dispersion of the wave vectors $k$. Thus, the laser beam paths were adapted in such a way as to accomplish phase matching on the sample.

The CARS radiation generated by the interaction of the laser light with sample molecules was collected in the forward direction, through a second microscope objective lens. After passing through a set of laser-light and scattered-light rejection filters, CARS spectra were recorded using a spectrometer—CCD–array detector combination. Spatial scanning of the sample was not by piezo-driven galvanometric mirrors—as is common in the SRL measurements described much earlier—but by moving the sample itself on a three-axis piezo-displacement stage.

The example image data shown in Figure 19.20 are for mouse fibroblasts (L929) cells. These are commercially available cloned cells (and are frequently used to test biomedical procedures); thus, they are also ideal for serving as a test bed for CARS imaging feasibility studies.

In the figure, three (color-coded) hyperspectral component CARS images are shown, which are associated with the molecular vibrational modes indicated in the representative CARS spectrum from an individual, spatial pixel position. These three components correspond to molecule-specific images of proteins (ring-stretch vibration by phenylalanine residues, at $\tilde{v} = 1002$ cm$^{-1}$—coded in red); lipids (CH$_2$ scissoring mode, at $\tilde{v} = 1446$ cm$^{-1}$—coded in green); and nucleic acids (the vibrational mode due to the purine rings of adenine and guanine, at $\tilde{v} \approx 1565 - 1590$ cm$^{-1}$—coded in blue). The signal intensities (indicated as usual by the color tint) are proportional to the concentration of each molecule; thus, quantitative evaluation of molecular concentrations can be performed as well.

The image at the very bottom of the figure is a composite of the three individual molecular subsets, visualizing unequivocally that indeed the cell core is dominated by nucleic acids. Note that for the exposure time for recording the full

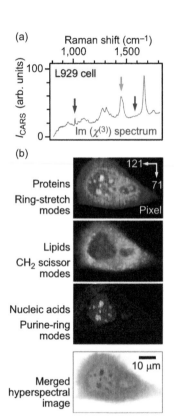

**Figure 19.20** Hyperspectral CARS imaging of living L929 cells. (a) Representative, denoised Im[$\chi^{(3)}$] spectrum. (b) False-color CARS images for the three individual vibrational modes, indicated by the colored arrows in the spectrum; the three images are merged into a hyperspectral representation. (Data adapted from Kano, H. et al., *J. Raman Spectrosc.* 47, no. 1 (2016): 116–123.)

CARS spectra at each pixel point was 50 ms. For the image area of 71 × 121 pixels shown in the figure, this equates to an acquisition time for the full hyperspectral image of ~430 s.

In the second example, an application to CARS imaging is given for quantitative, vibrational-band multiplexed visualization of metabolic activity in single, individual living yeast cells (see Okuno et al. 2010). In their study, the authors used more or less the same laser system configuration described in the previous example. Thus, the setup did not require scanning, neither of the laser nor the spectrometer wavelengths. The spectrometer resolution across the fingerprint range for molecular vibrational modes was sufficient to allow for multiplexed extraction of distinct vibrational bands from the congested CARS spectra. From using a microscope with a ×100 objective, the spatial resolution of the CARS imaging system was $\delta d_{x,y} \approx 0.5$ μm and $\delta d_z = 4.5$ μm in the lateral and axial directions, respectively.

Imaging of individual yeast cells was done by raster scanning the sample; each image is made up of 21 × 21 pixels (for imaging of the small yeast cells, this was sufficient, resulting in substantial speeding up of full-image acquisition). With an exposure time of 50 ms for each pixel, the overall measurement time adds up to ~22 s.

Multifrequency, label-free CARS Im[$\chi^{(3)}$] images were recorded for 12 individual vibrational frequencies, including the six annotated in the spectra of Figure 19.21; note that all images were obtained simultaneously (multiplexed) in one single scan of the sample. This method was applied to a time-resolved study of dynamical processes in a single living yeast cell.

The CARS and Raman spectra in the figure reveal that the vibrational band features are more or less the same, but that the band intensities differ significantly; the latter is expected by nature of the nonlinear CARS process. Some of the prominent bands that are vital for the interpretation of cell metabolism are

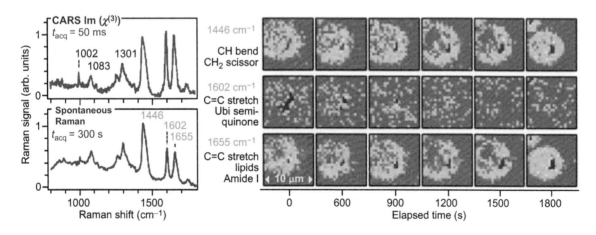

**Figure 19.21** Quantitative CARS molecular fingerprinting of single living cells. (Left) Typical CARS Im[$\chi^{(3)}$] spectrum a living yeast cell, after noise-reducing singular value decomposition analysis (top data panel), compared with a spontaneous Raman spectrum (bottom data panel); note the different data acquisition times. (Right) Selection of time-resolved CARS images for three different vibrational bands, extracted from the CARS spectra. Images were recorded every 12 s; the false-color intensity range is from blue (lowest values) to red (highest values). (Data adapted from Okuno, M. et al., *Angew. Chem. Int. Ed. Engl.* 49, no. 38 (2010): 6773-6777.)

annotated in the spectra (note that for clarity not all annotations are included in both data traces):

- The sharp band at $\tilde{v} = 1002$ cm$^{-1}$ originates from the phenylalanine residues in proteins.

- The bands at $\tilde{v} = 1083$ cm$^{-1}$ and $\tilde{v} = 1301$ cm$^{-1}$ originate from the phospho-lipids that constitute mitochondrial membranes.

- The feature at $\tilde{v} = 1446$ cm$^{-1}$ corresponds to the overlapping CH bend bands (or scissoring modes), originating from the $CH_2$ and $CH_3$ molecular groups.

- The band at $\tilde{v} = 1602$ cm$^{-1}$ originates exclusively from mitochondria; its intensity sharply reflects the metabolic activity of mitochondria. It is assigned to the in-phase C=C stretch mode of *ubisemiquinone* (a phospholipid), an electron carrier in the electron-transport chain in mitochondria. This band is often addressed as the "Raman spectroscopic signature of life."

- The band at $\tilde{v} = 1655$ cm$^{-1}$ is the superposition of the C=C stretching of lipid chains and the amide I mode of proteins.

Three series of time-resolved images at the Raman bands at $\tilde{v} = 1446$ cm$^{-1}$, $\tilde{v} = 1602$ cm$^{-1}$, and $\tilde{v} = 1655$ cm$^{-1}$ are in the right-hand part of Figure 19.21. The most striking feature observed in this sequence is that the intensity of the $\tilde{v} = 1602$ cm$^{-1}$ band gradually decreases with time; after about 20 minutes (=1800 s), it has almost disappeared, indicating that the metabolic activity in mitochondria has become minimal. The authors also observed that concomitantly with the disappearance of the $\tilde{v} = 1602$ cm$^{-1}$ band, the polyphosphate band at $\tilde{v} = 1160$ cm$^{-1}$ (not shown here), associated with a so-called dancing body (DB), suddenly appeared inside the vacuole at about 15 minutes and was as suddenly disrupted again after about 25 minutes. The interpretation of the latter was that all metabolic activity has ceased, and at this stage, the cell is dead.

These results indicated that irradiation of the living yeast cells with elevated laser power $P_L \sim 20$ mW significantly affected the metabolic activity of mitochondria, ultimately leading to cell death after about half an hour. The authors note that they did not observe this "killing-off" effect in their earlier studies in which they used lower laser power of $P_L \sim 4$ mW.

Regardless, the authors showed that by combining CARS microspectroscopy with maximum entropy evaluation methods, they were able to realize label-free and quantitative molecular mapping of single, living cells, generating vibrational-mode-specific images with improved time resolution over conventional spontaneous RM. These experiments clearly revealed that the loss of the Raman spectroscopic signature of life precedes the formation of a DB; this was interpreted that the loss of metabolic activity in mitochondria causes—through a yet unknown mitochondria–vacuolar cross talk—the formation of a DB in a vacuole. The authors note that a temporal evolution process could not be resolved using the much slower imaging based on spontaneous RM.

## 19.5 BREAKTHROUGHS AND THE CUTTING EDGE

Undoubtedly, Raman microscopic imaging and its enhanced/nonlinear derivatives have become the workhorses for imaging applications requiring both submicrometer spatial resolution and superb spectrochemical analytical prowess. Several quantum leaps in capabilities occurred over the years, normally associated with the improvement in measurement equipment or the introduction of novel methodologies. One particular exciting development was the proof-of-principle experiments by Ideguchi et al. (2013), who utilized a dual-frequency comb laser excitation source to realize hyperspectral CARS imaging, which in the future may well revolutionize certain applications in RM.

### 19.5.1 Breakthrough: Hyperspectral CARS imaging utilizing frequency combs

In their specific experiment Ideguchi et al. (2013) used two synchronized femtosecond-pulse duration frequency combs, providing an apodized resolution of ~4–10 cm$^{-1}$ for the 300 individual spectral elements. For proof of principle, the authors imaged a capillary plate (capillary diameter $d_{cap}$ = 25 μm), filled with a mixture of three organic liquids. An area of $45 \times 45$ μm$^2$ was scanned, with a step size of 1 μm (corresponding to 2025 pixels); for each pixel, complete CARS spectra were extracted from the recorded spectral interferograms and stored in the hyperspectral data cube.

In Figure 19.22, some representative results are shown, namely, a segment of the full (unaveraged) CARS spectrum, associated with a single pixel, and two selected spectrochemical maps (for nitromethane and toluene), as well as a (zero) background image, demonstrating the excellent signal-to-noise ratio achieved in this full spectral multiplex demonstration experiment (SNR ~ 10$^3$). It should be noted that the acquisition rate of 50 pixels·s$^{-1}$ (~40 s for the complete hyperspectral image cube) was much slower than what the sampling time of only <20 μs for an entire Fourier-transformed spectrum would suggest; basically this was limited by the particular refresh rate of the interferograms. The authors suggest some solutions how their image acquisition rates could be improved, possibly up to real-time video rates.

### 19.5.2 At the cutting edge: Superresolution Raman microscopy

Superresolution in microscopy, i.e., imaging resolutions below the diffraction limit, can now routinely be achieved utilizing fluorescence-based techniques

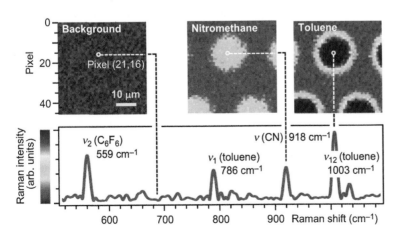

**Figure 19.22** Hyperspectral CARS imaging of a capillary plate, filled with a mixture of organic liquids, utilizing frequency combs. (Bottom) Dual-comb CARS spectrum from a single pixel; relevant vibrational bands are annotated. (Top) Spectrochemical maps retrieved from the hyperspectral data cube; the liquids are clearly confined to the capillary cores. (Data adapted from Ideguchi, T. et al., *Nat. Commun.* 5 (2013): article 3375.)

**Figure 19.23** Demonstration of superresolution RM. (a) Depiction of the subdiffraction limit Raman imaging concept. (b) Line scan (as part of a Raman image) across a diamond plate, showing clear improvement of resolution with the addition of the decoherence beam. (Data adapted from Silva, W.R. et al., *ACS Photon.* 3, no. 1 (2016): 79–86.)

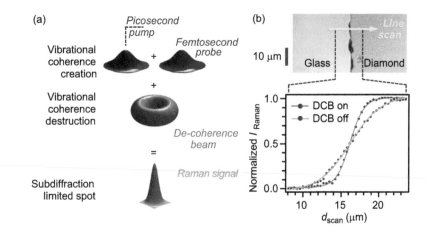

such as, e.g., stimulated emission depletion (STED) microscopy or photo-activated localization microscopy. However, in all these techniques, spectro-chemical specificity has been rather elusive; this may actually be at the brink of changing.

In a proof-of-principle experiment, Silva et al. (2016) demonstrated that as in fluorescence imaging, in Raman imaging, spatial resolution below the diffraction limit is also achievable. The authors implemented a far-field Raman spectro-scopic imaging technique that constitutes a sort of a hybrid of techniques, combining concepts from STED microscopy and femtosecond-stimulated Raman spectroscopy (fs-SRS; with picosecond-pump and femtosecond-Stokes laser beams). In particular, the stimulated Raman process generates vibrational-mode coherences, while the toroidally shaped "decoherence" pulse does elim-inate the Raman signal from the edges of the focal spot.

Vibrational-mode coherence is generated during the interaction of the molecule with the pump and probe laser pulses; the result is SRS, here observed in the form of SRG. Said coherences typically persist for about $10^2$–$10^3$ fs, depending on the vibrational dephasing time of the mode of interest. At some instance during this dephasing time, a third interaction with the pump laser radiation is "switched on"; this influences the system in such a way as to accelerate the destruction of the coherence, which leads to the elimination of the SRG signal. In this sense, fs-SRS constitutes a nonlinear four-wave mixing process, exploiting vibrational-mode coherences to generate SRG signals. As a consequence, the toroidally shaped decoherence beam reduces the central spot area, which can contribute to the SRG signal. This allows for subdiffraction limit imaging; as in STED, the spatial resolution improves with increasing decoherence laser pulse power. This concept is depicted in the upper part of Figure 19.23.

# CHAPTER 20

# Diffuse Optical Imaging

When a (laser) light beam impinges on a biological interface, the four processes described in the introductory chapter will occur: reflection, transmission, scattering, and absorption. However, other than the optical materials—for which the said processes can be described by simple formulae and for which scattering and absorption are normally minor—this is not longer true for biological tissue. In general, it is very inhomogeneous and consists of a multitude of constituents, of different relative composition, and normally it is very "opaque" to light (near-UV to near-IR). Quintessentially, scattering and absorption now become the dominant processes for light interaction. This means that exact ray tracing becomes next to impossible and no simple formulae for the interaction are available, in general. A simplified sketch of the light–tissue interaction scenario is provided in Figure 20.1.

The figure shows that while ray tracing conceptually might look feasible (see the accumulative pathways of the photons through the medium with scattering and absorbing particles), in practice this is normally rather difficult. However, since light beams are spatially extended, specifically compared to the scale of tissue particles, surface roughness and statistical inhomogeneities result in varying random paths, and imaging becomes increasingly "diffuse." As a consequence, imaging of "objects" within the tissue becomes blurred.

## 20.1 BASIC CONCEPTS

### 20.1.1 Scattering and absorption in biological tissue

Biological tissue is very heterogenic to scattering, i.e., it comprises a range of constituents of very different size, including cell nuclei, cell membranes, mitochondria, and whole cells, with dimensional range from 100 nm up to a few micrometers. Thus, scattering, in general, comprises a mix of Rayleigh (particles small with respect to the light wavelength) and Mie (particles comparable to or larger than the light wavelength) scattering. Overall, its effect is often described by a global scattering coefficient, $\mu_s = \rho_s \cdot \sigma_s$ (in cm$^{-1}$), associated with the depth progression into the tissue; here $\rho_s$ is the volume density of the scattering particles (in cm$^{-3}$) and $\sigma_s$ is the effective scattering cross section (in cm$^2$). This scattering coefficient $\mu_s$ is certainly wavelength dependent, and its measurement is rather difficult: it has to be made for very thin samples to avoid multiple scattering. But as a consequence, now heterogeneity of any tissue sample may become apparent. In general, one finds that tissue scattering coefficients are typically in the range of $\mu_s(\lambda = 1000 \text{ nm}) \cong 10 \text{ cm}^{-1}$ to $\mu_s(\lambda = 400 \text{ nm}) \cong 100 \text{ cm}^{-1}$.

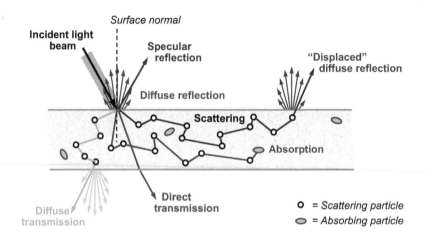

**Figure 20.1**  Interaction processes of photons with biological tissue material; of the extended incident beam (grey block area), only selected rays are traced for clarity. The diffuse "cones" at the various observation locations are the result of multiple ray propagation.

When it comes to absorption in tissue, once again, one has to consider a wide range of potential candidates (although the volume fraction of some may be rather small) and their wavelength dependence. Typically, biological tissue and its constituent cells comprise about 65–75% of water; a few percent of whole blood (of the order 3–7%); and the various molecular ingredients like lipids, proteins, hydrocarbons, and RNA/DNA. These exhibit absorption in different wavelength regions; some absorb stronger in the UV (like proteins and DNA), while others absorb substantially in the infrared (like water and lipids). Absorption spectra for the dominant (pure) components in tissue are summarized in Figure 20.2 in the interval 400–1000 nm.

As the composition of tissue types strongly varies, so does total absorption. For example, the overall absorption of just a few percent of blood in tissue would push the hemoglobin absorption down by nearly two orders of magnitude, leaving blood still dominant at short wavelengths; but now water starts to dominate toward the IR. The balanced sum of all contributions then leads to a "trough" in the absorption spectrum, or a spectral window for diagnosis, therapy, and imaging, the so-called NIR window.

For further details of how to tackle scattering and absorption properties of biological tissue, the reader may refer to the review by Jacques (2013).

## 20.1.2  What can we learn from diffuse optical imaging and spectroscopy?

Probably everybody has, as a child, followed the challenge "... if you shine a flashlight onto your hand, and look at it from the back, what can be seen?..." And invariably the answer may have followed more or less the same line "... a reddish glow and some blurry dark areas ..." The investigative conclusion therefore was that light can travel through a few centimeters of tissue and still be detected, and that hidden obstacles such as bones are crudely visible; see Figure 20.3. This then may lead to the question "can this light be used to see, or image, inside the body?" The great advances made over the last two decades in understanding of light migration through tissue, in the development of algorithms for light propagation in tissue, and subsequent spatial reconstruction of subsurface features, have unequivocally demonstrated that imaging with diffuse light—diffuse optical imaging (DOI)—is indeed possible.

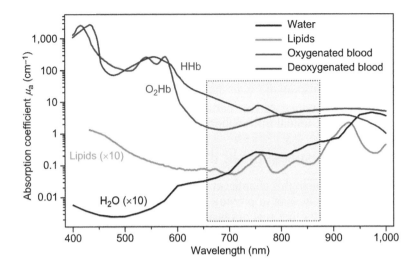

Figure 20.2 Absorption coefficients, as a function of wavelength, for (pure) dominant components of biological tissue: $O_2Hb$ = oxygenated blood (red trace); HHb = deoxygenated blood (blue trace); $H_2O$ = water (black trace); lipids = fatty tissue (green trace); note that the data for water and lipids are multiplied by ×10 for scaling purposes. The grayed area indicates the range of the optical NIR window. (Data adapted from those cited in Jacques, S.L., *Phys. Med. Biol.* 58, no. 11 (2013): R37–R61.)

In particular, much of the success of DOI could be linked to the knowledge from optical spectroscopy, which broadly stipulates that every "chromophore" in the investigated tissue absorbs light at specific wavelengths, which are often unique to an individual molecule. Referring to Figure 20.2, lipids exhibit a clear absorption peak at around 930 nm; and one finds that oxygenated and deoxygenated hemoglobin in whole blood have substantially different absorption profiles. Therefore, DOI gains significantly in strength through spectroscopy because of its ability to quantify molecular concentrations and dispositions (state and environment). By measuring the interaction of the stimulating photons, at multiple wavelengths, with the tissue during DOI studies, chromophore content and composition can be individually separated and analyzed. For example, using carefully selected laser wavelengths, one can easily quantify the relative composition of tissue blood, i.e., the ratio between oxy- and deoxyhemoglobin. And then, by using suitable reconstruction of the measured signals, one may be able to spatially map, e.g., differences between benign and malignant tissue regions.

Very crudely, DOI can be subdivided into two main categories: topography or tomography, although the distinction between these two techniques has become increasingly blurred. The term *optical topography* is, in general, used when referring to DOI implementations, which generate two-dimensional (2D) images of a plane parallel to the laser excitation source(s) and detector(s); only limited depth information is provided. The term *optical tomography* addresses techniques in which full three-dimensional (3D) images are reconstructed from measurement geometries with excitation sources and detectors widely spaced over the surface of the (3D) object. In the latter case, one often addresses the technique as diffuse optical tomography (DOT).

Nowadays, DOT is maturing into a widely applicable tool for biomedical diagnostics. DOT is perceived to be a painless, noninvasive, simple-to-use, and relatively cheap methodology that, in principle, can be performed safely on nearly any patient, at any time. This is in stark contrast to other, common imaging techniques based on CT and MRI scanning devices. In addition, in an apparatus that incorporates multiple laser sources and spectrally filtered detectors, spectroscopic information about the sample can be obtained, often in near real time.

Figure 20.3 Photograph of a hand illuminated from the back by a white light source.

For example, using DOT instrumentation, exogenous contrast agents are not required and metabolic information can be monitored continuously, unlike with positron emission tomography (PET). However, one should be clear that, in many cases, DOT may only be complementary to other imaging techniques. This stems from the inexorable fact that DOT provides only limited spatial resolution, which decreases substantially with increasing penetration depth. This is an unavoidable consequence of the nature of diffusely scattered light and the accumulative loss in spatial information with each scattering event.

### 20.1.3 Historical snapshots in the development of DOI

It is beyond the scope of this chapter to review the full historical development of DOI, and here we only wish to provide selected snapshots. It is probably fair to say that DOI found its first real applications in the late 1920s and early 1930s, although the idea of optical transillumination of tissue goes back about another 100 years. The applications in question were the diagnosis of breast tumors (Cutler 1929, 1931) and the *in vivo* measurement of blood oxygen saturation using arterial transillumination (Kramer 1935). But it was only in the 1980s that DOI started to boom, mainly due to the fact that time- and frequency-resolved (TD and FD) techniques were developed, which greatly benefited NIR spectroscopy. Specifically, these techniques allowed for NIR measurements with vastly improved sensitivity and specificity. The 1990s then saw the publication of hundreds of studies utilizing NIR spectroscopy and DOI (both 2D topographic and 3D tomographic). In parallel, with the clinical success of the DOI techniques, commercial instrumentation was developed and marketed by a number of companies, predominantly in Japan and the United States. Since the early 2000s, many more applications were developed, and evermore sophisticated and reliable instrumentation was marketed so that today DOI and its brethren can confidently be termed established diagnostic tools for biological tissue analysis, now even allowing near real-time monitoring of metabolic processes. The reader may consult some relevant reviews (e.g., Boas et al. 2002; Kalisz 2004; Gibson et al. 2005; Gibson and Dehghani 2009; Durduran et al. 2010) and textbooks (Wang and Wu 2007; Jiang 2010) for further insight.

## 20.2 BASIC IMPLEMENTATION AND EXPERIMENTAL METHODOLOGIES

When contemplating to set up a system for DOI, it is paramount to first select the best methodology for the problem at hand, and then assemble the necessary components for its realization. Over the past two decades, a vast range of problem-oriented DOI devices have been set up, tested, and applied to topographic and tomographic imaging problems of biomedical samples. Depending on the sample form and size, and the information desired from the measurements, different excitation/detection techniques have to be used, but by and large the conceptual setup is the same:

- The stimulating light from a (or more than one) light source is coupled to the sample at single or multiple point(s).

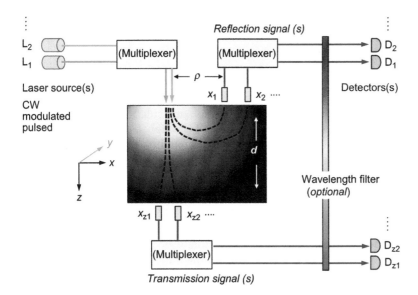

**Figure 20.4** Conceptual experimental implementation for DOI.

- Scattered light is collected at a single or multiple location(s), either in transmission or reflection mode.

- The detector signal(s) are processed and analyzed to finally generate topographical (2D) or tomographic (3D) maps of the sample.

This basic setup concept is depicted in Figure 20.4.

Here we briefly summarize the equipment components that are required to implement DOI in is various guises, namely continuous wave (CW), frequency domain (FD), and time domain (TD) variants; the three methodologies are outlined further below. Note, however, that we will not dwell on the actual spatial arrangement of sources and detectors; these depend crucially on the specific application, and representative examples will be described below.

## 20.2.1 Key equipment components for DOI

*Light sources.* The question as to which wavelength to use in imaging applications of biomedical tissue is of utmost importance, but not an easy one to answer. In general for tissue optics, the choice is bounded by the so-called "NIR window": at wavelengths below about 650 nm, absorption by hemoglobin limits the penetration into tissue; at wavelengths above about 850 nm, absorption by water dominates. Therefore, by and large, wavelengths within the 650–850 nm band are utilized. In the early periods of diffuse imaging, the actual wavelengths used in a given study were often *ad hoc*, simply dictated by the availability of a suitable (laser) light source. The first systematic evaluation of the optimal wavelengths for NIR imaging was carried out by Corlu et al. (2003), who tried to identify wavelength combinations that generated maximum uniqueness to distinguish between different chromophores. Today wavelengths are carefully preselected, in general, tailored to the specific problem under study; still, the actual choice may be dictated by available (cheap) light sources. In early DOI studies, broadband (white-light) sources like Xe lamps were used out of which relevant wavelength bands were filtered for wavelength-selective stimulus, or spectroscopic selection on the detector side was implemented. With the progress in semiconductor laser technology, nowadays mostly laser diodes (LDs) are utilized because of their ease of use, relatively low cost, wide selection of distinct wavelengths throughout the tissue NIR window, high power density within a narrow wavelength interval (up to 100 mW even in a single longitudinal mode), and extremely versatile temporal-variation capabilities (continuous operation, modulation up to multi-megahertz, and ultrashort pulses). Alternatively to LDs, light-emitting diodes (LEDs) at similar selected wavelengths are used, although their spectral bandwidth is much broader (20–50 nm rather than the ≤1 nm of LDs), and therefore their spectral power density (mW/nm) is much lower.

*Detectors.* Photon detectors for DOI have to be very sensitive since, in general, the initially high photon flux from the light source is severely reduced by the tissue scattering and absorption processes, sometimes down to a few individual photons per second. In general use is made of "amplified" photon detectors, which convert the rather small energy of the individual photon into much larger electrical signals, which are easy to capture by any subsequent electronics devices. Rather than simple photo diodes (PDs), which nevertheless have their merits in certain experiments, one uses avalanche PDs (APDs), photomultiplier tubes (PMTs), and microchannel plate (MCP) devices. These increase in this sequence in their amplification capabilities, providing photon-signal enhancements of up to $10^6$–$10^8$; therefore, many of these detector devices are sensitive enough to record (count) single photons, if so required. Increasingly, use is made too of charge-coupled devices (CCDs), which constitute 1D or 2D array devices. They have clear advantages over all the other detectors, being multiple-point rather than single-point detectors. On the other hand, in general, they suffer from "low speed," meaning that while they can accumulate sequences of few photon events, the repetition rate is limited by the rather lengthy array readout times.

*Signal processing electronics.* All signals from the photon detectors require, in general, post-acquisition treatment and analysis. Despite the fact that the small individual photon energy had normally been amplified by the detector device itself, the actual electrical signals are still rather small to be directly analyzed and displayed. Therefore, in general postdetector signal amplification is paramount; in particular, devices are required that can suppress spurious noise signals. The choice of electronics normally depends on the DOI methodology, which was utilized in the experiment, i.e., whether measurements were in the continuous, FD, or TD regime (for details on these, see further below). Even in the so-called CW experiments, the light source is amplitude-modulated at low frequencies, up to a few kilohertz or so, in order to be able to separate the scattered light associated with the stimulus from spurious other light sources whose frequency content is unrelated to the source modulation. In this case, simple low-noise DC/AC amplifiers and standard lock-in amplifiers are used. In the FD, the source is normally modulated at a high frequency (often up to a few hundreds of megahertz, i.e., RF frequencies). The scattered signals exhibit some phase shift with respect to the stimulus wave, due to the finite travel-time through the medium. Besides the amplitude reduction, this phase shift can be measured and then used for image reconstruction purposes. This means that both in-phase (for amplitude information) and phase-shifted (for phase-delay information) need to be measured. The instruments used for this are normally quadrature demodulators, which can directly cope with the RF-modulated waves, or dual-phase lock-in amplifier, for which the wave frequency has to be reduced by heterodyning to match it to the slower response of standard lock-in electronics. For TD measurements in which nanosecond duration scattering sequence related to a picosecond stimulus pulse, time-to-amplitude converters (TACs) or constant-fraction discriminators (CFDs) are used. Both types of devices measure the delay between the stimulus reference picosecond pulse and the arrival of photons at the detector, with time resolution of about 20–100 ps.

*Light delivery and collection coupling.* In general, the coupling between the light sources to and the detectors from the sample under investigation is afforded by optical fibers, which allow for flexible location and to reach difficult-to-reach positions. Both single-mode (SM) and multimode (MM) fibers are used. SM fibers have a core diameter of 3–6 µm (depending on the light wavelength) and supports—as the name suggests—only a single (transverse) mode, and the light travels only along the optical core axis. The light coupled into the (round) fiber face then emerges at the other (round) end with Gaussian $TEM_{00}$ beam profile. Also, such a fiber light source can be treated as a near-perfect point source. The drawback of SM fibers is that they are difficult to align, and normally one aims for rugged fiber coupling, e.g., via mechanically fixed FC connectors. In contrast, MM fibers have larger core diameters (>50 µm), as a consequence of light waves being allowed different beam paths along the fiber, as long as the total internal reflection condition is met. This multiple-pass transit of light and the associated temporal dispersion means that one now encounters the so-called "speckle" in the output light because of coherent interference. This effect may cancel out over longer data averaging times, but nevertheless may have to be considered more in the photon transport modeling since this MM fiber light source may now have to be treated as an extended rather than a point source. Finally, in multiple-source multiple-detector configurations, one will have to multiplex, or switch, between several source/detector fibers. The related optical switch may operate by (1) mechanical means, such as physically shifting an optical fiber to drive one or more alternative fibers; (2) electro-optic or magneto-optic effects; or (3) utilizing other methods. Probably the most elegant method is to use micro-optical electromechanical systems (MOEMS) optical switches, which are very common in telecommunications and have switching times of the order of milliseconds. The minor drawback has been that only telecommunications wavelengths in the 1300/1500 nm range were commercially available. However, since a few years also devices in the range 500–1000 nm have become available, for example, the *Thorlabs* 1×2 optical switches OSW12-633-SM, which can be used in the 600–800 nm range (Thorlabs 2014). The concept of MOEMS multiplexers is shown in Figure 20.5.

*Software.* For the understanding and interpretation of DOI, there are two key ingredients: (1) the modeling of the photon transport in tissue and (2) the reconstruction of images from diffuse scattering measurements; these will be discussed in more detail in Section 20.3. The former is normally tackled via Monte Carlo methods or diffusion theory, based on the radiation transfer equation; for the latter, finite element methods (FEMs) are normally utilized. Over the past 20 years, many groups have developed sophisticated theoretical modeling approaches and written bespoke computer codes, many of them based on Matlab™ because of its wide use and its relative ease of graphical presentation of data. Certain *ad hoc* development and implementations of computer codes for DOI are still widespread. On the other hand, a few elegant open-source software

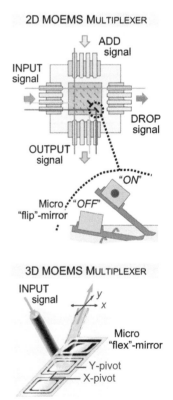

**Figure 20.5** Concept of light multiplexers based on MOEMS devices. (Top) 2D in-plane multiplexer utilizing microflip mirrors (their operation is shown in the magnified insert). (Bottom) 3D multiplexer utilizing x–y flex mirrors.

packages are now available, which may ease the burden of having to rewrite vast segments of computer code, which in principle is common to many applications in DOI. Such complete packages are—without claiming completeness—EIDORS (Adler and Lionheart 2006), NIRFAST (Jermyn et al. 2013), and TOAST++ (Schweiger and Arridge 2014). EIDORS is based on Matlab™ code, and its goal was "to provide free software algorithms for forward and inverse modeling for... Diffusion-based Optical Tomography (DOT), in medical and industrial settings." NIRFAST is also run within the Matlab™ environment. It is a FEM-based package for modeling NIR-light transport in tissue and includes (1) single wavelength absorption and scatter; (2) multiwavelength spectrally constrained models; and (3) fluorescence models. TOAST++ is written in C++ but also contains bindings for Matlab™ and Python™. It is a software suite for image reconstruction in DOT and contains a forward-solver module "using FEM for simulating the propagation of light in highly scattering, inhomogeneous biological tissues" and an inverse-solver module that "uses an iterative, model-based approach to reconstruct the unknown distributions of absorption and scattering coefficients in the volume of interest from boundary measurements of light transmission." All three packages incorporate the capabilities to adapt the solvers to novel sample properties and geometries.

A selection of the most common elements and devices and their key operating parameters are collated in Table 20.1.

**Table 20.1**   Essential Equipment Components for the Implementation of Diffuse Imaging Applications

| Device | Operating Parameters | Remarks |
|---|---|---|
| **Light sources** | | |
| White-light Xe-lamp | 650–1000 nm; 1–20 mW/nm | Continuum; required $\Delta\lambda$-filter |
| LEDs | 650–1000 nm; 0.1–2.0 mW/nm | Bandwidth $\Delta\lambda \sim$ 20–50 nm |
| LD | 650–1000 nm; 2–200 mW/nm | Bandwidth $\Delta\lambda$ = 0.01–2 nm |
| **Detectors** | | |
| PD | Responsivity 0.5 mA/mW; ×1 | Requires signal above DC noise |
| APD | Responsivity 0.5 mA/mW; $\times 10^2$ | Able to see single photons |
| PMT | Responsivity 0.5 mA/mW; $\times 10^6$ | Able to see single photons |
| MCP | Responsivity 0.5 mA/mW; $\times 10^6$ | Nanosecond-gating possible |
| CCD | Responsivity 0.5 mA/mW; $\times 10^6$ | Sensitive 1D/2D array detector |
| **Electronics** | | |
| Low-noise DC/AC amplifiers | Amplification $\times 10$–$10^6$; ≤10 MHz | DC noise often <10 nV/$\sqrt{\text{Hz}}$ |
| Photon counter | Event counter; rate <1 MHz | Analogue count rate ≤200 MHz |
| (Dual-phase) lock-in amplifier | Amplification $\times 1$–$10^5$; ≤1 MHz | Frequency-modulation detector |
| Quadrature demodulator | Frequency response >500 MHz | Frequency-modulation detector |
| Time-to-amplitude converter | ~50 ps resolution at 50 ns range | Start/stop arrival comparison |
| Time-to-amplitude converter | <100 ps resolution at 20 ns range | Up to 50 MHz count rate |
| **Light delivery/multiplexing** | | |
| SM fibers | Core diameter 6–8 μm | Can be treated as point source |
| MM fibers | Core diameter 25–250 μm | Not a point source; laser speckle |
| Wave multiplexers | $1 \times N$ or $N \times 1$ MEMS switches | Multiplexing at up to 1 kHz |
| **Software** | | |
| Photon-transport modeling | RTE and Monte Carlo codes | Mostly bespoke software |
| Image reconstruction | Various reverse-model coding | e.g., TOAST++ (GPL-license) |

For DOI of biological tissues, three main techniques are currently used; their main difference lies in the way the tissue is illuminated. These techniques are CW measurements, amplitude-modulated excitation for the frequency-resolved domain (FD), and pulsed excitation for the time-resolved domain (TD). The principle of these three methods is illustrated in Figure 20.6.

## 20.2.2 Experimental methodology 1: CW systems

Measuring of the intensity of light transmitted between two points on the surface of tissue is rather straightforward (and normally inexpensive); see the upper panels in Figure 20.6.

Still, such measurements contain a remarkable amount of useful information. CW light sources have been used for imaging purposes since the early nineteenth century (Bright 1831; Curling 1843; Cutler 1929), but it was only after LDs in the NIR wavelength range and avalanche photodiodes (APDs) or PMTs with ultra-high sensitivity—down to the counting of individual photons—became available that this method received considerable attention. A significant improvement of CW systems was achieved by using sources whose light intensity is modulated at low (a few kilohertz) frequency so that phase-locked detection could be employed to enhance the method's sensitivity.

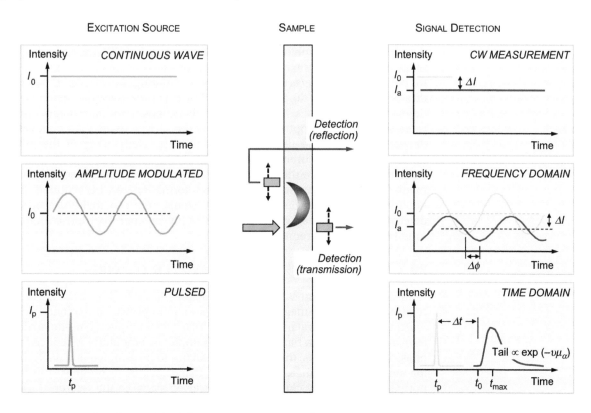

**Figure 20.6** Temporal evolution of the diffuse scattering response of tissue, with excitation and detection spatially displaced to each other. The lateral distance between the two positions may be variable. (Top panels) CW measurements, with intensity attenuation $\Delta I = I_0 - I_a$. (Central panels) Amplitude-modulated excitation; the transmitted wave (red color) has less amplitude and its mean intensity reduced by $\Delta I$, and a phase difference $\Delta\phi$. (Bottom panels) Pulsed excitation; the transmitted pulse is delayed by $\Delta t$ and temporally broadened, with the long-delay tail often exhibiting an exponential shape. For further details, see text.

CW techniques are now commonly used in the study of hemodynamic and oxygenation variation in human superficial tissues, and in the 2D mapping of cortical regions of the brain (optical topography; see Gibson et al. 2005). By keeping the separation between excitation and detection points small, the measured signals are relatively high. As a consequence, they can be acquired on a very short time scale, enabling to even monitor hemodynamic changes with characteristic responses as fast as a few tens of milliseconds.

As indicated in Figure 20.6, it is possible, and now rather common, to have more than a single source/detector pair but to attach an array of sources and detectors to the sample under investigation. Sources are either activated sequentially or intensity-modulated at different frequencies, so they can be operated simultaneously. However, for reasons of cost, frequently a single laser source is multiplexed sequentially to different optical fibers instead of using a number of distinct devices. Most current systems use multiple detectors to record signals continuously and in parallel. Signals from specific sources, which are operated simultaneously, are isolated either by using lock-in amplifiers or by Fourier transformation and appropriate filtration.

Based on this multisource/multidetector philosophy and with the advancement of near-IR diode laser and detector technology systems have been developed and are now commercially marketed for clinical applications, which can generate 3D tomographic images, for example, of the breast (see Grable et al. 2004).

While greatly successful, there remain a number of conceptual drawbacks despite the great technological advances. In particular, CW imaging relies on the measurement of diffuse light intensity at a single wavelength; this does not allow for decoupling the effects of absorption and scattering (see Arridge 1998). Further disadvantages of CW imaging are as follows: (1) the detected signal is greatly dependent on surface coupling and (2) the intensity measurements are much more sensitive to the optical properties of the surface tissue than to those of the deeper regions of the tissue. In order to circumvent many of the problems inherent to CW measurements, TD or FD methodologies have been developed. Both approaches have their merits, but note that while time-resolved measurements have the advantage that sample information can be acquired simultaneously at all frequencies, in the FD systems, one can make use of source/detector combinations, which are significantly cheaper than those required for time-resolved measurement systems.

## 20.2.3 Experimental methodology 2: FD systems

In these systems, a radio-frequency oscillator is used both to drive a laser diode (typically in the high-megahertz RF range) and to provide a reference signal. As illustrated in the central panels of Figure 20.6, when frequency-modulated light is transmitted through a biological tissue sample, its amplitude is reduced and its phase is affected. Thus, the detected signal has to be sampled and analyzed both for amplitude and phase information. Using lock-in techniques, the latter is normally achieved by comparing the received signal to the reference oscillator wave from the laser source driver. Note also that because of the electronic properties of lock-in devices, heterodyning is utilized to convert the RF to few-kilohertz waves prior to phase detection.

Since the turn of the millennium, a range of FD instruments have been developed for optical topography and tomography. In particular, noteworthy is the development of a hybrid CW/FD optical tomographic instrument (Culver et al. 2003a), which combines the speed and low cost of CW measurement components with the capability that absorption and scatter components can be separated. The said hybrid system makes use of amplitude-modulated LDs at four common NIR wavelengths (690, 750, 786, and 836 nm), which are multiplexed through a 5×9 optical fiber array. The diffusely reflected light is recorded via a number of fiber-coupled APD detectors, whose positions are interlaced with the source fiber array. The amplitude and phase of the APD signals are extracted by homodyning. Essentially the information deduced from the FD channels is combined with spatial CW interpretations in order to improve the functional tomography.

### 20.2.4 Experimental methodology 3: TD systems

Time-resolved techniques were initially developed in the belief that despite the predominant diffuse scattering, inherent to the light transmission through highly scattering media, the collection of minimally scattered photons—the so-called ballistic photons—could provide well-resolved images as in x-ray computed tomography (CT). This approach was soon proved to be impractical as ballistic photons can only be detected in very thin (a few millimeters) samples.

Nowadays, TD techniques are those that employ a short pulse of laser excitation light (normally $\tau_p < 200$ ps) and detectors that measure the time of flight of the photons transmitted through the sample, as illustrated in the lower panels of Figure 20.6. In general, this photon distribution is known as the temporal point light spread function (TPSF) and extends over a few nanoseconds for photons having travelled through several centimeters of biological tissue. In particular, advances in time-correlated single-photon counting (TCSPC) hardware, in conjunction with short-pulse LDs, have significantly reduced the cost and complexity of TD measurement systems. The principle of TCSPC is based on correlating the arrival time of a scattered photon by processing the scattered and reference light signals through a constant-fraction discriminator (CFD), an electronic processing unit that mimics the mathematical operation of finding a maximum of a pulse by finding its zero slope. This process eliminates timing jitter, and photon-arrival times can be determined with a resolution of 20–30 ps (Kalisz 2004; Wahl 2014).

Note that the actual shape of the recorded signal, i.e., its overall shift, duration, and slope of the signal tail provide crucial information about the properties of the investigated tissue—in particular, the absorption and scattering coefficients of distinct parts of the sample (see Figure 20.7). From these, (semi-)quantitative reconstruction of tissue composition in 2D or 3D is possible, in principle.

Commonly, two distinct implementations of TD measurements to DOI are encountered, namely transillumination techniques and tomographic approaches (see Figures 20.4 and 20.6). In the former case, the contact fibers from the laser source and to the detector are arranged on opposite sides of the tissue sample (normally a "slab"). These are scanned in two dimensions across each surface, and a 2D projection image is generated from these data. In the latter case, the laser source(s) and detector(s) are arranged suitably over the available surface of the tissue sample. In this way, multiple lines of sight are recorded across the entire volume, either simultaneously or sequentially. 3D images are then reconstructed using any of the techniques described in Section 20.4.

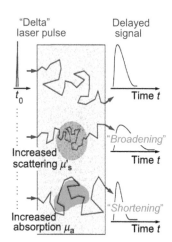

**Figure 20.7** Features in diffuse transmission influencing the shape of the time-broadened signal following a delta-shaped input light pulse. (Top) Homogeneously scattering tissue; (middle) inclusion with different scattering coefficient; (bottom) inclusion with strong absorption.

**Table 20.2** Comparison of Merits between the Three Experimental Methods of Diffuse Optical Measurements

| Method | Drawbacks | Advantages |
|---|---|---|
| CW | Penetration depth<br>Difficult to decouple<br> absorption and scattering | Sample rate<br>Cost<br>Instrumental simplicity |
| FD | Penetration depth | Sampling rate<br>Acceptable separation of<br> absorption and scattering |
| TD | Sampling rate<br>Cost<br>Instrument dimensions and<br> weight | Spatial resolution<br>Penetration depth<br>Accurate separation of<br> absorption and scattering |

### 20.2.5 Comparison between the three experimental methods

Overall, each of the three experimental techniques has its respective advantages and drawbacks, which are summarized in Table 20.2.

Essentially, the choice of the most appropriate modality depends on the desired specific information. In particular, over the last decade, the relative advantages and disadvantages of FD and TD optical imaging systems have been subject to major debate. It is certainly true that TD data provide the richest information in diffuse imaging applications (Schweiger and Arridge 1999). In particular, they allow for a rather accurate separation between absorption and scattering conditions, and spatial resolution even at large penetration depths is quite remarkable. But they are, in general, slow and costly. In contrast, FD systems are less costly, are relatively easy to set up, and use and provide temporal sampling times of just a few milliseconds. However, they suffer from relatively poor penetration depth. Moreover, for hospital applications, the inability of FD systems to identify light that invariably leaks around an object, but that is temporally correlated with the measurement, is detrimental to accurate diagnostics.

## 20.3 MODELING OF DIFFUSE SCATTERING AND IMAGE RECONSTRUCTION

### 20.3.1 Modeling light transport through tissue

Before one can contemplate diffuse optical image reconstruction, one requires some understanding on how to model photon migration in tissue. Today, this is typically modeled using either of the following three common approaches: (1) models based on numerical Monte Carlo simulations; (2) analytical methods based on the radiative transfer equation (RTE)—also addressed as the Boltzmann equation—and then utilizing diffusion approximations; or (3) hybrid models suitably combining both numerical and analytical aspects. It should be noted that the RTE is difficult to solve without introducing approximations. As a rule, computing solutions for the photon transport exploiting the diffusion equation in one form or another is more efficient but less accurate than using Monte Carlo simulations. For a full description of some or all aspects of these aforementioned methodologies, the reader may consult relevant reviews, for example, that of Arridge (1998) or the textbook by Wang and Wu (2007). Here, we only want to give a brief glimpse at the complexity of the problem, exemplified for the formulation of the RTE, which is conceptually easier to follow.

The RTE is a differential equation mathematically describing the transfer of energy as photons move inside a tissue sample, i.e., the flow of radiation energy through a small area element. It basically states that a beam of light on the one hand loses energy through divergence and extinction (which includes both absorption and spatial scattering), but on the other hand may gain energy from light sources in the medium and scattering directed toward the original beam. In its fundamental form, coherence and polarization properties as well as nonlinearities are neglected, but other optical properties like the refractive index, absorption and (elastic) scattering coefficients, and scattering anisotropies can be included; normally, these parameters are treated as time-invariant, but spatial variations are permissible. The notation of the RTE given here is that used by Wang and Wu (2007):

$$\frac{1}{\upsilon} \cdot \frac{\partial L(\vec{r}, \hat{s}, t)}{\partial t} = -\hat{s} \cdot \nabla L(\vec{r}, \hat{s}, t) \qquad \text{I}$$

$$- \mu_t L(\vec{r}, \hat{s}, t) \qquad \text{II}$$

$$+ \mu_s \int_{4\pi} L(\vec{r}, \hat{s}', t) P(\hat{s}', \hat{s}) d\Omega' \qquad \text{III}$$

$$+ S(\vec{r}, \hat{s}, t) \qquad \text{IV}$$

(20.1)

where $L(\vec{r}, \hat{s}, t)$ is the radiance in units of $W \cdot m^{-2} \cdot sr^{-1}$; $\mu_t = \mu_a + \mu_s$ is the extinction coefficient (made up from the coefficients for absorption and scattering); $P(\hat{s}', \hat{s})$ is a phase function, representing the probability that photons with propagation direction $\hat{s}'$ are scattered into a solid angle $d\Omega$ around $\hat{s}$, and $S(\vec{r}, \hat{s}, t)$ (also in units of $W \cdot m^{-2} \cdot sr^{-1}$) describes the (laser) excitation source; $\upsilon$ is the velocity of light within the medium. For the meaning of the individual variables, consult Figure 20.8.

Note that the four terms in Equation 20.1 are associated with I—energy balance at $r$ in direction $\hat{s}$; II—loss due to scattering and absorption; III—gain due to scattering at $r$ from $\hat{s}'$ to $\hat{s}$; and IV—source term.

By making appropriate assumptions about how the photons in the tissue medium propagate, the number of independent variables can be reduced, leading to a diffusion equation for the photon transport. In particular, one assumes that (1) there are few absorption events relative to scattering events and (2) in such a medium over the transport mean free path, the fractional change in current density is very small (this latter property is sometimes called "temporal broadening").

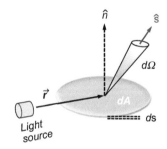

**Figure 20.8** Geometrical definitions for the photon energy flow through a differential area element $dA$ at position $r$ within a differential solid angle element $d\Omega$.

Recalling that the radiance, $L$, is related to the photon fluence rate, or intensity, $\Phi$, by

$$\Phi(\vec{r}, t) = \int_{4\pi} L(\vec{r}, \hat{s}, t)d\Omega \;\; (\text{in units of W} \cdot \text{cm}^{-2}),$$

and incorporating the above approximations, one arrives at the photon diffusion equation for the light intensity:

$$\frac{1}{\upsilon} \cdot \frac{\partial \Phi(\vec{r}, t)}{\partial t} = -\nabla \cdot [D(\vec{r}) \nabla \Phi(\vec{r}, t)] - \mu_a \Phi(\vec{r}, t) + S(\vec{r}, t). \qquad (20.2)$$

where $D(\vec{r})$ is the spatially variant diffusion coefficient (related to the absorption and reduced scattering coefficients). For selected configurations of boundaries (e.g., layers of tissue) and light sources (e.g., a point source), the diffusion equation may be solved by applying appropriate boundary conditions and a well-defined source term $S(\vec{r}, t)$.

It should be reiterated that the RTE, and with it the diffusion equation, are conservation equations, stating that the radiance for photons traveling from a given point in a given direction at a certain time is equal to the sum of all the mechanisms that increase minus those effects that reduce it. This approach has been used successfully to model light transport in a range of media, including diffusive media, and it can be formulated both in the TD and FD (Equation 20.1 is in the TD). However, it should be noted that the methodology has severe limitations: (1) it does not include wave effects (i.e., the light wavelength has to be much smaller than the dimensions of the object under study); and (2) it requires the refractive index to be constant in the medium. A few key publications by Arridge and coworkers, from the late 1990s (Hebden et al. 1997; Arridge and Hebden 1997; Arridge 1999), still serve as a useful introduction to the basic theoretical (and experimental) concepts. Progress since then has, by and large, had at its heart (1) how to develop more realistic and efficient models of light transport in media/tissue; and (2) how to solve the so-called "inverse problem" with increasing rigor. As a consequence, it is now rather common to include prior information of the structure, shape, and optical properties of the medium/tissue in both the modeling of tissue scattering and the image reconstruction.

Three key steps are required in the sequence to reconstruct an optical image from the diffuse scattering data:

- Stage 1: The transport of light in tissue must be modeled.

- Stage 2: The transport model is then used to predict the distribution and propagation of light in the probed sample. This so-called *forward problem* allows one to model the measurement and to generate a sensitivity matrix, which relates the measurement data to internal optical properties of the sample. This (explicit or implicit) matrix is frequently addressed as the "forward mapping Jacobian."

- Stage 3: The image is reconstructed by inverting the Jacobian from step 2, and thus the *inverse problem* is solved.

Referring back to Figure 20.1, it is clear that each measurement is sensitive to the whole volume, meaning that the forward problem constitutes a series of integrals over the entire volume, including a range of scattering, absorption, and re-emission probability parameters. As a consequence, the inverse problem is ill-posed (i.e., the Jacobian is underdetermined) and is nonlinear, and normally complex reconstruction techniques are required, which are demanding on computing power. The overall concept yielding reconstructed images from measurement data is shown in Figure 20.9.

## 20.3.2 The forward problem

A number of modeling approaches to the forward problem in DOI are being used, including analytical, statistical, and numerical modeling techniques. Below we briefly summarize the pros and cons for those techniques.

In *analytical techniques*, Green's functions are used to model the RTE analytically. Green's functions provide solutions when the light excitation source can be treated like a spatial and temporal $\delta$-function. From such "point" solutions, one can derive a solution for an extended source by convolution. Unfortunately, solutions only exist for simple homogeneous objects or media, which only include a single spherical perturbation; both are unrealistic approximations for most biological tissues. On the other hand, Green's function techniques are now frequently used to solve the forward problem, in which the geometry can be approximated as a slab or an infinite half-space (see Culver et al. 2003b).

In *statistical techniques*, the trajectories of individual photons through the medium are modeled; the inherent advantage is that the statistical (Poisson) error is incorporated into the model that is quite natural and elegant. In the context of diffuse optics, the most commonly used statistical technique is the *Monte Carlo method*, often regarded as the "gold standard" to which other techniques are compared. The photon trajectories are followed until they either escape from the sample under study or are absorbed (see Boas et al. 2002). An alternative approach is *random walk theory*, in which the photon migration is modeled as a series of space-propagation steps. For this, the sample is rastered into a discrete cubic lattice; the time steps may be discrete or continuous. Random walk theory is particularly suited to model measurements in the TD (see Chernomordik et al. 2000).

In complex scenarios—such as inhomogeneous distribution of optical properties in an arbitrary geometry—*numerical techniques* are required, in general. As a natural choice, models based on the *finite element method* (FEM) have evolved. It requires dividing the domain to be modeled into a finite element mesh. While this is a completely general technique, in principle, it is normally rather difficult to create a mesh of irregular objects with complex internal structure. The FEM modeling approach was first introduced into optical tomography by Arridge and coworkers (Arridge et al. 1993; Arridge and Schweiger 1995).

Overall, it has been recognized that for efficient and robust modeling of the forward problem in complex biological samples, like the human breast or head, one should attempt to reduce the general complexity of the problem. This can be done by including as much reliable "prior information" as possible. For example, anatomical information from complementary MRI images can be included; or one may replace the isotropic diffusion coefficients by diffusion tensors to account for anisotropic scattering of tissue with known anatomic anisotropy at the cellular level.

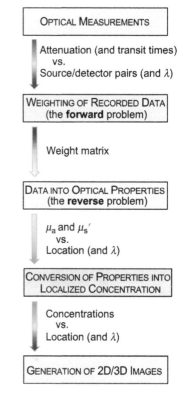

**Figure 20.9** Overall concept for image reconstruction from diffuse optical scattering data.

### 20.3.3 The reverse Problem—Principles of image reconstruction

The goal in DOI (DOT), be it 2D topography or 3D tomography, is to reconstruct the spatial distribution of optical/physiological properties at each volume element in the tissue sample from experimental measurement data (in general, fluence rates), recorded by detectors placed on the sample surface. As mentioned above, this reconstruction approach is typically called the *inverse problem*. Related to this is the *forward problem* in which one attempts to calculate the fluence rate at a specific tissue surface point, assuming a predetermined spatial distribution of the optical/physiological properties inside the tissue sample. Key to the success of the forward/inverse problem solution is the understanding of the photon transport in tissue. In the majority of practical approaches, a tractable mathematical basis for this are the aforementioned transport or diffusion equations. However, often these approximations are inadequate, and other model approaches are required. For a summary of recent modeling approaches, see Pogue et al. (2006) or Durduran et al. (2010).

In order to tackle the inverse problem, normally one first has to formulate a "specific" forward model based on a general forward model of diffuse scattering. For this, one simulates diffuse scattering data, $Y_D$, which correspond to the radiance $L(\vec{r}, \hat{s}, t)$, the fluence rate $\Phi(\vec{r}, t)$, or any other variables in approximate models to the RTE. These $Y_D$ data (which may be CW, TD, or FD data) are linked to the knowledge (or approximation) of the sources, $S_q$, and tissue-internal optical properties, $X_D$ (which may include the absorption coefficients, $\mu_a$, and effective scattering coefficients, $\mu'_s$), by a forward operator, $\hat{F}$:

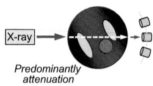

Object reconstruction: $\mu=[\hat{J}]^{-1} \cdot S$

Detected signal: $S=\hat{J} \cdot \mu$

X-ray

*Predominantly attenuation*

$$Y_D = \hat{F}(X_D). \tag{20.3}$$

To reconstruct an actual image from the measurement data, one needs to solve the inverse problem, i.e., one has to calculate the internal optical properties $X_D$ from the known source(s) $S_q$ and the measurement data $Y_{D,m}$:

Linear reconstruction problem

vs.

Non linear reconstruction problem

$$X_D = \hat{F}^{-1}(Y_{D,m}). \tag{20.4}$$

In general, this is a nonlinear problem for two reasons. Firstly, scattering in light propagation is the dominant process and thus makes any measurement of the fluence rate $\Phi(\vec{r}, t)$ sensitive to a relatively large, most likely inhomogeneous tissue volume. Secondly, the inverse problem in DOI is intrinsically nonlinear with respect to the actual optical properties of the tissue. This means that "simple" linear inversion algorithms, which are common in x-ray CT or PET, are normally inadequate for DOT, although for a few specific applications, the problem can be linearized. But, in general, computationally intensive methods have to be used to arrive at the inverse solution. The basic configurations and concepts relevant in linear and nonlinear image reconstruction are visualized in Figure 20.10.

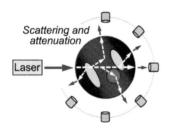

*Scattering and attenuation*

Laser

Detected signal: $S=\phi(\mu)+r$

Object reconstruction:
Iterative minimization
of error = $\|S - \phi(\mu)\|^2_{min}$

**Figure 20.10** Conceptual differences between linear (top) and nonlinear (bottom) imaging geometries.

*Linear image reconstruction.* A nominally nonlinear inverse problem can be linearized in the case that the actual (reconstructed) optical properties $X_D$ and the measured data $Y_D$ are close to an initial estimate $X_0$ and the simulated data

$Y_0$. One encounters this, for example, in "difference imaging," a method in which data are recorded before and after a small change in the optical properties. Then one can expand Equation 20.3 as a Taylor series:

$$Y_{D,m} = Y_0 + \hat{\boldsymbol{F}}'(X_0) \cdot (X_D - X_0) + \hat{\boldsymbol{F}}''(X_0) \cdot (X_D - X_0)^2 + \cdots. \qquad (20.5)$$

where $\hat{\boldsymbol{F}}'$ and $\hat{\boldsymbol{F}}''$ are the first- and second-order Fréchet derivatives of $\hat{\boldsymbol{F}}$, respectively. Note that the Fréchet derivative is a linear integral operator, which maps a function in the image space to a function in the data space; in numerical methods, these are often known as the Jacobian and Hessian matrices $\boldsymbol{J}$ and $\boldsymbol{H}$, respectively. Thus, by omitting the nonlinear, high-order terms in the Taylor expansion, one finds

$$\Delta Y_D = J(X_0) \cdot \Delta X_D, \qquad (20.6)$$

where $\Delta X_D = X_D - X_0$ and $\Delta Y_D = Y_D - Y_0$. From this, one can calculate $\Delta X_D$ by inverting the Jacobean $\boldsymbol{J}$. However, in general, $\boldsymbol{J}$ is underdetermined and the so-called "regularization" methods need to be applied. As the most common techniques, one encounters (1) truncated singular-value decomposition, (2) Tikhonov regularization, and (3) algebraic reconstruction; for details, see Arridge (1999, 2011), Gibson et al. (2005), and Dehghani et al. (2008). For example, a Tikhonov-type formulation of linear inverse problem can then be expressed as

$$\Delta X_D = \boldsymbol{J}^T \cdot \left( \boldsymbol{J} \cdot \boldsymbol{J}^T + \lambda \cdot \boldsymbol{I} \right)^{-1} \cdot \Delta Y_D. \qquad (20.7)$$

where $\boldsymbol{I}$ is the identity matrix and $\lambda$ is a regularization parameter; $\boldsymbol{J}^T$ is the transposed Jacobian. The Jacobian $\boldsymbol{J}$ itself is calculated using the forward model.

*Nonlinear image reconstruction.* In the majority of cases, the inverse problem cannot be solved using a linearizing approximation. The underlying assumption leading to Equations 20.5 and 20.6 relied on reconstructing the *difference* between two similar (measurement) states, i.e., it constitutes a *relative* measure. However, more often than not, reconstruction has to be accomplished from a single acquisition to obtain *absolute* values of the optical properties, meaning that now the full nonlinear problem must be solved. The absolute values of the unknown variable can be derived by iteratively minimizing the difference between the measured data and the simulated model data.

The common philosophy then is first to define an objective function, $O_D$, which represents the difference between the measured data $Y_{D,m}$ and data that are simulated using the forward model $\hat{\boldsymbol{F}}(X_D)$. If one surmises that $X_D$ represents the distribution of optical parameters, which minimizes $O_D$, then it can be treated as the model that best fits the data. Therefore, it may be taken to constitute the desired image. But because of the single-acquisition scenario, the problem is ill-posed and underdetermined, in general, and thus must be "regularized." Two regularization approaches are common, namely using the so-called L1- and L2-norms, which minimize the least absolute error (LAE) or the least-square error (LSE), respectively. In general, in image reconstruction problems, the L2-norm is preferred because among other features (1) it results always in one solution, and this is stable; and (2) it is computationally efficient due to exhibiting analytical solutions. Thus, a typical objective function for nonlinear image reconstruction is

$$O_D = \| Y_{D,m} - \hat{F}(X_D) \|^2 + \alpha \cdot \| \Pi \|^2, \qquad (20.8)$$

where $\alpha$ is a regularization parameter, and $\Pi$ represents any prior information of the target volume. It is increasingly common to include anatomical and other information (like the physics and the physiology of the problem) in $\Pi$.

Note that for the (simplistic) case $\Pi = I$, Equation 20.8 constitutes an iterative solution of Equation 20.7. Note also that Equation 20.8 may be extended to a weighted least-squares approach, which reduces the influence of noise and cross-talk both in the data and the resulting image—both are common problems in DOI. For further details on methodologies and algorithms, see Arridge (1999, 2011), Dunduran et al. (2010), Chaillat and Biros (2011), and Darne et al. (2014).

In this context, 3D DOI—also known as diffuse optical tomography (DOT)—is particularly demanding. This is linked to the fact that the inverse problem in DOT is highly ill-posed since, in the majority of practical applications, high sparsity of excitation/sensing sites is encountered, with respect to the complexity of the (normally biomedical) sample to be imaged; this makes reconstruction of high-quality images a major challenge. Because of the nature of sparsity in DOT, common matrix regularization methodologies have been utilized in high-quality DOT reconstruction. However, these conventional approaches require substantial programming efforts and computational power, and, moreover, lack certain selection criteria to optimize the regularization parameters. Here, a novel type of algorithm may come to the rescue, dubbed *dimensionality reduction-based optimization* DOT—or DRO-DOT; this approach has been proposed by Bhowmik et al. (2016). The key concept of the authors' approach is to reduce the dimensionality of the inverse DOT problem by reducing the number of unknowns in two linked, consecutive steps. First, a "low-resolution" support mask for a (nonzero) voxel set is generated based on the properties of the sensing matrix. Subsequently, a sparse image is reconstructed within the support framework generated in the first step. This was done by using $L_1$-minimization inside the recovered support mask only, with size selected smaller than the actual, full 3D imaging volume.

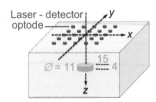

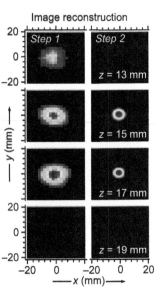

**Figure 20.11** 3D image reconstruction of a subsurface object measured with a sparse planar optode array. (Data adapted from Bhowmik, T. et al., *Sci. Rep.* 6 (2016): article 22242.)

The result of the dimensionality reduction in both steps is that the number of unknowns is reduced as a consequence, and in conjunction with their adaptive scheme to find the problem residuals via statistical interpretation of the $L_1$-regularization, the authors were able to substantially reduce the computational complexity, while at the same time being able to recover high-resolution images even when using only a limited number of optodes (excitation source/detector pairs). A conceptual example for the success of this approach is shown in Figure 20.11; here the shape and position of the cylinder "hidden" at some significant depth below the surface could be reproduced truthfully, only using a sparse planar arrangement of optodes.

With all the success of DOI imaging, it should still be kept in mind that also DOI may "suffer" from the general pitfalls in image reconstruction, which are depicted in Figure 20.12. In Figure 20.12a, four differently shaped, schematic 3D objects are shown imbedded in a (tissue) sample. It is clear from the two 2D projections shown (here in the $x$- and $z$-directions) that the information one can extract from the data does not necessarily result in a unique interpretation as to the

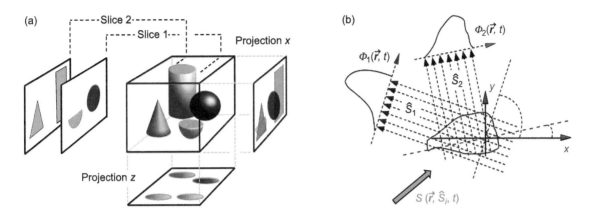

**Figure 20.12** Conceptual principles in diffuse optical image reconstruction. (a) 2D projections and image planes of 3D objects imbedded in a (tissue) sample; (b) photon flux distributions $\Phi(\vec{r}, t)$ at two detector sites, 1 and 2, associated with the scattered light from an arbitrarily shaped object, which is illuminated by a light source $S$ in direction $\vec{r}$.

nature and actual shape of the objects. Take for example the projection in the z-direction: all objects appear as circles, and one cannot deduce any further information on the shape of the objects other than they are round in one dimension. Additional projections or image slices through the sample volume are required to find the 3D solution for the inverse problem. In Figure 20.12b, the scattering into directions $\hat{s}_i$ from an odd-shaped object is shown, after illumination by a source $S(\vec{r}, \hat{s}_i, t)$; the scattered photon flux pattern $\Phi_i(\vec{r}, t)$ for two (arbitrarily) selected general directions $\hat{s}_1$ and $\hat{s}_2$ is shown, where a single-device detector is moved laterally by (small) amounts or a multiple-detector array records all positions simultaneously.

It is clear from the sketch that the resolution of the reconstructed image will become better the more the positions and directions are probed in individual measurements. Note also that, for the sake of simplicity, the trajectories of the photons traveling within the medium are shown as straight lines. However, in a heavily scattering medium, the photons follow more a "random-walk" pattern and contributions from scattering sites other than the designated target object contribute to the signal. From this, it is indeed clear that the image reconstruction procedure is as complex as described further above.

## 20.4 CLINICAL APPLICATIONS OF DOI AND SPECTROSCOPY

In general most applications of DOI, including tomographic and spectroscopic modalities, have been developed to address biomedical problems. The two main advantages that are exploited, and that are extremely attractive for medical diagnostics, are that (1) the method is noninvasive and (2) non-ionizing radiation is used. Probably the most common applications these days are the monitoring of blood flow and oxygenation (e.g., in pulse oximetry); the detection and monitoring of breast cancer; and the investigation of brain functions. These and a few other examples have been reviewed by Gibson et al. (2005), Huppert (2013), and Hillman (2007); also, one may consult the books by Jiang (2010) and Madsen (2013).

*Pulse oximetry* is a simple, noninvasive method of monitoring the concentration of hemoglobin, both in its oxygen-desaturated (Hb) and oxygen-saturated (HbO$_2$) forms, whose absorption spectra are different. Using probe laser sources at two wavelengths (usually at red and NIR wavelengths), the related diffuse optical responses at the two wavelengths allow one to calculate the proportion of hemoglobin that is oxygenated, and to follow dynamic processes (see Lopez Silva et al. 2003). The methodology is now quite mature, and commercial (handheld) pulse oximeters are in widespread use; the simplest instruments now cost less than $100, although these are normally equipped with LEDs rather than LDs.

*Breast cancer* now accounts for about one third of all new cancers diagnosed in women. Due to safety issues and shortcomings of current x-ray imaging methods in screening, other imaging modalities, such as DOT, are evolving for the purpose of breast cancer screening and treatment monitoring. In particular, its bio-marker-specific response when using multispectral probing proves to be rather advantageous. The methodology is discussed in more detail further below.

Clearly, the *mapping of (human) brain function* has revolutionized the field of neuroscience. Traditionally, PET or functional magnetic resonance imaging (fMRI) has been and is used in brain scanning. Optical neuroimaging based on *DOT* offers a noninvasive alternative, in principle, which avoids some of the disadvantages encountered in PET and fMRI. However, DOT has lacked the spatial resolution of the aforementioned techniques. But recently promising attempts have been reported in which a "high-density diffuse optical tomography imaging array that can map higher-order, distributed brain functions" was successfully tested (Eggebrecht et al. 2014). A few examples for imaging of the brain are given in Section 20.4.2.

## 20.4.1 DOT and spectroscopy of breast cancer

Breast cancer is one of the most abundant: types of cancer and, unless detected at an early stage, one of the leading causes of death for women. The most widely used preventive screening technique for early detection of this type of cancer is x-ray mammography; however, the predictive power of film-screen mammography has been rather poor (Liberman et al. 1998) and remains so, despite perpetual improvements in technology and more extensive screening of risk groups. Additional modalities like full-field digital mammography with the addition of digital breast tomosynthesis may be required to increase correct diagnosis and to reduce false positives (see, e.g., Michell et al. 2012). Alternative techniques currently under consideration for breast cancer detection include ultrasound imaging, CT, and magnetic resonance imaging (MRI). Of these, MRI is the most widely used and promising alternative for dense breast imaging. However, while highly sensitive, its specificity is only modest, and it is not yet utilized for routine breast analysis due to its high cost. Within the framework of a search for alternative technologies, NIR DOT slowly emerges as a noninvasive and low-cost technique for breast cancer detection; a typical implementation is shown in Figure 20.13.

A quite systematic early review on DOI imaging for breast cancer detection has been given by Leff et al. (2008). In particular, NIR spectroscopy and imaging are currently used to evaluate the concentrations of hemoglobin, water, and lipids, as well as optical scattering properties in normal and cancerous breast tissue (McBride et al. 2002). In this context, topographical examination of the breast has

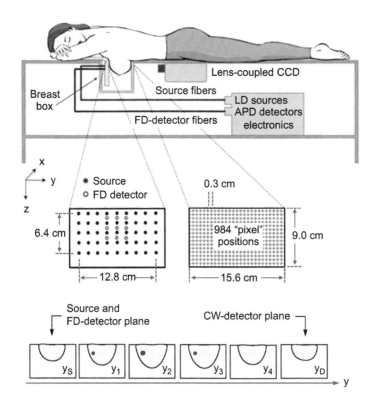

**Figure 20.13** Schematic of the "parallel-plate" diffuse optical tomography (DOT) instrument, used for imaging the female breast; for details see text. Adapted from Choe, R. et al., *J. Biomed. Opt.*14, no. 2 (2009): article 024020 reproduced with permission of SPIE and the authors.

received increasing attention, and both commercial and academic instruments have been developed; see, e.g., van der Mark et al. (2000), Hadjipanayis et al. (2011), and Fantini and Sassaroli (2012), and references therein. In most of the scanner configurations, the breast is either in contact with the source and detector optics (e.g., van der Mark et al. 2000), or immersed in an optical matching fluid (e.g., Culver et al. 2003a; Choe et al. 2009). A particularly promising development in instrumentation is that of a noninvasive line-scan, handheld probe, which determines the microvascular blood flow contrast in the breast in the *in vivo* detection and monitoring of tumors (Choe et al. 2014). A typical DOI setup for breast monitoring is that shown in Figure 20.13; this particular configuration is rather "patient-friendly," since the woman can lie down on a bed.

The DOI instrumentation is based on a hybrid FD/CW configuration (Choe et al. 2009). Four LDs (at 690, 750, 786, and 830 nm) were used for the FD measurements, whose amplitude is modulated, and which are switched between 45 optical fibers, arranged in a 9×5 array configuration. The nine FD-detection fibers are interlinked with the source fibers, arranged in a 3×3 pattern. Two additional LDs at 650 and 905 nm were used in CW mode (these latter two wavelengths helped to improve the separation of chromophore contributions). The breast tissue response to the excitation is measured in remission FD mode (by an array of APDs) and transmission CW mode (by a lens-coupled CCD array detector). For purposes of easier image reconstruction, the breast box is filled with an index-matching fluid. To illustrate the capability of such technique, Figure 20.14 depicts several DOT images generated in breast tissue.

The simultaneous use of the (suitably selected) four modulated LD wavelengths allows one to identify and map various biomolecule compounds; in the figure,

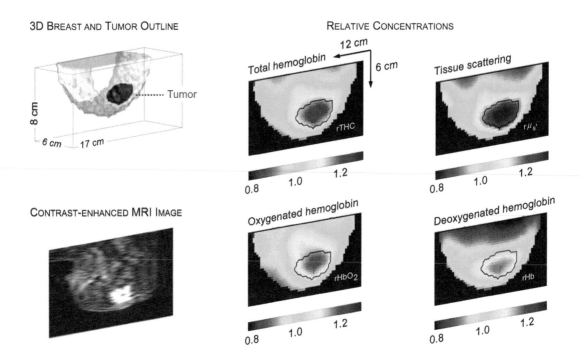

**Figure 20.14**  Multispectral DOT images of an invasive ductal carcinoma in a woman's breast, mapping relative hemoglobin concentrations and relative tissue scattering (the four right-side panels). For comparison, a DCE-MRI image slice containing the tumor center is shown (lower left). For further details, see text. (Data adapted from Choe, R. et al., *J. Biomed. Opt.* 14, no. 2 (2009): article 024020.)

the relative oxygenated (rHbO$_2$), deoxygenated (rHb), and total (rTHC) hemoglobin concentrations are shown, which all exhibit enhanced absorption in the tumor region (marked in its outline by the black line). In addition, a map of the relative tissue scattering coefficient, $\mu'_s$, is included (top-right data panel); this also mimics an increased response in the same region. A comparison with a dynamic, contrast-enhanced (DCE) MRI image (bottom left of the figure), taken to guide the analysis, is also included. Certainly, in the case shown, the tumor is clearly identifiable in both the DCE-MRI image and the bioparameters of the DOT map. Note that all DOT images represent slices through the same plane in the breast, containing the center of the tumor and oriented in caudal–cranial view. As a consequence of these and other investigations, there is now significant evidence that cancerous tissue is associated with higher water and hemoglobin concentrations, and a lower lipid concentration with respect to normal breast tissue (Fantini et al. 2012).

Finally, it should be mentioned that over the past few years, a technique called CT laser mammography (CTLM) has emerged as a valuable tool to visualize vascular structures, not only physiological blood vessels but also neovascularization (see Grable et al. 2004; Qi and Ye 2013). CTLM employs diode laser radiation (normally at $\lambda = 808$ nm), which is absorbed in blood pigments of physiological and pathological blood vessels and is able to display their distribution, offering new possibilities of breast disease diagnosis with demonstration of neovascularization. The technique is then able to recognize malignant tumor from benign lesion by imaging their neovascularization.

## 20.4.2 Diffuse optical topography and tomography of the brain

Near-infrared optical imaging of the brain is probably one of the most widely used applications of DOI, both in basic science and for clinical problems. Using light to image living tissue has significant benefits; in particular, it is based on non-ionizing radiation, and instruments are in general much less costly compared to other clinical imaging modalities. Furthermore, it is sensitive to functional changes (reflected in changes in absorption, fluorescence, and/or scatter) to monitor—even in real time—the concentrations of intrinsic chromophores oxy- and deoxyhemoglobin, as well as other important cytochromes and metabolites, which all have distinctive absorption or fluorescence characteristics. Finally, patients find their exposure to DOI instrumentation much less stressful and daunting, in general: the subjects undergoing functional near-infrared spectroscopy (fNIRS) imaging of brain activity may smile throughout because of their "comical headset." The major challenge of DOI, in general, and for imaging of the brain, in particular, is to overcome the effects of light scattering, which limit penetration depth and achievable imaging resolution. In this context, functional mapping of the human brain can adopt the modalities of optical topography and tomography (see Figure 20.15).

For example, optical topography can be used to measure hemodynamic changes occurring in the human cortex. Using widely spaced sources and detectors, light can be measured, which has passed through deeper regions of the brain, although the depth limit, in general, is of the order 2–3 cm, and the spatial resolution of the related tomographic reconstruction is rather poor. Sort of full 3D tomography has only been achieved in neonatal brains, since the baby's head is, in general, small enough, and its skull is still relatively translucent. The most exciting feature of any brain imaging is not necessarily the anatomical structure but the actual brain activity/functionality.

The traditional method for imaging brain functions is fMRI. When neurons are activated, they require energy—typically in the form of increased levels of sugar and oxygen. This is afforded by a change in blood flow, a process called hemodynamic response, by which blood releases oxygen to the activated area. This response leads to an increase in the concentration of oxyhemoglobin, $HbO_2$, and a corresponding decrease in the concentration of deoxyhemoglobin, HbR. Essentially, the difference in magnetic susceptibility between $HbO_2$ and HbR, i.e., oxygenated and deoxygenated blood, produces a magnetic signal that can be detected by an MRI scanner. Thus, the brain activity is reflected in the blood oxygen-level dependent (BOLD) signal.

Similarly, since the two chromophores $HbO_2$ and HbR have distinct optical and spectral properties, functional NIR spectroscopy (fNRS) can be employed for noninvasive investigations of brain functions by detecting changes in blood hemoglobin concentrations related to neural activity. Since the early 2000s, fNIRS has been successfully used to investigate brain activity in both adult and neonatal infants (Yamamoto et al. 1999; Gibson et al. 2005; Ferrari et al. 2012a,b).

Moreover, simultaneous measurement of cerebral oxygenation during brain activation by functional NIR spectroscopy and functional MRI demonstrated a clear linear relation between fNIRS and BOLD-fMRI measurements for hemoglobin changes over motor cortex in animals and humans (see Mehagnoul-Schipper et al. 2002; Eggebrecht et al. 2014). Another example in this comparative

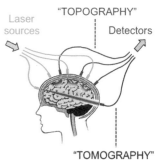

**Figure 20.15** Concept of optical topography and tomography of the brain. The photograph shows a typical "helmet" configuration for f-NIRS imaging of brain activity. (Reproduced from Jasinska and Petitto, *Dev. Cogn. Neurosci* 6, (2013): 87–101.)

context is that the evaluation of language lateralization based on brain imaging is an important tool in the diagnosis and treatment of epileptic patients and children. The fMRI method would require that subjects remain motionless during data acquisition, making the assessment of receptive and expressive language difficult for these groups of patients. In contrast, fNIRS has proven to be more tolerant to motion artifacts. Paquette et al. (2010) showed that the technique is indeed useful to assess receptive language in adults and language lateralization among children and epileptic patients.

Overall, the applications of optical topography are manifold, including the spatial identification of brain activity centers and functional response to stimuli. Without claiming to be comprehensive one finds fNIRS applications in/to neurosurgery; functional mapping of foci in epilepsy; the investigation of speech (communication, linguistics, language disability); sleep science (monitoring of dream phases); diagnosis of psychiatric diseases; Alzheimer's disease research; and so on (for a summary and relevant references, see, e.g., Koizumi et al. 2003). Two such applications are briefly touched upon below.

Franceschini et al. (2000) mapped cerebral oxy-/deoxyhemoglobin concentrations, measuring—in real time—the activated cortical area during voluntary hand tapping (see Figure 20.16). In the concentration maps for $\Delta$[Hb], and the response signals from individual detectors, one can clearly see how the motor cortex activation is induced by voluntary hand tapping: the maps show distinct differences for the tapping and rest phases. Note that in the original publication, a real-time time-lapse video is included, which impressively demonstrates the temporal changes of brain activity as a result of the physical stimulus.

Brain injury in infants is a major cause of permanent disability or death. Since optical imaging techniques are well suited to imaging infants as they are less sensitive to motion artifact than fMRI, several groups developed optical tomography instruments particularly suitable to investigate the infant's brain and its most important types of injury. *Interventricular hemorrhage* (IVH) constitutes one of the typical injures that may occur in premature infants, due to the weakness of their cerebral vasculature, which cannot resist fluctuations in blood pressure during birth. Instrumentation based on time-resolved NIR imaging to generate 3D images of the neonatal head has been developed in the early 2000s (see Schmidt et al. 2000; Gibson et al. 2005). Typical fNIRS images taken with such an instrument are shown in Figure 20.17. The images were obtained from a pair of twins, recorded consecutively on the same day. One of the twins was anatomically normal, while the other suffered from IVH; this had caused bleeding into the brain tissue. As can be seen, the images of blood volume and oxygenation saturation in the baby with the normal brain are nearly symmetrical about the midline and appear to show a decrease in blood volume in the white matter. In the brain image of the baby with IVH, the hemorrhage clearly shows up as an increase in blood volume and a decrease in oxygenation. For further details, see Gibson et al. (2005).

## 20.5 NONCLINICAL APPLICATIONS OF DOI AND SPECTROSCOPY

As shown in the previous section, DOI has evolved into a well-accepted medical diagnostics technique, with numerous commercial solutions now being

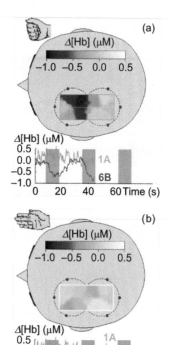

**Figure 20.16** Recording of the real-time brain response to tapping with the right hand. Two laser excitation (red dots)/detection (open white circles) fiber assemblies A and B are indicated. In the rectangular area (of dimension 4×9 cm²), the relative deoxyhemoglobin concentration of the imaged brain segment is shown. Temporal traces from two representative source–detector pairs (1A and 6B) are shown at the bottom; the blue bars represent the tapping periods of 10 s each. (a) Video snapshot during tapping period; (b) video snapshot during rest period. For further details, see text and Franceschini et al. (2000).

marketed. However, nonclinical/nonmedical use of DOI is far less common. Nevertheless, two interesting topics in which it is applied have emerged over the past few years: (1) in the characterization of some nanoparticle structures and features in catalysis and (2) for the analysis and monitoring of food stuff. Both exploit time-resolved reflectance spectroscopy (TRS) and hyperspectral imaging (HSI); the latter allows one to add chemical component mapping to spatial images.

Understanding the alignment of nanoparticle building blocks into ordered superstructures and their catalytic and/or (photo-) dynamic behavior constitutes key topics in modern colloid and material chemistry; here we only mention a few examples. During a multitechnique investigation of ZnO mesocrystals, Bian et al. (2014) showed that femtosecond TRS measurements helped to clarify some of the ultrafast interfacial charge transfer dynamics in these composites. Asahara et al. (2014) investigated the dynamics of photoreversible transitions in nano-granular $Ti_3O_5$: time-resolved diffuse reflection spectroscopy was conducted over the range from femtoseconds to microseconds in order to characterize its overall relaxation behavior.

One particularly interesting application of the diffuse optical spectroscopy and imaging consists of the assessment of (internal) fruit quality. In a sense, the problem may be classified as being similar to that described earlier, namely that of imaging the subsurface or interior of an "odd-shaped" biological object, like the breast or the head. The particular techniques frequently applied for fruit analysis are time-resolved reflectance spectroscopy (TRS), at one or a few excitation wavelengths, or multispectral imaging (MSI)/HSI exploiting a range of excitation and detection wavelengths. Traditionally, colorimetry (based on the use of visible light and its reflection from surfaces) has been employed to determine the food quality from the surface color of food samples, like whole fruits. For example, attenuated total reflectance spectroscopy (ATR) in the near-IR has been used to estimate the sugar content or firmness in fruits. However, a key limitation of ATR is that since the reflected light intensity depends on both light absorption and scattering, one cannot separate the effects of these two properties. This limitation can be circumvented to a large extent by using TRS (see Tijskens et al. 2007; Torricelli et al. 2008), and in addition, one may gain detailed information about the subsurface regions.

## 20.5.1 Single-point bulk measurements on fruits

The fundamental aspects of TRS are very much similar to DOI in the TD, as described in Section 20.2. A short light pulse (of picosecond or nanosecond duration) is injected into the medium to be analyzed. Due to the absorption and scattering of photons, the initial pulse is attenuated, broadened, and delayed. The technique measures the distribution of photons time of flight at a fixed source detector distance. In other words, it monitors the escape of light as reflectance versus time at a fixed source detector distance, $\rho$. The temporal profile of the measured TRS curve is analyzed using a solution of the diffusion equation for a semi-infinite homogeneous medium: $R(\rho, \mu_a, \mu'_s, t)$; for the actual functional form, see Toricelli et al. (2008). Basically, one finds that the wavelength-dependent absorption- and scattering-related contributions are proportional to $\exp(-A \cdot t)$ and $B \cdot t^{5/2} \cdot \exp(-C/t)$, respectively. The parameter $A$ incorporates the absorption coefficient $\mu_{a,\lambda}$, while $B$ and $C$ are functions of the reduced scattering coefficient $\mu'_{s,\lambda} = (1-g)\mu_{s,\lambda}$, where $g$ is the anisotropy factor for the medium. The different functional time behavior for absorption and scattering lead to the

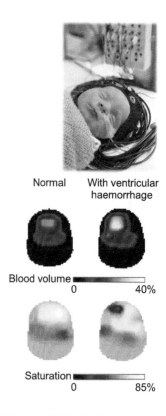

Normal     With ventricular haemorrhage

Blood volume 0 — 40%

Saturation 0 — 85%

**Figure 20.17** Optical tomography of the neonatal brain. The central column shows coronal slices through 3D images of blood volume and oxygen saturation in an anatomically normal (early-born) baby. The column on the right shows equivalent images, reconstructed from the first baby's identical twin, which had a left ventricular haemorrhage. The haemorrhage can be seen as an increase in blood volume and a decrease in oxygen saturation (adapted from Gibson, A.P. et al., *Phil. Trans. R. Soc. A* 367, no. 1900 (2009): 3055–3072; reproduced with permission. © IOP Publishing, all rights reserved.

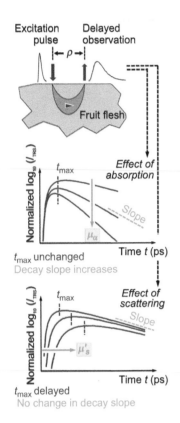

**Figure 20.18** Influence of the absorption and scattering coefficients on the TRS curves from nondestructive probing of fruits (homogeneous medium).

general observation that for a given scattering coefficient, the higher the absorption coefficient, the greater the decay of the TRS time profile, while for a given absorption coefficient, the maximum in the TRS curve is delayed as the scattering coefficient is increased, with no effect in the slope of the TRS decay. This means that the shape of the TRS curve is predominantly influenced by the absorption coefficient at early times and basically only depends on the scattering coefficient for later times (see Figure 20.18).

It is precisely this different influence that makes possible that the two optical properties, i.e., the absorption and scattering coefficients, can be extracted from fitting the experimental TRS curve to an appropriate theoretical model.

One of the most interesting findings of TRS investigations into the (ripening) status of fruit was that the postharvest time evolution of $\mu_{a,670}$ value follows a logistic curve and "occurs in all fruits with the same rate during ripening," but is shifted in time from fruit to fruit, in association with its specific degree of maturity (Torricelli et al. 2008). The absorption at $\lambda \sim 670$ nm corresponds to that of chlorophyll; the decrease in the amount of chlorophyll can be associated with the degree of maturity and ripening: the lower $\mu_{a,670}$ and hence the concentration of chlorophyll, the higher the maturity of the fruit. An example of such behavior is illustrated in Figure 20.19 where the firmness of *Spring Bright* apples is depicted as a function of the fruit biological times, for two storage temperatures after harvesting. The use of the fruit firmness is very common in this type of studies, since it has been shown that it is directly (linearly) correlated to the absorption coefficient. In this plot, each of the data was corrected for its *biological time-shift factor*, which indicates the time required to reach a reference stage of maturity (normally taken to be the midpoint of the logistic curve). Essentially, the biological time-shift factor takes into account the different state of maturity at the time of harvest, basically due to the individual age of the fruit at the tree.

## 20.5.2 Multipoint measurements on fruits yielding 2D images

Besides the possibility of simultaneously estimating the absorption and scattering coefficients, the main features of TRS are (1) its sensitivity to bulk rather than superficial optical properties and (2) its penetration depth of the order of the source–detector distance. Therefore, TRS offers the potential to sense internal fruit disorders, without having to cut it open. One such disorder, encountered particularly in pears and apples, is Brown Heart (BH) disorder; it frequently shows up after lengthy storage, even under carefully controlled conditions. The symptoms are rarely recognizable from the outside of the fruit and are visible only after cutting the fruit.

Measurements in which BH was successfully identified have been published, for example, by Eccher Zerbini et al. (2002) for pears and by Torricelli et al. (2008) and Vanoli et al. (2011) for apples. Particularly interesting is the work by Eccher Zerbini and coworkers who carried out multipoint TRS studies of fruit to determine the absorption and reduced scattering coefficients at two distinct NIR wavelengths (690 and 720 nm), trying firstly to identify the presence of BH disorder, its spatial location within the fruit, and its temporal evolution, and secondly to attempt to follow the ripening process (up to overripening) and the effect of bruises. The latter was deduced from the change of the reduced scattering coefficient $\mu'_{s,720}$ in the subpeal translucent fruit tissue, while the former was evaluated using the absorption coefficients $\mu_{a,720}$ and $\mu_{a,690}$. Since longer NIR

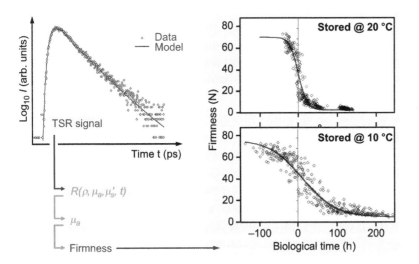

**Figure 20.19**  Change of the firmness of tested apples, as a function of the (normalised) biological time scale, which is related to the ripening potential prior to and after harvesting, for two different storage temperatures (data adapted from Torricelli, A. et al., *Sens. Instrumen. Food Qual.* 2, no. 2 (2008): 82–89; reproduced with permission of Springer). The data are based solely on measured $\mu_{a,670}$ values, derived from TRS-curves (an example is shown on the left).

wavelengths penetrate deeper into the fruit, a (normalized) density profile of the BH-affected area could be built up, and the spatial and temporal evolution of the defect area could be determined. Examples for this for two BH-affected pears are shown in Figure 20.20.

## 20.5.3  MSI and HSI of fruits

In extension to the examples described above, it might be desirable to probe the fruit (or food) sample at more than one or two selected excitation wavelengths to capture information about more than one biochemical compound, with spatial resolution. Here the techniques of MSI and HSI come into their own. In MSI, images are recorded for a handful of discrete, suitably selected excitation/observation wavelengths (related to the detection of the desired molecular compound). In HSI, the image is recorded for narrow spectral bands over a continuous spectral range, at each spatial "pixel." This allows one to execute postacquisition analysis to identify the biochemical specimen, and it has the advantage that one does not have to guess the relevant probe wavelength prior to an experiment. Recall that the outline for the two spectral imaging approaches was described in Chapter 17.

In recent years, MSI and HSI have emerged as new analytical tools in food quality and safety control (see, e.g., the reviews by Huang et al. 2014; Manley 2014). Both techniques have been developed for reflectance and transmittance modalities, in the visible and NIR, and were applied to the investigation of a wide range of different food samples, including fruits. For example, Kim et al. (2008) developed a prototype line-scan imaging system, capable of simultaneously acquiring a combination of multispectral reflectance and fluorescence from rapidly moving fruit, mounted on a commercial apple sorting machine. They demonstrated that they could detect defects and surface/subsurface contamination while the apples were sorted at a speed of 3–4 apples per second.

Traditionally, MSI/HSI systems incorporated tungsten-halogen light sources; but more recently, laser sources are being used, which offer the advantage of (1) higher spectral intensity and (2) short-pulse generation, so that faster acquisition times are possible, and fast biochemical processes can be traced with

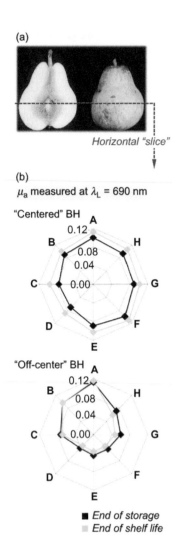

Figure **20.20** Identification of "brown-heart" disorder in pears by measuring the absorption coefficient at 690 nm using time-resolved reflection spectroscopy at selected positions around the pear's equator (indicated by the red line in the picture part). (Measurement data for two specimens adapted from Eccher Zerbini, P. et al., *Postharvest Biol. Technol.* 25, no. 1 (2002): 87–97.)

temporal resolution. For example, Lu and Peng (2007) set up a multispectral device with a set of four NIR LDs, arranged for "point" excitation (1 mm spot size) and 2D imaging (a few millimeters in diameter). Their measurements provided, in real time, information on the firmness of apples moving on a conveyor belt. Spinelli et al. (2013) set up a multispectral TRS imaging system, which incorporated a supercontinuum laser source and filter wheels for the selection of excitation and observation wavelength bands. With this system, they investigated the maturity of mango fruits, and they were able to determine the production rates of a few biochemicals; they could link the TRS measurements to the maturity status of the fruit.

## 20.6 BRIEF COMPARISON WITH OTHER MEDICAL IMAGING TECHNIQUES

As will have become clear from the previous two sections, DOI in its various guises has become a relatively mature methodology; in particular, it has become a potent diagnostic tool in biomedicine, with the most common applications found in the form of optical tomography of the skin, the breast, and the brain. The primary advantages of DOI over other (medical) imaging techniques are (1) its relative simplicity; (2) its reasonably low cost; (3) its "miniaturization" toward portable instruments; (4) its rapid data acquisition and evaluation, allowing for near real-time measurements; and (5) its spectroscopic capabilities, which allow to identify and trace specific molecular agents.

However, despite its great success in a wide range of studies and applications, in general, DOI cannot be used on its own, but one still requires supplementary measurements, including the measurements of basic optical properties. In addition, more often than not, preknowledge of structural, anatomical nature of the target sample is required to achieve reasonable image reconstruction from the measurements, because of its relatively poor spatial resolution.

As mentioned above, and as highlighted in most of the relevant literature, the predominant factor that reduces the image quality in DOI is the severe scatter in biological tissue. Over the past decade, the detrimental influence of scattering on image quality has been largely reduced, thanks to the tremendous advances in (1) measurement instrumentation (in particular, reliable diode laser sources, fast and sensitive photon detectors, and precise signal evaluation electronics); (2) refinement of experimental techniques (in particular, the introduction of multiwavelength, ultrafast TD methodologies and hybrid techniques); and (3) theory (better understanding and modeling of the photon transport in tissue and the development of efficient, elegant algorithms for the forward and inverse problems).

Probably the most promising hybrid technique for enhancing spatial resolution in DOI is the combination of the functional sensitivity of the NIR optical measurement with the superior spatial resolution of ultrasound, in the form of photoacoustic imaging. If implemented with care, anatomical and functional images with a spatial resolution of the order 200 μm may be achievable; see, e.g., the *in vivo* study of Wang et al. (2003) of small animal brains. On the other hand, it is widely accepted that DOI struggles to compete with "anatomical" imaging techniques in terms of spatial resolution. However, it offers distinct advantages when it comes to sensitivity to functional changes; moreover, it is—as mentioned above—cost-effective and small enough to be used at the bedside.

To put DOI in the context of anatomical imaging approaches, the four most relevant techniques (optical coherence tomography—OCT; x-ray CT—xCT; positron emission transaxial tomography—PET; MRI) are briefly summarized below, and some of their relative merits and disadvantages are outlined. All of them rely on the fact that image reconstruction, by and large, does not suffer from scattering processes but can make use of "ballistic," (nearly) straight-line probe/detection trajectories.

*Optical coherence tomography* may be seen as a derivative technique to DOI. It is an interferometric technique, typically employing NIR light (to penetrate sufficiently deep into the tissue). It is based on microscope-type optical configurations incorporating a Michelson interferometer; this allows for "confocal" gating in the lateral directions and "coherence" gating in the axial dimension (with the axial information being retrieved from the envelope of the Doppler-broadened carrier frequency). The light source is of low coherence and broad bandwidth light source. As usual for Michelson interferometer systems, the incoming light is split into two components, to propagate in a "sample arm" containing the tissue sample and a "reference arm" with (normally) a mirror. The conceptual setup of an OCT system is shown in Figure 20.21.

Two basic operational modes are utilized, namely *time-domain OCT* (or TD-OCT) and *frequency-domain OCT* (or FD-OCT); several variants have been developed over the years, which for specific applications increased the amount of useful information that could be extracted from the data, or allowed for speedier operation or increased resolution.

In *TD-OCT*, the reference mirror is translated longitudinally (in time), altering the path length. A property of low-coherence interferometry is that it only occurs when the path difference is less than the coherence length of the light source. Recall from optical theory that this interference is known as *auto-correlation* for a "symmetric" interferometer (both arms have the same reflectivity) or *cross-correlation* in the common, asymmetric case. The envelope of the interferogram modulation changes with the path length difference; its peak amplitude corresponds to matching path length in the two arms. This axial reflectivity profile—often called the *A-scan*—contains information about the spatial dimensions and location of structures within the sample. A cross-sectional 3D tomograph—or *B-scan*—is generated by laterally combining a series of axial depth (*A-*) scans. Note that the axial and lateral resolutions in OCT are decoupled: the former is equivalent to the coherence length of the light source, while the latter is a function of the optics.

In *FD-OCT*, the broadband interference is acquired with spectrally separated detectors. This can be achieved in two ways: (1) by encoding the optical frequency in time (in this case, the wavelength of the source is scanned) or (2) by utilizing a dispersive detector (e.g., a combination of a grating and a light detector array). As a consequence of the Fourier relation—recall the Wiener–Khintchine theorem, which mediates between the autocorrelation and the spectral power density—the depth profile can be derived directly from the Fourier transform of the acquired spectra without the need for any movement of the mirror in the reference arm (see Schmitt 1999). This dramatically improves the speed with which an image can be acquired; at the same time, the signal-to-noise ratio

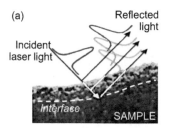

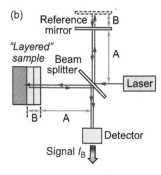

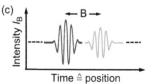

**Figure 20.21** Concept of OCT based on a Michelson interferometer and a broadband, low-coherence light source. (a) Reflection from different layers in a tissue sample; (b) schematic experimental setup; (c) temporal interference from different layers.

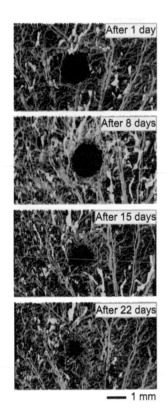

— 1 mm

**Figure 20.22** Microvasculature and lymphatic vessel response to a wound induced by a biopsy punch, over a period of three weeks. The lymphatic vessels are color-coded and overlaid on top of blood vessels. Data adapted from Yousefi, S. et al., *J. Biomed. Opt.* 18, no. 8 (2013): article 086004; reproduced with the permission of SPIE and the authors.

improves as well, being proportional to the number of detector elements. Note that the parallel detection at multiple wavelengths limits the scanning range; and that the full spectral bandwidth of the source determines the axial resolution.

In principle, OCT images are reconstructed more or less analogous to confocal microscopy, and depending on the actual light source, submicrometer resolutions have been demonstrated. On the downside, OCT is only sensible if the region to be imaged is relatively small (otherwise, lateral x–y scans become excessively time consuming), and if the feature to be imaged is not too far below the surface (i.e., the depth-range is limited).

Further information about the implementation of OCT, related image reconstruction, and a wide range of applications may be found in Drexler and Fujimoto (2015). Out of the wealth of applications of OCT, we show only one example. In their study of visualization of lymphatic vessels in response to an injury, Yousefi et al. (2013) monitored the dynamics and healing process of lymphatic and blood vessels over a period of a few weeks; selected snapshots of the process are shown in Figure 20.22.

*Magnetic resonance imaging (MRI)*, sometimes also addressed as nuclear MRI (NMRI) or magnetic resonance tomography (MRT), is used to form images of anatomic and/or physiologic features of the body. Most medical applications rely on detecting the RF signal emitted subsequent to the excitation of H-atoms in the tissue by strong, uniform, oscillating magnetic fields (the majority of commercial systems operate at 1.0–1.5 T), applied at the appropriate resonant frequency. The contrast between different tissue types—which is exploited to generate spatial 3D images—is determined by the rate at which excited H-atoms return to their equilibrium state. In order to increase the contrast, and hence the image quality, often exogenous contrast agents are introduced, which are normally administered intravenously or orally; consequently, MRI using contrast-enhancement agents can no longer be classified as being fully noninvasive. While only the H-atom is probed in MRI, nevertheless certain spectroscopic properties transpire; the proton resonance exhibits specific frequency shifts depending on its chemical environment, dictated by neighboring protons within a particular (organic) molecule. Therefore, some metabolites can be characterized by their unique H-atom chemical shifts (see Rosen and Lenkinski 2007). This form of metabolic MRI is usually addressed as magnetic resonance spectroscopy (MRS).

*X-ray computed tomography (xCT, or simply CT)* is based on the exposure of tissue to x-rays and generating 2D "shadowgrams" in modern instruments normally on phosphor-coated surfaces. In order to form 3D reconstructions of the tissue sample, spatial scanning and subsequent back-projection algorithms from the individual 2D "slices" are required But other than in DOI, which constitutes an ill-posed problem because of the severe scattering of the NIR photons, image reconstruction in xCT is never ill-posed because of the ballistic nature of the x-ray photon transmission through the tissue. While xCT is noninvasive, as are DOI and MRI, a major concern is that x-rays are ionizing radiation and thus have the potential to cause radiation damage and contribute to radiation-induced cancer. Also, contrast media containing iodine are routinely used in xCT, which have as main adverse effects anaphylactic reactions and nephrotoxicity (see Bettmann 2004).

*Positron emission transaxial tomography (PET)* involves the injection of radionuclides whose uptake is specific to the various metabolites in biological tissue.

As such, PET is an invasive method that also introduces potentially harmful radiation doses since quite a substantial amount of the radionuclide has to be (preferentially) bound to a metabolic substrate to provide sufficient contrast for meaningful imaging. PET plays an increasingly important role in the diagnosis and staging of malignant disease and in monitoring response to therapy, providing accurate anatomical localization of functional abnormalities. However, one encounters fundamental limits of spatial resolution; these are predominantly related to the combined effects of positron noncollinearity, positron range, and sampling geometry. Thus, effectively one deals with similar ill-posed image reconstruction problems as in DOI, although they are less severe.

Overall, one may summarize the comparative imaging capabilities of the aforementioned techniques in the following way. From the point of view of spatial resolution, DOI (able to resolve features of size ~xx mm) is inferior to the "traditional" techniques of PET (normally >1 mm), xCT (in the range 0.3–1.0 mm), and MRI (down to a few tens of micrometers), and the modern methodology of OCT (submicrometer resolution has been demonstrated). However, DOI comes into its own if molecular and functional image information is sought, due to its unique spectral resolution power; of the traditional methods, only MRI offers limited spectral capabilities.

## 20.7 BREAKTHROUGHS AND THE CUTTING EDGE

DOI has seen many ups and downs during the close-to-100 years of its use in biomedical analysis, diagnosis, and process monitoring—the many "breakthroughs" were more often than not followed by setbacks in acceptance of a particular methodology. Here a single snapshot out of the sequence of perceived breakthroughs is highlighted.

### 20.7.1 Breakthrough: DOI of brain activities

The fact that brain activity is associated with increased, local hemoglobin concentrations can be exploited by DOI in pinpointing said variations using a point array of near-IR laser excitation sources and light detectors. The feasibility of the approach was first demonstrated for motor-activity stimuli of the hand, revealing differences between ipsilateral and contralateral response (see Maki et al. 1995). The methodology evolved rapidly not only to follow motor stimuli but also to trace the brain's response to audio (see Peña et al. 2003) and visual stimuli, including covert "thought-reaction" to a visual request (see Eggebrecht et al. 2014).

The example shown in Figure 20.23 is related to Peña's study of neonatals and their response to speech. Here, the variation of hemoglobin concentration, $\Delta$[Hb], is mapped following a voice stimulus. The distribution and variation of hemoglobin in the infant's brain was recorded using a sparse 3×3 measurement array of fiber-coupled near-IR light sources and detectors (*Hitachi* ETG-100 OT). The response to repeated audio stimuli was statistically analyzed; the resulting topographical maps from the left hemisphere (LH) of the brain clearly reveal the difference in response, whether an audio (speech) stimulus was present or not— see the right part of the figure. Also, a significant difference in speech response was found between the left and right hemisphere (RH); the results for the latter

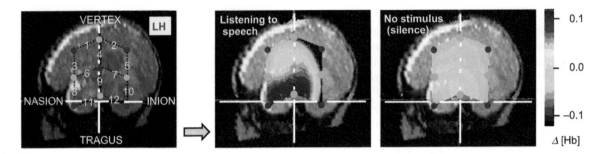

**Figure 20.23**   Language recognition of neonatals probed by optical topography. (Left) Positioning of the OT probes; ℓ red dots = light emitters; ℓ green dots = optical fibers to detectors. The numbers indicate adjacent emitter–detector pairs used in determination of changes in Hb concentration. (Right) Reconstruction maps of relative changes in Hb concentration, revealing the activity difference between periods of sound and silence. (Data adapted from Peña, M. et al., *Proc. Natl. Acad. Sci. USA* 100, no. 20 (2003): 11702–11705.)

are not shown here. This observed LH/RH imbalance confirmed, for the first time, that newborn infants process language in the left temporal lobe, as do adults.

### 20.7.2 At the cutting edge: Photoacoustic tomography–toward DOI with high spatial resolution

As stated in several parts of the present chapter, perhaps the main disadvantage of DOI derives from its poor spatial resolution, when compared to optical microscopy (which admittedly is only applicable to surface features). The primary cause for this resolution deficiency is the high degree of scattering of the diffusing photons. Consequently, the experimental data not only have a low signal-to-noise ratio but also carry, in general, insufficient information to reconstruct a clear tissue image.

An elegant solution to overcome the aforementioned difficulties was put forward by Wang and coworkers (Wang 2008), who developed a novel hybrid imaging technique—called *photoacoustic tomography* (PAT)—which integrates optical waves with ultrasound. In the said technique, light is absorbed by the biological tissue and converted to heat. As a result, an acoustic wave is produced due to thermoelastic expansion that is ultimately detected by an ultrasonic transducer. This constitutes the same physical phenomenon that is exploited in photoacoustic spectroscopy (see Chapter 7.5). Essentially, the technique combines rich optical contrast and high ultrasonic resolution, providing high-resolution tomographic images of regions undergoing optical absorption and, therefore, can be applied to obtain structural, functional, and molecular imaging *in vivo* and in real time.

Currently, one of the most used modalities of PAT is based on *photoacoustic microscopy* and this is, not surprisingly, abbreviated as PAM. This is further classified into *optical-resolution* PAM (OR-PAM) or *acoustic-resolution* PAM (AR-PAM), depending on which resolution is greater, optical or ultrasonic. In the first approximation, the lateral resolution for the two modalities are $R_{opt} \approx 0.51 \times (\lambda_0/NA_{opt})$ and $R_{acu} \approx 0.71 \times [v_S/(f_0 \cdot NA_{acu})]$, respectively. Using these expressions, for laser light at $\lambda_0 = 532$ nm and $NA_{opt} = 1.5$, one finds $R_{opt} \sim 200$ nm, while for ultrasound frequency, $f_0 = 50$ MHz and $NA_{acu} = 0.45$, and a typical velocity of sound in tissue $v_S = 1500$ m·s$^{-1}$, the resolution reduces to $R_{acu} \sim 50$ μm. In the

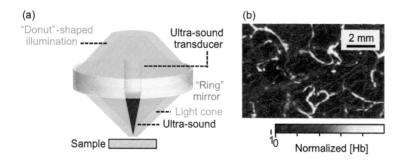

**Figure 20.24** Representative *in vivo* AR-PAM image of normalized total-hemoglobin concentration, [Hb], in a human palm. (a) Conceptual imaging setup. (b) AR-PAM image, scanned with step-size resolution of 20 μm. (Data adapted from Wang, L.V. and S. Wu, *Science* 335, no. 6072 (2012): 1458–1462.)

former case, imaging depths of $d \approx 100$ μm could be realized, while in the latter case, this increases to $d \approx 3$ mm; for further details, see Wang and Wu (2012). Note that these resolution values are scalable with relative ease by variation of the relevant parameters. This means that one can link microscopic biological structures (usually imaged by optical microscopy) to macroscopic structures (normally imaged using non-optical modalities, like x-ray CT). In general, correlating such microscopic and macroscopic images, obtained via different methodologies, can be challenging.

Based on hemoglobin absorption, an example of an AR-PAM anatomical image of a human palm is shown in Figure 20.24, revealing in particular subpapillary blood vessels. Note, that the (false-color) scale is that for the normalized total-hemoglobin concentration, [Hb].

# CHAPTER 21

# Imaging Based on Absorption and Ion Detection Methods

In this final chapter, two imaging techniques are being addressed that sometimes are seen as being outside the mainstream.

The first class is *imaging based on absorption spectroscopy detection methods.* Recall that absorption and laser-induced fluorescence spectroscopies are, in principle, related to each other, insofar as the latter may be seen as a detection modality for the absorbed photon, the difference being that absorption detection is against the background of the laser excitation field, while fluorescence photon detection is by and large background-free. Therefore, for reasons of contrast (but others as well), fluorescence imaging has had the upper hand. This, however, has been changing with technological advances and the development of novel absorption-specific measurement modalities, and advantage is now taken of the label-free nature of absorption spectroscopy in chemical imaging. This will be shown through relevant examples in Sections 21.1 and 21.2.

The second class of imaging techniques is based on *charged particle detection* (predominantly ions, but occasionally electrons as well). Charged particle detection normally is a "destructive" method, since the ionization—translation from the neutral state to the ionized state of the molecule (or atom)—removes the probed particle from the sample. Charged particle imaging has really found its niche in the probing and understanding of chemical collision (uni- and bimolecular) processes; this will be discussed in detail in Sections 21.3 and 21.4.

## 21.1 IMAGING EXPLOITING ABSORPTION SPECTROSCOPY: FROM THE MACRO- TO THE NANOSCALE

It is probably fair to say that with respect to laser spectroscopic imaging techniques, absorption spectroscopy has long been the poor relation in comparison with imaging modalities based on laser-induced fluorescence or Raman scattering. Generally, this can be linked to the lack of suitable experimental equipment, since both excitation laser sources and detectors—operating in the infrared (IR) and near-infrared (NIR), the wavelength regions of absorption by the fundamental vibrations of most molecules—were not easily available or cumbersome to handle (e.g., needing liquid nitrogen or even liquid helium cooling) for viable imaging solutions.

Over the past decade or so, this has changed dramatically, with the introduction of, e.g., easy-to-operate quantum-cascade laser (QCL) sources or the availability of efficient IR photodetector arrays, to name but two examples. With this change in available opportunities, absorption-based spectrochemical imaging has rapidly gained popularity, in particular for two reasons. Firstly, in contrast to many of the laser-induced fluorescence modalities, no sample additives or labeling of molecular sites is necessary, since molecular specificity is provided by the absorbing molecular vibrational transitions themselves. Secondly, while Raman spectroscopy also exhibits chemical specificity related to molecular vibrations, the Raman scattering process is orders of magnitude weaker than absorption; hence, absorption-based spectroscopic imaging should be much more sensitive, in principle.

As for all the other techniques, imaging based on absorption spectroscopy has also been developed in many guises, depending on the actual scientific problem and the availability of equipment, leading to a wide range of measurement modalities.

### 21.1.1 Experimental implementation of imaging exploiting absorption spectroscopy

Probably the most common/popular are (1) "standard" IR/NIR and the variant of terahertz chemical imaging, (2) photoacoustic imaging (PAI), and (3) photothermal imaging at the nanoscale. These are briefly outlined in the following sections, together with selected, relevant examples to highlight their pros and cons.

#### IR/NIR chemical imaging

Basically, like any other spectroscopic measurement procedure, absorption spectroscopy constitutes the conceptual chain laser light source → light delivery → light–matter interaction → detection of the interaction process → recorded spectrum. In imaging, the spatial dimension is added, resulting in a hyperspectral data cube (with a spectrum for each spatial pixel point). The concept is summarized in Figure 21.1a. As for the other spectroscopic techniques, the actual setup depends very much on the task and the available equipment. For example, to realize the molecular specificity, either the laser wavelength is tuned or the absorption response signal is spectrally dispersed or both; and the detection can be made with a single-element or an array detector, which means raster scanning the sample with a focused laser beam in the former case and fully illuminating it in the latter. In addition, various excitation and detection modalities have evolved into stand-alone application fields, such as the recording of spectra by Fourier-transform techniques (Fourier-transform infrared [FTIR] imaging) or the implementation of specific wavelength regimes outside the standard range (terahertz imaging); in particular, the latter has undergone rapid development and is now found in a wide range of practical applications (see, e.g., the review by Chan et al. 2007, or the recent book by Peiponen et al. 2013).

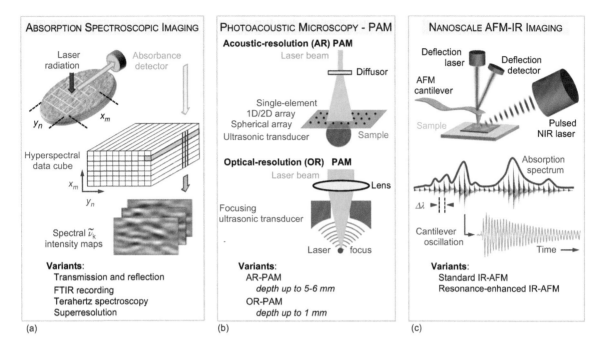

**Figure 21.1** Experimental realization: (a) general absorption spectroscopic imaging; (b) photoacoustic microscopy; (c) nanoscale IR atomic force microscopy.

## Photoacoustic imaging

While the spectral recording and imaging modalities just outlined rely on photodetection of the laser–sample interaction process, indirect detection methods based on the photothermal response of the sample (see Sections 7.5 and 8.6) have always been tremendously important in absorption spectroscopy when sample properties or geometrical constraints did not allow for direct measurement of the absorption process in transmission or reflection. The experimental setup is slightly different from standard IR/NIR chemical imaging, as shown in Figure 21.1b. Two main modalities are encountered in the general implementation of photoacoustic microscopy (PAM), namely, acoustic-resolution (AR-PAM) setups, with a variety of ultrasound detector geometries (single-element and multielement) and accessible depths within a sample of $d_z \approx 3\text{-}5$ mm, and optical-resolution (OR-PAM) setups, normally using a focusing ultrasound transducer with and accessible depths within a sample of up to $d_z \leq 1$ mm. A wealth of technical details and applications of PAI modalities can be found, e.g., in the reviews by Yao and Wand (2014) and Wang et al. (2016).

## IR/NIR imaging at the nanoscale

The spatial resolution of imaging based on absorption spectroscopy can be expanded to below the optical subdiffraction limit, exploiting the photothermal response of the sample on pulsed laser excitation and its local probing by atomic force microscopy (AFM) techniques. The technique is widely known as *infrared atomic force microscopy* (IR-AFM) or *infrared photoinduced force microscopy* (IR-PiFM); the conceptual setup is shown in Figure 21.1c. The thermal expansion following the absorption of laser radiation, tuned into a molecular vibrational band transition, is probed by the AFM tip, whose oscillation amplitude reflects the absorption strength and the local molecular concentration. Note that in IR-AFM/IR-PiFM, topographic and chemical information images can be recorded concurrently. Details of this approach and a multitude of application examples can be found, e.g., in the reviews of Muller et al. (2015) and Dazzi and Prater (2017).

### 21.1.2 IR/NIR chemical imaging

As an analytical technique, near-infrared chemical imaging (NIR-CI) is a technique that slowly has begun to replace traditional single-point NIR spectroscopy (providing only a bulk average spectrum to reflect the average composition of the sample). In addition to spectral information, NIR-CI simultaneously adds the spatial dimension to NIR spectroscopy, thus yielding the capability to acquire distributional information of constituents in the sample. As in many other imaging spectroscopic methodologies, a hyperspectral data cube is generated, which for each spatial measurement location contains the full spectral information.

Presently the most established technology for chemical imaging is based on FTIR imaging spectrometers. However, one of its drawbacks is that while complete (broadband) spectra can be recorded, for high-resolution applications, scanning the entire spectrum can be very time consuming. The advent of high-intensity, broadly tunable QCLs now potentially promises accelerated IR imaging. For example, Yeh et al. (2015) demonstrated the feasibility of IR absorption microscopic imaging based on a rapidly step-tunable QCL. It should be noted that the approach by Yeh and coworkers constitutes a fundamentally different methodology in comparison with FTIR imaging, basically constituting discrete-frequency IR spectral imaging.

In a certain way, chemical imaging is only the second step in the overall procedure, meaning that the spectral signature of each pixel has to be translated into chemical information, i.e., the concentration of individual chemical compounds in the sample. Then, the pixel-by-pixel chemical images will reflect the spatial distribution of components in the sample.

As an example, here an application of NIR-CI to the spatial concentration analysis of pharmaceutical tablets will be discussed, namely, that of chlorpheniramine maleate (CPM) tablets. CPM is a so-called $H_1$ receptor antagonist (antihistamine); clinically it is used, e.g., to alleviate symptoms of cold or to treat a variety of allergic diseases. Typically, CPM tablets contain 1 or 4 mg of the active pharmaceutical ingredient (API), corresponding to about 1–5% w/w in a tablet. In their experiment on CPM tablets, Xu et al. (2016) used NIR Fourier transform instrumentation, covering the wavelength wavenumber range $\tilde{\nu} \approx 4000$–$7800$ cm$^{-1}$, with spectral resolution $\Delta\tilde{\nu} = 16$ cm$^{-1}$. An area of $2 \times 2$ mm$^2$ of the sample tablets was raster scanned, with a spatial resolution of $25 \times 25$ μm$^2$, i.e., 6400 local sample points in a hyperspectral image data cube. These data were used to generate two-dimensional (2D) concentration images, using multivariate regression in the form of partial least squares (PLS); typical images for two particular sample tablets are shown in the upper part of Figure 21.2.

Since PLS is based on the relation between the measured spectral signals and calibrated reference data, a prediction/validation regression between the two quantities is required. Such a plot is shown in the bottom part of the figure.

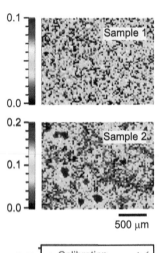

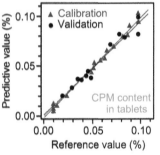

**Figure 21.2** Quantitative spatial distribution analysis in pharmaceutical tablets. (Top) 2D images of CPM concentration values in two tablet samples; images reconstructed from hyperspectral image cube using PLS. (Bottom) CPM concentration prediction versus reference results, used for concentration calibration of the images. (Data adapted from Xu, M. et al., *J. Innov. Opt. Health Sci.* 9, no. 6 (2006): article 1650002.)

Note that for their calibration, the authors produced 33 calibration batches of samples from four-ingredient pharmaceutical tablet formulation (content ranges: CPM—1–10%; pregelatinized starch [STA]—20–90%; microcrystalline cellulose—1–8%; and magnesium stearate—0.04%).

Based on their calibration set, the authors interpreted the measurement results for two aspects, namely, (1) average CPM concentration in their samples and (2) homogeneity of the CPM distribution throughout the sample. For the two samples 1 and 2 shown in Figure 21.2, they derived average CPM concentration values of 4.09% and 5.95%, respectively. Note that these are above the typical CPM content in commercially sold tablets, but it should be kept in mind that these particular experiments were performed for proof of principle only. In relation to the second aspect, it is already clear from a simple inspection by eye that sample 1 is substantially more homogeneous than sample 2, for which local concentrations well in excess of 10–15% are observed (the red areas in the image). Note, however, that for objective, quantitative assessment of the homogeneity in different samples, criteria such as the distributional homogeneity index may need to be incorporated in the analysis (see, e.g., Sacré et al. 2014).

Note also that, while the homogeneity of API distribution is not necessarily of high importance in a treatment with whole-tablet consumption, it certainly plays an important role in both medicine safety and efficacy (especially for small dose products) or sustained- and controlled-release products. Therefore, efficient, precise, and fast concentration imaging with the capability of multiple-ingredient recognition—as shown by the authors—can be invaluable in the screening of pharmaceutical tablets.

## 21.1.3 Detecting "hidden" structures using terahertz imaging

In Section 7.5 it was pointed out that the terahertz region of the spectrum, also called the submillimeter wave band, offers important advantages because many inter- and intramolecular resonances can be seen within its spectral window (typically in the range 0.1–10 THz). In addition, unlike IR radiation, terahertz waves can penetrate through the majority of "soft" materials, such as fabric, plastic, wood, and paper, and other solid materials; this feature opens up the possibility of nondestructive detection and imaging of concealed objects and materials.

For these reasons, it is not surprising that, aided by significant technical advances over recent years, terahertz imaging is gaining acceptance and popularity in a wide range of screening applications, last but not least in, e.g., homeland security.

### Terahertz imaging for weapon and explosive detection

Terahertz imaging systems can be both passive and active. *Passive imaging* systems are anomaly detectors, which exploit the difference in emissivity and reflectivity between humans and other materials. Thus, metallic, ceramic, or plastic objects hidden beneath their clothing will be distinguished from the human, as those objects emit and reflect ambient terahertz radiation differently. *Active imaging* systems are laser-based techniques, ranging from single-frequency up to broadband systems, which are suitable for both spectroscopy and imaging. Both types of systems are rather similar in their capability to detect

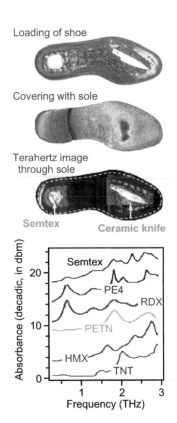

**Figure 21.3** Terahertz imaging for weapon and explosive detection. (Top) Visible and terahertz images of a shoe loaded with a ceramic knife and explosive (top) and covered by a sole (middle) and image using terahertz radiation (bottom). (Adapted from Leahy-Hoppa, M.R., "Terahertz for weapon and explosive detection." In F. Flammini (ed.), *Critical Infrastructure Security*, WIT Press, Southampton, UK, pp. 207–230, 2012.) (Bottom) Terahertz absorption spectra for (from top to bottom) Semtex H, Plastic Explosive No. 4 (PE4), RDX, pentaerythritol tetranitrate (PETN), octogen (HMX, related to RDX), and trinitrotoluene (TNT), as annotated; data traces are offset for clarity. (Data adapted from TeraView, Terahertz applications: Detection of explosives and materials characterization, 2016.)

solid materials under concealed conditions, but, in general, active systems achieve higher spatial resolution.

Figure 21.3 illustrates how hidden objects can be seen within a shoe, using terahertz time-domain spectroscopy (THz-TDS); concurrently, spectroscopic and imaging information can be obtained. In the upper image, one can see how a ceramic knife and a piece of explosive (Semtex) are loaded into hollowed parts of a shoe. In the center image, the cavities in the shoe, and the objects therein, are covered again with the sole; the thus hidden objects are not visible. Finally, the bottom image is a (contrast-enhanced) terahertz image, in which the explosive and the knife are clearly seen.

From the spectroscopy point of view, one of the most active areas of terahertz technology has been, and still is, dedicated to identifying the spectral signatures of explosives since, as shown in the bottom data panel of Figure 21.3, many of these compounds have unique spectral signatures in the terahertz spectral region. It is possible to not only distinguish different types of explosives but also to differentiate between members of a class of explosives. An example of the latter is evident in the second and third spectral traces in the figure: the plastic explosive PE4 contains the "parent" compound RDX (formally cyclo-trimethylene-trinitramine), confirmed by the typical, species-specific spectral feature at ~0.75 THz. Indeed, this particular terahertz signature of RDX has been used as a unique identifier in a variety of explosives, since other explosives do not contain the same signature (note that this spectral band is attributed to phonon modes within the crystalline structure).

These examples and others, not discussed here for brevity, indicate that nondestructive, standoff detection of objects and materials relevant for homeland security, based on terahertz technology, nowadays constitutes an area of intense fundamental and military research and related technological and procedural development. One already finds numerous commercial systems in a range of common applications, such as, e.g., weapons and explosive detection in airline passenger screening; of course, their use has to be safe for human exposure, and stringent safety guidelines need to be followed (for a summary, see, e.g., Berry et al. 2003).

### Time-gated terahertz spectral imaging

In the example just described, it was demonstrated that THz-TDS can be used for nondestructive detection, identification, and imaging of concealed objects and materials. THz-TDS can also detect *structural defects* in many solid materials, such as wooden objects, plastic components, or pharmaceutical products. But despite many, otherwise successful applications, THz-TDS is incapable of extracting occluded content from layers whose thicknesses and separations are comparable with the terahertz wavelength utilized. This limitation is a consequence of several factors, mainly related to spatial resolution, spectral contrast, and occlusion of deeper layers by the content from front layers.

Recently, Redo-Sanchez et al. (2016) introduced a time-gated spectral imaging variant that overcomes these difficulties; the authors demonstrated how occluded textual content can be extracted from a sample, such as, e.g., a closed book. They showed that for a test structure of pages with single-sided writing, content could be recovered down to a depth of nine pages without human supervision.

A schematic layout of the approach, with its measurement geometry, as well as demonstrative test results is shown in Figure 21.4. On the left, the conceptual

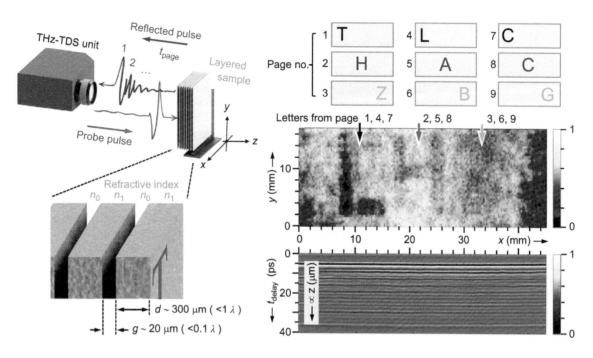

**Figure 21.4** Retrieving written content from pages in a closed book. (Left) Confocal terahertz time-domain (THz-TDS) measurement, configured for reflection geometry. (Right) Test sample of a nine-page stack; nine roman letters are written on the pages, with the indicated pattern (top). The data constitute a single time frame of the recorded terahertz electric field amplitude data cube (center); and an averaged $x$-$t$, equivalent to $x$-$z$ data slice (bottom); for further details, see text. (Data adapted from Redo-Sanchez, A. et al., *Nat. Commun.* 7 (2016): article 12665.)

setup of a confocal terahertz time-domain (THz-TDS) apparatus is shown, arranged in reflection geometry with respect to the sample allocated on an $x$-$y$-motorized stage. Moving the sample mechanically along the two axes, raster scanning along the sample plane is implemented. Note that the electric field of the incident terahertz pulses and the reflected signal (shown in blue and red, respectively) are bipolar; note also that the reflected signal comprises a series of dense reflections from the layered sample. These reflections provide sort of "time-of-flight" information for boundaries of the sampled pages along the axial $z$-axis. As indicated in the figure, each layer of the sample was about 300 μm thick, and the (nonuniform) gaps between the layers were ~20 μm, after pressing the pages together. Then, by time gating the response signal with respect to the terahertz probe, pulse information from distinct depths in the sample can be recovered.

In the right-hand part of the figure, example data are shown that demonstrate the capability of the time-gating technique to retrieve written content from pages in a stacked thin-sheet structure (a closed book). The authors tested the principle on a nine-page stack of sheets—with page 1 facing the terahertz measurement system—with the thickness properties indicated above and in the figure. They applied nine Roman letters (T, H, Z, L, A, B, C, C, and G—with the pattern as shown at the top of the figure); this meant three letters each were stacked on top of each other but occurring at different depths. A single time-gated THz-TDS frame from the recorded terahertz electric field amplitude data cube is shown at the center. It constitutes an $x$-$y$ lateral map, with its horizontal and vertical axes given in millimeters; the normalized field amplitude of the signal can be read off the scale bar (color coded, arbitrary units). The arrows indicate the locations (1) of occlusion (black arrow: T is occluding L) (2) of shadowing (blue arrow: shadow of H); and (3) of interlayer reflection-induced noise (green arrow), Note that the signal timing was roughly that for the depth of page 4 (with the letter L). In the bottom part, an $x$–$t$ (equivalent to $x$–$z$) cut through the terahertz data cube is shown; the image intensity indicated in the scale bar is the normalized field amplitude in arbitrary units. This cut clearly reveals the layered structure of the sample, which, however, becomes less distinct with increasing depth in the stacked sample.

The success of the method employed by the authors relies on several factors. First, statistical analysis of the terahertz electric field response needs to be undertaken, to lock the gated response into each layer position. Second, time-gated spectral *kurtosis averaging* is applied (subsequent to layer locking), to "tune" into the spectral images with the highest contrast of the selected layer. This is performed by using thin time slices ($\Delta t \sim 3$ ps) of the waveform around the position of the layer extracted from the initial statistical analysis. One should bear in mind that the kurtosis is created by the presence of two different reflective materials for each layer; in other words, the higher the contrast between paper with ink and blank paper for a certain frequency, the higher the kurtosis. Third, an extraction algorithm for shape composition has to be used that overcomes partial occlusion of the desired signal (for further details, see Redo-Sanchez et al. 2016).

With the experiment just described, the authors demonstrated that their proposed method of time-gated THz-TDS can be used to inspect densely layered samples, such as, e.g., coatings and polymer-based laminates, or objects of cultural value, such as, e.g., artworks, documents, and fragile medieval books, to name but a few.

### 21.1.4 IR imaging at the nanoscale

Correlating spatiochemical information with the material morphology, down to the nanostructure level, has been a challenge and sort of the holy grail for scientific researchers. Nanoscale patterns are not easily interrogated for their chemical content in real space (i.e., without placing the sample in vacuum), at least not for some of the existing, nondestructive techniques based on optical imaging or electron scattering.

New developments in scanning probe technology promise to alleviate this. Specifically, it is a relatively novel technique known as IR-PiFM that causes excitement in the research community. Its principle is based on the direct measurement of the photoinduced polarizability of the sample, in the near field, by recording the time-integrated force between the tip and the sample. By raster imaging at multiple IR wavelengths—corresponding to selected absorption peaks that are characteristic of individual chemical species—it was demonstrated that IR-PiFM has the capability to spatially map samples at the nanoscale, with chemical specificity. Descriptions of the principles and discussion of representative application examples can be found, e.g., in the reviews of Centrone (2015) or Dazzi and Prater (2017).

In short, the aforementioned photoinduced force is the result of a dipole–dipole force attraction between the imaging tip and the sample when the contact region is illuminated by monochromatic laser light. If one assumes that the (spatial) phase of the electric field of the illuminating laser radiation does not change at the tip–sample junction, then one can find a relatively straightforward relation between localized ($F_{loc}$) and nonlocalized ($F_{nonloc}$) forces, which make up the time-averaged photoinduced force $\langle F \rangle$ and the laser radiation field ($E$):

$$\langle F \rangle = F_{loc} + F_{nonloc}$$
$$\propto -z^{-4} \cdot \mathrm{Re}\{a_S \cdot a_T^*\} \cdot |E_z|^2 + \mathrm{Im}\{a_T\} \cdot |E_x|^2$$

(21.1)

Here, $a_S$ and $a_T$ stand for the polarizabilities of the sample and tip, respectively; $E_x$ and $E_z$ are the transverse $x$- and axial $z$-components of the incident laser field. Note that the dependences of $\langle F \rangle$ on $E_z$, and the $z^{-4}$ nature of $F_{loc}$, give rise to the highly localized force behavior in PiFM; this is the basic origin of its high spatial resolution. For further insight and details, see, e.g., Nowak et al. (2016) and references therein. Typically, in IR-PiFM, spatial resolution on the order of ~100 nm is achieved, although in a few experimental variations, even lower values have been reported.

To demonstrate high spatial resolution, in combination with chemical specificity, Nowak et al. (2016) imaged a variety of so-called block copolymer materials. These consist of two (or more) immiscible polymeric segments (blocks), joined by a covalent bond. When annealed, they undergo microphase separation, which generates nanometer-scale periodic structures. The experimental setup used by the authors was constructed along the same concept as shown in Figure 21.1, allowing for simultaneous measurement of the topography and the chemically selective image. The results shown in Figure 21.5 are for the copolymer compound poly(styrene-$b$-methyl-methacrylate)—in short, PS-$b$-PMMA.

Note that to image the PS or PMMA component, the excitation laser was tuned to their main absorption bands at $\tilde{v} = 1492\ \mathrm{cm}^{-1}$ or $\tilde{v} = 1733\ \mathrm{cm}^{-1}$, respectively. The two related IR-PiFM images clearly show the complementary composition character expected from the makeup of the sample compounds. It is also clear from the topographic (AFM) image that, while the same features are visible, their contrast is far inferior to that achieved in the IR-PiFM images. The lamella periodicity of about ~40 nm deduced from the images was very much consistent with the authors' expectation from their specimen manufacture procedure.

Although the example results shown here are for a specific copolymer material, IR-PiFM nanoscale imaging was applied by the authors, and others, to a variety of other chemically heterogeneous systems, including biomedical samples (see, e.g., Danielli et al. 2014). This demonstrates that IR-PiFM may serve as a powerful analytical tool for imaging at the nanometer scale, exhibiting chemical specificity, being nondestructive, and not requiring any additives or labeling to achieve this.

In a similar type of study, Tang et al. (2016) used IR-AFM to investigate the compositions of nanodomains of another copolymer material, the so-called *high-impact polypropylene* (HIPP). In HIPP alloys, "rubber" particles with multilayered core–shell structure are dispersed in the polypropylene (PP) matrix. The rubber particles (size submicrometers to a few micrometers) have a "rigid"

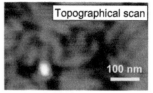

Topographical scan

100 nm

0.00 ▬▬▬▬ 3.20
Height (nm)

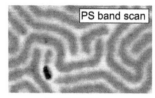

PS band scan

1.79 ▬▬▬▬ 2.86
PiFM signal (arb. units)

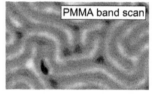

PMMA band scan

0.79 ▬▬▬▬ 1.76
PiFM signal (arb. units)

**Figure 21.5** Chemically selective PiFM imaging of PS-$b$-PMMA. (Top) Topography image. (Center and bottom) Fingerprint PiFM images at $\tilde{v} = 1492\ \mathrm{cm}^{-1}$ and $\tilde{v} = 1733\ \mathrm{cm}^{-1}$, corresponding to the absorption bands for PS and PMMA, respectively. Relative signal intensity range according to the bar scales. (Data adapted from Nowak, D. et al., *Sci. Adv.* 2, no. 3 (2016): article e1501571.)

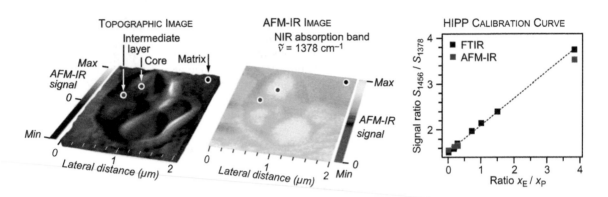

**Figure 21.6**   Images of HIPP nanodomains and their quantitative composition analysis. (Left) AFM topographic image. (Center) IR-AFM map of the methyl symmetric C–H bending mode at $\tilde{v} = 1378$ cm$^{-1}$. (Right) Calibration for quantitative composition analysis; FTIR reference data (black squares) and IR-AFM data from standard samples (red squares, representing the 1456/1378 cm$^{-1}$ absorption peak ratios). (Data adapted from Tang, F. et al., *Anal. Chem.* 88, no. 9 (2016): 4926–4930.)

core and a "soft" (rubbery) intermediate layer. In their study, the authors used an optical parametric oscillator laser system (OPO, tunable over the range $\tilde{v} = 900$–$2000$ cm$^{-1}$ and with spectral resolution $\Delta\tilde{v} = 4$ cm$^{-1}$) as the excitation source. The thermal expansion of the sample following absorption by specific molecular vibrational bands was measured with the AFM tip, raster scanning the sample area of interest.

The HIPP absorption spectra are dominated by the symmetric C–H bending of the methyl group located at $\tilde{v} = 1378$ cm$^{-1}$, characteristic of PP, and the symmetric C–H bending band of the methylene group (at $\tilde{v} = 1456$ cm$^{-1}$, characteristic of polyethylene, frequently abbreviated as PE). In Figure 21.6, some representative results are shown; the AFM topography and the IR-AFM spectral image for the PP response were recorded concurrently.

The authors extracted three key elements from these images. First, it is quite evident that the various phase domains (matrix, intermediate rubber layer, and core) correspond well to each other in the topographic and spectral images. Second, based on the relative intensity of the $\tilde{v} = 1378$ cm$^{-1}$ band throughout the image area, it appears that all three phases always contain significant amounts of PP (see the relative, color-coded intensity scale). Third, the cores of the imaged particles are quite bright (indicated by the yellow intensity code), suggesting that the content of PP in the cores is high. For the PE content, the authors found that (1) in the matrix it was close to zero and that (2) it was high in the rubbery intermediate layers, both as expected. These findings contradict the often-proposed models in the literature, predicting the (rigid) cores in the rubbery particles as PE-rich and not the inverse (see relevant reference in Tang et al., 2016).

Overall, the two studies discussed here demonstrate that novel measurement methodologies such as IR-AFM/IR-PiFM can be useful tools for assessing the structure and chemical composition of nanodomains in complex (polymeric) systems. Specifically, results can easily be quantified since they can reliably be cross calibrated against data obtained using complementary, other absorption measurement techniques.

## 21.2 IMAGING EXPLOITING ABSORPTION SPECTROSCOPY: SELECTED APPLICATIONS IN BIOLOGY AND MEDICINE

In Section 21.1, it was pointed out that NIR absorption microscopy and imaging was becoming a popular field of biomedical research and that a range of practical application was appearing. By and large, this is due to the fact that all related methodological implementations do not need any molecular marker and are thus label-free: absorption by vibrational modes of the very molecules of the sample is responsible for the signal response. This is indeed a beneficial property in biomedical imaging, specifically when it comes to real-time *in vivo* measurements. Here, only a few selected examples of the three most common imaging modalities can be discussed, namely, those of (1) FTIR-based imaging, (2) terahertz-spectroscopic imaging, and (3) PAM imaging.

### 21.2.1 Imaging based on FTIR methodologies

FTIR chemical imaging is emerging as an important technology for examining biomedical tissues; for example, it is establishing its usefulness in cancer pathology.

In traditional FTIR, single spectra from a sample are recorded, averaging the absorption information over a predetermined aperture size, i.e., averaging over a finite area of the sample (see Section 6.1 for the principles of FTIR). Evidently, for imaging applications, this approach is out of the question. Transforming the imaged area to smaller sizes utilizing optical means and then raster scanning the sample surface is possible, in principle, but for practical reasons, it is not a satisfactory solution: sequentially acquiring spectra from thousands of raster points would take an unacceptably long time to be practical for routine analysis. The advent of and technical advances in focal-plane array (FPA) IR-sensitive detectors, since the early 2000s, has alleviated the problem. Today's IR-FPA detectors have sizes of typically 128 × 128 active pixels, permitting concurrent individual recordings from >$10^4$ sample locations. This then constitutes a feasible approach to spatially resolved FTIR imaging and thus indeed would allow for developing FTIR methodologies into a potential tool for aiding in clinical disease diagnosis and assessment (see, e.g., Kazarian and Chan, 2013).

An example of the validity of the approach is shown in Figure 21.7, comprising results from a feasibility study on a sectional sample from rat brain (see Dorling and Baker 2013). In typical optical microscopy investigations, samples are stained to enhance contrast and highlight certain features of the sample. In the example here, so-called H&E staining was applied: hematoxylin (H) stains negatively charged nucleic acids, while eosin (E) stains protein-rich regions. The optical representation of the sample (top image in the figure) reveals only structural information, in principle; any further interpretation may have to rely on prior experience. From the FTIR hyperspectral data cube, a mosaic image (at $\tilde{\nu}$ = 3300 cm$^{-1}$) of a small area of the brain slice (1.4 × 1.4 mm$^2$, with spatial resolution of ~10 μm), more detailed (spectrochemical) information can be extracted. This type of spectral images at specific wavenumbers, related to distinct molecular vibrations, greatly helps in the interpretation of image. For example, the authors associate the intensity distribution observed in the image

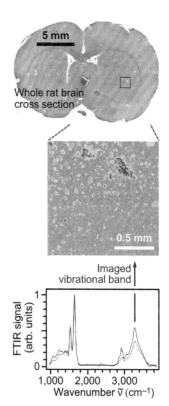

**Figure 21.7** (Top) Optical microscope image of a whole hematoxylin and eosin (H&E)-stained cross section of rat brain. (Center) FTIR spectral image (at $\tilde{\nu}$ = 3300 cm$^{-1}$) of the target area indicated in the optical image (resolution 128 × 128 pixels). (Bottom) Average FTIR spectra; red and blue data traces recorded at red and green locations in the spectral image. For details, see text. (Reprinted from *Cell* 31, no. 8, Dorling, K.M. and M.J. Baker, 437–438, Copyright 2013, with permission from Elsevier.)

(roughly, red for high intensity and green for low intensity) with different spatial orientation of the brain nerve fibers in the sample: the high-intensity areas are thought to originate from "end-on" oriented fibers, while the low-intensity areas are attributed to "other" orientations.

While rather successful in proof-of-principle research, for clinical approval and usefulness, any new method has to be robust and reproducible for reliable diagnostics. Therefore, it is paramount for the incorporation of the technique into clinical routine analysis that practical protocols are developed that guarantee optimal data acquisition, data evaluation algorithms, result classification, and validation against standards.

Bhargava (2007) illustrated the development of such a practical protocol for automated histopathology, exemplified for the histology of prostate tissue. In particular, he addressed how experimental parameters affect the prediction accuracy, as a function of varying spatial resolution, spectral resolution, and signal-to-noise ratio of the recorded data. But the author also points out, and demonstrates with examples, that for reliably relating pathological (or physiological) tissue states to FTIR chemical imaging data, not only are experimental reproducibility and precision important but correct data analysis, predictive modeling, and training also matter equally, to reach unsupervised classification.

Part of the aforementioned issues and arguments are highlighted in the data presented in Figure 21.8, for a prostate tissue sample; in the figure standard microscopy images of stained tissue samples are compared with a couple of spatiospectral slice images from the full hyperspectral data cube.

The microscope image reveals "structural" information typical in prostate tissue; these include, e.g., epithelial cells, which form foci of width 10–35 μm around the cross sections of ducts (these ducts appear as white areas in the image). In addition, three locations are marked in the microscope image that correspond to areas that are dominated by epithelium, stroma, and "stone" contributions (marked 1, 2, and 3, respectively); note that in the prostate histology that the

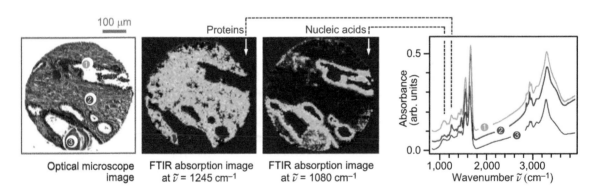

**Figure 21.8** Correspondence between conventionally stained microscopy and FTIR chemical imaging in pathology applications. (Left image) H&E-stained image of a prostate tissue section. (Center image) Spatial intensity distribution at $\tilde{v} = 1245$ cm$^{-1}$ (peak attributed to protein absorption), highlighting differences in the eosin-stained regions of the microscope image. (Right image) Spatial intensity distribution at $\tilde{v} = 1080$ cm$^{-1}$ (peak attributed to nucleic acids absorption), highlighting nuclei-rich epithelial cells, associated with hematoxylin-stained regions of the microscope image. (Spectral plots) Complete absorption spectra (vertically shifted for clarity), recorded at about the location indicated in the microscope image. (Data adapted from Bhargava, R., *Anal. Bioanal. Chem.* 389, no. 4 (2007): 1155–1169.)

author used for classification, these rank as relatively low in the ensemble of 10 indicators. Full FTIR spectra recorded at these locations are displayed in the data panel on the right of the figure. At first sight, they look rather similar, reflecting that various vibrational bands are common to several chemical compounds, but also exhibiting subtle differences, some of which are even recognizable by the naked eye. For two of the minor peaks—associated with protein absorption (at $\tilde{v} = 1245$ cm$^{-1}$) and nucleic acid absorption (at $\tilde{v} = 1080$ cm$^{-1}$)—spectral intensity map images are shown. Certainly, these reflect spatial features observed in the microscope image, but these univariate representations do not yet allow for unequivocal identification of actual cell types and their health status.

To determine the cell types and their locations within the image while providing quantitative measures of accuracy and statistical confidence for diagnostics, automated mathematical algorithms for multivariate analysis are required. Note that this latter step is not shown here; the spectral image examples only serve to demonstrate that individual molecular vibrations can be visualized (and quantified) with high spatial resolution—in the two images about $10 \times 10$ μm$^2$. The important issue here is that all this is possible without requiring an external marker (i.e., no staining or other labeling processes are required).

## 21.2.2 Imaging based on terahertz methodologies

There are various reasons why terahertz radiation has become increasingly popular in biomedical imaging applications. For one, the photon energy of terahertz light is far too low to cause ionization; thus, from this perspective, it is judged to be safe in its use on tissue and humans. Furthermore, the power levels used in most terahertz systems are in the order of a few microwatts, which is substantially lower than the terahertz radiation naturally emitted by the human body (on the order of 1 W). Thus, other than very localized at the probing site, thermal effects are negligible. And, finally, the spectrally sensitive interaction of terahertz radiation with tissue makes it possible, in principle, to identify and image molecule-specific features in the sample.

Terahertz imaging has been applied to a vast range of medical applications, ranging from the detection of cancers (breast, colon, skin, etc.), to monitoring of phenomena in the skin, to detecting the onset of dental caries, to name but a few. A wider insight into terahertz imaging for applications in biology and medicine and a thorough introduction into the underlying physical processes and practical instrumentational issues can be found, e.g., in the book by Yin et al. (2012). Here, the two examples described in the following sections can give only a small flavor of the versatility of biomedical terahertz imaging.

### *Terahertz dynamic imaging of skin drug absorption*

Transdermal drug delivery systems have a series of advantages, such as, e.g., better patient compliance, absence of the hepatic first-pass metabolic effects, and noninvasive drug delivery, to name but a few. However, despite these benefits, exact and reproducible data about the topical drug local distribution are required to fully understand skin absorption mechanisms. Although *in vitro* methods can provide valuable information—for example, on the rate of chemical penetration—in general they are unable to provide real-time information. Also, they often require a series of sequential steps, including, e.g., high-performance liquid chromatography analysis, which often make the methods time consuming and tedious.

Today terahertz radiation is increasingly used as an excellent alternative modality for investigating drug distribution and for providing information on drug permeation mechanisms; this is due to its spectroscopic and imaging abilities, and because *in vivo* real-time measurements can be performed.

As an example, Kim et al. (2012) used terahertz dynamic imaging to investigate, *in vivo*, the penetration and distribution of a topical drug, *ketoprofen*, dissolved in dimethyl sulfoxide (DMSO), and compared the results with those of *in vitro* skin absorption tests. A simplified scheme of the experimental technique is illustrated in the bottom right corner of Figure 21.9.

The samples were prepared from freshly excised full-thickness skin of hairless mice, which was subsequently placed on the receptor chamber, with the dermis side facing downward and the stratum corneum side facing upward. Terahertz reflection imaging data of the skin were recorded using THz-TDS equipment.

The terahertz pulses were focused onto the sample and the reflected terahertz response signals were then guided to the detector. In a typical imaging experiment, the signal time-domain waveform was acquired using a suitable optical delay between excitation and signal pulses; the sample was moved by an $x$–$y$-axes translational stage, in increments of $\Delta x(\Delta y) = 0.25$ mm, to obtain the pixel-by-pixel 2D images. One should bear in mind that the terahertz pulse is reflected back when it reaches an interface between media with different refractive indices; hence, reflections from different manifest themselves by distinct optical time delays (with reference to the excitation pulse), which can be used to obtain depth information.

In the top row of Figure 21.9 a sequence of terahertz dynamic reflection images of the drug-applied site is shown; note that the total scan time for an individual frame was ~8 minutes. The sequence of images recorded after ~1/4, ~1/2, 1, and 6 hours clearly reveal the time evolution of the drug penetration across the area of application (diameter approximately 5 mm). Although not directly evident

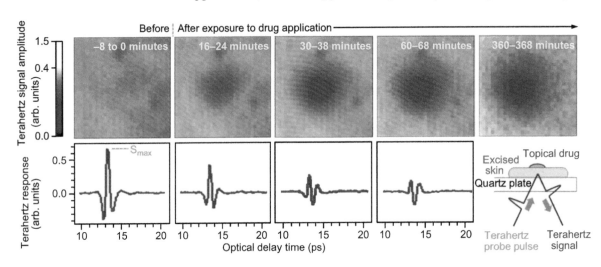

**Figure 21.9** Time-resolved terahertz imaging of dynamic drug absorption through skin. (Top row) Sequence of terahertz reflection images of the drug-applied site as a function of time (indicated within the images); image area ~ 8 × 8 mm². The scan time for individual images was 8 minutes. (Bottom row) Time-domain terahertz response signals, recorded at a center position of the related top row images. (Bottom right) Conceptual layout of the experiment. (Data adapted from Kim, K.W. et al., *Opt. Express* 20, no. 9 (2012): 9476–9484.)

from the limited number of frames shown here, the authors observed that the terahertz reflection signal from the drug-applied site gradually decreased up to about 1 hour after application and then remained constant until the end of the terahertz imaging experiment (in most cases 6 hours).

A closer inspection of the terahertz signal waveforms (bottom row in the figure), measured at about the center location of the related terahertz image, reveals the following. The maximum amplitude $S_{max}$ of the first peak (at a delay of about $\Delta t \sim 12.5$ ps) gradually decreases during the first hour after drug application, remaining approximately constant thereafter. With reference to the experimental setup sketch, this first peak is associated with the reflection from a boundary between the quartz plate and the dermis. The observed reduction in reflectivity means that the permeated DMSO reduces the difference of the refractive index between the quartz plate and the skin dermis and thus decreases the reflection of the terahertz pulse at the quartz plate-dermis interface. This explains the reduced terahertz signal in the images, across the drug-active area. The signal amplitude of the second, smaller peak (at a delay of about $\Delta t \approx 14$–$15$ ps) increases as time progresses. The authors attribute this part of the signal to the layer just above the quartz plate-dermis interface, i.e., to the reflection from the interface between the lower part of the skin dermis and the drug-permeated layer in the skin. It is also clear from the signal shapes that this second peak moves gradually closer in time to the first peak, indicating the increased penetration of the administered drug toward the dermis.

This study nicely demonstrates how terahertz imaging can be employed to monitor the spatial distribution and penetration of drug-applied sites.

### Terahertz imaging for early screening of diabetic foot syndrome

Most people with diabetes suffer from the so-called diabetic foot syndrome (DFS)—also known as "diabetic foot." This is characterized by a combination of microvascular and neurological deterioration, causing poor irrigation, loss of sensitivity in the patient's feet, and often ulceration or even amputation. Unfortunately, DFS is difficult to detect at an early stage and no direct predictive method exists, which would alert to the potential risk to the patient's quality of life and prevent costly medical treatments.

Recently, Hernandez-Cardoso et al. (2017) proposed a terahertz imaging method for early screening of DFS. The basic concept of the technique relies on the assumption that the skin dehydration of the feet of diabetics constitutes a key feature of the deterioration process. Consequently, terahertz radiation might be used to determine the water content of the tissue, at each point on the foot sole. This is because most of the foot ulcers experienced by diabetic patients occur in the greater toe, the heel area, and the metatarsal area (the region where the toes meet the rest of the foot).

Essentially, the experimental setup utilized by the authors comprises a terahertz time-domain spectrometer and a raster scanning system to produce the 2D images. Their terahertz system was based on an Er-doped fiber laser ($\lambda_L = 1550$ nm, average power $P_L = 120$ mW), which generated pulses of duration $\delta\tau_p = 90$ fs, at a repetition rate of PRF = 100 MHz. The experimental setup could be configured in transmission or reflection geometry. While the former was employed for monitoring the dielectric properties of dehydrated skin samples, the latter was used for the generation of the feet images.

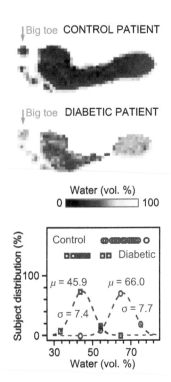

**Figure 21.10** Terahertz imaging and diagnosis of DFS. (Upper image) Terahertz image of the foot from a typical member of the control group. (Lower image) Terahertz image of the foot of a typical member of the diabetic group. Intensity scale: volumetric fraction of water content. (Bottom) Statistical distributions of the average water content across the greater toe for both the diabetic (squares) and control (circles) groups; the dashed lines represent Gaussian fits to the data. The mean and standard deviation of the fits are indicated in the plot. (Data adapted from Hernandez-Cardoso, G.G. et al., *Sci. Rep.* 7 (2017): article 42124.)

In the top half of Figure 21.10, terahertz images of the feet of typical control and diabetic subjects are shown; notice the remarkable difference between the two images, with the water content in the foot of the control subject significantly larger than that of the diabetic subject.

The authors carried out statistical analysis for the control (33 members) and diabetic (38 members) groups for different locations of the foot, namely, on the sole, the heel, and the greater toe, all yielding similar results (note that not all group members were included in the analysis process). For clarity, here only the statistical distributions for the average water content across the greater toe are displayed in the data graph (bottom part of the figure). The two distributions reveal clear, significant differences between the hydration of the feet of subjects in the control and diabetic groups; hence, the much lower water content may serve as an indicator of the deterioration of the feet of diabetics. The authors interpret their results as a proof of concept that terahertz imaging is indeed suitable, in principle, for early screening of diabetes foot syndrome; at the time of writing, the design of a future clinical trial is under way.

A last remark on terahertz imaging exploiting water content monitoring is worth making. In an approach slightly different from the direct measurement of water content to assess DFS, Son (2013) describes in his review a technique known as *terahertz molecular imaging* (TMI) for medical diagnostics; for example, TMI has been successfully applied to cancer diagnosis and nanoparticle drug delivery imaging. The technique makes use of nanoparticle probes, administered to the region of interest, to achieve greatly enhanced sensitivity when compared with conventional terahertz imaging. This comes about because of surface plasmons, which are induced around the nanoparticles. These induce a rise in temperature of cell water, which in turn change the temperature-dependent optical properties of water. Within the terahertz frequency range, these are large, and they can easily be differentially measured and imaged by terahertz waves.

## 21.2.3 Imaging based on photoacoustic methodologies

Over the last decade or so, PAI (in two dimensions) and photoacoustic tomography (in three dimensions) have drawn ever-increasing attention from a variety of disciplines, including chemistry, materials science, physics, and biomedicine. In particular, for applications in the last field, PAI is proving to be attractive because it is a label-free technique. This is a consequence of the wavelength-selective absorption of the constituent molecules themselves, at distinct, characteristic vibrational frequencies. On its own, this probably would not be enough to explain the increase in popularity, since in the very common technique of Raman spectroscopic imaging, the said molecular vibrations are also exploited. But favorable to PAI are two particular aspects, namely, (1) that the absorption process underlying PAI is substantially stronger than Raman scattering, so that it promises to be more sensitive, and (2) that the indirect detection of the absorption process via photoacoustic mechanisms does not require the setup of carefully optimized light collection paths.

In the following, only one particular example is given for imaging on the small scale, namely, PAM. Details about the methodology underlying PAM and a selection of relevant applications in biomedicine can be found, e.g., in Yao and Wang (2014).

*Scanning acoustic microscopy* (SAM) is a well-developed technique, with spatial resolution of about 1 μm; nowadays, SAM is widely used in biomedical applications such as, e.g., single-cell imaging, with contrast and resolution not available in ordinary optical microscopy. When short-pulse laser excitation is added to SAM, the signal response is in the range of ultrasound frequencies, and the technique is known as photoacoustic microscopy (PAM).

In PAM imaging, laser pulses of nanosecond duration, in the visible or NIR spectral region, are absorbed by molecules of biomedical importance, e.g., hemoglobin. As a result, broadband ultrasound waves are generated, which can be recorded by ultrasound receiver(s)/transducer(s) arranged over the tissue surface (see Figure 21.1 for the concept); using appropriate excitation and detection geometries, from the response signal data one can reconstruct 3D images of a sample, in principle.

Interestingly, PAM relies on the fact that the ultrasound waves are encoded with the optical properties of the tissue. Consequently, the limitations often encountered in purely optical imaging techniques—such as low penetration depth and poor spatial resolution due to the strong optical scattering of tissues—are avoided. Furthermore, PAM images preserve not only molecular contrast but also the spectral specificity of optical methods; this allows for recognition of anatomical features, which may be indistinguishable when using typical bioimaging modalities, such as ultrasound imaging. Clinical applications of PAM include (1) the assessment of skin, breast, and colon cancer; (2) cardiovascular disease; and (3) dermatological conditions, to name but three key examples. Note that the technique can also be applied to the investigation of single cells, using exogenous chromophores (such as, e.g., dyes or nanoparticles).

Recently Kolios and coworkers (Strohm et al. 2016) set up a combined acoustic/photoacoustic microscope to record (simultaneously) high-resolution ultrasound and photoacoustic images of (stained) neutrophils, lymphocytes, and monocytes from a blood smear. In particular, this approach provided the authors with a novel way of probing leukocyte structure and diseased cells at the single-cell level. For the generation of the photoacoustic images, the authors used an ultrashort-pulse laser, at $\lambda_L = 532$ nm, whose radiation was coupled into a single-mode fiber yielding a number of discrete output wavelengths via stimulated Raman scattering, in the wavelength range $\lambda_{SRS} \approx 532$–620 nm. The wavelength-selected light pulses were then focused onto the sample, using a ×20 microscope objective. A 1 GHz focusing transducer was coaligned with the laser spot, which allowed for reconstruction of ultrasound and photoacoustic images with micrometer spatial resolution.

Different cell types could be identified from the variations in contrast within the acoustic and photoacoustic images, a feature that provided a new way of probing, e.g., leukocyte structure. In Figure 21.11, acquired optical, ultrasound, and photoacoustic images of two of the most abundant leukocytes in blood, namely, neutrophils and monocytes, are collated; the top row contains the images of a monocyte, while in the bottom row the same sequence for a neutrophil is shown.

Looking at the images in the top row, one can see that (1) the stained monocyte exhibits a color profile that shows the cytoplasm as light purple and the nuclei as dark purple; (2) the same features appear with higher and lesser attenuation for the cytoplasm and the nucleus, respectively, but the image is fuzzier; and (3) the

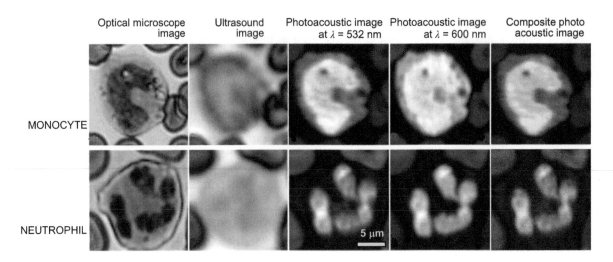

**Figure 21.11**  OR-PAM imaging of single cells of monocytes (top row) and neutrophils (bottom row). (Left to right) Optical microscopy image (for comparison of cell features), ultrasound image; PAM images at $\lambda$ = 532 nm and $\lambda$ = 600 nm; and a composite PAM image. (Data adapted from Strohm, E.M. et al., *Photoacoustics* 4, no. 1 (2016): 36–42.)

photoacoustic images are of similar sharpness as the optical image, even revealing additional features. In particular, the vacuoles, which are barely visible in the ultrasound image, are much easier to identify in the photoacoustic images, which exhibit even better contrast than the optical image.

In the bottom row—images of a neutrophil—the staining process was different, showing up the nuclear lobes as blue and the cytoplasm as pink. The general findings in the measurements were that (1) the acoustic attenuation through the surrounding red blood cells was more intense than through the neutrophils and (2) strong photoacoustic signals were obtained from the nucleus at both $\lambda$ = 532 nm and $\lambda$ = 600 nm, while the cytoplasm produces very weak signals. For image contrast and detail, the same holds as was discussed for the monocyte case.

Note that the last two images in the two rows constitute composite photoacoustic images, which were merged using the green channel (the $\lambda$ = 532 nm images) and the red channel (the $\lambda$ = 600 nm image). This enhanced some of the features, which were not easily noticed in the individual grayscale images; note that the yellow areas correspond to equal signal amplitude in both images. The clarity and the unprecedented resolution of the composite PAM images are impressive.

The main goal of this second example is to provide a glimpse of the state of the art in PAI, with the first demonstration of PAM measurements of single cells, with micrometer resolution.

Finally, it is worth noting that, as in all other laser spectroscopic microscopy measurement methodologies, tremendous efforts are always made to improve on the resolution of PAM.

Relatively recently, a novel PAM measurement modality was developed, dubbed *photoimprint photoacoustic microscopy* (PiPAM) by the authors (Yao et al. 2014). It relies on the nonlinear "bleaching" of the absorption of a molecular vibration by a first laser pulse and subsequent probing of the same location by a second laser pulse. The authors demonstrated a significant improvement in spatial resolution, specifically in the axial direction. Details of their experiment are given in Section 21.5.

At about very much the same time, Berer et al. (2015) began a particular research project, *superresolution photoacoustic microscopy by stimulated depletion*, with the aim to achieve sub-Abbe limit spatial resolution in PAM. Their proposed approach is akin to the superresolution laser fluorescence method of stimulated emission depletion (STED) microscopy, exploiting spatially dependent depletion of absorbing molecules exposed to strong, saturating laser radiations with tailored beam shape. At the time of writing, the ultimate aim in the project has not yet been reached, but the researchers are well on track, having already proven the saturation effects in two-laser two-wavelength test measurements (see, e.g., Langer and Berer 2016).

Besides this, two approaches for photoacoustic spectral imaging to breach the Abbe limit for the optical-frequency regime, i.e., for near-surface PAM, very recently it was successfully demonstrated that superresolution PAM is also feasible in the acoustic-frequency regime, i.e., for deep-tissue applications (see Chaigne et al. 2017).

In their proof-of-principle study for deep-tissue PAM, the authors surpassed the common acoustic diffraction limit in PAM by exploiting inherent temporal fluctuations in the photoacoustic signals, associated with sample dynamics (e.g., those induced, e.g., by the flow of absorbing red blood cells). This was achieved by modifying a conventional PAM system to include concepts from superresolution fluorescence fluctuation microscopy, specifically exploiting the statistical analysis of the acoustic signals from flowing acoustic emitters. Applying this technique to photoacoustic imaging of microfluidic phantom circuits, the authors demonstrated an improvement in resolution for deep-tissue PAM by a factor of ~1.6, close to the expected theoretical limit for their specific experiment.

## 21.3 CHARGED PARTICLE IMAGING: BASIC CONCEPTS AND IMPLEMENTATION

Toward the end of the 1980s, Chandler and Houston (1987) developed a new imaging technique for elucidating chemical reaction dynamics. They demonstrated the viability and advantages of the novel approach with an experiment, in which they imaged the spatial distribution of (laser-ionized) photodissociation products, produced by laser photolysis of methyl iodide; they referred to the method as to *photofragment ion imaging* or—if applied to the study of chemical reaction products—as *product reaction imaging*.

In ion imaging (in which the ions are generated using a range of laser ionization techniques), the speed and angular direction of the product are measured simultaneously for any state-selected product from the chemical process, in just one single measurement. This unique feature significantly changed the way of investigating and understanding many of the elementary processes in molecular reaction dynamics, such as, e.g., photodissociation, bimolecular reactions, or energy transfer, to cite but a few (see, e.g., Heck and Chandler 1995; or Ashfold et al. 2006). In this section, the basic concepts underlying this relatively new methodology will be described, together with the key ingredients of the experimental technique, placing emphasis on its advantages and limitations.

### 21.3.1 Basic concepts of unimolecular and bimolecular collisions

As is well known, in photodissociation studies the reference axis is the electric field vector of the excitation laser; the photofragment angular distribution exhibits axial symmetry with respect to it. In contrast, the reference axis in a bimolecular collision is normally taken to be the relative velocity vector of the colliding particles provided that the collision takes place without any external field applied. In this context, it is worth recalling that in the study of elementary chemical reactions, one needs to quantify the number of products scattered in a given direction per unit time, i.e., the so-called *differential cross section* (DCS), denoted in the center-of-mass (CM) frame by $d\sigma/d\Omega$, where $\sigma$ and $\Omega$ stand for the integral cross section and solid angle, respectively. This quantity refers to the effective area of the colliding particles that leads to scattering into the angular range $d\Omega = \sin\theta \cdot d\theta \cdot d\phi$.

If in addition one selects a particular quantum product state in the measurement of the DCS, this is then referred to as the *state-resolved DCS* (this was already briefly addressed in Section 16.3 in the discussion of resonance-enhanced multiphoton ionization [REMPI]). Obviously, the most detailed information on any collision process would be provided by so-called *state-to-state DCS*, i.e., the measurement of the DCS for a given initial state of the reagent into a given final state of the product.

#### Unimolecular collisions (photofragmentation dynamics)

The use of molecular beams and polarized laser light, coupled with time-of-flight (ToF) mass spectrometric measurements, allows one to probe the dynamics of photodissociation processes (see Levine and Bernstein 1987). The conceptual realization of such an experiment is shown in Figure 21.12: a molecular beam (for simplicity containing only diatomic molecules, AB) is crossed with a linearly polarized laser beam, to excite the electronic ground state of the molecule to an excited repulsive state that photodissociates into the two fragments, A and B.

In a typical photodissociation experiment, one then would measure $I_A(\theta)$, i.e., the intensity of fragments A, as a function of the angle $\theta$ (the angle between the detection direction)—the $z$-axis—and the electric field vector of the laser light. Note that a second laser beam is usually intersecting the photodissociation region, thus ionizing fragment A, to yield $A^+$ necessary for ion detection. Note also that no external field is applied, to spatially orient the molecule; thus, they will be randomly oriented inside the laser–beam interaction volume.

In the case that the excited dissociative state AB* lives long enough to rotate many times before falling apart, then the fragments would appear in all directions with the same intensity, and $I_A(\theta)$ would exhibit no angular dependence. If, in contrast, the dissociative state lives a only very short time, such that the bond breaks long before the molecule can even rotate, this scenario is normally described as *prompt dissociation*; as a result $I_A(\theta)$ would be nonisotropic.

From a conceptual point of view, one needs to recall that according to electric dipole transition, the probability $P$ that a molecule absorbs one photon is given by the relation $P \propto |\boldsymbol{\mu}\cdot\boldsymbol{E}|^2$; $\boldsymbol{\mu}$ and $\boldsymbol{E}$ are the transition moment and the electric field vector of the excitation laser light, respectively. Because of the scalar product of the two vectors in this proportionality expression, it can also be written as $P \propto \mu^2 \cdot E^2 \cdot \cos^2\alpha$, where $\alpha$ is the angle between $\boldsymbol{\mu}$ and $\boldsymbol{E}$.

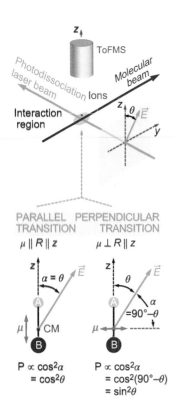

**Figure 21.12** Investigating photofragmentation dynamics. (Top) Basic experimental setup. ToFMS, ToF mass spectrometry. (Bottom) Schematic representation of the parallel and perpendicular transition probabilities for a diatomic molecule, including their respective angular dependence; for further details, see text.

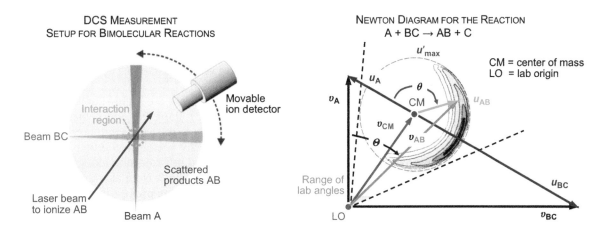

**Figure 21.13**   Concept of bimolecular (reactive) collisions. (Left) Typical experimental arrangement for measuring differential scattering cross sections. (Right) Newton diagram for the reactive collision A + BC → AB + C. The velocities are shown in the laboratory frame ($v_A$, $v_{BC}$, $v_{AB}$, $v_{CM}$) and the CM frame ($u_A$, $u_{BC}$, $u_{AB}$); the product AB scatters with angles $\Theta$ and $\theta$ in the laboratory and CM frames, respectively. The internal-state-dependent differential scattering of products is indicated by the circle with $u'_{max}$.

For a diatomic molecule, the transition moment can be only parallel or perpendicular to the internuclear axis, which gives rise to either parallel or perpendicular transitions with $\mu \| R$ or $\mu \perp R$, where $R$ is the internuclear distance vector. Thus, in the photodissociation of a randomly oriented sample of molecules, the fragment atoms will fly off either perpendicular or parallel to the polarization direction of the light field for a perpendicular or parallel transition, respectively. As a consequence, the expected fragment angular distribution $I_A(\theta)$ exhibits $\cos^2 \theta$ or $\sin^2 \theta$ behavior for a parallel or perpendicular transition, respectively (see the bottom part of Figure 21.12). In general, for common optical excitation, the normalized angular distribution of the ejected fragment(s) is given by (see, e.g., Zare 1972)

$$I(\theta) = (1 + \beta \cdot P_2(\cos \theta))/4\pi, \tag{21.2}$$

in which $\theta$ is the angle between the fragment velocity vector and the laser polarization direction, $P_2(\cos \theta)$ is a second-order Legendre polynomial and $\beta$ is the so-called *anisotropy parameter*; the said parameter is $\beta = 2$ or $\beta = -1$ for parallel or perpendicular transitions, respectively. In the case of mixed transitions, the value for $\beta$ lies between the two extremes. In fact, the parameter $\beta$ is associated with the vectorial character of the photodissociation reaction, namely, the so-called $E$-$\mu$-$v$ vector correlation, where $E$, $\mu$, and $v$ stand for the electric field vector of the photodissociation laser radiation, the transition moment vector, and the fragment recoil velocity vector, respectively. For this and other types of vector correlations in reaction dynamics, see, e.g., Houston (1987, 1995).

### Bimolecular reactive and nonreactive scattering

Bimolecular collisions of the type A + BC are the second main class of collisions; for experimental convenience in such collisions, the atom A and the molecule BC collide at right angles in the laboratory frame, constituting so-called crossed molecular-beam experiments (see the left-hand part of Figure 21.13). The related Newton diagram (also known as the velocity vector diagram) of the collision is shown on the right. Here the velocities in the laboratory, $v_A$ and $v_{BC}$, are decomposed into the velocity of the CM $v_{CM}$ and the velocities in the CM reference frame ($u_A$ and $u_{BC}$). In the CM reference frame, the (reaction) product AB has scattered with an angle $\theta$ and travels with velocity $u_{AB}$.

In ion imaging experiments, the relative velocity vector $v_{rel} \equiv u_A + u_{BC}$ is normally chosen as the reference axis (related to the CM reference frame). Then, two specific directions for the ionization laser beam are highly convenient, namely, (1) along the relative velocity vector or (2) along the CM velocity vector, i.e., $E \| v_{rel}$ or $E \| v_{CM}$, respectively. Note that, due to the axial symmetry of the angular scattering with respect to the relative velocity vector, in the first case the angle between the product velocity vector $u_{AB}$ and its projection along the laser beam axis would exactly replicate the scattering angle $\theta$. Note also that the maximum product velocity $u'_{max}$ in the bimolecular scattering process is associated with the energy balance between (potential) internal quantum excitation and maximal possible kinetic energy of the product molecule, i.e.,

$$u'_{max} \propto \left[ E_{kin,max}(AB) - E_{int}(AB) \right]^{1/2} \tag{21.3}$$

### 21.3.2 Newton sphere

To understand the basic principle underlying ion imaging out of the two cases discussed earlier (unimolecular and bimolecular collisions), the slightly simpler process of molecular photodissociation is addressed here, to exemplify the problem. Laser dissociation of a molecule AB creates the fragments A and B, i.e.,

$$AB + h\nu_{diss} \rightarrow (AB)^* \rightarrow A + B + E_T, \tag{21.4}$$

where $E_T$ stands for the total excess energy released as product translational energy, after subtracting the internal energy of the products A and B. Exploiting conservation of momentum and energy, this excess energy partitions as

$$E_{kin}(A) = (M_B/M_{AB}) \cdot E_T \quad \text{and} \quad E_{kin}(B) = (M_A/M_{AB}) \cdot E_T \tag{21.5}$$

Here $M_i$ stands for the mass of the $i$th particle and $E_{kin}(A)$ and $E_{kin}(B)$ are the kinetic energies of fragments A and B, respectively. Each photodissociation event yields two fragments, flying apart with equal momentum in opposite directions in the CM reference frame. When repeating the photolysis for a statistically relevant large number of times, the fragments will be distributed spherically in velocity space forming the so-called Newton (velocity) spheres, whose size is proportional to the fragment's speed (this can be deduced from the relations in Equation 21.5). Thus, for a fixed total energy $E_T$, one finds that the higher the internal energy of the nascent fragment, the lower its translational energy and, consequently, the smaller its Newton sphere radius.

In Figure 21.14, the two Newton spheres for fragments A and B are shown, for two events only in the upper part and for the full statistical distribution in the lower part. It is clear from the few-event representation that dissociation events generate particle pairs with identical speed fractions, but which are different in directions. Note that for the ideal-case scenario shown here, the parent molecule AB is always located at the same origin in space. Note also that the built-up surface pattern representing the Newton spheres resemble anisotropic patterns for scattering in forward/backward direction, with most of the events distributed near the vicinity of the "poles" (basically representing the aforementioned cos$^2$ $\theta$-type distribution).

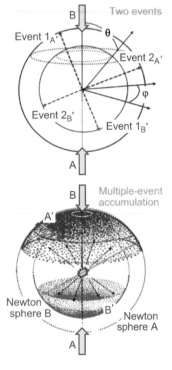

INELASTIC FORWARD-SCATTERING COLLISION A+B → A'+B'

**Figure 21.14** Newton spheres in the CM frame for inelastic scattering in the forward direction. By defining the energy of the collision, conservation of energy and momentum determines the velocity of the final products A and B (mass B > mass A), which recoil in opposite directions. (Top) Display of two events from A + B scattering. (Bottom) Multiple-event accumulation, generating a sphere surface pattern revealing the dynamics of the scattering process. In the CM frame, one finds cylindrical symmetry about the azimuthal angle $\varphi$. (Adapted from Brouard, M. et al., *Chem. Soc. Rev.* 43, no. 21 (2014): 7279–7294.)

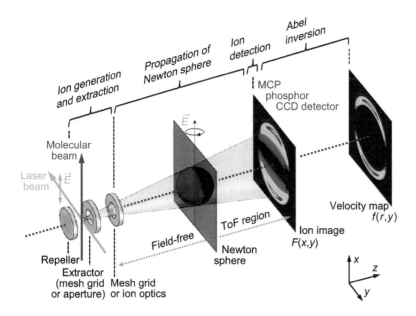

**Figure 21.15** Generalized, schematic setup of an ion imaging experiment, showing the different parts of the basic technique. In the ion extraction and steering, the mesh components are for standard ion imaging, while the nonmesh configurations are utilized in the velocity map imaging and slicing modalities. For further details, see text. MCP, microchannel plate; CCD, charge-coupled device.

### 21.3.3 Basic experimental setups

The aim in ion imaging is to measure the full distribution of the collision events, i.e., to image the Newton sphere as truthfully as possible. The overall conceptual setup for measuring molecular photodissociation, within a molecular-beam environment, is shown in Figure 21.15. Briefly, one or two laser beams, with short laser pulses, cross a molecular beam; with the first laser pulse, the molecule is dissociated, and with the second laser pulse, the fragment of interest is ionized. Note that often the two laser beams are counterpropagating, but configurations with the laser beams under a small mutual angle between them are also rather common; note that for clarity, only a single laser beam is shown in the figure. The laser photon–molecule interaction region is situated between a pair of repeller/extractor electrodes. A suitable electric field between these electrodes accelerates the ionized photofragment through a further electrode, at ground potential, into a field-free ToF drift region. The ionization laser pulse is normally timed to arrive with only a short delay (normally a few nanoseconds only) relative to the photodissociation laser pulse, to ensure that the products have not moved significantly from their position of generation. Note that the ionization process does not affect the recoil velocity of the photofragment(s), due to the small mass of the ejected electron.

The extracted (accelerated) ions fly along the detection axis inside the ToF tube in such a way that those with the same velocity are found on the surface of a Newton sphere whose radius changes with time. In other words, during their flight to the detector, the ions separate spatially, according to the speed and angular distribution. Finally the ions reach a 2D detector that consists of a dual microchannel plate (MCP), coupled to a phosphor screen. The electrons produced at the back of the MCP are accelerated onto a fast-reacting (i.e., very short phosphorescence lifetime) screen, whose emitted light is imaged typically by a charge-coupled device camera. The result is a 2D projection image of the Newton sphere (the velocity distribution) on the detection plane. Hence, the original three-dimensional (3D) velocity distribution can be reconstructed from the ion image in performing the so-called *inverse Abel transformation*, or by a fit to a model function.

It should be noted that the inverse Abel transform can be used only if an axis of cylindrical symmetry parallel to the imaging plane is well defined in the experimental arrangement. For example, in photodissociation studies, initiated by linearly polarized laser light, the velocity distribution is cylindrically symmetric around this polarization direction; this means that the inverse Abel transformation can be applied when the electric polarization vector is parallel to the imaging plane. In the case of double-resonance experiments, i.e., when photofragment polarization is encountered, both the photolysis and (ionizing) probe laser polarization directions must be parallel if the inverse Abel transform were to be applied.

While ion imaging began and very much thrived on unimolecular collision scenarios, studies of the dynamics of bimolecular reaction have become increasingly popular and are now routine in many research laboratories, a testimonial to which is the topical issue "Developments and Applications of Velocity Mapped Imaging Techniques" of *The Journal of Chemical Physics* (July 2017 issue), with an introductory perspective of the pioneers of the technique (Chandler et al. 2017). The application of ion imaging to bimolecular collisions requires combining crossed molecular-beam techniques and laser spectroscopic ionization with ion imaging. Laser radiation is employed to ionize (normally via a suitable REMPI scheme) a specific rotational–vibrational state of a nascent product formed in the crossed molecular-beam reaction. Imaging of the probe ions allows for the simultaneous measurement of the product angle and speed distributions, from which the state-resolved DCS can be determined (this will be discussed further in the following).

### 21.3.4 Methods for improving the resolution in ion imaging

#### Technique of velocity map imaging

The limit in energy resolution encountered in the original implementation of ion imaging ($\Delta E/E \sim 15$–20%) was improved upon by Eppink and Parker (1997). They developed a new type of ion lens extraction optic, based on open electrodes rather than the original mesh-covered electrodes (see further in the following); in addition, they introduced inhomogeneous extraction fields to focus ions with the same velocities, but distinct spatial locations, onto the same imaging plane. This development is commonly referred to as *velocity map imaging* (*VMI*); it improves the energy resolution by almost an order of magnitude and—not surprisingly—is nowadays used in the majority of studies that follow the dynamics of photodissociation and bimolecular reactions (see, e.g., Parker and Eppink, 2003).

The (historical) conventional experimental setups for ion imaging suffer various, important drawbacks due to the presence of grid electrodes. In particular, the presence of fine wire grids, used as electrodes to ensure parallel field lines, results in a "shaded" projection of the ionization volume onto the detector area, leading to a significant blurring of the image. In addition, the grid also produces ion deflection, distorting the ion paths to the detector. These problems are avoided in the VMI method, which, as mentioned earlier, does not use grid electrodes but open lens as illustrated in Figure 21.16.

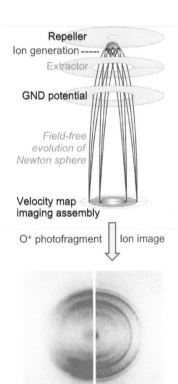

**Figure 21.16** Principle of ToF VMI. (Top) Conceptual setup, illustrating velocity mapping trajectories of a distribution of ions, comprising four velocities and two nonzero kinetic energy release distributions; for details see text. GND, ground potential with $U_{GND} = 0$ V. (Bottom) Example of VMI measurement, for the detection of $O^+$ in an $O_2$ photolysis experiment; imaging with a fine-mesh-grid electrode (left) and an open-aperture ion Einzel lens (right). (Data adapted from Eppink, A.T.J.B. and D.H. Parker, *Rev. Sci. Instrum.* 68, no. 9 (1997): 3477–3484.)

The use of open-electrode configurations ensures that all particles with the same initial velocity vector are mapped onto the same point on the detector, irrespective of their initial spatial position inside the ionization region. In their pioneering experiment, Parker and Eppink (1997) demonstrated the improvement in spatial resolution, associated with their new technology, in a photodissociation experiment of molecular oxygen. Dissociation of $O_2$ is subsequent to two-photon excitation (at $\lambda \approx 225$ nm) into the Rydberg level $3d\pi\Sigma^3_{1g}$ ($v = 2, N = 2$). In the lower part of Figure 21.16, a comparison of the $O^+$ ion images from the photodissociation processes is shown, using the conventional method of ion imaging with a grid electrode (left image) and the method of VMI (right image). While the image obtained with the grid electrode shows deflections induced by the grid and sharp details are blurred out, in the Einzel lens VMI image these defects are largely resolved; distinct concentric rings are observed, which were unambiguously assigned to several different dissociation/ionization pathways (see the detailed analysis by Parker and Eppink 1997).

A last comment regarding VMI is noteworthy. Despite the enhanced resolution of the VMI method, inverse Abel transform is still required to reconstruct the 3D velocity distribution; in the $O_2$ photolysis study of Eppink and Parker (1997), this was possible because the laser polarization (i.e., the axis of cylindrical symmetry) was along the vertical axis of the ion images.

### Technique of "slice" imaging

Just a few years after the introduction of VMI, Gebhardt et al. (2001) developed a further, novel modality of the already popular methods of ion imaging and velocity mapping; this has become known as *slice imaging*. Its great advantage is that the product angular and velocity distributions can be measured directly from the images, without the need for inverse Abel transformation. A further bonus of the new approach is that cylindrical symmetry, previously paramount for implementing Abel inversion, is no longer necessary but is still of benefit for ease of interpretation.

The basic idea behind Gebhardt and coworkers' approach is *delayed pulsed extraction* of the ions, following the generation of fragments in a photodissociation event, and positioning of the nascent products (the ion Newton sphere is first allowed to expand in a field-free environment). Therefore, because of the relatively large volume of ion origin, the subsequent time-delayed pulsed-electric-field extraction causes a substantial axial velocity spread in the ion cloud, so that on arrival at the detector, the ion packet (Newton sphere) is stretched in time by typically a few hundred nanoseconds. Then, by using a narrow gating pulse for intensified charge-coupled device (ICCD) image detector (typically on the order of 20-40 ns, which is equivalent to about 1/10 of the arrival time spread), one is able—with appropriate time relation between the ion extraction and the detector gating pulses—to image only the center slice of the ion packet.

This is shown schematically on the left of Figure 21.17a. This "center-slice" result is equivalent to that obtained with common ion imaging methods, which use inverse Abel-transform reconstruction. In a certain way, delayed-pulse ion slice imaging is analogous to time-lag focusing introduced by Wiley and McLaren (1955) in their pioneering paper on ToF mass spectrometry.

A couple of further advantages of this approach are worth noting: (1) the artificial noise that is introduced by the Abel transformation is eliminated and (2) the

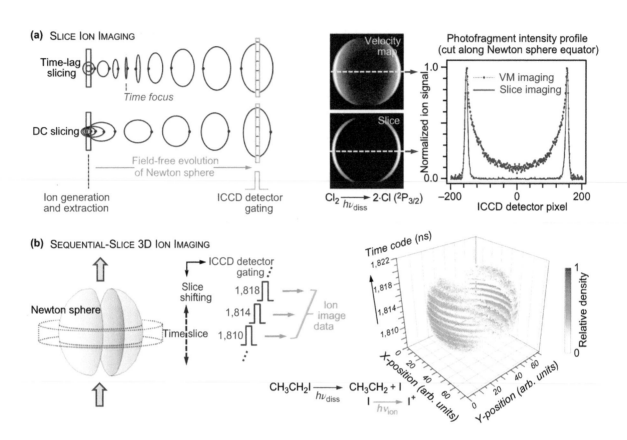

**Figure 21.17**  Method of slice ion imaging. (a) Key concept of evolution of the Newton sphere for time lag and direct-current (DC) slicing modalities (for details see text). On the right, a typical example of imaging of photofragments is shown (exemplified for Cl⁺), comparing VMI and slice images with each other; the improvement in resolution is evident from the intensity profiles along the equator of the images.(Data adapted from Gebhardt, C.R. et al., *Rev. Sci. Instrum.* 72, no. 10 (2001): 3848–3853.) (b) Sequential slicing of the Newton sphere, resulting in a series of images from which the full 3D distribution can be assembled (here exemplified for the photolysis of ethyl iodide). (Data adapted from Slater, C.S., *Studies of Photoinduced Molecular Dynamics Using a Fast Imaging Sensor*, 2016.)

energy resolution is not compromised but remains comparable with that achieved using the common VMI technique.

A representative example of the improvement that can be achieved by slice ion imaging is shown in the right part of Figure 21.17a, namely, for the ground-state spin-orbit component of the $^{35}$Cl atom photofragment, obtained from the photodissociation of $Cl_2$ at $\lambda = 355$ nm (Gebhardt et al. 2001). The upper ion image was obtained using the velocity mapping method, while the lower ion image is that obtained by using slice ion imaging. For a better comparison, the intensity cuts along the equator of the two images are depicted on the very right of the figure, revealing how the long (velocity) tail observed in the "squashed" full projection of the entire Newton sphere in VMI is eliminated from the slice image.

This is a clear advantage of the latter technique, in particular, when multiple photodissociation channels contribute to the signal. For example, one can imagine situations in which a very small interior peak would be hidden below the tail signal of the outer peak and, consequently, would not be recovered through the inverse Abel-transform mechanism. In contrast, with the inherently reduced noise and the tail-free presentation of the center velocity slice, such a small peak will become recognizable.

Besides the time-lag slicing method, a number of variants have been developed over the years; their various merits and pitfalls have been summarized, e.g., in a review article by Ashfold et al. (2006). Among said variants, so-called *DC time slicing* has become a rather widely used alternative (see Townsend et al. 2003, 2004b). Briefly, the goal in all slicing ion imaging variants is to achieve sufficiently wide spreads in the ion packet arrival time while maintaining VMI conditions.

In DC slicing, low electric fields are used for ion extraction; in addition, the arrival time spread is further influenced by using extra electrodes to extend the region along the ion flight path over which acceleration by the electrostatic lens takes place (for the concept see Figure 21.17a). Such configurations are relatively easy to implement, and the DC operating mode guarantees reasonably homogeneous ion lens behavior. Overall, the ion image size becomes less dependent on the position of ionization along the $Z$-axis; on the other hand, DC slicing rarely achieves the same degree of slicing as is possible with pulsed-field slicing.

Finally, in some cases, it might be interesting or important not only to record the information on the center slice of the Newton sphere ion distribution but also to have data available for the complete sphere available; this is specifically the case if no cylindrical symmetry (which is normally assumed in photofragmentation experiments) is prevalent. Such full 3D recording of the Newton sphere is now possible, by buildup from a sufficiently large series of thin time slices on arrival; this is afforded by the availability of very fast ICCD array detectors, with gating times as low as ~5 ns. The concept of 3D sliced ion image reconstruction of the Newton sphere is shown schematically in Figure 21.17b, together with example data from a photolysis experiment of ethyl iodide (see Slater 2016).

## 21.3.5 Measuring time and position: Direct 3D ion imaging

As stated a few times already, if the collision process under study lacks cylindrical symmetry, then common mathematical reconstruction methods (like, e.g., the inverse Abel transform) cannot be applied to recover the full 3D product velocity distribution from the 2D projection image of the Newton sphere. In addition to the aforementioned technique of slice imaging, several methodologies have been developed to overcome this difficulty. One of these is *direct 3D imaging*, a specific 3D photofragment imaging technique that allows for the simultaneous measurement of the complete velocity (speed and angle) distributions of the photodissociation products. This novel approach, developed by Chichinin et al. (2002), utilizes a so-called 2D *delay-line detector* (DLD), located directly behind the MCP element and replacing the phosphor screen plus ICCD image detector common to all other ion imaging implementations discussed thus far (see the conceptual representation in the left-hand part of Figure 21.18). This type of detector measures single events only, but for these, the two transverse velocity components ($v_x$ and $v_y$) of the nascent fragment are directly derived from the position of the impinging electron on 2D DLD plane; and from the measured arrival time, the longitudinal velocity component ($v_z$) can be extracted.

The concept behind delay-line detection of a charged particle event is rather simple and around for quite some time (see, e.g., Lampton et al. 1987): it is basically made up of two crossed electrical signal delay lines, one each in the $X$- and $Y$-directions. Each consists of a pair of wires, wound parallel to each other

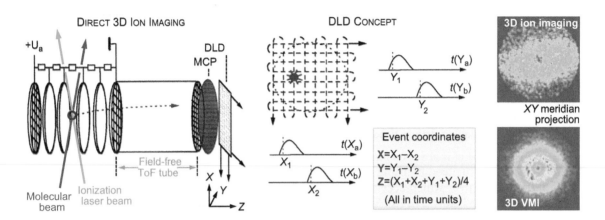

**Figure 21.18** Direct 3D ion imaging using a DLD. (Left) Common experimental setup. (Center) Concept of DLD detection, with correlation of actual 3D coordinate values to temporal–spatial signal information. (Right) Density map of a meridian plot of NO products emerging from the $N_2O \cdots N_2O$ intracluster reaction (see text), recorded by conventional and VMI (top and bottom images, respectively). (Reprinted with permission from Kauczok, S. et al., *Rev. Sci. Instrum.* 80, no. 8: article 083301, Copyright 2009. American Institute of Physics.)

in a zigzag pattern across the detector plane, with a small potential difference of ~30 V applied between the two wires of each pair. The crossed layer of the delay lines share the charge from each photoion event; therefore, one can localize the *X-Y* position of the event impact from the two delay times. Typical DLDs exhibit propagation speeds on the order 1-2 mm·ns$^{-1}$; combined with the time resolution of standard detector electronics (in general <0.1 ns), one can achieve spatial resolution of less than 100 µm. The principle of signal readout and its relation to the position on the detector, and the addition *Z*-component information, is briefly summarized in the central part of Figure 21.18; for further details, see Chichinin et al. (2002).

More recently, Kauczok et al. (2009) reported yet another imaging variant, namely, a hybrid technique in which they combined *velocity mapping* with *direct 3D imaging*. Essentially, in the new development, advantages of the improved spatial resolution seen in VMI were combined with the direct observation of all three components of the product velocity vector in 3D ion event imaging. The authors tested their novel approach to characterize several (elementary) unimolecular and bimolecular chemical reactions.

A representative example of the improved resolution is illustrated on the very right of Figure 21.18; here a comparison of NO products from the intracluster reaction

$$N_2O \cdots N_2O + h\nu \rightarrow N_2 + O(^1D) \cdots N_2O \rightarrow NO + NO + N_2 \qquad (21.6)$$

is shown: the top image stems from the results of conventional 3D ion imaging, while the bottom image constitutes the density map of a meridian plot of NO recorded by 3D VMI. A closer look at the figure reveals that whereas the distribution in the top panel is of *prolate* shape, that in the bottom panel is nearly *spherical*. The reason for the oblate shape is mainly a consequence of the projection of the length of the excitation volume, since this image was recorded without velocity mapping; it is clear that an evaluation of details visible in the image recorded in the hybrid mode hardly would have been possible in images generated in the conventional approach.

## 21.3.6 Product-pair correlation by ion imaging

Prior to the advent of ion imaging, the overwhelming number of bimolecular reaction processes studied for their full dynamics belong to the class of three-atom systems, in general symbolized by the reaction formula

$$A + BC \rightarrow AB + C. \tag{21.7}$$

For this type of reaction, the most detailed quantity that can be measured is the *state-to-state* DCS, also referred to as *state-to-state angular distribution*. But detailed DCS measurements have been successful for only a small number of "simple" reactions such as, for example, the exchange reaction $H + D_2 \rightarrow HD + D$ (see Section 16.2).

However, since the majority of uni- and bimolecular reactions involve more than three atoms, their study implies (1) increased complexity on the one hand and on the other hand (2) new challenges and opportunities to deeply understand molecular collision/reaction mechanisms.

One of the key examples of the latter is mode-selective chemistry in a four-atom reaction (as a demonstration for this, REMPI measurements for the particular reaction $O^* + N_2O$ were outlined in Section 16.2). On contemplating, e.g., the four-atom reaction

$$A + BCD(v, J) \rightarrow AB(v_1, J_1) + CD(v_2, J_2), \tag{21.8}$$

it is clear that product channel branching into AB and CD modes will most likely be heavily dependent on the distinct mode excitation in the reagent molecule BCD. In particular, correlation measurements of the formed products have proven to be rather powerful for elucidating the dynamics of reactions of the type given in Equation 21.8, and even more complex ones (involving a higher number of atoms). In particular, the approach of *state-resolved pair-correlated DSC* (also referred to, in short, as *product-pair correlation*) has become a powerful tool in this task; specifically, ion imaging has played a major role in its advancement.

The key idea behind (ion) imaging in product-pair correlation measurements for reactions of the type shown in Equation 21.8 is based on state-selective REMPI detection of one of the reaction products, for example, the molecule AB. The product AB is imaged with high-resolution VMI, such that not only can its quantum state be assigned, but also its translational energy can be evaluated with high precision. Then, by conservation of energy and momentum, the maximum velocities of the coproduct CD, in different quantum states, can be derived using the following conditions:

$$\text{Energy conservation}: \; \Delta E = E_{kin}(AB) + E_{kin}(CD) + E_{int}(AB) + E_{int}(CD) \tag{21.9a}$$

$$\text{Momentum conservation} \quad M_{AB} \cdot u_{AB} = M_{CD} \cdot u_{CD} \tag{21.9b}$$

Here $\Delta E$ is the total energy available to products (a quantity that is normally known for a given experiment configuration), which includes (1) the collision energy of the reagents, (2) the reaction exothermicity, and (3) the internal energy of the reactants. The internal energy of the product molecule AB, $E_{int}(AB)$, is known from the REMPI scheme used in the (ion) detection; the kinetic energy

$E_{kin}(AB)$ is extracted from the VMI map. Then, the translational energy of the coproduct molecule CD, $E_{kin}(CD)$, can be deduced from applying the condition in Equation 21.9b; thus, with the knowledge of the rest of the energies, the remaining internal energy, $E_{int}(CD)$, can be calculated from Equation 21.9a. Note that $M_i$ and $u_i$ are the mass and velocity of the $i$th particle in the CM frame.

The outcome of this procedure is that the full angular distribution of a given coproduct pair can be resolved; the results are often labeled as *(i,j) product-pair correlation*, where $i$ and $j$ stand for the quantum states of the coproduct molecules AB and CD. This type of ion imaging-related information can reveal detailed information about the chemical event, which might not be accessible by other means. A representative example is given in Section 21.4.

# 21.4 CHARGED PARTICLE IMAGING: SELECTED EXAMPLES FOR ION AND ELECTRON IMAGING

This section is dedicated to ion (and electron) imaging applications in unimolecular reactions and bimolecular (reactive and nonreactive) collisions; these are the scientific areas in which the technique has had and still has a major impact.

The first realization of ion imaging dates back to the late 1980s; it was demonstrated by Chandler and Houston (1987) for a (unimolecular) photodissociation experiment of methyl iodide ($CH_3I$); for some details, see also Section 21.5. Since those early days, technological developments and novel measurement variants have greatly improved the temporal and spatial resolution of ion image generation and interpretation, and although charged particle imaging has by and large remained in the arena of investigation of fundamental chemical processes, its results increasingly become important in other research fields as well.

Without claiming exhaustive coverage, representative examples for some of the key applications of charged particle imaging are discussed in the following sections.

## 21.4.1 Photodissociation with oriented molecules

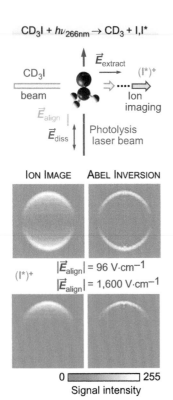

$CD_3I + h\nu_{266nm} \rightarrow CD_3 + I, I^*$

$\vec{E}_{extract}$

CD$_3$I beam

$(I^*)^+$

Ion imaging

$\vec{E}_{align}$

$\vec{E}_{diss}$

Photolysis laser beam

ION IMAGE     ABEL INVERSION

$(I^*)^+$

$|\vec{E}_{align}| = 96\ V\cdot cm^{-1}$

$|\vec{E}_{align}| = 1{,}600\ V\cdot cm^{-1}$

0 ▬▬▬▬▬ 255
Signal intensity

**Figure 21.19** Photofragmentation of state-selected and oriented molecules. (Top) Experimental concept, with orientation of the molecule by the field $E_{align}$, dissociation by the laser field $E_{diss}$, and extraction of the fragment ions by $E_{extract}$. (Bottom) 2D ion images of the recoil distribution of the $I^*(^2P_{1/2})$ fragments, from the photodissociation of state-selected $|J,K,M_J\rangle = |1,1,1\rangle$ and oriented CD$_3$I parent molecules, for two different orientation field strengths; the right data panels constitute the Abel-inverted images. (Data adapted from Janssen, M.H.M. et al., *J. Phys. Chem. A* 101, no. 41 (1997): 7605–7613.)

Ten years after the first realization of ion imaging, an advanced photodissociation imaging technique was developed by Stolte and coworkers, in which they combined state-selected and oriented molecular-beam methods with ion imaging detection (Janssen et al. 1997); the new approach was illustrated for the photodissociation of CD$_3$I. In the said experiments, the authors recorded 2D images of the recoil distribution of $I^*(^2P_{1/2})$ photofragments.

In a first step, electric-hexapole (JKM) quantum-state selection was employed, to create a single-state molecular beam of the symmetric-top molecule CD$_3$I; the hexapole focusing voltage was set as to the selection of the quantum state $|J,K,M_J\rangle = |1,1,1\rangle$. This state-selected molecule was then oriented in space by an external electric field. Subsequently, these state-selected and oriented (parent) molecules were photodissociated by photons of wavelength $\lambda = 266$ nm, yielding the iodide atom fragments $I(^2P_{3/2})$ and $I^*(^2P_{1/2})$. Finally, the excited fragment $I^*$ $(^2P_{1/2})$ was ionized by using a (2+1) REMPI scheme, and detected via ion imaging (see the concept representation in the upper part of Figure 21.19; for further details, see Janssen et al. 1997).

The images shown in the lower part of the figure are ion image projections of the angular recoil distribution of the $I^*(^2P_{1/2})$ fragments onto the 2D detector plane. The images displayed here were taken at different orientation field strengths, namely, for weak orientation at $E = 96$ V·cm$^{-1}$ and for strong orientation at $E = 1600$ V·cm$^{-1}$. Note that the orientation field was aligned along the vertical axis, being parallel to the linear polarization of the dissociation laser, and that the parent molecules were oriented with the I atom end of the molecule in the upward direction. Inspecting the low-field image (only minor spatial orientation is expected), indeed one observes a distribution exhibiting two projected "polar" caps along the direction of the polarization of the photolysis laser; this feature is typical of a parallel-type photon transition, in a randomly oriented ensemble of molecules. As soon as a higher electric field is applied across the photolysis region, the state-selected $CD_3I$ molecules become (partially) oriented, with the iodine pointing in the upward direction. The ensuing asymmetry in the recoil distribution is clearly manifested in the observed image.

As outlined earlier on in Section 21.3, in their most general form, ion images are squashed 2D projections of the complete 3D Newton sphere distribution. Using inverse Abel transformation, one can reconstruct an (equatorial) cut through the cylindrically symmetric 3D distribution (see the right-side images in Figure 21.19). Then, rotating this reconstructed image around the vertical axis recoups the full Newton sphere, from which one can extract the fragment recoil intensity. The authors did this, as a function of the angle between the space-fixed direction of the DC orientation field and the recoil direction. The recoil distribution images of the photofragment, for the orientation of the parent molecule with a strong electric field, clearly mimic the initial orientation. From this, the authors concluded that the photodissociation process was very fast and axial, meaning that the dissociation process could be described as "prompt," on a timescale of <100 fs; this then means that no rotation had yet occurred, thus maintaining the very directional ejection of the $I^*(^2P_{1/2})$ fragment in the direction of the orientation axis of the $CD_3I$ molecule.

In a more recent experiment, Nakamura et al. (2017) investigated the photo-dissociation dynamics of the more complex asymmetric-top molecule 2-bromobutane:

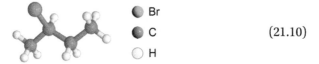

$$\begin{array}{ll} \bullet & Br \\ \bullet & C \\ \bigcirc & H \end{array} \qquad (21.10)$$

Note that the molecule is asymmetric with respect to the bromine substitution at the 2-position, and possesses two stereoisomers (R) and (S), with the latter shown here in the presentation in Equation 21.10. Note also that as a consequence of its asymmetry, 2-bromobutane is a chiral molecule (chirality is briefly discussed in Section 12.2); this aspect is not addressed here. In the authors' experiment, the molecules were also spatially oriented by means of a hexapole-field state selector, favoring the C–Br bond axis (analogous to the C–I bond axis, as in the $CD_3I$ case just discussed earlier). The molecular beam entered the photodissociation region along the ToF axis for the photofragment ions; it was crossed at a right angle by a linearly polarized laser beam, whose polarization was tilted at an angle of 45° with respect to the ion detector surface. This particular arrangement allowed for determining the mixing ratio of perpendicular and parallel transitions, when tuning the wavelength of the photolysis laser across a range of ro-vibrational resonances. The angular distributions for the two fine-structure components of the bromine atom photofragment, associated with different dissociation laser wavelengths, were acquired by velocity-map ion imaging.

Overall, the authors demonstrated that ion imaging of the photodissociation process of oriented molecules provides insight into molecular stereodynamics, even when congested rotational state structures and multiple transitions are involved in the photolysis process.

### 21.4.2 Imaging of the pair-correlated fragment channels in photodissociation

In Section 21.3, ion pair correlation was briefly discussed. To demonstrate the power of this approach, i.e., to unequivocally determine the full photo-fragmentation dynamics from mapping only one of the products, here the photodissociation of isocyanic acid (HNCO) is presented as a typical example:

$$HNCO(S_1) \xrightarrow{h\nu_{201\ nm}} NH(a^1\Delta) + CO(X^1\Sigma^+) \tag{21.11}$$

In their experiment, Zhang et al. (2014) utilized slice VMI to map the dynamics of the product fragment $NH(a^1\Delta)$; for the ionization step, a (2+1) REMPI scheme was applied. By tuning the photodissociation laser wavelength suitably ($\lambda \approx$ 201 nm), a range or rotational states ($J = 2, ..., 9$) could be probed for the two vibrational levels $NH(a^1\Delta|v = 0, 1)$. In Figure 21.20, some (raw data) ion images are shown for selected ro-vibrational levels of $NH(a^1\Delta)$. Note that in all measurements, the polarization vector of the dissociation laser was in the vertical direction in the images. The product angular distribution exhibits $\sin^2 \theta$-type dependence, which—together with the laser polarization direction—is indicative of a direct dissociation process for the N–C bond. In all ion images, well-resolved

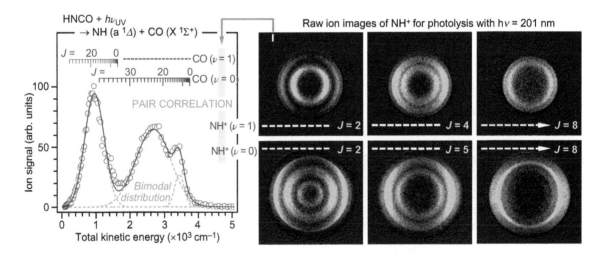

**Figure 21.20** Pair-correlated fragment channels in the photodissociation of HNCO. (Right) Raw ion images of NH $(a^1\Delta|v = 0, 1;J)$ products formed as a result of the photodissociation of HNCO, at wavelength $\lambda \approx 201$ nm. The ring features correspond to ro-vibrational states of the coproduct CO $(v')$. (Left) Pair correlation of the coproducts NH $(a^1\Delta|v = 0, 1;J)$ and $CO(X^1\Sigma^+|v';J')$; for details, see text. (Data adapted from Zhang, Z. et al., *J. Phys. Chem. A* 118, no. 13 (2014): 2413–2418.)

anisotropic rings were observed; these structures could be assigned to the vibrational states of the CO coproduct in the binary dissociation process channel $NH(a\Delta) + CO(X\Sigma^+)$.

Exploiting the energy and momentum conservation relation in Equations 21.9a and 21.9b, discussed in Section 21.3, these $NH^+$ ion images allowed the correlation to be determined between individual NH $(a^1\Delta|v;J)$ states and the CO coproduct ro-vibrational state distributions. The results, shown in the left-hand data panel of Figure 21.20, reveal two things. First, with increasing $J$, some inner rings disappear, and the strongest rings become sharper; this indicates a clear pair correlation between coproducts NH $(a^1\Delta|v = 0;J)$ and $CO(X\,{}^1\Sigma^+|v)$. Second, a clear bimodal rotational distribution is observed for $CO(X\,{}^1\Sigma^+|v = 0)$. The authors attributed this to two distinctive pathways for the HNCO dissociation in the excited $S_1$-state, related to the two stable isomers of the parent molecule (*trans*- and *cis*-HNCO).

### 21.4.3 Nonreactive scattering: Energy transfer in bimolecular collisions

A further representative example of the power of the velocity imaging technique in molecular-beam scattering is outlined in the following paragraphs, addressing the study of the rotational excitation of molecules following their (nonreactive) collision with inert gas atoms. Two specific examples will be discussed, namely, the collision systems Ar + NO and He + NO.

The first measurement of state-resolved DCS, combining crossed molecular-beam scattering and ion imaging, were carried out by Houston and coworkers for the system Ar + NO (see Bontuyan et al. 1993). The Newton diagram for the inelastic collision,

$$Ar(E_{kin}) + NO(X^2\Pi_{1/2}|v = 0; J'' = 0.5) \rightarrow Ar(E'_{kin}) + NO(X^2\Pi_{1/2}|v = 0; J'), \quad (21.12)$$

is illustrated on the left-hand side of Figure 21.21. The precollision laboratory and CM velocity vectors of the Ar and NO beams—$v_{Ar}$, $v_{NO}$ and $u_{Ar}$, $u_{NO}$, respectively—are indicated; so too are the CM velocity vector $v_{CM}$ and the postcollision CM velocity for NO, $u'_{NO}$. Scattered NO molecules are expected on the Newton sphere, indicated by the dotted-line circle; it represents the sphere on which NO products would be found if their speed were not changed by the collision. By convention, in the CM frame, the forward direction (scattering angle $\theta = 0°$) is indicated by the $u_{NO}$ vector; a NO molecule, scattered at an angle $\theta$, with its CM velocity $u'_{NO}$, is also shown.

The ion images (uncorrected raw data) nicely show that the scattering process in Equation 21.12 is strongly dependent on the final $J'$ value of the scattered NO molecule. As the final rotational level increases, the scattering distribution shifts from forward scattering through a more sideward direction to backward scattering. Interestingly, for $NO(v' = 0, J' = 18.5)$, two so-called rainbow peaks are observed. Such double-rainbows had been predicted for scattering of atoms from heteronuclear molecules, but they had not been directly observed in angular distributions prior to this imaging experiment for the Ar + NO scattering system.

The experiment just discussed relied on the assumption that using a rotationally cold molecular beam ($T_{rot}$ only a few kelvins), most of the rotational NO population was in the $J = 0.5$ rotational state. However, even if this were the only state

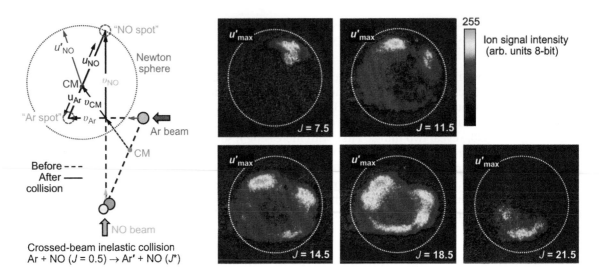

**Figure 21.21** Energy transfer in the bimolecular collision system Ar + NO. (Left) Newton diagram for the crossed-beam collision system. (Right) Ion images (raw data) for selected, final rotational levels $J'$ of the scattered NO molecule; the potential maximum product velocity $u'_{max}$ in the CM frame is indicated. (Data adapted from Bontuyan, L.S. et al., *J. Phys. Chem.* 97, no. 24 (1993): 6342–6350.)

populated, it would comprise the two near-degenerate $\Lambda$-doublet states, whose scattering properties cannot be resolved unless it were possible to prepare the molecular beam in a specific $J$- and $\Lambda$-state.

The first measurement of state-selective DCS for inelastic scattering from a *single quantum state* was carried out in collaboration between two research groups, for the He + NO system (Gijsbertsen et al. 2005). Using an optimized hexapole state selector, the authors prepared NO(X $^2\Pi_{1/2}|v = 0; J'' = 0.5$) in its single, upper $\Lambda$-doublet state ($\bar{\Omega} = 1/2$), with symmetry index $\varepsilon = -1$; the amount of molecules in other quantum states was substantially less than 1%. Scattering of this single-quantum state-selected molecule with He atoms was carried out in a crossed-beam arrangement, in very much the same fashion as the Ar + NO scattering experiment described earlier. The scattering scenario thus is

$$\mathrm{He}(E_{kin}) + \mathrm{NO}\left(\mathrm{X}^2\Pi_{1/2}|v = 0; J'' = 0.5; \bar{\Omega} = 1/2; \varepsilon = -1\right)$$
$$\rightarrow \mathrm{He}\left(E'_{kin}\right) + \mathrm{NO}\left(\mathrm{X}^2\Pi_{1/2}|v = 0; J'\bar{\Omega}; \varepsilon'\right)$$

(21.13)

Individual scattering product quantum states of NO were then detected using a resonant (1+1′) REMPI ionization scheme, in combination with velocity-mapped ion imaging. The use of the doubly resonant (1+1′) scheme allowed for the exact matching of transition between single quantum levels. Scattering ion images were recorded for all rotational transition branches, with all combinations of ($\bar{\Omega}' = 1/2, 3/2; \varepsilon' = -1, +1$). Selected ion images for state-to-state quantum scattering in the system He + NO are shown in Figure 21.22 for the rotational $R_{21}$-branch with ($\bar{\Omega}' = 1/2; \varepsilon' = -1$).

Looking at the images, in conjunction with the schematic Newton diagram (on the left of the figure), the overall scattering patterns observed in the images confirm expectation: for low-$J$ rotational quantum states, forward scattering dominates and then more sideways scattering is observed, while, finally, for high-$J$ rotational quantum states, the scattering is predominantly backward. Note that

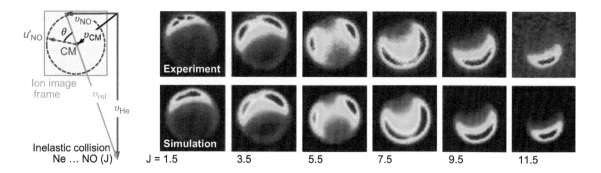

**Figure 21.22**  State-to-state scattering measurements in the single-quantum state-selected He + NO system, using (1+1′) REMPI ionization and VMI ion imaging. (Left) Newton diagram for the scattering system. (Right) Ion images (raw data) for selected *J*-states, for the rotational transition branch $R_{21}$ with ($\bar{\Omega}' = 1/2; \varepsilon' = -1$). The lower row of images constitutes simulations, which incorporate blurring as well as velocity and alignment corrections. (Data adapted from Gijsbertsen, H. et al., *J. Chem. Phys.* 123, no. 22 (2005): article 224305.)

the experimental results are fully replicated by theoretical model calculations (lower row of images in the figure).

Although not discussed here in further detail, it should be noted that the authors also found that scattering into $\bar{\Omega}' = 3/2$ states (spin-orbit changing collisions) exhibited more pronounced forward scattering than spin-orbit conserving collisions with $\bar{\Omega}' = 1/2$. This finding is relevant insofar as this outcome cannot be predicted simply from the energy splitting between the spin-orbit components. In general, at this level of detailed results and interpretation, only a quantum mechanical treatment will lead to complete understanding of the underlying collision dynamics.

## 21.4.4 Reactive scattering: Bimolecular reactions

The first study of a neutral bimolecular reaction using a crossed-beam arrangement and (2+1) REMPI detection of an atomic product was carried out for the exchange reaction $H + D_2 \rightarrow HD + D$ (Kitsopoulos et al., 1993). This study derived differential reaction cross sections, summed over all ro-vibrational states of the HD product, from the ion images of the product D atom.

Since this pioneering study, ion imaging detection of products attracted considerable attention, and with some new technical developments, unprecedented resolution was reached. An example of this is a study carried out by Toomes and Kitsopoulos (2003) for the reaction of Cl atoms with *n*-butane, using (2+1) REMPI ionization and VMI ion imaging for the reaction product HCl($v' = 0; J'$) in the scattering process

$$n - C_4H_{10} + Cl \rightarrow HCl\left(v'; J'\right) + C_4H_9 \qquad (21.14)$$

Two product ion images for HCl($v' = 0; J = 1$ and $J = 5$) are shown in Figure 21.23; the CM frame reagent velocity vector $\boldsymbol{u}_{Cl}$ and a product velocity vector $\boldsymbol{u}_{HCl}$, together with its scattering angle $\theta$, are indicated.

The two images clearly reveal a strong propensity for scattering in the forward direction for $J' = 1$ that is substantially reduced for the $J' = 5$ state.

The number of bimolecular reaction systems that has been investigated using ion imaging techniques, in one form or other, is simply too large to cover here in greater detail. Only a few key aspects will be noted that proved to be important in the evolution of the technique for applications in reaction dynamics.

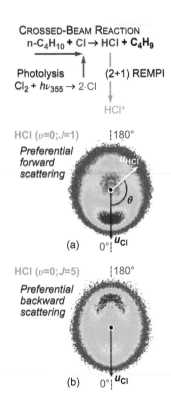

**Figure 21.23** Ion imaging of the molecular product HCl in the bimolecular reactive scattering of Cl atoms with $n$-butane. The product rotational states $HCl(v' = 0; J' = 1)$ (image a) and $HCl(v' = 0; J' = 5)$ (image b) exhibit strong forward- and backward-scattering behaviors, respectively. (Data adapted from Toomes, R.L. and T.N. Kitsopoulos, *Phys. Chem. Chem. Phys.* 5, no. 12 (2003): 2481–2483.)

One of the main restrictions is often imposed by the low sensitivity inherent to the two-photon transition nature of (2+1) REMPI schemes. An elegant way to circumvent this problem is to utilize one-photon photoionization at very short wavelengths. To this end, tunable synchrotron radiation, in conjunction with conventional crossed-beam configuration, was used to detect, e.g., hydrocarbon reaction radicals. Alternately, high-energy excimer lasers, such as, e.g., the fluorine laser operating at $\lambda = 157$ nm, have been incorporated in the studies of crossed-beam reactions, detecting the ionized reaction products via ion VMI.

An example of the latter is a study of the crossed molecular-beam reaction of Cl atoms with polyatomic hydrocarbons (Joalland et al. 2014); the authors implemented "soft" one-photon ionization of the radical product HCl at $\lambda_{VUV} = 157$ nm and then used DC slice imaging to derive full velocity–flux contour maps of the radical product ion, measuring both angular and energy distributions. The particular hydrocarbon reagents investigated by the authors were alkanes and alkenes (saturated and unsaturated hydrocarbon molecules, respectively). The authors found that in the reaction of Cl atoms with alkanes, only direct reactive encounters, i.e., hydrogen abstraction, persevered. However, in the Cl atom reaction with unsaturated hydrocarbons, the overall dynamics was significantly modified by the presence of the C=C double bond; in particular, at low collision energies, complex-mediated addition/elimination reactions seemed to dominate the dynamics. In view of these findings, the authors concluded that only a so-called *roaming* mechanism could provide the pathway to HCl elimination; the said type of reaction mechanism was first suggested by Townsend et al. (2004a).

The direct and roaming mechanisms by which the Cl atom binds with an H atom and escapes as HCl is conceptually illustrated in the center picture of Figure 21.24.

The roaming excursion of the Cl atom from its strongly bound adduct location is depicted on the right, for the particular case of

$$Cl + C_4H_8 \underset{\text{Cl roaming}}{\rightarrow} HCl + C_4H_7 \tag{21.15a}$$

On its roaming path, the Cl atom unlinks from the C=C double bond, and—in principle—is about to be eliminated (i.e., returning to reactants), but instead it wanders in the direction of the allylic site of the alkane molecule, where it picks up an H- atom from a methyl group, thus forming the parting product HCl. For comparison, the direct mechanism of HCl formation in the reaction

$$Cl + C_4H_{10} \underset{\text{H abstraction}}{\rightarrow} HCl + C_4H_9 \tag{21.15b}$$

is shown in the left half of the picture.

The roaming mechanism manifests itself in the shape of the product angular distribution maps. Examples of angular ion flux plots are included at the bottom of the picture; clearly, the predominance of forward scattering in the direct mechanism (left plot) is partially lost in the roaming mechanism (right plot).

## 21.4.5 Product-pair correlation in bimolecular reactions

Much earlier in this section, an example of ion imaging and product-pair correlation was given describing the results from a unimolecular photodissociation experiment. The technique allows one to deduce the complete scattering product distributions from evaluating full, measured ion images from only one product

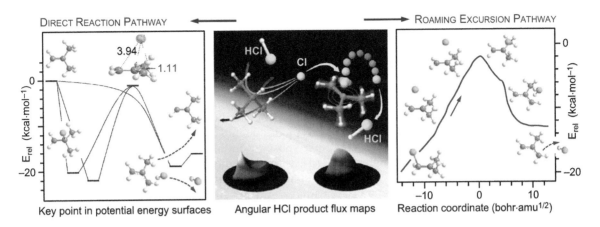

**Figure 21.24** Ion imaging of the formation of HCl in the reaction of Cl atoms with alkanes and alkenes. The conceptual reaction pathways of direct abstraction and roaming are illustrated in the center. The energetics of H abstraction in the $Cl + C_4H_{10}$ reaction are shown on the left, together with relevant steric presentations of the molecular complexes; the energetics of H capture by the roaming Cl atom is shown on the right, again with steric configurations based on theoretical modeling. (Data adapted from Joalland, B. et al., *Phys. Chem. A* 118, no. 40 (2014): 9281–9295.)

under the constraint of energy and momentum conservation. As was already pointed out in Section 21.3, not only can the technique be applied to the investigation of unimolecular collisions but it can also be expanded to the bimolecular collision case as well. As a representative example, here the reaction

$$Cl + CD_3H(v) \rightarrow HCl(v') + CD_3(v'') \tag{21.16}$$

is discussed with reference to the experiments carried out by Yan et al. (2007). The authors measured, by precise tuning of translational energies and exciting vibrational modes of the reagent molecule, the influence of vibration and translation on the relative efficiencies for scattering into specific product states and their differential scattering cross sections.

At this point it is worthwhile noting that while the influence of vibrational excitation on chemical reaction dynamics is well understood for triatomic reactions of the type $A + BC \rightarrow AB + C$; this is not the case for larger systems. For them, the presence of multiple vibrational modes makes the prediction of the general behavior rather difficult, despite the fact that a few studies seem to indicate possible, selective vibrational enhancement.

In their experiment, Yan and coworkers crossed two pulsed, seeded supersonic beams of Cl and $CD_3H$; the relative collision energy could be tuned by changing the intersection angle of the two beams. The vibrational ground-state product $CD_3(v' = 0)$ was probed (ionized) via a (2+1) REMPI scheme, using a pulsed ultraviolet (UV) laser source tuned to wavelengths near $\lambda = 333$ nm. The ions were measured using the technique of time-sliced velocity ion imaging, to map the recoil vector distribution of the reaction fragment $CD_3$. In addition to the aforementioned selection of translational energy, the authors also prepared the reagent molecule $CD_3H$ with vibrational mode excitation. Two specific modes were utilized, namely, (1) the C–H stretch mode, excited by using an IR laser, and (2) the bend mode of the molecule, excited by heating the pulsed beam valve.

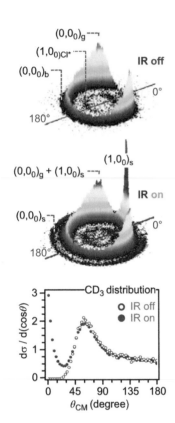

**Figure 21.25** Product-pair correlation ion mapping for the reaction $Cl + CD_3H(v) \rightarrow HCl(v') + CD_3(v'')$. (Top) 3D ion flux maps for the product $CD_3$ $(v'' = 0)$, without and with excitation of $CD_3H(v_s = 1)$ by IR photons; for the annotations, see text. (Bottom) DCS of the product $CD_3$ (scattering angle $\theta_{CM}$ in the CM frame). (Data adapted from Yan, S. et al., *Science* 316, no. 5832 (2007): 1723–1726.)

As an example of results from this study, 3D ion flux maps of the raw ion images of the probed $CD_3(v' = 0)$ products are shown in Figure 21.25, with and without IR excitation of the reagent $CD_3H(v_{CH \text{ stretch}})$, for a relative collision energy of $E_c = 8.9$ kcal·mol$^{-1}$.

Invoking energy and momentum conservation according to Equations 21.9a and 21.9b, the ringlike features in each image could be assigned to product pairs. These are labeled as follows: the numbers in the parentheses denote (from left to right) the vibrational mode quanta in the products $CD_3(v'')$ and $HCl(v')$, respectively. The outer subscript indicates the vibrational state of the reactant parent molecule $CD_3H$, with "g" indicating the ground vibrational state $v = 0$, while "s" and "b" stand for the C–H stretch excited state with $v_s = 1$ and for the bend excited state with $v_b = 1$, respectively.

In the bottom part of the figure, plots of the DCS $d\sigma/d(\cos\theta)$ for the product molecule $CD_3$ are shown as a function of the scattering angle $\theta_{cm}$ in the CM frame, both with and without vibrational excitation of the parent molecule $CD_3H$ by IR laser radiation.

Both data representation—ion flux maps and DCS plots—reveal that without IR laser excitation (i.e., $CD_3H$ is in its vibrational ground state), the reaction is dominated by sideways scattering. If C–H stretch excitation of the reagent is present, then a sharp forward-scattering peak appears in the inner ring (also very evident in the DCS plot); using product-pair correlation along the conditions in Equations 21.9a and 219b, the authors concluded that much of the vibrational excitation of the reagent is channeled into vibrational excitation of the coproduct, $HCl(v' = 1)$.

In addition to the evident, very strong features, the presence of further broad-scattering products in outer rings is observed. The authors also observed (1) that C–H stretch excitation is no more effective than an equivalent amount of translational energy in raising the overall reaction efficiency and (2) that $CD_3$ bend excitation is, on a small scale, more effective. These experiments demonstrated that indeed vibrational excitation does have a strong impact on product state and angular distributions.

## 21.4.6 Imaging the motion of electrons across semiconductor heterojunctions

It is well known that in solar cells, photons excite high-energy electrons in one material, and that subsequent fast charge transfer takes place to low-energy states in a neighboring material, thus providing electric power in a circuit. Indeed, this electron transfer through semiconductor heterojunctions constitutes the basic process not only in solar cells but also in all kinds of modern electronic devices. In the last decades, (1) the advances in solar cell technology promised green energy; (2) miniature transistors are powering computers; and (3) light-emitting diodes have become the linchpin for efficient, energy-saving light sources. These few examples highlight how revolutionary semiconductor technologies have transformed our society.

From the experimental point of view, the study of these phenomena has triggered the development of new techniques for observing electron dynamics, with high spatial and time resolution, to improve our understanding of these processes and, thus, open the way for better device design and performance. In general, the most common technique for this is *time-resolved photoemission electron microscopy* (TR-PEEM), which combines ultrafast optics with photoelectron excitation, for imaging electron densities (and their evolution) with high spatiotemporal resolution (see, *e.g.*, Fukumoto et al. 2014). TR-PEEM constitutes a perfect combination since optical techniques—although limited in spatial resolution—provide excellent time resolution, on the one hand, and on the other hand, electron microscopy—although limited in temporal resolution—provides excellent spatial resolution.

It was only in 2016 at the Okinawa Institute of Science and Technology that researchers combined femtosecond-laser pump–probe techniques with spectroscopic photoemission electron microscopy to image—for the first time—the motion of photoexcited electrons from high-energy to low-energy states in a type II 2D InSe/GaAs heterostructure, shortly after photoexcitation (see Man et al. 2016, 2017). The researchers were able to record "movies" of the fundamental operating phenomena in optoelectronic semiconductor devices. A conceptual view of the experimental setup is shown on the left of Figure 21.26.

To image the motion of photoexcited electrons in the InSe/GaAs sample in their experiment—as a function of space, time, and energy—the authors used an NIR femtosecond-laser source ($\lambda = 800$ nm/$h\nu = 1.55$ eV), which provided the "pump" pulse to phototransfer electrons into the conduction band of the InSe/GaAs sample. The photoexcited electrons were liberated from the sample into free space by a time-delayed "probe" UV laser pulse ($\lambda = 266$ nm). A photoemission electron microscope is then used that capture (image) the spatial and energy redistribution of the photoexcited electrons as a function of time after photoexcitation; a time-lag sequence (with picosecond resolution) of such images can even be combined into a movie of the evolution from the initial to the final electron density distribution in the sample. Essentially, the authors' measure-

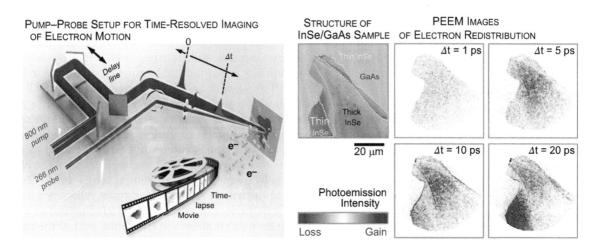

**Figure 21.26**  Imaging of the electron transfer across a 2D InSe/GaAs heterostructure. (Left) Schematic concept of the TR-PEEM setup. (Center) InSe/GaAs heterojunction sample with labels indicating different thicknesses of InSe and the GaAs substrate. (Right) Electron transport in the InSe/GaAs heterostructure as a function of time; for additional details, see text. (Adapted from Man, M.K.L. et al., *Nat. Nanotech.* 12, no. 1 (2016): 36–40; Man, M.K.L. et al., "Imaging electron motion in 2D semiconductor heterojunctions." In CLEO/QELS Fundamental Science 2017 Conference, CLEO_QELS-2017-FTh4F.2. Reproduced with permission.)

ments revealed a high nonequilibrium distribution of photoexcited carriers in space and energy right after the instant of the photoexcitation.

For illustration of the electron dynamics, a few snapshots from the image sequence are shown in the right-hand part of Figure 21.26, for different pump–probe laser pulse delays, revealing the charge redistribution across the heterojunction; note that the images were recorded in energy-integrated mode. Note also that an image taken at delay $\Delta t = 0.5$ ps, serving as the "background," was subtracted from all longer-delay images.

Visual inspection of the images already reveals that for few picosecond delays, an increased presence of electrons (denoted by red) in all parts of the InSe flake is noticeable, as well as electron depletion (blue) in GaAs. Once the delays become longer (longer than about 10 ps), the electrons continue to accumulate in thicker regions of InSe while simultaneously they begin to deplete in thin regions, similar to GaAs. Eventually, for delays of about 100 ps and longer (not shown here), the entire sample becomes depleted of electrons. As one of the concluding remarks in the publication of Man et al.'s (2017) exciting results, the authors state "understanding the formation of internal fields and how charge separation and transport happen in a microscopic scale will help the development of future semiconductor devices that operate at the ultrafast timescale."

## 21.5 BREAKTHROUGHS AND CUTTING EDGE

To follow the motion of individual atoms or molecules, with quantum-state selectivity, inelastic (nonreactive and reactive) scattering has long been a key feature in chemical reaction dynamics, but experimental realization had long been tedious and rather time consuming. This all changed with the invention of the technique of ion imaging, which allows for the simultaneous recording of the full DCS distribution in three dimensions.

### 21.5.1 Breakthrough: First ion imaging experiment

The first demonstration experiment was that of the photodissociation of methyl iodide ($CH_3I$), carried out by Chandler and Houston (1987). Photolysis photons at wavelength $\lambda = 266$ nm initially yields neutral fragments, according to $CH_3I + h\nu_{266\ nm} \rightarrow CH_3 + I(^2P_{1/2})/I^*(^2P_{1/2})$, with an I* fragmentation fraction of 0.8. The authors utilized a second laser (at wavelength $\lambda \approx 330$ nm) to interrogate the methyl radical in its vibrational ground state, exploiting a (2+1) REMPI scheme via the resonant intermediate two-photon transition $2p\,^2\tilde{A}_2''(v = 0) \rightarrow 3p\,^2\tilde{A}_2''(v = 0)$. The two laser beams in the experiment were counterpropagating, as shown schematically in the top of Figure 21.27. The $CH_3^+$ ions were then extracted from the photolysis region by an electric field; after passing through a field-free ToF region, they were detected, with spatial resolution and time gating, by a detector assembly comprising a MCP, a phosphor screen, and a 2D intensified Reticon array. A typical ion image (raw data) is shown in the center part of the figure; a flux intensity plot derived from the ion image is shown below it.

The ion image allows for immediate interpretation of the ionization process. First, the dissociation laser is aligned along the north–south axis in the figure, with its polarization vector on the image plane. Since the transition moment responsible for the photodissociation is known to be parallel to the C–I bond,

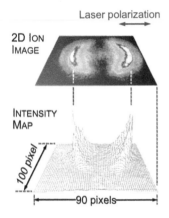

**Figure 21.27** First ion imaging experiment: the photolysis of $CH_3I$. (Top) Concept of the experiment, with counterpropagating photolysis (at 266 nm) and ionization laser (at ~330 nm) beams. (Center) Ion image of the spatial distribution of the $CH_3(v = 0)$ fragment (2D projection of the 3D Newton sphere. (Data adapted from Houston. P.L., *Acc. Chem. Res.* 28, no. 11 (1995): 453–460.) (Bottom) Product flux intensity map, derived from the ion image. (Data adapted from Chandler, D.W. and P.L. Houston, *J. Chem. Phys.* 87, no. 2 (1987): 1445–1447.)

therefore the observed intensity distribution in the 2D image of the $CH_3^+$ ion fragment is consistent with such a parallel transition, with its characteristic $\cos^2$ $\theta$-type distribution. Second, the anisotropic intensity distribution suggests that excited methyl iodide dissociates more rapidly than it rotates. Third, the $CH_3$ fragment energy is sharply defined by the photolysis energy balance; consequently, it should recoil from the methyl iodide CM with nearly singular speed. This is indeed reflected in the observed, relatively sharp image projection of the Newton sphere.

This illustrates the power and great advantages of ion imaging in the study of unimolecular dissociation processes. In particular, one finds that (1) if the experiment is designed in such a way that the symmetry axis of the velocity distribution is oriented parallel to the face of the imaging detector, one single image is all that is required to uniquely define the 3D angular distribution, and (2) the multiplexing nature of measuring all angles at once simplifies the experiment and significantly reduces the measurement time.

## 21.5.2 At the cutting edge: PAM—toward label-free superresolution imaging

For the majority of laser-based imaging techniques, at some stage the scientific and practical drive has been to beat the optical diffraction limit. For example, in laser-induced fluorescence microscopy, superresolution imaging down to spatial resolutions of about 20 nm has been demonstrated (using the method of interferometric photoactivation and localization microscopy; see Section 18.4). However, all "traditional" superresolution techniques rely on doping the sample with suitable label molecules. The first efforts now bear fruit to utilize PAM techniques, based on laser absorption spectroscopy, to implement label-free superresolution imaging.

Quite recently, Wang and coworkers at the Washington University of St. Louis, Missouri, United State, developed the novel PAM measurement modality PiPAM (see Yao et al. 2014). Their approach is akin to the superresolution laser fluorescence method of STED microscopy (see Section 18.3).

A first excitation (pump) laser pulse inhomogeneously bleaches the absorbers inside the excitation light spot, depending on the local excitation intensity. As a consequence, a second (probe) laser pulse encounters a greater reduction in

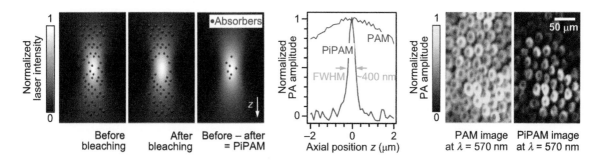

**Figure 21.28** Toward label-free subdiffraction PAM: the addition of photoimprint—PiPAM. (Left) Conceptual principle of PiPAM. (Center) Comparison of the depth resolutions in standard PAM and PiPAM. (Right) PAM and PiPAM images of red blood cells, recorded at an absorption wavelength of $\lambda_{abs}$ = 570 nm. (Data adapted from Yao, J. et al., *Phys. Rev. Lett.* 112, no. 1 (2014): article 014302.)

absorption in the center of the excitation spot than in the periphery. Thus, the difference between the photoacoustic responses for two consecutive laser pulses nonlinearly incorporates an absorption reduction distribution, which sharpens the center of the focus; this is true for both the lateral and axial directions. Exploiting PiPAM, the authors demonstrated superresolution at visible laser wavelengths on the order of <100 nm in the lateral direction and ~400 nm in the axial direction. An example of the latter case is shown in Figure 21.28. It is noteworthy that, in contrast to STED, the PiPAM images shown here are for a label-free sample and required only a singular pump–probe laser wavelength.

# Bibliography

**Note**: All online sources were checked, prior to going to print, for their integrity and accessibility, on January 27 2018.

## Chapter 1    Introduction

*Journal articles*

Dixit, G., O. Vendrell and R. Santra. "Imaging electronic quantum motion with light." *PNAS* 109, no. 29 (2012): 11636–11640. DOI: 10.1073/pnas.1202226109.

Gemma Frisius, R. "De radio astronomico et geométrico." Antwerp: apud. Greg. Brontium (1545).

Newton, I. "New theory about light and colors." *Phil. Trans. Royal Soc.* 80 (19 February 1671): 3075–3087.

Young, T. "On the theory of light and colours." *Phil. Trans. Royal. Soc. Lond.* 92 (1802): 12–48.

## Chapter 2    Interaction of light with matter

*Books and chapters in books*

Atkins, P.V. *Molecular Quantum Mechanics*, 5th edition. Oxford, UK: Oxford University Press (2010).

Bohren, C.F. and D.R. Huffman. *Absorption and Scattering of Light by Small Particles*, 2nd edition. Heidelberg, Germany: Wiley-VCH (2009).

Demtröder, W. *Atoms, Molecules and Photons: An Introduction to Atomic, Molecular and Quantum Physics*, 2nd edition. Berlin, Germany: Springer (2010).

Hollas, J.M. *Modern Spectroscopy*, 4th edition. Chichester, UK: John Wiley & Sons (2004).

Schiller, C. *The Adventure of Physics—Vol. 4: The Quantum of Change.* München, Germany: CreateSpace Independent Publishing Platform (2014). [Free eBook edition 27.06 available from http://www.motionmountain.net.]

van de Hulst, H.C. *Light Scattering by Small Particles*. New York, NY, USA: John Wiley & Sons (1957). [Reprint 2003, eBook edition 2012.]

*Journal articles*

Berke, H. "Chemistry in ancient times: The development of blue and purple pigments." *Angew. Chem. Int. Ed.* 41, no. 14 (2002): 2483–2487. DOI: 10.1002/1521-3773(20020715)41:14<2483 ::AID-ANIE2483>3.0.CO;2-U

Caruthers, J.W. "On Raleigh and Mie scattering." *Proc. Meet. Acoust. (POMA)* 14, no. 1 (2011): article 070001. DOI: 10.1121/1.3664646.

Cox, A.J., A.J. DeWeerd and J. Linden. "An experiment to measure Mie and Rayleigh total scattering cross sections." *Am. J. Phys.* 70, no. 6 (2002): 620–625. DOI: 10.1119/1.1466815.

Einstein, A. "Quantentheorie der Strahlung (On the quantum theory of radiation)." *Phys. Z.* 18 (1917): 121–128.

Gonzálvez, A.G. and A. González Ureña. "Transmission resonance Raman spectroscopy: Experimental results versus theoretical model calculations." *Appl. Spectrosc.* 66, no. 10 (2012): 1163–1170. DOI: 10.1366/12-06632.

Juster, N.J. "Color and chemical constitution." *J. Chem. Edu.* 39, no. 11 (1962): 596–601. DOI: 10.1021/ed039p596.

Kolwas, M. "Scattering of light on droplets and spherical objects: 100 years of Mie scattering." *Comp. Meth. Sci. Technol.* SI, no. 2 (2010): 107–113. DOI: 10.12921/cmst.2010.SI.02.107-113.

Levoni, C., M. Cervino, R. Guzzi and F. Torricella. "Atmospheric aerosol optical properties: A database of radiative characteristics for different components and glasses." *Appl. Opt.* 36, no. 30 (1997): 8031–8041. DOI: 10.1364/AO.36.008031.

Mie, G. "Beiträge zur Optik trüber Medien speziell kolloidaler Goldlösungen (Contributions to the optics of diffuse media, especially colloid metal solutions)." *Ann. Phys.* 25, no. 3 (1908): 377–445.

Pike, A.W.G., D.L. Hoffmann, M. García-Diez, P.B. Pettitt, J. Alcolea, R. De Balbín, C. González-Sainz, C. de las Heras, J.A. Lasheras, R. Montes and J. Zilhão. "U-series dating of Paleolithic art in 11 caves in Spain." *Science* 336, no. 6087 (2012): 1409–1413. DOI: 10.1126/science.1219957.

Rezus, Y.L.A., S.G. Walt, R. Lettow, A. Renn, G. Zumofen, S. Götzinger and V. Sandoghdar. "Single-photon spectroscopy of a single molecule." *Phys. Rev. Lett.* 108, no. 9 (2012): article 093601. DOI: 10.1103/PhysRevLett.108.093601.

Strutt, J.W. (Lord Rayleigh). "On the light from the sky, its polarization and colour." *Philos. Mag.* 41, no. 271 (1871): 107–120 and 274–279. DOI: 10.1080/14786447108640452 and 10.1080/14786447108640479.

Weisskopf, V.F. "How light interacts with matter." *Sci. Am.* 219, no. 3 (1968): 60–71. DOI: 10.1038/scientificamerican0968-60.

Yodh, A. and B. Chance. "Spectroscopy and imaging with diffusing light." *Phys. Today* 48, no. 3 (1995): 34–40. DOI: 10.1063/1.881445.

*Other sources*

Hoppe, E.O. "Einstein playing the violin" (1921). **Access**: https://i.pinimg.com/originals/4a/3d/a6/4a3da6912f8460cde548c7a5ac61f3ee.jpg.

Rammssos. "Reproduction of a bison of the cave of Altamira" (2008). **Access:** http://commons.wikimedia.org/wiki/File:AltamiraBison.jpg.

## Chapter 3  The basics of lasers

*Books and chapters in books*

Diels, J.C. and W. Rudolf. *Ultrashort Laser Pulse Phenomena*, 2nd edition. Burlington, MA, USA: Academic Press (2006).

Eichhorn, M. *Laser Physics: From Principles to Practical Work in the Lab*. Heidelberg, Germany: Springer (2014).

Hodgson, N. and H. Weber. *Laser Resonators and Beam Propagation: Fundamentals, Advanced Concepts and Applications*. Springer Series in Optical Sciences, vol. 108. Heidelberg, Germany: Springer (2005).

Renk, K.F. *Basics of Laser Physics*. Heidelberg, Germany: Springer (2012).

Rulliere, C. (ed.). *Femtosecond Laser Pulses: Principles and Experiments*, 2nd edition. Heidelberg, Germany: Springer (2007).

Siegman, A.E. *Lasers*. Herndon, VA, USA: University Science Books (1990).

Silfvast, W.T. *Laser Fundamentals*, 2nd edition. Cambridge, UK: Cambridge University Press (2008).

Smith, F.G., T.A. King and D. Wilkins. *Optics and Photonics: An Introduction*, 2nd edition. Chichester, UK: Wiley-Blackwell (2007).

Thyagarajan, K. and A. Ghatak. *Lasers: Fundamentals and Applications*, 2nd edition. Heidelberg, Germany: Springer (2011).

*Journal articles*

Barnes, N.P. and B.M. Walsh. "Relaxation oscillation suppression in Nd:YAG lasers using intra-resonator harmonic generation." In *Conference on Lasers and Electro-Optics/Quantum Electronics and Laser Science Conference*, Technical Digest (Optical Society of America, 2003): paper CFM5.

Cerjan, A., Y.D. Chong and A.D. Stone. "Steady-state *ab initio* laser theory for complex gain media." *Opt. Express* 23, no. 5 (2015): 6455–6477. DOI: 10.1364/OE.23.006455.

Daraei, A., S.M. Izadyar and N. Chenarani. "Simulation and analysis of carrier dynamics in the InAs/GaAs quantum dot laser, based upon rate equations." *Opt. Photon. J.* 3 (2013): 112–116. DOI: 10.4236/opj.2013.31018.

Demirkhanyan, G.G. "Intensities of inter-Stark transitions in YAG-Yb$^{3+}$ crystals." *Laser Phys.* 16, no. 7 (2006): 1054–1057. DOI: 10.1134/S1054660X0607005X.

Di Domenico, G., S. Schilt and P. Thomann. "Simple approach to the relation between laser frequency noise and laser line shape." *Appl. Opt.* 49, no. 25 (2010): 4801–4807. DOI: 10.1364 /AO.49.004801.

Feng, S. and H.G. Winful. "Physical origin of the Gouy phase shift." *Opt. Lett.* 26, no. 8 (2001): 485–487. DOI: 10.1364/OL.26.000485.

Fork, R.L., O.E. Martinez and J.P. Gordon. "Negative dispersion using pairs of prisms." *Opt. Lett.* 9, no. 5 (1984): 150–152. DOI: 10.1364/OL.9.000150.

Fox, A.G. and T. Li. "Resonant modes in a Maser interferometer." *Bell Sys. Tech. J.* 40, no. 2 (1961): 453–488. DOI: 10.1002/j.1538-7305.1961.tb01625.x.

Haken, H. "A nonlinear theory of laser noise and coherence. I & II." *Z. Physik* 181, no. 1 (1964): 96–124 and *Z. Physik* 182, no. 4 (1965): 346–359. DOI: 10.1007/BF01383921 and 10.1007 /BF01383115.

Hirata, S., T. Akatsuka, Y. Ohtake and A. Morinaga. "Sub-hertz-linewidth diode laser stabilized to an ultralow-drift high-finesse optical cavity." *Appl. Phys. Express* 7, no. 2 (2014): article 022705. DOI: 10.7567/APEX.7.022705.

Keller, U., D.A.B. Miller, G.D. Boyd, T.H. Chiu, J.F. Ferguson and M.T. Asom. "Solid-state low-loss intra-cavity saturable absorber for Nd:YLF lasers: An anti-resonant semiconductor Fabry-Perot saturable absorber." *Opt. Lett.* 17, no. 7 (1992): 505–507. DOI: 10.1364/OL.17 .000505.

Kuizenga, D.J. and A.E. Siegman. "FM and AM mode locking of the homogeneous laser—Part I: Theory & Part II: Experimental results in a Nd:YAG laser with internal FM modulation." *IEEE J. Quantum Electron.* 6, no. 11 (1970): 694–708 and 709-715. DOI: 10.1109/JQE.1970.1076343 and 10.1109/JQE.1970.1076344.

Lamb, W.E. Jr. "Theory of an optical Maser." *Phys. Rev.* 134, no. 6A (1964): A1429–A1450. DOI: 10.1103/PhysRev.134.A1429.

Oron, R., S. Blit, N. Davidson, A.A. Friesem, Z. Bomzon and E. Hasman. "The formation of laser beams with pure azimuthal or radial polarization." *Appl. Phys. Lett.* 77, no. 21 (2000): 3322–3324. DOI: 10.1063/1.1327271.

Pan, L., I. Utkin and R. Fedosejevs. "Two-wavelength ytterbium-doped fiber laser with sustained relaxation oscillation." *Appl. Opt.* 48, no. 29 (2009): 5484–5489. DOI: 10.1364/AO.48 .005484.

Saito, Y., M. Kobayashi, D. Hiraga, K. Fujita, S. Kawano, N.I. Smith, Y. Inouye and S. Kawata. "z-Polarization sensitive detection in micro-Raman spectroscopy by radially polarized incident light." *J. Raman Spectrosc.* 39, no. 11 (2008): 1643–1648. DOI: 10.1002/jrs.1953.

Saliba, S.D. and R.E. Scholten. "Linewidths below 100 kHz with external cavity diode lasers." *Appl. Opt.* 48, no. 36 (2009): 6961–6966. DOI: 10.1364/AO.48.006961.

Schawlow, A.L. and C.H. Townes. "Infrared and optical masers." *Phys. Rev.* 112, no. 6 (1958): 1940–1949. DOI: 10.1103/PhysRev.112.1940.

Spence, D.E., P.N. Kean and W. Sibbett. "60-fsec pulse generation from a self-mode-locked Ti: sapphire laser." *Opt. Lett.* 16, no. 1 (1991): 42–44. DOI: 10.1364/OL.16.000042.

Thirugnanasambandam, M.P., Y. Senatsky and K. Ueda. "Generation of radially and azimuthally polarized beams in Yb:YAG laser with intra-cavity lens and birefringent crystal." *Opt. Express* 19, no. 3 (2011): 1905–1914. DOI: 10.1364/OE.19.001905.

Türeci, H.E., A.D. Stone and B. Collier. "Self-consistent multimode lasing theory for complex or random lasing media." *Phys. Rev. A* 74, no. 4 (2006): article 043822. DOI: 10.1103/PhysRevA.74 .043822.

Wang, X.H., M. Xu, H.W. Ren and Q.H. Wang. "A polarization converter array using a twisted-azimuthal liquid crystal in cylindrical polymer cavities." *Opt. Express* 21, no. 13 (2013): 16222–16230. DOI: 10.1364/OE.21.016222.

Zhan, Q. "Trapping metallic Rayleigh particles with radial polarization." *Opt. Express* 12, no. 15 (2004): 3377–3382. DOI: 10.1364/OPEX.12.003377.

*Other sources*    Brosson, P. "Transient response of a semiconductor laser." Wolfram Demonstrations Project (2005). **Access**: http://demonstrations.wolfram.com/TransientResponseOfASemiconductor Laser/.

Newport Technology and Applications Center. "Prism compressor for ultra-short laser pulses." Application Note 29, Irvine, CA, USA: Newport Corp. (2006). **Access**: http://www.newport.com /images/webDocuments-EN/images/12243.pdf.

"*VirtualLab*™" with Laser resonator toolbox: Flexible Eigenmode analysis of laser resonators." Jena, Germany: LightTrans GmbH (2015). [Latest version of *VirtualLab*™ 5.11, http://www .lighttrans.com, released 17/12/2014.]

## Chapter 4    Lasers sources based on gaseous, liquid, or solid-state active media

*Books and chapters in books*

Alfano, R.R. *The Super-continuum Laser Source: Fundamentals with Updated References*, 2nd edition. Heidelberg, Germany: Springer (2006).

Binh, L.N. and N.Q. Ngo. *Ultra-Fast Fiber Lasers: Principles and Applications with MATLAB®  Models*. Boca Raton, FL, USA: CRC Press (2010).

Demtröder, W. *Laser Spectroscopy: Vol. 1: Basic Principles and Laser Spectroscopy: Vol. 2: Experimental Techniques*, 4th edition. Heidelberg, Germany: Springer (2008).

Digonnet, M.J.F. (ed.). *Rare-Earth-Doped Fiber Lasers and Amplifiers*, 2nd edition. New York, NY, USA: Marcel Dekker (2001).

Duarte, F.J. and L.W. Hillman (eds.). *Dye Laser Principles: With Applications*. San Diego, CA, USA: Academic Press (1990).

Dudley, J.M. and J.R. Taylor (eds.). *Super-continuum Generation in Optical Fibers*. Cambridge, UK: Cambridge University Press (2010).

Okhotnikov, O.G. (ed.). *Fiber Lasers*. Weinheim, Germany: Wiley-VCH (2012).

Paschotta, R. *Field Guide to Optical Fiber Technology*. SPIE Press Field Guides, vol. FG16. Bellingham, WA, USA: SPIE (2010).

Raineri, F., A. Bazin and R. Raj. "Optically pumped semiconductor photonic crystal lasers." Chapter 2 in R.M. De La Rue, S. Yu and J.M. Lourtioz (eds.). *Compact Semiconductor Lasers*. Weinheim, Germany: Wiley-VCH (2014). DOI: 10.1002/9783527655342.ch2.

Schäfer, F.P. (ed.). *Dye Lasers*, 3rd enlarged and revised edition. Heidelberg, Germany: Springer (1990).

Silfvast, W.T. *Laser Fundamentals*, 2nd edition. Cambridge, UK: Cambridge University Press (2008).

Ter-Mikirtychev, V. *Fundamentals of Fiber Lasers and Fiber Amplifiers*. Springer Series in Optical Sciences, vol. 181. Heidelberg, Germany: Springer (2014).

Zagumennyi, A.I., V.A. Mikhailov and I.A. Shcherbakov. "Rare earth ion lasers—$Nd^{3+}$." In C. Webb and J. Jones (eds.). *Handbook of Laser Technology and Applications, Vol. 2: Laser Design and Laser Applications*. Bristol, UK: IOP Publishing (2004): pp. 353–382.

*Journal articles*

Asher, S.A., R.W. Bormett, X.G. Chen, D.H. Lemmon, N. Cho, P. Peterson, M. Arrigoni, L. Spinelli and J. Cannon. "UV resonance Raman spectroscopy using a new cw laser source: Convenience and experimental simplicity." *Appl. Spectrosc.* 47, no. 5 (1993): 628–633. DOI: 10 .1366/0003702934067225.

Bethune, D.S. "Dye cell design for high-power low-divergence excimer-pumped dye lasers." *Appl. Opt.* 20, no. 11 (1981): 1897–1899. DOI: 10.1364/AO.20.001897.

Bridges, W.B. "Laser oscillation in singly ionized argon in the visible spectrum." *Appl. Phys. Lett.* 4, no. 7 (1964): 128–130. DOI: 10.1063/1.1753995.

Chen, K.K., S. Alam, J.H.V. Price, J.R. Hayes, D. Lin, A. Malinowski, C. Codemard, D. Ghosh, M. Pal, S.K. Bhadra and D.J. Richardson. "Picosecond fiber MOPA pumped supercontinuum source with 39 W output power." *Opt Express* 18, no. 6 (2010): 5426–5432. DOI: 10.1364/OE.18.005426.

Chernikov, S.V., Y. Zhu, J.R. Taylor and V. Gapontsev. "Super-continuum self-Q-switched ytterbium fiber laser." *Opt. Lett.* 22, no. 5 (1997): 298–300. DOI: 10.1364/OL.22.000298.

Corless, J.D., J.A. West, J. Bromage and C.R. Stroud Jr. "Pulsed single-mode dye laser for coherent control experiments." *Rev. Sci. Instrum.* 68, no. 6 (1997): 2259–2265. DOI: 10.1063/1 .1148056.

Cumberland, B.A., J.C. Travers, S.V. Popov and J.R. Taylor. "Toward visible cw-pumped super-continua." *Opt. Lett.* 33, no. 18 (2008): 2122–2124. DOI: 10.1364/OL.33.00212.

Dekker, S.A., A.C. Judge, R. Pant, I. Gris-Sánchez, J.C. Knight, C.M. de Sterke and B.J. Eggleton. "Highly-efficient, octave spanning soliton self-frequency shift using a specialized photonic crystal fiber with low OH loss." *Opt. Express* 19, no. 18 (2011): 17766–17773. DOI: 10.1364/OE .19.017766.

Dudley, J., G. Genty and S. Coen. "Super-continuum generation in photonic crystal fiber." *Rev. Mod. Phys.* 78, no. 4 (2006): 1135–1184. DOI: 10.1103/RevModPhys.78.1135.

Duling, I.N. III. "All-fiber ring soliton laser mode locked with a nonlinear mirror." *Opt. Lett.* 16, no. 8 (1991): 539–541. DOI: 10.1364/OL.16.000539.

Durfee, C.G., T. Storz, J. Garlick, S. Hill, J.A. Squier, M. Kirchner, G. Taft, K. Shea, H. Kapteyn, M. Murnane and S. Backus. "Direct diode-pumped Kerr-lens mode-locked Ti:sapphire laser." *Opt. Express* 20, no. 13 (2012): 13677–13683. DOI: 10.1364/OE.20.013677.

Ell, R., U. Morgner, F.X. Kärtner, J.G. Fujimoto, E.P. Ippen, V. Scheuer, G. Angelow, T. Tschudi, M.J. Lederer, A. Boiko and B. Luther-Davies. "Generation of 5-fs pulses and octave-spanning spectra directly from a Ti:sapphire laser." *Opt. Lett.* 26, no. 6 (2001): 373–375. DOI: 10.1364/OL.26.000373.

Georgiev, D., V.P. Gapontsev, A.G. Dronov, M.Y. Vyatkin, A.B. Rulkov, S.V. Popov and J.R. Taylor. "Watts-level frequency doubling of a narrow line linearly polarized Raman fiber laser to 589nm." *Opt. Express* 13, no. 18 (2005): 6772–6776. DOI: 10.1364/OPEX.13.006772.

Gerginov, V., C.E. Tanner, S.A. Diddams, A. Bartels and L. Hollberg. "High-resolution spectroscopy with a femtosecond laser frequency comb." *Opt. Lett.* 30, no. 13 (2005): 1734–1736. DOI: 10.1364/OL.30.001734.

Geusic, J.E., H.M. Marcos and L.G. van Uitert. "Laser oscillations in Nd-doped yttrium aluminum, yttrium gallium and gadolinium garnets." *Appl. Phys. Lett.* 4, no. 10 (1964): 182–184. DOI: 10.1063/1.1753928.

Hercher, M. "Tunable single-mode operation of gas lasers using intra-cavity tilted etalons." *Appl. Opt.* 8, no. 6 (1969): 1103–1106. DOI: 10.1364/AO.8.001103.

Hill, K.O., B.S. Kawasaki and D.C. Johnson. "Low-threshold cw Raman laser." *Appl. Phys. Lett.* 29, no. 3 (1976): 181–183. DOI: 10.1063/1.89016.

Huang, S., Y. Feng, J. Dong, A. Shirakawa, M. Musha and K. Ueda. "1083 nm single frequency ytterbium doped fiber laser." *Laser Phys. Lett.* 2, no. 10 (2005): 498–501. DOI: 10.1002/lapl.200510032.

Javan, A., W.R. Bennett and D.R. Herriott. "Population inversion and continuous optical maser oscillation in a gas discharge containing a He–Ne mixture." *Phys. Rev. Lett.* 6, no. 3 (1961): 106–110. DOI: 10.1103/PhysRevLett.6.106.

Jeong, Y., J.K. Sahu, D.N. Payne and J. Nilsson. "Ytterbium-doped large-core fiber laser with 1.36 kW continuous-wave output power." *Opt. Express* 12, no. 25 (2004): 6088–6092. DOI: 10.1364/OPEX.12.006088.

Kitahara, T. "Device for simultaneous multiple-line mode-locking of an argon-ion laser." *Opt. Lett.* 12, no. 11 (1987): 873–875. DOI: 10.1364/OL.12.000873.

Knight, J.C. "Photonic crystal fibers and fiber lasers." *J. Opt. Soc. Am. B* 24, no. 8 (2007): 1661–1668. DOI: 10.1364/JOSAB.24.001661.

Krause, M., S. Cierullies, H. Renner and E. Brinkmeyer. "Design of widely tunable Raman fibre lasers supported by switchable FBG resonators." *Electron. Lett.* 39, no. 25 (2003): 1795–1797. DOI: 10.1049/el:20031175.

Ling, W.J., S.G. Zhang, M.X. Zhang, Z. Dong, K. Li, Y.Y. Zuo, X.H. Guo and Y.L. Jia. "High power continuous-wave actively mode-locked diode-pumped Nd:YAG laser." *Chin. Phys. Lett* 27, no. 11 (2010): article 114202. DOI: 10.1088/0256-307X/27/11/114202.

Ma, J., M. Du and Q. Chen. "A new absorption band of Ti-doped $Al_2O_3$ crystals after thermal annealing and $\gamma$-irradiation." *Z. Naturforsch.* 61A, no. 10–11 (2006): 564–568.

Maiman, T.H. "Stimulated optical radiation in ruby." *Nature* 187, no. 4736 (1960): 493–494. DOI: 10.1038/187493a0.

Morgner, U., F.X. Kärtner, S.H. Cho, Y. Chen, H.A. Haus, J.G. Fujimoto, E.P. Ippen, V. Scheuer, G. Angelow and T. Tschudi. "Sub-two cycle pulses from a Kerr-lens mode-locked Ti:sapphire laser," *Opt. Lett.* 24, no. 6 (1999): 411–413. DOI: 10.1364/OL.24.00041.

Moulton, P.F. "Advances in tunable transition metal lasers." *Appl. Phys. B* 28, no. 2–3 (1982): 233. DOI: 10.1007/BF00697850. See also: Moulton, P. "Ti-doped sapphire: tunable solid-state laser." *Opt. News* 8, no. 6 (1982): 9. DOI: 10.1364/ON.8.6.000009.

Po, H., E. Snitzer, L. Tumminelli, F. Hakimi, N.M. Chu and T. Haw. "Doubly-clad high-brightness Nd-fiber laser-pumped by GaAlAs phased array." In *Optical Fiber Communication Conference*, Vol. 5 of the 1989 OSA Technical Digest Series (Optical Society of America, 1989), paper PD7.

Ranka, J.K., R.S. Windeler and A.J. Stentz. "Optical properties of high-delta air–silica microstructure optical fibers." *Opt. Lett.* 25, no. 11 (2000): 796–798. DOI: 10.1364/OL.25.000796.

Roth, P.W., A.J. Maclean, D. Burns and A.J. Kemp. "Directly diode-laser-pumped Ti:sapphire laser." *Opt. Lett.* 34, no. 21 (2009): 3334–3336. DOI: 10.1364/OL.34.003334.

Royon, R., J. L'hermite, L. Sarger and E. Cormier. "High power, continuous-wave ytterbium-doped fiber laser tunable from 976 to 1120 nm." *Opt. Express* 15, no. 20 (2007): 12882–12889. DOI: 10.1364/OE.21.013818.

Russell, J.A., D.P. Pacheco and H.R. Aldag. "Efficient tunable near-infrared solid-state dye laser with good beam quality" *Proc. SPIE* 5707 (2005): 217–226. DOI: 10.1117/12.601019.

Samanta, G.K., S. Chaitanya Kumar, K. Devi and M. Ebrahim-Zadeh. "High-power, continuous-wave Ti:sapphire laser pumped by fiber-laser green source at 532 nm." *Opt. Las. Eng.* 50, no. 2 (2012): 215–219. DOI: 10.1016/j.optlaseng.2011.09.001.

Schäfer, F.P., W. Schmidt and J. Volze. "Organic dye solution laser." *Appl. Phys. Lett.* 9, no. 8 (1966): 306–309. DOI: 10.1063/1.1754762.

Schliesser, A., N. Picqué and T.W. Hänsch. "Mid-infrared frequency combs." *Nature Photon.* 6, no. 7 (2012): 440–444. DOI: 10.1038/nphoton.2012.142.

Snitzer, E. "Optical MASER action of $Nd^{3+}$ in a barium Crown Glass." *Phys. Rev. Lett.* 7, no. 12 (1961a): 444–446. DOI: 10.1103/PhysRevLett.7.444.

Snitzer, E. "Proposed fiber cavities for optical masers." *J. Appl. Phys.* 32, no. 1 (1961b): 36–39. DOI: 10.1063/1.1735955.

Sorokin, P.P. and J.R. Lankard. "Stimulated emission observed from an organic dye, chloro-aluminum phtalocyanine." *IBM J. Res. Dev.* 10, no. 2 (1966): 162–163. DOI: 10.1147/rd.102 .0162.

Spence, D.E., P.N. Kean and W. Sibbett. "Sub-100fs pulse generation from a self-mode-locked titanium:sapphire laser." Paper CPDP10 – CLEO'90; OSA Technical Digest Series (1990): 619–620. See also: Spence, D.E., P.N. Kean and W. Sibbett. "60-fsec pulse generation from a self-mode-locked Ti:sapphire laser." *Opt. Lett.* 16, no. 1 (1991): 42–44. DOI: 10.1364/OL.16.000042.

Spühler, G.J., R. Paschotta, U. Keller, M. Moser, M.J.P. Dymott, D. Kopf, J. Meyer, K.J. Weingarten, J.D. Kmetec, J. Alexander and G. Truong. "Diode-pumped passively mode-locked Nd:YAG laser with 10-W average power in a diffraction-limited beam." *Opt. Lett.* 24, no. 8 (1999): 528–530. DOI: 10.1364/OL.24.000528.

Stone, J.M. and J.C. Knight. "Visibly 'white' light generation in uniform photonic crystal fiber using a microchip laser." *Opt. Express* 16, no. 4 (2008): 2670–2675. DOI: 10.1364/OE.16.002670.

Supradeepa, V.R., J.W. Nicholson, C.E. Headley, M.F. Yan, B. Palsdottir and D. Jakobsen. "A high efficiency architecture for cascaded Raman fiber lasers." *Opt. Express* 21, no. 6 (2013): 7148–7155. DOI: 10.1364/OE.21.007148.

Wadsworth, W.C., J.C. Knight, W.H. Reeves, P.S. Russell and J. Arriaga. "$Yb^{3+}$-doped photonic crystal fibre laser." *Electron. Lett.* 36, no. 17 (2000): 1452–1454. DOI: 10.1049/el:20000942.

Yang, B., X. Liu, X. Wang, J. Zhang, L. Hu and L. Zhang. "Compositional dependence of room-temperature Stark splitting of $Yb^{3+}$ in several popular glass systems." *Opt. Lett.* 39, no. 7 (2014): 1772–1774. DOI: 10.1364/OL.39.001772.

Zhu, X. and N. Peyghambarian. "High-power ZBLAN glass fiber lasers: Review and prospect." *Adv. OptoElectron.* 2010 (2010): article 501956. DOI: 10.1155/2010/501956.

*Other sources*

Brackmann, U. *Lambdachrome® : Laser Dyes*, 3rd edition. Göttingen, Germany: Lambda Physik AG (2000).

Nelson, A. Video: "Maiman's first laser light shines again." *SPIE Newsroom*, May 20, 2010. DOI: 10.1117/2.3201005.04. **Access:** http://spie.org/x40717.xml.

SnakeCreek lasers. "MicroGreen™ APC Series." Friendsville, PA, USA: SnakeCreek Lasers (2013). **Access:** http://www.snakecreeklasers.com/products/laserheads.aspx.

**Chapter 5**  **Laser sources based on semiconductor media and nonlinear optic phenomena**

*Books and chapters in books*

Cerulla, G. and C. Manzoni. "Solid-state ultrafast optical parametric amplifiers." Chapter 11 in A. Sennaroglu (ed.). *Solid-State Lasers and Applications*. Boca Raton, FL, USA: CRC Press (2007): pp. 437–472.

Demtröder, W. *Laser Spectroscopy: Vol. 1: Basic Principles and Laser Spectroscopy: Vol. 2: Experimental Techniques*, 4th edition. Heidelberg, Germany: Springer (2008).

Erneux, T. and P. Glorieux. "Optical parametric oscillator." Chapter 12 in *Laser Dynamics*. Cambridge, UK: Cambridge University Press (2010): pp. 294–335.

Faist, J. *Quantum Cascade Lasers*. Oxford, UK: Oxford University Press (2013).

Guha, S. and L.P. Gonzalez. *Laser Beam Propagation in Non-Linear Optical Media*. Boca Raton, FL, USA: CRC Press (2014).

Numai, T. *Fundamentals of Semiconductor Lasers*, 2nd edition. Springer Series in Optical Sciences, vol. 93. Heidelberg, Germany: Springer (2014).

Ross, I.N. "Optical parametric amplification techniques." In T. Brabec (ed.), *Strong Field Laser Physics*. Springer Series in Optical Sciences, vol. 134. Heidelberg, Germany: Springer (2009): pp. 35–59.

Silfvast, W.T. *Laser Fundamentals*, 2nd edition. Cambridge, UK: Cambridge University Press (2008).

Suhara, T. *Semiconductor Laser Fundamentals*. New York, NY, USA: Marcel Dekker (2004).

Thyagarajan, K. and A. Ghatak. "Optical parametric oscillators." Chapter 14 in *Lasers— Fundamentals and Applications*, 2nd edition. Heidelberg, Germany: Springer (2011): pp. 363–385.

Yao, J. and Y. Wang. *Nonlinear Optics and Solid-State Lasers: Advanced Concepts, Tuning-Fundamentals and Applications*. Springer Series in Optical Sciences, vol. 164. Heidelberg, Germany: Springer (2012).

Zhang, J.Y., J.Y. Huang and Y.R. Shen. *Optical Parametric Generation and Amplification*. Boca Raton, FL, USA: CRC Press (1995).

*Journal articles*   Alferov, Z.I., V.M. Andreev, E.L. Portnoi and M.K. Trukan. "AlAs-GaAs heterojunction injection lasers with a low room-temperature threshold." *Fiz. Tekh. Poluprovodn.* 3, no. 9 (1969): 1328–1332 [in English: *Sov. Phys. Semicond.* **3**, no. 9 (1970): 1107–1110].

Alferov, Z.I., V.M. Andreev, D.Z. Garbazov, Y.V. Zhilyaev, E.P. Morozov, E.L. Portnoi and V.G. Trofim. "Investigation of the influence of the AlAs-GaAs heterostructure parameters on the laser threshold current and the realization of continuous emission at room temperature." *Fiz. Tekh. Poluprovodn.* 4, no. 9 (1970): 1826–1829 [in English: *Sov. Phys. Semicond.* **4**, no. 9 (1971): 1573–1575].

Arslanov, D.D., M. Spunei, J. Mandon, S.M. Cristescu, S.T. Persijn and F.J.M. Harren. "Continuous-wave optical parametric oscillator based infrared spectroscopy for sensitive molecular gas sensing." *Laser Photon. Rev.* 7, no. 2 (2013): 188–206. DOI: 10.1002/lpor.201100036

Bass, M., P.A. Franken, A.E. Hill, C.W. Peters and G. Weinreich. "Optical mixing." *Phys. Rev. Lett.* 8, no. 1 (1962): 18. DOI: 10.1103/PhysRevLett.8.18.

Batchko, R.G., D.R. Weise, T. Plettner, G.D. Miller, M.M. Fejer and R.L. Byer. "Continuous-wave 532-nm-pumped singly resonant optical parametric oscillator based on periodically poled lithium niobate." *Opt. Lett.* 23, no. 3 (1998): 168–170. DOI: 10.1364/OL.23.000168.

Beck, M., D. Hofstetter, T. Aellen, J. Faist, U. Oesterle, M. Ilegems, E. Gini and H. Melchior. "Continuous wave operation of a mid-infrared semiconductor laser at room temperature." *Science* 295, no. 5553 (2002): 301–305. DOI: 10.1126/science.1066408.

Cerulla, G. and S. De Silvestri. "Ultrafast optical parametric amplifiers." *Rev. Sci. Instrum.* 74, no. 1 (2003): article 1523642. DOI: 10.1063/1.1523642.

Chen, M.C., C. Mancuso, C. Hernández-García, F. Dollar, B. Galloway, D. Popmintchev, P.C. Huang, B. Walker, L. Plaja, A.A. Jaroń-Becker, A. Becker, M.M. Murnane, H.C. Kapteyn and T. Popmintchev. "Generation of bright isolated attosecond soft X-ray pulses driven by multi-cycle mid-infrared lasers." *Proc. Natl. Acad. Sci.* 111, no. 23 (2014): E2361–E2367. DOI: 10.1073/pnas.1407421111.

Faist, J., F. Capasso. D.L. Sivco, C. Sirtori, A.L. Hutchinson and A.Y. Cho. "Quantum cascade laser." *Science* 264, no. 5158 (1994): 553–556. DOI: 10.1126/science.264.5158.553.

Fallahi, M., L. Fan, Y. Kaneda and C. Hessenius. "5-W yellow laser by intra-cavity frequency doubling of high-power vertical-external-cavity surface-emitting laser." *IEEE Photon. Technol. Lett.* 20, no. 20 (2008): 1700–1702. DOI: 10.1109/LPT.2008.2003413.

Franken, P., A. Hill, C. Peters and G. Weinreich. "Generation of optical harmonics." *Phys. Rev. Lett.* 7, no. 4 (1961): 118–119. DOI: 10.1103/PhysRevLett.7.118.

Giordmaine, J.A. and R.C. Miller, "Tunable coherent parametric oscillation in LiNbO$_3$ at optical frequencies." *Phys. Rev. Lett.* 14, no. 24 (1965): 973–975. DOI: 10.1103/PhysRevLett.14.973.

Gladyshev, A.V., M.I. Belovolov, S.A. Vasiliev, E.M. Dianov, O.I. Medvedkov, A.I. Nadezhdinskii, O.V. Ershov, A.G. Beresin, V.P. Duraev and E.T. Nedelin. "Tunable single-frequency diode laser at wavelength λ=1.65μm for methane concentration measurements." *Spectrochim. Acta A* 60, no. 14 (2004): 3337–3340. DOI: 10.1016/j.saa.2003.12.055.

Guina, M., A. Härkönen, V.M. Korpijärvi, T. Leinonen and S. Suomalainen. "Semiconductor disk lasers: Recent advances in generation of yellow-orange and mid-IR radiation." *Adv. Opt. Technol.* 2012 (2012): article 265010. DOI: 10.1155/2012/265010.

Hall, R.N., G.E. Fenner, J.D. Kingsley, T.J. Soltys and R.O. Carlson. "Coherent light emission from GaAs junctions." *Phys. Rev. Lett.* 9, no. 9 (1962): 366–369. DOI: 10.1103/PhysRevLett.9.366.

Hayashi, I., M. Panish and P. Foy. "A low-threshold room-temperature injection laser." *IEEE J. Quant. Electron.* 5, no. 4 (1969): 211–212. DOI: 10.1109/JQE.1969.1075759.

Kumar, S.C., A. Esteban-Martin, T. Ideguchi, M. Yan, S. Holzner, T.W. Hänsch, N. Picqué and M. Ebrahim-Zadeh. "Few-cycle, broadband, mid-infrared optical parametric oscillator pumped by a 20-fs Ti:sapphire laser." *Laser Photon. Rev.* 8, no. 5 (2014): L86–L91. DOI: 10.1002/lpor.201400091.

Kuznetsov, M., F. Hakimi, R. Sprague and A. Mooradian. "High-power (>0.5W CW) diode-pumped vertical-external-cavity surface-emitting semiconductor lasers with circular $TEM_{00}$ beams." *IEEE Photon. Technol. Lett.* 9, no. 8 (1997): 1063–1065. DOI: 10.1109/68.605500.

Liebel, M., C. Schnedermann and P. Kukura. "Sub-10-fs pulses tunable from 480 to 980 nm from a NOPA pumped by an Yb:KGW source." *Opt. Lett.* 39, no. 14 (2014): 4112–4115. DOI: 10.1364/OL.39.004112.

Lucas-Leclin, G., D. Paboeuf, P. Georges, J. Holm, P. Andersen, B. Sumpf and G. Erbert. "Wavelength stabilization of extended-cavity tapered lasers with volume Bragg gratings." *Appl. Phys. B* 91, no. 3–4 (2008): 493–498. DOI: 10.1007/s00340-008-3034-2.

McPherson, A., G.G.H. Jara, U. Johann, T.S. Luk, I.A. McIntyre, K. Boyer and C. K. Rhodes. "Studies of multiphoton production of vacuum-ultraviolet radiation in the rare gases." *J. Opt. Soc. Am. B* 4, no. 4 (1987): 595–601. DOI: 10.1364/JOSAB.4.000595.

Mes, J. M. Leblans and W. Hogervorst. "Single-longitudinal-mode optical parametric oscillator for spectroscopic applications." *Opt. Lett.* 27, no. 16 (2002): 1442–1444. DOI: 10.1364/OL.27.001442.

Nathan, M.I., W.P. Dumke, G. Burns, F.H. Dill and G. Lasher. "Stimulated emission of radiation from GaAs p-n junctions." *Appl. Phys. Lett.* 1, no. 3 (1962): 62–64. DOI: 10.1063/1.1777371.

Ou, Z.Y., S.F. Pereira, E.S. Polzik and H.J. Kimble. "85% efficiency for cw frequency doubling from 1.08 to 0.54 μm." *Opt. Lett.* 17, no. 9 (1992): 640-642. DOI: 10.1364/OL.17.000640.

Popov, Y.M. "Historical works at Lebedev Institute on injection lasers." *Semicond. Sci. Technol.* 27, no. 9 (2012): article 090203. DOI: 10.1088/0268-1242/27/9/090203.

Pupeza, I., M. Högner, J. Weitenberg, S. Holzberger, D. Esser, T. Eidam, J. Limpert, A. Tünnermann, E. Fill and V. S. Yakovlev. "Cavity-enhanced high-harmonic generation with spatially tailored driving fields." *Phys. Rev. Lett.* 112, no. 10 (2014): article 103902. DOI: 10.1103/PhysRevLett.112.103902.

Smith, R.G., J.E. Geusic, H.J. Levinstein, J.J. Rubin, S. Singh and L.G. Van Uitert. "Continuous optical parametric oscillation in $Ba_2NaNb_5O_{15}$." *Appl. Phys. Lett.* 12, no. 9 (1968): 308–309. DOI: 10.1063/1.1652004.

Steckman, G.J., W. Liu, R. Platz, D. Schroeder, C. Moser and F. Havermeyer. "Volume holographic grating wavelength stabilized laser diodes." *IEEE J. Sel. Topics Quantum Electron.* 13, no. 3 (2007): 672–678. DOI: 10.1109/JSTQE.2007.896060.

Wysocki, G. and D. Weidmann. "Molecular dispersion spectroscopy for chemical sensing using chirped mid-infrared quantum cascade laser." *Opt. Express*, 18, no. 25 (2010): 26123–26140. DOI: 10.1364/OE.18.026123

Yang, L., N. Ye, B. Wang, D. Zhou, X. An, J. Bian, J. Pan, L. Zhao and W. Wang. "Monolithic integration of widely tunable sampled grating DBR laser with tilted semiconductor optical amplifier." *J. Semicon.* 31, no. 7 (2010): article 074003. DOI: 10.1088/1674-4926/31/7/074003.

Ziegelberger, G. "ICNIRP guidelines on limits of exposure to laser radiation of wavelengths between 180nm and 1000μm." *Health Phys.* 105, no. 3 (2013): 271–295. DOI: 10.1097/HP.0b013e3182983fd4

*Other sources*   "ANSI Z136 Standards: The foundation of a successful laser safety program." Orlando, FL: Laser Institute of America (2014). **Access**: https://www.lia.org/publications/ansi.

Block Engineering. "Mini-QCL™: Widely-tunable mid-IR OEM laser module." Marlborough, MA, USA: Block Engineering (2017). **Access**: http://www.blockeng.com/products/miniqcl.pdf.

LaserVision. "Laser safety products," 22nd edition. Fürth, Germany: Laservision GmbH (2014). **Access**: http://www.uvex-laservision.de/fileadmin/user_upload/Download/laservision_edition22.pdf.

"LaserSafe PC—Fast and accurate laser safety calculations," v.5. Bournemouth, UK: Lasermet Ltd/Schaumburg, IL, USA: Lasemet Inc. (2015). **Access**: http://www.lasermet.com/laser-safety-software.php.

"The Evaluator—Laser safety hazard analysis system." Orlando, FL, USA: Laser Institute of America (2014). **Access**: http://www.lasersafetyevaluator.org/.

"Safety of laser products—Part 1: Equipment classification and requirements." IEC 60825-1, 3rd edition. Geneva: International Electro-technical Commission (2014). **Access**: http://webstore.iec.ch/Webstore/webstore.nsf/standards+ed/IEC%2060825-1%20Ed.%203.0?Open Document.

"XUUS—The eXtreme Ultraviolet Ultrafast Source." Boulder, CO, USA: KMLabs (2009). **Access**: http://www.kmlabs.com/content/high-harmonic-generation.

## Chapter 6    Common spectroscopic and imaging detection techniques

*Books and chapters in books*   Chang, C.I. *Hyperspectral Imaging: Techniques for Spectral Detection and Classification.* Heidelberg, Germany: Springer (2003).

de Hoffmann, E. and V. Stroobant. *Mass Spectrometry: Principles and Applications*, 3rd edition. Chichester, UK: John Wiley & Sons (2007).

Demtröder, W. "Spectroscopic instrumentation." Chapter 4 in *Laser Spectroscopy, Vol. 1: Basic Principles*. Heidelberg, Germany: Springer (2008): pp. 92–233.

Griffiths, P.R. and J.A. de Haseth. *Fourier Transform Infrared Spectrometry*, 2nd edition. Chichester, UK: John Wiley & Sons (2007).

Gross, J.H. *Mass Spectrometry*. Heidelberg, Germany: Springer (2011).

Kauppinen, J. and J. Partanen. *Fourier Transform in Spectroscopy*. Weinheim, Germany: Wiley-VCH Verlag (2001).

Michaelian, K.H. *Photoacoustic IR Spectroscopy: Instrumentation, Applications and Data Analysis*, 2nd edition. Weinheim, Germany: Wiley-VCH Verlag (2010).

Smith, B.C. *Fundamentals of Fourier Transform Infrared Spectroscopy*, 2nd edition. Boca Raton, FL, USA: CRC Press (2011).

Throck Watson, J. and O.D. Sparkman. *Introduction to Mass Spectrometry: Instrumentation, Applications, and Strategies for Data Interpretation*, 4th edition. Chichester, UK: John Wiley & Sons (2007).

Timothy, J.G. "Microchannel plates for photon detection and imaging in space." Chapter 22 in M.C.E. Huber, A. Pauluhn, J.L. Culhane, J.G. Timothy, K. Wilhelm and A. Zehnder (eds.). *Observing Photons in Space*. New York, NY, USA: Springer (2013): pp. 391–421. DOI: 10.1007/978-1-4614-7804-1_22.

*Journal articles*   Amelio, G.F., M.F. Tompsett and G.E. Smith. "Experimental verification of the charge coupled device concept." *Bell Sys. Tech. J.* 49, no. 4 (1970): 593–600. DOI: 10.1002/j.1538-7305.1970.tb01791.x.

Bajaj, J. "HgCdTe infrared detectors and focal plane arrays." In Proc. 1998 Conf. Optoelectronic and Microelectronic Materials and Devices. Piscataway, NJ, USA: IEEE (1999): 23–31. DOI: 10.1109/COMMAD.1998.791581.

Beard, P. "Biomedical photoacoustic imaging." *Interface Focus* 1, no. 4 (2011): 602–631. DOI: 10.1098/rsfs.2011.0028.

Beck, J., M. Woodall, R. Scritchfield, M. Ohlson, L. Wood, P. Mitra and J. Robinson. "Gated IR imaging with 128×128 HgCdTe electron Avalanche photodiode FPA." *J. Electron. Mat.* 37, no. 9 (2008): 1334–1343. DOI: 10.1007/s11664-008-0433-4

Benzer, S. "Excess-defect germanium contacts." *Phys. Rev.* 72, no. 12 (1947): 1267–1268. DOI: 10.1103/PhysRev.72.1267.

Bodkin, A., A. Sheinis, A. Norton, J. Daly, S. Beaven and J. Weinheimer. "Snapshot hyperspectral imaging: the hyperpixel array camera." *Proc. SPIE.* 7334 (2009): 73340H1–73340H11. DOI: 10.1117/12.818929.

Boyle, W.S. and G.E. Smith. "Charge-coupled semiconductor devices." *Bell Sys. Tech. J.* 49, no. 4 (1970): 587–593. DOI: 10.1002/j.1538-7305.1970.tb01790.x.

Chandler, D.W. and P.L. Houston. "Velocity and internal state distributions by two-dimensional imaging of products detected by multiphoton ionization." *J. Chem. Phys.* 87, no. 2 (1987): 1445–1447. DOI: 10.1063/1.453276.

Fassel, V.A. (chairman, IUPAC, Analytical Chemistry Division, Commission on Spectrochemical and Other Optical Procedures for Analysis). "Nomenclature, symbols, units and their usage in spectrochemical analysis—II. Data interpretation." *Pure Appl. Chem.* 45, no. 2 (1976): 99–103. DOI: 10.1351/pac197645020099.

Haisch, C. "Photoacoustic spectroscopy for analytical measurements." *Meas. Sci. Technol.* 23, no. 1 (2012): article 012001. DOI: 10.1088/0957-0233/23/1/012001.

Hajireza, P., A. Forbrich and R.J. Zemp. "Multifocus optical-resolution photoacoustic microscopy using stimulated Raman scattering and chromatic aberration." *Opt. Lett.* 38, no. 15 (2013): 2711–2713. DOI: 10.1364/OL.38.002711.

Han, X., W. Du, R. Yu, C. Pan and Z.L. Wang. "Piezo-phototronic enhanced UV sensing based on a nanowire photodetector array." *Adv. Mater.* 27, no. 48 (2015): 7963–7969. DOI: 10.1002/adma.201502579.

Harrison, G.R., J.E. Archer and J. Camus. "A fixed-focus broad-range Echelle spectrograph of high speed and resolving power." *J. Opt. Soc. Am.* 42, no. 10 (1952): 706–712. DOI: 10.1364/JOSA.42.000706.

Hayden, O., R. Agarwal and C.M. Lieber. "Nanoscale avalanche photodiodes for highly sensitive and spatially resolved photon detection." *Nature Mater.* 5, no. 5 (2006): 352–356. DOI: 10.1038/nmat1635.

Heck, A.J.R. and D.W. Chandler. "Imaging techniques for the study of chemical reaction dynamics." *Ann. Rev. Phys. Chem.* 46 (1995): 335–372. DOI: 10.1146/annurev.pc.46.100195.002003.

Kind, H., H. Yan, B. Messer, M. Law and P. Yang. "Nanowire ultraviolet photodetectors and optical switches." *Adv. Mater.* 14, no. 2 (2002): 158–160. DOI:10.1002/1521-4095(20020116)14:2<158::AID-ADMA158>3.0.CO;2-W.

Michels, W.C. and N.L. Curtis. "A pentode lock-in amplifier of high frequency selectivity." *Rev. Sci. Instrum.* 12, no. 9 (1941): 444–447. DOI: 10.1063/1.1769919.

Natarajan, C.M., M.G. Tanner and R.H. Hadfield. "Superconducting nanowire single-photon detectors: physics and applications." *Supercond. Sci. Technol.* 25, no. 6 (2012): article 063001. DOI: 10.1088/0953-2048/25/6/063001.

Owen, K. "The impact of volume phase holographic filters and gratings on the development of Raman instrumentation." *J. Chem. Edu.* 84, no. 1 (2007): 61–66. DOI: 10.1021/ed084p61.

Patimisco, P., G. Scamarcio, F.K. Tittel and V. Spagnolo. "Quartz-enhanced photoacoustic spectroscopy: A review." *Sensors* 14, no. 4 (2014): 6165–6206. DOI:10.3390/s140406165.

Rosencwaig, A. and A. Gersho. "Theory of the photoacoustic effect with solids." *J. Appl. Phys.* 47, no. 1 (1976): 64–69. DOI: 10.1063/1.322296.

Schippers, W., M. Köhring, S. Böttger, U. Willer, G. Flachenecker and W. Schade. "Simultaneous detection of Raman- and collision-induced molecular rotations of O and N via femtosecond multi-pulses in combination with quartz-enhanced photoacoustic spectroscopy." *Appl. Phys. B* 116, no. 1 (2014): 53–60. DOI: 10.1007/s00340-013-5647-3.

Smith, G.E. "Nobel Lecture: The invention and early history of the CCD." *Rev. Mod. Phys.* 82, no. 3 (2010): 2307–2312. DOI: 10.1103/RevModPhys.82.2307.

Soci, C., A. Zhang, B. Xiang, S.A. Dayeh, D.P.R. Aplin, J. Park, X.Y. Bao, Y.H. Lo and D. Wang. "ZnO nanowire UV photodetectors with high internal gain." *Nano Lett.* 7, no. 4 (2007): 1003–1009. DOI: 10.1021/nl070111x.

Spencer, C.L., V. Watson and M. Hippler. "Trace gas detection of molecular hydrogen $H_2$ by photoacoustic stimulated Raman spectroscopy (PARS)." *Analyst* 137, no. 6 (2012): 1384–1388. DOI: 10.1039/c2an15990b.

Talmi, Y. and R.W. Simpson. "Self-scanned photodiode array: A multichannel spectrometric detector." *Appl. Opt.* 19, no. 9 (1980): 1401–1414. DOI: 10.1364/AO.19.001401.

Tompsett, M.F., G.F. Amelio and G.E. Smith. "Charge-coupled 8-bit shift register." *Appl. Phys. Lett.* 17, no. 3 (1970): 111–115. DOI: 10.1063/1.1653327.

Tompsett, M.F., N.J. Murray Hill, G.F. Amelio, W.J. Bertram Jr., R.R. Buckley, W.J. McNamara, J.C. Mikkelsen Jr. and D.D. Sealer. "Charge-coupled imaging devices: Experimental results." *IEEE Trans. Electron. Dev.* 18, no. 11 (1971): 992–996. DOI: 10.1109/T-ED.1971.17321.

Voigtman, E. "Limits of detection and decision, Parts 1–4." *Spectrochim. Acta B* 63, no. 2 (2008): 115–128, 129–141, 142–153, 154–165. DOI: 10.1016/j.sab.2007.11.014.

Wang, P., J.R. Rajian and J.X. Cheng. "Spectroscopic imaging of deep tissue through photoacoustic detection of molecular vibration." *J. Phys. Chem. Lett.* 4, no. 13 (2013): 2177–2185. DOI: 10.1021/jz400559a.

Wester, R. "Velocity map imaging of ion–molecule reactions." *Phys. Chem. Chem. Phys.* 16, no. 2 (2014): 396–405. DOI: 10.1039/C3CP53405G.

Yakovleva, V.V., H.F. Zhang, G.D. Noojin, M.L. Denton, R.J. Thomas and M.O. Scully. "Stimulated Raman photoacoustic imaging." *PNAS* 107, no. 47 (2010): 20335–20339. DOI: 10.1073/pnas.1012432107.

Zworykin, H. and J.A. Rajchman. "The electrostatic electron multiplier." *Proc. IRE* 27, no. 9 (1939) 558–566. DOI: 10.1109/JRPROC.1939.228753.

*Other sources*    EMCCD Forum. "Fundamentals of EMCCD—A tutorial." Belfast, UK: Andor Technology plc (2005). **Access**: http://www.EMCCD.com/What_is_EMCCD/EMCCD_tutorial.

EMCCD Forum. "Unravelling sensitivity, signal-to-noise and dynamic range—EMCCD vs CC." Belfast, UK: Andor Technology plc (2005). **Access**: http://www.EMCCD.com/What_is_EMCCD/unraveling_sensitivity.

Hamamatsu, Electron Tube Division. "Image intensifiers." Iwata City, Japan: Hamamatsu Photonics K.K. (2009). **Access**: https://www.hamamatsu.com/resources/pdf/etd/II_TII0004E.pdf.

Hamamatsu Electron Tube Division. "MCP assembly," Technical Information TMCP9002E01. Iwata City, Japan: Hamamatsu K.K. (2006). **Access**: http://www.triumf.ca/sites/default/files/Hamamatsu%20MCP%20guide.pdf.

Hamamatsu, Solid State Division. "Opto-semiconductor handbook." Hamamatsu City, Japan: Hamamatsu Photonics K.K. (2014). **Access**: https://www.hamamatsu-news.de/hamamatsu_optosemiconductor_handbook/.

Hakamata, T. (chief editor). "Photomultiplier tubes: Basics and applications," 3rd edition Hamamatsu City, Japan: Hamamatsu Photonics K.K. (2007). **Access**: https://www.hamamatsu.com/resources/pdf/etd/PMT_handbook_v3aE.pdf.

PAR model HR-8. "New PAR lock-in amplifier measures signals in the presence of noise by cross-correlation." *IEEE Spectrum* 2, no. 12 (1965): 3. [Advertisement.]

## Chapter 7    Absorption spectroscopy and its implementation

*Books and chapters in books*    Buldyreva, J., N. Lavrentieva and V. Starikov. *Collisional Line Broadening and Shifting of Atmospheric Gases: A Practical Guide for Line Shape Modelling by Current Semi-Classical Approaches.* London, UK: Imperial College Press (2011).

Demtröder, W. *Laser Spectroscopy, Vol. 1: Basic Principles* and *Laser Spectroscopy, Vol. 2: Experimental Techniques.* Heidelberg, Germany: Springer (2008).

Levenson, M.D. "Saturation spectroscopy" and "Multi-photon absorption." Chapters 3 and 5 in *Introduction to Non-linear Laser Spectroscopy.* New York, NY, USA: Academic Press (1982).

Michaelian, K.H. *Photoacoustic IR Spectroscopy: Instrumentation, Applications and Data Analysis,* 2nd edition. Weinheim, Germany: Wiley-VCH Verlag (2010). DOI: 10.1002/9783527633197.

Peiponen, K.E., A. Zeitler and M. Kuwata-Gonokami (eds.). *Terahertz spectroscopy and imaging.* Springer Series in Optical Sciences, vol. 171. Heidelberg, Germany: Springer Verlag (2013).

Ye, J. and S. Cundiff (eds.). *Femtosecond Optical Frequency Comb: Principle, Operation and Applications.* New York, NY, USA: Springer (2005).

*Journal articles*    Ahmadi, M., B.X.R. Alves, C.J. Baker, W. Bertsche, E. Butler, A. Capra, C. Carruth, C.L. Cesar, M. Charlton, S. Cohen, R. Collister, S. Eriksson, A. Evans, N. Evetts, J. Fajans, T. Friesen, M.C.

Fujiwara, D.R. Gill, A. Gutierrez, J.S. Hangst, W.N. Hardy, M.E. Hayden, C.A. Isaac, A. Ishida, M.A. Johnson, S.A. Jones, S. Jonsell, L. Kurchaninov, N. Madsen, M. Mathers, D. Maxwell, J.T.K. McKenna, S. Menary, J.M. Michan, T. Momose, J.J. Munich, P. Nolan, K. Olchanski, A. Olin, P. Pusa, C.Ø. Rasmussen, F. Robicheaux, R.L. Sacramento, M. Sameed, E. Sarid, D.M. Silveira, S. Stracka, G. Stutter, C. So, T.D. Tharp, J.E. Thompson, R.I. Thompson, D.P. van der Werf and J.S. Wurtele. "Observation of the 1S–2S transition in trapped antihydrogen". *Nature* 541 (2017): 506–5115. doi: 10.1038/nature21040.

Amamou, H., B. Ferhat and A. Bois. "Calculation of the Voigt function in the region of very small values of the parameter $a$ where the calculation is notoriously difficult." *Am. J. Anal. Chem.* 4, no. 12 (2013): 725–731. DOI: 10.4236/ajac.2013.412087.

Ambartsumyan, R.V., Y.A. Gorohkov, V.S. Letokhov and G.N. Makarov. "Interaction of $SF_6$ molecules with a powerful infrared laser pulse and the separation of sulfur isotopes." *Sov. Phys. JETP* 42, no. 6 (1975): 993–1000. [Russian original version: *Zh. Eksp. Teor. Fiz.* 69, no. 6 (1975): 1956–1970.]

Antognini, A., F. Nez, K. Schuhmann, F.D. Amaro, F. Biraben, J.M.R. Cardoso, D.S. Covita, A. Dax, S. Dhawan, M. Diepold, L.M.P. Fernandes, A. Giesen, A.L. Gouvea, T. Graf, T.W. Hänsch, P. Indelicato, L. Julien, C.Y. Kao, P. Knowles, F. Kottmann, E.O. Le Bigot, Y.W. Liu, J.A.M. Lopes, L. Ludhova, C.M.B. Monteiro, F. Mulhauser, T. Nebel, P. Rabinowitz, J.M.F. dos Santos, L.A. Schaller, C. Schwob, D. Taqqu, J.C.A. Veloso, J. Vogelsang and R. Pohl. "Proton structure from the measurement of 2S–2P transition frequencies of muonic hydrogen." *Science* 339, no. 6118 (2013): 417–420. DOI: 10.1126/science.1230016.

Beyer, A., C.G. Parthey, N. Kolachevsky, J. Alnis, K. Khabarova, R. Pohl, E. Peters, D.C. Yost, A. Matveev, K. Predehl, S. Droste, T. Wilken, R. Holzwarth, T.W. Hänsch, M. Abgrall, D. Rovera, C. Salomon, P. Laurent and T. Udem. "Precision spectroscopy of atomic hydrogen." *J. Phys. Conf. Ser.* 467 (2013): article 012003. DOI: 10.1088/1742-6596/467/1/012003.

Biraben, F., B. Cagnac and G. Grynberg. "Experimental evidence of two-photon transition without Doppler broadening." *Phys. Rev. Lett.* 32, no. 12 (1974): 643–645. DOI: 10.1103/PhysRevLett.32.64.

Biraben, F., M. Bassini and B. Cagnac. "Line-shapes in Doppler-free two-photon spectroscopy: The effect of finite transit time." *J. Phys. France* 40, no. 5 (1979): 645–655. DOI: jphys:01979004005044500.

Cerez, P. and S.J. Bennett. *Appl. Opt.* 18, no. 7 (1979): 1079–1083. DOI: 10.1364/AO.18.001079.

Coddington, I., N. Newbury and W. Swann. "Dual-comb spectroscopy." *Optica* 3, no. 4 (2016): 414–426. DOI: 10.1364/OPTICA.3.000414.

Costain, C.C. "The use of saturation dip absorption in microwave spectroscopy and in microwave frequency stabilization." *Can. J. Phys.* 47, no. 21 (1969): 2431–2433. DOI: 10.1139/p69-299.

Cubillas, A.M., J.M. Lazaro, O.M. Conde, M.N. Petrovich and J.M. Lopez-Higuera. "Gas sensor based on photonic crystal fibres in the $2v_3$ and $v_2+2v_3$ vibrational bands of methane." *Sensors* 2009, no. 9 (2009): 6261–6272. DOI: 0.3390/s90806261.

de Rosa, M., C. Corsi, M. Gabrysch and F. d'Amato. "Collisional broadening and shift of lines in the $2v_1+2v_2+v_3$ band of $CO_2$." *J. Quant. Spectrosc. Rad. Trans.* 61, no. 1 (1997): 97–104. DOI: 10.1016/S0022-4073(97)00207-0.

Deng, Y., Q. Sun and J. Yu. "On-line calibration for linear time-base error correction of terahertz spectrometers with echo pulses." *Metrologia* 51, no. 1 (2014): 18–24. DOI: 10.1088/0026-1394/51/1/18.

Eckstein, J.N., A.I. Ferguson and T.W. Hänsch. "High-resolution two-photon spectroscopy with picosecond light pulses." *Phys. Rev. Lett.* 40, no. 13 (1978): 847–850. DOI: 10.1103/PhysRevLett.40.847.

Engeln, R., G. Berden, R. Peeters and G. Meijer. "Cavity-enhanced absorption and cavity-enhanced magnetic rotation spectroscopy." *Rev. Sci. Instr.* 69, no. 11 (1998): 3763–3769. DOI: 10.1063/1.1149176.

Foltynowicz, A., P. Masłowski, T. Ban, F. Adler, K.C. Cossel, T.C. Briles and J. Ye. "Optical frequency comb spectroscopy." *Faraday Discuss.* 150 (2011): 23–31. DOI: 10.1039/C1FD00005E.

Galtier, S., H. Fleurbaey, S. Thomas, L. Julien, F. Biraben and F. Nez. "Progress in spectroscopy of the 1S–3S transition in hydrogen." *J. Phys. Chem. Ref. Data* 44, no. 3 (2015): article 031201. DOI: 10.1063/1.4922255.

Goeppert-Mayer, M. "Über Elementarakte mit zwei Quantensprüngen." *Annal. Phys.* 9, no. 3 (1931): 273–295. DOI: 10.1002/andp.19314010303.

Goto, M. and S. Moriata. "Self-reversal in hydrogen Lyman-α line profile." *Plasma Fusion Res.* 5, special issue 2 (2010): article S2089. DOI: 10.1585/pfr.5.S2089

Hänsch, T.W., M.D. Levenson and A.L. Schawlow. "Complete hyperfine structure of a molecular iodine line." *Phys. Rev. Lett.* 26, no. 16 (1971): 946–949. DOI: 10.1103/PhysRevLett.26.946.

Hänsch, T.W., I.S. Shahin and A.L. Schawlow. "Optical resolution of the Lamb shift in atomic hydrogen by laser saturation spectroscopy." *Nature* 235, no. 5333 (1972): 63–65. DOI: 10.1038/physci235063a0.

Hänsch, T.W., K.C. Harvey, G. Miesel and A.L. Schawlow. "Two-photon spectroscopy of Na 3s-4d without Doppler broadening using a cw dye laser." *Opt. Commun.* 11, no. 1 (1974): 50–53. DOI: 10.1016/0030-4018(74)90331-9.

Hermann, H., J. Kruse, A. Piel, E.W. Weber and V. Helbig. "Charged and neutral particle shift and broadening of resolved Balmer-β fine structure lines at low electron densities." *Plasma Sources Sci. Technol.* 2, no. 3 (1993): 214–218.

Huet, N., S. Krins, P. Dubé and T. Bastin. "Hyperfine-structure splitting of the 716 nm R(90)3-10 molecular iodine transition." *J. Opt. Soc. Am. B* 30, no. 5 (2013): 1317-1321. DOI: 10.1364/JOSAB.30.001317.

Hult, J., R.S. Watt and C.F. Kaminski. "High bandwidth absorption spectroscopy with a dispersed supercontinuum source." *Opt. Express* 15, no. 18 (2007): 11385–11395. DOI: 10.1364/OE.15.011385

Jacques, V., B. Hingant, A. Allafort, M. Pigeard and J.F. Roch. "Nonlinear spectroscopy of rubidium: an undergraduate experiment." *Eur. J. Phys.* 30, no. 5 (2009) 921–934. DOI: 10.1088/0143-0807/30/5/001.

Jepsen, P.U., D.G. Cooke and M. Koch. "Terahertz spectroscopy and imaging—Modern techniques and applications." *Laser Photon. Rev.* 5, no. 1 (2011): 124–166. DOI: 10.1002/lpor.201000011.

Juvells, I., A. Carnicer, J. Ferré-Borrull, E. Martín-Badosa and M. Montes-Usategui. "Understanding the concept of resolving power in the Fabry-Perot interferometer using a digital simulation." *Eur. J. Phys.* 27, no. 5 (2006): 1111–1119. DOI: 10.1088/0143-0807/27/5/010

Kassi, S., A. Campargue, K. Pachucki and J. Komasa. "The absorption spectrum of $D_2$: Ultrasensitive cavity ring-down spectroscopy of the (2-0) band near 1.7μm and accurate *ab initio* line list up to 24000cm$^{-1}$." *J. Chem. Phys.* 136, no. 18 (2012): article 184309. DOI: 10.1063/1.4707708.

Kosterev, A., G. Wysocki, Y. Bakhirkin, S. So, R. Lewicki, M. Fraser, F. Tittel and R.F. Curl. "Application of quantum cascade lasers to trace gas analysis." *Appl. Phys. B.* 90, no. 2 (2008): 165–176. DOI: 10.1007/s00340-007-2846-9.

Laity, G., A. Fierro, J. Dickens, A. Neuber and K. Frank. "Simultaneous measurement of nitrogen and hydrogen dissociation from vacuum ultraviolet self-absorption spectroscopy in a developing low temperature plasma at atmospheric pressure." *Appl. Phys. Lett.* 102, no. 18 (2013): article 184104. DOI: 10.1063/1.4804369.

Lamb Jr., W.E. "Theory of an optical maser." *Phys. Rev.* 134, no. 6A (1964): A1429–A1450. DOI: 10.1103/PhysRev.134.A1429.

Levenson, M.D. and N. Bloembergen. "Observation of two-photon absorption without Doppler broadening on the 3S–5S transition in sodium vapor." *Phys. Rev. Lett.* 32, no. 12 (1974): 645–648. DOI: 10.1103/PhysRevLett.32.645.

Mandon, J., E. Sorokin, I.T. Sorokina, G. Guelachvili and N. Picqué. "Supercontinua for high-resolution absorption multiplex infrared spectroscopy." *Opt. Lett.* 33, no. 3 (2008): 285–287. DOI: 10.1364/OL.33.000285.

Mathi, P., P.K. Sahoo, D.N. Joshi, M.N. Deo, A.K. Nayak, S.K. Sarkar and V. Parthasarathy. "A novel two-stage process for laser enrichment of sulphur isotopes." *J. Photochem. Photobiol. A: Chem.* 194, no. 2-3 (2008): 344–350. DOI: 10.1016/j.jphotochem.2007.08.034.

Matveev, A., C.G. Parthey, K. Predehl, J. Alnis, A. Beyer, R. Holzwarth, T. Udem, T. Wilken, N. Kolachevsky, M. Abgrall, D. Rovera, C. Salomon, P. Laurent, G. Grosche, O. Terra, T. Legero, H. Schnatz, S. Weyers, B. Altschul and T.W. Hänsch. "Precision measurement of the hydrogen 1S-2S frequency via a 920-km fiber link." *Phys. Rev. Lett.* 110, no. 23 (2013): article 230801. DOI: 10.1103/PhysRevLett.110.230801.

McFarlane, R.A., W.R. Bennett, Jr. and W.E. Lamb. "Single-mode tuning dip in the power output of an He–Ne optical maser." *Appl. Phys. Lett.* 2, no. 10 (1963): 189-191. DOI: 10.1063/1.1753727.

Mohr, P.J., B.N. Taylor and D.B. Newell. "CODATA recommended values of the fundamental physical constants: 2010." *Rev. Mod. Phys.* 84, no. 4 (2012): 1527-1606. DOI: 10.1103/RevModPhys.84.1527.

O'Keefe, A. "Integrated cavity output analysis of ultra-weak absorption." *Chem. Phys. Lett.* 293, no. 5-6 (1998): 331–336. DOI: 10.1016/S0009-2614(98)00785-4.

O'Keefe, A. and D.A.G. Deacon. "Cavity ring-down optical spectrometer for absorption measurements using pulsed laser sources." *Rev. Sci. Instrum*, 59, no. 12 (1988): 2544–2551. DOI: 10.1063/1.1139895.

Patimisco, P., G. Scamarcio, F.K. Tittel and V. Spagnolo. "Quartz-enhanced photoacoustic spectroscopy: A review." *Sensors* 14, no. 4 (2014): 6165–6206. DOI:10.3390/s140406165.

Pawar, A.Y., D.D. Sonawane, K.B. Erande and D.V. Derle. "Terahertz technology and its applications." *Drug Invention Today* 5, no. 2 (2013): 157–163. DOI: 10.1016/j.dit.2013.03.009.

Pearman, C.P., C.S. Adams, S.G. Cox, P.F. Griffin, D.A. Smith and I.G. Hughes. "Polarization spectroscopy of a closed atomic transition: Applications to laser frequency locking." *J. Phys. B: At. Mol. Opt. Phys.* 35, no. 24 (2002): 5141–5151. DOI: 10.1088/0953-4075/35/24/315.

Pritchard, D., J. Apt and T.W. Ducas. "Fine structure of Na $4d^2$ $D$ using high-resolution two-photon spectroscopy." *Phys. Rev. Lett.* 32, no. 12 (1974): 641–643. DOI: 10.1103/PhysRevLett .32.641.

Robert, C. "Simple, stable, and compact multiple-reflection optical cell for very long optical paths." *Appl. Opt.* 46, no. 22 (2007): 5408–5418. DOI: 10.1364/AO.46.005408.

Ryan, R.E., L.A. Westling and H.J. Metcalf. "Two-photon spectroscopy in rubidium with a diode laser." *J. Opt. Soc. Am. B* 10, no. 9 (1993): 1643–1648. DOI: 10.1364/JOSAB.10.001643.

Salter, R., J. Chu and M. Hippler. "Cavity-enhanced Raman spectroscopy with optical feedback cw diode lasers for gas phase analysis and spectroscopy." *Analyst* 137, no. 20 (2012): 4669–4676. DOI: 10.1039/c2an35722d.

Schneidenbach, H. and S. Franke. "Basic concepts of temperature determination from self-reversed spectral lines." *J. Phys. D: Appl. Phys.* 41, no. 14 (2008): article 144016. DOI: 10.1088 /0022-3727/41/14/144016.

Sheng, D., A. Pérez Galván and L.A. Orozco. "Lifetime measurements of the 5d states of rubidium." *Phys. Rev. A* 78, no. 6 (2008): article 062506. DOI: 10.1103/PhysRevA.78.062506.

Smith, P.W. and T.W. Hänsch. "Cross-relaxation effects in the saturation of the 6328-Å neon-laser line." *Phys. Rev. Lett.* 26, no. 13 (1971): 740–744. DOI: 10.1103/PhysRevLett.26.740.

Strohm, E.M., M.J. Moore and M.C. Kolios. "Single cell photoacoustic microscopy: A review." *IEEE J. Sel. Topics Quant. Electron.* 22, no. 3 (2016): article 6801215. DOI: 10.1109/JSTQE.2015 .2497323.

Vasilenko, L.S., V.P. Chebotayev and A.V. Shishaev. "Line shape of two-photon absorption in a standing-wave field in a gas." *JETP Lett.* 12, no. 3 (1970): 113–116.

Werblinski, T., B. Lämmlein, F.J.T. Huber, L. Zigan and S. Will. "Supercontinuum high-speed cavity-enhanced absorption spectroscopy for sensitive multispecies detection." *Opt. Lett.* 41, no. 10 (2016): 2322–2325. DOI: 10.1364/OL.41.00232.

Wieman, C. and T.W. Hänsch. "Doppler-free laser polarization spectroscopy." *Phys. Rev. Lett.* 36, no. 20 (1976): 1170–1173. DOI: 10.1103/PhysRevLett.36.1170.

Zeller, W., L. Naehle, P. Fuchs, F. Gerschuetz, L. Hildebrandt and J. Koeth. "DFB lasers between 760 nm and 16 μm for sensing applications." *Sensors* 10, no. 4 (2010): 2492–2510. DOI: 10.3390 /s100402492.

*Other sources*

HITRAN. "The HITRAN data base." Cambridge, MA, USA: The Harvard-Smithonian Center for Astrophysics (2012). **Access**: https://www.cfa.harvard.edu/hitran/.

Znamenáček, J. (maintainer). "IUPAC compendium of chemical terminology—the Gold book," release 2.3.3. Research Triangle Park, NC, USA: International Union of Pure and Applied Chemistry (2014). **Access**: http://goldbook.iupac.org/index.html.

## Chapter 8    Selected applications of absorption spectroscopy

*Books and chapters in books*

de Martinis, D., T. Koyama and C. Chang (eds.). *One Rotten Apple Spoils the Whole Barrel: The Plant Hormone Ethylene, the Small Molecule and Its Complexity.* Lausanne, Switzerland: Frontiers Media (2015). DOI: 10.3389/978-2-88919-623-4.

Harren, F.J.M., G. Cotti, J. Oomens and S.L. Hekkert. "Photoacoustic spectroscopy in trace gas monitoring." In R.A. Meyers (ed.). *Encyclopedia of Analytical Chemistry.* Chichester, UK: John Wiley & Sons (2012): pp. 2203–2226.

Son, J.H. (ed.). *Terahertz Biomedical Science and Technology.* Boca Raton, FL, USA: CRC Press (2014).

Ye, J. and S.T. Cundiff (eds.). *Femtosecond Optical Frequency Comb: Principle, Operation, and Applications.* Norwell, MA, USA: Kluwer Academic Publishers (2005).

*Journal articles*

Adler, F., P. Masłowski, A. Foltynowicz, K.C. Cossel, T.C. Briles, I. Hartl and J. Ye. "Mid-infrared Fourier transform spectroscopy with a broadband frequency comb." *Opt. Express* 18, no. 21 (2010): 21861–21872. DOI: 10.1364/OE.18.021861.

Auston, D.H. "Picosecond optoelectronic switching and gating in silicon." *Appl. Phys. Lett.* 26, no. 3 (1975): 101–104. DOI: 10.1063/1.88079.

Bartlome, R., A. Feltrin and C. Ballif. "Infrared laser-based monitoring of the silane dissociation during deposition of silicon thin films." *Appl. Phys. Lett.* 94, no. 20 (2009): article 201501. DOI: 10.1063/1.3141520.

Bell, A.G. "On the production and reproduction of sound by light." *Am. J. Sci. – Ser. 3* 20, no. 118 (1880): 305–324. DOI: 10.2475/ajs.s3-20.118.305.

Bolshov, M.A., Y.A. Kuritsyn and Y.V. Romanovskii. "Tunable diode laser spectroscopy as a technique for combustion diagnostics." *Spectrochim. Acta B* 106, no. 4 (2015): 45–66. DOI: 10.1016/j.sab.2015.01.010.

Castro-Camus, E. and M.B. Johnston. "Conformational changes of photoactive yellow protein monitored by terahertz spectroscopy." *Chem. Phys. Lett.* 455, no. 4–6 (2008): 289–292. DOI: 10.1016/j.cplett.2008.02.084.

Chen, J., Y. Chen, H. Zhao, G.J. Bastiaans and X.C. Zhang. "Absorption coefficients of selected explosives and related compounds in the range of 0.1–2.8 THz." *Opt. Express* 15, no. 19 (2007): 12060–12067. DOI: 10.1364/OE.15.012060.

Coddington, I., N. Newbury and W. Swann. "Dual-comb spectroscopy." *Optica* 3, no. 4 (2016): 414–426. DOI: 10.1364/OPTICA.3.000414.

Cremer, J.W., K.M. Thaler, C. Haisch and R. Signorell. "Photo-acoustics of single laser-trapped nano-droplets for the direct observation of nano-focusing in aerosol photo-kinetics." *Nature Commun.* 7 (2016): article 10941. DOI: 0.1038/ncomms10941.

Curl, R.F., F. Capasso, C. Gmachl, A.A. Kosterev, B. McManus, R. Lewicki, M. Pusharsky, G. Wysocki and F.K. Tittel. "Quantum cascade lasers in chemical physics." *Chem. Phys. Lett.* 487, no. 1–3 (2010): 1–18. DOI: 10.1016/j.cplett.2009.12.073.

Diddams, S.A., L. Hollberg and V. Mbele. "Molecular fingerprinting with the resolved modes of a femtosecond laser frequency comb." *Nature* 445, no. 7128 (2007): 627–630. DOI: 10.1038/nature05524.

Eckstein, J.N., A.I. Ferguson and T.W. Hänsch. "High-resolution two-photon spectroscopy with picosecond light pulse." *Phys. Rev. Lett.* 40, no. 13 (1978): 847–850. DOI: 10.1103/PhysRevLett.40.847.

Fan, W.H., A. Burnett, P.C. Upadhya, J. Cunningham, E.H. Linfield and A.G. Davies. "Far-infrared spectroscopic characterization of explosives for security applications using broadband Terahertz time-domain spectroscopy." *Appl. Spectrosc.* 61, no. 6 (2007): 638–643. DOI: 10.1366/000370207781269701.

Fischer, B.M., M. Walther and P.U. Jepsen. "Far-infrared vibrational modes of DNA components studied by terahertz time domain spectroscopy." *Phys. Med. Biol.* 47, no. 21 (2002): 3807–3814. DOI: 10.1088/0031-9155/47/21/319.

Fischer, M., N. Kolachevsky, M. Zimmermann, R. Holzwarth, T. Udem and T.W. Hänsch. "New limits on the drift of fundamental constants from laboratory measurements." *Phys. Rev. Lett.* 92, no. 23 (2004): article 230802. DOI: 10.1103/PhysRevLett.92.230802.

Gohle, C., B. Stein, A. Schliesser, T. Udem and T.W. Hänsch. "Frequency-comb Vernier spectroscopy for broadband, high-resolution, high-sensitivity absorption and dispersion spectra." *Phys. Rev. Lett.* 99, no. 26 (2007): article 263902. DOI: 10.1103/PhysRevLett.99.263902.

Grischkowsky, D., S. Keiding, M.P. van Exter and C. Fattinger. "Far-infrared time-domain spectroscopy with terahertz beams of dielectrics and semiconductors." *J. Opt. Soc. Am. B* 7, no. 10 (1990): 2006–2015. DOI: 10.1364/JOSAB.7.002006.

Harren, F.J.M., F.G.C. Bijnen, J. Reuss, L.A.C.J. Voesenek and C.W.P.M. Blom. "Sensitive intracavity photoacoustic measurements with a $CO_2$ waveguide laser,' *Appl. Phys. B* 50, no. 2 (1990): 137–144. DOI: 10.1007/BF00331909.

Hébert, N.B., S.K. Scholten, R.T. White, J. Genest, A.N. Luiten and J.D. Anstie. "A quantitative mode-resolved frequency comb spectrometer." *Opt. Express* 23, no. 11 (2015): 13991–14001. DOI: 10.1364/OE.23.013991.

Hippler, M., C. Mohr, K.A. Keen and E.D. McNaghten. "Cavity-enhanced resonant photoacoustic spectroscopy with optical feedback cw diode lasers: A novel technique for ultra-trace gas analysis and high-resolution spectroscopy." *J. Chem. Phys.* 133, no. 4 (2010): article 044308. DOI: 10.1063/1.3461061.

Hoshina, H., T. Seta, T. Iwamoto, I. Hosako, C. Otani and Y. Kasai. "Precise measurement of pressure broadening parameters for water vapor with a terahertz time-domain spectrometer." *J. Quant. Spectrosc. Radiat. Transfer* 109, no. 12 (2008): 2303–2314. DOI: 10.1016/j.jqsrt.2008.03.005.

Ideguchi, T., A. Poisson, G. Guelachvili, N. Picqué and T.W. Hänsch. "Adaptive real-time dual-comb spectroscopy." *Nature Commun.* 5 (2014): article 3375. DOI: 10.1038/ncomms4375.

Ikeda, Y., Y. Ishihara, T. Moriwaki, E. Kato and K. Terada. "A novel analytical method for pharmaceutical polymorphs by Terahertz spectroscopy and the optimization of crystal form at the discovery stage." *Chem. Pharma. Bull.* 58, no. 1 (2010): 76–81. DOI: 10.1248/cpb.58.76.

International Union of Pure and Applied Chemistry (IUPAC), Analytical Chemistry Division, Commission on Spectrochemical and Other Optical Procedures for Analysis. "Nomenclature, symbols, units and their usage in spectrochemical analysis—II: Data interpretation." *Pure Appl. Chem.* 45, no. 2 (1976): 99–103. DOI: 10.1351/pac197645020099.

Johnson, P.R. and J.R. Ecker. "The ethylene gas signal transduction pathway: A molecular perspective." *Annu. Rev. Genet.* 32 (1998): 227–254. DOI: 10.1146/annurev.genet.32.1.227.

Kachanov, A.A., V.R. Mironenko and I.K. Pashkovich. "Quantum threshold of the sensitivity of an intra-cavity traveling-wave laser spectrometer." *Sov. J. Quantum Electron.* 19, no. 1 (1989): 95–98.

Kassi, S., A. Campargue, K. Pachucki and J. Komasa. "The absorption spectrum of $D_2$: Ultra-sensitive cavity ring down spectroscopy of the (2–0) band near 1.7 μm and accurate ab initio line list up to 24 000 cm$^{-1}$." *J. Chem. Phys.* 136, no. 18 (2012): article 184309. DOI: 10.1063/1.4707708.

Köhler, R., A. Tredicucci, F. Beltram, H.E. Beere, E.H. Linfield, A.G. Davies, D.A. Ritchie, R.C. Iotti and F. Rossi. "Terahertz semiconductor-heterostructure laser." *Nature.* 417, no. 6885 (2002): 156–159. DOI:10.1038/417156a.

Langridge, J.M., T. Laurila, R.S. Watt, R.L. Jones, C.F. Kaminski and J. Hult. "Cavity enhanced absorption spectroscopy of multiple trace gas species using a supercontinuum radiation source." *Opt. Express* 16, no. 14 (2008): 10178–10187. DOI: 10.1364/OE.16.010178.

Li, J., B. Yu, W. Zhao and W. Chen. "A review of signal enhancement and noise reduction techniques for tunable diode laser absorption spectroscopy." *Appl. Spectrosc. Rev.* 49, no. 8 (2014): 666–691. DOI: 10.1080/05704928.2014.903376.

Lukashevskaya, A.A., O.V. Naumenko, D. Mondelain, S. Kassi and A. Campargue. "High sensitivity cavity-ringdown spectroscopy of the $3\nu_1+3\nu_2+\nu_3$ band of $NO_2$ near 7587 cm$^{-1}$." *J. Quant. Spectrosc. Radiat. Transfer* 177 (2016) 225–233. DOI: 10.1016/j.jqsrt.2015.12.017.

Lytkine, A., W. Jäger and J. Tulip. "Frequency tuning of long-wavelength VCSELs." *Spectrochim. Acta A* 63, no. 5 (2006): 940–946. DOI: 10.1016/j.saa.2005.11.004

Mandon, J., G. Guelachvili and N. Picqué. "Fourier transform spectroscopy with a laser frequency comb." *Nature Photon.* 3, no. 2 (2009): 99–102. DOI: 10.1038/nphoton.2008.293.

Matveev, A., C.G. Parthey, K. Predehl, J. Alnis, A. Beyer, R. Holzwarth, T. Udem, T. Wilken, N. Kolachevsky, M. Abgrall, D. Rovera, C. Salomon, P. Laurent, G. Grosche, O. Terra, T. Legero, H. Schnatz, S. Weyers, B. Altschul and T.W. Hänsch. "Precision measurement of the hydrogen 1S-2S frequency via a 920-km fiber link." *Phys. Rev. Lett.* 110, no. 23 (2013): article 230801. DOI: 10.1103/PhysRevLett.110.230801.

Naumenko, O.V., B.A. Voronin, F. Mazzotti, J. Tennyson and A. Campargue. "Intracavity laser absorption spectroscopy of HDO between 12 145 and 13 160 cm$^{-1}$." *J. Mol. Spectrosc.* 248, no. 2 (2008): 122–133. DOI: 10.1016/j.jms.2007.12.005.

Nelson, D.D., J.B. McManus, S.C. Herndon, J.H. Shorter, M.S. Zahniser, S. Blaser, L. Hvozdara, A. Muller, M. Giovannini and J. Faist. "Characterization of a near-room-temperature, continuous-wave quantum cascade laser for long-term, unattended monitoring of nitric oxide in the atmosphere." *Opt. Lett.* 31, no. 13 (2006): 2012–2014. DOI: 10.1364/OL.31.002012.

Niering, M., R. Holzwarth, J. Reichert, P. Pokasov, T. Udem, M. Weitz, T.W. Hänsch, P. Lemonde, G. Santarelli, M. Abgrall, P. Laurent, C. Salomon and A. Clairon. "Measurement of the hydrogen 1S-2S transition frequency by phase coherent comparison with a microwave cesium fountain clock." *Phys. Rev. Lett.* 84, no. 24 (2000): 5496–5499. DOI: 10.1103/PhysRevLett.84.5496.

Nwaboh, J.A., O. Werhahn and D. Schiel. "Measurement of CO amount fractions using a pulsed quantum-cascade laser operated in the intra-pulse mode." *Appl. Phys. B* 103, no. 4 (2011): 947–957. DOI: 10.1007/s00340-010-4322-1.

Pakhomycheva, L.N., E.A. Sviridenkov, A.F. Suchkov, L.V. Titova, and S.S. Churilov. "Line structure of generation spectra of lasers with inhomogeneous broadening of the amplification line." *J. Exp. Theor. Phys. Lett.* 12, no. 1 (1970): 43–45.

Parthey, C.G., A. Matveev, J. Alnis, B. Bernhardt, A. Beyer, R. Holzwarth, A. Maistrou, R. Pohl, K. Predehl, T. Udem, T. Wilken, N. Kolachevsky, M. Abgrall, D. Rovera, C. Salomon, P. Laurent and T.W. Hänsch. "Improved measurement of the hydrogen 1S–2S transition frequency." *Phys. Rev. Lett.* 107, no. 20 (2011): article 203001. DOI: 10.1103/PhysRevLett.107. 203001.

Patimisco, P., G. Scamarcio, F.K. Tittel and V. Spagnolo. "Quartz-enhanced photo-acoustic spectroscopy: A review." *Sensors* 14, no. 4 (2014): 6165–6206. DOI: 10.3390/s140406165.

Pawar, A.Y., D.D. Sonawane, K.B. Erande and D.V. Derle. "Terahertz technology and its applications." *Drug Invent. Today* 5, no. 2 (2013): 157–163. DOI: 10.1016/j.dit.2013.03.009.

Pickwell-MacPherson, E. and V.P. Wallace. "Terahertz pulsed imaging: A potential medical imaging modality." *Photodiagnosis Photodyn. Ther.* 6, no. 2 (2009): 128–134. DOI: 10.1016/j .pdpdt.2009.07.002.

Rieker, G.B., J.B. Jeffries and R.K. Hanson. "Calibration-free wavelength-modulation spectroscopy for measurements of gas temperature and concentration in harsh environments. *Appl. Opt.* 48, no. 29 (2009): 5546–5560. DOI: 10.1364/AO.48.005546.

Romanini, D. and K.K. Lehmann. "Ring-down cavity absorption spectroscopy of the very weak HCN overtone bands with six, seven, and eight stretching quanta." *J. Chem. Phys.* 99, no. 9 (1993): 6287–6301. DOI: 10.1063/1.465866.

Rothman, L.S., I.E. Gordon, Y. Babikov, A. Barbe, D.C. Benner, P.F. Bernath, M. Birk, L. Bizzocchi, V. Boudon, L.R. Brown, A. Campargue, K. Chance, E.A. Cohen, L.H. Coudert, V.M. Devi, B.J. Drouin, A. Fayt, J.-M. Flaud, R.R. Gamache, J.J. Harrison, J.-M. Hartmann, C. Hill, J.T. Hodges, D. Jacquemart, A. Jolly, J. Lamouroux, R.J. Le Roy, G. Li, D.A. Long, O.M. Lyulin, C.J. Mackie, S.T. Massie, S. Mikhailenko, H.S.P. Müller, O.V. Naumenko, A.V. Nikitin, J. Orphal, V. Perevalov, A. Perrin, E.R. Polovtseva, C. Richard, M.A.H. Smith, E. Starikova, K. Sung, S. Tashkun, J. Tennyson, G.C. Toon, G. Tyuterev and G. Wagner. "The HITRAN2012 molecular spectroscopic database." *J. Quant. Spectrosc. Radiat. Transfer* 130 (2013): 4–50. DOI: 10.1016/j .jqsrt.2013.07.002.

Santamaria, L., V. Di Sarno, P. De Natale, M. De Rosa, M. Inguscio, S. Mosca, I. Ricciardi, D. Calonico, F. Levie and P. Maddaloni. "Comb-assisted cavity ring-down spectroscopy of a buffer-gas-cooled molecular beam." *Phys. Chem. Chem. Phys.* 18, no. 25 (2016): 16715–16720. DOI: 10 .1039/c6cp02163h.

Scholten, S.K., J.D. Anstie, N.B. Hébert, R.T. White, J. Genest and A.N. Luiten. "Complex direct comb spectroscopy with a virtually imaged phased array." *Opt. Lett.* 41, no. 6 (2016): 1277–1280. DOI: 10.1364/OL.41.001277.

Sepman, A., Y. Ögren, M. Gullberg and H. Wiinikka. "Development of TDLAS sensor for diagnostics of CO, $H_2O$ and soot concentrations in reactor core of pilot-scale gasifier." *Appl. Phys. B* 122, no. 2 (2016): article 29. DOI: 10.1007/s00340-016-6319-x.

Shirasaki, M. "Virtually imaged phased array." *Fujitsu Sci. Tech. J.* 35, no. 1 (1999): 113–125.

Shrivastava, A. and V.B. Gupta. "Methods for the determination of limit of detection and limit of quantitation of the analytical methods." *Chron. Young Scientists* 2, no. 1 (2011): 21–25. DOI:10.4103/2229-5186.79345.

Sur, R., K. Sun, J.B. Jeffries, J.G. Socha and R.K. Hanson. "Scanned-wavelength-modulation-spectroscopy sensor for CO, $CO_2$, $CH_4$ and $H_2O$ in a high-pressure engineering-scale transport-reactor coal gasifier." *Fuel* 150 (2015): 102–111. DOI: 10.1016/j.fuel.2015.02.003.

Thain, S.C., F. van den Bussche, L.J.J. Laarhoven, M.J. Dowson-Day, Z.Y. Wang, E.M. Tobin, F.J.M. Harren, A.J. Milla and D. van der Straeten, "Circadian rhythms of ethylene emission in *Arabidopsis*." *Plant Physiol.* 136, no. 3 (2004): 3751–3761. DOI: 10.1104/pp.104.042523.

Tsai, T. and G. Wysocki. "External-cavity quantum cascade lasers with fast wavelength scanning." *Appl. Phys. B* 100, no. 2 (2010): 243–251. DOI: 10.1007/s00340-009-3865-5.

Udem, T., R. Holzwarth and T.W. Hänsch. "Optical frequency metrology." *Nature* 416, no. 6877 (2002): 233–237. DOI: 10.1038/416233a.

van Helden, J.H., D. Lopatik, A. Nave, N. Lang, P.B. Davies and J. Röpcke. "High resolution spectroscopy of silane with an external-cavity quantum cascade laser: Absolute line strengths of the $v_3$ fundamental band at 4.6µm." *J. Quant. Spectrosc. Radiat. Transfer* 151 (2015): 287–294. DOI: 10.1016/j.jqsrt.2014.10.016.

Venables, D.S., T. Gherman, J. Orphal, J.C. Wenger and A.A. Ruth. "High sensitivity *in situ* monitoring of $NO_3$ in an atmospheric simulation chamber using incoherent broadband cavity-enhanced absorption spectroscopy." *Environ. Sci. Technol.* 40, no. 21 (2006): 6758–6763. DOI: 10.1021/es061076j.

Walther, M., B. Fischer, M. Schall, H. Helm and P.U. Jepsen. "Far-infrared vibrational spectra of all-*trans*, 9-*cis* and 13-*cis* retinal measured by THz time-domain spectroscopy." *Chem. Phys. Lett.* 332, no. 3–4 (2000) 389–395. DOI: 10.1016/S0009-2614(00)01271-9.

Wang, C., Z. Gong, Y.L. Pan and G. Videen. "Optical trap-cavity ring-down spectroscopy as a single-aerosol-particle-scope." *Appl. Phys. Lett.* 107, no. 24 (2015): article 241903. DOI: 10.1063/1.4937467

Werblinski, T., B. Lämmlein, F.J.T. Huber, L. Zigan and S. Will. "Supercontinuum high-speed cavity-enhanced absorption spectroscopy for sensitive multispecies detection." *Opt. Lett.* 41, no. 10 (2016): 2322–2325. DOI: 10.1364/OL.41.002322.

Wu, S., P. Dupre and T.A. Miller. "High-resolution IR cavity ring-down spectroscopy of jet-cooled free radicals and other species." *Phys. Chem. Chem. Phys.* 8, no. 14 (2006): 1682–1689. DOI: 10.1039/b518279d.

Yamaguchi, S., Y. Fukushi, O. Kubota, T. Itsuji, T. Ouchi and S. Yamamoto. "Brain tumor imaging of rat fresh tissue using terahertz spectroscopy." *Sci. Rep.* 6 (2016): article 30124. DOI: 10.1038/srep30124.

Yoo, J., N. Traina, M. Halloran and T. Lee. "Minute concentration measurements of simple hydrocarbon species using supercontinuum laser absorption spectroscopy." *Appl. Spectrosc.* 70, no. 6 (2016): 1063–1071. DOI: 10.1177/0003702816641563.

## Chapter 9    Fluorescence spectroscopy and its implementation

*Books and chapters in books*

Gilbert, A. and J.E. Baggott. *Essentials of Molecular Photochemistry*. Oxford, UK: Blackwell Scientific Publications (1991).

Harvey, E.N. *A History of Luminescence from the Earliest Times until 1900*. Philadelphia, PA, USA: The American Philosophical Society (1957). [Reprint: Mineola, NY, USA: Dover Publications (2005)].

Jameson, D.M. *Introduction to Fluorescence*. Boca Raton, FL, USA: CRC Press (2014).

Lakowicz, J.R. *Principles of Fluorescence Spectroscopy*, 3rd edition. New York, NY, USA: Springer Science + Business Media (2006).

Turro, N.J., V. Ramamurthy and J.C. Scaiano. *Modern Molecular Photochemistry of Organic Molecules*. Sausalito, CA, USA: University Science Books (2010). [Note: This is an updated revision of Turro, N.J. *Modern Molecular Photochemistry*. San Francisco, CA, USA: Benjamin Cummings (1978)].

*Journal articles*

Beer, M. and H.C. Longuet-Higgins. "Anomalous light emission of Azulene." *J. Chem. Phys.* 23, no. 8 (1955): 1390–1391. DOI: 10.1063/1.1742314.

Braslavsky, S.E. "Glossary of terms used in photochemistry, 3rd edition (IUPAC recommendations 2006)." *Pure Appl. Chem.* 79, no. 3 (2007): 293–465. DOI: 10.1351/pac200779030293.

Byron, C.M. and T.C. Werner. "Experiments in synchronous fluorescence spectroscopy for the undergraduate instrumental chemistry course." *J. Chem. Educ.* 68, no. 5 (1991): 433–436 DOI: 10.1021/ed068p433.

Castro, A. and E.B. Shera. "Single-molecule detection: applications to ultrasensitive biochemical analysis." *Appl. Opt.* 34, no. 18 (1995): 3218–3222. DOI: 10.1364/AO.34.003218.

Demtröder, W., M. McClintock and R.N. Zare. "Spectroscopy of $Na_2$ using laser-induced fluorescence." *J. Chem. Phys.* 51, no. 12 (1969): 5495–5508. DOI: 10.1063/1.1671977.

Ezekiel, S. and R. Weiss. "Laser-induced fluorescence in a molecular beam of iodine." *Phys. Rev. Lett.* 20, no. 3 (1968): 91–93. DOI: 10.1103/PhysRevLett.20.91.

Förster, T. "Energiewanderung und Fluoreszenz." *Naturwissenschaften* 33, no. 6 (1946): 166–175. [Translated into English by K. Suhling: Förster, T. "Energy migration and fluorescence." *J. Biomed. Opt.* 17, no. 1 (2012): 011002. DOI: 10.1117/1.JBO.17.1.011002.]

Gaviola, E. "Die Abklingungszeiten der Fluoreszenz von Farbstofflösungen." *Annal. Phys.* 386, no. 23 (1926): 681–710. DOI: 10.1002/andp.19263862304.

Herschel, J.F.W. "On a case of superficial colour presented by a homogeneous liquid internally colourless." *Phil. Trans. R. Soc. London.* 135 (1845): 143–145. DOI: 10.1098/rstl.1845.0004.

Ishikawa-Ankerhold, H.C., R. Ankerhold and G.P.C. Drummen. "Advanced fluorescence microscopy techniques—FRAP, FLIP, FLAP, FRET and FLIM." *Molecules* 17 (2012): 4047–4132. DOI: 10.3390/molecules17044047.

Jablonski, A. "Efficiency of anti-Stokes fluorescence in dyes." *Nature* 131, no. 3319 (1933): 839–840. DOI: 10.1038/131839b0.

Jablonski, A. "Über den Mechanismus der Photolumineszenz von Farbstoffphosphoren." *Z. Phys.* 94, no. 1–2 (1935): 38–46. DOI: 10.1007/BF01330795.

Kasha, M. "Characterization of electronic transitions in complex molecules." *Discuss. Faraday Soc.* 9 (1950): 14–19. DOI: 10.1039/DF9500900014.

Kastrup, L. and S.W. Hell. "Absolute optical cross section of individual fluorescent molecules." *Angew. Chem. Int. Ed.* 43, no. 48 (2004): 6646–6649. DOI: 10.1002/anie.200461337.

Kaye, T.G., A.R. Falk, M. Pittman, P.C. Sereno, L.D. Martin, D.A. Burnham, E. Gong, X. Xu and Y. Wang. "Laser-stimulated fluorescence in paleontology." *PLoS One* 10, no. 5 (2015): e0125923. DOI: 10.1371/journal.pone.0125923.

Lippert, E. "Spektroskopische bestimmung des dipolmomentes aromatischer verbindungen im ersten angeregten singulettzustand." *Z. Elektrochem. Ber. Bunsenges. Phys. Chem.* 61, no. 8 (1957): 962–975. DOI: 10.1002/bbpc.19570610819.

Moerner, W.E. "A dozen years of single-molecule spectroscopy in physics, chemistry and biophysics." *J. Phys. Chem. B* 106, no. 5 (2002): 910–927. DOI: 10.1021/jp012992g.

Moerner, W.E. "Examining nano-environments in solids on the scale of a single, isolated impurity molecule." *Science* 265, no. 5168 (1994): 46–53. DOI: 10.1126/science.265.5168.46.

Moerner, W.E. and L. Kador. "Optical detection and spectroscopy of single molecules in a solid." *Phys. Rev. Lett.* 62, no. 21 (1989): 2535–2538. DOI: 10.1103/PhysRevLett.62.2535.

Moerner, W.E. and M. Orrit. "Illuminating single molecules in condensed matter." *Science* 283, no. 5408 (1999): 1670–1676. DOI: 10.1126/science.283.5408.1670.

Orrit, M. and J. Bernard. "Single pentacene molecules detected by fluorescence excitation in a p-terphenyl crystal." *Phys. Rev. Lett.* 65, no. 21 (1990): 2716–2719. DOI: 10.1103/PhysRevLett.65.2716.

Owicki, J.C. "Fluorescence polarization and anisotropy in high throughput screening: Perspectives and primer." *J. Biomol. Screening* 5, no. 5 (2000): 297–306. DOI: 10.1177/108705710000500501.

Perrin, M.F. "Polarization de la lumière de fluorescence. Vie moyenne de molécules dans létat excité." *J. Phys. Radium* 7, no. 12 (1926): 390–401. DOI: 10.1051/jphysrad:01926007012039000.

Sandanayaka, A.S.D., K. Yoshida, T. Matsushima and C. Adachi. "Exciton quenching behavior of thermally-activated delayed fluorescence molecules by charge carriers." *J. Phys. Chem. C* 119, no. 14 (2015): 7631–7636. DOI: 10.1021/acs.jpcc.5b01314.

Shimomura, O., F.H. Johnson and Y. Saiga. "Extraction, purification and properties of aequorin, a bioluminescent protein from the luminous hydromedusan, Aequorea." *J. Cell. Comp. Physiol.* 59, no. 3 (1962): 223–239. DOI: 10.1002/jcp.1030590302.

Sreenivasan, V.K.A., A.V. Zvyagin and E.M. Goldys. "Luminescent nanoparticles and their applications in the life sciences." *J. Phys. Condens. Matter* 25, no. 19 (2013): 194101. DOI: 10.1088/0953-8984/25/19/194101.

Stern, O. and M. Volmer. "Über die Abklingungszeit der Fluoreszenz." *Phys. Zeitschrift* 20 (1919): 183–188.

Stokes, G.G. "On the change of refrangibility of light: No. I." *Phil. Trans. R. Soc. Lond.* 142 (1852): 463–562. DOI: 10.1098/rstl.1852.0022.

Stokes, G.G. "On the change of refrangibility of light: No. II." *Phil. Trans. R. Soc. Lond.* 143 (1853): 385–396. DOI: 10.1098/rstl.1853.0016.

Strickler, S.J. and R.A. Berg. "Relationship between absorption intensity and fluorescence lifetime of molecules." *J. Chem. Phys.* 37, no. 4 (1962): 814–820. DOI: 10.1063/1.1733166.

Suhling, K., L.M. Hirvonen, J.A. Levitt, P.H. Chung, C. Tregidgo, A. Le Marois, D.A. Rusakov, K. Zheng, S. Ameer-Beg, S. Poland, S. Coelho, R. Henderson and N. Krstajic. "Fluorescence lifetime imaging (FLIM): Basic concepts and some recent developments." *Med. Photonics* 27, no. 1 (2015): 3–40. DOI: 10.1016/j.medpho.2014.12.001.

Tango, W.J., J.K. Link and R.N. Zare. "$K_2$ fluorescence excited by the 6328 Å He-Ne laser line." *Bull. Am. Phys. Soc.* 12, FC12 (1967): 1147.

Tango, W.J., J.K. Link and R.N. Zare. "Spectroscopy of $K_2$ using laser-induced fluorescence." *J. Chem. Phys.* 49, no. 10 (1968): 4264–4268. DOI: 10.1063/1.1669869.

Valeur, B. and M.N. Berberan-Santos. "A brief history of fluorescence and phosphorescence before the emergence of quantum theory." *J. Chem. Edu.* 88, no. 6 (2010): 731–738. DOI: 10.1021/ed100182h.

## Chapter 10      Selected applications of laser-induced fluorescence spectroscopy

*Books and chapters in books*      Campargue, R. (ed.). *Atomic and Molecular Beams: The State of the Art 2000*. Heidelberg, Germany: Springer Verlag (2001).

Cox, R.A. "Atmospheric chemistry." In M.J. Pilling and I.W.M. Smith (eds.). *Atmospheric Chemistry in Modern Gas Kinetics*. Oxford, UK: Blackwell Scientific Publications (1987): pp. 262–283.

Crosley, D.R. "Laser fluorescence detection of atmospheric hydroxil radicals." In J.R. Barker (ed.). *Problems and Progress in Atmospheric Chemistry*. River Edge, NJ, USA: World Scientific Publishing (1995): pp. 256–317.

Jameson, D.M. *Introduction to Fluorescence*. Boca Raton, FL, USA: CRC Press (2014).

Lakowicz, J.R. *Principles of Fluorescence Spectroscopy*, 3rd edition. New York, NY, USA: Springer (2006).

Lee, Y.T. "Reactive scattering I: Non-optical methods." In G. Scoles, D. Bassi, U. Buck and D.C. Laine (eds.). *Atomic and Molecular Beam Methods, Vol. 1*. Oxford, UK: Oxford University Press (1988): pp. 553–569.

Levine, R.D. and R.B. Bernstein. *Molecular Reaction Dynamics and Chemical Reactivity*. Oxford, UK: Oxford University Press (1987).

Moore, J.H. Jr., C.C. Davis and M.A. Coplan. *Building Scientific Apparatus: A Practical Guide to Design and Construction*. Reading, MA, USA: Addison-Wesley (1983): pp. 307–316.

Telle, H.H., A. Gonzalez Ureña and R.J. Donovan. *Laser Chemistry: Spectroscopy, Dynamics and Applications*. Chichester, UK: John Wiley & Sons (2007).

*Journal articles*      Bloss, W.J., J.D. Lee, D.E. Heard, R.A. Salmon, S.J.B. Bauguitte, H.K. Roscoe and A.E. Jones. "Observations of OH and $HO_2$ radicals in coastal Antarctica." *Atmos. Chem. Phys.* 7, no. 16 (2007): 4171-4185. DOI: 10.5194/acp-7-4171-2007.

Czar, M.F., A. Zarrine-Afsar, F. Zosel, I. Konig, R.A. Jockusch, D. Nettels, B. Wunderlich and B. Schuler. "Gas-phase FRET-efficiency measurements to probe the conformation of mass-selected proteins." *Anal. Chem.* 87, no. 15 (2015): 7559-7565. DOI: 10.1021/acs.analchem .5b01591.

Dantus, M., M.J. Rosker and A.H. Zewail. "Femtosecond real-time probing of reactions, II. The dissociation reaction of ICN." *J. Chem. Phys.* 89, no. 10 (1988): 6128-6140. DOI: 10.1063/1 .455428.

Gratton, E. and M. Limkeman. "A continuously variable frequency cross-correlation phase fluorometer with picosecond resolution." *Biophys. J.* 44, no. 3 (1983): 315-324. DOI: 10.1016 /S0006-3495(83)84305-7.

Hard, T.M., R.J. O'Brien, T.B. Cook and G.A. Tsongas. "Interference suppression in HO fluorescence detection." *Appl. Opt.* 18, no. 19 (1979): 3216-3217. DOI: 10.1364/AO.18.003216.

Heard, D.E. and M.J. Pilling. "Measurement of OH and $HO_2$ in the troposphere." *Chem. Rev.* 103, no. 12 (2003): 5163-5198. DOI: 10.1021/cr020522s.

L'Hermite, J.M., G. Rahmat and R. Vetter. "The $Cs(7P) + H_2 \rightarrow CsH + H$ reaction, I: Angular scattering measurements by Doppler analysis." *J. Chem. Phys.* 93, no. 1 (1990): 434-444. DOI: 10.1063/1.459543.

Maiti, S., U. Haupts and W.W. Webb. "Fluorescence correlation spectroscopy: Diagnostics for sparse molecules." *Proc. Natl. Acad. Sci. USA* 94, no. 22 (1997): 11753-11757. DOI: 10.1073 /pnas.94.22.11753.

Murphy, E.J., J.H. Brophy, G.S. Arnold, W.L. Dimpfl and J.L. Kinsey. "Velocity and angular distributions of reactive collisions from Fourier transform Doppler spectroscopy: First experimental results." *J. Chem. Phys.* 70, no. 12 (1979): 5910-5911. DOI: 10.1063/1.437422.

Parkin, D.M., P. Pisani and J. Ferlay. "Estimates of the worldwide incidence of 25 major cancers in 1990." *Int. J. Cancer* 80, no. 6 (1999): 827-841. DOI: 10.1002/(SICI)1097-0215(19990315) 80:6<827::AID-IJC6>3.0.CO;2-P

Rosker, M.J., M. Dantus and A.H. Zewail. "Femtosecond real-time probing of reactions. I. The technique." *J. Chem. Phys.* 89, no. 10 (1988): 6113-6127. DOI: 10.1063/1.455427.

Schultz, A., H.W. Cruse and R.N. Zare. "Laser-induced fluorescence: A method to measure the internal state distribution of reaction products." *J. Chem. Phys.* 57, no. 3 (1972): 1354-1355. DOI: 10.1063/1.1678401.

Sinclair, W.E. and D.W. Pratt. "Structure and vibrational dynamics of aniline and aniline–Ar from high resolution electronic spectroscopy in the gas phase." *J. Chem. Phys.* 105, no. 18 (1996): 7942–7956. DOI: 10.1063/1.472710.

Sun, Y., R.N. Day and A. Periasamy. "Investigating protein–protein interactions in living cells using fluorescence lifetime imaging microscopy." *Nat. Protoc.* 6, no. 9 (2011): 1324–1340. DOI: 10.1038/nprot.2011.364.

Szmacinsky, H.K. and Q. Chang. "Micro- and sub-nanosecond lifetime measurements using a UV light-emitting diode." *Appl. Spectrosc.* 54, no. 1 (2000): 106–109. DOI: 10.1366/0003702001948187.

Vo-Dinh, T., M. Panjehpour, B.F. Overholt and P. Buckley III. "Laser-induced differential fluorescence for cancer diagnosis without biopsy." *Appl. Spectrosc.* 51, no. 1 (1997): 58–63. DOI: 10.1366/0003702971938768.

Zare, R.N. "My life with LIF: A personal account of developing laser-induced fluorescence." *Annu. Rev. Anal. Chem.* 5 (2012): 1–14. DOI: 10.1146/annurev-anchem-062011-143148.

*Other sources*
Cancer statistics center—A service of cancer.org. Atlanta, GA, USA: American Cancer Society (2016). **Access**: https://cancerstatisticscenter.cancer.org/#/.

Fluorophores.org. "Database of fluorescent dyes, properties and applications." **Access**: http://www.fluorophores.tugraz.at.

Heard, D. "Fluorescence assay by gas expansion (FAGE)." Atmospheric Chemistry and Astrochemistry, Department of Chemistry, University of Leeds, UK (2011). **Access**: http://www.chem.leeds.ac.uk/fage/the-fage-technique.html.

Wilhelm, S. "Confocal laser scanning microscopy." Göttingen, Germany: Carl Zeiss Microscopy GmbH (2011).

## Chapter 11    Raman spectroscopy and its implementation

*Books and chapters in books*
Asher, S.A. "Ultraviolet Raman spectroscopy." In J.M. Chalmers and P.R. Griffiths (eds.). *Handbook of Vibrational Spectroscopy*. Chichester, UK: John Wiley & Sons (2002).

Eesley, G.L. *Coherent Raman Spectroscopy*. Oxford, UK: Pergamon Press (2013).

Ferraro, J.R., K. Nakamoto and C.W. Brown. *Introductory Raman Spectroscopy*, 2nd edition. San Diego, CA, USA: Academic Press (2003).

Ghomi, M. (ed.). "Applications of Raman spectroscopy to biology." *Advances in Biomedical Spectroscopy*, vol. 5. Amsterdam, Netherlands: IOS Publishing (2012).

Herzberg, G. *Molecular Spectra and Molecular Structure. II: Infrared and Raman Spectra of Polyatomic Molecules*. New York, NY, USA: van Nostrand Reinhold (1945).

Koningstein, J.A. "Laser Raman spectroscopy." In H.A. Szymanski (ed.). *Raman Spectroscopy*. Heidelberg, Germany: Springer (1967): pp. 82–100 DOI: 10.1007/978-1-4684-3024-0_3.

Larkin, P. *Infrared and Raman Spectroscopy; Principles and Spectral Interpretation*. Amsterdam, Netherlands: Elsevier (2011).

Le Ru, E. and P. Etchegoin. *Principles of Surface-Enhanced Raman Spectroscopy and Related Plasmonic Effects*. Amsterdam, Netherlands: Elsevier Science (2008).

Lewis, I.R. and H.G.M. Edwards (eds.). *Handbook of Raman Spectroscopy: From the Research Laboratory to the Process Line*. New York, NY, USA: Marcel Dekker (2001).

Long, D.A. *The Raman Effect: A Unified Treatment of the Theory of Raman Scattering by Molecules*. Chichester, UK: John Wiley & Sons (2002).

McCreery, R.L. *Raman Spectroscopy for Chemical Analysis*. Chichester, UK: John Wiley & Sons (2005).

Placzek, G. "Rayleigh-Streuung und Raman-Effekt." In E. Marx (ed.). *Quantenmechanik der Materie und Strahlung; Handbuch der Radiologie',* vol. 6.2. Leipzig, Germany: Akademische Verlagsgesellschaft (1934): pp. 209–374. [Translation: A. Werbin. "The Rayleigh and Raman Scattering." UCRL, Trans 526, Livermore, CA, USA (1959).]

Prochazka, M. *Surface-Enhanced Raman Spectroscopy: Bioanalytical, Biomolecular and Medical Applications*. Cham, Switzerland: Springer International (2016).

Smith, E. and G. Dent. *Modern Raman Spectroscopy—A Practical Approach*. Chichester, UK: John Wiley & Sons (2005).

Vandenabeele, P. *Practical Raman Spectroscopy: An Introduction*. Chichester, UK: John Wiley & Sons (2013).

Zhang, S.L. *Raman Spectroscopy and Its Application in Nanostructures.* Chichester, UK: John Wiley & Sons (2012).

*Journal articles*

Albert, S., S. Bauerecker, V. Boudon, LR. Brown, J.-P. Champion, M. Loëte, A. Nikitin and M. Quack. "Global analysis of the high resolution infrared spectrum of methane $^{12}CH_4$ in the region from 0 to 4800 $cm^{-1}$." *Chem. Phys.* 356, no. 1–3 (2009): 131–146. DOI: 10.1016/j.chemphys.2008.10.019.

Ashok, P.C. and K. Dholakia "Microfluidic Raman spectroscopy for bio-chemical sensing and analysis." In W., Fritzsche and J. Popp. *Optical Nano- and Microsystems for Bioanalytics.* Springer Series on Chemical Sensors and Biosensors, vol. 10 (2012): pp. 247–268. DOI: 10.1007/978-3-642-25498-7_9.

Chandra, S., A. Compaan and E. Wiener-Avnear. "Phase matching in coherent anti-stokes Raman scattering." *J. Raman Spectrosc.* 10, no. 1 (1981): 103–105. DOI: 10.1002/jrs.1250100118.

Ciddor, P.E. "Refractive index of air: new equations for the visible and near infrared." *Appl. Opt.* 35, no. 9 (1996): 1566–1573. DOI: 10.1364/AO.35.001566.

De Luca, A.C., K. Dholakia and M. Mazilu. "Modulated Raman spectroscopy for enhanced cancer diagnosis at the cellular level." *Sensors* 15, no. 6 (2015): 13680–13704. DOI: 10.3390/s150613680.

Fodor, S.P.A., R.P. Rava, T.R. Hays and T.G. Spiro. "Ultraviolet resonance Raman spectroscopy of the nucleotides with 266-, 240-, 218-, and 200-nm pulsed laser excitation." *J. Am. Chem. Soc.* 107, no. 6 (1985): 1520–1529. DOI: 10.1021/ja00292a012.

James, T.M., S. Rupp and H.H. Telle. "Trace gas and dynamic process monitoring by Raman spectroscopy in metal-coated hollow glass fibres." *Anal. Methods* 7, no. 6 (2015): 2568–2576. DOI: 10.1039/c4ay02597k.

James, T.M., M. Schlösser, R.J. Lewis, S. Fischer, B. Bornschein and H.H. Telle. "Automated quantitative spectroscopic analysis combining background subtraction, cosmic ray removal, and peak fitting." *Appl. Spectrosc.* 67, no. 8 (2013): 949–959. DOI: 10.1366/12-06766.

Lee, W. and W.R. Lempert. "Enhancement of spectral purity of injection-seeded Ti:sapphire laser by cavity locking and stimulated Brillouin scattering." *Appl. Opt.* 42, no. 21 (2003): 4320–4326. DOI: 10.1364/AO.42.004320.

Li, J.F., Y.F. Huang, S. Duan, R. Pang, D.Y. Wu, B. Ren, X. Xu and Z.Q. Tian. "SERS and DFT study of water on metal cathodes of silver, gold and platinum nanoparticles." *Phys. Chem. Chem. Phys.* 12, no. 10 (2010): 2493–2502. DOI: 10.1039/b919266b.

Lin, K., X.G. Zhou; S.L Liu and Y. Luo. "Identification of free OH and its implication on structural changes of liquid water." *Chin. J. Chem. Phys.* 26, no. 2 (2013): 121–126. DOI: 10.1063/1674-0068/26/02/121-126.

Matousek, P., M. Towrie and A.W. Parker. "Fluorescence background suppression in Raman spectroscopy using combined Kerr gated and shifted excitation Raman difference techniques." *J. Raman Spectrosc.* 33, no. 4 (2002): 238–242. DOI: 10.1002/jrs.840.

Nafie, L.A. "Recent advances in linear and nonlinear Raman spectroscopy."

Part I. *J. Raman Spectrosc.* 38, no. 12 (2007): 1538–1553. DOI: 10.1002/jrs.1902.

Part II. *J. Raman Spectrosc.* 39, no. 12 (2008): 1710–1725. DOI: 10.1002/jrs.2171.

Part III. *J. Raman Spectrosc.* 40, no.12 (2009): 1766–1779. DOI: 10.1002/jrs.2555.

Part IV. *J. Raman Spectrosc.* 41, no. 12 (2010): 1566–1586. DOI: 10.1002/jrs.2859.

Part V. *J. Raman Spectrosc.* 42, no. 12 (2011): 2049–2068. DOI: 10.1002/jrs.3115.

Part VI. *J. Raman Spectrosc.* 43, no. 12 (2012): 1845–1863. DOI: 10.1002/jrs.4221.

Part VII. *J. Raman Spectrosc.* 44, no. 12 (2013): 1629–1648. DOI: 10.1002/jrs.4417.

Part VIII. *J. Raman Spectrosc.* 45, no. 11–12 (2014): 1326–1346. DOI: 10.1002/jrs.4619.

Newton, H., L.L. Walkup, N. Whiting, L. West, J. Carriere, F. Havermeyer, L. Ho, P. Morris, B.M. Goodson and M.J. Barlow. "Comparative study of *in situ* $N_2$ rotational Raman spectroscopy methods for probing energy thermalisation processes during spin-exchange optical pumping." *Appl. Phys. B* 115, no. 2 (2014): 167–172. DOI: 10.1007/s00340-013-5588-x.

Placzek, G. and E. Teller. "Die Rotationsstruktur der Ramanbanden mehratomiger Moleküle." *Z. Phys.* 81, no. 3–4 (1933): 209–258. DOI: 10.1007/BF01338366.

Porto, S.P.S. and D.L. Wood. "Ruby optical maser as a Raman source." *J. Opt. Soc. Am.* 52, no. 3 (1962): 251–252. DOI: 10.1364/JOSA.52.000251.

Raman, C.V. and K.S. Krishnan. "A new type of secondary radiation." *Nature* 121, no. 3048 (1928): 501–502. DOI: 10.1038/121501c0.

Rogalski, A. "Progress in focal plane array technologies." *Prog. Quant. Electron.* 36, no. 2–3 (2012): 342–473. DOI: 10.1016/j.pquantelec.2012.07.001.

Salter, R., J. Chua and M. Hippler. "Cavity-enhanced Raman spectroscopy with optical feedback cw diode lasers for gas phase analysis and spectroscopy." *Analyst* 137, no. 20 (2012): 4669–4676. DOI: 10.1039/C2AN35722D.

Schippers, W., E. Gershnabel, J. Burgmeier, O. Katz, U. Willer, I. S. Averbukh, Y. Silberberg and W. Schade. "Stimulated Raman rotational photo-acoustic spectroscopy using a quartz tuning fork and femtosecond excitation." *Appl. Phys. B* 105, no. 2 (2011): 203–211. DOI: 10.1007/s00340-011 -4725-7.

Siebert, D.R., G.A. West and J.J. Barrett. "Gaseous trace analysis using pulsed photo-acoustic Raman spectroscopy." *Appl. Opt.* 19, no. 1 (1980): 53–60. DOI: 10.1364/AO.19.000053.

Smekal, A. "Zur Quantentheorie der Dispersion." *Naturwissenschaften* 11, no. 43 (1923): 873–875. DOI: 10.1007/BF01576902.

Wang, D., W. Guo, J. Hu, F. Liu, L. Chen, S. Du and Z. Tang. "Estimating atomic sizes with Raman spectroscopy." *Nature Sci. Rep.* 3 (2013): article 1486. DOI: 10.1038/srep01486.

Weber, A. and S.P.S. Porto. "HeNe laser as a light source for high-resolution Raman spectroscopy." *J. Opt. Soc. Am.* 55, no. 8 (1965): 1033–1034. DOI: 10.1364/JOSA.55.001033.

Yakovleva, V.V., H.F. Zhang, G.D. Noojin, M.L. Denton, R.J. Thomas and M.O. Scully. "Stimulated Raman photo-acoustic imaging." *PNAS* 107, no. 47 (2010): 20335–20339. DOI: 10.1073 /pnas.1012432107.

Zhou, Q. and T. Kim. "Review of microfluidic approaches for surface-enhanced Raman scattering." *Sens. Actuators B Chem.* 227 (2016): 504–514. DOI: 10.1016/j.snb.2015.12.069.

*Other sources*

Al-Marashi, J.F.M. and H.H. Telle. "Raman spectra of *S. epidermidis*." Unpublished results (2013).

Paschotta, R. "*RP Fiber Power* – Simulation and Design Software for Fiber Optics, Amplifiers and Fiber Lasers." Bad Durrheim, Germany: RP Photonics Consulting GmbH (2015).

## Chapter 12  Linear Raman spectroscopy

*Books and chapters in books*

Herzberg, G. *Molecular Spectra and Molecular Structure—Vol II: Infrared and Raman of Polyatomic Molecules.* New York, NY, USA: Van Nostrand Reinold (1945). [Reprint: Malabar, FL, USA: Krieger Publishing (1990)].

Long, D.A. *The Raman Effect—A Unified Treatment of the Theory of Raman Scattering by Molecules.* Chichester, UK: John Wiley & Sons (2002).

Placzek, G. "Rayleigh-Streuung und Raman-Effekt." In E. Marx (ed.). *Handbuch der Radiologie* 6, no. 2. Leipzig, Germany: Akademische Verlags-Gesellschaft (1934): pp. 205–374. [Translation: The Rayleigh and Raman Scattering. University of California Radiation Laboratory (UCRL), Trans 526(L) (1962).]

Turner, D.D. and D.N. Whiteman. "Remote Raman spectroscopy: Profiling water vapor and aerosols in the troposphere using Raman LIDARs." In J.M. Chalmers and P.R. Griffiths (eds.). *Handbook of Vibrational Spectroscopy*, vol. 4. Chichester, UK: John Wiley & Sons (2002): pp. 2857–2878.

Vaskova, H. and M. Buckova. "Measuring and identification of oils." In N. Mastorakis, K. Psarris, G. Vachtsevanos, P. Dondon, V. Mladenov, A. Bulucea, I. Rudas and O. Martin (eds.). *Latest Trends on Systems—I; Recent Advances in Electrical Engineering Series* 37. Montclair, NJ, USA: Institute for Natural Sciences and Engineering–INASE (2014): pp. 211–215.

*Journal articles*

Baritaux, J.C., A.C. Simon, E. Schultz, C. Emain, P. Laurent and J.M. Dinten. "A study on identification of bacteria in environmental samples using single-cell Raman spectroscopy: feasibility and reference libraries." *Environ. Sci. Pollut. Res.* 23, no. 9 (2016): 8184–8191. DOI: 10.1007/s11356-015-5953-x.

Barron, L.D. "The development of biomolecular Raman optical activity spectroscopy." *Biomed. Spectrosc. Imaging* 4, no. 3 (2015): 223–253. DOI 10.3233/BSI-150113.

Barron, L.D., M.P. Bogaard and A.D. Buckingham. "Raman scattering of circularly polarized light by optically active molecules." *J. Am. Chem. Soc.* 95, no. 2 (1973): 603–605. DOI: 10.1021/ja00783a058.

Barron, L.D., A.R. Gargaro and Z.Q. Wen. "Vibrational Raman optical activity of peptides and proteins." *J. Chem. Soc. Chem. Commun.* 1990, no. 15 (1990): 1034–1036. DOI: 10.1039/C39900001034.

Barron, L.D., L. Hecht, E.W. Blanch and A.F. Bell. "Solution structure and dynamics of biomolecules from Raman optical activity." *Prog. Biophys. Mol. Biol.* 73, no. 1 (2000): 1–49. DOI: 10.1016/S0079-6107(99)00017-6.

Birch, K.P. and M.J. Downs. "An updated Edlén equation for the refractive index of air." *Metrologia* 30, no. 3 (1993): 155–162. DOI: 10.1088/0026-1394/30/3/004.

Blacksberg, J., E. Alerstam, Y. Maruyama, C.J. Cochrane and G.R. Rossman. "Miniaturized time-resolved Raman spectrometer for planetary science based on a fast single photon avalanche diode detector array." *Appl. Opt.* 55, no. 4 (2016): 739–748. DOI: 10.1364/AO.55.000739.

Carriere, J.T.A., F. Havermeyer and R.A. Heyler. "THz-Raman spectroscopy for explosives, chemical and biological detection." *Proc. SPIE* 8710 (2013): article 87100M. DOI: 10.1117/12.2018095.

Chakraborty, T. and S.N. Rai. "Depolarization ratio and correlation between the relative intensity data and the abundance ratio of various isotopes of liquid carbon tetrachloride at room temperature." *Spectrochim. Acta A* 62, no. 1–3 (2005): 438–445. DOI: 10.1016/j.saa.2005.01.012.

Chen, K., T. Wu, H. Wei, X. Wu and Y. Li. "High spectral specificity of local chemical components characterization with multi-channel shift-excitation Raman spectroscopy." *Sci. Rep.* 5 (2015): article 13952. DOI: 10.1038/srep13952.

Choquette, S.J., E.S. Etz, W.S. Hurst, D.H. Blackburn and S.D. Leigh. "Relative intensity correction of Raman spectrometers: NIST SRMs 2241 through 2243 for 785 nm, 532 nm, and 488 nm/514.5 nm excitation." *Appl. Spectrosc.* 61, no. 2 (2007): 117–129. DOI: 10.1366/000370207779947585.

Cooper, J.B., M. Abdelkader and K.L. Wise. "Sequentially shifted excitation Raman spectroscopy: Novel algorithm and instrumentation for fluorescence-free Raman spectroscopy in spectral space." *Appl. Spectrosc.* 67, no. 8 (2013): 973–984. DOI: 10.1366/12-06852.

de Gelder, J., K. de Gussem, P. Vandenabeele and L. Moens. "Reference database of Raman spectra of biological molecules." *J. Raman Spectrosc.* 38, no. 9 (2007): 1133–1147. DOI: 10.1002/jrs.1734.

Dunstan, P.R., T.G.G. Maffeïs, M.P. Ackland, G.T. Owen and S.P. Wilks. "The correlation of electronic properties with nanoscale morphological variations measured by SPM on semiconductor devices." *J. Phys. Cond. Matter* 15, no. 42 (2004): S3095–S3112. DOI: 10.1088/0953-8984/15/42/008.

Ferralis, N., E.D. Matys, A.H. Knoll, C. Hallmann and R.E. Summons. "Rapid, direct and non-destructive assessment of fossil organic matter via microRaman spectroscopy." *Carbon* 108 (2016): 440–449. DOI: 10.1016/j.carbon.2016.07.039.

Fischer, S., M. Sturm, M. Schlösser, B. Bornschein, G. Drexlin, F. Priester, R.J. Lewis and H.H. Telle. "Monitoring of tritium purity during long-term circulation in the KATRIN test experiment LOOPINO using laser Raman spectroscopy." *Fusion Sci. Technol.* 60, no. 3 (2011): 925–930.

Fuest, F., R.S. Barlow, G. Magnotti, A. Dreizler, I.W. Ekoto and J.A. Sutton. "Quantitative acetylene measurements in laminar and turbulent flames using 1D Raman/Rayleigh scattering." *Combust. Flame* 162, no. 5 (2015): 2248–2255. DOI: 10.1016/j.combustflame.2015.01.021.

Glebov, A.L., O. Mokhun, A. Rapaport, S. Vergnole, V. Smirnov and L.B. Glebov. "Volume Bragg Gratings as ultra-narrow and multiband optical filters." *Proc. SPIE* 8428 (2012): article 84280C. DOI: 10.1117/12.923575.

Herranz, J. and B.P. Stoicheff. "High-resolution Raman spectroscopy of gases: Part XVI. The $\nu_3$ Raman band of methane." *J. Mol. Spectrosc.* 10, no. 1-6 (1963): 448–483. DOI: 10.1016/0022-2852(63)90190-5.

Hutchinson, I.B., R. Ingley, H.G.M. Edwards, L. Harris, M. McHugh, C. Malherbe and J. Parnell. "Raman spectroscopy on Mars: Identification of geological and bio-geological signatures in Martian analogues using miniaturized Raman spectrometers." *Phil. Trans. R. Soc. A* 372, no. 2030 (2014): article 20140204. DOI: 10.1098/rsta.2014.0204.

James, T.M., M. Schlösser, S. Fischer, M. Sturm, B. Bornschein, R.J. Lewisa and H.H. Telle. "Accurate depolarization ratio measurements for all diatomic hydrogen isotopologues." *J. Raman Spectrosc.* 44, no. 6 (2013a): 857–865. DOI: 10.1002/jrs.4283.

James, T.M., M. Schlösser, R.J. Lewis, S. Fischer, B. Bornschein and H.H. Telle. "Automated quantitative spectroscopic analysis combining background subtraction, cosmic ray removal, and peak fitting." *Appl. Spectrosc.* 67, no. 8 (2013b): 949–959. DOI: 10.1366/12-06766.

Jehlička, J., H.G.M. Edwards and P. Vítek. "Assessment of Raman spectroscopy as a tool for the non-destructive identification of organic minerals and biomolecules for Mars studies." *Planet. Space Sci.* 57, no. 5 (2009): 606–613. DOI: 10.1016/j.pss.2008.05.005.

Jennings, D.E., A. Weber and J.W. Brault. "Raman spectroscopy of gases with a Fourier transform spectrometer: The spectrum of $D_2$." *Appl. Opt.* 25, no. 2 (1986): 284–290. DOI: 10.1364/AO.25.000284.

Jones, W.J. "High-resolution Raman spectroscopy of gases and the determination of molecular bond lengths." *Can. J. Phys.* 78, no. 5–6 (2000): 327–390. DOI: 10.1139/p00-041.

Kessler, J., J. Kapitán and P. Bouř. "First-principles predictions of vibrational Raman optical activity of globular proteins." *J. Phys. Chem. Lett.* 6, no. 16 (2015): 3314–3319. DOI: 10.1021/acs.jpclett.5b01500.

Kojima, J.J. and D.G. Fischer. "Multi-scalar analyses of high-pressure swirl-stabilized combustion via single-shot dual-SBG Raman spectroscopy." *Combust. Sci. Technol.* 185, no. 12 (2013): 1735–1761. DOI: 10.1080/00102202.2013.832231.

Lin, K., X. Zhou, Y. Luo and S. Liu. "The microscopic structure of liquid methanol from Raman spectroscopy." *J. Phys. Chem. B* 114, no. 10 (2010): 3567–3573. DOI: 10.1021/jp9121968.

Loeffen, P.W., G. Maskall, S. Bonthron, M. Bloomfield, C. Tombling and P. Matousek. "The performance of spatially offset Raman spectroscopy for liquid explosive detection." *Proc. SPIE* 9995 (2016): article 99950D. DOI: 10.1117/12.2241535.

Malka, D., G. Berkovic, Y. Hammer and Z. Zalevsky. "Super-resolved Raman spectroscopy." *Spectrosc. Lett.* 46, no. 4 (2013): 307–313. DOI: 10.1080/00387010.2012.728553.

Matousek, P., I.P. Clark, E.R.C. Draper, M.D. Morris, A.E. Goodship, N. Everall, M. Towrie, W.F. Finney and A.W. Parker. "Sub-surface probing in diffusely scattering media using spatially offset Raman spectroscopy." *Appl. Spectrosc.* 59, no. 4 (2005): 393–400. DOI: 10.1366/0003702053641450.

Medders, G.R. and F. Paesani. "Infrared and Raman spectroscopy of liquid water through "First-Principles" many-body molecular dynamics." *J. Chem. Theory Comput.* 11, no. 3 (2015): 1145–1154. DOI: 10.1021/ct501131j.

Milani, A., A. Lucotti, V. Russo, M. Tommasini, F. Cataldo, A. Li Bassi and C.S.J. Casari. "Charge transfer and vibrational structure of sp-hybridized carbon atomic wires probed by surface-enhanced Raman spectroscopy." *J. Phys. Chem. C* 115, no. 26 (2011): 12836–12843. DOI: 10.1021/jp203682c.

Milani, A., M. Tommasini, V. Russo, A. Li Bassi, A. Lucotti, F. Cataldo and C.S. Casari. "Raman spectroscopy as a tool to investigate the structure and electronic properties of carbon-atom wires." *Beilstein J. Nanotechnol.* 6 (2015): 480–491. DOI: 10.3762/bjnano.6.49.

Pérez, F.R. and J. Martinez-Frias. "Raman spectroscopy goes to Mars." *Spectrosc. Europe* 18, no. 1 (2006): 18–21.

Placzek, G. and E. Teller. "Die Rotationsstruktur der Ramanbanden mehratomiger Moleküle." *Z. Physik* 81, no. 3 (1933): 209–258. DOI: 10.1007/BF01338366.

Sakai, T., D.N. Whiteman, F. Russo, D.D. Turner, I. Veselovskii, S.H. Melfi, T. Nagai and Y. Mano. "Liquid water cloud measurements using the Raman LIDAR technique: Current understanding and future research needs." *J. Atmos. Oceanic Technol.* 30, no. 7 (2013): 1337–1353. DOI: 10.1175/JTECH-D-12-00099.1.

Samek, O., J.F.M. Al-Marashi and H.H. Telle. "The potential of Raman spectroscopy for the identification of biofilm formation by *Staphylococcus epidermidis*." *Laser Phys. Lett.* 7, no. 5 (2010): 378–383. DOI: 10.1002/lapl.200910154.

Samek, O., P. Zemánek, A. Jonáš and H.H. Telle. "Characterization of oil-producing microalgae using Raman spectroscopy." *Laser Phys. Lett.* 8, no. 10 (2011): 701–709. DOI 10.1002/lapl.201110060.

Sasaki, K., O. Tanaike and H. Konno. "Distinction of Jarosite-group compounds by Raman spectroscopy." *Can. Mineral.* 36, no. 5 (1998): 1225–1235.

Schlösser, M., S. Rupp, H. Seitz, S. Fischer, B. Bornschein, T.M. James and H.H. Telle. "Accurate calibration of the laser Raman system for the Karlsruhe Tritium Neutrino experiment." *J. Mol. Struct.* 1044 (2013): 61–66. DOI: 10.1016/j.molstruc.2012.11.022.

Schlösser, M., S. Rupp, T. Brunst and T.M. James. "Relative intensity correction of Raman systems with National Institute of Standards and Technology Standard Reference Material

2242 in 90°-scattering geometry. *Appl. Spectrosc.* 69, no. 5 (2015): 597–607. DOI: 10.1366 /14-07748.

Sil, S. and S. Umapathy. "Raman spectroscopy explores molecular structural signatures of hidden materials in depth: Universal multiple-angle Raman spectroscopy." *Sci. Rep.* 4 (2014): article 5308. DOI: 10.1038/srep05308.

Stöckel, S., J. Kirchhoff, U. Neugebauer, P. Rösch and J. Popp. "The application of Raman spectroscopy for the detection and identification of microorganisms." *J. Raman Spectrosc.* 47, no. 1 (2016): 89–109. DOI: 10.1002/jrs.4844.

Stoicheff, B.P., C. Cumming, G.E. St. John and H.L. Welsh. "Rotational structure of the $v_3$ Raman band of methane." *J. Chem. Phys.* 20, no. 3 (1952): 498–506. DOI: 10.1063/1.1700446.

Strola, S.A., J.C. Baritaux, E. Schultz, A.C. Simon, C. Allier, I. Espagnon, D. Jary and J.M. Dinten. "Single bacteria identification by Raman spectroscopy." *J. Biomed. Opt.* 19, no. 11 (2014): article 111610. DOI: 10.1117/1.JBO.19.11.111610.

Tan, K.M., I. Barman, N.C. Dingari, G.P. Singh, T.F. Chia and W.L. Tok. "Toward the development of Raman spectroscopy as a non-perturbative online monitoring tool for gasoline adulteration." *Anal. Chem.* 85, no. 3 (2013): 1846–1851. DOI: 10.1021/ac3032349.

Teboul, V., J.L. Godet and Y. Le Duff. "Collection angle dependence of the depolarization ratio in light-scattering experiments." *Appl. Spectrosc.* 46, no. 3 (1992): 476–478. DOI: 10.1366 /0003702924125285.

Wu, S., X. Song, B. Liu, G. Dai, J. Liu, K. Zhang, S. Qin, D. Hua, F. Gao and L. Liu. "Mobile multi-wavelength polarization Raman LIDAR for water vapor, cloud and aerosol measurement." *Opt. Express* 23, no. 26 (2015): 33870–33892. DOI: 10.1364/OE.23.033870.

Yu, Y., K. Lin, X. Zhou, H. Wang, S. Liu and X. Ma. "Precise measurement of the depolarization ratio from photoacoustic Raman spectroscopy." *J. Raman Spectrosc.* 38, no. 9 (2007): 1206–1211. DOI: 10.1002/jrs.1754.

Zhang, X., X.F. Qiao, W. Shi, J.Bin. Wu, D.S. Jiang and P.H. Tan. "Phonon and Raman scattering of two-dimensional transition metal dichalcogenides from monolayer, multilayer to bulk material." *Chem. Soc. Rev.* 44, no. 9 (2015): 2757–2785. DOI: 10.1039/C4CS00282B.

Zhu, F., N.W. Isaacs, L. Hecht and L.D. Barron. "Raman optical activity: A tool for protein structure analysis." *Structure* 13, no. 10 (2005): 1409–1419. DOI: 10.1016/j.str.2005.07.009.

*Other sources*

ASTM E1840. "Standard guide for Raman shift standards for spectrometer calibration." West Conshohocken, PA, USA: ASTM International. Active Standard ASTM E1840 - 96 (2014). DOI: 10.1520/E1840-96R14. **Access**: https://www.astm.org/Standards/E1840.htm.

Mittelholz, A., M. Maloney and G.R. Osinski. "The use of Raman spectroscopy for the 2015 CanMars analogue mission." Presentation at 47th Lunar and Planetary Science Conference. The Woodlands, TX, USA (March 21–25, 2016). Contribution no. 1903, p. 1578. **Access**: http:// adsabs.harvard.edu/abs/2016LPI....47.1578M.

Morris, G.A. "Methane formation in tritium gas exposed to stainless steel." Livermore, CA, USA: Lawrence Livermore Laboratory Report, UCRL-52262 (1977). **Access**: http://www.iaea.org/inis /collection/NCLCollectionStore/_Public/08/346/8346052.pdf

Muray, A.J., R.L. Teitzel and K. Muray. "A stabilized HBLED suitable as calibration standard." Presentation at CIE Conference Light and Lighting Conference with Special Emphasis on LEDs and Solid State Lighting. Budapest, Hungary (May 27–29, 2009). **Access**: http://www .precisionphotometrics.com/pdf-files/CIE_Conference_Paper-HBLED_Standards.pdf.

NIST Standard Reference Database 78. "Atomic transition database, version 5." Gaithersburg, MD, USA: NIST Physical Measurement Laboratory (2016). **Access**: https://www.nist.gov/pml /atomic-spectra-database.

Sacher Lasertechnik GmbH, Marburg, Germany (2016). **Access** to Raman spectral data of gasoline: https://www.sacher-laser.com/applications/overview/raman_spectroscopy/gas_oil.html.

*SpecTools*—an OpenSource software package for Automated quantitative spectroscopic analysis combining background subtraction, cosmic-ray removal and peak fitting, programmed in LabVIEW (2012). **Access**: http://spectools.sourceforge.net.

## Chapter 13    Enhancement techniques in Raman spectroscopy

*Books and chapters in books*

Long, D.A. *The Raman Effect: A Unified Treatment of the Theory of Raman Scattering by Molecules.* Chichester, UK: John Wiley & Sons (2002).

*Journal articles*   Abdullah, H.H., R.M. Kubba and M. Shanshal. "Vibration frequencies shifts of naphthalene and anthracene as caused by different molecular charges." *Z. Naturforsch.* 58a, no. 11 (2003): 645–655.

Albrecht, A.C. "On the theory of Raman intensities." *J. Chem. Phys.* 34, no. 5 (1961): 1476–1484. DOI: 10.1063/1.170103.

Albrecht, M.G. and J.A. Creighton. "Anomalously intense Raman spectra of pyridine at a silver electrode." *J. Am. Chem. Soc.* 99, no. 15 (1977): 5215–5217. DOI: 10.1021/ja00457a071.

Altkorn, R., M. Duval Malinski, R.P. van Duyne and I. Koev. "Intensity considerations in liquid core optical fiber Raman spectroscopy." *Appl. Spectrosc.* 55, no. 4 (2001): 373–381. DOI: 10.1366/0003702011951939.

Asher, S.A. "Ultraviolet resonance Raman spectrometry for detection and speciation of trace polycyclic aromatic hydrocarbons." *Anal. Chem.* 56, no. 4 (1984): 720–724. DOI: 10.1021/ac00268a029.

Buric, M.P., K.P. Chen, J. Falk and S.D. Woodruff. "Enhanced spontaneous Raman scattering and gas composition analysis using a Photonic Crystal Fiber." *Appl. Opt.* 47, no. 23 (2008): 4255–4261. DOI: 10.1364/AO.47.004255.

Buric, M.P. K. Chen, J. Falk, R. Velez and S. Woodruff. "Raman sensing of fuel gases using a reflective coating capillary optical fiber." *Proc. SPIE* 7316 (2009): article 731608. DOI: 10.1117/12.818746.

Buric, M.P., B.T. Chorpening, J.C. Mullen, S.D. Woodruff and J.A. Ranelli. "Field tests of the Raman gas composition sensor." In Proc. Future of Instrumentation International Workshop (FIIW) 2012. Gatlinburg, TN, USA: IEEE (2012). DOI: 10.1109/FIIW.2012.6378319.

Cabalo, J.B., S.K. Saikin, E.D. Emmons, D. Rappoport and A. Aspuru-Guzik. "State-by-state investigation of destructive interference in resonance Raman spectra of neutral tyrosine and the tyrosinate anion with the simplified sum-over-states approach." *J. Phys. Chem. A* 118, no. 41 (2014): 9675–9686. DOI: 10.1021/jp506948h.

Chow, K.K., M. Short, S. Lam, A. McWilliams and H. Zeng. "A Raman cell based on hollow core photonic crystal fiber for human breath analysis." *Med. Phys.* 41, no. 9 (2014): article 092701. DOI: 10.1118/1.4892381.

Conde, J., C. Bao, D. Cui, P.V. Baptista and F. Tian. "Antibody–drug gold nano-antennas with Raman spectroscopic fingerprints for *in vivo* tumour theranostics." *J. Control. Release* 183 (2014): 87–93. DOI: 10.1016/j.jconrel.2014.03.045.

Courtillot, I., J. Morville, V. Motto-Ros and D. Romanini. "Sub-ppb $NO_2$ detection by optical feedback cavity-enhanced absorption spectroscopy with a blue diode laser." *Appl. Phys. B* 85, no. 2–3 (2006): 407–412. DOI: 10.1007/s00340-006-2354-3.

Dinish, U.S., G. Balasundaram, Y.T. Chang and M. Olivo. "Actively targeted *in vivo* multiplex detection of intrinsic cancer biomarkers using biocompatible SERS nano-tags." *Sci Rep.* 4 (2014): article 4075. DOI: 10.1038/srep04075.

Eaglesfield, C.C. "Optical pipeline." *J. Inst Elec. Eng.* 8, no. 85 (1962): 34–36. DOI: 10.1049/jiee-3.1962.0019.

Eiler, J.M. "Clumped-isotope geochemistry—The study of naturally-occurring, multiply-substituted isotopologues." *Earth Planet. Sci. Lett.* 262, no. 3–4 (2007): 309–327. DOI: 10.1016/j.epsl.2007.08.020.

Fenner, W.R., H.A. Hyatt, J.M. Kellam and S.P.S. Porto. "Raman cross section of some simple gases." *J. Opt. Soc. Am.* 63, no. 1 (1973): 73–77. DOI: 10.1364/JOSA.63.000073.

Fluhr, J.W., P. Caspers, J.A. van der Pol, H. Richter, W. Sterry, J. Lademann and M.E. Darvin. "Kinetics of carotenoid distribution in human skin in vivo after exogenous stress: Disinfectant and wIRA-induced carotenoid depletion recovers from outside to inside." *J. Biomed. Opt.* 16, no. 3 (2011): article 035002. DOI: 10.1117/1.3555183.

Ghosh, M., L. Wang and S.A. Asher. "Deep-ultraviolet resonance Raman excitation profiles of $NH_4NO_3$, PETN, TNT, HMX and RDX." *Appl. Spectrosc.* 66, no. 9 (2012): 1013–1021. DOI: 10.1366/12-06626.

Gonzálvez, A.G., N.L. Martínez, H.H. Telle and Á. González Ureña. "Monitoring LED-induced carotenoid increase in grapes by transmission resonance Raman spectroscopy." *Chem. Phys. Lett.* 559 (2013): 26–29. DOI: 10.1016/j.cplett.2012.12.054.

Hanf, S., T. Bögözi, R. Keiner, T. Frosch and J. Popp. "Fast and highly sensitive fiber-enhanced Raman spectroscopic monitoring of molecular $H_2$ and $CH_4$ for point-of-care diagnosis of

*Malabsorption Disorders* in exhaled human breath." *Anal. Chem.* 87, no. 2 (2015): 982–988. DOI: 10.1021/ac503450y.

Hanf, H., R. Keiner, D. Yan, J. Popp and T. Frosch. "Fiber-enhanced Raman multigas spectroscopy: A versatile tool for environmental gas sensing and breath analysis." *Anal. Chem.* 86, no. 11 (2014): 5278–5285. DOI: 10.1021/ac404162w.

Harmon, P.A. and S.A. Asher. "Differentiation between resonance Raman scattering and single vibrational level fluorescence 5100 cm$^{-1}$ above the $^1B_{2u}$ origin in benzene vapor by means of excitation profiles." *J. Chem. Phys.* 88, no. 5 (1988): 2925–2938. doi: 10.1063/1.453985.

He, L., T. Chen and T.P. Labuza. "Recovery and quantitative detection of thiabendazole on apples using a surface swab capture method followed by surface-enhanced Raman spectroscopy." *Food Chem.* 148 (2014): 42–46. DOI: 10.1016/j.foodchem.2013.10.023.

Hickman, R.S. and L. Liang. "Intra-cavity laser Raman spectroscopy using a commercial laser." *Appl. Spectrosc.* 27, no. 6 (1973): 425–427. DOI: 10.1366/000370273774333074.

Hildebrandt, P.G., R.A. Copeland, T.G. Spiro, J. Otlewski, M. Laskowski Jr. and F.G. Pendergast. "Tyrosine hydrogen-bonding and environmental effects in proteins probed by UV resonance Raman spectroscopy." *Biochemistry* 27, no. 15 (1988): 5426–5433. DOI: 10.1021/bi00415a007.

Hippler, M. "Cavity-enhanced Raman spectroscopy of natural gas with optical feedback cw-diode lasers." *Anal. Chem.* 87, no. 15 (2015): 7803–7809. DOI: 10.1021/acs.analchem.5b01462.

Hu, S., I.K. Morris, J.P. Singh, K.M. Smith and T.G. Spiro. "Complete assignment of cytochrome c resonance Raman spectra via enzymic reconstitution with isotopically labeled hemes." *J. Am. Chem. Soc.* 115, no. 26 (1993): 12446–12458. DOI: 10.1021/ja00079a028.

James, T.M., S. Ruppa and H.H. Telle. "Trace gas and dynamic process monitoring by Raman spectroscopy in metal-coated hollow glass fibres." *Anal. Methods* 7, no. 6 (2015): 2568–2576. DOI: 10.1039/c4ay02597k.

Jochum, T., L. Rahal, R.J. Suckert, J. Poppa and T. Frosch. "All-in-one: A versatile gas sensor based on fiber enhanced Raman spectroscopy for monitoring postharvest fruit conservation and ripening." *Analyst* 141, no. 6 (2016): 2023–2029. DOI: 10.1039/c5an02120k.

Jones, C.M., T.A. Naim, M. Ludwig, J. Murtaugh, P.L. Flaugh, J.M. Dudik, C.R. Johnson and S.A. Asher. "Analytical applications of ultraviolet resonance Raman spectroscopy." *Trends Anal. Chem.* 4, no. 3 (1985): 75–80. DOI: 10.1016/0165-9936(85)87089-8.

Khetani, A., J. Riordon, V. Tiwari, A. Momenpour, M. Godin and H. Anis. "Hollow core photonic crystal fiber as a reusable Raman biosensor." *Opt. Express* 21, no. 10 (2013): 12340–12350. DOI:10.1364/OE.21.012340.

Manfred, K.M., G.A.D. Ritchie, N. Lang, J. Röpcke and J.H. van Helden. "Optical feedback cavity-enhanced absorption spectroscopy with a 3.24 μm interband cascade laser." *Appl. Phys. Lett.* 106, no. 22 (2015): article 221106. DOI: 10.1063/1.4922149.

Martin, D., A. González Gonzalvez, R. Mateos Medina and A. González Ureña. "Modeling tomato ripening based on carotenoid Raman spectroscopy: Experimental versus kinetic model." *Appl. Spectrosc.* 71, no. 6 (2017): 1310–1320. DOI: 10.1177/0003702816681012.

Matsuura, Y., R. Kasahara, T. Katagiri and M. Miyagi. "Hollow infrared fibers fabricated by glass-drawing technique." *Opt. Express* 10, no. 12 (2002): 488–492. DOI: 10.1364/OE.10.000488.

Measor, P., L. Seballos, D.L. Yin, J.Z. Zhang, E.J. Lunt, A.R. Hawkins and H. Schmidt. "On-chip surface-enhanced Raman scattering detection using integrated liquid-core waveguides." *Appl. Phys. Lett.* 90, no. 21 (2007): article 211107. DOI: 10.1063/1.2742287.

Miller, J.J., R.L. Brooks and J.L. Hunt. "Raman spectrum of solid hydrogen deuteride." *Phys. Rev. B* 47, no. 22 (1993): 14886–14897. DOI: 10.1103/PhysRevB.47.14886.

Müller, C., L. David, V. Chiş and S.C. Pînzaru. "Detection of thiabendazole applied on citrus fruits and bananas using surface enhanced Raman scattering." *Food Chem.* 145 (2014): 814–820. DOI: 10.1016/j.foodchem.2013.08.136.

Osawa, M., Y. Kato, T. Watanabe, M. Miyagi, A. Abe, M. Aizawa and S. Onodera. "Fabrication of fluorocarbon polymer-coated silver hollow-glass waveguides for the infrared by the liquid-phase coating method." *Opt. Laser Technol.* 27, no. 6 (1995): 393–396. DOI: 10.1016/0030-3992(95)00041-0.

Persichetti, G., G. Testa and R. Bernini Institute. "Opto-fluidic jet waveguide enhanced Raman spectroscopy." *Sens. Actuators B Chem.* 207, Part A (2015): 732–739. DOI: 10.1016/j.snb.2014.10.060.

Qi, D. and A.J. Berger. "Correction method for absorption-dependent signal enhancement by a liquid-core optical fiber." *Appl. Opt.* 45, no. 3 (2006): 489–494. DOI: 10.1364/AO.45.000489.

Qi, D. and A.J. Berger. "Chemical concentration measurement in blood serum and urine samples using liquid-core optical fiber Raman spectroscopy." *Appl. Opt.* 46, no. 10 (2007): 1726-1734. DOI: 10.1364/AO.46.001726.

Rupp, S., A. Off, H. Seitz-Moskaliuk, T.M. James and H.H. Telle. "Improving the detection limit in a capillary Raman system for *in situ* gas analysis by means of fluorescence reduction." *Sensors* 15, no. 9 (2015): 23110-23125. DOI:10.3390/s150923110.

Salter, R., J. Chu and M. Hippler. "Cavity-enhanced Raman spectroscopy with optical feedback cw diode lasers for gas phase analysis and spectroscopy." *Analyst* 137, no. 20 (2012): 4669-4676. DOI: 10.1039/c2an35722d.

Spiro, T.G. and T.C. Strekas. "Resonance Raman spectra of hemoglobin and cytochrome c: Inverse polarization and vibronic scattering." *Proc. Nat. Acad. Sci. USA* 69, no. 9 (1972): 2622-2626. PMCID: PMC427002.

Taylor, D.J., M. Glugla and R.D. Penzhorn. "Enhanced Raman sensitivity using an actively stabilized external resonator." *Rev. Sci. Instrum.* 72, no. 4 (2001): 1970-1975. DOI: 10.1063/1.1353190.

Thorstensen, J., K.H. Haugholt, A. Ferber, K.A.H. Bakke and J. Tschudi. "Low-cost resonant cavity Raman gas probe for multi-gas detection." *J. Eur. Opt. Soc. – RP* 9 (2014): article 14054. DOI: 10.2971/jeos.2014.14054.

Walrafen, G.E. and J. Stone. "Intensification of spontaneous Raman spectra by use of liquid core optical fibers." *Appl. Spectrosc.* 26, no. 6 (1972): 585-589. DOI: 10.1366/000370272774351688.

Ziegler, L.D. "Rotational Raman excitation profiles of symmetric tops: Sub-picosecond rotation-dependent lifetimes in the Ã-state of ammonia." *J. Chem. Phys.* 86, no. 4 (1987): 1703-1704. DOI: 10.1063/1.452169

*Other sources*

Coursey, J.S., D.J. Schwab, J.J. Tsai and R.A. Dragoset. "Atomic weights and isotopic composition with relative atomic masses." Gaithersburg, MD, USA: NIST Physical Measurement Laboratory (2013). **Access**: http://www.nist.gov/pml/data/comp.cfm. Date created: 23 August 2009; last updated: 15 May 2013. [Last accessed on 10 February 2016.]

Kaye & Laby Online. "Tables of Physical and Chemical Constants: 3.1.4 Composition of the Earth's Atmosphere," v2.0—updated 16 October 2012. Teddington, UK: National Physics Laboratory. **Access**: http://www.kayelaby.npl.co.uk. [Last accessed on 10 February 2016.]

Mitchell, J.R. "Diode laser pumped Raman gas analysis system with reflective hollow tube gas cell." United States Patent 5,521,703 (28 May 1996).

NKT Photonics. "Hollow Core Photonic Bandgap Fiber HC-532-02." Birkenrod, Denmark: NKT Photonics A/S (2015). **Access**: http://www.nktphotonics.com/wp-content/uploads/2015/01/HC-532.pdf. [Last accessed on 17 March 2016.]

Opto-Knowledge. "VNIR Hollow Fiber Optics." Torrance, CA, USA: Opto-Knowledge Systems Inc (2015). **Access**: www.optoknowledge.com/documents/fliers/Flyer_OKSI_HollowVNIR_2015_03_V1.pdf

Schlösser, M. T.M. James, M. Ojo Rabe and H.H. Telle. The presented data are from a systematic study of a V-shape CERS setup, utilizing SLM and MLM laser diodes at $\lambda_L \approx 660$nm. Measurement campaign at Tritium Laboratory Karlsruhe, Karlsruhe Institute of Technology (KIT), and at Instituto Pluridisciplinar, Universidad Complutense Madrid (2013-2014).

## Chapter 14     Nonlinear Raman spectroscopy

*Books and chapters in books*

Aroca, R. *Surface-Enhanced-Vibrational Spectroscopy*. Chichester, UK: John Wiley & Sons (2006).

Kneipp, K., M. Moskovits and H. Kneipp (eds.). *Surface-Enhanced Raman Scattering: Physics and Applications*. Heidelberg, Germany: Springer (2006).

Le Ru, E. and P. Etchegoin. *Principles of Surface-Enhanced Raman Spectroscopy and Related Plasmonic Effects*. Amsterdam, Netherlands: Elsevier Science (2009).

Ozaki, Y., K. Kneipp, and R. Aroca (eds.). *Frontiers of Surface-Enhanced Raman Scattering: Single Nano-particles and Single Cells*. Chichester, UK: John Wiley & Sons (2014).

Potma, E.O. and S. Mukamel. "Theory of coherent Raman scattering." In J.X. Cheng and X.S. Xie (eds.). *Coherent Raman Scattering Microscopy*. Boca Raton, FL, USA: CRC Press (2013).

Procházka, M. *Surface-Enhanced Raman Spectroscopy: Bioanalytical, Biomolecular and Medical Applications*. Heidelberg, Germany: Springer (2016).

Schlucker, S. (ed.). *Surface-Enhanced Raman Spectroscopy: Analytical, Biophysical and Life Science Applications*. Chichester, UK: John Wiley & Sons (2010).

Ziegler, L.D. "Hyper-Raman spectroscopy: Theory and instrumentation." In J.M. Chalmers and P.R. Griffiths (eds.). *Handbook of Vibrational Spectroscopy*. Chichester, UK: Wiley (2001). DOI: 10.1002/0470027320.s0412.

*Journal articles*     Albrecht, M.G. and J.A. Creighton. "Anomalously intense Raman spectra of pyridine at a silver electrode." *J. Am. Chem. Soc.* 99, no. 15 (1977): 5215–5217. DOI: 10.1021/ja00457a071.

Alden, A., H. Edner and S. Svanberg. "Coherent anti-Stokes Raman spectroscopy (CARS) applied in combustion probing." *Phys. Scripta* 27, no. 1 (1983): 29–38. DOI: 10.1088/0031-8949/27/1/004.

Anderson, M.S. "Locally enhanced Raman spectroscopy with an atomic force microscope (AFM-TERS)." *Appl. Phys. Lett.* 76, no. 21 (2000): 3130–3133. DOI:10.1063/1.126546.

Chiang, N., N. Jiang, D.V. Chulhai, E.A. Pozzi, M.C. Hersam, L. Jensen, T. Seideman and R.P. Van Duyne. "Molecular-resolution interrogation of a porphyrin monolayer by ultrahigh vacuum tip-enhanced Raman and fluorescence spectroscopy." *Nano Lett.* 15, no. 6 (2015): 4114–4120. DOI: 10.1021/acs.nanolett.5b01225.

Chung, Y.C. and L.D. Ziegler. "The vibronic theory of resonance hyper-Raman scattering." *J. Chem. Phys.* 88, no. 12 (1988): 7287–7294. DOI: 10.1063/1.454339.

Crampton, K.T., A. Zeytunyan, A.S. Fast, F.T. Ladani, A. Alfonso-Garcia, M. Banik, S. Yampolsky, D.A. Fishman, E.O. Potma and V.A. Apkarian. "Ultrafast coherent Raman scattering at plasmonic nano-junctions." *J. Phys. Chem. C* 120, no. 37 (2016): 20943–20953. DOI: 10.1021/acs.jpcc.6b02760.

Dantus, M., R.M. Bowman and A.H. Zewail. "Femtosecond laser observations of molecular vibration and rotation." *Nature* 343, no. 6260 (1990): 737–739. DOI: 10.1038/343737a0.

Dieringer, J.A., R.B. Lettan II, K.A. Scheidt and R.P. van Duyne. "A frequency domain existence proof of single-molecule surface-enhanced Raman spectroscopy." *J. Am. Chem. Soc.* 129, no. 51 (2007): 16249–16256. DOI: 10.1021/ja077243c.

Dieringer, J.A., A.D. McFarland, N.C. Shah, D.A. Stuart, A.V. Whitney, C.R. Yonzon, M.A. Young, X. Zhang and R.P. Van Duyne. "Surface enhanced Raman spectroscopy: New materials, concepts, characterization tools, and applications." *Faraday Discuss.* 132 (2006): 9–26. DOI: 10.1039/B513431P.

Ding, S.Y., J. Yi, J,F. Li, B. Ren, D.Y. Wu, R. Panneerselvam and Z.Q. Tian. "Nanostructure-based plasmon-enhanced Raman spectroscopy for surface analysis of materials." *Nature Rev. Mats.* 1 (2016): article 16021. DOI: 10.1038/natrevmats.2016.21.

Doménech, J.L. and M. Cueto. "Sensitivity enhancement in high resolution stimulated Raman spectroscopy of gases with hollow-core photonic crystal fibers." *Opt. Lett.* 38, no. 20 (2013): 4074–4077. DOI: 10.1364/OL.38.004074.

El-Diasty, F. "Coherent anti-Stokes Raman scattering: Spectroscopy and microscopy." *Vibr. Spectrosc.* 55, no. 1 (2011): 1–37. DOI: 10.1016/j.vibspec.2010.09.00.

Fabelinsky, V.I., B.B. Krynetsky, L.A. Kulevsky, V.A. Mishin, A.M. Prokhorov, A.D. Savel'ev and V.V. Smirnov. "High resolution cw cars spectroscopy of the Q-branch of the $v_2$-band in $C_2H_2$." *Opt. Commun.* 20, no. 3 (1977): 389–391. DOI:10.1016/0030-4018(77)90211-5

Fickenscher, M. and A. Laubereau. "High-precision femtosecond CARS of simple liquids." *J. Raman Spectrosc.* 21, no. 12 (1990): 857–861. DOI: 10.1002/jrs.1250211215

Fleischmann, M., P.J. Hendra and A.J. McQuillan. "Raman spectra of pyridine adsorbed at a silver electrode." *Chem. Phys. Lett.* 26, no. 2 (1974): 163–166. DOI: 10.1016/0009-2614(74)85388-1.

Gustafson, K.E., J.C. McDaniel and R.L. Byer. "High-resolution continuous-wave coherent anti-Stokes Raman spectroscopy in a supersonic jet." *Opt. Lett.* 7, no. 9 (1982): 434–436. DOI: 10.1364/OL.7.000434.

He, L., T. Chen and T.P. Labuza. "Recovery and quantitative detection of thiabendazole on apples using a surface swab capture method followed by surface-enhanced Raman spectroscopy." *Food Chem.* 148 (2014): 42–46. DOI: 10.1016/j.foodchem.2013.10.023.

Jeanmaire, D.L. and R.P. van Duyne. "Surface Raman electrochemistry, Part I. Heterocyclic, aromatic and aliphatic amines adsorbed on the anodized silver electrode." *J. Electroanal. Chem.* 84, no. 1 (1977): 1–20. DOI: 10.1016/S0022-0728(77)80224-6.

Kerker, M. "Electromagnetic model for surface-enhanced Raman scattering (SERS) on metal colloids." *Acc. Chem. Res.* 17, no. 8 (1984): 271–277. DOI: 10.1021/ar00104a002.

Kim, H.M., H. Kim, I. Young, S.M. Jin and Y.D. Suh. "Time-gated pre-resonant femtosecond stimulated Raman spectroscopy of diethylthiatricarbocyanine iodide." *Phys. Chem. Chem. Phys.* 16, no. 11 (2014): 5312–5318. DOI: 10.1039/c3cp54870h.

Kneipp, K., A.S. Haka, H. Kneipp, K. Badizadegan, N. Yoshizawa, C. Boone, K.E. Shafer-Peltier, J.T. Motz, R.R. Dasari and M.S. Feld. "Surface-enhanced Raman spectroscopy in single living cells using gold nanoparticles." *Appl. Spectrosc.* 56, no. 2 (2002): 150–154. DOI: 10.1366/0003702021954557.

Kneipp, K., Y. Wang, H. Kneipp, L.T. Perelman, I. Itzkan, R.R. Dasari and M.S. Feld. "Single molecule detection using surface-enhanced Raman scattering (SERS)." *Phys. Rev. Lett.* 78, no. 9 (1997): 1667–1670. DOI: 10.1103/PhysRevLett.78.1667.

Knutsen, K.P., B.M. Messer, R.M. Onorato and R.J. Saykally. "Chirped coherent anti-Stokes Raman scattering for high spectral resolution spectroscopy and chemically selective imaging." *J. Phys. Chem. B* 110, no. 12 (2006): 5854–5864. DOI: 10.1021/jp052416a.

Kukura, P., D.W. McCamant and R.A. Mathies. "Femtosecond stimulated Raman spectroscopy." *Annu. Rev. Phys. Chem.* 58 (2007): 461–488. DOI: 10.1146/annurev.physchem.58.032806.104456.

Kulatilaka, W.D., H.U. Stauffer, J.R. Gord and S. Roy. "One-dimensional single-shot thermometry in flames using femtosecond-CARS line imaging." *Opt. Lett.* 36, no. 21 (2011): 4182–4184. DOI: 10.1364/OL.36.004182.

Kumar, N., S. Mignuzzi, W. Su and D. Roy. "Tip-enhanced Raman spectroscopy: Principles and applications." *Eur. Phys. J. Tech. Instrum.* 2, no. 1 (2015): article 9. DOI: 10.1140/epjti/s40485-015-0019-5.

Lang, T. and M. Motzkus. "Single-shot femtosecond coherent anti-Stokes Raman-scattering thermometry." *J. Opt. Soc. Am. B* 19, no. 2 (2002): 340–344. DOI: 10.1364/JOSAB.19.000340.

Langelüddecke, L., P. Singh and V. Deckert. "Exploring the nanoscale: Fifteen years of tip-enhanced Raman spectroscopy." *Appl. Spectrosc.* 69, no. 12 (2015): 1357–1371. DOI: 10.1366/15-08014.

Litorja, M., C.L. Haynes, A.J. Haes, T.R. Jensen and R.P. Van Duyne. "Surface-enhanced Raman scattering detected temperature programmed desorption: Optical properties, nano-structure, and stability of silver film over $SiO_2$ nano-sphere surfaces." *J. Phys. Chem. B* 105, no. 29 (2001): 6907–6915. DOI: 10.1021/jp010333y.

Maker, P.D. and R.W. Terhune. "Study of optical effects due to an induced polarization third order in the electric field strength." *Phys. Rev.* 137, no. 3a (1965): A801–A817. DOI: 10.1103/PhysRev.137.A801.

McCamant, D.W., P. Kukura and R.A. Mathies. "Femtosecond time-resolved stimulated Raman spectroscopy: Application to the ultrafast internal conversion in β-carotene." *J. Phys. Chem. A* 107, no. 40 (2003): 8208–8214. DOI: 10.1021/jp030147n.

McFarland, A.D., M.A. Young, J.A. Dieringer and R.P. Van Duyne. "Wavelength-scanned surface-enhanced Raman excitation spectroscopy." *J. Phys. Chem. B* 109, no. 22 (2005): 11279–11285. DOI: 10.1021/jp050508u.

Milojevich, C.B., B.K. Mandrell, H.K. Turley, V. Iberi, M.D. Best and J.P. Camden. "Surface-enhanced hyper-Raman scattering from single molecules." *J. Phys. Chem. Lett.* 4, no. 20 (2013): 3420–3423. DOI: 10.1021/jz4017415.

Müller, C., L. David, V. Chiş and S.C. Pînzaru. "Detection of thiabendazole applied on citrus fruits and bananas using surface enhanced Raman scattering." *Food Chem.* 145 (2014): 814–820. DOI: 10.1016/j.foodchem.2013.08.136.

Myers Kelley, A. "Hyper-Raman scattering by molecular vibrations." *Ann. Rev. Phys. Chem.* 61 (2010): 41–61. DOI: 10.1146/annurev.physchem.012809.103347.

Neddersen, J.P., S.A. Mounter, J.M. Bostick and C.K. Johnson. "Non-resonant hyper-Raman and hyper-Rayleigh scattering in benzene and pyridine." *J. Chem. Phys.* 90, no. 9 (1989): 4719–4726. DOI: 10.1063/1.456592.

Nie, S. and S.R. Emory. "Probing single molecules and single nanoparticles by surface-enhanced Raman scattering." *Science* 275, no. 5303 (1997): 1102–1106. DOI: 10.1126/science.275.5303.1102.

Owyoung, A. "High-resolution cw stimulated Raman spectroscopy in molecular hydrogen." *Opt. Lett.* 2, no. 4 (1978): 91–93. DOI: 10.1364/OL.2.000091.

Polavarapu, L., A. La Porta, S.M. Novikov, M. Coronado-Puchau and L.M. Liz-Marzán. "Pen-on-Paper approach toward the design of universal surface-enhanced Raman scattering substrates." *Small* 10, no. 15 (2014): 3065–3071. DOI: 10.1002/smll.201400438.

Radziuk, D. and H. Moehwald. "Prospects for plasmonic hot spots in single molecule SERS towards the chemical imaging of live cells." *Phys. Chem. Chem. Phys.* 17, no. 33 (2015): article 21072. DOI: 10.1039/c4cp04946b.

Roy, S., J.R. Gord and A.K. Patnaik. "Recent advances in coherent anti-Stokes Raman scattering spectroscopy: Fundamental developments and applications in reacting flows." *Prog. Energy Combust. Sci.* 36, no. 2 (2010): 280–306. DOI: 0.1016/j.pecs.2009.11.001.

Schmitt, M., G. Knopp, A. Materny and W. Kiefer. "Femtosecond time-resolved coherent anti-Stokes Raman scattering for the simultaneous study of ultrafast ground and excited state dynamics: Iodine vapour." *Chem. Phys. Lett.* 270, no. 1–2 (1997): 9–15. DOI: 10.1016/S0009-2614(97)00347-3.

Shoute, L.C.T., M. Blanchard-Desce and A. Myers Kelley. "Tunable resonance hyper-Raman spectroscopy of second-order nonlinear optical chromophores." *J. Chem. Phys.* 121, no. 15 (2004): 7045–7048. DOI: 10.1063/1.1806131.

Simmons Jr, P.D., H.K. Turley, D.W. Silverstein, L. Jensen and J.P. Camden. "Surface-enhanced spectroscopy for higher-order light scattering: A combined experimental and theoretical study of second hyper-Raman scattering." *J. Phys. Chem. Lett.* 6, no. 24 (2015): 5067–5071. DOI: 10.1021/acs.jpclett.5b02342.

Sonntag, M.D., E.A. Pozzi, N. Jiang, M.C. Hersam and R.P. van Duyne. "Recent advances in tip-enhanced Raman spectroscopy." *J. Phys. Chem. Lett.* 5, no. 18 (2014): 3125–3130. DOI: 10.1021/jz5015746.

Stadler, J., T. Schmid and R. Zenobi. "Developments in and practical guidelines for tip-enhanced Raman spectroscopy." *Nanoscale* 4, no. 6 (2012): 1856–1870. DOI: 10.1039/C1NR11143D.

Stiles, P.L., J.A. Dieringer, N.C. Shah and R.P. van Duyne. "Surface-enhanced Raman spectroscopy." *Annu. Rev. Anal. Chem.* 1 (2008): 601–626. DOI: 10.1146/annurev.anchem.1.031207.112814.

Stöckle, R.M., Y.D. Suh, V. Deckert and R. Zenobi. "Nanoscale chemical analysis by tip-enhanced Raman spectroscopy." *Chem. Phys. Lett.* 318, no. 1–3 (2000): 131–136. DOI: 10.1016/S0009-2614(99)01451-7.

Stolen, R.H., E.P. Ippen and A.R. Tynes. "Raman oscillation in glass optical waveguides." *Appl. Phys. Lett.* 20, no. 2 (1972): 62–64. DOI: 10.1063/1.1654046.

Terhune, R.W., P.D. Maker and C.M. Savage. "Measurements of nonlinear light scattering." *Phys. Rev. Lett.* 14, no. 17 (1965): 681–685. DOI: 10.1103/PhysRevLett.14.681.

Tolles, W.M., J.W. Nibler, J.R. McDonald and A.B. Harvey. "A review of the theory and application of coherent anti-Stokes Raman spectroscopy (CARS)." *Appl. Spectrosc.* 31, no. 4 (1977): 253–271. DOI: 10.1366/000370277774463625.

Tuesta, A.D., A. Bhuiyan, R.P. Lucht and T.S. Fisher. "$H_2$ mole fraction measurements in a microwave plasma using coherent anti-Stokes Raman scattering spectroscopy." *J. Micro Nano-Manuf.* 4, no. 1 (2015): article 011005. DOI: 10.1115/1.4031916.

Wang, A.X. and X. Kong. "Review of recent progress of plasmonic materials and nano-structures for surface-enhanced Raman scattering." *Materials* 8, no. 6 (2015): 3024–3052. DOI: 10.3390/ma8063024.

Wang, Y. and J. Irudayaraj. "Surface-enhanced Raman spectroscopy at single-molecule scale and its implications in biology." *Phil. Trans. R. Soc. B* 368, no. 1611 (2013): article 20120026. DOI: 10.1098/rstb.2012.0026.

Yampolsky, S., D.A. Fishman, S. Dey, E. Hulkko, M. Banik, E.O. Potma and V.A. Apkarian. "Seeing a single molecule vibrate through time-resolved coherent anti-Stokes Raman scattering." *Nature Photon.* 8, no. 8 (2014): 650–656. DOI: 10.1038/nphoton.2014.143.

Yang, S., X. Dai, B. Boschitsch Stogin and T.S. Wong. "Ultrasensitive surface-enhanced Raman scattering detection in common fluids." *Proc. Natl. Acad. Sci. USA* 113, no. 2 (2016): 268–273. DOI: 10.1073/pnas.1518980113.

Zhang, R., Y. Zhang, Z.C. Dong, S. Jiang, C. Zhang, L.G. Chen, L. Zhang, Y. Liao, J. Aizpurua, Y. Luo, J. L. Yang and J.G. Hou. "Chemical mapping of a single molecule by plasmon-enhanced Raman scattering." *Nature* 498, no. 7452 (2013): 82–86. DOI: 10.1038/nature12151.

Zrimsek, A.B., N.L. Wong and R.P. van Duyne. "Single molecule surface-enhanced Raman spectroscopy: A critical analysis of the bi-analyte versus isotopologue proof." *J. Phys. Chem. C* 120, no. 9 (2016): 5133–5142. DOI: 10.1021/acs.jpcc.6b00606.

## Chapter 15    Laser-induced breakdown spectroscopy

*Books and articles in books*    Cremers, D.A. and L.J. Radziemski. *Handbook of Laser-Induced Breakdown Spectroscopy.* Chichester, UK: John Wiley & Sons (2006).

Griem, H.R. *Plasma Spectroscopy.* New York, NY, USA: McGraw-Hill (1964).

Griem, H.R. *Spectral Line Broadening by Plasmas.* New York, NY, USA: Academic Press (1974).

Harilal, S.S., J.R. Freeman, P.K. Diwakar and A. Hassanei. "Femtosecond laser ablation: Fundamentals and applications." In S. Musazzi and U. Perini (eds.). *Laser-Induced Breakdown Spectroscopy: Theory and Application.* Berlin, Germany: Springer Verlag (2014): pp. 143–168. DOI: 10.1007/978-3-642-45085-3_6.

Harris, D. C. *Quantitative Chemical Analysis*, 9th edition. New York, NY, USA: W.H. Freeman & Co. (2015).

McWhirter, R.W.P. "Spectral intensities." In R.H. Huddlestone and S.L. Leonard (eds.). *Plasma Diagnostic Techniques.* New York, NY, USA: Academic Press (1965): pp. 201–264.

Miziolek, A.W., V. Palleschi and I. Schechter. *Laser-Induced Breakdown Spectroscopy (LIBS): Fundamentals and Applications.* Cambridge, UK: Cambridge University Press (2006).

Moros, J., F.J. Fortes, J.M. Vadillo and J.J. Laserna. "LIBS detection of explosives in traces." In S. Musazzi and U. Perini (eds.). *Laser-Induced Breakdown Spectroscopy: Theory and Application.* Berlin, Germany: Springer Verlag (2014): pp. 349–376. DOI: 10.1007/978-3-642-45085 -3_13.

Musazzi, S. and U. Perini (eds.). *Laser-Induced Breakdown Spectroscopy: Theory and Application.* Berlin, Germany: Springer Verlag (2014).

Noll, R. *Laser-Induced Breakdown Spectroscopy, Fundamentals and Applications.* Heidelberg, Germany: Springer Verlag (2012).

*Journal articles*    Beddows, D.C.S., O. Samek, M. Liška and H.H. Telle. "Single-pulse laser-induced breakdown spectroscopy of samples submerged in water using a single-fibre light delivery system." *Spectrochim Acta B* 57, no. 6 (2002): 1461–1471. DOI: 10.1016/S0584-8547(02)00083-6.

Bescos, B. and A. Gonzalez Ureña. "Laser chemical analysis of metallic elements in aluminum samples." *J. Laser Appl.* 7, no. 1 (1995): 47–50. DOI: 10.2351/1.4745371.

Cáceres, J.O., J. Tornero López, H.H. Telle and A. González Ureña. "Quantitative analysis of trace metal ions in ice using laser induced breakdown spectroscopy." *Spectrochim. Acta B* 56, no. 6 (2001): 831–838. DOI: 10.1016/S0584-8547(01)00173-2.

Cahoon, E.M. and J. R. Almirall. "Quantitative analysis of liquids from aerosols and microdrops using laser-induced breakdown spectroscopy." *Anal. Chem.* 84, no. 5 (2012): 2239–2244. DOI: 10.1021/ac202834j.

Chichkov, B.N. C. Momma, S. Nolte, F. von Alvensleben and A. Tunnermann. "Femto-second, picosecond and nanosecond laser ablation of solids." *Appl. Phys. A* 63, no. 2 (1996): 109–115. DOI: 10.1007/BF01567637.

Ciucci, A., M. Corsi, V. Palleschi, S. Rastelli, A. Salvetti and E. Tognoni. "New procedure for quantitative elemental analysis by laser-induced plasma spectroscopy." *Appl. Spectrosc.* 53, no. 8 (1999): 960–964. DOI: 10.1366/0003702991947612.

Davies, C.M., H.H. Telle, D.J. Montgomery and R.E. Corbett. "Quantitative analysis using remote laser-induced breakdown spectroscopy (LIBS)." *Spectrochim. Acta B* 50, no. 9 (1995): 1059–1075. DOI: 10.1016/0584-8547(95)01314-5.

DeLucia Jr, F.C., A.C. Samuels, R.S. Harmon, R.A. Walters, K.L. McNesby, A. LaPointe, R.J. Winkel Jr. and A.W. Miziolek. "Laser-induced breakdown spectroscopy (LIBS): A promising versatile chemical sensor technology for hazardous material detection." *IEEE Sens. J.* 5, no. 4 (2005): 681–689. DOI: 10.1109/JSEN.2005.848151.

Fortes F.J. and J.J. Laserna. "The development of fieldable laser-induced breakdown spectrometer: No limits on the horizon." *Spectrochim. Acta B* 65, no. 12 (2010): 975–990. DOI: 10.1016/j.sab.2010.11.009.

Fortes, F.J., J. Moros, P. Lucena, L.M. Cabalín and J.J. Laserna. "Laser-induced breakdown spectroscopy." *Anal. Chem.* 85, no. 2 (2013): 640–669. DOI: 10.1021/ac303220r.

Gimenez, Y., B. Busser, F. Trichard, A. Kulesza, J.M. Laurent, V. Zaun, F. Lux, J.M. Benoit, G. Panczer, P. Dugourd, O. Tillement, F. Pelascini, L. Sancey and V. Motto-Ros. "3D imaging of nanoparticle distribution in biological tissue by laser-induced breakdown spectroscopy." *Sci. Rep.* 6 (2016): article 29936. DOI: 10.1038/srep29936.

Gottfried, L.J., F.C. DeLucia Jr., C.A. Munson and A.W. Miziolek. "Double-pulse standoff laser-induced breakdown spectroscopy for versatile hazardous materials detection." *Spectrochim. Acta B* 62, no. 12 (2007): 1405–1411. DOI: 10.1016/j.sab.2007.10.039.

Gurevich, E.L. and R. Hergenroder. "Femtosecond laser-induced breakdown spectroscopy: Physics, applications and perspectives." *Appl. Spectrosc.* 61, no. 10 (2007): 233A–242A. DOI: 10.1366/000370207782217824.

Hahn, D.W. and N. Omenetto. "Laser-induced breakdown spectroscopy (LIBS), part I: Review of basic diagnostics and plasma–particle interactions: Still-challenging issues within the analytical plasma community." *Appl. Spectrosc.* 64, no. 12 (2010): 335–366. DOI: 10.1366/000370210793561691.

Lazic, V. and S. Jovićević. "Laser induced breakdown spectroscopy inside liquids: Processes and analytical aspects." *Spectrochim. Acta B* 101 (2014b): 288–311. DOI: 10.1016/j.sab.2014.09.006.

Margetic, V., A. Pakulev, A. Stockhaus, M. Bolshov, K. Niemax and R. Hergenröder. "A comparison of nanosecond and femtosecond laser-induced plasma spectroscopy of brass samples." *Spectrochim. Acta B* 55, no. 11 (2000): 1771–1785. DOI: 10.1016/S0584-8547(00)00275-5.

Matsumoto, A., A. Tamura, K. Fukami, Y.H. Ogata and T. Sakka. "Single-pulse underwater laser-induced breakdown spectroscopy with non-gated detection scheme." *Anal. Chem.* 85, no. 8 (2013): 3807–3811. DOI: 10.1021/ac400319v.

Miziolek, A.W. "Progress in fieldable laser induced breakdown spectroscopy (LIBS)." *Proc. SPIE* 8374 (2012): article 837402. DOI: 10.1117/12.919492

Moros, J.J., Lorenzo, P. Lucena, L. M. Tobaria and J.J. Laserna. "Simultaneous Raman spectroscopy—Laser-induced breakdown spectroscopy for instant standoff analysis of explosives using a mobile integrated sensor platform." *Anal. Chem.* 82, no. 4 (2010b): 1389–1400. DOI: 10.1021/ac902470v.

Noll, R., C. Fricke-Begemann, M. Brunk, S. Connemann, C. Meinhardt, M. Scharunb, V. Sturm, J.Makowe and C. Gehlen. "Laser-induced breakdown spectroscopy expands into industrial applications." *Spectrochim. Acta B* 93 (2014): 41–51. DOI: 10.1016/j.sab.2014.02.001.

Ohba, H., M. Saeki, I.Wakaida, R. Tanabe and Y. Ito. "Effect of liquid-sheet thickness on detection sensitivity for laser-induced breakdown spectroscopy of aqueous solution." *Opt. Express* 22, no. 20 (2014): 24478–24490. DOI: 10.1364/OE.22.024478.

Pasquini, C., J. Cortez, L.M.C. Silva and F.B. Gonzaga. "Laser-induced breakdown spectroscopy." *J. Braz. Chem. Soc.* 18, no. 3 (2007): 463–512. DOI: 10.1590/S0103-50532007000300002.

Piñon, V., M.P. Mateo and G. Nicolas. "Laser-induced breakdown spectroscopy for chemical mapping of materials." *Appl. Spectrosc. Rev.* 48, no. 5 (2013): 357–383. DOI: 10.1080/05704928.2012.717569.

Rifai, K., S. Laville, F. Vidal, M. Sabsabi and M. Chaker. "Quantitative analysis of metallic traces in water-based liquids by UV-IR double-pulse laser-induced breakdown spectroscopy." *J. Anal. At. Spectrom.* 27, no. 2 (2012): 276–283. DOI: 10.1039/C1JA10178A.

Runge, E.F., R.W. Minck and F.R. Bryan. "Spectrochemical analysis using a pulsed laser source." *Spectrochim. Acta* 20, no. 4 (1964): 733–736. DOI: 10.1016/0371-1951(64)80070-9.

Sallé, B., J. Lacour, E. Vors, P. Fichet, S. Maurice, D.A. Cremers and R.C. Wiens. "Laser-induced breakdown spectroscopy for Mars surface analysis: Capabilities at stand-off distances and detection of chlorine and sulfur elements." *Spectrochim. Acta B* 59, no. 9 (2004): 1413–1422. DOI: 10.1016/j.sab.2004.06.006.

Samek, O., D.C.S. Beddows, J. Kaiser, S.V. Kukhlevsky, M. Liška, H.H. Telle and J. Young. "Application of laser-induced breakdown spectroscopy to in situ analysis of liquid samples." *Opt. Eng.* 39, no. 8 (2000): 2248–2262. DOI: 10.1117/1.1304855.

Samek, O., D.C.S. Beddows, H.H. Telle, J. Kaiser, M. Liška, J.O. Cáceres and A. Gonzáles Ureña. "Quantitative laser-induced breakdown spectroscopy analysis of calcified tissue samples." *Spectrochim Acta B* 56, no. 6 (2001): 865–875. DOI: 10.1016/S0584-8547(01)00198-7.

Wiens, R.C., S. Maurice, B. Barraclough *et al.* (for the MSL-ChemCam collaboration). "The ChemCam instrument suite on the Mars Science Laboratory (MSL) rover: Body unit and combined system tests." *Space Sci. Rev.* 170, no. 1 (2012): 167–227. doi: 10.1007/s11214-012-9902-4.

Yamamoto, K.Y., D.A. Cremers, M.J. Ferris and L.E. Foster. "Detection of metals in the environment using a portable laser-induced breakdown spectroscopy instrument." *Appl. Spectrosc.* 50, no. 2 (1996): 222–233. DOI: 10.1366/0003702963906519.

Zorba, V., X. Mao and R.E. Russo. "Ultrafast laser induced breakdown spectroscopy for high spatial resolution chemical analysis." *Spectrochim. Acta B* 66, no. 2 (2011): 189–192. DOI: 10.1016/j.sab.2010.12.008.

*Other sources*

Clausen, J. and Z. Courville. "Ice-core analysis in a polar environment using laser-induced breakdown spectroscopy (LIBS)." Poster presented at 13th Polar Technology Conference, U.S. Naval Academy, Annapolis, MD, USA (April 2–4, 2013). DOI: 10.13140/2.1.4190.6566. **Access**: http://polar.sri.com/polarpower.org/PTC/2013_pdf/PTC_2013_Clausen.pdf.

Gottfried, J.L. and F.C. De Lucia Jr. *Laser-induced breakdown spectroscopy: Capabilities and applications.* Report ARL-TR-5238. Aberdeen Proving Ground, MD, USA: Army Research Laboratory (2010).

Kramida, A., Y. Ralchenko, J. Reader and NIST ASD Team. NIST Atomic Spectra Database (version 5.4, online). Gaithersburg, MD, USA: National Institute of Standards and Technology-NIST (2016). **Access**: http://physics.nist.gov/asd. [Last accessed on 14 May 2017.]

LIBSCAN 25+. *Fully portable, battery-powered modular LIBS system.* Skipton, UK: Applied Photonics Ltd (2015).

NASA-JPL–NASA's Mars Science Laboratory (MSL) mission: *"Curiosity" rover. "Chemistry & Camera (ChemCam)."* Pasadena, CA, USA: Jet Propulsion Laboratory, California Institute of Technology (2012). **Access**: https://mars.nasa.gov/msl/mission/instruments/spectrometers/chemcam/.

## Chapter 16 Laser ionization techniques

*Books and articles in books*

Schlag, E.W. *ZEKE Spectroscopy.* Cambridge, UK: Cambridge University Press (1998).

Turner, D.W., C. Baker, A.D. Baker and C.R. Brundle. *Modern Photoelectron Spectroscopy.* London, UK: Wiley-Blackwell (1970).

*Journal articles*

Bean, B.D., F. Fernández-Alonso and R.N. Zare. "Distribution of rovibrational product states for the "prompt" reaction H + $D_2$(v=0 | j=0-4) $\rightarrow$ HD(v'=1,2 | j')+D near 1.6 eV collision energy." *J. Phys. Chem. A* 105, no. 11 (2001): 2228–2233. DOI: 10.1021/jp0027288.

Beyer, M. and F. Merkt. "Observation and calculation of the quasi-bound rovibrational levels of the electronic grounds of $H_2^+$." *Phys. Rev. Lett.* 116, no. 9 (2016a): article 093001. DOI: 10.1103/PhysRevLett.116.093001.

Beyer, M. and F. Merkt. "Structure and dynamics $H_2^+$ of near the dissociation threshold: A combined experimental and computational investigation." *J. Mol. Spectrosc.* 330 (2016b): 147–157. doi: 10.1016/j.jms.2016.08.001.

Blais, N.C. and D.G. Truhlar. "The H + $D_2$ reaction: Quasiclassical simulation of nascent HD rovibrational state distributions under experimentally probed high-energy conditions." *Chem. Phys. Lett.* 162, no. 6 (1989): 503–510. DOI: 10.1016/0009-2614(89)87015-0.

Boesl, U. "Laser mass spectrometry for environmental and industrial chemical trace analysis." *J. Mass Spectrom.* 35, no. 3 (2000): 289–304. DOI: 10.1002/(SICI)1096-9888(200003)35:3<289::AID-JMS960>3.0.CO;2-Y.

Braun, J.E. and H.J. Neusser. "Threshold photoionization in time-of-flight mass spectrometry." *Mass. Spectrom. Rev.* 21, no. 1 (2002): 16–36. DOI: 10.1002/mas.10014.

Brinkhaus, S.G., J.B. Thiäner and C. Achten. "Ultra-high sensitive PAH analysis of certified reference materials and environmental samples by GC-APLI-MS." *Anal. Bioanal. Chem.* 409, no. 11 (2017): 2801–2812. DOI: 10.1007/s00216-017-0224-y.

Chupka, W.A. "Factors affecting lifetimes and resolution of Rydberg states observed in zero-electron-kinetic-energy spectroscopy." *J. Chem. Phys.* 98, no. 9 (1993): 4520–4530. doi: 10.1063/1.474712.

Cockett, M.C.R. "Photoelectron spectroscopy without photoelectrons: Twenty years of ZEKE spectroscopy." *Chem. Soc. Rev.* 34 (2005): 935–948. DOI: 10.1039/B505794A.

Cockett, M.C.R., D.A. Beattie, N.A. Macleod, K.P. Lawley, T. Ridley and R.J. Donovan. "The spectroscopy of Rydberg to Rydberg transitions in $I_2$ and $Br_2$ investigated by vibrationally induced autoionization." *Phys. Chem. Chem. Phys.* 4, no. 8 (2002): 1419–1424. DOI: 10.1039/B109560A.

Cockett, M.C.R., J.G. Goode, K.P. Lawley and R.J. Donovan. "Zero kinetic energy photoelectron spectroscopy of Rydberg excited molecular iodine." *J. Chem. Phys.* 102, no. 13 (1995): 5226–5234. doi: 10.1063/1.469248.

Cockett, M.C.R. and M.J. Watkins. "Hydrogenic Rydberg states of molecular van der Waals complexes: Resolved Rydberg spectroscopy of DABCO-$N_2$." *Phys. Rev. Lett.* 92, no. 4 (2004): article 43001. DOI: 10.1103/PhysRevLett.92.043001.

Dessent, C.E.H., S.R. Haine and K. Muller-Dethlefs. "A new detection scheme for synchronous, high resolution ZEKE and MATI spectroscopy demonstrated on the phenol≡Ar complex." *Chem. Phys. Lett.* 315, no. 1–2 (1999): 103–108. DOI: 10.1016/S0009-2614(99)01193-8.

Dixit, S.N., D. L. Lynch and V. McKoy. "Rotational branching ratios in (1+1) resonant-enhanced multiphoton ionization of NO via the $A^2\Sigma^+$state." *Phys. Rev. A* 32, no. 2 (1985): 1267–1270. DOI: 10.1103/PhysRevA.32.1267.

Dong, F., S.H. Lee and K. Liu. "Reactive excitation functions for F + p-$H_2$/n-$H_2$/$D_2$ and the vibrational branching for F+HD." *J. Chem. Phys.* 113, no. 9 (2000): 3633–3640. DOI: 10.1063/1.1287840.

Dong, W., C. Xiao, T. Wang, D. Dai, X. Yang and D.H. Zhang. "Transition-state spectroscopy of partial wave resonances in the F + HD Reaction." *Science* 327, no. 5972 (2010): 1501–1502. DOI: 10.1126/science.1185694.

Fischer, I., R. Linder and K. Müller-Dethlefs. "State-to-state photoionization dynamics probed by zero kinetic energy (ZEKE) photoelectron spectroscopy." *J. Chem. Soc. Faraday Trans.* 90, no. 17 (1994): 2425–2442. DOI: 10.1039/FT9949002425.

Ford, M., R. Linder and K. Müller-Dethlefs. "Fully rotationally resolved ZEKE photoelectron spectroscopy of $C_6H_6$ and $C_6D_6$: Photoionization dynamics and geometry of the benzene cation." *Mol. Phys.* 101, no. 4–5 (2003): 705–716. DOI: 10.1080/0026897021000054916.

Gasmi, K., R.M. Al-Tuwirqi, S. Skowronek, H.H. Telle and A. González Ureña. "Rotationally resolved (1+1') resonance-enhanced multiphoton ionisation (REMPI) of CaR (R=H,D) in supersonic beams: CaR X $^2\Sigma^+$ (v"= 0) $\rightarrow$ CaR B $^2\Sigma^+$ (v' = 1,0) $\rightarrow$ CaR$^+$ X $^1\Sigma^+$." *J. Phys. Chem. A* 107, no. 50 (2003): 10960–10968. DOI: 10.1021/jp036351s.

Goldberg, N.T., K. Koszinowski, A.E. Pomerantz and R.N. Zare. "Doppler-free ion imaging of hydrogen molecules produced in bimolecular reactions." *Chem. Phys. Lett.* 433, no. 4–6 (2007): 439–443. DOI: 10.1016/j.cplett.2006.11.073.

Groß, A., S. Wilke and M. Scheffler. "Six-dimensional quantum dynamics of adsorption and desorption of $H_2$ at Pd(100): No need for a molecular precursor adsorption state." *Surf. Sci.* 357–358 (1996): 614–618. DOI: 10.1016/0039-6028(96)00232-4.

Große Brinkhaus, S.G., J.B. Thiäner and C. Achten. "Ultra-high sensitive PAH analysis of certified reference materials and environmental samples by GC-APLI-MS." *Anal. Bioanal. Chem.* 409, no. 11 (2017): 2801–2812. DOI: 10.1007/s00216-017-0224-y.

Haines, S.R., C.E.H. Dessent and K. Müller-Dethlefs. "Is the phenol≡Ar complex van der Waals or hydrogen-bonded? A REMPI and ZEKE spectroscopic study." *J. Electron Spectrosc. Rel. Phenom.* 108, no. 1–3 (2000): 1–11. DOI: 10.1016/S0368-2048(00)00140-7.

Heger, H.J., R. Zimmermann, R. Dorfner, M. Beckmann, H. Griebel, A. Kettrup and U. Boesl. "On-line emission analysis of polycyclic aromatic hydrocarbons down to pptv concentration levels in the flue gas of an incineration pilot plant with a mobile resonance-enhanced multiphoton ionization time-of-flight mass spectrometer." *Anal. Chem.* 71, no. 1 (1999): 46–57. DOI: 10.1021/ac980611y.

Held, A. and E.W. Schlag. "Zero kinetic energy spectroscopy." *Acc. Chem. Res.* 31, no. 8 (1998): 467–473. DOI: 10.1021/ar9702987.

Hippler, M. and J. Pfab. "Detection and probing of nitric oxide (NO) by two-colour laser photoionisation (REMPI) spectroscopy on the A← X transition." *Chem. Phys. Lett.* 243, no. 5–6 (1995): 500–505. DOI: 10.1016/0009-2614(95)00870-A.

Jacobs, D.C., R.J. Madix and R.N. Zare. "Reduction of 1+1 resonance enhanced MPI spectra to population distributions: Application to the NO A $^2\Sigma^+$ - X $^2\Pi$ system." *J. Chem. Phys.* 85, no. 10 (1986): 5469–5479. DOI: 10.1063/1.451557.

Jahn, H.A. and Teller, E. "Stability of polyatomic molecules in degenerate electronic states. I-orbital degeneracy." *Proc. R. Soc. London A* 161, no. 905 (1937): 220–235. DOI: 10.1098/rspa.1937.0142.

Johnson, P.M. "Multiphoton ionization spectroscopy: A new state of benzene." *J. Chem. Phys.* 62, no. 11 (1975): 4562–4563. DOI: 10.1063/1.430366.

Johnson, P.M. "The multiphoton ionization spectrum of benzene." *J. Chem. Phys.* 64, no. 10 (1976): 4143–4148. DOI: 10.1063/1.431983.

Keil, M., H.G. Krämer, A. Kudell, M.A. Baig, J. Zhu, W. Demtröder and W. Meyer. "Rovibrational structures of the pseudorotating lithium trimer $^{21}Li_3$: Rotationally resolved spectroscopy and *ab initio* calculations of the A$^2$E″ ← X$^2$E′ system." *J. Chem. Phys.* 113, no. 17 (2000): 7414–7431. DOI: 10.1063/1.1308091.

Lai, L.H., D.C. Che and K. Liu. "Photo-dissociative pathways of $C_2H_2$ at 121.6 nm revealed by a Doppler-selected time-of-flight (a 3-D Mapping) technique." *J. Phys. Chem.* 100, no. 16 (1996): 6376–6380. DOI: 10.1021/jp953672y.

Lee, S.H. and K. Liu. "Exploring insertion reaction dynamics: A case study of S($^1$D) + D$_2$ → SD + D." *J. Phys. Chem. A* 102, no. 45 (1998): 8637–8640. DOI: 10.1021/jp983220w.

Lee, S.H., F. Dong and K. Liu. "A crossed-beam study of the F + HD → HF + D reaction: The resonance-mediated channel." *J. Chem. Phys.* 125, no. 13 (2006): article 133106. DOI: 10.1063/1.2217374.

Marinero, E.E., C.T. Rettner and R.N. Zare. "H + D$_2$ reaction dynamics: Determination of the product state distribution at a collision energy of 1.3 eV." *J. Chem. Phys.* 80, no. 9 (1984): 4142–4156. DOI: 10.1063/1.447242.

Mitscherling, C., C. Maul and K.H. Gericke. "Ultra-sensitive detection of nitric oxide isotopologues." *Phys. Scr.* 80, no. 4 (2009): article 048122. DOI: 10.1088/0031-8949/80/04/048122.

Montero, C., J.B. Jiménez, J.M. Orea, A. González Ureña, S.M. Cristescu, S. te Lintel Hekkert and F.J.M. Harren. "*trans*-Resveratrol and grape disease resistance: A dynamical study by high-resolution laser-based techniques." *Plant Physiol.* 131, no. 1 (2003): 129–138. DOI: 10.1104/pp.010074.

Moss, R.E. "Calculations for the vibration-rotation levels of H$_2^+$ in its ground and first excited electronic states." *Mol. Phys.* 80, no. 6 (1993): 1541–1554. DOI: 10.1080/00268979300103211.

Müller-Dethlefs, K., M. Sander and E.W. Schlag. "Two-color photoionization resonance spectroscopy of NO: Complete separation of rotational levels of NO$^+$ at the ionization threshold." *Chem. Phys. Lett.* 112, no. 4 (1984): 291–294. DOI: 10.1016/0009-2614(84)85743-7.

Müller-Dethlefs, K. and E.W. Schlag. "Chemical applications of zero kinetic energy (ZEKE) photoelectron spectroscopy." *Angew. Chem. Int. Ed.* 37, no. 10 (1998): 1346–1374. DOI: 10.1002/(SICI)1521-3773(19980605)37:10<1346::AID-ANIE1346>3.0.CO;2-H.

Nishide, T. and T. Suzuki. "Photodissociation of nitrous oxide revisited by high-resolution photofragment imaging: Energy partitioning." *J. Phys. Chem. A* 108, no. 39 (2004): 7863–7870. DOI: 10.1021/jp048966a.

Orea, J.M., C. Montero, J.B. Jiménez and A. González Ureña. "Analysis of trans-resveratrol by laser desorption coupled with resonant ionisation spectrometry: Application to *trans*-resveratrol content in vine leaves and grape skin." *Anal. Chem.* 73, no. 24 (2001): 5921–5929. DOI: 10.1021/ac010439p.

Park, G., B.C. Krüger, S. Meyer, A.M. Wodtke and T. Schäfer. "A (1 +1') resonance-enhanced multiphoton ionization scheme for rotationally state-selective detection of formaldehyde via the allowed Ã $^1$A$_2$ ← $\tilde{X}$ $^1$A$_1$ (4$^1_0$) transition." *Phys. Chem. Chem. Phys.* 18, no. 32 (2016): article 22355. DOI: 10.1039/c6cp03833f.

Park, H. and R. Zare. "Photoionization dynamics of the NO A $^2\Sigma^+$ state deduced from energy- and angle-resolved photoelectron spectroscopy." *J.Chem. Phys.* 99, no. 9 (1993): 6537–6544. DOI: 10.1063/1.465845.

Pogrebnya S.K., J. Palma, D.C. Clary and J. Echave. "Quantum scattering and quasi-classical trajectory calculations for the H$_2$+OH → H$_2$O+H reaction on a new potential surface." *Phys. Chem. Chem. Phys.* 2, no. 4 (2000): 693–700. DOI: 10.1039/A908080E.

Radischat, C., O. Sippula, B. Stengel, S. Klingbeil, M. Sklorz, R. Rabe, T. Streibel, H. Harndorf and R. Zimmermann. "Real-time analysis of organic compounds in ship engine aerosol emissions using resonance-enhanced multiphoton ionisation and proton transfer mass spectrometry." *Anal. Bioanal. Chem.* 407, no. 20 (2015): 5939–5951. DOI: 10.1007/s00216-015-8465-0.

Reiser, G., W. Habenicht, K. Müller-Dethlefs and E.W. Schlag. "The ionization energy of nitric oxide." *Chem. Phys. Lett.* 152, no. 2–3 (1988): 119–123. DOI: 10.1016/0009-2614(88)87340-8.

Ridley, T., D.A. Beattie, M.C.R. Cockett, R.J. Donovan and K.P. Lawley. "A re-analysis of the vacuum ultraviolet absorption spectrum of $I_2$, $Br_2$ and ICl using ionization energies determined from their ZEKE-PFI photoelectron spectra." *Phys. Chem. Chem. Phys.* 4, no. 8 (2002): 1398–1411. DOI: 10.1039/B109555M.

Rinnen, K.D., M.A. Buntine, D.A.V. Kliner, R. N. Zare and W.M. Huo. "Quantitative determination of $H_2$, HD and $D_2$ internal-state distributions by (2+1) resonance-enhanced multiphoton ionization." *J. Chem. Phys.* 95, no. 1 (1991): 214–225. DOI: 10.1063/1.461478.

Rinnen, K.D., D.A.V. Kliner and R.N. Zare. "The H + $O_2$ reaction: Quantum-state distributions at collision energies of 1.3 and 0.55 eV." *J. Chem. Phys.* 91, no. 12 (1989): 7514–7529. DOI: 10.1063/1.457275.

Sander, M., L.A. Chewter, K. Müller-Dethlefs and E.W. Schlag. "High-resolution zero-kinetic-energy photoelectron spectroscopy of nitric oxide." *Phys. Rev. A* 36, no. 9 (1987): 4543–4546. DOI: 10.1103/PhysRevA.36.4543.

Schmidt, S., M.F. Appel, R.M. Garnica, R.N. Schindler and T. Benter. "Atmospheric pressure laser ionization: An analytical technique for highly selective detection of ultralow concentrations in the gas phase." *Anal. Chem.* 71, no. 17 (1999): 3721–3729. DOI: 10.1021/ac9901900.

Schnieder, L., W. Meier, K.H. Welge, M.N R. Ashfold and C.M. Western. "Photo-dissociation dynamics of $H_2S$ at 121.6 nm and a determination of the potential energy function of SH (A $^2\Sigma^+$)." *J. Chem. Phys.* 92, no. 12 (1990): 7027–7037. DOI: 10.1063/1.458243.

Schnieder, L., K. Seekamp-Rahn, J. Borkowski, E. Wrede, F.J. Aoiz, L. Bañares, M.J. D'Mello, V.J. Herrero, V.S. Rabanos and R.E. Wyatt. "Experimental studies and theoretical predictions for the $H+D_2 \rightarrow HD+D$ reaction." *Science* 269, no. 5221 (1995): 207–210. DOI: 10.1126/science.269.5221.207.

Schnieder, L., K. Seekamp-Rahn, E. Wrede and K.H. Welge. "Experimental determination of quantum state resolved differential cross sections for the hydrogen exchange reaction $H+D_2 \rightarrow HD+D$." *J. Chem. Phys.* 107, no. 16 (1997): 6175–6195. DOI: 10.1063/1.474283.

Shea, D.A., L. Goodman and M.G. White. "Acetone n-radical cation internal rotation spectrum: The torsional potential surface." *J. Chem. Phys.* 112, no. 6 (2000): 2762–2768. DOI: 10.1063/1.480850.

Shinkai, M., S. Suzuki, A. Miyashita, H. Koboyashi, T. Okubo and Y. Ishigatsubo. "Analysis of exhaled nitric oxide by the helium bolus method." *Chest* 121, no. 6 (2002): 1847–1852. DOI: 10.1378/chest.121.6.1847.

Short, L.C., R. Frey and T. Benter. "Real-time analysis of exhaled breath via resonance-enhanced multiphoton ionization mass spectrometry with a medium pressure laser ionization source: Observed nitric oxide profile." *Appl. Spectrosc.* 60, no. 2 (2006): 217–222. DOI: 10.1366/000370206776023241.

Signorell, R. and F. Merkt. "General symmetry selection rules for the photoionization of polyatomic molecules." *Mol. Phys.* 92, no. 5 (1997): 793–804. DOI: 10.1080/00268979709482151.

Skowronek, S., R. Pereira and A.G. Ureña. "Spectroscopy and dynamics of excited harpooning reactions: The photo-depletion action spectrum of the Ba···$FCH_3$ complex." *J. Phys. Chem. A* 101, no. 41 (1997): 7468–7475. DOI: 10.1021/jp970385g.

Stert, V., P. Farmanara, H.H. Ritze, W. Radloff, K. Gasmi and A.G. Ureña. "Femtosecond time-resolved electron spectroscopy of the intra-cluster reaction in Ba···$FCH_3$." *Chem. Phys. Lett.* 337, no. 4–6 (2001): 299–305. DOI: 10.1016/S0009-2614(01)00218-4.

Strazisar, B.R., C. Lin and H.F. Davis. "Mode-specific energy disposal in the four-atom reaction $OH+D_2 \rightarrow HOD+D$." *Science* 290, no. 5493 (2000): 958–961. DOI: 10.1126/science.290.5493.958.

Streibel, T., K. Hafner, F. Mühlberger, T. Adam and R. Zimmermann. "Resonance-enhanced multiphoton ionization Time-of-Flight mass spectrometry for detection of nitrogen containing aliphatic and aromatic compounds: Resonance-enhanced multiphoton ionization spectroscopic investigation and on-line analytical application." *Appl. Spectrosc.* 60, no. 1 (2006): 72–79. DOI: 10.1366/000370206775382767.

Swain, S. "A simple model for multiphoton ionisation." *J. Phys. B* 12, no. 19 (1979): 3201–3228. DOI: 10.1088/0022-3700/12/19/007.

Tokel, O., J. Chen, C.K. Ulrich and P.L. Houston. "$O(^1D)$ + $N_2O$ reaction: NO vibrational and rotational distribution." *J. Phys. Chem. A* 114, no. 42 (2010): 11292–11297. DOI: 10.1021/jp1042377.

Weickhardt, C., R. Zimmermann, K. W. Schramm, U. Boesl and E.W. Schlag. "Laser mass spectrometry of the di-, tri- and tetrachlorobenzenes: Isomer-selective ionization and detection." *Rapid Commun. Mass Spectrom.* 8, no. 5 (1994): 381–384. DOI: 10.1002/rcm.1290080508.

Wetzig, D., M. Rutkowski and H. Zacharias. "Vibrational and rotational population distribution of $D_2$ associatively desorbing from Pd (100)." *Phys. Rev. B* 63, no. 20 (2001): article 205412. DOI: 10.1103/PhysRevB.63.205412.

Whetten, R.L., K.J. Fu and E.R. Grant. "Ultraviolet two-photon spectroscopy of benzene: A new *gerade* Rydberg series and evidence for the 1 $^1E_{2g}$ valence state." *J. Chem. Phys.* 79, no. 6 (1983): 2626–2640. DOI: 10.1063/1.446159.

Wilke S. and M. Scheffler. "Potential-energy surface for $H_2$ dissociation over Pd(100)." *Phys. Rev. B* 53, no. 8 (1996): 4926–4932. DOI: 10.1103/PhysRevB.53.4926.

Xie, J, and R.N. Zare. "Selection rules for the photoionization of diatomic molecules." *J. Chem. Phys.* 93, no. 5 (1990): 3033–3038. DOI: 10.1063/1.458837.

Yang, X. "Probing state-to-state reaction dynamics using H-atom Rydberg tagging time-of-flight spectroscopy." *Phys. Chem. Chem. Phys.* 13, no. 18 (2011): 8112–8121. DOI: 10.1039/C1CP00005E.

Zacharias, H., R. Schrniedl and K.H. Welge. "State selective step-wise photoionization of NO with mass spectroscopic ion detection." *Appl. Phys.* 21, no. 2 (1980): 127–133. DOI: 10.1007/BF00900674.

Zhu, L. and P. Johnson. "Mass analyzed threshold ionization spectroscopy." *J. Chem. Phys.* 94, no. 6 (1991): 5769–5771. DOI: 10.1063/1.460460.

Zimmermann, R. "Laser ionization mass spectrometry for on-line analysis of complex gas mixtures and combustion effluents." *Anal. Bioanal. Chem.* 381, no. 1 (2005): 57–60. DOI: 10.1007/s00216-004-2886-5.

## Chapter 17    Basic concepts of laser imaging

*Books and chapters in books*

Chang, C.I. *Hyperspectral Data Processing: Algorithm Design and Analysis*. Chichester, UK: John Wiley & Sons (2013).

Cheeseman, P., B. Kanefsky, R. Kraft, J. Stutz and R. Hanson. "Super-resolved surface reconstruction from multiple images." In G.R. Heidbreder (ed.), *Fundamental Theories of Physics, Vol. 62—Maximum Entropy and Bayesian Methods*. Dordrecht, Netherlands: Springer Science & Business Media B.V. (1996): pp. 293–308. DOI: 10.1007/978-94-015-8729-7_23.

Cristobal, G., P. Schelkens and H. Thienpont (eds.). *Optical and Digital Image Processing: Fundamentals and Applications*. Weinheim, Germany: Wiley-VCH Verlag (2011).

Diaspro, A. and M.A.M.J. van Zandvoort (eds.). "Super-resolution-imaging-in-biomedicine." Boca Raton FL, USA: CRC Press (2017).

Milanfar, P. (ed.). "Super-resolution imaging." Boca Raton FL, USA: CRC Press (2011).

Nguyen, R.M.H., D.K. Prasad and M.S. Brown. "Training-based spectral reconstruction from a single RGB image." In D. Fleet, T. Pajdla, B. Schiele and T. Tuytelaars (eds.). *Lecture Notes in Computer Science*, vol. 8695. Cham, Switzerland: Springer International Publishing AG (2014): pp. 186–201. DOI: 10.1007/978-3-319-10584-0_13.

Pawley, J.B. "Points, pixels, and gray levels: Digitizing image data." Chapter 4 in J.B. Pawley (ed.), *Handbook of Biological Confocal Microscopy*, 3rd edition. New York, NY, USA: Springer Science and Business Media (2006).

Pratt, W.K. *Introduction to Digital Image Processing*. Boca Raton, FL, USA: CRC Press (2014).

*Journal articles*

Blom, H. and J. Widengren. "Stimulated emission depletion microscopy." *Chem. Rev.* 117, no. 11 (2017): 7377–7427. DOI: 10.1021/acs.chemrev.6b00653.

Brede, N. and M. Lakadamyali. "GraspJ: an open source, real-time analysis package for super-resolution imaging." *Opt. Nanoscopy* 1 (2012): article 11. doi: 10.1186/2192-2853-1-11. Software accessible via: https://omictools.com/gpu-run-analysis-for-storm-and-palm-tool or https://github.com/isman7/graspj.

Cao, X., T. Yue, X. Lin, S. Lin, X. Yuan, Q. Dai, L. Carin and D.J. Brady. "Computational Snapshot Multispectral Cameras: Toward dynamic capture of the spectral world." *IEEE Signal Process. Mag.* 33, no. 5 (2016): 95–108. DOI: 10.1109/MSP.2016.2582378.

Chin, M.S., B.B. Freniere, Y.C. Lo, J.H. Saleeby, S.P. Baker, H.M. Strom, R.A. Ignotz, J.F. Lalikos and T.J. Fitzgerald. "Hyperspectral imaging for early detection of oxygenation and perfusion changes in irradiated skin." *J. Biomed. Opt.* 17, no. 2 (2012): article 026010. DOI: 10.1117/1.JBO.17.2.026010.

Hagan, I.M. and R.E. Palazzo. "Warming up at the poles—Workshop on centrosomes and spindle pole bodies." *Eur. Mol. Biol. Org. Rep.* 7, no. 4 (2006): 364–371. DOI: 10.1038/sj.embor.7400660.

Hagen, N., R.T. Kester, L. Gao and T.S. Tkaczyk. "Snapshot advantage: a review of the light collection improvement for parallel high-dimensional measurement systems." *Opt. Eng.* 51, no. 11 (2012): article 111702. DOI: 10.1117/1.OE.51.11.111702.

Hardie, R.C., K.J. Barnard, J.G. Bognar, E.E. Armstrong and E.A. Watson. "High-resolution image reconstruction from a sequence of rotated and translated frames and its application to an infrared imaging system." *Opt. Eng.* 37, no. 1 (1998): 247–260. DOI: 10.1117/1.601623.

Hell, S.W. and J. Wichmann. "Breaking the diffraction resolution limit by stimulated emission: stimulated-emission-depletion fluorescence microscopy." *Opt. Lett.* 19, no. 11 (1994): 780–782. doi: 10.1364/OL.19.000780.

Henriques, R., M. Lelek, E.F. Fornasiero, F. Valtorta, C. Zimmer and M.M. Mhlanga. "QuickPALM: 3D real-time photo-activation nanoscopy image processing in ImageJ." *Nature Meth.* 7, no. 5 (2010): 339–340. DOI: 10.1038/nmeth0510-339. Software accessible via: https://code.google.com/archive/p/quickpalm/.

Huff, J. "The Airy-scan detector from ZEISS: confocal imaging with improved signal-to-noise ratio and super-resolution." *Nature Meth.* 12, no. 12 (2015): Application Notes, i–ii.

Hugelier, S. J.J. de Rooi, R. Bernex, S. Duwé, O. Devos, M. Sliwa, P. Dedecker, P.H.C. Eilers and C. Ruckebusch. "Sparse deconvolution of high-density super-resolution images." *Sci. Rep.* 6 (2016): article 21413. DOI: 10.1038/srep21413.

Jordan, J., E. Angelopoulou and A. Maier. "A novel framework for interactive visualization and analysis of hyperspectral image data." *J. Elec. Comp. Eng.* 2016 (2016): article 2635124. DOI: 10.1155/2016/2635124.

Kim, M.S., A.M. Lefcourt and Y.R. Chen. "Multispectral laser-induced fluorescence imaging system for large biological samples." *Appl. Opt.* 42, no. 19 (2003): 3927–3934. DOI: 10.1364/AO.42.003927.

Křížek, P., T. Lukeš, M. Ovesný, K. Fliegel and G.M. Hagen. "SIMToolbox: a MATLAB toolbox for structured illumination fluorescence microscopy." *Bioinform.* 32, no. 2 (2016): 318–320. DOI: 10.1093/bioinformatics/btv576. Software accessible via: http://mmtg.fel.cvut.cz/SIMToolbox.

Laine, R.F., G.S. Kaminski-Schierle, S. van de Linde and C.F Kaminski. "From single-molecule spectroscopy to super-resolution imaging of the neuron: a review." *Meth. Appl. Fluoresc.* 4, no. 2 (2016): article 022004. DOI: 10.1088/2050-6120/4/2/022004.

Lau, L., Y.L. Lee, S.J. Sahl, T. Stearns and W.E. Moerner. "STED microscopy with optimized labeling density reveals 9-fold arrangement of a centriole protein." *Biophys. J.* 102, no. 12 (2012): 2926–2935. DOI: 10.1016/j.bpj.2012.05.015.

Lawo, S., M. Hasegan, G.D. Gupta and L. Pelletier. "Sub-diffraction imaging of centrosomes reveals higher-order organizational features of pericentriolar material." *Nature Cell Biol.* 14, no. 11 (2012): 1148–1158. DOI: 10.1038/ncb2591.

Li, Z., J. Suo, X. Hu, C. Deng, J. Fan and Q. Dai. "Efficient single-pixel multispectral imaging via non-mechanical spatio-spectral modulation." *Sci. Rep.* 7 (2017): article 41435. DOI: 10.1038/srep41435.

Mennella, V., B. Keszthelyi, K.L. McDonald, B. Chhun, F. Kan, G.C. Rogers, B. Huang and D.A. Agard. "Sub-diffraction-resolution fluorescence microscopy reveals a domain of the centrosome critical for pericentriolar material organization." *Nature Cell Biol.* 14, no. 11 (2012): 1159–1168. DOI: 10.1038/ncb2597.

Müller, M., V. Mönkemöller, S. Hennig, W. Hñbner and T. Huser. "Open-source image reconstruction of super-resolution structured illumination microscopy data in ImageJ." *Nature Commun.* 7 (2016): article10980. DOI: 10.1038/ncomms10980. Software accessible via: https://github.com/fairSIM.

Omasa, K., F. Hosoi and A. Konishi. "3D lidar imaging for detecting and understanding plant responses and canopy structure." *J. Exp. Bot.* 58, no. 4 (2007): 881–898. DOI: 10.1093/jxb/erl142.

Park, J.H., S.W. Lee, E.S. Lee and J.Y. Lee. "A method for super-resolved CARS microscopy with structured illumination in two dimensions." *Opt. Express* 22, no. 8 (2014): 9854–9870. DOI: 10.1364/OE.22.009854.

Rieger, B., R.P.J. Nieuwenhuizen and S. Stallinga. "Image processing and analysis for single-molecule localization microscopy." *IEEE Signal Process. Mag.* 32, no. 1 (2015): 49–57. DOI: 10.1109/MSP.2014.2354094.

Saxena, M., G. Eluru and S.S. Gorthi. "Structured illumination microscopy." *Adv. Opt. Photon.* 7, no. 2 (2015): 241–275. DOI: 10.1364/AOP.7.000241.

Sitter, D.N. Jr. and Gelbart, A. "Laser-induced fluorescence imaging of the ocean bottom." *Opt. Eng.* 40, no. 8 (2001): 1545–1553. DOI: 10.1117/1.1385510.

Su, Y., Q. Zhang, Z. Gao, X. Xu and X. Wu. "Fourier-based interpolation bias prediction in digital image correlation." *Opt. Express* 23, no. 15 (2015): 19242–19260. DOI: 10.1364/OE.23 .019242.

Waithe, D., M.P. Clausen, E. Sezgin and C. Eggeling. "FoCuS-point: software for STED fluorescence correlation and time-gated single photon counting." *Bioinformatics* 32, no. 6 (2016): 958–960. DOI: 10.1093/bioinformatics/btv687. Software accessible via: https://github.com /dwaithe/FCS_point_correlator.

Wolter, S., A. Löschberger, T. Holm, S. Aufmkolk, M.C. Dabauvalle, S. van de Linde and M. Sauer. "rapidSTORM: accurate, fast open-source software for localization microscopy." *Nature Meth.* 9, no. 11 (2012): 1040–1041. DOI: 10.1038/nmeth.2224. Software accessible via: https://omictools.com/rapidstorm-tool.

Zhang, L., Q. Yuan, H. Shen and P. Li. "Multiframe image super-resolution adapted with local spatial information." *J. Opt. Soc. Am. A* 28, no. 3 (2011): 381–390. DOI: 10.1364/JOSAA.28.000381.

*Other sources*

Huff, J., W. Bathe, R. Netz, T. Anhut and K. Weisshart. "The Airy-scan detector from ZEISS: Confocal imaging with improved signal-to-noise ratio and super-resolution." *Technical Note EN-41-013-105.* Jena, Germany: Carl Zeiss Microscopy GmbH (2015). **Access**: https:// applications.zeiss.com/C125792900358A3F/0/BF16BECDC08849E1C1257ED60029AF8D /$FILE/EN_41_013_105_wp_Airyscan-detector.pdf.

Messina, E. "Standards for visual acuity." Gaithersburg, MD, USA: National Institute for Standards and Technology (2006): report document NIST-12930. **Access**: https://www.nist.gov /sites/default/files/documents/el/isd/ks/Visual_Acuity_Standards_1.pdf.

SMLM Software Benchmarking—Comprehensive review of the single-molecule localization microscopy (SMLM, PALM, STORM) software packages. Lausanne, Switzerland: Biomedical Imaging Group, Ecole Polytechnique Fédérale de Lausanne (EPFL) (2016). **Access**: http:// bigwww.epfl.ch/smlm/#&panel1-1.

Thurman, S.T. "OTF estimation using a Siemens Star target." In *Imaging and Applied Optics, OSA Technical Digest* (CD). Washington, DC, USA: Optical Society of America (2011): paper IMC5. **Access**: https://www.osapublishing.org/abstract.cfm?URI=ISA-2011-IMC5.

Weisshart, K. "The basic principle of Airy-scanning." *Technical Note EN_41_013_084.* Jena, Germany: Carl Zeiss Microscopy GmbH (2014). **Access**: http://forum.sci.ccny.cuny.edu/cores /microscopy-imaging/confocal-microscopy/documents/Basic-Principle-Airyscan.pdf.

Woese, C.R. "Confocal, Airy-scan and structured illumination super-resolution microscopy." Lunch with the Core series. Urbana, IL, USA: Inst. Genomic Biology, University of Illinois (2015). **Access**: https://www.igb.illinois.edu/sites/default/files/upload/core/PDF/Lunch%20with%20the %20core-Shiv%202015.pdf.

## Chapter 18    Laser-induced fluorescence imaging

*Books and chapters in books*

Patwardhan, S.P., W.J. Akers and D.S. Bloch. "Fluorescence molecular imaging: Microscopic to macroscopic." In A.P. Dhawan, H.K. Huang and D. Kim (eds.). *Principles and Advanced Methods in Medical Imaging and Image Analysis.* Singapore: World Scientific. (2008): pp. 311–336. DOI: 10.1142/9789812814807_0013.

*Journal articles*

Abbe, E. "Beiträge zur Theorie des Mikroskops und der micrkoskopischen Wahrnehmung." *Arch. Mikrosk. Anat.* 9, no. 1 (1873): 413–468. DOI: 10.1007/BF02956173.

Arridge, S.R. "Optical tomography in medical imaging." *Inverse Probl.* 15, no. 2 (1999): R41–R93.

Balzarotti, F., Y. Eilers, K.C. Gwosch, A.H. Gynnå, V. Westphal, F.D. Stefani, J. Elf, S.W. Hell. Nanometer resolution imaging and tracking of fluorescent molecules with minimal photon fluxes." *Science* 355, no. 6325 (2017): 606–612. DOI: 10.1126/science.aak9913.

Bates, M., T.R. Blosser and X. Zhuang. "Short-range spectroscopic ruler based on a single-molecule optical switch." *Phys. Rev. Lett.* 94, no. 10 (2005): article 108101. DOI: 10.1103 /PhysRevLett.94.108101

Berning, S., K.I. Willig, H. Steffens, P. Dibaj and S.W. Hell. "Nanoscopy in a living mouse brain." *Science* 335, no. 6068 (2012): 551. DOI: 10.1126/science.1215369.

Betzig, E., G.H. Patterson, R. Sougrat, O.W. Lindwasser, S. Olenych, J.S. Moerner Bonifacino, M.W. Davidson, J. Lippincott-Schwartz and H.F. Hess. "Imaging intra-cellular fluorescent proteins at nanometer resolution." *Science* 313, no. 5793 (2006): 1642–1645. DOI: 10.1126/science.1127344.

Blomberg, S., J. Zhou, J. Gustafson, J. Zetterberg and E. Lundgren. "2D and 3D imaging of the gas phase close to an operating model catalyst by planar laser induced fluorescence." *Phys. Condens. Matter* 28, no. 45 (2016): article 453002. DOI: 10.1088/0953-8984/28/45/453002.

Brown, T.A., A.N. Tkachuk, G. Shtengel, B.G. Kopek, D.F. Bogenhagen, H.F. Hess and D.A. Clayton. "Super-resolution fluorescence imaging of mitochondrial nucleoids reveals their spatial range, limits, and membrane interaction." *Mol. Cell Biol.* 31, no. 24 (2011): 4994–5010. DOI: 10.1128/MCB.05694-11.

Chalfie, M., Y. Tu, G. Euskirchen, W.W. Ward and D.C. Prasher. "Green fluorescent protein as a marker for gene expression." *Science* 263, no. 5148 (1994): 802–805. DOI: 0.1126/science.8303295.

Cho, K.Y., A. Satija, T.L. Pourpoint, S.F. Son and R.P. Lucht. "High-repetition-rate three-dimensional OH imaging using scanned planar laser-induced fluorescence system for multiphase combustion." *Appl. Opt.* 53, no. 3 (2014): 316–326. DOI: 10.1364/AO.53.000316.

Cohn, L.C. "Transillumination of breasts." *J. Am. Med. Assoc.*, 93, no. 21 (1929): 1671. DOI: 10.1001/jama.1929.02710210065031.

Deliolanis, N., T. Lasser, D. Hyde, A. Soubret, J. Ripoll and V.S. Ntziachristos. "Free-space fluorescence molecular tomography utilizing 360° geometry projections." *Opt. Lett.* 32, no. 4 (2007): 382–384. DOI: 10.1364/OL.32.000382.

Dickson, R.M, A.B. Cubitt, R.Y. Tsien and W.E. Moerner. "On/off blinking and switching behaviour of single molecules of green fluorescent protein." *Nature* 388, no. 6640 (1997): 355–358. DOI: 10.1038/41048.

Gibson, A.P., J.C. Hebden and S.R. Arridge. "Recent advances in diffuse optical imaging." *Phys. Med. Biol.* 50, no. 4 (2005): R1–R43. DOI: 10.1088/0031-9155/50/4/R01.

Göttfert, F., C.A. Wurm, V. Mueller, S. Berning, V.C. Cordes, A. Honigmann and S.W. Hell. "Coaligned dual-channel STED nanoscopy and molecular diffusion analysis at 20 nm resolution." *Biophys. J.* 105, no. 1 (2013): L01–L03. DOI: 10.1016/j.bpj.2013.05.029.

Gustafsson, M.G.L. "Surpassing the lateral resolution limit by a factor of two using structured illumination microscopy." *J. Microsc.* 198, no. 2 (2000): 82–87. DOI: 10.1046/j.1365-2818.2000.00710.x.

Gustafsson, M.G.L. "Nonlinear structured-illumination microscopy: Wide-field fluorescence imaging with theoretically unlimited resolution." *PNAS* 102, no. 37 (2005): 13081–13086. DOI: 10.1073/pnas.0406877102.

Hell, S.W. "Toward fluorescence nanoscopy." *Nature Biotechnol.* 21, no. 11 (2003): 1347–1355. DOI: 10.1038/nbt895.

Hell, S.W. and J. Wichmann. "Breaking the diffraction resolution Limit by stimulated emission: Stimulated emission-depletion fluorescence microscopy." *Opt. Lett.* 19, no. 11 (1994): 780–782. DOI: 10.1364/OL.19.000780.

Hess, S.T, T.P.K. Girirajan and M.D. Mason. "Ultra-high resolution imaging by fluorescence photo-activation localization microscopy." *Biophys. J.* 91, no. 11 (2006): 4258–4272. DOI: 10.1529/biophysj.106.091116.

Hoffman, R.M. "The multiple uses of fluorescent proteins to visualize cancer *in vivo*." *Nat Rev Cancer* 5, no. 10 (2005): 796–806. DOI: 10.1038/nrc1717.

Hoffman, R.M. "Application of GFP imaging in cancer." *Lab. Invest.* 95, no. 4 (2015): 432–452. DOI: 10.1038/labinvest.2014.154.

Huang, B., H. Babcock and X. Zhuang. "Breaking the diffraction barrier: super-resolution imaging of cells." *Cell* 143, no. 7 (2010): 1047–1058. DOI: 10.1016/j.cell.2010.12.002.

Huang, B., M. Bates and X. Zhuang. "Super-resolution fluorescence microscopy." *Annu. Rev. Biochem.* 78 (2009): 993–10165. DOI: 10.1146/annurev.biochem.77.061906.092014.

Izeddin, I., C.G. Specht, M.l. Lelek, X. Darzacq, A. Triller, C. Zimmer and M. Dahan. "Super-resolution dynamic imaging of dendritic spines using a low-affinity photo-convertible actin probe." *PLoS ONE* 6, no. 1 (2011): article: e15611. DOI: 10.1371/journal.pone.0015611.

Jacques, S.L. and B.W. Pogue. "Tutorial on diffuse light transport." *J. Biomed. Opt.* 13, no. 4 (2008): article 041302. DOI: 10.1117/1.2967535.

Kaminski, C.F., J. Hult and M. Aldén. "High repetition rate planar laser induced fluorescence of OH in a turbulent non-premixed flame." *Appl. Phys. B* 68, no. 4 (1999): 757–760. DOI: 10.1007/s003400050700.

Klar, T.A., S. Jakobs, M. Dyba, A. Egner and S.W. Hell. "Fluorescence microscopy with diffraction resolution barrier broken by stimulated emission." *PNAS* 97, no. 15 (2000): 8206–8210. DOI: 10.1073/pnas.97.15.8206.

Lal, A., C. Shan and P. Xi. "Structured illumination microscopy image reconstruction algorithm." *IEEE J. Sel. Top. Quant. Electron.* 22, no. 4 (2016): article 6803414. DOI: 10.1109/JSTQE.2016.2521542.

Leung, B.O. and K.C. Chou. "Review of super-resolution fluorescence microscopy for biology." *Appl. Spectros.* 65, no. 9 (2011): 967–980. DOI: 10.1366/11-06398.

Ntziachristos, V. "Fluorescence molecular imaging." *Annu. Rev. Biomed. Eng.* 8 (2006): 1–33. DOI: 10.1146/annurev.bioeng.8.061505.095831.

Ntziachristos, V. "Optical imaging of molecular signatures in pulmonary inflammation." *Proc. Am. Thorac. Soc.* 6, no. 5 (2009): 416–418. DOI: 10.1513/pats.200901-003AW.

Ntziachristos, V., C. Bremer and R. Weissleder. "Fluorescence imaging with near-infrared light: new technological advances that enable in vivo molecular imaging." *Eur. Radiol.* 13, no. 1 (2003): 195–208. DOI: 10.1007/s00330-002-1524-x.

Ntziachristos, V., G. Turner, J. Dunham, S. Windsor, A. Soubret, J. Ripoll and H.A. Shihet. "Planar fluorescence imaging using normalized data." *J. Biomed. Opt.* 10, no. 6 (2005): article 064007. DOI: 10.1117/1.2136148.

Orain, M., F. Grisch, E. Jourdanneau, B. Rossow, C. Guin and B. Trétout. "Simultaneous measurements of equivalence ratio and flame structure in multipoint injectors using PLIF." *C. R. Mécanique* 337, no. 6–7 (2009): 373–384. DOI: 10.1016/j.crme.2009.06.019.

Patterson, G.H. and J. Lippincott-Schwartz. "A photo-activatable GFP for selective photo-labeling of proteins and cells." *Science* 297, no. 5588 (2002): 1873–1877. DOI: 10.1126/science.1074952

Rust, M.J., M. Bates and X. Zhuang. "Sub-diffraction-limit imaging by stochastic optical reconstruction microscopy (STORM)." *Nature Meth.* 3, no. 10 (2006): 793–795. DOI: 10.1038/nmeth929.

Saxena, M., G. Eluru and S.S. Gorthi. "Structured illumination microscopy." *Adv. Opt. Photon.* 7, no. 2 (2015): 241–275. DOI: 10.1364/AOP.7.000241.

Schulz, C. and V. Sick. "Tracer-LIF diagnostics: Quantitative measurement of fuel concentration, temperature and fuel/air ratio in practical combustion systems." *Prog. Energy Combust. Sci.* 31, no. 1 (2005): 75–121. DOI: 10.1016/j.pecs.2004.08.002.

Shcherbakova, D.M., P. Sengupta, J. Lippincott-Schwartz and V.V. Verkhusha. "Photocontrollable fluorescent proteins for super-resolution imaging." *Ann. Rev. Biophys.* 43, (2014): 303–329. DOI: 10.1146/annurev-biophys-051013-022836.

Shtengel, G., J.A. Galbraith, C.G. Galbraith, J. Lippincott-Schwartz, J.M. Gillette, S. Manley, R. Sougrat, C.M. Waterman, P. Kanchanawong, M.W. Davidson, R.D. Fetter and H.F. Hess. "Interferometric fluorescent super-resolution microscopy resolves 3D cellular ultrastructure." *PNAS* 106, no. 9 (2009): 3125–3130. DOI: 10.1073/pnas.0813131106.

Takehara, H., O. Kazutaka, M. Haruta, T. Noda, K. Sasagawa, T. Tokuda and J. Ohta. "On-chip cell analysis platform: Implementation of contact fluorescence microscopy in microfluidic chips." *AIP Adv.* 7, no. 9 (2017): article 095213. DOI: 10.1063/1.4986872.

Testa, I., N.T. Urban, S. Jakobs, C. Eggeling, K.I. Willig and S.W. Hell. "Nanoscopy of living brain slices with low light levels." *Neuron* 75, no. 6 (2012): 992–1000. DOI: 10.1016/j.neuron.2012.07.028.

Vintersten, K., C. Monetti, M. Gertsenstein, P. Zhang, L. Laszlo, S. Biechele and A. Nagy. "Mouse in red: Red fluorescent protein expression in mouse ES cells, embryos, and adult animals." *Genesis* 40, no. 4 (2004): 241–246. DOI: 10.1002/gene.20095.

Wang, Y. and P. Kanchanawong. "Three-dimensional super-resolution microscopy of F-actin filaments by interferometric photo-activated localization microscopy (iPALM)." *J. Vis. Exp.* 118 (2016): article e54774. DOI: 10.3791/54774.

Xu, K., G. Zhong and X. Zhuang. "Actin, spectrin, and associated proteins form a periodic cytoskeletal structure in axons." *Science* 339, no. 6118 (2013): 452–456. DOI: 10.1126/science.1232251.

Yamanaka, M., N.I. Smith and K. Fujita. "Introduction to super-resolution microscopy." *Microscopy* 63, no. 3 (2014): 177–192. DOI: 10.1093/jmicro/dfu007.

Yang, M., J. Reynoso, M. Bouvet and R.M. Hoffman. "A transgenic red fluorescent protein-expressing nude mouse for color-coded imaging of the tumor micro-environment." *J. Cell Biochem.* 106, no. 2 (2009): 279–284. DOI: 10.1002/jcb.21999.

Zacharakis, G., H. Kambara, H. Shih, J. Ripoll, J. Grimm. "Volumetric tomography of fluorescent proteins through small animals *in-vivo.*" *PNAS* 102, no. 51 (2005): 18252–18257. DOI: 10.1073/pnas.0504628102.

Zacharakis, G., J. Ripoll, R. Weissleder and V. Ntziachristos. "Fluorescent protein tomography scanner for small animal imaging." *IEEE Trans. Med. Imaging* 24, no. 7 (2005): 878–885. DOI: 10.1109/TMI.2004.843254.

*Other sources*

The Carl Zeiss Microscopy Online Campus. "Super-resolution microscopy." In *Education in microscopy and digital imaging;* providing review articles, interactive tutorials and reference libraries (2017). **Access:** http://zeiss-campus.magnet.fsu.edu/articles/superresolution/index.html.

Zhuang, X. "*Comparison of conventional (upper) and 3D STORM (lower) images of actin in the axons of neurons* (STORM Image gallery)." Harvard University, Cambridge, MA, USA **Access:** http://zhuang.harvard.edu/storm_images.html.

## Chapter 19    Raman imaging and microscopy

*Books and chapters in books*

Cheng, J.X. and X.S. Xie. *Coherent Raman Scattering Microscopy.* Boca Raton, FL, USA: CRC Press (2013).

Dieing, T., O. Hollricher and J. Toporski (eds.). *Confocal Raman Microscopy.* Springer Series in Optical Sciences, vol. 158. Heidelberg, Germany: Springer Verlag (2011).

Freudiger, C. and X.S. Xie. "Stimulated Raman scattering microscopy." Chapter 4 in J.X. Cheng and X.S. Xie (eds.). *Coherent Raman Scattering Microscopy.* Boca Raton, FL, USA: CRC Press (2013): pp. 99–120.

Fries, M. and A. Steele. "Raman spectroscopy and confocal Raman imaging in mineralogy and petrography." In T. Dieing, O. Hollricher and J. Toporski (eds.). *Confocal Raman Microscopy.* Springer Series in Optical Sciences, vol. 158. Heidelberg, Germany: Springer Verlag (2011): pp. 111–135. DOI: 10.1007/978-3-642-12522-5_6.

Salzer, R. and H.W. Siesler (eds.). *Infrared and Raman Spectroscopic Imaging*, 2nd edition. Chichester, UK: John Wiley & Sons (2014).

Zoubir, A. (ed.). *Raman Imaging: Techniques and Applications.* Springer Series in Optical Sciences, vol. 168. Heidelberg, Germany: Springer Verlag (2012).

*Journal articles*

Chen, C., N. Hayazawa and S. Kawata. "A 1.7 nm resolution chemical analysis of carbon nanotubes by tip-enhanced Raman imaging in the ambient." *Nature Commun.* 5 (2014): article 3312. DOI: 10.1038/ncomms4312.

Conde, J., C. Bao, D. Cui, P.V. Baptista and F. Tian. "Antibody-drug gold nano-antennas with Raman spectroscopic fingerprints for in vivo tumour theranostics." *J. Control. Release* 183 (2014): 87–93. DOI: 10.1016/j.jconrel.2014.03.045.

Day, J.P.R., K.F. Domke, G. Rago, H. Kano, H. Hamaguchi, E.M. Vartiainen and M. Bonn. "Quantitative coherent anti-Stokes Raman scattering (CARS) microscopy." *J. Phys. Chem. B* 115, no. 24 (2011): 7713–7725. DOI: 101021/jp200606e.

Deng, S., L. Liu, Z. Liu, Z. Shen, G. Li and Y. He. "Line-scanning Raman imaging spectroscopy for detection of fingerprints." *Appl. Opt.* 51 no. 17 (2012): 3701–3706. DOI: 10.1364/AO.51.003701.

Dinish, U.S., G. Balasundaram, Y.T. Chang and M. Olivo. "Actively targeted *in vivo* multiplex detection of intrinsic cancer biomarkers using biocompatible SERS nano-tags." *Sci Rep.* 4 (2014): article 4075. DOI: 10.1038/srep04075.

Downes, A. and A. Elfick. "Raman spectroscopy and related techniques in biomedicine." *Sensors* 10, no. 3 (2010): 1871–1889. DOI: 10.3390/s100301871.

Guo, L., J.A. Jackman, H.H. Yang, P. Chen, N.J. Cho and D.H. Kim. "Strategies for enhancing the sensitivity of plasmonic nano-sensors." *Nano Today* 10, no. 2 (2015): 213–239. DOI: 10.1016/j.nantod.2015.02.007.

Ideguchi, T., S. Holzner, B. Bernhardt1, G. Guelachvili, N. Picqué and T.W. Hänsch. "Coherent Raman spectro-imaging with laser frequency combs." *Nature* 502, no. 7471 (2013): 355–358. DOI: 10.1038/nature12607.

Kann, B., H.L. Offerhaus, M. Windbergs and C. Otto. "Raman microscopy for cellular investigations: From single cell imaging to drug carrier uptake visualization." *Adv. Drug Deliv. Rev.* 89 (2015): 71–90. DOI: 10.1016/j.addr.2015.02.006.

Kano, H., H. Segawa, M. Okuno, P. Leproux and V. Couderc. "Hyperspectral coherent Raman imaging—Principle, theory, instrumentation, and applications to life sciences." *J. Raman Spectrosc.* 47, no. 1 (2016): 116–123. DOI: 10.1002/jrs.4853.

Kawata, S., T. Ichimura, A. Taguchi and Y. Kumamoto. "Nano-Raman scattering microscopy: Resolution and enhancement." *Chem. Rev.* 117 (2017): 4983–5001. DOI: 101021/acs.chemrev.6b00560.

Keshava, N. and J.F. Mustard. "Spectral unmixing." *IEEE Signal Proc. Mag.* 19, no. 1 (2002): 44–57. DOI: 10.1109/79.974727.

Klein, K., A.M. Gigler, T. Aschenbrenner, R. Monetti, W. Bunk, F. Jamitzky, G. Morfill, R.W. Stark and J. Schlegel. "Label-free live-cell imaging with confocal Raman microscopy." *Biophys. J.* 102, no. 2 (2012): 360–368. DOI: 10.1016/j.bpj.2011.12.027.

Kumar, K., S. Mignuzzi, W. Su and D. Roy. "Tip-enhanced Raman spectroscopy: Principles and applications." *EPJ Techn. Instrum.* 2 (2015): article 9. doi: 10.1140/epjti/s40485-015-0019-5.

Langelüddecke, L., P. Singh and V. Deckert. "Exploring the nanoscale: Fifteen years of tip-enhanced Raman spectroscopy." *Appl. Spectrosc.* 69, no. 12 (2015): 1357–1371. DOI: 10.1366/15-08014.

Li, X., R. Zhou, Y. Xu, X. Wei and Y. He. "Spectral unmixing combined with Raman imaging, a preferable analytic technique for molecule visualization." *Appl. Spectrosc. Rev.* 52, no. 5 (2017): 417–438. DOI: 10.1080/05704928.2016.1226183.

Liao, C.S., M.N. Slipchenko, P. Wang, J. Li, S.Y. Lee, R.A Oglesbee and J.X. Cheng. "Microsecond scale vibrational spectroscopic imaging by multiplex stimulated Raman scattering microscopy." *Light Sci. Appl.* 4 (2015): article e265. DOI: 10.1038/lsa.2015.38.

Ling, Z. and A. Wang. "Spatial distributions of secondary minerals in the Martian meteorite MIL 03346,168 determined by Raman spectroscopic imaging." *J. Geophys. Res. Planets* 120, no. 6 (2015): 1141–1159. DOI: 10.1002/2015JE004805.

Majzner, K., K. Kochan, N. Kachamakova-Trojanowska, E. Maslak, S. Chlopicki and M. Baranska. "Raman imaging providing insights into chemical composition of lipid droplets of different size and origin: In hepatocytes and endothelium." *Anal. Chem.* 86, no. 13 (2014): 6666–6674. DOI: 10.1021/ac501395g.

Matthäus, C., C. Krafft, B. Dietzek, B.R. Brehm, S. Lorkowski and J. Popp. "Non-invasive imaging of intracellular lipid metabolism in macrophages by Raman microscopy in combination with stable isotopic labeling." *Anal. Chem.* 84, no. 20 (2012): 8549–8556. DOI: 10.1021/ac3012347.

Min, W., C.W. Freudiger, S. Lu and X.S. Xie. "Coherent nonlinear optical imaging: Beyond fluorescence microscopy." *Annu. Rev. Phys. Chem.* 62 (2011): 507–530. DOI: 10.1146/annurev.physchem.012809.103512.

Nandakumar, P., A. Kovalev and A. Volkmer. "Vibrational imaging based on stimulated Raman scattering microscopy." *New J. Phys.* 11, no. 3 (2009): article 033026. DOI: 10.1088/1367-2630/11/3/033026.

Noothalapati, H., T. Sasaki, T. Kainol, M. Kawamukai, M. Ando, H. Hamaguchi and T. Yamamoto. "Label-free chemical imaging of fungal spore walls by Raman microscopy and multivariate curve resolution analysis." *Sci. Rep.* 6 (2016): article 27789. DOI: 10.1038/srep27789.

Okuno, M., H. Kano, P. Leproux, V. Couderc, J.P.R. Day, M. Bonn and H. Hamaguchi. "Quantitative CARS molecular fingerprinting of single living cells with the use of the maximum entropy method." *Angew. Chem. Int. Ed. Engl.* 49, no. 38 (2010): 6773–6777. DOI: 10.1002/anie.201001560.

Opilik, L., T. Schmid and R. Zenobi. "Modern Raman imaging: Vibrational spectroscopy on the micrometer and nanometer scales." *Annu. Rev. Anal. Chem.* 6 (2013): 379–398. DOI: 10.1146/annurev-anchem-062012-092646.

Palonpon, A.F., M. Sodeoka and K. Fujita. "Molecular imaging of live cells by Raman microscopy." *Curr. Opin. Chem. Biol.* 17, no. 4 (2013a): 708–715. DOI: 10.1016/j.cbpa.2013.05.021.

Palonpon, A.F., J. Ando, H. Yamakoshi, K. Dodo, M. Sodeoka, S. Kawata and K. Fujita. "Raman and SERS microscopy for molecular imaging of live cells." *Nat. Protoc.* 8, no. 4 (2013b): 677–692. DOI: 10.1038/nprot.2013.030.

Rodriguez, R.D., E. Sheremet, T. Deckert-Gaudig, C. Chaneac, M. Hietschold, V. Deckert and D.R.T. Zahn. "Surface- and tip-enhanced Raman spectroscopy reveals spin-waves in iron oxide nanoparticles." *Nanoscale* 7, no. 21 (2015): 9545–9551. DOI: 10.1039/c5nr01277e.

Rusciano, G., G. Zito, R. Isticato, T. Sirec, E. Ricca, E. Bailo and A. Sasso. "Nanoscale chemical imaging of bacillus subtilis spores by combining tip-enhanced Raman scattering and advanced statistical tools." *ACS Nano* 8, no. 12 (2014): 12300–12309. DOI: 10.1021/nn504595k.

Saar, B.G., C.W. Freudiger, J. Reichman, C.M. Stanley, G.R. Holtom and X.S. Xie. "Video-rate molecular imaging *in vivo* with stimulated Raman scattering." *Science* 330, no. 6009 (2010): 1368–1370. DOI: 10.1126/science.1197236.

Schie, I.W. and T. Huser. "Methods and applications of Raman micro-spectroscopy to single-cell analysis." *Appl Spectrosc.* 67, no. 8 (2013): 813–828. DOI: 10.1366/12-06971.

Schmid, T., L. Opilik, C. Blum and R. Zenobi. "Nanoscale chemical imaging using tip-enhanced Raman spectroscopy: A critical review." *Angew. Chem. Int. Ed.* 52, no. 23 (2013): 5940–5954. DOI: 10.1002/anie.201203849.

Shi, X., N. Coca-López, J. Janik and A. Hartschuh. "Advances in tip-enhanced near-field Raman microscopy using nanoantennas." *Chem. Rev.* 117 (2017): 4945–4960. DOI: 10.1021/acs.chemrev.6b00640.

Silva, W.R., C.T. Graefe and R.R. Frontiera. "Toward label-free super-resolution microscopy." *ACS Photon.* 3, no. 1 (2016): 79–86. DOI: 10.1021/acsphotonics.5b00467.

Tipping, W.J., M. Lee, A. Serrels, V.G. Brunton and A.N. Hulme. "Stimulated Raman scattering microscopy: An emerging tool for drug discovery." *Chem. Soc. Rev.* 45, no. 8 (2016): 2075–2089. DOI: 10.1039/c5cs00693g.

Touzalin, T., A.L. Dauphin, S. Joiret, I.T. Lucas and E. Maisonhaute. "Tip-enhanced Raman spectroscopy imaging of opaque samples in organic liquid." *Phys. Chem. Chem. Phys.* 18, no. 23 (2016): 15510–15513. DOI: 10.1039/c6cp02596j.

Verma, P. "Tip-enhanced Raman spectroscopy: Technique and recent advances." *Chem. Rev.* 117 (2017): 6447–6466. DOI: 10.1021/acs.chemrev.6b00821.

Volkmer, A. "Vibrational imaging and micro-spectroscopies based on coherent anti-Stokes Raman scattering microscopy." *J. Phys. D: Appl. Phys.* 38, no. 5 (2005): R59–R81. DOI: 10.1088/0022-3727/38/5/R01.

Wei, L. Y. Yu, Y. Shen, M.C. Wang and W. Min. "Vibrational imaging of newly synthesized proteins in live cells by stimulated Raman scattering microscopy." *PNAS* 110, no. 28 (2013): 11226–11231. DOI: 10.1073/pnas.1303768110.

Winterhalder, M.J. and A. Zumbusch. "Beyond the borders—Biomedical applications of non-linear Raman microscopy." *Adv. Drug Deliv. Rev.* 89 (2015): 135–144. DOI: 10.1016/j.addr.2015.04.024.

Xu, Q., W. Liu, L. Li, F. Zhou, J. Zhou and Y. Tian. "Ratiometric SERS imaging and selective biosensing of nitric oxide in live cells based on trisoctahedral gold nanostructures." *Chem. Commun.* 53, no. 11 (2017): 1880–1883. DOI: 10.1039/c6cc09563a.

Zhang, R., Y. Zhang, Z.C. Dong, S. Jiang, C. Zhang, L.G. Chen, L. Zhang, Y. Liao, J. Aizpurua, Y. Luo, J.L. Yang and J.G. Hou. "Chemical mapping of a single molecule by plasmon-enhanced Raman scattering." *Nature* 498, no. 7452 (2013): 82–86. DOI: 10.1038/nature12151.

Zhang, Y., H. Hong, D.V. Myklejord and W. Cai. "Molecular imaging with SERS-active nanoparticles." *Small* 7, no. 23 (2011): 3261–3269. DOI: 10.1002/smll.201100597.

Zumbusch, A., G.R. Holtom and X.S. Xie. "Three-dimensional vibrational imaging by coherent anti-Stokes Raman scattering." *Phys. Rev. Lett.* 82, no. 20 (1999): 4142–4145. DOI: 10.1103/PhysRevLett.82.4142.

## Chapter 20    Diffuse optical imaging

*Books and chapters in books*

Bright, R. *Diseases of the Brain and Nervous System*, vol. 2. London, UK: Longman and Co. (1831).

Curling, T.B. *A Practical Treatise on the Diseases of the Testis and of the Spermatic Cord and Scrotum.* London, UK: Samuel Highley (1843): pp 125–181.

Drexler, W. and J.G. Fujimoto (eds.). *Optical Coherence Tomography: Technology and Applications* (3-volume set), 2nd edition. Heidelberg, Germany: Springer (2015).

Huppert, T.J. "History of diffuse optical spectroscopy of human tissue." In S.J. Madsen (ed.), 'Optical methods and instrumentation in brain imaging and therapy'. Bioanalysis, vol. 3. Heidelberg, Germany: Springer (2013), pp. 23–56. DOI: 10.1007/978-1-4614-4978-2.

Jiang, H. *Diffuse Optical Tomography: Principles and Applications.* London, UK: CRC Press, Taylor & Francis Group (2010).

Madsen, S.J. (ed.). *Optical Methods and Instrumentation in Brain Imaging and Therapy.* Series Bioanalysis, vol. 3. Heidelberg, Germany: Springer (2013).

Torricelli A., M. Vanoli, M. Leitner, A. Nemeth, N.N.D. Trong, B. Nicolai and W. Saeys. "Chapter 5—Optical coherence tomography, space-resolved reflectance spectroscopy and time-resolved reflectance spectroscopy: principles and applications to food microstructures." In V.J. Morris and K. Groves (eds.), *Food Microstructures: Microscopy, Measurement and Modelling*. Cambridge, UK: Woodhead Publishing (2013): pp. 150–155.

Wang, L.V. and H. Wu. *Biomedical Optics*. Chichester, UK: Wiley (2007).

*Journal articles*     Adler, A. and W.R.B Lionheart. "Uses and abuses of EIDORS: An extensible software base for EIT." *Physiol. Meas.* 27, no. 5 (2006): S25–S42. doi: 10.1088/0967-3334/27/5/S06. Download of software (last update 29 May 2013): http://eidors3d.sourceforge.net/download.shtml. [Last accessed on 2 September 2014.]

Arridge, S.R. "Optical tomography in medical imaging." *Inverse Problems* 15, no. 2 (1999): R41–R93. DOI: 10.1088/0266-5611/15/2/022.

Arridge, S.R. "Methods in diffuse optical imaging." *Phil. Trans. R. Soc. A* 369, no. 1955 (2011): 4558–4576. DOI: 10.1098/rsta.2011.0311.

Arridge, S.R. and J.C. Hebden. "Optical imaging in medicine: II. Modelling and reconstruction." *Phys. Med. Biol.* 42, no. 5 (1997): 841–853. DOI: 10.1088/0031-9155/42/5/008.

Arridge, S.R. and W.R.B. Lionheart. "Non-uniqueness in diffusion-based optical tomography." *Opt. Lett.* 23, no. 11 (1998): 882–884. DOI: 10.1364/OL.23.000882

Arridge, S.R. and M. Schweiger. "Photon measurement density functions. Part II: Finite element method calculations." *Appl. Opt.* 34, no. 34 (1995): 8026–8037. DOI: 10.1364/AO.34.008026.

Arridge, S.R., M. Schweiger, M. Hiraoka, and D.T. Delpy. "Finite element approach for modelling photon transport in tissue." *Med. Phys.* 20, no. 2 (1993): 299–309. DOI: 10.1118/1.597069.

Asahara, A., H. Watanabe, H. Tokoro, S.I. Ohkoshi and T. Suemoto. "Ultrafast dynamics of photo-induced semiconductor-to-metal transition in the optical switching nano-oxide $Ti_3O_5$." *Phys. Rev. B* 90, no. 1 (2014): article 014303. DOI: 10.1103/PhysRevB.90.014303.

Bettmann, M.A. "Frequently asked questions: Iodinated contrast agents." *Radiographics* 24, no. suppl-1 (2004): S3–S10. DOI: 10.1148/rg.24si045519.

Bhowmik, T., H. Liu, Z. Ye and S. Oraintara. "Dimensionality reduction based optimization algorithm for sparse 3-D image reconstruction in diffuse optical tomography." *Sci. Rep.* 6 (2016): article 22242. DOI: 10.1038/srep22242.

Bian, Z., T. Tachikawa, P. Zhang, M. Fujitsuka and T. Majima. "A nanocomposite superstructure of metal oxide with effective charge transfer interfaces." *Nature Commun.* 5 (2014): article 3038. DOI: 10.1038/ncomms4038.

Boas, D.A., J.P. Culver, J.J. Stott, and A.K. Dunn. "Three dimensional Monte Carlo code for photon migration through complex heterogeneous media including the adult human head." *Opt. Express* 10, no. 3 (2002): 159–169. DOI: 10.1364/OE.10.000159.

Chaillat, S. and G. Biros. "A fast and adaptive algorithm for the inverse medium problem with multiple frequencies and multiple sources for the time-harmonic wave equation." In G. De Roeck, G. Degrande, G. Lombaert and G. Müller (eds.). *Proc. 8th Int. Conf. on Structural Dynamics—EURODYN 2011*. Volos, Greece: The European Association for Structural Dynamics (2011): pp. 2535–2542.

Chernomordik, V., D.W. Hattery, A.H. Gandjbakhche, A. Pifferi, P. Taroni, A. Torricelli, G. Valentini and R. Cubeddu. "Quantification by random walk of the optical parameters of nonlocalized abnormalities embedded within tissue-like phantoms." *Opt. Lett.* 25, no. 13 (2000): 951–953. DOI: 10.1364/OL.25.000951.

Choe, R., S.D. Konecky, A. Corlu, K. Lee, T. Durduran, D.R. Busch, S. Pathak, B.J. Czerniecki, J. Tchou, D.L. Fraker, A. DeMichele, B. Chance, S.R. Arridge, M. Schweiger, J.P. Culver, M.D. Schnall, M.E. Putt, M.A. Rosen and A.G. Yodh. "Differentiation of benign and malignant breast tumors by *in-vivo* three-dimensional parallel-plate diffuse optical tomography." *J. Biomed. Opt.* 14, no. 2 (2009): article 024020. DOI:10.1117/1.3103325.

Choe, R., M.E. Putt, P.M. Carlile, T. Durduran, J.M. Giammarco, D.R. Busch, K.W. Jung, B.J. Czerniecki, J. Tchou, M.D. Feldman, C. Mies, M.A. Rosen, M.D. Schnall, A. DeMichele and A.G. Yodh. "Optically measured microvascular blood flow contrast of malignant breast tumors." *PLoS ONE* 9, no. 6 (2014): e99683. DOI:10.1371/journal.pone.0099683.

Corlu, A., T. Durduran, R. Choe, M. Schweiger, E.M.C. Hillman, S.R. Arridge and A.G. Yodh. "Uniqueness and wavelength optimization in continuous-wave multispectral diffuse optical tomography." *Opt. Lett.* 28, no. 23 (2003): 2339–2431. DOI: 10.1364/OL.28.002339.

Culver, J.P., R. Choe, M.J. Holboke, L. Zubkov, T. Durduran, A. Slemp, V. Ntziachristos, B. Chance and A.G. Yodh. "Three-dimensional diffuse optical tomography in the parallel plane transmission geometry: Evaluation of a hybrid frequency domain/continuous wave clinical system for breast imaging." *Med. Phys.* 30, no. 2 (2003a): 235–247. DOI: 10.1118/1 .1534109.

Culver, J.P., A.M. Siegel, J.J. Stott and D.A. Boas. "Volumetric diffuse optical tomography of brain activity." *Opt. Lett.* 28, no. 21 (2003b): 2061–2063. DOI: 10.1364/OL.28.002061.

Cutler, M. "Transillumination as an aid in the diagnosis of breast lesions." *Surg. Gynecol. Obstet.* 48 (1929): 721–728.

Cutler, M. "Transillumination of the breast." *Ann. Surgery.* 93, no. 1 (1931): 223–234.

Darne, C., Y. Lu and E.M. Sevick-Muraca. "Small animal fluorescence and bioluminescence tomography: A review of approaches, algorithms and technology update." *Phys. Med. Biol.* 59, no. 1 (2014): R1–R64. DOI:10.1088/0031-9155/59/1/R1.

Dehghani, H., M.E. Eames, P.K. Yalavarthy, S. C. Davis, S. Srinivasan, C.M. Carpenter, B.W. Pogue and K.D. Paulsen. "Near infrared optical tomography using NIRFAST: Algorithm for numerical model and image reconstruction." *Commun. Numer. Meth. Eng.* 25, no. 6 (2008): 711–732. DOI: 10.1002/cnm.1162.

Durduran, T., R. Choe, W.B. Baker and A.G. Yodh. "Diffuse optics for tissue monitoring and tomography." *Rep. Prog. Phys.* 73, no.7 (2010): 076701. DOI: 10.1088/0034-4885/73/7/076701.

Eccher Zerbini, P., M. Grassi, R. Cubeddu, A. Pifferi and A. Torricelli. "Non-destructive detection of brown heart in pears by time-resolved reflectance spectroscopy." *Postharvest Biol. Technol.* 25, no.1 (2002): 87–97. DOI: 10.1016/S0925-5214(01)00150-8.

Eggebrecht, A.T., S.L. Ferradal, A. Robichaux-Viehoever, M.S. Hassanpour, H. Dehghani, A.Z. Snyder, T. Hershey and J.P. Culver. "Mapping distributed brain function and networks with diffuse optical tomography." *Nature Photon.* 8, no. 6 (2014): 448–454. DOI: 10.1038/NPHOTON .2014.107.

Fantini, S. and A. Sassaroli. "Near-Infrared mammography for breast cancer detection with intrinsic contrast." *Ann. Biomed. Eng.* 42, no. 2 (2012): 398–407. DOI: 10.1007/s10439-011-0404-4.

Ferrari, M. K., H. Norris and M.G. Sowa. "Medical near infrared spectroscopy 35 years after the discovery." *J. Near Infrared Spectrosc.* 20, no. 1 (2012a): vii–ix. DOI: 10.1255/jnirs.982.

Ferrari, M. and V. Quaresima. "A brief review on the history of human functional near-infrared spectroscopy (fNIRS) development and fields of application." *NeuroImage* 63, no. 2 (2012b): 921–935. DOI: 10.1016/j.neuroimage.2012.03.049.

Franceschini, M.A., V. Toronov, M.E. Filiaci, E. Gratton and S. Fantini. "On-line optical imaging of the human brain with 160-ms temporal resolution," *Opt. Express* 6, no. 3 (2000): 49–57. DOI: 10.1364/OE.6.000049.

Gibson, A. and H. Dehghani. "Diffuse optical imaging." *Phil. Trans. R. Soc. A* 367, no. 1900 (2009): 3055–3072. DOI: 10.1098/rsta.2009.0080.

Gibson, A.P., J.C. Hebden and S.R. Arridge. "Recent advances in diffuse optical imaging." *Phys. Med. Biol.* 50 (2005): R1–R43. DOI:10.1088/0031-9155/50/4/R01.

Grable, R.J., D.P. Rohler, and S. Kla. "Optical tomography breast imaging." *Proc. SPIE* 2979 "Optical Tomography and Spectroscopy of Tissue: Theory, Instrumentation, Model, and Human Studies II" (2004): 197–210. DOI: 10.1117/12.280266.

Hadjipanayis G.C., H. Jiang, D.W. Roberts and L. Yang. "Current and future clinical applications for optical imaging of cancer: From intraoperative surgical guidance to cancer screening." *Semin Oncol.* 38, no. 1 (2011): 109–118. DOI: 10.1053/j.seminoncol.2010.11.008.

Hebden, J.C., S.R. Arridge and D.T. Delpy. "Optical imaging in medicine: I. Experimental techniques." *Phys. Med. Biol.* 42, no. 5 (1997): 825–840. DOI: 10.1088/0031-9155/42/5/007.

Hillman, E.M.C. "Optical brain imaging in vivo: Techniques and applications from animal to man." *J. Biomed. Opt.* 12, no. 5 (2007): article 051402. DOI:10.1117/1.2789693.

Huang, H. L. Liu and M.O. Ngadi. "Recent developments in hyperspectral imaging for assessment of food quality and safety." *Sensors* 14, no. 4 (2014): 7248–7276. DOI: 10.3390 /s140407248.

Jacques, S.L. "Optical properties of biological tissues: A review." *Phys. Med. Biol.* 58, no. 11 (2013): R37–R61. DOI:10.1088/0031-9155/58/11/R37.

Jasinska, K.K. and L.A. Petitto. "How age of bilingual exposure can change the neural systems for language in the developing brain: A functional near infrared spectroscopy investigation of syntactic processing in monolingual and bilingual children." *Dev. Cogn. Neurosci.* 6 (2013): 87–101. DOI: 10.1016/j.dcn.201306005.

Jermyn, M., H. Ghadyani, M.A. Mastanduno, W. Turner, S.C. Davis, H. Dehghani and B.W. Pogue. "Fast segmentation and high-quality three-dimensional volume mesh creation from medical images for diffuse optical tomography." *J. Biomed. Opt.* 18, no. 8 (2013): article 086007.

doi: 10.1117/1.JBO.18.8.086007. Download of software (last update 13 November 2013): https://code.google.com/p/nirfast/.

Kalisz, J. "Review of methods for time interval measurements with picosecond resolution." *Metrologia* 4, (2004): 17–32. DOI: 10.1088/0026-1394/41/1/004.

Kim, M.S., K. Lee, K. Chao, A.M. Lefcourt, W. Jun and D.E. Chan. "Multispectral line-scan imaging system for simultaneous fluorescence and reflectance measurements of apples: Multitask apple inspection system." *Sens. Instrumen. Food Qual.* 2, no. 2 (2008): 123–129. DOI: 10.1007/s11694-008-9045-1.

Koizumi, H., T. Yamamoto, A. Maki, Y. Yamashita, H. Sato, H. Kawaguchi and N. Ichikawa. "Optical topography: Practical problems and new applications." *Appl. Opt.* 42, no. 16 (2003): 3054–3062. DOI: 10.1364/AO.42.003054.

Kramer, K. "Ein Verfahren zur laufenden Messung des Sauerstoffgehaltes im strömenden Blute an uneröffneten Gefäßen." *Z. Biol.* 96 (1935): 61–75.

Leff, D.R., O.J. Warren, L.C. Enfield, A. Gibson, T. Athanasiou, D.K. Patten, J. Hebden, G.Z. Yang and A. Darzi. "Diffuse optical imaging of the healthy and diseased breast: A systematic review." *Breast Cancer Res. Treat.* 108, no. 1 (2008): 9–22. DOI: 10.1007/s10549-007-9582-z.

Liberman, L., T.L. Feng, D.D. Dershaw, E.A. Morris and A.F. Abramson. "US-guided core breast biopsy: Use and cost-effectiveness." *Radiology* 208, no. 3 (1998): 717–723. DOI: 10.1148/radiology.208.3.9722851.

Lopez Silva, S.M., M.L. Dotor Castilla and J.P. Silveira Martin. "Near-infrared transmittance pulse oximetry with laser diodes." *J. Biomed. Opt.* 8, no. 3 (2003): 525–533. DOI: 10.1117/1.1578495.

Lu, R. and Y. Peng. "Development of a multispectral imaging prototype for real-time detection of apple fruit firmness." *Opt. Eng.* 46, no. 12 (2007): article 123201. DOI: 10.1117/1.2818812.

Maki, A., Y. Yamashita, Y. Ito, E. Watanabe, Y. Mayanagi and H. Koizumi. "Spatial and temporal analysis of human motor activity using non-invasive NIR topography." *Med. Phys.* 22, no. 12 (1995): 1997–2005. DOI: 10.1118/1.597496.

Manley, M. "Near-infrared spectroscopy and hyperspectral imaging: non-destructive analysis of biological materials." *Chem Soc Rev.* 43, no. 24 (2014): 8200–8214. DOI: 10.1039/c4cs00062e.

McBride, T.O., B.W. Pogue, S. Poplack, S. Soho, W.A. Wells, S. Jiang, U.L. Österberg and K.D. Paulsen. "Multispectral near-infrared tomography: a case study in compensating for water and lipid content in hemoglobin imaging of the breast." *J. Biomed. Opt.* 7, no. 1 (2002): 72–79. DOI:10.1117/1.1428290.

Mehagnoul-Schipper, D.J., B.F.W. van der Kallen, W.N.J.M. Colier, M.C. van der Sluijs, L.J.Th.O. van Erning, H.O.M. Thijssen, B. Oeseburg, W.H.L. Hoefnagels and R.M.M. Jansen. "Simultaneous measurements of cerebral oxygenation changes during brain activation by near-infrared spectroscopy and functional magnetic resonance imaging in healthy young and elderly subjects." *Human Brain Mapping* 16, no. 1 (2002): 14–23. DOI: 10.1002/hbm.10026.

Michell, M.A., A. Iqbal, R.K. Wasan, D.R. Evans, C. Peacock, C.P. Lawinski, A. Douiri, R. Wilson and P. Whelehan. "A comparison of the accuracy of film-screen mammography, full-field digital mammography, and digital breast tomosynthesis." *Clin. Radiology* 67, no. 10 (2012): 976–981. DOI: 10.1016/j.crad.2012.03.009.

Paquette, N., B. Gonzalez-Frankenberger, P. Vannasing, J. Tremblay, O. Florea, R. Beland, F. Lepore and M. Lassonde. "Lateralization of receptive language function using near infrared spectroscopy." *Neurosci. Med.* 1, no. 2 (2010): 64–70. DOI: 10.4236/nm.2010.12010.

Peña, M., A. Maki, D. Kovačić, G. Dehaene-Lambertz, H. Koizumi, F. Bouquet and J. Mehler. "Sounds and silence: An optical topography study of language recognition at birth." *Proc. Natl. Acad. Sci. USA* 100, no. 20 (2003): 11702–11705. DOI: 10.1073/pnas.1934290100.

Pogue, B.W., S.C. Davis, X. Song, B.A. Brooksby, H. Dehghani and K.D. Paulsen. "Image analysis methods for diffuse optical tomography." *J. Biomed. Opt,* 11, no. 3 (2006): article 033001. DOI: 10.1117/1.2209908.

Qi, J. and Z. Ye. "CTLM as an adjunct to mammography in the diagnosis of patients with dense breast." *Clin. Imaging* 37, no. 2 (2013): 289–294. DOI: 10.1016/j.clinimag.2012.05.003.

Rosen, Y., R.E. Lenkinski. "Recent advances in magnetic resonance neuro-spectroscopy." *Neurotherapeutics* 4, no. 3 (2007): 330–345. DOI: 10.1016/j.nurt.2007.04.009.

Schmidt, F.E.W., M.E. Fry, E.M.C. Hillman, J.C. Hebden and D.T. Delpy. "A 32-channel time-resolved instrument for medical optical tomography." *Rev. Sci. Instrum.* 71, no.1 (2000): 256–265. DOI: 10.1063/1.1150191.

Schmitt, J.M. "Optical coherence tomography (OCT): A review." *IEEE J. Sel. Topics Quantum Electron.* 5, no. 4 (1999): 1205–1215. DOI: 10.1109/2944.796348.

Schweiger, M. and S.R. Arridge. "Applications of temporal filters to time-resolved data in optical tomography." *Phys. Med. Biol.* 44, no. 7 (1999): 1699–1717. DOI:10.1088/0031-9155/44 /7/310.

Schweiger, M. and S.R. Arridge. "The Toast++ software suite for forward and inverse modeling in optical tomography." *J. Biomed. Opt.* 19, no. 4 (2014): 040801. doi: 10.1117/1.JBO.19.4 .040801. Download of software (last update 22 July 2014): http://sourceforge.net/projects /toastpp/.

Spinelli, L., A. Rizzolo, M. Vanoli, M. Grassi, P. Eccher Zerbini, R. Meirelles de Azevedo Pimentel and A. Torricelli. "Nondestructive assessment of fruit biological age in Brazilian mangoes by time-resolved reflectance spectroscopy in the 540–900 nm spectral range."

Tijskens, L.M.M., P. Eccher Zerbini, R.E. Schouten, M. Vanoli, S. Jacob, M. Grassi, R. Cubeddud, L. Spinelli and A. Torricelli. "Assessing harvest maturity in nectarines." *Postharvest Biol. Tec.* 45, no. 2 (2007): 204–213. DOI: 10.1016/j.postharvbio.2007.01.014.

Torricelli, A., L. Spinelli, D. Contini, M. Vanoli, A. Rizzol and P. Eccher Zerbini. "Time resolved reflectance spectroscopy for non-destructive assessment of food quality." *Sens. Instrumen. Food Qual.* 2, no. 2 (2008): 82–89. DOI: 10.1007/s11694-008-9036-2.

Vanoli, M., A. Rizzolo, M. Grassi, A. Farina, A. Pifferi, L. Spinelli, B.E. Verlinden and A. Torricelli. "Non destructive detection of brown heart in 'Braeburn' apples by time-resolved reflectance spectroscopy." *Proc. Food Sci.* 1 (2011): 413–420. DOI: 10.1016/j.profoo.2011.09.064.

Wang, L.V. "Prospects of photo-acoustic tomography." *Med. Phys.* 35, no. 12 (2008): 5758–5767. DOI: 10.1118/1.3013698.

Wang, L.V. and S. Wu. "Photo-acoustic tomography: In vivo imaging from organelles to organs." *Science* 335, no. 6072 (2012): 1458–1462. DOI: 10.1126/science.1216210.

Wang, X., Y. Pang, G. Ku, X. Xie, G. Stoica, and L.V. Wang. "Non-invasive laser-induced photo-acoustic tomography for structural and functional *in vivo* imaging of the brain." *Nature Biotech.* 21, no. 7 (2003): 803–806. DOI: 10.1038/nbt839.

Yamamoto, T., Y. Yamashita, H. Yoshizawa, A. Maki, M. Iwata, E. Watanabe and H. Koizumi. "Noninvasive measurement of language function by using optical topography." *Proc. SPIE* 3597, (1999): 230–237. DOI: 10.1117/12.356812.

Yousefi, S., J. Qin, Z. Zhi and R.K. Wang. "Label-free optical lymphangiography: Development of an automatic segmentation method applied to optical coherence tomography to visualize lymphatic vessels using Hessian filters." *J. Biomed. Opt.* 18, no. 8 (2013): article 086004. DOI: 10.1117/1.JBO.18.8.086004.

*Other sources*

Austin, T. (2014). Picture of Integrated optical-EEG cap on infant. **Access**: http://www .neuroscience.cam.ac.uk/directory/profile.php?topunaustin.

Inside Food Symposium, Leuven, Belgium (9–12 April 2013). **Access**: http://www.insidefood .eu/INSIDEFOOD_WEB/UK/WORD/proceedings/027P.pdf.

Thorlabs Inc. *MEMS Fiber-Optic Switches.* Online catalogue (2014). **Access**: http://www .thorlabs.de/newgrouppage9.cfm?objectgroup_id=1553 or http://www.thorlabs.de/thorproduct .cfm?partnumber=OSW12-633-SM.

Toscano, J. (2014). Photograph of head helmet for brain DOT. **Access**: http://publish.illinois. edu/jtoscano/fast-diffuse-optical-imaging/.

van der Mark, M.B., G.W. 't Hooft, J. Chen, R. Richards-Kortum and B. Chance. "Clinical study of the female breast using spectroscopy diffuse optical tomography." Conference Biomedical Optical Spectroscopy and Diagnostics, Miami Beach, FL, USA (2000). Paper PD.6. DOI: 10.1364 /BOSD.2000.PD6.

Wahl, M. "Time-Correlated Single Photon Counting." PicoQuant Technical Note 7253 (2014). **Access**: http://www.picoquant.com/images/uploads/page/files/7253/technote_tcspc .pdf.

## Chapter 21    Imaging based on absorption and ion detection methods

*Books and chapters in books*

Leahy-Hoppa, M.R. "Terahertz for weapon and explosive detection." In F. Flammini (ed.), *Critical Infrastructure Security*, WIT Transactions on state-of-the-art in science and engineering, vol. *54*. Southampton, UK: WIT Press (2012): pp. 207–230.

Levine, R.D. and R.B. Bernstein. *Molecular Reaction Dynamics and Chemical Reactivity.* New York, NY, USA: Oxford University Press (1987).

Parker, D.H. and A.T.J.B. Eppink. "Velocity mapping in applications in molecular dynamics and experimental aspects." In B.J. Whitaker (ed.), *Imaging in Molecular Dynamics: Technology and Applications—A Users Guide.* Cambridge, UK: Cambridge University Press (2003): pp. 20–112.

Peiponen, K.I., J.A. Zeitler and M. Kuwata-Gonokami (eds.). *Terahertz Spectroscopy and Imaging.* Springer Series in Optical Sciences, vol. 171. Heidelberg, Germany: Springer Verlag (2013).

Slater, C.S. *Studies of Photo-Induced Molecular Dynamics Using a Fast Imaging Sensor.* Cham, Switzerland: Springer International (2016).

Yin, X., B.W.H. Ng and D. Abbott. *Terahertz Imaging for Biomedical Applications: Pattern Recognition and Tomographic Reconstruction.* New York, NY, USA: Springer (2012).

*Journal articles*

Ashfold, M.N.R., N.H. Nahler, A.J. Orr-Ewing, O.P.J. Vieuxmaire, R.L. Toomes, T.N. Kitsopoulos, I.A. Garcia, D.A. Chestakov, S.M. Wuc and D.H. Parker. "Imaging the dynamics of gas phase reactions." *Phys. Chem. Chem. Phys.* 6, no. 1 (2006): 26–53. DOI: 10.1039/B509304J.

Berry, E., G.C. Walker, A.J. Fitzgerald, N.N. Zinov'ev, M. Chamberlain, S.W. Smye, M.R. Miles and M.A. Smith. "Do *in vivo* terahertz imaging systems comply with safety guidelines?" *J. Laser Appl.* 15, no. 3 (2003): 192–198. DOI: 10.2351/1.1585079.

Bhargava, R. "Towards a practical Fourier transform infrared chemical imaging protocol for cancer histopathology." *Anal. Bioanal. Chem.* 389, no. 4 (2007): 1155–1169. DOI: DOI 10.1007/s00216-007-1511-9.

Bontuyan, L.S. A.G. Suits, and P.L. Houston and B.J. Whitaker. "State-resolved differential cross sections for crossed-beam Ar-NO inelastic scattering by direct ion imaging." *J. Phys. Chem.* 97, no. 24 (1993): 6342–6350. DOI: 10.1021/j100126a006.

Brouard, M., D.H. Parker and S.Y.T. van de Meerakker. "Taming molecular collisions using electric and magnetic fields." *Chem. Soc. Rev.* 43, no. 21 (2014): 7279–7294. DOI: 10.1039/c4cs00150h.

Centrone, A. "Infrared imaging and spectroscopy beyond the diffraction limit." *Annu. Rev. Anal. Chem.* 8, no. 1 (2015): 101–126. DOI: 10.1146/annurev-anchem-071114-040435.

Chaigne, T., B. Arnal, S. Vilov, E. Bossy and O. Katz. "Super-resolution photoacoustic imaging via flow-induced absorption fluctuations." *Optica.* 4, no. 11 (2017): 1397–1404. DOI: 10.1364/OPTICA.4.001397.

Chan, W.L., J. Deibel and D.M Mittleman. "Imaging with terahertz radiation." *Rep. Prog. Phys.* 70, no. 8 (2007): 1325–1379. DOI: 10.1088/0034-4885/70/8/R02.

Chandler, D.W. and P.L. Houston. "Two-dimensional imaging of state-selected photo-dissociation products detected by multi-photon ionisation." *J. Chem. Phys.* 87, no. 2 (1987): 1445–1447. DOI: 10.1063/1.453276.

Chandler, D.W., P.L. Houston and D.H. Parker. "Perspective: Advanced particle imaging." *J. Chem. Phys.* 147, no. 1 (2017): article 013601 (2017). DOI: 10.1063/1.4983623. Introductory article to the special issue topic 'Developments and applications of velocity mapped imaging techniques'.

Chichinin, A.I., T.S. Einfeld, C. Maul and K.H. Gericke. "Three-dimensional imaging technique for direct observation of the complete velocity distribution of state-selected photo-dissociation products." *Rev. Sci. Instrum.* 73, no. 4 (2002): 1856–1865. DOI: 10.1063/1.1453505.

Danielli, A., K. Maslov, A. Garcia-Uribe, A.M. Winkler, C. Li, L. Wang, Y. Chen, G.W. Dorn II and L.V. Wang. "Label-free photoacoustic nanoscopy." *J. Biomed. Opt.* 19, no. 8 (2014): article 086006. DOI: 10.1117/1.JBO.19.8.086006.

Dazzi, A. and C.B. Prater. "AFM-IR: Technology and applications in nanoscale infrared spectroscopy and chemical imaging." *Chem. Rev.* 117, no. 7 (2017): 5146–5173. DOI: 10.1021/acs.chemrev.6b00448.

Dorling, K.M. and M.J. Baker. "Rapid FTIR chemical imaging: highlighting FPA detectors." *Cell* 31, no. 8 (2013): 437–438. DOI: 10.1016/j.tibtech.2013.05.008.

Eppink, A.T.J.B. and D.H. Parker. "Velocity map imaging of ions and electrons using electrostatic lenses: Application in photoelectron and photofragment ion imaging of molecular oxygen." *Rev. Sci. Instrum.* 68, no. 9 (1997): 3477–3484. DOI: 10.1063/1.1148310.

Fukumoto, K., K. Onda, Y. Yamada, T. Matsuk, T. Mukuta1, S.I. Tanaka and S.Y. Koshihara. "Femtosecond time-resolved photoemission electron microscopy for spatiotemporal imaging of photogenerated carrier dynamics in semiconductors." *Rev. Sci. Instrum.* 85, no. 8 (2014): article 083705. DOI: 10.1063/1.4893484.

Gebhardt, C.R., T.P. Rakitzis, P.C. Samartzis, V. Ladopoulos and T.N. Kitsopoulos. "Slice imaging: A new approach to ion imaging and velocity mapping." *Rev. Sci. Instrum.* 72, no. 10 (2001): 3848–3853. DOI: 10.1063/1.1403010.

Gijsbertsen, A., H. Linnartz, G. Rus, A.E. Wiskerke, S. Stolte, D.W. Chandler and J. Kłos. "Differential cross sections for collisions of hexapole state-selected NO with He." *J. Chem. Phys.* 123, no. 22 (2005): article 224305. DOI: 10.1063/1.2126969.

Heck, A.J.R. and D.W. Chandler. "Imaging techniques for the study of chemical reaction dynamics." *Annu. Rev. Phys. Chem.* 16 (1995): 335–372. DOI: 10.1146/annurev.pc.46.100195.002003.

Hernandez-Cardoso, G.G., S.C. Rojas-Landeros, M. Alfaro-Gomez1, A.I. Hernandez-Serrano, I. Salas-Gutierrez, E. Lemus-Bedolla, A.R. Castillo-Guzman, H.L. Lopez-Lemus and E. Castro-Camus. "Terahertz imaging for early screening of diabetic foot syndrome: A proof of concept." *Sci. Rep.* 7 (2017): article 42124. DOI: 10.1038/srep42124.

Houston, P.L. "Vector correlations in photo-dissociation dynamics." *J. Phys. Chem.* 91, no. 21 (1987): 5388–5397. DOI: 10.1021/j100305a003.

Houston, P.L. "Snapshots of chemistry: Product imaging of molecular reactions." *Acc. Chem. Res.* 28, no. 11 (1995): 453–460. DOI: 10.1021/ar00059a003.

Janssen, M.H.M., J.W.G. Mastenbroek and S. Stolte. "Imaging of oriented molecules." *J. Phys. Chem. A* 101, no. 41 (1997): 7605–7613. DOI: 10.1021/jp971159+.

Joalland. B., Y. Shi, A.D. Estillore, A. Kamasah, A.M. Mebel and A.G. Suit. "Dynamics of chlorine atom reactions with hydrocarbons: Insights from imaging the radical product in crossed beams." *Phys. Chem. A* 118, no. 40 (2014): 9281–9295. DOI: 10.1021/jp504804n.

Kauczok, S., N. Gödecke, A.I. Chichinin, M. Veckenstedt, C. Maul and K.H. Gericke. Three-dimensional velocity map imaging: Setup and resolution improvement compared to three-dimensional ion imaging." *Rev. Sci. Instrum.* 80, no. 8 (2009): article 083301. DOI: 10.1063/1.3186734.

Kazarian, S.G. and K.L.A. Chan. "ATR-FTIR spectroscopic imaging: recent advances and applications to biological systems." *Analyst* 138, no. 7 (2013): 1940–1951. DOI: 10.1039/C3AN36865C.

Kim, K.W., K.S. Kim, H. Kim, S.H. Lee, J.H. Park, J.H. Han, S.H. Seok, J. Park, Y.S. Choi, Y.I. Kim, J.K. Han and J.H. Son. "Terahertz dynamic imaging of skin drug absorption." *Opt. Express* 20, no. 9 (2012): 9476–9484. DOI: 10.1364/OE.20.009476.

Kitsopoulos, T.N., M.A. Buntine, D.P. Baldwin, R.N. Zare and D.W. Chandler. "Reaction product imaging—The H+D$_2$ reaction." *Science* 260, no. 5114 (1993): 1605–1610. DOI: 10.1126/science .260.5114.1605.

Lampton, M., O. Siegmund and R. Raffanti. "Delay line anodes for microchannel-plate spectrometers." *Rev. Sci. Instrum.* 58, no. 12 (1987): 2298–2305. DOI: 10.1063/1.1139341.

Langer, G. and T. Berer. "Nonlinear photoacoustic and fluorescence microscopy using a modulated laser diode." *Proc. SPIE* 9708 (2016): article 97082K. DOI: 10.1117/12.2213027.

Liu, K. "Crossed-beam studies of neutral reactions: State-specific differential cross sections." *Annu. Rev. Phys. Chem.* 52 (2001): 139–164. DOI: 10.1146/annurev.physchem.52.1.139.

Man, M.K.L., A. Margiolakis, S. Deckoff-Jones, T. Harada, E.L. Wong, M.B.M. Krishna, J. Madéo, A. Winchester, S. Lei, R. Vajtai, P.M. Ajayan and K.M. Dani. "Imaging the motion of electrons across semiconductor heterojunctions." *Nature Nanotech.* 12, no. 1 (2016): 36–40. DOI: 10.1038 /NNANO.2016.183.

Muller, E.A., B. Pollard and M.B. Raschke. "Infrared chemical nano-imaging: Accessing structure, coupling, and dynamics on molecular length scales." *J. Phys. Chem. Lett.* 6, no. 7 (2015): 1275–1284. DOI: 10.1021/acs.jpclett.5b00108.

Nakamura, N., S. Yang, K.C. Lin, T. Kasa, D.C. Che, A. Lombardi, F. Palazzetti and V. Aquilanti. "Stereodirectional images of molecules oriented by a variable-voltage hexapolar field: Fragmentation channels of 2-bromobutane electronically excited at two photolysis wavelengths." *J. Chem. Phys.* 147, no. 1 (2017): article 013917. DOI: 10.1063/1.4981025.

Nowak, D., W. Morrison, H.K. Wickramasinghe, J. Jahng, E. Potma, L. Wan, R. Ruiz, T.R. Albrecht, K. Schmidt, J. Frommer, D.P. Sanders and S. Park. "Nanoscale chemical imaging by photo-induced force microscopy." *Sci Adv.* 2, no. 3 (2016): article e1501571. DOI: 10.1126/sciadv.1501571.

Parker, D.H. and A.T.J.B. Eppink. "Photoelectron and photofragment velocity map imaging of state-selected molecular oxygen dissociation/ionization dynamics." *J. Chem. Phys.* 107, no. 7 (1997): 2357–2362. DOI: 10.1063/1.474624.

Redo-Sanchez, A., B. Heshmat, A. Aghasi, S. Naqvi, M. Zhang, J. Romberg and R. Raskar. "Terahertz time-gated spectral imaging for content extraction through layered structures." *Nature Commun.* 7 (2016): article 12665. DOI: 10.1038/ncomms12665.

Sacré, P.J., P. Lebrun, P.F. Chavez, C.D. Bleye, L. Netchacovitch, E. Rozet, R. Klinkenberg, B. Streel, P. Hubert and E. Ziemons. "A new criterion to assess distributional homogeneity in hyperspectral images of solid pharmaceutical dosage forms." *Anal. Chim. Acta* 818 (2014): 7–14. DOI: 10.1016/j.aca.2014.02.014.

Son, J.K. "Principle and applications of terahertz molecular imaging." *Nanotechnol.* 24, no. 21 (2013): article 214001. DOI: 10.1088/0957-4484/24/21/214001.

Strohm, E.M., M.J. Moore and M.C. Kolios. "High resolution ultrasound and photoacoustic imaging of single cells." *Photoacoustics* 4, no. 1 (2016): 36–42. DOI: 10.1016/j.pacs.2016.01.001.

Tang, F., P. Bao and Z. Su. "Analysis of nano-domain composition in high-impact polypropylene by atomic force microscopy-infrared." *Anal. Chem.* 88, no. 9 (2016): 4926–4930. DOI: 10.1021/acs.analchem.6b00798.

Toomes, R.L. and T.N. Kitsopoulos. "Rotationally resolved reaction product imaging using crossed molecular beams." *Phys. Chem. Chem. Phys.* 5, no. 12 (2003): 2481–2483. DOI: 10.1039/B303166G.

Townsend, D., S.A. Lahankar, S.K. Lee, S.D. Chambreau, A.G. Suits, X. Zhang, J. Rheinecker, L.B. Harding and J.M. Bowman. "The roaming atom: Straying from the reaction path in formaldehyde decomposition." *Science* 306, no. 5699 (2004a): 1158–1161. DOI: 10.1126/science.1104386.

Townsend, D., S.K. Lee and A.G. Suits. "Orbital polarization from DC slice imaging: S($^1$D) alignment in the photodissociation of ethylene sulfide." *Chem. Phys.* 301, no. 2–3 (2004b): 197–208. DOI: 10.1016/j.chemphys.2003.10.020.

Townsend, D., M.P. Minitti and A.G. Suits. "Direct current slice imaging." *Rev. Sci. Instrum.* 74, no. 4 (2003): 2530–2539. DOI: 10.1063/1.1544053.

Wang, D., Y. Wu and J. Xia. "Review on photoacoustic imaging of the brain using nanoprobes." *Neurophotonics* 3, no. 1 (2016): article 010901. DOI: 10.1117/1.NPh.3.1.010901].

Wiley, W.C. and I.H. McLaren. "Time-of-flight mass spectrometer with improved resolution." *Rev. Sci. Instrum.* 26, no. 12 (1955): 1150–1157. DOI: 10.1063/1.1715212.

Xu, M., L. Zhou, Q. Zhang, Z. Wu, X. Shi and Y. Qiao. "Near-infrared chemical imaging for quantitative analysis of chlorpheniramine maleate and distribution homogeneity assessment in pharmaceutical formulations." *J. Innov. Opt. Health Sci.* 9, no. 6 (2016): article 1650002. DOI: 10.1142/S1793545816500024.

Yan, S., Y.T. Wu, B. Zhang, X.F. Yue and K. Liu. "Do vibrational excitations of CHD$_3$ preferentially promote reactivity toward the chlorine atom?" *Science* 316, no. 5832 (2007): 1723–1726. DOI: 10.1126/science.1142313.

Yao, J. and L.V. Wang. "Sensitivity of photoacoustic microscopy." *Photoacoustics* 2, no. 2 (2014): 87–101. DOI: 10.1016/j.pacs.2014.04.002.

Yao, J., L. Wang, C. Li, C. Zhang and L.V. Wang. "Photoimprint photoacoustic microscopy for three-dimensional label-free subdiffraction imaging." *Phys. Rev. Lett.* 112, no. 1 (2014): article 014302. DOI: 10.1103/PhysRevLett.112.014302

Yeh, K., S. Kenkel, J.N. Liu and R. Bhargava. "Fast infrared chemical imaging with a quantum cascade laser." *Anal. Chem.* 87, no. 1 (2015): 485–493. DOI: 10.1021/ac5027513.

Zare, R.N. "Photoejection dynamics (1)." *Mol. Photochem.* 4, no. 1 (1972): 1–37.

Zhang, Z., Z. Chen, C. Huang, Y. Chen, D. Dai, D.H. Parker and X. Yang. "Imaging the pair-correlated HNCO photodissociation: The NH(a $^1\Delta$) + CO(X $^1\Sigma^+$) channel." *J. Phys. Chem. A* 118, no. 13 (2014): 2413–2418. DOI: 10.1021/jp500625m.

Other sources     Berer, T, T.A. Klar, G. Langer and B. Buchegger. "SuRePAM—Super-resolution photo-acoustic microscopy by stimulated depletion." Austrian Science Fund (FWF) research project P 27839-N36/06/2015–05/2018 (2015). Linz, Austria: Research Center for Non-Destructive Testing GmbH (RECENDT GmbH). **Access**: http://www.recendt.at/1730_DEU_HTML.php?g_currMenuName=1223.

Man, M.K.L., S. Deckoff-Jones, T. Harada, E.L. Wong, A. Margiolakis, M.B.M. Krishna, J. Madéo, A. Winchester, S. Lei, R. Vajtai, P.M. Ajayan and K.M. Dani. "Imaging electron motion in 2D semiconductor heterojunctions." CLEO/QELS Fundamental Science 2017 Conference, San Jose, CA, USA (May 14–19, 2017). Conference paper FTh4F.2. **Access**: https://www.osapublishing.org/abstract.cfm?uri=CLEO_QELS-2017-FTh4F.2.

TeraView. "Terahertz applications: Detection of explosives and materials characterization." Cambridge, UK: TeraView Ltd. (2016). **Access**: http://www.teraview.com/applications/homeland-security-defense-industry/explosive-detection.html

# Index

Printed and bound by CPI Group (UK) Ltd, Croydon, CR0 4YY

01/11/2024

01782601-0020